AutoCAD LT® 2000:

A Problem-Solving Approach

Sham Tickoo

Professor
Department of Manufacturing Engineering Technologies
And Supervision
Purdue University Calumet
Hammond, Indiana

Thomson Learning™

Africa • Australia • Canada • Denmark • Japan • Mexico • New Zealand
Philippines • Puerto Rico • Singapore • Spain • United Kingdom • United States

NOTICE TO THE READER

Autodesk Press Staff
Executive Director: Alar Elken
Executive Editor: Sandy Clark
Aquisitions Editor: Michael Kopf
Developmental Editor: John Fisher
Executive Marketing Manager: Maura Theriault
Executive Production Manager: Mary Ellen Black
Production Coordinator: Jennifer Gaines
Art and Design Coordinator: Mary Beth Vought
Marketing Coordinator: Paula Collins
Technology Project Manager: Tom Smith
Editorial Assistant: Jasmine Hartman

Cover illustration by Scott Keidong's Image Enterprises. AutoCAD® image © 2000, Autodesk.

Printed in the United States of America
1 2 3 4 5 6 7 8 9 10 XXX 05 04 03 02 01 00

For more information, contact
Autodesk Press
3 Columbia Circle, Box 15-015
Albany, New York USA 12212-15015;
or find us on the World Wide Web at http://www.autodeskpress.com

Library of Congress Cataloging-in-Publication Data

Tickoo, Sham.
AutoCAD LT 2000: a problem solving approach/Sham Tickoo.
p. cm.
ISBN 0-7668-2095-5
1. Computer graphics. 2. AutoCAD. 3. Computer-aided design. I. Title

T385 .T52432 2000
620'.0042'02855369--dc21

00-23793

Table of Contents

Chapter 3: Drawing Aids

Chapter 4: Editing Commands

Chapter 5: Controlling Drawing Display and Creating Text

Chapter 12: Blocks

Advanced AutoCAD LT

Chapter 13: Defining Block Attributes

Chapter 14: External References

Chapter 15: Plotting Drawings and Draw Commands

Chapter 19: Isometric Drawing

Chapter 20: User Coordinate System

Chapter 21: Drawing and Viewing 3D Objects

Chapter 22: Model Space Viewports, Paper Space Viewports, and Layouts

Customizing AutoCAD LT

Chapter 23: Template Drawings

Chapter 24: Script Files and Slide Shows

Chapter 25: Creating Linetypes and Hatch Patterns

Chapter 26: Customizing the ACLT.PGP File, DIESEL: A String Expression Language

Chapter 27: Pull-down, Shortcut, and Partial Menus and Customizing Toolbars

Chapter 28: Tablet Menus

Chapter 29: Image Tile, Button and Auxilliary Menus

Chapter 30: AutoCAD LT on the Internet

Appendices

Index

Preface

AutoCAD LT, developed by Autodesk Inc., is the most popular PC-CAD system available in the market. AutoCAD LT 2000 is a Windows based application used to generate various kinds of drawings. AutoCAD LT's drafting system is a two-dimensional CAD application with many graphical user interface tools. This package is very useful for architects, engineers, draftsmen for creating drawings manufacturing parts, site plans and fine tuning design details accurately and quickly. AutoCAD LT has also provided facilities that allow users to customize AutoCAD LT to make it more efficient and therefore increase their productivity.

This book contains a detailed explanation of AutoCAD LT 2000 commands and how to use them in solving drafting and design problems. The book also unravels the customizing power of AutoCAD LT. Every AutoCAD LT command and customizing technique is thoroughly explained with examples and illustrations that make it easy to understand their function and application. At the end of each topic, there are examples that illustrate the function of the command and how it can be used in the drawing. When you are done reading this book, you will be able to use AutoCAD LT commands to make a drawing, create text, make and insert symbols, dimension a drawing, create 3D objects and solid models, write script files, define linetypes and hatch patterns, write your own menus, customize the toolbars, customize the status line using DIESEL, and edit the Program Parameter file (ACLT.PGP).

The book also covers basic drafting and design concepts — such as orthographic projections, dimensioning principles, sectioning, auxiliary views, and assembly drawings — that provide you with the essential drafting skills you need to solve drawing problems with AutoCAD LT. In the process, you will discover some new applications of AutoCAD LT that are unique and might have a significant effect on your drawings. You will also get a better idea of why AutoCAD LT has become such a popular software package and an international standard in PC-CAD. Please refer to the following table for conventions used in this text.

Convention	Example
• Command names are capitalized and bold.	The **MOVE** command
• A key icon appears when you should respond by pressing the ENTER or RETURN key.	

Convention	Example
• Command sequences are indented. Responses are indicated by boldface. Directions are indicated by italics. Comments are enclosed in parentheses.	Command: **MOVE** Select object: **G** Enter group name: *Enter a group name (the group name is group1)*
• The command selection from the toolbars, menus, and Command prompt are enclosed in a shaded box.	**Toolbar:** Draw>Arc **Pull-down:** Draw>Arc **Command:** ARC or A
• AutoCAD 2000 features are indicated by an asterisk symbol at the end of the feature.	AutoCAD Design Center*
• Project exercises are designated by the project number and the part number of the project.	Project Exercise 1-3

Author's Web Sites

For Faculty: Please contact the author at **stickoo@calumet.purdue.edu** to access the Web Site that contains the following:

1. **Power Point presentations, programs, and drawings used in this textbook.**
2. **Syllabus, chapter objectives and hints, and questions with answers for every chapter.**

For Students:You can download drawing-exercises, tutorials, programs, and special topics by accessing authors Web Site at **http://www.calumet.purdue.edu/public/mets/tickoo/index.html**.

DEDICATION

To teachers, who make it possible to disseminate knowledge
to enlighten the young and curious minds
of our future generations

To students, who are dedicated to learning new technologies
and making the world a better place to live

Thanks

To the faculty and students of the MET department of
Purdue University Calumet for their cooperation

To Renu Muthoo, Anupama Pande, and Santosh Tickoo
of CADCIM Technologies
for their valuable help

Chapter 1

Getting Started

Learning Objectives

After completing this chapter, you will be able to:

- *Invoke AutoCAD LT commands from the menu, digitizer, command line, or toolbars.*
- *Understand the functioning of dialog boxes in AutoCAD LT.*
- *Draw lines using the* ***LINE*** *command and its options.*
- *Understand different coordinate systems used in AutoCAD LT.*
- *Use the* ***ERASE*** *and* ***MOVE*** *commands.*
- *Create selection sets using Window and Crossing options.*
- *Draw circles using different options of the* ***CIRCLE*** *command.*
- *Use* ***REDRAW***, ***ZOOM***, *and* ***PAN*** *commands.*
- *Plot drawings.*
- *Save the work using different file-saving commands.*
- *Open an existing file and start a new drawing.*
- *Use the* ***Options*** *dialog box to specify the settings.*
- *Recall and edit command line.*
- *Use Learning Assistance and AutoCAD LT's help.*

STARTING AUTOCAD LT

When you turn on your computer, the operating system (Windows 95, Windows NT or Windows 98) is automatically loaded (Figure 1-1). This will display the Windows screen with various application icons. You can load AutoCAD LT by double-clicking on the AutoCAD LT 2000 icon. You can also load AutoCAD LT from the task bar by choosing the **Start** button at the bottom left corner of the screen (default position) to display the menu. Choose **Programs** to display the program folders. Now, choose the **AutoCAD LT 2000** folder to display AutoCAD LT programs and then choose **AutoCAD LT 2000** to start AutoCAD LT.

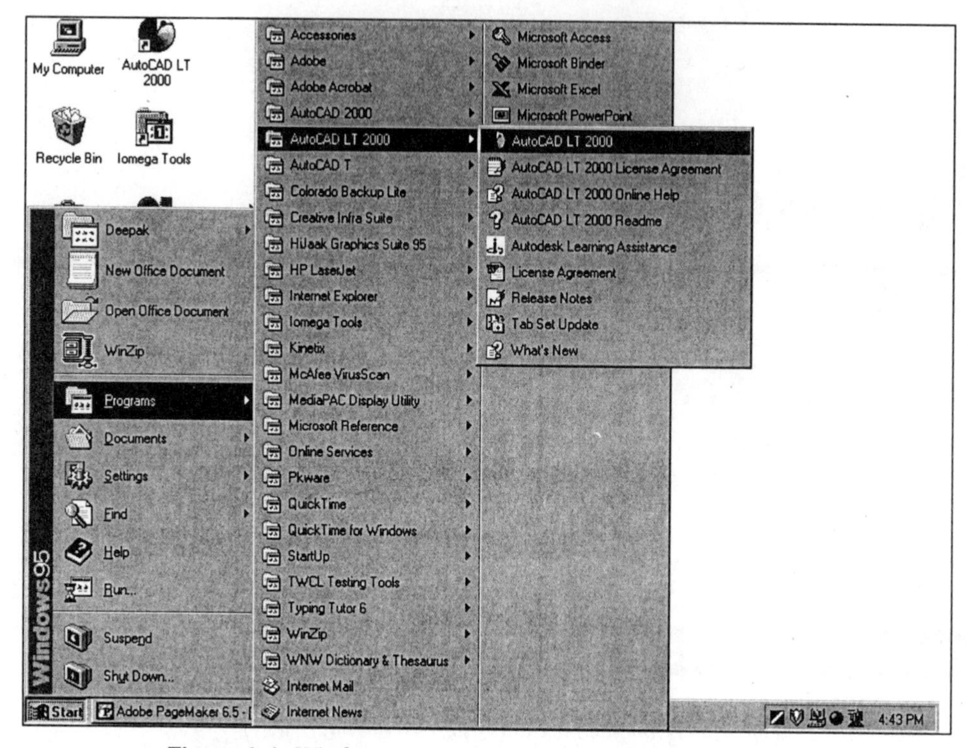

Figure 1-1 Windows screen with taskbar and application icons

STARTING A DRAWING

When you start AutoCAD LT, AutoCAD LT displays the **Startup** dialog box on the screen (Figure 1-2). The dialog box is disabled if you clear the **Show Startup dialog** check box in the **Startup** dialog box. To turn on the display of the dialog box, Choose **Options** from the **Tools** menu. Choose the **System** tab and under the **General Options** area, select **Show Startup dialog** check box and choose **OK**. The **Startup** dialog box provides the following four options for starting a new drawing.

Open a Drawing

When you choose the **Open a Drawing** button (Figure 1-3), a list of the four most recently opened drawings is displayed for you to select from. The **Browse** button displays the **Select File** dialog box which allows you to look for another file. The dialog box is disabled if the **FILEDIA** system variable is set to 0.

Start from Scratch

When you choose the **Start from Scratch** button (Figure 1-4), AutoCAD LT starts a new drawing that contains the default AutoCAD LT setup for English (aclt.dwt) or Metric drawings (acltiso.dwt). For example, if you select the English default setting, the limits are 12x9, text height is 0.20, dimensions and linetype scale factors are 1.

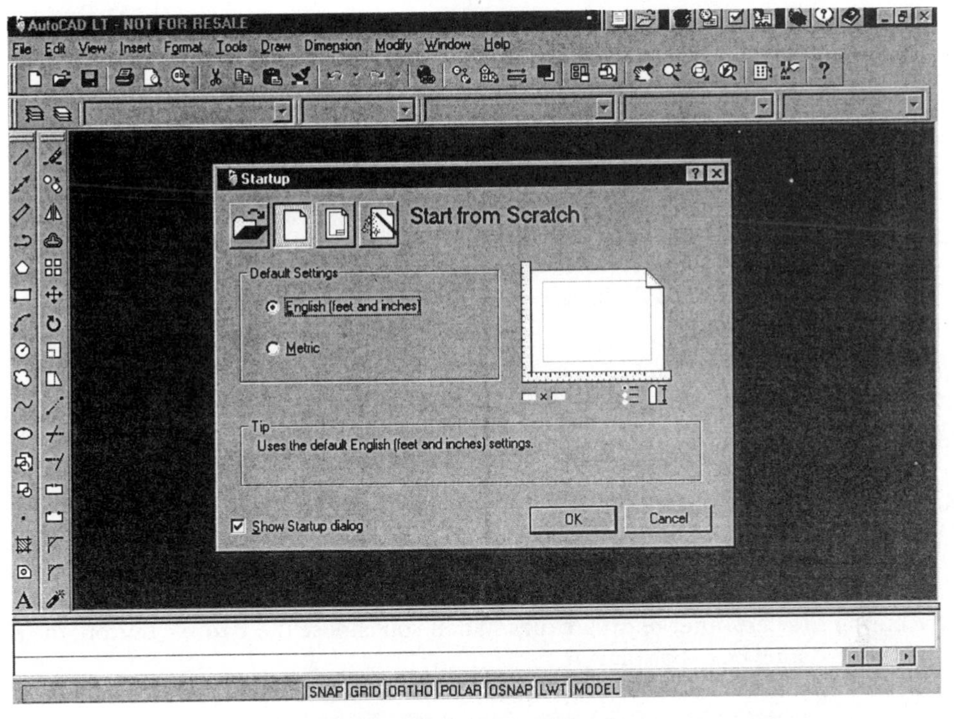

***Figure 1-2 Startup** dialog box*

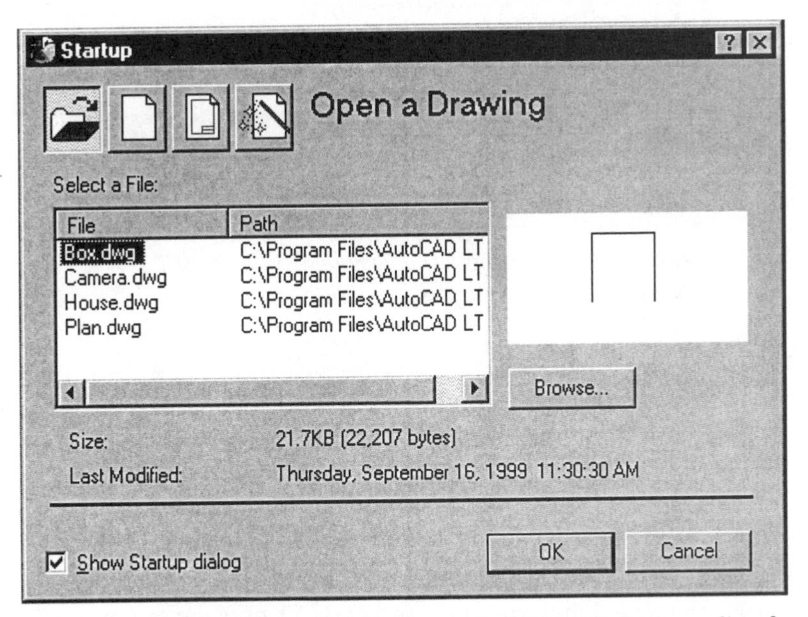

***Figure 1-3** When using Open a Drawing, AutoCAD LT displays a list of four most recently opened drawings.*

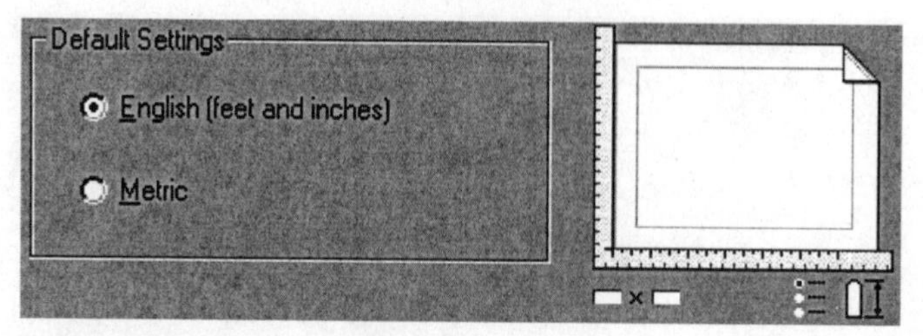

Figure 1-4 When using Start from Scratch, the new drawing contains the default settings.

Use a Template

When you choose the **Use a Template** button in the **Startup** dialog box, AutoCAD LT displays a list of templates supplied with AutoCAD LT (Figure 1-5). The default template file is aclt.dwt or acltiso.dwt, depending on the installation. If you use the template file, the new drawing will have the same setting as the template file. You can also define your own template files that are customized to your requirements (See Chapter 23, Template Drawings). To differentiate the template files from the drawing files, the template files have **.dwt** extension whereas the drawing files have **.dwg** extension. Any drawing files can be saved as a template file. You can use the **Browse** button to select other template files. When you choose the **Browse** button, the **Select a template file** dialog box is displayed.

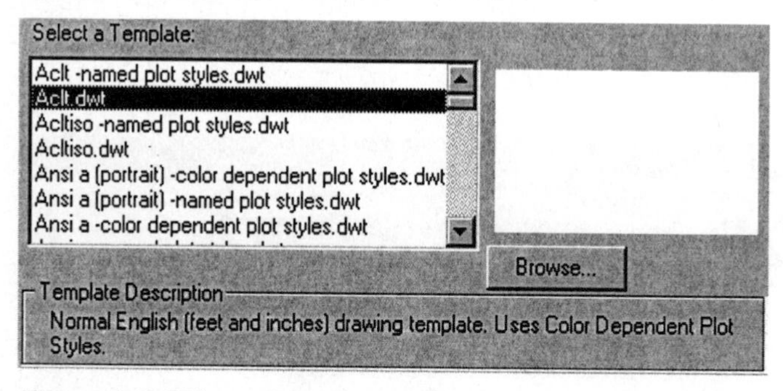

Figure 1-5 When using Use a Template, AutoCAD LT displays a list of template files.

Use a Wizard

When you choose the **Use a Wizard** button, you have the choice of using **Quick Setup** or **Advanced Setup**. In the **Quick Setup**, you specify the units, width, and length of the work area. In the **Advanced Setup** you specify units, angle, angle measure, angle direction, and area.

INVOKING COMMANDS IN AUTOCAD LT

This chapter assumes that the software is loaded on your system and that you are familiar with the hardware you are using. When you start AutoCAD LT and you are in the drawing editor, you need to invoke AutoCAD LT commands to perform any operation. For example, if you want to draw a line, first you have to invoke the **LINE** command, and then you define the start point and endpoint of the line. Similarly, if you want to erase objects, you must enter the

ERASE command, and then select the objects you want to erase. AutoCAD LT has provided the following methods to enter or select commands:

Keyboard **Menu** **Toolbar** **Digitizing tablet**

Keyboard

You can invoke any AutoCAD LT command at the keyboard by typing the command name at the Command prompt, and then pressing ENTER or the SPACEBAR. Before you enter a command, make sure the Command prompt is displayed as the last line in the command prompt area. If the Command prompt is not displayed, you must cancel the existing command by pressing ESC (escape) on the keyboard. The following example shows how to invoke the **LINE** command from the keyboard:

Command: **LINE** [Enter]

Menu

You can also select commands from the menu. The **menu bar** that displays the menu bar titles is at the top of the screen. As you move the pointing device sideways, different menu bar titles are highlighted. You can choose the desired item by pressing the pick button of your pointing device. Once the item is selected, the corresponding menu is displayed directly under the title. You can invoke a command from the menu by pressing the pick button of your pointing device. Some of the menu items in the menu display an arrow on the right side, which indicates that the menu item has a cascading menu. You can display the cascading menu by selecting the menu item or by just moving the arrow pointer to the right of that item. You can then select any item in the cascading menu by highlighting the item or command and pressing the pick button of your pointing device, or simply by pressing the stylus (pen) down if you are using a stylus with your digitizer. For example, if you want to draw an ellipse using the Center option, choose **Draw** from the menu bar, choose **Ellipse** from the **Draw** menu, then choose **Center** from the cascading menu (Figure 1-6). In this text, this command selection sequence will be referenced as Choose **Draw** > **Ellipse** > **Center.**

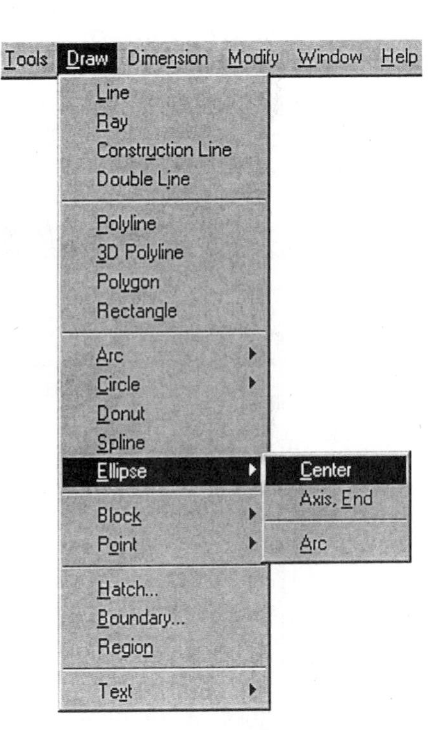

***Figure 1-6** Invoking the **ELLIPSE** command from the **Draw** menu.*

Toolbar

In Windows, the toolbar is an easy and convenient way to invoke a command. For example, you can invoke the **LINE** command by choosing the **Line** button (the upper left tool, Figure 1-7) in the **Draw** toolbar. When you select a command from the toolbar, the Command prompts are displayed in the Command prompt area.

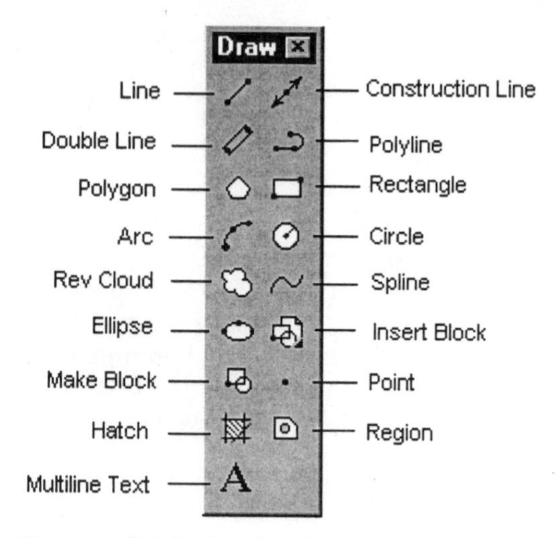

*Figure 1-7 Selecting the **LINE** command from the **Draw** toolbar*

Displaying Toolbars. The toolbars can be displayed by selecting the respective check boxes in the **Toolbars** dialog box (Figure 1-8). The **Toolbars** dialog box can be invoked by selecting **Toolbars** in the **View** menu. Each toolbar contains a group of tools representing different AutoCAD LT commands. When you move the cursor over the tools of a toolbar, the tool-tip (name of command) is displayed below the tool on which the cursor is resting. Once you locate the desired tool, the command associated with that tool can be invoked by choosing the tool.

The toolbars can be moved anywhere on the screen by placing the cursor on the title bar area and then dragging it to the desired location. You must hold the pick button down while dragging. You can also change the shape of the toolbars by placing the cursor anywhere on the border of

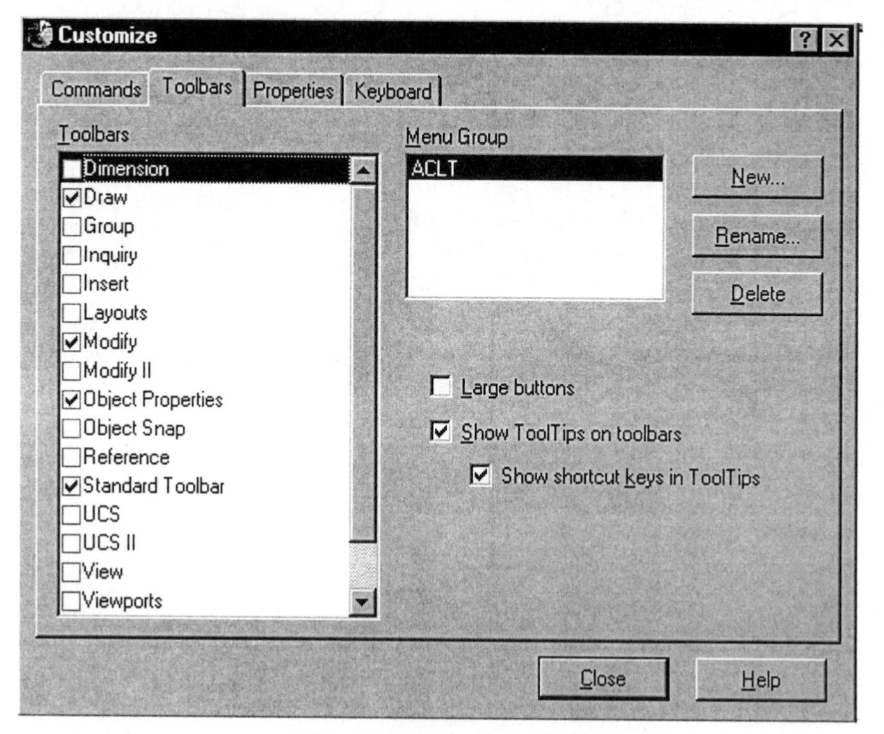

*Figure 1-8 **Toolbars** dialog box*

the toolbar and then dragging it in the desired direction (Figures 1-9 and 1-10). You can also customize toolbars to meet your requirements (See Chapter 27, Pull-down, Shortcut, and Partial Menus and Customizing Toolbars).

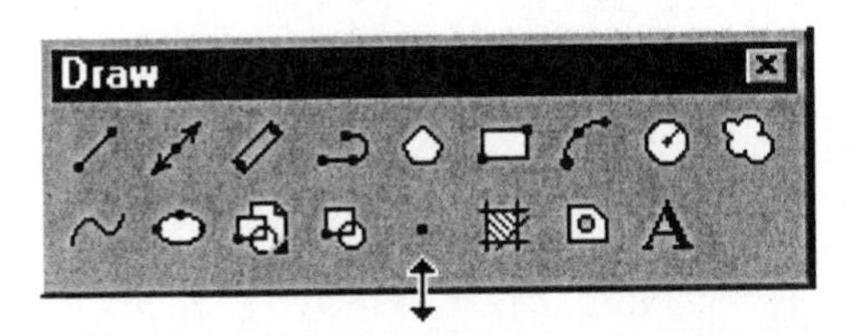

*Figure 1-9 Reshaping the **Draw** toolbar*

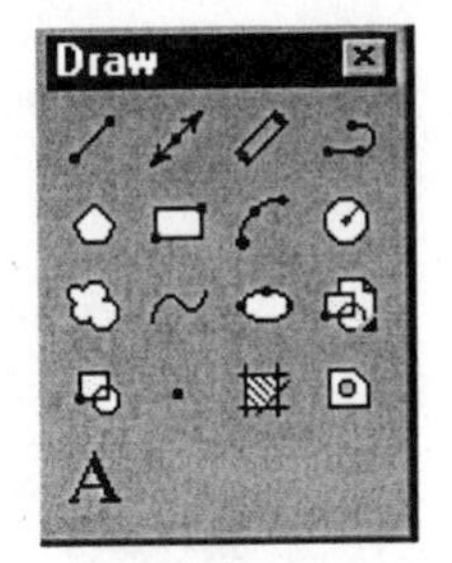

*Figure 1-10 **Draw** toolbar reshaped*

Digitizing Tablet

In the digitizing tablet, the commands are selected from the template that is secured on the surface of the tablet. To use the tablet menu you need a digitizing tablet and a pointing device. You also need a tablet template that contains AutoCAD LT commands arranged in various groups for easy identification. To invoke a command from the digitizer, move your pointing device so that the crosshairs of the pointing device (or the tip of the stylus) are directly over the cell that contains the command you want to invoke. Now press the pick button of the pointing device, or simply press down the stylus, and AutoCAD LT will automatically invoke that command.

AUTOCAD LT DIALOG BOXES

A dialog box can consist of a dialog label, toggle buttons, radio buttons, edit boxes, slider bars, image boxes, and a box that encloses these components. These components are also referred to as TILES. Some of the components of a dialog box are shown in Figure 1-11. Some of the AutoCAD LT dialog boxes have number of tabs. Choosing these tabs gives you the options in the dialog box related to these tabs. On the upper right corner of the dialog boxes you will find a **close button (X)** which is used to close the dialog boxes. Also in some of the dialog boxes you will find a **question mark** button (?). When you choose this button, the question mark gets attached to the cursor and starts moving along with it. Now, when you drop it on any topic in the dialog box, AutoCAD LT displays the help related to that topic. Also provided with the dialog boxes is the **HELP** button. Choosing this button displays the help topics related to that particular dialog box.

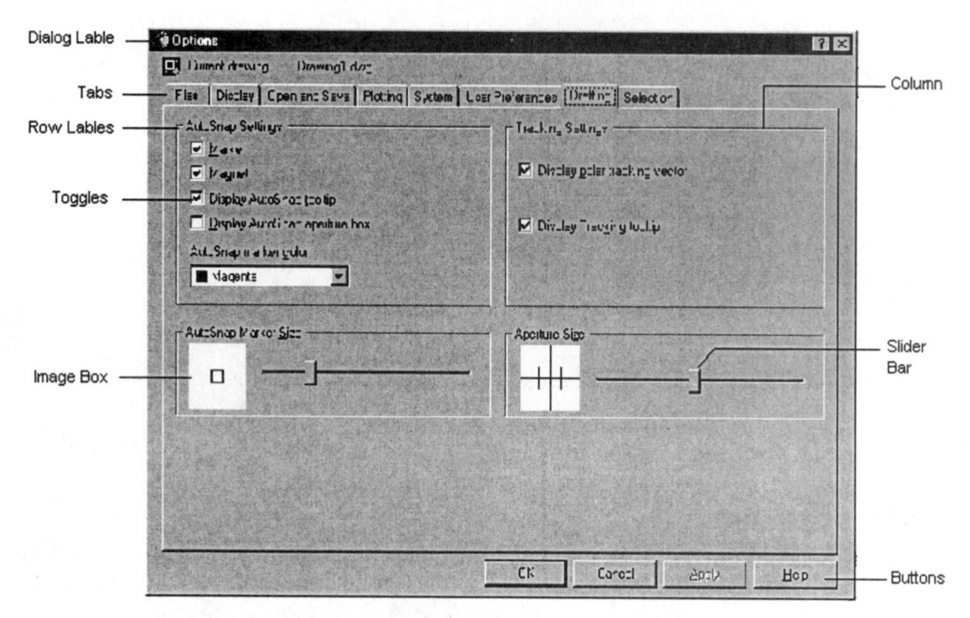

Figure 1-11 *Components of a dialog box*

DRAWING LINES IN AUTOCAD LT

Toolbar:	Draw > Line
Menu:	Draw > Line
Command:	LINE

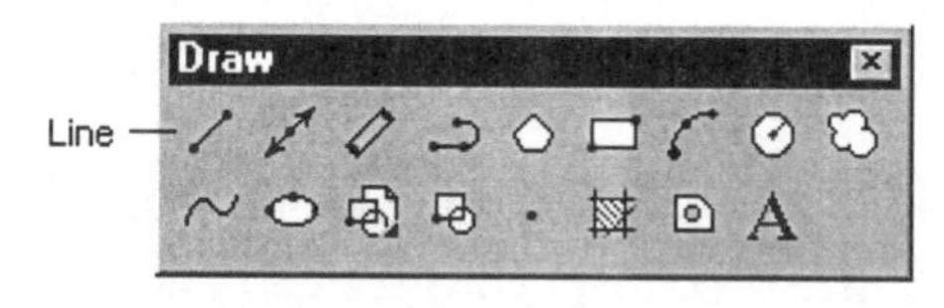

*Figure 1-12 Invoking the **LINE** command from the **Draw** toolbar*

The most fundamental object in a drawing is the line. A line can be drawn between any two points by using AutoCAD LT's **LINE** command. Once you have invoked the **LINE** command, the next prompt, **Specify first point:** requires you to specify the starting point of the line. After the first point is selected, AutoCAD LT will prompt you to enter the second point at the **Specify next point or [Undo]:** prompt. When you select the second point of the line, AutoCAD LT will again display the prompt **Specify next point or [Undo]:**. At this point you may continue to select points or terminate the **LINE** command by pressing ENTER, ESC, or the SPACEBAR. You can also right-click to display the shortcut menu from where you can choose **Enter** or **Cancel** options to exit from the **LINE** command. After terminating the **LINE** command, AutoCAD LT will again display the Command prompt. The prompt sequence for drawing Figure 1-13 is as follows:

Command: **LINE** [Enter]
Specify first point: *Move the cursor (mouse) and left-click to specify the first point.*
Specify next point or [Undo]: *Move the cursor and left-click to specify the second point.*
Specify next point or [Undo]: *Specify the third point.*
Specify next point or [Close/Undo]: [Enter] *(Press ENTER to terminate the **LINE** command.)*

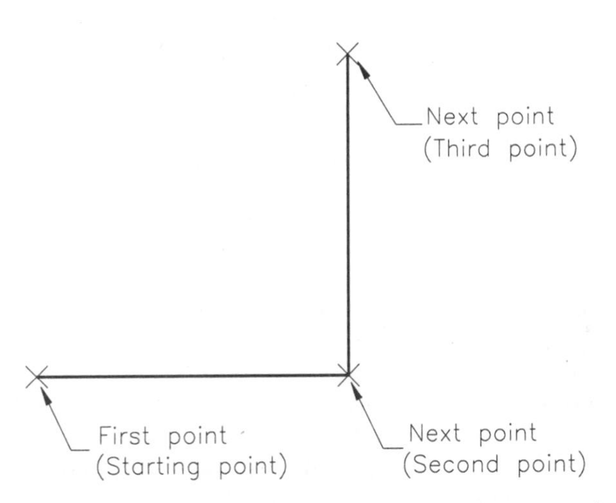

***Figure 1-13** Drawing lines using the **LINE** command*

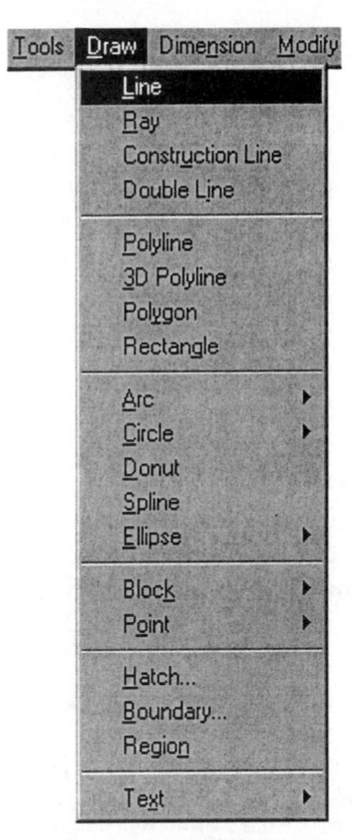

***Figure 1-14** Invoking the **LINE** command from the **Draw** menu*

Note

*To clear the graphics area (drawing screen) to gain space to work out the exercises and the examples, type **ERASE** at the Command prompt and press ENTER. The screen crosshairs will change into a box called a pick box and AutoCAD LT will prompt you to select objects. You can select the object by positioning the pick box anywhere on the object and pressing the pick button of the pointing device. Once you have finished selecting the objects, press ENTER to terminate the **ERASE** command and the objects you selected will be erased. If you enter **All** at the **Select objects:** prompt, AutoCAD LT will erase all objects from the screen. (See **ERASE** command "Erasing Objects" discussed later in this chapter). You can use the **U** (undo) command to undo the last command.*

Command: **ERASE** [Enter]
Select objects: *Select objects.* *(Select objects using the pick box.)*
Select objects: [Enter]
Command: **R** [Enter] *(Type **R** for redraw.)*
Command: **ERASE** [Enter]
Select objects: **ALL** [Enter]
Select objects: [Enter]
Command: **U** [Enter] *(**U** command will undo the last command.)*

The **LINE** command has the following three options:

Continue **Close** **Undo**

The Continue Option

After exiting from the **LINE** command, you may want to draw another line starting from the point where the previous line ended. In such cases you can use the **Continue** option. This option enables you to grab the endpoint of the previous line and to continue drawing the line from that point (Figure 1-15). The following is the prompt sequence for the **Continue** option:

Command: **LINE** [Enter]
Specify first point: *Pick first point of the line.*
Specify next point or [Undo]: *Pick second point.*
Specify next point or [Undo]: [Enter]
Command: **LINE** [Enter] *(Or press ENTER to repeat the command.)*
Specify first point: [Enter] *(Press ENTER)*
Specify next point or [Undo]: *Pick second point of the second line (Third point in Figure 1-15).*
Specify next point or [Undo]: [Enter]

You can also type the @ symbol to start the line from the **last point**. For example, if you draw a line and then immediately repeat the **LINE** command, the @ will snap to the endpoint of the last line. The **Continue** option will snap to the endpoint of the last line or arc, even if other points have been defined after the line was drawn.

Command: **LINE** [Enter]
Specify first point: *Pick first point of the line.*
Specify next point or [Undo]: *Pick second point.*
Specify next point or [Undo]: [Enter]
Command: **LINE** [Enter] *(**L** is command alias for **LINE**)*
Specify first point: @ [Enter] *(Continues the drawing line from the last point.)*
Specify next point or [Undo]: *Pick second point of the second line.*
Specify next point or [Undo]: [Enter]

The Close Option

The **Close** option can be used to join the current point with the initial point of the first line when two or more lines are **drawn in continuation**. Apart from closing the objects it also at the same time comes out of the line command which means it will not ask you **Specify next point or [Undo]:** again. For example, this option can be used when an open figure needs one more line to close it and make a polygon (a polygon is a closed figure with at least three sides, for example, a triangle or rectangle). The following is the prompt sequence for the **Close** option (Figure 1-16):

Command: **LINE** [Enter]
Specify first point: *Pick first point.*
Specify next point or [Undo]: *Pick second point.*
Specify next point or [Undo]: *Pick third point.*
Specify next point or [Close/Undo]: *Pick fourth point.*
Specify next point or [Close/Undo]: **C** [Enter] *(Joins the fourth point with the first point.)*

You can also choose the **Close** option from the shortcut menu, which appears when you right-click in the drawing area.

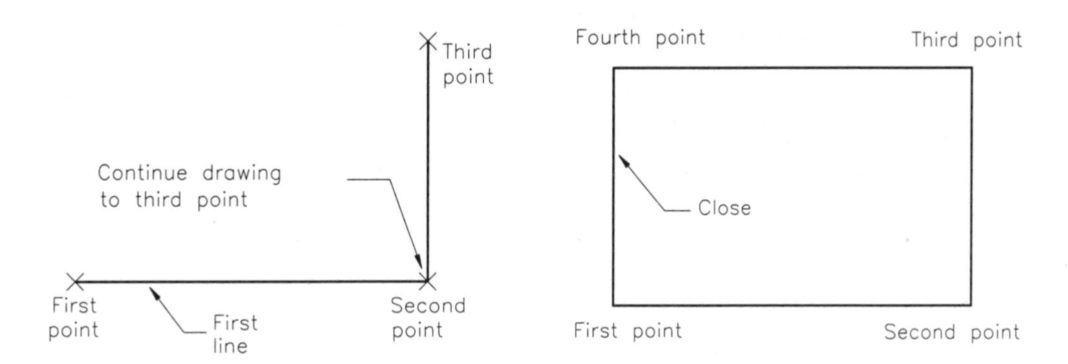

*Figure 1-15 Using the **Continue** option with the **LINE** command*

*Figure 1-16 Using the **Close** option with the **LINE** command*

The Undo Option

If you draw a line, and then realize that you made an error, you can remove the line using the **Undo** option. If you need to remove more than one line, you can use this option multiple times and go as far back as you want. In this option, you can type **Undo** (or just **U**) at the **Specify next point or [Undo]** prompt. You can also right-click to display the shortcut menu which gives you the **Undo** option. The following example illustrates the use of the **Undo** option (Figure 1-17):

Command: **LINE** [Enter]
Specify first point: *Pick first point* *(Point 1 in Figure 1-17).*
Specify next point or [Undo]: *Pick second point (Point 2).*
Specify next point or [Undo]: *Pick third point.*
Specify next point or [Close/Undo]: *Pick fourth point.*
Specify next point or [Close/Undo]: **U** [Enter] *(Removes last line from Point 3 to Point 4.)*

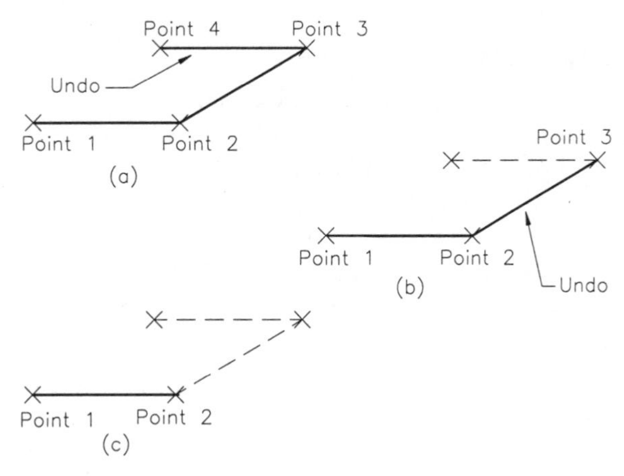

*Figure 1-17 Removing lines using the **Undo** option of the **LINE** command*

Specify next point or [Close/Undo]: **U** Enter *(Removes next line from Point 2 to Point 3.)*
Specify next point or [Close/Undo]: Enter

COORDINATE SYSTEMS

To specify a point in a plane, we take two mutually perpendicular lines as references. The horizontal line is called the **X axis**, and the vertical line is called the **Y axis.** The point of intersection of these two axes is called the **origin**. The X and Y axes divide the XY plane into four parts, generally known as quadrants. The X coordinate measures the horizontal distance from the origin (how far left or right) on the X axis. The Y coordinate measures the vertical distance from the origin (how far up or down) on the Y axis. The origin has the coordinate values of X = 0, Y = 0. The origin is taken as the reference for locating any point in the XY plane. The X coordinate is positive if measured to the right of the origin and negative if measured to the left of the origin. The Y coordinate is positive if measured above the origin and negative if measured below the origin. This method of specifying points is called the **Cartesian coordinate system** (Figure 1-18). In AutoCAD LT, the default origin is located at the lower left corner of the graphics area of the screen. AutoCAD LT uses the following coordinate systems to locate a point in an XY plane:

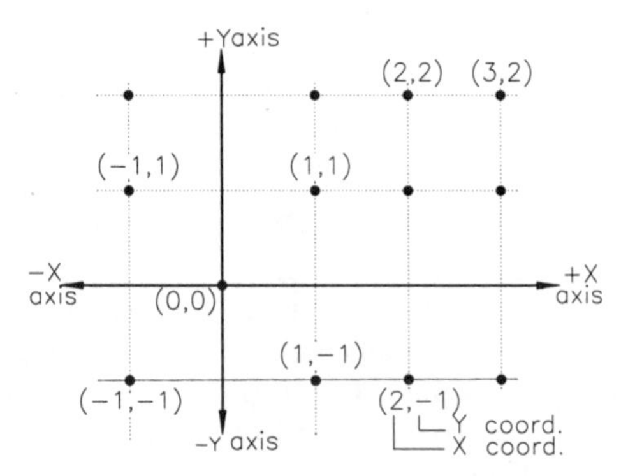

Figure 1-18 *Cartesian coordinate system*

Absolute coordinates
Relative coordinates
Polar coordinates
Direct distance entry

Absolute Coordinates

In the absolute coordinate system the points are located with respect to the origin (0,0) by specifying their exact X and Y coordinates from (0,0). For example, a point with X = 4 and Y = 3 is measured 4 units horizontally (displacement along the X axis) and 3 units vertically (displacement along the Y axis) from the origin (Figure 1-19).

In AutoCAD LT, the absolute coordinates are specified by entering X and Y coordinates, separated by a comma. The following example illustrates the use of absolute coordinates (Figure 1-20):

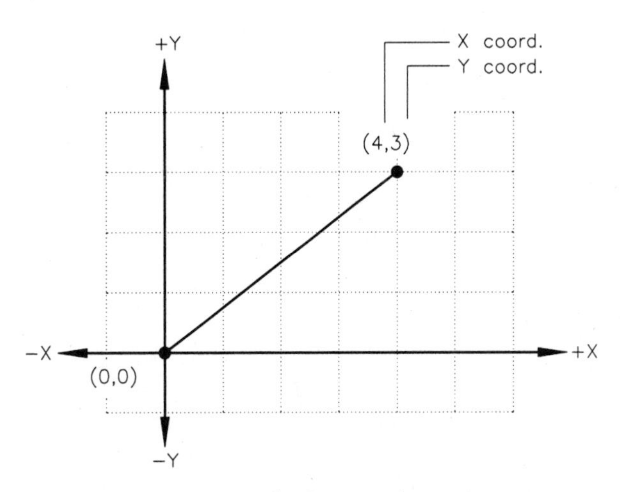

Figure 1-19 *Absolute coordinate system*

Command: **LINE** [Enter]
Specify first point: **1,1** [Enter]
(X = 1 and Y = 1.)
Specify next point or [Undo]: **4,1** [Enter]
(X = 4 and Y = 1.)
Specify next point or [Undo]: **4,3** [Enter]
Specify next point or [Close /Undo]: **1,3** [Enter]
Specify next point or [Close/Undo]: **1,1** [Enter]
Specify next point or [Close/Undo]: [Enter]

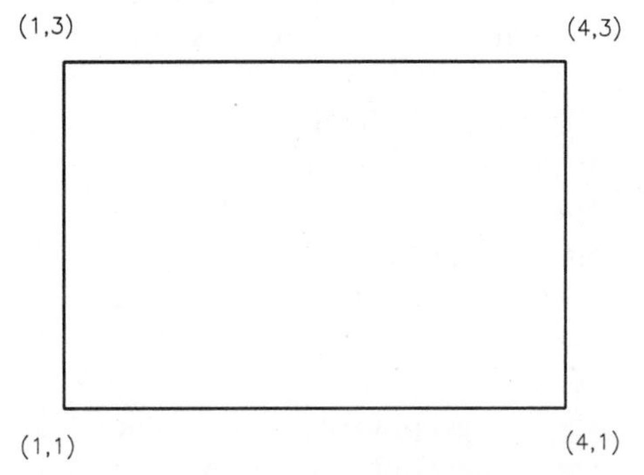

Figure 1-20 *Drawing lines using absolute coordinates*

Example 1

For Figure 1-21, enter the absolute coordinates of the points in the given table. Then draw the figure using absolute coordinates.

Point	Coordinates	Point	Coordinates
1	3,1	5	5,2
2	3,6	6	6,3
3	4,6	7	7,3
4	4,2	8	7,1

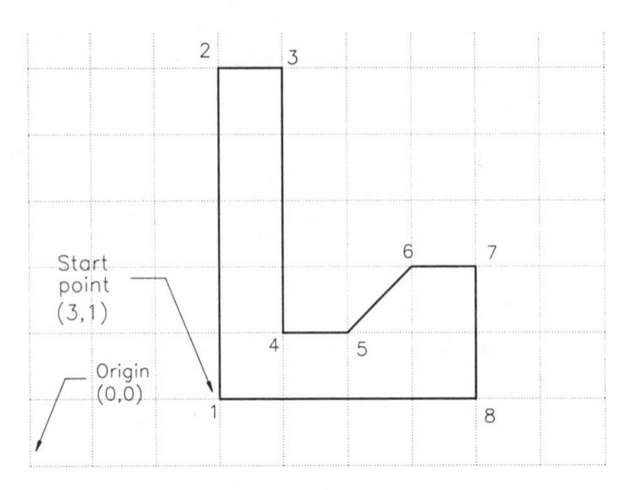

***Figure 1-21** Drawing a figure using absolute coordinates*

Once the coordinates of the points are known you can draw the figure by using AutoCAD LT's **LINE** command. The prompt sequence is:

Command: **LINE** Enter
Specify first point: **3,1** Enter (*Start point.*)
Specify next point or [Undo]: **3,6** Enter
Specify next point or [Undo]: **4,6** Enter
Specify next point or [Close/Undo]: **4,2** Enter
Specify next point or [Close/Undo]: **5,2** Enter
Specify next point or [Close/Undo]: **6,3** Enter
Specify next point or [Close/Undo]: **7,3** Enter
Specify next point or [Close/Undo]: **7,1** Enter
Specify next point or [Close/Undo]: **3,1** Enter
Specify next point or [Close/Undo]: Enter

Exercise 1 *General*

For Figure 1-22, enter the absolute coordinates of the points in the given table, then use those coordinates to draw that same figure.

Point	Coordinates	Point	Coordinates
1	2, 1	6	________
2	________	7	________
3	________	8	________
4	________	9	________
5	________		

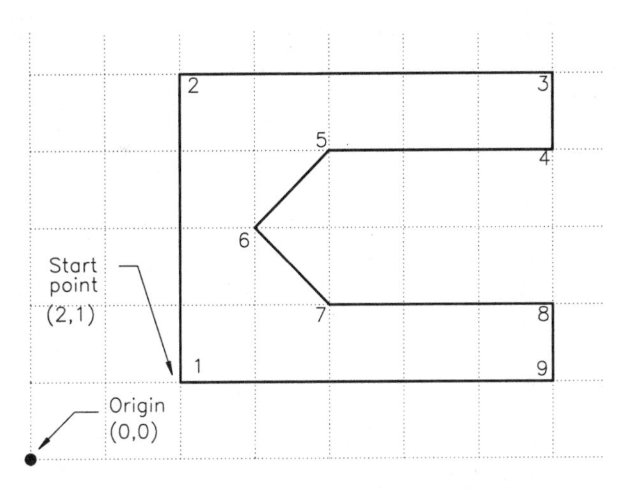

Figure 1-22 Drawing for Exercise 1

Relative Coordinates

In the relative coordinate system, the displacements along the X and Y axes (DX and DY) are measured with reference to the previous point rather than to the origin. In AutoCAD LT the relative coordinate system is designated by the symbol @ and it should precede any relative entry. The following prompt sequence illustrates the use of the relative coordinate system to draw a rectangle that has the lower left corner at point (1,1). The length of the rectangle is 4 units and the width is 3 units (Figure 1-23).

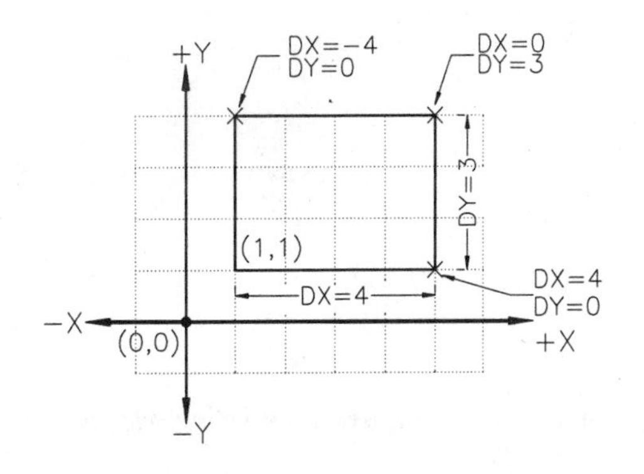

Figure 1-23 Drawing lines using relative coordinates

Command: **LINE** [Enter]
Specify first point: **1,1** [Enter] *(Start point.)*
Specify next point or [Undo]: **@4,0** [Enter] *(Second point DX = 4, DY = 0.)*
Specify next point or [Undo]: **@0,3** [Enter] *(Third point DX = 0, DY = 3.)*
Specify next point or [Close/Undo]: **@-4,0** [Enter] *(Fourth point DX = -4, DY = 0.)*
Specify next point or [Close/Undo]: **@0,-3** [Enter] *(Start point DX = 0, DY = -3.)*
Specify next point or [Close/Undo]: [Enter]

Sign Convention. As just mentioned, in the relative coordinate system the displacements along the X and Y axes are measured with respect to the previous point. Imagine a horizontal line and a vertical line passing through the previous point so that you get four quadrants. If the new point is located in the first quadrant, the displacements DX and DY are both positive. If the new point is located in the third quadrant, the displacements DX and DY are both negative. In other words up or right are positive and down or left are negative.

Example 2

Draw Figure 1-24 using relative coordinates of the points given in the table below.

Point	Coordinates	Point	Coordinates
1	3,1	8	@-1,-1
2	@4,0	9	@-1,1
3	@0,1	10	@-1,0
4	@-1,0	11	@0,-2
5	@1,1	12	@1,-1
6	@0,2	13	@-1,0
7	@-1,0	14	@0,-1

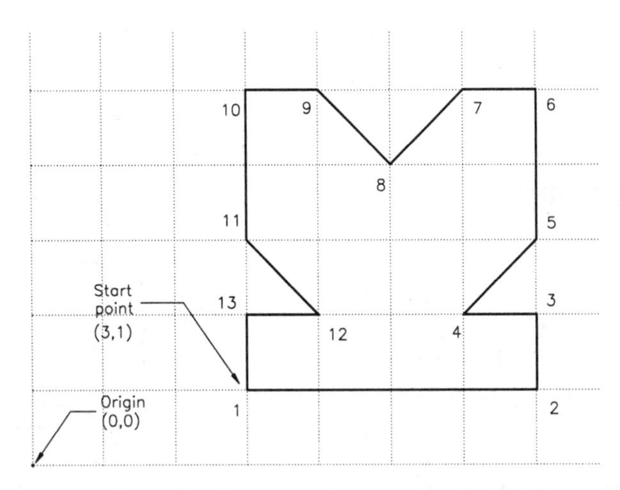

***Figure 1-24** Using relative coordinates with the **LINE** command*

Once you know the coordinates of the points, you can draw the figure by using AutoCAD LT's **LINE** command and entering the coordinates of the points.

Command: **LINE** [Enter]
Specify first point: **3,1** [Enter] *(Start point.)*
Specify next point or [Undo]: **@4,0** [Enter]
Specify next point or [Undo]: **@0,1** [Enter]
Specify next point or [Close/Undo]: **@-1,0** [Enter]
Specify next point or [Close/Undo]: **@1,1** [Enter]
Specify next point or [Close/Undo]: **@0,2** [Enter]
Specify next point or [Close/Undo]: **@-1,0** [Enter]
Specify next point or [Close/Undo]: **@-1,-1** [Enter]
Specify next point or [Close/Undo]: **@-1,1** [Enter]
Specify next point or [Close/Undo]: **@-1,0** [Enter]
Specify next point or [Close/Undo]: **@0,-2** [Enter]
Specify next point or [Close/Undo]: **@1,-1** [Enter]
Specify next point or [Close/Undo]: **@-1,0** [Enter]

Specify next point or [Close/Undo]: **@0,-1** [Enter]
Specify next point or [Close/Undo]: [Enter]

Exercise 2 *General*

For Figure 1-25, enter the relative coordinates of the points in the given table, then use these coordinates to draw the figure.

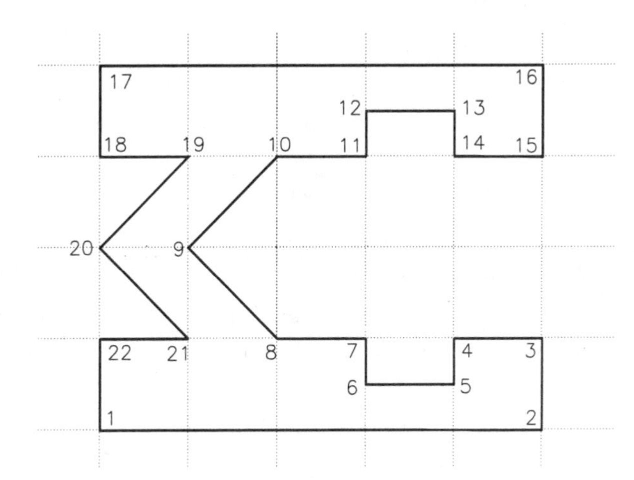

***Figure 1-25** Drawing for Exercise 2*

Point	Coordinates	Point	Coordinates
1	2, 1	12	________
2	________	13	________
3	________	14	________
4	________	15	________
5	________	16	________
6	________	17	________
7	________	18	________
8	________	19	________
9	________	20	________
10	________	21	________
11	________	22	________

Relative Polar Coordinates

In the polar coordinate system, a point can be located by defining both the distance of the point from the current point and the angle that the line between the two points makes with the positive X axis. The prompt sequence to draw a line from a point at 1,1 to a point at a distance of 5 units from the point (1,1), and at an angle of 30 degrees to the X axis, is (Figure 1-26):

Command: **LINE** [Enter]
Specify first point: **1,1** [Enter]
Specify next point or [Undo]: **@5<30** [Enter]

Sign Convention. In the polar coordinate system the angle is measured from the horizontal axis (3 o' clock) as the zero degree baseline. Also, the angle is positive if measured in a counterclockwise direction and negative if measured in a clockwise direction. Here we assume that the default setup of angle measurement has not been changed. For more information about changing the default setup, see "Setting Units" in Chapter 3.

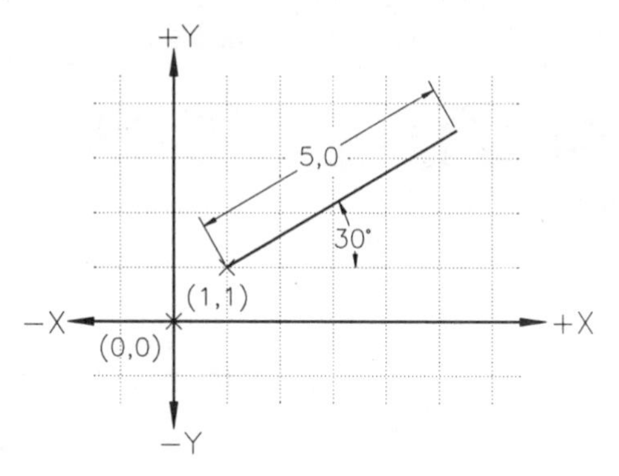

Figure 1-26 Drawing a line using polar coordinates

Example 3

For Figure 1-27, enter the polar coordinates of each point in the table, then generate the drawing. Use absolute coordinates for the start point (1.5, 1.75). The dimensions are shown in the drawing.

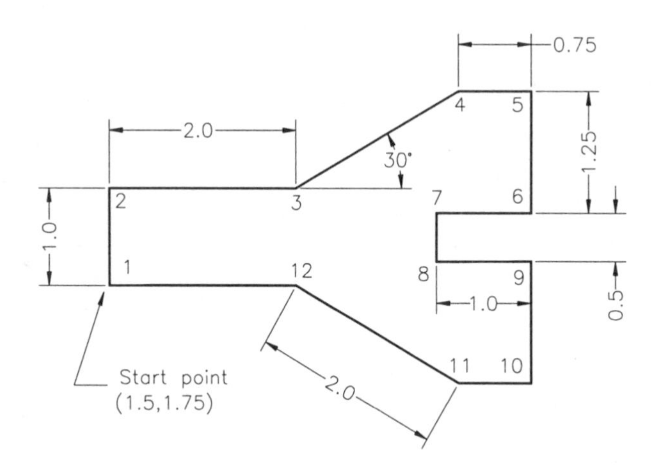

Figure 1-27 Drawing for Example 3

Point	Coordinates	Point	Coordinates
1	1.5,1.75	7	@1.0<180
2	@1.0<90	8	@0.5<270
3	@2.0<0	9	@1.0<0
4	@2.0<30	10	@1.25<270
5	@0.75<0	11	@0.75<180
6	@1.25<-90 (or <270)	12	@2.0<150

Once you know the coordinates of the points, you can generate the drawing by using AutoCAD LT's **LINE** command and entering the coordinates of the points.

Command: **LINE** [Enter]
Specify first point: **1.5,1.75** [Enter] *(Start point.)*
Specify next point or [Undo]: **@1<90** [Enter]
Specify next point or [Undo]: **@2.0<0** [Enter]
Specify next point or [Close/Undo]: **@2<30** [Enter]
Specify next point or [Close/Undo]: **@0.75<0** [Enter]
Specify next point or [Close/Undo]: **@1.25<-90** [Enter]
Specify next point or [Close/Undo]: **@1.0<180** [Enter]
Specify next point or [Close/Undo]: **@0.5<270** [Enter]
Specify next point or [Close/Undo]: **@1.0<0** [Enter]
Specify next point or [Close/Undo]: **@1.25<270** [Enter]
Specify next point or [Close/Undo]: **@0.75<180** [Enter]
Specify next point or [Close/Undo]: **@2.0<150** [Enter]
Specify next point or [Close/Undo]: **C** [Enter] *(Joins the last point with the first point.)*

Exercise 3 *General*

Draw the object shown in Figure 1-28 using the absolute, relative, and polar coordinate systems to locate the points. Do not draw the dimensions; they are for reference only. Assume the missing dimensions.

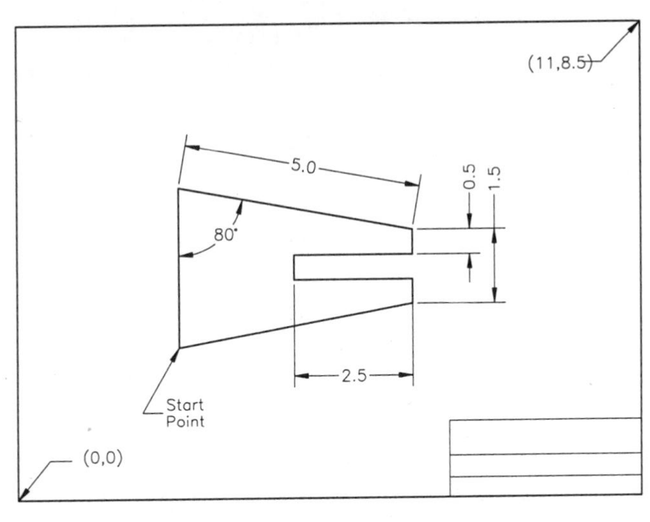

Figure 1-28 *Drawing for Exercise 3*

Direct Distance Entry

You can draw a line by specifying the length of the line and its direction, using the Direct Distance Entry (Figure 1-29). The direction is determined by the position of the cursor, and the length of the line is entered from the keyboard. If Ortho is on, you can draw lines along the X or Y axis by specifying the length of line and positioning the cursor along ortho direction. You can also use it with other draw commands like **RECTANGLE**. You can also use the Direct Distance Entry with polar tracking and **SNAPANG.** For example, if **SNAPANG** is 45 degrees and ortho is off, you can draw a line at 45 or 135 degrees direction by positioning the cursor and entering the distance from the keyboard. Similarly, if the polar tracking is on, you can position the cursor at the predefined angles and then enter the length of the line from the keyboard.

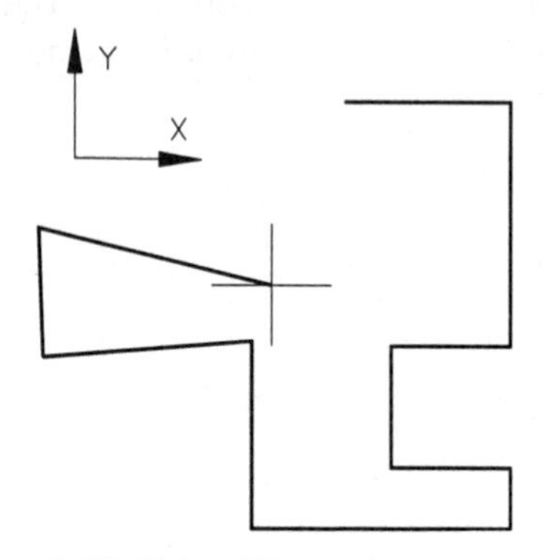

Figure 1-29 *Using Direct Distance Entry to draw lines*

Command: **LINE** [Enter]
Specify first point: *Start point.*
Specify next point or [Undo]: *Position the cursor and then enter distance.*
Specify next point or [Undo]: *Position the cursor and then enter distance.*

Example 4

In this example you will draw the object as shown in Figure 1-30, using Direct Distance Entry. The starting point is 2,1.

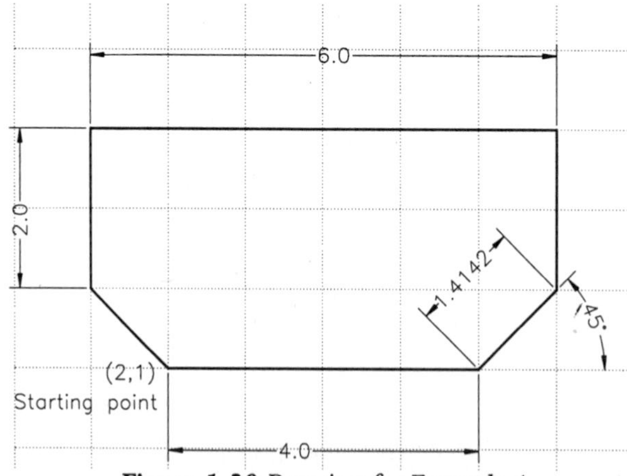

Figure 1-30 *Drawing for Example 4*

Before you invoke the **LINE** command, you should turn on polar tracking. This will make it easier to specify the direction of lines. To turn polar tracking on, choose the **Polar Tracking** button in the status bar. You can also turn polar tracking on or off while you are in a command.

As you move the cursor, AutoCAD LT displays a dotted line when the position of the cursor matches one of the predefined angles for polar tracking. The following is the Command prompt sequence for drawing the object in Figure 1-30:

Command: **LINE** [Enter]
Specify first point: **2,1** [Enter]
Specify next point or [Close/Undo]: **4** [Enter] *(Move the cursor horizontally and enter the length of the line, 4, from the keyboard.)*
Specify next point or [Close/Undo]: **1.4142** [Enter] *(Position the cursor in a 45-degree direction and enter 1.4142.)*
Specify next point or [Close/Undo]: **2** [Enter] *Move the cursor up vertically, then enter 2.*
Specify next point or [Close/Undo]: **6** [Enter] *Move the cursor left horizontally then enter 6.*
Specify next point or [Close/Undo]: **2** [Enter] *Move the cursor down vertically then enter 2.*
Specify next point or [Close/Undo]: **C** [Enter]

Exercise 4 — *General*

Use the direct distance entry method to draw a parallelogram. The base of the parallelogram = 4 units, side = 2.25 units, and the angle = 45 degrees. Draw the same parallelogram using absolute, relative, and polar coordinates. Note the differences and the advantage of using Direct Distance Entry.

ERASING OBJECTS

Toolbar:	Modify > Erase
Menu:	Modify > Erase
Command:	ERASE

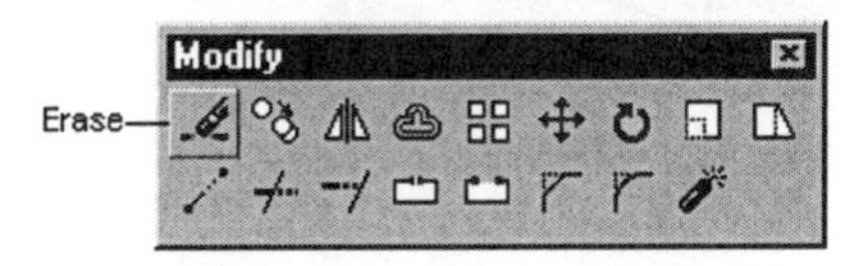

***Figure 1-31** Invoking the **ERASE** command from the **Modify** toolbar*

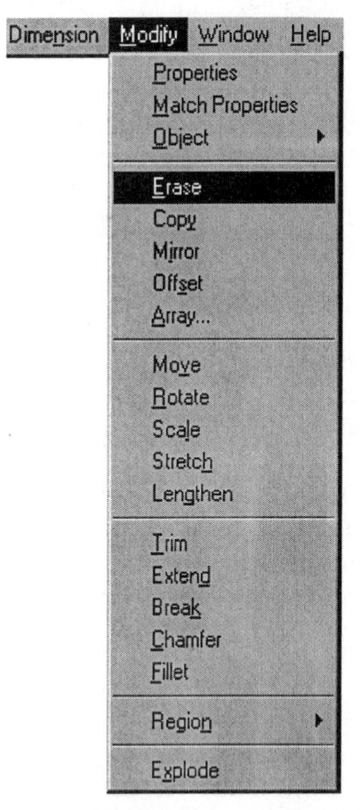

***Figure 1-32** Invoking the **ERASE** command from the **Modify** menu*

After drawing some objects you may want to erase some of them from the screen. To erase you can use AutoCAD LT's **ERASE** command. This command is used exactly the same way as an eraser is used in manual drafting to remove unwanted information. When you invoke the **ERASE** command, a small box known as the pick box replaces the screen cursor. To erase an object, move the **pick box** so it touches the object. You can select the object by pressing the pick button of your pointing device (Figure 1-33). AutoCAD LT confirms the selection by changing the selected objects into dashed lines, and the **Select objects** prompt returns. You can either continue selecting objects or press ENTER to terminate the object selection process and erase the selected objects. If you are entering the command from the keyboard, you can type **E** or **ERASE**. The following is the prompt sequence:

Command: **ERASE** [Enter]
Select objects: *Select first object.*
Select objects: *Select second object.*
Select objects: [Enter]

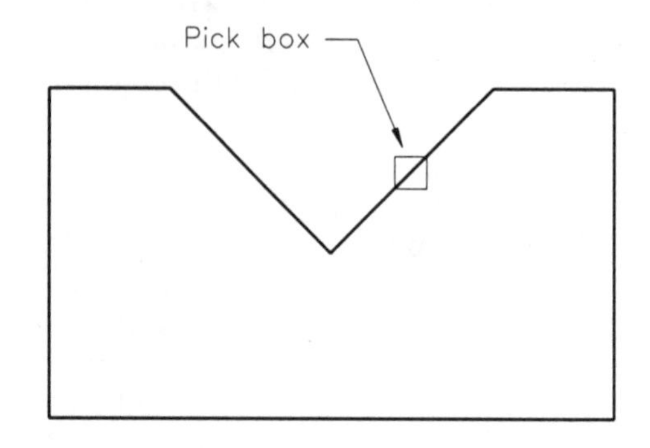

Figure 1-33 *Selecting objects by positioning the pick box at the top of the object and then pressing the pick button on the pointing device*

If you enter All at the **Select objects** prompt, AutoCAD LT will erase all objects in the drawing, even if the objects are outside the screen display area.

Command: **ERASE** [Enter]
Select objects: **All** [Enter]

You can also first select the objects to be erased from the drawing and then right-click on the drawing area to display the shortcut menu. From here, you can choose the **Erase** option.

CANCELING AND UNDOING A COMMAND

If you are in a command and you want to cancel or get out of that command, press the ESC (Escape) key on the keyboard.

Command: **ERASE** [Enter]
Select objects: *Press ESC (Escape) to cancel the command.*

Similarly, sometimes you unintentionally erase some object from the screen. When you discover such an error, you can correct it by restoring the erased object by means of the **OOPS** command. The **OOPS** command restores objects that have been accidentally erased by the previous **ERASE** command, Figure 1-34. You can also use the **U** (Undo) command to undo the last command.

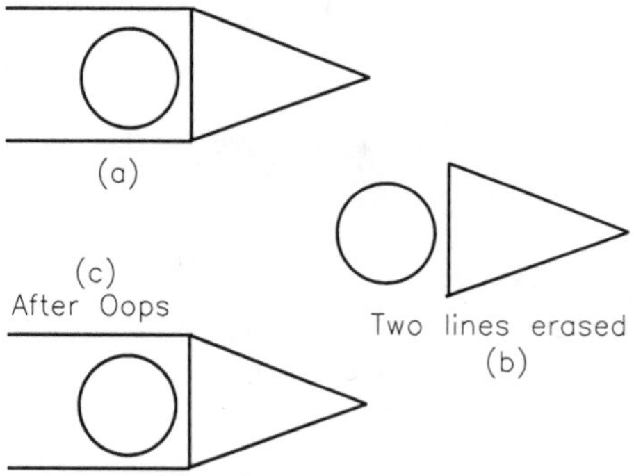

(a)
(c) After Oops
Two lines erased (b)

Figure 1-34 *Use of the* ***OOPS*** *command*

Command: **OOPS** [Enter] *(Restores the last erased objects.)*
Command: **U** [Enter] *(Undoes the last command.)*

CREATING SELECTION SETS

One of the ways to select objects is to select them individually, which can be time-consuming if you have a number of objects to edit. This problem can be solved by creating a selection set that enables you to select several objects at a time. The selection set options can be used with those commands that require object selection, such as **MOVE** and **ERASE**. There are many selection options for creating a selection set, like **All, Last,** and **Add**. At this point we will explore two options: **Window** and **Crossing.** The remaining options are discussed in the next chapter.

The Window Option

This option is used to select an object or group of objects by drawing a box or window around them. The objects to be selected should be completely enclosed within the window; those objects that lie partially inside the boundaries of the window are not selected. You can select the **Window** option by typing W at the **Select objects:** prompt. AutoCAD LT will prompt you to select the two opposite corners of the window. After selecting the first corner, you can select the other corner by moving the cursor to the desired position and specifying the particular point. As you move the cursor, a box or window is displayed that changes in size as you move the cursor. The objects selected by the **Window** option are displayed as dashed objects, i.e. highlighted (Figure 1-35). The following prompt sequence illustrates the use of the **Window** option with the **ERASE** command:

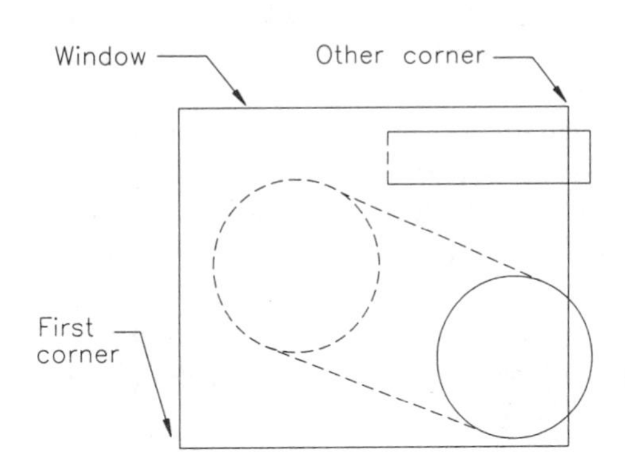

Figure 1-35 *Selecting objects using the* ***Window*** *option*

Command: **ERASE** [Enter]
Select objects: **W** [Enter]
Specify first corner: *Select the first corner.*
Specify opposite corner: *Select the second corner.*
Select objects: [Enter]

You can also select the **Window** option by selecting a blank point on the screen at the **Select objects:** prompt. This is automatically taken as the first corner of the window. Dragging the cursor to the right will display a window. After getting all the objects to be selected inside this window, you can specify the other corner with your pointing device. The objects that are completely enclosed within the window will be selected and highlighted. The following is the prompt sequence for automatic window selection with the **ERASE** command:

Command: **ERASE** [Enter]
Select objects: *Select a blank point as the first corner of the window.*
Specify opposite corner: *Drag the cursor to the right to select the other corner of the window.*
Select objects: [Enter]

The Crossing Option

This option is used to select an object or group of objects by creating a box or window around them. The objects to be selected should be touching the window boundaries or completely enclosed within the window. You can invoke the **Crossing** option by entering **C** at the **Select objects:** prompt. After you choose the **Crossing** option, AutoCAD LT prompts you to select the first corner at the **Specify first corner** prompt. Once you have selected the first corner, a box or window made of dashed lines is displayed. By moving the cursor you can change the size of the crossing box, hence putting the objects to be selected within (or touching) the box. Here you can select the other corner. The objects selected by the **Crossing** option are highlighted by displaying them as dashed objects (Figure 1-36). The following prompt sequence illustrates the use of the **Crossing** option with the **ERASE** command:

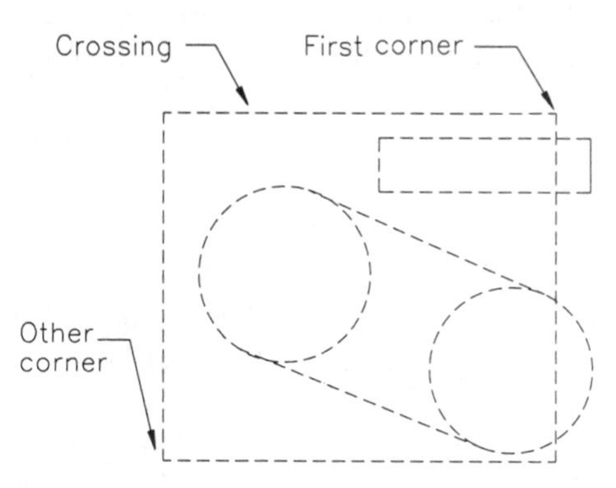

Figure 1-36 *Selecting objects using the Crossing option*

Command: **ERASE** [Enter]
Select objects: **C** [Enter]
Specify first corner: *Select the first corner of the crossing window.*
Specify opposite corner: *Select the other corner of the crossing window.*
Select objects: [Enter]

You can also select the **Crossing** option automatically by selecting a blank point on the screen at the **Select objects:** prompt and dragging the cursor to the left. The blank point you selected becomes the first corner of the crossing window and AutoCAD LT will then prompt you to select the other corner. As you move the cursor, a box or window made of dashed lines is displayed. The objects that are touching or completely enclosed within the window will be selected. The objects selected by the **Crossing** option are highlighted by being displayed as dashed objects. The prompt sequence for automatic crossing selection is:

Command: **ERASE** [Enter]
Select objects: *Select a blank point as the first corner of the crossing window.*
Specify opposite corner: *Drag the cursor to the left to select the other corner of the crossing window.*
Select objects: [Enter]

MOVE COMMAND

Toolbar:	Modify > Move
Menu:	Modify > Move
Command:	MOVE

Sometimes objects may not be located where they should be. In such situations you can use the **MOVE** command. This command lets you move the object (or objects) from their present location to a new one. This shifting does

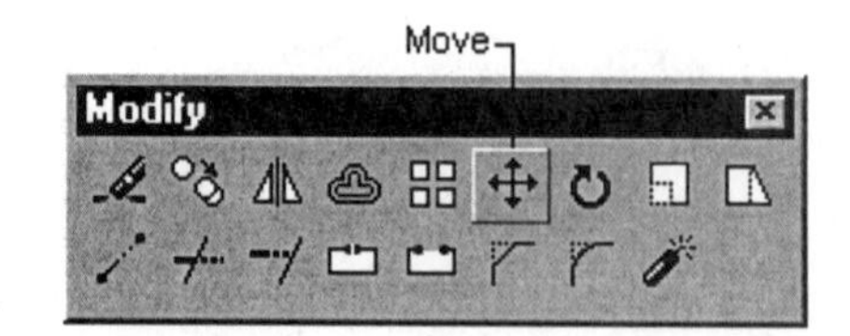

Figure 1-37 *Invoking the **MOVE** command from the **Modify** toolbar*

not change the size or orientation of the objects. After you enter this command, AutoCAD LT will prompt you to select the objects to be moved. You can select the objects individually or use any selection techniques discussed earlier (Window, Crossing, etc.). Next, AutoCAD LT prompts you for the base point. This is any point on or next to the object. It is better to select a point on the object, a corner (if the object has one), or the center of a circle. The next prompt asks you for a second point of displacement. This is the location where you want to move the object. The selected objects are moved from the specified base point to the second point of displacement (Figure 1-39). The prompt sequence is as follows:

> Command: **MOVE** Enter
> Select objects: *Choose objects individually or use the selection set.*
> Select objects: Enter
> Specify base point or displacement: *Specify any point on or near the object.*
> Specify second point of displacement or <use first point as displacement>: *Select the new location by specifying a point on the screen.*

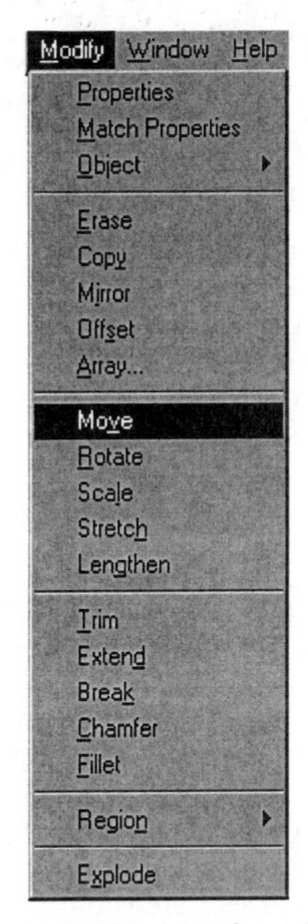

***Figure 1-38** Invoking the **MOVE** command from the **Modify** menu*

We can also move the selected objects by using the short cut menu by clicking the right button. If null response is given at **Specify second point of displacement or <use first point as displacement>** prompt by pressing enter, AutoCAD LT interprets the first point as relative value of displacement in X,Y, and Z directions respectively.

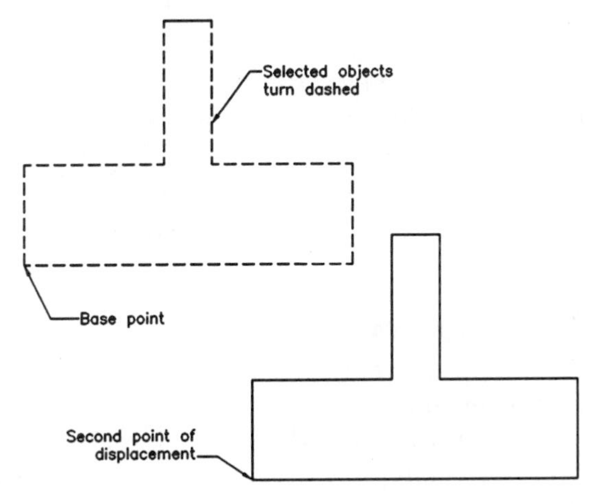

***Figure 1-39** Moving an object using the **MOVE** command*

DRAWING CIRCLES

Toolbar:	Draw > Circle
Menu:	Draw > Circle
Command:	CIRCLE

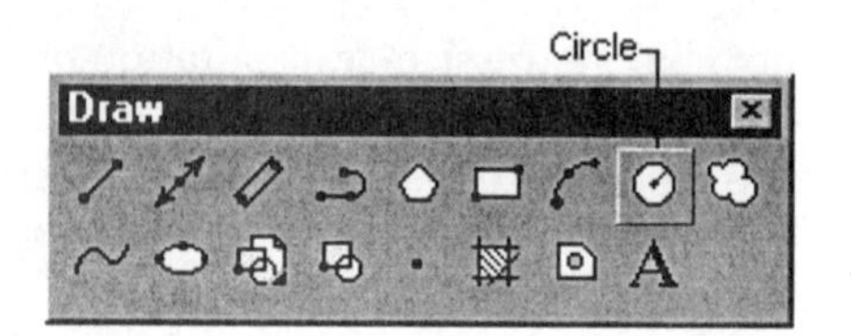

Figure 1-40 Invoking the ***CIRCLE*** *command from the* ***Draw*** *toolbar*

To draw a circle you can use the AutoCAD LT's **CIRCLE** command. The following is the prompt sequence for the **CIRCLE** command:

Command: **CIRCLE** [Enter]
Specify center point for circle or [3P/2P/Ttr(tan tan radius)]:

The different options of the **CIRCLE** command follow.

The Center and Radius Option

In this option you can draw a circle by defining the center and the radius of the circle, Figure 1-42. After entering the **CIRCLE** command, AutoCAD LT will prompt you to enter the center of the circle, which can be selected by specifying a point on the screen or by entering the coordinates of the center point. Next, you will be prompted to enter the radius of the circle. Here you can accept the default value, enter a new value, or select a point on the circumference of the circle to specify the radius. The following is the prompt sequence for drawing a circle with a center at 3,2 and a radius of 1 unit:

Command: **CIRCLE** [Enter]
Specify center point for circle or [3P/2P/Ttr(tan tan radius): **3,2** [Enter]
Specify radius of circle or [Diameter] <current>: **1** [Enter] (or @1<0, or @1,0)

Note

You can also set the radius by assigning a value to the ***CIRCLERAD*** *system variable. The value you assign becomes the default value for radius.*

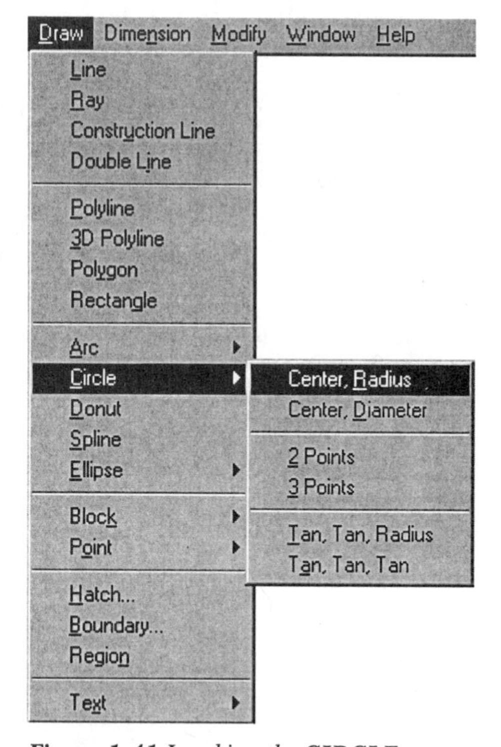

Figure 1-41 Invoking the ***CIRCLE*** *command from the* ***Draw*** *menu*

The Center and Diameter Option

In this option you can draw a circle by defining the center and diameter of the circle. After invoking the **CIRCLE** command, AutoCAD LT prompts you to enter the center of the circle, which can be selected by specifying a point on the screen or by entering the coordinates of the center point. Next, you will be prompted to enter the radius of the circle. At this prompt enter D. After this you will be prompted to enter the diameter of the circle. For entering the diameter you can accept the default value, enter a new value, or drag the circle to the desired diameter and select a point. If you use a menu option to select the **CIRCLE** command with the **Diameter** option, the menu automatically enters the **Diameter** option and prompts for the diameter after you specify the center. The following is the prompt sequence for drawing a circle with the center at (2,3) and a diameter of 2 units (Figure 1-43):

Command: **CIRCLE** Enter
Specify center point for circle or [3P/2P/Ttr(tan tan radius): **2,3** Enter
Specify radius of circle or [Diameter] <current>: **D** Enter
Specify diameter of circle <current>: **2** Enter

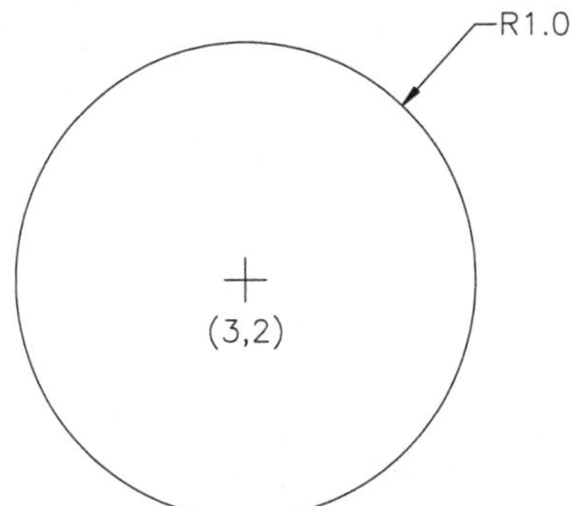

Figure 1-42 *Drawing a circle using the Center and Radius option*

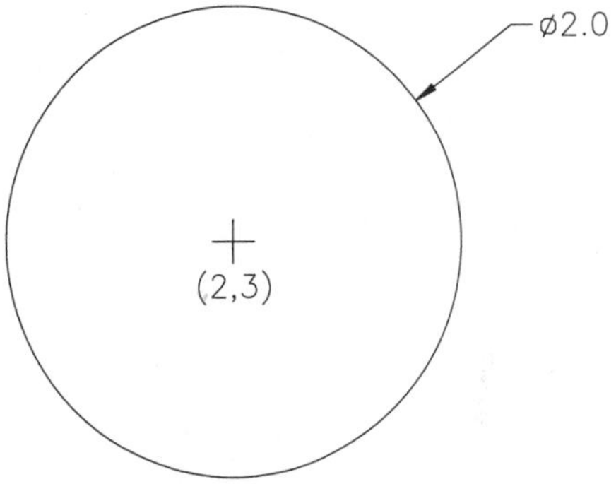

Figure 1-43 *Drawing a circle using the Center and Diameter option*

The Two-Point Option

You can also draw a circle using the **Two-Point** option. In this option AutoCAD LT lets you draw the circle by specifying the two endpoints of the circle's diameter. For example, if you want to draw a circle that passes through the points (1,1) and (2,1), you can use the **CIRCLE** command with **2P** option, as shown in the following example (Figure 1-44):

Command: **CIRCLE** Enter
Specify center point for circle or [3P/2P/Ttr(tan tan radius): **2P** Enter
Specify first end point of circle's diameter: **1,1** Enter
Second end point of circle's diameter: **2,1** Enter *(You can also use polar or relative coordinates.)*

The Three-Point Option

For drawing a circle, you can also use the **Three-Point** option by defining three points on the circumference of the circle. The three points may be entered in any order. To draw a circle that passes through the points 3,3, 3,1 and 4,2 (Figure 1-45), the prompt sequence is:

Command: **CIRCLE** Enter
Specify center point for circle or [3P/2P/Ttr(tan tan radius)]: **3P** Enter
Specify first point on circle: **3,3** Enter
Specify second point on circle: **3,1** Enter
Specify third point on circle: **4,2** Enter

You can also use **relative coordinates** to define the points:

Command: **CIRCLE** [Enter]
Specify center point for circle or [3P/2P/Ttr(tan tan radius)]: **3P** [Enter]
Specify first point on circle: **3,3** [Enter]
Specify second point on circle: **@0,-2** [Enter]
Specify third point on circle: **@1,1** [Enter]

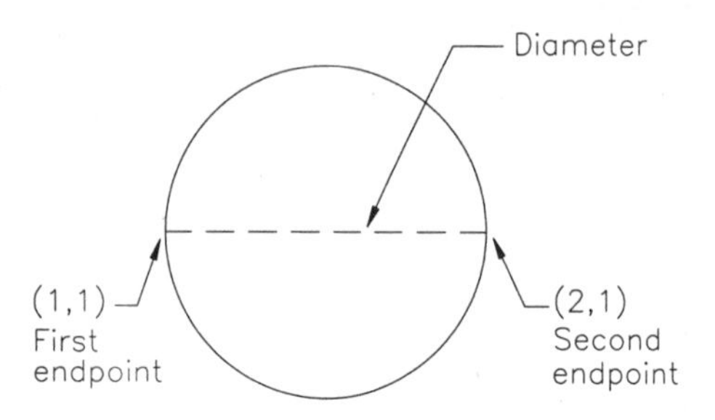

Figure 1-44 Drawing a Circle using the Two-Point option

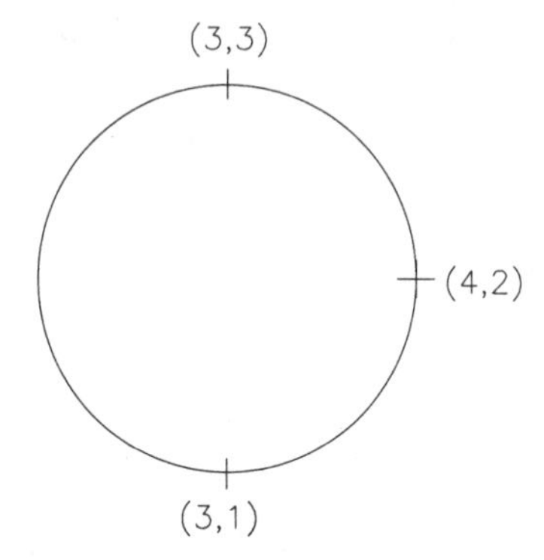

Figure 1-45 Drawing a Circle using the Three-Point option

The Tangent Tangent Radius Option

A tangent is an object (line, circle, or arc) that contacts the circumference of a circle at only one point. In this option AutoCAD LT uses the Tangent object snap to locate two tangent points on the selected objects that are to be tangents to the circle. Then you have to specify the radius of the circle. The prompt sequence for drawing a circle using the **Ttr** option:

Command: **CIRCLE** [Enter]
Specify center point for circle or [3P/2P/Ttr(tan tan radius)]: **T** [Enter]
Specify point on object for first tangent of circle: *Select first line, circle, or arc.*
Specify point on object for second tangent of circle: *Select second line, circle, or arc.*
Specify radius of circle <current>: **0.75** [Enter]

In Figures 1-46 through 1-49, the dotted circles represent the circles that are drawn by using the **Ttr** option. The circle AutoCAD LT actually draws depends on how you select the objects that are to be tangent to the new circle. The figures show the effect of selecting different points on the objects. The dashed circles represent the circles that are drawn using the **Ttr** option. If you specify too small or large a radius, you may get unexpected results or a "Circle does not exist" prompt.

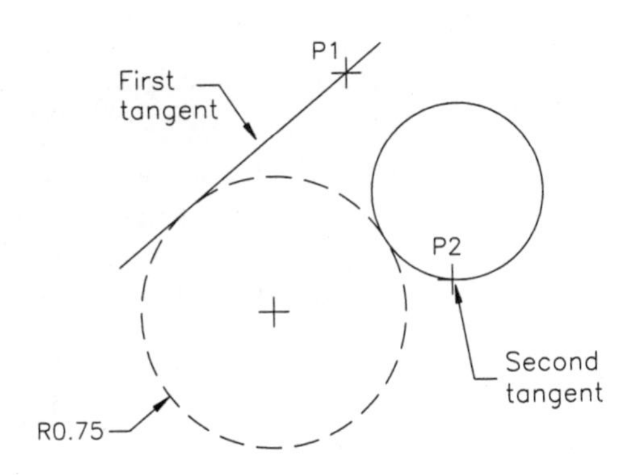

***Figure 1-46** Tangent, tangent, radius (Ttr) option*

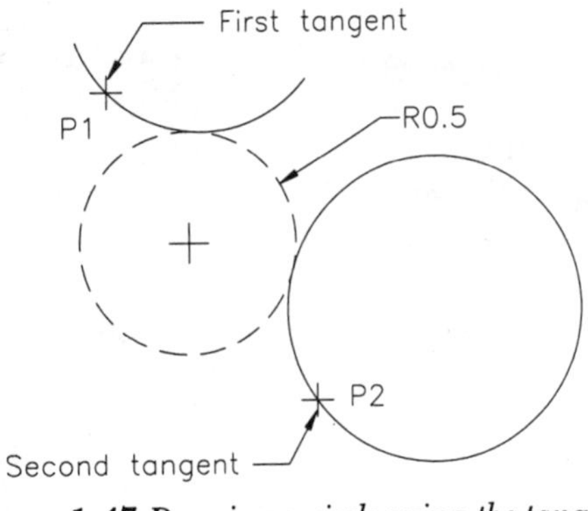

***Figure 1-47** Drawing a circle using the tangent, tangent, radius (Ttr) option*

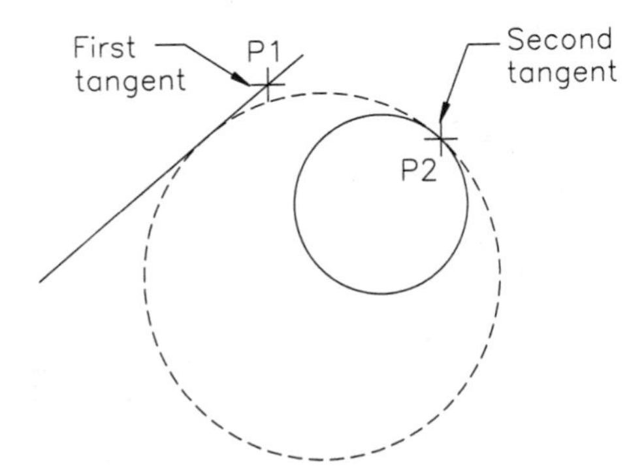

***Figure 1-48** Tangent, tangent, radius (Ttr) option*

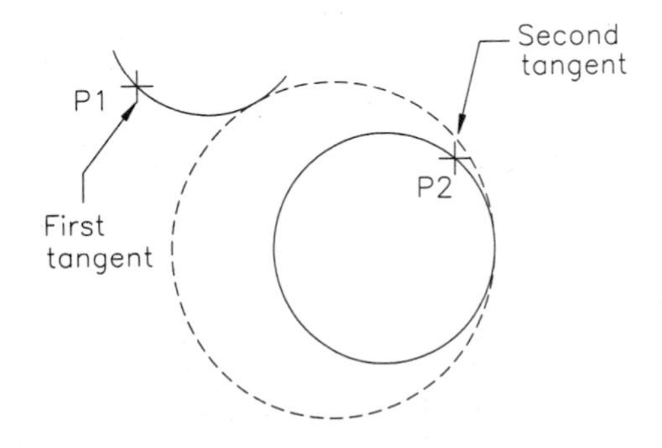

***Figure 1-49** Tangent, tangent, radius (Ttr) option*

The Tangent, Tangent, Tangent Option

You can invoke this option from the menu bar. This option is a modification of the **Three-Point** option. In this option AutoCAD LT uses the Tangent osnap to locate three points on three selected objects to which the circle is drawn tangent. The following is the prompt sequence for drawing a circle using the **Tan, Tan, Tan** option (Figure 1-50):

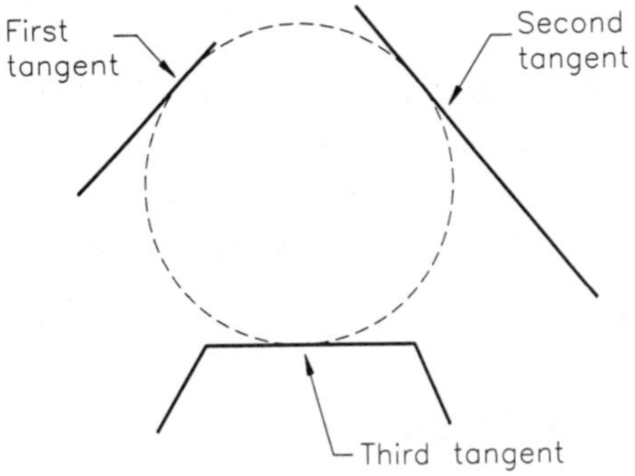

Command: **CIRCLE** [Enter]
Specify center point for circle or [3P/2P/Ttr (tan tan radius): *Select Tan, Tan, Tan option from the **Draw** menu.*
_ 3P Specify first point on circle: _tan to *Select the first object.*
Specify second point on circle: _tan to *Select the second object.*

***Figure 1-50** Drawing a circle using the Tan, Tan, Tan option*

Specify third point on circle: _tan to *Select the third object.*

Exercise 5 — *Mechanical*

Generate Figure 1-51 using different options of the **LINE** and **CIRCLE** commands. Use absolute, relative, or polar coordinates for drawing the triangle. The vertices of the triangle will be used as the center of the circles. The circles can be drawn using the **Center and Radius**, **Center and Diameter**, or **TTT** option. (Height of triangle = 2.25 x sin 60 = 1.949.) Do not draw the dimensions; they are for reference only.

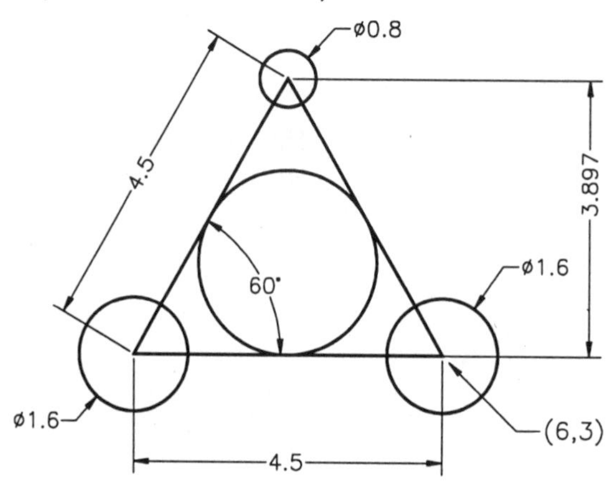

Figure 1-51 *Drawing for Exercise 5*

BASIC DISPLAY COMMANDS

Drawing in AutoCAD LT is much simpler than manual drafting in many ways. Sometimes while drawing, it is very difficult to see and alter minute details. In AutoCAD LT, you can overcome this problem by viewing only a specific portion of the drawing. This is done using the **ZOOM** command, which lets you enlarge or reduce the size of the drawing displayed on the screen. We will introduce here some of the drawing display commands, like **REDRAW**, **ZOOM**, and **PAN**. A detailed explanation of these commands and other display options appears in Chapter 5, Controlling Drawing Display and Creating Text.

REDRAW Command

The **REDRAW** command redraws the screen, thereby removing the cross marks that appear when a point is specified on the screen. These marks, known as blip marks or **blips**, indicate the points you have selected (points specified).

Command: **REDRAW** [Enter]

ZOOM Command

Toolbar:	Zoom toolbar, Standard > Zoom flyout
Menu:	View > Zoom
Command:	ZOOM

The **ZOOM** command enlarges or reduces the view of the drawing on screen, but it does not affect the actual size of drawing. After the **ZOOM** command has been invoked, different options can be used to obtain desired display. If you use a menu, it issues the appropriate option at the initial **ZOOM** prompt. The following is the prompt sequence of the **ZOOM** command and some of the **ZOOM** options:

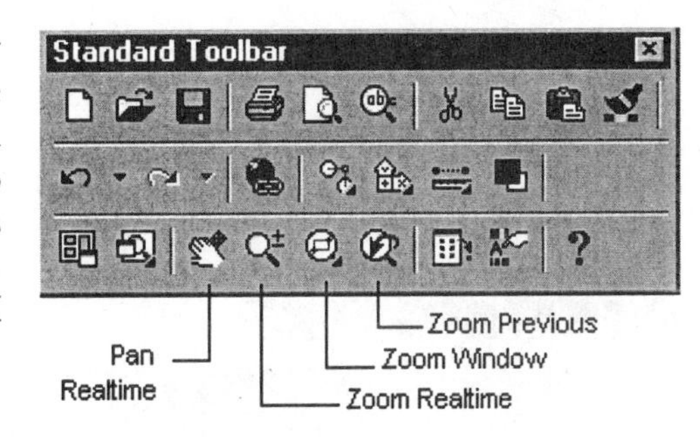

Figure 1-52 Selecting ***Zoom*** *options from the* ***Standard*** *toolbar*

Command: **ZOOM** Enter
Specify corner of window, enter a scale factor (nX or nXP), or [All/Center/Dynamic/Extents/Previous/Scale/Window]<realtime>:

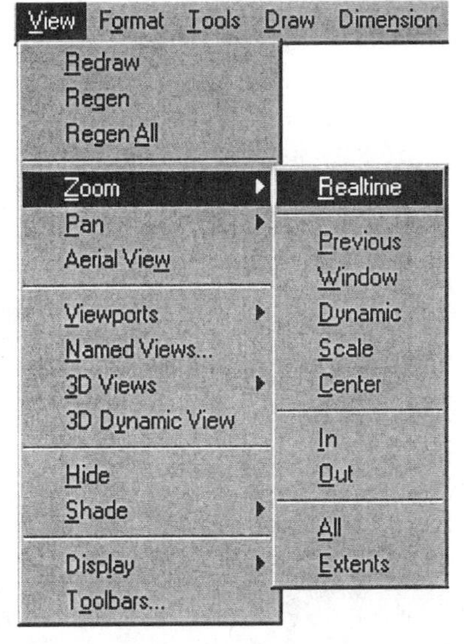

Figure 1-53 Invoking the ***ZOOM*** *command from the* ***View*** *menu*

Realtime Zooming

You can use the **Realtime Zoom** to zoom in and zoom out interactively. To zoom in, invoke the command, then hold the pick button down and move the cursor up. If you want to zoom in further, bring the cursor down, Specify a point and move the cursor up. Similarly, to zoom out, hold the pick button down and move the cursor down. Realtime zoom is the default setting for the **ZOOM** command. Pressing ENTER after entering the **ZOOM** command automatically invokes the realtime zoom.

Window Option

This is the most commonly used option of the **ZOOM** command. It lets you specify the area you want to zoom in on by letting you specify two opposite corners of a rectangular window. The center of the specified window becomes the center of the new display screen. The area inside the window is magnified in size to fill the display as completely as possible. The points can be specified by selecting them with the help of the pointing device or by entering their coordinates.

Previous Option

While working on a complex drawing, you may need to zoom in on a portion of the drawing to edit some minute details. When you have completed the editing you may want to return to the previous view. This can be done using the **Previous** option of the **ZOOM** command. AutoCAD LT remembers the last ten views which can be recalled by using the **Previous** option.

Panning in Realtime

You can use the **Realtime Pan** to pan the drawing interactively. To pan a drawing, invoke the command and then hold the pick button down and move the cursor in any direction. When you select the realtime pan, AutoCAD LT displays an image of a hand indicating that you are in PAN mode. When in realtime zoom or pan, you can right-click to display the shortcut menu, Figure 1-54. You can use the shortcut menu to select **Zoom**, **Pan**, **Exit**, or other zoom options.

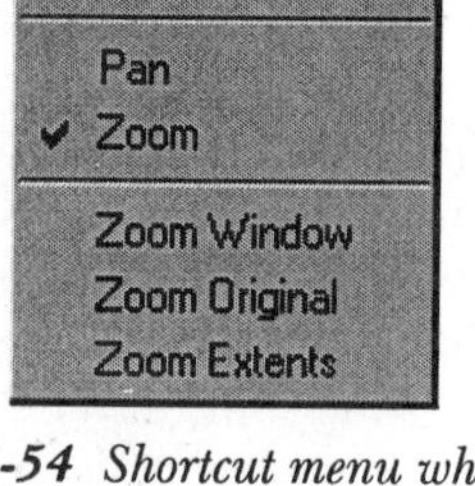

*Figure 1-54 Shortcut menu when in realtime **ZOOM** and **PAN** commands*

CREATING TEXT

Menu:	Draw > Text > Single-Line Text
Command:	TEXT

The **TEXT** command lets you write several lines of text on a drawing. The prompt sequence is:

Command: **TEXT** [Enter]
Current text style: "current" Text height: current
Specify start point of text or [Justify/Style]: *Specify the starting point of text.*
Specify height <current>: *Enter the text height.*
Specify rotation angle of text <0>:
Enter text: *Enter the text.*
Enter text: *Enter the second line of text.*
Enter text: [Enter]

You can use the BACKSPACE key to edit the text on the screen while you are writing the text. The text commands are discussed in detail in Chapter 5.

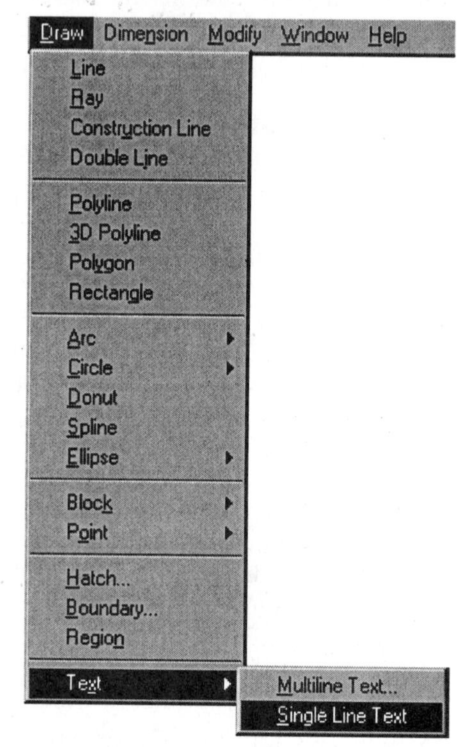

*Figure 1-55 Invoking the **TEXT** command from the **Draw** menu*

PLOTTING DRAWINGS*

Toolbar:	Standard toolbar > Plot
Menu:	File > Plot
Command:	PLOT

Drawings can be plotted by using the **PLOT** command. AutoCAD LT will display the **Plot** dialog box (Figure 1-58) when you enter or select the **PLOT** command.

Command: **PLOT** [Enter]

The values in this dialog box are the ones that were set during the configuring of AutoCAD LT. If the displayed values conform to your requirements, you can start plotting without making any changes. If necessary, you can make changes in the default values according to your plotting requirements.

Basic Plotting

In this section you will learn how to set up the basic plotting parameters. Later, you will learn about the advance option, that allows you to plot according to your plot drawing specifications. Basic plotting involves selecting the correct output device (plotter), specifying the area to plot, selecting paper size, specifying the plot origin, orientation, and the plot scale.

Example 5

You will plot the drawing of Example 4 (Figure 1-56) using the **Window** option to select the area to plot. Assume that AutoCAD LT is configured for two output devices, System Printer and HP 7475A.

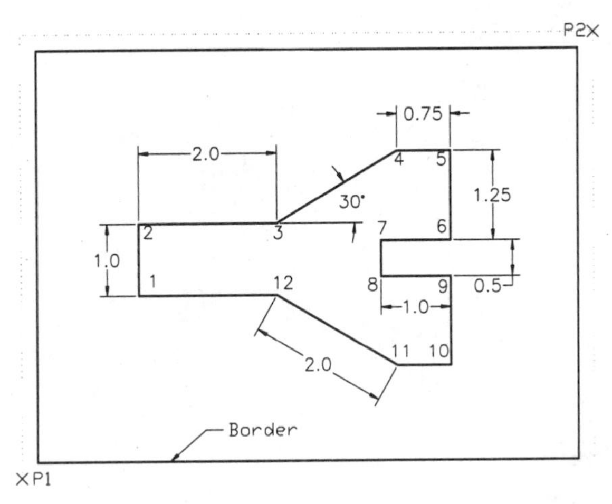

Figure 1-56 Drawing for Example 5

Step 1

Invoke the **Plot** dialog box from the **Standard** toolbar, or the **File** menu (Choose **Plot**) or by entering **PLOT** at the Command prompt.

Step 2

Choose the **Plot Device** tab, Figure 1-57. In the **Plotter Configuration** area, select the name of the device by choosing the drop-down arrow button in the **Name:** edit box, to display the drop-down list of the currently configured devices. Related information about the selected device is displayed below this box. For example, if you have configured AutoCAD LT for two output devices, System Printer and HP 7475A, AutoCAD LT will display these names in the **Name:** drop-down list. If you want to plot the drawing on the System Printer, select System Printer (if it is not already selected).

Step 3

Now choose the **Plot Settings** tab and here, choose the **Window** button (located in the **Plot area**) in the **Plot** dialog box. You can enter the first and second corners of the area you want to plot by specifying points on the screen. When you choose the **Window** button, the dialog box will temporarily disappear and the drawing will appear on the screen. Now, select the first and second corners (Points P1 and P2), specifying the plot area (the area you want to plot). Once you have defined the two corners, the **Plot** dialog box will reappear.

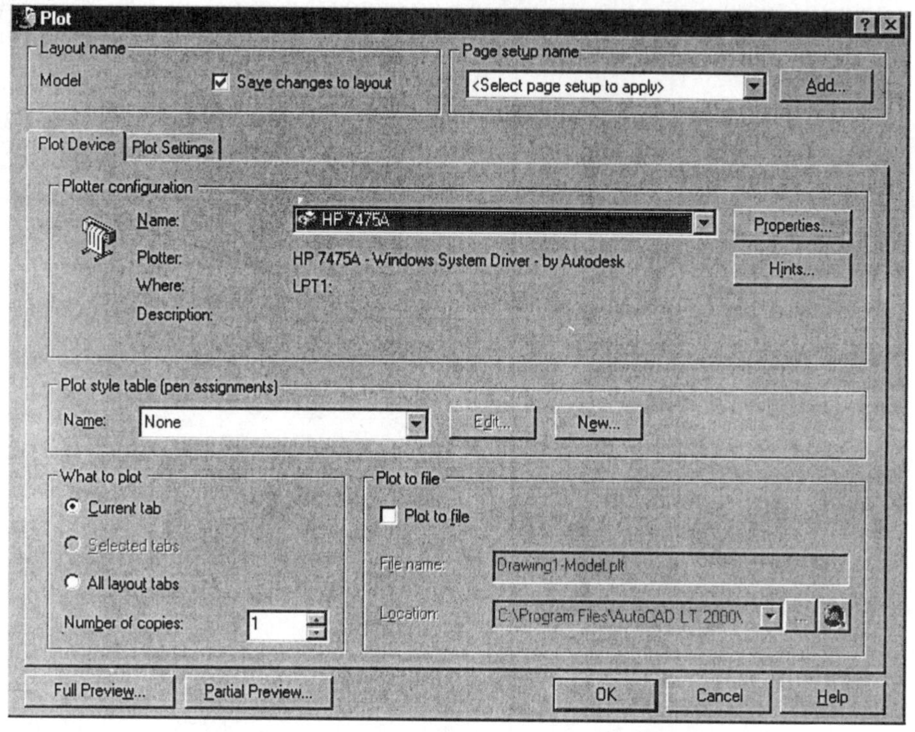

***Figure 1-57 Plot** dialog box with the **Plot Device** tab*

Step 4

To set the size for the plot, select from the **Paper size:** edit box drop-down list in the **Paper size and paper units** area (Figure 1-58), which lists all the plotting sizes that the present plotter can support. You can select any one of the sizes listed in the dialog box or specify a size (width and height) of your own through the **Plotter Manager**. (This option is discussed later in Chapter 15, Plotting Drawings and Draw Commands.) Once you select a size, you can also select the orientation by choosing either the **Landscape** or **Portrait** radio buttons under **Drawing orientation**. The sections in the **Plot** dialog box pertaining to paper size and orientation are automatically revised to reflect the new paper size and orientation. In this example you will specify Paper size A (8.5 by 11 inches).

Step 5

You can also modify values for **Plot offset**, the default values for X and Y are 0. Similarly, you can enter values for **Plot scale**. Choose the **Scale:** drop-down arrow button to display the various scale factors. From this list you can select a scale factor you want to use. For example if you select the scale factor 1/4"=1'-0", the **Custom:** edit boxes show 1 inch= 48 drawing units. If you want the drawing to be plotted so that it fits on the specified sheet of paper, select the **Scaled to Fit** option. When you select this option, AutoCAD LT will determine the scale factor and display the scale factor in the **Custom:** edit boxes. In this example, you will plot the drawing so that it fits on 8.5 x 11 paper. Therefore, select the **Scaled to Fit** option and notice the change in the **Custom:** (Inches = Drawing Units) edit boxes. Selecting the **Custom** option, allows you to enter your own values.

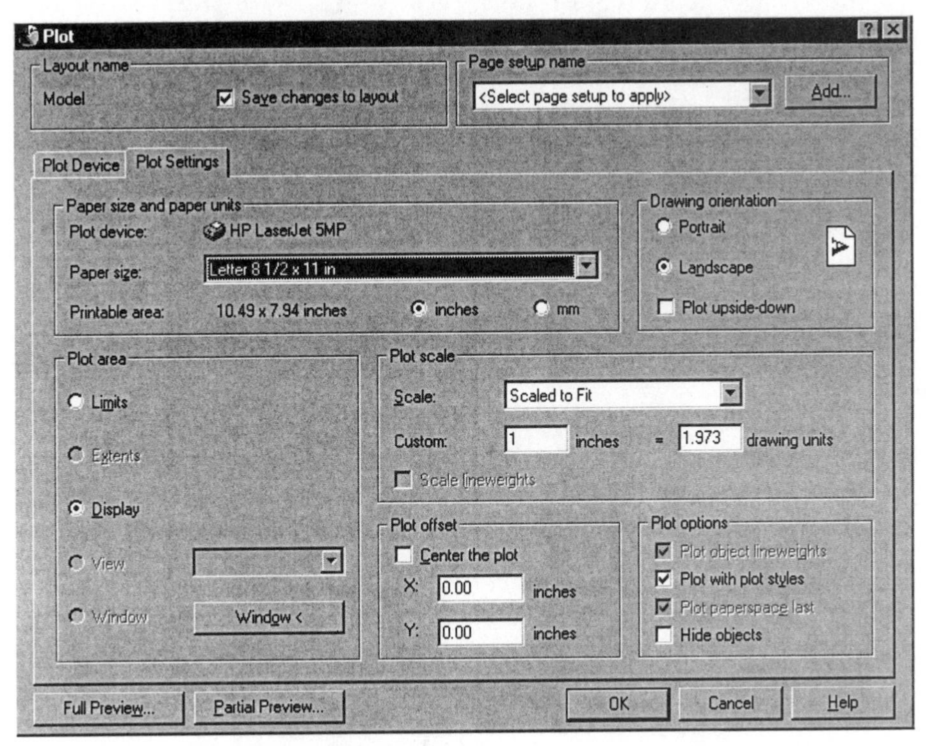

***Figure 1-58 Plot** dialog box with **Plot Settings** tab*

Step 6

You can view the plot on the specified paper size before actually plotting it by choosing the **Preview** buttons in the **Plot** dialog box. This way you can save time and stationery. AutoCAD LT provides two types of Plot Previews, partial and full. To generate partial preview of a plot, choose the **Partial Preview** button. The **Partial Plot Preview** dialog box is displayed (Figure 1-59). To display the drawing just as it would be plotted on the paper, choose the **Full Preview** button. **Full Preview** takes more time than **Partial Preview** because the drawing is regenerated. Once regeneration is complete, the preview image is displayed on the screen, Figure 1-60. In place of the cursor a realtime zoom icon is displayed. You can hold the pick button of your pointing device and then move it up to zoom into the preview image and move the cursor down to zoom out of the preview image. To exit the preview image, press the ESC or ENTER key. You can also right-click to display the shortcut menu where you can choose **Exit** to exit the preview image.

Step 7

If the plot preview is satisfactory and you want to plot the drawing, choose the **OK** button in the **Plot** dialog box. You can also choose **Plot** from the shortcut menu displayed in the previous step. AutoCAD LT will plot the drawing on the specified plotter.

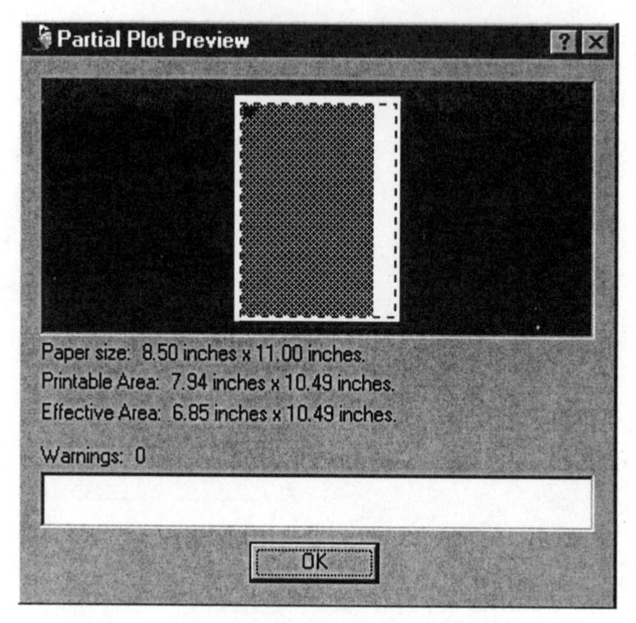

***Figure 1-59 Partial Plot Preview** dialog*

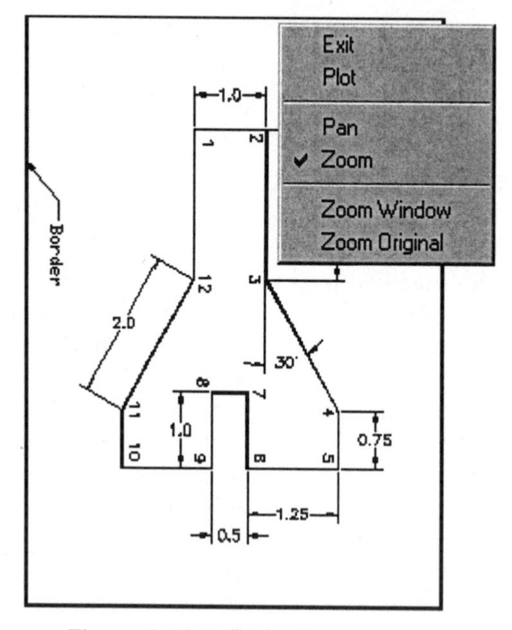

***Figure 1-60** Full Plot Preview*

SAVING YOUR WORK

Toolbar:	Standard > Save
Menu:	File >Save or Save As
Command:	SAVE , SAVEAS . QSAVE

In AutoCAD LT or any computer system, you must save your work before you exit from the drawing editor or turn the system off. Also, it is recommended that you save your drawings after regular time intervals. In case of a power failure, a serious editing error, or some other problem, all work saved prior to the problem will be retained. The commands that AutoCAD LT has provided to save the work can be entered at the Command prompt or by selecting the appropriate command from the menu bar (Choose **File**) (Figure 1-61).

AutoCAD LT has provided the following commands that let you save your work on the hard disk of the computer or on the floppy diskette:

SAVE **SAVEAS** **QSAVE**

The **SAVE**, **SAVEAS**, and **QSAVE** commands allow you to save your drawing by writing it to a permanent storage device, such as a hard drive, or on a diskette in the A or B drive. When you invoke the **SAVEAS** command or the **SAVE** command (**SAVE** command can be invoked from the Command prompt only. **SAVE** in the **File** menu and with the **Standard** toolbar invoke the **QSAVE** command). The **Save Drawing As** dialog box (Figure 1-62) is displayed. In this dialog box, you are supposed to enter the file name in which the drawing will be saved. The **SAVEAS** command works in the same way as the **SAVE** command, but in addition to saving the drawing it sets the name of the current drawing to the file name you specify. The **QSAVE** command saves the current named drawing without asking you to enter a file name, thus allowing you to do a quick save. If the current drawing is unnamed, **QSAVE** acts like **SAVEAS** and will prompt you to enter the file name in the **Save Drawing As** dialog box.

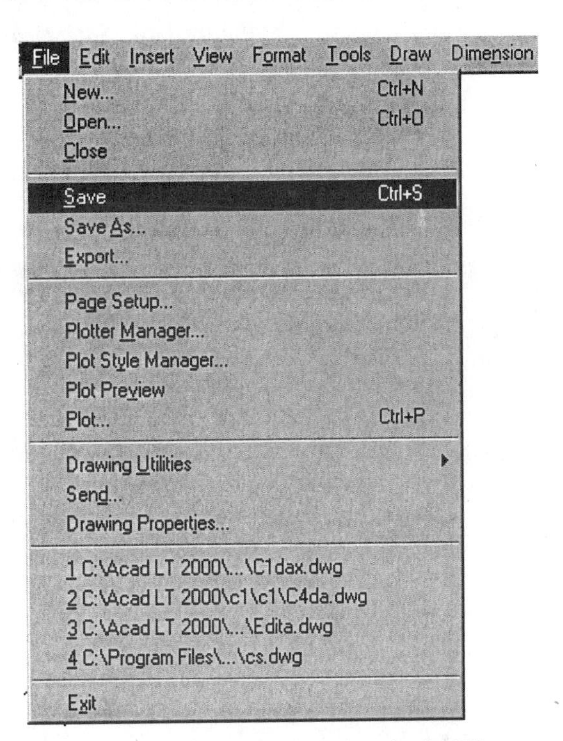

***Figure 1-61** Different Save options in the **File** menu*

The **Save Drawing As** dialog box displays the information related to the drawing files on your system. To save your work, enter the name of the drawing in the **File Name:** edit box or select a file from the list box. The drawing extension **.dwg** is not required as it is the default (AutoCAD LT drawing) and is displayed in the **Save as type:** list box. You can change the directory through the **Save in:** drop-down list. If you want to save the drawing on the A drive, enter **a:house**, and AutoCAD LT will write the information to the diskette in the A drive. After entering the name, choose the **Save** button.

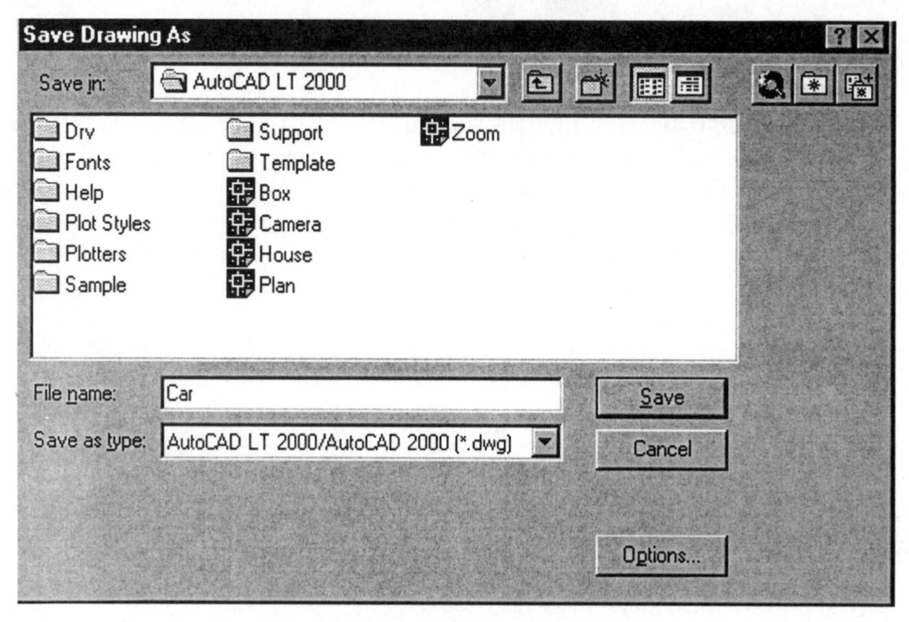

***Figure 1-62 Save Drawing As** dialog box*

Note

*If you want to save a drawing on the A or B drive, make sure the diskette you are using to save the drawings is formatted. Other features of the **Save Drawing As** dialog box are described near the end of this chapter.*

OPENING AN EXISTING FILE

Toolbar:	Standard toolbar > Open
Menu:	File > Open
Command:	OPEN

To open an existing file in the drawing editor, you can use the **OPEN** command. When you invoke the **OPEN** command, AutoCAD LT displays the **Select File** dialog box (Figure 1-63). You can enter the name of the drawing file you want to open in the **File Name:** edit box, or you can select the name of the drawing from the list box.

Command: **OPEN** [Enter]

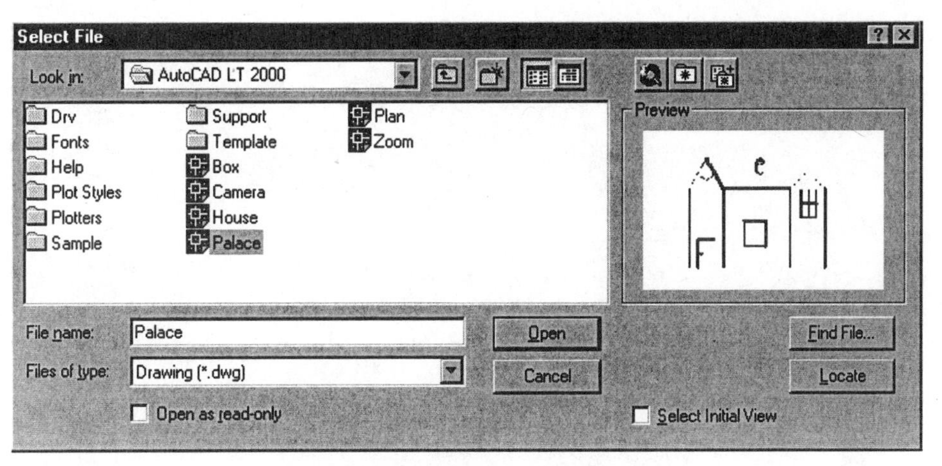

*Figure 1-63 **Select File** dialog box*

STARTING A NEW DRAWING

Toolbar:	Standard toolbar > New
Menu:	File > New
Command:	NEW

When you run AutoCAD LT, and you are in the drawing editor, the drawing you create is named **Drawing 1.dwg** by default. You can save this drawing under a different name by using the **SAVE** or **SAVEAS** command. You can either specify the file name first and then begin work or do it before exiting. The **NEW** command is used to create a new drawing while you are in an editing session.

Command: **NEW** Enter

After you invoke the **NEW** command, AutoCAD LT will display the **Create New Drawing** dialog box (Figure 1-64). Choose **OK** to accept the **Start from Scratch** option (default). A new drawing having the default AutoCAD LT setup is started.

QUITTING A DRAWING

Menu:	File > Exit
Command:	QUIT

You can exit from the drawing editor by using the **QUIT** command. If the changes you have made to the drawing have been saved, AutoCAD LT allows you to exit without saving them again. In case the drawing has not been saved, it allows you to save the work first through a dialog box. This box gives you an option to discard the current drawing or changes made to it. It also gives you an option to cancel the command.

Note

*The following sections describe in detail the **Save Drawing As** and **Select File** dialog boxes. If you are using AutoCAD LT for the first time, you can skip this portion and come back to it later.*

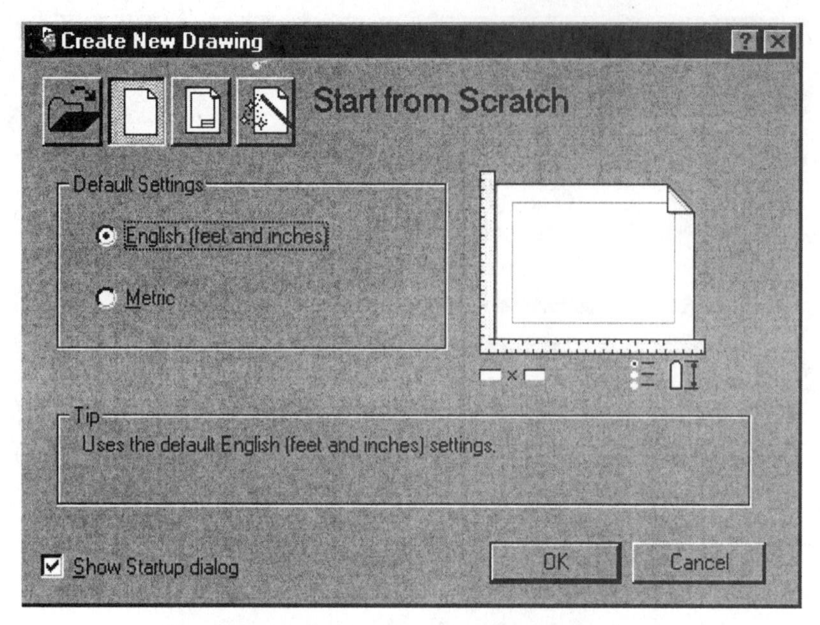

Figure 1-64 ***Create New Drawing*** *dialog box*

SAVE DRAWING AS DIALOG BOX

Save as type: List Boxes

The **Save as type:** drop-down list (Figure 1-65) is used to specify the drawing format in which you want to save the file. For example, to save the file as an AutoCAD LT 2000 drawing file, select **AutoCAD LT 2000** Drawing from the drop-down list.

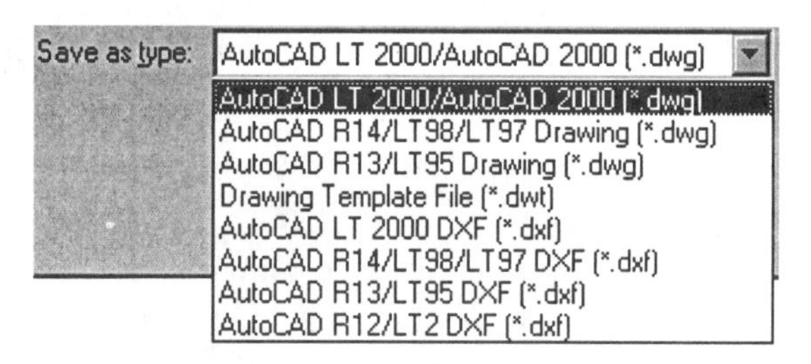

Figure 1-65 Save as type list box

File name Edit Box

The **File name:** edit box is used to enter the name of the file you want the drawing to be saved as. This can be done by typing the file name or selecting it from the list box. If you select the file name you want from the list box, the name you select automatically appears in the **File name:** edit box. If you have already assigned a name to the drawing, the current drawing name is taken as default. If the drawing is unnamed, the default name **Drawing1** is displayed in the **File Name:** edit box.

Save in: List Box

The current drive and path information is listed in the **Save in:** drop-down list (Figure 1-66). AutoCAD LT will initially save the drawing in the default directory, but if you want to save the drawing in a different directory, you have to specify the path. For example, if you want to save the present drawing under the file name **PALACE** in the C1 subdirectory, choose the arrow button in the **Save in:** list box to display the drop-down list and select C:. When you select C: all directories in C drive will be listed in the file list box. Double-click on AutoCAD LT 2000 or select AutoCAD LT 2000 or choose the **Open** button to display its directories. Again double-click on C1 or select C1 and choose the **Open** button to display drawing names in the file list box. Select PALACE from the list, if it is already listed there, or enter it in the **File name:** edit box and then choose the **Save** button. Your drawing (Palace) will be saved in the C1 folder (C:\Acad LT 2000\C1\PALACE.dwg).

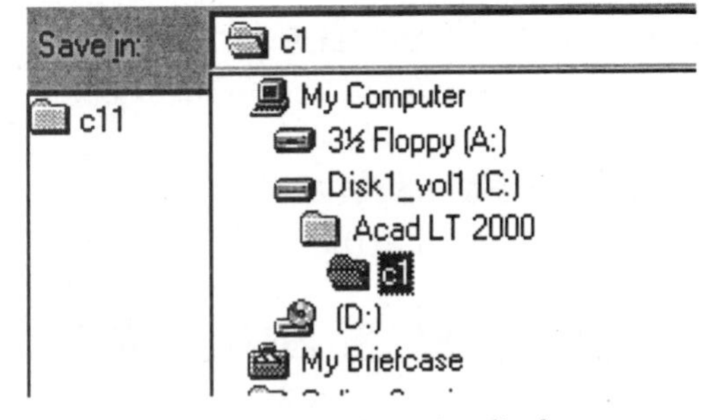

Figure 1-66 Save in: *list box*

List and Detail Buttons

If you choose the **List** button (Figure 1-67), all files present in the current directory will be listed in the file list box. If you choose the **Detail** button, it will display detailed information about the files (size, type, date, and time of modification).

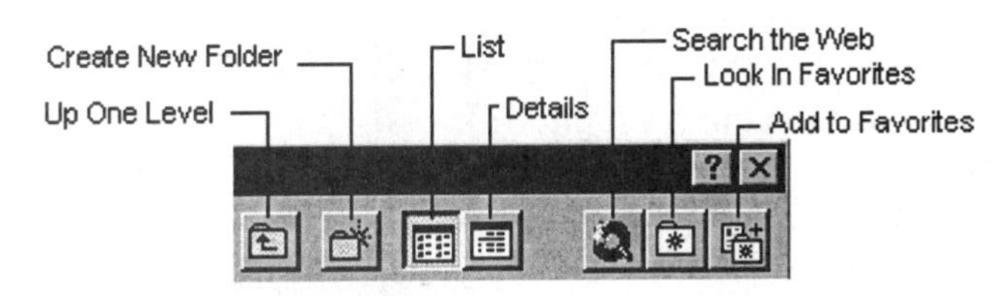

Figure 1-67 Save in: *list box*

Folder Button

If you choose the **Create New Folder** button, AutoCAD LT creates a new directory under the name **New Folder**. The new folder is displayed in the file list box. You can accept the name or change it to your requirement.

Up One Level Button

The **Up One Level** button displays the directories which are up by one level.

Options Button

The **Options** button displays the **Saveas Options** dialog box where you can save the proxy images of custom objects. It has the **DWG Options** and **DXF Options** tabs.

Search the Web

It displays the **Browse the Web** dialog box that enables you to access and store AutoCAD LT files on the internet.

Look in Favorites

Sets the search path to look in the system's **Favorites** folder.

Add to Favorites

Creates a shortcut to the selected file or folder and adds it to the systems **Favorites** folder.

SELECT FILE DIALOG BOX

When you enter the **OPEN** command or choose **Open** from the **File** menu or the **Standard** toolbar, AutoCAD LT displays the **Select File** dialog box. This box is similar to the **Save Drawing As** dialog box (described earlier, Figure 1-67), except the **Preview box, Locate, Partial Open** and **Find File** buttons, **Select Initial View** and **Open as read-only** check boxes.

Select Initial View

A view is defined as the way you look at an object. The **Select Initial View** option allows you to specify the view you want to load initially when AutoCAD LT loads the drawing. This option will work if the drawing has saved views. You can save a desired view by using AutoCAD LT's **VIEW** command (see "**VIEW** Command," Chapter 5). If the drawing has no saved views, selecting this option will load the last view. If you select the **Select Initial View** check box and then the **OK** button, AutoCAD LT will display the **Select Initial View** dialog box (Figure 1-68). You can select the view name from this dialog box, and AutoCAD LT will load the drawing with the selected view displayed.

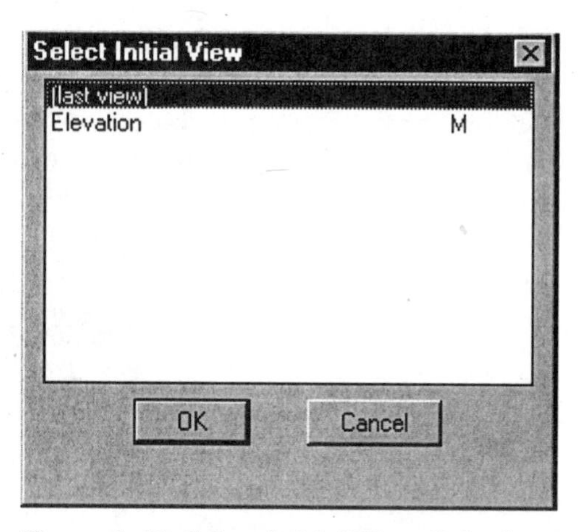

Figure 1-68 Select Initial View dialog box.

Open as read-only

If you want to view a drawing without altering it, you must select the **Open as read-only** check box. In other words, read only protects the drawing file from changes. AutoCAD LT does not prevent you from editing the drawing, but if you try to save the opened drawing to the original file name, AutoCAD LT warns you that the drawing file is **write protected.** However, you can save the edited drawing to a file with a different file name.

Find File

Select this button to display the **Browse/Search** dialog box (Figure 1-69). This allows you to search files in different drives and subdirectories.

Browse

The file name of the drawing that is currently selected is displayed in the **File Name:** drop-down list. The bitmap images of the drawings in the current directory are displayed in the **Preview** box. The drawing display size can be made small, medium, or large by selecting the desired

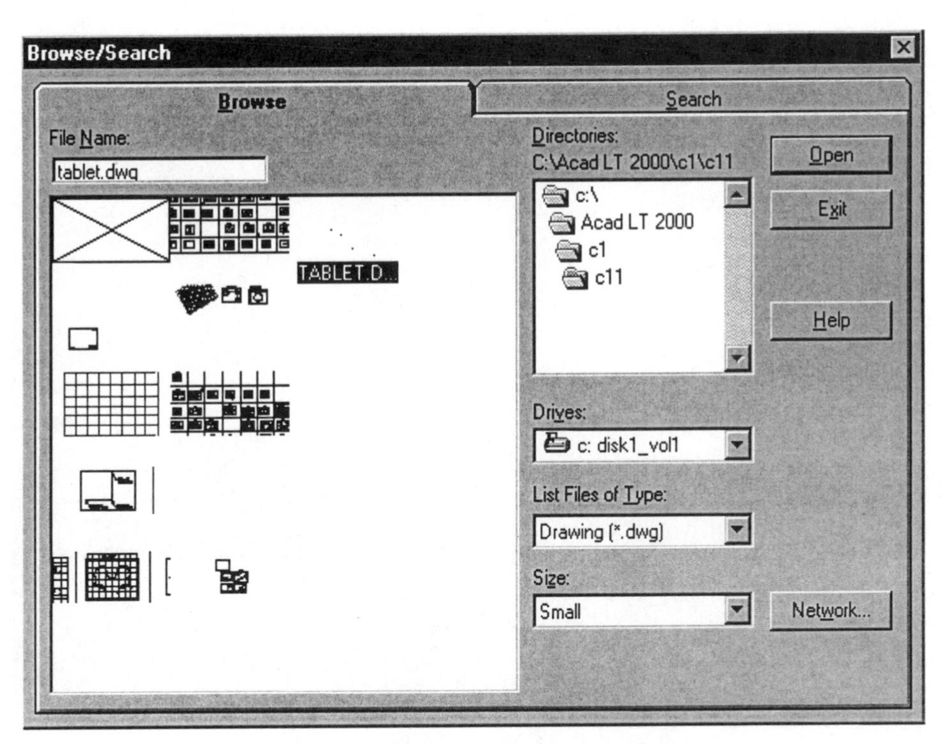

Figure 1-69 ***Browse/Search*** *dialog box*

option from the **Size:** drop-down list. The **Drives:** drop-down list and **Directories:** list box display the different drives and directories available.

Search

To activate the **Search** dialog box (Figure 1-70), choose the **Search** tab. To start the search routine, choose the **Search** button. Searching for files depends on the file type or creation date. As the search continues, the search progress is updated in the **Preview** box. After the search is completed, the bitmap images and the file names of those files that fulfill the search criteria are displayed in the list box. The total number of these files is also displayed. The **Search Pattern** allows you to search for files having a specific pattern of specified file type. From the **Files types** drop-down list you can select the desired file type for searching. In the **Date Filter** area you can specify a certain time and date so that only those files are searched that have been created before or after the particular time and date. The time and date can be specified in the **Time** and **Date** text boxes in the **Date Filter** area. In the **Search Location** area you can specify the drives and paths for searching. From the **Drives** drop-down list you can specify the drive for searching. The **All Drives** drop-down list displays all the fixed drives. In the **Path** edit box you can specify the directories to be searched. The **Edit** button opens the **Edit Path** dialog box which is used to edit a path.

Locate

The **Locate** button enables you to search a specified File by using the search path specified in the **Options** dialog box.

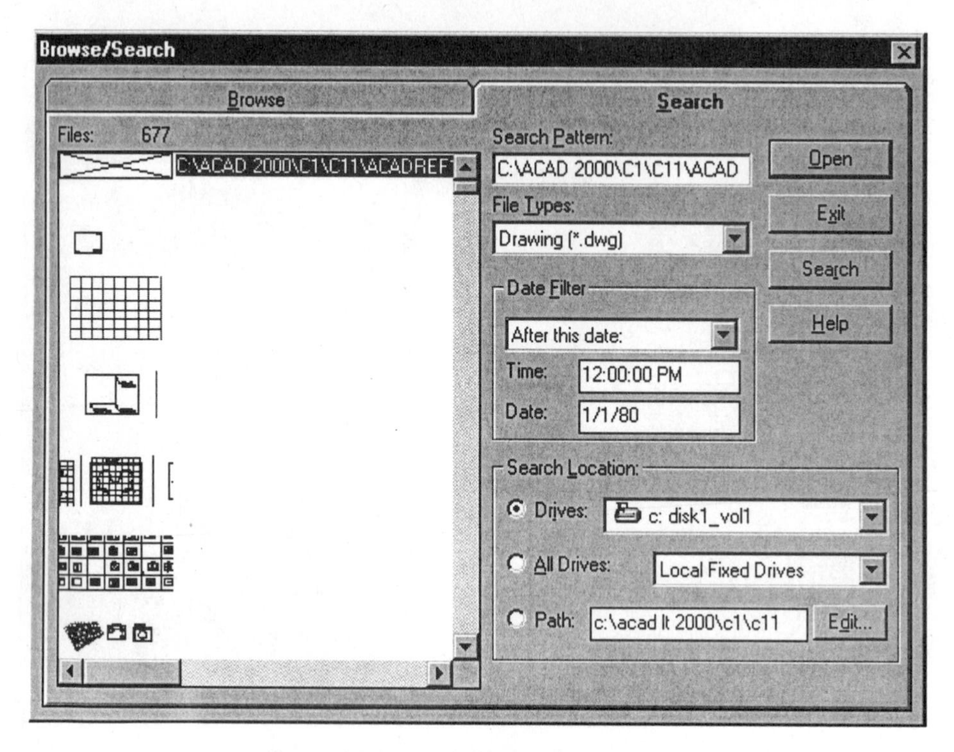

*Figure 1-70 **Browse/Search** dialog box*

Preview

If you select a file name from the **File Name:** list box, the bitmap image of that particular file is displayed in the Preview box.

AUTOMATIC TIMED SAVE

AutoCAD LT allows you to save your work automatically at specific intervals. To change the time intervals you can use the system variable **SAVETIME**. You can also change the time intervals from the **Options** dialog box (**Open and Save** tab), which can be invoked from the **Tools** menu. Depending on the power supply, hardware, and type of drawings, you should decide on an appropriate time and assign that time to this variable. AutoCAD LT saves the drawing under the file name **AUTO.SV$.** The extension of the auto-save file is **.SV$.**

Command: **SAVETIME** Enter
Enter new value for SAVETIME <120>: *Enter time, in minutes.*

CREATION OF BACKUP FILES

If the drawing file already exists and you use **SAVE** or **SAVEAS** commands to update the current drawing, AutoCAD LT creates a backup file. AutoCAD LT takes the previous copy of the drawing and changes it from a file type **.DWG** to **.BAK**, and the updated drawing is saved as a drawing file with the **.DWG** extension. For example, if the name of the drawing is **MYPROJ.DWG**, AutoCAD LT will change it to **MYPROJ.BAK** and save the current drawing as **MYPROJ.DWG.**

OPTIONS DIALOG BOX*

You can use the **Options** dialog box (Figure 1-71) to change the settings that affect the drawing environment or AutoCAD LT interface. For example, you can use the **Options** dialog box to display scroll bars or specify the support directories that contain the files you need. The Current profile: name and the Current drawing: name are displayed on the top, above the tabs.

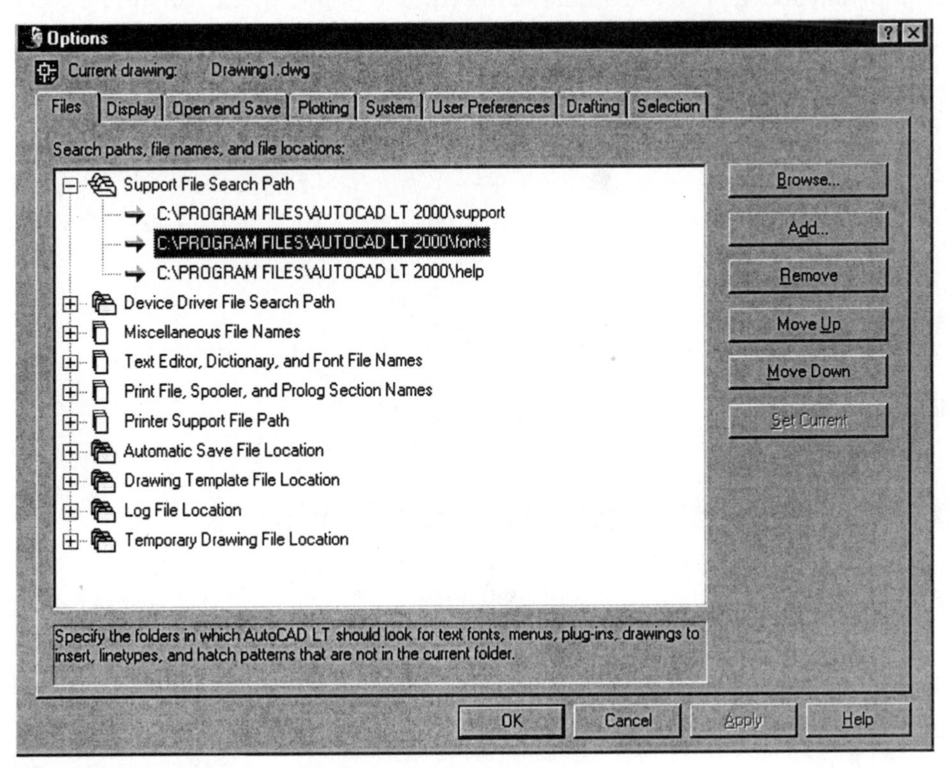

Figure 1-71 ***Options*** *dialog box*

The Options Dialog Box Tabs

Files

Stores the directories in which AutoCAD LT looks for driver, menu, support, and other files. It uses three icons: folder, paper stack, and file cabinet. The folder icon is for search path, the paper stack icon is for files, and the file cabinet icon is for a specific folder.

Display

Controls drawing and window settings like scroll bar display. For example, if you don't want to display the scroll bars, invoke the **Options** dialog box by selecting **Options** in the **Tools** menu. In the **Options** dialog box choose the **Display** tab and then clear the check box located to the left of the **Display scroll bars in drawing window** under the **Window Elements** area. Similarly, you can change the color of the graphics window background. It also allows you to modify the Display resolution, Display performance, display of Layout elements and also change the cross hair size.

Open and Save

Controls parameters related to opening and saving of files in AutoCAD LT like the Automatic Save feature.

Plotting

Controls parameters related to plotting of drawings in AutoCAD LT like the Default output device.

System

Contains AutoCAD LT system settings options like Pointing device settings options.

User Preferences

Controls settings that depend on the way the user prefers working on AutoCAD LT, like the Right-click Customization.

Drafting

Controls settings like the Autosnap and Tracking Settings and the Aperture Size.

Selection

Controls settings related to the methods of Object Selection like the Grips.

COMMAND LINE RECALL AND EDITING

You can recall a command by using the up and down arrow key in the History Window (Command prompt area). As you press the arrow key, previously entered commands and the Command prompt entries are displayed at the Command prompt. Once the desired command is displayed at the Command prompt, you can execute the command by pressing the ENTER key. You can also edit the command or the values displayed in the History Window.

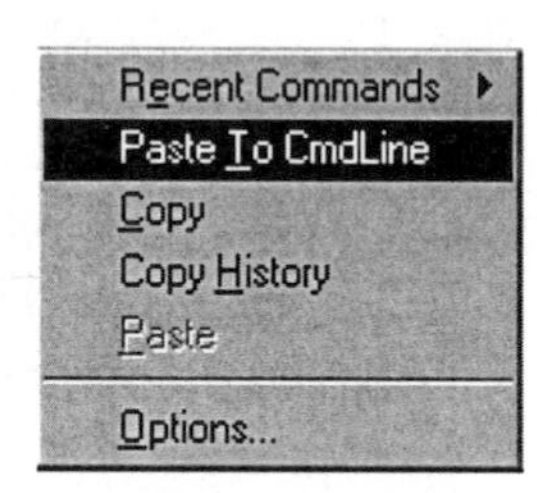

***Figure 1-72** The shortcut menu, displayed on right-clicking on the command prompt area*

The commands and command prompt entries that are displayed in the History Window can be selected, copied, and pasted in the command line. To copy and paste, select the lines and then right-click in the History Window to display the shortcut menu (Figure 1-72). In the shortcut menu select copy and then paste the selected lines in the command line. After the lines are pasted, you can edit them.

AUTOCAD LT'S HELP

You can get the on-line help and documentation on the working of AutoCAD LT 2000 commands from the **Help** menu (Figure 1-73).

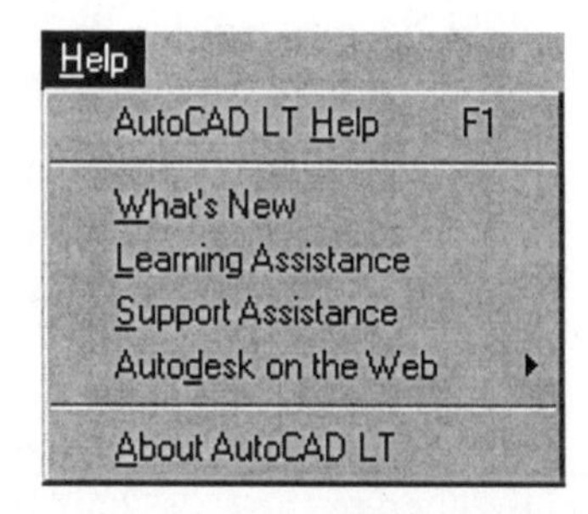

***Figure 1-73 Help** menu*

AutoCAD LT Help Topics

Toolbar:	Standard > Help
Menu:	Help > AutoCAD LT Help
Command:	HELP

Invoking **HELP** displays the **AutoCAD LT 2000 Help System** dialog box (Figure 1-74). You can use this dialog box to access help on different topics and commands. It has three tabs: **Contents**, **Index** and **Search**, which display the corresponding help topics. If you are in the middle of a command and require help regarding it, choosing the **Help** button, displays information about that particular command.

Contents

This tab displays the help topics that are organized by categories pertaining to different sections of AutoCAD LT. To select a category, double-click on the corresponding book icon. The topics associated with that category are displayed in a window. Select a topic and then double-click; AutoCAD LT displays information about the selected topic or command. In this display, if you select items that are underlined and blue in color, a step-by-step tutorial on that topic is displayed.

Index

This tab displays the complete index (search keywords) in an alphabetical order. To display information about an item or command, type the item (word) or command name in the edit box (Figure 1-74). With each letter entered the listing keeps on changing in the list area, displaying the possible topics. When you enter the word and if AutoCAD LT finds that word, it is automatically highlighted in the list area. Choose the **Display** button to display information about it. If the topic selected is a command and has more than one explanation, then the **Topic Found** dialog box is displayed. Select the topic and choose the **Display** button to display the information about that particular topic.

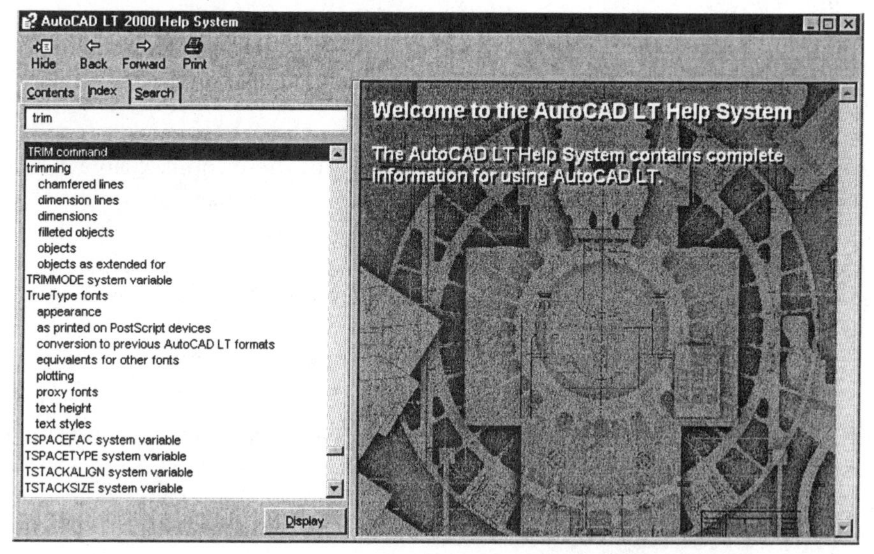

***Figure 1-74 AutoCAD LT 2000 Help Systems** dialog box (**Index** tab)*

Search

This tab creates a word list based on all the keywords present in the on-line help files. When you type a word you are looking for and choose the **Start Search** button, a list of matching words appears in a window below to narrow down your search.

What's New

This option gives you an interactive list of all the new features in AutoCAD LT 2000. You can choose this option from the **Help** menu that displays the **What's New** screen with a list of new features. When you choose a topic, a brief description of the feature improvement is displayed. Choose a feature to display an explanation related to that topic. You can choose a particular topic by positioning your cursor on it (cursor changes into a pointing hand icon), and then pressing the pick button of the pointing device. An explanation of the selected topic is displayed. You can exit the **What's New** by selecting the **Close** (**X**) button.

Learning Assistance

This is a multimedia learning option that provides you with an on-line environment where you can learn the AutoCAD LT software application interactively. You can choose this option from the **Help** menu. Choosing this option displays an **Insert Disc** dialog box. On inserting the disc and after specifying the path for it, choose **OK** to display the **Autodesk Learning Assistance** screen (Figure 1-75). This screen is divided into two areas, the one on the left displays the various topics to select from. When you choose a heading, the related items under that particular topic are displayed on the right-hand side. When you choose an item, AutoCAD LT displays the related explanation. On the top right-hand side of this screen is the **Play Video** icon. Choose this icon to give you an animated working of the selected command or feature. The animation is also supported by voice (audio) which explains the working interactively with animation. You can exit the animation by selecting **Back** from **Go** menu. You can exit from the Learning Assistance by choosing **Exit** from the **File** menu or by selecting the **Close** (**X**) button.

Support Assistance

This help option displays the Autodesk Support Assistance screen having six main headings: **Search for a solution**, **Download**, **What's Hot**, **More Resources**, **Phone Support** and **Getting the Most From Support Assistance**. Choosing any one of these topics displays further sub topics to choose from with their brief explanations. You can choose any of these to look for their respective descriptions and related topics. To go to search or Up one level, you choose the **Search** or **Up one level** buttons, provided on the screen. This option gives you all the Technical Support information in a question and answer format. It also provides information about other help resources.

Autodesk on the Web

This utility connects you to the various AutoCAD LT/ Autodesk web pages and sites through the Microsoft Internet Explorer.

About AutoCAD LT

This option gives you information about the Release, Serial Number, and Licensed To and

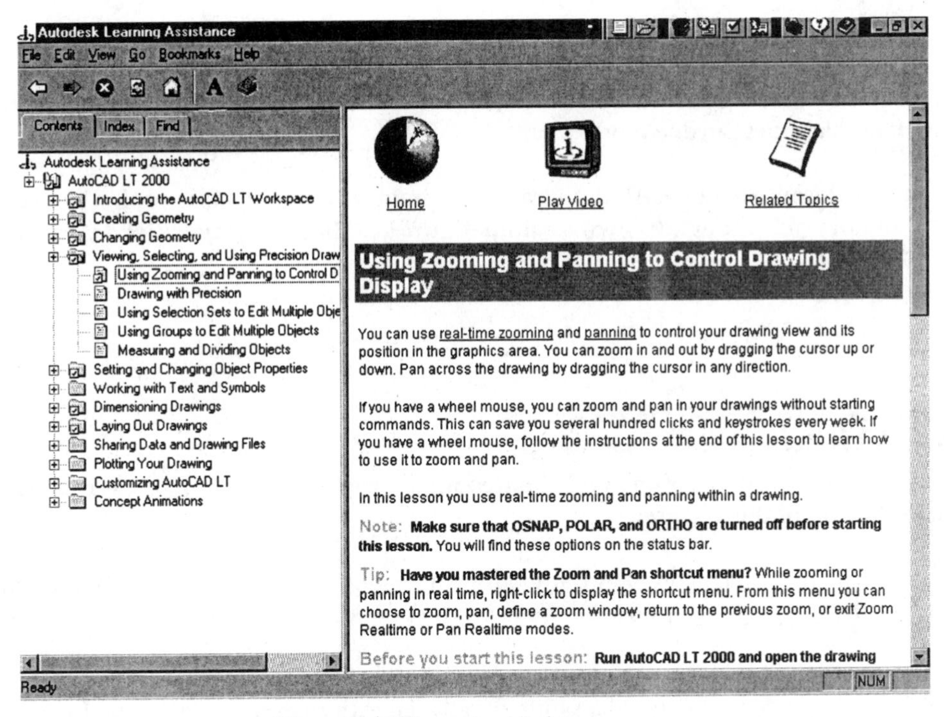

Figure 1-75 *Learning Assistance screen*

also the legal description about AutoCAD LT.

ADDITIONAL HELP RESOURCES

1. You can get help for a command while working by pressing the F1 key. The **AutoCAD LT 2000 Help System** dialog box containing information about the command is displayed. You can exit the dialog box and continue with the command.

2. You can get help about a certain dialog box by choosing the **Help** button in that dialog box.

3. Some of the dialog boxes have a **question mark** (?) button at the top-right corner just adjacent to the **close** button. When you choose this button the ? gets attached to the cursor. You can then drop it on any item in the dialog box to display information about that particular item.

4. Autodesk has provided several resources that you can use to get assistance with your AutoCAD LT questions. You can also use these resources to get information on Autodesk products, product updates, and other services provided by Autodesk. The following is a list of some of the resources:

   a. Autodesk Web Site **http://www.autodesk.com**
   b. AutoCAD LT 2000 Web Site: **http://www.autodesk.com/AutoCAD LT**
   c. Autodesk Fax Information System: **(415) 446-1919**
   d. AutoCAD LT Technical Assistance Web Site **autodesk.com/support**.

e. AutoCAD LT Discussion Groups Web Site **autodesk.com/support/discsgrp/acad.htm**.

5. You can also get help by contacting the author, Sham Tickoo, at **stickoo@calumet.purdue.edu**.

6. You can download AutoCAD LT drawings, programs, and special topics by accessing the authors Web Site at **http://www.calumet.purdue.edu/public/mets/tickoo/index.html**.

Self-Evaluation Test

Answer the following questions and then compare your answers to the correct answers given at the end of this chapter.

1. If you exit the **LINE** command after drawing a line and then select **LINE** command again, the __________ option can be used to draw a line from the endpoint of the last line.

2. The __________coordinates of a point are located with reference to the previous point.

3. To draw a line from point (2,2) to another point at a distance of 3 units and an angle of 60 degrees, you will have the prompt sequence:
   Command: __________
   Specify first point: __________
   Specify next point or [Undo]: __________

4. When using the **Window** option in the **ERASE** command, the objects to be erased must __________lie within the window.

5. The prompt sequence for drawing a circle with its center at (4,5) and a diameter of 3 units is:
   Command: __________
   Specify center point or [3P/2P/Ttr(tan tan radius)]<Center point>:__________
   Specify radius of circle or [Diameter]<current>: __________
   Specify diameter of circle <current>: __________

6. The __________option of the **CIRCLE** command can be used to draw a circle, if you want the circle to be tangent to two previously drawn objects.

7. You can erase a previously drawn line using the __________option of the **LINE** command.

8. The __________system variable can be used to change the time interval for automatic save.

9. When you select the **ERASE** command, a small box known as the __________ replaces the screen cursor.

10. The __________ command is used to plot a drawing.

11. By choosing the __________ button in the **Plot** dialog box, you can specify the section of the drawing to be plotted with the help of a __________.

Review Questions

1. The prompt sequence for the **LINE** command is:

   Command: _______
   ____
   ____

2. The __________ option can be used to draw the last side of a polygon.

3. In the two mutually intersecting lines for the X and Y coordinates, the X coordinate value is calculated along the __________ line, and the Y coordinate value is calculated along the __________ line.

4. In the __________ coordinate system, the points are located with respect to origin 0,0.

5. In the relative coordinate system, if you want a displacement of 6 units along the Y axis and 0 displacement along the X axis, you should enter __________ at the **Specify next point or [Undo]:** prompt.

6. In the polar coordinate system, you need to specify the __________ of the point from the __________ point and the __________ it makes with the positive X axis.

7. The **ERASE** command can be selected from the __________ screen menu or by entering __________ at the Command prompt.

8. When AutoCAD LT puts you in the drawing editor, the __________ command is used to create a new drawing .

9. The **NEW** command can be invoked from the __________ menu or by entering __________ at the Command prompt.

10. The prompt sequence for drawing a circle with (2,2), (4,5), and (7,1) as three points on its circumference is:

Command: _________
Specify center point for circle or [3P/2P/Ttr(tan tan radius)]: _________
Specify first point on circle: _________
Specify second point on circle: _________
Specify third point on circle: _________

11. To open or load an existing file into the drawing editor, you can use the _________ command.

12. A file name for a new drawing can also be specified from the command line if the _________ system variable is set to 0.

13. When you enter the **SAVEAS** command, AutoCAD LT displays the standard file dialog box entitled _________

14. What is the difference between the **Window** option and the **Crossing** option?

15. Explain the function of various boxes in the **Select File** dialog box.

16. Explain briefly the function of the **Create New Drawing** dialog box.

17. How will you create a new file with the name **DRAW** using the command line and not through the dialog box?

18. Explain the differences between the **SAVE**, **SAVEAS**, and **QSAVE** commands.

19. Explain the functions of various components of the **Save Drawing As** dialog box.

20. You can view the plot on the specified paper size before actually plotting it by choosing the _________ button in the **Plot** dialog box.

Exercises

Exercise 6 — *Mechanical*

Use the following relative and absolute coordinate values in the **LINE** command to draw the object.

Point	Coordinates	Point	Coordinates
1	3.0, 3.0	5	@3.0,5.0
2	@3,0	6	@3,0
3	@-1.5,3.0	7	@-1.5,-3
4	@-1.5,-3.0	8	@-1.5,3

Exercise 7 — *Mechanical*

For Figure 1-76, enter the relative coordinates of the points in the given table, then use these coordinates to draw the figure.

1	3.0, 1.0	9	________
2	________	10	________
3	________	11	________
4	________	12	________
5	________	13	________
6	________	14	________
7	________	15	________
8	________	16	________

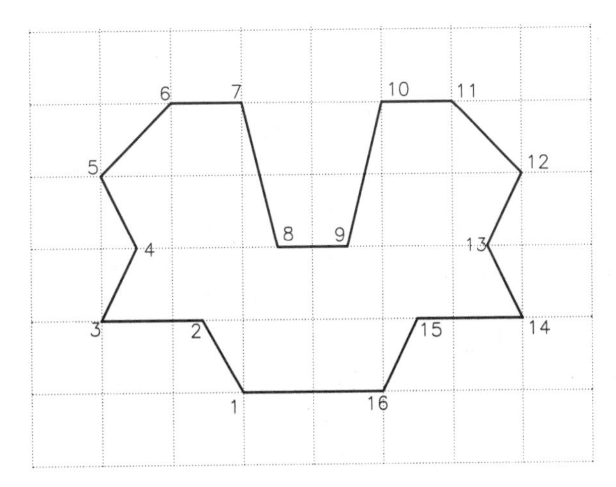

Figure 1-76 *Drawing for Exercise 7*

Exercise 8 *Mechanical*

For Figure 1-77, enter the polar coordinates of the points in the given table. Then use those coordinates to draw the figure. Do not draw the dimensions.

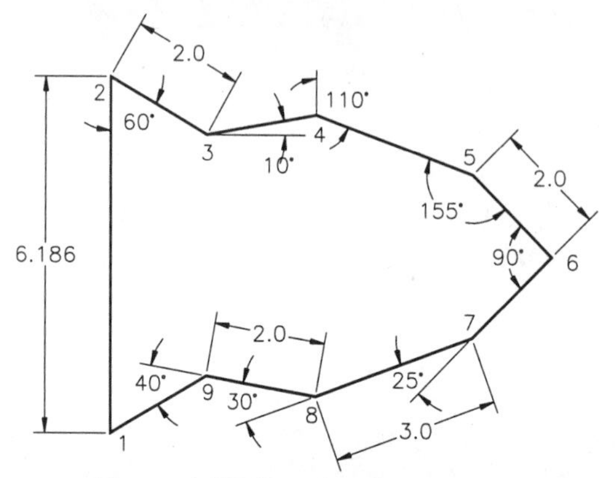

***Figure 1-77** Drawing for Exercise 8*

Point	Coordinates	Point	Coordinates
1	1.0, 1.0	6	__________
2	__________	7	__________
3	__________	8	__________
4	__________	9	__________
5	__________		

Exercise 9 *Mechanical*

Generate the drawing in Figure 1-78, using the absolute, relative, or polar coordinate system. Draw according to the dimensions shown in the figure, but do not draw the dimensions.

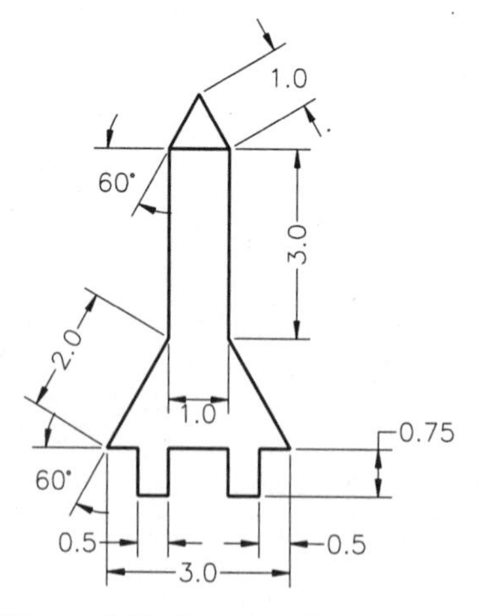

***Figure 1-78** Drawing for Exercise 9*

Exercise 10 *Mechanical*

Draw Figure 1-79, using the **LINE** command and the **Ttr** option of the **CIRCLE** command.

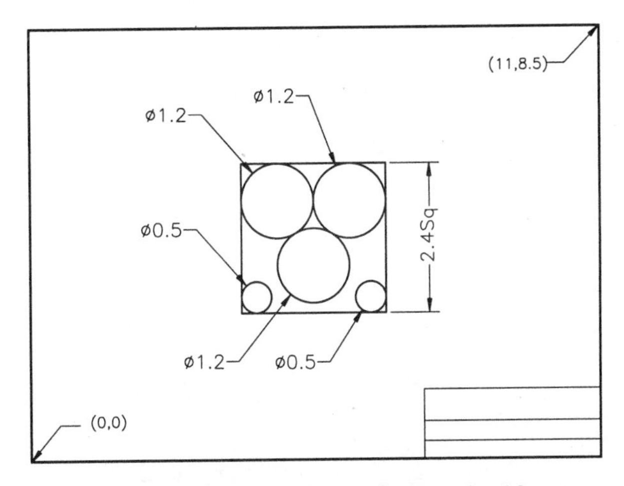

Figure 1-79 *Drawing for Exercise 10*

Exercise 11 *Architectural*

Create the drawing in Figure 1-80, using the pointing device to select points. The drawing you generate should resemble the one in the figure.

Figure 1-80 *Drawing for Exercise 11*

Exercise 12 — General

Draw the object shown in Figure 1-81, using various options of the **CIRCLE** command.

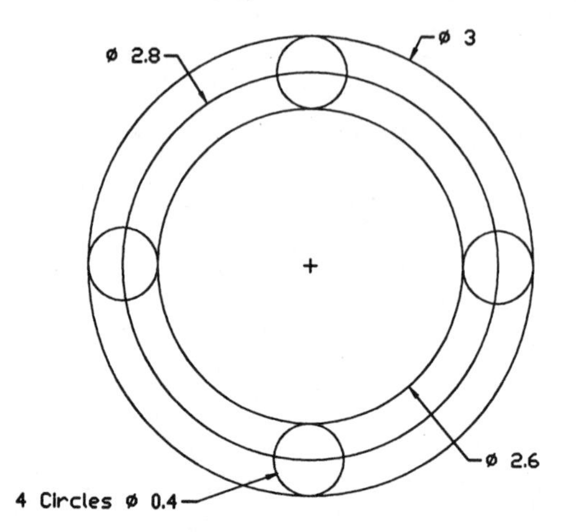

Figure 1-81 *Drawing for Exercise 12*

Problem-Solving Exercise 1 — Mechanical

Draw the object shown in Figure 1-82, using the **LINE** and **CIRCLE** commands. In this exercise only the diameters of the circles are given. In order to draw the lines and small circles (R0.6), you need to find the coordinate points for the lines and the center points of the circles. For example, if the center of concentric circles is at 5,3.5, then the X coordinate of the lower left corner of the rectangle is 5.0 - 2.4=2.6.

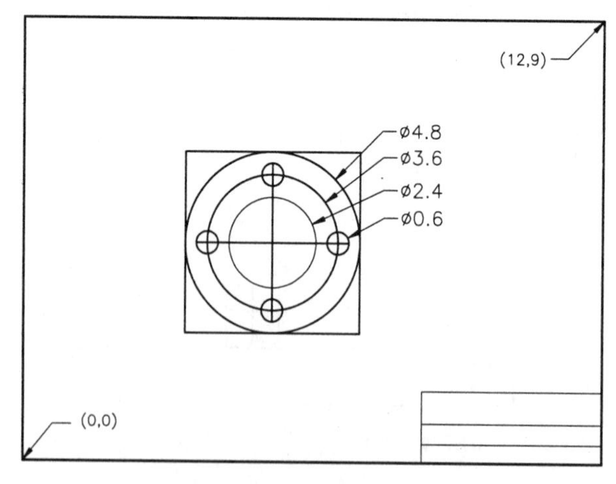

Figure 1-82 *Drawing for Problem-Solving Exercise 1*

Problem-Solving Exercise 2 — *Mechanical*

Draw Figure 1-83, using various options of the **CIRCLE** and **LINE** commands. In this exercise you have to find coordinate points for drawing lines and circles. Also, you need to determine the best and easiest method to draw the 1.7 diameter circles along the outermost circle.

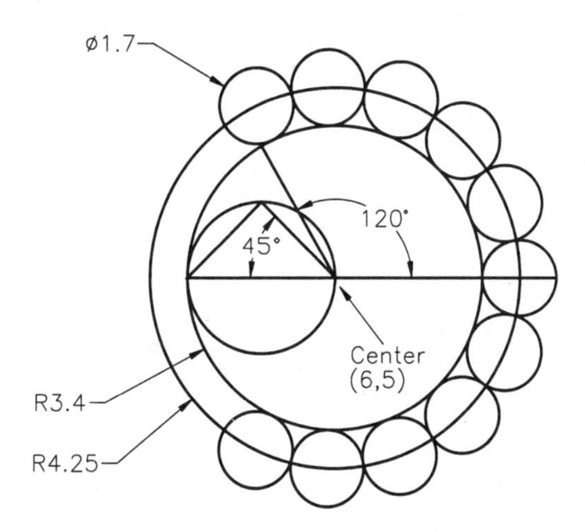

Figure 1-83 *Drawing for Problem-Solving Exercise 2*

Project Exercise 1-1 — *Mechanical*

Generate the drawing in Figure 1-84, using the absolute, relative, or polar coordinate system. Draw according to the dimensions shown in the figure, but do not draw the dimensions. Save the drawing as BENCH1-2. (See Chapters 2, 6, 7, 19, and 23 for other drawings of Project Exercise 1.)

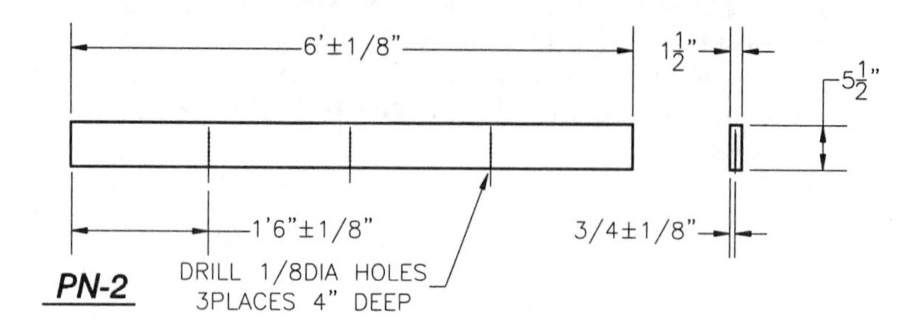

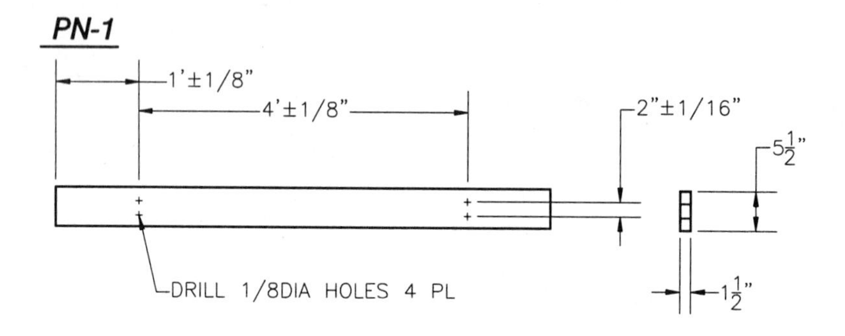

Figure 1-84 *Drawing for Project Exercise 1-1*

Answers to Self-Evaluation Test

1 - Continue, **2** - Relative, **3** - **LINE** / 2,2 / @3<60, **4** - completely, **5** - **CIRCLE** / 4,5 / D / 3, **6** - Tangent tangent radius, **7** - Undo, **8** - **SAVETIME**, **9** - pick box, **10** - **PLOT**, **11**- window, window

Chapter 2

Draw Commands

Learning Objectives

After completing this chapter, you will be able to:

- *Draw arcs using different options.*
- *Draw rectangles, ellipses, and elliptical arcs.*
- *Draw polygons like hexagons and pentagons.*
- *Draw polylines, and doughnuts.*
- *Draw points and change point style and point size.*

DRAWING ARCS

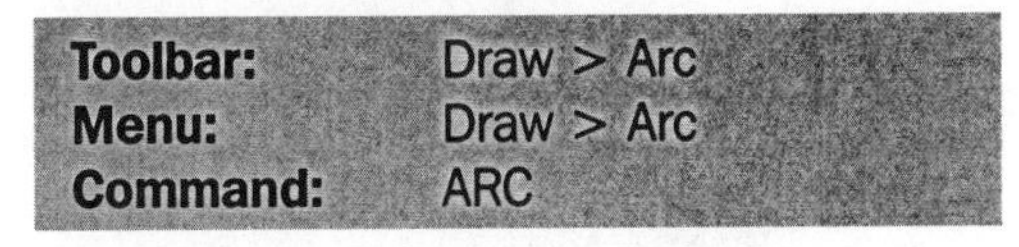

Toolbar:	Draw > Arc
Menu:	Draw > Arc
Command:	ARC

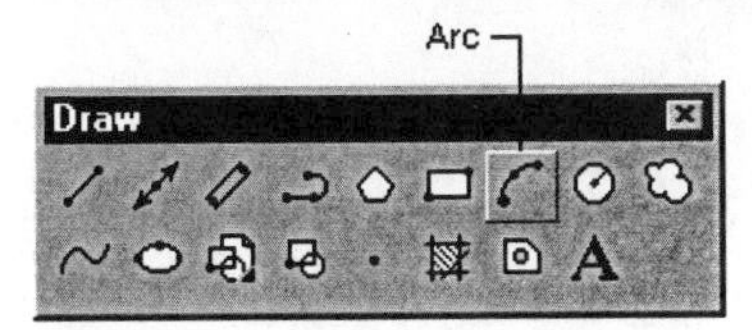

***Figure 2-1** Invoking the **ARC** command from the **Draw** toolbar*

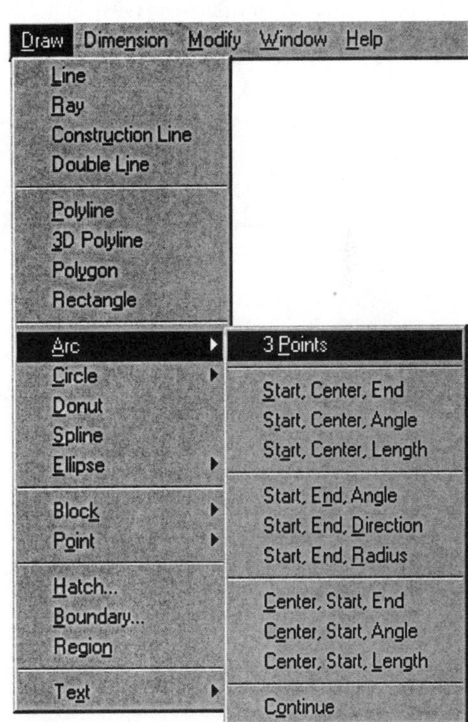

***Figure 2-2** Invoking the **ARC** command from the **Draw** menu*

An arc is defined as a part of a circle; it can be drawn using the **ARC** command. An arc can be drawn in **11** distinct ways using the options listed under the **ARC** command. The default method for drawing an arc is the **3 Points** option. Other options can be invoked by entering the appropriate letter to select an option. If you have set the **DRAGMODE** variable to Auto (Default Value), the last parameter to be specified in any arc generation is automatically dragged into the relevant location.

The 3 Points Option

When you enter ARC at the Command prompt, you automatically get into the **3 Points** option. The **3 Points** option requires the start point, the second point, and the endpoint of the arc (Figure 2-3 and Figure 2-4). The arc can be drawn in a clockwise or counterclockwise direction by dragging the arc with the cursor. The following is the prompt sequence to draw an arc with a start point at (2,2), second point at (3,3), and an endpoint at (3,4). (You can also specify the points by moving the cursor and then specifying points on the screen.)

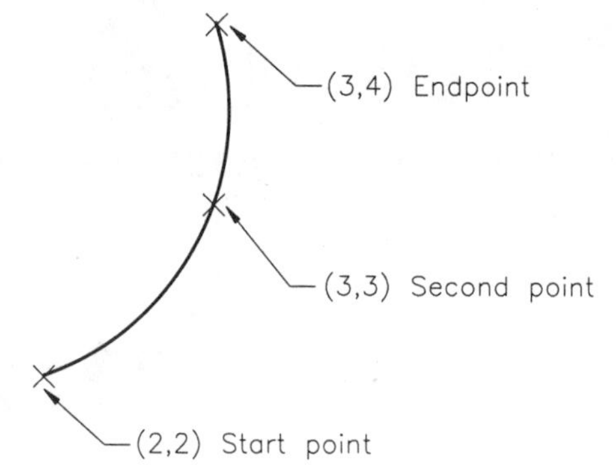

Figure 2-3 Drawing an arc using the 3 Points option

Command: **ARC** Enter
Specify start point of arc or [CEnter]: **2,2** Enter
Specify second point of arc or [CEnter/ENd]: **3,3** Enter
Specify end point of arc: **3,4** Enter

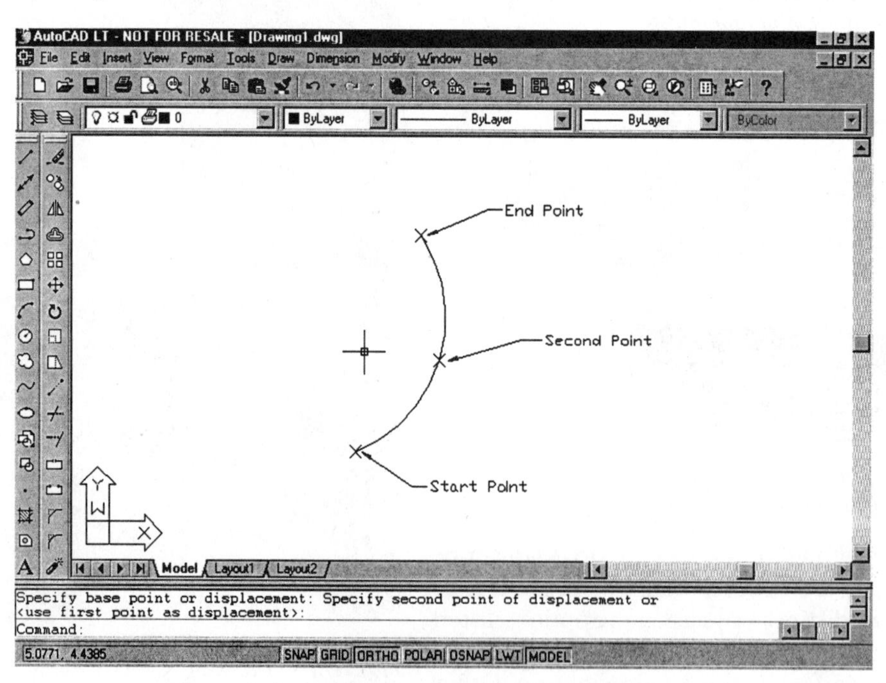

Figure 2-4 Drawing an arc using the 3 Points option

Exercise 1 *General*

Draw several arcs using the **3 Points** option. The points can be selected by entering coordinates or by specifying points on the screen. Also, try to create a circle by drawing two separate arcs and by drawing a single arc. Notice the limitations of the **ARC** command.

The Start, Center, End Option

This option is slightly different from the **3 points** option. In this option, instead of entering the second point, you enter the center of the arc. Choose this option when you know the start point, endpoint, and center point of the arc. The arc is drawn in a counterclockwise direction from the start point to the endpoint around the specified center. The endpoint specified need not be on the arc and is used only to calculate the angle at which the arc ends. The radius of the arc is determined by the distance between the center point and the start point. The prompt sequence for drawing an arc with a start point of (3,2), center point of (2,2), and endpoint of (2,3.5) (Figure 2-5) is as follows:

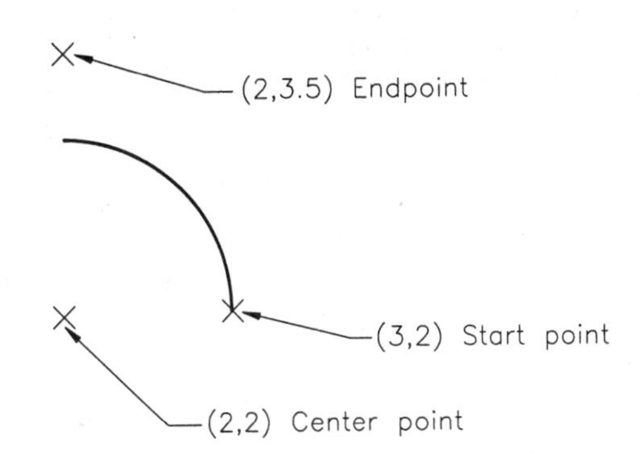

***Figure 2-5** Drawing an arc using the Start, Center, End option*

Command: **ARC** [Enter]
Specify start point of arc or [CEnter]: **3,2** [Enter]
Specify second point of arc or [CEnter/ENd]: **CE** [Enter]
Specify center point of arc: **2,2** [Enter]
Specify end point of the arc or [Angle/ chord Length]: **2,3.5** [Enter]

The Start, Center, Angle Option

This option is the best choice if you know the **included angle** of the arc. The included angle is the angle formed by the start point and the endpoint of the arc with the specified center. This option draws an arc in a counterclockwise direction with the specified center and start point spanning the indicated angle (Figure 2-6). If the specified angle is negative, the arc is drawn in a clockwise direction (Figure 2-7).

The prompt sequence for drawing an arc with center at (2,2), a start point of (3,2), and an included angle of 60 degrees (Figure 2-6) is:

Command: **ARC** [Enter]
Specify start point of arc or [CEnter]: **3,2** [Enter]
Specify second point of arc or [CEnter/ENd]: **CE** [Enter]
Specify center point of arc: **2,2** [Enter]
Specify end point of the arc or [Angle/ chord Length]: **A** [Enter]
Specify included angle: **60** [Enter]

Chapter 2

You can draw arcs with negative angle values in the Start, Center, Included Angle (St,C,Ang) option by entering "-" (negative sign) followed by the angle values of your requirement at the **Specify included angle** prompt (Figure 2-7). This is illustrated by the following prompt sequence:

Command: **ARC** [Enter]
Specify start point of arc or [CEnter]: **4,3** [Enter]
Specify second point of arc or [CEnter/ENd]: **CE** [Enter]
Specify center point of arc: **3,3** [Enter]
Specify end point of the arc or [Angle/ chord Length]: **A** [Enter]
Specify included angle: **-180** [Enter]

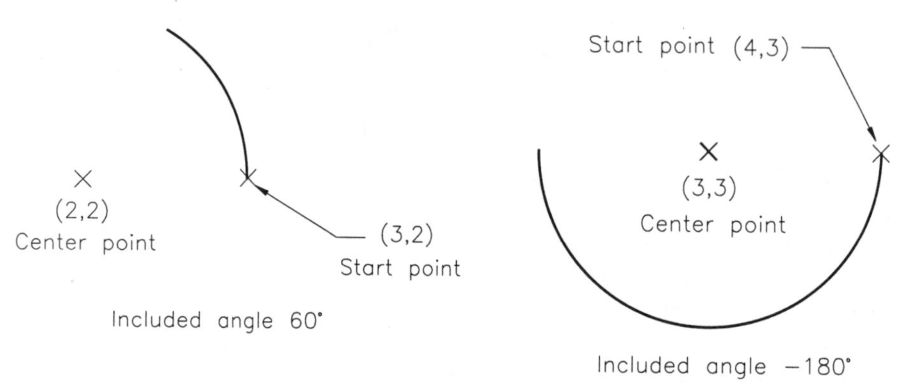

Figure 2-6 *Drawing an arc using the Start, Center, Angle option*

Figure 2-7 *Drawing an arc using a negative angle in the Start, Center, Angle option*

Exercise 2 — *Mechanical*

a. Draw an arc using the **St, C, Ang** option. The start point is (6,3), the center point is (3,3), and the angle is 240 degrees.

b. Make the drawing shown in Figure 2-8. The distance between the dotted lines is 1.0 units. Create the radii by using the arc command options as indicated in the drawing.

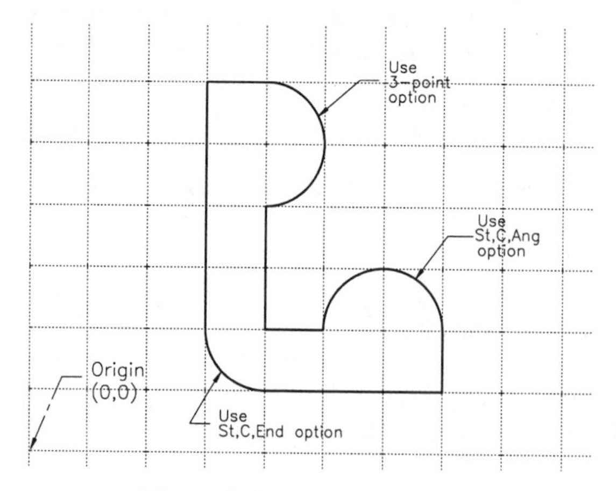

Figure 2-8 *Drawing for Exercise 2(b)*

The Start, Center, Length Option

In this option you are required to specify the start point, center point, and length of chord. A **chord** is defined as the straight line connecting the start point and the endpoint of an arc. The chord length needs to be specified so that AutoCAD LT can calculate the ending angle. Identical start, center, and chord length specifications can be used to define four different arcs. AutoCAD LT settles this problem by always drawing this type of arc counterclockwise from the start point. Therefore, a positive chord length gives the smallest possible arc with that length. This is known as the minor arc. The minor arc is less than 180 degrees. A negative value for chord length results in the largest possible arc, also known as the major arc. The chord length can be determined by using the standard chord length tables or using mathematical relation (L = 2*Sqrt [h(2r-h)]). For example, an arc of radius 1 unit, with an included angle of 30 degrees, has a chord length of 0.51764 units. The prompt sequence for drawing an arc that has a start point of (3,1), center of (2,2) and chord length of (2) (Figure 2-9) is:

Command: **ARC** Enter
Specify start point of arc or [CEnter]: **3,1** Enter
Specify second point of arc or [CEnter/ENd]: **CE** Enter
Specify center point of arc: **2,2** Enter
Specify end point of the arc or [Angle/ chord Length]: **L** Enter
Specify length of chord: **2** Enter

You can draw the major arc by defining the length of the chord as negative (Figure 2-10). In this case the arc with a start point of (3,1), a center point of (2,2), and a negative chord length of (-2) is drawn with the following prompt sequence:

Command: **ARC** Enter
Specify start point of arc or [CEnter]: **3,1** Enter
Specify second point of arc or [CEnter/ENd]: **CE** Enter
Specify center point of arc: **2,2** Enter
Specify end point of the arc or [Angle/ chord Length]: **L** Enter
Specify length of chord: **-2** Enter

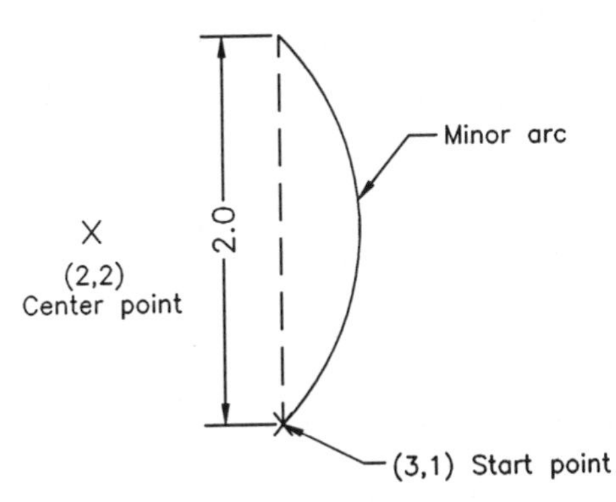

Figure 2-9 Drawing an arc using the Start, Center, Length option

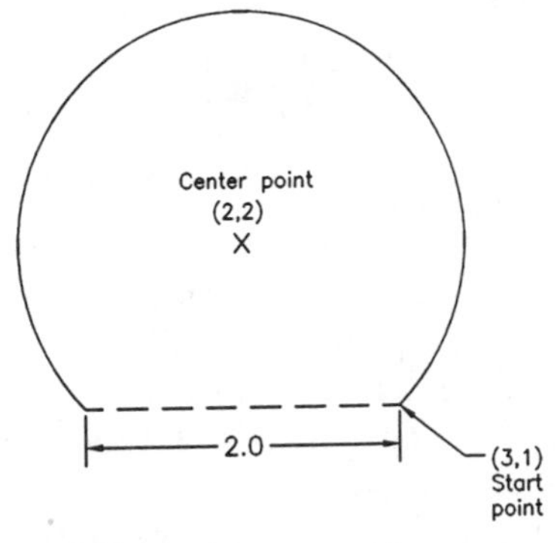

Figure 2-10 Drawing an arc using a negative chord length value in the Start, Center, Length option

Exercise 3 *General*

Draw a minor arc with the center point at (3,4), start point at (4,2), and chord length of 4 units.

The Start, End, Angle Option

With this option you can draw an arc by specifying the start point of the arc, the endpoint, and the included angle. A positive included angle value draws an arc in a counterclockwise direction from the start point to the endpoint, spanning the included angle; a negative included angle value draws the arc in a clockwise direction. The prompt sequence for drawing an arc with a start point of (3,2), endpoint of (2,4), and included angle of 120 degrees (Figure 2-11) is:

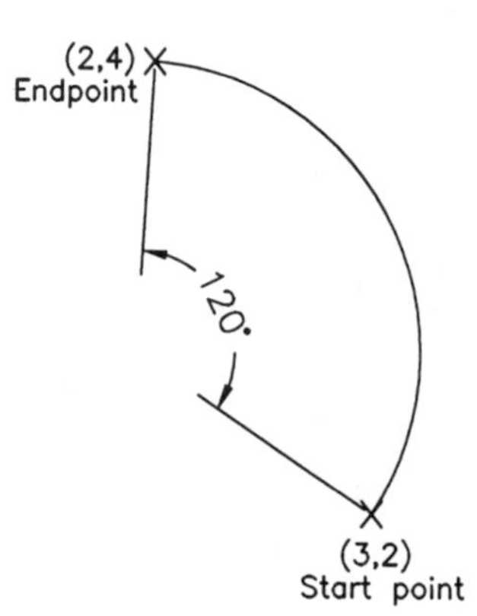

Figure 2-11 Drawing an arc using the Start,End, Angle option

Command: **ARC** [Enter]
Specify start point of arc or [CEnter]: **3,2** [Enter]
Specify second point of arc or [CEnter/ENd]: **EN** [Enter]
Specify end point of arc: **2,4** [Enter]
Specify center point of arc or [Angle/Direction/Radius]: **A** [Enter]
Specify included angle: **120** [Enter]

The Start, End, Direction Option

In this option you can draw an arc by specifying the start point, endpoint, and starting direction of the arc, in degrees. In other words, the arc starts in the direction you specify (the start of the arc is established tangent to the direction you specify). This option can be used to draw a major or minor arc, in a clockwise or counterclockwise direction. The size and position of the arc are determined by the distance between the start point and endpoint and the direction specified. To illustrate the positive direction option (Figure 2-12), the prompt sequence for an arc having a start point of (4,3), endpoint of (3,5), and direction of 90 degrees is:

Command: **ARC** [Enter]
Specify start point of arc or [CEnter]: **4,3** [Enter]
Specify second point of arc or [CEnter/ENd]: **EN** [Enter]
Specify end point of arc: **3,5** [Enter]
Specify center point of arc or [Angle/Direction/Radius]: **D** [Enter]
Specify tangent direction for the start of arc: **90** [Enter]

To illustrate the option using a negative direction degree specification (Figure 2-13), the prompt sequence for an arc having a start point of (4,3), endpoint of (3,4), and direction of -90 degrees is:

Command: **ARC** [Enter]
Specify start point of arc or [CEnter]: **4,3** [Enter]
Specify second point of arc or [CEnter/ENd]: **EN** [Enter]
Specify end point of arc: **3,4** [Enter]
Specify center point of arc or [Angle/Direction/Radius]: **D** [Enter]
Specify tangent direction for the start of arc: **- 90** [Enter]

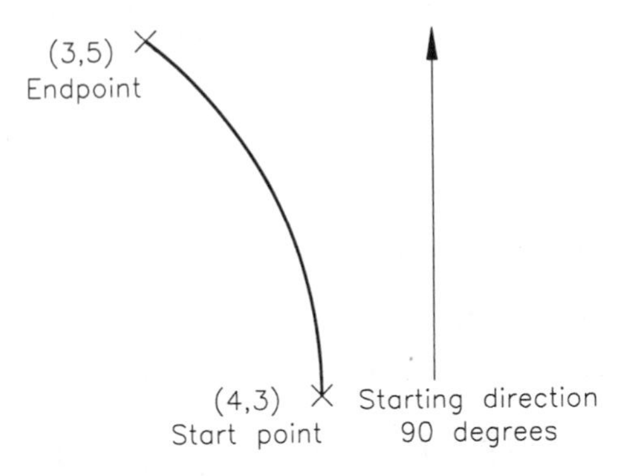

Figure 2-12 Drawing an arc using the Start, End, Direction option

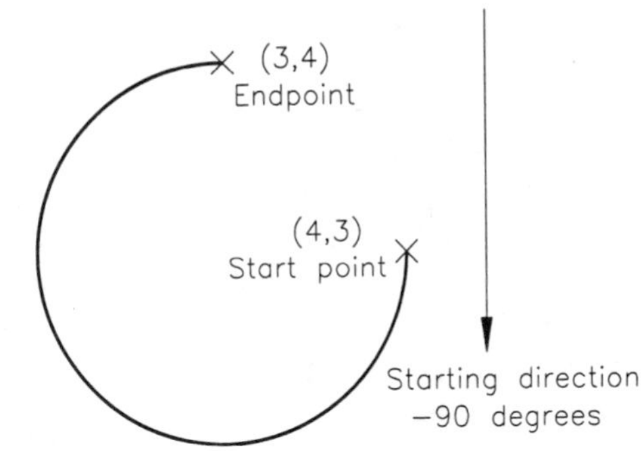

Figure 2-13 Drawing an arc using a negative direction in the Start, End, Direction option

Exercise 4 — *Mechanical*

a. Specify the directions and the coordinates of two arcs in such a way that they form a circular figure.

b. Make the drawing shown in Figure 2-14. Create the radii by using the **ARC** command options indicated in the drawing. (Use the @ symbol to snap to the previous point.) Example: **Specify startpoint of arc or[CEnter]:** @

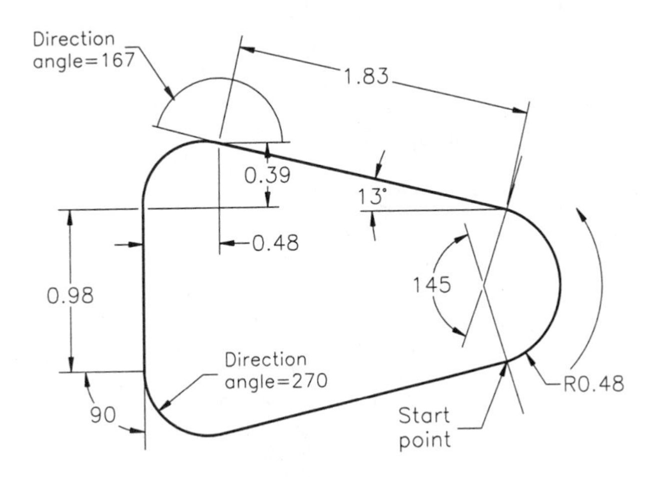

Figure 2-14 Drawing for Exercise 4(b)

The Start, End, Radius Option

This option is used when you know the start point, endpoint, and radius of the arc. The same values for the three variables (start point, endpoint, and radius) can result in four different arcs. AutoCAD LT resolves this by always drawing this type of arc in a counterclockwise direction from the start point. Hence, a positive radius value results in a **minor arc** (smallest arc between the start point and the endpoint) Figure 2-15(b), while a negative radius value results in a **major arc** (the largest arc between two endpoints) Figure 2-15(a). The prompt sequence to draw a major arc having a start point of (3,3), endpoint of (2,5), and radius of -2,Figure 2-15(a), is

Command: **ARC** Enter
Specify start point of arc or [CEnter]: **3,3** Enter
Specify second point of arc or [CEnter/ENd]: **EN** Enter
Specify end point of arc: **2,5** Enter
Specify center point of arc or [Angle/Direction/Radius]: **R** Enter
Specify radius of arc: **-2** Enter

The prompt sequence to draw a minor arc having its start point at (3,3), endpoint at (2,5), and radius as 2 (Figure 2-15b) is:

Command: **ARC** Enter
Specify start point of arc or [CEnter]: **3,3** Enter
Specify second point of arc or [CEnter/ENd]: **EN** Enter
Specify end point of arc: **2,5** Enter
Specify center point of arc or [Angle/Direction/Radius]: **R** Enter
Specify radius of arc: **2** Enter

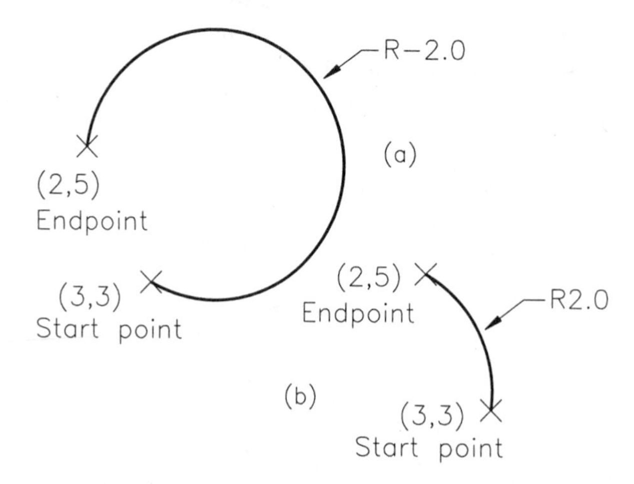

Figure 2-15 *Drawing an arc using the Start, End, Radius option*

The Center, Start, End Option

The **Center, Start, End** option is a modification of the **Start, Center, End** option. Use this option whenever it is easier to start drawing an arc by establishing the center first. Here the arc is always drawn in a counterclockwise direction from the start point to the endpoint, around the specified center. The prompt sequence for drawing an arc that has a center point at (3,3), start point at (5,3), and endpoint at (3,5) (Figure 2-16) is:

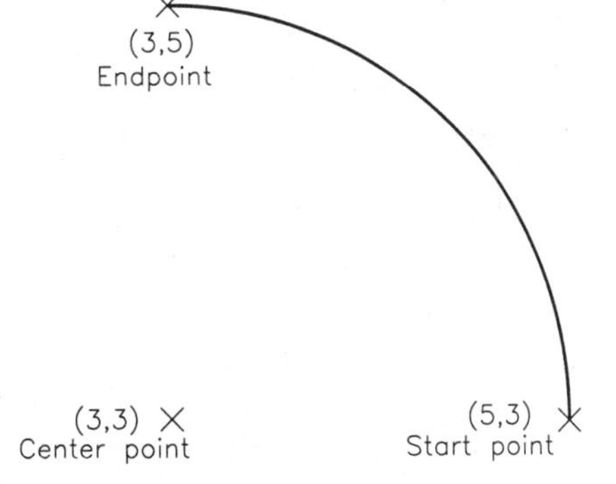

Figure 2-16 Drawing an arc using the Center, Start, End option

Command: **ARC** [Enter]
Specify start point of arc or [CEnter]: **C** [Enter]
Specify center point of arc: **3,3** [Enter]
Specify start point of arc: **5,3** [Enter]
Specify end point of arc or [Angle/chord Length]: **3,5** [Enter]

The Center, Start, Angle Option

This option is a variation of the **Start, Center, Angle** option. Use this option whenever it is easier to draw an arc by establishing the center first. The prompt sequence for drawing an arc that has a center point at (4,5), start point at (5,4), and included angle of 120 degrees (Figure 2-17) is:

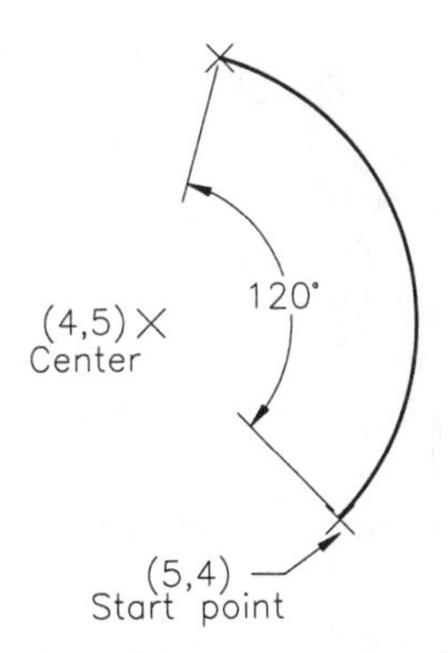

Figure 2-17 Drawing an arc using the Center, Start, Angle option

Command: **ARC** [Enter]
Specify start point of arc or [CEnter]: **C** [Enter]
Specify center point of arc: **4,5** [Enter]
Specify start point of arc: **5,4** [Enter]
Specify end point of arc or [Angle/chord Length]: **A** [Enter]
Specify included angle: **120** [Enter]

The Center, Start, Length Option

The **Center, Start, Length** option is a modification of the **Start, Center, Length** option. This option is used whenever it is easier to draw an arc by establishing the center first. The prompt sequence for drawing an arc that has a center point at (2,2), start point at (4,3), and length of chord of 3 (Figure 2-18) is:

Command: **ARC** [Enter]
Specify start point of arc or [CEnter]: **CE** [Enter]
Specify center point of arc: **2,2** [Enter]
Specify start point of arc: **4,3** [Enter]
Specify end point of arc or [Angle/chord Length]: **L** [Enter]
Specify length of chord: **3** [Enter]

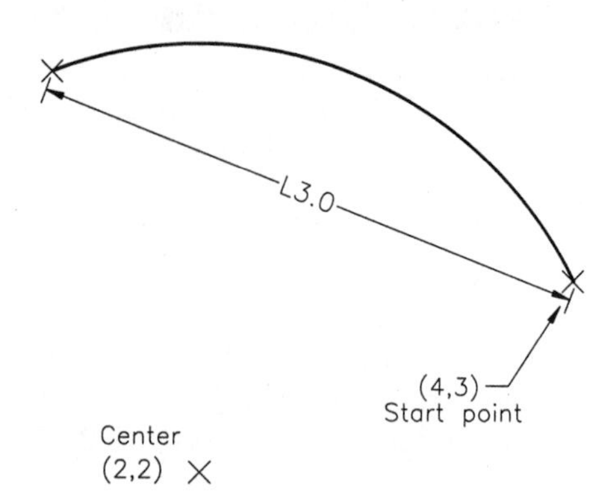

Figure 2-18 *Drawing an arc using the Center, Start, Length option*

Continue Option

With this option you can continue drawing an arc from a previously drawn arc or line. This option resembles the **Start, End, Direction** option in that if you do not specify a start point but just press ENTER, or you select the **Continue** option (**Draw** menu), then the start point and direction of the arc will be taken from the endpoint and ending direction of the previous line or arc drawn on the current screen. When this option is used to draw arcs, each successive arc is tangent to the previous one. Most often this option is used to draw arcs tangent to a previously drawn line.

The prompt sequence to draw an arc tangent to an earlier drawn line using the **Continue** option (Figure 2-19) is:

Command: **LINE** [Enter]
Specify first point: **2,2** [Enter]
Specify next point or [Undo]: **4,3** [Enter]
Specify next point or [Undo]: [Enter]
Command: **ARC** [Enter]
Specify start point of arc or [CEnter]: *Select the* **Continue** *option or press* ENTER.
Specify endpoint of arc : **4,5** [Enter]

The prompt sequence to draw an arc continued from a previously drawn arc (Figure 2-20) is:

Command: **ARC** [Enter]
Specify start point of arc or [CEnter]: **2,2** [Enter]
Specify second point of arc or [CEnter/ENd]: **EN** [Enter]
Specify endpoint of arc : **3,4** [Enter]
Specify center point of arc or [Angle/Direction/Radius]: **R** [Enter]
Specify radius of arc: **2** [Enter]
Command: **ARC** [Enter]
Specify start point of arc or [CEnter]: [Enter]
Endpoint: **5,4** [Enter]

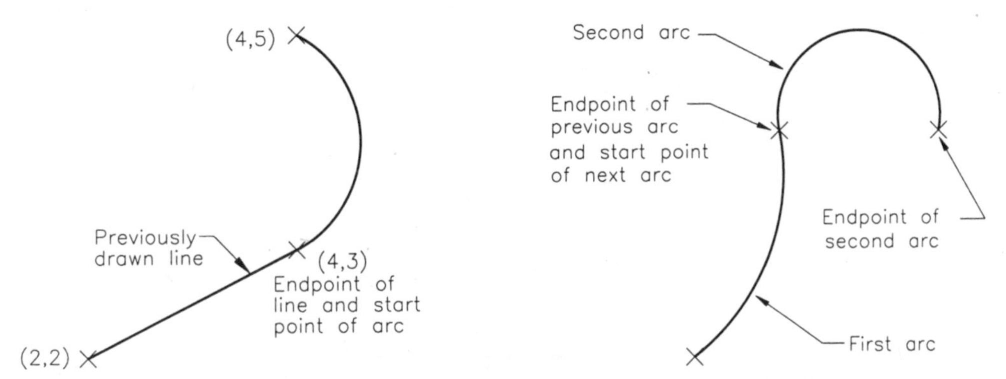

Figure 2-19 Drawing an arc using the Continue option

Figure 2-20 Drawing an arc using the Continue option

Exercise 5 *Graphics*

a. Use the **Center, Start, Angle** and the**Continue** options to draw the figures shown in Figure 2-21.

b. Make the drawing shown in Figure 2-22. The distance between the dotted lines is 1.0 units. Create the radii as indictated in the drawing by using the **ARC** command options.

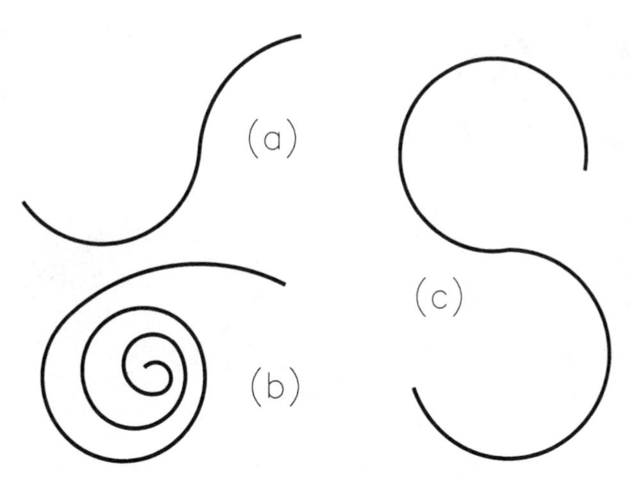

Figure 2-21 Drawing for Exercise 5(a)

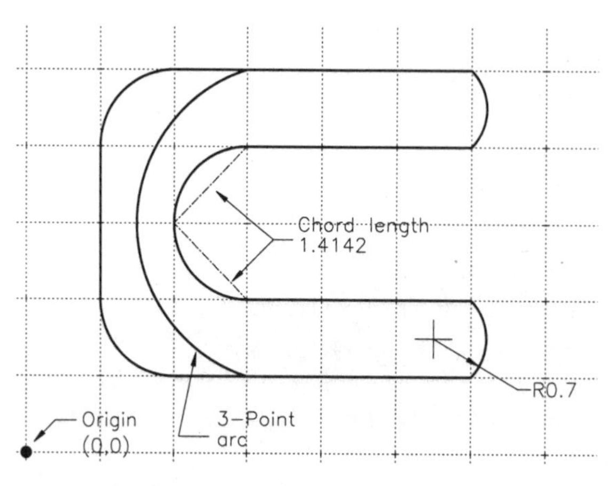

Figure 2-22 Drawing for Exercise 5(b)

DRAWING RECTANGLES

Toolbar:	Draw > Rectangle
Menu:	Draw > Rectangle
Command:	RECTANG

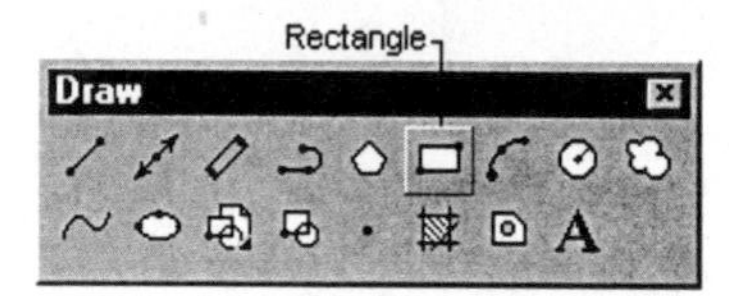

*Figure 2-23 Invoking the **RECTANG** command from the **Draw** toolbar*

A rectangle can be drawn using the **RECTANG** command. After invoking the **RECTANG** command, you are prompted to specify the first corner of the rectangle at the **Specify first corner point**: prompt. Here you can enter the coordinates of the first corner or specify the desired point with the pointing device. The first corner can be any one of the four corners. Then you are prompted to enter the coordinates or specify the other corner at the **Specify other corner point**: prompt. This corner is taken as the corner diagonally opposite the first corner. The prompt sequence for drawing a rectangle with (3,3) as its lower left corner coordinate and (6,5) as its upper right corner (Figure 2-25) is:

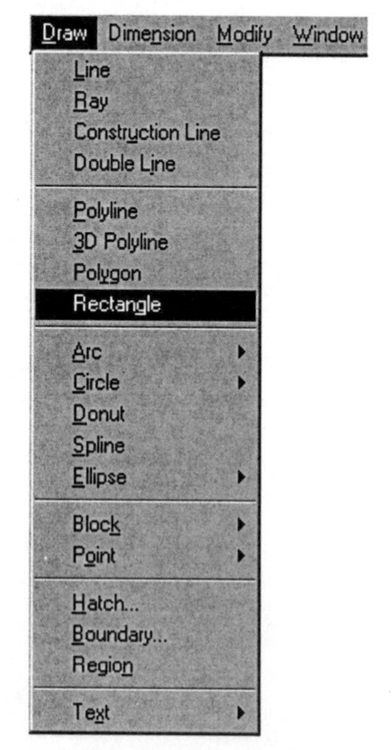

*Figure 2-25 Invoking the **RECTANG** command from the **Draw** menu*

Command: **RECTANG** Enter
Specify first corner point or [Chamfer/Elevation/Fillet/Thickness/Width]:
3,3 Enter *(Lower left corner location.)*
Specify other corner point: **6,5** Enter
(Upper right corner location.)

(6,5)
Second corner
(3,3) First corner

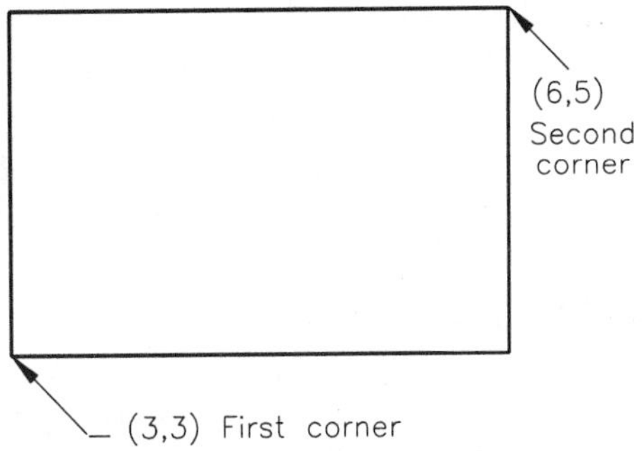

Figure 2-25 *Drawing a rectangle using the* ***RECTANG*** *command*

You can also specify the first corner and drag the cursor to specify the other corner. The **RECTANG** command has the following options:

Chamfer

The **Chamfer** option creates a chamfer by specifying the chamfer distances (Figure 2-26).

Command: **RECTANG** or **RECTANGLE** Enter
Specify first corner point or [Chamfer/Elevation/Fillet/Thickness/Width]: **C** Enter
Specify first chamfer distance for rectangles <0.0000>: *Enter a value.*
Specify second chamfer distance for rectangles <0.0000>: *Enter a value.*
Specify first corner point or [Chamfer/Elevation/Fillet/Thickness/Width]: *Select a point as lower left corner location.*
Specify other corner point : *Select a point as upper right corner location.*

Fillet

The **Fillet** option allows you to create a filleted rectangle by specifying the fillet radius (Figure 2-27).

Specify first corner point or [Chamfer/Elevation/Fillet/Thickness/Width]: **F** Enter
Specify fillet radius for rectangles <0.0000>: *Enter a value.*

Now, if you draw a rectangle it will be filleted provided the length and width of the rectangle is equal to or greater than twice the value of the specified fillet. Otherwise, AutoCAD LT will draw a rectangle without fillet.

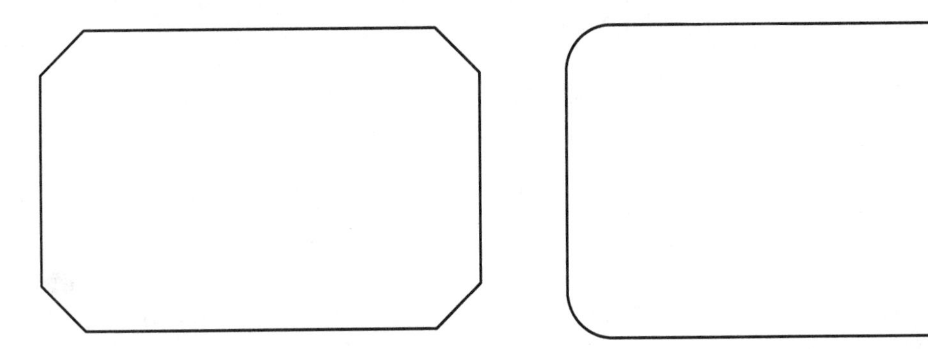

Figure 2-26 *Drawing a rectangle with chamfer*

Figure 2-27 *Drawing a rectangle with fillet*

Width

The **Width** option allows you to control the line width of the rectangle by specifying the width (Figure 2-28).

Specify first corner point or [Chamfer/Elevation/Fillet/Thickness/Width]: **W** Enter
Specify lines width for rectangles <0.0000>: *Enter a value.*

Thickness

The **Thickness** option allows you to draw a rectangle that is extruded in the Z direction by the specified value of thickness. For example, if you draw a rectangle with thickness of 2 units, you will get a rectangular box whose height is 2 units (Figure 2-29).

Specify first corner point or [Chamfer/Elevation/Fillet/Thickness/Width]: **T** Enter
Specify thickness for rectangles <0.0000>: *Enter a value.*

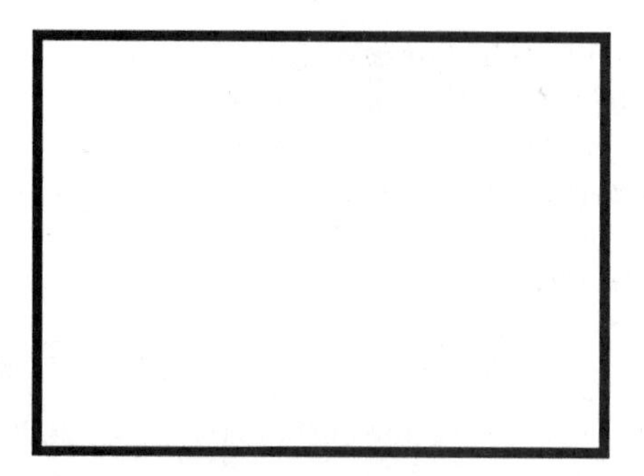

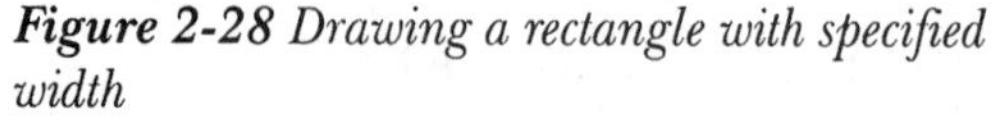

Figure 2-28 *Drawing a rectangle with specified width*

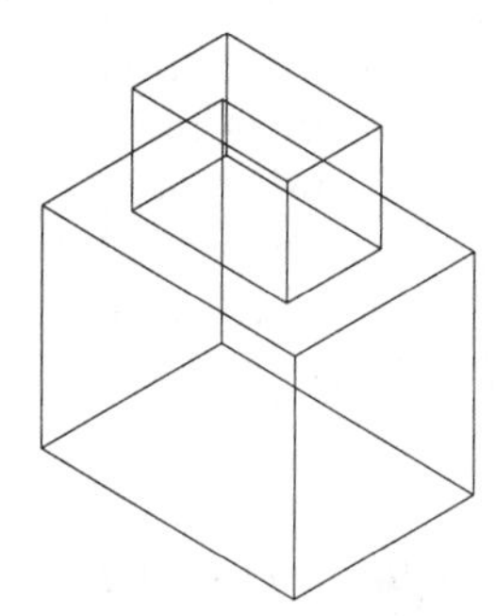

Figure 2-29 *Drawing rectangles with thickness and elevation specified*

Elevation

The **Elevation** option allows you to draw a rectangle at a specified distance from the XY plane along the Z axis. For example, if the elevation is 2 units, the rectangle will be drawn two units above the XY plane. If the thickness of the rectangle is 1 unit, you will get a rectangular box of 1 unit height located 2 units above the XY plane (Figure 2-29).

Specify first corner point or [Chamfer/Elevation/Fillet/Thickness/Width]: **E** Enter
Specify elevation for rectangles <0.0000>: *Enter a value.*

To view the objects in 3D space, change the viewpoint (Select **View, 3D Viewpoint > SE Isometric**).

Note

The value you enter for Fillet, Width, Elevation, and Thickness becomes the current value for the subsequent rectangle command. Therefore, you must reset the values if they are different from the current values. The thickness of rectangle is always controlled by its thickness settings.

The rectangle generated on the screen is treated as a single object. Hence, the individual sides can be edited only after the rectangle has been exploded using the ***EXPLODE*** *command.*

Exercise 6 — *General*

Draw a rectangle 4 units long, 3 units high, and with its first corner at (1,1). Inside the first rectangle draw another rectangle whose sides are 0.5 units.

DRAWING ELLIPSES

Toolbar:	Draw > Ellipse
Menu:	Draw > Ellipse
Command:	ELLIPSE

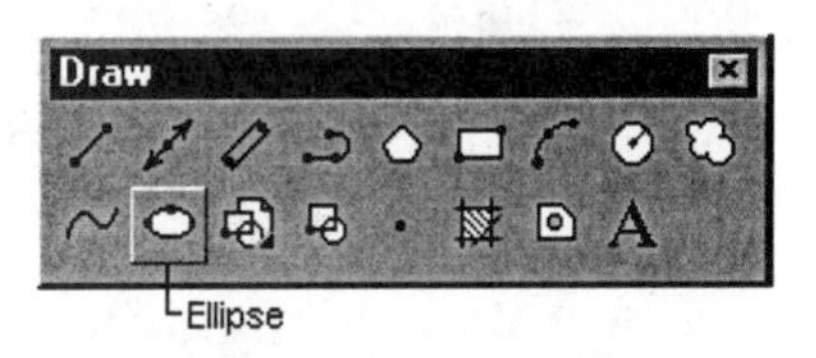

Figure 2-30 *Invoking the* ***ELLIPSE*** *command from the* ***Draw*** *toolbar*

If a circle is observed from an angle, the shape seen is called an **Ellipse**. An ellipse can be created using various options listed within the **ELLIPSE** command. If the **PELLIPSE** is set to 0 (default), AutoCAD LT creates a true ellipse, also known as NURBS-based (Non-Uniform Rational Bezier Spline) ellipse. The true ellipse has a center and quadrant points. If you select it, the grips will be displayed at the center and the quadrant points of the ellipse. If you move one of the grips located on the perimeter of the ellipse, the major or minor axis will change, which changes the size of the ellipse, as shown in Figure 2-32(d).

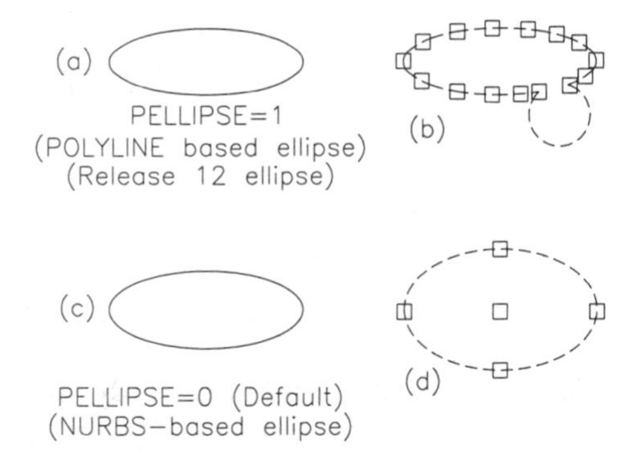

Figure 2-32 *Drawing polyline and NURBS-based ellipses*

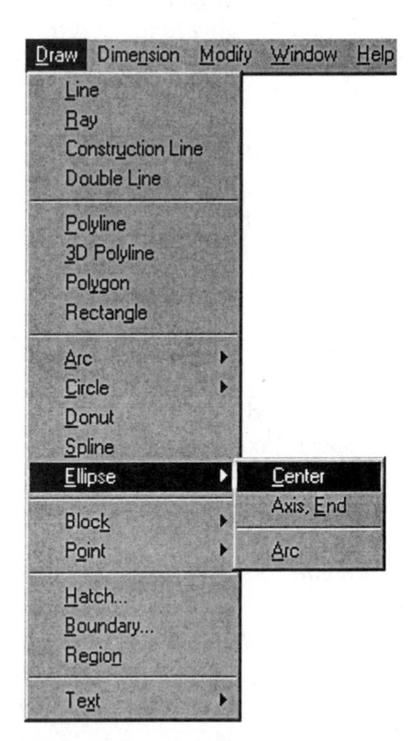

Figure 2-31 *Invoking the* ***ELLIPSE*** *command from the* ***Draw*** *menu*

Note

Up to AutoCAD LT 2.0, ellipses were based on polylines. They were made of multiple polyarcs and as a result, it was difficult to edit an ellipse. For example, if you select an ellipse, the grips will be displayed at the endpoints of each polyarc. If you move a vertex point, you get the shape shown in Figure 2-32(b). Also, you cannot snap to the center or the quadrant points of a polyline-based ellipse. In AutoCAD LT 2000, you can still draw the polyline-based ellipse by setting the value of ***PELLIPSE*** *system variable to 1.*

Once you invoke the **ELLIPSE** command, AutoCAD LT will acknowledge with the prompt **Specify axis endpoint of ellipse or [Arc/Center]** or **Specify axis endpoint of ellipse or [Arc/Center/Isocircle]** (if isometric snap is on). The response to this prompt depends on the option you want to choose. The different options are explained next.

Drawing an Ellipse Using the Axis and Eccentricity Option

In this option you draw an ellipse by specifying one of its axes and its eccentricity. To use this option, acknowledge the **Specify axis endpoint of ellipse or [Arc/Center]**: prompt by specifying a point, either by specifying a point using a pointing device or by entering its coordinates. This is the first endpoint of one axis of the ellipse. AutoCAD LT will then respond with the prompt **Specify other endpoint of axis:**. Here, specify the other endpoint of the axis. The angle at which the ellipse is drawn depends on the angle made by these two axis endpoints. Your response to the next prompt determines whether the axis is the **major axis** or the **minor axis.**

The next prompt is **Specify distance to other axis or [Rotation]:**. If you specify a distance, it is presumed as half the length of the second axis. You can also specify a point. The distance from this point to the midpoint of the first axis is again taken as half the length of this axis. The ellipse will pass through the selected point only if it is perpendicular to the midpoint of the first axis. To visually analyze the distance between the selected point and the midpoint of the first axis, AutoCAD LT appends an elastic line to the crosshairs, with one end fixed at the midpoint of the first axis. You can also drag the point, dynamically specifying half of the other axis distance. This helps you to visualize the ellipse. The prompt sequence for drawing an ellipse with one axis endpoint is located at (3,3), the other at (6,3), and the distance of the other axis being 1 (Figure 2-33) is:

Command: **ELLIPSE** Enter
Specify axis endpoint of ellipse or [Arc/Center]: **3,3** Enter
Specify other endpoint of axis : **6,3** Enter
Specify distance to other axis or [Rotation]: **1** Enter *(You can also enter @1<90 or @1<0)*

Another example for drawing an ellipse (Figure 2-34) using the **Axis and Eccentricity** option is illustrated by the following prompt sequence:

Command: **ELLIPSE** Enter
Specify axis endpoint of ellipse or [Arc/Center]: **3,3** Enter
Specify other endpoint of axis : **4,2** Enter
Specify distance to other axis or [Rotation]: **2** Enter

If you enter ROTATION or R at the **Specify distance to other axis or [Rotation]**: prompt, the first axis specified is automatically taken as the major axis of the ellipse. The next prompt is **Specify rotation around major axis:**. The major axis is taken as the diameter line of the circle, and the rotation takes place around this diameter line into the third dimension. The ellipse is formed when AutoCAD LT projects this rotated circle into the drawing plane. You can enter the rotation angle value in the range of 0 to 89.4 degrees only, because an angle value greater than 89.4 degrees changes the circle into a line. Instead of entering a definite angle value at the **Specify Rotation around major axis:** prompt, you can specify a point relative to the midpoint

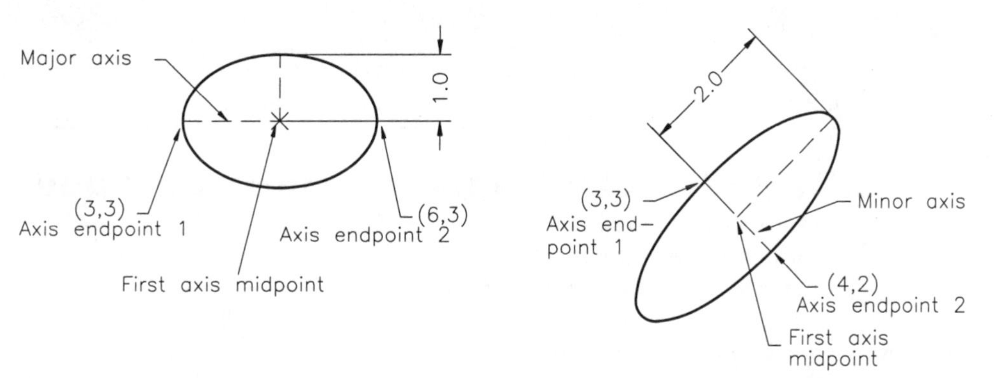

Figure 2-33 *Drawing an ellipse using the Axis and Eccentricity option*

Figure 2-34 *Drawing an ellipse using the Axis and Eccentricity option*

of the major axis. This point can be dragged to specify the ellipse dynamically. The following is the prompt sequence for a rotation of 0 degrees around the major axis, as shown in Figure 2-35(a).

Command: **ELLIPSE** [Enter]
Specify axis endpoint of ellipse or [Arc/Center]: *Select point (P1)* [Enter]
Specify other endpoint of axis : *Select another point (P2)* [Enter]
Specify distance to other axis or [Rotation]: **R** [Enter]
Specify rotation around major axis: **0** [Enter]

The Figure 2-35 also shows rotations of 45 degrees, 60 degrees, and 89.4 degrees.

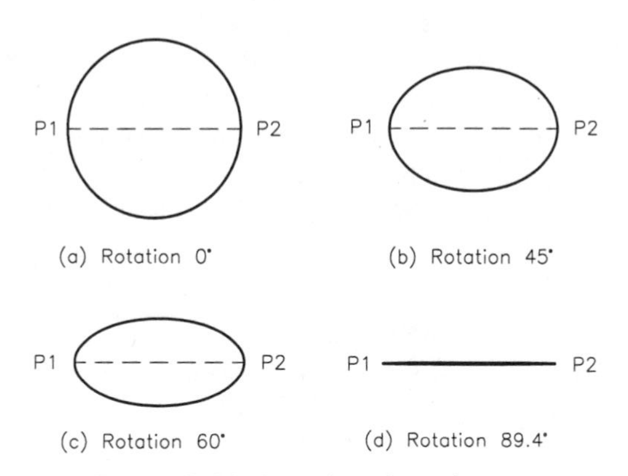

Figure 2-35 *Rotation about the major axis*

Note

*The **Isocircle** option is not available if the **Isometric** suboption in the **Style** option of the **SNAP** command is set to Off. In this section the **SNAP** mode **Isometric** option is not being used.*

*The **Arc** option is not available if you have set the **PELLIPSE** system variable to 1.*

Exercise 7 — General

Draw an ellipse whose major axis is 4 units and whose rotation around this axis is 60 degrees. Draw another ellipse, whose rotation around the major axis is 15 degrees.

Drawing Ellipse Using the Center and Two Axes Option

In this option you can construct an ellipse by specifying the center point, the endpoint of one axis, and the length of the other axis. The only difference between this method and the ellipse by axis and eccentricity method is that instead of specifying the second endpoint of the first axis, the center of the ellipse is specified. The center of an ellipse is defined as the point of intersection of the major and minor axes. In this option the first axis need not be the major axis. For example, to draw an ellipse with center at (4,4), axis endpoint at (6,4), and length of the other axis as 2 units (Figure 2-36), the command sequence is:

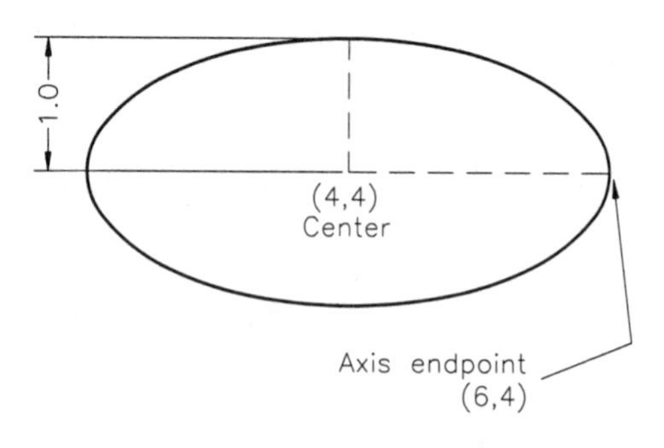

***Figure 2-36** Drawing an ellipse using the Center option*

Command: **ELLIPSE** Enter
Specify axis endpoint of ellipse or [Arc/Center]: **C** Enter
Specify center of ellipse: **4,4** Enter
Specify end point of axis: **6,4** Enter
Specify distance to other axis or [Rotation]: **1** Enter

Instead of entering the distance, you can enter ROTATION or R at the **Specify distance to other axis [Rotation]:** prompt. This takes the first axis specified as the major axis. The next prompt, **Specify rotation around major axis:**, prompts you to enter the rotation angle value. The rotation takes place around the major axis, which is taken as the diameter line of the circle. The rotation angle values should range from 0 to 89.4 degrees.

Drawing Elliptical Arcs

You can use the **Arc** option of the **ELLIPSE** command to draw an elliptical arc. When you enter the **ELLIPSE** command and select the **Arc** option, AutoCAD LT will prompt you to enter information about the geometry of the ellipse and the arc limits. You can define the arc limits by using the following options:

1. Start and End angle of the arc
2. Start and Included angle of the arc
3. Specifying Start and End parameters

The angles are measured from the first point and in a counterclockwise direction if AutoCAD LT's default setup has not been changed. The following example illustrates the use of these three options.

Example 1

Draw the following elliptical arcs as shown in Figure 2-37:

a. Start angle = -45, end angle = 135
b. Start angle = -45, included angle = 225
c. Start parameter = @1,0, end parameter = @1<225

Specifying Start and End Angle of the Arc [Figure 2-37(a)]

Command: **ELLIPSE**
Specify axis endpoint of ellipse or [Arc/Center]: **A** Enter
Specify axis endpoint of elliptical arc or [Center]: *Select the first endpoint.*
Specify other endpoint of axis : *Select the second point.*
Specify distance to other axis or [Rotation]: *Select a point or enter a distance.*
Specify start angle or [Parameter]: **-45** Enter
Specify end angle or [Parameter/Included angle]: **135** Enter *(Angle where arc ends.)*

Specifying Start and Included Angle of the Arc [Figure 2-37(b)]

Command: **ELLIPSE**
Specify axis endpoint of ellipse or [Arc/Center]: **A** Enter
Specify axis endpoint of elliptical arc or [Center]: *Select the first endpoint.*
Specify other endpoint of axis : *Select the second point.*
Specify distance to other axis or [Rotation]: *Select a point or enter a distance.*
Specify start angle or [Parameter]: **-45** Enter
Specify end angle or [Parameter/Included angle]: **I** Enter
Specify included angle for arc<current>: 225 Enter *(Included angle.)*

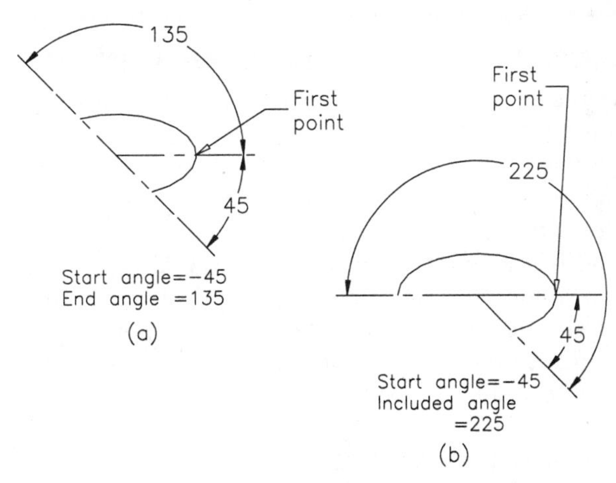

Figure 2-37 *Drawing elliptical arcs*

Chapter 2

Specifying Start and End Parameters (Figure 2-38)

Command: **ELLIPSE**
Specify axis endpoint of ellipse or [Arc/Center]: **A** Enter
Specify axis endpoint of elliptical arc or [Center]:*Select the first endpoint*
Specify other endpoint of axis : *Select the second endpoint.*
Specify distance to other axis or [Rotation]: *Select a point or enter a distance.*
Specify start angle or [Parameter]: **P**
Specify start parameter or [Angle]: **@1,0**
Specify end parameter or angle[Angle/Included angle]: **@1<225**

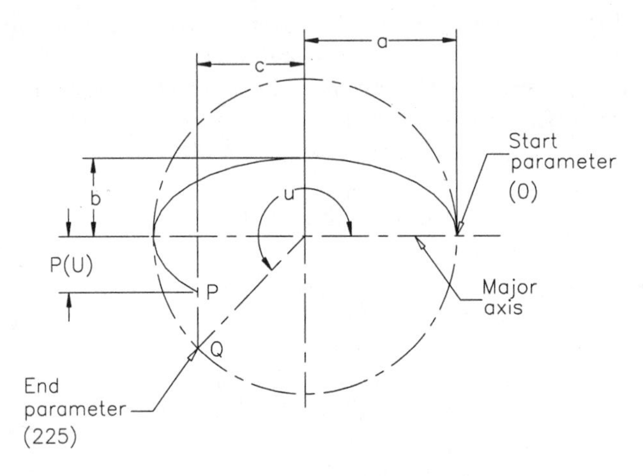

***Figure 2-38** Drawing an elliptical arc by specifying the start and end parameters*

Calculating Parameters for an Elliptical Arc

The start and end parameters of an elliptical arc are determined by specifying a point on the circle whose diameter is equal to the major diameter of the ellipse as shown in Figure 2-39. In this drawing, the major axis of the ellipse is 2.0 and the minor axis is 1.0. The diameter of the circle is 2.0. To determine the start and end parameters of the elliptical arc, you must specify the points on the circle. In the example, the start parameter is @1,0 and the end parameter is @1<225. Once you specify the points on the circle, AutoCAD LT will project these points on the major axis and determine the endpoint of the elliptical arc. In the Figure 2-39, Q is the end parameter of the elliptical arc. AutoCAD LT projects point Q on the major axis and locates intersection point P, which is the endpoint of the elliptical arc. The coordinates of point P can be calculated by using the following equations:

The equation of an ellipse with center as origin is
x^2/a^2 + y^2/b^2 = 1
In parametric form x = a * cos(u)
y = b * sin(u)
For the example a = 1
b = 0.5

Therefore x = 1 * cos(225) = -0.707
y = 0.5 * sin(225) = -0.353

The coordinates of point P are (-0.707, -0.353) with respect to the center of the ellipse.

Note: v = atan(b/a*tan(u)) = end angle
v = atan(0.5/1*tan(225)) = 206.56o

Also e = 1-b^2/a^2)^.5 = eccentricity
e = 1-.5^2/1^2)^.5 = .866
r = x^2 + y^2)^.5
r = .707^2 + .353^2)^.5 = 0.790

or using the polar equation r = b/(1 - e^2 * cos(v)^2)^.5
r = .5/(1 - .866^2 * cos(206.56)^2)^.5
r = 0.790

Exercise 8 — *Mechanical*

a. Construct an ellipse with center at (2,3), axis endpoint at (4,6), and the other axis endpoint a distance of 0.75 units from the midpoint of the first axis.

b. Make the drawing as shown in Figure 2-39. The distance between the dotted lines is 1.0 units. Create the elliptical arcs using **ELLIPSE** command options.

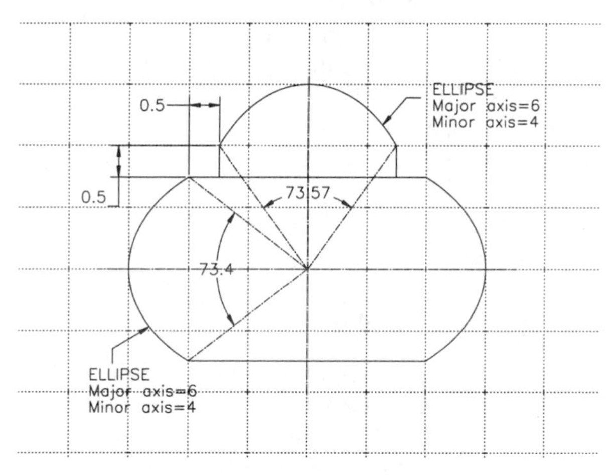

***Figure 2-39** Drawing for Exercise 8(b)*

DRAWING REGULAR POLYGONS

Toolbar:	Draw > Polygon
Menu:	Draw > Polygon
Command:	POLYGON

A **regular polygon** is a closed geometric figure with equal sides and equal angles. The number of sides varies from **3** to **1024**. For example, a triangle is a three-sided polygon and a pentagon is a five-sided polygon. In AutoCAD LT, the **POLYGON** command is used to draw regular two-dimensional polygons. The characteristics of a polygon drawn in AutoCAD LT are those

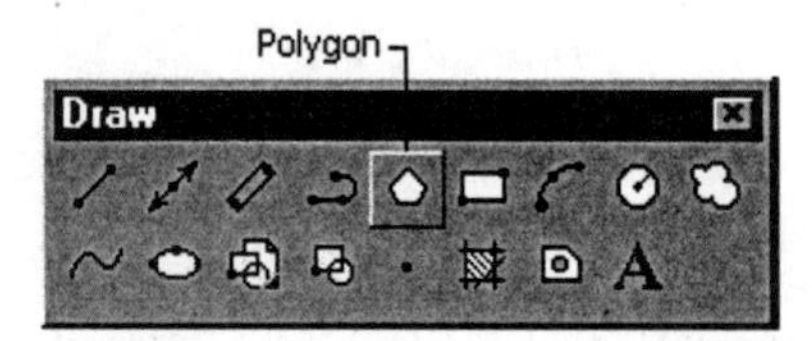

*Figure 2-40 Invoking the **POLYGON** command from the **Draw** toolbar*

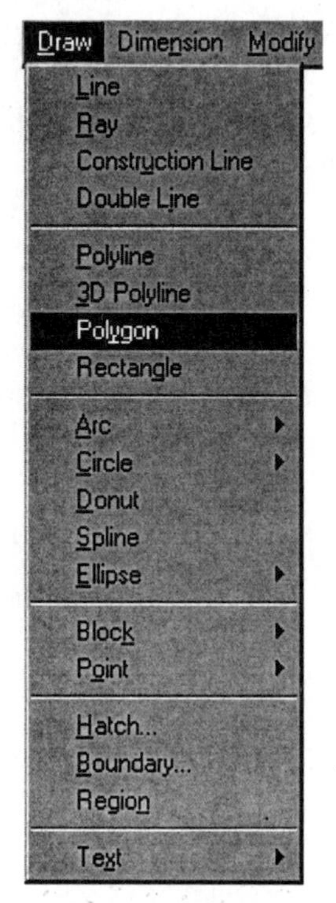

*Figure 2-41 Invoking the **POLYGON** command from the **Draw** menu*

of a closed polyline having 0 width. You can change the width of the polyline forming the polygon. The prompt sequence is:

Command: **POLYGON** [Enter]
Enter number of sides <4>:

Once you invoke the **POLYGON** command, it prompts you to enter the number of sides. The number of sides determines the type of polygon (for example, 6 sides defines a hexagon). The default value for the number of sides is **4**. You can change the number of sides to your requirement (in the range of 3 to 1024) and then the new value becomes the default. You can also set a different default value for the number of sides by using the **POLYSIDES** system variable. For example, if you want the default for the number of sides to be **3**, the prompt sequence is:

Command: **POLYSIDES** [Enter]
Enter new value for POLYSIDES <4>: **3** [Enter]

The Center of Polygon Option

The default option prompts you to select a point that is taken as the center point of the polygon. The next prompt is **Enter an option[Inscribed in circle/Circumscribed about circle]<I>:**. A polygon is said to be inscribed when it is drawn inside an imaginary circle and its vertices (corners) touch the circle (Figure 2-42). Likewise, a polygon is **circumscribed** when it is drawn outside the imaginary circle and the sides of the polygon are tangent to the circle (midpoint of each side of the polygon will lie on the circle) (Figure 2-43). If you want to have an inscribed polygon, enter **I** at the prompt. The next prompt issued is **Specify radius of circle:**. Here you are required to specify the radius of the circle on which all the vertices of the polygon will lie. Once you specify the radius, a polygon will be generated. If you want to select the circumscribed option, enter **C** at the prompt **Enter an option[Inscribed in circle/ Circumscribed about circle]<I>:**. After this enter the radius of the circle. The inscribed or circumscribed circle is not drawn on the screen. The radius of the circle can be dynamically dragged instead of a numerical value entered. The prompt sequence for drawing an inscribed octagon with center at (4,4), and a radius of 1.5 units (Figure 2-42) is:

Command: **POLYGON** Enter
Enter number of sides<4>: **8** Enter
Specify center of polygon or [Edge]: **4,4** Enter
Enter an option[Inscribed in circle/Circumscribed about circle]<I>: **I** Enter
Specify radius of circle: **1.5** Enter

The prompt sequence for drawing a circumscribed pentagon (polygon with five sides) with center at (9,4) and a radius of 1.5 units (Figure 2-43) is:

Command: **POLYGON** Enter
Enter number of sides<4>: **5** Enter
Specify center of polygon or [Edge]: **9,4** Enter
Enter an option[Inscribed in circle/Circumscribed about circle]<I>: **C** Enter
Specify radius of circle: **1.5** Enter

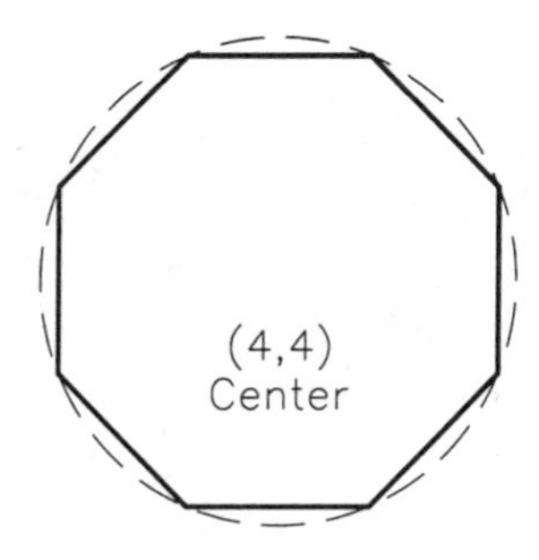

Figure 2-42 Drawing an inscribed polygon using the Center of Polygon option

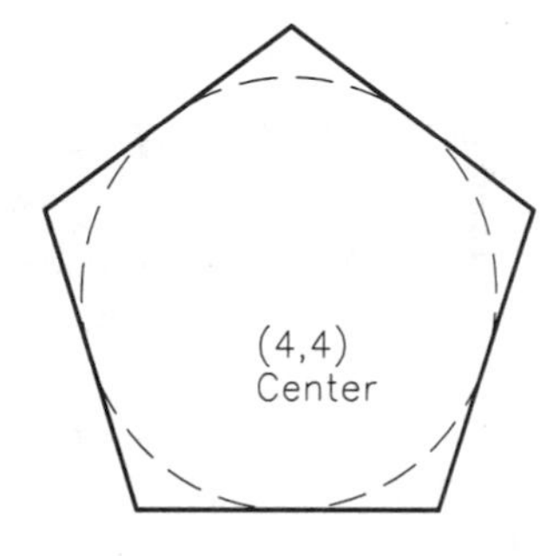

Figure 2-43 Drawing a circumscribed polygon using the Center of Polygon option

Note

If you select a point to specify the radius of an inscribed polygon, one of the vertices is positioned on the selected point. In the case of circumscribed polygons, the midpoint of an edge is placed on the point you have specified. In this manner you can specify the size and rotation of the polygon.

In the case of numerical specification of the radius, the bottom edge of the polygon is rotated by the prevalent snap rotation angle.

Exercise 9 *General*

Draw a circumscribed polygon of eight sides. The polygon should be drawn by the **Center of Polygon** method.

The Edge Option

The other method for drawing a polygon is to select the **Edge** option. This can be done by entering EDGE or E at the **Specify center of polygon or [Edge]:** prompt. The next two prompts issued are **Specify first endpoint of edge:** and **Specify second endpoint of edge:**. Here you need to specify the two endpoints of an edge of the polygon. The polygon is drawn in a counterclockwise direction, with the two points entered defining its first edge. To draw a hexagon (six-sided polygon) using the **Edge** option, with the first endpoint of the edge at (2,4) and the second endpoint of the edge at (2,2.5) (Figure 2-44), the following will be the prompt sequence:

Command: **POLYGON** Enter
Enter number of sides<4>: **6** Enter
Specify center of polygon or [Edge]: **E** Enter
Specify first endpoint of edge: **2,4** Enter
Specify second endpoint of edge: **2,2.5** Enter

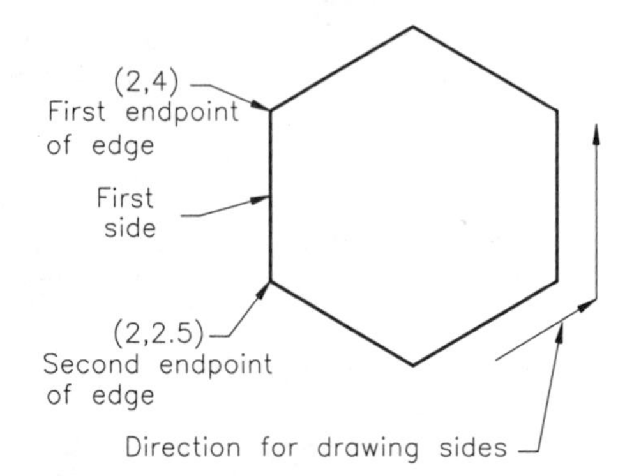

Figure 2-44 *Drawing a polygon (hexagon) using the Edge option*

Exercise 10 *General*

Draw a polygon with 10 sides using the **Edge** option and an elliptical arc as shown in Figure 2-45. Let the first endpoint of the edge be at (7,1) and the second endpoint be at (8,2).

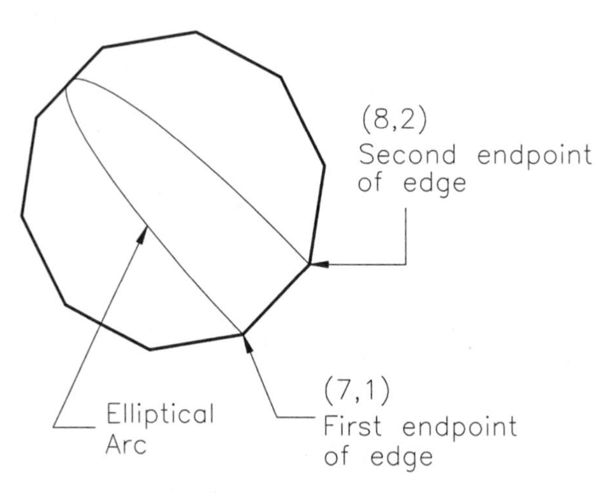

Figure 2-45 *Polygon and elliptical arc for Exercise 10*

DRAWING POLYLINES

Toolbar:	Draw > Polyline
Menu:	Draw > Polyline
Command:	PLINE

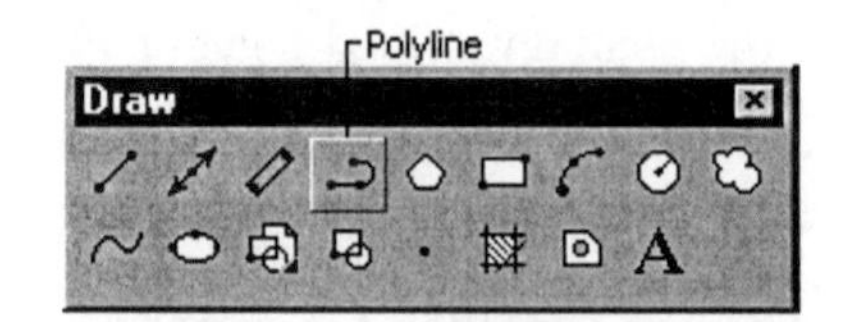

Figure 2-46 *Invoking the **Polyline** command from the **Draw** toolbar*

A **polyline** is a line that can have different characteristics. The term POLYLINE can be broken into two parts: POLY and LINE. POLY means "many". This signifies that a polyline can have many features. Some of the features of polylines are:

1. Polylines are thick lines having a desired width.
2. Polylines are very flexible and can be used to draw any shape, such as a filled circle or a doughnut.
3. Polylines can be used to draw objects in any linetype (for example, hidden linetype).
4. Advanced editing commands can be used to edit polylines (for example, **PEDIT** command).
5. A single polyline object can be formed by joining polylines and polyarcs of different thickness.
6. It is easy to determine the area or perimeter of a polyline feature. Also, it is easy to offset when drawing walls.

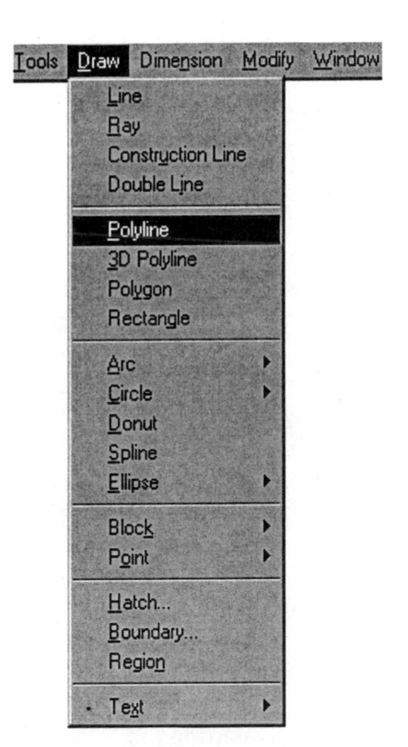

Figure 2-48 *Invoking the* ***Polyline*** *command from the* ***Draw*** *menu*

The command to draw a polyline is **PLINE**. The **PLINE** command functions fundamentally like the **LINE** command, except that additional options are provided and all the segments of the polyline form a single object. After invoking the **PLINE** command, the next prompt is:

Specify start point: *Specify the starting point or enter its coordinates.*
Current line width is nn.n.

Current line width is nn.n is displayed automatically, which indicates that the polyline drawn will have nn.n width. If you want the polyline to have a different width, invoke the **Width** option at the next prompt and then set the polyline width. The next prompt is:

Specify next point or [Arc/Close/Halfwidth/Length/Undo/Width]: *Specify next point or enter an option.*

The options that can be invoked at this prompt, depending on your requirement, are:

Next Point of Line

This option is maintained as the default and is used to specify the next point of the current polyline segment. If additional polyline segments are added to the first polyline, AutoCAD LT automatically makes the endpoint of the first polyline segment the start point of the next polyline segment. The prompt sequence is:

Command: **PLINE** Enter
Specify start point: *Specify the starting point of the polyline.*
Current line width is 0.0000.
Specify next point or [Arc/Close/Halfwidth/Length/Undo/Width]: *Specify the endpoint of the first polyline segment.*
Specify next point or [Arc/Close/Halfwidth/Length/Undo/Width]: *Specify the endpoint of*

the second polyline segment, or press ENTER to exit the command.

Width

You can change the current polyline width by entering W (**Width** option) at the last prompt. Then you are prompted for the starting width and the ending width of the polyline.

Specify next point or [Arc/Close/Halfwidth/Length/Undo/Width]: **W** Enter
Specify starting width <current>: *Specify the starting width.*
Specify ending width <starting width>: *Specify the ending width.*

The starting width value is taken as the default value for the ending width. Hence, to have a uniform polyline you need to give a null response (press ENTER) at the **Specify ending width <starting width>** prompt. The start point and endpoint of the polyline are located at the center of the line width. To draw a uniform polyline (Figure 2-48) with a width of 0.25 units, a start point at (4,5), an endpoint at (5,5), and the next endpoint at (3,3), the following will be the prompt sequence:

Command: **PLINE** Enter
Specify start point: **4,5** Enter
Current line-width is 0.0000
Specify next point or [Arc/Close/Halfwidth/Length/Undo/Width]: **W** Enter
Specify starting width <current>: **0.25** Enter
Specify ending width <0.25>: Enter
Specify next point or [Arc/Close/Halfwidth/Length/Undo/Width]: **5,5** Enter
Specify next point or [Arc/Close/Halfwidth/Length/Undo/Width]: **3,3** Enter
Specify next point or [Arc/Close/Halfwidth/Length/Undo/Width]: Enter

You can get a tapered polyline by entering two different values at the starting width and the ending width prompts. To draw a tapered polyline (Figure 2-49) with a starting width of 0.5 units and an ending width of 0.15 units, a start point at (2,4), and an endpoint at (5,4), the prompt sequence is:

Command: **PLINE** Enter
Specify start point: **2,4** Enter
Current line-width is 0.0000
Specify next point or [Arc/Close/Halfwidth/Length/Undo/Width]: **W** Enter
Specify starting width <0.0000>: **0.50** Enter
Specify ending width <0.50>: **0.15** Enter
Specify next point or [Arc/Close/Halfwidth/Length/Undo/Width]: **5,4** Enter
Specify next point or [Arc/Close/Halfwidth/Length/Undo/Width]: Enter

Close

This option closes the polyline by drawing a polyline segment from the most recent endpoint to the initial start point, and on doing so exits from the **PLINE** command. This option can be invoked by entering Close or C at the following prompt:

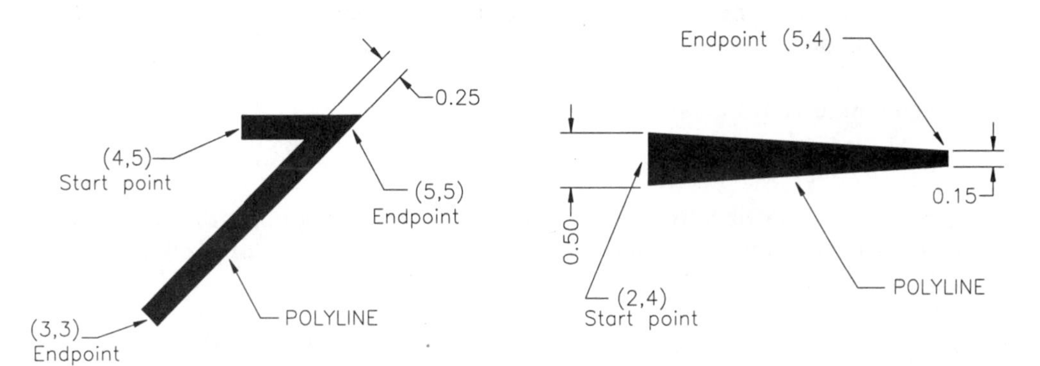

Figure 2-48 *Drawing a uniform polyline using the **PLINE** command*

Figure 2-49 *Drawing a tapered polyline using the **PLINE** command*

Specify next point or [Arc/Close/Halfwidth/Length/Undo/Width]: **C** Enter

The width of the closing segment can be changed by using the **Width/Halfwidth** option before invoking the **Close** option.

Halfwidth

With this option you can specify the starting and ending halfwidth of a polyline. This halfwidth distance is equal to half of the actual width of the polyline. This option can be invoked by entering Halfwidth or H at the following prompt:

Specify next point or [Arc/Close/Halfwidth/Length/Undo/Width]: **H** Enter
Specify starting half-width <0.0000>: **0.12** Enter *(Specify desired starting halfwidth.)*
Specify ending half-width <0.1200>: **0.05** Enter *(Specify desired ending halfwidth.)*

Length

This option prompts you to enter the length of a new polyline segment. The new polyline segment will be the length you have entered. It will be drawn at the same angle as the last polyline segment or tangent to the previous polyarc segment. This option can be invoked by entering Length or L at the following prompt:

Specify next point or [Arc/Close/Halfwidth/Length/Undo/Width]: **L** Enter
Specify length of line: *Specify the desired length of the Pline.*

Undo

This option erases the most recently drawn polyline segment. This option can be invoked by entering Undo or U at the following prompt:

Specify next point or [Arc/Close/Halfwidth/Length/Undo/Width]: **UNDO** or **U** Enter

You can use this option repeatedly until you reach the start point of the first polyline segment. Further use of **Undo** option evokes this message:

All segments already undone.

Arc

This option is used to switch from drawing polylines to drawing polyarcs, and provides you the options associated with drawing polyarcs. The **Arc** option can be invoked by entering Arc or A at the following prompt:

Specify next point or [Arc/Close/Halfwidth/Length/Undo/Width]: **A** Enter

The next prompt is:

Specify end point of arc or [Angle/CEnter/CLose/Direction/Halfwidth/Line/Radius/Second pt/Undo/Width]: *Enter an option.*

By default the arc segment is drawn tangent to the previous segment of the polyline. The direction of the previous line, arc, or polyline segment is default for polyarc. The preceding prompt contains options associated with the PLINE Arc. The detailed explanation of each of these options follows:

Angle. This option prompts you to enter the included angle for the arc. If you enter a positive angle, the arc is drawn in a counterclockwise direction from the start point to the endpoint. If the angle specified is negative, the arc is drawn in a clockwise direction. The prompt issued for this option is:

Specify included angle: *Specify the included angle.*

The next prompt is:

Specify endpoint of arc or [Center/Radius]:

Center refers to the center of the arc segment, Radius refers to the radius of the arc, and Endpoint draws the arc.

CEnter. This option prompts you to specify the center of the arc to be drawn. As mentioned before, usually the arc segment is drawn so that it is tangent to the previous polyline segment; in such cases AutoCAD LT determines the center of the arc automatically. Hence, the **CEnter** option provides the freedom to choose the center of the arc segment. The **CEnter** option can be invoked by entering **CE** at the **Specify end point of arc or [Angle/CEnter/CLose/Direction/Halfwidth/Line/Radius/Second pt/Undo/Width]:** prompt. Once you specify the center point, AutoCAD LT issues the following prompt:

Specify endpoint of arc or [Angle/Length]:

Angle refers to the included angle, Length refers to the length of the chord, and Endpoint refers to the endpoint of the arc.

CLose. This option closes the polyline by drawing a polyarc segment from the previous endpoint to the initial start point, and on doing so exits from the **PLINE** command. The **CLose** option can be invoked by entering **CL**.

Direction. Usually the arc drawn with the **PLINE** command is tangent to the previous polyline segment. In other words, the starting direction of the arc is the ending direction of the previous segment. The Direction option allows you to specify the tangent direction of your choice for the arc segment to be drawn. The next prompt is:

Specify tangent direction for the start point of arc: *Specify the direction.*

You can also specify the direction by specifying a point. AutoCAD LT takes it as a direction from the starting point. Once the direction is specified, AutoCAD LT prompts:

Specify endpoint of arc: *Specify the endpoint of arc.*

Halfwidth. This option is the same as for the **Line** option and prompts you to specify the starting and ending halfwidth of the arc segment.

Line. This option takes you back to the Line mode. You can draw polylines only in Line mode.

Radius. This option prompts you to specify the **radius** of the arc segment. The prompt sequence is:

Specify radius of arc: *Specify the radius of the arc segment.*
Specify endpoint of arc or [Angle]:

If you specify a point, the arc segment is drawn (Figure 2-50). If you enter an angle, you will have to specify the angle and the direction of the chord at the **Specify included angle** and **Specify direction of chord for arc<current>** prompts, respectively.

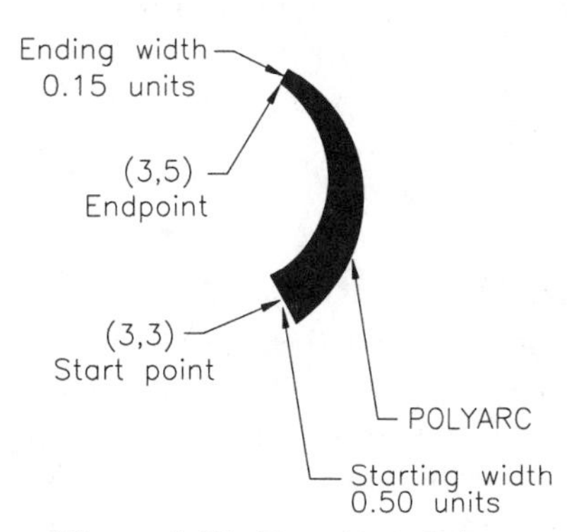

Figure 2-50 *Drawing a Polyarc*

Second pt. This option selects the second point of an arc in the **three-point** arc option. The prompt sequence is:

Specify second point of arc: *Specify the second point on the arc.*
Specify endpoint of arc: *Specify the third point on the arc.*

Undo. This option reverses the changes made in the previously drawn segment.

Width. This option prompts you to enter the width of the arc segment. To draw a tapered arc segment you can enter different values at the starting width and ending width prompts. The prompt sequence is identical to that for the polyline. Also, a specified point on a polyline refers to the midpoint on its width.

Endpoint of arc. This option is maintained as the default and prompts you to specify the endpoint of the current arc segment. The following is the prompt sequence for drawing an arc with start point at (3,3), endpoint at (3,5), starting width of 0.50 units, and ending width of 0.15 units (Figure 2-50):

Command: **PLINE** [Enter]
Specify start point: **3,3** [Enter]
Current line-width is 0.0000
Specify nextpoint or [Arc/Close/Halfwidth/Length/Undo/Width]: **A** [Enter]
Specify endpoint of arc or [Angle/CEnter/CLose/Direction/Halfwidth/Line/Radius/Second pt/Undo/Width]: **W** [Enter]
Specify starting width <current>: **0.50** [Enter]
Specify ending width <0.50>: **0.15** [Enter]
Specify endpoint of arc or [Angle/CEnter/CLose/Direction/Halfwidth/Line/Radius/Second pt/Undo/Width]: **3,5** [Enter]
Specify endpoint of arc or [Angle/CEnter/CLose/Direction/Halfwidth/Line/Radius/Second pt/Undo/Width]: [Enter]

Note

*The **FILL** or **FILLMODE** applies to polylines and hatch patterns and the change is effective on regeneration. If **FILL** is on or if **FILLMODE** is 1, the polylines are drawn filled and the hatch patterns are visible. Also **PLINEGEN** controls the linetype pattern between the vertex points of a 2D polyline. A value of 0 centers the linetype for each polyline segment and 1 makes them continuous.*

Optimized Polylines

The optimized polylines also known as **lightweight polylines** are created when you use the **PLINE** command (an AutoCAD LT feature since AutoCAD LT 97). The optimized polylines are similar to regular 2D polylines in functionality, but the data base format of an optimized polylines is different from a 2D polyline. In case of an optimized polyline the vertices are not stored as separate entities, but as a single object with an array of information. This feature results in reduced object and file size.

If you use the **PEDIT** command on a optimized polyline and use spline fit or curve fit options, the polyline automatically becomes a regular 2D polyline. Also, when you load a AutoCAD LT 2.0 or earlier drawing, AutoCAD LT automatically converts the polylines into optimized polylines. Commands like **POLYGON, DONUT, ELLIPSE, PEDIT, BOUNDARY, RECTANG** create optimized polylines. To see the difference between these two polylines; draw a polyline consisting of several segments and then use the **LIST** command to list the information about the object. Now, use the **PEDIT** command to change this polyline into a splined polyline and then use the **LIST** command again. Notice the difference in the database associated with these polylines.

CONVERT Command

You can use the **CONVERT** command to manually change a 2D polyline to an optimized (lightweight) polyline. You can also use this command to change an AutoCAD LT 97 associative hatch pattern to AutoCAD LT 2000 hatch object, except the solid-fill hatches.

Command: **CONVERT** [Enter]
Enter type of objects to optimize [Hatch/Polylines/All]: *Enter an option.*
Enter object selection preference [Select/All] <All>: *Press Enter or S to Select.*

Exercise 12 *General*

Draw the objects shown in Figure 2-51 and Figure 2-52. Approximate the width of different polylines.

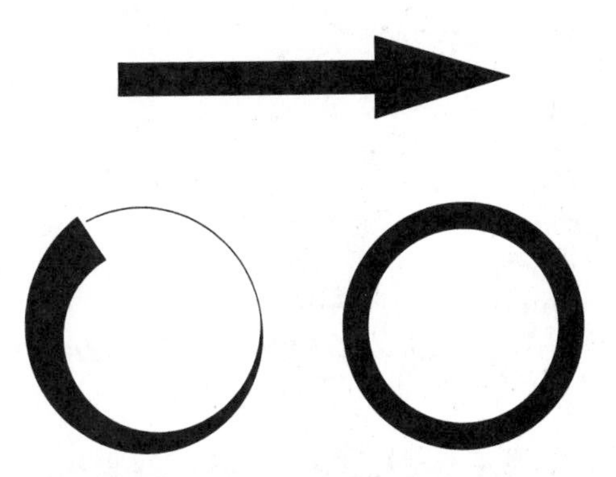

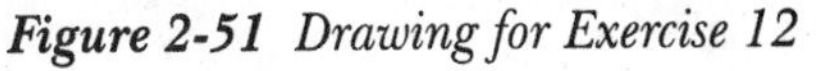

Figure 2-51 *Drawing for Exercise 12*

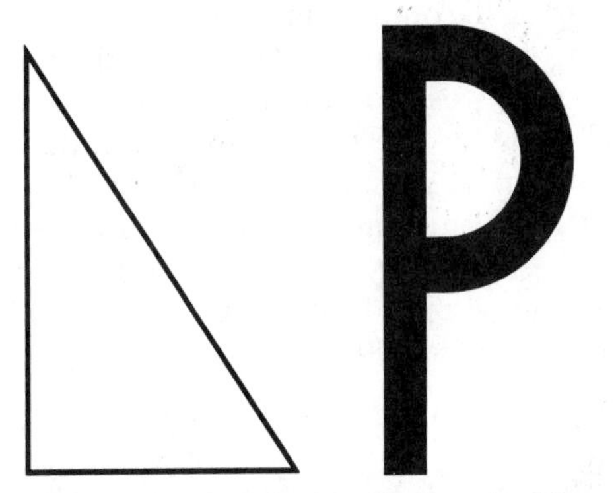

Figure 2-52 *Drawing for Exercise 12*

DRAWING DOUGHNUTS

Menu:	Draw > Donut
Command:	DONUT

In AutoCAD LT the **DONUT** command is issued to draw an object that looks like a filled circle ring called a donut. Actually, AutoCAD LT's donuts are made of two semicircular polyarcs having a certain width. Hence the **DONUT** command allows you to draw a thick circle. The donuts can have any inside and outside diameters. If the **FILLMODE** is off, the donuts look like circles (if the inside diameter is zero) or concentric circles (if the inside diameter is not zero). You can also select a point for the center of the donut anywhere on the screen with the help of a pointing device. The prompt sequence for drawing donuts is:

Command: **DONUT** [Enter]
Specify inside diameter of donut <current>: *Specify the inner diameter of the donut.*
Specify outside diameter of donut <current>: *Specify the outer diameter of the donut.*
Specify center of donut or <exit>: *Specify the center of the donut.*
Specify center of donut or <exit>: *Specify the center of the donut to draw more donuts of previous specifications or give a null response to exit.*

The defaults for the inside and outside diameters are the respective diameters of the most recent donut drawn. The values for the inside and outside diameters are saved in the **DONUTID** and **DONUTOD** system variables. You can specify a new diameter of your choice by entering a numeric value or by specifying two points to indicate the diameter. A solid-filled circle is drawn by specifying the inside diameter as zero (**FILLMODE** is on). Once the diameter specification is completed, the donuts are formed at the crosshairs and can be placed anywhere on the screen. The location at which you want the donuts to be drawn has to be specified at the **Specify center of donut or <exit>:** prompt. You can enter the coordinates of the point at that prompt or specify the point by dragging the center point. Once you have specified center of the donut, AutoCAD LT repeats the **Specify center of donut or <exit>:** prompt. As you go on specifying the locations for the center point, donuts with specified diameters at specified locations are drawn. To end the **DONUT** command, give a null response to this prompt by pressing ENTER. Since donuts are circular polylines, they can be edited with **PEDIT** command or any other editing command that can be used to edit polylines.

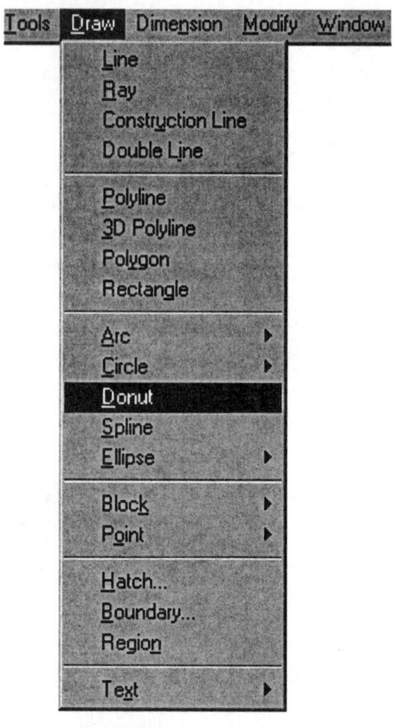

Figure 2-54 *Invoking the* ***DONUT*** *command from the* ***Draw*** *menu*

Example 2

If you want to draw an unfilled doughnut with an inside diameter of 0.75 units, an outside diameter of 2.0 units, and centered at (2,2) (Figure 2-54), the following is the prompt sequence:

Command: **FILLMODE** Enter
New value for FILLMODE <1>: **0** Enter
Command: **DONUT** Enter
Specify inside diameter of donut<0.5000>: **0.75** Enter
Specify outside diameter of donut <1.000>: **2** Enter
Specify center of donut or <exit>: **2,2** Enter
Specify center of donut or <exit>: Enter

The following is the prompt sequence for drawing a filled doughnut with an inside diameter of 0.5 units, outside diameter of 2.0 units, centered at a specified point (Figure 2-55).

Command: **FILLMODE** Enter
Enter new value for FILLMODE <1>: **1** Enter
Command: **DONUT** Enter
Specify inside diameter of donut<0.5000>: **0.50** Enter
Specify outside diameter of donut <1.000>: **2** Enter
Specify center of donut or <exit>: *Specify a point.*
Specify center of donut or <exit>: Enter

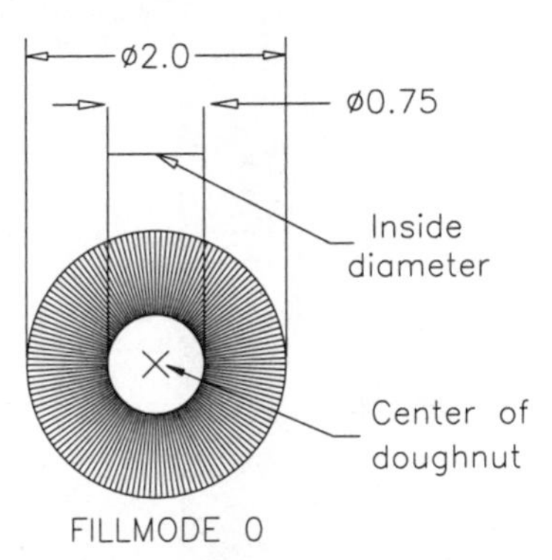

*Figure 2-54 Drawing an unfilled doughnut using the **DONUT** command*

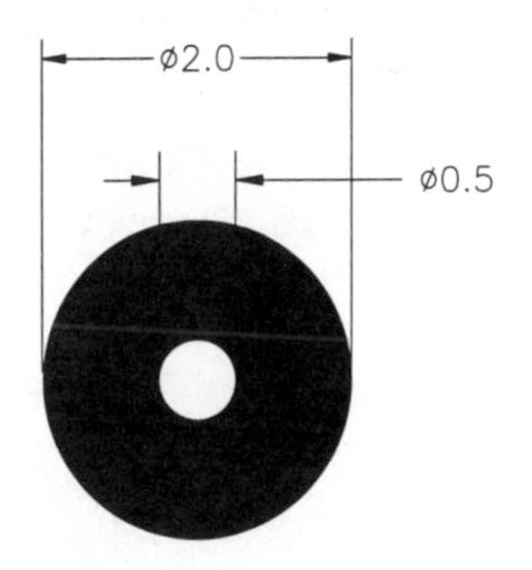

*Figure 2-55 Drawing a filled doughnut using the **DONUT** command*

To draw a solid-filled doughnut with an outside diameter of 2.0 units (Figure 2-56), the following is the prompt sequence:

Command: **DONUT** [Enter]
Specify inside diameter of donut <0.5000>: **0** [Enter]
Specify outside diameter of donut <1.000>: **2** [Enter]
Specify center of donut or <exit>: *Specify a point.*
Specify center of donut or <exit>: [Enter]

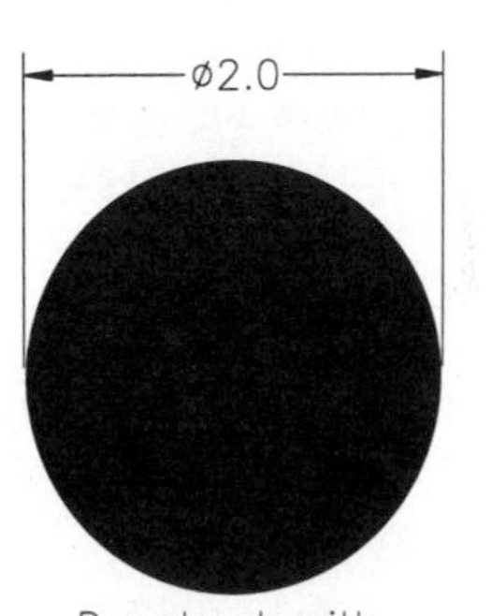

Figure 2-56 Solid-filled doughnut

DRAWING POINTS

Toolbar:	Draw > Point
Menu:	Draw > Point
Command:	POINT

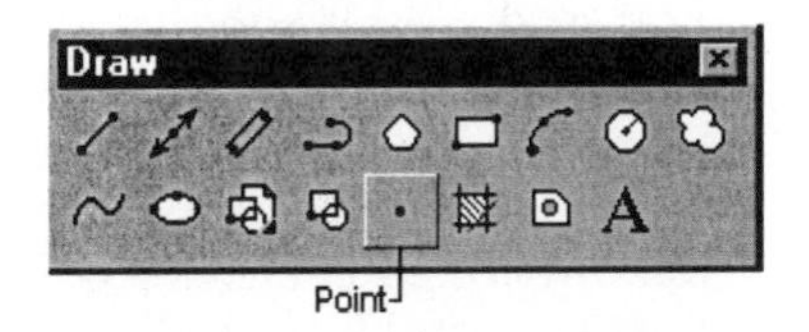

*Figure 2-57 Invoking the **POINT** command from the **Draw** toolbar*

The point is the basic drawing object. Points are invaluable in building a drawing file. To draw a point anywhere on the screen, AutoCAD LT provides the **POINT** command.

Command: **POINT** [Enter]
Current pointmodes: PDMODE=n PDSIZE=n.nn
Specify a point: *Specify the location where you want to plot the point.*

When a point is drawn, a mark appears on the screen. This mark, known as a **blip**, is the construction marker for the point. The blip mark is cleared once the screen is redrawn with the **REDRAW** command and a point of the specified point type is left on the screen. If you invoke the **POINT** by entering POINT at the Command prompt, you can draw only one point in a single point command. On the other hand, if you invoke the **POINT** command from the toolbar or the menu, you can draw as many points as you desire in a single command with **Multiple Point** option. In this case you can exit from the **POINT** command by pressing ESC.

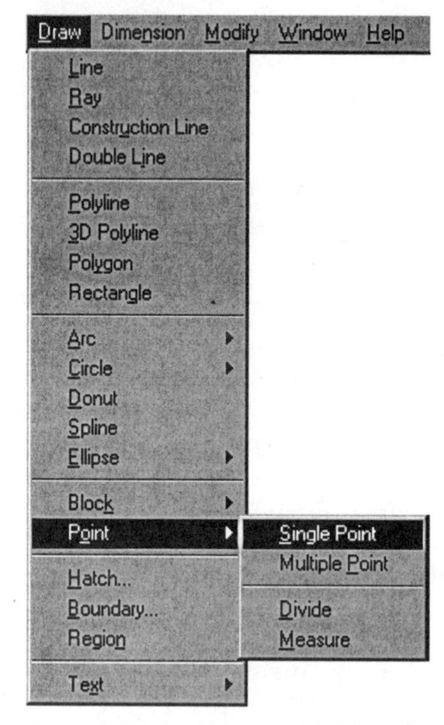

*Figure 2-58 Invoking the **POINT** command from the **Draw** menu*

Changing the Point Type

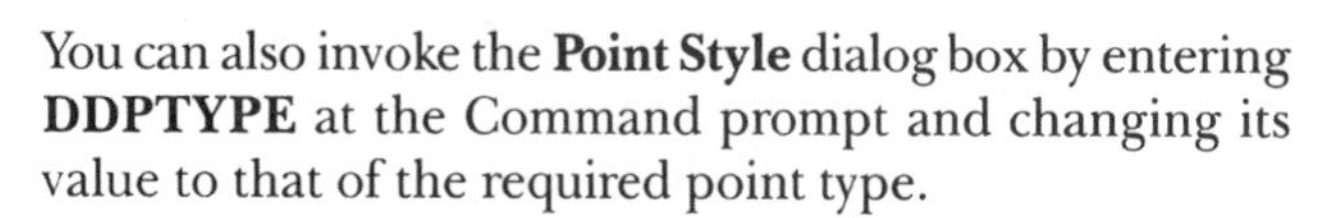

Menu:	Format > Point Style
Command:	DDPTYPE

The type of point drawn is controlled by the **PDMODE** (Point Display MODE) system variable. The point type can be set either from the **Point Style** dialog box or by entering **PDMODE** at the Command prompt. There are twenty combinations of point types. The **Point Style** dialog box can be accessed from the format menu (Select **Format** > **Point Style**). You can choose a point style in this dialog box by clicking your pointing device on the point style of your choice. A box is formed around that particular point style to acknowledge the selection made. Next, choose the **OK** button. Now all the points will be drawn in the selected style until you change it to a new style.

You can also invoke the **Point Style** dialog box by entering **DDPTYPE** at the Command prompt and changing its value to that of the required point type.

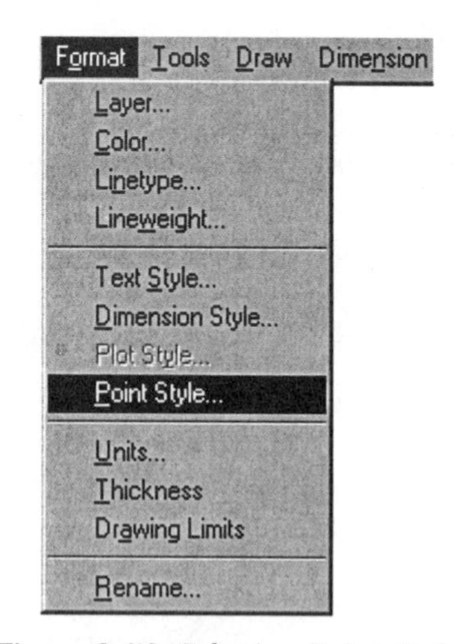

*Figure 2-59 Selecting **Point Style** from the **Format** menu*

Command: **PDMODE** [Enter]
Enter new value for PDMODE <current>: *Enter the new value.*
Command: **POINT** [Enter]
Specify a point: *Select a point.*

The **PDMODE** values for different point types are:

1. A value of 0 is the default for the **PDMODE** variable and generates a dot at the specified point.

2. A value of 1 for the **PDMODE** variable generates nothing at the specified point.

3. A value of 2 for the **PDMODE** variable generates plus mark (+) through the specified point.

4. A value of 3 for the **PDMODE** variable generates cross mark (X) through the specified point.

5. A value of 4 for the **PDMODE** variable generates a vertical line in the upward direction from the specified point.

6. When you add **32** to the **PDMODE** values of 0 to 4, a circle is generated around the symbol obtained from the original **PDMODE** value. For example, to draw a point having a cross mark and a circle around it, the value of the **PDMODE** variable will be **3 + 32 = 35.** Similarly, you can generate a square around the symbol with the **PDMODE** of value 0 to 4 by adding **64** to the original **PDMODE** value. For example, to draw a point having a plus mark and a square around it, the value of the **PDMODE** variable will be **2 + 64 = 66**. You can also generate a square and a circle around the symbol with the **PDMODE** of value 0 to 4 by adding **96** to the original **PDMODE** value. For example, to draw a point having a dot mark and a circle and a square around it, the value of the **PDMODE** variable will be **0 + 96 = 96.**

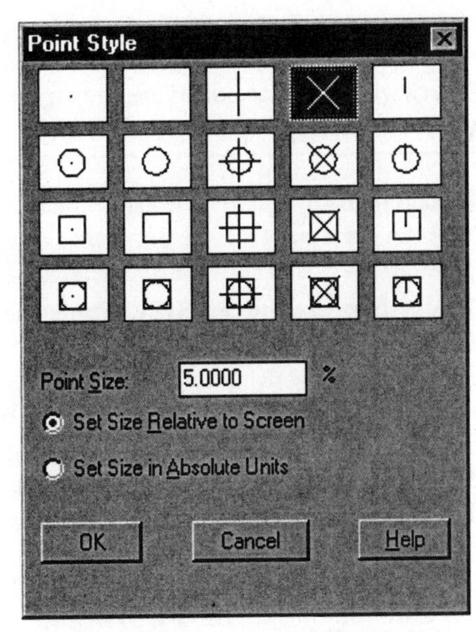

***Figure 2-61 Point Style** dialog box*

When the value of the **PDMODE** is changed or a new point type is selected from the dialog box, all of the previously drawn points retain their styles until the drawing is regenerated (Figure 2-26). After REGEN, all of the points on the drawing are drawn in the shape designated by the current value of the **PDMODE** variable. The **REGEN** command does the calculations for all the objects on the screen and draws them according to the new values obtained upon calculation.

Pdmode value	Point style	Pdmode value	Point style
0		64+0=64	
1		64+1=65	
2		64+2=66	
3		64+3=67	
4		64+4=68	
32+0=32		96+0=96	
32+1=33		96+1=97	
32+2=34		96+2=98	
32+3=35		96+3=99	
32+4=36		96+4=100	

***Figure 2-62** Different point types*

Exercise 13 *General*

Check what types of points are drawn for each value of the **PDMODE** variable. Use the **REGEN** command to regenerate the drawing, and notice the change in the previously drawn points.

Changing the Point Size

Command: PDSIZE

The system variable **PDSIZE** (Point Display SIZE) governs the size of the point (except for the **PDMODE** values of 0 and 1). The size of a point can be set from the **Point Style** dialog box (Figure 2-60) by entering the desired point size in the **Point Size** edit box. The point size can also be set by entering PDSIZE at the Command prompt and then changing its value to a new one (Figure 2-63). A value of 0 for the **PDSIZE** variable generates the point at 5 percent of the graphics area height. A positive value for **PDSIZE** defines an absolute size for the point. This can also be specified by selecting the **Set Size in Absolute Units** radio button in the **Point Style** dialog box. If **PDSIZE** is negative or if the **Set Size Relative to Screen** radio button is selected in the dialog box, the size is taken as a percentage of the viewport size, and as such the appearance (size) of the point is not altered by the use of the **ZOOM** command (the **ZOOM** command should perform regeneration). For example, a setting of 5 makes the point 5 units high; a setting of -5 makes the point 5 percent of the current drawing area. The prompt sequence for changing the size of the point is:

Figure 2-63 *Changing point size using the* ***PDSIZE*** *variable*

Command: **PDSIZE** [Enter]
Enter new value for PDSIZE <0.000>: **1** [Enter]

Command: **POINT** [Enter]
Specify a point: *Select a point.*

PDMODE and **PDSIZE** values can also be changed through the **SETVAR** command.

Exercise 14 *General*

a. Try various combinations of the **PDMODE** and **PDSIZE** variables.

b. Check the difference between the points generated from negative values of **PDSIZE** and points generated from positive values of **PDSIZE**. Use the **ZOOM** command to zoom in, if needed.

Self-Evaluation Test

Answer the following questions and then compare your answers to the correct answers given at the end of this chapter.

1. Different options for drawing arcs can be invoked by entering the **ARC** command and then selecting an appropriate letter. (T/F)

2. A negative value for chord length in the **St, C, Len** option results in the largest possible arc, also known as the major arc. (T/F)

3. If you do not specify a start point but just press ENTER or select the **Continue** option, the start point and direction of the arc will be taken from the endpoint and ending direction of the previous line or arc drawn on the current screen. (T/F)

4. If you select a point to specify the radius of a circumscribed polygon, one of the vertices is positioned on the selected point. In the case of inscribed polygons, the midpoint of an edge is placed on the point you have specified. (T/F)

5. When the value of **PDMODE** is changed, all the previously drawn points also change to the type designated by the **PDMODE** variable when the drawing is regenerated. (T/F)

6. Fill in the command, entries, and operations required to draw an arc with start point at (3,3), center at (2,3), and an included angle of 80 degrees.
   Command: ________
   Specify start of arc or [CEnter]: ________
   Specify second point of arc or [CEnter/ENd]: ________
   Specify center point of arc: ________
   Specify endpoint of arc or [Angle/chord Length]: ________
   Specify included angle: ________

7. Fill in the command, entries, and operations required to draw an ellipse with first and second axis endpoints are at (4,5) and (7,5), respectively, and the distance of the other axis 1 unit.
   Command: ________
   Specify axis endpoint of ellipse[Arc/Center]: ________
   Specify other endpoint of axis : ________
   Specify distance to other axis or [Rotation]: ________

8. Fill in the command, entries, and operations needed to erase the most recently drawn object in a figure comprising many objects.
   Command: ________
   Select objects: ________
   1 found
   Select objects: ________

9. The system variable ________ governs the size of the point.

10. The maximum number of sides a polygon can have in AutoCAD LT is ________ .

11. The **Fill** option can be accessed through the ________ command or the ________ system variable.

12. The values for the inside and outside diameters of a doughnut are saved in the ________ and ________ system variables.

13. Summarize the various ways of drawing an ellipse. ________________________________
____.

14. What is the function of the **FILLMODE** system variable?________________________________
____.

15. Define inscribed and circumscribed polygons. ________________________________
____.

Review Questions

1. The default method of drawing an arc is the **three-point** option. (T/F)

2. An arc can be drawn only in a counterclockwise direction. (T/F)

3. In the **St, C, Ang** option, if the specified angle is negative, the arc is drawn in a counterclockwise direction. (T/F)

4. In the **St, C, Len** option, the arc is drawn in a counterclockwise fashion from the start point. (T/F)

5. In the **St, C, Len** option, positive chord length gives the largest possible arc (known as the major arc) with that length. (T/F)

6. A unit radius arc with an included angle of 45 degrees has a chord length of 0.765 units. (T/F)

7. In the **St, E, Rad** option, a positive radius value results in a major arc, and a negative radius value results in a minor arc. (T/F)

8. In the **St, E, Dir** option, the start of the arc is established tangent to the specified direction. (T/F)

9. Using the **RECTANG** command, the rectangle generated on the screen is treated as a combination of different objects; hence, individual sides can be edited immediately. (T/F)

10. The **ELLIPSE** command draws an ellipse as a composition of small arc segments forming a polyline. (T/F)

11. The characteristics of a polygon drawn in AutoCAD LT are those of a closed polyline having zero width and no tangent specification. (T/F)

12. If the **FILL** mode is off, only the donut outlines are drawn. (T/F)

13. Regeneration or redrawing takes the same amount of time for both non-solid-filled plines and solid-filled plines. (T/F)

14. Polylines can be used with any type of line. (T/F)

15. Doughnuts in AutoCAD LT are circular polylines. (T/F)

16. After regeneration, all the points on the drawing are drawn in the shape designated by the current value of the **PDMODE** variable. (T/F)

17. The system variable **PDSIZE** controls the size of the point except for the **PDMODE** values of 0 and 1. (T/F)

18. Though a selection set is being formed using the **Last** option, only one object is selected, even if you use the **Last** option a number of times. (T/F)

19. Fill in the command, entries, and operations required to draw an arc tangent to the end point of a previously drawn arc.

    Command: _______
    Specify start point of arc or [CEnter]: _______
    Specify second point of arc [CEnter/End]: _______

20. Fill in the command, entries, and operations required to draw an ellipse whose first and second axis endpoints are at (2,3) and (5,3) respectively, with a rotation of 45 degrees around the major axis.

    Command: _______
    Specify axis endpoint of ellipse or [Arc/Center]: _______
    Specify other endpoint of axis: _______
    Specify distance to other axis or [Rotation]: _______
    Specify rotation around major axis: _______

21. Give the command, entries, and operations required to draw an ellipse by the **Center and Two Axes** method. The center of the ellipse is located at (5,6), the axis endpoint at (6,6), and the width of the other axis is 2 units.

Command: ______
Specify axis endpoint of ellipse or [Arc/Center]: ______
Specify center of ellipse: ______
Specify endpoint of axis: ______
Specify distance to other axis or [Rotation]: ______

22. Fill in the command, entries, and operations needed to draw a pentagon with the **Edge** option, where the first and second endpoints of the edge are at (4,2) and (3,3), respectively.

    Command: ______
    Enter number of sides <4>: ______
    Specify center of polygon or [Edge]>: ______
    First endpoint of edge: ______
    Second endpoint of edge: ______

23. To save time, you can turn on the **FILL** mode when the drawing has taken the final shape and then use the ________ command to fill the plines in the drawing.

24. The ________ option of the **PLINE** command closes the polyline by drawing a polyline segment from the most recent endpoint to the initial start point.

25. The ________ option of the **PLINE** command is used to switch from drawing polylines to drawing polyarcs, and provides you the options associated with the drawing of polyarcs.

26. The type of point drawn is controlled by the ________ system variable.

27. You can have ________ combinations of point types.

28. A value of zero for the **PDSIZE** variable generates the point at ________ percent of the graphics area height.

29. A positive value for **PDSIZE** defines ________ size for the point.

30. Explain the difference between an optimized polyline and a 2D polyline. ____________
    ____.

31. What is the function of **CONVERT** command? ______________________________
    ____.

32. You cannot draw a filleted rectangle using the **RECTANG** command. (T/F)

Exercises

Exercise 15 *Mechanical*

Make the drawing shown in Figure 2-64. The distance between the dotted lines is 1.0 units. Create the radii by using appropriate **ARC** command options.

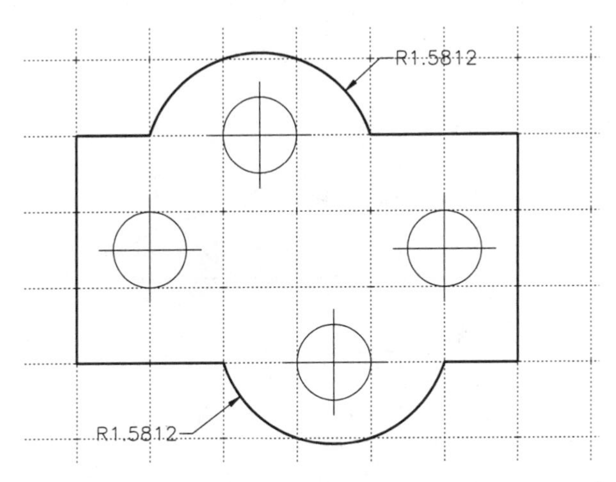

Figure 2-64 *Drawing for Exercise 15*

Exercise 16 *Graphics*

Make the drawing shown in Figure 2-65. The distance between the dotted lines is 1.0 units. Create the radii using appropriate **ARC** command options.

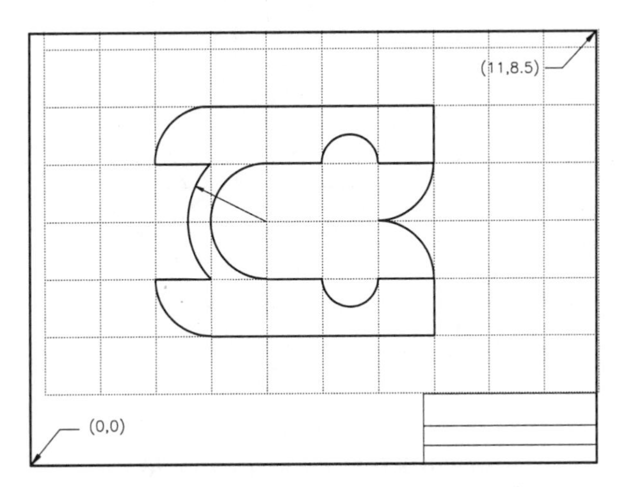

Figure 2-65 *Drawing for Exercise 16*

Exercise 17 *Mechanical*

Make the drawing shown in Figure 2-66. The distance between the dotted lines is 0.5 units. Create the ellipses using the **ELLIPSE** command.

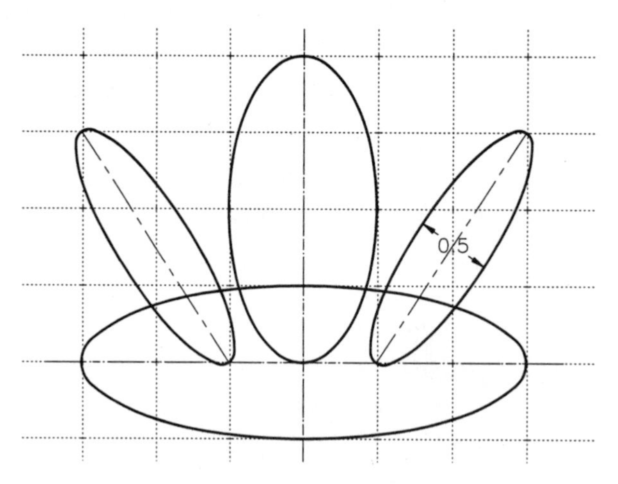

***Figure 2-66** Drawing for Exercise 17*

Exercise 18 *Mechanical*

Use the appropriate options of the **CIRCLE, LINE,** and **ARC** commands to generate the drawing in Figure 2-67. The dimensions are as shown in the figure. Use the **POINT** command to draw the center points of the circles.

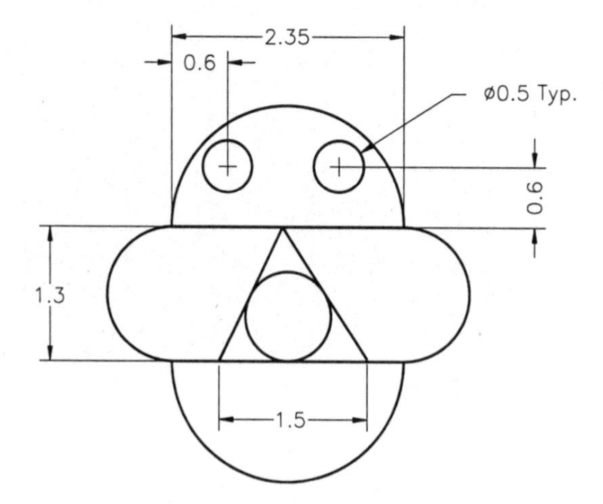

***Figure 2-67** Drawing for Exercise 18*

Exercise 19 *Mechanical*

Make the drawing shown in Figure 2-68. You must use only the **LINE**, **CIRCLE**, and **ARC** command options. Use the **POINT** command to draw the center points of the circles and arcs.

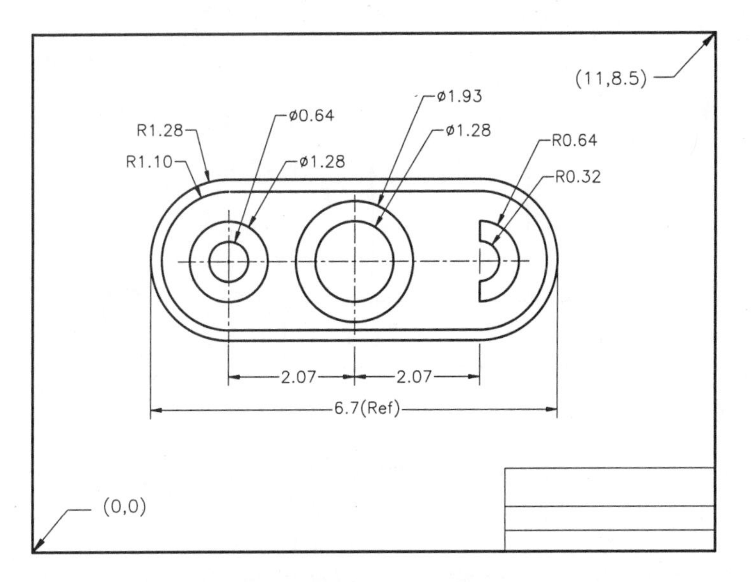

Figure 2-68 Drawing for Exercise 19

Exercise 20 *Piping*

Draw the following Figure 2-69, using the **RECTANGLE** command.

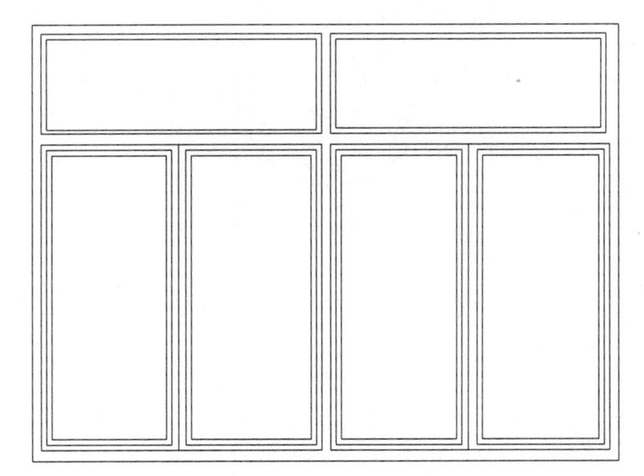

Figure 2-69 Drawing for Exercise 20

Exercise 21 *Mechanical*

Draw the following Figure 2-70 using the **LINE**, **CIRCLE**, **PLINE**, and **ARC** commands.

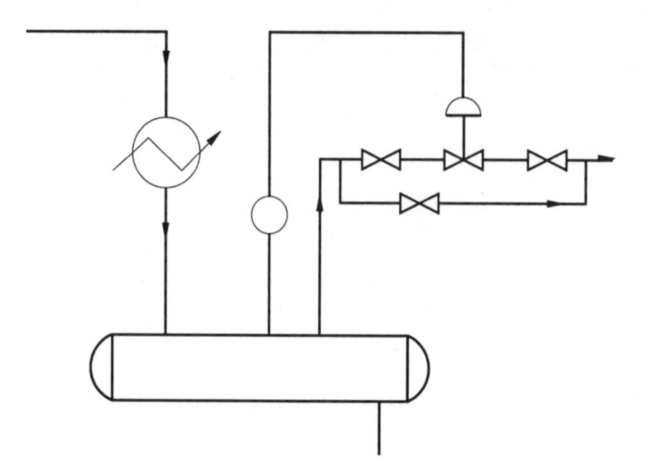

Figure 2-70 *Drawing for Exercise 21*

Problem Solving Exercise 1 *Mechanical*

Draw Figure 2-71 using only the **LINE**, **CIRCLE**, and **ARC** commands or their options.

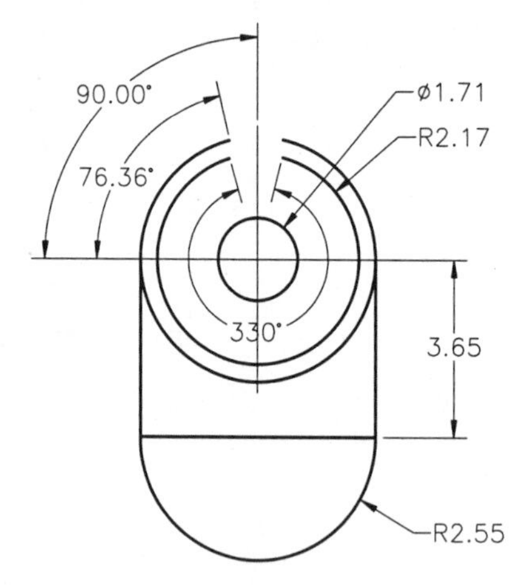

Figure 2-71 *Drawing for Problem Solving Exercise 1*

Problem-Solving Exercise 2 — *Mechanical*

Draw Figure 2-72 using the **POLYGON** and **CIRCLE** commands. The drawing must be created according to the given dimensions. Use the **POINT** command to draw the centerlines going through the circles.

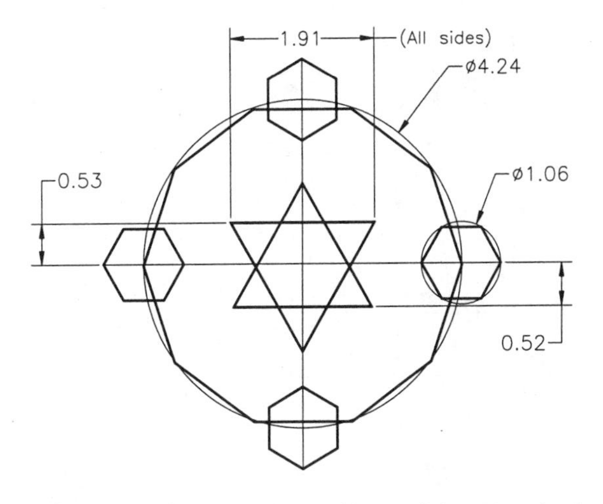

Figure 2-72 *Drawing for Problem Solving Exercise 2*

Project Exercise 1-2 — *Mechanical*

Generate the drawing in Figure 2-73 according to the dimensions shown in the figure, but do not draw the dimensions. Save the drawing as BENCH3-4. (See Chapters 1, 6, 7, and 19 for other drawings of Project Exercise 1).

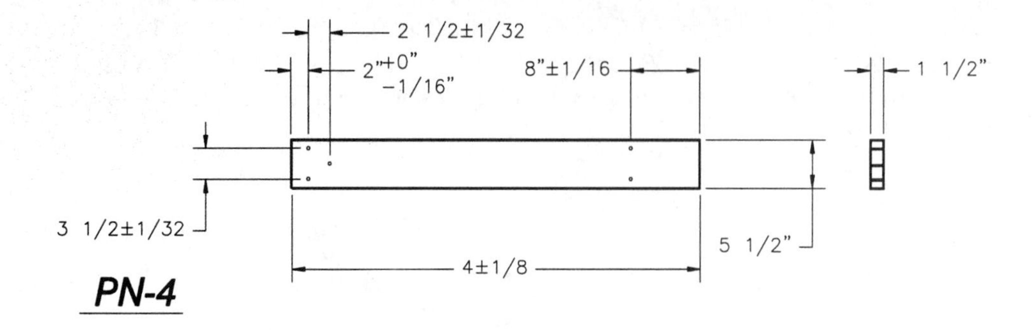

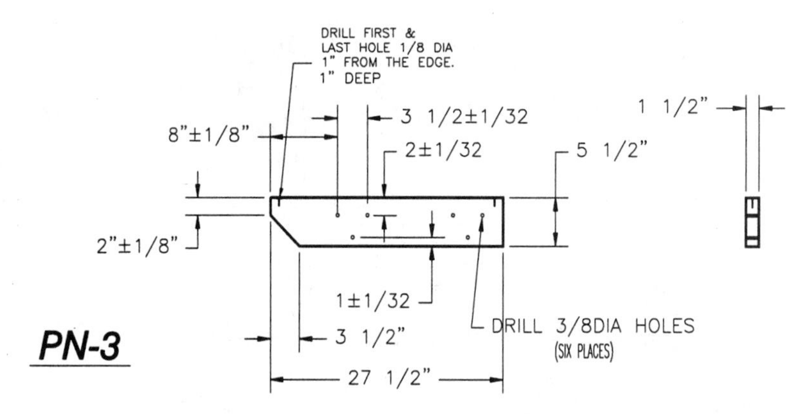

Figure 2-73 Drawing for Project Exercise 1-2

Answers to Self-Evaluation Test.

1 - T, **2** - T, **3** - T, **4** - F, **5** - T, **6** - ARC, 3,3 / C, 2,3 / A / 80, **7** - ELLIPSE / 4,5 / 7,5 / 1, 8 - ERASE / LAST, Enter, **9** - PDSIZE, **10** - 1024, **11** - FILL / FILLMODE, **12** - DONUTID / DONUTOD, **13** - NURBS Ellipse, **14** -Turns the fill on and also controls the hatch pattern visibility, **15** -An inscribed polygon is drawn inside the circle and a circumscribed polygon is drawn outside the circle.

Chapter 3

Drawing Aids

Learning Objectives

After completing this chapter, you will be able to:

- *Set up units using the **Drawing Units** dialog box and the **UNITS** command.*
- *Set up and determine limits for a given drawing.*
- *Determine limits for engineering, architectural, and metric drawings.*
- *Set up layers and assign colors and linetypes to them.*
- *Set up Grid, Snap, and Ortho modes based on the drawing requirements.*
- *Use Object Snaps and understand their applications.*
- *Combine Object Snap modes and set up running object snap modes.*
- *Use AutoTracking, current, and global linetype scaling.*
- *Determine **LTSCALE** factor for plotting.*

In this chapter you will learn about the drawing setup. This involves several factors that can affect the quality and accuracy of your drawing. This chapter contains a detailed description of how to set up units, limits, and layers. You will also learn about some of the drawing aids, like Grid, Snap, and Ortho. These aids will help you to draw accurately and quickly.

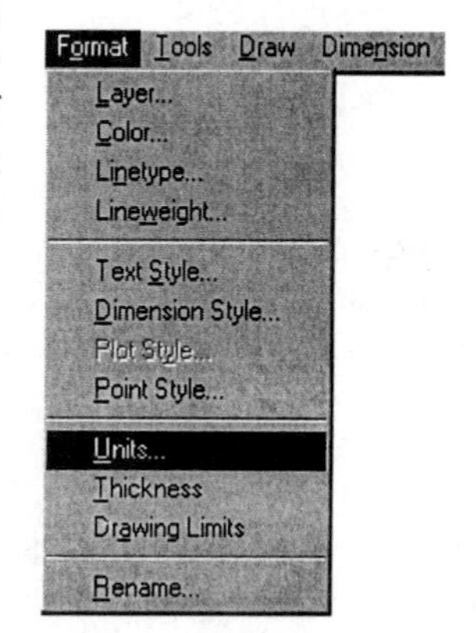

***Figure 3-1** Invoking the **UNITS** command from the **Format** menu*

SETTING UNITS

Using the Drawing Units Dialog Box

Menu:	Format > Units
Command:	UNITS

The **UNITS** command is used to select a format for the units of distance and angle measurement. You can use the

Drawing Units dialog box to set the units. In the **Drawing Units** dialog box (Figure 3-2), you can select a desired format of units or angles from the drop-down list displayed on choosing the down arrow button to the right of the **Type:** edit box. You can also specify the precision of units and angles from the **Precision:** drop-down list. (Figure 3-3).

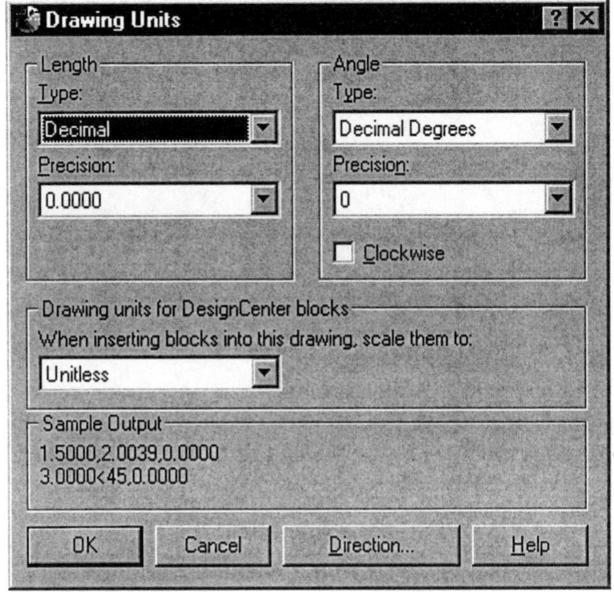

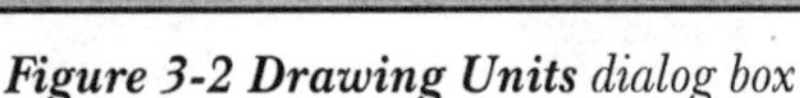

*Figure 3-2 **Drawing Units** dialog box*

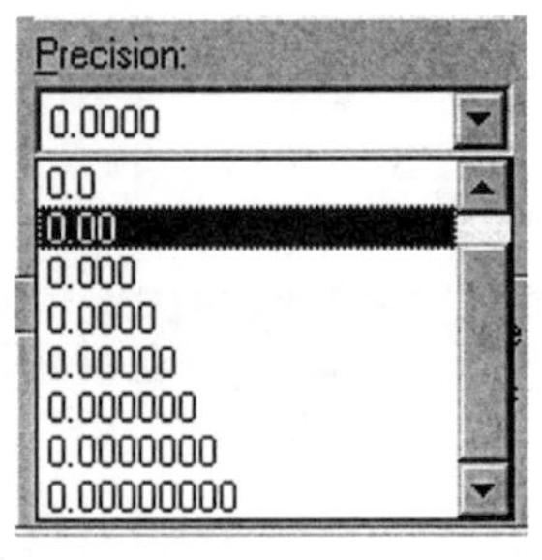

*Figure 3-3 Specifying Precision from the **Drawing Units** dialog box*

Specifying Units. You can use the **Drawing Units** dialog box to specify the units you want in the drawing.You can select one of the following five formats:

1. Architectural
2. Decimal
3. Engineering
4. Fractional
5. Scientific

If you select the Scientific, Decimal, or Fractional format, you can enter the distances or coordinates in any of these three formats, but not in engineering or architectural units. In the following example, the units variously are set as decimal, scientific, fractional, and decimal and fractional to enter the coordinates of different points:

Command: **LINE** Enter
Specify from point: **1.75,0.75** Enter *(Decimal.)*
Specify next point or [Undo]: **1.75E+01, 3.5E+00** Enter *(Scientific.)*
Specify next point or [Undo]: **10-3/8,8-3/4** Enter *(Fractional.)*
Specify next point or [Close/Undo]: **0.5,17/4** Enter *(Decimal and fractional.)*

If you choose the Engineering or Architectural format, you can enter the distances or coordinates in any of the five formats. In the following example, the units are set as architectural; hence, different formats are used to enter the coordinates of points:

Command: **LINE** Enter
Specify first point: **1-3/4,3/4** Enter *(Fractional.)*
Specify next point or [Undo]: **1'1-3/4",3-1/4** Enter *(Architectural.)*
Specify next point or [Undo]: **0'10.375,0'8.75** Enter *(Engineering.)*
Specify next point or [Close/Undo]: **0.5,4-1/4"** Enter *(Decimal and engineering.)*

Note

The inch symbol (") is optional. For example, 1'1-3/4" is the same as 1'1-3/4, and 3/4" is the same as 3/4.
You cannot use the feet (') or inch (") symbols if you have selected Scientific, Decimal, or Fractional unit formats.

Specifying Angle. You can select one of the following five angle measuring systems:

1. Decimal degrees
2. Deg/min/sec
3. Grads
4. Radians
5. Surveyor's units

If you select any of the first four measuring systems, you can enter the angle in the Decimal, Degrees/minutes/seconds, Grads, or Radians system, but you cannot enter the angle in Surveyor's units. However, if you select Surveyor's units, you can enter the angles in any of the five systems. If you enter an angle value without any indication of measuring system, it is taken in the current system. To enter the value in another system, use the appropriate suffixes and symbols, such as r (Radians), d (Degrees), g (Grads), or the others shown in the following examples. In the following example, the system of angle measure is Surveyor's units and different systems of angle measure are used to define the angle of the line:

Command: **LINE** Enter
Specify first point: **3,3** Enter
Specify next point or [Undo]: **@3<45.5** Enter *(Decimal degrees.)*
Specify next point or [Undo]: **@3<90d30'45"** Enter *(Degrees/min/sec.)*
Specify next point or [Close/Undo]: **@3<75g** Enter *(Grads.)*
Specify next point or [Close/Undo]: **@3<N45d30'E** Enter *(Surveyor's units.)*

In **Surveyor's units** you must specify the bearing angle that the line makes with the north-south direction (Figure 3-4). For example, if you want to define an angle of 60 degrees with north, in the Surveyor's units the angle will be specified as N60dE. Similarly, you can specify angles like S50dE, S50dW, N75dW, as shown in Figure 3-5. You cannot specify an angle that exceeds 90 degrees (N120E). The angles can also be specified in **radians** or **grads** for example, 180 degrees is equal to **PI** (3.14159) radians. You can convert degrees into radians or radians into degrees using the following equations:

radians = degrees x 3.14159/180
degrees = radians x 180/3.14159

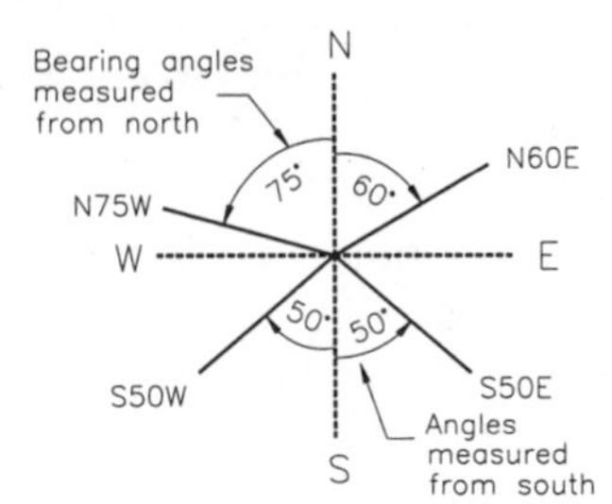

Figure 3-4 Specifying angles in Surveyor's units

Grads are generally used in land surveys. There are 400 grads or 360 degrees in a circle. A 90-degree angle is equal to 100 grads.

In AutoCAD LT, by default the angles are positive if measured in the counterclockwise direction (Figure 3-5) and the angles are measured with positive X axis (East, Figure 3-6). The angles are negative if measured clockwise. If you want the angles measured as positive in the clockwise direction, select the **Clockwise** check box. Then the positive angles will be measured in clockwise direction and the negative angles in counterclockwise direction.

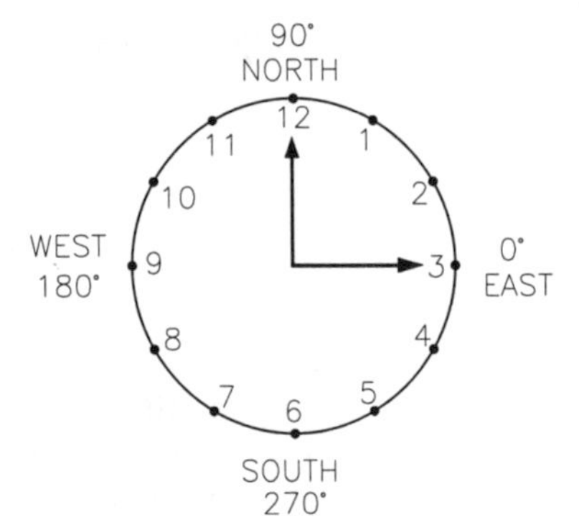

Figure 3-5 N,S,E,W directions

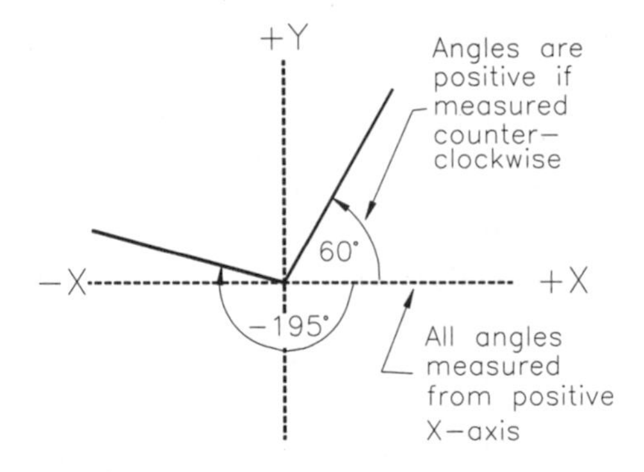

Figure 3-6 Measuring angles counterclockwise from the positive X axis (default)

Choosing the **Direction** button in the **Drawing Units** dialog box displays the **Direction Control** dialog box, which gives you an option of selecting the setting for direction of the Base Angle (Figure 3-7). If you select the **Other** option, you can set your own direction for the Base Angle by either entering a value in the **Angle:** edit box or choosing the **Pick an angle** button to specify the angle on the screen.

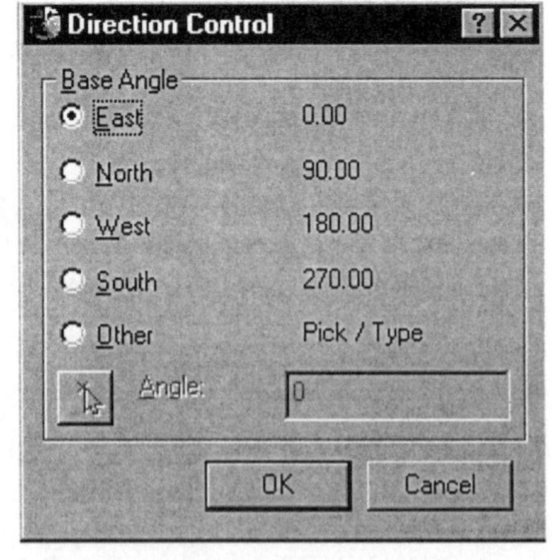

***Figure 3-7** Selecting Direction from the **Units Control** dialog box*

Choosing the down arrow button in the **When inserting blocks in this drawing, scale them to:** edit box, for **DesignCenter blocks**, you are allowed to choose from a list of units. The default value is Unitless.

The **Sample Output area** shows an example of the current format of the units and angles.

Using the -UNITS Command

You can also set up units by entering **-UNITS** at Command prompt. The prompt sequence is:

Command: **-UNITS** Enter

The AutoCAD LT Text Window is displayed, which shows the following five report formats with the examples:

Report formats:	(Examples)
1. Scientific	1.55E+01
2. Decimal	15.50
3. Engineering	1'-3.50"
4. Architectural	1'-3 1/2"
5. Fractional	15 1/2

With the exception of the Engineering and Architectural formats, these formats can be used with any basic units of measurement. For example, Decimal mode is perfect for metric units as well as decimal English units.

Enter choice, 1 to 5 <default>:

AutoCAD LT is now waiting for you to select a format of your choice. After the selection is made, AutoCAD LT will prompt you to select the precision required, depending on the selected format. For selection 1, 2, or 3, the next prompt says:

Enter number of digits to right of decimal point (0 to 8) <default>:

If the default value is acceptable to you, press ENTER; otherwise, enter your choice. For example, the number 2.1250 has four digits to the right of the decimal point; therefore, you will enter 4 at this prompt. For choices 4 and 5 the prompt says:

Enter denominator of smallest fraction to display
(1, 2, 4, 8, 16, 32, 64, 128, or 256) <default>:

You can now enter your choice or choose default by pressing ENTER. For example, if you want the distances or coordinates measured up to 1/64 (2-1/64), enter 64. After you enter the format and precision for distances and coordinates, the **-UNITS** command proceeds to angles and allows you to select any one of the five listed systems of angle measurement.

Systems of angle measure:	(Examples)
1. Decimal degrees	45.0000
2. Degrees/minutes/seconds	45d0'0"
3. Grads	50.0000g
4. Radians	0.7854r
5. Surveyor's units	N 45d0'0" E

Enter choice, 1 to 5 <default>:

Enter number of fractional places for display of angles (0 to 8) <default>:

After you select the system of angle measure and precision to display angles, you must select a direction for angle measurement. The prompt says:

Direction for angle 0:
East 3 o'clock = 0
North 12 o'clock = 90
West 9 o'clock = 180
South 6 o'clock = 270
Enter direction for angle <0> <current>:

You can now select any starting direction to measure angles. In the default mode, angle measurement always starts with 0 degrees corresponding to East (Figure 3-6). You can enter a new direction or press ENTER to select 0 degrees. The next prompt says:

Measure angles clockwise? [Yes/No] <current>:

If you want angles measured clockwise, enter Y. If you enter N, then a positive angle is measured in a counterclockwise direction and a negative angle is measured clockwise. This is the last prompt for setting the units, and it is followed by the Command prompt.

Example 1 *General*

In this example we will set the units for a drawing according to the following specifications and then draw Figure 3-9:

1. Set **UNITS** to fractional, with the denominator of the smallest fraction equal to 32.
2. Set the angular measurement to Surveyor's units, with the number of fractional places for display of angles equal to zero.
3. Set the direction to 90 degrees (north) and the direction of measurement of angles to clockwise (angles measured positive in clockwise direction) (Figure 3-8).

Using the Drawing Units dialog box

1. Invoke the **Drawing Units** dialog box by entering UNITS at the Command prompt or by choosing **Units** from the **Format** menu.

   Command: **UNITS** Enter

2. Select **Fractional units** from the **Type:** drop-down list in the **Length** area. From the **Precision** drop-down list, in the **Length** area, select **0 1/32** if it is not already selected.

3. In the **Angle** area of the dialog box, select **Surveyor's Units** from the **Type:** drop-down list. From the **Precision** drop-down list, in the **Angle** area, select **N 0d E** if it is not already selected. Also, select the **Clockwise** check box to set clockwise angle measurement as positive.

4. Choose the **Direction** button to display the **Direction Control** dialog box.

5. Select the **North** radio button. Choose the **OK** button to exit the **Direction Control** dialog box.

6. Choose the **OK** button to exit the **Drawing Units** dialog box.

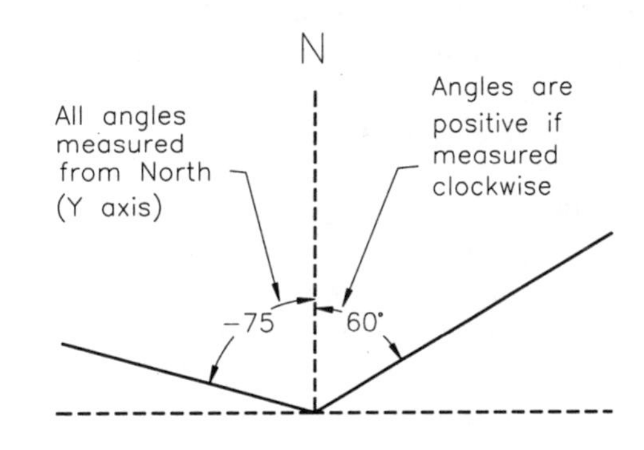

***Figure 3-8** Angles measured from north (Y axis)*

7. With the units set, draw Figure 3-9 using polar coordinates. Here the units are fractional and the **angles are measured from north** (90-degree axis). Also, the angles are measured as positive in a clockwise direction and **negative** in a counterclockwise direction.

Command: **LINE** Enter
Specify first point: **2,2** Enter
Specify next point or [Undo]: **@2.0<0** Enter
Specify next point or [Undo]: **@2.0<60** Enter
Specify next point or [Close/Undo]: **@1<180** Enter
Specify next point or [Close/Undo]: **@1<90** Enter
Specify next point or [Close/Undo]: **@1<180** Enter
Specify next point or [Close/Undo]: **@2.0<60** Enter
Specify next point or [Close/Undo]: **@0.5<90** Enter
Specify next point or [Close/Undo]: **@2.0<180** Enter
Specify next point or [Close/Undo]: **C** Enter

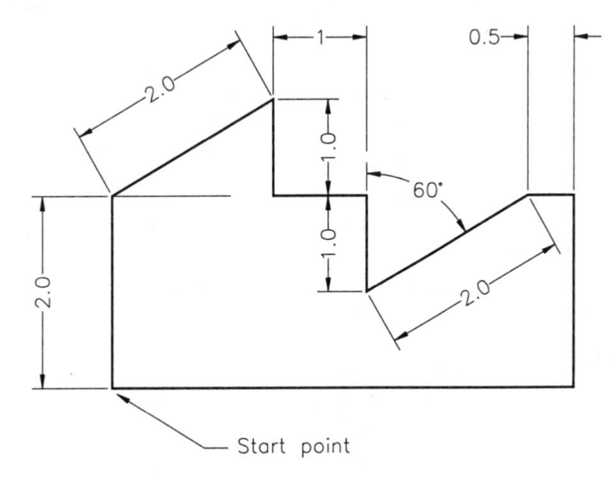

***Figure 3-9** Drawing for Example 1*

Chapter 3

Using the command line

1. Invoke the **UNITS** command and select the appropriate options as shown below:

Command:-**UNITS** Enter

Report formats:	(Examples)
1. Scientific	1.55E+01
2. Decimal	15.50
3. Engineering	1'-3.50"
4. Architectural	1'-3 1/2"
5. Fractional	15 1/2

With the exception of Engineering and Architectural formats, these formats can be used with any basic units of measurement. For example, Decimal mode is perfect for metric units as well as decimal English units.

Enter choice, 1 to 5 <default>: **5** Enter

Enter denominator of smallest fraction to display
(1,2,4,8,16,32, 64, 128, or 256) <default>: **32** Enter

Systems of angle measure:	(Examples)
1. Decimal degrees	45.0000
2. Degrees/minutes/seconds	45d0'0"
3. Grads	50.0000g
4. Radians	0.7854r
5. Surveyor's units	N 45d0'0" E

Enter choice, 1 to 5 <default>: **5** Enter
Enter number of fractional places for display of angles (0 to 8) <default>: **0** Enter

Direction for angle N 90d E

East	3 o'clock	= N 90d E
North	12 o'clock	= N 0d W
West	9 o'clock	= S 90d W
South	6 o'clock	= S 0d E

Enter direction for angle <0> <current>: **N** Enter

Measure angles clockwise? [Yes/No] <current>: **Y** Enter

Forcing Default Angles

When you define the direction by specifying the angle, the output of the angle depends on the following (Figure 3-10):

Angular units
Angle direction
Angle base

For example, if you are using the AutoCAD LT default setting, <70 represents an angle of 70 decimal degrees from the positive X axis, measured counterclockwise. The decimal degrees represent angular units, X axis represents the angle base, and counterclockwise represents the angle direction. If you have changed the default settings for measuring angles, it might be confusing to enter the angles. AutoCAD LT lets you bypass the current settings by entering << or <<< before the angle. If you enter << before the angle, AutoCAD LT will bypass the current angle settings and use the angle as decimal degrees from the default angle base and default angle direction (angles are referenced from the positive X axis and are positive if measured counterclockwise). If you enter <<< in front of the angle, AutoCAD LT will use current angular units, but it will bypass the current settings for angle base and angle direction and use the default angle setting (angles are referenced from the positive X axis and are positive if measured counterclockwise).

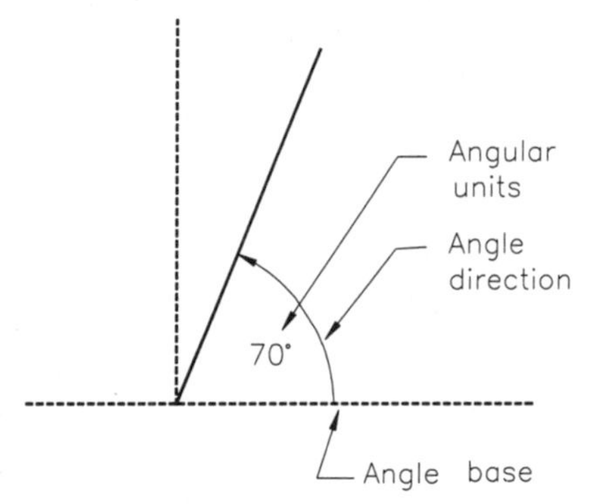

***Figure 3-10** Default angular units, direction, and base*

Assume that you have changed the current settings and made the system of angle measure radians with two places of precision: angle base north (the Y axis), and the direction clockwise. Now, if you enter <1.04 or <1.04r, all current settings will be considered and you will get an angle of 1.04 radians, measured in clockwise direction from the positive Y axis, Figure 3-11(a). If you enter <<60, AutoCAD LT will bypass the current settings and reference the angle in degrees from the positive X axis, measuring 60 degrees in a counterclockwise direction, Figure 3-11(b). If you enter <<<1.04r, the current angular units will be used,but will bypass the current angle base and angle direction. Hence, the angle will be referenced from the positive X axis, measuring 1.04 radians in a counterclockwise direction, Figure 3-11(c).

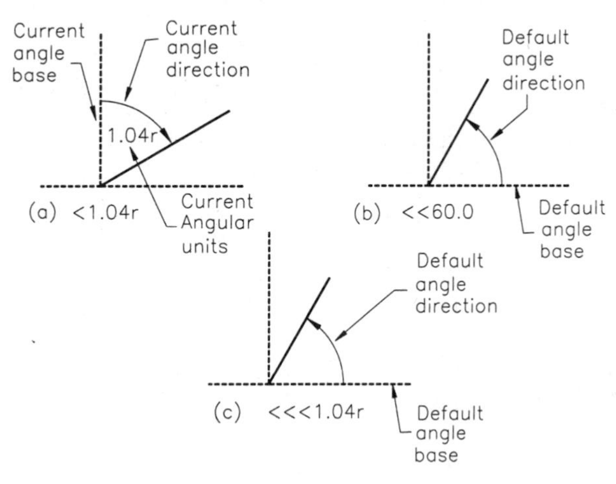

***Figure 3-11** Forcing default angles*

The effect of using the open angle brackets is summarized in the following table.

Angle prefix	Angular units	Angle direction	Angle base	Example
<	current	current	current	<1.04r
<<	degrees	counterclockwise	default	<<60.0
<<<	current	counterclockwise	default	<<<1.04r

You can also override the current angle units by entering the appropriate suffix, like 60d for degrees.

LIMITS COMMAND

Menu:	Format > Drawing Limits
Command:	LIMITS

When you start AutoCAD LT, the default limits are 12.00, 9.00. You can use the **LIMITS** command to set up new limits. The following is the prompt sequence of the **LIMITS** command for setting the limits of 24,18:

Command: **LIMITS** Enter
Specify lower left corner or [ON/OFF]<current>: **0,0** Enter
Specify upper right corner <current>: **24,18** Enter

At the preceding two prompts you are required to specify the lower left corner and the upper right corner of the sheet. Normally you choose (0,0) as the lower left corner, but you can enter any other point. If the sheet size is 24 x 18, enter 24,18 as the coordinates of the upper right corner.

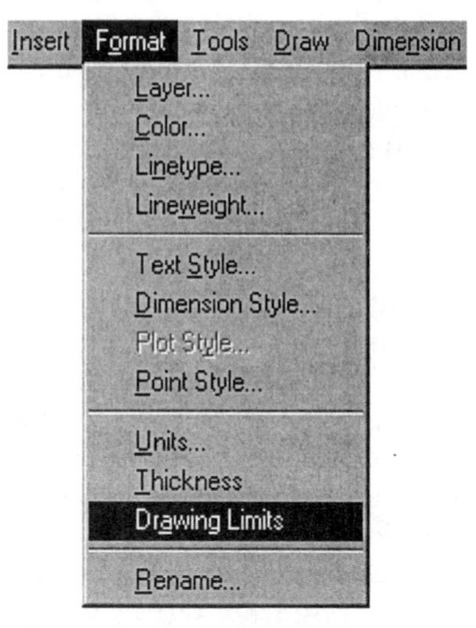

*Figure 3-12 Choosing **Drawing Limits** from the **Format** menu*

Setting Limits

In AutoCAD LT the drawings must be drawn full scale and, therefore, limits are needed to size up a drawing area. The limits of the drawing area are usually determined by the following factors:

1. The actual size of the drawing.
2. The space needed for putting down the dimensions, notes, bill of materials, and other necessary details.
3. The space between different views so that the drawing does not look cluttered.
4. The space for the border and a title block, if any.

To get a good idea of how to set up limits, it is always better to draw a rough sketch of the drawing to help calculate the area needed. For example, if an object has a front view size of 5 x 5, a side view size of 3 x 5, and a top view size of 5 x 3, the limits should be set so they can accommodate the drawing and everything associated with it. In Figure 3-13, the space between the front and side views is 4 units and between the front and top views is 3 units. Also, the space between the border and the drawing is 5 units on the left, 5 units on the right, 3 units at the bottom, and 2 units at the top. (The space between the views and between the border line and the drawing depends on the drawing.)

After you know the sizes of different views and have determined the space required between views, between the border and the drawing, and between the borderline and the edges of the paper, you can calculate the space you need as follows:

Space along (X axis) = 1 + 5 + 5 + 4 + 3 + 5 + 1 = 24
Space along (Y axis) = 1 + 3 + 5 + 3 + 3 + 2 + 1 = 18

Thus, the space or the work area you need for the drawing is 24 x18. Once you have determined the space you need, select the sheet size that can accomodate your drawing. In the above case, you will select D size (34 x 22) sheet. Therefore the actual drawing limits are 34, 22.

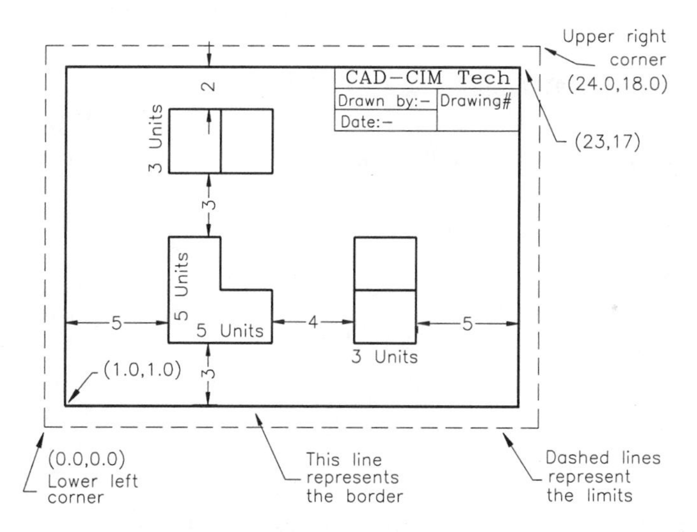

Figure 3-13 Setting limits in a drawing

Standard Sheet Sizes

When you make a drawing, you might want to plot the drawing to get a hard copy. Several standard sheet sizes are available to plot your drawing. Although in AutoCAD LT you can select any work area, it is recommended that you select the work area based on the sheet size you will be using to plot the drawing. The sheet size is the deciding factor for determining the limits (work area), text size (**TEXTSIZE**), dimensioning scale factor (**DIMSCALE**), linetype scale factor (**LTSCALE**) and other drawing-related parameters. The following tables list standard sheet sizes and the corresponding drawing limits for different scale factors:

Standard U.S. Size

Letter size	Sheet size	Limits (1:1)	Limits (1:4)	Limits (1/4"=1')
A	8.5 x 11	8.5,11	34,44	34',44'
B	11 x 17	11,17	44,68	44',68'
C	17 x 22	17,22	68,88	68',88'
D	22 x 34	22,34	88,136	88',136'
E	34 x 44	34,44	136,176	136'x176'

International Size

Letter size	Sheet size	Limits (1:1)	Limits (1:20)
A4	210 x 297	210,297	4200,5940
A3	297 x 420	297,420	5940,8400
A2	420 x 594	420,597	8400,11940
A1	595 x 841	595,841	11940,16820
A0	841 x 1189	841,1189	16820,23780

Limits for Architectural Drawings

Most architectural drawings are drawn at a scale of 1/4" = 1', 1/8" = 1', or 1/16" = 1'. You must set the limits accordingly. The following example illustrates how to calculate the limits in architectural drawings:

Given
Sheet size = 24 x 18
Scale is 1/4" = 1'

Calculate limits
Scale is 1/4" = 1'
  or 1/4" = 12"
  or 1" = 48"
X limit = 24 x 48
  = 1152" or 1152 Units
  = 96'
Y limit = 18 x 48
  = 864" or 864 Units
  = 72'

Thus, the scale factor is 48 and the limits are 1152", 864", or 96',72'.

Example 2 *Architectural*

In this example you will calculate limits and determine an appropriate drawing scale factor for Figure 3-14. The drawing is to be plotted on a 12" x 9" sheet.

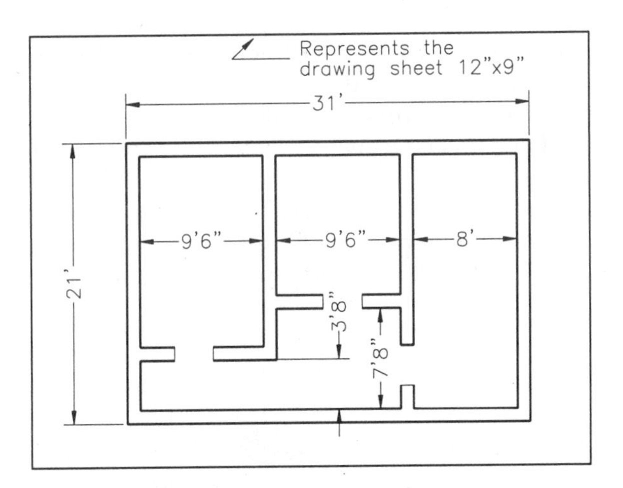

Figure 3-14 Drawing for Example 2

The scale factor can be calculated as follows:

Given or known
Overall length of the drawing = 31'
Length of the sheet = 12"
Approximate space between the drawing and the edges of the paper = 2"

Calculate scale factor
To calculate the scale factor, you have to try different scales until you find one that satisfies the given conditions. After some experience you will find this fairly easy to do. For this example, assume a scale factor of 1/4" = 1'.

Scale factor 1/4" = 1'
or 1" = 4'
Thus, a line 31' long will be = 31'/4' = 7.75" on paper. Similarly, a line 21' long = 21'/4' = 5.25".

Approximate space between the drawing and the edges of paper = 2"
Therefore, total length of the sheet = 7.75 + 2 + 2 = 11.75"
Similarly, total width of the sheet = 5.25 + 2 + 2 = 9.25"

Because you selected the scale 1/4" = 1', the drawing will definitely fit on the given sheet of paper (12" x 9"). Therefore, the scale for this drawing is 1/4" = 1'.

Calculate limits
Scale factor = 1" = 48" or 1" = 4'
The length of the sheet is 12"
Therefore, X limit = 12 x 4' = 48'
Also, Y limit = 9 x 4' = 36'

Limits for Metric Drawings

When the drawing units are metric, you must use **standard metric size sheets** or calculate the limits in millimeters (mm). For example, if the sheet size you decide to use is 24 x 18, the limits after conversion to the metric system will be 609.6,457.2 (multiply length and width by 25.4). You can round these numbers to the nearest whole numbers 610,457. Note that metric drawings do not require any special setup, except for the limits. Metric drawings are like any other drawings that use decimal units. As with architectural drawings, you can draw metric drawings to a scale. For example, if the scale is 1:20 you must calculate the limits accordingly. The following example illustrates how to calculate the limits for metric drawings.

Given
Sheet size = 24" x 18"
Scale = 1 : 20

Calculate limits
Scale is 1 : 20
Therefore, scale factor = 20
X limit = 24 x 25.4 x 20 = 12192 units
Y limits = 18 x 25.4 x 20 = 9144 units

Thus, the limits are 12192 and 9144.

Exercise 1 *Mechanical*

Set the units of the drawing according to the following specifications and then make the drawing shown in Figure 3-15 (Leave a space of 3 to 5 units around the drawing for dimensioning and title block). The space between the dotted lines is 1 unit.

1. Set **UNITS** to decimal units, with two digits to the right of the decimal point.
2. Set the angular measurement to decimal degrees, with the number of fractional places for display of angles equal to 1.
3. Set the direction to 0 degrees (east) and the direction of measurement of angles to counterclockwise (angles measured positive in counterclockwise direction)

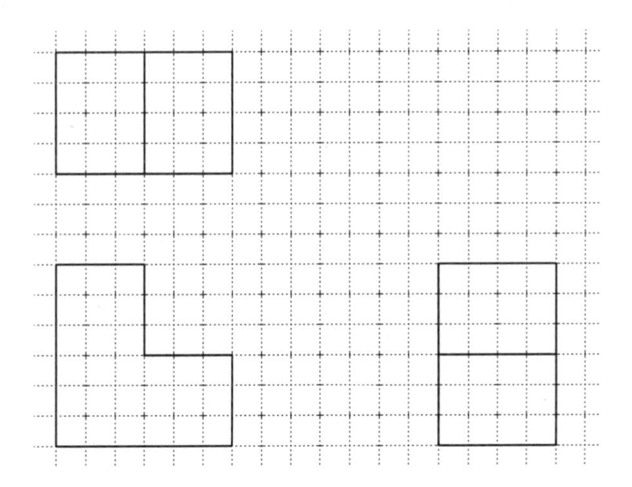

Figure 3-15 Drawing for Exercise 1

LAYERS

The concept of layers can be explained by using the concept of overlays in manual drafting. In manual drafting, different details of the drawing can be drawn on different sheets of transparent paper, or overlays. Each overlay is perfectly aligned with the others, and when all of them are placed on top of each other you can reproduce the entire drawing. As shown in Figure 3-16, the object lines have been drawn in the first overlay and the dimensions in the second overlay. You can place these overlays on top of each other and get a combined look at the drawing.

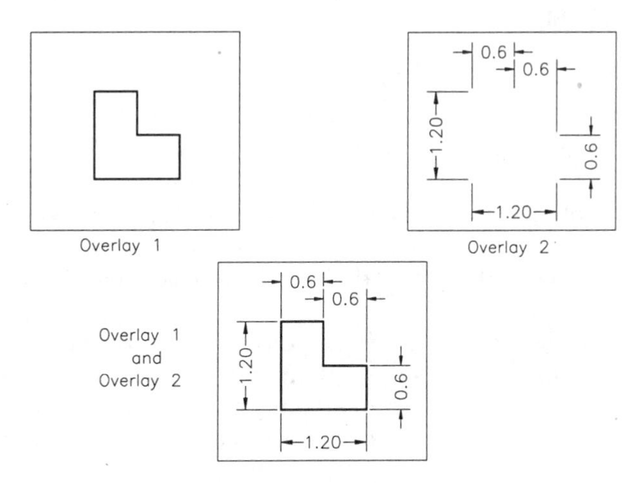

Figure 3-16 Drawing lines and dimensions in different overlays

In AutoCAD LT, instead of using overlays you use layers. Each layer is assigned a name. You can also assign a color and linetype to these layers. For example, in Figure 3-17 the object lines have been drawn in the OBJECT layer and the dimensions have been drawn in the DIM layer. The object lines will be red because the red color has been assigned to the OBJECT layer.

Similarly, the dimension lines will be green because the DIM layer has been assigned green color. You can display all the layers or display the layers individually or in any combination

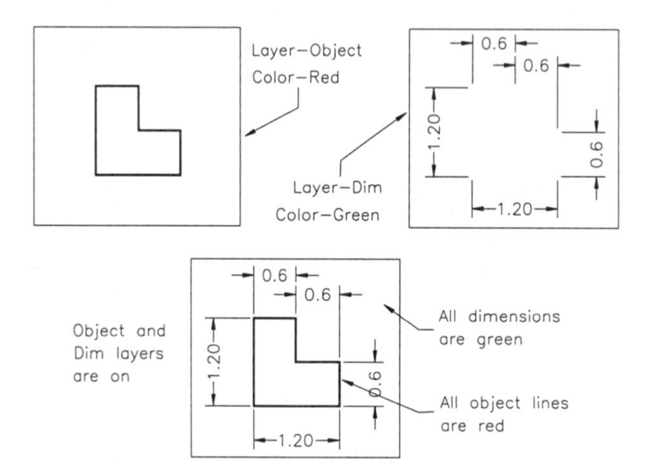

Figure 3-17 Drawing lines and dimensions in different layers

Advantages of Layers

1. Each layer can be assigned a different color. Assigning a particular color to a group of objects is very important for plotting. For example, if all object lines are red, at the time of plotting you can assign the red color to a slot (pen) that has the desired tip width (e.g., medium). Similarly, if the dimensions are green, you can assign the green color to another slot (pen) that has a thin tip. By assigning different colors to different layers you can control the width of the lines when the drawing is plotted. You can also make a layer plottable or nonplottable.

2. The layers are useful for some editing operations. For example, if you want to erase all dimensions in a drawing, you can freeze all layers except the dimension layer and then erase all dimensions by using the **Crossing** option to select objects.

3. You can turn a layer off or freeze a layer that you do not want to be displayed or plotted.

4. You can lock a layer, which will prevent the user from accidently editing the objects in it.

5. The colors help you to distinguish different groups of objects. For example, in architectural drafting, the plans for foundation, floors, plumbing, electrical, and heating systems may all be made in different layers. In electronics drafting and in PCB (printed circuit board), the design of each level of a multilevel circuit board can be drawn on separate layer.

LAYER PROPERTIES MANAGER DIALOG BOX*

Toolbar:	Object Properties > Layers
Menu:	Format > Layer
Command:	LAYER

You can use the **Layer Properties Manager** dialog box (Figure 3-19) to perform the functions associated with the **LAYER** command. For example, you can create new layers, assign colors, assign linetypes, or perform any operation that is shown in the dialog box. Using this dialog box for layers is efficient and provides a convenient way to use various options. You can also perform some layer functions like freeze, thaw, lock, unlock, etc., directly from the **Object properties** toolbar (Figure 3-20). If you have already created some layers, they will be listed with their current status in the dialog box as shown in Figure 3-19.

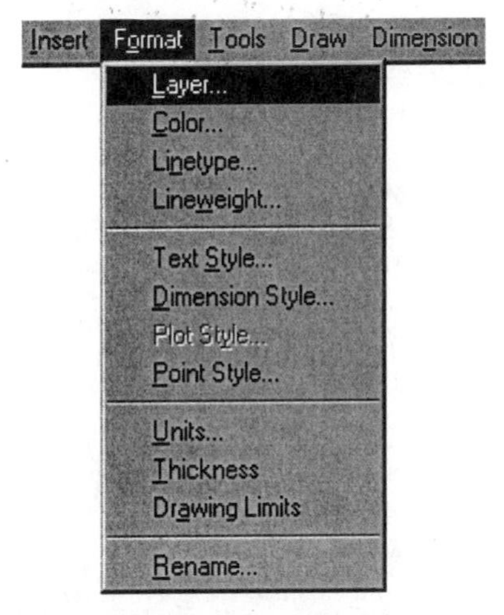

*Figure 3-18 Invoking the **LAYER** command from the **Format** menu*

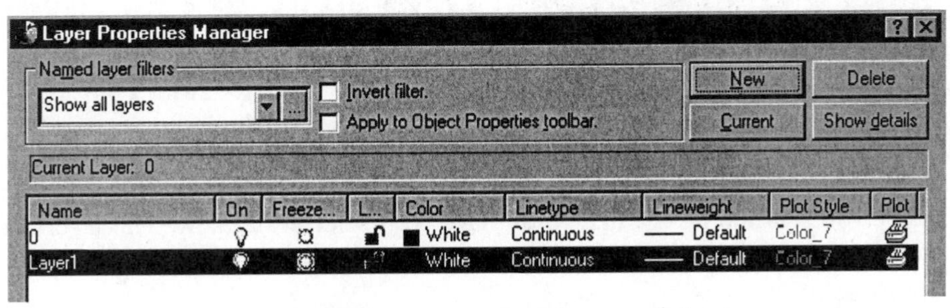

***Figure 3-19 Layer Properties Manager** dialog box*

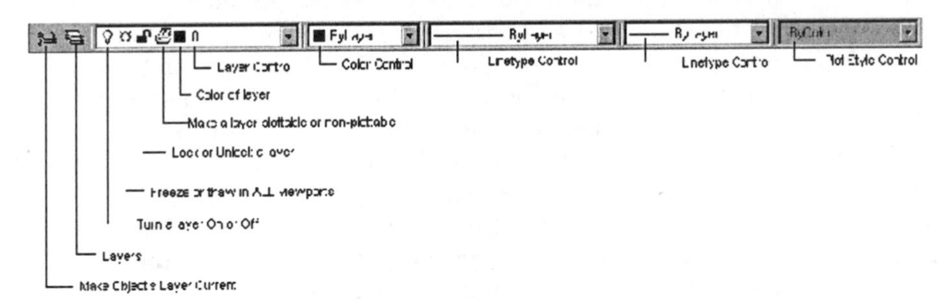

***Figure 3-20 Object Properties** toolbar*

Creating New Layers

If you want to create new layers, choose the **New** button. A new layer with name Layer1 and having the properties of the selected (highlighted) layer is created and listed in the dialog box just below the highlighted layer. You can change or edit the name by selecting it and then entering a new name. If no layer is selected, the new layer is placed at the end of the layers list and has the properties of the default layer 0. If more than one layer is selected, the new layer is placed below the last highlighted layer and has its properties. Also, right-clicking anywhere in the Layers list area of the **Layer Properties Manager** dialog box, displays a shortcut menu that gives you an option to create a new layer. You can right-click a layer whose properties you want to use in the new layer and then select **New Layer** from the shortcut menu.

Layer Names

1. A layer name can be upto 256 characters long,including letters (a-z),numbers (0-9) special characters ($ _ -) and spaces. Any combination of lowercase and uppercase letters can be used while naming a layer.
2. The layers should be named to help the user identify the contents of the layer. For example, if the layer name is HATCH, a user can easily recognize the layer and its contents. On the other hand, if the layer name is X261, it is hard to identify the contents of the layer.
3. Layer names should be short, but should also convey the meaning.

Note

If you exchange drawings with or provide drawings to consultants or others, it is very important that you standardize and coordinate layer names and other layer settings.

Assigning a Linetype, Color, Lineweight* or Plot Style* to a Layer

To assign a new linetype to a layer, select the current linetype displayed with a particular layer in the **Layer Properties Manager** dialog box. When you select the linetype, AutoCAD LT will display the **Select Linetype** dialog box (Figure 3-21), which displays the linetypes that are defined and loaded on your system. Select the new linetype and then choose the **OK** button. The linetype you selected is assigned to the layer you selected initially. The layers are, by default, assigned continuous linetype and white color.

If the linetypes have not been loaded on your system, choose the **Load** button in the **Select Linetype** dialog box. This displays the **Load or Reload Linetypes** dialog box (Figure 3-22), which displays all linetypes in the **ACLT.LIN** file. In this dialog box you can select individual linetypes or a number of linetypes by holding the SHIFT or CTRL key on the keyboard and then selecting the linetypes. If you right-click, AutoCAD LT displays the shortcut menu that you can use to select all linetypes. Then by choosing the **OK** button, the selected linetypes are loaded. You can also use the **LINETYPE** command to display the **Linetype Manager** dialog box which is similar to the **Layer Properties Manager** dialog box and allows you to load, delete linetypes, set filters or show details of linetypes. The **Linetype Manager** dialog box (Figure 3-23) can also be loaded through the **Format** menu by choosing **Linetype**. To use the command line to load the linetypes, the prompts are as follows:

Command: **-LINETYPE** Enter
Current line type: "ByLayer"
Enter an option [?/Create/Load/Set]: **L** Enter
Enter linetype(s) to load: *Enter name(s) of linetypes to load.*

The **Select Linetype File** dialog box appears that enables you to load the required linetypes from the specific files you need.

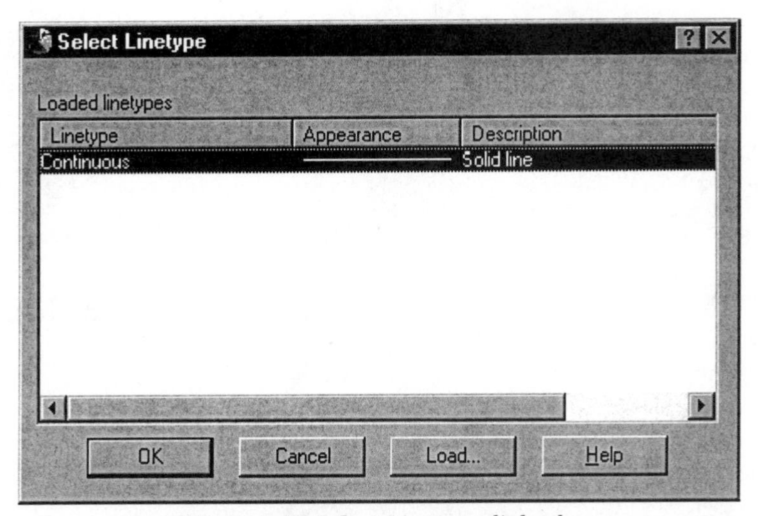

Figure 3-21 ***Select Linetype*** *dialog box*

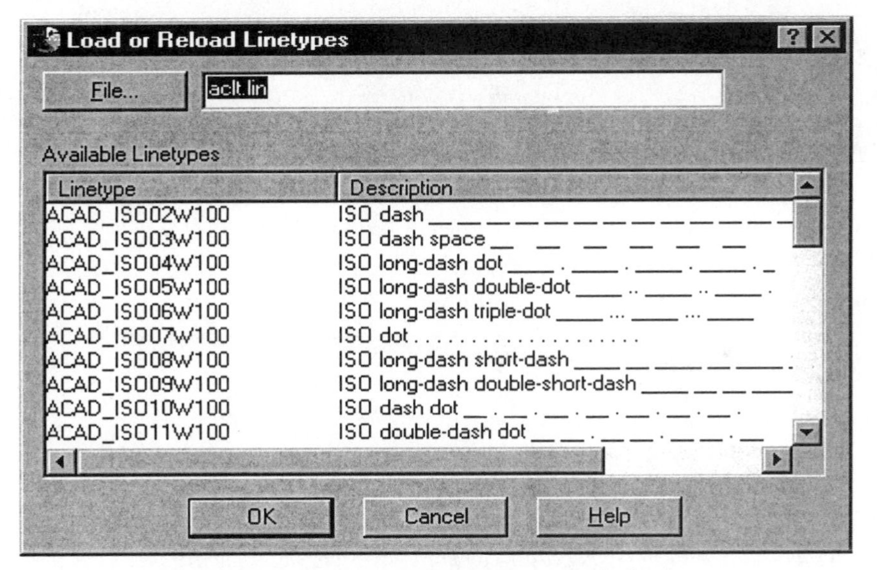

Figure 3-22 ***Load or Reload Linetypes*** *dialog box*

To assign a color, select the color icon in a particular layer; AutoCAD LT will display a **Select Color** dialog box. Select a desired color and then choose the **OK** button. The color you selected will be assigned to the selected layer. The number of colors is determined by your graphics card and monitor. Most color systems support eight or more colors. If your system allows it, you may choose a color number between 0 and 255 (256 colors). The following are the first seven standard colors:

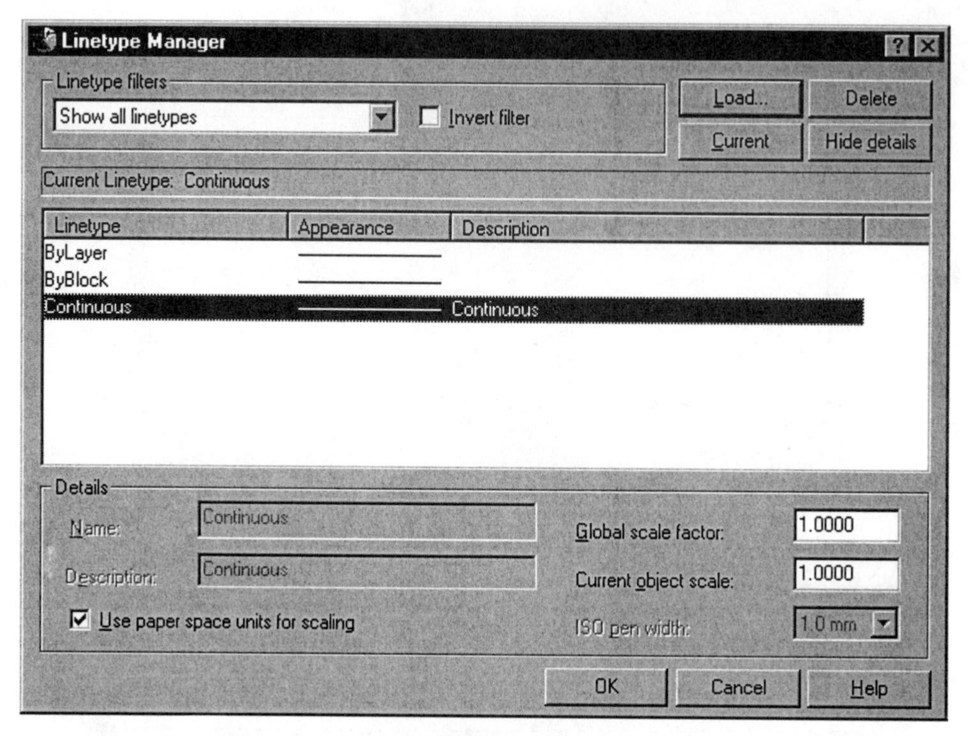

Figure 3-23 Linetype Manager *dialog box*

Color number	Color name	Color number	Color name
1	Red	5	Blue
2	Yellow	6	Magenta
3	Green	7	White
4	Cyan		

Note

Using nonstandard colors may cause compatibility problems if the drawings are used on other systems with different colors. On a light background, the color "white" appears black. Certain colors are hard to see on light backgrounds and others are hard to see on dark backgrounds, so you may have to use different colors than those specified in some examples and exercises.

To assign a Lineweight to a layer, select the layer and then click on the Lineweight associated with it; the **Lineweight** dialog box appears. Select a Lineweight from the **Lineweights:** list. Choose **OK** to return to the **Layer Properties Manager** dialog box.

Plot Style is a group of property settings like color, linetype, lineweight that can be assigned to a layer. The default Plot Style is **Normal** in which the color, linetype, and lineweight are BYLAYER. To assign a Plot Style to a layer, select the layer and then click on its plot style; the **Select Plot Style** dialog box appears where you can select a specific Plot Style from the **Plot**

Styles: list of available Plot Styles. Plot Styles have to be created before you can use them. Choose **OK** to return to the **Layer Properties Manager** dialog box.

Making a Layer Current

To make a layer current, select the name of desired layer and then choose the **Current** button in the dialog box. AutoCAD LT will display the name of the selected layer next to **Current Layer:** display box. Choose **OK** to exit the dialog box. You can also make a layer current by double-clicking on it in the layer list box of the **Layer Properties Manager** dialog box. Right-clicking a layer in the layer list box displays a shortcut menu which gives you an option to make the selected layer current. Only one layer can be made current and it is the layer where new objects will be drawn. You can also make a layer current by selecting it from the **Layer Control** drop-down list of the **Object Properties** toolbar. Choosing the **Make Object's Layer Current** button from the **Object Properties** toolbar prompts you to select the object whose layer you wish to make current and the layer associated with that object will be made current.

Controlling Display of Layers

You can control the display of the layers by selecting the **Turn a layer On or Off**, **Freeze or thaw in ALL viewports, Lock or Unlock a layer, Make layer plottable or non-plottable** toggle icons in the list box of any particular layer.

Turn a layer On or Off. With the **Turn a layer On or Off** toggle icon you can turn the layers on or off. The layers that are turned on are displayed and can be plotted. The layers that are turned off are not displayed and cannot be plotted. If you turn the current layer off, AutoCAD LT will display a warning informing you that the current drawing layer has been turned off.

Freeze or thaw in ALL viewports. While working on a drawing, if you do not want to see certain layers you can also use the **Freeze or thaw in ALL viewports** toggle icon to freeze the layers. The Thaw option negates the effect of the Freeze option, and the frozen layers are restored to normal. For example, while editing a drawing you may not want the dimensions displayed on the screen. To avoid this, you can freeze the DIM layer. The frozen layers are invisible and cannot be plotted. The difference between the **Off** option and the **Freeze** option is that the frozen layers are not calculated by the computer while regenerating the drawing. This saves time. The current layer cannot be frozen.

Active or New VP Freeze. When the **TILEMODE** is turned off (see Chapter 21, Drawing and Viewing 3D Objects) you can freeze or thaw the selected layers in the active floating viewport by selecting the **Active VP Freeze** icon for the selected layers. The frozen layers will still be visible in other viewports. If you want to freeze some layers in the new floating viewports, then select the **New VP Freeze** toggle icon for the selected layers. AutoCAD LT will freeze the layers in subsequently created new viewports without affecting the viewports that already exist. (Paper space is discussed in Chapter 22.) Also, check the **VPLAYER** command for selectively freezing the layers in viewports. The widths of the column headings in the **Layer Properties Manager** dialog box can be decreased or increased, by positioning the cursor between the column headings and then holding down the pick button of your pointing device and dragging the cursor to the right or the left.

LOck or Unlock a layer. While working on a drawing, if you do not want to accidentally edit some objects on a particular layer but still need to have them visible, you can use the **LOck or Unlock a layer** toggle icon to lock the layers. When a layer is locked you can still use the objects in the locked layer for Object Snaps and inquiry commands like **LIST**. You can also make the locked layer the current layer and draw objects on it. The locked layers are plotted. The **Unlock** option negates the **LOck** option and allows you to edit objects on the layers previously locked.

Make layer plottable or nonplottable*. If you do not wish to plot a particular layer, for example, construction lines, you can use the **Make layer plottable or nonplottable** toggle icon to make the layer nonplottable. The construction lines will not be plotted.

Deleting Layers

You can delete a layer by selecting the layer and then choosing the **Delete** button. To delete a layer it is necessary that the layer should not contain any objects. You cannot delete the Layers 0 and defpoints, a current layer and a Xref dependent layer.

You can also control the display of layers by first selecting a layer and then choosing the **Show details** button in **Layer Properties Manager** dialog box. This displays the layer details box that displays the properties associated with the selected layer. Using this box, you can change the name, color, linetype, lineweight, and other properties of the selected layer. Note that you cannot change the name of 0 layer. Once you have chosen the **Show details** button, the **Hide details** button is available.

Selective Display of Layers

If you have a limited number of layers it is easy to scan through them. However, if you have many layers it is sometimes difficult to search through the layers. To solve this problem you can use the drop-down list in the **Named layer filters** area of the **Layer Properties Manager** dialog box to display the layers selectively. If you want to display only those layers that are **XRef dependent**, you can select this option in the drop-down list. You can also define filter specification by choosing the **[...]** button to display the **Named layers filters** dialog box (Figure 3-24). When loaded, initially this dialog box contains the default specifications. If you want to list only those layers that are red in color, enter red in the **Color:** edit box, then give this filter a name in the **Filter name:** edit box and then choose the **Add** button. This filter gets added to the list in the **Named layer filters** area . Now, when you select this named filter from the drop-down list in the Named layer filters area, AutoCAD LT will display the layers that are red in color in the **Layer Properties Manager** dialog box. If you select **Show all layers** in the Show drop-down list, all layers will be displayed. If you again want to display only those layers that are red in color, select the name of the specific filter from the drop-down list. As the specifications are already set in the **Named Layer Filters** dialog box, all the layers that are red in color are displayed.

In the **Layer Properties Manager** dialog box, selecting the **Invert filter** check box, inverts the filter that you have selected. For example, if you have selected the **Show all layers**, none of the layers will be displayed. If you wish to apply the current layer filter to the Layer Control list in the **Object Properties** toolbar, select the **Apply to Object Properties toolbar** check box. When you do so, the tooltip indicates that the specified filter has been applied and only the filtered layers are displayed in the Layer Control drop-down list. When you right-click in the layers list box, AutoCAD LT displays a shortcut menu on the screen that you can use to select all layers, clear all layers, create new layer, or selectively display layers.

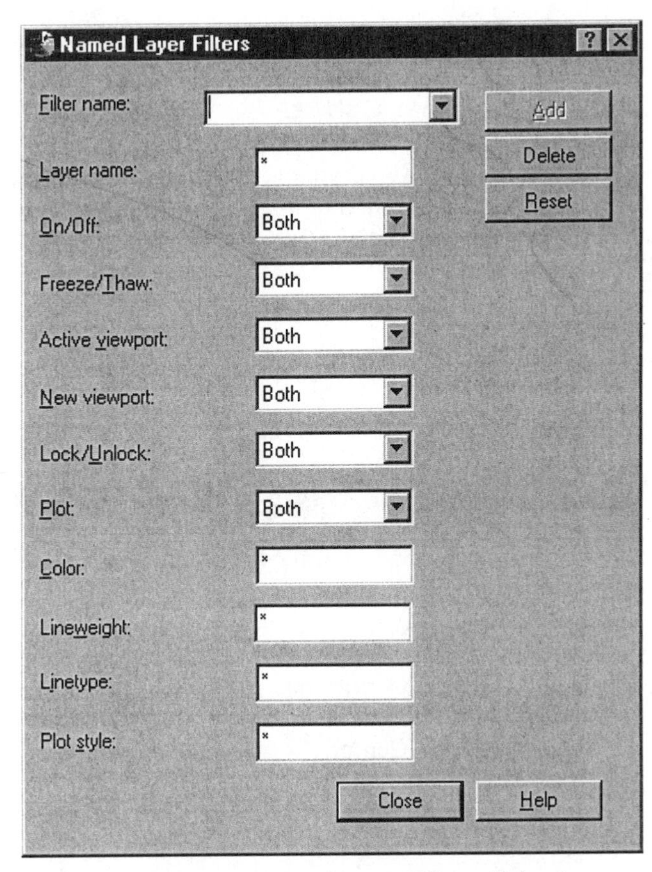

*Figure 3-24 **Named Layer Filters** dialog box*

Note

*You can also access the Layer options from the **Object Properties** toolbar (Figure 3-20) by selecting the appropriate icons.*

A frozen layer cannot be made current, but a layer that is turned off can be made current. However, neither can be plotted.

Setting Layers from the Command Line

You can also set layers from the command line by entering **-LAYER** at the Command prompt. The following is the prompt sequence:

Command: **-LAYER** [Enter]
Enter an option
[?/Make/Set/New/ON/OFF/Color/Ltype/LWeight/Plot/Pstyle/Freeze/Thaw/LOck/Unlock:

You can perform any function on the layer by entering the appropriate option at this prompt. For example, if you want to create two new layers, DIM and CEN, the prompt sequence is:

Command: **-LAYER** Enter
Enter an option
[?/Make/Set/New/ON/OFF/Color/Ltype/LWeight/Plot/Pstyle/Freeze/Thaw/LOck/Unlock:
N Enter
Enter name list for new layer(s): **DIM, CEN** Enter

You can use the **Make** option of the **LAYER** command to simultaneously create a layer (unless it already exists) and set it current. If your current layer is VIEW and you want to create a new layer, OBJECT, and also make it the current, the prompt sequence is:

Command: **-LAYER** Enter
Enter an option
[?/Make/Set/New/ON/OFF/Color/Ltype/Lweight/Plot/Pstyle/Freeze/Thaw/LOck/Unlock:
M Enter
Enter name for new layer (becomes the current layer) <VIEW>: **OBJECT** Enter

Example 3 *General*

Set up four layers with the following linetypes and colors. Then make the drawing (without dimensions) as shown in Figure 3-25.

Layer name	Color	Linetype
Obj	Red	Continuous
Hid	Yellow	Hidden
Cen	Green	Center
Dim	Blue	Continuous

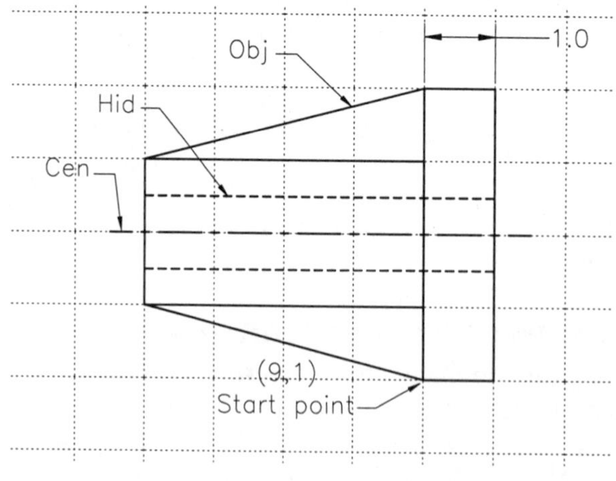

Figure 3-25 *Drawing for Example 3*

In this example, assume that the limits and units are already set. Before drawing the lines, you need to create layers and assign colors and linetypes to these layers. Also, depending on the objects that you want to draw, you need to set that layer as current. In this exercise you will create the layers using the **Layer Properties Manager** dialog box and the **-LAYER** command.

Using the dialog box

1. Choose the **Layers** button in the **Object Properties** toolbar to display the **Layer Properties Manager** dialog box. The default layer 0 with default properties is displayed in the list box.

2. Choose the **New** button. AutoCAD LT automatically creates a new layer (Layer1) having the default properties and displays it in the list box. Change its name by entering OBJ in place of Layer1.

3. Now choose the color icon to display the **Select Color** dialog box. Select color Red and then choose **OK**. Red color is assigned to layer **Obj**.

Using the command prompt

4. Now, use the **-LAYER** command to create the layer **Hid** and then assign to it the hidden linetype and yellow color.

   Command: **-LAYER** [Enter]
   Enter an option
   [?/Make/Set/New/ON/OFF/Color/Ltype/LWeight/Plot/Pstyle/Freeze/Thaw/LOck/Unlock]:
   N [Enter]
   Enter name list for new layer(s) name: **Hid** [Enter]
   Enter an option
   [?/Make/Set/New/ON/OFF/Color/Ltype/LWeight/Plot/Pstyle/Freeze/Thaw/LOck/Unlock]:
   C [Enter]
   Enter color name or number (1-255): **Yellow** or **Y** [Enter]
   Enter name list of layer(s) for color 2 (yellow) <0>: **Hid** [Enter]
   Enter an option
   [?/Make/Set/New/ON/OFF/Color/Ltype/LWeight/Plot/Pstyle/Freeze/Thaw/LOck/Unlock]:
   L [Enter]
   Enter loaded linetype name or [?] <CONTINUOUS>: **Hidden** [Enter]
   Enter name list for of layer(s) for linetype "HIDDEN" <0>: **Hid** [Enter]
   Enter an option
   [?/Make/Set/New/ON/OFF/Color/Ltype/LWeight/Plot/Pstyle/Freeze/Thaw/LOck/Unlock]:
   [Enter]

5. You can create the other two layers (CEN and DIM) using either the dialog box or the command prompt.

6. Enter **LAYER** at the Command prompt to invoke the **Layer Properties Manager** dialog box. Select **Obj** layer and then choose the **Current** button to make the **Obj** layer current. Choose the **OK** button to exit the dialog box.

7. Using the **LINE** command, draw the object lines in layer **Obj**.

   Command: **LINE** Enter
   Specify first point: **9,1** Enter
   Specify next point or [Undo]: **9,9** Enter
   Specify next point or [Undo]: **11,9** Enter
   Specify next point or [Close/Undo]: **11,1** Enter
   Specify next point or [Close/Undo]: **9,1** Enter
   Specify next point or [Close/Undo]: **1,3** Enter
   Specify next point or [Close/Undo]: **1,7** Enter
   Specify next point or [Close/Undo]: **9,9** Enter
   Specify next point or [Close/Undo]: Enter

   Command: **LINE** Enter
   Specify first point: **1,3** Enter
   Specify next point or [Undo]: **9,3** Enter
   Specify next point or [Undo]: Enter

   Command: **LINE** Enter
   Specify first point: **1,7** Enter
   Specify next point or [Undo]: **9,7** Enter
   Specify next point or [Undo]: Enter

8. Now, make the **Hid** layer current using the Command prompt and then draw the hidden lines of the drawing in the Hid layer.

   Command: **-LAYER** Enter
   Enter an option
   [?/Make/Set/New/ON/OFF/Color/Ltype/LWeight/Plot/Pstyle/Freeze/Thaw/LOck/Unlock]:
   S Enter
   Enter layer name to make current <OBJ>: **Hid** Enter
   Enter an option
   [?/Make/Set/New/ON/OFF/Color/Ltype/LWeight/Plot/Pstyle/Freeze/Thaw/LOck/Unlock]:
   Enter

   Command: **LINE** Enter
   Specify first point: **1,4** Enter
   Specify next point or [Undo]: **11,4** Enter
   Specify next point or [Undo]: Enter

   Command: **LINE** Enter
   Specify first point: **1,6** Enter
   Specify next point or [Undo]: **11,6** Enter
   Specify next point or [Undo]: Enter

9. Enter **LAYER** at the Command prompt or select **Layer** in the **Format** menu to invoke the **Layer Properties Manager** dialog box. Select Cen layer and then choose the **Current**

button to make the Cen layer current. Choose the **OK** button to exit the dialog box.

Command: **LINE** Enter
Specify first point: **0,5** Enter
Specify next point or [Undo]: **12,5** Enter
Specify next point or [Undo]: Enter

Exercise 2 *Mechanical*

Set up layers with the following linetypes and colors. Then make the drawing (without dimensions) as shown in Figure 3-26. The distance between the dotted lines is 1 unit.

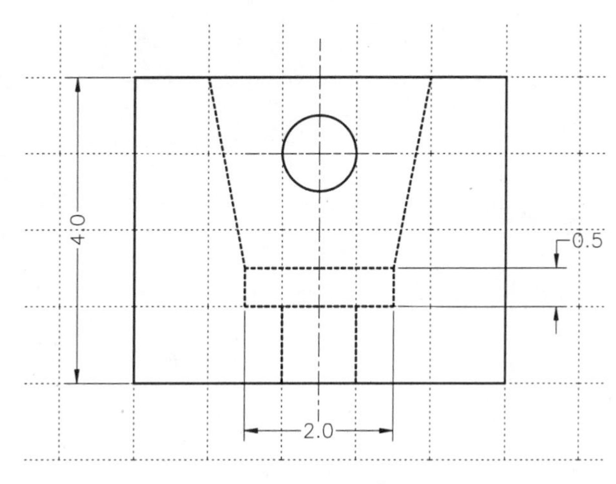

Figure 3-26 *Drawing for Exercise 2*

Layer name	**Color**	**Linetype**
Object	Red	Continuous
Hidden	Yellow	Hidden
Center	Green	Center
Dimension	Blue	Continuous

DRAFTING SETTINGS DIALOG BOX*

Menu: Tools > Drafting Settings
Command: DSETTINGS

You can use the **Drafting Settings** dialog box to set drawing modes such as Grid, Snap, Object Snap, and Polar tracking. You can also right-click on the **Snap**, **Grid**, **Polar**, **LWT**, or **osnap** buttons on the Status bar to display a shortcut menu and select **Settings** to display the **Drafting Settings** dialog box. This dialog box operates under three main tabs: **Snap and Grid**, **Object Snap,** and **Polar Tracking**. In the **Drafting Settings** dialog box (Figure 3-27), the Grid and Snap can be turned on or off by selecting their **Snap On** or **Grid On** check boxes. You can set the grid or snap spacing by entering either the same or different values in their respective **Snap/Grid X Spacing** and **Snap/Grid Y Spacing** edit boxes. When you exit the grid

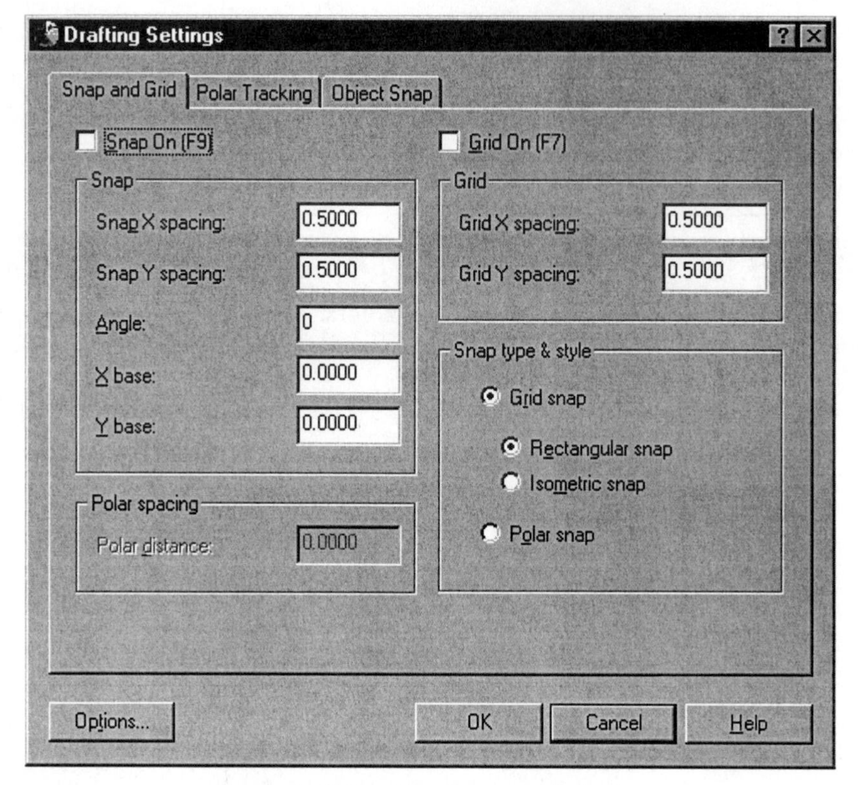

Figure 3-27 ***Drafting Settings*** *dialog box*

or snap X spacing edit boxes, the corresponding Y Spacing value is automatically set to match the X spacing value. Therefore, if you want different X and Y spacing values, you need to set the X spacing first, then set the Y spacing.

Setting Grid

The grid lines are lines of dots on the screen at predefined spacing (Figure 3-28). These dotted lines act as a graph that can be used as reference lines in a drawing. You can change the distance between the grid dots as per your requirement. The grid pattern appears within the drawing limits, which helps to define the working area. The grid also gives you a sense of the size of the drawing objects.

Grid On (F7): Turning the Grid On or Off. You can turn the grid on/off and change the grid spacing in the **Drafting Settings** dialog box (Figure 3-27). You can also turn grid on or off by choosing the **GRID** button in the Status bar (Figure 3-38). When a grid is turned on after having been off, the grid is set to the previous grid spacing. The function key F7 acts as a toggle key for turning the grid on or off.

Grid X Spacing and Grid Y Spacing. The **Grid X Spacing** and **Grid Y Spacing** edit boxes are used to define a desired grid spacing along the X and Y axis. For example, to set the grid spacing to 0.5 units, enter 0.5 in the **X Spacing** and **Y Spacing** edit boxes (Figure 3-29). You can also enter different values for horizontal and vertical grid spacing (Figure 3-30).

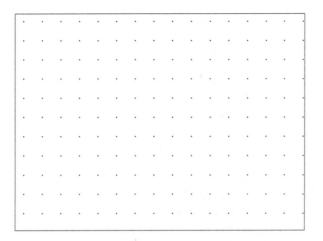

Figure 3-28 Grid lines

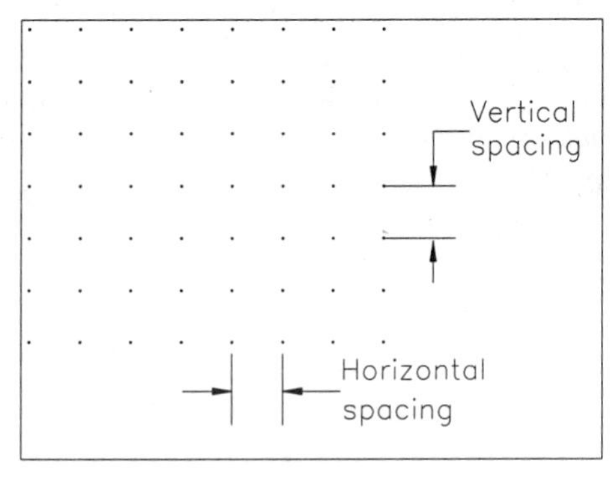

Figure 3-29 Controlling grid spacing

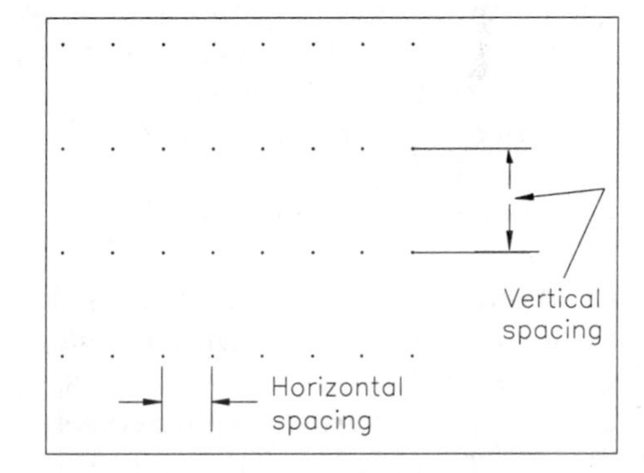

Figure 3-30 Creating unequal grid spacing

You can also use the **GRID** command to set the grid. The following is the prompt sequence for the **GRID** command:

Command: **GRID** Enter
Specify grid spacing (X) or [ON/OFF/Snap/Aspect] <0>: **0.5** Enter

The **Aspect** option is used to assign a different value to the horizontal and vertical grid spacing. The grid and the snap grid (discussed in the next section) are independent of each other. However, you can automatically display the grid lines at the same resolution as that of the Snap grid by using the **Snap** option. When you use this option, AutoCAD LT will automatically change the grid spacing to zero and display the grid lines at the same resolution as set for Snap.

Note

*In the **GRID** command, to specify the grid spacing as a multiple of the Snap spacing, enter X after the value (2X). In the **Drafting Settings** dialog box if the grid spacing is specified as zero, it automatically adjusts to equal the Snap resolution.*

Setting Snap

The snap is used to set increments for cursor movement. While moving the cursor it is sometimes difficult to position a point accurately. The **SNAP** command allows you to set up an invisible grid (Figure 3-31) that allows the cursor to move in fixed increments from one snap point to another. The snap points are the points where the invisible snap lines intersect. The snap spacing is independent of the grid spacing, so the two can have equal or different values. You generally set snap to an increment of the grid setting, for example, Snap=2 and Grid=10.

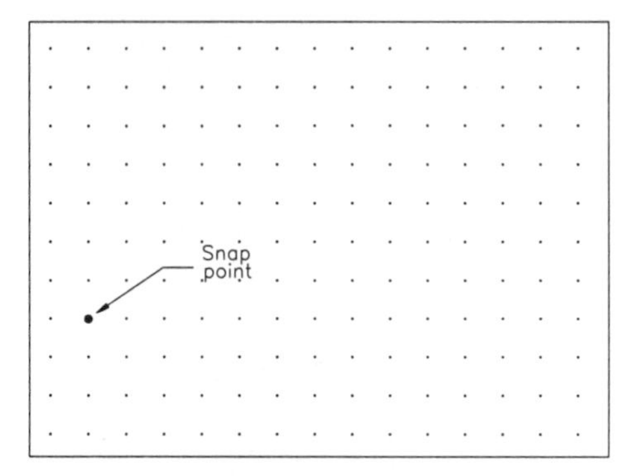

Figure 3-31 *Invisible Snap grid*

On: Turning the Snap On or Off. You can turn the snap on/off and set the snap spacing from the **Drafting Settings** dialog box (Figure 3-27). As in the case of GRID, the **Snap On** check box selection turns the invisible snap grid on or off and moves the cursor by increments. If the SNAP is off, the cursor will not move by increments. It will move freely as you move the pointing device. When you turn the snap off, AutoCAD LT remembers the value of the snap, and this value is restored when you turn the snap back on. You can also turn snap on or off by choosing the **SNAP** button in the Status bar (Figure 3-38) or using the function key F9 as a toggle key for turning snap on or off.

Snap Angle. The **Angle:** edit box is used to rotate the snap grid through an angle (Figure 3-32). Normally, the snap grid has horizontal and vertical lines, but sometimes you need the snap grid at an angle. For example, when you are drawing an auxiliary view (a drawing view that is at an angle to other views of the drawing), it is more useful to have the snap grid at an angle. If the rotation angle is positive, the grid rotates in a counterclockwise direction; if the rotation angle is negative, it rotates in a clockwise direction.

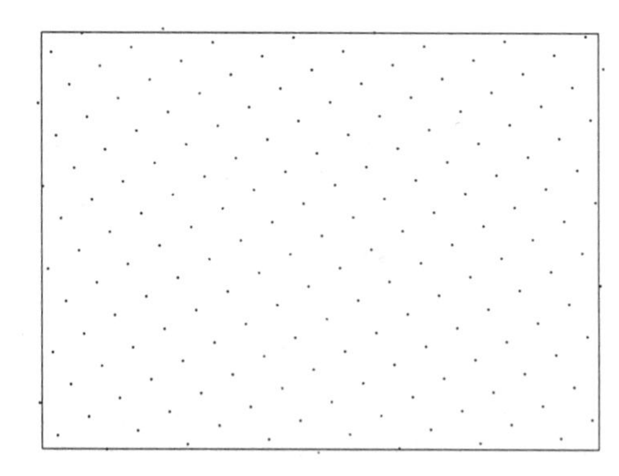

Figure 3-32 *Snap grid rotated 30 degrees*

X Base and Y Base. The **X base** and **Y base** edit boxes can be used to define the snap and grid origin point in the current viewport relative to current UCS. The value is stored in the **SNAPBASE** system variable.

You can also use the **SNAP** command to set the snap. The following is the Command prompt sequence for the **SNAP** command:

Command: **SNAP** [Enter]
Specify snap spacing or [ON/OFF/Aspect/Rotate/Style/Type] <current>: **A** [Enter]

Specify horizontal spacing <current>: **0.5** [Enter]
Specify vertical spacing <current>: **1** [Enter]

Command: **SNAP** [Enter]
Specify snap spacing or [ON/OFF/Aspect/Rotate/Style/Type] <current>: **R** [Enter]
Specify base point <0,0>: **2,2** [Enter] (New base point.)
Specify rotation angle <0>: **30** [Enter]

Note

It is generally preferable to rotate the UCS rather than rotating the snap grid. See Chapter 20.

Isometric Snap/Grid

You can choose the **Isometric snap** radio button to set the snap grid to isometric mode. The default is off (standard). The isometric mode is used to make isometric drawings. In isometric drawings, the isometric axes are at angles of 30, 90, and 150 degrees. The Isometric Snap/Grid enables you to display the grid lines along these axes (Figure 3-33). You can also use the **Style** option of the **SNAP** command to set the snap and grid to isometric mode.

Figure 3-33 *Isometric snap grid*

Command: **SNAP** [Enter]
Specify snap spacing or [ON/OFF/Aspect/Rotate/Style/Type] <current>: **S** [Enter]
Enter snap grid style [Standard/Isometric] <current>: **I** [Enter]
Specify vertical spacing <current>: **1.5** [Enter]

Once you set the isometric snap, AutoCAD LT automatically changes the cursor to align with the isometric axis. You can adjust the cursor orientation when you are working on the left, top, or right plane of the drawing by using the F5 key or holding down the CTRL key and then pressing the E key to cycle the cursor through different isometric planes (left, right, top).

Polar Spacing* controls polar settings. When you choose the **Polar Snap** radio button, the **Polar distance** edit box is highlighted. The **Polar distance** edit box, allows you to give a value to the Polar Snap. If this value is zero, it assumes the same value as the snap X spacing.

Snap Type and Style*. There are two types of Snap types, **Polar** snap and **Grid** snap. **Grid** snap snaps along the grid and is either of **Rectangular** style or **Isometric** style. **Polar** snap snaps along Polar alignment angles. These angles can be set in the **Polar Tracking** tab of the **Drafting Settings** dialog box. If you select the **Polar snap** radio button in the **Snap type and style** area of the **Snap and Grid** tab of the **Drafting Settings** dialog box and if Polar or Object tracking is on, the cursor will snap along the set Polar tracking angles, relative to the last point selected or aquired. The snap type is also controlled by the **SNAPTYPE** system variable.

Example 4 — *General*

Draw the auxiliary view of the object whose front view is shown in Figure 3-34. The thickness of the plate is 4 units; the length of the incline's face is 12 units. The following are the three different ways to draw the auxiliary view.

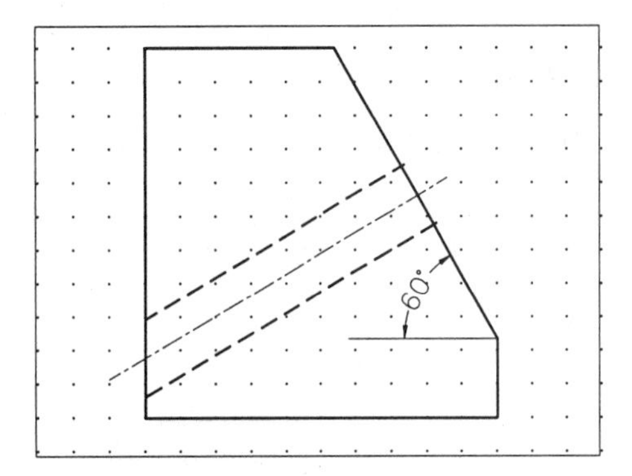

Figure 3-34 *Front view of the plate for Example 4*

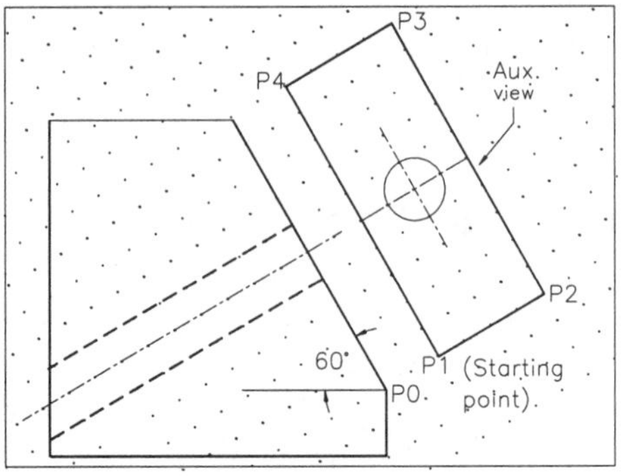

Figure 3-35 *Auxiliary view using rotated snap grid*

Rotating the Snap Grid

Command: **SNAP** Enter
Specify snap spacing or [ON/OFF/Aspect/Rotate/Style/Type] <current>: **R** Enter
Specify base point <current>: *Select point P0* *(Use Object snap.)*
Specify rotation angle <current>: **30** Enter

Command: **LINE** Enter
Specify first point: *Select point (P1).*
Specify next point or [Undo]: *Move the cursor 4 units right and select point (P2).*
Specify next point or [Undo]: *Move the cursor 10 units up and select point (P3).*
Specify next point or [Close/Undo]: *Move the cursor 4 units left and select point (P4).*
Specify next point or [Close/Undo]: **C** Enter

Using Polar Coordinates

Command: **LINE** Enter
Specify first point: *Select the starting point P1.*
Specify next point or [Undo]: **@4<30** Enter
Specify next point or [Undo]: **@10<120** Enter
Specify next point or [Close/Undo]: **@4<210** Enter
Specify next point or [Close/Undo]: **C** Enter

Note that the angle measurements do not change even though you have rotated the snap. The angle is still measured with reference to the positive X axis.

Rotating UCS Icon

When you rotate the UCS by 30 degrees, the X axis and Y axis also rotate through the same angle and the **UCSICON** will automatically align with the rotated snap grid (See Figure 3-35). (See the **UCS** command for a detailed explanation.)

Command: **UCS** Enter
Current UCS name:*WORLD*
Enter an option [New/Move/orthoGraphic/Prev/Restore/Save/Del/Apply/?/
World]<World>: **N** Enter
Specify origin of new UCS or [ZAxis/3Point/OBject/Face/View/X/Y/Z]<0,0,0>: **Z**
Specify rotation angle about Z axis <0>: **30** Enter

Command: **LINE** Enter
Specify first point: *Select the starting point P1.*
Specify next point or [Undo]: **@4<0** Enter
Specify next point or [Undo]: **@10<90** Enter
Specify next point or [Close/Undo]: **@4<180** Enter
Specify next point or [Close/Undo]: **C** Enter

Note that the angle measurements have changed to match the alignment of the rotated UCS.

Setting Ortho Mode

You can turn the Ortho mode on or off by choosing the **ORTHO** button in the Status bar (Figure 3-38). The **ORTHO** mode allows you to draw lines at right angles only. Whenever you are using the pointing device to specify the next point, the movement of the rubber-band line connected to the cursor is either horizontal (parallel to the X axis) or vertical (parallel to the Y axis). If you want to draw a line in the Ortho mode, specify the starting point at the **Specify first point:** prompt. To specify the second point, move the cursor with the pointing device and specify a desired point. The line drawn will be either vertical or horizontal, depending on the direction in which you moved the cursor (Figure 3-36 and Figure 3-37). You can also use the **ORTHO** command to turn the Ortho mode on or off. You can also use the function key F8 to turn the Ortho mode on or off.

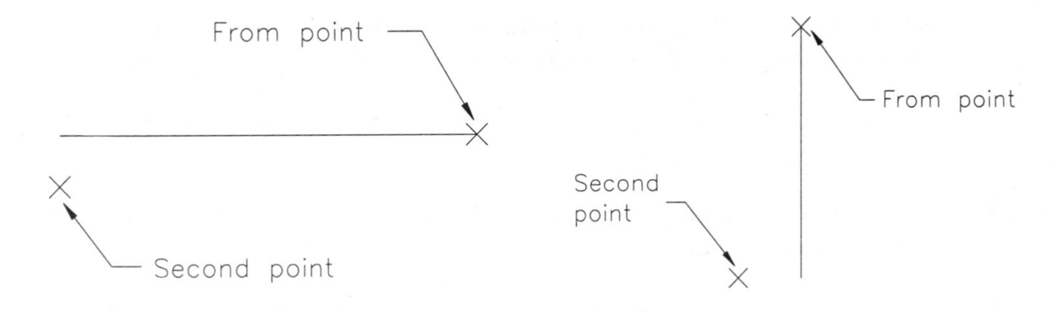

***Figure 3-36** Drawing a horizontal line using the Ortho mode*

***Figure 3-37** Drawing a vertical line using the Ortho mode*

Command: **ORTHO** [Enter]
Enter mode [ON/OFF] <current>: *Enter ON or OFF.*

STATUS BAR

When you are in AutoCAD LT, displayed is a status bar at the bottom of the graphics screen (Figure 3-38). This bar contains some useful information and tools that will make it easy to change the status of some AutoCAD LT functions. To change the status, you must choose the buttons. For example, if you want to display grid lines on the screen, choose the **GRID** button. Similarly, if you want to switch to paper space, choose **MODEL**. The status bar contains the following information:

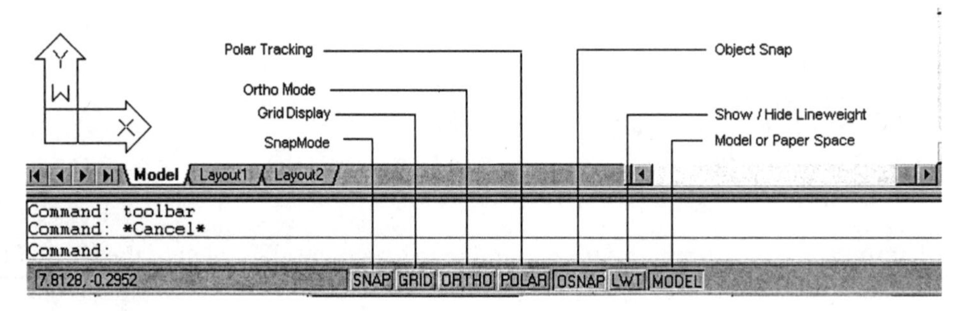

Figure 3-38 *Default Status bar display*

Coordinate Display. You can also use the function key F6 as a toggle key to turn the display of coordinates on or off. The **coordinates** information displayed in the Status bar can be static or dynamic. The **COORDS** system variable controls the type of display of the coordinates. If the value of the **COORDS** variable is set to 0, the coordinate display is static, i.e., the coordinate values displayed in the status line change only when you specify a point . However, if the value of the **COORDS** variable is set to 1 or 2, coordinate display is dynamic (default setting), AutoCAD LT constantly displays the absolute coordinates of the graphics cursor with respect to the UCS origin. AutoCAD LT can also display the polar coordinates (length<angle) if you are in an AutoCAD LT command and the **COORDS** system variable is set to 2.

SNAP. If **Snap Mode** is on, the **SNAP** button is displayed as pressed in the Status bar; otherwise, it is not. You can also use the function key F9 as a toggle key to turn snap off or on.

GRID. If **Grid Display** is on, grid lines are displayed on the screen. The function key F7 can also be used to turn on or off the grid display.

ORTHO. If **Ortho Mode** is on, the **ORTHO** button is displayed as pressed on the status bar. You can use the F8 key to turn on or off the **Ortho Mode**.

POLAR*. Choosing the **POLAR** button in the Status bar turns on the **Polar Tracking**. You can also use the function key F10. Turning the Polar Tracking on, automatically turns off the Ortho mode.

OSNAP. If **Object Snap** is on, the **OSNAP** button is displayed as pressed on the Status bar. You can also use the F3 key to turn the object snap on or off. When object snap is on, you can use the running object snaps. If OSNAP is off, the running Object Snaps are temporarily disabled. The status of OSNAP (Off or On) does not prevent you from using regular object snaps.

LWT. Choosing this button on the Staus bar, allows you to **Show/Hide Lineweight**. If the **LWT** button is not pressed, the display of Lineweight is turned off.

MODEL and PAPER Space. AutoCAD LT displays MODEL in the status line when you are working in the model space. If you are working in the paper space, AutoCAD LT will display PAPER in place of MODEL.

OBJECT SNAPS

Toolbar:	Object Snap or Standard > Object Snap flyout

Object snaps are one of the most useful features of AutoCAD LT. They improve your performance and the accuracy of your drawing and make drafting much simpler than it normally would be. Object snaps can also be invoked from the shortcut menu (Figure 3-39), which can be accessed by holding down the SHIFT key on the keyboard and then pressing the ENTER button on your pointing device. You can also invoke the different object snaps from the **Standard** toolbar by choosing the **Object Snap** button and holding the Pick button of your pointing device to display the flyout having the object snap buttons (Figure 3-40). The term object snap refers to the cursor's ability to snap exactly to a geometric point on an object. The advantage of using object snaps is that you do not have to specify an exact point. For example,if you want to place a point at the midpoint of a line,you may not be able to specify the exact point. Using the MIDpoint Object snap, all you do is move the cursor somewhere on the object. You will notice a marker (called the **AutoSnap** marker, in the form of a geometric shape) is automatically displayed at the middle point (snap point). You can click to place a point at the position of the **marker**. You can also have a target box attached to the cursor when any object snap is invoked. The object snaps recognize only the objects that are visible on the screen, which include the objects on locked layers. The objects on layers that are turned off or frozen are not visible, so they cannot be used for object snaps. The following are the Object Snap modes available in AutoCAD LT:

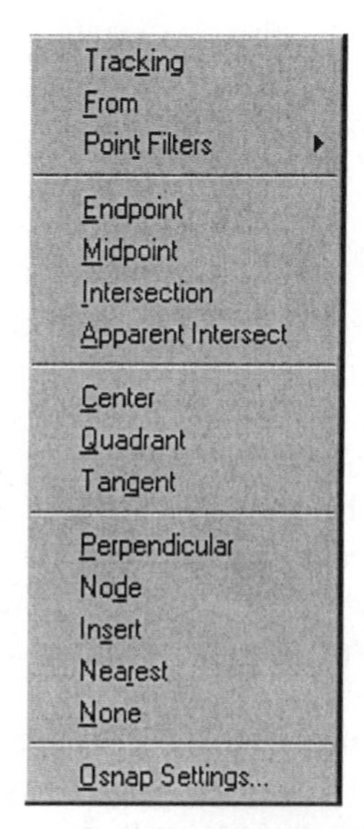

Figure 3-39 *Selecting object snap modes from the shortcut menu*

NEArest	CENter	TANgent	ENDpoint
QUAdrant	NONe	PERpendicular	MIDpoint
INTersection	INSertion	NODe	From
APParent Intersection			

Chapter 3

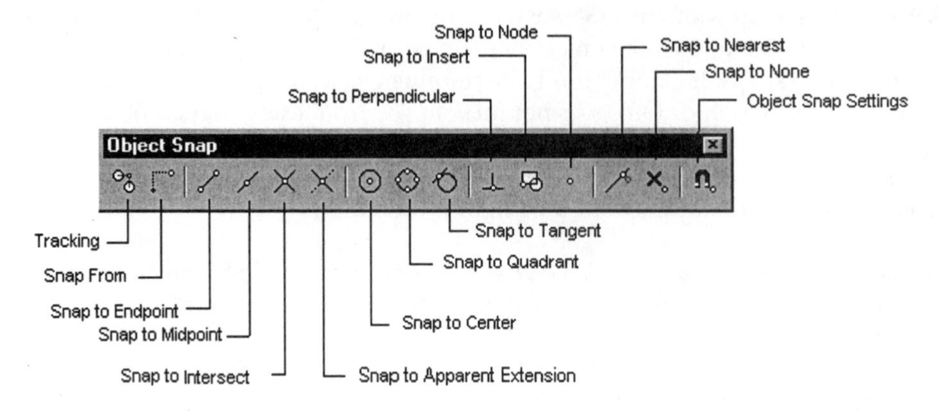

Figure 3-40 ***Object Snap*** *toolbar*

You can also invoke an object snap by entering an abbreviation in the command line.

AutoSnap

Menu:	Tools > Options
Command:	OPTIONS

The AutoSnap feature enables you to lock to geometric points of the objects. As you move the target box over the object, AutoCAD LT displays the geometric marker corresponding to the shapes shown in the **Object Snap** tab of the **Drafting Settings** dialog box. You can change the AutoSnap settings in the **Options** dialog box, under the **Drafting** tab. The **Options** dialog box can be invoked from the **Tools** menu, or by entering **OPTIONS** at the Command prompt. When you choose the **Drafting** tab in the **Options** dialog box, the AutoSnap options are displayed (Figure 3-41). You can select the **Marker** check box to activate the marker. You can also activate the Magnet and AutoSnap Tooltip by selecting their respecive check boxes. The **AutoSnap tooltip** displays a flag that describes the name of the object snap that AutoCAD LT has detected. The **Magnet** automatically moves and locks to the closest snap point. Similarly, if you select the **Display AutoSnap aperture box**, the target box is added to the screen crosshairs when you invoke an object snap. You can change the size of the marker by moving the **AutoSnap marker size** slider bar. You can also change the color of the markers through the **AutoSnap Marker Color:** list box. The size of the aperture box can be changed by moving the **Aperture size** slider bar in the **Options** (**Drafting tab**) dialog box. The size of the aperture is measured in PIXELS, short for picture elements. Picture elements are dots that make up the screen picture. The aperture size can also be changed using the **APERTURE** command. In AutoCAD LT the default value for the aperture size is 10 pixels. The display of the Marker and the AutoSnap tooltip is controlled by the **AUTOSNAP** system variable. The default value of this variable is 63 and is saved with the drawing. The following are the bit values for **AUTOSNAP**:

Bit Values	Function
0	Turns off the Marker, AutoSnap tooltip, and Magnet
1	Turns on the AutoSnap Marker
2	Turns on the AutoSnap tooltips

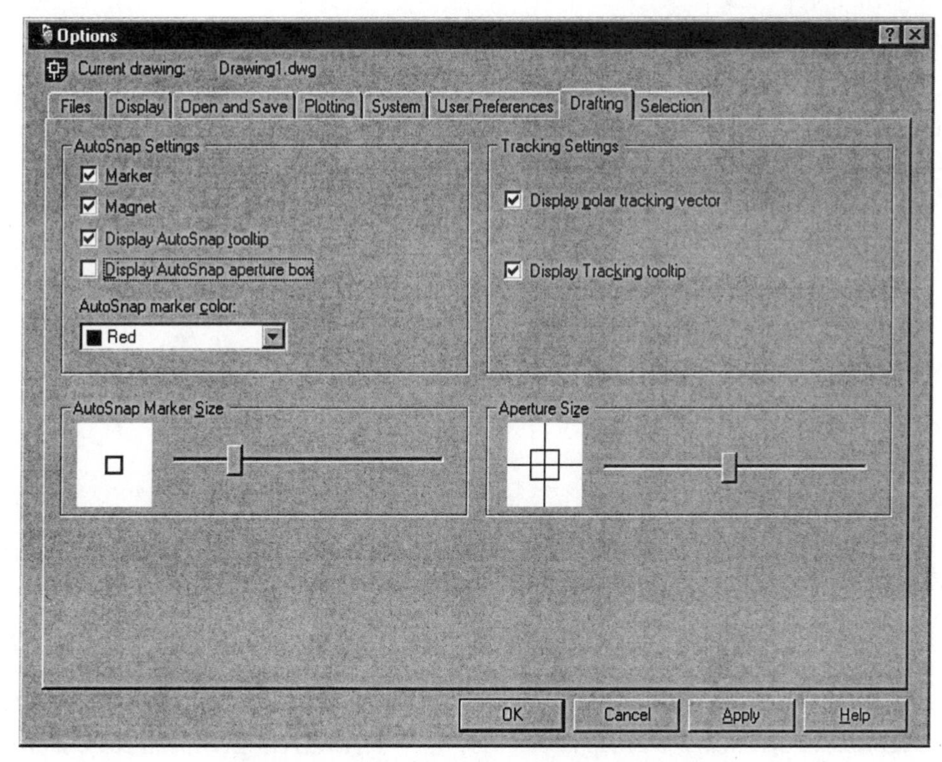

Figure 3-41 ***Drafting*** *tab of* ***Options*** *dialog box*

Bit Values	**Function**
4	Turns on the AutoSnap Magnet
8	Turns on polar tracking
32	Turns on the tooltips for polar tracking

NEArest

The **NEArest** Object Snap mode selects a point on an object (line, arc, circle, ellipse) that is visually closest to the graphics cursor (crosshairs). To use this mode, enter the command, then choose the NEArest object snap. Move the crosshairs near the intended point on the object so as to display the marker at the desired point and then select the object. AutoCAD LT will grab a point on the line where the marker was displayed. The following is the prompt sequence for drawing a line from a point on a line (Figure 3-42):

Command: **LINE** [Enter]
Specify first point: **NEA** [Enter] *(NEArest object snap.)*
to *Select a point near an existing object.*
Specify next point or [Undo]: *Select endpoint of the line.*

ENDpoint

The **ENDpoint** Object Snap mode snaps to the closest endpoint of a line or an arc. To use this Object Snap mode, select or enter the ENDpoint, move the cursor (crosshairs)

anywhere close to the endpoint of the object. The marker will be displayed at the endpoint, and then click to specify that point. AutoCAD LT will grab the endpoint of the object. If there are several objects near the cursor crosshairs, AutoCAD LT will grab the endpoint of the object that is closest to the crosshairs. The following is the prompt sequence for drawing a line from the endpoint of a line (Figure 3-43):

Command: **LINE** [Enter]
Specify first point: **END** [Enter] *(ENDpoint object snap.)*
of *Move the target box and select the line.*
Specify next point or [Undo]: *Select the endpoint of the line.*

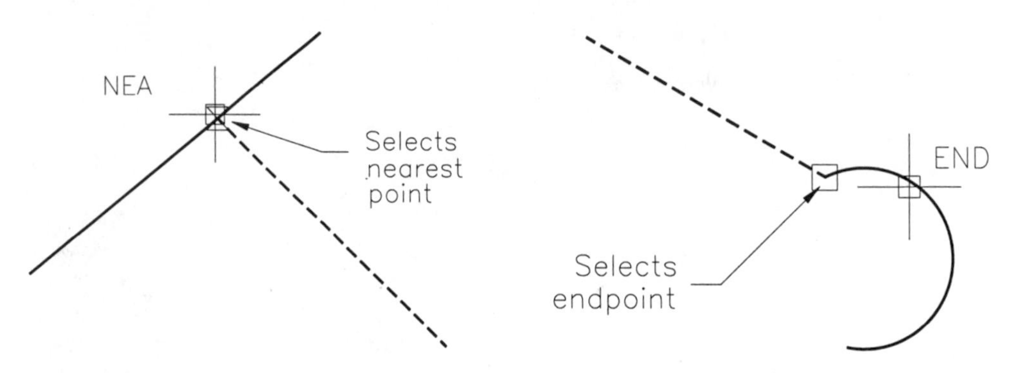

Figure 3-42 *NEArest object snap mode* ***Figure 3-43*** *ENDpoint object snap mode*

MIDpoint

The **MIDpoint** object snap mode snaps to the midpoint of a line or an arc. To use this object snap mode, select or enter MIDpoint and select the object anywhere. AutoCAD LT will grab the midpoint of the object. The following is the prompt sequence for drawing a line to the midpoint of a line (Figure 3-44):

Command: **LINE** [Enter]
Specify first point: *Select the starting point of the line.*
Specify next point or [Undo]: **MID** [Enter] *(MIDpoint object snap.)*
of *Move the cursor and select the original line.*

TANgent

The **TANgent** object snap allows you to draw a tangent to or from an existing ellipse, circle, or arc. To use this object snap, the cursor should be placed on the circumference of the circle or arc to select it. The following is the prompt sequence for drawing a line tangent to a circle (Figure 3-45):

Command: **LINE** [Enter]
Specify first point: *Select the starting point of the line.*
Specify next point or [Undo]: **TAN** [Enter] *(TANgent object snap)*
to *Move the cursor and select the circle.*

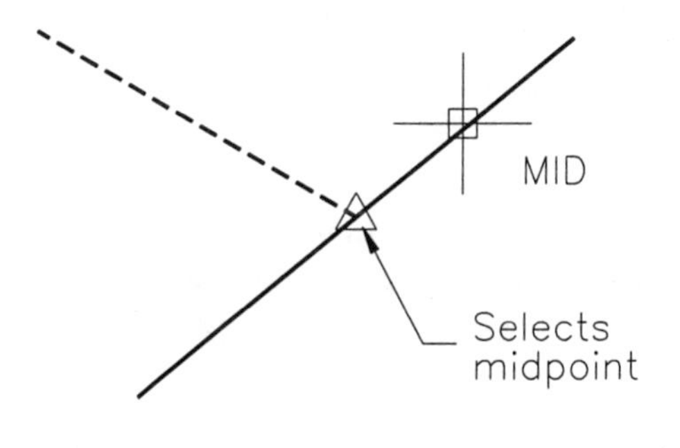

Figure 3-44 MIDpoint object snap mode

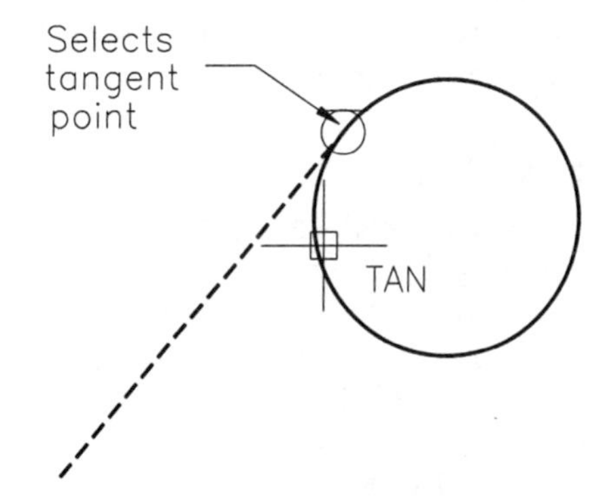

Figure 3-45 TANgent object snap mode

Figure 3-46 shows the use of NEArest, ENDpoint, MIDpoint, and TANgent object snaps.

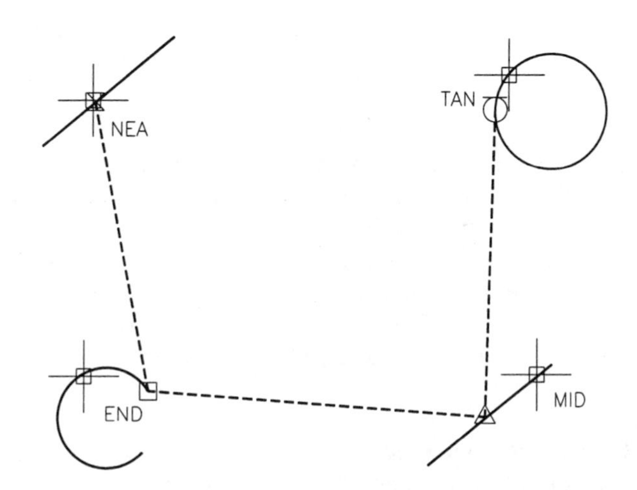

Figure 3-46 Using NEArest, ENDpoint, MIDpoint, and TANgent object snaps

CENter

The **CENter** object snap mode allows you to snap to the center point of an ellipse, a circle, or an arc. After selecting this option, you must point to the visible part of the circumference of a circle or arc. The following is the prompt sequence for drawing a line from the center of a circle (Figure 3-47):

Command: **LINE** Enter
Specify first point: **CEN** Enter

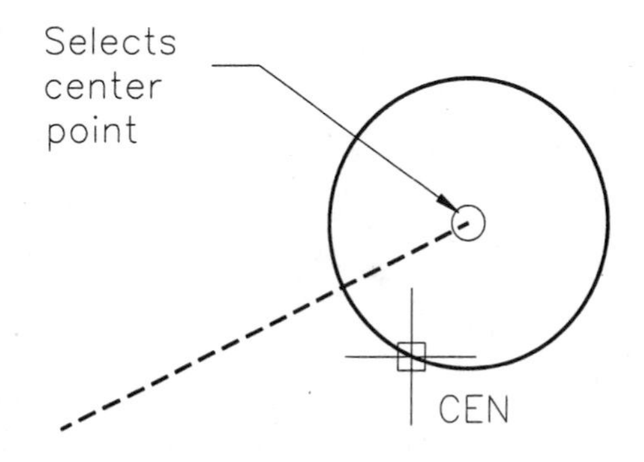

Figure 3-47 CENter object snap mode

(CENter object snap.)
of *Move the cursor and select the circle.*
Specify next point or [Undo]: *Select the endpoint of the line.*

INTersection

The **INTersection** object snap mode is used if you need to snap to a point where two or more lines, circles, ellipse, or arcs intersect. To use this object snap, move the cursor close to the desired intersection so that the intersection is within the target box, and then specify that point. The following is the prompt sequence for drawing a line from an intersection (Figure 3-48):

Command: **LINE** Enter
Specify first point: **INT** Enter *(INTersection object snap.)*
of *Position the cursor near the intersection and select it.*
Specify next point or [Undo]: *Select the endpoint of the line.*

If you select only one object with INTersection object snap, AutoCAD LT prompts **and** for selection of another object.

QUAdrant

The **QUAdrant** object snap mode is used when you need to snap to a quadrant point of an ellipse, arc, or circle. A circle has four quadrants, and each quadrant subtends an angle of 90 degrees. The quadrant points are located at 0-, 90-, 180-, 270-degree positions. If the circle is inserted as a block (see Chapter 12) that is rotated, the quadrant points are also rotated by the same amount. To use this object snap, position the cursor on the circle or arc closest to the desired quadrant. The prompt sequence for drawing a line from the third quadrant of a circle (Figures 3-49, 3-50, and 3-51) is:

Command: **LINE** Enter
Specify first point: **QUA** Enter *(QUAdrant object snap.)*
of *Move the cursor close to the third quadrant of the circle and select it.*
Specify next point or [Undo]: *Select the endpoint of the line.*

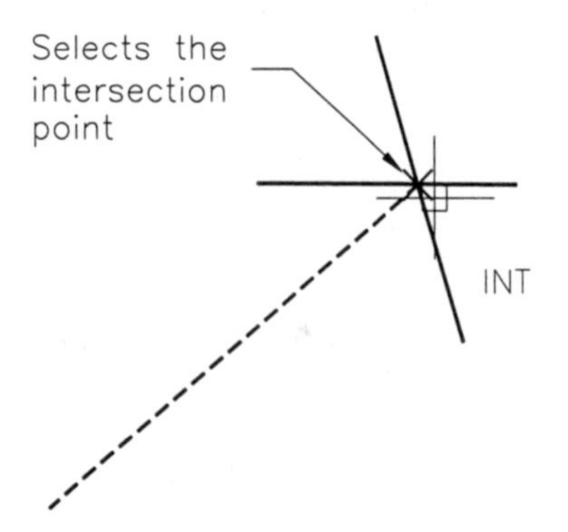

Figure 3-48 *INTersection object snap mode*

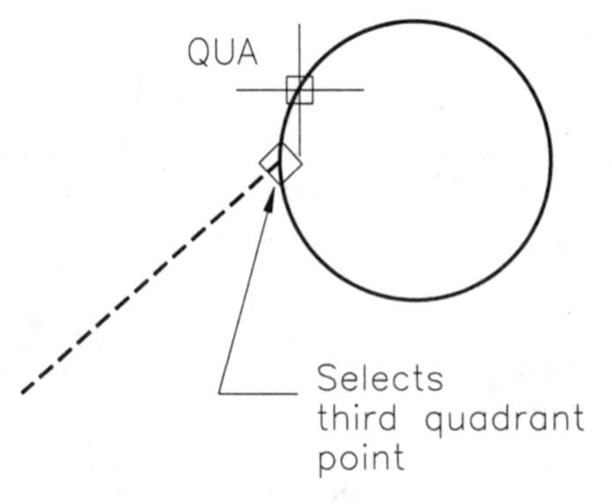

Figure 3-49 *QUAdrant object snap mode*

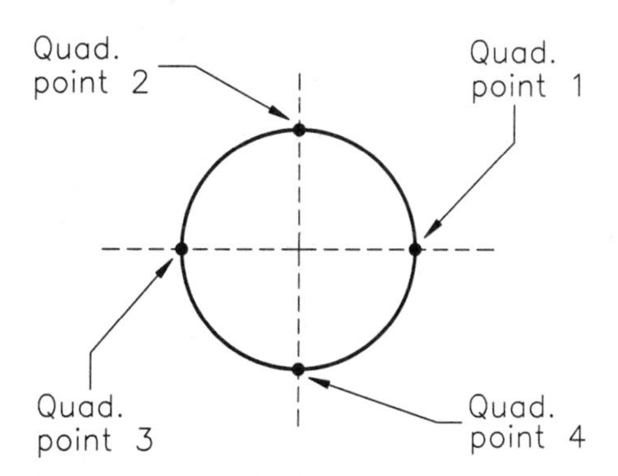

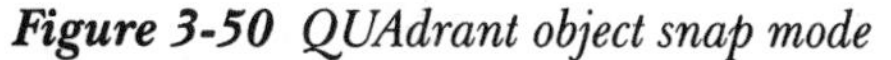
Figure 3-50 *QUAdrant object snap mode*

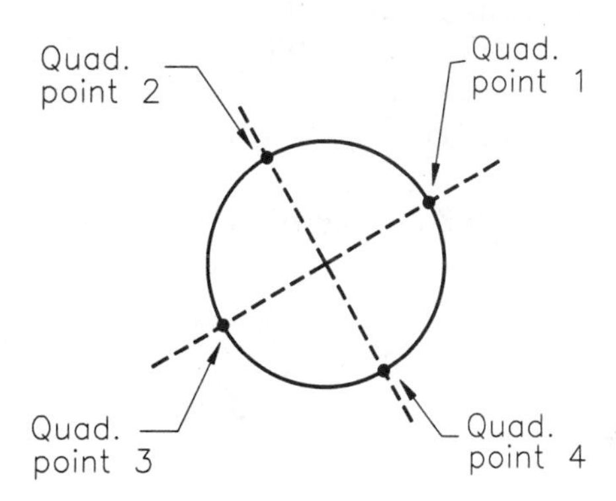

Figure 3-51 *Four quadrants of a circle*

PERpendicular

The **PERpendicular** object snap mode is used to draw a line perpendicular to or from another line, or normal to or from an arc or circle, or to an ellipse. When you use this mode and select an object, AutoCAD LT calculates the point on the selected object so that the previously selected point is perpendicular to the line. The object can be selected by positioning the cursor anywhere on the line. The following is the prompt sequence for drawing a line that is perpendicular to a given line (Figure 3-52):

Command: **LINE** [Enter]
Specify first point: *Select the starting point of the line.*
Specify next point or [Undo]: **PER** [Enter] *(PERpendicular object snap.)*
to *Select the line on which you want to draw perpendicular.*

When you select the line first, the rubber-band feature of the line is disabled. The line will appear only after the second point is selected. The prompt sequence for drawing a line perpendicular from a given line (Figure 3-53) is:

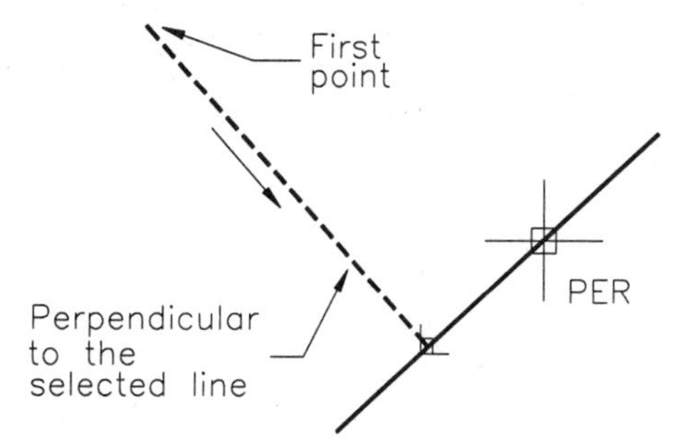

Figure 3-52 *Selecting the start point, then the perpendicular snap*

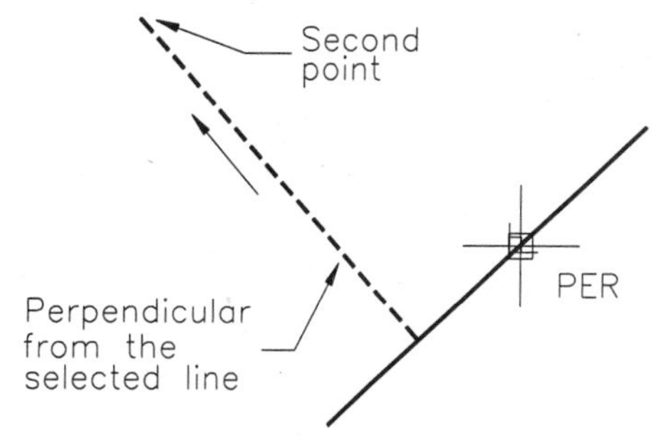

Figure 3-53 *Selecting the perpendicular snap first*

Command: **LINE** [Enter]
Specify first point: **PER** [Enter] *(PERpendicular object snap.)*
to *Select the line on which you want to draw perpendicular.*
Specify next point or [Undo]: *Select the endpoint of the line.*

In the Figure 3-54, shows the use of the CENter, INTersection, and PERpendicular object snaps.

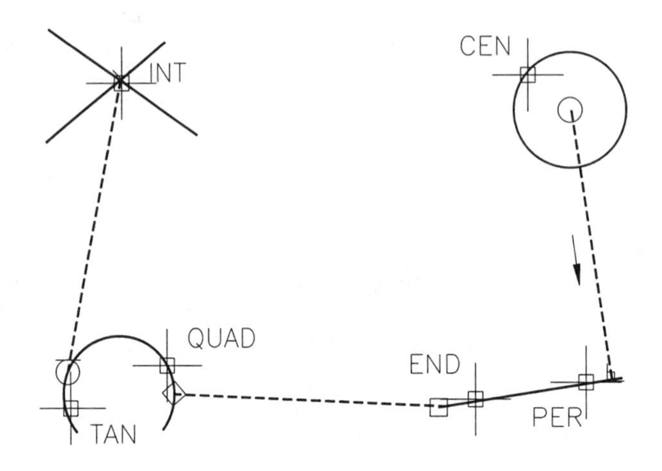

Figure 3-54 *Using CENter, Intersection, and PERpendicular object snap modes*

Exercise 3 *Graphics*

In Figure 3-55, P1 and P2 are the center points of the top and bottom arcs. The space between the dotted lines is 1 unit.

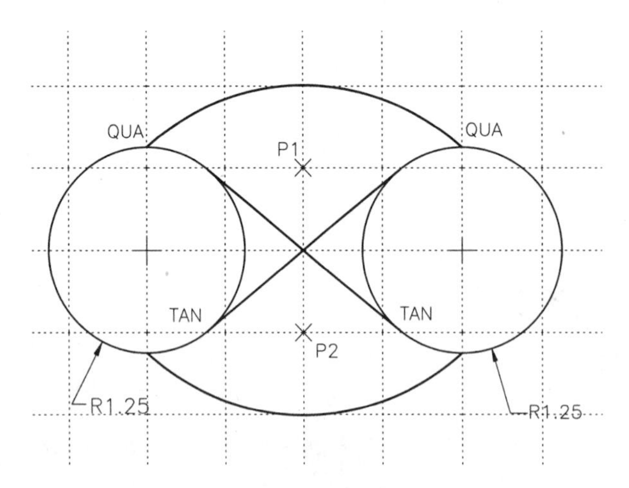

Figure 3-55 *Drawing for Exercise 3*

NODe

The **NODe** object snap can be used to snap to a point object drawn, using the **POINT** command or placed while using the **DIVIDE** or **MEASURE** commands. In Figure 3-56, three points have been drawn using the AutoCAD LT **POINT** command. You can snap to these points by using NODe snap mode, as shown on the following example, Figure 3-57:

Command: **LINE** [Enter]
Specify first point: **NOD** [Enter] *(NODe object snap.)*
of *Select point P1.*
Specify next point or [Undo]: **NOD** [Enter]
of *Select point P2.*
Specify next point or [Undo]: NOD [Enter]
of *Select point P3.*

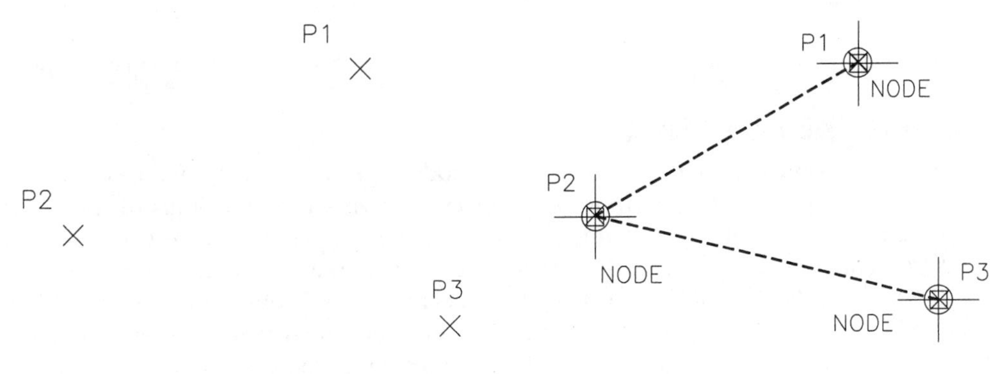

Figure 3-56 *Point objects* ***Figure 3-57*** *Using NODe object snap*

INSertion

The **INSertion** object snap mode is used to snap to the insertion point of a text, shape, block, attribute, or attribute definition. In Figure 3-58, the text **WELCOME** is left-justified and the text **AutoCAD** is center-justified. The point with respect to which the text is justified is the insertion point of that text string. If you want to snap to these insertion points or the insertion point of a block, use the INSertion object snap mode. The prompt sequence is:

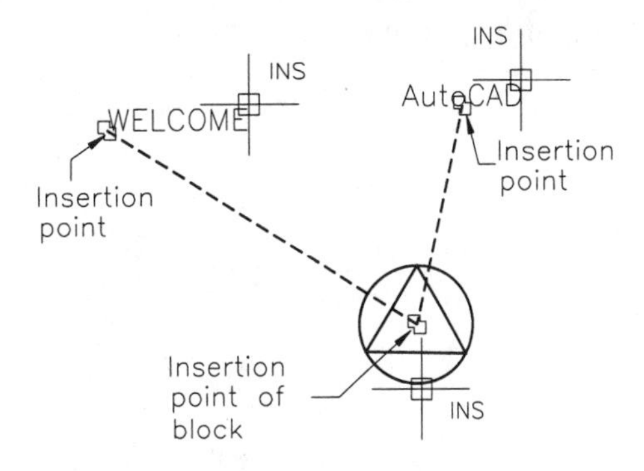

Figure 3-58 *INSertion object snap mode*

Command: **LINE** [Enter]
Specify first point: **INS** [Enter] *(INSert object snap.)*
of *Select WELCOME text.*
Specify next point or [Undo]: **INS** [Enter]
of *Select the block.*
Specify next point or [Undo]: **INS** [Enter]
of *Select AutoCAD LT text.*

NONe

The **NONe** object snap mode turns off any running object snap (see the section "Running Object Snap Mode" that follows.) for one point. The following example illustrates the use of this Object Snap mode:

Command: **-OSNAP** Enter
Enter list of object snap modes: **MID, CEN**

This sets the object snaps to Mid and Cen. You can disable those using the None object snap.

Command: **LINE** Enter
Specify first point: *Specify a point.*
Specify next point or [Undo]: **NONe** Enter *(Turns off the running object snap and allows you to select any point.)*

Once you have selected a point it continues the command with the specified object snaps.

APParent Intersection

The **APParent Intersection** object snap mode is similar to the INTersection snap mode, except that this mode selects projected or visual intersections of lines, arcs, circles, or ellipse (Figure 3-59 and Figure 3-60). The projected intersections are those intersections that are not present on the screen, but are imaginary ones that can be formed if line, or arcs are extended. Sometimes two objects appear to intersect one another in the current view, but in 3D space the two objects do not actually intersect. The APParent Intersection snap mode also selects such visual intersections. The prompt sequence is:

Command: **LINE** Enter
From point: **APP** *(APParent Intersection object snap.)*
of *Select first object.*
and *Select second object.*
To point: *Select the endpoint of the line.*

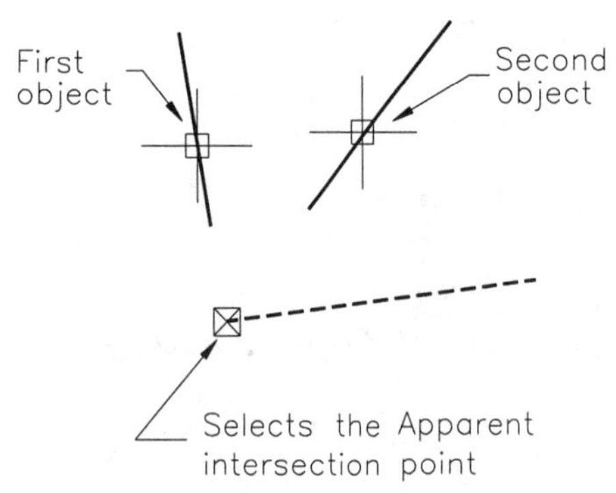

Figure 3-59 *Apparent Intersection Snap mode*

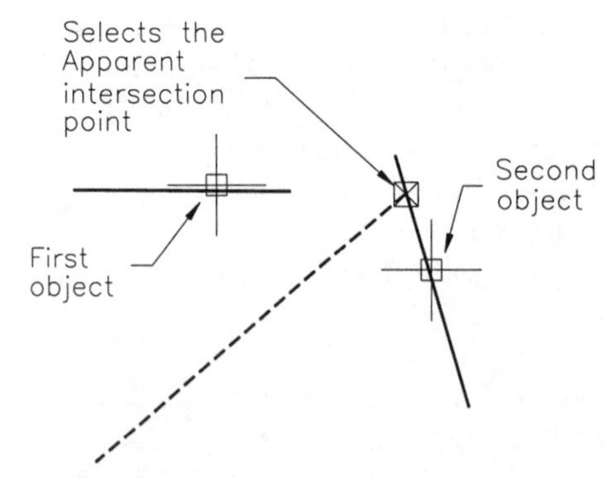

Figure 3-60 *APParent Intersection Snap mode*

From

The **From** object snap can be used to locate a point relative to a given point (Figure 3-61). For example, if you want to locate a point that is 2.5 units up and 1.5 units right from the endpoint of a given line, you can use the From object snap mode. The prompt sequence for the From object snap is:

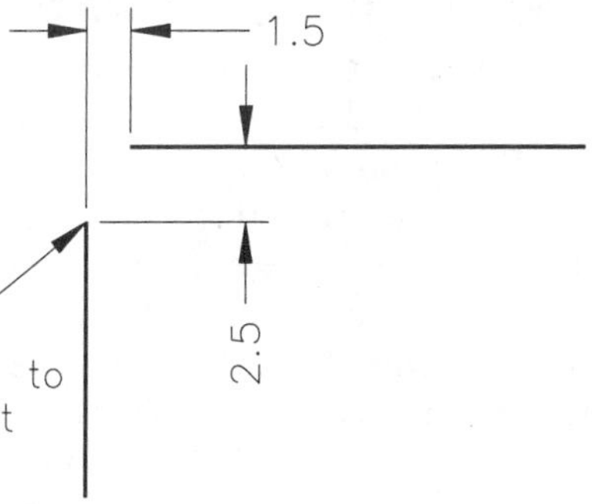

Figure 3-61 Using the From object snap to locate a point

Command: **LINE**
Specify first point: **From** *(or select Snap From from the **Object Snap** toolbar.)*
_from Base point: **End** *(or select Snap to Endpoint from the **Object Snap** toolbar and specify the endpoint of the given line.)*
of <Offset>: **@1.5,2.5**

Note
The From object snap cannot be used as running object snap.

Combining Object Snap Modes

You can also combine the snaps by separating the snap modes with a comma. AutoCAD LT will search for the specified modes and grab the point on the object closest to the point where the object is selected. The prompt sequence for using the MIDpoint and ENDpoint snaps is:

Command: **LINE** [Enter]
Specify first point: **MID, END** [Enter] *(MIDpoint or ENDpoint object snap.)*
to *Select the object.*

In this example, we have defined two object snap modes, MIDpoint and ENDpoint. The point that it will grab depends on where you select the line. If you select the line at a point that is closer to the midpoint, AutoCAD LT will snap to the midpoint (Figure 3-62). If you select the line closer to the endpoint of the line, the line will snap to the endpoint (Figure 3-63).

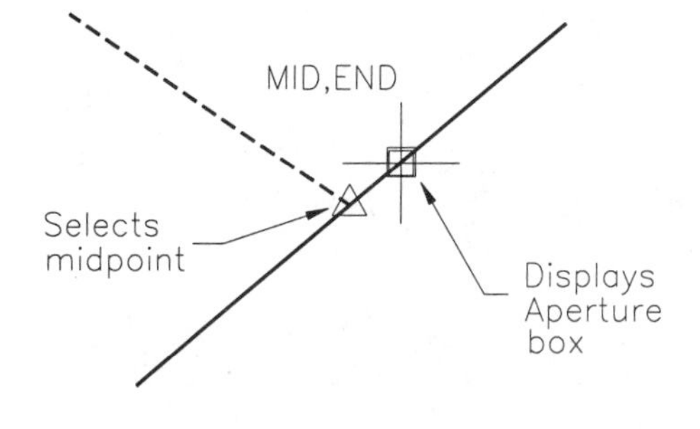

***Figure 3-62** Using MIDpoint, ENDpoint Snaps (Selecting midpoint)*

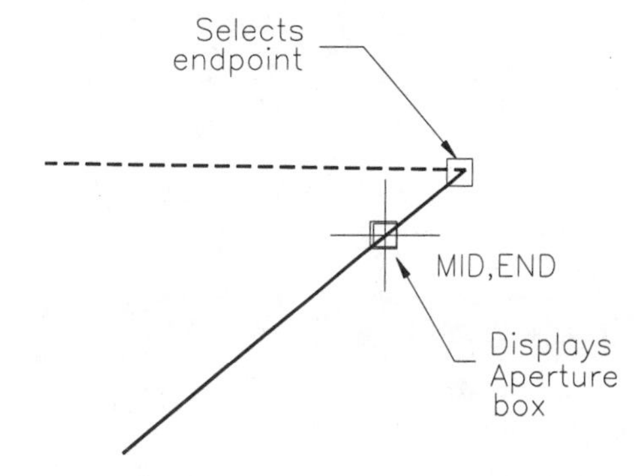

***Figure 3-63** Using MIDpoint, ENDpoint Snaps (Selecting endpoint)*

Cycling Through Snaps and Keyboard Entry Priority

AutoCAD LT displays the geometric marker corresponding to the shapes shown in the **Object Snap** tab of the **Drafting Settings** dialog box. You can use the **TAB** key to cycle through the snaps. For example, if you have a circle with an intersecting rectangle as shown in Figure 3-64 and you want to snap to one of the geometric points on the circle, you can use the **TAB** key to cycle through the geometric points. The geometric points for circle are center point, quadrant points, and in this example, the intersection points of the rectangle and circle. To snap to one of these points, the first thing you need to do is to set the snaps (center, quadrant, and intersection object snaps) in the **Drafting Settings** dialog box. After entering a command, when you drag the cursor over the objects, AutoSnap displays a marker and a SnapTip. You can cycle through the snap points available for an object by pressing the TAB key. For example, if you press the TAB key while the aperture box is on the circle, AutoSnap will display the center, quadrant, and intersection points. By selecting a point you can snap to one of these points.

The **OSNAPCOORD** system variable determines the priority between the keyboard entry and object snap. By selecting the **Running Object Snap** radio button, you can also set this variable through the **User Preferences** tab of the **Options** dialog box. For example, if intersection object snap is on and **OSNAPCOORD** is set to 1 and you enter the coordinates of the end-point of a line very close to the intersection point (Figure 3-65), the line will not snap to the intersection point. If **OSNAPCOORD** is set to 0, the line will snap to the intersection point because running object snap has the priority.

Note

In the discussions of object snaps, "line" generally includes xlines, rays, and polyline segments, and "arc" generally includes polyarc segments.

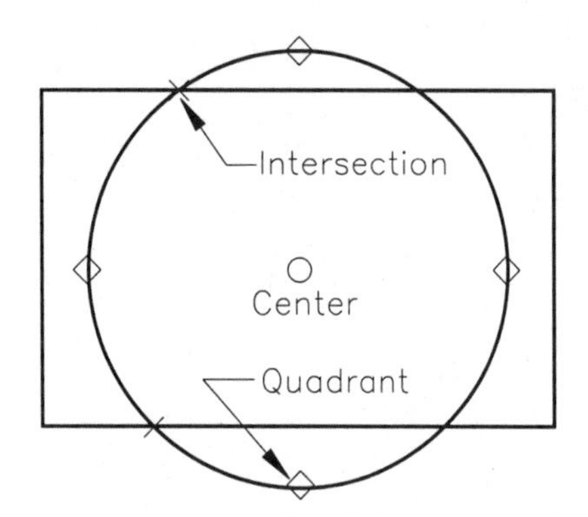

Figure 3-64 *Using TAB key to cycle through the Snap*

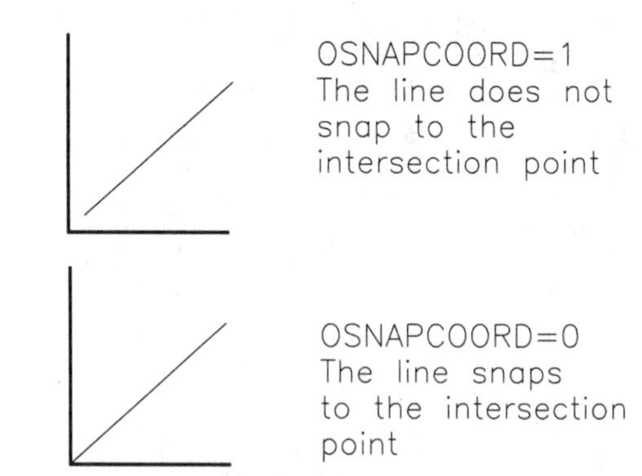

Figure 3-65 *Using OSNAPCOORD system variable to set keyboard entry priority*

Exercise 4 — *General*

Make the drawing in Figure 3-66. The space between the dotted lines is 1 unit.

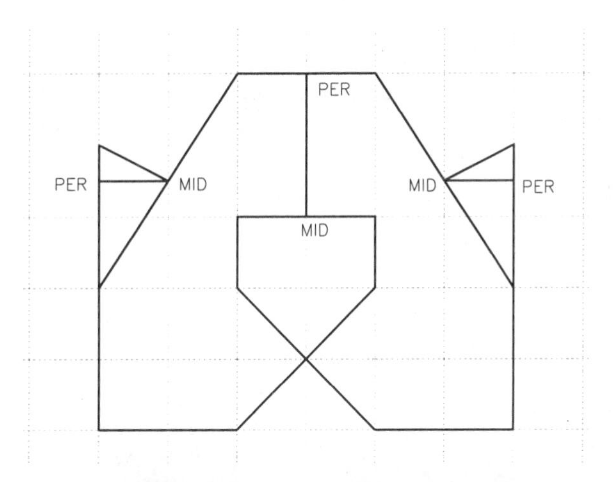

Figure 3-66 *Drawing for Exercise 4*

RUNNING OBJECT SNAP MODE

Toolbar:	Object Snap > Object Snap Settings
Menu:	Tools > Drafting Settings
Command:	OSNAP or DSETTINGS

In the previous sections you have learned how to use the object snaps to snap to different points of an object. One of the drawbacks of these object snaps is that you have to select them every time you use them, even if it is the same snap mode. This problem can be solved by using **running object snaps**. The Running osnap can be invoked from the **Drafting Settings** dialog box. To invoke the **Drafting Settings** dialog box, choose **Drafting Settings** from the **Tools** menu. You can also choose the **Object Snap Settings** button from the **Object Snap** toolbar. Right-clicking on the **Osnap** button in the Status bar displays a shortcut menu from which you can choose a setting to display the **Drafting Settings** dialog box. In the **Drafting Settings** dialog box, choose the **Object Snap** tab and set the running object snap modes by selecting the check boxes next to the snap modes. For example, if you want to set ENDpoint as the running object Snap mode, select the **ENDpoint** check box and then the **OK** button in the dialog box. You can also set it through the Command prompt:

Command: **-OSNAP** [Enter]
Current Osnap modes :
Enter list of object snap modes: **END** [Enter]

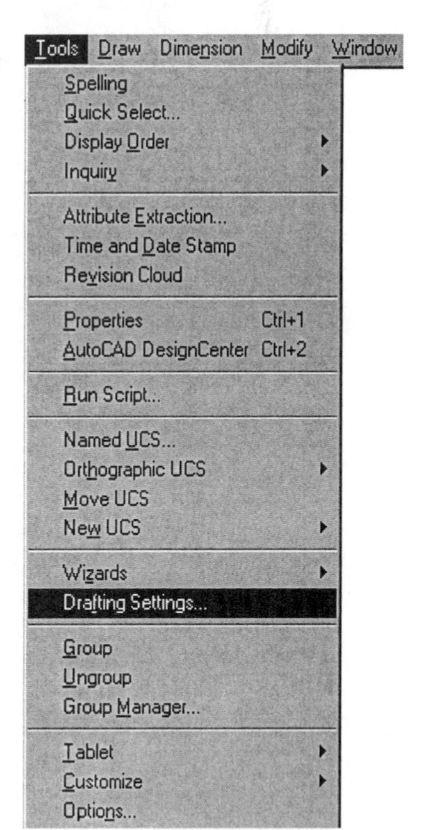

Figure 3-67 *Selecting* ***Drafting Settings*** *from the* ***Tools*** *menu*

Chapter 3

Once you set the running object snap mode, you are automatically in that mode and the marker is displayed when you move the crosshairs over the snap points. If you had selected a combination of modes, AutoCAD LT selects the mode that is closest to the screen crosshairs. The running object snap mode can be turned on or off, without losing the object snap settings, by choosing the **OSNAP** button in the Status bar located at the bottom of the screen. You can also accomplish this by pressing the function key F3 or CTRL+F keys. If no running osnaps are set, the **Drafting Settings** dialog box is displayed on choosing the **OSNAP** button in the Status bar (Figure 3-68).

Note

When you use a temporary object snap and fail to specify a suitable point, AutoCAD LT issues an Invalid point prompt and reissues the pending command's prompt. However, with a running object snap mode, if you fail to specify a suitable point, AutoCAD LT accepts the point specified without snapping to an object.

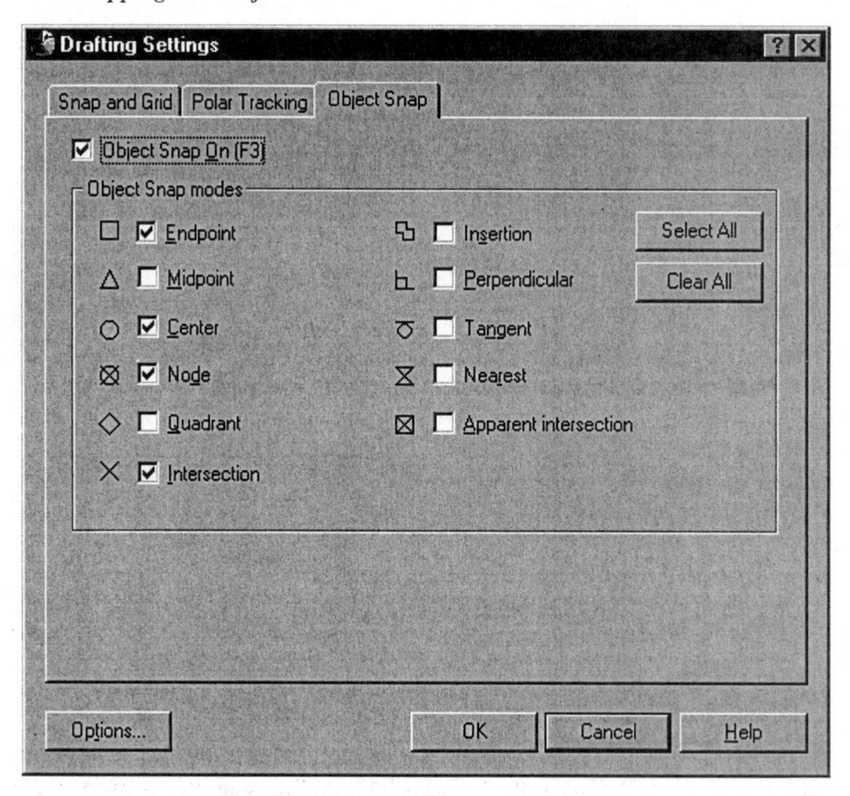

***Figure 3-68 Drafting Settings** dialog box*

Overriding the Running Snap

When you select the running object snaps, all other object snap modes are ignored unless you select another object snap mode. Once you select a different osnap mode, the running OSNAP mode is temporarily overruled. After the operation has been performed, the running OSNAP mode goes into effect again. If you want to discontinue the current running object snap modes totally, choose the **Clear all** button in the **Drafting Settings** dialog box. If you want to temporarily disable the running object snap, choose the **OSnap** button in the Status bar (Figure 3-38).

If you are overriding the running OSNAP modes for a point selection and no point is found to satisfy the override object snap mode, AutoCAD LT displays a message to this effect. For example, if you specify an override object snap mode of CENter and no circle, ellipse, or arc is found at that location, AutoCAD LT will display the message "No Center found for specified point. Point or option keyword required."

FUNCTION AND CONTROL KEYS

You can also us the function and control keys to change the status of coordinate display, Snap, Ortho, Osnap, tablet, isometric planes, running Object Snap, Grid, and Polar tracking. The following is a list of function and control keys:

F1	Help	F7	Grid On/Off **(Ctrl+G)**
F2	Graphics Screen/Text Window	F8	Ortho On/Off **(Ctrl+L)**
F3	Running Osnap On/Off **(Ctrl+F)**	F9	Snap On/Off **(Ctrl+B)**
F4	Tablet mode On/Off **(Ctrl+T)**	F10	Polar tracking On/Off
F5	Isoplane top/right/left **(Ctrl+E)**		
F6	Coordinate display On/Off **(Ctrl+D)**		

USING AUTOTRACKING*

When using AutoTracking, the cursor moves along temporary paths to locate key points in a drawing. It can be used to locate points with respect to other points or objects in the drawing. The Autotracking option in AutoCAD LT is the **Polar Tracking** option. Polar Tracking can be selected by choosing the **Polar** button in the Status bar, by using the function key F10 or by selecting the **Polar Tracking On** check box in the **Polar Tracking** tab of the **Drafting Settings** dialog box (Figure 3-69).

Polar Tracking constrains the movement of the cursor along a path that is based on the Polar Angle Settings. For example, if the **Increment Angle:** edit box value is set to 5 degrees in the **Polar Angle Settings** area of the **Polar Tracking** tab of the **Drafting Settings** dialog box, the cursor will move along alignment paths that are multiples of 5 degrees and a tooltip will display a distance and angle. Selecting the **Additional angles** check box and choosing the **New** button allows you to add a new increment angle value in the **Increment Angle:** drop-down list. Polar Tracking is on only when the Ortho mode is off.

The direction of the path is determined by the motion of the cursor or the point you select on an object. For example, if you want to draw a circle whose center is located in line with the center of two existing circles, you can use Autotracking as follows:

Command: **CIRCLE**
Specify center point for Circle or [3P/2P/Ttr (tan, tan, radius)]: **TK** [Enter]
First tracking point: **CEN** *(or Pause the cursor briefly at the center of the first circle to attain the point and then move it along the horizontal alignment path.)*

Next point (Press ENTER to end tracking) :**CEN** *(or Place the cursor briefly at the center of the second circle and then move it along the vertical alignment path.)*
Next point (Press ENTER to end tracking): Enter *(Select the intersection of the two alignment paths)*
Specify radius of circle or [Diameter] <current>: Enter radius or specify a point.

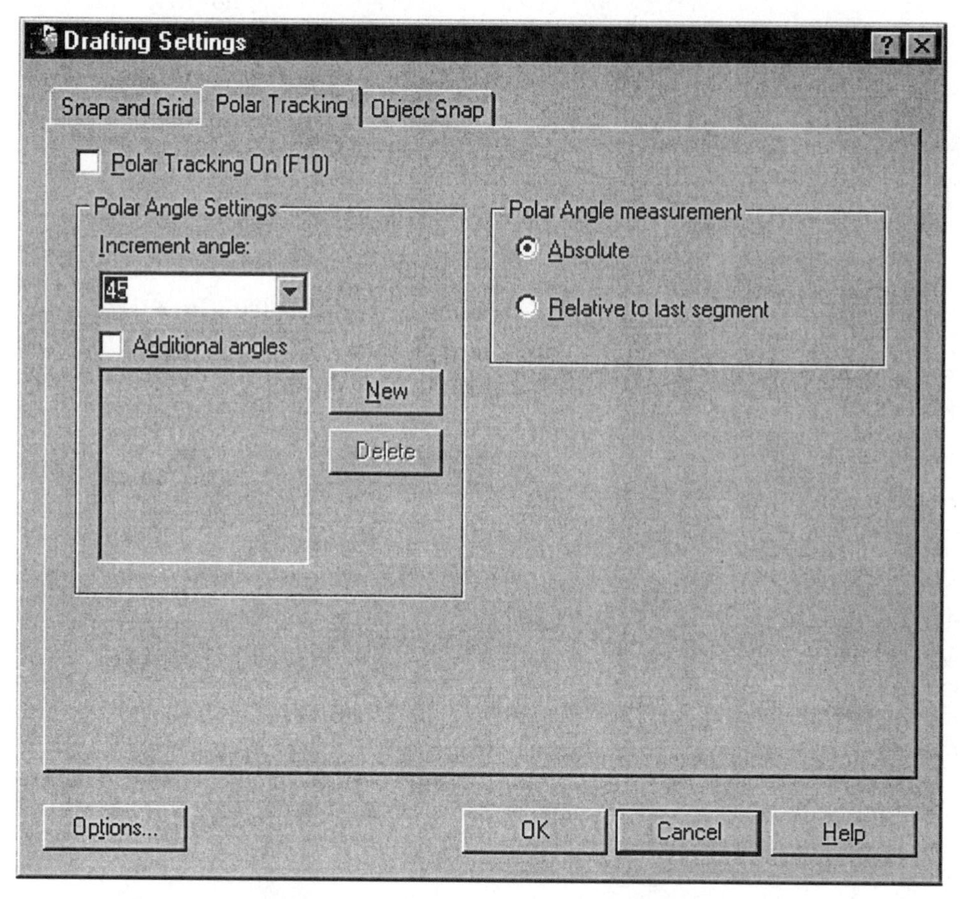

***Figure 3-69 Polar Tracking** tab of the **Drafting Settings** dialog box*

Similarly, you can use Autotracking in combination with the midpoint object snap to locate the center of the rectangle and then draw a circle (Figure 3-70 and Figure 3-71).

GLOBAL AND CURRENT LINETYPE SCALING

The **LTSCALE** system variable controls the global scale factor of the lines in a drawing. For example, if **LTSCALE** is set to 2, all lines in the drawing will be affected by a factor of 2. Like **LTSCALE**, the **CELTSCALE** system variable controls the linetype scaling. The difference is that **CELTSCALE** determines the current linetype scaling. For example, if you set **CELTSCALE** to 0.5, all lines drawn after setting the new value for **CELTSCALE** will have the linetype scaling factor of 0.5. The value is retained in the **CELTSCALE** system variable. Line (a) in Figure 3-72 is drawn with a **CELTSCALE** factor of 1; line (b) is drawn with a **CELTSCALE**

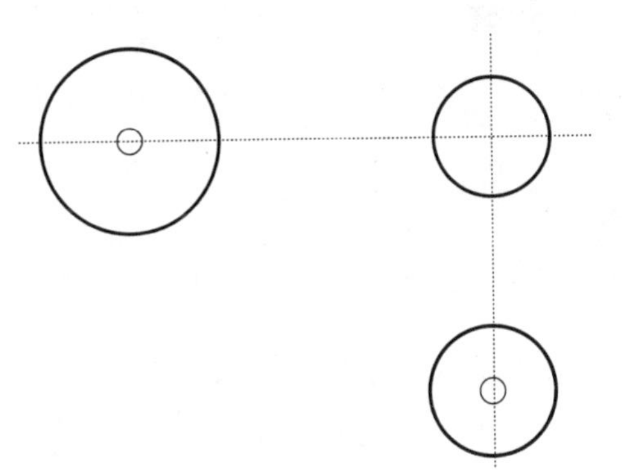

Figure 3-70 Using AutoTracking to locate a point (Center of circle)

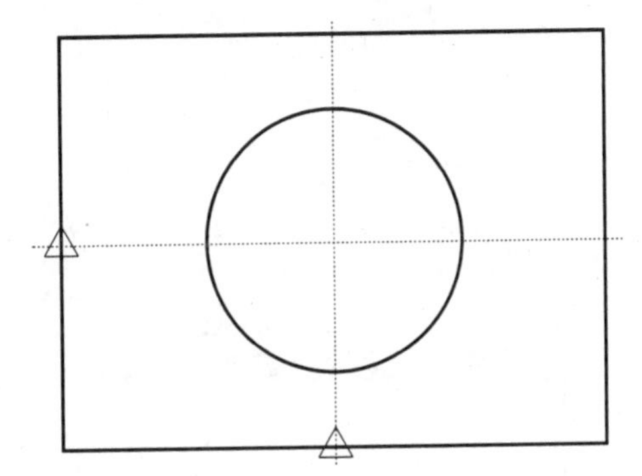

Figure 3-71 Using AutoTracking to locate a point (MIDpoint of rectangle)

factor of 0.5. The length of the dash is reduced by a factor of 0.5 when **CELTSCALE** is 0.5. The net scale factor is equal to the product of **CELTSCALE** and **LTSCALE**. Figure 3-72(c) shows a line that is drawn with **LTSCALE** of 2 and CELTSCALE of 0.25. The net scale factor = LTSCALE x **CELTSCALE** = 2 x 0.25 = 0.5. You can also change the global and current scale factors by entering the **LINETYPE** command at the Command prompt.and the **Linetype Manager** dialog box (Figure 3-73). If you choose the **Details** button, the properties associated with the selected linetype are displayed. You can change the values according to your drawing requirements.

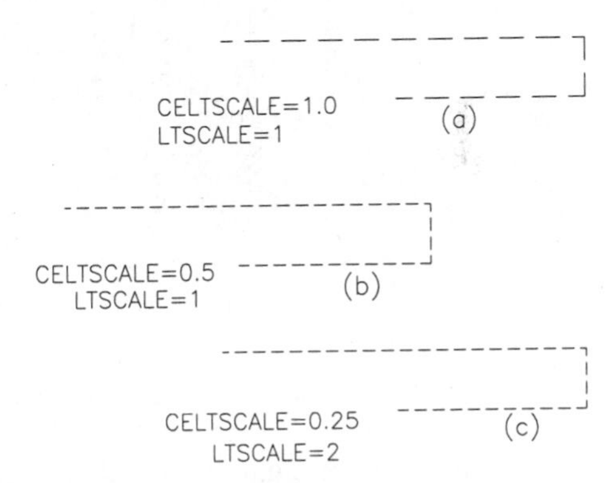

*Figure 3-72 Using **CELTSCALE** to control current linetype scaling*

LTSCALE FACTOR FOR PLOTTING

The **LTSCALE** factor for plotting depends on the size of the sheet you are using to plot the drawing. For example, if the limits are 48 by 36, the drawing scale is 1:1, and you want to plot the drawing on a 48" by 36" size sheet, then the **LTSCALE** factor is 1. If you check the specification of the Hidden linetype in the **ACAD.LIN** file, the length of each dash is 0.25. Hence, when you plot a drawing with 1:1 scale, the length of each dash in a hidden line is 0.25.

However, if the drawing scale is 1/8" = 1' and you want to plot the drawing on 48" by 36" paper, the **LTSCALE** factor must 8 x 12 = 96. The length of each dash in the hidden line will increase by a factor of 96, because the **LTSCALE** factor is 96. Therefore, the length of each dash will be (0.25 x 96 = 24) units. At the time of plotting, the scale factor for plotting must be 1:96 to plot the 384' by 288' drawing on 48" by 36" paper. Each dash of the hidden line that was 24" long on the drawing will be 24/96 = 0.25" long when plotted. Similarly, if the desired text size on the paper is 1/8", the text height in the drawing must be 1/8 x 96 = 12".

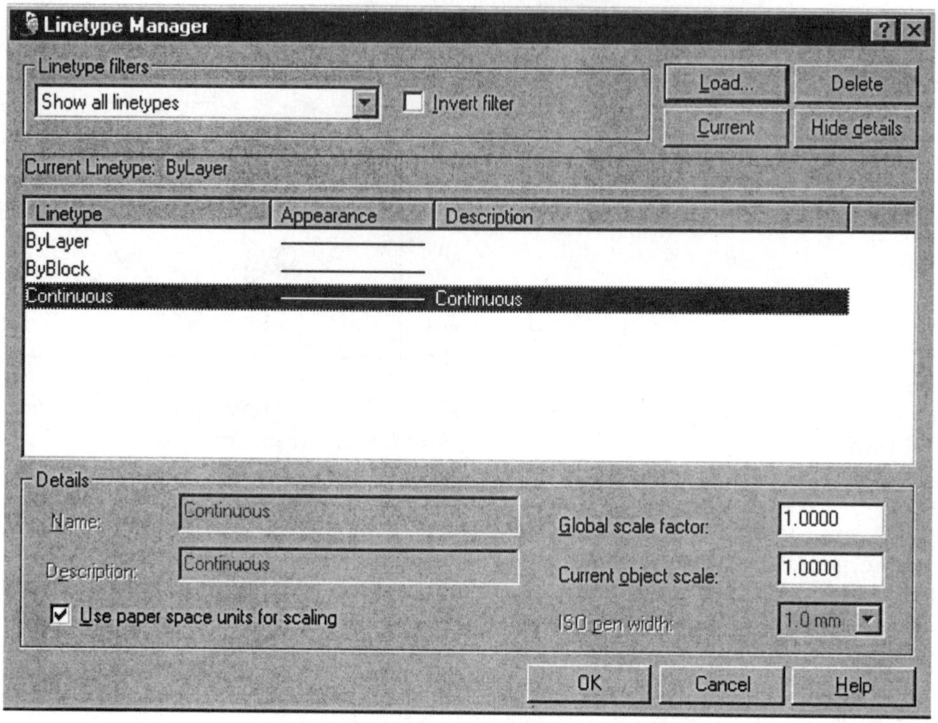

***Figure 3-73 Linetype Manager** dialog box*

Ltscale Factor for PLOTTING = Drawing Scale

Sometimes your plotter may not be able to plot a 48" by 36" drawing, or you might like to decrease the size of the plot so that the drawing fits within a specified area. To get the correct dash lengths for hidden, center, or other lines, you must adjust the **LTSCALE** factor. For example, if you want to plot the previously mentioned drawing in a 45" by 34" area, the correction factor is:

Correction factor = 48/45
= 1.0666

New LTSCALE factor = **LTSCALE** factor x Correction factor
= 96 x 1.0666
= 102.4

New Ltscale Factor for PLOTTING = Drawing Scale x Correction Factor

Note

*If you change the **LTSCALE** factor, all lines in the drawing are affected by the new ratio.*

Changing Linetyype Scale Using the PROPERTIES Command*

Toolbar:	Standard > Properties
Menu:	Modify > Properties
Command:	PROPERTIES

You can also change the current linetype scale by using the **PROPERTIES** command. You can invoke this command by choosing the **Properties** button from the **Standard Toolbar** (Figure 3-75). AutoCAD LT will display the **Properties** dialog box, Figure 3-74. All properties of the selected objects are displayed in the **Properties** dialog box. To change the current linetype scale, you should select the object first and then invoke the **PROPERTIES** command to display the **Properties** dialog box (Figure 3-74). In this dialog box, select the **Categorized** tab and locate the **Linetype scale** edit box. Enter the new linetype scale and then close the dialog box. The linetype scale of the selected objects is changed to the new value you entered.

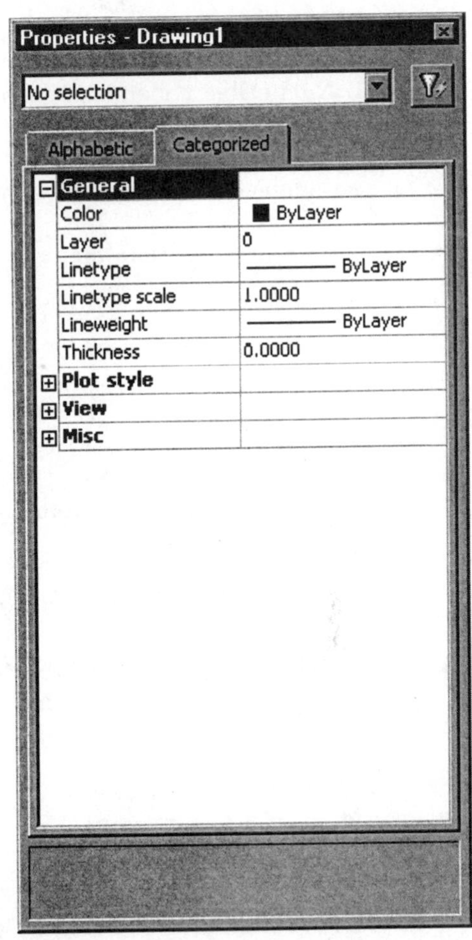

*Figure 3-74 **Properties** dialog box (**Categorized** tab)*

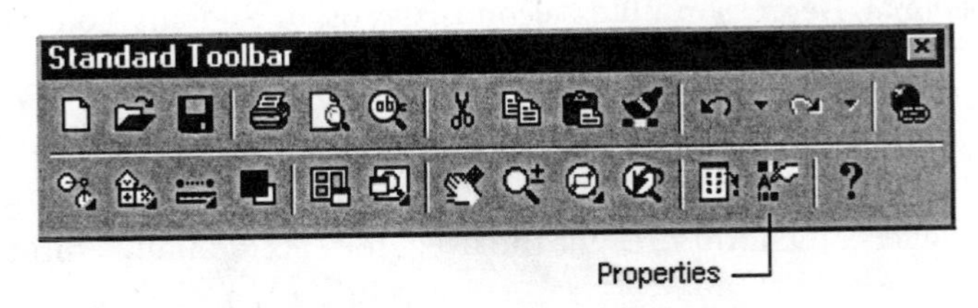

*Figure 3-75 Selecting Properties from the **Standard** toolbar*

Self-Evaluation Test

Answer the following questions and then compare your answers to the correct answers given at the end of this chapter.

1. In the default mode, the measurement of angles always starts with 90 degrees corresponding to north. (T/F)

2. The layers that are turned on are displayed on the screen and cannot be plotted. (T/F)

3. The TANgent Object Snap mode allows you to draw a tangent to or from an existing line. (T/F)

4. You can combine object snaps by separating the snap modes with a comma. (T/F)

5. The **INSertion** object snap mode is used to snap to the insertion point of a block only. (T/F)

6. In Surveyor's units you are required to specify the bearing angle the line makes with the __________ .

7. The layers are, by default, assigned __________ linetype and white color.

8. You can display the grid lines at the same resolution as that of the Snap by using the __________.

9. The __________ object snap can be used to snap to a point object.

10. You can load and display the **Layer Properties Manager** dialog box by entering __________ at the Command prompt.

Review Questions

1. You can use the decimal units mode for metric units as well as decimal English units. (T/F)

2. You cannot use the feet (') or inch (") symbols if you have selected the Scientific, Decimal, or Fractional unit format. (T/F)

3. If you select Decimal, Degrees/minutes/seconds, Grads, or Radians, you can enter an angle in any of the five measuring systems. (T/F)

4. Assigning a color to a group of objects is very important for plotting. (T/F)

5. You can lock a layer, which will prevent the user from accidentally editing the objects in that layer. (T/F)

6. The **Set** option of the **-LAYER** command allows you to create a new layer and simultaneously make it the current layer. (T/F)

7. When a layer is locked, you cannot use the objects in the locked layer for Osnaps. (T/F)

8. When the tile mode is turned off, you can freeze the selected layers in the current model space viewport or in the paper space environment by selecting **Active VP Freeze**. (T/F)

9. To specify the grid spacing as a multiple of the snap spacing in the **GRID** command, enter X after the value (2X). (T/F)

10. If the grid spacing is specified as 1, it automatically adjusts to the snap resolution. (T/F)

11. When you are drawing an auxiliary view (a drawing view that is at an angle to other views of the drawing), it is useful to have the UCS snap grid at an angle. (T/F)

12. The **CENter** Object Snap mode allows you to snap to the center point of a circle or an arc. (T/F)

13. To use the **QUAdrant** object snap, position the target box anywhere on the circle or arc. (T/F)

14. In AutoCAD LT the default is: Angles measured in the _________ direction are positive.

15. If you enter _________ before the angle, AutoCAD LT will bypass the current angle settings and use the angle as decimal degrees from the default angle base and default angle direction.

16. If you enter _________ in front of the angle, AutoCAD LT will use current angular units, but it will bypass the current settings of angle base and angle direction and use the default angle, base, and direction settings.

17. You can also load the **Drawing Units** dialog box by entering _________ at the AutoCAD LT Command prompt.

18. A layer name can be up to _________ characters long, including letters, numbers, and special characters.

19. The difference between the **Off** option and the **Freeze** option is that the frozen layers are not _________ by the computer while regenerating the drawing.

20. You can use the _________ option of the **Layer Properties Manager** dialog box to selectively display the layers.

21. You can use the **Style** option to set the snap grid to a standard or _________

22. The _________ object snap option stops searching as soon as AutoCAD LT finds a point on the object qualifying for the specified object snap.

Exercises

Exercise 5 *Architecture*

Set the units for a drawing according to the following specifications:

1. Set the **UNITS** to architectural, with the denominator of the smallest fraction equal to 16.
2. Set the angular measurement to degrees/minutes/seconds, with the number of fractional places for display of angles equal to 2.
3. Set the direction to 0 degrees (east) and the direction of measurement of angles to counterclockwise (angles measured positive in a counterclockwise direction).

Based on Figure 3-76, determine and set the limits of the drawing. Also, set an appropriate value for Grid and Snap. The scale for this drawing is 1/4" = 1'. Leave enough space around the drawing for dimensioning and the title block. (HINT: Scale factor = 48; sheet size required is 12 x 9; therefore, the limits are 12 x 48, 9 x 48 = 576, 432. Use athe **ZOOM** command and then select the **All** option to display the new limits.)

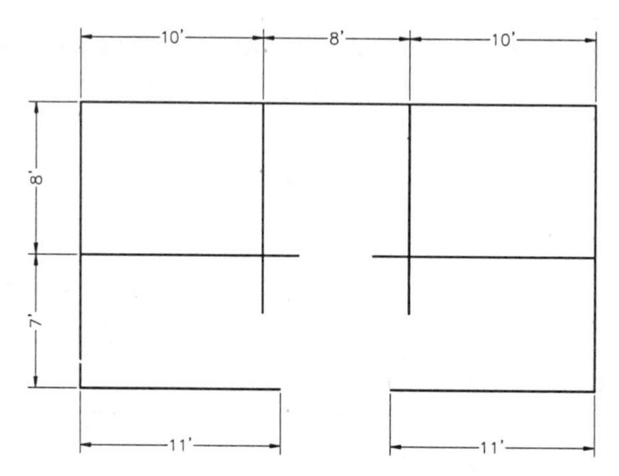

Figure 3-76 *Drawing for Exercise 5*

Exercise 6 *General*

Set up four layers with the following linetypes and colors. Then make the drawing (without dimensions) as shown in Figure 3-77.

Layer name	**Color**	**Linetype**
Object	Red	Continuous
Hidden	Yellow	Hidden
Center	Green	Center

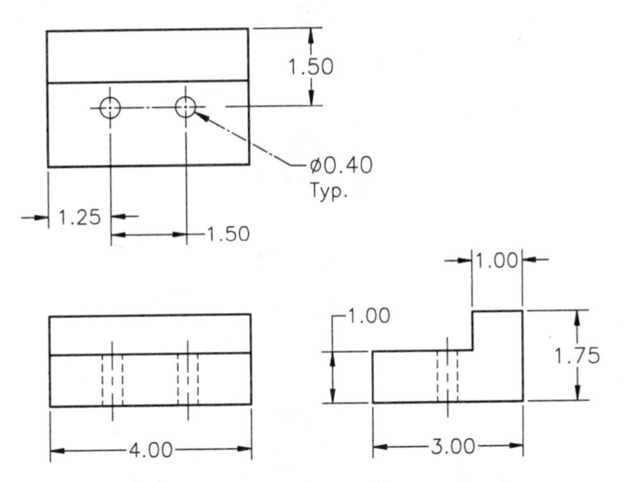

Figure 3-77 Drawing for Exercise 6

Exercise 7 — *Mechanical*

Set up a drawing with the following layers, linetypes, and colors.Then make the drawing as shown in Figure 3-78. The distance between the dotted lines is 1 unit.

Layer name	Color	Linetype
Object	Red	Continuous
Hidden	Yellow	Hidden

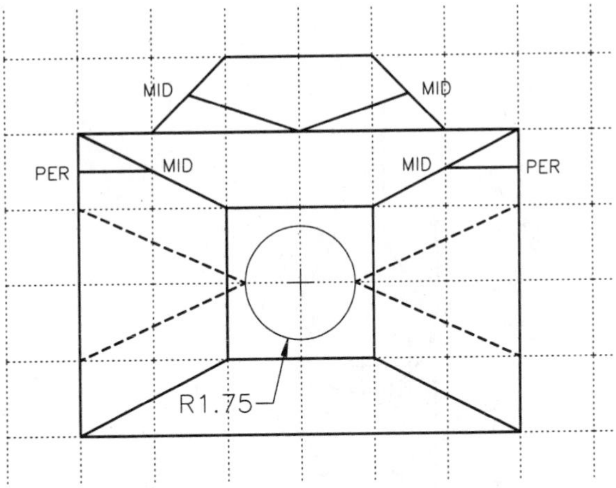

Figure 3-78 Drawing for Exercise 7

Exercise 8 — *Mechanical*

Set up a drawing with the following layers, linetypes, and colors. Then make the drawing (without dimensions) as shown in Figure 3-79.

Layer name	**Color**	**Linetype**
Object	Red	Continuous
Center	Green	Center

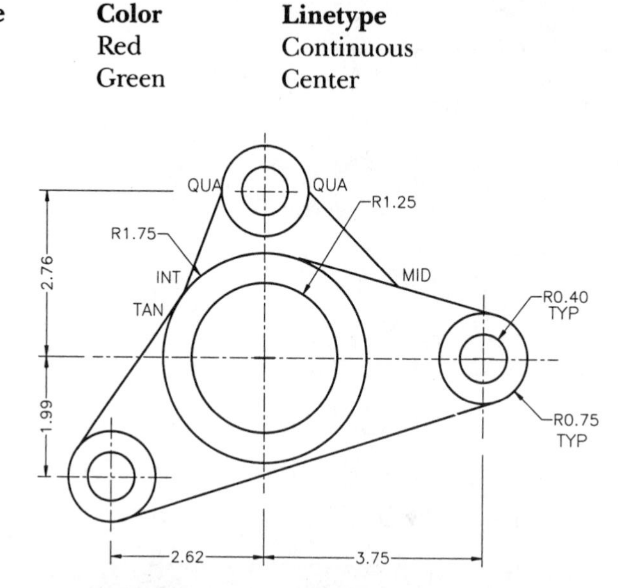

Figure 3-79 Drawing for Exercise 8

Problem-Solving Exercise 1 — *Mechanical*

Make the drawing shown in Figure 3-80. To create the arcs you need to calculate their start points and endpoints. Also, you need to figure out the start point for the lines (lines inside the arcs) so that the lines can be drawn with the existing dimensions.

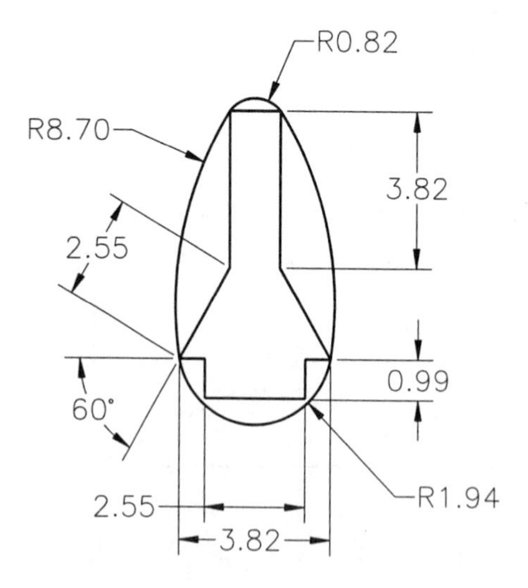

Figure 3-80 Drawing for Problem-Solving Exercise 1

Problem-Solving Exercise 2 — *Mechanical*

Based on Figure 3-81, determine and set the limits and units of the drawing. Also, set an appropriate value for Grid and Snap. The scale for this drawing is 1:10. Leave enough space around the drawing for dimensioning and the title block. (HINT: Scale factor = 10; sheet size required is 600 x 450; therefore, the limits are 600 x 10, 450 x 10 = 6000, 4500. Also, change the LTSCALE to 500 so that the hidden lines are displayed as hidden lines on screen and use the **ZOOM** command with **All** option to display the new limits.)

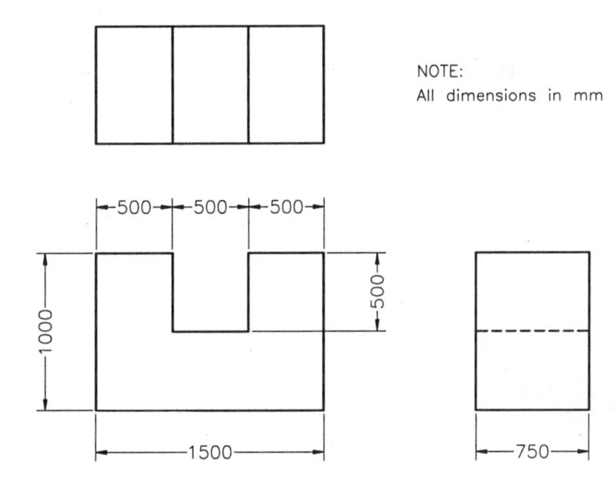

Figure 3-81 *Drawing for Problem-Solving Exercise 2*

Answers to Self-Evaluation Test

1 - F, **2** - F, **3** - F, **4** - T, **5** - F, **6** - north-south direction, **7** - continuous, **8** - Snap option, **9** - NODe, **10** - **LAYER**

Chapter 4

Editing Commands

Learning Objectives

After completing this chapter, you will be able to:

- *Create selection sets using various object selection options.*
- *Make copies of existing objects using the **COPY** command.*
- *Copying objects with base point using the **COPYBASE** command.*
- *Use the **OFFSET** and **BREAK** commands.*
- *Fillet and chamfer objects using the **FILLET** and **CHAMFER** commands.*
- *Cut and extend objects using the **TRIM** and **EXTEND** commands.*
- *Stretch objects using the **STRETCH** command.*
- *Create polar and rectangular arrays using the **ARRAY** command.*
- *Use the **ROTATE** and **MIRROR** commands.*
- *Scale objects using the **SCALE** command.*
- *Lengthen objects like line, arc, spline using **LENGTHEN** command.*
- *Use the **MEASURE** and **DIVIDE** commands.*
- *Use the **PROPERTIES** command to change object properties.*
- *Use the **QSELECT** command for quick selection of objects.*

CREATING A SELECTION SET

In Chapter 1, we discussed two options of the selection set (Window, Crossing). In this chapter you will learn additional selection set options you can use to select objects. The following options are explained here.

Last	CPolygon	Add	Undo	Previous
Fence	BOX	SIngle	ALL	Group
AUto	WPolygon	Remove	Multiple	

Note

The quickest and easiest way to enter most selection options is by typing their required (shown in uppercase) one or more character abbreviations.

Last

Last is the most convenient option if you want to select the most recently drawn object that is visible on the screen. Although a selection set is being formed using the **Last** option, only one object is selected even if you use the Last option a number of times. You can use the Last selection option with any command that requires selection of objects (e.g., **COPY**, **MOVE**, **ERASE**). After entering the particular command at the Command prompt, enter LAST or L at the **Select objects**: prompt.The most recently drawn object on the screen will be selected and highlighted.

Exercise 1 — *General*

Using the **LINE** command, draw Figure 4-1(a). Then use the **ERASE** command with the **Last** option to erase the three most recently drawn lines to obtain Figure 4-1(d).

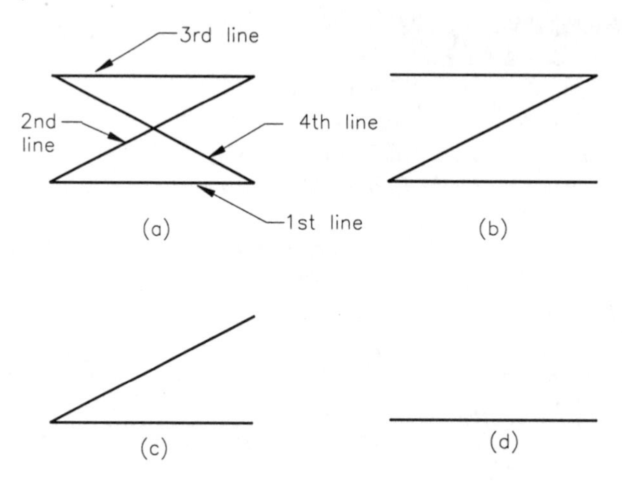

Figure 4-1 *Erasing objects using the* ***Last*** *selection option*

Previous

The Previous selection option automatically selects the objects in the most recently created selection set. To invoke this option you can enter **PREVIOUS** or **P** at the **Select objects:** prompt. AutoCAD LT 2000 saves the previous selection set and lets you select it again by using the Previous option. In other words, with the help of the Previous option you can edit the previous set without reselecting its objects individually. Another advantage of the Previous selection option is that you need not remember the objects if more than one editing operation has to be carried out on the same set of objects. For example, if you want to copy a number of objects and then move them, you can use the **Previous** option to select the same group of objects with the **MOVE** command. The prompt sequence will be as follows:

Command: **COPY** [Enter]
Select objects: *Select the objects.*
Select objects: [Enter]
Specify base point or displacement, or [Multiple]: *Specify the base point.*
Specify second point of displacement or <use first point as displacement>: *Specify the point for displacement.*

Command: **MOVE** [Enter]
Select objects: **P** [Enter]
found

The **Previous** option does not work with some editing commands, like **STRETCH**. A previous selection set is cleared by the various deletion operations and the commands associated with them, like **UNDO**. You cannot select the objects in model space and then use same selection set in paper space, or vice versa, because AutoCAD LT notes the space (paper space or model space) in which the individual selection set is obtained.

WPolygon

This option is similar to the **Window** option, except that in this option you can define a window that consists of an irregular polygon. You can specify the selection area by specifying points around the object you want to select (Figure 4-2). In other words, the objects to be selected should be completely enclosed within the polygon. The polygon is formed as you specify the points and can take any shape except one in which it intersects itself. The last segment of the polygon is automatically drawn to close the polygon. The polygon can be created by specifying the coordinates of the points or by specifying the points with the help of a pointing device. With the **Undo** option, the most recently specified WPolygon point can be undone. To use the **WPolygon** option with object selection commands (like **ERASE**, **MOVE**, **COPY**), first enter the particular command at the Command prompt, and then enter WP at the **Select objects:** prompt. The prompt sequence for the **WPolygon** option is:

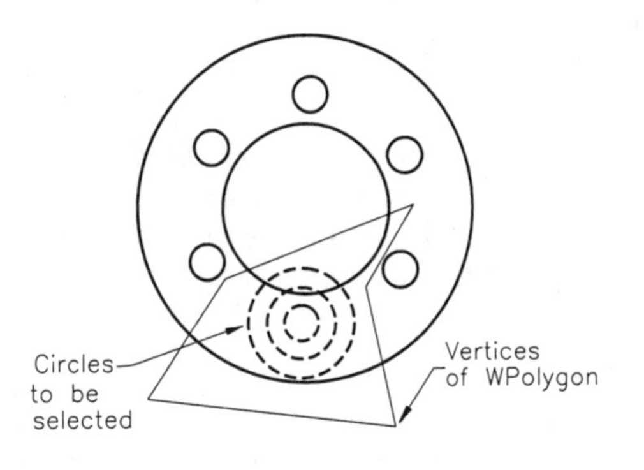

Figure 4-2 *Selecting objects using the* ***WPolygon*** *option*

Select objects: **WP** [Enter]
First polygon point: *Specify the first point.*
Specify endpoint of line or [Undo]: *Specify the second point.*
Specify endpoint of line or [Undo]: *Specify the third point.*
Specify endpoint of line or [Undo]: *Specify the fourth point.*
Specify endpoint of line or [Undo]: [Enter] *(Press ENTER after specifying the last point of polygon.)*

Exercise 2 — General

Draw a number of objects on the screen, and then erase some of them using the WPolygon to select the objects you want to erase.

CPolygon

This method of selection is similar to the **WPolygon** method except that **CPolygon** also selects those objects that are not completely enclosed within the polygon, but are touching the polygon boundaries. In other words, if a portion of an object is lying inside the polygon, the particular object is also selected in addition to those objects that are completely enclosed within the polygon (Figure 4-3). **CPolygon** is formed as you specify the points. The points can be specified at the Command line or by specifying points with a pointing device. Just as in the **WPolygon** option, the crossing polygon can take any shape except the one in which it intersects itself. Also, the last segment is drawn automatically, so the CPolygon is closed at all times. The prompt sequence for the **CPolygon** option is:

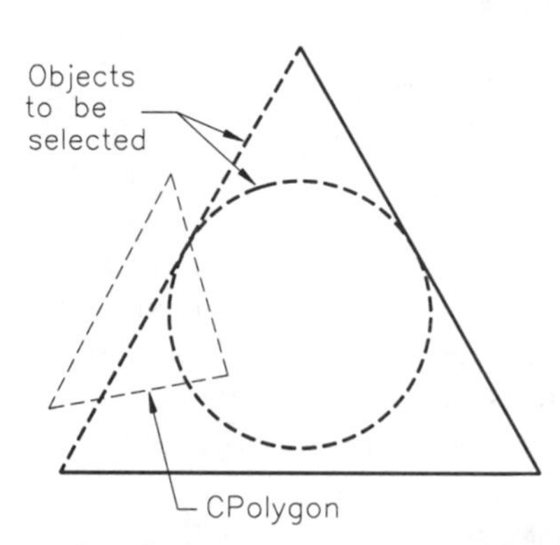

Figure 4-3 *Selecting objects using the* ***CPolygon*** *option*

Select objects: **CP** [Enter]
First polygon point: *Specify the first point.*
Specify endpoint of line or [Undo]: *Specify the second point.*
Specify endpoint of line or [Undo]: *Specify the third point.*
Specify endpoint of line or [Undo]: [Enter] *(When finished specifying the last point of the polygon.)*

Remove

The **Remove** option is used to remove an object from the selection set (but not from the drawing). After selecting many objects from a drawing by any selection method, there may be need for removing some of the objects from the selection set. You can remove the objects by selecting the objects using the pointing device with the SHIFT key depressed, when system variable **PICKADD** is set to 1 (default). The following prompt sequence illustrates the use of removing objects by using the SHIFT key depressed in the **ERASE** command (Figure 4-4):

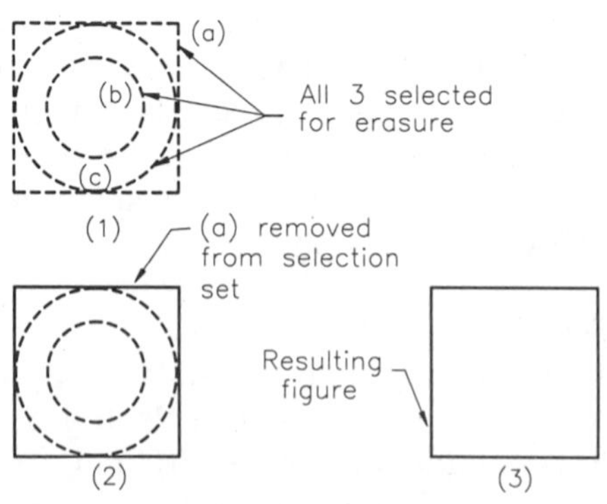

Figure 4-4 *Using the* ***Remove*** *option*

Command: **ERASE** [Enter]
Select objects: *Select objects (a), (b), and (c).*

Now if do not want to erase object (a), you can do the same by depressing the SHIFT key and selecting object (a) with the pointing device. AutoCAD LT will display the following message:

Select objects: 1 found, 1 removed, 2 total

Add

The **Add** option can be used to add objects to the selection set. When you begin creating a selection set, you are in **Add** mode. After you create a selection set by any selection method, you can add more objects by simply selecting them by the pointing device, when system variable **PICKADD** is set to 1(default). When system variable **PICKADD** is set to 0, to add objects to the selection set, you need to select objects by pointing device with SHIFT key depressed.

ALL

The **ALL** selection option is used to select all the objects on the drawing screen. Objects that are in the "OFF" layers are not selected with the **ALL** option. You can use this selection option with **ERASE** or **MOVE** or any other command that requires object selection. After invoking the command, the ALL option can be entered at the **Select objects:** prompt. Once you enter this option, all the objects drawn on the screen will be highlighted (dashed). For example, if there are four objects on the screen and you want to erase all of them, the prompt sequence is:

Command: **ERASE** [Enter]
Select objects: **ALL** [Enter]
4 found
Select objects: [Enter]

You can use this option in combination with other selection options. To illustrate this, consider there are five objects on the drawing screen and you want to erase three of them. After invoking the **ERASE** command you can enter **ALL** at the **Select objects:** prompt, then press the SHIFT key and select the two particular objects to remove from the selection set. Hence, the remaining three objects are erased.

Fence

In the **Fence** option, a selection set is created by drawing an open polyline fence through the objects to be selected. Any object touched by the fence polyline is selected (Figure 4-5). This mode of selection is like the **CPolygon** option, except that the last line of the selection polygon is not closed and objects that do not touch are not selected. The selection fence can be created by entering the coordinates at the Command line or by specifying the points with the pointing device. More flexibility for selection is provided, because the fence can intersect itself. The Undo option can be used to undo the most recently selected fence point. Like the other selection options, this option is also used with commands that need object selection. The prompt sequence is:

Select objects: **F** [Enter]
First fence point: *Specify the first point.*
Specify endpoint of line or [Undo]: *Specify the second point.*
Specify endpoint of line or [Undo]: *Specify the third point.*
Specify endpoint of line or [Undo]: *Specify the fourth point.*
Specify endpoint of line or [Undo]: [Enter] *(Press ENTER.)*

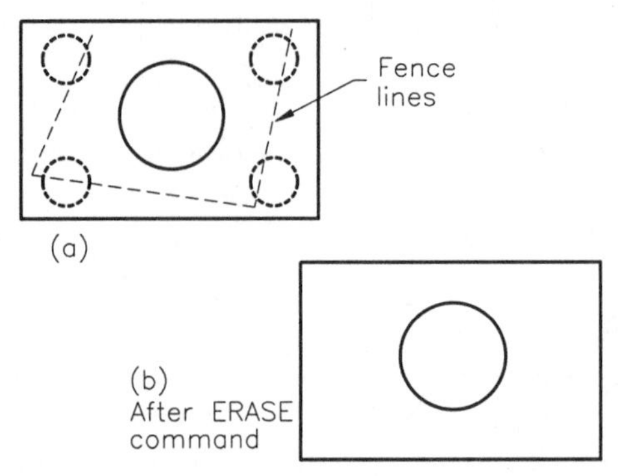

Figure 4-5 *Erasing objects using the* ***Fence*** *option*

Group

The **Group** option enables you to select a group of objects by their group name. You can create a group and assign a name to it with the help of the **GROUP** command (see Chapter 16). Once a group has been created, you can select the group using the **Group** option for editing purposes. This makes the object selection process easier and faster, for a set of objects is selected by entering just the group name. The prompt sequence is:

Command: **MOVE** [Enter]
Select objects: **G** [Enter]
Enter group name: *Enter the name of the predefined group you want to select.*
4 found
Select objects: [Enter]

Exercise 3 — *General*

Draw six circles and select all of them to erase using the **ERASE** command with the **ALL** option. Now change the selection set contents by removing alternate circles from the selection set by holding the SHIFT key down and selecting objects so that alternate circles are erased.

BOX

When system variable **PICKAUTO** is set to 1 (default), the **BOX** selection option is used to select objects inside a rectangle. After you enter BOX at the **Select objects:** prompt, you are required to specify the two corners of a rectangle at the **First corner:** and the **Other corner:** prompts. If you specify the first corner on the right and the second corner on the left, the BOX is equivalent to the Crossing selection option; hence, it also selects those objects that are touching the rectangle boundaries in addition to those that are completely enclosed within the rectangle. If you specify the first corner on the left and the second corner on the right, this option is equivalent to the Window option and selects only those objects that are completely enclosed within the rectangle. The prompt sequence is:

Select objects: **BOX** [Enter]
Specify first corner: *Specify a point.*
Specify Opposite corner: *Specify opposite corner point of the box.*

AUto

The **AUto** option is used to establish automatic selection. You can select a single object by selecting that object, as well as select a number of objects, by creating a window or a crossing. If you select a single object, it is selected; if you specify a point in the blank area, you are automatically in the BOX selection option and the point you have specified becomes the first corner of the box. Auto and Add are default selections.

Multiple

When you enter **M** (Multiple) at the **Select objects:** prompt, you can select multiple objects at a single **Select objects:** prompt without the objects being highlighted. Once you give a null response to the **Select objects:** prompt, all the selected objects are highlighted together. Multiple commands can also be used effectively to select two intersecting objects if the intersection point is selected twice.

Undo

This option removes the most recently selected object from the selection set.

SIngle

When you enter SI (SIngle) at the **Select objects:** prompt, the selection takes place in the SIngle selection mode. The **Select objects:** prompt repeats until you make the selection. Once you select an object or a number of objects using a Window or Crossing option, the Select objects: prompt is not repeated and AutoCAD LT proceeds with the command for which the selection is made.

Command: **ERASE** [Enter]
Select objects: **SI** [Enter]
Select objects: *Select the object for erasing. The selected object is erased.*

You can also create a selection set with help of **SELECT** command. The prompt sequence is:

Command: **SELECT** [Enter]
Select objects: *Use any selection method.*

EDITING COMMANDS

To use AutoCAD LT effectively, you need to know the editing commands and how to use them. In this section you will learn about the editing commands. These commands can be invoked from the toolbar, or menu, or can be entered at the Command: prompt. We are going to discuss the following editing commands (**MOVE**, **ERASE**, and **OOPS** commands were discussed in Chapter 1):

MOVE	SCALE	EXTEND	MIRROR
COPY	FILLET	STRETCH	BREAK
OFFSET	CHAMFER	LENGTHEN	MEASURE
ROTATE	TRIM	ARRAY	DIVIDE

COPY COMMAND

Toolbar:	Modify > Copy Object
Menu:	Modify > Copy
Command:	COPY

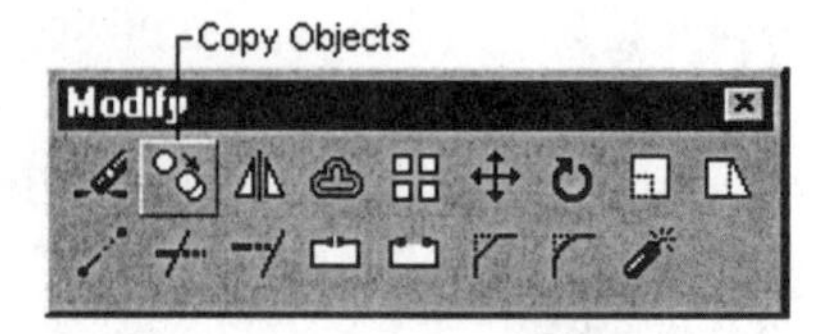

Figure 4-6 Invoking the ***COPY*** *command from the* ***Modify*** *toolbar*

The **COPY** command is used to copy an existing object. This command is similar to the **MOVE** command, in the sense that it makes copies of the selected objects and places them at specified locations, but the originals are left intact. In this command also, you need to select the objects and then specify the base point. Then you are required to specify the second point, where you want the objects to be copied (see Figure 4-7). The prompt sequence is:

Command: **COPY** [Enter]
Select objects: *Select the objects to copy.*
Select objects: [Enter]
Specify base point or displacement, or [Multiple]: *Specify the base point.*
Specify second point or displacement, or <use first point as displacement>: *Specify a new position on the screen using the pointing device or entering coordinates.*

Multiple Copies

This option of the **COPY** command is used to make multiple copies of the same object (Figure 4-8). To use this option, select the **Multiple** option or enter **M** at the **Specify base point or displacement, or [Multiple]:** prompt. Next, AutoCAD LT prompts you to enter the second point. When you select a point, a copy is placed at that point, and the prompt is automatically repeated until you press ENTER to terminate the **COPY** command. The prompt sequence for the **COPY** command with the **Multiple** option is:

Specify base point or displacement, or [Multiple]: **M** [Enter]
Specify base point: *Specify the base point.*
Specify second point of displacement or <use first point as displacement>: *Specify a point for placement.*

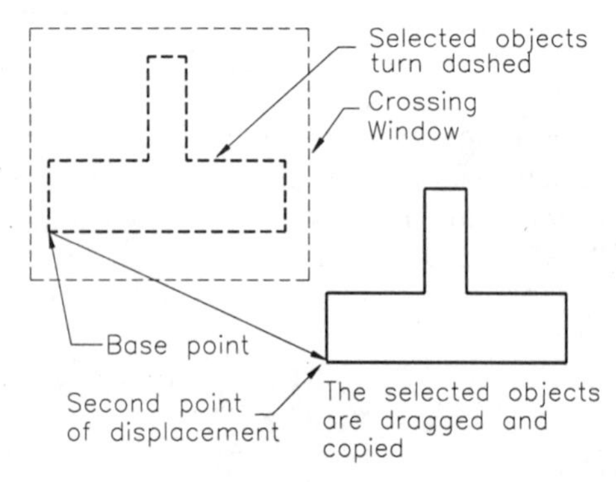

Figure 4-7 Using the ***COPY*** *command*

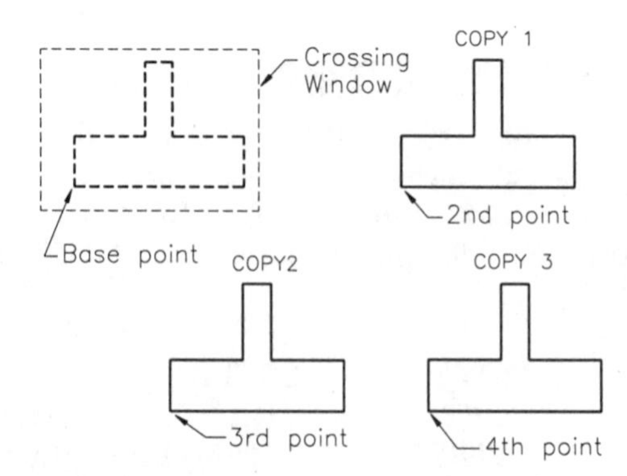

Figure 4-8 Making multiple copies using the ***COPY*** *command*

Specify second point of displacement or <use first point as displacement>: *Specify another point for placement.*
Specify second point of displacement or <use first point as displacement>: *Specify another point for placement.*
Specify second point of displacement or <use first point as displacement>: [Enter]

COPYBASE COMMAND*

Menu:	Edit > Copy with Basepoint
Command:	COPYBASE

The command **COPYBASE** is used to specify the basepoint of objects to be copied. This is a new feature in AutoCAD LT 2000 that can be very useful while pasting an object very precisely in the same diagram or into another diagram. The **COPYBASE** command can also be invoked from the shortcut menu by right-clicking in the drawing area and choosing **Copy with Base Point**. The command sequence is as follows:

Command: **COPYBASE** [Enter]
Specify base point: *Specify the base point of the object.*
Select objects: *Select the object to be copied.*

After copying objects with the **COPYBASE** command , you can **paste** the objects at the desired place with greater accuracy.

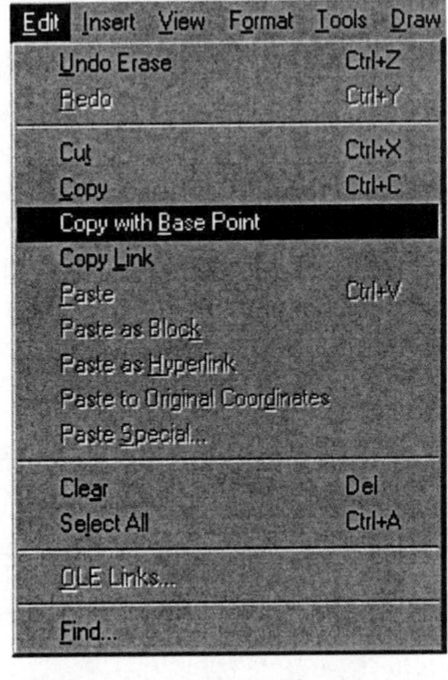

***Figure 4-9** Invoking the **Copy with Base Point** from the shortcut menu*

PASTEBLOCK COMMAND*

Menu:	Edit > Paste as Block
Command:	PASTEBLOCK

PASTEBLOCK command pastes a copied block into another drawing. You can also invoke the **PASTEBLOCK** command from the shortcut menu by right-clicking in the drawing area and choosing **Paste as Block**.

PASTEORIG COMMAND*

Menu:	Edit > Paste to Original Coordinates
Command:	PASTEORIG

PASTEORIG command pastes a copied object into a new drawing using the coordinates from the original drawing. You can also invoke **PASTEORIG** command from the shortcut menu by right-clicking in the drawing area and choosing **Paste to Original Coordinates**. **PASTEORIG** is available only when the Clipboard contains AutoCAD LT data from a drawing other than the current drawing.

Exercise 4 *General*

a. Draw a circle and use the **COPY** command to make a single copy of the circle or any other figure you have drawn.
b. Make four copies of the figure drawn in (a) using the Multiple (**M**) option.
c. Draw a rectangle and make a single copy of the rectangle using the **COPYBASE** command and **Paste** commands.

OFFSET COMMAND

Toolbar:	Modify > Offset
Menu:	Modify > Offset
Command:	OFFSET

If you want to draw parallel lines, polylines, concentric circles, arcs, curves, etc., you can use the **OFFSET** command (Figure 4-10). This command creates another object that is similar to the selected one. When offsetting an object you can specify the offset distance and the side to offset, or you can specify a point through which you want to offset the selected object. Depending on the side to offset, you can create smaller or larger circles, ellipses, and arcs. If the offset side is toward the inner side of the perimeter, the arc, ellipse, or circle will be smaller than the original. The prompt sequence is:

Command: **OFFSET** Enter
Specify offset distance or [Through] <current>:

At this last prompt you can enter either the **Offset distance** or the **Through** option (Figure 4-11). The offset distance can be specified by entering a value or by specifying two points with the pointing device. AutoCAD LT will measure the distance between these two points and use it as the offset distance. The prompt sequence for the **Offset distance** option is:

Specify offset distance or [Through] <current>: *Specify two points or enter a value.*
Select object to offset or <exit>: *Select the object to offset.*
Specify point on side to offset: *Specify the side for offsetting.*

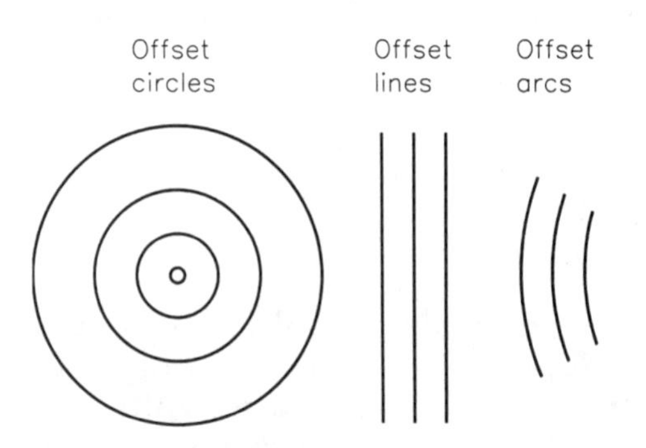

***Figure 4-10** Using the **OFFSET** command*

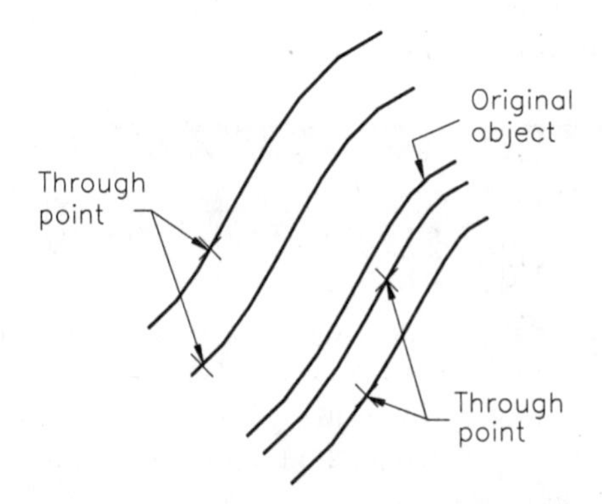

***Figure 4-11** Using the **Through** option in the **OFFSET** command*

Select object to offset or <exit>: *Select another object or press ENTER to complete the command.*

In the case of the **Through** option, you do not have to specify a distance; you simply specify an offset point after selecting. The offset object is created at the specified point. The prompt sequence is:

Specify offset distance or [Through] <current>: **T** Enter
Select object to offset or <exit>: *Select the object.*
Specify through point: *Specify the offset point.*

The offset distance is stored in the **OFFSETDIST** system variable. A negative value indicates OFFSET is set to the Through option. You can offset lines, arcs, 2D polylines, xlines, circles, ellipses, elliptical arcs rays, and planar splines. If you try to offset objects other than these, the message **Cannot offset that object** is displayed.

Exercise 5 — Graphics

1. Draw five concentric circles using the **OFFSET** command.
2. Use the **OFFSET** edit command to draw Figures 4-12(a) and 4-12(b).

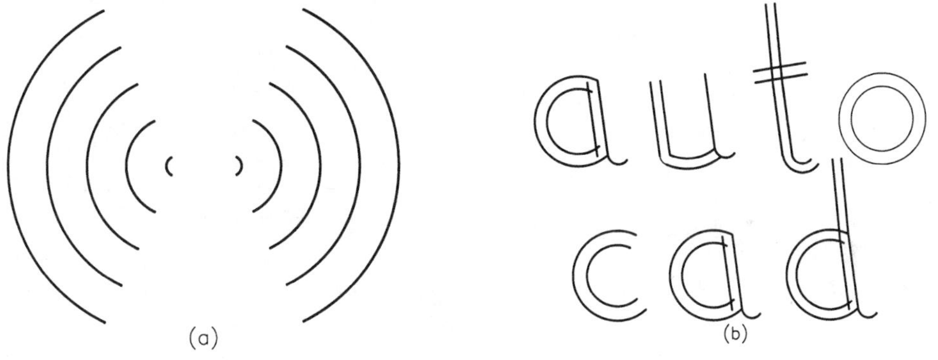

Figure 4-12 *Drawing for Exercise 5*

ROTATE COMMAND

Toolbar:	Modify > Rotate
Menu:	Modify > Rotate
Command:	ROTATE

Sometimes when making drawings you may need to rotate an object or a group of objects. You can accomplish this by using the **ROTATE** command. When you invoke this command, AutoCAD LT will prompt you to select the objects and the base point about which the selected objects will be rotated. You should be careful when selecting the base point as it is easy to get confused if the base point is not located on a known object. After you specify the base point you are required to enter a rotation angle. Positive angles produce a counterclockwise rotation; negative angles produce a clockwise rotation (Figure 4-13). The

ROTATE command can also invoked from the shortcut menu by selecting the object and clicking the right button in the drawing area and choosing **Rotate**. The prompt sequence is:

Command: **ROTATE** [Enter]
Current positive angle in UCS: ANGDIR=*current* ANGBASE=*current*
Select objects: *Select the objects for rotation.*
Select objects: [Enter]
Specify base point: *Specify a base point on or near the object.*
Specify rotation angle or [Reference]: *Enter a positive or negative rotation angle, or specify a point.*

If you need to rotate objects with respect to a known angle, you can do this in two different ways using the **Reference** option. The first way is by specifying the known angle as the reference angle, followed by the proposed angle to which the objects will be rotated (Figure 4-14). Here the object is first rotated clockwise from the X axis, to the reference angle. Then the object is rotated to the new angle from this reference position in a counterclockwise direction. The prompt sequence is:

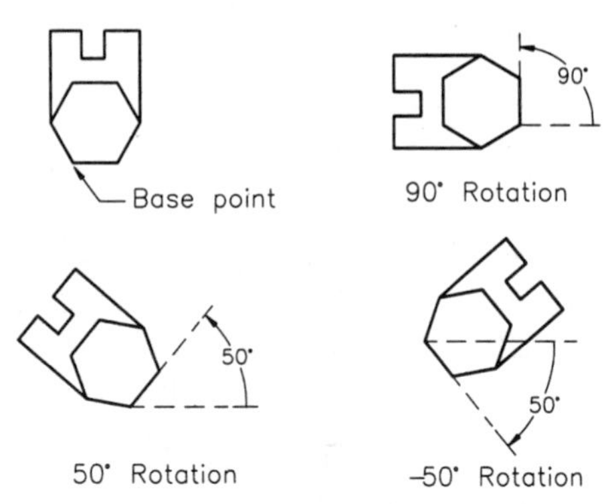

Figure 4-13 Rotation of objects with different rotation angles

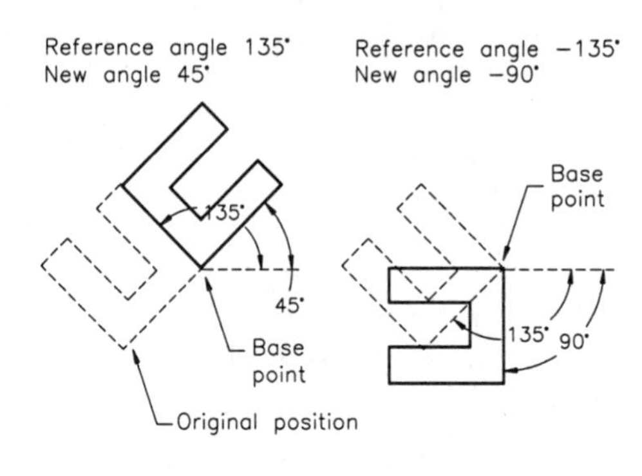

*Figure 4-14 Rotation using the **Reference** Angle option*

Command: **ROTATE** [Enter]
Current positive angle in UCS: ANGDIR=*current* ANGBASE=*current*
Select objects: *Select the objects for rotation.*
Select objects: [Enter]
Specify base point: *Specify the base point.*
Specify rotation angle or [Reference]: **R** [Enter]
Specify the reference angle <0>: **10** [Enter]
Specify the new angle: **90** [Enter]

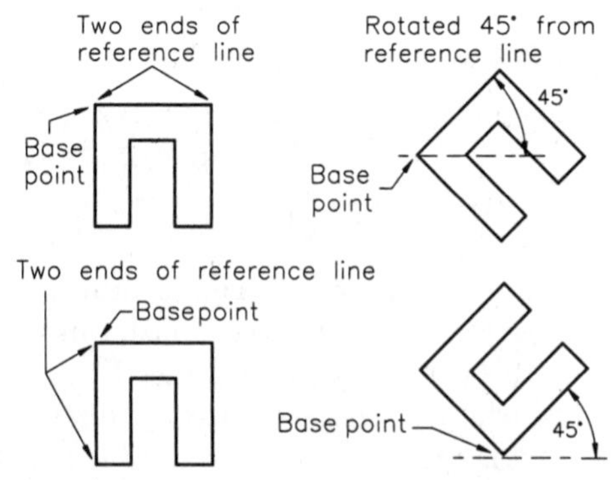

*Figure 4-15 Rotation using the **Reference** Line option*

The other method is to specify two points to indicate a reference line on the object; then specify a new angle to rotate the object with respect to this line (Figure 4-15). The direction of rotation depends on the point you choose first on the reference line. The prompt sequence is:

Command: **ROTATE** [Enter]
Current positive angle in UCS: ANGDIR=*current* ANGBASE=*current*
Select objects: *Select the object for rotation.*
Select objects: [Enter]
Specify base point: *Specify the base point.*
Specify rotation angle or [Reference]: **R** [Enter]
Specify the reference angle <0>: *Specify the first point on the reference line.*
Specify second point: *Specify the second point on the reference line.*
Specify new angle: **45** [Enter]

SCALE COMMAND

Toolbar:	Modify > Scale
Menu:	Modify > Scale.
Command:	SCALE

Many times you will need to change the size of objects in a drawing. You can do this with the **SCALE** command. This command enlarges or shrinks the selected object in the same ratio for the X and Y dimensions about a base point. Application of the identical scale factor to the X and Y dimensions ensures that the shape of the objects being scaled do not change. This is a useful and timesaving editing command because instead of redrawing objects to the required size, you can scale the objects with a single **SCALE** command. Another advantage of this command is that if you have already put the dimensions on the drawing, they will also change according to the new scale. You can also invoke the **SCALE** command from the shortcut menu by right-clicking in the drawing area and choosing **Scale**. The prompt sequence is:

Command: **SCALE** [Enter]
Select objects: *Select objects to be scaled.*
Select objects: [Enter]
Specify base point: *Specify the base point, preferably a known point.*
Specify scale factor or [Reference]: **0.75** [Enter]

When a drawing is rescaled, a scale factor is used to change the size of selected objects around a chosen base point (Figure 4-16). The chosen base point remains fixed, whereas everything around it may increase or reduce in size according to the scale factor.

To reduce the size of an object, the scale factor should be less than 1 and to increase the size of an object, the scale factor should be greater than 1. You can enter a scale factor or select two points to specify a distance as a factor. When you select two points to specify a distance as factor, the first point should be on the referenced object.

Sometimes it is time-consuming to calculate the relative scale factor. In such cases you can scale the object by specifying a desired size in relation to the existing size (a known dimension).

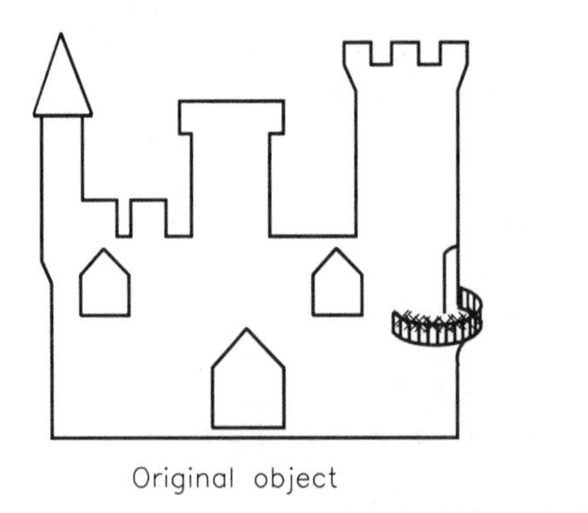

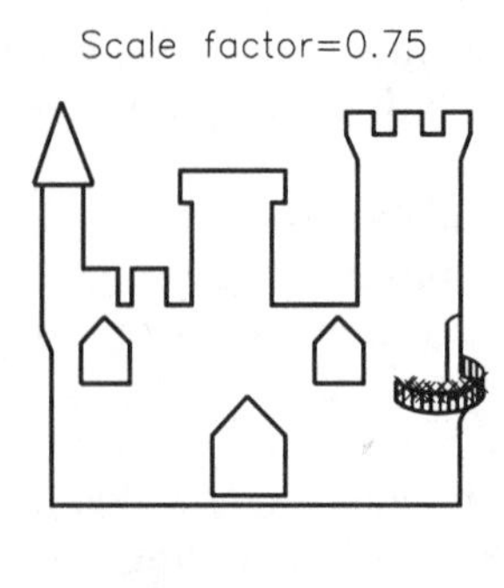

***Figure 4-16** Using the Scale factor option in the **SCALE** command*

In other words you can use a **reference length**. To do this, type **R** at the **Specify scale factor or [Reference]** prompt. Then, you either specify two points to specify the length or enter a length. At the next prompt, enter the length relative to the reference length. For example, if a line is **2.5** units long and you want the length of the line to be **1.00** units, then instead of calculating the relative scale factor, you use the **Reference** option (Figure 4-17). This is illustrated as follows:

Command: **SCALE** [Enter]
Select objects: *Select the object.*
Select objects: [Enter]
Specify base point: *Specify the base point.*
Specify scale factor or [Reference]: **R** [Enter]
Specify reference length <1>: **2.5** [Enter]
Specify new length: **1.0** [Enter]

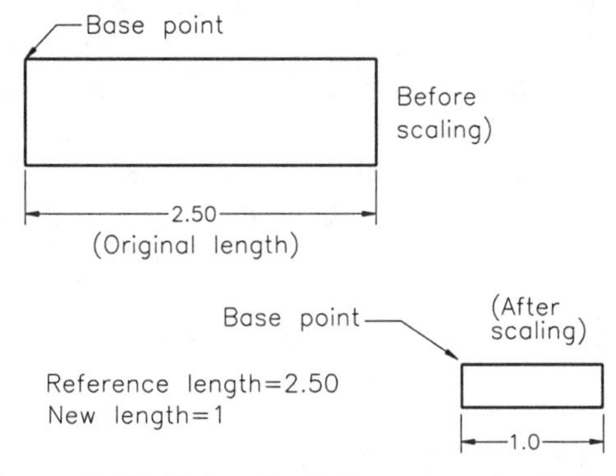

***Figure 4-17** Using the **Reference** option in the **SCALE** command*

If you have not used the required drawing units for a drawing, you can use the **Reference** option of the **SCALE** command to correct the error. Select the entire drawing with the help of the All selection option. Specify the **Reference option**, then select the endpoints of the object whose desired length you know. Specify the new length, and all objects in the drawing will be rescaled automatically to the desired size.

FILLET COMMAND

Toolbar:	Modify > Fillet
Menu:	Modify > Fillet
Command:	FILLET

The **FILLET** command is used to create smooth round arcs to connect two objects. In mechanical drafting, inside rounded corners are known as FILLETS and outside ones are known as ROUNDS, but in AutoCAD LT all rounded corners are referred to as fillets. The **FILLET** command helps you form round corners between any two lines by asking you to identify the two lines. This can be done by selecting them with the cursor. A fillet can be drawn between two intersecting parallel lines as well as nonintersecting and nonparallel lines, arcs, polylines, xlines, rays, splines, circles, and true ellipses. The radius of the arc to create the fillet has to be specified. The prompt sequence is:

Command: **FILLET** Enter
Current Settings: Mode = TRIM, Radius = 0.5000
Select first object or [Polyline/Radius/Trim]:

Select First Object Option

This is the default method to fillet two objects. As the name implies, it prompts for the first object required for filleting. The prompt sequence is:

Command: **FILLET** Enter
Current Settings: Mode = TRIM, Radius = 0.5000
Select first object or [Polyline/Radius/Trim]: *Specify first object.*
Select second object: *Select second object.*

You can select lines with the Window, Crossing, or Last option, but to avoid unexpected results it is safer to select by picking objects individually. Also, selection by picking objects is necessary in the case of arcs and circles that have the possibility of more than one fillet. They are filleted

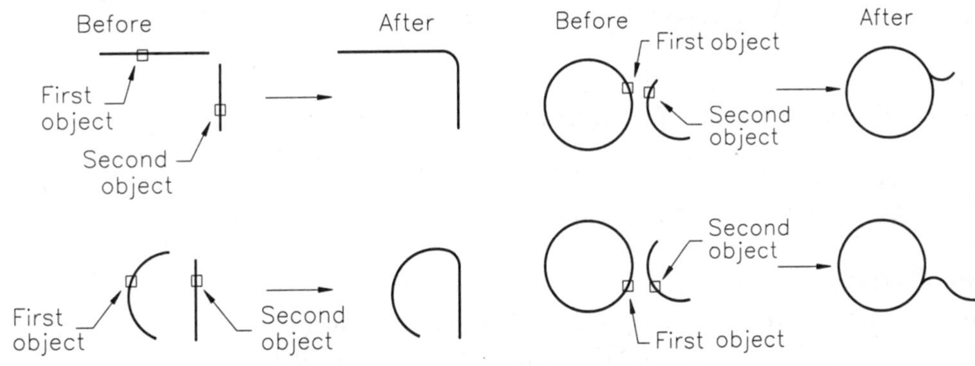

Figure 4-18 *Using the* ***FILLET*** *command*

Figure 4-19 *Using the* ***FILLET*** *command on circles and arcs*

closest to the select points (Figures 4-18 and 4-19).
The **FILLET** command can also be used to cap the ends of two parallel lines (Figure 4-20). The cap is a semicircle whose radius is equal to half the distance between the two parallel lines. The cap distance is calculated automatically when you select the two parallel lines for filleting.

Command: **FILLET** Enter
Current Settings: Mode = TRIM, Radius = 0.5000
Select first object or [Polyline/Radius/Trim]: *Select the first parallel line.*
Select second object: *Select the second parallel line.*

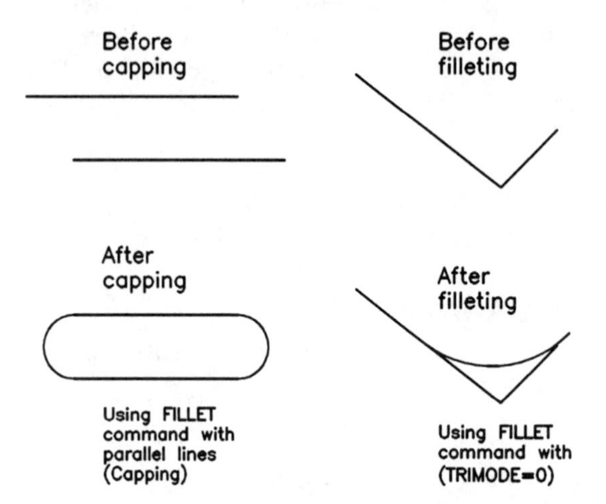

*Figure 4-20 Capping parallel lines with the **FILLET** command, and filleting without cutting the geometry.*

Radius Option

The fillet you create depends on the radius distance you specify. The default radius is 0.5000. You can enter a distance or two points. The new radius you enter becomes the default and remains in effect until changed. Fillet with zero radius creates sharp corners and is used to clean up lines at corners (in case they overlap or have a gap). The prompt sequence is:

Select first object or [Polyline/Radius/Trim]: **R** Enter
Specify fillet radius <current>: *Enter a fillet radius or press ENTER to accept current value.*

Note

***FILLETRAD** system variable controls and stores the current fillet radius.*

Trim Option

If this option is set to Trim, the selected objects are either trimmed or extended to the fillet arc endpoints. If set to NoTrim they are left intact. The prompt sequence is:

Select first object or [Polyline/Radius/Trim]: **T** Enter
Specify trimmode option [Trim/No trim] <current>: *Enter* **T** *to trim the edges,* **N** *to leave them intact.*

Setting TRIMMODE. **TRIMMODE** is a system variable that eliminates any size restriction on the **FILLET** command. By setting **TRIMMODE** to 0, you can create a fillet of any size without actually cutting the existing geometry. Also, there is no restriction on the fillet radius; the fillet radius can be larger than one or both objects that are being filleted. The value of the **TRIMMODE** system variable can also be set by entering **TRIMMODE** at the Command prompt:

Command: **TRIMMODE** Enter

Enter new value for TRIMMODE <1>: **0**

Note

TRIMMODE = 0 *Fillet or chamfer without cutting the existing geometry.*
TRIMMODE = 1 *Extend or trim the geometry.*

*When you enter the **FILLET** command, AutoCAD LT displays the current* **TRIMMODE** *and the current fillet radius.*

Polyline Option

Using this option you can fillet a single polyline (Figure 4-21). If the polyline is selected with the Window, Crossing, or Last option, the most recent vertex is filleted. If you select the P option, AutoCAD LT prompts you to select a polyline, and then all its vertices are filleted. If the selected polyline is not closed, then the beginning corner is not filleted. The prompt sequence is:

Command: **FILLET** [Enter]
Current Settings: Mode= *current*, Radius= *current*
Select first object or [Polyline/Radius/Trim]: **P** [Enter]
Select 2D polyline: *Select the polyline.*

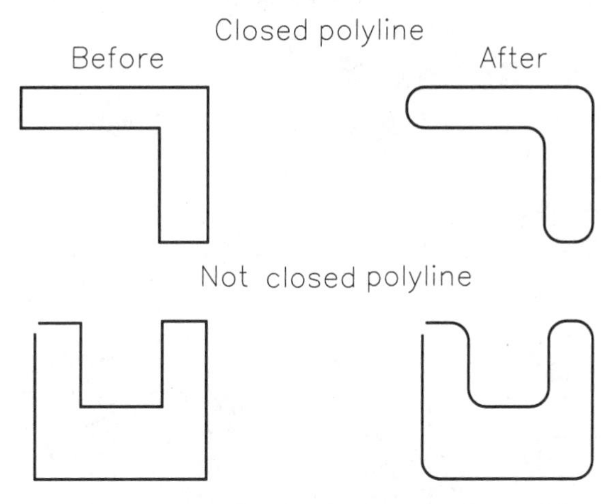

***Figure 4-21** Using the **FILLET** command on polylines*

Filleting Objects with Different UCS

The fillet command will also fillet objects that are not in the current UCS plane. To create a fillet for such objects, AutoCAD LT will automatically change the UCS transparently so that it can generate a fillet between the selected objects.

Note

*The **FILLET** command can fillet between a polyline and other objects, or between two adjacent segments of a single polyline, but it cannot fillet between two separate polylines. You can select* *the segments to fillet, or enter C or W and use Crossing or Window selection. If you select more than two segments with a Crossing or Window selection, the two segments found nearest the polyline's first endpoint will be filleted.*

CHAMFER COMMAND

Toolbar:	Modify > Chamfer
Menu:	Modify > Chamfer
Command:	CHAMFER

In drafting, the **CHAMFER** is defined as the taper provided on a surface. Sometimes the chamfer is used to avoid a sharp corner. In AutoCAD LT a chamfer is any angled corner of a drawing. A beveled line connects two separate objects to create a chamfer. The size of a chamfer depends on its distance from the corner. If a chamfer is equidistant from the corner in both directions, it is a 45-degree chamfer. A chamfer can be drawn between two lines that may or may not intersect. This command also works on a single polyline. The prompt sequence is:

Command: **CHAMFER** [Enter]
(TRIM mode) Current chamfer Dist1 = 0.5000, Dist2 = 0.5000
Select first line or [Polyline/Distances/Angle/Trim/Method]:

The next prompts displayed are dependent on the option you choose at this last prompt.

Select First Line Option

In this option you need to select two nonparallel objects so that they are joined with a beveled line. The prompt sequence is:

Command: **CHAMFER** [Enter]
(TRIM mode) Current chamfer Dist1 = 0.5000, Dist2 = 0.5000

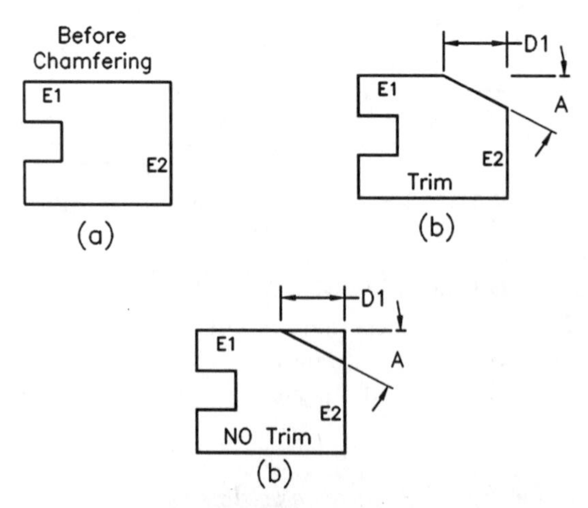

Figure 4-22 *Using the **CHAMFER** command to create a chamfer*

Select first line or [Polyline/Distance/Angle/Trim/Method]: *Specify the first line.*
Select second line: *Select the second line.*

Distances Option

Under this option you can enter the chamfer distance. To do this, type **D** at the **Select first line or [Polyline/Distance/Angle/Trim/Method]:** prompt. Next, enter the first and the second chamfer distances. The default value for the chamfer distance is 0.5. The new chamfer distances remain in effect until you change them. Instead of entering the distance values, you can specify two points to indicate each distance (Figure 4-23). The prompt sequence is:

Select first line or [Polyline/Distance/Angle/Trim/Method]: **D** Enter
Specify first chamfer distance <current>: *Enter a distance or specify two points.*
Specify second chamfer distance <current>: *Enter a distance or specify two points.*

The first and second chamfer distances are stored in the **CHAMFERA** and **CHAMFERB** system variables.

Note

*If Dist 1 and Dist 2 are set to zero, the **CHAMFER** command will extend or trim the selected lines so that they end at the same point.*

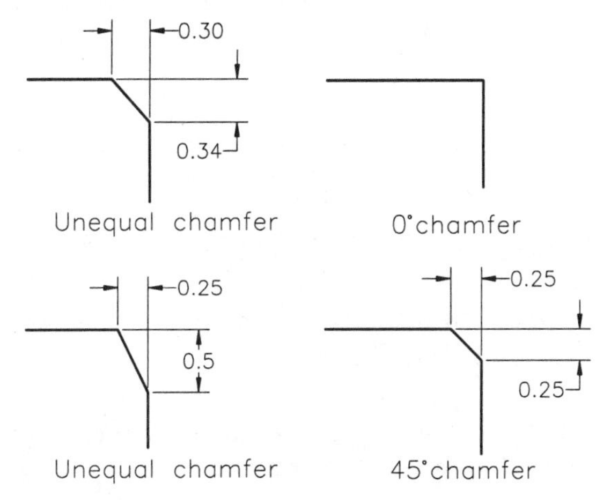

Figure 4-23 *Different types of chamfers*

Polyline Option

You can use the **CHAMFER** command to chamfer all corners of a closed or open polyline (Figure 4-24). To do this, select the **Polyline** option after entering the command. Next, select the polyline. In case of a closed polyline, all the corners of the polyline are chamfered to the set distance values. Sometimes the polyline may appear closed. But if the **Close** option was not used to create it, it may not be. In this case, the beginning corner is not chamfered. The prompt sequence is:

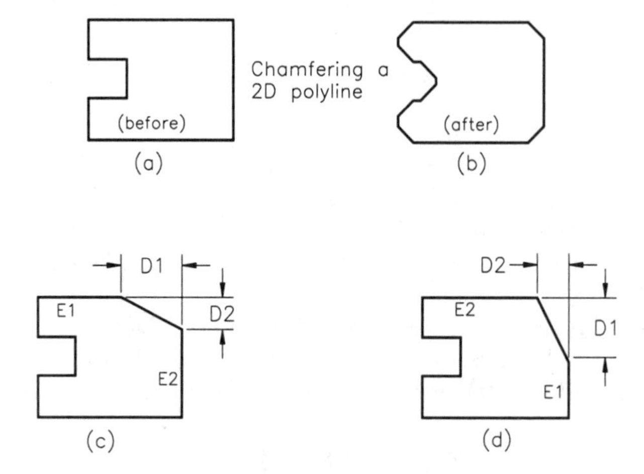

Figure 4-24 *Using the **CHAMFER** command to create a chamfer between two lines or polyline segments*

Command: **CHAMFER** Enter
(TRIM mode) Current chamfer Dist1 = 0.5000, Dist2 = 0.5000
Select first line or [Polyline/Distance/Angle/Trim/Method]: **P** Enter
Select 2D polyline: *Select the polyline.*

Angle Option

The **Angle** option is used to set the chamfer distance by specifying an angle and a distance. The prompt sequence is:

Select first line or [Polyline/Distance/Angle/Trim/Method]: **A** Enter
Specify chamfer length on the first line <current>: *Specify a length.*
Specify chamfer angle from the first line <current>: *Specify an angle.*

Trim Option

Depending on this option, the selected objects are either trimmed extended to the endpoints of the chamfer line or left intact. The prompt sequence is:

Select first line or [Polyline/Distance/Angle/Trim/Method]: **T** Enter
Enter Trim mode option [Trim/No Trim] <current>:

Method Option

By using this option you can choose between the Distance option and the Angle option using the current distance or angle settings. The prompt sequence is:

Select first line or [Polyline/Distance/Angle/Trim/Method]: **M** Enter
Enter trim method [Distance/Angle] <current>: *Enter* **D** *for Distance option,* **A** *for Angle option.*

Note

*If you set the **TRIMMODE** system variable to 1 (Default value), the objects will be trimmed or extended after they are chamfered and filleted. If **TRIMMODE** is set to zero, the objects are left untrimmed.*

Setting the CHAMFER System Variables

The chamfer modes, distances, length, and angle can also be set by using the following system variables:

CHAMMODE = 0	Distance/Distance (default)
CHAMMODE = 1	Length/Angle
CHAMFERA	Sets first chamfer distance on the first selected line (default = 0.5000)
CHAMFERB	Sets second chamfer distance on the second selected line (default = 0.5000)
CHAMFERC	Sets the chamfer length (default = 1.0000)
CHAMFERD	Sets the chamfer angle from the first line (default = 0)

TRIM COMMAND

Toolbar: Modify > Trim
Menu: Modify > Trim
Command: TRIM

You may need to trim existing objects of a drawing. Breaking individual objects takes time if you are working on a complex drawing with many objects. The **TRIM** command trims objects that extend beyond a required point of intersection. With this command you must select the cutting edges or boundaries first. There can be more than one cutting edge and you can use any selection method to select them. After the cutting edge or edges are selected, you must select each object to be trimmed. If you want to break a single object into two by deleting a portion, you can select two cutting edges on the same object and then select the object to trim between those two cutting edges. An object can be both a cutting edge and an object to trim. You can trim lines, circles, arcs, polylines, splines, ellipses, xlines, and rays. The prompt sequence is:

Command: **TRIM** [Enter]
Current settings:Projection=current Edge = current
Select cutting edges...
Select objects: *Select the cutting edges.*
Select objects: [Enter]
Select object to trim or [Project/Edge/Undo]:

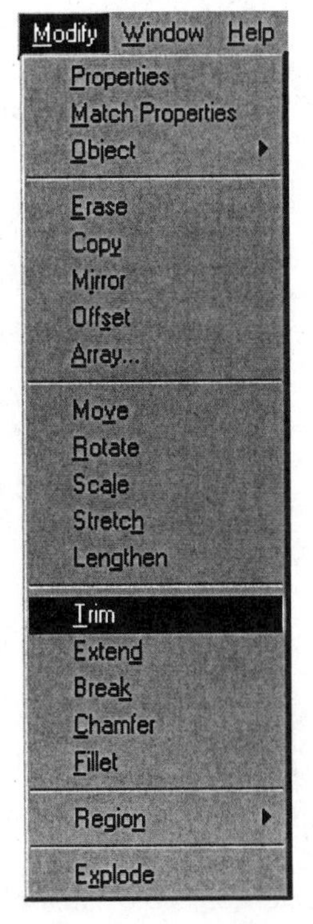

***Figure 4-25** Invoking the **TRIM** command from the **Modify** menu*

Select Object to Trim Option

Here you have to specify the objects you want to trim. This

Chapter 4

prompt is repeated until you press ENTER. This way you can select several objects with a single **TRIM** command (Figure 4-26). The prompt sequence is:

Command: **TRIM** [Enter]
Current settings: Projection=current Edge=current
Select cutting edges...
Select objects: *Select the first cutting edge.*
Select objects: *Select the second cutting edge.*
Select objects: [Enter]
Select object to trim or [Project/Edge/Undo]: *Select the first object to trim.*
Select object to trim or [Project/Edge/Undo]: *Select the second object to trim.*
Select object to trim or [Project/Edge/Undo]: [Enter]

Edge Option

This option is used whenever you want to trim those objects that do not intersect the cutting edges, but would intersect if the cutting edges were extended and intersected the selected objects in 3D space, (Figure 4-27). The prompt sequence is:

Command: **TRIM** [Enter]
Current settings: Projection=current Edge=current
Select cutting edges...
Select objects: *Select the cutting edge.*
Select objects: [Enter]
Select object to trim or [Project/Edge/Undo]: **E** [Enter]
Enter an implied edge extension mode [Extend/No extend] <current>: **E** [Enter]
Select object to trim or [Project/Edge/Undo]: *Select the object to trim.*

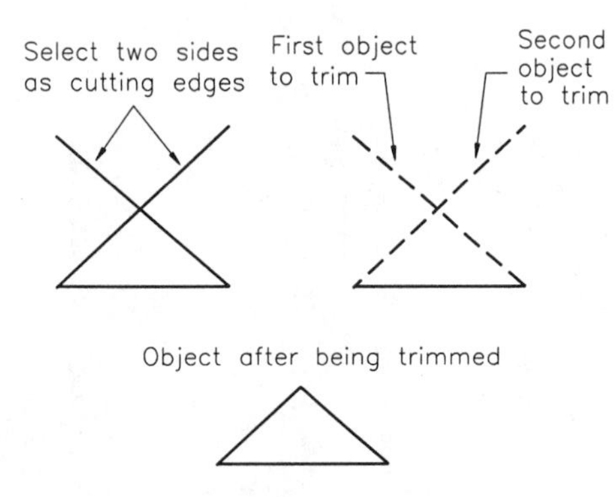

*Figure 4-26 Using the **TRIM** command*

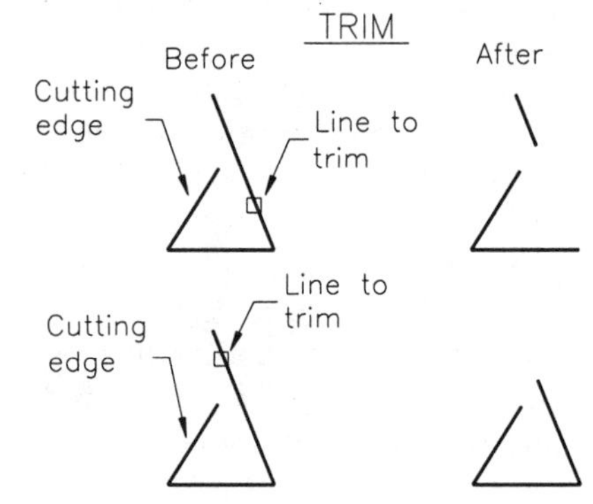

*Figure 4-27 Trimming an object using the **Edge** option (Extend)*

Project Option

In this option you can use Project mode while trimming objects. The prompt sequence is:

Select object to trim or [Project/Edge/Undo]: **P** [Enter]
Enter an projection option [None/Ucs/View] <current>:

The **None** option is used whenever the objects to trim intersect the cutting edges in 3D space. If you want to trim those objects that do not intersect the cutting edges in 3D space, but do visually appear to intersect in a particular UCS or the current view, use the **UCS** or **View** options. The **UCS** option projects the objects to the XY plane of the current UCS, while the **View** option projects the objects to the current view direction (trims to their apparent visual intersections).

Undo Option

If you want to remove the previous change created by the **TRIM** command, enter **U** at the **Select object to trim or [Project/Edge/Undo]:** prompt.

Exercise 6 — *Mechanical*

Draw the top illustration in Figure 4-28. Then use the **FILLET**, **CHAMFER**, and **TRIM** commands to obtain the figure shown in the bottom of Figure 4-28. (Set the SNAP = 0.05; assume the missing dimensions.)

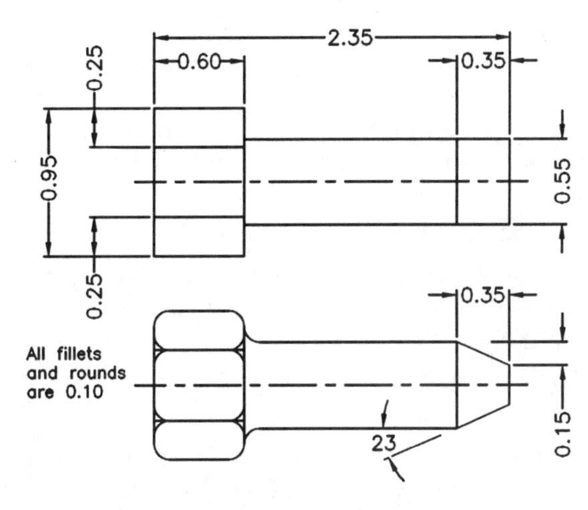

Figure 4-28 *Drawing for Exercise 6*

EXTEND COMMAND

Toolbar:	Modify > Extend
Menu:	Modify > Extend
Command:	EXTEND

This command may be considered the opposite of the **TRIM** command. In the **TRIM** command you trim objects; in the **EXTEND** command you can lengthen or extend lines, polylines, rays, and arcs to meet other objects. This command does not extend closed polylines. The command format is similar to that of the **TRIM** command. You are required to select the boundary edges first. The boundary edges are those objects that the selected lines or arcs extend to meet. These edges can be lines, polylines, circles, arcs, ellipses, xlines, rays, splines, text, or even viewports. If a polyline is selected as the boundary edge, the line extends to the polyline's center. The following is the prompt sequence:

Command: **EXTEND** Enter
Current settings: Projection=current Edge=current
Select boundary edges ...
Select objects: *Select the boundary edges.*
Select objects: Enter
Select object to extend or [Project/Edge/Undo]:

Project Option

In this option you can use projection mode while trimming objects. The prompt sequence is:

Command: **EXTEND** Enter
Current settings: Projection=current Edge=current
Select boundary edges ...
Select objects: *Select the boundary edges.*
Select objects: Enter
Select object to extend or [Project/Edge/Undo]: **P** Enter
Enter a projection option [None/UCS/View] <current>:

The **None** option is used whenever the objects to be extended intersect with the boundary edge in 3D space. If you want to extend those objects that do not intersect the boundary edge in 3D space, use the **UCS** or **View** options. The **UCS** option projects the objects to the XY plane of the current UCS, while the **View** option projects the objects to the current view.

Select Object to Extend Option

Here you have to specify the object you want to extend to the particular boundary (Figure 4-29). This prompt is repeated until you press ENTER. This way you can select a number of objects in a single **EXTEND** command.

Edge Option

You can use this option whenever you want to extend objects that do not actually intersect the boundary edge, but would intersect its edge if the boundary edge were extended (Figure 4-30). If you enter **E** at the prompt, the selected object is extended to the implied boundary edge. If you enter **N** at the prompt, only those objects that would actually intersect the real boundary edge are extended (the default). The prompt sequence is:

Command: **EXTEND** Enter
Current settings: Projection=current Edge=current
Select boundary edges ...
Select objects: *Select the boundary edge.*
Select objects: Enter
Select object to extend or [Project/Edge/Undo]: **E** Enter
Enter an implied edge extension mode [Extend/No extend] <current>: **E** Enter
Select object to extend or [Project/Edge/Undo]: *Select the line to extend.*

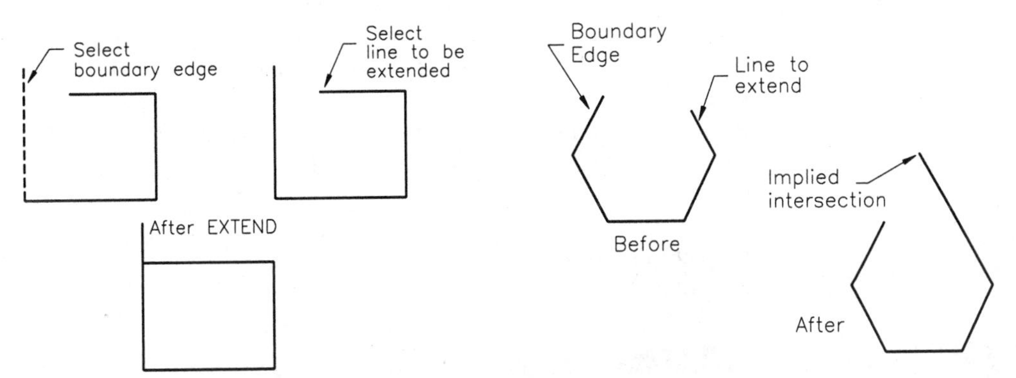

*Figure 4-29 Using the **EXTEND** command*

*Figure 4-30 Extending an object using the **Edge** option*

Undo Option

If you want to remove the previous change created by the **EXTEND** command, enter **U** at the **Select object to extend or [Project/Edge/Undo]** prompt.

Trimming and Extending with Text, Region, or Spline

The **TRIM** and **EXTEND** commands can be used with text, regions, or splines as edges (Figure 4-31). This makes the **TRIM** and **EXTEND** commands two of the most useful editing commands in AutoCAD LT. The **TRIM** and **EXTEND** commands can also be used with arcs, elliptical arcs, splines, ellipses, 3D Pline, rays, and lines. See Figure 4-32.

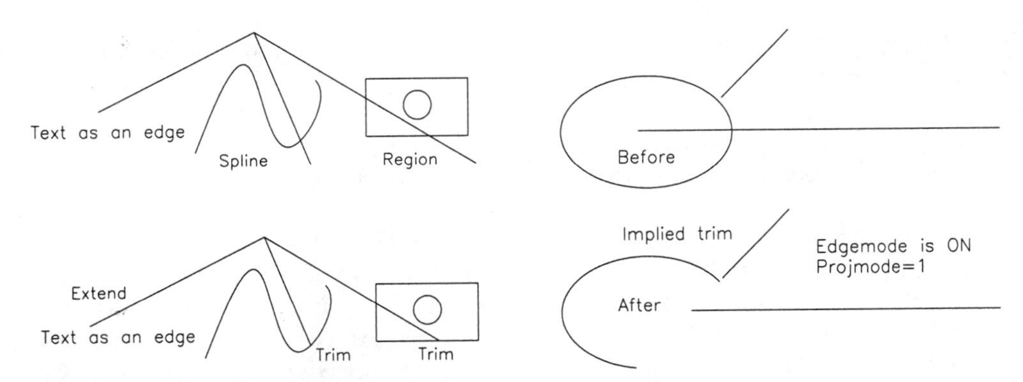

*Figure 4-31 Using the **TRIM** and **EXTEND** commands with text, spline, and region*

*Figure 4-32 Using the **Edge** option to do an implied trim*

The system variables **PROJMODE** and **EDGEMODE** determine how the **TRIM** and **EXTEND** commands are executed. The following is the list of the values that can be assigned to these variables:

Value	PROJMODE	EDGEMODE
0	True 3D mode	Use regular edge without extension (default)
1	Project to current UCS XY plane (default)	Extend the edge to natural boundary
2	Project to current view plane	

STRETCH COMMAND

Toolbar:	Modify > Stretch
Menu:	Modify > Stretch
Command:	STRETCH

This command can be used to stretch objects, altering selected portions of the objects. With this command you can lengthen objects, shorten them, and alter their shapes (see Figure 4-33). You must use a Crossing, or CPolygon selection to specify the objects to stretch. The prompt sequence is:

Command: **STRETCH** Enter
Select objects to stretch by crossing-window or crossing-polygon ...
Select objects: *Select the objects using crossing window or crossing polygon.*
Select objects: Enter

After selecting the objects, you have to specify the point of displacement. You should select only that portion of the object that needs stretching.

Specify base point or displacement: *Specify the base point.*
Specify second point of displacement: *Specify the displacement point.*

You normally use a Crossing or CPolygon selection with **STRETCH** because if you use a Window, those objects that are partially inside the window are not selected, and the objects selected because they are fully within the window are moved, not stretched as it is the property of **STRETCH** command that if the object lies completely inside the Window (or even Crossing), then **STRETCH** command acts like a simple **MOVE** command. The object selection and stretch specification process of **STRETCH** is a little unusual. You are actually specifying two things. First, you are selecting objects, and second, you are specifying the portions of those selected objects to be stretched. You can use a Crossing or CPolygon selection to simultaneously specify both, or you can select objects by any method, and then use any window or crossing specification to specify what parts of those objects to stretch. Objects or portions of selected objects which are completely within the window or crossing specification are moved. If selected objects cross the window or crossing specification, their defining points within the window or crossing specification are moved, their defining points outside the window or crossing specification remain fixed, and the parts crossing the window or crossing specification are stretched. Only the last window or crossing specification made determines what is stretched or moved. Figure 4-33 illustrates using a crossing selection to simultaneously select the two angled lines and specify that their right ends will be stretched. Alternatively, you could select the lines by any method and then use a Window selection (which would not actually select anything) to specify that their right ends will be stretched.

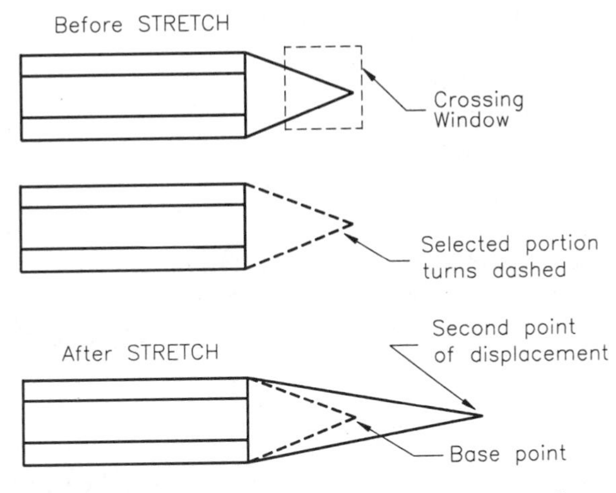

***Figure 4-33** Using the **STRETCH** command*

LENGTHEN COMMAND

Toolbar:	Modify > Lengthen
Menu:	Modify > Lengthen
Command:	LENGTHEN

Like the **TRIM** and **EXTEND** commands, the **LENGTHEN** command can be used to extend or shorten lines, polylines, elliptical arcs, and arcs. The **LENGTHEN** command has several options that allow you to change the length of objects by dynamically dragging the object endpoint, entering the delta value, entering the percentage value, or entering the total length of the object. The **LENGTHEN** command also allows the repeated selection of objects for editing. The **LENGTHEN** command has no effect on closed objects, like circles. The prompt sequence is:

Command: **LENGTHEN** (Enter)
Select an object or [DElta/Percent/Total/DYnamic]:

<Select object>

This is the default option, which returns the current length or the angle of the selected object. If the object is a line, AutoCAD LT returns only the length. However, if the selected object is an arc, AutoCAD LT returns the length and the angle.

DElta

The **DElta** option is used to increase or decrease the length or angle of an object by defining the delta distance or delta angle. The delta value can be entered by entering a numerical value or by specifying two points. A positive value will increase (Extend) the length of the selected object and a negative value will decrease the length (Trim). The following is the command prompt sequence for decreasing the angle of an arc by 30 degrees (Figure 4-34):

Command: **LENGTHEN** [Enter]
Select an object or [DElta/Percent/Total/DYnamic]: **DE** [Enter]
Enter delta length or [Angle] <current>: **A** [Enter]
Enter delta angle <current>: **-30** [Enter]
Select object to change or [Undo]: *Select object.*
Select object to change or [Undo]: [Enter]

Percent

The **Percent** option is used to extend or trim an object by defining the change as a percentage of the original length of the angle. For example, a positive number of 150 will increase the length by 50 percent and a positive number of 75 will decrease the length by 25 percent of the original value. (Negative values are not allowed).

Total

The **Total** option is used to extend or trim an object by defining the new total length or angle. For example, if you enter a total length of 1.25, AutoCAD LT will automatically increase or decrease the length of the object so that the new length of the object is 1.25. The value can be entered by entering a numerical value or by specifying two points. The object is shortened or lengthened with respect to the endpoint that is closest to the selection point. The selection point is determined by where the object was selected. The prompt sequence in Total option is:

Specify total length or [Angle] <current>: *Specify the length or enter* **A** *for angle.*

If you enter a positive value in the above prompt then the length of the selected objects is changed accordingly. You can change the angle of an arc by entering A and giving the new value.

DYnamic

The **DYnamic** option allows you to dynamically change the length or angle of an object by specifying one of the endpoints and dragging it to a new location. The other end of the object stays fixed and is not affected by dragging. The angle of lines, radius of arcs, and shape of elliptical arcs are unaffected.

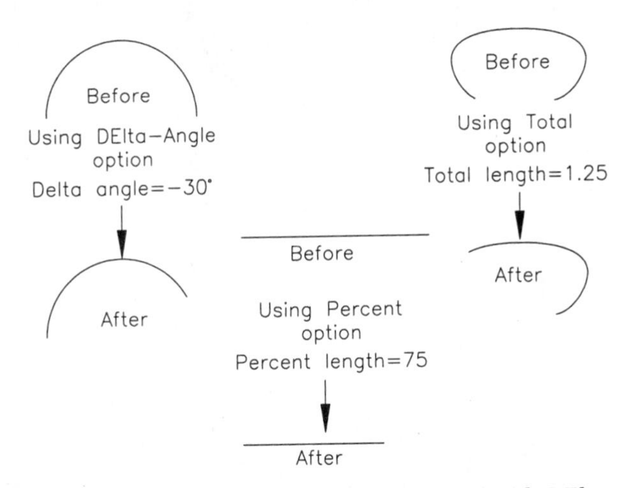

*Figure 4-34 Using the **LENGTHEN** command with DElta, Percent, and Total options*

ARRAY COMMAND*

Toolbar:	Modify > Array
Menu:	Modify > Array
Command:	ARRAY

In some drawings you may need to specify an object multiple times in a rectangular or circular arrangement. For example, suppose you have to draw six chairs around a table. This job can be accomplished by drawing each chair separately or by using the **COPY** command to make multiple copies of the chair. You could also draw one chair and then, with the help of the **ARRAY** command, create the other five. This method is more efficient and less time-consuming. The **ARRAY** command invokes the Array dialog box that allows you to make multiple copies of selected objects in a **rectangular** or **polar** fashion. Each resulting element of the array can be controlled separately.

Rectangular Array

A rectangular array is formed by making copies of the selected object along the X and Y axes (along rows and columns). The command allows you to enter the number of rows and columns in the **ARRAY** dialog box, Figure 4-35.

Array Dialog Box (Rectangular Option)

As you select the **Rectangular Array** radio button, the Array dialog box gives you the following options:

Select Objects

This button is used to select the object(s) you want to array in the drawing. As you choose the **Select Objects** button, the **ARRAY** dialog box is temporarily closed so that you can select the object(s) to be arrayed in the drawing.

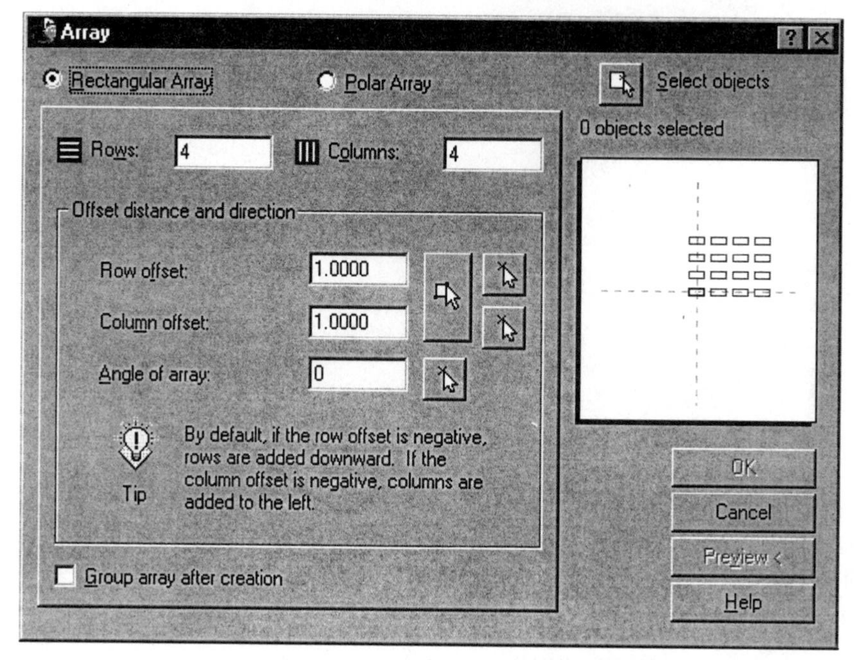

Figure 4-35 ***Array*** *dialog box (Rectangular option)*

Rows

This option allows you to enter the number of rows in the array (Figure 4-36). You can enter the required number of rows in the **Rows** edit box. The default value for this is 4.

Columns

This option allows you to enter the number of columns in the array. The required number of columns can be entered in the **Columns** edit box. The default value for this is 4.

Note

You can not have an object arrayed with one row and one column at the same time. If the number of rows is one , then the number of columns have to be more than one, and if the number of columns is one, then the number of rows have to be more than one.

Offset distance and direction

Row offset

This option allows you to enter the distance between the rows. The value can be entered in the **Row offset** edit box. Default value for this is 1. If the value entered in the **Row offset** edit box is negative, then the rows are added in downward direction, Figure 4-37.

Column offset

This option allows you to enter the distance between columns. The value of distance between columns can be entered in the **Column offset** edit box. The default value of this is also 1. Negative value in the **Column offset** edit box adds the columns to the left of the original object selected.

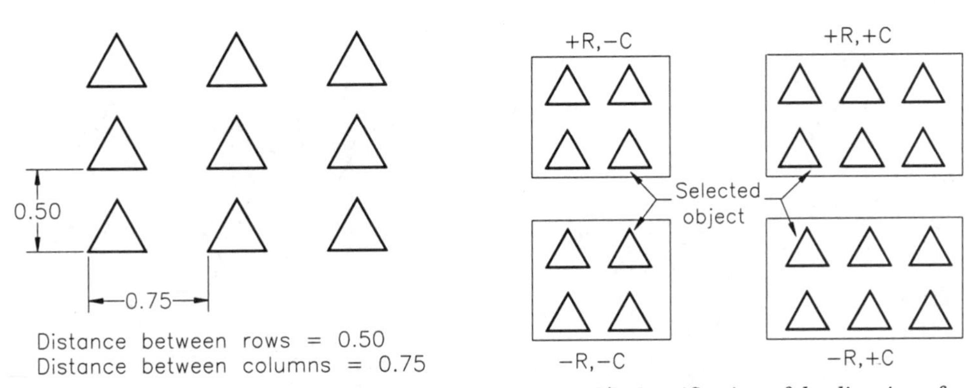

***Figure 4-36** Rectangular array with row and column distance*

***Figure 4-37** Specification of the direction of arrays*

Pick Row Offset

Choosing this button temporarily closes the **ARRAY** dialog box so that you can enter the distance between the rows by using two points on the screen.

Pick Column Offset

As you choose this button the **ARRAY** dialog box is temporarily closed so that you can enter the value of distance between columns using two points on the screen.

Pick Both Offset

As you choose this button, the **ARRAY** dialog box is temporarily closed so that you can specify distance between both rows and columns at the same time by specifing two diagonal points of a rectangle (Unit cell) using your pointing device (Mouse). The horizontal distance between the diagonal points is taken as the distance between columns and the vertical distance between the diagonal point is taken as the distance between the rows, Figure 4-38 . The command sequence is:

> Specify unit cell: *Specify first corner.*
> Other corner: *Specify other corner.*

Angle of array

This option allows you to enter the angle of rotation for the rows and columns at which they are arranged in the **Array** command, Figure 4-39.

Pick Angle of Array

As you choose this button the **Array** dialog box is temporarily closed so that you can select the angle of rotation using two points on the screen.

Note

*The angle measurement convention can be changed with the help of the **UNITS** command. These settings can also be changed using the **ANGBASE** and **ANGDIR** System variables.*

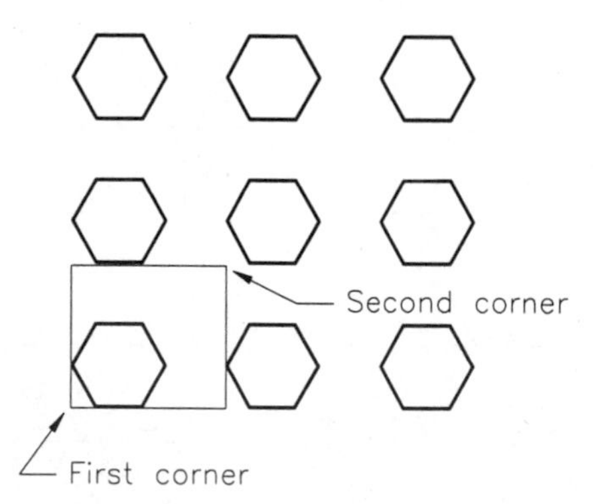

Figure 4-38 Rectangular array selecting unit cell distance

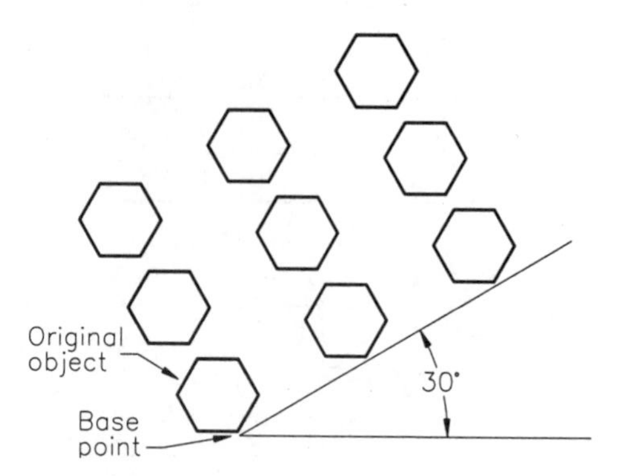

Figure 4-39 Rotated rectangular array

Preview

This button helps you to have a preview of the arrayed objects depending upon the settings and values you have entered in the **Array** dialog box. Along with the preview of array, another dialog box, Figure 4-40, is displayed on the screen with options like **Accept**, **Modify**, and **Cancel.**

Choosing the **Accept** button completes the **ARRAY** command and you can see the arrayed objects on the screen. Choosing the **Modify** button brings back the **ARRAY** dialog box so that you can make the required modification in the **Array** dialog box. Choosing the **Cancel** button cancels the **Array** command and you have to repeat the complete command in case you want to array the object.

*Figure 4-40 **Preview** dialog box in **Array** command*

Group array after creation

This option groups all of the objects that are the result of the **Array** command in one group. This means all the objects that are the result of the **Array** command will be selected upon selecting one of the objects among them. If this check box is cleared, each of the objects will be a different entity and can be modified independently.

Polar Array

A polar array is an arrangement of objects around a point in a circular pattern. You can use the **Polar** option of the **ARRAY** command dialog box to construct this kind of array, Figure 4-41.

Array Dialog Box (Polar Option)

As you select **Polar Array** radio button of the **ARRAY** dialog box, the following options appears in the dialog box:

Select objects

Selecting this option temporarily closes the **ARRAY** dialog box so that you can select the object(s) to be arrayed from the drawing.

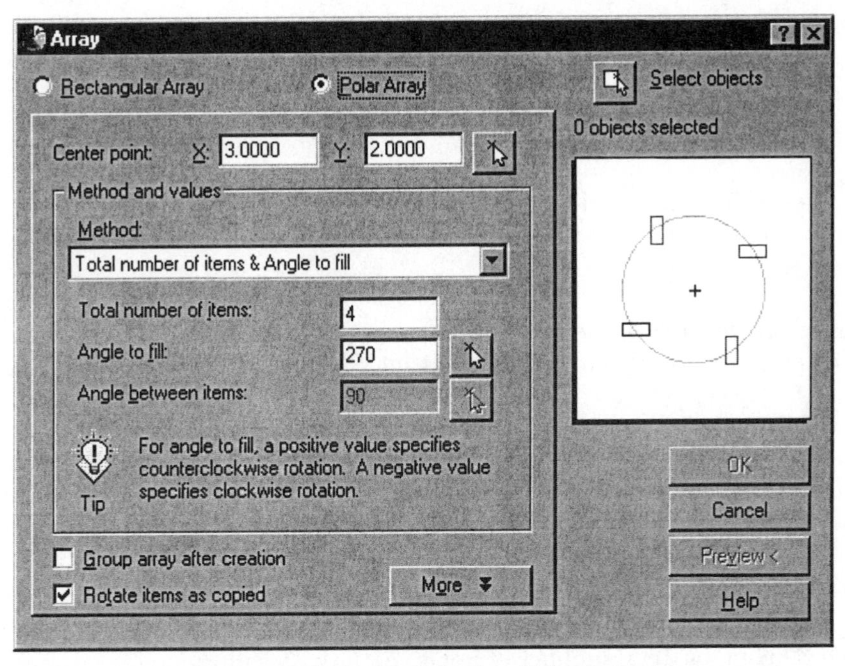

Figure 4-41 ***Array*** *dialog box (Polar option)*

Center Point

This option allows you to enter the center point of the imaginary circle about which the objects will be arrayed. You can enter the values X and Y coordinates of the center point in the **X** and **Y** edit box respectively.

Pick Center Point

Choosing this option temporarily closes the **Array** dialog box so that you can select the center point of array in the drawing using your pointing device.

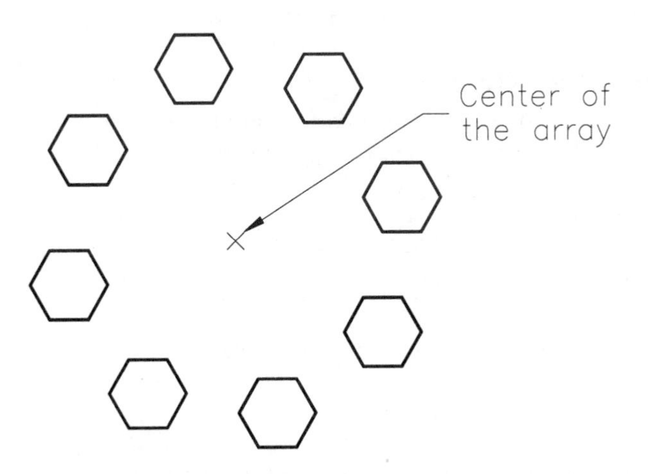

Figure 4-42 Using the ***Polar*** *option to create a circular array*

Method and values Area

Method

This option allows you to select the method by which you want to array the objects. There are three methods from which you can select any one to array the objects. These methods are:

Total number of items & Angle to fill: In this option you have to specify the total number of items you want in the array and the total angle to fill between the first object and the last object.

Total number of items & Angle between items: In this option you have to enter the total number of items in the array and the included angle between any two adjacent objects.

Angle to fill & Angle between items: This option allows you to enter the total angle to fill and the included angle between any two of the adjacent items. In this option AutoCAD LT itself calculates the number of items that will fit into the specified total angle to fill, and at the same time maintain the included angle between the two adjacent items you have specified.

Total number of items

This option allows you to enter the total number of items you want in the array. You can enter the desired value in the **Total number of items** edit box. Default value for this is 4.

Angle to fill

This option allows you to enter the value of the angle to fill between all of the items. The default value for this is 360.

Pick Angle to fill

Choosing this button temporarily closes the **ARRAY** dialog box so the you can define angle to fill in the drawing using your pointing device.

Angle between items

This option allows you to enter the included angle between any two adjacent items.

Pick Angle between Items

As you choose this button the **Array** dialog box is temporarily closed so that you can define the included angle between the two adjacent items using your pointing device in the drawing.

Note

Positive value of Angle to fill and Angle between the items specifies the rotation in a counterclockwise direction and negative values specifies the rotation in a clockwise direction.

Zero (0) value of the Angle to fill or Angle between the items is not allowed.

Group array after creation

This option groups all of the objects that are the result of array into an unnamed group so that a complete group is selected upon selecting any of the objects from the group. If this check box is cleared, all of the resultant objects will be different entities.

Rotate items as copied

This option gives you a choice that whether you want the objects to be rotated as they are copied during array. If this check box is not cleared, a particular face of all the replicated objects will face towards the center point of array as shown in Figure 4-43. If this check box is cleared, the replicated objects remain in the same orientation as that of actual object as shown in Figure 4-44.

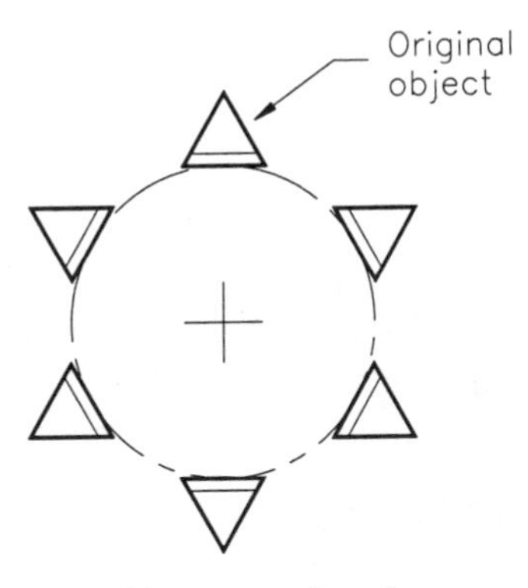

*Figure 4-43 Objects rotated as they are arrayed in the **ARRAY** command*

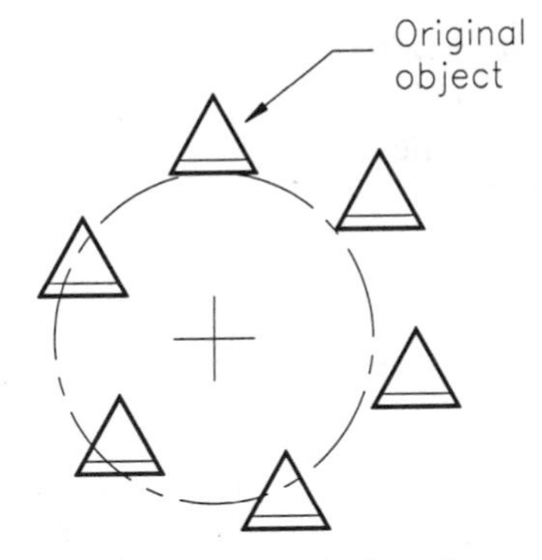

*Figure 4-44 Objects not rotated as they are arrayed in the **ARRAY** command*

More

As you choose the **More** button the **Array** dialog box expands and displays the **Object base point** area, Figure 4-45. Upon choosing the **More** button, it gets converted into the **Less** button. When you choose this button, the expanded portion of the **Array** dialog box is contracted and the original **Array** dialog box is displayed back. The **Object base point** area gives you following options:

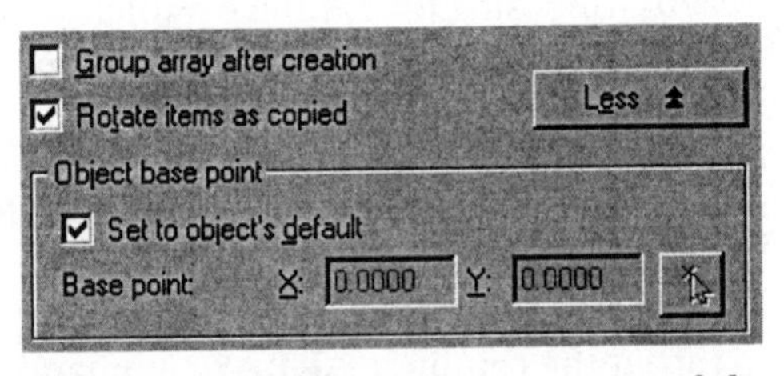

*Figure 4-45 Object base point area of the **Array** dialog box*

Set to object's default

This option uses the default base point of the objects to position the arrayed objects. The point used depends upon the object you have selected to array. If you clear the **Set to object's default** check box, you can manually set the base point using the **X** and **Y** edit boxes. The following table gives the details of the object type and their respective default base point.

Object Type	**Defualt base point**
Arc, Circle, Ellipse	Center point
Polygon, Rectangle	First corner
Donut, Line, Polyline, 3D Polyline, Ray, Spline	Starting point
Block, Paragraph text, Single line text	Insertion point
Construction lines	Mid point
Region	Grip point

Base point

This option allows you to define the base point of object which is used to calculate the rotation of the array. To array the objects in polar fashion, AutoCAD LT calculates the distance from center point of array to a reference point, which is generally the base point, on the last object you have selected. You can enter the values in the **X** and **Y** edit boxes. This option is activated only if the **Set to object default** is cleared.

Pick Base Point

Choosing this button temporarily closes the **ARRAY** dialog box so that you can define the base point in the drawing.

Preview

Choosing this button temporarily closes the **ARRAY** dialog box so that you can have a preview of the arrayed objects. It shows another dialog box, Figure 4-46, giving you the following options:

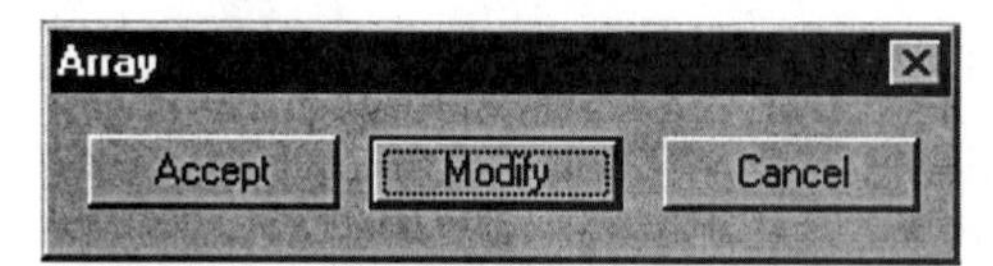

Figure 4-46 *Preview dialog box in the Polar Array option*

Accept. Choosing this button completes the **ARRAY** command and arrange the objects based on the values and settings you have given in the **ARRAY** dialog box.

Modify. Choosing this button brings back the **ARRAY** dialog box so that you can make the required modification in the values you have entered in the **ARRAY** dialog box.

Cancel. Choosing this button terminates the **ARRAY** command here itself and you will have to repeat the complete **ARRAY** command in case you have to array any object.

-ARRAY Command

You can use the **-ARRAY** command to array an object using the command line. The following is the prompt sequence:

Command: **-ARRAY** Enter
Select objects: *Select objects to copy.*
Select objects: Enter
Enter type of array [Rectangular/Polar] <last>:

The option that was used in the previous **-ARRAY** command is set as current and the "<>" is placed around it. The function and results obtained by using the **-ARRAY** command vary, depending upon which type of array you want to generate: **Rectangular** or **Polar**. The prompt sequence for the rectangular array is as follows:

Command: **-ARRAY** Enter
Select objects: *Select object to be arrayed.*
Select objects: Enter
Enter the type of array [Rectangular or Polar array]<R>: **R** Enter
Enter the number of rows (---) <1>: *Enter the value.*
Enter the number of columns (|||) <1>: *Enter the value.*
Enter the distance between rows or specify unit cell (---): *Enter the value of distance between rows or specify first corner for unit cell.*
Specify the distance between columns(|||): *Enter the value of distance between columns.*

The prompt sequence for **Polar Array** is:

Command: **-ARRAY** Enter
Select objects: *Select object to be arrayed.*
Select objects: Enter
Enter type of array [Rectangular or Polar array]<R>: **P** Enter
Specify center point of array: *Specify a point.*
Enter the number of items in the array: *Enter the value.*
Specify the angle to fill (+=ccw, -=cw)<360>: *Enter the value of angle to fill.*
Rotate arrayed objects [Yes/No] <Y>: *Enter the choice.*

Note

You have the option of specifying any two of the three parameters that follows the prompt ***Specify center point of array:****. For the first prompt,* ***Specify the number of items:****, enter the total number of items in the array, including the original item. If you give a null response to this written prompt, you will have to specify the angle to fill as well as the angle between the items. The next prompt sequence is* ***Specify the angle to fill (+=CCW,-=CW)<360>:****. The letters in the parentheses stand for the counterclockwise and clockwise directions. As is clear from the prompt, a positive angle generates an array in a counterclockwise direction, and a negative angle generates it in a clockwise direction. The array is completely defined if you specify the number of items and the angle to fill. If you supply only one of the two parameters, AutoCAD LT prompts you for the angle between the items in the array at the prompt* ***Angle between items:****. You should specify the direction of the array at this prompt. If you have not previously specified the number of items, AutoCAD LT will calculate them automatically. To construct the array, AutoCAD LT calculates the distance from the array's center point to a point of reference on the last object selected. The reference point applied depends on the object.*

MIRROR COMMAND

Toolbar:	Modify > Mirror
Menu:	Modify > Mirror
Command:	MIRROR

The **MIRROR** command creates a mirror copy of the selected objects; the objects can be mirrored at any angle. This command is helpful in drawing symmetrical figures. When you invoke this command, AutoCAD LT will prompt you to select the objects and then the mirror line.

After you select the objects to be mirrored, AutoCAD LT prompts you to enter the beginning point and endpoint of a mirror line (the imaginary line about which objects are reflected). You can specify the endpoints of the mirror line by specifying the points or by entering their coordinates. The mirror line can be specified at any angle. After the first endpoint has been selected, AutoCAD LT displays the selected objects as they would appear in a mirror. Next, you need to specify the second endpoint of the mirror line. Once this is accomplished, AutoCAD LT prompts you to specify whether you want to retain the original figure (Figure 4-47) or delete it and just keep the mirror image of the figure (Figure 4-48). The prompt sequence is:

Command: **MIRROR** [Enter]
Select objects: *Select the objects to be mirrored.*
Select objects: [Enter]
Specify first point of mirror line: *Specify the first endpoint.*
Specify second point of mirror line: *Specify the second endpoint.*
Delete source objects ? [Yes/No] <N>: *Enter* Y *for deletion,* N *for retaining the previous objects.*

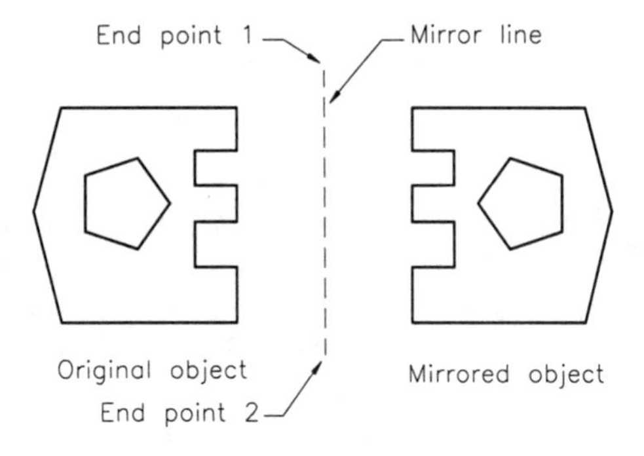

Figure 4-47 *Reflecting an object using the* ***MIRROR*** *command*

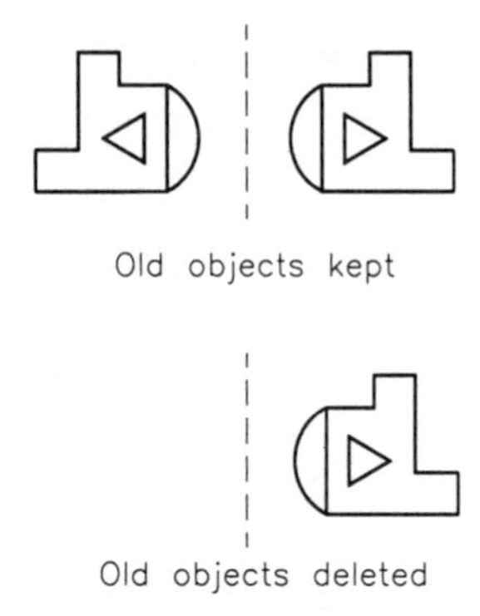

Figure 4-48 *Retaining and deleting old objects using the* ***MIRROR*** *command*

Text Mirroring

By default, the **MIRROR** command reverses all the objects, including texts and dimensions. But you may not want the text reversed (written backward). In such a situation, you should use the system variable **MIRRTEXT**. The **MIRRTEXT** variable has the following two values (Figure 4-49):

1 = Text is reversed in relation to the original object. This is the default value.

0 = Inhibits the text from being reversed with respect to the original object.

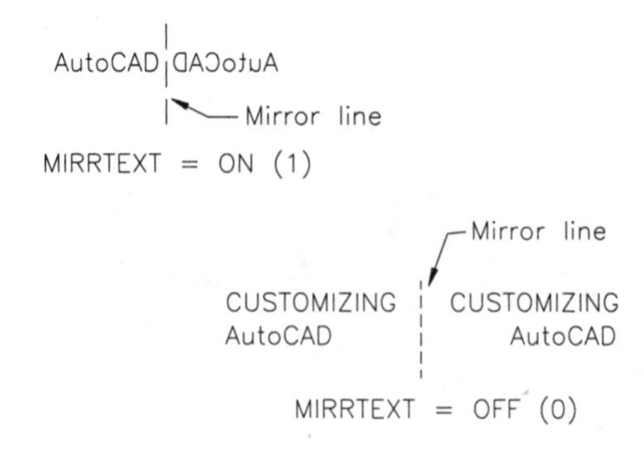

Figure 4-49 *Using the* ***MIRRTEXT*** *variable to mirror the text*

Hence, if you want the existing object to be mirrored, but at the same time you want the text to be readable and not reversed, set the **MIRRTEXT** variable to 0, then use the **MIRROR** command.

Command: **MIRRTEXT** Enter
Enter new value for MIRRTEXT <1>: **0** Enter
Command: **MIRROR** Enter
Select objects: *Select the objects.*
Select objects: Enter
Specify first point of mirror line: *Specify the first point of the mirror line.*
Specify second point of mirror line: *Specify the second point.*
Delete source objects?[Yes/No] <N>: *Enter* **Y** *to delete,* **N** *to retain the original objects.*

BREAK COMMAND

Toolbar:	Modify > Break
Menu:	Modify > Break
Command:	BREAK

The **BREAK** command cuts existing objects into two or erases portions of the objects. This command can be used to remove a part of the selected objects or to break objects like lines, arcs, circles, ellipses, xlines, rays, splines, and polylines. After entering this command, AutoCAD LT prompts you to select the object to be broken, and then the breakpoints. The four different options for using the **BREAK** command are discussed next.

1 Point Option

Using this option, you can break the object into two parts. Here, the selection point is taken as the breakpoint, and the object is broken there. The prompt sequence is:

Command: **BREAK** Enter
Select object: *Select the object to be broken.*
Enter second break point or [First point]: @ Enter

1 Point Select Option

Using this option, you can break the object into two parts, but you are allowed to specify a different breakpoint. The selection point does not affect the breakpoint. The prompt sequence is:

Command: **BREAK** Enter
Select object: *Select the object to be broken.*
Enter second break point or [First point]: **F** Enter
Specify first break point: *Specify a new breakpoint.*
Specify second break point: @ Enter

2 Points Option

Here you are allowed to break an object between two selected points. The point at which you select the object becomes your first breakpoint and then you are prompted to enter the second breakpoint. The prompt sequence is:

Command: **BREAK** Enter
Select object: *Select the object to be broken.*
Enter second break point or [First point]: *Specify the second breakpoint.*

The object is broken between those two points and the in-between portion of the object is removed (Figure 4-50).

2 Points Select Option

This option is similar to the 2 Points option; the only difference is that instead of making the selection point as the first breakpoint, you are allowed to specify a new first point. The prompt sequence is:

Command: **BREAK** Enter
Select object: *Select the object to be broken.*
Enter second break point or [First point]: **F** Enter
Specify first break point: *Specify a new breakpoint.*
Specify second break point: *Specify second point.*

If you need to work on arcs or circles, make sure that you work in a counterclockwise direction, or you may end up cutting the wrong part. In this case the second point should be selected in a counterclockwise direction with respect to the first one (Figure 4-51). You can use the 2 Points and 2 Points Select options to break an object into two without removing a portion in between. This can be achieved by specifying the same point on the object as the first and the second break points. If you specify the first break point on the line and the second break point beyond the end of the line, one complete end starting from the first break point will be removed. The extrusion direction of the selected object need not be parallel to the Z axis of the UCS.

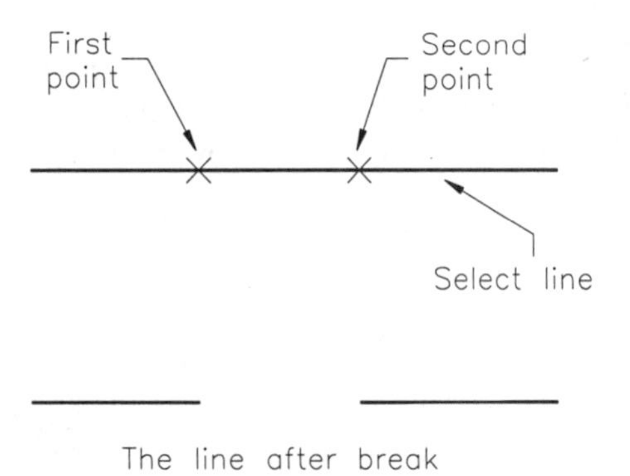

Figure 4-50 Using the ***BREAK*** *command on a line*

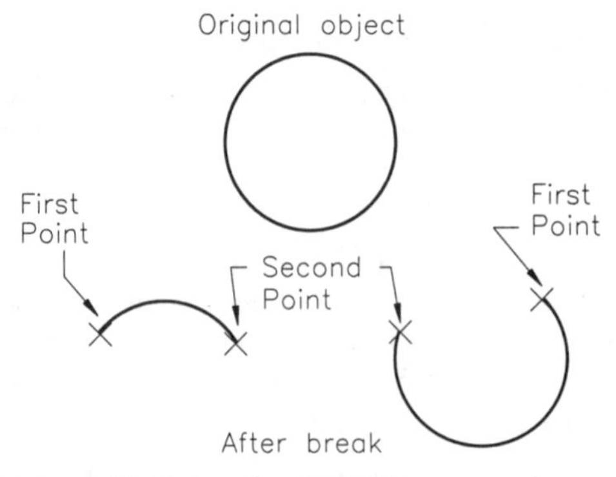

Figure 4-51 Using the ***BREAK*** *command on arcs and circles*

Exercise 7 *General*

Break a line at five different points and then erase the alternate segments. Next, draw a circle and break it into four equal parts.

MEASURE COMMAND

Menu: Draw > Point > Measure
Command: MEASURE

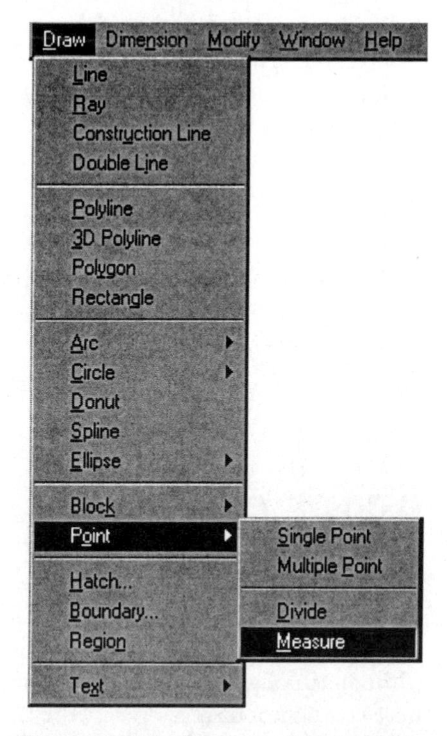

***Figure 4-52** Invoking the **MEASURE** command from the **Draw** menu*

While drawing, you may need to segment an object at fixed distances without actually dividing it. You can use the **MEASURE** command to do so. This command places points (nodes) on the given object at a specified distance. The shape of these points is determined by the **PDMODE** system variable and the size is determined by the **PDSIZE** variable. The **MEASURE** command starts measuring the object from the endpoint closest to where the object is selected. When a circle is to be measured, an angle from the center is formed that is equal to the Snap rotation angle. This angle becomes the starting point of measurement, and the markers are placed at equal intervals in counterclockwise direction.

This command goes on placing markers at equal intervals of the specified distance without considering whether the last segment is the same distance or not. Instead of entering a value, you can also select two points that will be taken as the distance. In the following example a line and a circle is measured at 0.80 unit distance (Figure 4-53). The Snap rotation angle is zero degrees. The **PDMODE** variable is set to 3 so that X marks are placed as markers.

The prompt sequence is:

Command: **MEASURE** [Enter]
Select object to measure: *Select the object to be measured.*
Specify length of segment or [Block]: 0.**80** [Enter]

You can also place blocks as markers (Figure 4-54), but the block must already be defined within the drawing. You can align these blocks with the object to be measured. The prompt sequence is:

Command: **MEASURE** [Enter]
Select object to measure: *Select the object to be measured.*

Specify length of segment or [Block]: **B** Enter
Enter name of block to insert: *Enter the name of the block.*
Align block with object? [Yes/No] <Y>: *Enter* Y *to align,* N *to not align.*
Specify length of segment: *Enter distance.*

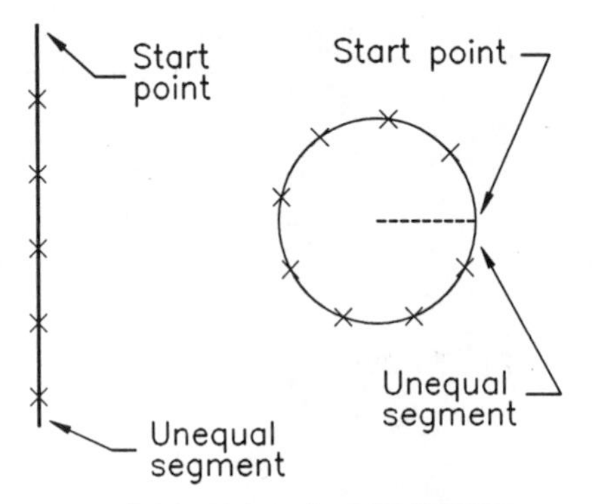

Figure 4-53 *Using the **MEASURE** command*

Figure 4-54 *Using blocks with the **MEASURE** command*

DIVIDE COMMAND

Menu:	Draw >Point > Divide
Command:	DIVIDE

The **DIVIDE** command is used to divide an object into a number of segments of equal length without actually breaking it. This command is similar to the **MEASURE** command except that here you do not have to specify the distance. The **DIVIDE** command calculates the full length of the object and places markers at equal intervals. This makes the last interval equal to the rest of the intervals. If you want a line to be divided, first enter the **DIVIDE** command and select the object to be divided. After this you enter the number of divisions or segments. The number of divisions entered can range from 2 to 32,767. See Figure 4-55. The prompt sequence for dividing a line and circle into five equal parts is:

Command: **DIVIDE** Enter
Select object to divide: *Select the object you want to divide.*
Enter number of segments or [Block]: **5** Enter

You can also place blocks as markers, but the block must be defined within the drawing. See Figure 4-56. You can align these blocks with the object to be measured. The prompt sequence is:

Command: **DIVIDE** Enter
Select object to divide: *Select the object to be measured.*
Enter number of segments or [Block]: **B** Enter
Enter name of block to insert: *Enter the name of the block.*
Align block with object? [Yes/No] <Y>: *Enter Y to align, N to not align.*
Enter number of segments: *Enter the number of segments.*

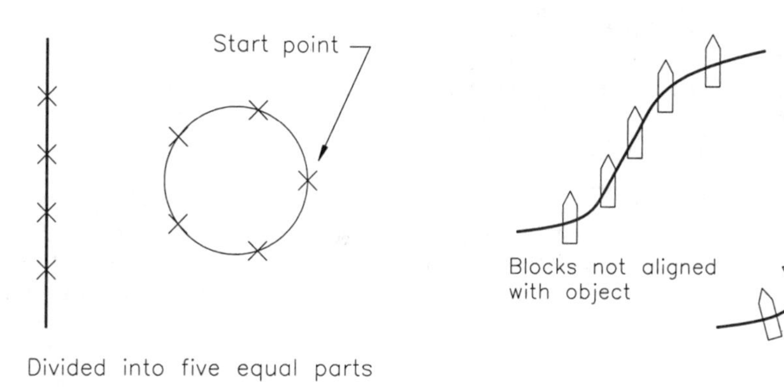

*Figure 4-55 Using the **DIVIDE** command*

*Figure 4-56 Using blocks with the **DIVIDE** command*

Note

*The size and shape of points placed by the **DIVIDE** and **MEASURE** commands are controlled by the **PDSIZE** and **PDMODE** system variables.*

Exercise 8 — *General*

a. Use the **MEASURE** command to divide a circle into parts with a of length 0.75" and divide a line into 10 equal segments (set **PDMODE** variable to 3).

b. Draw the Figure 4-57. Use the **DIVIDE** command to divide the circle and use the NODE object snap to select the points.

Figure 4-57 Drawing for Exercise 8

MATCHING PROPERTIES

Toolbar:	Standard > Match Properties
Menu:	Modify > Match Properties
Command:	MATCHPROP

The **MATCHPROP** or **PAINTER** command can be used to change some properties like color, layer, linetype, and linetype scale of the selected objects to the source object. When you invoke this command, AutoCAD LT will prompt you to select the source object and then the destination objects. The properties of the destination objects will be changed to that of the source object.

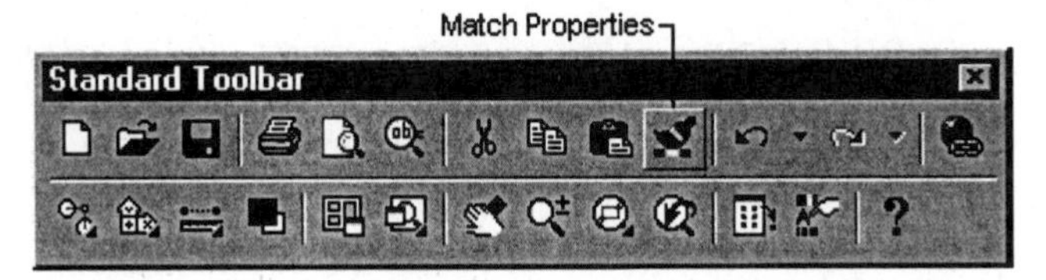

Figure 4-58 Invoking ***Match Properties*** *from the* ***Standard*** *toolbar*

Command: **MATCHPROP** [Enter]
Select Source Object: *Select the source object.*
Current active settings: Color Layer ltype Ltscale Lineweight Thickness Plotstyle Text Dim Hatch
Select Destination Object(s) or [Settings]: *Select the objects whose properties you want to change.*

When you select the Setting option, AutoCAD LT displays the **Properties settings** dialog box (Figure 4-59). The properties displayed in this dialog box are those of the source object. You can use this dialog box to change the properties that are copied from source to destination objects. The following table lists the properties associated with different types of objects:

Object	Color & layer	Linetype & Linetype Scale	Line-weight	Thick-ness	Text	Dime-nsion	Hatch	Plot style
Arc	x	x	x	x				x
AttDef	x			x	x			x
Body	x	x						x
Circle	x	x	x	x				x
Dimension	x	x				x		x
Ellipse	x	x	x					x
Hatch	x		x			+	x	
Image	x	x						x
Insert	x	x	x					x
Leader	x	x	x			x		x

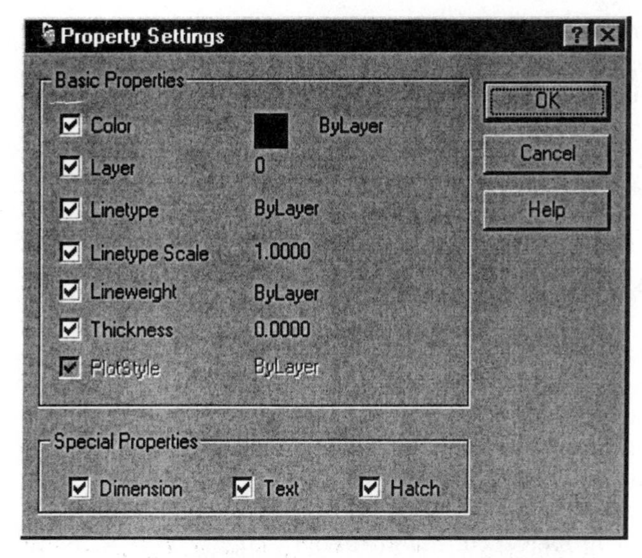

Figure 4-59 Properties Settings *dialog box*

Line	x	x		x	x			x	
Mtext	x						x		x
OLE object									
Point	x			x	x			x	
2D polyline	x	x	x		x			x	
3D polyline	x	x	x					x	
Ray	x	x	x					x	
Region	x	x	x		x			x	
2D solid	x	x	x					x	
Spline	x	x	x					x	
Text	x	x			x	x		x	
Tolerance	x	x	x				x	x	
Viewport	x			x				x	
Xline	x	x	x					x	
Xref	x	x	x					x	
Zombie	x	x	x					x	

PROPERTIES COMMAND*

Toolbar:	Standard > Properties
Menu:	Modify > Properties
	Tools > Properties
Command:	PROPERTIES

You can also modify an object by using the **PROPERTIES** command. When you invoke this command, AutoCAD LT will display the **Properties** Window on the screen (Figure 4-60). Similarly, if you select text, the Properties Window is displayed. For example, if you select an object in the drawing area, the properties of that object will be displayed in the Properties Window. The contents of the Properties Window changes according to the objects selected.

The **Properties** Window can also be invoked from shortcut menu by right-clicking in the drawing area and choosing **Properties** from the menu.

The **Properties Window** shows the common properties of the multiple objects if more than one object is selected. To change properties of the selected objects, you can click in the cell next to the name of the property and change the values manually or you can choose from the available options in the drop-down list if one is available. You can cycle through the options by double-clicking in the property cell. You can also choose the **Quick Select** button to create a selection set that either includes or excludes all the objects matching the criteria specified by you.

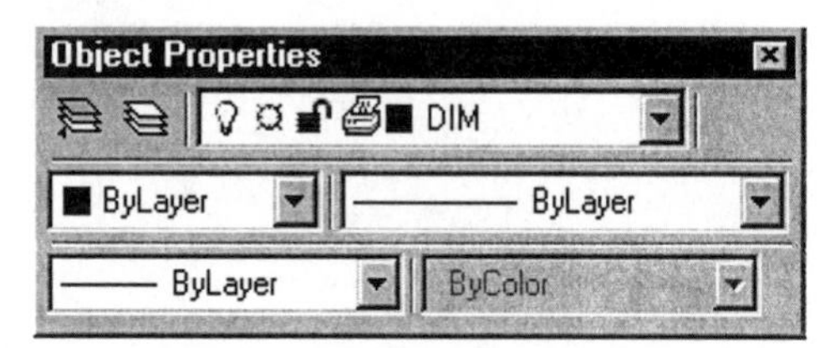

Figure 4-60 *Selecting Properties from the* ***Object Properties*** *toolbar*

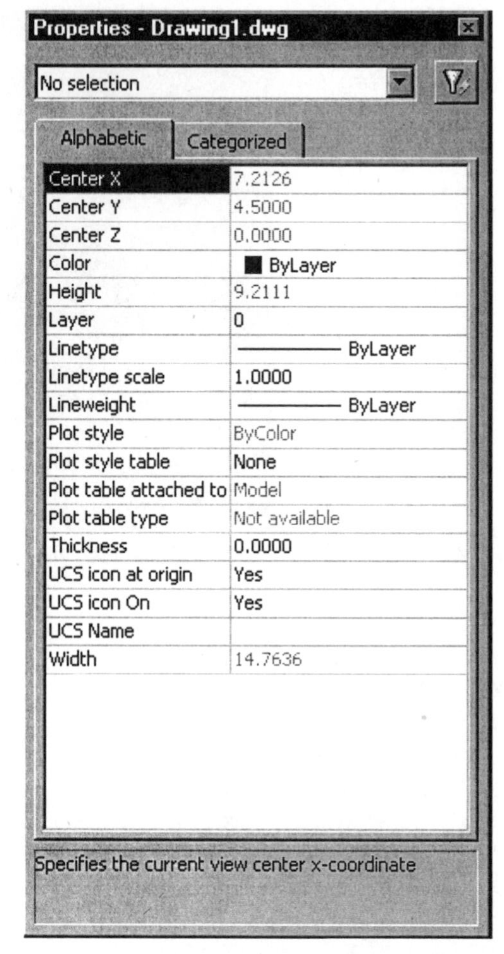

Figure 4-61 *Using the* ***Properties*** *command to edit (Objects Properties Window)*

QSELECT COMMAND*

Menu: Tools > Quick Select
Command: QSELECT

Quick Select is a new feature of AutoCAD LT 2000. This creates a new selection set which will either include or exclude all objects that matches specified object type and property criteria. The **QSELECT** command can be applied to the entire drawing or existing selection set. If a drawing is partially opened, QSELECT does not consider the objects that are not loaded. The **QSELECT** command can be invoked by choosing the **Quick Select** button in the **Properties** Window. In the shortcut menu, the **QSELECT** command can be invoked by choosing **Quick Select**. The prompt sequence is as follows:

Command: **QSELECT** [Enter]

AutoCAD LT 2000 displays the **Quick Select** dialog box (Figure 4-62).

The **Quick Select** dialog box specifies the object filtering criteria and creates a selection set from that criteria.

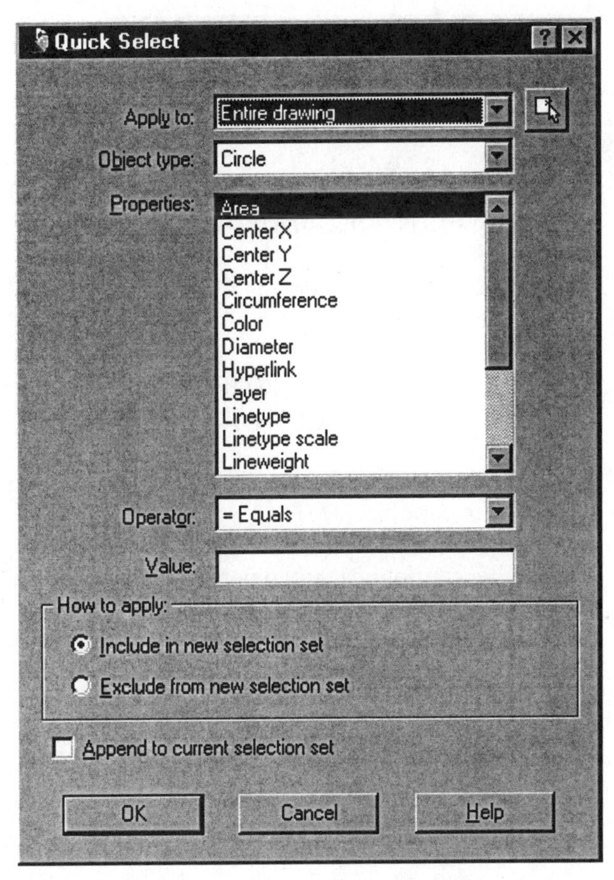

Figure 4-62 Quick Select dialog box

Apply To. Apply to specifies whether to apply the filtering criteria to the entire drawing or to the current selection set. If there is some existing current selection set, Current Selection is the default value. Otherwise entire drawing is the default value.

Object Type. This specifies the type of object to be filtered. It lists all the available object types and if some objects are selected, it lists all selected object types. Multiple is the default setting.

Properties. It specifies the properties to be filtered.

Operator. It specifies the range of the filter for the chosen property:

- Equals =
- Does not Equal < >
- Greater than >
- Less than <

Value. It specifies the property value of the filter. If values are known, then it becomes a list of the available values from which you can select a value. Otherwise you can enter a value.

How to Apply. It gives you choice of inclusion or exclusion from the new selection set. You can choose Include in new selection set for creating a new selection set only composed of those objects that conform to the filtering criteria. The other option creates a new selection set of objects that does not conform to the filtering criteria.

Append to Current Selection Set. This creates a cumulative selection set by using multiple uses of the Quick Select. It specifies whether **QSELECT** replaces or is appended to the current selection set.

Quick Select supports custom objects (objects that are created by some other applications) and their properties. If custom objects has other properties than AutoCAD LT 2000, then the source application of the object should be running in order for the properties to be available by the **QSELECT**.

Self-Evaluation Test

Answer the following questions and then compare your answers to the correct answers given at the end of this chapter.

1. When you shift a group of objects using the **MOVE** command, the size and orientation of these objects is changed. (T/F)

2. The **COPY** command makes copies of the selected object, leaving the original object intact. (T/F)

3. Depending on the side to offset, you can create smaller or larger circles, ellipses, and arcs with the **OFFSET** command. (T/F)

4. With the **BREAK** command, when you select an object, the selection point becomes the firstbreak point. (T/F)

5. A fillet cannot be created between two parallel and nonintersecting lines. (T/F)

6. The __________ command prunes objects that extend beyond a required point of intersection.

7. The offset distance is stored in the __________ system variable.

8. Instead of specifying the scale factor, you can use the **Reference** option to scale an object. (T/F)

9. If the **MIRRTEXT** system variable is set to a value of 1, the mirrored text is not reversed with respect to the original object. (T/F)

10. A Rectangular array can be rotated. (T/F)

11. There are two types of arrays: __________ and __________.

12. The __________ option of the __________ command is used to draw an array in which the objects of the array are placed in a circular pattern around a point.

Review Questions

1. In the case of the **Through** option of the **OFFSET** command, you do not have to specify a distance; you simply have to specify an offset point. (T/F)

2. With the **BREAK** command, you cannot break an object in two without removing a portion in between. (T/F)

3. With the **FILLET** command, the extrusion direction of the selected object must be parallel

to the Z axis of the UCS. (T/F)

4. With the **FILLET** command, there can be more than one fillet for arcs and circles. (T/F)

5. The _________ command helps you form round corners between any two lines by asking you to identify the two lines.

6. The _________ command does not work on closed lines or polylines.

7. If an entire selected object is within a window or crossing specification, the _________ command works like the **MOVE** command.

8. The _________ option of the **EXTEND** command is used to extend objects to the implied boundary.

9. With the **FILLET** command, using the **Polyline** option, if the selected polyline is not closed, the _________ corner is not filleted.

10. When the chamfer distance is zero, the chamfer created is in the form of a _________.

11. Compare the commands **EXTEND** and **STRETCH**.______________________________
____

12. Describe the commands **FILLET** and **CHAMFER**.______________________________
____

13. Explain how you can move an object from one position to another.__________________
____

14. AutoCAD LT saves the previous selection set and lets you select it again by using the _________ selection option.

15. The **R** (**Remove**) option is used to shift from Add mode to _________ .

16. The _________ command restores objects that have been erased by the previous **ERASE** command.

17. Objects on frozen or locked layers are selected by the **All** selection option. (T/F)

18. The **Remove** selection option removes the selected object from the drawing. (T/F)

19. When you use the **SIngle** selection option with the **ERASE** command, you can select either a single object or multiple objects. (T/F)

20. The _________ selection option is used to select objects by touching the objects to be selected with the selection line.

21. The __________ variable is used to set the default for the number of sides in a polygon.

22. In the case of numerical specification of a radius, the bottom edge of the polygon is rotated by the prevalent __________ angle in the current UCS.

23. The __________ selection option is used if you want to select the most recently drawn object on the screen.

24. The __________ selection option enables you to select all the objects on the drawing screen.

25. With the **ARRAY** command, if you specify __________ column distance, the columns are added to the left.

26. You can use the __________ option of the __________ command to rotate an object that has been rotated previously with respect to a known angle.

27. What is the purpose of the **ARRAY** command? ______________________________

____.

28. Explain the difference between a polar array and a rectangular array. ______________

____.

29. Before creating a rectangular array, what values should you know? ______________

____.

30. Define a unit cell. ____

____.

31. Suppose an object is 2 inches wide and you want a rectangular array with 1-inch spacing between objects. What should the column distance be? ______________________

____.

32. Explain how you can create a rotated rectangular array. ______________________

____.

Exercises

Exercise 9 — Mechanical

Draw the object shown in Figure 4-63 and save the drawing. Assume the missing dimensions.

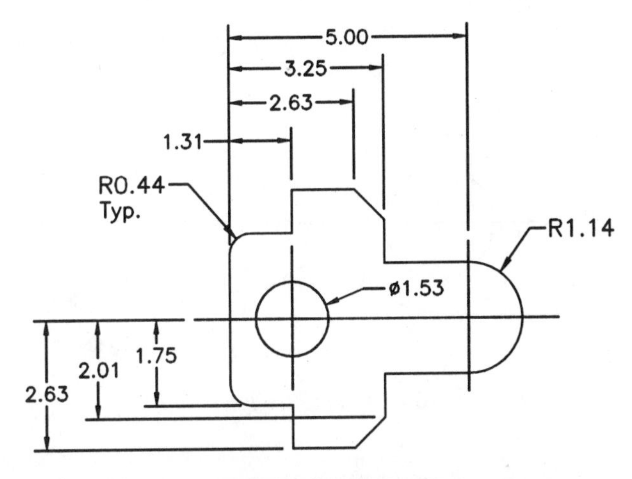

Figure 4-63 *Drawing for Exercise 9*

Exercise 10 — Mechanical

Make the drawing shown in Figure 4-64 and save the drawing. You can use the **ARRAY** command to make copies of the bolt hole details.

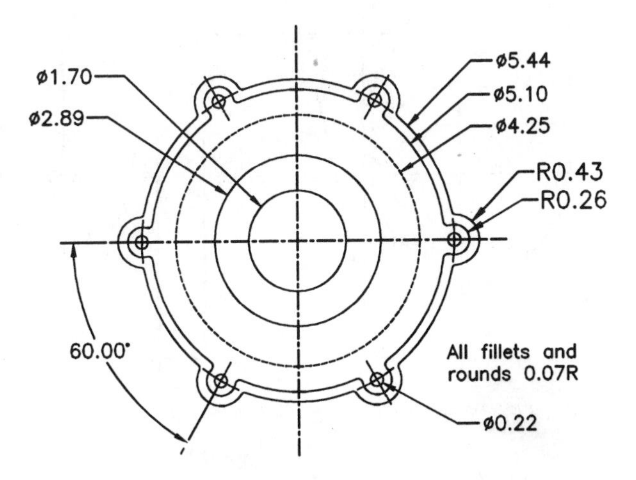

Figure 4-64 *Drawing for Exercise 10*

Exercise 11 — Mechanical

Make the drawing shown in Figure 4-65 and save the drawing.

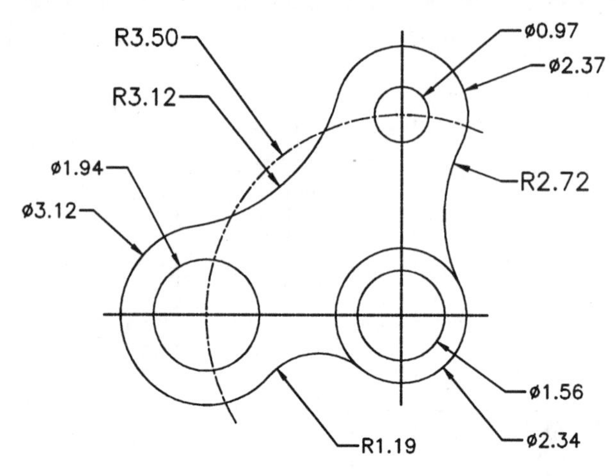

Figure 4-65 Drawing for Exercise 11

Exercise 12 — Mechanical

Make the drawing shown in Figure 4-66 and save the drawing. Assume the missing dimensions.

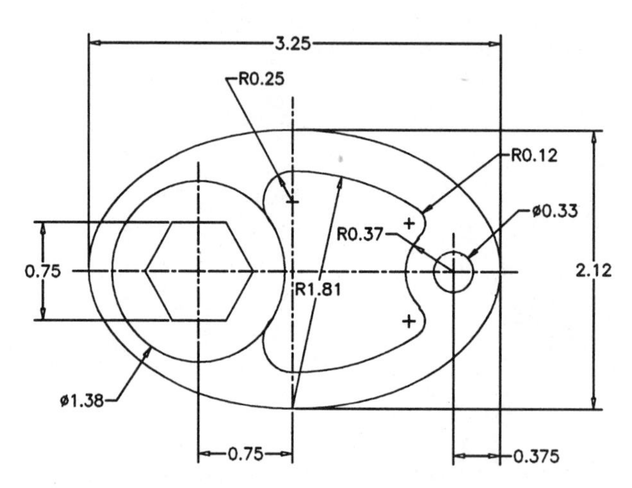

Figure 4-66 Drawing for Exercise 12

Problem-Solving Exercise 1 — *Mechanical*

Draw Figure 4-67 using the draw and edit commands. Use the **MIRROR** command to mirror the shape 9 units across the Y axis so that the distance between two center points is 9 units. Mirror the shape across the X axis and then reduce the mirrored shape by 75 percent. Join ends to complete the shape of the open end spanner. Save the file. Assume the missing dimensions.

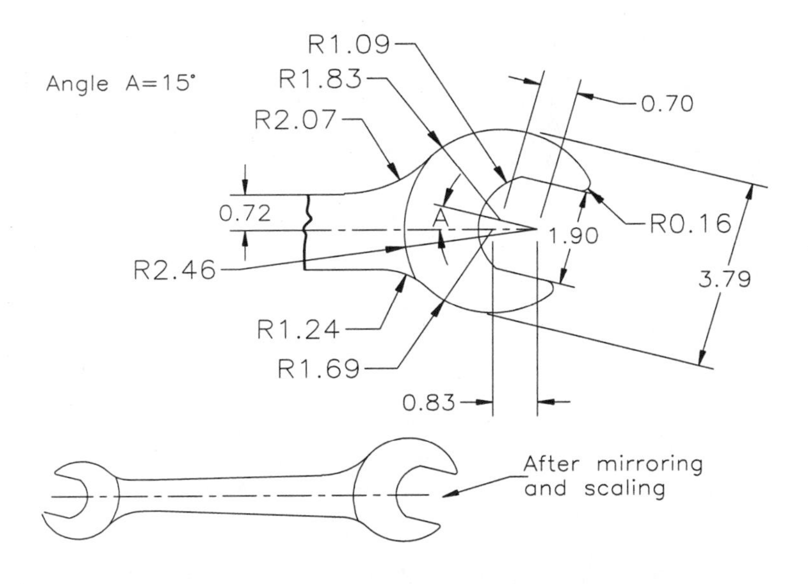

Figure 4-67 Partial drawing for Problem-Solving Exercise 1

Problem Solving Exercise 2 — *Architectural*

Draw the floor plan as shown in Figure 4-68 using AutoCAD LT commands that you have learned in Chapters 1 through 4. Assume the missing dimensions or measure the relative dimensions from the given drawing (the drawing is not to scale).

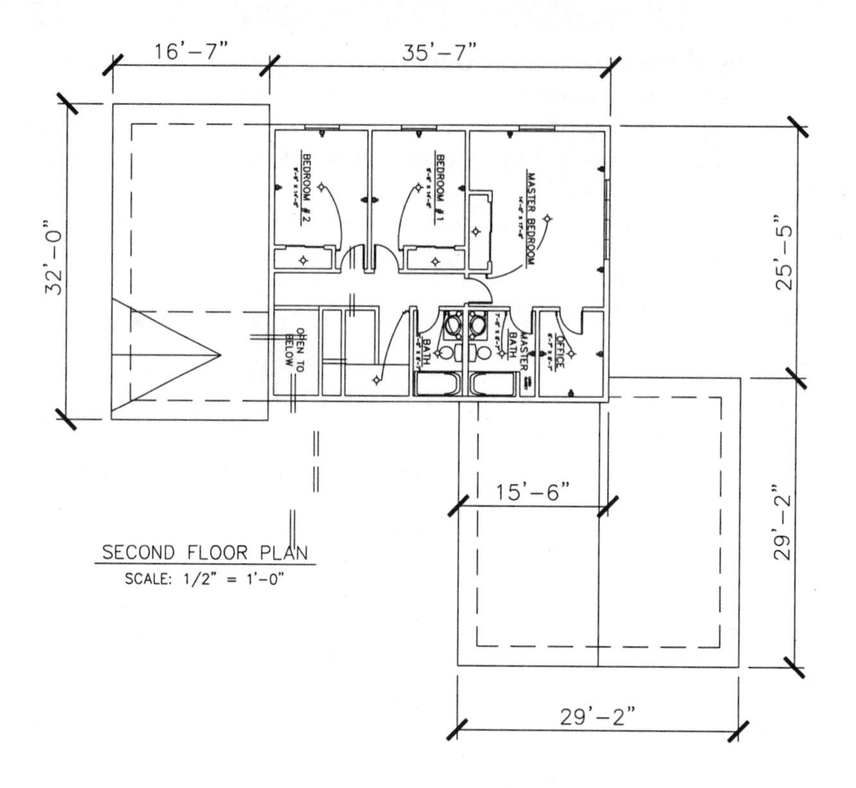

Figure4-68 Drawing for Problem-Solving Exercise 2

.Answers to Self-Evaluation Test

1 - F, **2** - T, **3** - T, **4** - T, **5** - , **6** - **TRIM**, **7** - **OFFSETDIST**, **8** - T, **9** - F, **10** - T, **11** - **Rectangular, Polar**, **12** - **Polar, ARRAY**

Chapter 5

Controlling Drawing Display and Creating Text

Learning Objectives

After completing this chapter, you will be able to:

- *Use the **REDRAW** and **REGEN** commands.*
- *Use the **ZOOM** command and its options.*
- *Understand the **PAN** and **VIEW** commands.*
- *Use AutoCAD LT text command **TEXT**.*
- *Edit text using the **DDEDIT** command.*
- *Draw special characters in AutoCAD LT.*
- *Create paragraph text using the **MTEXT** command.*
- *Edit text using the **DDEDIT** and **PROPERTIES** commands.*
- *Substitute fonts and specify alternate default fonts.*
- *Create text styles using the **STYLE** command.*
- *Determine text height and check spelling.*
- *Format paragraph text.*

BASIC DISPLAY OPTIONS

Drawing in AutoCAD LT is much simpler than manual drafting in many ways. Sometimes while drawing, it is very difficult to see and alter minute details. In AutoCAD LT , you can overcome this problem by viewing only a specific portion of the drawing. For example, if you want to display a part of the drawing on a larger area, you can use the **ZOOM** command, which lets you enlarge or reduce the size of the drawing displayed on the screen. Similarly, you can use **REGEN** command to regenerate the drawing and **REDRAW** to refresh the screen. In this chapter you will learn some of the drawing display commands, like **REDRAW**, **REGEN**,

PAN, **ZOOM**, and **VIEW**. These commands can also be used in the transparent mode. Transparent commands are commands that can be used while another command is in progress. Once you have completed the process involved with a transparent command, AutoCAD LT automatically returns you to the command with which you were working before you invoked the transparent command.

REDRAW COMMAND

Menu:	View > Redraw
Command:	REDRAW

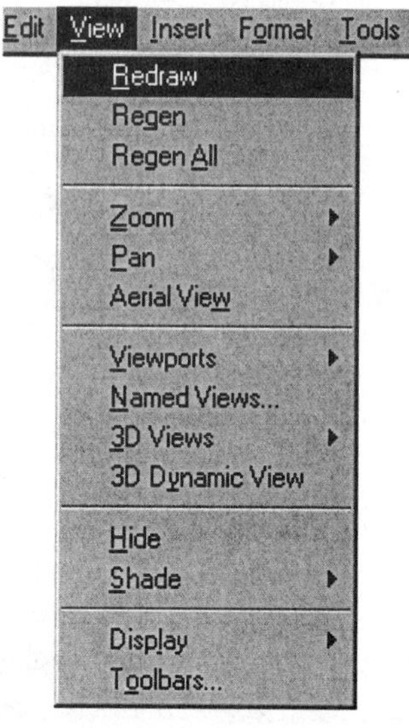

*Figure 5-1 Invoking the **REDRAW** command from the **View** menu*

This command redraws the screen, thereby removing the small cross marks that appear when a point is specified on the screen. These marks, known as blip marks or **blips**, indicate the points you have selected (points picked). The blip mark is not treated as an element of the drawing. In AutoCAD LT, several commands redraw the screen automatically (for example, when a grid is turned off), but it is sometimes useful to redraw the screen explicitly. Redrawing is used to remove blips and fill empty spaces left on the screen after objects have been deleted from the drawing. It also redraws the objects that do not display on the screen as a result of editing some other object. In AutoCAD LT, the **REDRAW** command can also be used in the transparent mode. Use of the **REDRAW** command does not involve a prompt sequence; instead, the redrawing process takes place without any prompting for information. You can also invoke this command from the tablet menu if you have the tablet facility. The prompt sequence is:

Command: **REDRAW** [Enter]

If the command is to be entered while you are working inside another command, type an apostrophe in front of the command. The apostrophe appended to a command indicates that the command is to be used as a transparent command (Command: **'REDRAW** [Enter]).

As with any other command, pressing ESC stops the command from taking effect and takes you back to the Command prompt. The **REDRAW** command affects only the current viewport.

While working on complex drawings it may be better to set the blipmode off instead of using the **REDRAW** command to clear blips. This can be done by using the **BLIPMODE** command. The prompt sequence is:

Command: **BLIPMODE** [Enter]
Enter mode [ON/OFF]<ON>: **OFF** [Enter]

REGEN COMMAND

Menu:	View > Regen
Command:	REGEN

The **REGEN command** makes AutoCAD LT regenerate the entire drawing to update it. The need for regeneration usually occurs when you change certain aspects of the drawing. All the objects in the drawing are recalculated and the current viewport is redrawn. This is a longer process than **REDRAW** and is seldom needed. One of the advantages of this command is that the drawing is refined by smoothing out circles and arcs. To use this command, enter **REGEN** at the Command prompt:

Command: **REGEN** [Enter]

AutoCAD LT displays the message **Regenerating model** while it regenerates the drawing. The **REGEN** command affects only the current viewport. If you have more than one viewport, you can use the **REGENALL** command to regenerate all the viewports. The **REGEN** command can be aborted by pressing ESC. This saves time if you are going to use another command that causes automatic regeneration.

Note

*Under certain conditions, the **ZOOM** and **PAN** commands automatically regenerate the drawing. Some other commands also perform regenerations under certain conditions.*

ZOOM COMMAND

Toolbar:	Zoom Standard > Zoom flyout
Menu:	View>Zoom
Command :	ZOOM

Creating drawings on the screen would not be of much use if you could not magnify the drawing view to work on minute details. Getting close to or away from the drawing is the function of the **ZOOM** command. In other words, this command enlarges or reduces the view of the drawing on the screen, but it does not affect the actual size of the objects. In this way the **ZOOM** command functions like the zoom lens on a camera. When you magnify the apparent size of a section of the drawing, you see that area in greater detail. On the other hand, if you reduce the apparent size of the drawing, you see a larger area.

The ability to zoom in, or magnify, has been helpful in creating the miniscule circuits used in the electronics and computer industries. This is one of the most frequently used commands. Also, this command can be used transparently, which means that it can be used while working in other commands. When you are working with some other command, the **ZOOM** command can be invoked from the shortcut menu by right-clicking in the drawing area and choosing **Zoom** from the menu (Figure 5-3). This command has several options and can be used in a number of ways. After the **ZOOM** command has been entered, different options are listed:

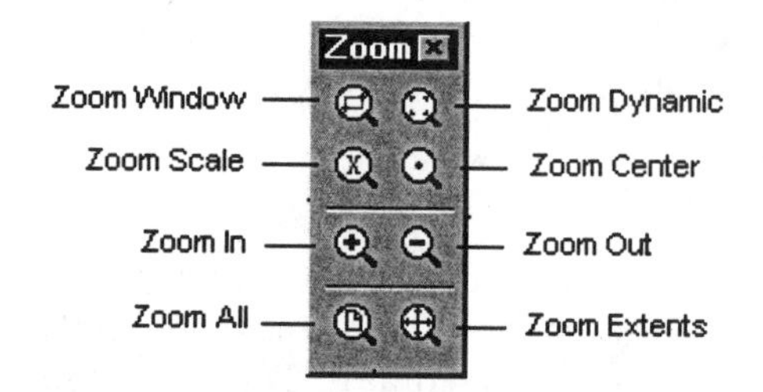

Figure 5-2 Selecting ***Zoom*** *options from the* ***Zoom*** *toolbar*

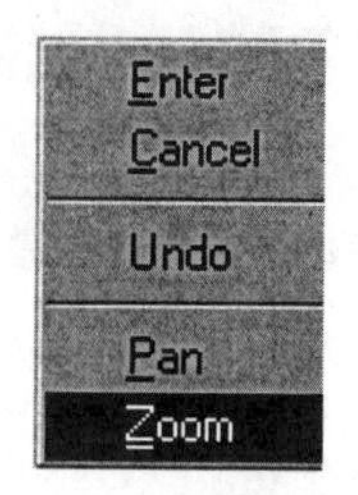

Figure 5-3 Invoking the ***ZOOM*** *command from the* ***shortcut*** *menu, when in* ***LINE*** *Command*

Command: **ZOOM** [Enter]
Specify corner of window, enter a scale factor (nX or nXP) , or
[All/Center/Dynamic/Extents/Previous/Scale/Window] <real time>:

Realtime Zooming

You can use the **Realtime Zoom** to zoom in and zoom out interactively. To zoom in, invoke the command, then hold the pick button down and move the cursor up. If you want to zoom in further, bring the cursor down. Specify a point and move the cursor up. Similarly, to zoom out, hold the pick button down and move the cursor down. If you move the cursor vertically up from the midpoint of the screen to the top of the window, the drawing is magnified by 100% (Zoom in 2x magnification). Similarly, if you move the cursor vertically down from the midpoint of the screen to the bottom of the window, the drawing display is reduced 100% (Zoom out 0.5x magnification). Realtime zoom is the default setting for the **ZOOM** command. Pressing ENTER after entering the **ZOOM** command automatically invokes the realtime zoom.

When you use the realtime zoom, AutoCAD LT displays a plus sign (+) and a minus sign (-) with a magnifying glass. When you reach the zoom out limit, AutoCAD LT does not display the minus sign (-) with the magnifying glass. Similarly, when you reach the zoom in limit, AutoCAD LT does not display the plus sign (+) with the cursor. To exit the realtime zoom, press Enter, Esc, or select Exit from the cursor menu.

All Option

This option of the **ZOOM** command zooms to the limits of the drawing if all the objects in the drawing are within the specified limits of the drawing area (Figure 5-4) If the objects are drawn beyond the limits of the drawing,the ALL option zooms to the extents of object. Hence, with the help of the ALL option, you can view the entire drawing in the current viewport, Figure 5-5.

Center Option

This option lets you define a new display window by specifying its center point (Figures 5-6 and 5-7). Here, you are required to enter the **center** and the **height** of the subsequent screen display. If you press ENTER instead of entering a new center point, the center of the view will remain unchanged. Instead of entering a height, you can enter the

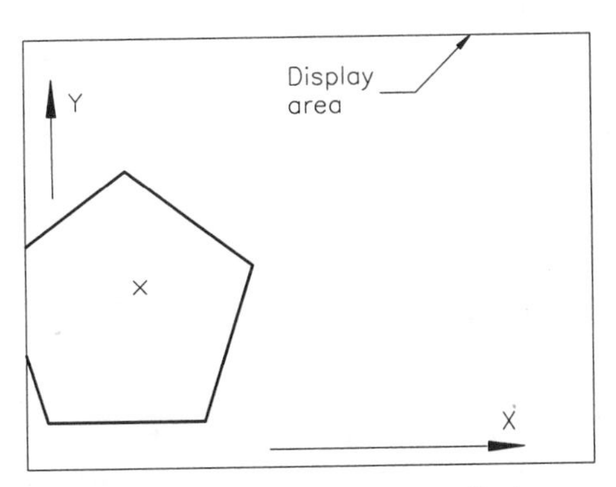

Figure 5-4 Drawing showing limits

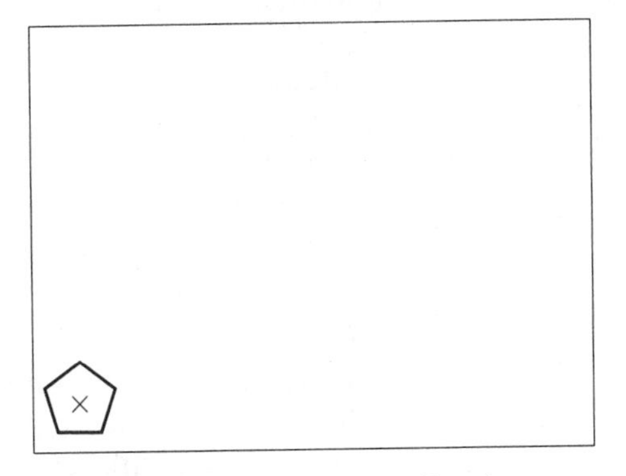

Figure 5-5 The Zoom All option

magnification factor by typing a number. If you press ENTER at the height prompt, or if the height you enter is the same as the current height, magnification does not take place. For example, if the current height is 2.7645 and you press ENTER at the **Enter magnification or height <2.7645>:** prompt, magnification will not take place. The smaller the value, the greater the enlargement of the image. You can also enter a number followed by **X**. This indicates the change in magnification, not as an absolute value, but as a value relative to the current screen. The following is the prompt sequence:

Command: **ZOOM** Enter
Specify corner of window, enter a scale factor (nX or nXP), or
[All/Center/Dynamic/Extents/Previous/Scale/Window] <realtime>: **C** Enter
Specify center point: *Specify a center point.*
Enter magnification or height <current>: **5X** Enter

In Figure 5-7, the current magnification height is 5, which magnified the display 5 times. If you enter a value of 2, the size (height and width) of the display area changes to 2 x 2, that will change the size of the object in graphics window. In Figure 5-8, since the diameter of the circle

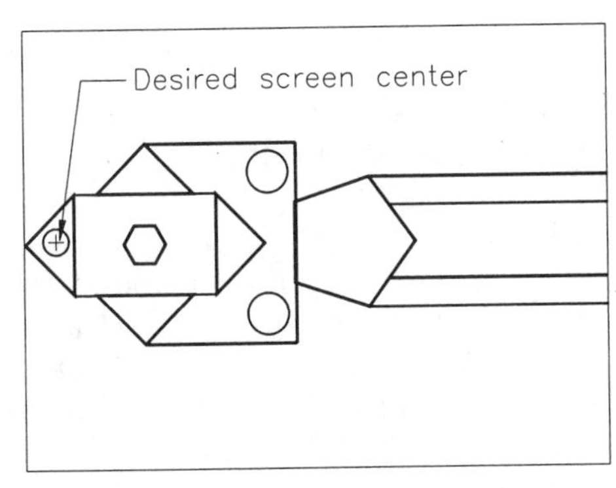

Figure 5-6 Drawing before using the ***ZOOM Center*** *option*

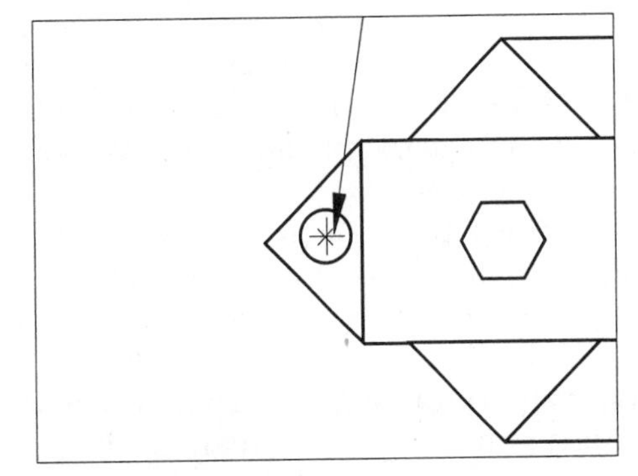

Figure 5-7 Drawing after using the ***ZOOM Center*** *option*

is 2 units, it will fit in the 2 x 2 window. The prompt sequence is:

Command: **ZOOM** [Enter]
Specify corner of window,enter a scale factor (nX or nXP), or
[All/Center/Dynamic/Extents/Previous/Scale/Window]<realtime>: **C** [Enter]
Center point: *Select the center of the circle.*
Magnification or Height <5.0>:**2** [Enter]

Extents Option

As the name indicates, this option lets you zoom to the extents of the drawing. The extents of the drawing comprise the area that has the drawings in it. The rest of the empty area is neglected. With this option, all the objects in the drawing are magnified to the largest possible display (Figure 5-9).

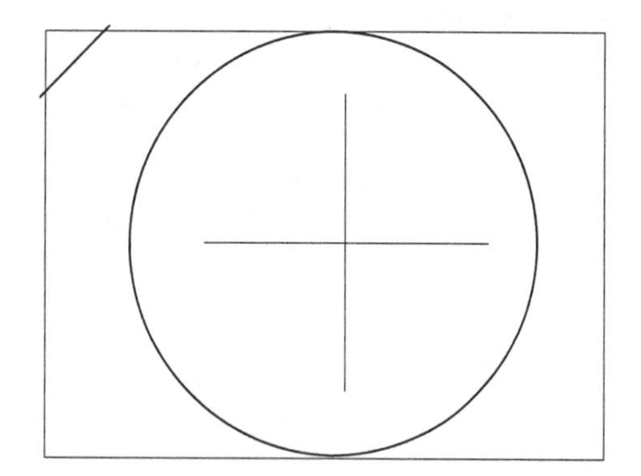

Figure 5-8 *Drawing after using the* ***ZOOM Center*** *option*

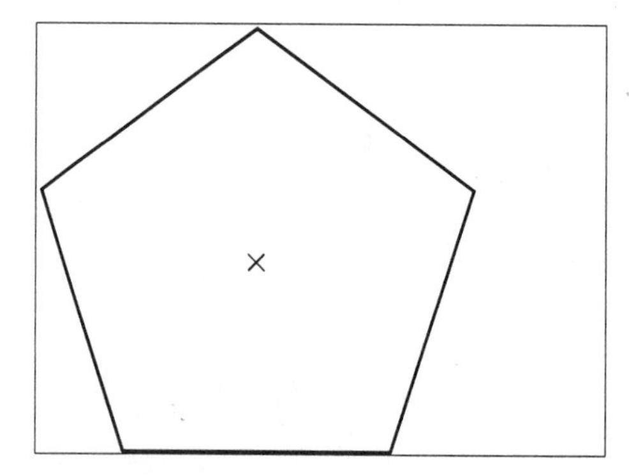

Figure 5-9 *The* ***ZOOM Extents*** *option*

Dynamic Option

This option displays the portion of the drawing that you have already specified. The prompt sequence for using this option of the **ZOOM** command is:

Command: **ZOOM** [Enter]
Specify corner of window,enter a scale factor (nX or nXP), or
[All/Center/Dynamic/Extents/Previous/Scale/Window]<realtime>: **D** [Enter]

You can then specify the area you want to be displayed by manipulating a view box representing your viewport. This option lets you enlarge or shrink the view box and move it around. When you have the view box in the proper position and size, the current viewport is cleared by AutoCAD LT and a special view selection screen is displayed. This special screen comprises information regarding the current view as well as available views. In a color display, the different viewing windows are very easy to distinguish because of their different colors, but in a monochrome monitor, they can be distinguished by their shape.

Blue Dashed Box Representing Drawing Extents. Drawing extents are represented by a dashed blue box (Figure 5-10), which constitutes the larger of the drawing limits or the actual area occupied by the drawing.

Green Dashed Box Representing the Current View. A green dotted box is formed to represent the area that the current viewport comprises when the **Dynamic** option of the **ZOOM** command is invoked (Figure 5-11).

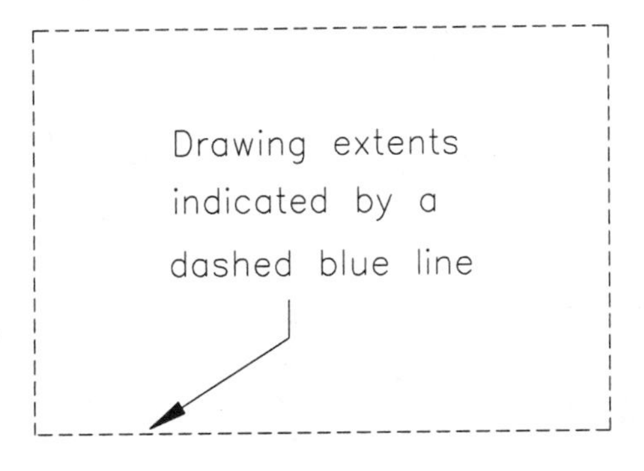

Figure 5-10 Box representing drawing extents

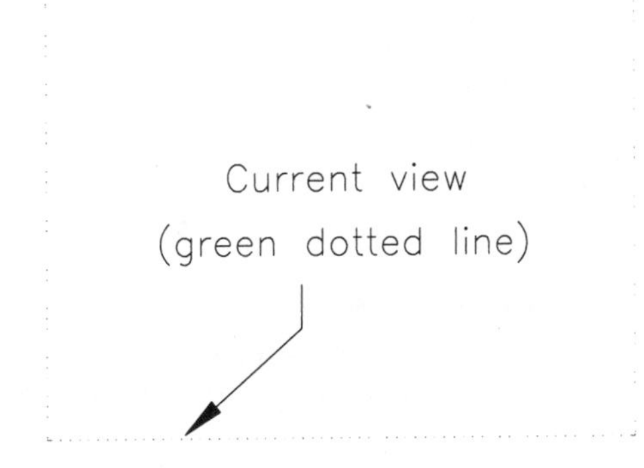

Figure 5-11 Representation of the current view

Panning View Box (X in the center). A view box initially of the same size as the current view box (the green or magenta box) is displayed with an X in the center (Figure 5-12). You can move this box with the help of your pointing device. This box, known as the **panning view box**, helps you to find the center point of the zoomed display you want. When you have found the center, you press the pick button to make the zooming view box appear.

Zooming View Box (arrow on the right side). After you press the pick button in the center of the panning view box, the X in the center of the view box is replaced by an arrow pointing to the right edge of the box. This **zooming view box** (Figure 5-13) indicates the ZOOM mode. You can now increase or decrease the area of this box according to the area you want to zoom into. To shrink the box, move the pointer to the left; to increase it, move the pointer to the right. The top, right, and bottom sides of the zooming view box move as you move the pointer, but the left side remains fixed, with the zoom base point at the midpoint of the left side. When you have the zooming view box in the desired size for your zoom display, press ENTER to

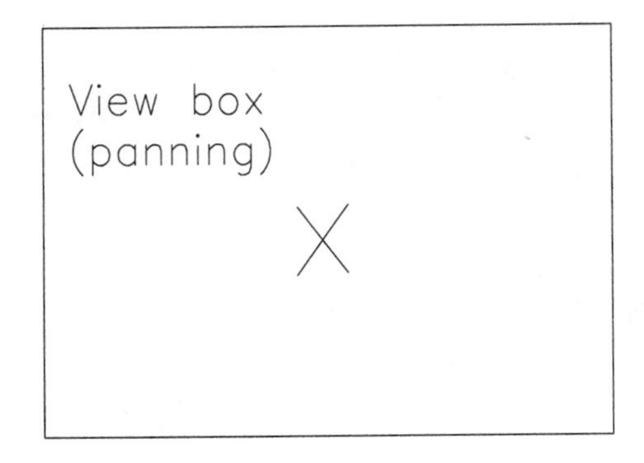

Figure 5-12 The panning view box

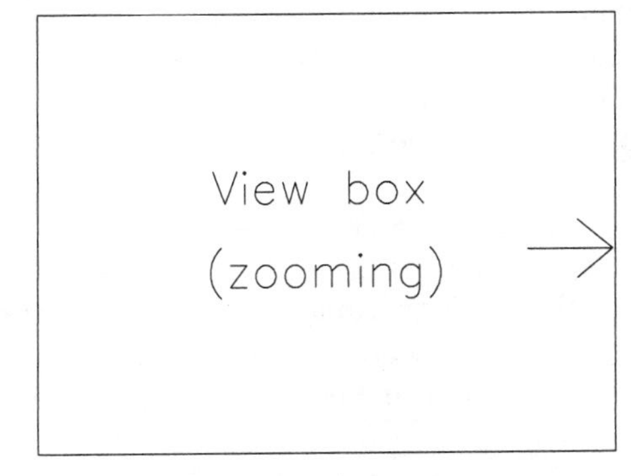

Figure 5-13 The zooming view box

complete the command and zoom into the desired area of the drawing. Before pressing ENTER, if you want to change the position of the zooming view box, click the pick button of your pointing device to make the panning view box reappear. After repositioning, press ENTER.

Previous Option

While working on a complex drawing, you may need to zoom in on a portion of the drawing to edit some minute details. Once the editing is over you may want to return to the previous view. This can be done using the **Previous** option of the **ZOOM** command. Without this option it would be very tedious to zoom back to previous views. AutoCAD LT saves the view specification of the current viewport whenever it is being altered by any of the ZOOM options or by the PAN, VIEW Restore, **DVIEW** or **PLAN** commands (which are discussed later).The prompt sequence for this option is:

Command: **ZOOM** Enter
Specify corner of window, enter a scale factor (nX or nXP), or
[All/Center/Dynamic/Extents/Previous/Scale/Window]<realtime>: **P** Enter

Successive **ZOOM, P** commands can restore upto ten previous views. **VIEW** here refers to the area of the drawing defined by its display extents. If you erase some objects and then issue a **ZOOM** Previous command, the previous view is restored, but the erased objects are not.

Window Option

This is the most commonly used option of the **ZOOM** command. It lets you specify the area you want to zoom in on, by letting you specify two opposite corners of a rectangular window. The center of the specified window becomes the center of the new display screen. The area inside the window is magnified or reduced in size to fill the display as completely as possible. The points can be specified either by selecting them with the help of the pointing device or by entering their coordinates. The prompt sequence is:

Command: **ZOOM** Enter
Specify corner of window, enter a scale factor (nX or nXP), or
[All/Center/Dynamic/Extents/Previous/Scale/Window]<realtime>: *Specify a point.*
Specify opposite corner: *Specify another point.*

Whenever the **ZOOM** command is invoked, the window method is one of two default options. This is illustrated by the previous prompt sequence, where you can specify the two corner points of the window without invoking any option of the **ZOOM** command. The **Window** option can also be used by entering **W**. In this case the prompt sequence is:

Command: **ZOOM** Enter
Specify corner of window, enter a scale factor (nX or nXP), or
[All/Center/Dynamic/Extents/Previous/Scale/Window]<realtime>: **W** Enter
Specify first corner: *Specify a point.*
Specify opposite corner: *Specify another point.*

Scale Option

The Scale option is used in the following ways.

Scale: Relative to Full View. This option of the **ZOOM** command lets you magnify or reduce the size of a drawing according to a scale factor, Figure 5-14. A scale factor equal to 1 displays an area equal in size to the area defined by the established limits. This may not display the entire drawing if the previous view was not centered on the limits or if you have drawn outside the limits. To get a magnification relative to the full view, you can enter any other number. For example, you can type 4 if you want the displayed image to be enlarged four times. If you want to decrease the magnification relative to the full view, you need to enter a number that is less than 1.

In Figure 5-15, the image size decreased because the scale factor is less than 1. In other words, the image size is half of the full view because the scale factor is 0.5. The prompt sequence is:

Command: **ZOOM** [Enter]
Specify corner of window, enter a scale factor (nX or nXP), or
[All/Center/Dynamic/Extents/Previous/Scale/Window]<realtime>: **S** [Enter]
Enter a scale factor (nX or nXP): **0.5** [Enter]

*Figure 5-14 Drawing before **ZOOM Scale** option*

*Figure 5-15 Drawing after **ZOOM Scale** option*

Scale: Relative to Current View. The second way to scale is with respect to the current view (Figure 5-16). In this case, instead of entering only a number, enter a number followed by an **X**. The scale is calculated with reference to the current view. For example, if you enter **0.25X**, each object in the drawing will be displayed at one-fourth (1/4) of its current size. The following example increases the display magnification by a factor of 2 relative to its current value (Figure 5-17):

Command: **ZOOM** [Enter]
Specify corner of window, enter a scale factor (nX or nXP), or
[All/Center/Dynamic/Extents/Previous/Scale/Window]<realtime>: **2X** [Enter]

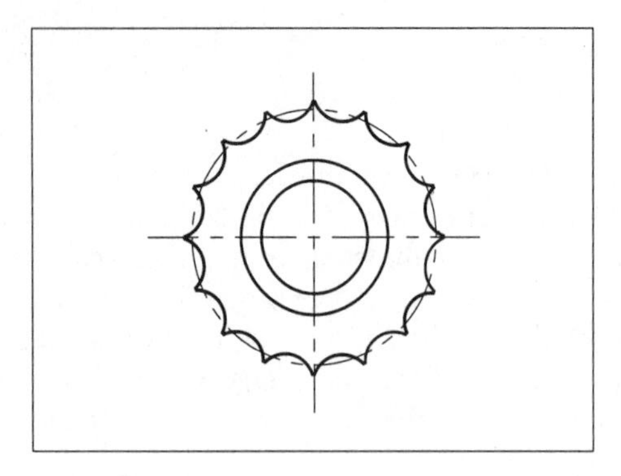

***Figure 5-16** Drawing before **ZOOM Scale**(X) option*

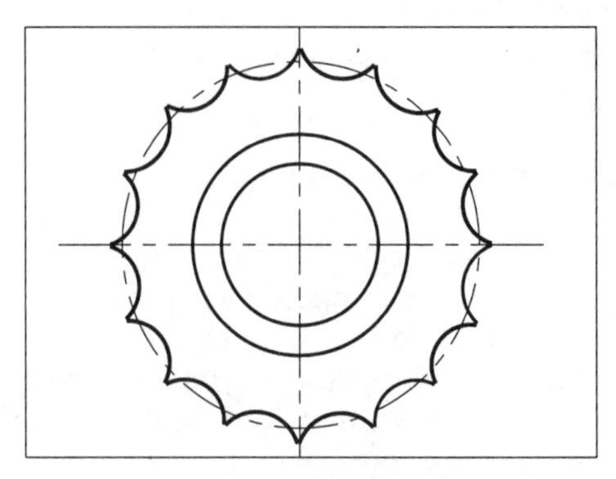

***Figure 5-17** Drawing after **ZOOM Scale** (X) option*

Scale: Relative to Paper Space Units. The third method of scaling is with respect to paper space. You can use paper space in a variety of ways and for various reasons. For example, you can array and plot various views of your model in paper space. To scale each view relative to paper space units, you can use the **ZOOM XP** option. Each view can have an individual scale. The drawing view can be at any scale of your choice in a model space viewport. For example, to display a model space at one-fourth (1/4) the size of the paper space units, the prompt sequence is:

Command: **ZOOM** Enter
Specify corner of window, enter a scale factor (nX or nXP), or
[All/Center/Dynamic/Extents/Previous/Scale/Window]<realtime>: **1/4XP** Enter

Note

For a better understanding of this topic, refer to "Model Space and Paper Space" in Chapter 22.

Zoom In and Out

You can also zoom into the drawing using the **In** option, which doubles the image size.

Similarly, you can use the **Out** option to decrease the size of the image by half. To invoke these options from the command line, enter **ZOOM 2X** for the **In** option or **ZOOM .5X** for the **Out** option at the Command prompt.

Exercise 1 — *Mechanical*

Draw Figure 5-18 according to the given dimensions. Use the **ZOOM** command to get a bigger view of the drawing. Do not dimension the drawing.

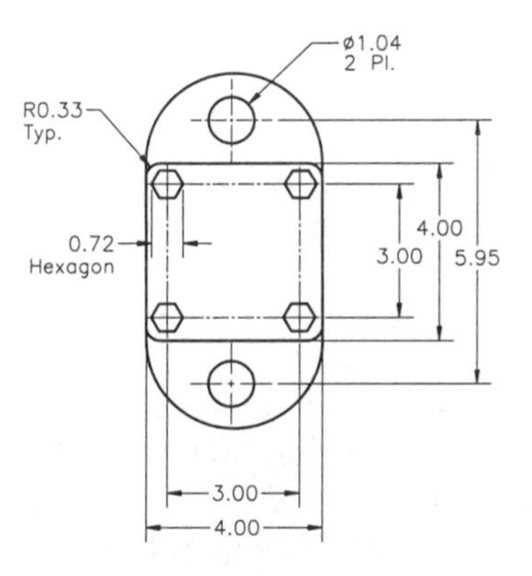

Figure 5-18 Drawing for Exercise 1

PAN REALTIME COMMAND

Toolbar:	Standard > Pan
Menu:	View > Pan
Command:	PAN

You may want to view or draw on a particular area outside the current viewport. You can do this using the **PAN** command. If done manually, this would be like holding one corner of the drawing and dragging it across the screen. The **PAN** command allows you to bring into view portions of the drawing that are outside the current viewport. This is done without changing the magnification of the drawing. The effect of this command can also be illustrated by imagining that you are looking at a big drawing through a window known as the **display window** that allows you to slide the drawing right, left, up, and down to bring the part you want to view inside this window. You can invoke the **PAN** command from the shortcut menu also.

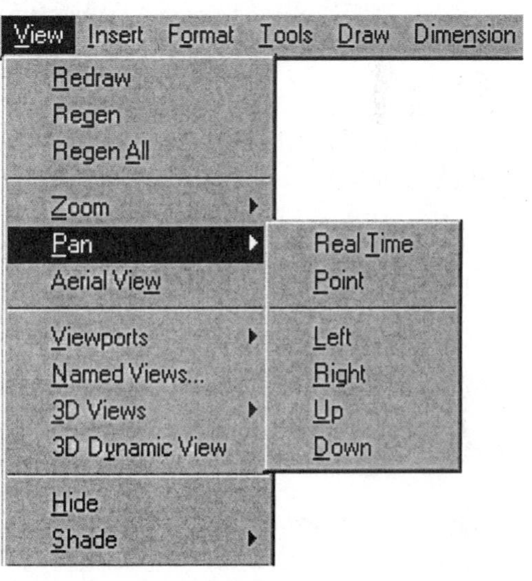

*Figure 5-19 Invoking the **PAN** command from the **View** menu*

Panning in Realtime

You can use the **Realtime PAN** to pan the drawing interactively. To pan a drawing, invoke the command and then hold the pick button down and move the cursor in any direction. When you select the realtime pan, AutoCAD LT displays an image of a hand indicating that you are in PAN mode.

Realtime pan is the default setting for the **PAN** command. Pressing ENTER after entering the **PAN** command automatically invokes the realtime pan. To exit the realtime pan, press ENTER, ESC, or choose **Exit** from the shortcut menu.

There are different menu items and buttons predefined for the **PAN** command, which can be used to pan the drawing in a particular direction. These items can be invoked from the menu.

Point. In this option you are required to specify the displacement. To do this you need to specify in what direction to move the drawing and by what distance. You can give the displacement either by entering the coordinates of the points or by specifying the coordinates by using a pointing device. The coordinates can be entered in two ways. One way is to specify a single coordinate pair. In this case, AutoCAD LT takes it as a relative displacement of the drawing with respect to the screen. For example, in the following case the **PAN** command would shift the displayed portion of the drawing 2 units to the right and 2 units up.

Command: *Specify the **Point** item from the **View** menu.*
Specify base point or displacement: **2,2** [Enter]
Specify second point: [Enter]

In the second case, you can specify two coordinate pairs. AutoCAD LT computes the displacement from the first point to the second. Here, displacement is calculated between point (3,3) and point (5,5).

Command: *Specify the Point item from the View menu.*
Specify base point or displacement: **3,3** [Enter] *(Or specify a point.)*
Specify second point: **5,5** [Enter] *(Or specify a point.)*

Left. Moves the drawing so that some of the left portion of the screen is brought into view.

Right. Moves the drawing so that some of the right portion of the screen is brought into view.

Up. Moves the drawing toward the bottom so that some of the top portion of the screen is brought into view.

Down. Moves the drawing toward the top so that some of the bottom portion of the screen is brought into view.

CREATING VIEWS

While working on a drawing, you may frequently be working with the **ZOOM** and **PAN** commands, and you may need to work on a particular drawing view (some portion of the drawing) more often than others. Instead of wasting time by recalling your zooms and pans and selecting up the same area from the screen over and over again, you can store the view under a name and restore the view using the name you have given it. The **View dialog box** (Figure 5-21) is used to save the current view under a name so that you can restore (display) it later. It does not save any drawing object data, only the view parameters needed to redisplay that portion of the drawing.

View Dialog Box

Toolbar:	Viewpoint > Named Views
	Standard > Named Views
Menu:	View > Named Views
Command:	View

***Figure 5-20** Invoking Named Views from the **View** toolbar*

You can save and restore the views from the **View** dialog box (Figure 5-21). This dialog box is very useful when you are saving and restoring many view names. With this dialog box you can name the current view or restore some other view. The list box shows all the existing named views. The following are the different options in the **View** dialog box:

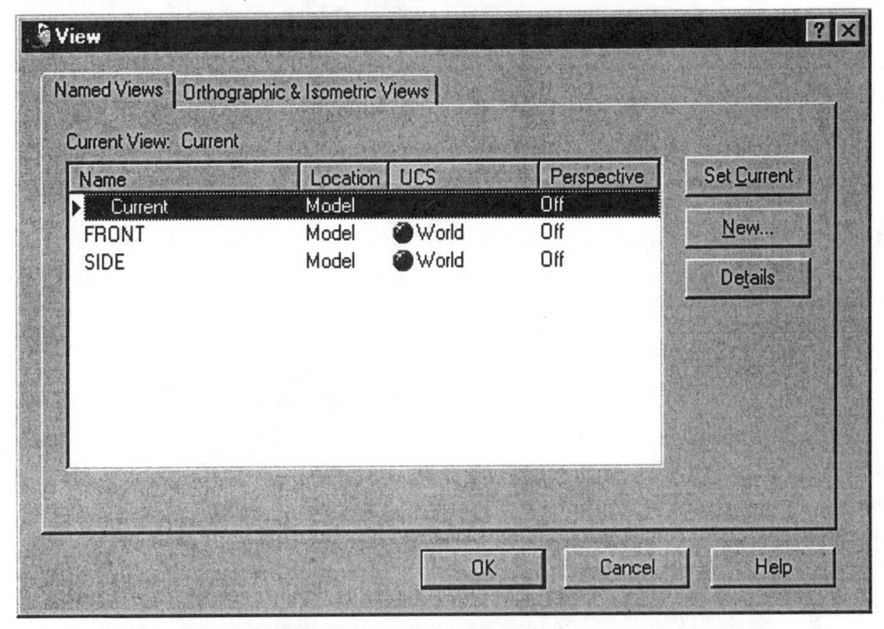

***Figure 5-21** **View** dialog box*

Current Views List Box

The Views list box displays a list of the named views in the drawing. The list appears with the names of all saved views and the space in which each was defined (as discussed in Chapter 21).

New

The **New** button allows you to create a new view and save it by giving it a name. When you choose the **New** button, the **New View** dialog box (Figure 5-22) is displayed. Enter a name for the view in the **View name** edit box. If you want to save the current view, then activate the **Current display** radio button. If you want to save a rectangular portion of the current drawing as a view (without first zooming in on that area), then choose the **Define window** radio button to activate it. You can specify two points on the screen to describe the window by choosing the **define view window** button or you can enter the X and Y coordinates in the **Specify first corner** and **Specify other corner** in the command lines. After specifying the view, choose the **OK** button to save the named view.

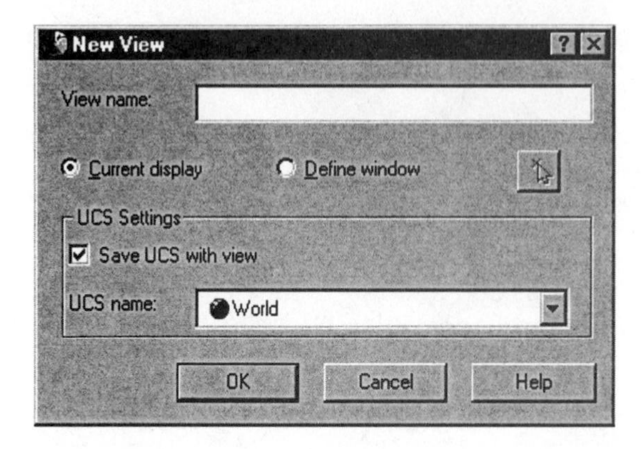

Figure 5-22 ***New View*** *dialog box*

Set Current

The **Set Current** button allows you to replace the current viewport by the view you specify AutoCAD LT uses the center point and magnification of each saved view and executes a **ZOOM Center** with this information when a view is restored.

Note

Refer to Chapter 22. Also, if ***TILEMODE*** *is on, you cannot restore a Paper space view.*

Details

You can also see the description of the general parameters of a view by selecting the particular view and then choosing the **Details** button (Figure 5-23).

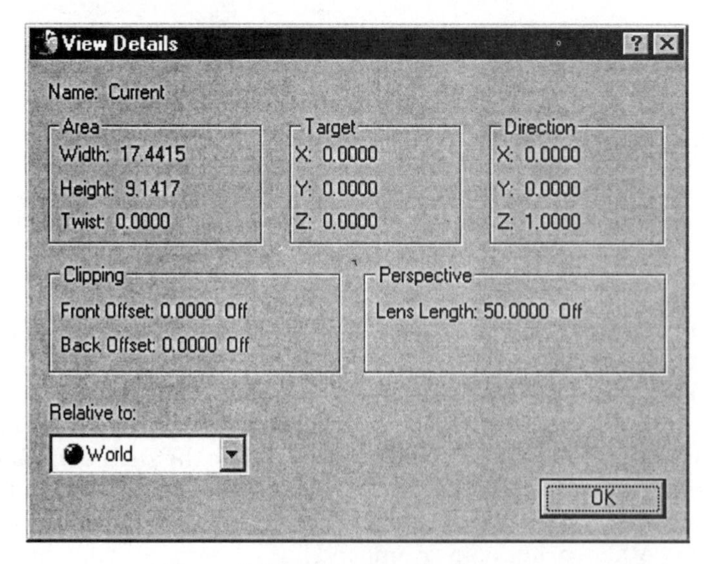

Figure 5-23 ***View Details*** *dialog box*

Using the Command Prompt

You can also use the **-VIEW** command to work with views at the Command: prompt.

Command: **-VIEW** [Enter]
Enter an option [?/Orthographic/Delete/Restore/Save/Ucs/Window]:

You can use the different options to save, restore, delete, or list the views. If you try to restore a Model space while working in Paper space, AutoCAD LT automatically switches to Model space, and vice versa. In this case AutoCAD LT will prompt you further as follows:

Enter view name to restore: *Select viewport for restoring*

You can select the viewport you want by selecting its border. That particular viewport must be on and active. The restored viewport also becomes the current one. You can use the Window option to define the view; the prompt sequence is:

Command: **-VIEW** [Enter]
Enter an option [?/Orthographic/Delete/Restore/Save/Ucs/Window]: **W** [Enter]
Enter view name to save: *Enter the name.*
Specify first corner: *Specify a point.*
Specify other corner: *Specify another point.*

AERIAL VIEW

Menu: View > Aerial View
Command: DSVIEWER

As a navigational tool, AutoCAD LT provides you with the option of opening another drawing display window along with the graphics screen window you are working on. This window, called the **Aerial View** window, can be used to view the entire drawing and select those portions you want to quickly zoom or pan. The AutoCAD LT graphics screen window resets itself to display the portion of the drawing you have selected for Zoom or Pan in the Aerial View window. You can keep the Aerial View window open as you work on the graphics screen, or minimize it so that it stays on the screen as an button which can be restored when required (Figure 5-24). As with all windows, you can resize or move it. The Aerial View window can also be invoked when you are in the midst of any command other than the **DVIEW** command.

The Aerial View window has two menus, View and Options. It also has a toolbar containing zoom in, zoom out, and zoom global options. The following is a description of the available options in the Aerial View:

Toolbar Buttons

The Aerial View window has three buttons: **Zoom In, Zoom Out,** and **Zoom Global** (Figure 5-25).

Zoom In. This option leads to the magnification by a factor of **2** which is centered on the current view box.

Zoom Out. This option leads to the reduction by **half** centered on the current view box.

Zoom Global. The Global option displays the complete drawing in the Aerial Viewport.

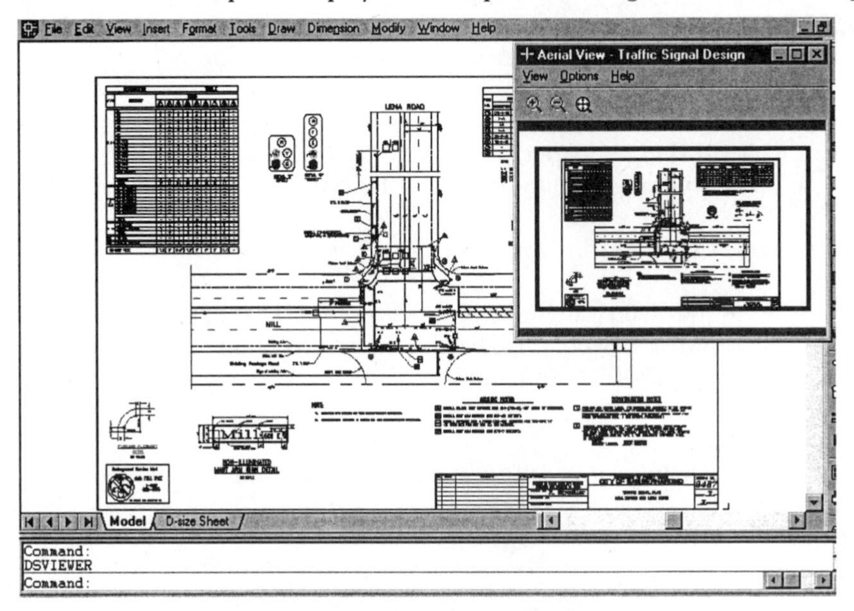

Figure 5-24 AutoCAD LT graphics screen with the Aerial View window

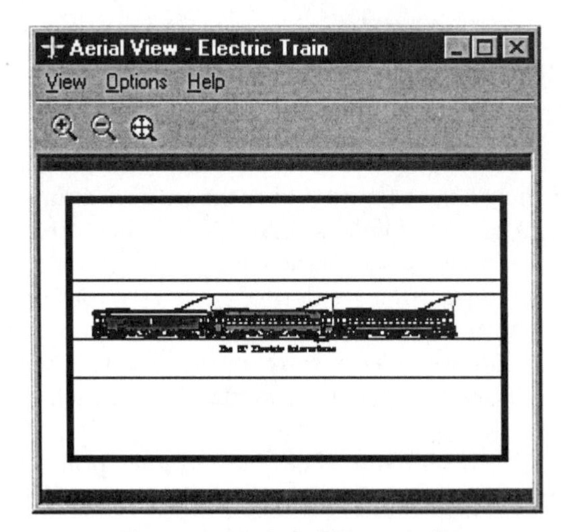

Figure 5-25 Aerial View window

Menus

The menus available in the menu title bar are **View** and **Options**. They are described below:

View. The **View** menu contains **Zoom In**, **Zoom Out**, and **Global options**. The **Zoom Out** and **Zoom In** options magnify or reduce the display, and the **Global option** acts like Zoom Extents in the Aerial View window.

Options. The **Options** menu has the following three options:

Auto Viewport. When you are working with viewports, you may need to change the view in the Aerial View window to display the current viewport view. You can achieve this by selecting the Auto Viewport option from the **Options** menu. If you are working in multiple viewports and you zoom or pan in the Aerial Viewport window, only the view in the current viewport is affected.

Dynamic Update. When you make any changes in the current drawing, the View box is updated simultaneously in the Aerial View window if you have selected the Dynamic Update option. If Dynamic Update is unselected, the drawing is not updated simultaneously. The drawing in the Aerial View window is updated only when you move the cursor in the Aerial View window or invoke any of the menus or the toolbar buttons.

Realtime Zoom. This option if on, updates the drawing area when you are zooming in Aerial view in realtime.

CREATING TEXT

In manual drafting, lettering is accomplished by hand using a lettering device, pen, or pencil. This is a very time-consuming and tedious job. Computer-aided drafting has made this process extremely simple. Engineering drawings invoke certain standards to be followed in connection with the placement of a text in a drawing. In this section, you will learn how text can be added in a drawing by using the **TEXT** and **MTEXT** commands. The following is a list of some of the functions associated with the **TEXT** and **MTEXT** commands that are described in this section:

1. Using the **TEXT** and **MTEXT** commands to create text.
2. Changing text styles using the **STYLE** command.
3. Drawing special symbols using the proper control characteristics.
4. Editing single line and paragraph text.

TEXT COMMAND

Menu:	Draw > Text > Single Line Text
Command:	TEXT

The **TEXT** command lets you write text on a drawing. This command allow you to delete what has been typed by using the BACKSPACE key. They also let you enter multiple lines in one command. The **TEXT** command display a line after you enter the start point, height, and the rotation angle. This line identifies the start point and the size of the text height entered. The characters appear on the screen as you enter them. When you press Enter after typing a line, the cursor automatically places itself at the start of the next line and repeats the prompt **Enter text:**. You can end the command by giving null response (pressing the ENTER key) to the **Enter text:** prompt. This command can be canceled by pressing ESC. If you do this, the entire text entered during this particular command sequence is erased.

The screen crosshairs can be moved irrespective of the cursor line for the text. If you specify a point, this command will complete the current line of the text and move the cursor line to the point you selected. This cursor line can be moved and placed anywhere on the screen; hence, multiple lines of text can be entered at any desired locations on the screen with a single **TEXT** command. By pressing BACKSPACE you can delete one character to the left of the current position of the cursor box. Even if you have entered several lines of the text, you can use Backspace and go on deleting until you reach the start point of the first line entered.

This command can be used with the text alignment modes, although it is most useful in the case of left-justified texts. Irrespective of the **Justify** option chosen, the text is first left-aligned at the selected point. After the **TEXT** command ends, the text is momentarily erased from the screen and regenerated with the requested alignment. For example, if you use the **Middle** option, the text will first appear on the screen as left-justified, starting at the point you designated as the middle point. After you end the command and press Enter, it is regenerated with the proper alignment.

Command: **TEXT** [Enter]
Specify start point of text or [Justify/Style]: *Specify the start point.*
Specify height <0.25>: **0.15** [Enter]
Specify rotation angle of text <0>: [Enter]
Enter text: *Enter first line of the text.*
Enter text: *Enter second line of the text.*
Enter text: [Enter] *Press ENTER to terminate the text command.*

Start Point Option

This is the default and the most commonly used option in the **TEXT** command. By specifying a start point, the text is left-justified along its baseline starting from the location of the starting point. Specifying a start point does let you **left-justify** your text; however, before AutoCAD LT draws the text, it needs some more parameters. AutoCAD LT must know the text height, the rotation angle for the baseline, and the text string to be drawn. It prompts you for all this information. The prompt sequence is:

Specify height <default>:
Specify rotation angle of text <default>
Enter text:

The **Specify height:** prompt determines the distance by which the text extends above the baseline, measured by the capital letters. This distance is specified in drawing units. You can specify the text height by specifying two points or entering a value. In the case of a null response, the default height, that is, the height used for the previous text drawn in the same style will be used.

Next comes the **Specify rotation angle of text:** prompt. It determines the angle at which the text line will be drawn. The default value of the rotation angle is **0 degrees** (3 o'clock, or east), and in this case the text is drawn horizontally from the specified start point. The rotation angle is measured in a counterclockwise direction. The last angle specified becomes the current rotation angle, and if you give a null response the last angle specified will be used as default. You can also specify the rotation angle by specifying a point. The text is drawn upside down if a point is specified at a location to the left of the start point.

As a response to the **Enter text:** prompt, enter the text string. Spaces are allowed between words. After entering the text, press ENTER. The following example illustrates these prompts:

Command: **TEXT** [Enter]
Specify start point of text or [Justify/Style]: **1,1** [Enter]
Specify height <0.25>:**0.15** [Enter]
Specify rotation angle of text <0>: [Enter]
Enter text: *Enter the text string.*

Justify Option

AutoCAD LT offers different options to align the text. **Alignment** refers to the layout of the text. The main text alignment modes are **left, center,** and **right**. You can align a text using a combination of modes; e.g., top/middle/baseline/bottom and left/center/right (Figure 5-26). **Top** refers to the line along which lie the top points of the capital letters; **Baseline** refers to the line along which lie their bases. Letters with descenders (such as p, g, y) dip below the baseline to the bottom.

When the **Justify** option is invoked, the user can place text in one of the fourteen various alignment types by selecting the desired alignment option. The orientation of the text style determines the command interaction for Text Justify. (Text styles and fonts are discussed later in this chapter). For now, assume that the text style orientation is horizontal. The prompt sequence using this option of the **TEXT** command is:

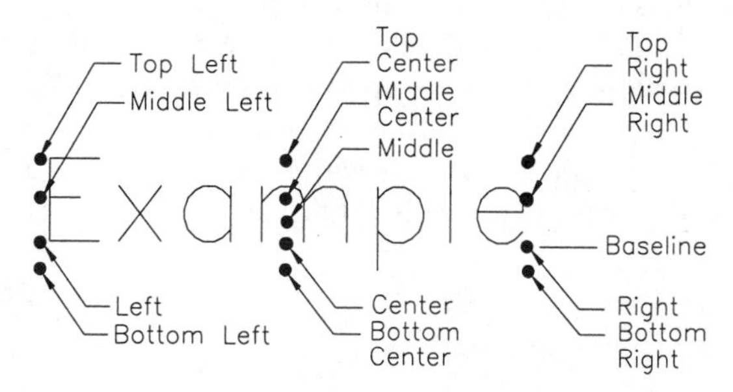

Figure 5-26 *Text alignment positions*

Command: **TEXT** [Enter]
Specify start point of text or [Justify/style]: **J** [Enter]
Enter an option [Align/Fit/Center/Middle/Right/TL/TC/TR/ML/MC/MR/BL/BC/ BR]: *Select any of these options.*

If the text style is vertically oriented (refer to the "**STYLE** Command" section later in this chapter), the prompt sequence is as follows:

Command: **TEXT** [Enter]
Specify start point of text or [Justify/Style]: **J** [Enter]
Enter an option [Align/Center/Middle/Right]:

If you know what justification you want, you can enter it directly at the **Specify start point of text or [Justify/Style]:** prompt instead of first entering J to display the justification prompt. If you need to specify a style as well as a justification, you must specify the style first. The available text justification options with alignment, the corresponding abbreviations, and the text style orientations (horizontal or vertical) are as follows:

Alignment	**Abbreviation**	**Orientation**
Align	A	Horizontal/Vertical
Fit	F	Horizontal
Center	C	Horizontal/Vertical
Middle	M	Horizontal/Vertical
Right	R	Horizontal/Vertical
Top/left	TL or Tleft	Horizontal
Top/center	TC or Tcenter	Horizontal
Top/right	TR or Tright	Horizontal
Middle/left	ML or Mleft	Horizontal
Middle/center	MC or Mcenter	Horizontal
Middle/right	MR or Mright	Horizontal
Bottom/left	BL or Bleft	Horizontal
Bottom/center	BC or Bcenter	Horizontal
Bottom/right	BR or Bright	Horizontal

Align Option. This option is invoked under the Justify option. Here, enter A at the **Enter an option[Align/Fit/Center/Middle/Right/TL/TC/TR/ML/MC/MR/BL/BC/BR]:** prompt. In this option the text string is written between two points (Figure 5-27). You must specify the two points that act as the endpoints of the baseline. The two points may be specified horizontally or at an angle. AutoCAD LT adjusts the text width (compresses or expands) so that it fits between the two points. The text height is also changed, depending on the distance between

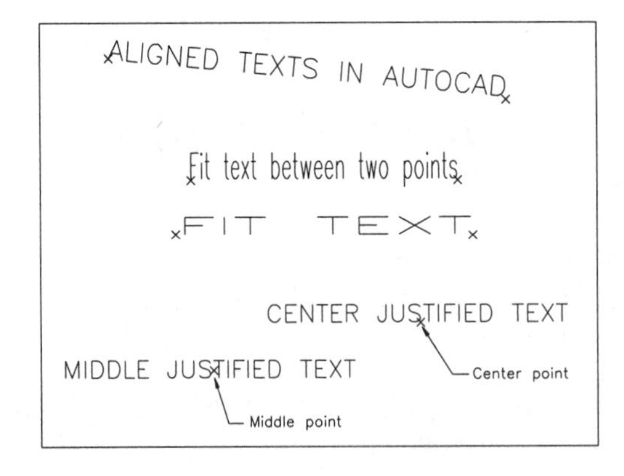

Figure 5-27 *Writing the text using* ***Align, Fit, Center,*** *and* ***Middle*** *options*

points and number of letters.

> Command: **TEXT** Enter
> Specify start point of text or [Justify/Style]: **J** Enter
> Enter an option [Align/Fit/Center/Middle/Right/TL/TC/TR/ML/MC/MR/BL/BC/BR]: **A** Enter
> Specify first endpoint of text baseline: *Specify a point.*
> Specify second endpoint of text baseline: *Specify a point.*
> Enter text: *Enter the text string.*

Fit Option. This option is very similar to the previous one. The only difference is that in this case you select the text height, and it does not vary according to the distance between the two points. AutoCAD LT adjusts the letter width to fit the text between the two given points, but the height remains constant (Figure 5-27). The Fit option is not accessible for vertically oriented text styles. If you try the Fit option on the vertical text style, you will notice that the text string does not appear in the prompt. The prompt sequence is:

> Command: **TEXT** Enter
> Specify start point of text or[Justify/Style]: **J** Enter
> Enter an option [Align/Fit/Center/Middle/Right/TL/TC/TR/ML/MC/MR/BL/BC/BR]: **F** Enter
> Specify first endpoint of text baseline: *Specify a point.*
> Specify second endpoint of text baseline: *Specify a point.*
> Specify height<current>: *Enter the height.*
> Enter text: *Enter the text.*

Note

*You do not need to select the **Justify** option (J) for selecting the text justification. You can enter the text justification by directly entering justification when AutoCAD LT prompts "Specify start point of text or [Justify/Style]:"*

Center Option. You can use this option to select the midpoint of the baseline for the text. This option can be invoked by entering **Justify** and then **Center** or **C**. After you select or specify the center point, you must enter the letter height and the rotation angle (Figure 5-27). The prompt sequence is:

> Command: **TEXT** Enter
> Specify start point of text or [Justify/Style]: **C** Enter
> Specify center point of text: *Specify a point.*
> Specify height<current>: **0.15** Enter
> Specify rotation angle of text<0>: Enter
> Enter text: **CENTER JUSTIFIED TEXT** Enter

Middle Option. Using this option you can center text not only horizontally, as with the previous option, but also vertically. In other words, you can specify the middle point of the text string (Figure 5-27). You can alter the text height and the angle of rotation to your requirement. The prompt sequence is:

Command: **TEXT** [Enter]
Specify start point of text or[Justify/Style]: **M** [Enter]
Specify middle point of text : *Specify a point.*
Specify height<current>: **0.15** [Enter]
Specify rotation angle of text<0>: [Enter]
Enter text: **MIDDLE JUSTIFIED TEXT** [Enter]

Right Option. This option is similar to the default left-justified Start point option. The only difference is that the text string is aligned with the lower right corner (the endpoint you specify); i.e., the text is **right-justified** (Figure 5-28). The prompt sequence is:

Command:**TEXT** [Enter]
Specify start point of text or [Justify/Style]: **R** [Enter]
Specify right endpoint of text baseline: *Specify a point.*
Specify height<current>: **0.15** [Enter]
Specify rotation angle of text<0>: [Enter]
Enter text: **RIGHT JUSTIFIED TEXT** [Enter]

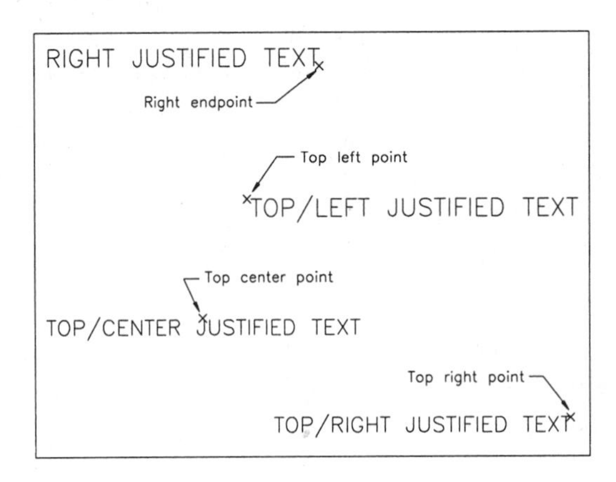

***Figure 5-28** Writing text using the Right, Top-Left, Top-Center, and Top-Right options*

TL Option. The full form of this abbreviation has already been written. In this option the text string is justified from the **top left** (Figure 5-28). The command prompt sequence is:

Command: **TEXT** [Enter]
Specify start point of text or[Justify/Style]: **TL** [Enter]
Specify top-left point of text : *Specify a point.*
Specify height<current>: **0.15** [Enter]
Specify rotation angle of text <0>: [Enter]
Enter text: **TOP/LEFT JUSTIFIED TEXT** [Enter]

Note

The rest of the text alignment options are similar to those just discussed, and you can try them on your own. The prompt sequence is almost the same as those given for the previous examples.

Style Option

With this option you can specify another existing text style. Different text styles can have different text fonts, heights, obliquing angles, and other features. This option can be invoked by entering TEXT and then S at the next prompt. The prompt sequence is:

Command: **TEXT** Enter
Specify start point of text or [Justify/Style]: **S** Enter
Enter style name or [?] <current>:

If you want to work in the previous text style, just press ENTER at the last prompt. If you want to activate another text style, enter the name of the style at the last prompt. You can also choose from a list of available text styles, which can be displayed by entering ?. After you enter ?, the next prompt is:

Enter text style(s) to list <*>:

Press ENTER to display the available text style names and the details of the current styles and commands in the **AutoCAD LT Text Window**. The prompt sequence in the AutoCAD LT Text Window is as follows:

Specify start point of text or [Justify/Style]: S Enter
Enter style name or [?] <current>: *Enter a style name and press F2 to return to the screen.*

Note

With the help of the **Style** *option of the* **TEXT** *command, you can select a text style from an existing list. If you want to create a new style, make use of the* **STYLE** *command, which is explained later in this chapter.*

DRAWING SPECIAL CHARACTERS

In almost all drafting applications, you need to draw **special characters** (symbols) in the normal text and in the dimension text. For example, you may want to draw the degree symbol (°) or the diameter symbol (ø), or you may want to underscore or overscore some text. This can be achieved with the appropriate sequence of control characters (control code). For each symbol, the control sequence starts with a percent sign written twice (%%). The character immediately following the double percent sign depicts the symbol. The control sequences for some of the symbols are:

Control sequence	Special character
%%c	Diameter symbol (ø)
%%d	Degree symbol (°)
%%p	Plus/minus tolerance symbol (ñ)
%%o	Toggle for overscore mode on/off
%%u	Toggle for underscore mode on/off
%%%	Single percent sign (%)

For example, if you want to draw **25° Celsius**, you need to enter 25%%dCelsius. If you enter

43.0%%c, you get **43.0ø** on the drawing screen. To underscore (underline) text, use the **%%u** control sequence followed by the text to be underscored. For example, to underscore the text: **UNDERSCORED TEXT IN AUTOCAD LT**, enter **%%uUNDERSCORED TEXT IN AUTOCAD LT** at the prompt asking for text to be entered. To underscore and overscore a text string, include **%%u%%o** at the text string.

Note

*The special characters %%O and %%U act as toggles. For example, if you enter "**This %%Utoggles%%U the underscore**", the word **toggles** will be underscored (toggles).*

None of these codes will be translated in the **TEXT** command until this command is complete. For example, to draw the degree symbol, you can enter **%%d.** As you are entering these symbols, they will appear as %%d on the screen. After you have completed the command and pressed ENTER, the code %%d will be replaced by the degree symbol (ø).

You may be wondering why a percent sign should have a control sequence when a percent sign can easily be entered at the keyboard by pressing the percent (%) key. The reason is that sometimes a percent symbol is immediately followed (without a space) by a control sequence. In this case, the **%%%** control sequence is needed to draw a single percent symbol. To make the concept clear, assume you want to draw **67%±3.5.** Try drawing this text string by entering **67%%%p3.5.** The result will be **67%p3.5,** which is wrong. Now enter **67%%%%%p3.5** and notice the result on the screen. Here you obtain the correct text string on the screen, i.e., **67%±3.5**. If there were a space between the 67% and ±3.5, you could enter 67% a%%p3.5 and the result would be **67% ±3.5.** Figure 5-29 illustrates some special characters.

In addition to the control sequences shown earlier, you can use the %%nnn control sequence to draw special characters. The nnn can take a value in the range of 1 to 126. For example, to draw the & symbol, enter the text string **%%038**.

Exercise 2 *General*

Draw the text on the screen as shown in Figure 5-30. Use the text justification options shown in the drawing. The text height is 0.25 units.

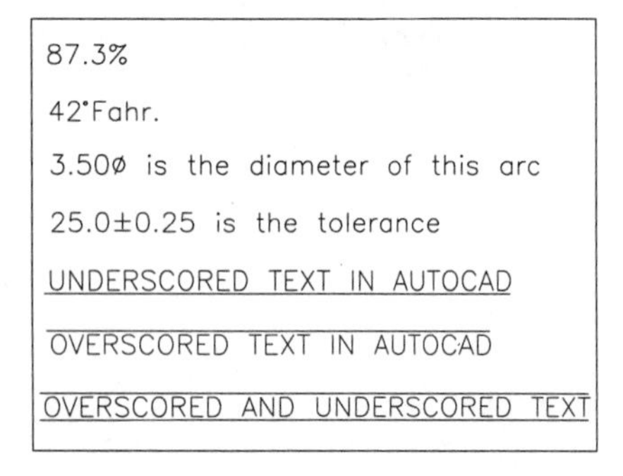

***Figure 5-29** Special characters*

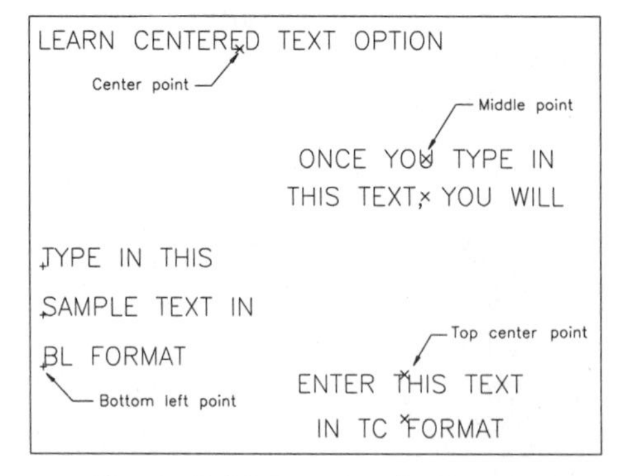

***Figure 5-30** Drawing for Exercise 2*

CREATING PARAGRAPH TEXT (MTEXT COMMAND)

Toolbar:	Draw > Multiline Text
Menu:	Draw > Text > Multiline Text
Command:	MTEXT

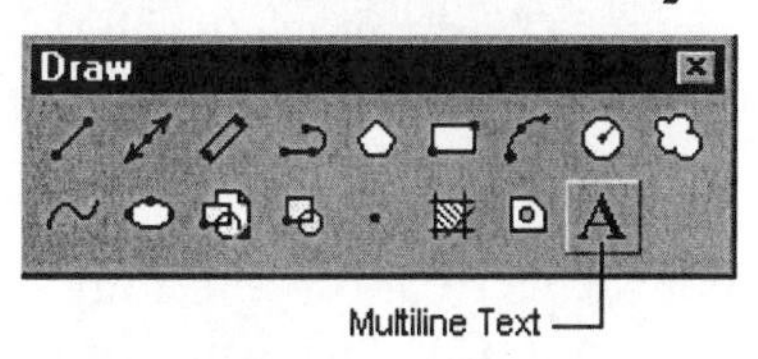

Figure 5-31 Invoking ***Multiline Text*** *from the* ***Draw*** *toolbar*

You can use the **MTEXT** command to write a paragraph text whose width can be specified by defining two corners of the text boundary or by entering a width, using coordinate entry. The text created by the **MTEXT** command is a single object regardless of the number of lines it contains. After specifying the width, you have to enter the text in the **Multiline Text Editor** dialog box (Figure 5-32).The following is the prompt sequence:

Command: **MTEXT** [Enter]
Current text style: STANDARD. Text height: 0.2000
Specify first corner: *Select a point to specify first corner.*
Specify opposite corner or [Height/Justify/Line spacing/Rotation/Style/Width]: *Select an option or select a point to specify other corner.*

After selecting the first corner, or you can drag the pointing device so that a box that shows the location and size of the paragraph text is formed. An arrow is displayed within the boundary, which indicates the direction of the text flow. When you define the text boundary, it does not mean that the text paragraph will fit within the defined boundary. AutoCAD LT only uses the width of the defined boundary as the width of the text paragraph. The height of the text boundary has no effect on the text paragraph. Once you have defined the boundary of the paragraph text, AutoCAD LT will display the **Multiline Text Editor** dialog box (Figure 5-32).

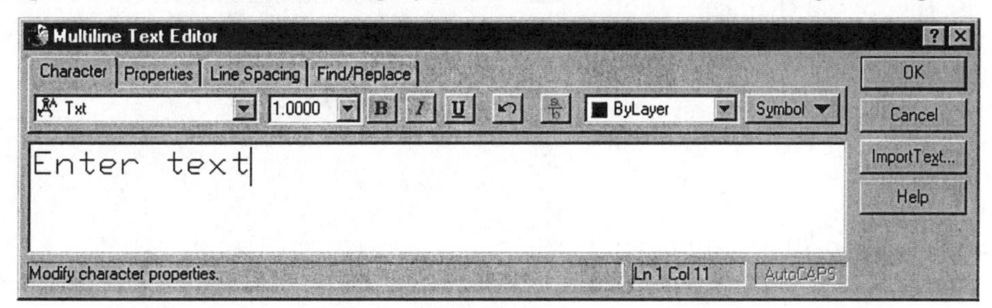

Figure 5-32 Multiline Text Editor *dialog box (Character tab)*

In addition to a Text Box, **Multiline Text Editor** dialog box has four tabs: **Character, Properties, Line Spacing,** and **Find/Replace**. Following is the description of the tabs and options available in each tab.

Text Box

The Text Box area displays the text you enter. The width of the active text area is determined by the specified width of paragraph text. The initial height of the text box is approximately equal to three lines of text. You can increase the size of the dialog box by dragging the edges of the box. You can also use the scroll bar to move up or down to display the text.

Character Tab

From the **Character** tab (default) you can use the following options that you can apply to the text.

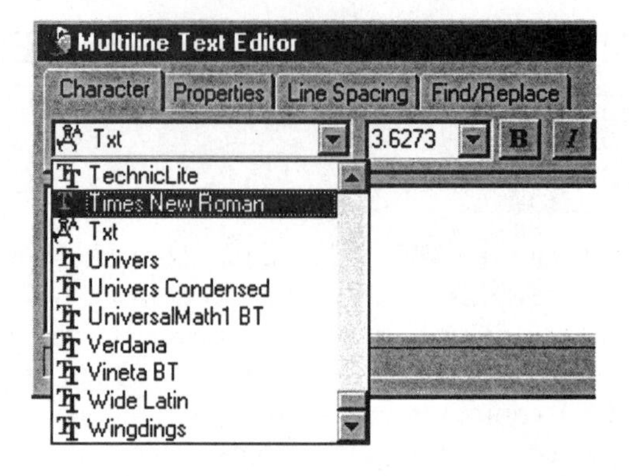

Figure 5-33 Font drop-down list

Font. If you choose the font drop-down arrow, AutoCAD LT will display a list of different fonts available (Figure 5-33). You can select any font to apply to the text.

Font Height. You can enter the new value of the text height in the Text Height edit box to change the text height of the entire paragraph text. Once you specify the height, AutoCAD LT retains that value unless you change it. The **MTEXT** height does not affect the size (**TEXTSIZE** system variable) specified for **TEXT** command.

Bold, Italic, Underline. You can use the appropriate tool icons to make the selected text boldface, italics, or create underlined text. Boldface and italics are not supported by SHX fonts.

Text Color. You can choose the color for the text from the Text color list box. You can also choose the color from the **Select Color** dialog box which is displayed by selecting **Other** in the list box.

Stacking Text. To stack the text you must use the special characters / and ^. The character / stacks the text above a line and the character ^ stacks the text without a line. After you enter the text with the special character and select the text, the stacking button is highlighted. The stacked text is displayed equal to 70 percent of the actual height. If you enter two numbers separated by / or ^, AutoCAD LT displays the **AutoStack Properties** dialog box (Figure 5-34). You can use this dialog box to control the stacking properties.

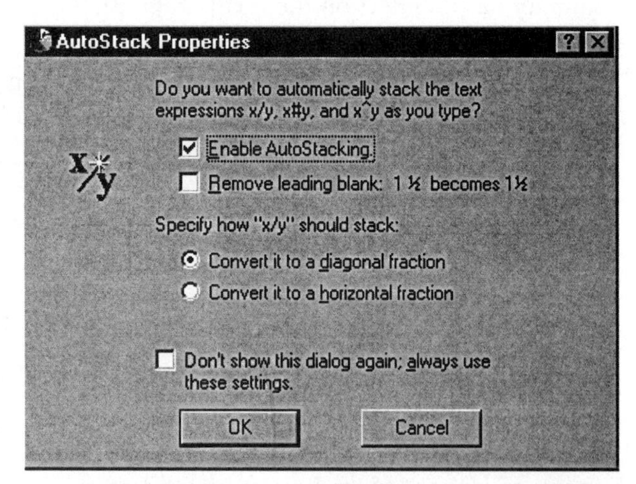

Figure 5-34 ***AutoStack Properties*** *dialog box*

Symbol. You can select the special characters from the Symbol menu that you can insert in the text. You can select Others to display Character Map of other special characters. To insert these characters, select the characters you want to copy and then choose the **Select** button. Once you are done selecting the special characters, choose the **Copy** button and then close the dialog box. In the text box of the Multiline Text Editor, position the cursor where you want to insert the special characters and right-click to display the shortcut menu. Select **Paste** to insert the selected special characters in the MultiLine Text Editor.

Properties Tab

From the **Properties** tab (Figure 5-35) you can use the following options that can be applied to the text.

***Figure 5-35 Multiline Text Editor** dialog box (Properties tab)*

Style. You can select the drop-down arrow to display the predefined styles and select a new text style. The text style that you want to use must be predefined.

Justification. You can choose the drop-down arrow to display the predefined text justifications and then choose a new justification from the list box. The justification is used to control the justification of the text paragraph. For example, if the text justification is bottom-right (BR), the text paragraph will spill to the left and above the insertion point, regardless of how you define the width of the paragraph. Figure 5-36 shows different text justifications for **MTEXT**. You can also choose the justification options from the command line.

Command: **-MTEXT**
Current text style: STANDARD. Text height: 0.2000
Specify first corner: *Select a point to specify first corner.*
Specify opposite corner or [Height/Justify/Line spacing/Rotation/Style/Width]: **J**
Enter justification [TL/TC/TR/ML/MC/MR/BL/BC/BR]<current>: *Select an option.*

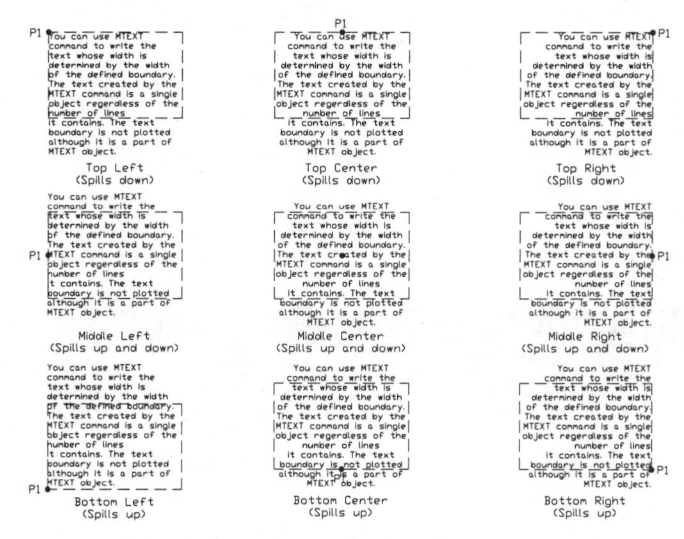

***Figure 5-36** Text justifications for **MTEXT**. P1 is the text insertion point.*

Width. You can enter a new width for the paragraph text or select the width from the previously defined values. The (no wrap) option is the same as entering a width of 0.

Rotation. You can choose a rotation for the text from the drop-down list box. The angles are in 15 degree increments. The rotation angle specifies the rotation of the text. For example, if the rotation angle is 15 degrees, the text will rotate 15 degrees counterclockwise.

Line Spacing Tab

From the **Line Spacing** tab (Figure 5-37) you can use the following options that you can apply to change the text.

Figure 5-37 ***Multiline Text Editor*** *dialog box (Line Spacing tab)*

Line Spacing. This has two options. **Atleast** is the default setting in AutoCAD LT where spacing is based upon the height of the largest character in the line, while the **Exactly** option forces the line spacing for all the lines to be uniform, Figure 5-38. This can be used effectively for making the tables.

Single[1.0x]. This option determines the vertical distance between the two lines in multiline text, Figure 5-38. Single gap is the default setting. Single line spacing is equal to 1.66 times the height of the text. The available options include Single (1x), 1.5 lines (1.5x), and Double (2x). You can also enter an absolute value for line spacing in the range of 0.0833 (0.25x) to 1.333 (4x) by entering a value in the edit box.

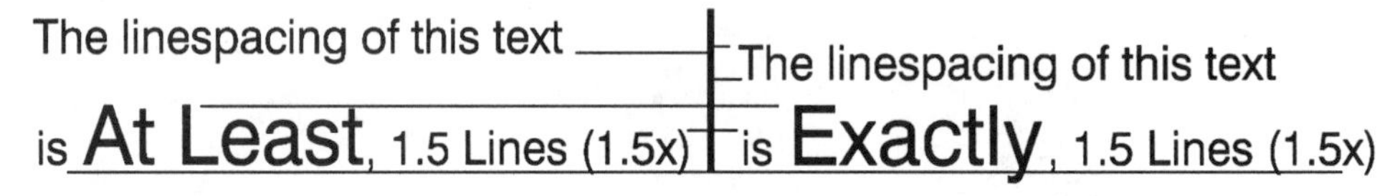

Figure 5-38 Specifying linespacing in multiline text

Find/Replace Tab

From the **Find/Replace** tab (Figure 5-39) you can use the following options that you can apply to the text.

Figure 5-39 ***Multiline Text Editor*** *dialog box (Find/Replace tab)*

Find. In this edit box, you can define the text string to be searched.

Replace With. If you want some text to be replaced, enter new text for replacement in this edit box.

Replace button. Select this icon to replace the highlighted text with the text in the **Replace With** edit box.

Match Case. Activate this check box if you want to find the text, only if all characters in text object are identical to the text in the Find box.

Whole Word. Activate this check box if you want to match the text in the Find box, only if it is a single word.

Import Text

When you choose this option, AutoCAD LT displays the **Open** dialog box. In this dialog box you can select any text file you want to import in the **Multiline Text Editor** dialog box. The imported text is displayed in the text area. Note that only ASCII or RTF files are interpreted properly.

Example 1

In this example you will use the **Multiline Text Editor** dialog box to write the following text on the screen.

For long, complex entries, create multiline text using the MTEXT option. The angle is 10 degrees. Diameter = 1/2" and Length = 32 1/2".

The font of the text is **Swis721 BT,** text height is 0.20, color Red, and written at an angle of 10 degrees with Middle-Left justification. Make the word "multiline" bold, underline the text "multiline text", and make the word "angle" italic. The line spacing type and line spacing between the lines are AtLeast and 1.5x respectively. Use the symbol for degrees and replace the word "option" with "command".

1. The first step is to enter the **MTEXT** command at the command prompt. You can also invoke this command from the **Draw** menu or the **Draw** toolbar. After invoking the command, specify two corners on the screen to define the paragraph text boundary. The prompt sequence is :

   Command: **MTEXT** Enter
   Current text style: STANDARD. Text height: 0.1000
   Specify first corner: *Select a point to specify first corner.*
   Specify opposite corner or [Height/Justify/Rotation/ Style/ Width]: *Select a point to specify other corner.*
   The **Multiline Text Editor** dialog box is displayed with the Character tab activated (default).

2. Choose the arrow in the **Font** list box to display the drop-down list. Select **Swis721 BT** True Type font from the list.

3. Enter 0.20 in the **Font height** edit box.

4. Choose the **Color** icon to display the list box and select Red from the list.

5. Now enter the text in the Text Editor (Figure 5-40) and to put the degrees symbol, choose the Symbol to display the menu and select **Degrees.** When you type 1/2 and then press the SPACEBAR, AutoCAD LT displays the **AutoStack Properties** dialog box. Select **Convert it to a diagonal fraction**, if it is not already selected, and then choose **OK**.

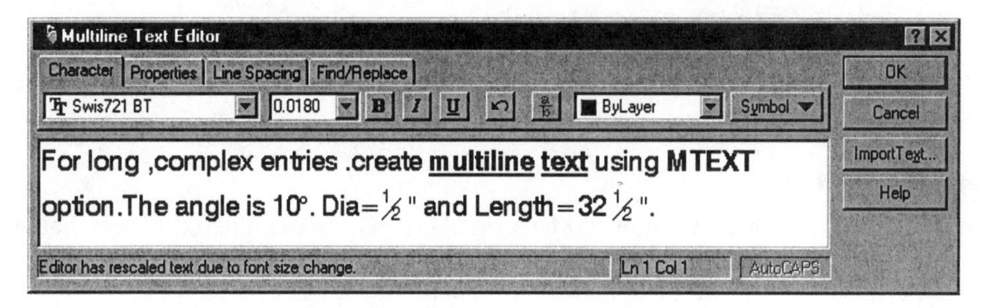

***Figure 5-40 Multiline Text Editor** dialog box (**Character** tab)*

6. Double-click on the word "multiline" to select it and then choose the **Underline** button.

7. Similarly, highlight the word "text" to select it and then again select the **Underline** button to underline it.

8. Now highlight on the word "multiline" (or pick and drag to select the text) and then choose the **Bold** button to make it boldface.

9. Highlight on the word "angle" (or pick and drag to select the text) and then choose the **Italic** button, Figure 5-40.

10. Select the **Properties** tab to display its options.

11. From the **Justification** drop-down list choose **ML**, and from the **Rotation** drop-down list choose **10**.

12. Select the **Line Spacing** tab to change the line spacing.

13. In the **line spacing type** drop-down list choose the **AtLeast** option and in the **spacing** drop-down list choose the **1.5Lines(1.5x)** option.

14. Now select the **Find/Replace** tab (Figure 5-41) to display its different options.

15. In the **Find** edit box, enter **option** and in the **Replace** edit box, enter **command**.

16. Choose the **Find** button. AutoCAD LT finds the word "option", Figure 5-41. Choose the **Replace** button to replace **option** by **command**.

17. Choose the **OK** button to exit the dialog box and the text is displayed in the selected width on the screen, Figure 5-42.

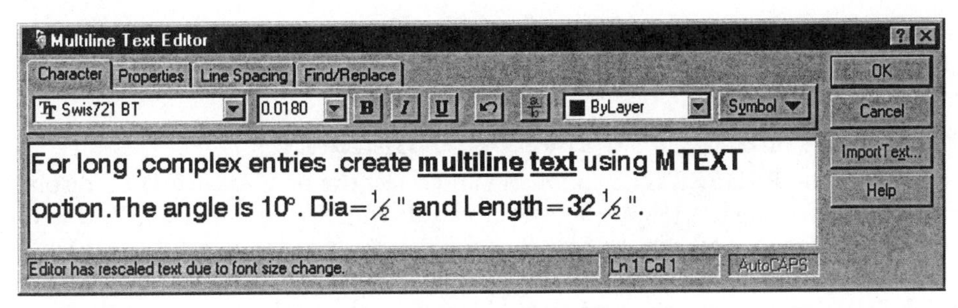

Figure 5-41 ***Multiline Text Editor** dialog box (Find/Replace tab)*

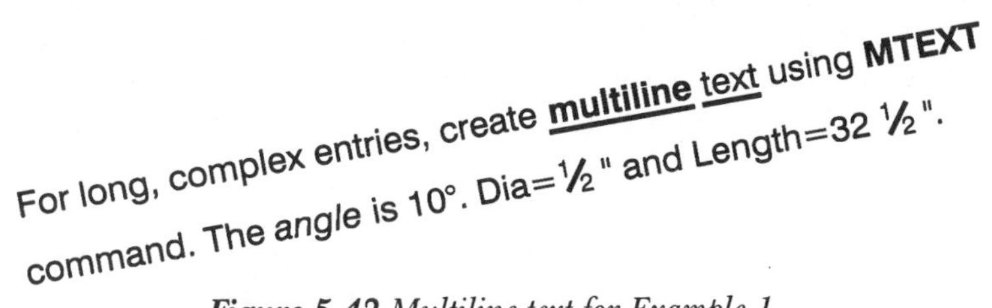

Figure 5-42 *Multiline text for Example 1*

EDITING TEXT

The contents of **MTEXT**, **TEXT**, and **DTEXT** objects can be edited by using the **DDEDIT** and **PROPERTIES** commands. You can also use the AutoCAD LT editing commands, like **MOVE**, **ERASE**, **ROTATE**, **COPY**, **MIRROR**, and **GRIPS** with any text object.

Editing Text Using DDEDIT Command

Toolbar:	Modify II > Edit Text
Menu:	Modify > Object > Text
Command:	DDEDIT

You can use the **DDEDIT** command to edit the text. If you select **TEXT** object, AutoCAD LT displays the **Edit Text** dialog box (Figure 5-43), where the selected text is displayed in the **Text:** edit box. You can make the changes and then choose the **OK** button. If you select the paragraph text, AutoCAD LT displays the text in the **Multiline Text Editor** dialog box. You can make the changes using the different options in the dialog box. The text in the dialog box can be selected by double-clicking on the word to select a word, by holding

Figure 5-43 *Using the **Edit Text** dialog box to edit text*

down the pick button of the pointing device and then dragging the cursor, or by triple-clicking on the text to select the entire line or paragraph.

Command: **DDEDIT** [Enter]
Select an annotation object or [Undo]: *Select a text object.*

Editing Text Using Properties Command

When you enter the **PROPERTIES** command and select the text, AutoCAD LT displays the **Object Properties** dialog box (Figure 5-44). You can also invoke the **PROPERTIES** command by choosing **Properties** button in the **Standard** toolbar. You need to select the text which you want to edit. To select the text, choose **Quick Select** in the **Object Properties** dialog box. In the **Quick Select** dialog box, choose **Select objects** and then select the text. Choose **OK** to return to **Object Properties** dialog box. Move the slider bar down till you see Text and Contents. If the text is a **TEXT** object, you can make the changes in the **Contents** edit box If you are editing multiline text, you must choose the **Full editor** button in the **Content** edit box, AutoCAD LT automatically switches to the **Multiline Text Editor** dialog box, where you can make changes to the paragraph text.

Note

*The **PROPERTIES** command has replaced the **DDMODIFY** command. If you invoke the **DDMODIFY** command, AutoCAD LT displays the **Object Properties** dialog box. You can also edit multiline text by using the **MTPROP** command. After entering the command, you are required to select the Mtext for editing. Once you select the Mtext, the **Multiline Text Editor** dialog box is displayed. This command cannot be used to edit the text created by the **TEXT** command.*

Properties - Drawing1

MText

Alphabetic | Categorized

General	
Color	ByLayer
Layer	0
Linetype	ByLayer
Linetype scale	1.0000
Plot style	ByColor
Lineweight	ByLayer
Hyperlink	
Text	
Contents	autocadlt 2000
Style	Standard
Justify	Top left
Direction	By style
Width	13.7174

Specifies the text string of the Mtext

*Figure 5-44 **Properties** dialog box*

Using Shortcut Menu

You can edit the text by using the **Shortcut** menu. Select the text object and right-click in the drawing area, AutoCAD LT displays the **Shortcut** menu, Figure 5-45. From the **Shortcut** menu select **Text Edit** or **Mtext Edit**, depending on if the selected text is a paragraph text created by **MTEXT** command or text created by **TEXT** or **DTEXT** commands. If you select **Mtext Edit** from the **Shortcut** menu, **Multiline Text Editor** dialog box is displayed. If you select Text Edit, **Edit Text** dialog box is displayed.

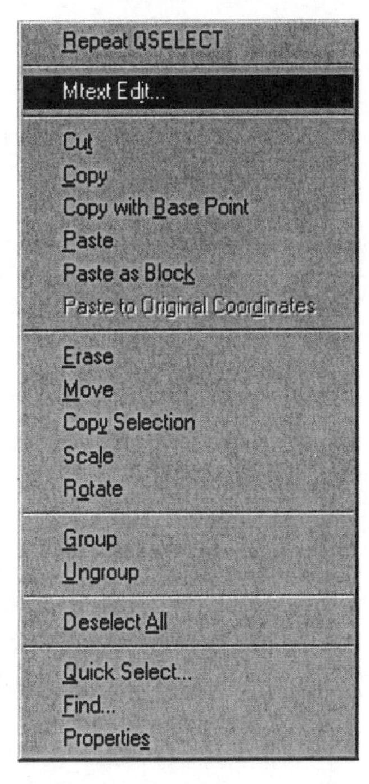

Figure 5-45 Shortcut menu when paragraph text is selected

SUBSTITUTING FONTS

AutoCAD LT provides the facility to designate the fonts that you want to substitute for other fonts used in the drawing. The information about font mapping is specified in the font mapping file (ACAD.FMP). The font mapping has the following advantages:

1. You can specify a font to be used when AutoCAD LT cannot find a font used in the drawing.
2. You can enforce the use of a particular font in your drawings. If you load a drawing that uses different fonts, you can use the font mapping to substitute the desired font for the fonts used in the drawing.
3. You can use .shx fonts when creating or editing a drawing. When you are done and ready to plot the drawing, you can substitute other fonts for the .shx fonts.

The Font mapping file is an ASCII file with FMP extension containing one font mapping per line. The format on the line is:

Base name of the font file;Name of the substitute font with extension (ttf, shx, etc.)

For example, if you want to substitute the ROMANC font for SWISS.TTF; the entry is:

SWISS;ROMANC.SHX

You can enter this line in the ACAD.FMP file or create a new file. You may use the **FONTMAP** system variable or use the **Options** dialog box to specify the new font map file. To specify a font mapping table in the **Options** dialog box, choose **Options** from the **Tools** menu to display the **Options** dialog box. Choose the **File** tab and click on the **plus** sign next to **Text Editor, Dictionary**, and **Font File Names**. Now, click on the plus sign next to **Font Mapping File** to display the path and the name of the font mapping file. Double click on the file to display the **Select a File** dialog box. Select the new font mapping file and exit the **Options** dialog box. At the Command prompt enter **REGEN** to convert the existing text font to the font as specified in the new font mapping file.

Command: **FONTMAP** Enter
Enter new value for FONTMAP, or . for none <"c:\ Program Files\AutoCAD LT 2000\ACLT.jmp">: *Enter the name of the font mapping file.*

The following file is a partial listing of ACAD.FMP file with the new font mapping line added (swiss;romanc.shx).

swiss;romanc.shx
cibt;CITYB___.TTF
cobt;COUNB___.TTF
eur;EURR____.TTF
euro;EURRO___.TTF
par;PANROMAN.TTF
rom;ROMANTIC.TTF
romb;ROMAB___.TTF
romi;ROMAI___.TTF
sas;SANSS___.TTF
sasb;SANSSB__.TTF
sasbo;SANSSBO_.TTF
saso;SANSSO__.TTF
suf;SUPEF___.TTF

Note

The text styles that were created using the PostScript fonts are substituted with an equivalent TrueType font and plotted using the substituted font.

SPECIFYING AN ALTERNATE DEFAULT FONT

When you open a drawing file that specifies a font file that is not on your system or is not specified in the font mapping file, AutoCAD LT, by default, substitutes the simplex.shx font file. You can specify a different font file in the **Options** dialog box or by changing the **FONTALT** system variable.

Command: **FONTALT** Enter
New value for FONTALT, or . for none <"simplex.shx">: *Enter the font file name.*

STYLE COMMAND

Menu:	Format > Text Style
Command:	STYLE

You can change the current text styles in the **TEXT** command using the **Style** option, but to create new text styles and modify the existing style you need the **STYLE** command. When you enter **STYLE** at the Command prompt, or select **Text Style** in the **Format** menu, the **Text Style** dialog box (Figure 5-46) is displayed.

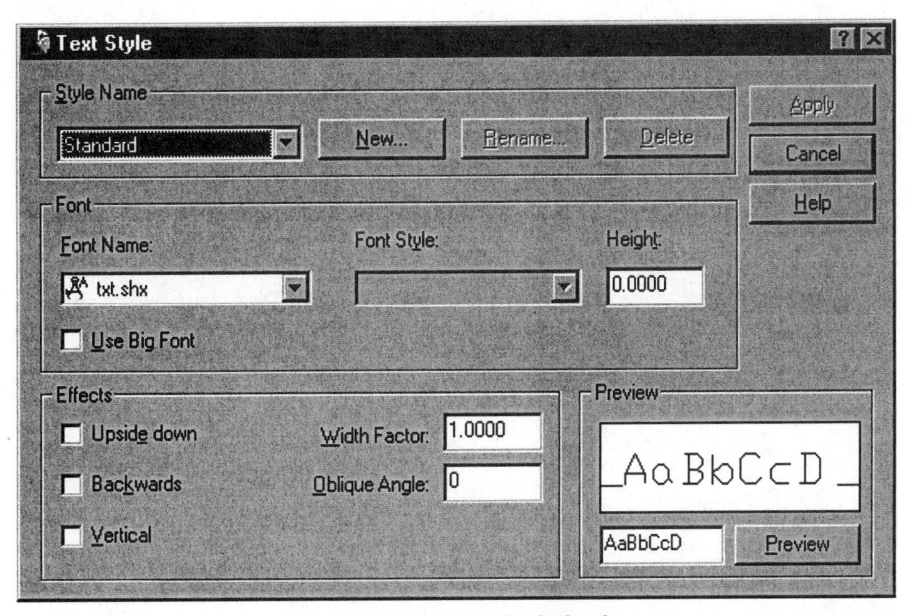

Figure 5-46 Text style dialog box

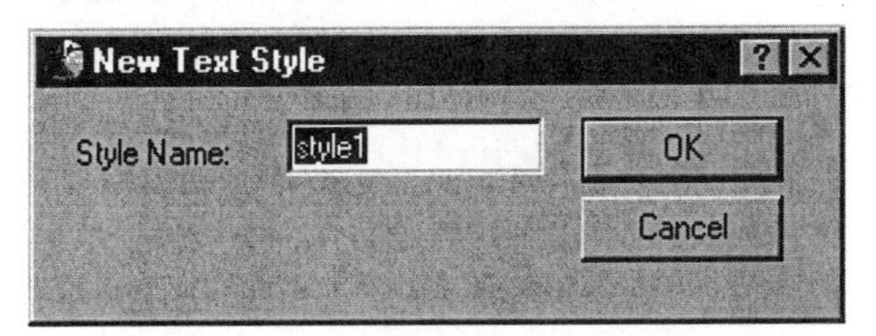

Figure 5-47 New Text Style dialog box

In the Style Name edit box the default style (STANDARD) will be displayed. For creating a new style, choose the **New** button to display the **New Text Style** dialog box (Figure 5-47) and enter the name of the style you want to create. A new style having the entered name and the properties present in the **Text Style** dialog box will be created. To modify this style, select the style name from the list box and then change the different settings by entering new values in the appropriate boxes. You can change the font by selecting a new font from the Font drop-down list. Similarly, you can change the text height, width, and oblique angle. If you specify the height of the text in the dialog box, AutoCAD LT will not prompt you to enter the text height when you use the **TEXT** command. The text will be created using the height specified in the text style. If you want AutoCAD LT to prompt you for the text height, specify 0 text height in the dialog box. For **Width factor,** 1 is the default. If you want the letters expanded, give a width factor greater than 1; for compressed letters, give a width factor less than 1. Similarly, for the **Oblique angle**, 0 is the default. If you want the slant of the letters toward the right, the value should be greater than 0; to slant the letters toward the left, the value should be less than 0. You can also force the text to be written upside down (Figure 5-48), backwards and vertically by checking their respective check boxes. As you make the changes, you can see their effect in the **Preview** box. After making the desired changes, choose the **Apply** button and the **Close** button to exit the dialog box. You can also use the **-STYLE** command to define a new test style from the command line.

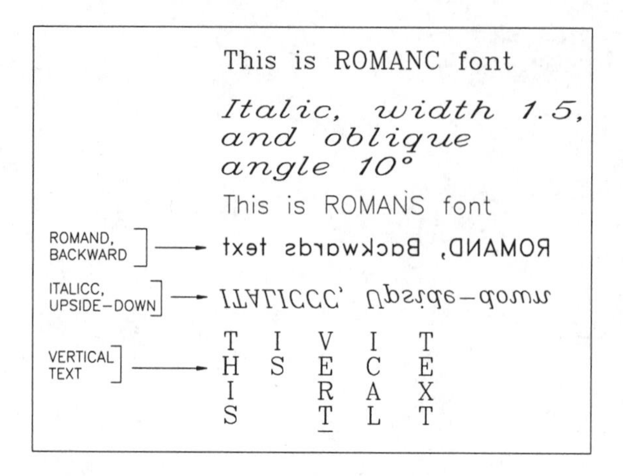

Figure 5-48 *Specifying different features to text style files*

DETERMINING TEXT HEIGHT

The actual text height is equal to the product of the **scale factor** and the **plotted text height**. Hence, scale factors are important numbers for plotting the text at the correct height. This factor is a reciprocal of the drawing plot scale. For example, if you plot a drawing at a scale of 1/4 = 1, you calculate the scale factor for text height as follows:

1/4" = 1" (that is, the scale factor is 4)

The scale factor for an architectural drawing that is to be plotted at a scale of 1/4" = 1'0" is calculated as:

1/4" = 1'0", or 1/4" = 12", or 1 = 48

Therefore, in this case, the scale factor is 48.

For a civil engineering drawing with a scale 1"= 50', the scale factor is as follows:

*1" = 50', or 1" = 50 * 12", or 1 = 600*

Therefore, the scale factor is 600.

Next, calculate the height of the AutoCAD LT text. If it is a full-scale drawing (1=1) and the text is to be plotted at 1/8" (0.125), it should be drawn at that height. However, in a civil engineering drawing a text drawn 1/8" high will look like a dot. This is because the scale for a civil engineering drawing is 1"= 50', which means that the drawing you are working on is 600 times larger. To draw a normal text height, multiply the text height by 600. Now the height will be:

0.125" x 600 = 75

Similarly, in an architectural drawing, which has a scale factor of 48, a text that is to be 1/8" high on paper must be drawn 6 units high, as shown in the following calculation:

0.125 x 48 = 6.0

It is very important to evaluate scale factors and text heights before you begin a drawing. It would be even better to include the text height in your prototype drawing by assigning the value to the **TEXTSIZE** system variable.

CHECKING SPELLING

Menu:	Tools > Spelling
Command:	SPELL

You can check the spelling of text (text generated by the **TEXT** or **MTEXT** commands) by using the **SPELL** command. The prompt sequence is:

Command: **SPELL** Enter
Select object: *Select the text for spell check or enter ALL to select all text objects.*

If no misspelled words are found in the selected text, then AutoCAD LT displays a message. If the spelling is not correct for any word in the selected text, AutoCAD LT displays the **Check Spelling** dialog box (Figure 5-49). The misspelled word is displayed under **Current word**, and correctly spelled alternate words are listed in the **Suggestions**: box. The dialog box also displays the misspelled word with the surrounding text in the **Context** box. You may select a word from the list, ignore the correction, and continue with the spell check, or accept the change. To add the listed word in the custom dictionary, choose the **Add** button. The dictionary can be changed from the **Check Spelling** dialog box or by specifying the name in the **DCTMAIN** or **DCTCUST** system variables. Dictionary must be specified for spell check.

Note

*You can also rename and use the Word'97 dictionary. Choose the **Change Dictionaries** button to display the **Change Dictionaries** dialog box and enter the new dictionary name with the .cus extension.*

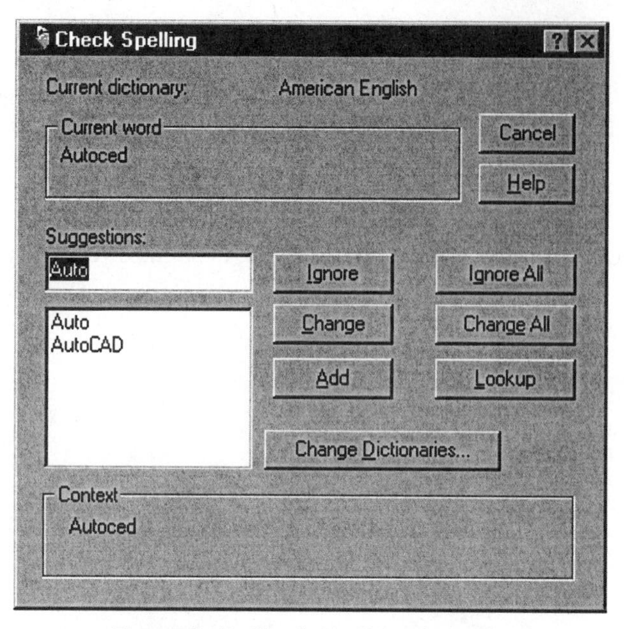

Figure 5-49 ***Check Spelling*** *dialog box*

FORMATTING PARAGRAPH TEXT

The text can be formatted by entering formatting codes in the text. To enter a paragraph text, you can use the **MTEXT** command at the command line. With the formatting codes you can underline or overline a text string, create stacked text, or insert nonbreaking space between two words. You can also use the formatting codes to change the color, font, text height, obliquing angle, or width of the text.

To obtain the paragraph text shown in Figure 5-50, use the following formatting codes in the **MTEXT** command using the command line. The prompt sequence is:

Command: **-MTEXT**
Specify first corner: *Select a point to specify first corner.*
Specify opposite corner or [Height/Justify/Rotation/Style/Width]: *Select a point to specify opposite corner.*
MText:**{\H0.15{\FROMANC;{\LNOTE}}}**
MText: **{\FROMANC;Hardness}unless**
MText: **otherwise specified**
Mtext: **{\FROMANC;60RC}\S+3.0^-2.0**
MText:

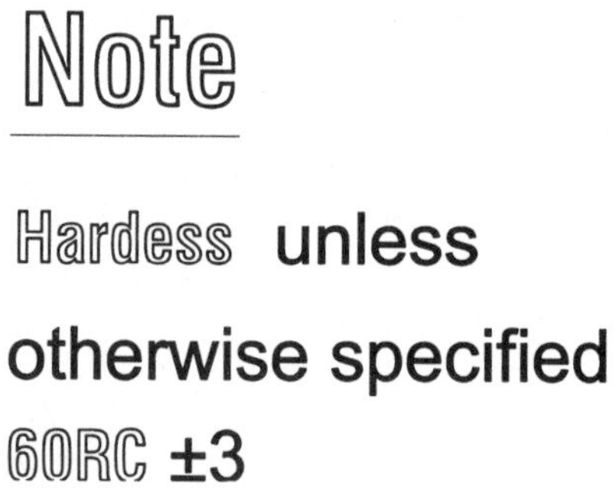

Note

Hardess unless

otherwise specified

60RC ±3

Figure 5-50 *Using TrueType font with TEXTQLTY set to 100*

Figure 5-51 lists the formatting codes for paragragh text.

\O...\o	Turns overline on and off	Turns \Ooverline\o on and off	Turns overline on and off
\L...\l	Turns underline on and off	Turns \Lunderline\l on and off	Turns underline on and off
\~	Inserts a nonbreaking space	Keeps the\~words together	Keeps the words together
\\	Inserts a backslash	Inserts \\ a backslash	Inserts \ a backslash
\{...\}	Inserts an opening and closing brace	This is \{bracketed\} word	The {bracketed} word
\Cvalue;	Changes to the specified color	Change \C1; the color	Change the color
\File name;	Changes to the specified font file	Chnage \Fromanc; this word	Change this word
\Hvalue;	Changes to the specified text height	Change \H0.15; this word	Change this word
\S...^...	Stacks the subsequent text at the \ or ^ symbol	2.005\S+0.001^-0.001	2.005$^{+0.001}_{-0.001}$
\Tvalue;	Adjusts the space between characters from .75 to 4 times	\T2;TRACKING	T R A C K I N G
\Qangle;	Changes obliquing angle	\Q15;OBLIQUE TEXT	*OBLIQUE TEXT*
\Wvalue;	Changes width factor to produce wide text	\W2;WIDE LETTERS	WIDE LETTERS
\P	Ends paragraph	First paragraph\PSecond paragraph	First paragraph Second paragraph

***Figure 5-51** Formatting codes for paragraph text*

Note

Use the curly braces if you want to apply the format codes only to the text within the braces. The curly braces can be nested up to eight levels deep.

TEXT QUALITY AND TEXT FILL

AutoCAD LT supports **TrueType fonts**. You can use your own TrueType fonts by adding them to the Fonts directory. You can also keep your fonts in a separate directory, in which case you must specify the location of your fonts directory in the AutoCAD LT search path.

The resolution and text fill of the TrueType font text is controlled by the **TEXTFILL** and **TEXTQLTY** system variables. If **TEXTFILL** is set to 1, the text will be filled. If the value is set to 0, the text will not be filled. On the screen the text will appear filled, but when it is plotted the text will not be filled. The **TEXTQLTY** variable controls the quality of the TrueType font text. The value of this variable can range from 0 to 100. The default value is 50, which gives a resolution of 300 dpi (dots per inch). If the value is set to 100, the text will be drawn at 600 dpi. The higher the resolution, the more time it takes to regenerate or plot the drawing.

FINDING AND REPLACING TEXT

Toolbar:	Standard Toolbar > Find and Replace
Menu:	Edit > Find
Command:	FIND

You can use the **FIND** command to find and replace the text. The text could be a line text created by the **TEXT** and **DTEXT** commands, paragraph text created by the **MTEXT** command, dimension annotation text, block attribute value, hyperlinks, or hyperlink description. When you invoke this command, AutoCAD LT displays the **Find and Replace** dialog box, Figure 5-52. You can use this dialog box to perform the following functions:

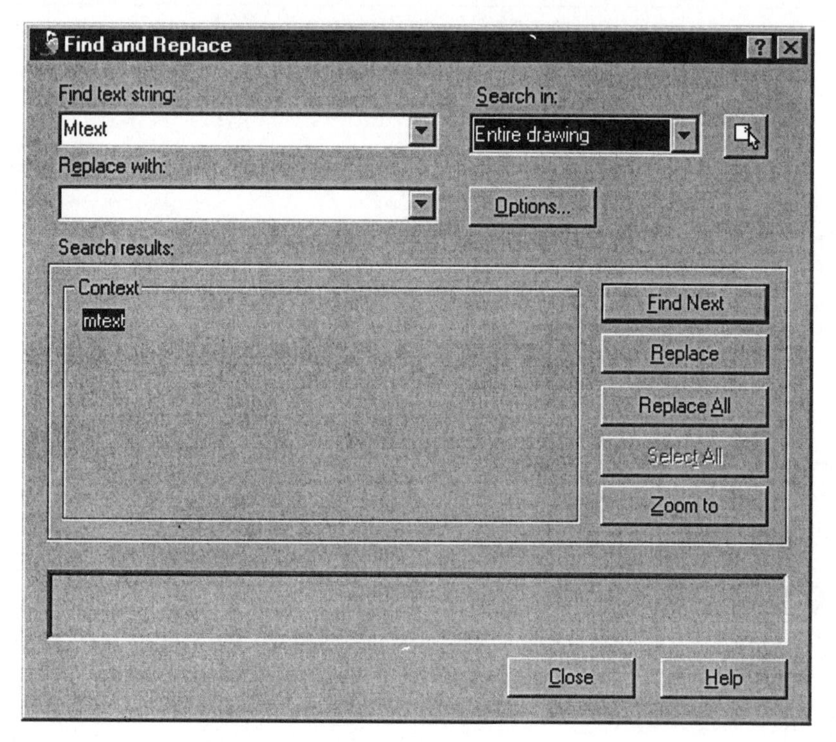

*Figure 5-52 **Find and Replace** dialog box*

Find Text

To find the text, enter the text you want to find in the **Find text string** edit box. You can do the search in entire drawing or confine the search to selected text. To select text, choose the **Select Objects** button: AutoCAD LT temporarily switches to drawing window. Once you are done selecting text, the **Find and Replace** dialog box reappears. In the **Search in** drop-down list you can specify if you want to search the entire drawing or current selection. If you choose the Options button, AutoCAD LT displays the **Find and Replace Options** dialog box, Figure 5-53. In this dialog box you can specify whether to find the whole word and whether to match the case of the specified text. To find the text, choose the **Find** button. AutoCAD LT displays the found text with the surrounding text in the Context area. To find the next occurrence of the text choose the **Find Next** button.

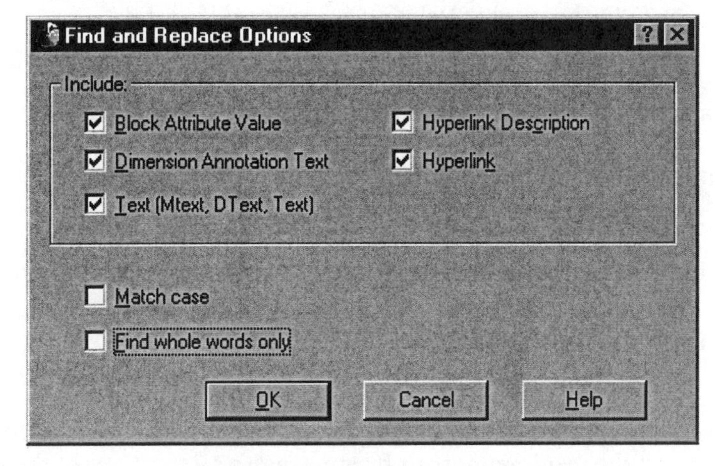

Figure 5-53 Find and Replace Options *dialog box*

Replacing Text

If you want to replace the specified text with new text, enter the new text in the **Replace with** edit box. Now, if you choose the **Replace** button, only the found text is replaced. If you choose the **Replace All** button, all occurrences of the specified text will be replaced with the new text.

Self-Evaluation Test

Answer the following questions and then compare your answers to the answers given at the end of this chapter.

1. Transparent commands cannot be used while another command is in progress. (T/F)

2. The virtual screen contains the last regeneration of the graphics database. (T/F)

3. An absolute magnification value is specified by a value followed by X. (T/F)

4. Regeneration does not involve redrawing action. (T/F)

5. The **TEXT** command does not allow you to enter multiple lines of text with a single **TEXT** command. (T/F)

6. Give the prompt sequence to create a view of the current view under name **VIEW1**.
   Command: __________
   Enter an option [?/Orthographic/Delete/Restore/Save/Window]: __________
   Enter view name to save: __________

7. The scale factor for scaling the text is a reciprocal of the __________.

8. The four main text alignment modes are__________, __________, __________, and __________.

9. To scale a view relative to paper space units, the **ZOOM** __________ option is used.

10. While using the **ZOOM Previous** option, you can go back up to __________ views for a viewport.

11. You can use __________to write a paragraph text whose width can be specified by defining __________of the text boundary or by entering the __________of the paragraph.

12 The **Justify** option is used to control the __________of paragraph text.

13. AutoCAD LT supports TrueType fonts. (T/F).

Review Questions

1. Blip marks are a part of the drawing. (T/F)

2. The **REDRAW** command can be used as a transparent command. (T/F)

3. After completion of a transparent command, AutoCAD LT returns you to the Command prompt. (T/F)

4. If the **BLIPMODE** variable is set to On, blip marks do not appear on the screen. (T/F)

5. Drawn objects are recalculated after a **REGEN** command. (T/F)

6. Drawn objects are recalculated after a **REDRAW** command. (T/F)

7. With the **ZOOM** command, the actual size of the object changes. (T/F)

8. You can use the **ZOOM** command an indefinite number of times on a drawing. (T/F)

9. A relative magnification value is specified by a value followed by X. (T/F)

10. The **TEXT** command does not allow you to see the text on the screen as you type it. (T/F)

11. Give the prompt sequence to zoom in on the center of a drawing with a magnification of 2.
    Command: __________
    Specify corner of window, enter a scale factor (nX or nXP), or
    [All/ Center/ Dynamic/ Extents/ Previous/ Scale / Window] <Realtime>: __________
    Specify center point: __________
    Enter magnification or height<current>: __________

12. By specifying the start point you need, your text is __________-justified along its __________.

13. The __________of a text determines how far above the baseline the capital letters can extend.

14. The __________determines the orientation of the text baseline with respect to the start point.

15. You can view the entire drawing (even if it is beyond limits) with the help of the __________ option.

16. In the **ZOOM** Window option, the area inside the window is __________to completely __________ the current viewport.

17. Explain the difference between the **REDRAW** and **REGEN** commands.__________

18. What are blips and what commands make them disappear from the screen?__________

19. Name the command that allows you to change the display of blips. __________

20. Explain the difference between the **Extents** and the **All** options of the **ZOOM** command.__________

21. When during the drawing process should you use the **ZOOM** command? __________

22. Name the view boxes that are displayed during the **ZOOM** Dynamic command, and explain the significance of each one. __________

23. What is the **PAN** command and how does it work? __________

24. Give the letter you must enter at the prompt after entering the **Justify** option for the **TEXT** command for:
    a. Top/left justified text
    b. Right justified text

25. What command do you use to see the text on the screen as it is typed?

26. Calculate the AutoCAD LT text height for a text to be plotted 1/8" using a half scale; i.e., 1" = 2" scale.

27. Find the AutoCAD LT text height for a text to be plotted 1/8" high using a scale of 1/16" = 1'-0".

28. What is the function of the **REGENALL** and **REDRAWALL** commands? ____________
____

29. The text created by the **MTEXT** command is a single object regardless of the number of lines it contains. (T/F)

30. The text boundary of paragraph text is also plotted when you plot the drawing. (T/F).

31. The contents of an **MTEXT** object (paragraph text) can be edited by using the _________ and the _________ commands.

32. When using the **MTEXT** command, the special characters _________ and _________ are used to stack the text.

Exercises

Exercise 3 *General*

Draw the text on the screen as shown in Figure 5-54(a). Use the text justification that will produce the text as shown in the drawing . Assume a value for text height. Use the text related commands to change the text as shown in Figure 5-54(b).

Figure 5-54(a) Drawing for Exercise 3

Figure 5-54(b) Drawing for Exercise 3 (After changing the text)

Exercise 4 General

Draw the text on the screen shown in Figure 5-55. You must first define text style files with the attributes as shown in the drawing. The text height is 0.25 units. Do not dimension the drawing.

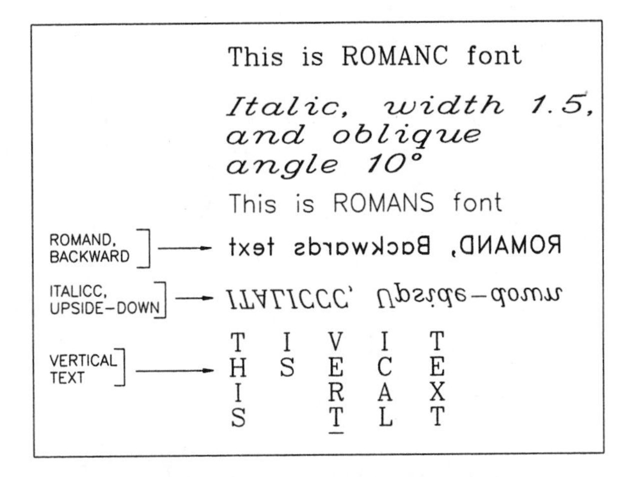

Figure 5-55 Drawing for Exercise 4

Exercise 5 Graphics

Draw Figure 5-56 using the **MIRROR** command to duplicate the features that are symmetrical. Also, use display commands to facilitate the drawing process. Do not dimension the drawing.

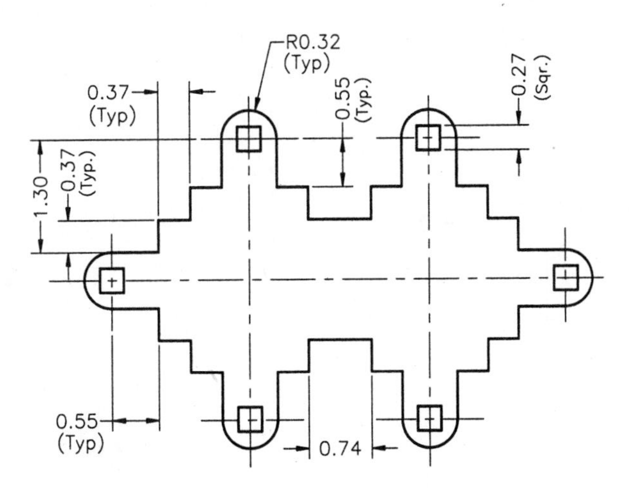

Figure 5-56 Drawing for Exercise 5

Exercise 6

Mechanical

Draw Figure 5-57 making use of the draw, edit, and display commands.

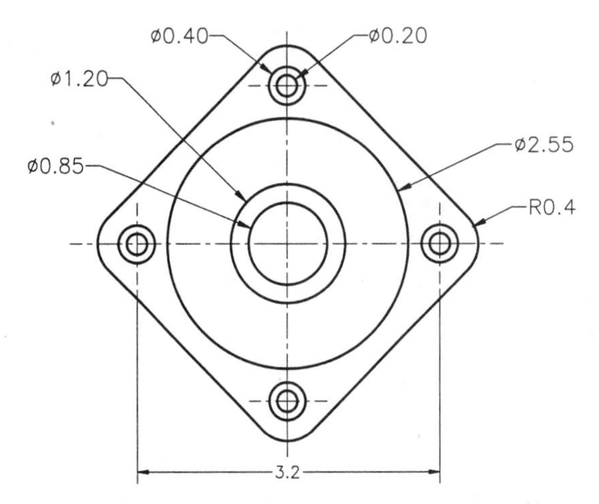

Figure 5-57 *Drawing for Exercise 6*

Exercise 7

Mechanical

Draw Figure 5-58 using the commands studied in this chapter to the maximum advantage. Do not dimension the drawing.

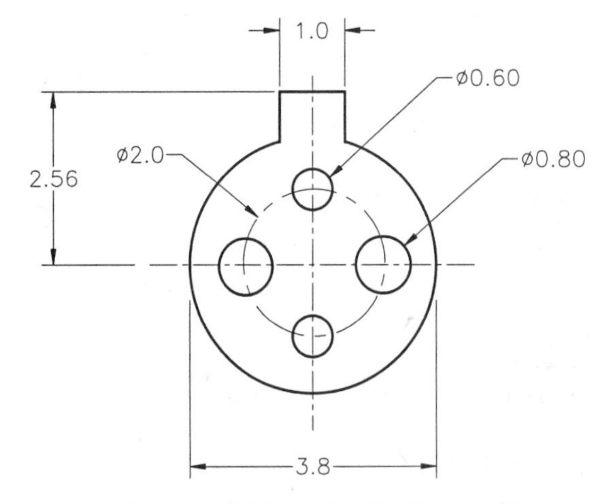

Figure 5-58 *Drawing for Exercise 7*

Exercise 8 *Architectural*

Make the drawing shown Figure 5-59. Assume the distance between the dotted lines.

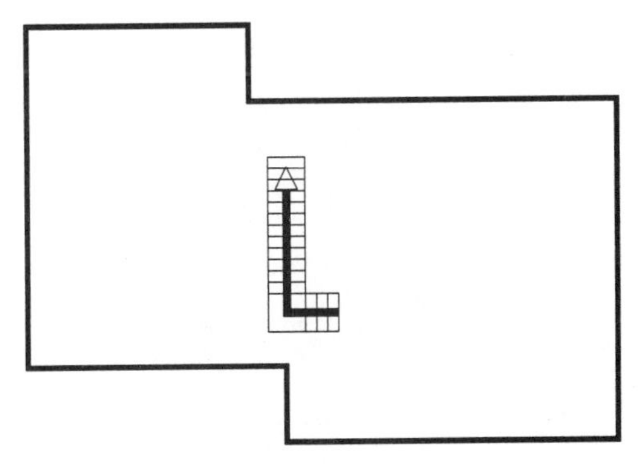

***Figure 5-59** Drawing for Exercise 8*

Exercise 9 *Architectural*

Make the drawing shown in Figure 5-60. Assume the distance between the dotted lines.

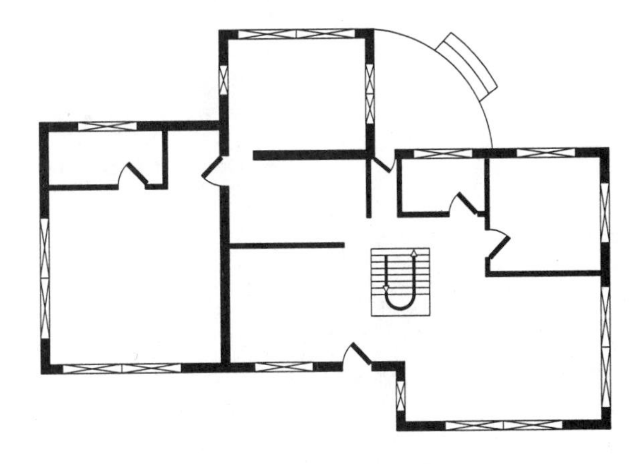

***Figure 5-60** Drawing for Exercise 9*

Answers to Self-Evaluation Test

1 - F, **2** - T, **3** - F, **4** - F, **5** - T, **6** - **VIEW**/Save/VIEW1, **7** - drawing scale, **8** - Align/Center/Middle/ Left, **9** - XP, **10** - 10, **11** - **MTEXT**, two corners, Width, **12** - justification, **13** - T

Chapter 6

Basic Dimensioning

Learning Objectives

After completing this chapter, you will be able to:
- *Understand the need for dimensioning in drawings.*
- *Understand the fundamental dimensioning terms.*
- *Understand associative dimensioning.*
- *Select dimensioning commands in AutoCAD LT.*
- *Create linear, aligned, rotated, baseline, and continue dimensions.*
- *Create angular, radial, diameter, and ordinate dimensions.*
- *Create center marks and centerlines.*
- *Use the **QLEADER** command to attach annotation to an object.*
- *Use the **LEADER** command.*

NEED FOR DIMENSIONING

To make designs more informative and practical, the drawing must convey more than just the graphic picture of the product. To manufacture an object, the drawing must contain size descriptions such as the length, width, height, angle, radius, diameter, and location of features. All of this information is added to the drawing with the help of **dimensioning**. Some drawings also require information about tolerances with the size of features. All of this information conveyed through dimensioning is vital and often just as important as the drawing itself. With the advances in computer-aided design/drafting and computer-aided manufacturing, it has become mandatory to draw the part to actual size so that the dimensions reflect the actual size of the features. At times it may not be necessary to draw the object the same size as the actual object would be when manufactured, but it is absolutely essential that the dimensions be accurate. Incorrect dimensions will lead to manufacturing errors.

By dimensioning, you are not only giving the size of a part, you are also giving a series of instructions to a machinist, an engineer, or an architect. The way the part is positioned in a machine, the sequence of machining operations, and the location of different features of the part depend on how you dimension the part. For example, the number of decimal places in a dimension (2.000) determines the type of machine that will be used to do that machining operation. The machining cost of such an operation is significantly higher than for a dimension that has only one digit after the decimal (2.0). Similarly, whether a part is to be forged or cast, the radii of the edges and the tolerance you provide to these dimensions determine the cost of the product, the number of defective parts, and the number of parts you get from a single die.

DIMENSIONING IN AUTOCAD LT

The objects that can be dimensioned in AutoCAD LT range from straight lines to arcs. The dimensioning commands provided by AutoCAD LT can be classified into four categories:

Dimension Drawing Commands
Dimension Style Commands
Dimension Editing Commands
Dimension Utility Commands

While dimensioning an object, AutoCAD LT automatically calculates the length of the object or the distance between two specified points. Also, settings like a gap between the dimension text and the dimension line, the space between two consecutive dimension lines, arrow size, and text size are maintained and used when the dimensions are being generated for a particular drawing. The generation of arrows, lines (dimension lines, extension lines), and other objects that form a dimension are automatically performed by AutoCAD LT to save the user's time. This also results in uniform drawings. However, you can override the default measurements computed by AutoCAD LT and change the settings of various standard values. The modification of dimensioning standards can be achieved through the dimension variables.

The dimensioning functions offered by AutoCAD LT provide you with extreme flexibility in dimensioning by letting you dimension various objects in a variety of ways. This is of great help because different industries, like architectural, mechanical, civil, or electrical, have different standards for the placement of dimensions.

FUNDAMENTAL DIMENSIONING TERMS

Before studying AutoCAD LT's dimensioning commands, it is important to know and understand various dimensioning terms that are common to linear, angular, radius, diameter, and ordinate dimensioning. Figure 6-1 shows some dimensioning objects.

Dimension Line

The **dimension line** indicates which distance or angle is being measured. Usually this line has arrows at both ends, and dimension text is placed along the dimension line. By default dimension line is drawn between the extension lines (Figure 6-2). If dimension line does not fit inside, two short lines with arrows pointing inward are drawn outside the extension lines. The dimension line for angular dimensions (which are used to dimension angles) is an arc. You can control positioning and various other features of the dimension lines by setting the dimensioning system variables. (The dimensioning system variables are discussed in Chapter 8.)

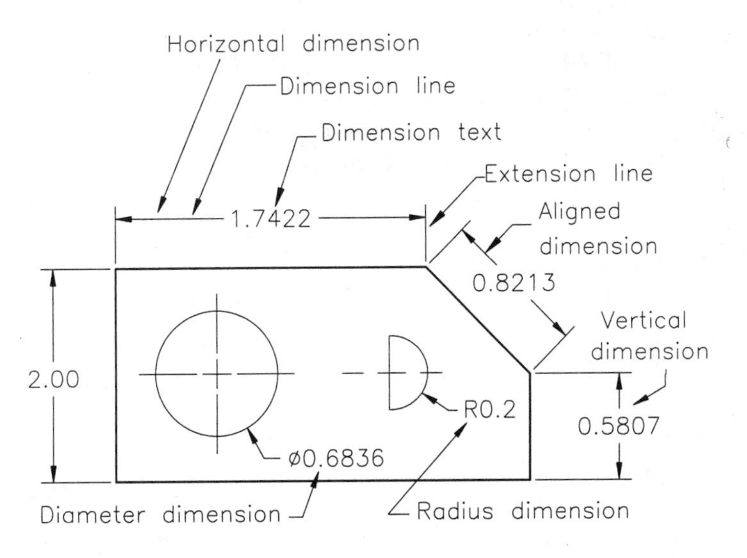

Figure 6-1 AutoCAD LT's dimensioning terms

Dimension Text

Dimension text is a text string that reflects the actual measurement (dimension value) between the selected points as calculated by AutoCAD LT. You can accept the value that AutoCAD LT returns or enter your own value. In case you use the default text, AutoCAD LT can be supplied with instructions to append the tolerances to it. Also, you can attach prefixes or suffixes of your choice to the dimension text.

Arrowheads

An **arrowhead** is a symbol used at the end of a dimension line (where dimension lines meet the extension lines). Arrowheads are also called **terminators** because they signify the end of the dimension line. Since the drafting standards differ from company to company, AutoCAD LT allows you to draw arrows, tick marks, closed arrows, open arrows, dots, right angle arrows, or user-defined blocks (Figure 6-2). The user-defined blocks at the two ends of the dimension line can be customized to your requirements. The size of the arrows, tick marks, user blocks, and so on, can be regulated by using the dimension variables.

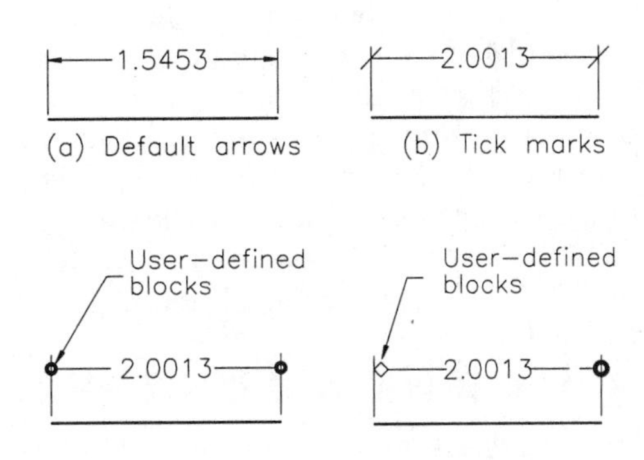

Figure 6-2 Using arrows, tick marks, and user-defined blocks

Extension Lines

Extension lines are drawn from the object measured to the dimension line (Figure 6-3). These lines are also called **witness lines.** Extension lines are used in linear and angular dimensioning. Generally, extension lines are drawn perpendicular to the dimension line. However, you can make extension lines incline at an angle by using the **DIMEDIT** command (**Oblique** option) or by selecting **Dimension Edit** from the **Dimension** toolbar. AutoCAD LT also allows you to suppress either one or both extension lines in a dimension (Figure 6-4). Other aspects of the extension line can be controlled by using the dimension variables (these variables are discussed in Chapter 8).

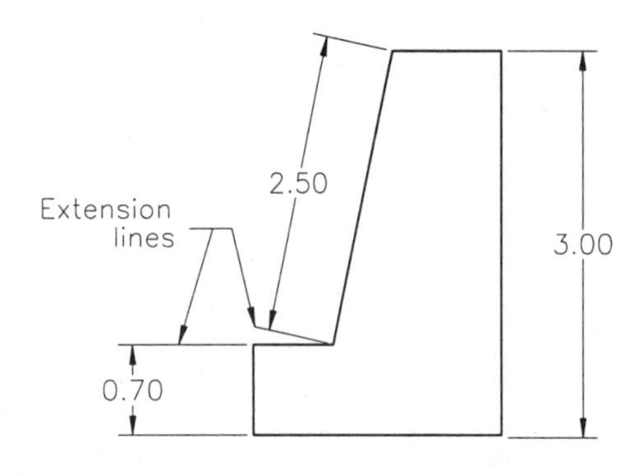

Figure 6-3 Extension lines

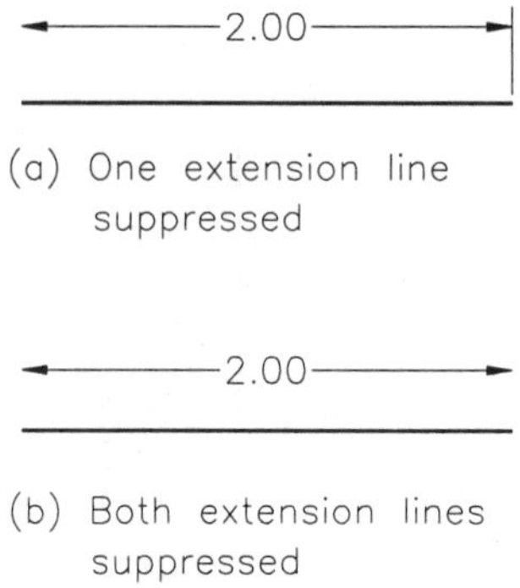

2.00

(a) One extension line suppressed

2.00

(b) Both extension lines suppressed

Figure 6-4 Extension line suppression

Leader

A **leader** is a line that stretches from the dimension text to the object being dimensioned. Sometimes the text for dimensioning and other annotations does not adjust properly near the object. In such cases, you can use a leader and place the text at the end of the leader line. For example, the circle shown in Figure 6-5 has a keyway slot that is too small to be dimensioned. In this situation, a leader can be drawn from the text to the keyway feature. Also, a leader can be used to attach annotations such as part numbers, notes, and instructions to an object.

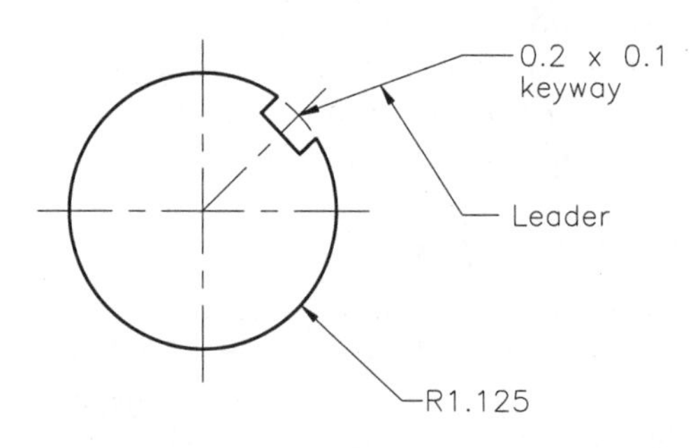

Figure 6-5 Leader used to attach annotation

Center Mark and Centerlines

The **center mark** is a cross mark that identifies the center point of a circle or an arc. Centerlines are mutually perpendicular lines passing through the center of the circle/arc and intersecting the circumference of the circle/arc. A center mark or the centerlines are automatically drawn when you dimension a circle or arc (see Figure 6-6). The length of the center mark and the extension of the centerline beyond the circumference of the circle is determined by the value assigned to the **DIMCEN** dimension variable or the value assigned in the **Center Mark for Circle** in the **Nem/Modify/Override/Dimension Style** dialog box.

Alternate units

With the help of **alternate units** you can generate dimensions for two systems of measurement at the same time (Figure 6-7). For example, if the dimensions are in inches, you can use the alternate units dimensioning facility to append metric dimensions to the dimensions (controlling the alternate units through the dimension variables are discussed in Chapter 8).

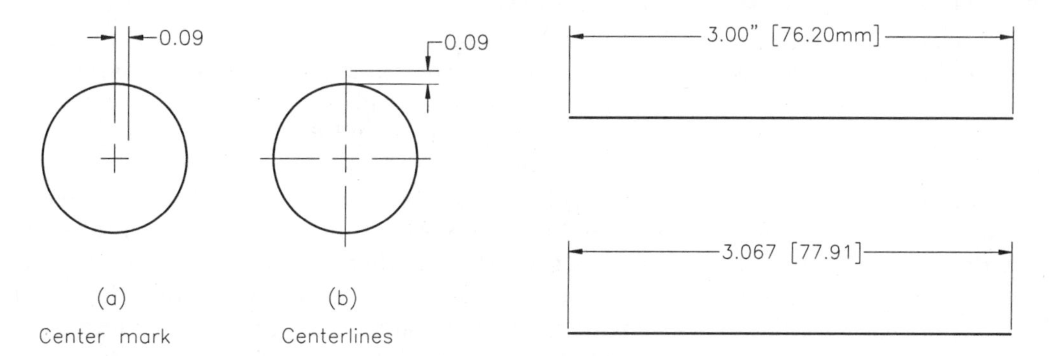

Figure 6-6 *Center mark and centerlines*

Figure 6-7 *Using alternate units dimensioning*

Tolerances

Tolerance is the amount by which the actual dimension can vary (Figure 6-8). AutoCAD LT can attach the plus/minus tolerances to the dimension text (actual measurement computed by AutoCAD LT). This is also known as **deviation tolerance**. The plus and minus tolerance that you specify can be the same or different. You can use the dimension variables to control the tolerance feature (these variables are discussed in Chapter 8).

Limits

Instead of appending the tolerances to the dimension text, you can apply the tolerances to the measurement itself (Figure 6-9). Once you define the tolerances, AutoCAD LT will automatically calculate the upper and lower **limits** of the dimension. These values are then displayed as a dimension text.

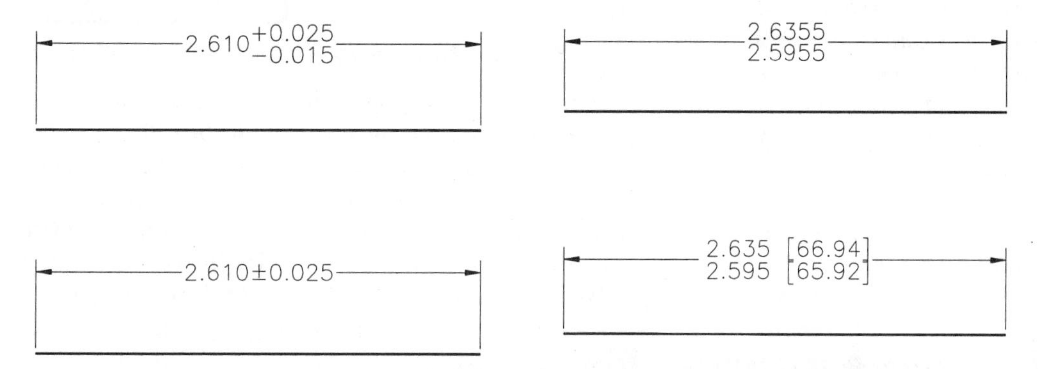

Figure 6-8 *Using tolerances with dimensions*

Figure 6-9 *Using limits dimensioning*

For example, if the actual dimension as computed by AutoCAD LT is 2.610 units and the tolerance values are +0.025 and -0.015, the upper and lower limits are 2.635 and 2.595. After calculating the limits, AutoCAD LT will display them as dimension text, as shown in Figure 6-9. The dimension variables that control the limits are discussed in Chapter 8.

ASSOCIATIVE DIMENSIONS

Associative dimensioning is a method of dimensioning in which the dimension is associated with the object that is dimensioned. In other words, the dimension is influenced by the changes in the size of the object. AutoCAD LT dimensions are not truly associative, but are related to the objects being dimensioned by definition points on the DEFPOINTS layer. To cause the dimension to be associatively modified, these definition points must be adjusted along with the object being changed. If, for example, you use the **SCALE** command to change an object's size and select the object, any associative dimensions will not be modified. If you select the object and its defpoints (using the Crossing selection method), then the dimension will be modified. In associative dimensioning, the items constituting a dimension (like dimension lines, arrows, extension lines, and dimension text) are drawn as a single object. If the associative dimension is disabled, the dimension lines, arrows, extension lines, and the dimension text are drawn as separate objects. In that case, you can edit the dimension lines, extension lines, text, and arrowheads as individual objects. You can also use the **EXPLODE** command to split an associative dimension into individual items (just as in the case when an associative dimension is disabled).

The dimensioning variable **DIMASO** controls associativity of dimensions. Its default setting is on. When **DIMASO** is set on and you dimension and edit the object (e.g., trimming or stretching), the dimensions associated with that object also change (Figure 6-10). Also, the appearance of associative dimensions can be preserved when they are edited by commands such as **STRETCH** and **DIMTEDIT**. For example, a vertical associative dimension is retained as a vertical dimension even after an editing operation. The associative dimension is always generated with the same dimension variable settings as defined in the dimension style. Another advantage of associative dimensioning is that you can edit dimension objects as single objects with various commands. The dimensioning commands that are affected by associative dimensioning are **Linear**, **Aligned**, **Rotated**, **Radius**, **Diameter**, and **Angular** commands. The **Center** and **Leader** commands are not affected by associative dimensioning. The associative dimension is always generated with same dimension variable settings as defined in the creation of the dimension style. The present linetype and the text style are used for the drawing of text.

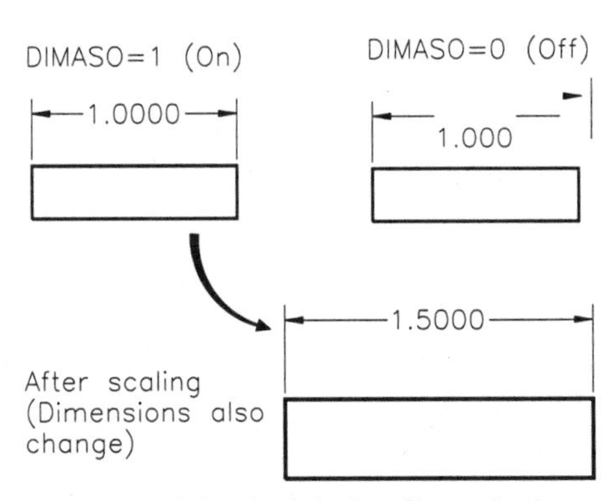

***Figure 6-10** Associative dimensioning*

Updating Associative Dimensions

As mentioned for this automatic revision of the associative dimension objects, the **definition points of the dimension object must be contained in the selection set** formed for the editing.

Therefore, becoming familiar with the definition points of different types of dimensions will make it easier to edit the dimensions.

DEFINITION POINTS

Definition points are the points drawn at the positions used to generate an associative dimension object. The definition points are used by the associative dimensions to control their updating and rescaling. AutoCAD LT draws these points on a special layer called DEFPOINTS. These points are not plotted by the plotter because AutoCAD LT does not plot any object on the DEFPOINTS layer. If you explode an associative dimension (which is as good as turning **DIMASO** off), the definition points are converted to point objects on the DEFPOINTS layer. In Figure 6-11, the small circles indicate the definition points for different objects.

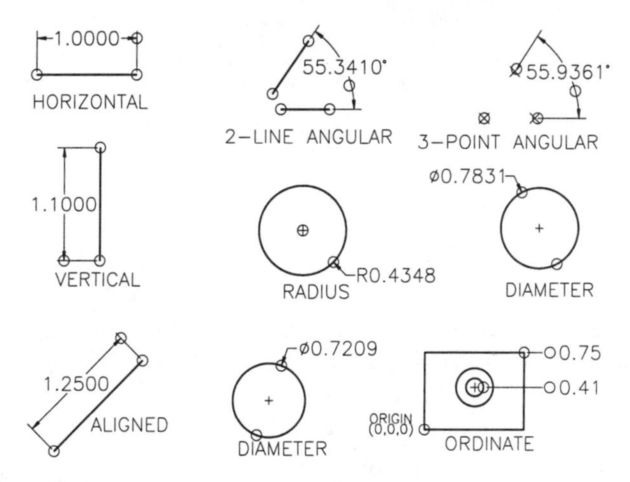

Figure 6-11 *Definition points of linear, angular, and ordinate dimensions*

The definition points for linear associative dimensions are the points used to specify the extension lines and the point of intersection of the first extension line and the dimension line. The definition points for the angular dimension are the endpoints of the lines used to specify the dimension and the point used to specify the dimension line arc. For example, for three-point angular dimension, the definition points are the extension line endpoints, the angle vertex, and the point used to specify the dimension line arc.

The definition points for the radius dimension are the center point of the circle or arc, and the point where the arrow touches the object. The definition points for the diameter dimension are the points where the arrows touch the circle. The definition points for the ordinate dimension are the UCS origin, the feature location, and the leader endpoint.

Note

In addition to the definition points just mentioned, the middle point of the dimension text serves as a definition point for all types of dimensions.

SELECTING DIMENSIONING COMMANDS

Using the Toolbar and the Dimension Menu

You can select the dimension commands from the **Dimension** toolbar by choosing the desired dimension button (Figure 6-12), or from the **Dimension** menu (Figure 6-13). The **Dimension** toolbar can also be displayed by right-clicking on any toolbar and choosing **Dimensions** from the shortcut menu.

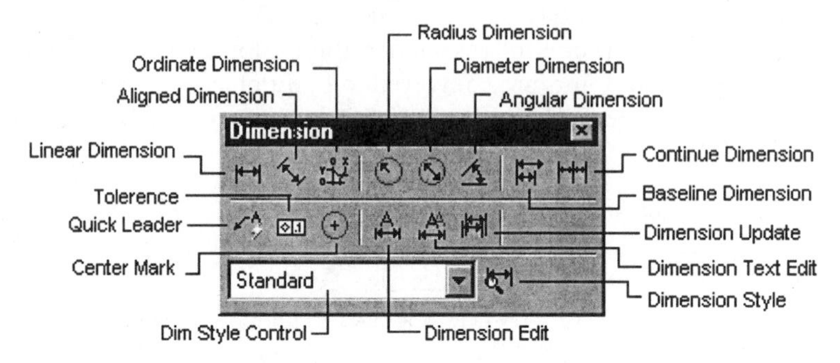

***Figure 6-12** Selecting dimension commands from the **Dimension** toolbar*

Using the Command Line

Using Dimensioning Commands. You can use the dimensioning commands at the Command prompt without using the **DIM** Command. For example, if you want to draw the linear dimension, the **DIMLINEAR** command can be entered directly at the Command prompt.

Command: **DIMLINEAR** [Enter]
Specify first extension line origin or <select object>: *Select a point or press ENTER.*
Specify second extension line origin: *Select second point.*
Specify dimension line location or
[Mtext/Text/Angle/Horizontal/Vertical/Rotated]: *Select a point to locate the position of the dimension.*
Command: *(After you have finished dimensioning, AutoCAD LT returns to the Command prompt.)*

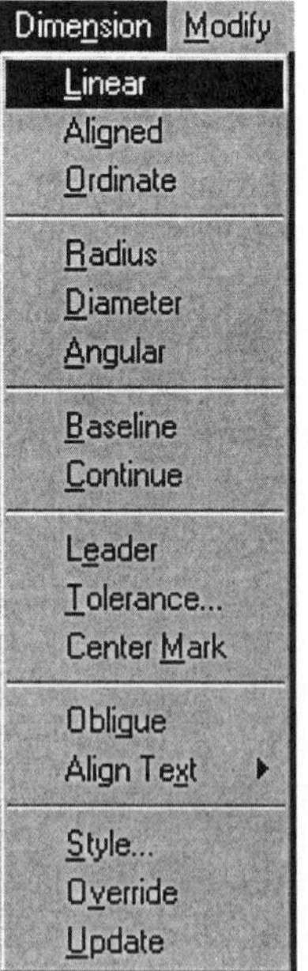

***Figure 6-13** Selecting dimensions from the **Dimension** menu*

DIM and DIM1 Commands. Since dimensioning has several options, it also has its own command mode. The **DIM** command keeps you in the dimension mode, and the **Dim:** prompt is repeated after each dimensioning command until you exit the dimension mode to return to the normal AutoCAD LT Command prompt. To exit the dimension mode, enter EXIT (or just E) at the **Dim** prompt. You can also exit by pressing ESC, or by pressing the third button of the digitizer puck. The previous command will be repeated if you press the SPACEBAR or ENTER at the **Dim** prompt. In the dimension mode, it is not possible to execute the normal set of AutoCAD LT commands, except function keys, object snap overrides, control key combinations, transparent commands, dialog boxes, and menus.

Command: **DIM** [Enter]
Dim: **Hor** [Enter]
Specify first extension line origin or <select object>: *Select a point or press ENTER.*
Specify second extension line origin: *Select the second point.*
Specify dimension line location or [MText/Text/Angle]: *Select a point to locate the position of the dimension.*
Enter dimension text <default>: *Press ENTER to accept the default dimension.*
Dim: *(After you have finished dimensioning, AutoCAD LT returns to the* ***Dim:*** *prompt.)*

The **DIM1** command is similar to the **DIM** command. The only difference is that **DIM1** lets you execute a single dimension command and then automatically takes you back to the normal Command prompt.

Command: **DIM1** [Enter]
Dim: **Hor** [Enter]
Specify first extension line origin or <select object>: *Select a point or press ENTER.*
Specify second extension line origin: *Select the second point.*
Specify dimension line location or [MText/Text/Angle]: *Select a point to locate the position of the dimension.*
Enter dimension text <default>: *Press ENTER to accept the default dimension.*
Command: *(After you are done dimensioning, AutoCAD LT returns Command prompt.)*

AutoCAD LT has provided the following five fundamental dimensioning types (Figure 6-14):

Linear dimensioning **Diameter dimensioning** **Radius dimensioning**
Angular dimensioning **Ordinate dimensioning**

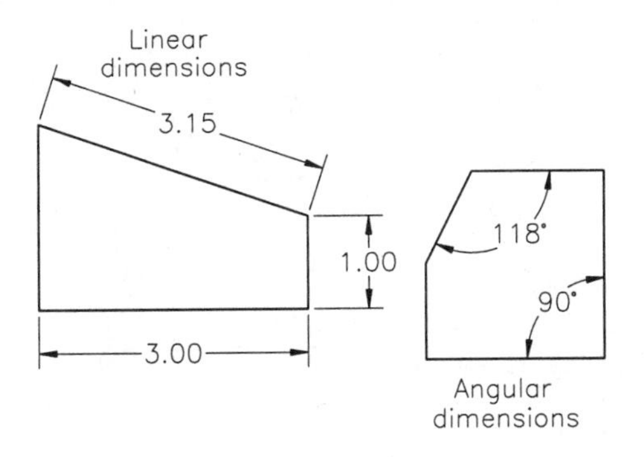

Figure 6-14(a) Linear and angular dimensions

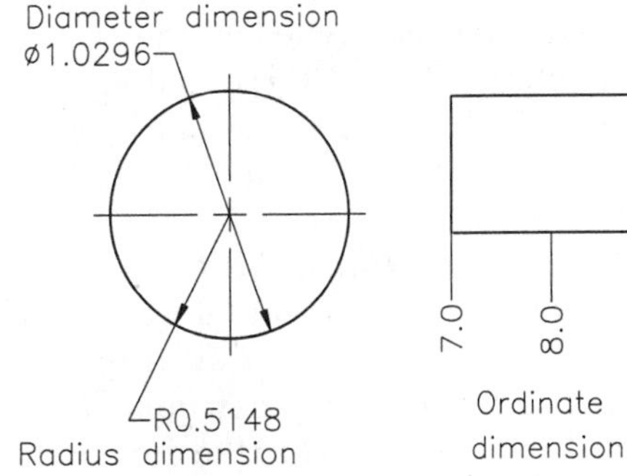

Figure 6-14(b) Radius, diameter, and ordinate dimensions

Note

*The **DIMDEC** variable sets the number of decimal places for the value of primary dimension and the **DIMADEC** variable for angular dimensions. For example, if **DIMDEC** is set to 3, AutoCAD LT will display the decimal dimension up to three decimal places (2.037).*

LINEAR DIMENSIONING

Toolbar:	Dimension > Linear Dimension
Menu:	Dimension > Linear
Command:	DIMLINEAR

Linear dimensioning applies to those dimensioning commands that measure the distance between two points. The points can be any two points in the space, the endpoints of an arc or line, or any set of points that can be identified. To achieve accuracy, selecting points must be done with the help of object snaps or by selecting an object to dimension. Linear dimensions include **Horizontal** and **Vertical** dimensioning. When you select a dimensioning command, AutoCAD LT will prompt you to enter information about the dimension. The dimension is not drawn until you respond to all of the dimensioning prompts.

Command: **DIMLINEAR** Enter
Specify first extension line origin or <select object>: Enter
Select object to dimension: *Select the object.*
Specify dimension line location or
[Mtext/Text/Angle/Horizontal/Vertical/Rotated]: *Select a point to locate the position of the dimension.*

Instead of selecting the object, you can also select the two endpoints of the line that you want to dimension (Figure 6-15). Usually the points on the object are selected by using the **object snaps** (endpoints, intersection, center, etc.). The prompt sequence is as follows:

Command: **DIMLINEAR** Enter
Specify first extension line origin or <select object>: *Select a point.*
Specify second extension line origin: *Select second point.*
Specify dimension line location or
[Mtext/Text/Angle/Horizontal/Vertical/Rotated]: *Select a point to locate the position of the dimension.*

When using the **DIMLINEAR** command, you can obtain the horizontal or vertical dimension by simply defining the appropriate dimension location point. If you select a point above or below the dimension, AutoCAD LT creates a horizontal dimension. If you select a point that is on the left or right of the dimension, AutoCAD LT creates a vertical dimension through that point.

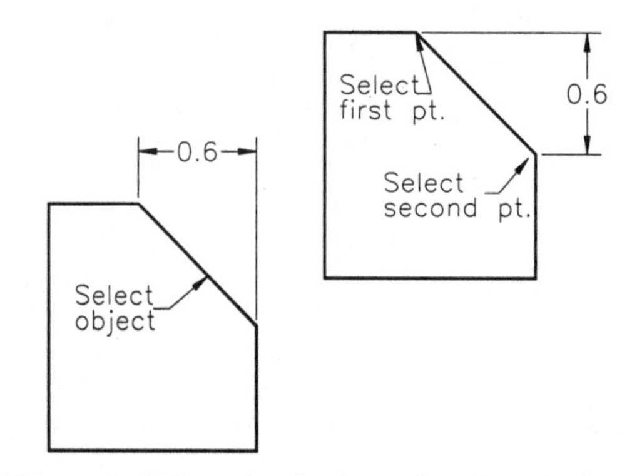

Figure 6-15 *Drawing horizontal and vertical dimension*

DIMLINEAR Options

Mtext Option. The **Mtext** option allows you to override the default dimension text and also change the font, height, etc., using the **Multiline Text Editor** dialog box. When you enter **M** at the **Specify dimension line location or [Mtext/Text/Angle/Horizontal/Vertical/Rotated]** prompt, the **Multiline Text Editor** dialog box is displayed. By default, it includes the <> code to represent the measured associative dimension text. You can change the text by entering a new text and deleting <> code. You can also use the different options of the dialog box (explained in Chapter 5) and then choose **OK**. However, if you override the default dimensions, the dimensional associativity of the **dimension text** is lost and AutoCAD LT will not recalculate the dimension when the object is scaled.

Command: **DIMLINEAR** Enter
Specify first extension line origin or <select object>: *Specify a point.*
Specify second extension line origin: *Specify second point.*
Specify dimension line location or
[Mtext/Text/Angle/Horizontal/Vertical/Rotated]: **M** *(Enter dimension text in **Mutiline Text Editor** dialog box; choose **OK**.)*
Specify dimension line location or
[Mtext/Text/Angle/Horizontal/Vertical/Rotated]: *Specify a point to locate the position of the dimension.*

Text Option. This option also allows you to override the default dimension. See Figure 6-16.

Command: **DIMLINEAR** Enter
Specify first extension line origin or <select object>: *Select a point.*
Specify second extension line origin: *Select second point.*
Specify dimension line location or
[Mtext/Text/Angle/Horizontal/Vertical/Rotated]: **T**
Enter dimension text <Current>: *Enter new text.*
Specify dimension line location or
[Mtext/Text/Angle/Horizontal/Vertical/Rotated]: *Select a point to locate the position of the dimension.*

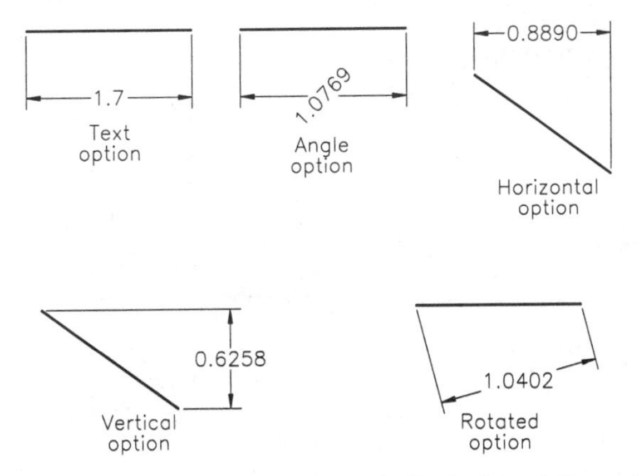

Figure 6-16 *Text, Angle, Horizontal, Vertical, and Rotated options*

Angle Option. This option lets you change the angle of the dimension text.

Horizontal Option. This option lets you create a horizontal dimension regardless of where you specify the dimension location.

Vertical Option. This option lets you create a vertical dimension regardless of where you specify the dimension location.

Rotated Option. This option lets you create a dimension that is rotated at a specified angle.

Note

If you override the default dimensions, the dimensional associativity of the dimension text is lost and AutoCAD LT will not recalculate the dimension when the object is scaled. .

Example 1

In this example, you will use the **DIMLINEAR** command to dimension a line by selecting the object and by specifying the first and second extension line origins (Figure 6-17). The following is the command prompt sequence:

Selecting the Object

1. Enter the **DIMLINEAR** command at the Command prompt. You can also select it from the **Dimension** menu or the toolbar.

   Command: **DIMLINEAR** [Enter]

2. At the next prompt, press ENTER to select the line. The endpoints of the line are taken as the origins for the extension line.

   Specify first extension line origin or <select object>: [Enter]
   Select object to dimension: *Select the line.*

3. At the next prompt, specify the location for the dimension line. This information is used by AutoCAD LT to set the location to place the dimension line. With the **Mtext** option, you can override the default dimension text.

   Specify dimension line location or [Mtext/Text/Angle/Horizontal/Vertical/Rotated]: **M** [Enter]

4. When you select the Mtext option, the **Multiline Text Editor** dialog box (Figure 6-18) is displayed. Enter the dimension text in the dialog box and choose the **OK** button to exit.

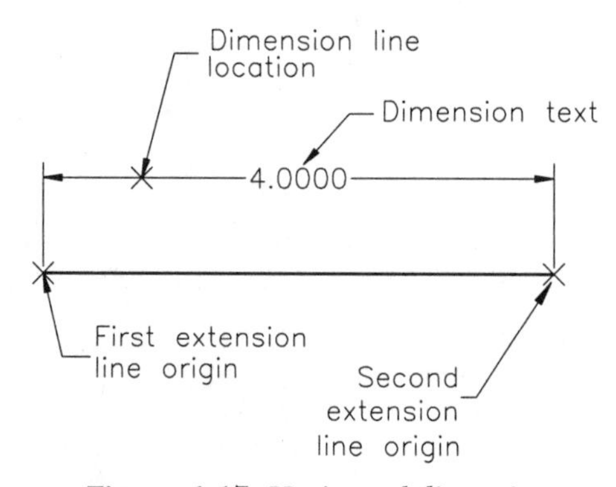

Figure 6-17 *Horizontal dimension*

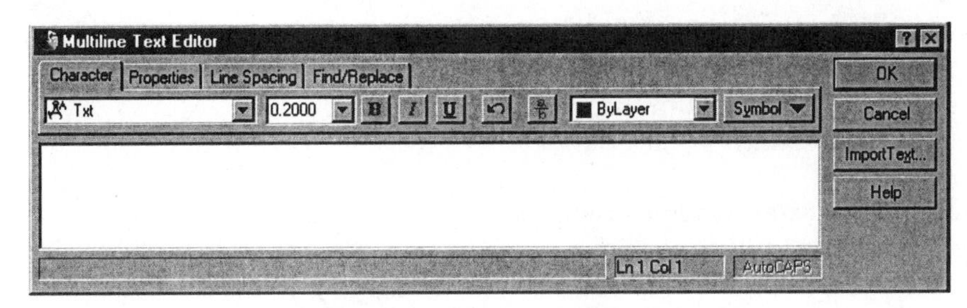

*Figure 6-18 **Multiline Text Editor** dialog box*

Specify dimension line location or
[Mtext/Text/Angle/Horizontal/Vertical/Rotated]: *Specify a point.*

Specifying Extension Line Origins

In case you want to specify the two endpoints (extension line origin) on the line, the Command prompt sequence is as follows:

Command: **DIMLINEAR** Enter
Specify first extension line origin or <select object>: *Select the first endpoint of the line using the Endpoint object snap.*
Specify second extension line origin: *Select the second endpoint of the line using the Endpoint object snap.*
Specify dimension line location or
[Mtext/Text/Angle/Horizontal/Vertical/Rotated]: *Specify the location for the dimension line.*

ALIGNED DIMENSIONING

Toolbar:	Dimension > Aligned Dimension
Menu:	Dimension > Aligned
Command:	DIMALIGNED

You may want to dimension an object that is inclined at an angle, not parallel to the X axis or Y axis. In such a situation you can use **aligned dimensioning**. With the help of aligned dimensioning, you can measure the true distance between the two points. In horizontal or vertical dimensioning, you can only measure the distance from the first extension line origin to the second extension line origin along the horizontal or vertical axis, respectively. The working of the **ALIGNED** dimension command is similar to that of the other linear dimensioning commands. The dimension created with the **ALIGNED** command is **parallel to the object being dimensioned.**

Command: **DIMALIGNED** Enter
or
Command: **DIM** Enter
Dim: **ALIGNED** Enter

Figure 6-19 illustrates aligned dimensioning. The prompt sequence for aligned dimensioning is same as that for linear dimensions. Only difference is that you have to select **Aligned dimension** option. Following is the prompt sequence for the Aligned dimensioning command:

Command: **DIM** [Enter]
Dim: **ALIGNED** [Enter]
Specify first extension line origin or <select object>: *Select the first extension line origin.*
Specify second extension line origin: *Select the second extension line origin.*
Specify dimension line location or [MText/Text/Angle]: *Specify the dimension line location.*
Enter Dimension text <measured>: *Enter new text to override the distance computed by AutoCAD LT or press ENTER to retain the default dimension text.*

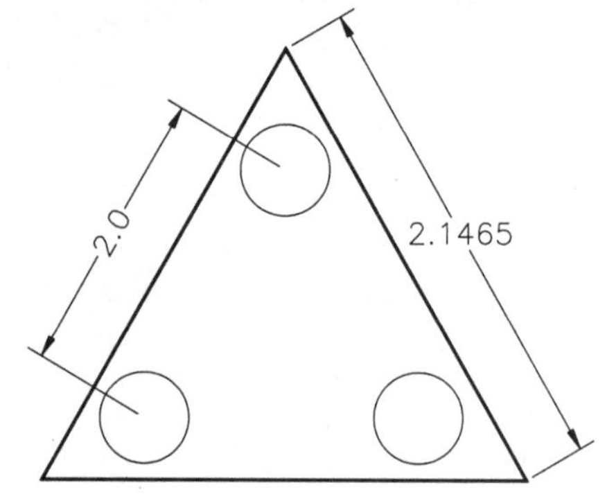

Figure 6-19 *Aligned dimensioning*

You can also respond to the **First extension line origin or <select object>** prompt by pressing ENTER. AutoCAD LT will prompt you to select the object to dimension. Once you select the object, AutoCAD LT will automatically align the dimension with the selected object.

Note

If you invoke this command from the ***Dimension*** *menu, enter* ***T*** *at the* ***Specify dimension line location or [MText/Text/Angle]:*** *prompt.*

Exercise 1 *Mechanical*

Draw Figure 6-20, and then use the **DIMLINEAR** and **DIMALIGNED** commands to dimension the part. The distance between the dotted lines is 0.5 units. The dimensions should be upto 2 decimal places. To get dimensions up to 2 decimal places, enter DIMDEC at Command prompt and then enter 2. (For more information about dimension variable, see Chapter 8)

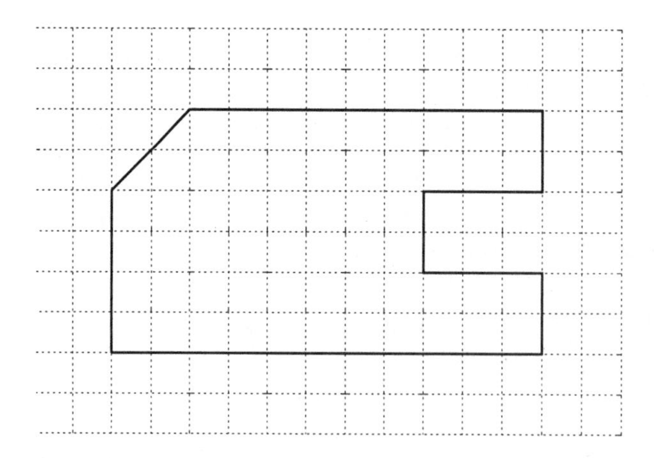

Figure 6-20 *Using the* ***DIMLINEAR*** *and* ***DIMALIGNED*** *commands to dimension the part*

ROTATED DIMENSIONING

Rotated dimensioning is used when you want to place the dimension line at an angle (if you do not want to align the dimension line with the extension line origins selected) (Figure 6-21). The **ROTATED** dimension option will prompt you to specify the dimension line angle. You can invoke this command by using the **DIMLINEAR** (**Rotate option**) or by entering **ROTATED** at the **Dim:** command prompt. The prompt sequence is:

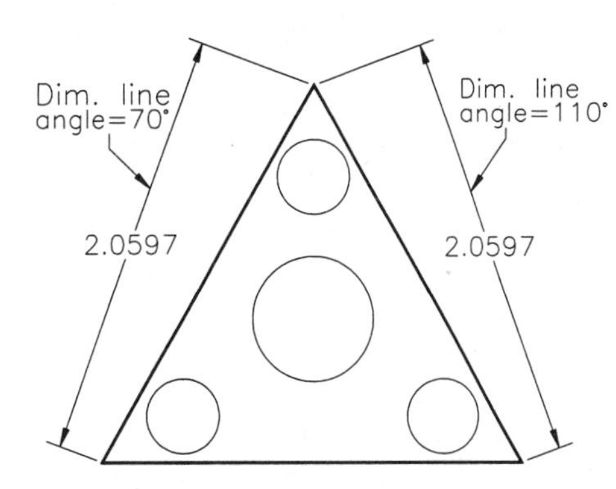

Figure 6-21 *Rotated dimensioning*

Dim: **ROTATED** [Enter]
Specify angle of dimension line <0>:
110 [Enter]
Specify first extension line origin or <select object>: *Select the lower right corner of the triangle.*
Specify second extension line origin: *Select the top corner.*
Specify dimension line location or [MText/Text/Angle]: *Select the location for the dimension line.*
Enter dimension text <2.0597>: [Enter]

Note

You can draw horizontal and vertical dimensioning by specifying the rotation angle of 0 degrees for horizontal dimensioning and 90 degrees for vertical dimensioning.

BASELINE DIMENSIONING

Toolbar:	Dimension > Baseline Dimension
Menu:	Dimension > Baseline
Command:	DIMBASELINE

Sometimes in manufacturing you may want to locate different points and features of a part with reference to a fixed point (base point or reference point). This can be accomplished by using **BASELINE dimensioning** (Figure 6-22). With this command you can continue a linear dimension from the first extension line origin of the first dimension. The new dimension line is automatically offset by a fixed amount to avoid drawing a dimension line and dimension text on top of the previous dimension. There must already be a linear, ordinate, or angular associative dimension to use the Baseline or

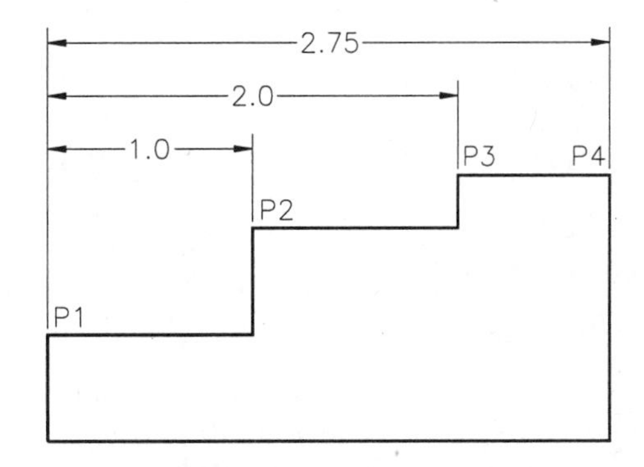

Figure 6-22 *Baseline dimensioning*

Continue dimensions. The prompt sequence for Baseline dimension is:

Command: **DIMLINEAR** Enter
Specify first extension line origin or <select object>: *Select left corner (P1). (Use Endpoint object snap.)*
Specify second extension line origin: *Select the origin of the second extension line (P2).*
Specify dimension line location or
[Mtext/Text/Angle/Horizontal/Vertical/Rotated]: *Select the location of the first dimension line.*
Dimension text = **1.0000** Enter

Command: **DIMBASELINE** Enter
Specify a second extension line origin or [Undo/Select]<Select>: *Select the origin of the next extension line (P3).*
Dimension text = **2.0000** Enter
Specify a second extension line origin or [Undo/Select]<Select>:

When you use the **DIMBASELINE** command, you cannot change the default dimension text. The **DIM** mode commands allows you to override the default dimension text.

Command: **DIM** Enter
Dim: **HOR** Enter
Specify first extension line origin or <select object>: *Select left corner (P1). (Use Endpoint object snap.)*
Specify second extension line origin: *Select the origin of the second extension line (P2).*
Specify dimension line location or [MText/Text/Angle]: **T** Enter
Enter dimension text <1.0000>: **1.0** Enter
Specify dimension line location or [MText/Text/Angle]: *Select the dimension line location.*
Dim: **BASELINE (or BAS)** Enter
Specify a second extension line origin or [Select]<Select>: *Select the origin of the next extension line (P3).*
Enter dimension text <2.0000>: **2.0** Enter
Dim: **BAS** Enter
Specify a second extension line origin or [Select]<Select>: *Select the origin of next extension line (P4).*
Enter dimension text <3.000>: **2.75** Enter

The next dimension line is automatically spaced and drawn by AutoCAD LT.

CONTINUE DIMENSIONING

Toolbar:	Dimension > Continue Dimension
Menu:	Dimension > Continue
Command:	DIMCONTINUE

With this command you can continue a linear dimension from the second extension line of the previous dimension. This is also called as **chained** or **incremental dimensioning**. The **DIM** command should be used if you want to change the text (Figure 6-23).

Command: **DIM** [Enter]
Dim: **HOR** [Enter]
Specify first extension line origin or <select object>: *Select left corner (P1). (Use Endpoint object snap.)*
Specify second extension line origin: *Select the origin of the second extension line (P2).*
Specify dimension line location or [MText/Text/Angle]: **T** [Enter]
Enter dimension text <1.0000>: **1.0** [Enter]
Specify dimension line location or [MText/Text/Angle]: *Select the dimension line location.*
Dim: **CONTINUE** [Enter]
Specify a second extension line origin or [Select]<Select>: *Select the origin of the next extension line (P3).*
Enter dimension text <1.0000>: [Enter]
Dim: **CONTINUE** [Enter]
Specify a second extension line origin or [Select]<Select>: *Select the origin of the next extension line (P4).*
Enter dimension text <0.7200>: **0.7500** [Enter]

The default base (first extension line) for the dimensions created with the **CONTINUE** command is the previous dimension's second extension line. You can override the default by pressing ENTER at the **Specify a second extension line origin or [Select]<Select>** prompt, and then specifying the other dimension. The extension line origin nearest to the selection point is used as the origin for the first extension line.

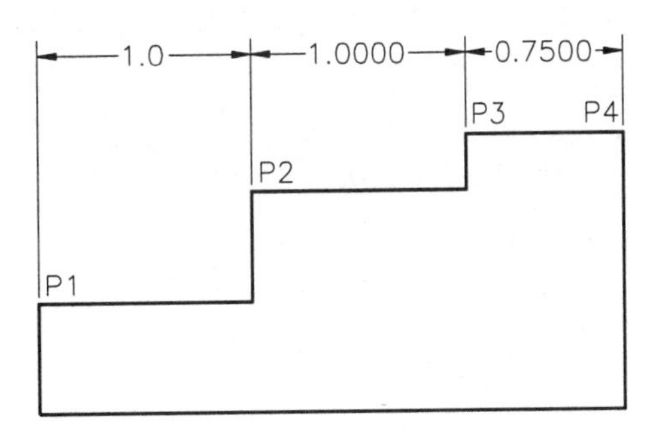

Figure 6-23 Continue dimensioning

Note

When you use the ***DIMCONTINUE*** *command, AutoCAD LT does not let you change the default dimension text. Use the* ***DIM*** *command if you want to override default dimension text.*

There must already be a linear, angular, or ordinate dimension for the ***CONTINUE*** *or* ***BASELINE*** *dimension commands.*

Exercise 2 — *General*

Make the drawing in Figure 6-24. Then use the **BASELINE** command to dimension the top half and **CONTINUE** to dimension the bottom half. The distance between the dotted lines is 0.5 units.

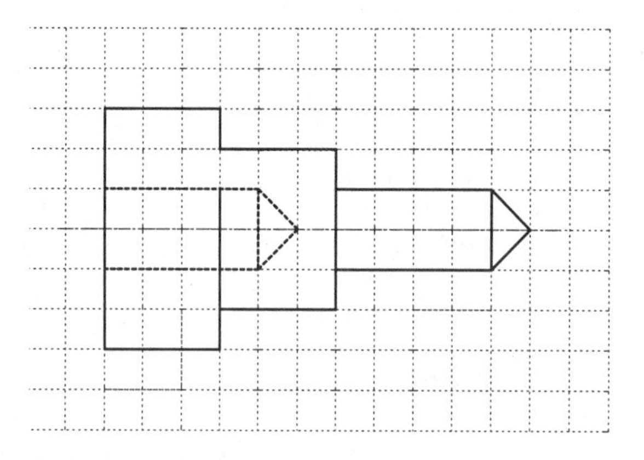

Figure 6-24 *Drawing for Exercise 2*

ANGULAR DIMENSIONING

Toolbar:	Dimension > Angular Dimension
Menu:	Dimension > Angular
Command:	DIMANGULAR

Angular dimensioning is used when you want to dimension an angle. This command generates a dimension arc (dimension line in the shape of an arc with arrowheads at both ends) to indicate the angle between two nonparallel lines. This command can also be used to dimension the vertex and two other points, a circle with another point, or the angle of an arc. For every set of points there exists one acute angle and one obtuse angle (inner and outer angles). If you specify the dimension arc location between the two points, you will get the acute angle; if you specify it outside the two points, you will get the obtuse angle. Figure 6-25 shows the four ways to dimension two nonparallel lines.

Command: **DIMANGULAR** [Enter]
Select arc, circle, line, or <specify vertex>:

Dimensioning the Angle Between Two Nonparallel Lines

The angle between two nonparallel lines or two straight line segments of a polyline can be dimensioned with the **DIMANGULAR** dimensioning command. The vertex of the angle is taken as the point of intersection of the two lines. The location of the extension lines and dimension arc is determined by how you specify the dimension arc location.

The following example illustrates the dimensioning of two nonparallel lines using the **DIMANGULAR** command:

Command: **DIMANGULAR** [Enter]
Select arc, circle, line, or <specify vertex>: *Select the first line.*
Select second line: *Select the second line.*

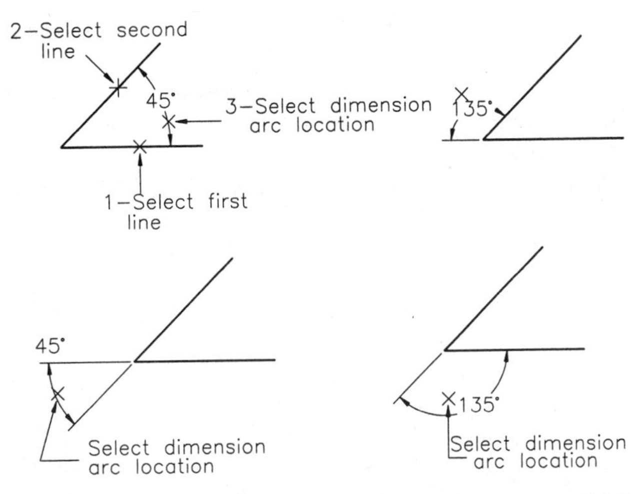

Figure 6-25 *Angular dimensioning between two nonparallel lines*

Specify dimension arc line location or [Mtext/Text/Angle]: **M** *(Enter the new value in the* ***Multiline Text Editor*** *dialog box.)*
Specify dimension arc line location or [Mtext/Text/Angle]: *Specify the dimension arc location.*

Dimensioning the Angle of an Arc

Angular dimensioning can also be used to dimension the angle of an arc. In this case, the center point of the arc is taken as the vertex and the two endpoints of the arc are used as the extension line origin points for the extension lines (Figure 6-26). The following example illustrates the dimensioning of an arc using the **DIMANGULAR** command. The prompt sequence is as follows:

Command: **DIMANGULAR** [Enter]
Select arc, circle, line, or <specify vertex>: *Select the arc.*
Specify dimension arc line location or [Mtext/Text/Angle]: **T** [Enter].
Enter dimension text <current value>: *Press ENTER or specify a new value.*
Specify dimension arc line location or [Mtext/Text/Angle]: *Specify a location for arc line.*

Angular Dimensioning of Circles

The angular feature associated with the circle can be dimensioned by selecting a circle object at the **Select arc, circle, line, or <specify vertex>** prompt. The center of the selected circle is used as the vertex of the angle. The first point selected (when the circle is selected for angular dimensioning) is used as the origin of the first extension line. In a similar manner, the second point selected is taken as the origin of the second extension line (Figure 6-27). The following is the prompt sequence for dimensioning a circle:

Command: **DIMANGULAR** [Enter]
Select arc, circle, line, or <specify vertex>: *Select the circle at the point where you want the first extension line.*
Specify second angle endpoint: *Select the second point on or away from the circle.*
Specify dimension arc line location or [Mtext/Text/Angle]: **M** *(Enter the new value in the* ***Multiline Text Editor*** *dialog box.)*
Specify dimension arc line location or [Mtext/Text/Angle]: *Select the location for the dimension line.*

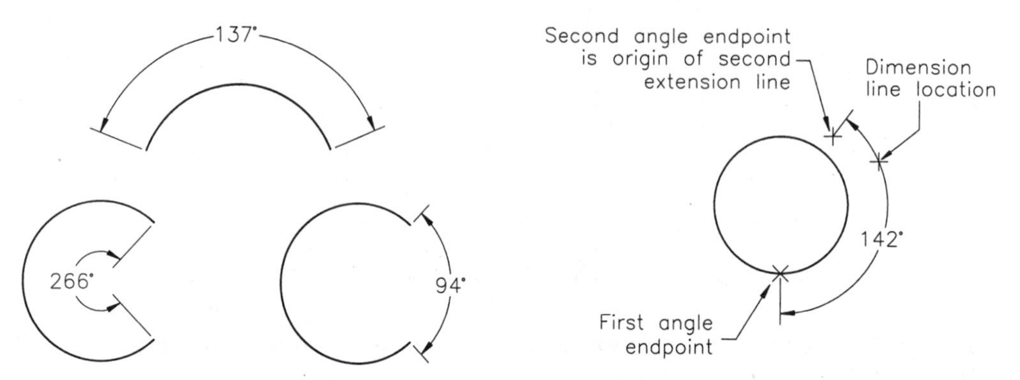

Figure 6-26 *Angular dimensioning of arcs* ***Figure 6-27*** *Angular dimensioning of circles*

Angular Dimensioning Based on Three Points

If you press ENTER at the **Select arc, circle, line, or <specify vertex>:** prompt, AutoCAD LT allows you to select three points to create an angular dimension. The first point is vertex point, and other two points are first and second angle endpoints of the angle (Figure 6-28). The coordinate specifications of the first and second angle endpoints must not be identical. However, the angle vertex coordinates and one of the angle endpoint coordinates can be identical. The following example illustrates angular dimensioning by defining three points:

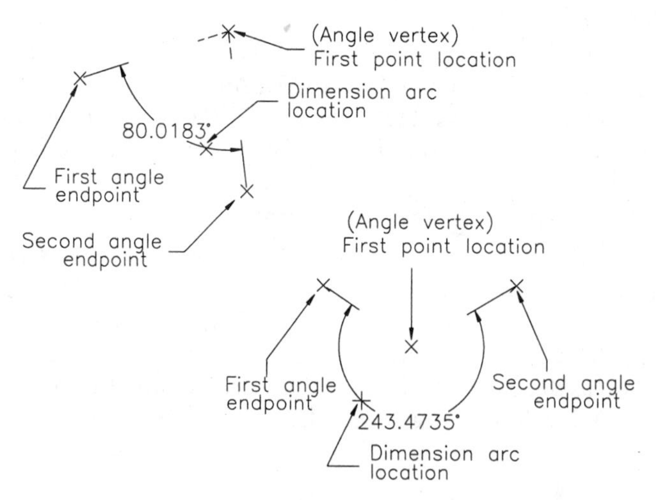

Figure 6-28 *Angular dimensioning for three points*

Command: **DIMANGULAR** [Enter]
Select arc, circle, line, or <specify vertex>: *Press ENTER.*
Specify angle vertex: *Specify the first point, vertex.*
Specify first angle endpoint: *Specify the second point.*
Specify second angle endpoint: *Specify the third point.*
Specify dimension arc line location or [Mtext/Text/Angle]: *Select the location for the dimension line.*

Note

*If you use the **DIMANGULAR** command, you cannot specify the text location. With the **DIMANGULAR** command, AutoCAD LT positions the dimensioning text automatically. If you want to position the dimension text yourself, use the **DIM, ANGULAR** command.*

Exercise 3 *Mechanical*

Make the drawing in Figure 6-29. Then use the **DIMANGULAR** command to dimension all angles of the part. The distance between the dotted lines is 0.5 units.

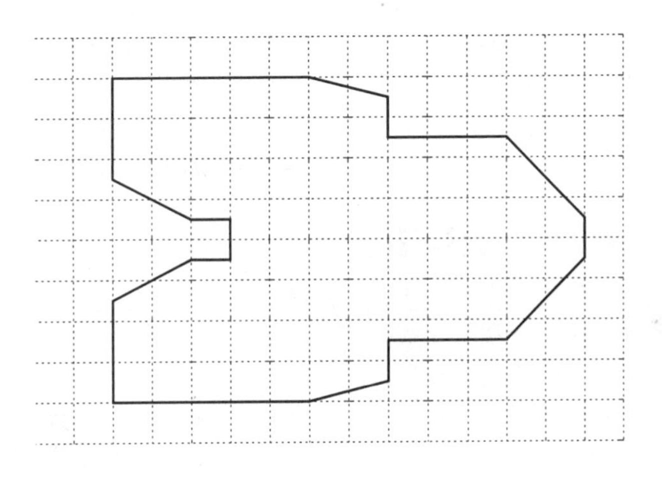

Figure 6-29 *Drawing for Exercise 3*

DIAMETER DIMENSIONING

Toolbar:	Dimension > Diameter Dimension
Pull-down:	Dimension > Diameter
Command:	DIMDIAMETER

Diameter dimensioning is used to dimension a circle; you can also use it to dimension an arc. Here, the measurement is done between two diametrically opposite points on the circumference of the circle or arc (Figure 6-30). The dimension text generated by AutoCAD LT commences with the ø symbol, to indicate a diameter dimension. The following is the prompt sequence for dimensioning a circle:

Command: **DIMDIAMETER** Enter
Select arc or circle: *Select an arc or circle by selecting a point anywhere on its circumference.*
Dimension text= Current
Specify dimension line location or [Mtext/Text/Angle]: *Specify a point to position the dimension.*

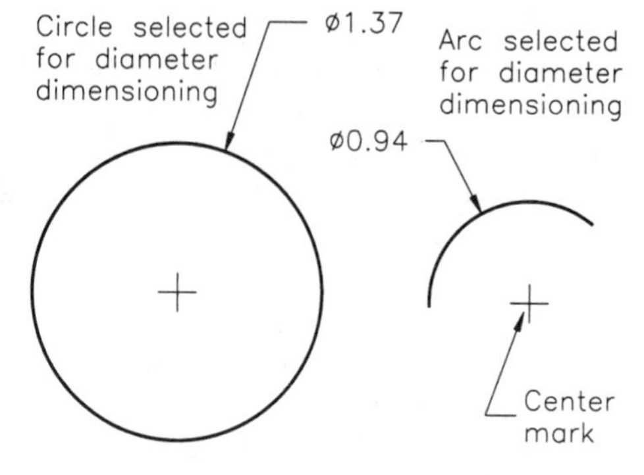

Figure 6-30 Diameter dimensioning

If you want to override the default value of the dimension text, use the Mtext or Text option.

Command: **DIMDIAMETER** Enter
Select arc or circle: *Select an arc or circle.*
Dimension text= Current
Specify dimension line location or [Mtext/Text/Angle]: **T** Enter
Enter dimension text <current>: **%%C2.5** Enter
Specify dimension line location or [Mtext/Text/Angle]: *Move the cursor and specify a point.*

The control sequence %%C is used to obtain the diameter symbol Ø. It is followed by the dimension text you want to appear in the diameter dimension.

Note

The control sequence %%d can be used to generate the degree symbol "o" (45 °).

RADIUS DIMENSIONING

Toolbar:	Dimension > Radius Dimension
Menu:	Dimension > Radius
Command:	DIMRADIUS

Radius dimensioning is used to dimension a circle or an arc (Figure 6-31). Radius and diameter dimensioning are similar; the only difference is that instead of diameter line, a radius line is drawn (half of the diameter line) which is measured from center to any point on the circumference. The dimension text generated by AutoCAD LT is preceded by the letter **R** to indicate a radius dimension. If you want to use the default dimension text (dimension text generated automatically by AutoCAD LT), simply specify a point to position dimension at the **Specify dimension line location or [Mtext/Text/Angle]** prompt. You can also enter a new value or specify a prefix or suffix, or suppress the entire text by entering a blank space following the **Enter dimension text <current>** prompt. A center mark for circle/arc is drawn automatically, provided

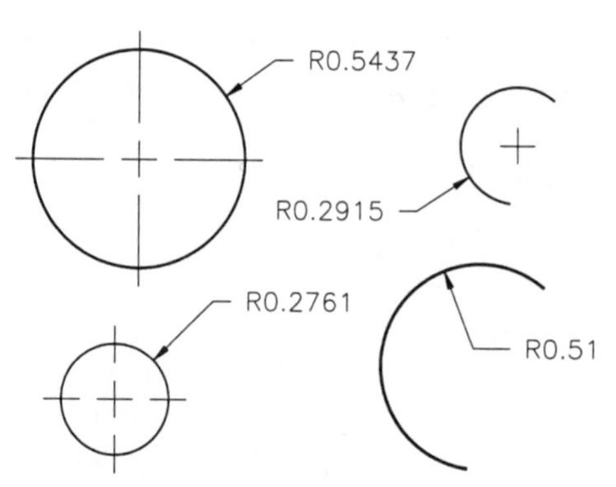

Figure 6-31 Radius dimensioning

the center mark value controlled by the **DIMCEN** variable is not 0.

Command: **DIMRADIUS** [Enter]
Select arc or circle: *Select the object you want to dimension.*
Dimension text= Current
Specify dimension line location or [Mtext/Text/Angle]: *Move the cursor and select a point.*

If you want to override the default value of the dimension text, use the Text option, as follows:

Command: **DIMRADIUS** [Enter]
Select arc or circle: *Select an arc or circle.*
Dimension text= current
Specify dimension line location or [Mtext/Text/Angle]: **T** [Enter]
Enter dimension text <current>: **R0.25** [Enter]
Specify dimension line location or [Mtext/Text/Angle]: *Move the cursor and select a point.*

GENERATING CENTER MARKS AND CENTERLINES

Toolbar:	Dimension > Center Mark
MENU:	Dimension > Center Mark
Command:	DIMCENTER

When circles or arcs are dimensioned with the **DIMRADIUS** or **DIMDIAMETER** command, a small mark known as a center mark may be drawn at the center of the circle/arc (Figure 6-32). Sometimes you may want to mark the center of a circle or an arc without using these dimensioning commands. This can be achieved by entering **DIMCENTER** at the Command prompt or **CENTER** (or **CEN**) at the **Dim:** prompt. You can control the size of the center mark by changing the size in the **Dimension Style Manager** dialog box or by using the **DIMCEN** variable. When the value of **DIMCEN** is negative, the centerlines are drawn (Figure 6-33).

Command: **DIMCENTER** [Enter]
Select arc or circle: *Select arc or circle.*

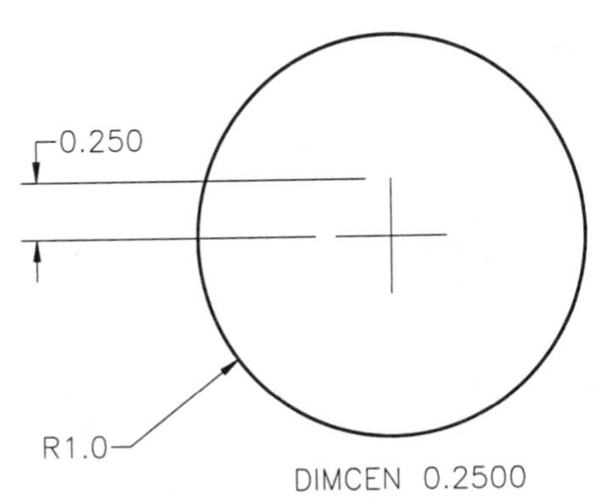

Figure 6-32 *Using the **DIMCEN** variable to control the size of the center mark*

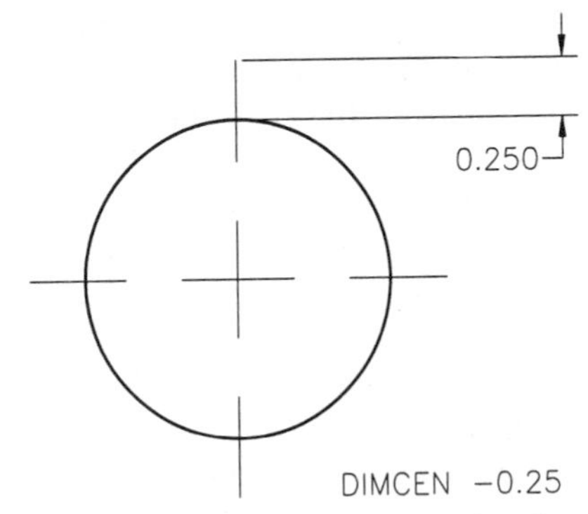

Figure 6-33 *Using a negative value for **DIMCEN***

Note

*The center marks created by **DIMCENTER** or **DIM, CENTER** are lines, not associative dimensioning objects, and they have an explicit linetype.*

Exercise 4 *Mechanical*

Make the drawing in Figure 6-34. Then use the **DIMRADIUS** and **DIMDIAMETER** commands to dimension the part. Use the **DIMCENTER** command to draw the centerlines through the circles. The distance between the dotted lines is 0.5 units.

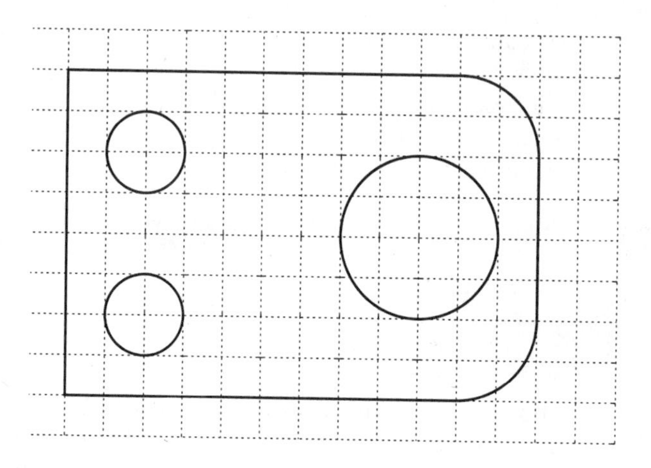

***Figure 6-34** Using the **DIMRADIUS** and **DIMDIAMETER** commands to dimension a part*

ORDINATE DIMENSIONING

Toolbar:	Dimension > Ordinate
Menu:	Dimension > Ordinate
Command:	DIMORDINATE

Ordinate dimensioning is also known as **arrowless** dimensioning because no arrowheads are drawn in this type of dimensioning. Ordinate dimensioning is also called datum dimensioning because all dimensions are related to a common base point. The current UCS (user coordinate system) origin becomes the reference or the base point for ordinate dimensioning. With ordinate dimensioning you can determine the X or Y displacement of a selected point from the current UCS origin. To locate the UCS, you can use the **UCS** command.

In ordinate dimensioning, AutoCAD LT automatically places the dimension text (X or Y coordinate value) and the leader line along the X or Y axis (Figure 6-35). Since the ordinate dimensioning pertains to either the X coordinate or the Y coordinate, you should keep ORTHO on. When ORTHO is off, the leader line is automatically given a bend when you select the second leader line point that is offset from the first point. This allows you to generate offsets and avoid overlapping text on closely spaced dimensions. In ordinate dimensioning, only one extension line (leader line) is drawn.

The leader line for an X coordinate value will be drawn perpendicular to the X axis, and the leader line for a Y coordinate value will be drawn perpendicular to the Y axis. Since you cannot override this, the leader line drawn perpendicular to the X axis will have the dimension text aligned with the leader line. The dimension text is the X datum of the selected point. The leader line drawn perpendicular to the Y axis will have the dimension text, which is the Y datum of the selected point, aligned with the leader line. Any other alignment specification for the dimension text is nullified. Hence, changes in Text Alignment in the **New/Modify/Override Dimension Style Manager** dialog box (**DIMTIH** and **DIMTOH** variables) have no effect on the alignment of dimension text. You can specify the coordinate value you want to dimension at the **Specify leader endpoint or [Xdatum/Ydatum/MText/Text/Angle]** prompt.

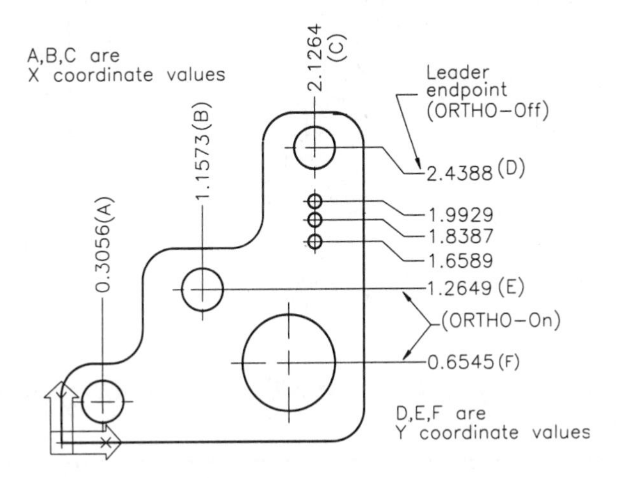

Figure 6-35 *Ordinate dimensioning*

If you select or enter a point, AutoCAD LT checks the difference between the feature location and the leader endpoint. If the difference between the X coordinates is greater, the dimension measures the Y coordinate; otherwise, the X coordinate is measured. In this manner AutoCAD LT determines whether it is an X or Y type of ordinate dimension. However, if you enter Y instead of specifying a point, AutoCAD LT will dimension the Y coordinate of the selected feature. Similarly, if you enter X, AutoCAD LT will dimension the X coordinate of the selected point. The following is the prompt sequence for ordinate dimensioning:

Command: **DIMORDINATE** [Enter]
Specify feature location: *Select a point on an object.*
Specify leader endpoint or [Xdatum/Ydatum/Mtext/Text/Angle]: *Enter the endpoint of the leader.*

You can override the text using the **Mtext** or the **Text** options.

Command: **DIMORDINATE** [Enter]
Specify feature location: *Select a point on an object.*

Specify leader endpoint or [Xdatum/Ydatum/Mtext/Text/Angle]: **M** *(Enter the new value in the **Multiline Text Editor** dialog box.)*
Specify leader endpoint or [Xdatum/Ydatum/Mtext/Text/Angle]: *Specify the endpoint of the leader line.*

Exercise 5 *Mechanical*

Make the drawing in Figure 6-36. Then use the **DIMORDINATE** command to dimension the part. The distance between the dotted lines is 0.5 units.

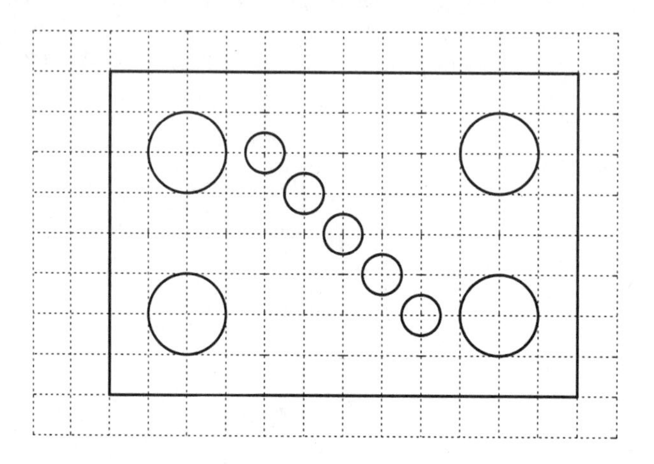

Figure 6-36 *Using the **DIMORDINATE** command to dimension the part*

DRAWING LEADERS

The leader line is used to attach annotations to an object or when user wants to show a dimension without using another dimensioning command. Sometimes leaders for circles or arcs are so complicated that you need to construct a leader of your own rather than using the leader generated by radius or diameter dimensioning (**DIMDIAMETER** or **DIMRADIUS** command). A leader drawn by using the **LEADER** or **QLEADER** command creates the arrow and the leader lines as a single object. The text is created as a separate object. These commands can create multiline annotations and offers several options like using Mtext or copying existing annotations.

QLEADER Command

Toolbar:	Dimension > Quick Leader
Menu:	Dimension > Leader
Command:	QLEADER

The **QLEADER** command creates a leader and annotation. Choosing **Leader** from the **Dimension** menu and choosing the **Quick Leader** button from the **Dimension** toolbar, executes the **QLEADER** command. You can customize the leader and annotation by selecting the Settings option at the **Specify first leader point, or**

[Settings]<Settings>: prompt. The following is the Command prompt sequence for the **QLEADER** command:

Command: **QLEADER** Enter
Specify first leader point, or [Settings]<Settings>:*Specify the start point of the leader.*
Specify next point: *Specify end point of the leader.*
Specify next point: *Specify next point.*
Specify text width <current>: *Enter text width of multiline text.*
Enter first line of annotation text <MText>: *Press ENTER; AutoCAD LT displays the **Multiline Text Editor** dialog box. Enter text in the dialog box and then choose **OK** to exit.*

Press ENTER at the **Specify first leader point or [Settings]<Settings>:** prompt; the **Leader Settings** dialog box is displayed (Figure 6-37). The **Leader settings** dialog box gives you a number of options for the leader line and the text attached to it. It has the following tabs:

Annotation Tab

Gives you various options to control annotation features like Annotation Type, Mtext Options, and, Annotation Reuse.

Mtext. When selected, AutoCAD LT uses the **Multiline Text Editor** to create annotation. Three Mtext options are available.

Prompt for width. Selecting this check box, allows you to specify width of mtext annotation.

Always left justify. Left justifies the mtext annotation in all situations. Selecting this check box makes the **Prompt for width** option unavailable.

Frame Text. Selecting this check box draws a box around the mtext annotation.

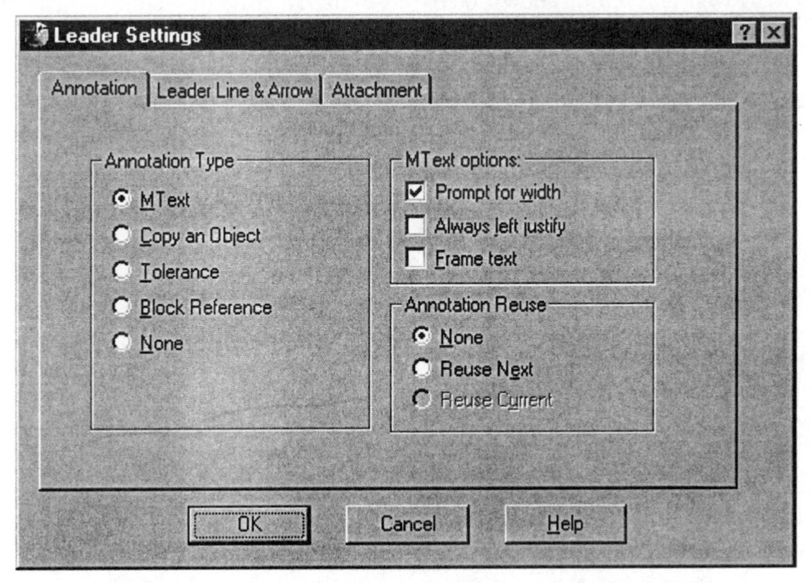

*Figure 6-37 **Leader Settings** dialog box, **Annotation** tab*

Copy an Object. The copy option allows you to copy an existing annotation object (like mtext, single line text, tolerance, or block) and attach it at the end of the leader. For example, if you have a text string in the drawing that you want to place at the end of the leader, you can use the **Copy an Object** option to place it at the end of the leader.

Tolerance. When you select the **Tolerance** option, AutoCAD LT displays the **Geometric Tolerance** dialog box on the screen. Specify the tolerance in the dialog box and choose OK to exit. AutoCAD LT will place the specified Geometric Tolerance with feature control frame at the end of the leader (Figure 6-38).

Block. The **Block** option allows you to insert a predefined block at the end of the leader. When you select this option, AutoCAD LT will prompt you to enter the block name, and insertion point.

None. Creates a leader without placing any annotation at the end of the leader.

Annotation Reuse. This option allows you to reuse the annotation.

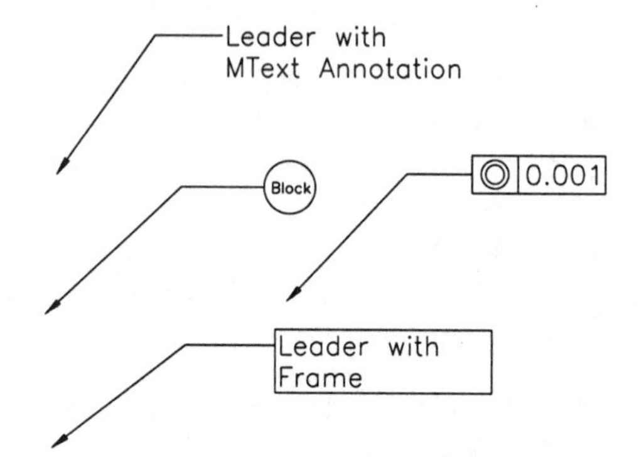

***Figure 6-38** Using the **QLEADER** command to draw a leader with MText, Block, Tolerance, and MText with Frame annotations*

None. When selected, AutoCAD LT does not reuse the leader annotation.

Reuse Next. This option allows you to reuse the annotation that you are going to create next for all subsequent leaders.

Reuse Current. This option allows you to reuse current annotation for all subsequent leaders.

Leader Line & Arrow Tab

Leader Line. Gives the options for the leader line type like straight or spline (Figure 6-39). The **Spline** option draws a spline through the specified leader point and the **Straight** option creates straight lines.

Number of Points. With this option you can specify the number of points you can specify in a leader. For example, if it is 3, AutoCAD LT will prompt you for annotation after you specify two leader points. First point is the leader start point.

Arrowheads. Allows you to define a leader arrowhead from the **Arrowhead** drop-down list. The arrowhead available here are the same as the one available for dimensioning. You can also use the user defined arrows by selecting **User Arrow** from the drop-down list.

Angle Constraint. This option allows you to set the angle of the first and the second leader segments. For example, you can set the first segment to 45 degrees and the second to horizontal to get a leader line at 45 degrees with horizontal extension.

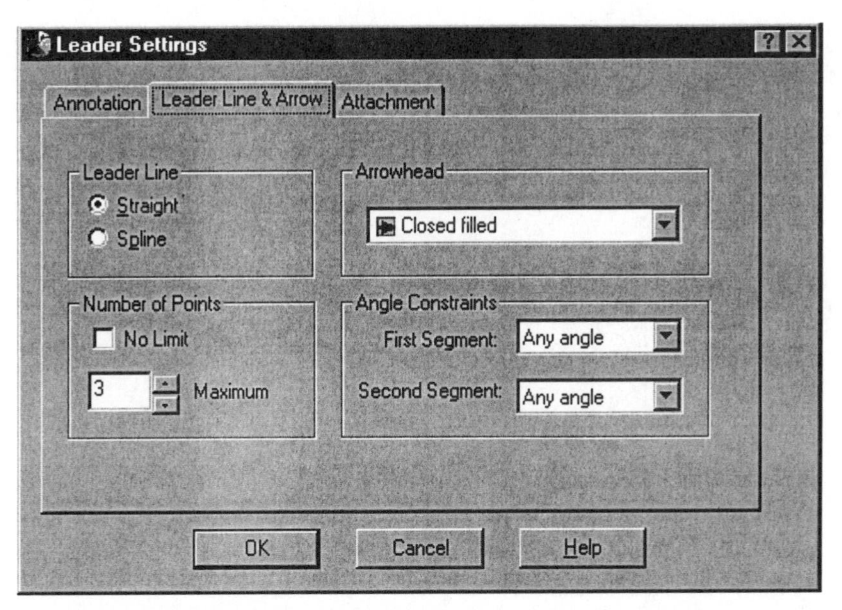

Figure 6-39 ***Leader Settings*** *dialog box,* ***Leader Line & Arrow*** *tab*

Attachment Tab

The **Attachment** tab (Figure 6-40) is available, only if you have selected **MText** in the **Annotation** tab.This tab allows you various options for attaching the Multi-line Text to the leader. It has two columns, **Text on left side** and **Text on right side**. Each column has different justifications. If you draw a leader from right to left, AutoCAD LT uses the settings under Text on left side. Similarly, if you draw a leader from left to right, AutoCAD LT uses the settings as specified under Text on right side.

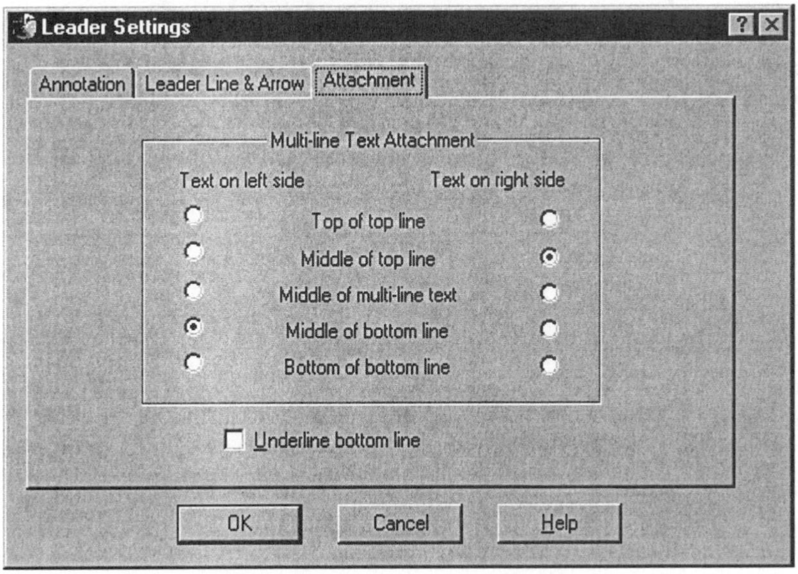

Figure 6-40 ***Leader settings*** *dialog box,* ***Attachment*** *tab*

Exercise 6 — *Mechanical*

Make the drawing in Figure 6-41. Then use the **QLEADER** command to dimension the part as shown. The distance between the dotted lines is 0.5 units.

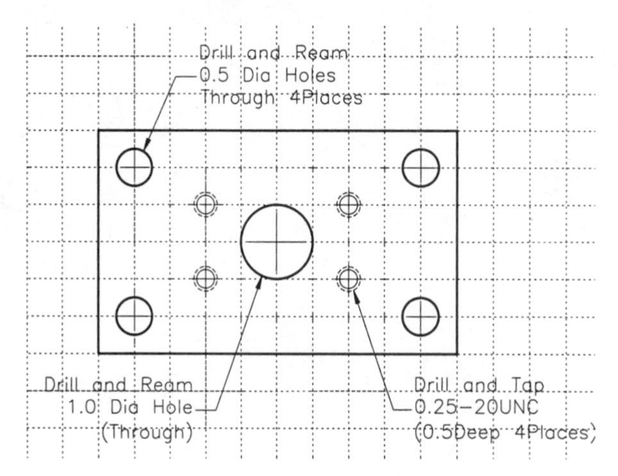

***Figure 6-41** Using the **QLEADER** command to dimension the part*

Using the LEADER Command

The **LEADER** command draws a line that extends from the object being dimensioned to the dimension text (Figure 6-42). The following is prompt sequence of the **LEADER** command:

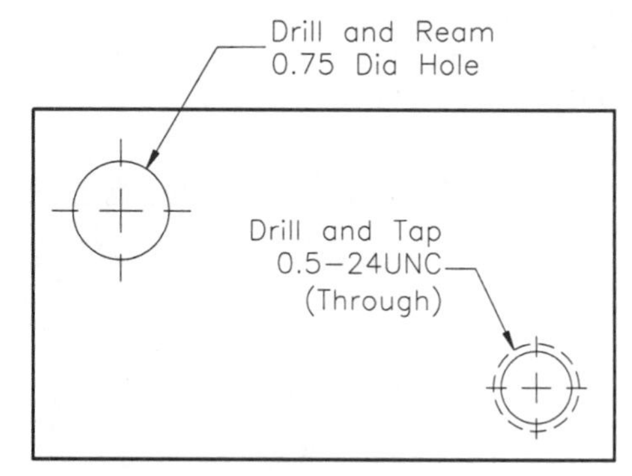

***Figure 6-42** Using the **LEADER** command to draw leaders*

Command: **LEADER** [Enter]
Specify leader start point: *Specify the starting point of the leader.*
Specify next point: *Specify the endpoint of the leader.*
Specify next point or [Annotation/Format/Undo]<Annotation>: *Press ENTER.*
Enter first line of annotation text or <options>: *Enter annotation text (for example, Drill and Ream).*
Enter the next line of annotation text: *Enter annotation text (for example, 0.75 Dia Hole).*
Enter the next line of annotation text: *Press ENTER when done entering text.*

Annotation Options

You can also use the **Copy**, **Tolerance**, **Block**, or **Mtext** options for annotations. The following is the description of the **Copy** option.

Copy. The **Copy** option allows you to copy an existing annotation object and attach it at the end of the leader. For example, if you have a text string in the drawing that you want to place at the end of the leader, you can use the **Copy** option to place it at the end of the leader. The following is the prompt sequence for copying an existing annotation:

Command: **LEADER** [Enter]
Specify leader start point: *Specify the starting point of the leader.*
Specify next point: *Specify the endpoint of the leader.*
Specify next point or [Annotation/Format/Undo]<Annotation>: *Press ENTER.*
Enter first line of annotation text or <options>: *Press ENTER.*
Enter an annotation option [Tolerance/Copy/Block/None/MText]<Mtext>: **C** [Enter]
Select an object to copy: *Select an existing annotation object.*

(Dimension text)
Leader start
Splined Leader
Straight line Leader
Leader start

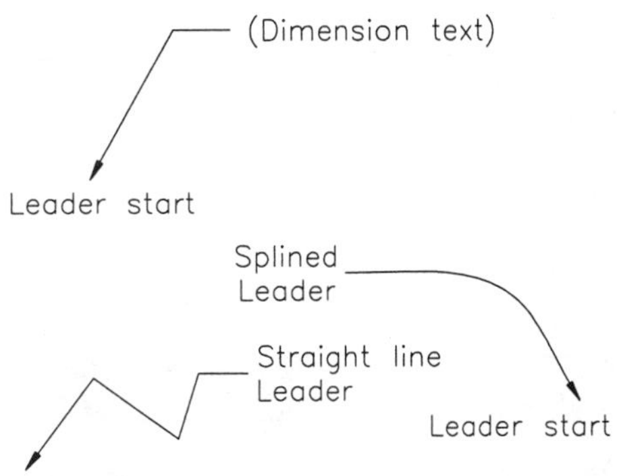

Figure 6-43 *Using the* ***LEADER*** *command to draw splined and straight leaders*

Other options like **Tolerance**, **Block**, **None**, and **MText** are similar to **QLEADER** options, discussed earlier under the **Annotation Tab** heading.

Format Options

If you select the **Format** option, you can control the shape of the leader and Arrow (Figure 6-43). The available options like **Spline**, **Straight**, **Arrow**, and **None** are similar to **QLEADER** options, discussed earlier under **Leader Line & Arrow Tab** heading.

Using Leader with the DIM Command

You can also draw a leader by using the **Leader** option of the **DIM** command. The **DIM, LEADER** command creates the arrow, leader lines, and the text as separate objects. This command has the feature of defaulting to the most recently measured dimension. Once you invoke the **DIM, LEADER** command and specify the first point, the prompt sequence is similar to that of the **LINE** command. The start point of the leader should be specified at the point closest to the object being dimensioned. After drawing the leader, enter a new dimension text or keep the default one.

Command: **DIM** [Enter]
Dim: **Leader** (or **L**) [Enter]
Leader start: *Specify the starting point of the leader.*
To point: *Specify the endpoint of the leader.*
To point: *Specify the next point.*
To point: *Press ENTER.*
Dimension text <current>: *Enter dimension text.*

The value between the angle brackets is the current value that is the measurement of the most recently dimensioned object. If you want to retain the default text, press ENTER at the

Dimension text <current>: prompt. You can enter text of your choice, specify a prefix/suffix, or suppress the text. The text can be suppressed by pressing the SPACEBAR, and then pressing ENTER at the **Dimension text <current>:** prompt. An arrow is drawn at the start point of the leader segment if the length of the segment is greater than two arrow lengths. If the length of the line segment is less than or equal to two arrow lengths, only a line is drawn.

Self-Evaluation Test

Answer the following questions, and then compare your answers to the answers given at the end of this chapter.

1. With the **DIM1** command, more than one dimensioning operation can be performed at a time. (T/F)
2. You can exit the dimension mode by entering E at the **Dim:** prompt. (T/F)
3. Only arrow marks can be generated at the two ends of the dimension line. (T/F)
4. You can specify dimension text of your own or accept the measured value computed by AutoCAD LT. (T/F)
5. Appending suffixes to dimension text is not possible. (T/F)
6. Appending prefixes to dimension text is not possible. (T/F)
7. The size of the dimension arrows can be regulated. (T/F)
8. With alternate units dimensioning, the dimension text generated is for a single system of measurement. (T/F)
9. When using tolerances in dimensioning, if the tolerance values are equal, the tolerances are drawn one over the other. (T/F)
10. Limits are appended to the dimension text. (T/F)

Review Questions

1. Fill in the command and entries required to dimension an inclined line length of 3.95 units. You must change the dimension text from 3.9500 to 4.0. The dimension line must be aligned with the line being dimensioned.

   Command: __________

Specify first extension line origin or <select object>: _________
Specify second extension line origin: _________
Specify dimension line location or [Mtext/Text/Angle]: _________
Enter dimension text <3.9500>: _________
Specify dimension line location or [Mtext/Text/Angle]: _________

2. Fill in the relevant information to dimension an arc. The dimension text should be R0.750. The radius of the arc is 0.75 units. After completing the dimensioning, return to the Command mode.
   Command: _________
   Select arc or circle: _________
   Dimension text = current
   Specify dimension line location or [Mtext/Text/Angle]: _________

3. Fill in the relevant information to dimension the diameter of a circle. The radius of the circle is 2.0 units. The dimension text should have a Ø symbol.
   Command: _________
   Select arc or circle: _________
   Specify dimension line location or [Mtext/Text/Angle]: _________

4. Fill in the blanks with the command and entries required to dimension the angle by defining three points.
   Command: _________
   Select arc, circle, line, or <specify vertex>: _________
   Specify angle vertex: _________
   Specify first angle endpoint: _________
   Specify second angle endpoint: _________
   Specify dimension arc line location or [Mtext/Text/Angle]: _________

5. Give the command and entries required to dimension the angle of an arc.
   Command: _________
   Select arc, circle, line, or <specify vertex>:_________
   Specify dimension arc line location or [Mtext/Text/Angle]: _________

6. Fill in the command and other entries required to draw the center mark for an existing circle/arc.
   Command: _________
   Select arc or circle: _________

7. Why is dimensioning needed?

8. Specify two ways to return to the Command prompt from the **Dim:** prompt (dimensioning mode).

9. What are the differences between the **DIM** and **DIM1** commands?

10. Explain briefly the five fundamental dimensioning types provided by AutoCAD LT. List

the different dimensioning options available in each of them.

11. Explain briefly dimension line, dimension text, and dimension arrows.

12. Describe extension lines and leaders.

13. Explain the term associative dimensioning.

14. Which dimensioning type generates the coordinate value of X or Y of a location in a drawing?

15. In associative dimensioning, the dimensioning objects are drawn as individual items. (T/F)

16. In aligned dimensioning, the dimension text, by default, is aligned with the object being dimensioned. (T/F)

17. Only inner angles (acute angles) can be dimensioned with angular dimensioning. (T/F)

18. In addition to the most recently drawn dimension (the default base dimension), you can use any other linear dimension as the base dimension. (T/F)

19. In continued dimensions, the base for successive continued dimensions is the base dimension's first extension line. (T/F)

20. Horizontal dimensions measure displacement along the Y axis. (T/F)

21. Vertical dimensions measure displacement along the X axis. (T/F)

22. In rotated dimensioning, the dimension line is always aligned with the object being dimensioned. (T/F)

23. The rotated dimension can be specified to get the effect of horizontal or vertical dimension. (T/F)

24. The center point of arc/circle is taken as the vertex angle in the angular dimensioning of the arc/circle. (T/F)

25. The **DIMORDINATE** command is used to dimension the X coordinate of an object only. (T/F)

Exercises

Exercise 7 *Mechanical*

Draw the object shown in Figure 6-44, and then dimension it. Save the figure drawn in the DIMEXR7 drawing file.

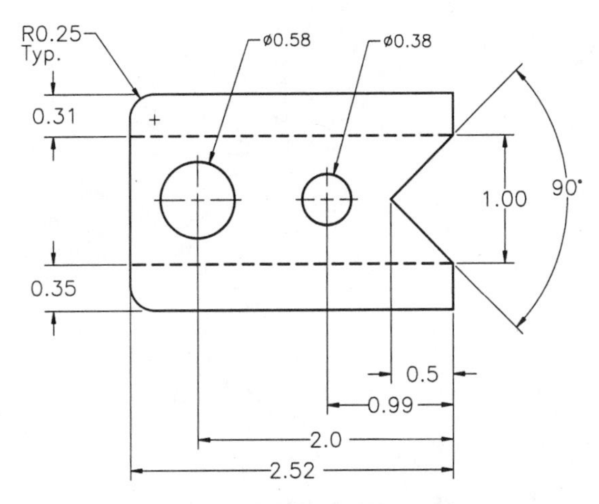

***Figure 6-44** Drawing for Exercise 7*

Exercise 8 *Mechanical*

Draw and dimension the object shown in Figure 6-45. The UCS origin is the lower left corner of the object. Save the drawing as DIMEXR8.

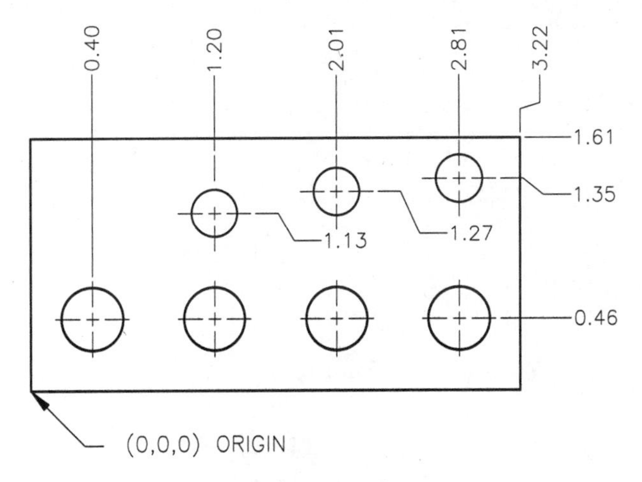

***Figure 6-45** Drawing for Exercise 8*

Exercise 9 Mechanical

Draw the object shown in Figure 6-46, then dimension it. Save the drawing as DIMEXR9.

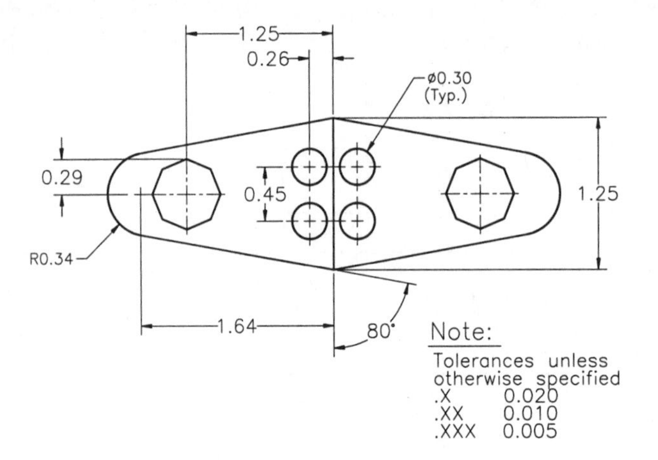

Figure 6-46 *Drawing for Exercise 9*

Exercise 10 Mechanical

Draw the object shown in Figure 6-47, then dimension it. Save the drawing as DIMEXR10.

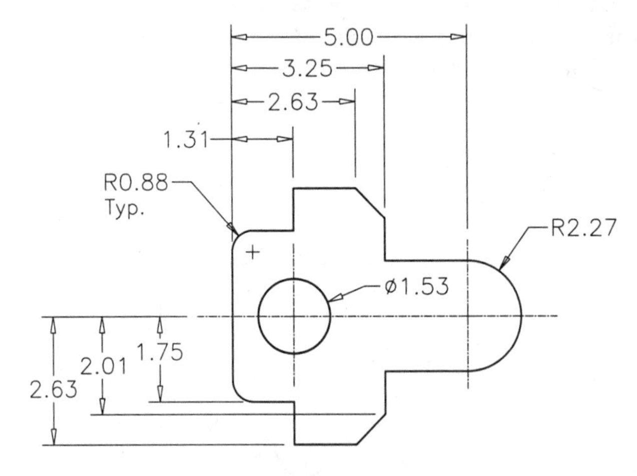

Figure 6-47 *Drawing for Exercise 10*

Exercise 11 *Mechanical*

Draw and dimension the object shown in Figure 6-48 and save it as DIMEXR11. You can use the **ARRAY** command to facilitate drawing the figure.

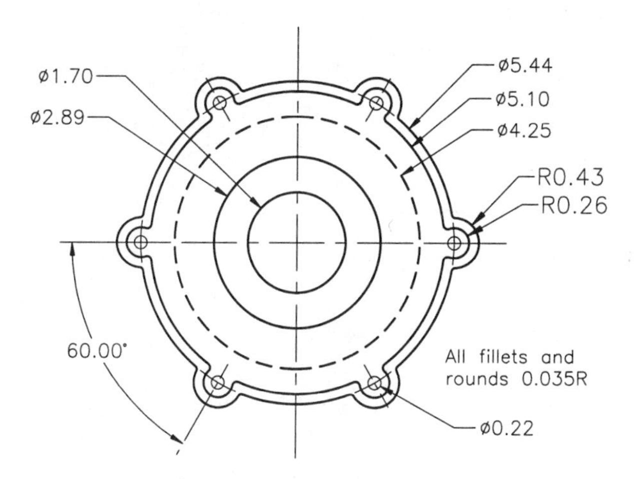

***Figure 6-48** Drawing for Exercise 11*

Exercise 12 *Mechanical*

Draw and dimension Figure 6-49 and save it as DIMEXR12.

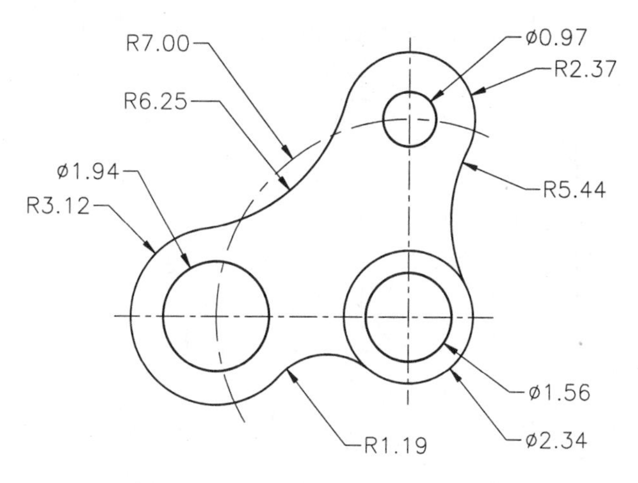

***Figure 6-49** Drawing for Exercise 12*

Exercise 13 — Mechanical

Draw and dimension Figure 6-50 and save it as DIMEXR13

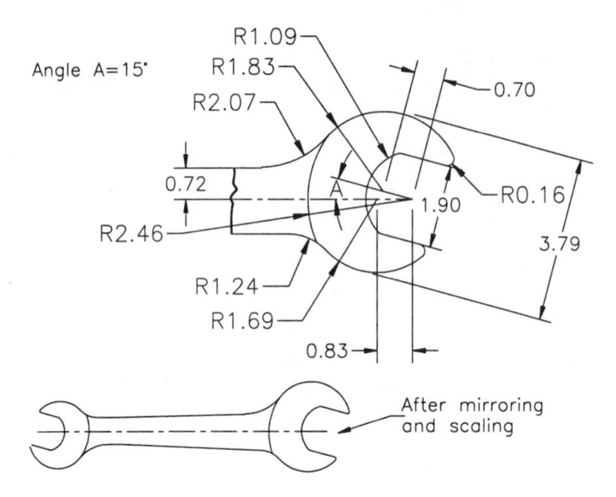

Figure 6-50 *Drawing for Exercise 13*

Exercise 14 — Mechanical

Draw Figure 6-51 and dimension it. Save the drawing as DIMEXR14.

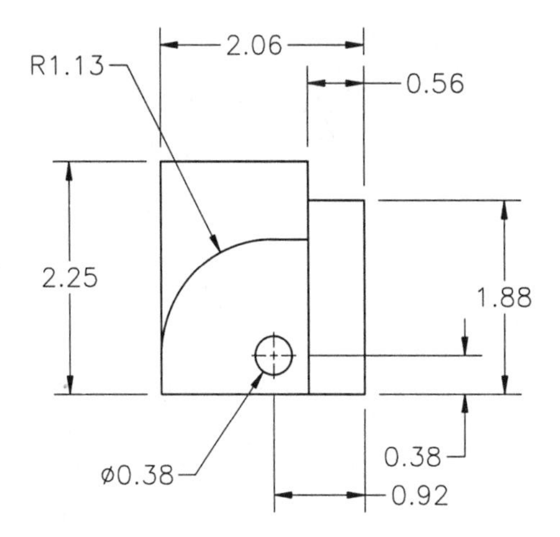

Figure 6-51 *Drawing for Exercise 14*

Exercise 15 *Mechanical*

Draw Figure 6-52 and dimension it. After you have drawn one of the doughnuts, use the **ARRAY** command to obtain other doughnuts. Save the drawing as DIMEXR15.

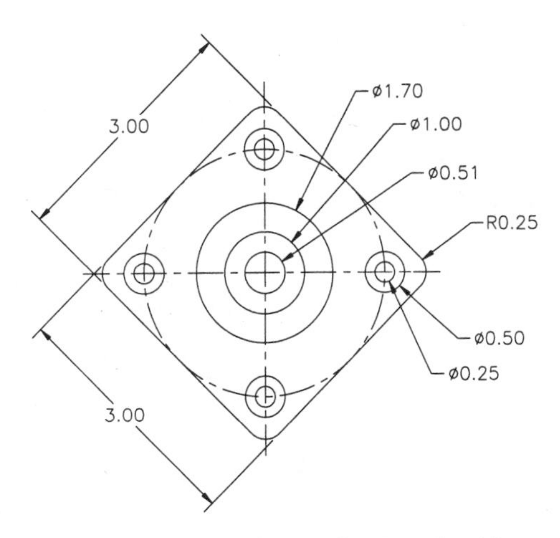

Figure 6-52 *Drawing for Exercise 15*

Exercise 16 *Mechanical*

Draw Figure 6-53 and dimension it as shown in the drawing. Use the **BASELINE** dimension command to dimension the drawing.

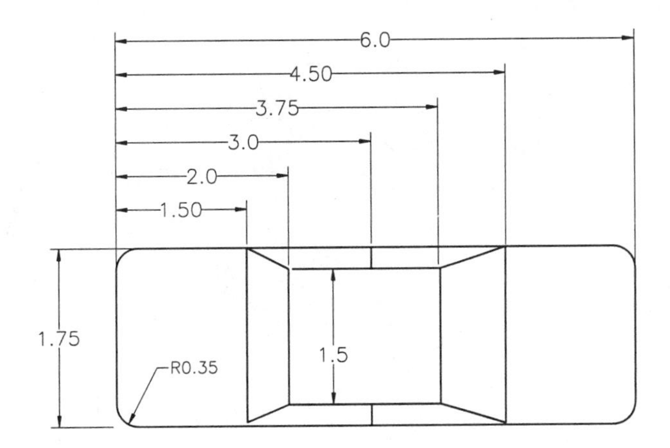

Figure 6-53 *Drawing for Exercise 16*

Problem-Solving Exercise 1 — *Mechanical*

Draw Figure 6-54 and dimension it exactly as shown in the drawing. The **FILLET** command should be used where needed. Save the drawing as DIMPSE1.

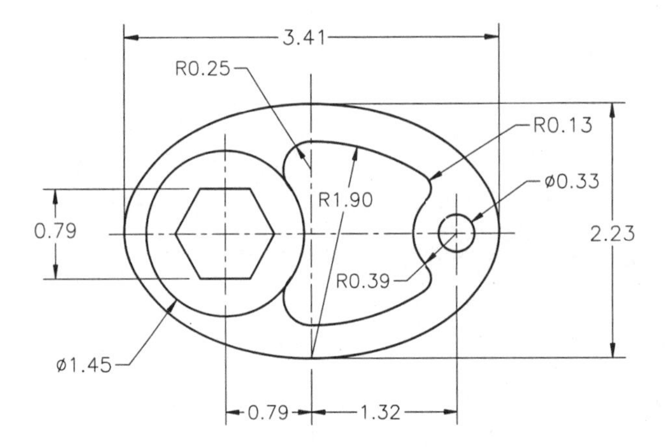

Figure 6-54 *Drawing for Problem-Solving Exercise 1*

Problem-Solving Exercise 2 — *Mechanical*

Draw Figure 6-55 and then dimension it exactly as shown in the drawing. Save the drawing as DIMPSE2.

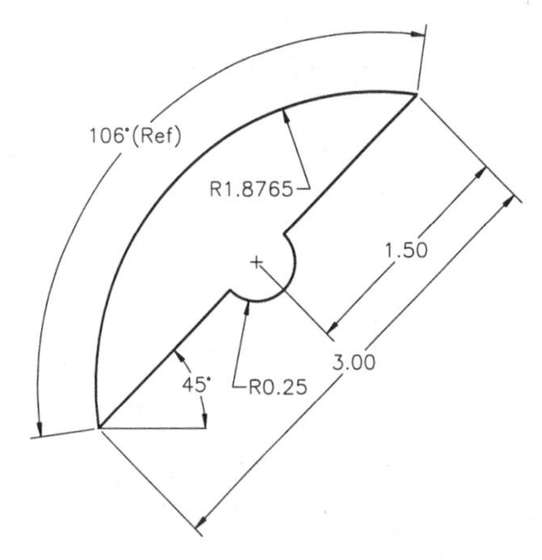

Figure 6-55 *Drawing for Problem-Solving Exercise 2*

Problem Solving Exercise 3 *Mechanical*

Draw Figure 6-56 and then dimension it exactly as shown in the drawing. Save the drawing as DIMPSE3.

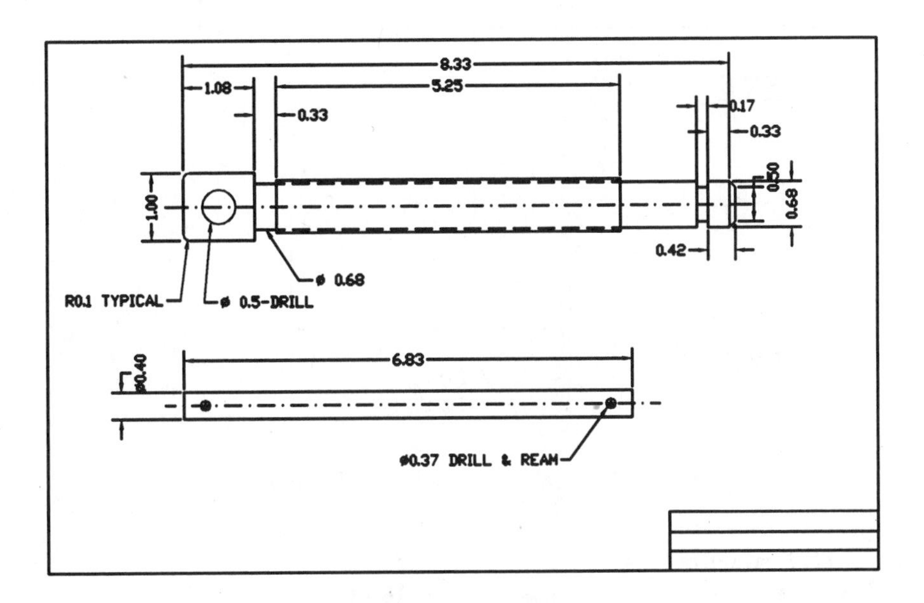

Project Exercise 1-3 *Mechanical*

Generate the drawing in Figure 6-57 according to the dimensions shown in the figure. Save the drawing as BENCH5-6. (See Chapters 1, 2, 7, 19, and 23 for other drawings of Project Exercise 1.)

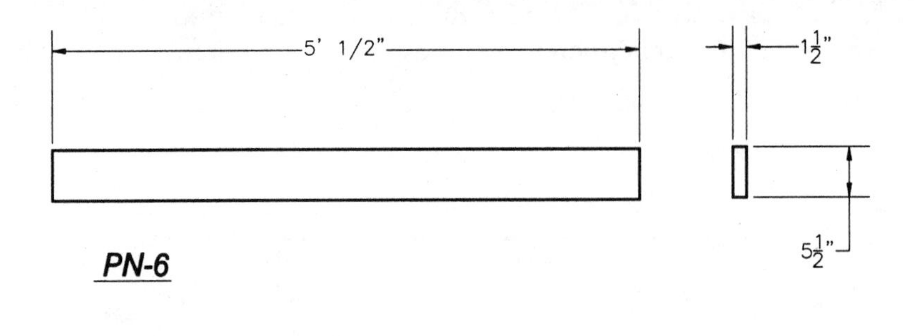

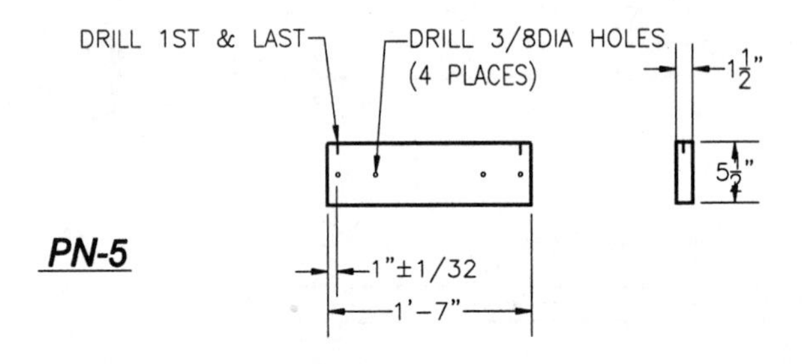

Figure 6-57 *Drawing for Project Exercise 1-3*

Answers to Self-Evaluation Test

1 - F, **2** - T, **3** - F, **4** - T, **5** - F, **6** - F, **7** - T, **8** - F, **9** - F, **10** - F

Chapter 7

Editing Dimensions

Learning Objectives

After completing this chapter, you will be able to:

- *Edit dimensions.*
- *Stretch, extend, and trim dimensions.*
- *Use the **DIMEDIT** and **DIMTEDIT** command options to edit dimensions.*
- *Update dimensions using the **DIM**, Update and **DIMSTYLE**, Apply commands.*
- *Use the **PROPERTIES** command to edit dimensions.*
- *Dimension in model space and paper space.*

EDITING DIMENSIONS

For editing dimensions, AutoCAD LT has provided some special editing commands that work with dimensions. These editing commands can be used to define new dimension text, return to home text, create oblique dimensions, and rotate and update the dimension text. You can also use the **TRIM**, **STRETCH**, and **EXTEND** commands to edit the dimensions. But unless you include the dimensioned objects in the edit selection set, objects do not change automatically. The properties of the dimensioned objects can also be changed using the **Properties** Window or **Dimension Style Manager**.

Stretching Dimensions

To stretch a dimension object, appropriate definition points must be included in the selection crossing box of the **STRETCH** command. Because the middle point of the dimension text is a definition point for all types of dimensions, you can use the **STRETCH** command to move the dimension text to any location you want. You can also choose **STRETCH** command from the **MODIFY** menu or **MODIFY** toolbar. If you stretch the dimension text but do not want the dimension line broken, the gap in the dimension line gets filled automatically. The dimension type remains the same after stretching. When editing, the definition points of the dimension being edited must be included in the selection crossing box. The aligned dimension adjusts itself so that it is aligned with the object being dimensioned; the dimension is automatically calculated as well. The vertical dimension maintains itself as a vertical dimension and measures the vertical distance between the base and the top of the triangle even though the line it dimensions may no longer be a vertical line. The following example illustrates the use of the **STRETCH** command in stretching a dimension.

Example 1

In this example you will stretch the objects and the dimension to a new location. The drawing in Figure 7-1(a) should already be drawn on the screen.

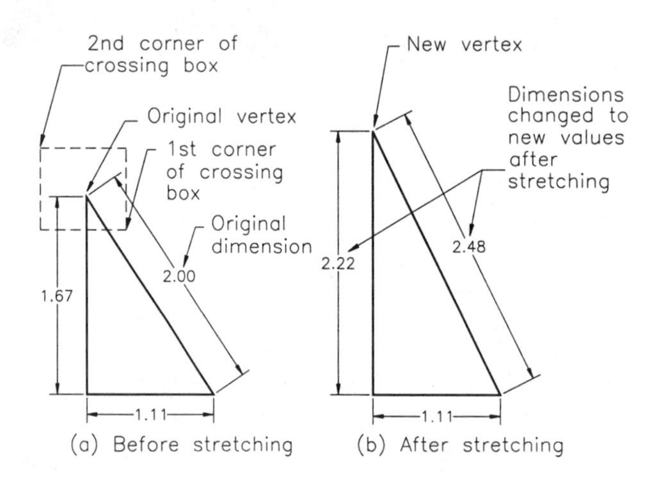

***Figure 7-1** Using the **STRETCH** command to stretch the dimensions*

1. At the AutoCAD LT Command prompt, enter the **STRETCH** command. At the next prompt, use the crossing option to select the objects and dimensions you want to stretch.

   Command: **STRETCH** [Enter]
   Select objects to stretch by crossing-window or crossing-polygon...
   Select objects: **C** [Enter]
   Specify first corner: *Select the first corner of the crossing box.*
   Specify opposite corner: *Select the second corner of the crossing box.*
   Select objects: [Enter]

2. Next, AutoCAD LT will prompt you to enter the base point and the second point of displacement. When you enter this information, the objects and the dimensions will be stretched as shown in Figure 7-1(b).

   Specify base point or displacement: *Specify the base point.*
   Specify second point of displacement: *Specify the point where you want to stretch the objects.*

Exercise 1 — Mechanical

The two dimensions in Figure 7-2(a) are too close. Fix the drawing by stretching the dimension as shown in Figure 7-2(b).

1. Stretch the outer dimension to the right so that there is some distance between the two dimensions.
2. Stretch the dimension text of the outer dimension so that the dimension text is staggered (lower than the first dimension).

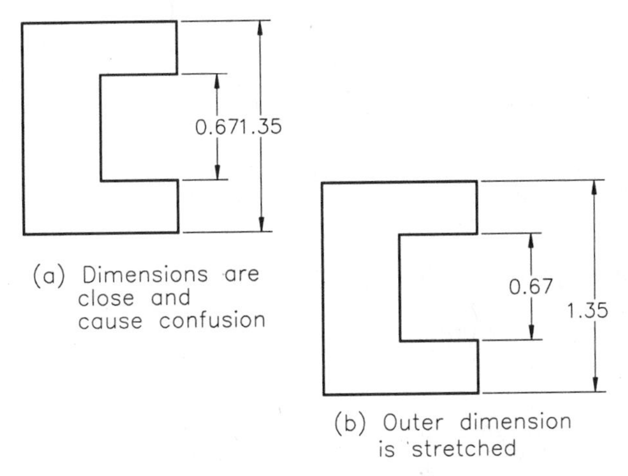

***Figure 7-2** Drawing for Exercise 1, stretching dimensions*

Trimming and Extending Dimensions

Trimming and extending operations can be carried out with all types of linear dimensions (horizontal, vertical, aligned, rotated) and the ordinate dimension. AutoCAD LT trims or extends a linear dimension between the extension line definition points and the object used as a boundary or trimming edge. To extend or trim an ordinate dimension, AutoCAD LT moves the feature location (location of the dimensioned coordinate) to the boundary edge. To retain the original ordinate value, the boundary edge to which the feature location point is moved is orthogonal to the measured ordinate. In both cases, the imaginary line drawn between the two extension line definition points is trimmed or extended by AutoCAD LT , and the dimension is adjusted automatically (Figure 7-3). The prompt sequence for extending the dimension is:

Command: **EXTEND** [Enter]
Current Settings: Projection=Current Edge=Current
Select boundary edges...
Select objects: *Select the upper tip of the polygon.*
Select objects: *Press ENTER.*
Select object to extend or [Project/Edge/Undo]: *Select the dimension.*

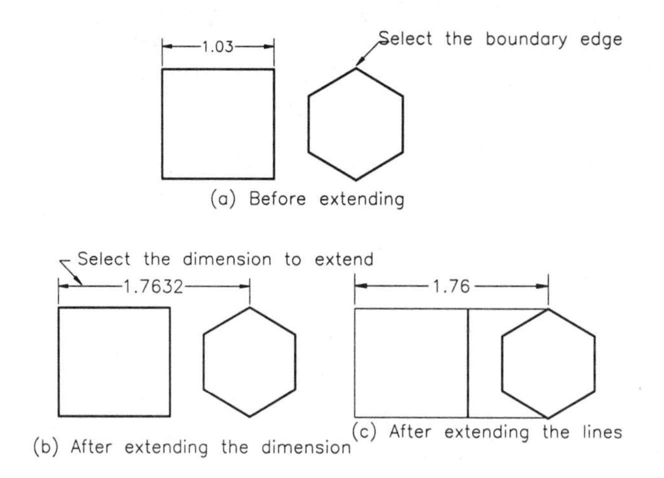

***Figure 7-3** Using the **EXTEND** command to extend dimensions*

Exercise 2 — *General*

Use the **TRIM** command to trim the dimension at the top of Figure 7-4(a) so it looks like Figure 7-4(b).

1. Make the drawing and dimension it as shown in Figure 7-4(a). Assume the dimensions where necessary.
2. Stretch the dimension as shown in Figure 7-4(b). You can also trim the dimension by setting the **EDGEMODE** system variable to 1, and then using the **TRIM** command. In this case, you do not need to draw a line to trim.

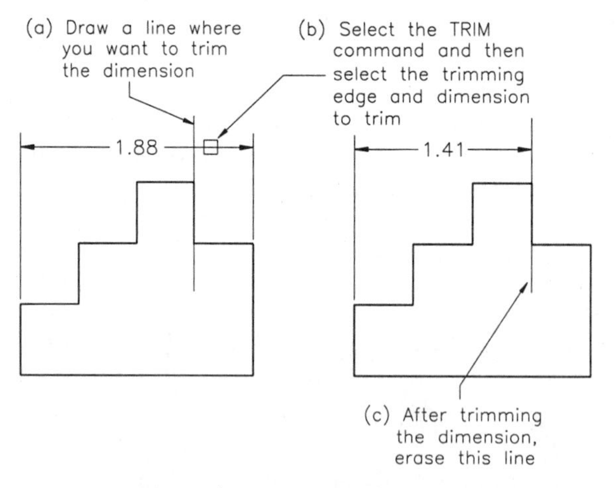

***Figure 7-4** Drawing for Exercise 2*

EDITING DIMENSIONS (DIMEDIT Command)

Toolbar:	Dimension > Dimension Edit
Command:	DIMEDIT

The dimension text can be edited by using the **DIMEDIT** command. This command has four options: New, Rotate, Home, and Oblique. The prompt sequence for the **DIMEDIT** command is:

Command: **DIMEDIT** [Enter]
Enter type of dimension editing (Home/New/Rotate/Oblique) <Home>: *Enter an option.*

New

You can use the **New** option to replace the existing dimension Mtext with a new text string.

Command: **DIMEDIT** [Enter]
Enter type of dimension editing (Home/New/Rotate/Oblique)
<Home>: **N** [Enter]

The **Multiline Text Editor** dialog box is displayed (Figure 7-6). Enter the text in the Text Editor and then choose the **OK** button. The following prompt is displayed in the command line:

Select objects: *Select the dimensions.*
Select objects: *Press ENTER.*

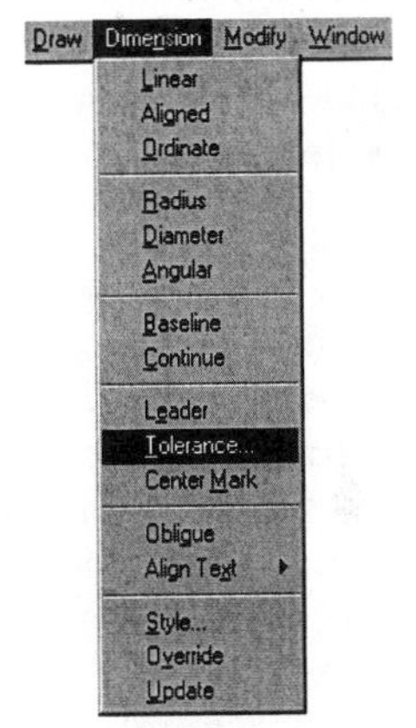

***Figure 7-5** Invoking **DIMEDIT** from the toolbar*

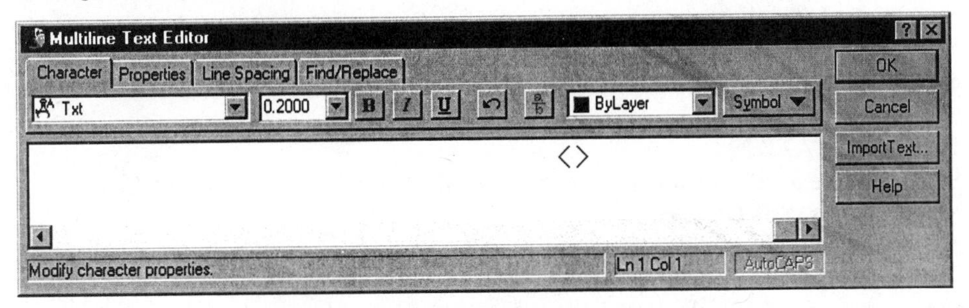

***Figure 7-6 Multiline Text Editor** dialog box*

In case a null response is given at the **Enter dimension text <current>:** prompt, durng dimensioning the actual dimension text is taken as the text. The <> prefix or suffix facility can also be used when you enter the new text string. In this case, the prefix/suffix is appended before or after the dimension measurement that is placed instead of the <> characters. For example, if you enter Ref-Dim before <> and mm after <> (Ref-Dim<>mm), AutoCAD LT will attach Ref-Dim as prefix and mm as suffix (Ref-Dim1.1257mm) (see Figure 7-7).

If information about the dimension style is available on the selected dimension, it is used to redraw the dimension, or the prevailing variable settings are used for the redrawing process.

You can also use the **DIM**, **NEWTEXT** commands to replace an existing text with simple text.

Command: **DIM** [Enter]
Dim: **NEWTEXT** [Enter]

The **Multiline Text Editor** dialog box is displayed where you can enter new text and then choose the **OK** button. The following prompt is displayed in the command line:

Select objects: *Select the dimensions.*
Select objects: *Press ENTER.*

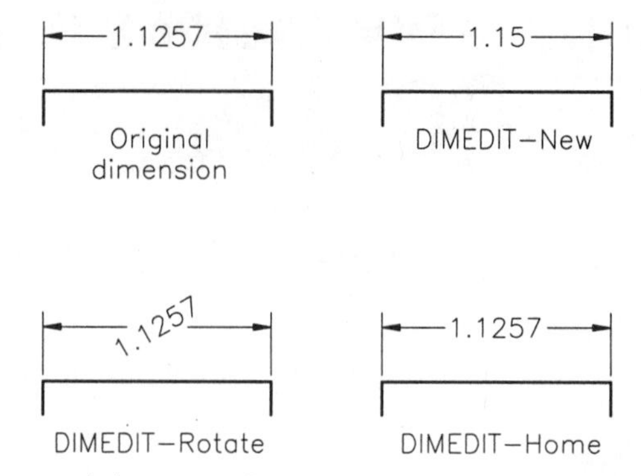

Figure 7-7 *Using the* ***DIMEDIT*** *command to edit dimensions*

Rotate

Using the **Rotate** option, you can position the dimension text at an angle you specify. The angle can be specified by entering its value at the **Specify angle for dimension text:** prompt or by specifying two points at the required angle. You will notice that the text rotates around its middle point. When the dimension text alignment is set to orient text horizontally, the dimension text is aligned with the dimension line. With this command, you can change the orientation (angle) of the dimension text of any number of associative dimensions. As an alternate method for rotating the text, use the **TROTATE** command at the **Dim:** prompt.

Command: **DIMEDIT** [Enter]
Enter type of dimension editing (Home/New/Rotate/Oblique) <Home>: **R** [Enter]
Specify angle for dimension text: *Enter text angle.*
Select objects: *Select the dimensions.*
Select objects: *Press ENTER.*

Dim: **TROTATE** [Enter]
Specify angle for dimension text: *Enter text angle.*
Select objects: *Select the dimensions.*
Select objects: *Press ENTER.*

An angle value of zero for the rotation angle of text places the text in its default orientation. If the text alignment is set to orient text horizontally, the dimension text is aligned with the dimension line.

Home

The Home option restores the text of an associative dimension to its original (home/default) location if the position of the text has been changed using the **STRETCH** or **DIMEDIT** command. For an alternate method use **HOMETEXT** at the **Dim:** prompt.

Command: **DIMEDIT** [Enter]
Enter type of dimension editing (Home/New/Rotate/Oblique) <Home>: **H** [Enter]

Select objects: *Select the dimension.*
Select objects: *Press ENTER.*

Dim: **HOMETEXT** [Enter]
Select objects: *Select the dimension object.*
Select objects: *Press ENTER.*

Oblique

In the linear dimensions, extension lines are drawn perpendicular to the dimension line. The **Oblique** option bends the linear associative dimensions. It draws extension lines at an oblique angle (Figure 7-8). This option is particularly important when creating isometric dimensions and can be used to resolve conflicting situations due to the overlapping of extension lines with other objects. Making an existing dimension oblique by specifying an angle oblique to it does not affect the generation of new linear dimensions. The oblique angle is maintained even after performing most editing operations. (See Chapter 19 for details about how to use this option.) The alternate method is using **OBLIQUE** at the **Dim**: prompt.

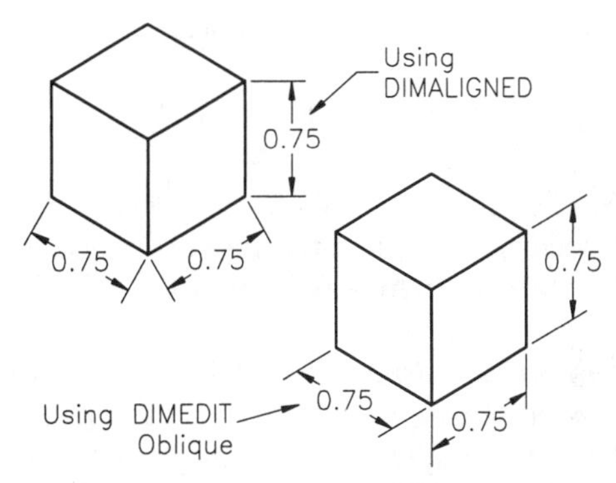

Figure 7-8 *Using **DIMEDIT-Oblique** option to edit dimensions*

Command: **DIMEDIT** [Enter]
Enter type of dimension editing (Home/New/Rotate/Oblique) Home>: **O**[Enter]
Select objects: *Select the dimension.*
Select objects: *Press ENTER.*
Enter obliquing angle (press ENTER for none): *Enter an angle.*

Dim: **OBLIQUE** [Enter]
Select objects: *Select the dimensions.*
Select objects: *Press ENTER.*
Enter obliquing angle (press ENTER for none): *Enter an angle.*

EDITING DIMENSION TEXT (DIMTEDIT Command)

Toolbar:	Dimension > Dimension Text Edit
Menu:	Dimension > Align Text
Command:	DIMTEDIT

The dimension text can also be edited by using the **DIMTEDIT** command (Figure 7-10). This command is used to edit the placement and orientation of a single existing associative dimension. The practical application of this command can be, for example, in cases where dimension text of two or more dimensions are too close together, resulting in confusion. In such cases, the

DIMTEDIT command is invoked to move the dimension text to some other location so that there is no confusion. This command has five options: **Left**, **Right**, **Center**, **Home**, and **Angle**.

Command: **DIMTEDIT** [Enter]
Select dimension: *Select a dimension.*
Specify new location for dimension text or [Left/Right/Center/Home/Angle): *Enter an option.*

As an alternate method, enter **TEDIT** at the **Dim:** prompt.

Command: **DIM** [Enter]
Dim: **TEDIT** [Enter]
Select dimension: *Select a dimension.*
Specify new location for dimension text or [Left/Right/Center/Home/Angle): *Enter an option.*

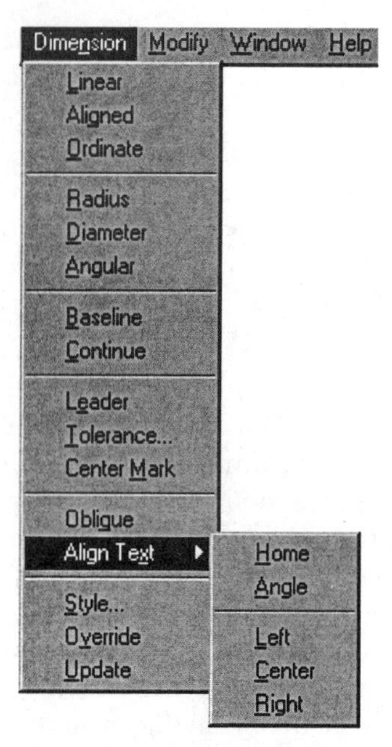

***Figure 7-9** Invoking **Align Text** from the **Dimension** menu*

If the **DIMSHO** variable is on, you will notice that as you move the cursor, the dimension text of the selected associative dimension is dragged with the cursor. In this manner, you can see where to place the dimension text while you are dynamically dragging the text. Another advantage of having **DIMSHO** on is that the dimension updates dynamically as you drag it. The initial value of **DIMSHO** is on. If **DIMSHO** is off, the dimension text and the extension lines are not dragged with the cursor; hence, it is difficult to see the position of the text as you move the cursor. The splits in the dimension line are calculated while keeping track of the new location of the dimension text. **DIMSHO** is not stored in dimension style. Another method of moving text (automatically) is by using any of the **Left**, **Right**, **Center**, **Home**, or **Angle** options provided in the **Specify new location for dimension text or [Left/Right/Center/Home/Angle)**: prompt. The effects of these options are described next (Figure 7-10).

Left

With this option, you can left-justify the dimension text along the dimension line. The vertical placement setting determines the position of the dimension text. The horizontally aligned text is moved to the left and the vertically aligned text is moved down. This option can be used only with the linear, diameter, and radius dimensions.

Right

With this option, you can right-justify the dimension text along the dimension line. As in the case of the Left option, the vertical placement setting determines the position of the dimension text. The horizontally aligned text is moved to the right, and the vertically aligned text is moved up. This option can be used only with the linear, diameter, and radius dimensions.

Center*

With this option you can center-justify the dimension text for linear, and aligned dimensions. The vertical setting controls the vertical position of the dimension text.

Home

This option works similarly to the **HOMETEXT** command. With the use of the **Home** option, the dimension text of an associative dimension is restored (moved) to its original (home/default) location if the position of the text has been changed.

Angle

With the Angle option, you can position the dimension text at the angle you specify. This option works similarly to the **TROTATE** command. The angle can be specified by entering its value at the **Specify angle for dimension text**: prompt or by specifying two points at the required angle. You will notice that the text rotates around its middle point. If the dimension text alignment is set to Orient Text Horizontally, the dimension text is aligned with the dimension line. If information about the dimension style is available on the selected dimension, AutoCAD LT uses it to redraw the dimension, or the prevailing dimension variable settings are used for the redrawing process. Entering 0 degree angle changes the text to its default orientation.

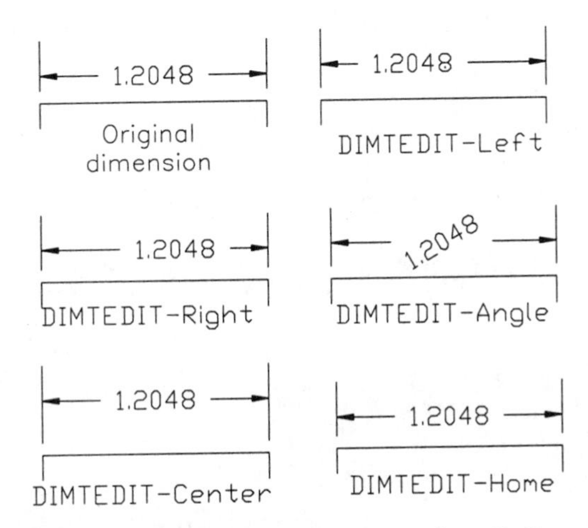

Figure 7-10 *Using the **DIMTEDIT** command to edit dimension text*

UPDATING DIMENSIONS

Toolbar:	Dimension > Dimension Update
Dim:	UPDATE

The **Update** option of the **DIM** command regenerates (updates) prevailing associative dimension objects (like arrows, text height) using current settings for the dimension variables, dimension style, text style, and units.

Command: **DIM** [Enter]
Dim: **UPDATE** [Enter]
Select objects: *Select the dimension objects.*

The dimensions can also be updated by using the **Apply** option of the **-DIMSTYLE** command. (For details, see Chapter 8.)

Command: **-DIMSTYLE** [Enter]
Current dimension style: Standard
Enter a dimension style option
[Save/Restore/STatus/Variables/Apply/?] <Restore>: **A** [Enter]
Select objects: *Select the dimension objects.*

EDITING DIMENSIONS WITH GRIPS

You can also edit dimensions by using GRIP editing modes. GRIP editing is the easiest and quickest way to edit dimensions. You can perform the following operations with GRIPS (see Chapter 10 for details):

1. Position the text anywhere **along** the dimension line. You cannot move the text and position it above or below the dimension line.
2. Stretch a dimension to change the spacing between the dimension line and the object line.
3. Stretch the dimension along the length. When you stretch a dimension, the dimension text automatically changes.
4. Move, rotate, copy, or mirror the dimensions.
5. Relocate a dimension origin.
6. Change properties like color, layer, linetype, and linetype scale.
7. Load Web browser (if any *Universal Resource Locator* is associated with the object)

EDITING DIMENSIONS USING THE PROPERTIES COMMAND*

Toolbar:	Standard > Properties
Menu:	Modify > Properties
Command:	PROPERTIES

You can also modify a dimension or leader created with the **LEADER** command by using the **PROPERTIES** command (Figure 7-11). When you choose this command, AutoCAD LT will display the **Properties** Window. All the properties of the selected objects are displayed in the **Properties** Window.

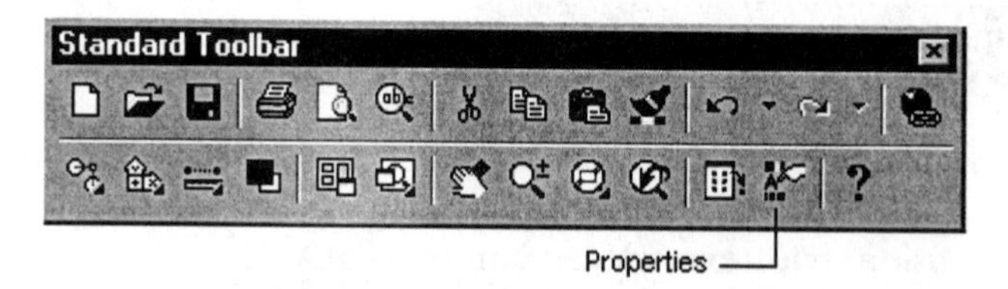

Figure 7-11 *Selecting Properties from the* ***Standard*** *toolbar*

Properties Window (DIMENSION)

You can use the **Properties** Window (Figure 7-12) to change the properties of a dimension, change the dimension text style, or change geometry, format, and annotation-related features of the selected dimension. The changes takes place dynamically in the drawing. The **Properties** Window is classified **Alphabetic** as well as **Category** under the respective tabs. The different categories for modification of dimensions are as follows:

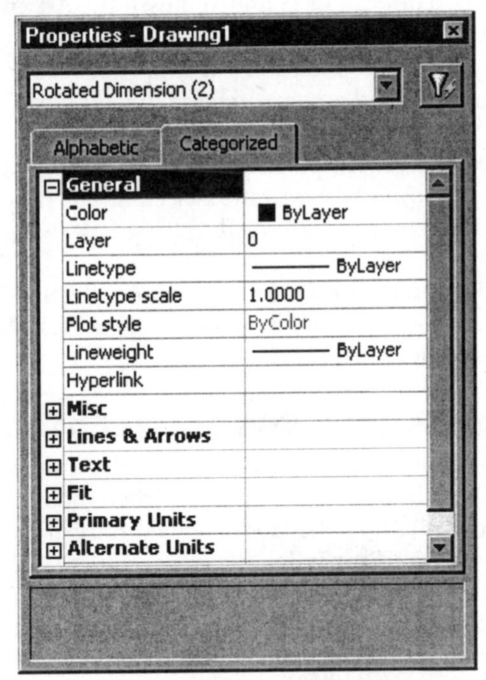

Figure 7-12 Properties Window (Dimension)

General. In the general category, the various parameters displayed are **Color**, **Layer**, **Linetype**, **Linetype scale**, **Plot style**, **Lineweight** and **Hyperlink** with their current values. For example, if you want to change the color of the selected object, then select Color property and choose the required color from the drop-down list. Similarly, Layer, Plot style, Linetype and Lineweight can be changed from the respective drop-down lists. The Linetype scale can be changed manually at the corresponding cell.

Misc. This category displays the dimension style by name (for **DIMSTYLE** system variable use **SETVAR**). You can change the dimension style from the drop-down list for the selected dimension object.

Lines and arrows. The various parameters of the lines and arrows in the dimension objects like arrowhead size, type, arrow lineweight, etc., can be changed in this category.

Text. The different parameters that control the text in the dimension object like text color, text height, vertical position text offset, etc., can be changed in this category.

Fit. In the fit category, the various parameters are **Dim line forced**, **Dim line inside**, **Dim scale overall**, **Fit**, **Text inside**, and **Text movement**. All the parameters can be changed by the drop-down list except Dim scale overall (which can be changed manually).

Primary Units. In the primary units category, the various parameters displayed are **Decimal separator**, **Dim prefix**, **Dim suffix**, **Dim roundoff**, **Dim scale linear**, **Dim units**, **Suppress leading zeroes**, **Suppress trailing zeroes**, **Suppress zero feet**, **Suppress zero inches** and **Precision**. Among the above properties, Dim units, Suppress leading zeroes, Suppress trailing zeroes, Suppress zero feet, Suppress zero inches and Precision properties can be changed with the corresponding drop-down lists and the other parameters can be changed manually.

Alternate Units. In the alternate unit category, there are various parameters for the alternate units. They can be changed only if the **Alt enabled** parameter is **on**. The parameters like **Alt format**, **Alt precision**, **Alt suppress leading zeroes**, and **Alt suppress trailing zeroes**, **Alt suppress zero feet**, and **Alt suppress zero inches** can be changed from the respective drop-down lists and others can be changed manually.

Tolerances. The parameters of this category can be changed only if Tolerances display parameter has some mode of tolerance selected. The various parameters are available corresponding to the mode of tolerance selected.

Tolerances and Text. This category displays the **Tolerance text height** which shows the scale factor of tolerance text as set by **DIMTXT (DIMTFAC** system variable) and can be changed manually.

Properties Window (LEADER)*

The **Properties Window** for **Leader** can be invoked by selecting a Leader and then choosing **Properties** from the **Modify** menu. You can also invoke the **Properties Window** (Figure 7-13) from the shortcut menu by right-clicking in the drawing area and choosing **Properties**. The various properties under Properties Window (Leader) are described as follows:

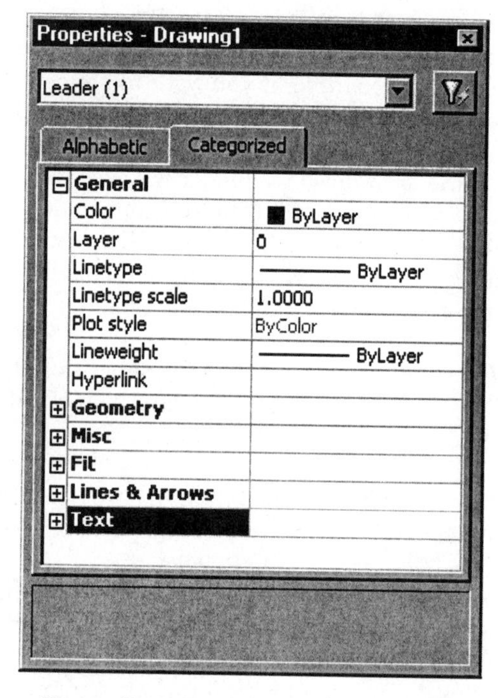

Figure 7-13 *Properties Window (Leader)*

General. The parameters in the general category are the same as those discussed in the previous section (**Properties Window** for dimensions).

Geometry. This category displays the coordinates of the Leader. The various parameters under this category are **Vertex**, **Vertex X**, **Vertex Y**, and **Vertex Z**. You can choose any vertex of the leader and change its coordinates.

Misc. This category displays the **Dim Style** and **Type** of the Leader. You can change the style name and the type of the Leader by using these properties.

Fit. This category displays the **Dim scale overall** property that specifies the overall scale factor applied to size, distances, or the offsets of the Leader.

Lines and Arrows. Lines and Arrows displays the different specifications of the arrowheads and line types for the Leader.

Text. Text category displays the **Text offset** to the dimension line and the vertical position (**Text pos vert**) of the dimension text and can be changed accordingly.

Example 2

Use the **PROPERTIES** command to modify the dimensions in Figure 7-14 so that it matches Figure 7-15.

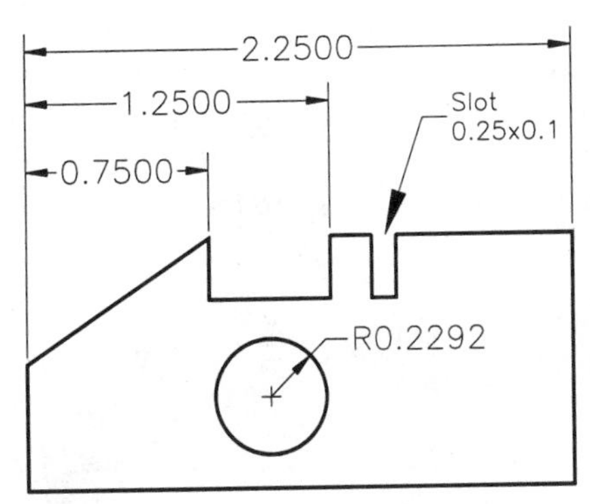

Figure 7-14 Drawing for Example 2

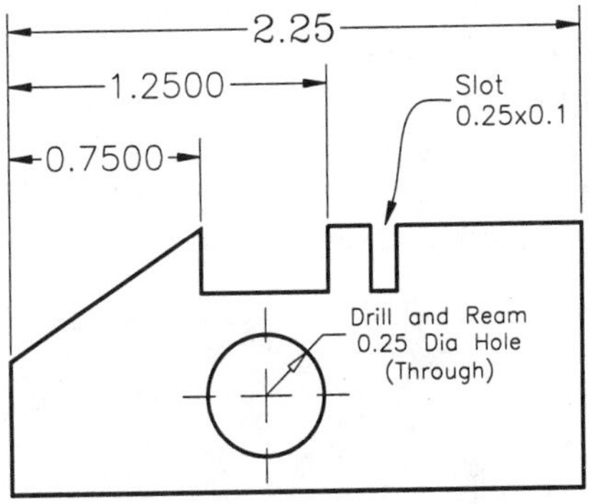

Figure 7-15 Drawing after editing the dimensions

1. Enter the **PROPERTIES** command at the Command prompt, or select it from the **Modify** menu or the toolbar.

   Command: **PROPERTIES** [Enter]

   AutoCAD LT displays the **Properties Window**.
   Select the dimension 2.2500.
2. To change the number of decimal places, choose the **Precision** drop-down list from the **Primary Units** category and choose (0.00). The number of decimal places will change dynamically in the drawing.
3. To change the text style, choose the **Text style** drop-down list under **Text** category and then choose the **ROMANC** text style. The changes will take place dynamically.
4. To edit the leader, select the leader and enter the **PROPERTIES** command. AutoCAD LT will display the **Properties Window (Leader)**. Choose **Arrow size** under the **Lines and Arrows** category. Change the arrowhead size to (0.09).
5. In the **Misc** category, from the **Type** drop-down list, choose the **Spline with arrow** option. The changes will take place dynamically in the drawing.
6. To edit the radius text, choose **Text override** parameter under Text category and in the adjacent cell enter **\P{Drill and Ream} \P{0.25 Dia Hole} \P{(Through)}**. Reduce the height of the text by changing the value in **Text height** parameter to (0.06).
7. In the **Lines and arrows** category, choose the **Closed** type of arrowhead from the **Arrow** drop-down list.

Now exit the **Properties Window**.

MODEL SPACE AND PAPER SPACE DIMENSIONING

Dimensioning objects can be drawn in model space or paper space. If the drawings are in model space, associative dimensions should also be created in model space. If the drawings are in model space and the associative dimensions are in paper space, the dimensions will not change when you perform such editing operations as stretching, trimming, and extending, or such display operations as zoom and pan in the model space viewport. The definition points of a dimension are located in the space where the drawing is drawn. You can check for **Paper Space Scaling** under **Scale for Dimension Features** in the **Fit** tab of **Modify**, **New**, or **Override** dialog boxes in the **Dimension Style Manager** dialog box, depends on whether you want to modify the present style or you want to create a new style (see Figure 7-16). Choose **OK/OK** to exit from both the dialog boxes. AutoCAD LT calculates a scale factor that is compatible with the model space and the paper space viewports. Choose **Update** from the **Dimension** menu and select the dimension objects for updating.

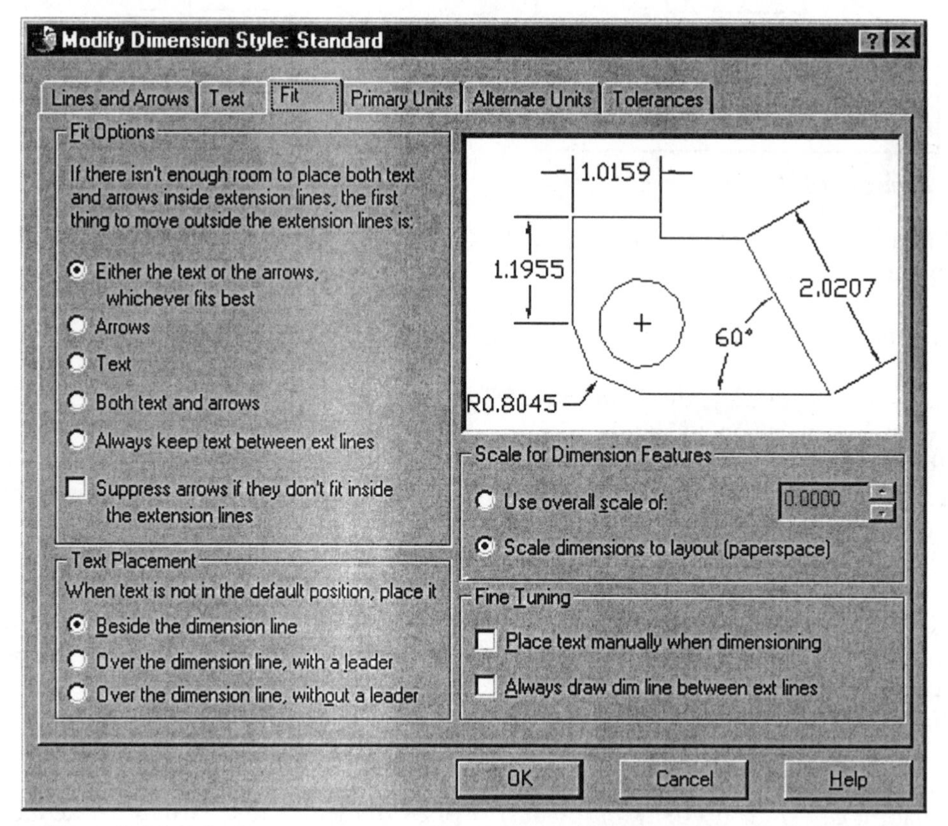

***Figure 7-16** Selecting paper space scaling in the **Modify Dimension Style** dialog box*

The drawing shown (Figure 7-17) uses paper space scaling. The main drawing and detail drawings are located in different floating viewports (paper space). The zoom scale factors for these viewports are different: 0.75XP, 1.0XP, and 1.5XP, respectively. When you use paper scaling, AutoCAD LT automatically calculates the scale factor for dimensioning so that the dimensions are uniform in all the floating viewports (model space viewports).

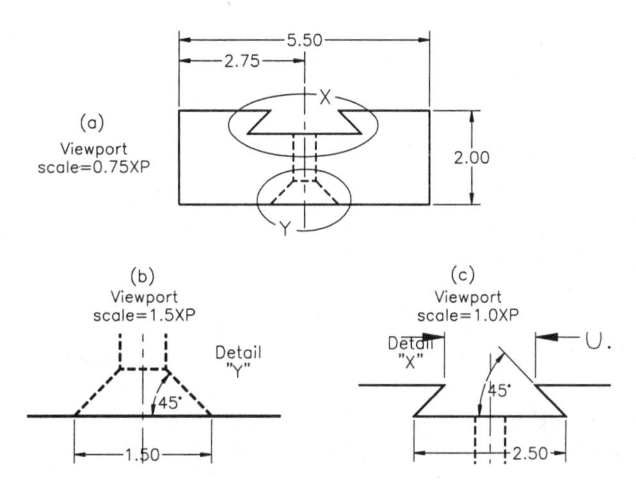

***Figure 7-17** Dimensioning in paper model space viewports using paper space scaling or setting **DIMSCALE** to 0*

Self-Evaluation

Answer the following questions, and then compare your answers to the correct answers given at the end of this chapter.

1. Explain the use and working of the **STRETCH** command.______________________________

2. Explain the use and working of the **TRIM** command on dimensions. ___________________

3. In associative dimensioning, the items constituting a dimension (such as dimension lines, arrows, leaders, extension lines, and dimension text) are drawn as a **single object.** (T/F)

4. If associative dimensioning is disabled, the dimension lines, arrows, leaders, extension lines, and dimension text are drawn as **independent** objects. (T/F)

5. The **DIMASO** dimensioning variable governs the creation of associative dimensioning. (T/F)

6. In associative dimensioning, you cannot edit the dimension objects as a single object with various commands. (T/F)

7. The **horizontal, vertical, aligned,** and **rotated** dimensions cannot be associative dimensions. (T/F)

8. Only the **CENTER** and **LEADER** commands are affected by associative dimensioning. (T/F)

9. The **EXPLODE** command can be used to break an associative dimension into individual items. (T/F)

Review Questions

1. To stretch a dimension object, appropriate definition points must be included in the ________ of the **STRETCH** command.

2. Trimming and extending operations can be carried out with all types of linear (horizontal, vertical, aligned, rotated) dimensions and with the ordinate dimension. (T/F)

3. To extend or trim an ordinate dimension, AutoCAD LT moves the feature location (location of the dimensioned coordinate) to the boundary edge. (T/F)

4. With the ________ or ________ commands you can edit the dimension text of associative dimensions.

5. The ________ or ________ command is particularly important for creating isometric dimensions and is applicable in resolving conflicting situations due to overlapping of extension lines with other objects.

6. The ________ command is used to edit the placement and orientation of a single existing associative dimension.

7. The ________ or ________ command restores the text of an associative dimension to its original (home/default) location if the position of the text has been changed by the **STRETCH**, **TEDIT**, or **TROTATE** command.

8. The ________ command regenerates (updates) prevailing associative dimension objects (like arrows, text height) using the current settings for the dimension variables, dimension style, text style, and units.

9. If the drawings are in model space, you should create associative dimensions in ________

10. The ________ variable can be automatically adjusted to the zoom scale factor of the model space viewport.

11. Explain the function of the **EXPLODE** command in dimensioning. ____________
____.

12. Explain the term **definition points.** ____________________
____.

13. Explain when to use the **EXTEND** command and how it works with dimensions. ________
____.

14. Explain the use and working of the **PROPERTIES** command for editing dimensions. __
____.

15. The **Properties** Window is classified in__________ as well as ___________ under the respective tabs.

Exercises

Exercise 3 *Mechanical*

1. Make the drawing as shown in Figure 7-18. Assume the dimensions where necessary.
2. Dimension the drawing as shown in Figure 7-18.
3. Edit the dimensions as shown in Figure 7-19, using the **STRETCH**, **EXTEND**, and **TRIM** commands.

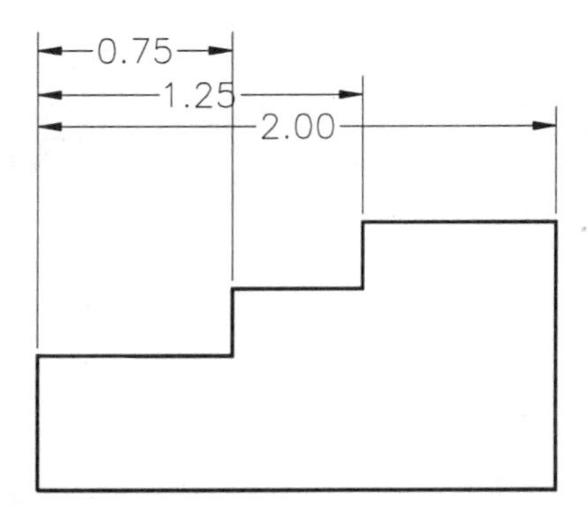

Figure 7-18 *Drawing for Exercise 3, before editing dimensions*

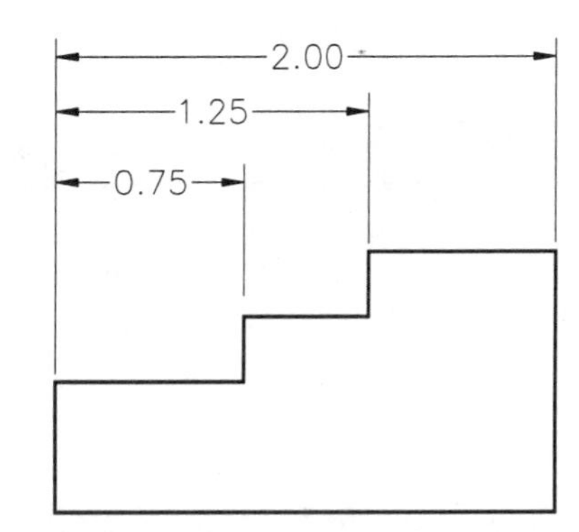

Figure 7-19 *Drawing for Exercise 3, after editing dimensions*

Exercise 4 *Mechanical*

1. Make the drawing as shown in Figure 7-20 (a). Assume the dimensions where necessary.
2. Dimension the drawing and edit them as shown in Figure 7-20(b).

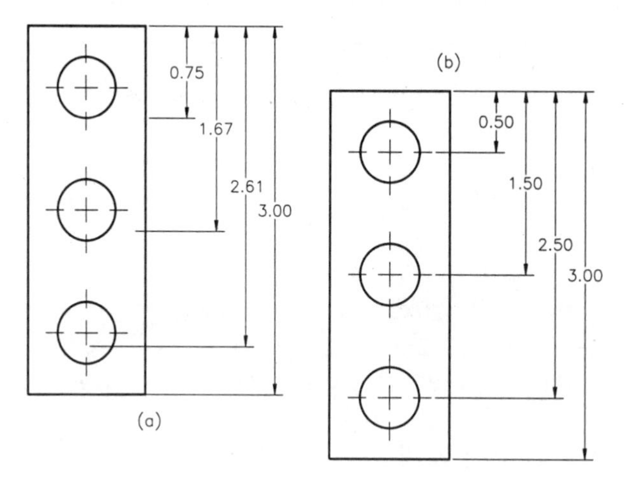

Figure 7-20 *Drawing for Exercise 4, showing dimensions (a) before and (b) after editing dimensions*

Exercise 5 — *Architectural*

1. Make the drawing as shown in Figure 7-21. Assume the dimensions where necessary.
2. Dimension, and then edit the dimension as shown.

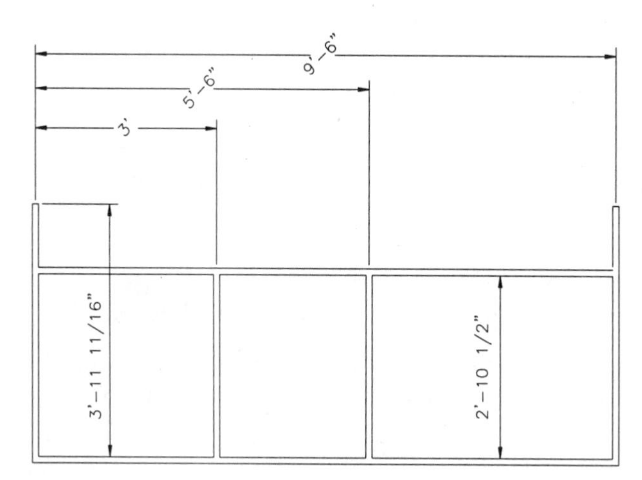

Figure 7-21 *Drawing for Exercise 5, editing dimensions*

Project Exercise 1-4 *Mechanical*

Generate the drawing in Figure 7-22 according to the dimensions shown in the figure. Save the drawing as BENCH7-8. (See Chapters 1, 2, 6, 19, and 23 for other drawings of Project Exercise 1.)

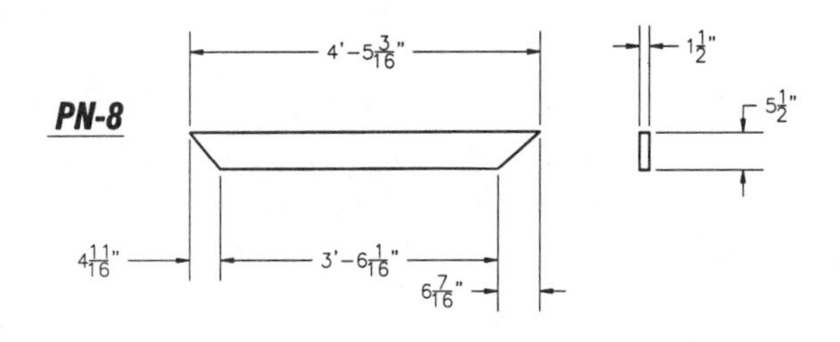

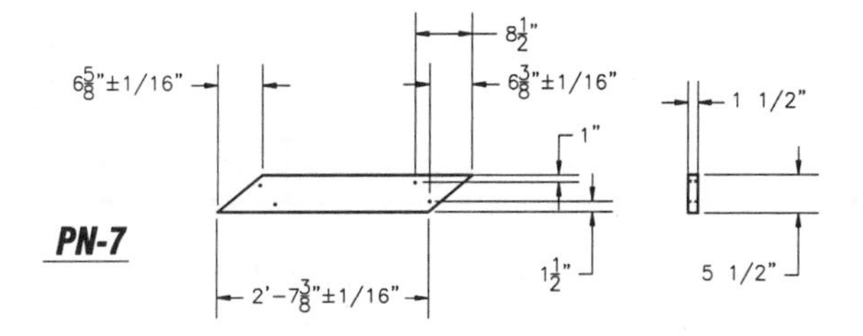

Figure 7-22 Drawing for Project Exercise 1-4

Project Exercise 1-5 *Mechanical*

Generate the assembly drawing with part numbers and the bill of materials as shown in Figure 7-23. Save the drawing as BENCHASM. (See Chapters 1, 2, 6, 19, and 23 for other drawings of Project Exercise 1.)

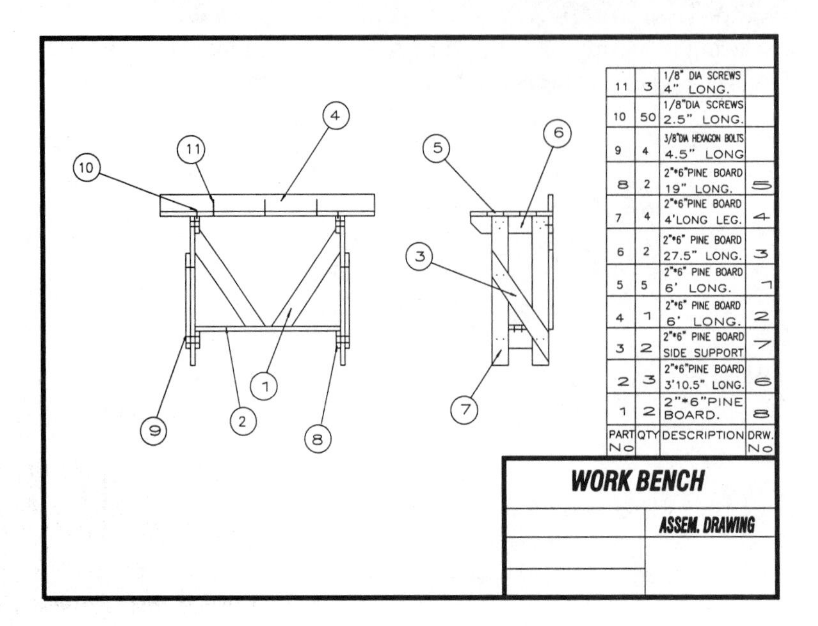

PART No	QTY	DESCRIPTION	DRW. No
11	3	1/8" DIA SCREWS 4" LONG.	
10	50	1/8"DIA SCREWS 2.5" LONG.	
9	4	3/8"DIA HEXAGON BOLTS 4.5" LONG	
8	2	2"*6"PINE BOARD 19" LONG.	5
7	4	2"*6"PINE BOARD 4'LONG LEG.	4
6	2	2"*6" PINE BOARD 27.5" LONG.	3
5	5	2"*6" PINE BOARD 6' LONG.	1
4	1	2"*6" PINE BOARD 6' LONG.	2
3	2	2"*6" PINE BOARD SIDE SUPPORT	7
2	3	2"*6"PINE BOARD 3'10.5" LONG.	6
1	2	2"*6"PINE BOARD.	8

Figure 7-23 Drawing for Project Exercise 1-5

Answers to Self-Evaluation Test

1 - The **STRETCH** command can be used to stretch the associative dimension. For stretching a dimension object, appropriate definition points must be included in the selection crossing box of the **STRETCH** command. **2** - The trimming operations can be carried out with all types of linear (horizontal, vertical, aligned, rotated) dimensions and the ordinate dimension. AutoCAD LT trims a linear dimension between the extension line definition points and the object used as trimming edge. For extending or trimming an ordinate dimension, AutoCAD LT moves the feature location (location of the dimensioned coordinate) to the boundary edge. **3** - T, **4** - T, **5** - T, **6** - F, **7** - F, **8** - F, **9** - T

Chapter 8

Dimension Styles and Dimensioning System Variables

Learning Objectives

After completing this chapter, you will be able to:

- *Use styles and variables to control dimensions.*
- *Create dimensioning styles using dialog boxes and the command line.*
- *Set dimension variables using the different tabs of the* ***New, Modify****, and* ***Override Dimension Style*** *dialog boxes and the command line.*
- *Set other dimension variables that are not in dialog boxes.*
- *Understand dimension style families and how to apply them in dimensioning.*
- *Use dimension style overrides.*
- *Compare and list dimension styles.*
- *Import externally referenced dimension styles.*

USING STYLES AND VARIABLES TO CONTROL DIMENSIONS

In AutoCAD LT, the appearance of dimensions on the drawing screen and the manner in which they are saved in the drawing database are controlled by a set of dimension variables. The dimensioning commands use these variables as arguments. The variables that control the appearance of the dimensions can be managed with dimension styles. You can use the **Dimension Style Manager** dialog box to control dimension styles and dimension variables through a set of dialog boxes. You can also do this by entering relevant dimensioning variable names at the **Command:** or **Dim**: prompt.

CREATING AND RESTORING DIMENSION STYLES

Using the Dialog Box

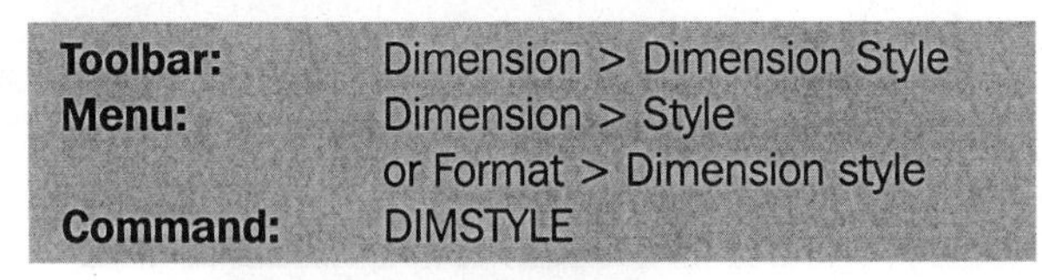

Toolbar:	Dimension > Dimension Style
Menu:	Dimension > Style
	or Format > Dimension style
Command:	DIMSTYLE

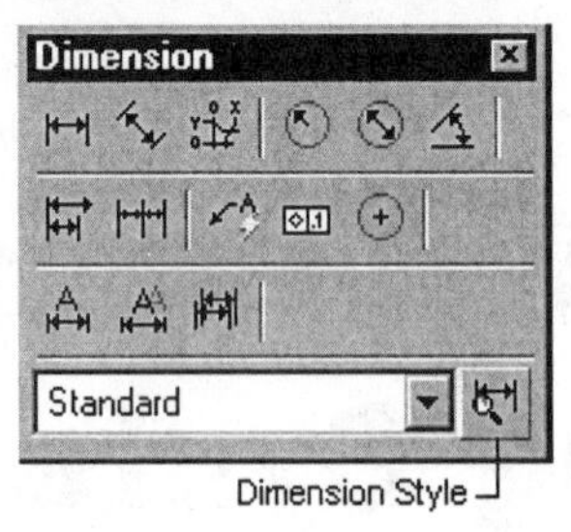

*Figure 8-1 Choosing **Dimension Style** from the **Dimension** toolbar*

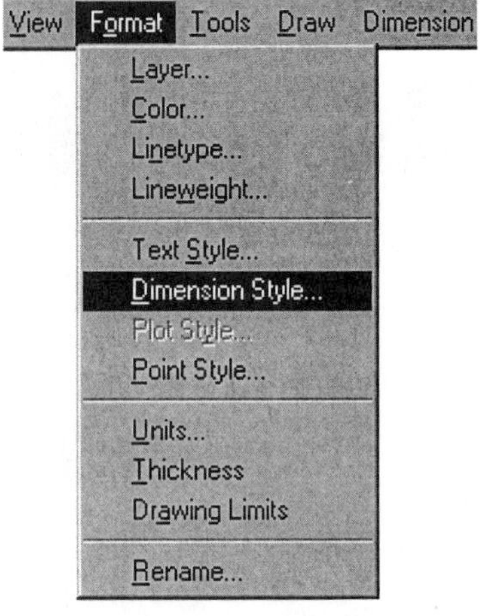

*Figure 8-2 Choosing **Dimension Style** from the **Format** menu*

Dimension style controls the appearance and positioning of dimensions. If the default dimensioning style (STANDARD) does not meet your requirements, you can select another dimensioning style or create one that does. The default dimension style file name is **STANDARD**. Dimension styles can be created by using the **Dimension Style Manager** dialog box (Figure 8-3) which appears on choosing **Style** from the **Dimension** menu or choosing

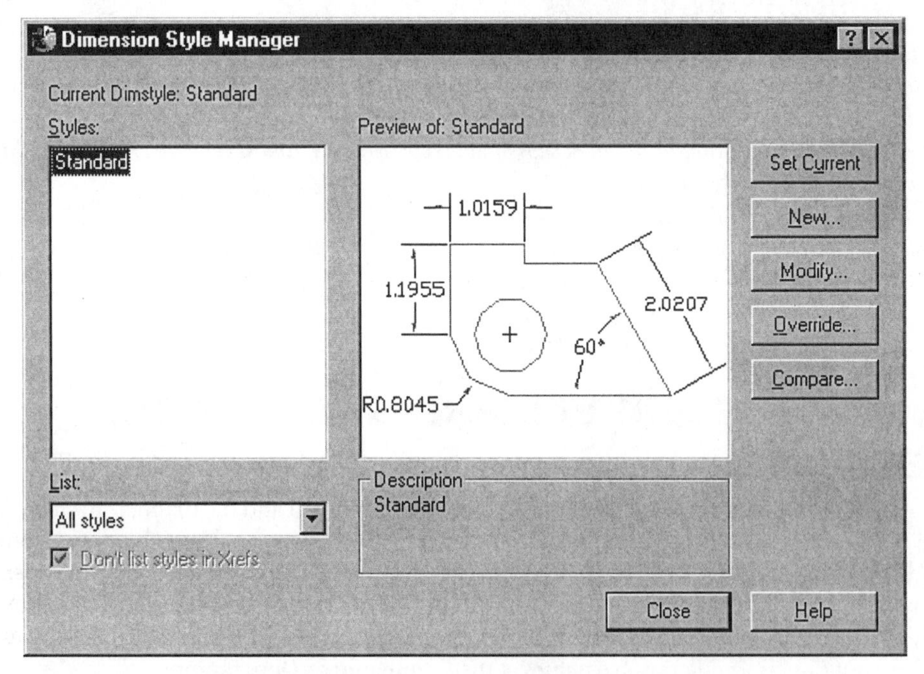

*Figure 8-3 **Dimension Style Manager** dialog box*

Dimension Style from the **Format** menu. You can also invoke the **Dimension Style Manager** dialog box by choosing the **Dimension Style** button in the **Dimension** toolbar.

In the **Dimension Style Manager** dialog box, choose the **New** button to display the **Create New Dimension Style** dialog box (Figure 8-4). Enter the dimension style name in the **New Style Name** text box and then select a style you want to base your style on from the **Start With** drop-down list. The **Use for** drop-down list, allows you to select the dimension type to which you want to apply the new dimension style. For example if you wish to use the new style for only the diameter dimension, you can select **Diameter dimension** from the **Use for** drop-down list. Choose the **Continue** button to display the **New Dimension Style** dialog box (Figure 8-4) where you can define the new style.

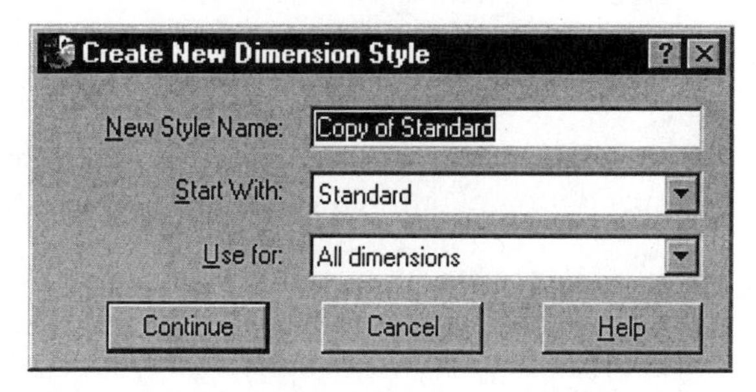

***Figure 8-4 Create New Dimension Style** dialog box*

In the **Dimension Style Manager** dialog box, the current dimension style name is shown in front of **Current dimstyle:** and is also shown highlighted in the **Styles:** list box. A brief description of the current style (it's differences from the default settings) is also displayed in the **Description** box. The **Dimension Style Manager** dialog box also has a **Preview of:** window which displays a preview of the current dimension style. A style can be made current (restored) by selecting the name of the dimension style you want to make current from the list of defined dimension styles and choosing the **Set Current** button or by double-clicking on the style name you want to make current in the list box. The drop-down list in the **Dimension** toolbar also displays the dimension styles. Selecting a Dimension style from this list also sets it current.The list of dimension styles displayed in the **Styles** list box is dependent on the option selected from the **List:** drop down list. If you select the **Styles in use** option, only the dimension styles in use will be listed in the **Style** list box. If you right-click a style in the **Style** list box, a shortcut menu is displayed which gives you the options to **Set current**, **Rename** or **Delete** a dimension style. Selecting the **Don't list styles in Xrefs** check box does not list the names of Xref styles in the **Styles:** list box. Choosing the **Modify** button displays the **Modify Dimension Style:** dialog box where you can modify an existing style. Choosing the **Override** button, displays the **Override Current Style** dialog box where you can define overrides to an existing style (discussed later in this chapter). Both these dialog boxes along with the **New Dimension Style:** dialog box have identical properties. Choosing the **Compare** button displays the **Compare Dimension Styles** dialog box (also discussed later in this chapter) which allows you to compare two existing styles.

Using the Command Line

You can also create dimension styles from the command line by entering **-DIMSTYLE** at the Command prompt or by entering **SAVE** at the **Dim:** prompt:

Command: **-DIMSTYLE** [Enter]
Current dimension style: *Displays the current style name.*
Enter a dimension style option

[Save/Restore/STatus/Variables/Apply/?] <Restore>: **SAVE** [Enter]
Enter name for new dimension style or [?]: *Enter the name of the dimension style.*
Command: **DIM** [Enter]
Dim: **SAVE** [Enter]
Enter name for new dimension style or [?]: *Enter the name of the dimension style.*

You can also restore the dimensions from the command line by entering **-DIMSTYLE** at the Command prompt or by entering **RESTORE** at the **Dim**: prompt.

Command: **-DIMSTYLE** [Enter]
Current dimension style: *Displays the current style name.*
Enter a dimension Style option
[Save/Restore/STatus/Variables/Apply/?] <Restore>: **RESTORE** [Enter]
Current dimension style: *Displays the current dimension style.*
Enter a dimension style name, [?] or <select dimension>: *Enter the dimension style name.*

Command: **DIM** [Enter]
DIM: **RESTORE** [Enter]
Current dimension style: *AutoCAD LT displays the current style name.*
Enter dimension style name, [?] or <select dimension>: *Enter the dimension style name.*

You can also select a dimension style by specifying a dimension on a drawing. This can be accomplished by pressing ENTER (null response) at the prompt **Enter a dimension style name, [?] or < select dimension>**. This way, you can select the dimension style without knowing the name of the dimension style. With the help of dimension styles, you can easily create and save groups of settings for as many types of dimensions as you require. Styles can be created to support almost any standards, including ANSI, DIN, and architectural.

NEW DIMENSION STYLE DIALOG BOX*

The **New Dimension Style** dialog box (Figure 8-5) can be used to specify the dimensioning attributes (variables) that affect the various properties of the dimensions. It has the following tabs.

Lines and Arrows Tab

The **Lines and Arrows** tab (Figure 8-5) can be used to specify the dimensioning attributes (variables) that affect the format of the dimension lines and arrows; that is, the appearance and behavior of the dimension lines, extension lines, arrowheads, and center marks. If the settings of the dimension variables have not been altered in the current editing session, the settings displayed in the dialog box are the default settings.

Dimension Line

Color. Dimension arrowheads have the same color as the dimension line because arrows constitute a part of the dimension line. You can establish a color for the dimension line and the dimension arrows. The color number or the special color label is stored in the **DIMCLRD** variable. The default color label for the dimension line is BYBLOCK. You can specify the color of the dimension line by selecting from the **Color** drop-down list. You can also choose **Other** from the **Color** drop-down list to display the **Select Color** dialog box where you can choose a specific color.

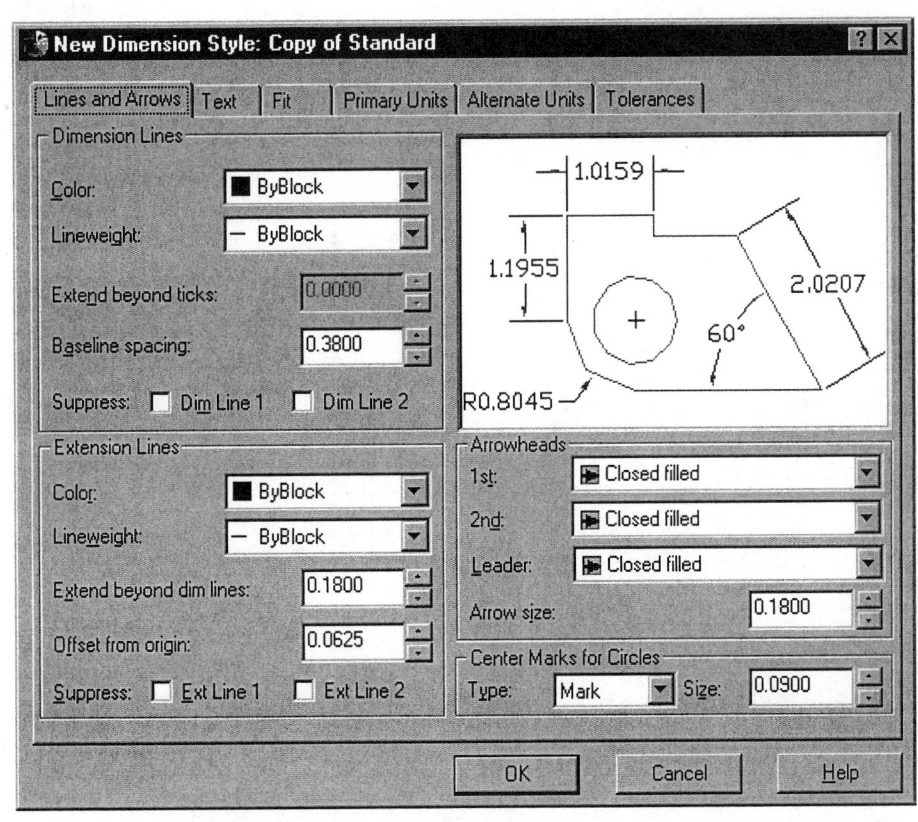

*Figure 8-5 **Lines and Arrows** tab of the **New Dimension Style** dialog box*

Lineweight. You can specify Lineweight for the dimension line by selecting from the **Lineweight** drop-down list. This value is stored in the **DIMLWD** variable. The default value is BYBLOCK.

Extend beyond ticks. The **Extend beyond ticks:** (Oblique tick extension) edit box is used to specify the distance by which the dimension line will extend beyond the extension line. The **Extend beyond ticks** edit box can be used only when you have selected the oblique, Architectural tick or any such arrowhead type in the **1st**, **2nd** drop-down lists in the **Arrowheads** area. The extension value entered in the **Extend beyond ticks** edit box gets stored in the **DIMDLE** variable. By default, this edit box is disabled because the oblique arrowhead type is not selected.

Baseline Spacing. The **Baseline Spacing:** (Baseline Increment) edit box is used to control the dimension line increment (gap between successive dimension lines) for the continuation of a linear dimension drawn with the **DIMBaseline** command (Figure 8-6). You can specify the dimension line increment to your requirement by entering the desired value in the **Baseline Spacing:** edit box. Also, when you are creating continued dimensions with the **DIMContinue** command, the contents of a **Baseline Spacing** edit box specify the offset distance for the successive dimension lines, if needed to avoid drawing over the previous dimension line. The default value displayed in the **Baseline Spacing** edit box is 0.38 units. The spacing (baseline

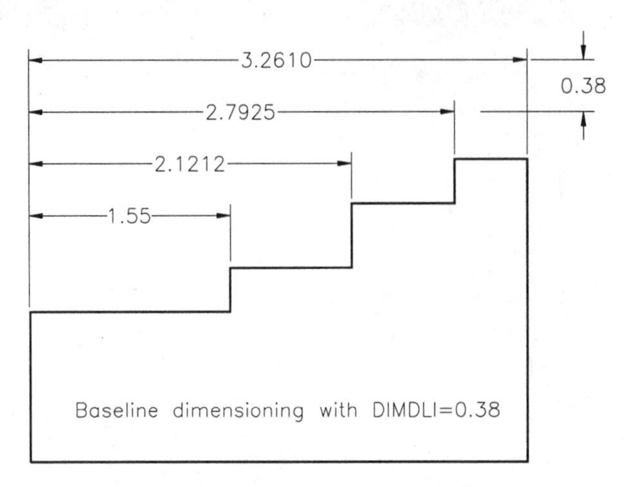

Figure 8-6 ***DIMDLI****, baseline increment*

increment) value is stored in the **DIMDLI** variable.

Suppress. The **Suppress:** check boxes control the drawing of the first and second dimension lines. By default, both dimension lines will be drawn. You can suppress one or both dimension lines by selecting their corresponding check boxes. The values of these check boxes are stored in the **DIMSD1** and **DIMSD2** variables.

Note

The first and second dimension lines are determined by how you select the extension line origins. If the first extension line origin is on the right, the first dimension line is also on the right. The initial value of ***DIMSD1*** *and* ***DIMSD2*** *is off.*

Extension Line

Color. In the **Color:** edit box, you can examine or modify the current extension line color (Figure 8-7). The default extension line color is BYBLOCK. You can change the color by assigning a new color to the extension lines. For example, if you want the extension line color to be red, select red from the **Color** drop-down list or select the **Other** option from the drop-down list. On selecting this option, the **Select Color** dialog box is displayed where you can choose the desired color. The color number or the color label is saved in the **DIMCLRE** variable.

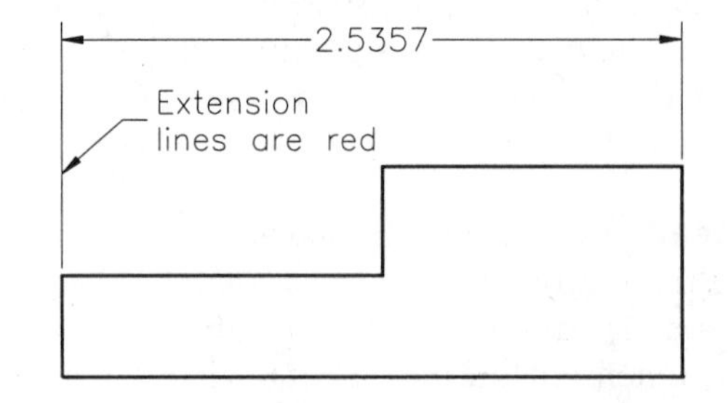

Figure 8-7 Changing extension line color, ***DIMCLRE***

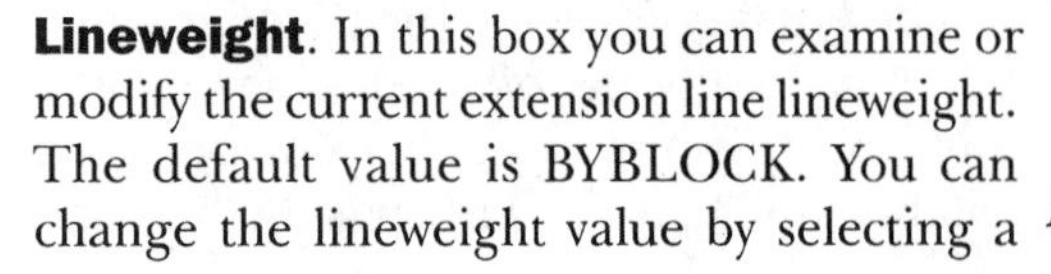

Lineweight. In this box you can examine or modify the current extension line lineweight. The default value is BYBLOCK. You can change the lineweight value by selecting a new value from the **Lineweight:** drop-down list. The value for lineweight is stored in the **DIMLWE** variable.

Extend beyond dim lines. It is the distance the extension line should extend past the dimension line. You can change the extension line offset by entering the desired distance value in the **Extend beyond dim lines:** edit box (Figure 8-8). The value of this edit box is stored in the **DIMEXE** variable. The default value for extension distance is 0.1800 units.

Offset from origin. This edit box displays the distance value the extension lines are offset from the extension line origins you specify. You may need to override this setting for specific dimensions when dimensioning curves and angled lines. You may have noticed that a small space exists between the origin points you specify and the start of the extension line (Figure 8-9). This space is due to the specified offset. You can specify an offset distance of your choice by entering it in this box. AutoCAD LT stores this value in the **DIMEXO** variable;. The default value for this distance is 0.0625.

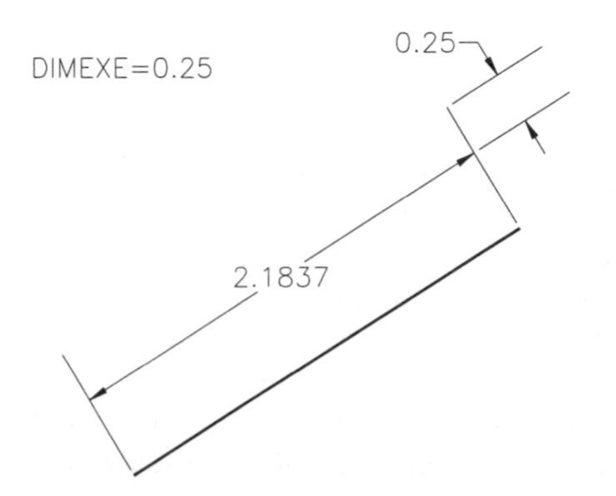

*Figure 8-8 Extension above line, **DIMEXE***

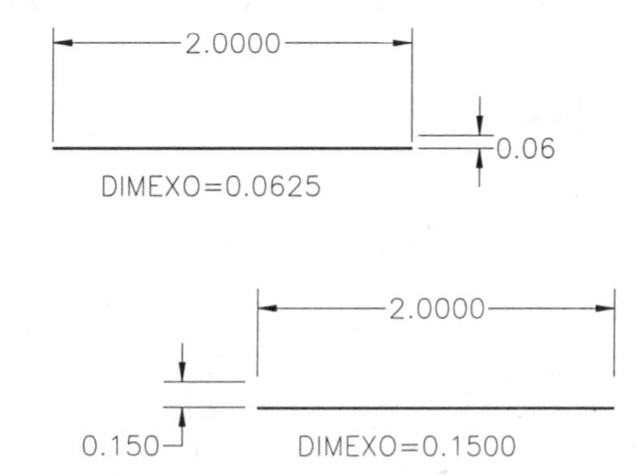

*Figure 8-9 Origin offset, **DIMEXO***

Chapter 8

Suppress. The **Suppress**: check boxes control the display of the extension lines. By default, both extension lines will be drawn. You can suppress one or both extension lines by selecting the corresponding check boxes (Figure 8-10). The values of these check boxes are stored in the **DIMSE1** and **DIMSE2** variables (Figure 8-11).

Note

*The first and second extension lines are determined by how you select the extension line origins. If the first extension line origin is on the right, the first extension line is also on the right. The initial value of **DIMSE1** and **DIMSE2** is off.*

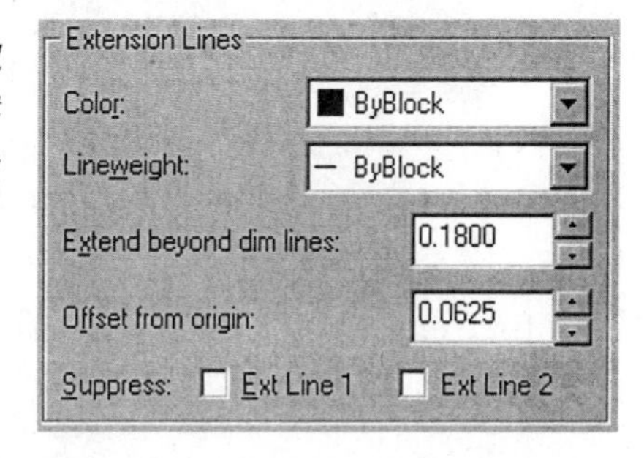

***Figure 8-10 Extension Lines** area of the **New Dimension Style** dialog box*

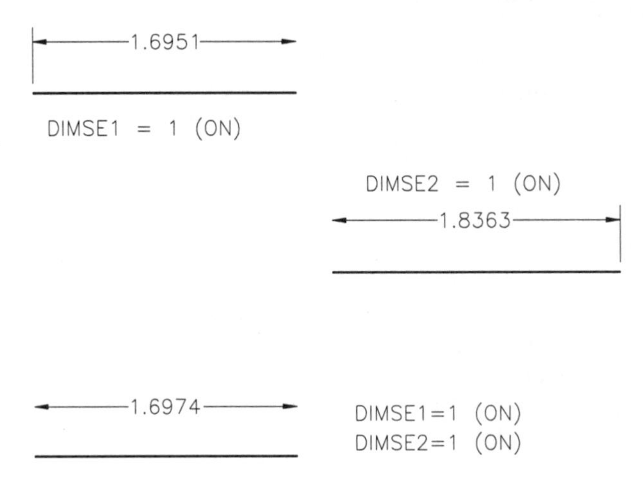

*Figure 8-11 Visibility of extension lines **DIMSE1** and **DIMSE2***

Arrowheads

1st: and 2nd: drop-down lists. When you create a dimension, AutoCAD LT draws the terminator symbols at the two ends of the dimension line. These terminator symbols, generally referred to as **arrowhead types**, represent the beginning and end of a dimension (see Figure 8-12). AutoCAD LT has provided nineteen standard termination symbols you can apply at each end of the dimension line. In addition to these, you can create your own arrows or terminator symbols. By default, the same arrowhead type is applied at both ends of the dimension line. If you select the first arrowhead, it is automatically applied to the second by default (see Figure 8-13). However, if you want to specify a different arrowhead at the second dimension line endpoint, you must select the desired arrowhead type or **User arrow** from the **2nd** drop-down list.

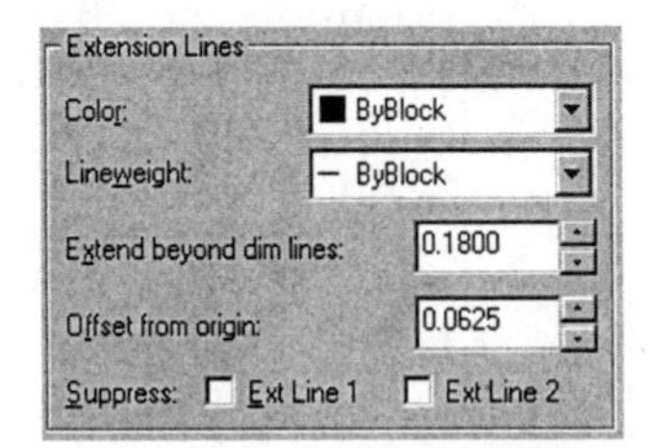

*Figure 8-12 Arrowheads area of the **Lines & Arrows** tab of the **New Dimension Style** dialog box*

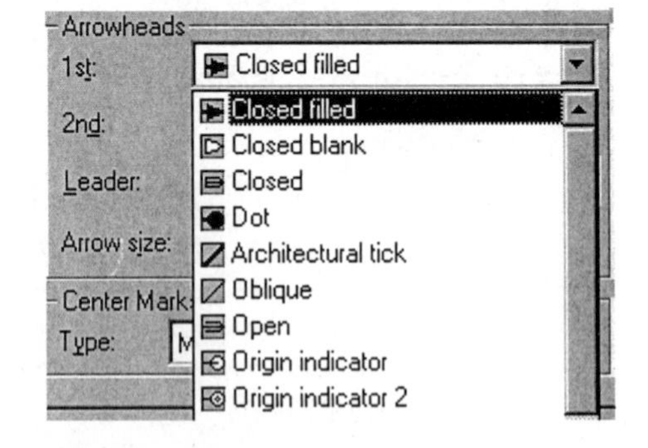

*Figure 8-13 Arrowheads **1st:** drop-down list*

The first endpoint of the dimension line is the intersection point of the first extension line and the dimension line. The first extension line is determined by the first extension line origin. However, in angular dimensioning the second endpoint is located in a counterclockwise direction from the first point, regardless of how the points were selected when creating the angular dimension. The specified arrowhead types are selected from the **1st** and **2nd** drop-down lists. The first arrowhead type is saved in the **DIMBLK1** system variable, and the second arrowhead type is saved in the **DIMBLK2** system variable. As mentioned earlier, by default, arrows are drawn at the two endpoints of the dimension line. If you specify a different block from the arrowhead, the name of the block is stored in the **DIMBLK** system variable (see Chapter 12 for information on creating blocks).

User Arrow. If you want to specify that, instead of the standard arrows, a user-defined block be drawn at the ends of the dimension line, select the **User Arrow...** from the **1st:** or **2nd:** arrowheads drop-down list to display the **Select Custom Arrow Block** dialog box. Select the name of the predefined block name from the **Select from Drawing Blocks:** drop-down list and then choose the **OK** button. The size of the block is determined by the value stored in the **Arrow size:** edit box. The name of the block you select is stored in the **DIMBLK1** or **DIMBLK2** variable.

Arrow size. Arrowheads and terminator symbols are drawn according to the size specified in the **Arrow size**: edit box. The default value is 0.18 units. This value is stored in the **DIMASZ** system variable. By default, **DIMTSZ** is set to 0 and **DIMASZ** to 0.18. See Figures 8-14 and 8-15.

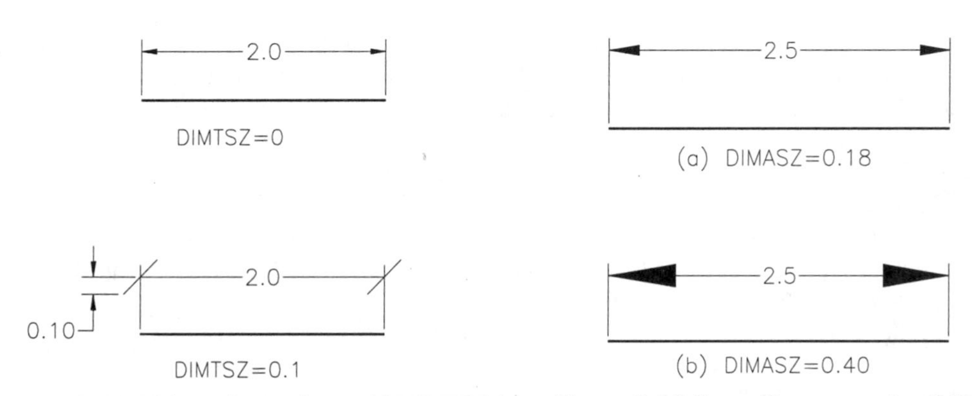

*Figure 8-14 Tick marks are drawn if **DIMTSZ** is not 0*

*Figure 8-15 Controlling arrow size, **DIMASZ***

Leader. The **Leader** drop-down list displays arrowhead types for the Leader arrow. Here also you can either select an option from the drop-down list or choose **User Arrow** which allows you to define and use a user-defined arrowhead type.

Creating Custom Arrowheads

As mentioned earlier, you can replace the default arrowheads at the end of the dimension lines with a user arrow (user-defined block). Figure 8-15, shows a block comprising a custom arrow. You can use this arrow to define a block and then use it as a user arrow. When you create a dimension, the custom arrow will automatically appear at the endpoint of the dimension line, replacing the standard arrow.

Considerations for Creating an Arrowhead Block

1. When you create a block for an arrowhead, the arrowhead block must be created in a 1 x 1 box. AutoCAD LT automatically scales the block's X and Y scale factors to arrowhead size x overall scale. The **DIMASZ** variable controls the length of the arrowhead. For example, if **DIMASZ** is set to 0.25, the length of the arrow will be 0.25 units. Also, if the length of the arrow is not 1 unit, it will leave a gap between the dimension line and the arrowhead block.
2. The arrowhead must be drawn as it would appear on the right-hand side of the dimension line.
3. The insertion point of the arrowhead block must be the point that will coincide with the extension line.

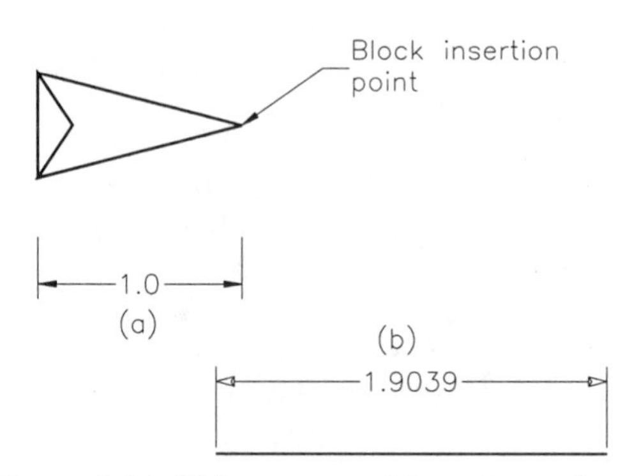

Figure 8-16 *Using arrows, tick marks, and user-defined blocks*

To create an arrowhead block, use the **BLOCK** command. The **Block Definition** dialog box is displayed on the screen. Enter a name for the block in **Name:** edit box, say **MYARROW**. Choose the **Pick point** button to select insertion base point on the screen. Once you have selected the insertion base point on screen, you are returned back to the **Block Definition** dialog box. Now, choose the **Select objects** button to select objects on the screen. Select the arrow and press ENTER. In the **Block Definition** dialog box, choose **OK**. Now, when you select the **User Arrow** option from the Arrowheads drop-down lists, the **Select from Drawing Blocks:** drop-down list displays the drawing block name you just created. You can select **MYARROW** from the drop-down list and when you create a dimension, the user-defined arrow is displayed at the end of the dimension line. The following is the prompt sequence for creating a block using the command line:

Command: **-BLOCK** Enter
Enter block name or [?]: *Give a name; for example, MYARROW.*
Specify insertion base point: *Select the endpoint as shown in figure 8-16.*
Select objects: *Select the arrow.*
Select objects: *Press ENTER.*

Center Marks for Circles

This area deals with options that control the appearance of Center Marks and Centerlines in Radius and Diameter dimensioning. When using the **DIMDIAMETER** and **DIMRADIUS** commands, Center Marks are drawn only if dimensions are placed outside the circle.

Type. The **Type:** drop-down list allows you to select the type of Center Mark you wish to use, such as **None**, **Mark**, or **Line**.

Mark. This is the default option and the size of the Center mark is stored as a positive value in the **DIMCEN** variable (Figure 8-17).

Line. If you want to draw centerlines for a circle or an arc, select **Line** from the **Type** drop-down list. The value in the **Size:** edit box determines the size of the centerlines, as shown in Figure 8-18. The value you enter in the **Size:** edit box is stored as a negative value in the **DIMCEN** variable. In the case of radius and diameter dimensioning, centerlines are drawn only when the dimension line is located outside the circle or arc. The default setting is a positive value, resulting in the generation of dimensions without centerlines.

None. If you select **None**, the center marks are not drawn and AutoCAD LT automatically disables the **Size** edit box.

If you use the **DIMCEN** command, a positive value will create a center mark, and a negative value will create a centerline. If the value is 0, AutoCAD LT does not create center marks or centerlines.

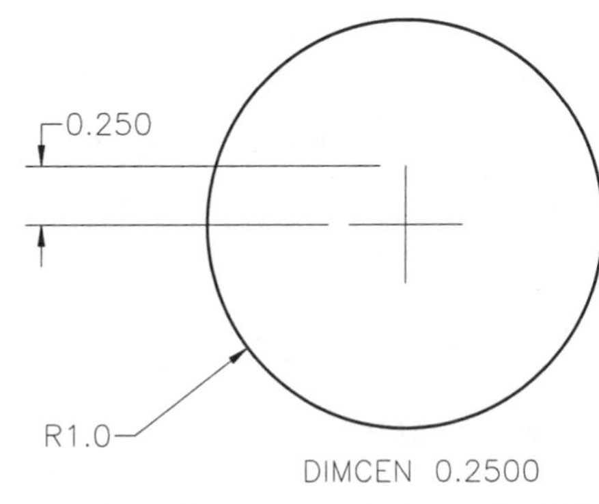

*Figure 8-17 Center mark size, **DIMCEN***

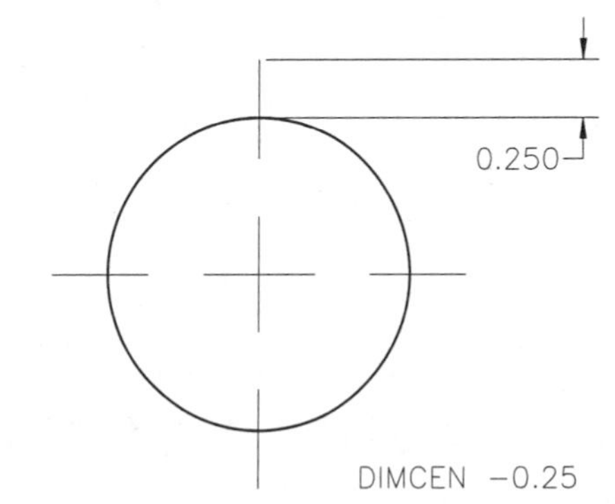

*Figure 8-18 Mark with center lines, (negative **DIMCEN**)*

Size. The **Size:** edit box in the **Center Marks for Circles** area of the **New, Modify, Override Dimension Style** dialog box displays the current size of the center marks (Figure 8-19). The center marks are created by using the Center, Diameter, and Radius dimensioning commands. The center marks are also created if you use the **DIMCENTER** command. In the case of radius and diameter dimensioning, the center mark is drawn only if the dimension line is located outside the circle or arc. You can specify the size of the center mark by entering the required value in the **Size:** edit box. If you do not want a center mark, enter 0 in the edit box or, better, select **None** in the **Type:** drop-down list. This value is stored in the **DIMCEN** variable. The default value of **DIMCEN** variable is 0.09.

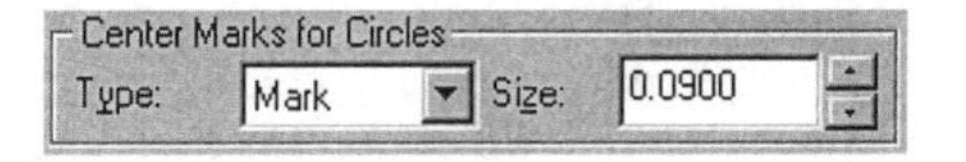

*Figure 8-19 The Center Marks for Circles area of the **New Dimension Style** dialog box*

Note

*Unlike specifying a negative value for the **DIMCEN** variable, you cannot enter a negative value in the **Size:** edit box. Selecting **Line** from the **Type:** drop-down list, automatically treats the value in the **Size:** edit box as the size for the centerlines and sets **DIMCEN** to the negative of the value shown.*

Exercise 1 *Mechanical*

Draw Figure 8-20. Then set the values in the **Lines and Arrows** tab of the **New Dimension Style** dialog box to dimension the drawing as shown in this figure. (Baseline Spacing = 0.25, Extension beyond dimension line = 0.10, Offset from Origin = 0.05, Arrowhead size= 0.09.) Assume the missing dimensions.

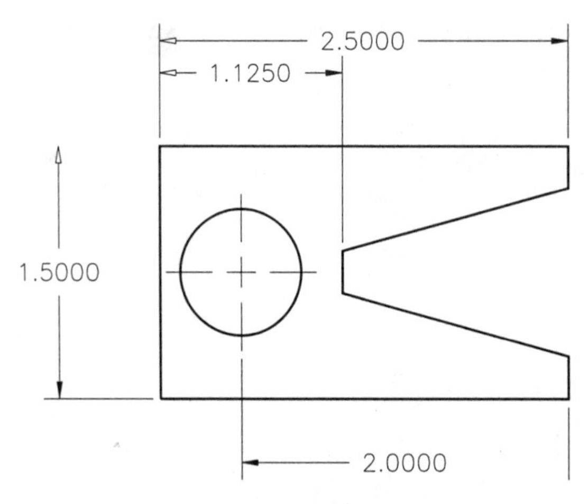

Figure 8-20 *Drawing for Exercise 1*

CONTROLLING DIMENSION TEXT FORMAT

Text Tab

You can control the dimension format through the **Text** tab of the **New Dimension Style** dialog box (Figure 8-21). In the **Text** tab, you can control the placement, appearance, horizontal and vertical alignment of dimension text. For example, you can force AutoCAD LT to align the dimension text along the dimension line. You can also force the dimension text to be displayed at the top of the dimension line. You can save the settings in a dimension style file for future use. The **New Dimension Style** dialog box has a Preview window that updates dynamically to display the text placement as the settings are changed. Individual items of the **Text** tab and the related dimension variables are described next.

Text Appearance

The **Text Appearance** area of the **New Dimension Style** dialog box lets you specify the text style, text height, text format and color of the dimension text. These options are described next (Figure 8-22).

Text Style. The **Text Style** drop-down list, displays the names of the predefined text styles. From this list you can select the style name that you want to use for dimensioning. You must define the text style before you can use it in dimensioning (see "**STYLE** command" in Chapter 5). Choosing the [**...**] button, displays the **Text Style** dialog box that allows you to create a new or modify an existing Text Style. The value of this setting is stored in the **DIMTXSTY** system variable. The change in dimension text style does not affect the text style

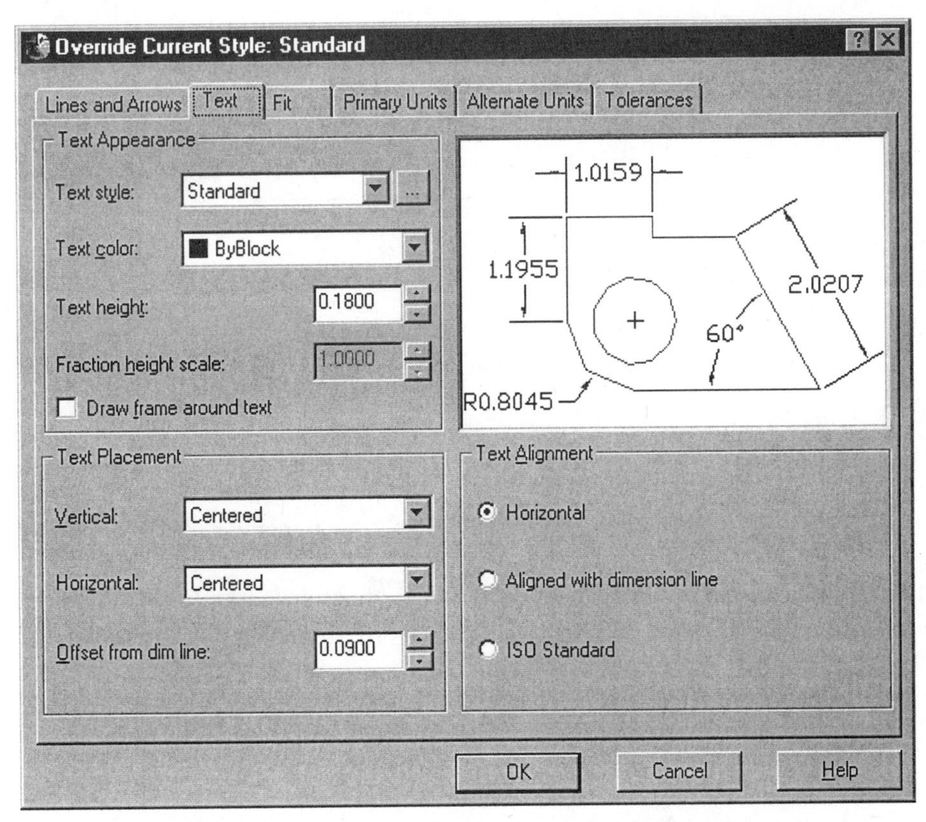

*Figure 8-21 The **Text** tab of the **New Dimension Style** dialog box*

you are using to draw other text in the drawing.

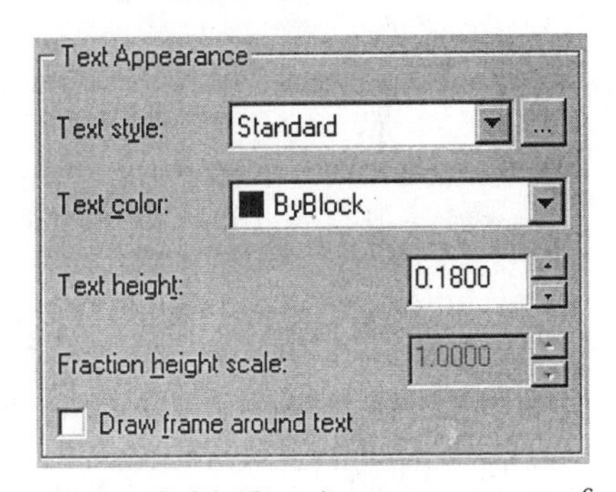

*Figure 8-22 Text Appearance area of the **Text** tab*

Text Color. A color can be assigned to the dimension text by selecting it from the **Text Color** drop-down list. The color number or special color label is held in the **DIMCLRT** variable. The default color label is BYBLOCK. If you choose the **Other** option from the **Text Color** drop-down list, the **Select Color** dialog box is displayed where you can choose a specific color.

Text Height. You can customize the height of the dimension text (Figure 8-23) by entering the required text height in the **Text height** edit box. You can change the dimension text height only when the current text style does not have a fixed height. In other words, the text height specified in the **STYLE** command should be zero because a predefined text height (specified in the **STYLE** command) overrides any other setting for the dimension text height. The value in the **Text height** edit box is stored in the **DIMTXT** variable. The default text height is 0.1800 units.

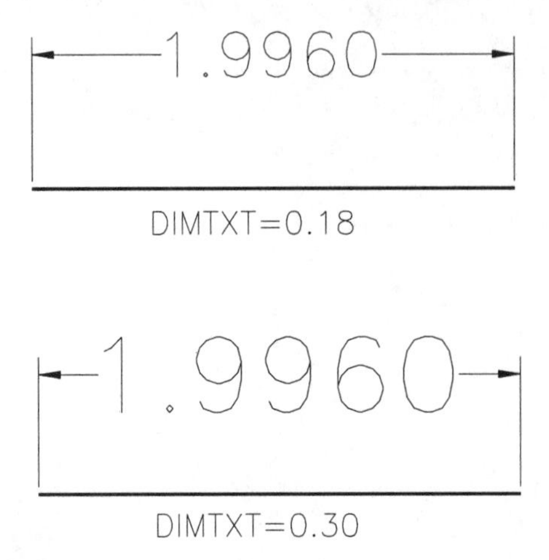

Figure 8-23 Dimension text height, DIMTXT

Fraction height scale. The **Fraction height scale:** edit box allows you to set the scale of the fractional units in relation to the Dimension text height. This value is stored in the **DIMTFAC** variable.

Draw frame around text. Selecting the **Draw frame around text** check box, draws a frame around the dimension text. This value is stored as a negative value in the **DIMGAP** system variable.

Text Placement

This area of the **Text** tab gives you options controlling the text placement. These options are discussed below:

Vertical. The **Vertical:** drop-down list displays the options that control the vertical placement of the dimension text. The current setting is highlighted. Controlling the vertical placement of the dimension text is possible only when the dimension text is drawn in its normal (default) location. This setting is stored in the **DIMTAD** system variable. The vertical text placement options are described next (Figure 8-24).

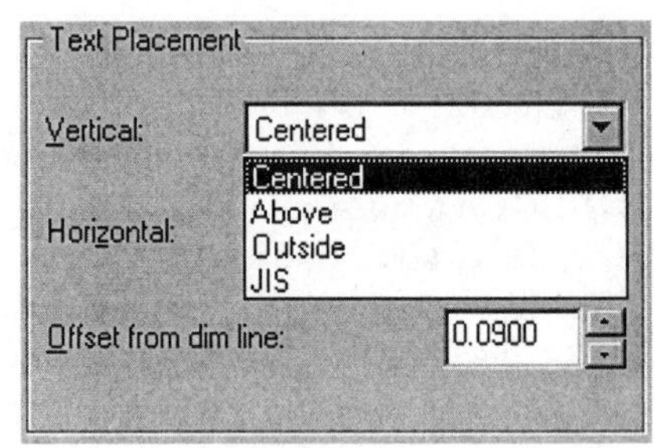

Figure 8-24 Controlling vertical placement

Centered. If this option is selected, the dimension text gets centered on the dimension line in such a way that the dimension line is split to allow for placement of the text. Selecting this option turns **DIMTAD** off. In this setting, **DIMTAD = 0** (Figure 8-25). If the **1st** or **2nd Extension Line** option is selected in the **Horizontal:** drop-down list, this centered setting will center the text on the extension line, not on the dimension line.

Above. If this option is selected, the dimension text is placed above the dimension line, except when the dimension line is not horizontal and the dimension text inside the extension lines is horizontal (**DIMTIH = 1**). The distance of the dimension text from the dimension line is controlled by the **DIMGAP** value. This results in an unbroken solid dimension line being drawn under the dimension text. In this setting, **DIMTAD=1**.

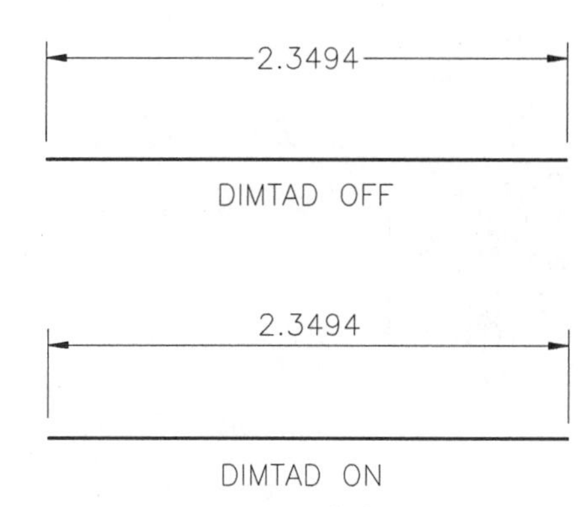

Figure 8-25 *Using Vertical placement to place text, Centered and Above a dimension line,* ***DIMTAD***

Outside. This option places the dimension text on the side of the dimension line. In this setting, **DIMTAD=2**.

JIS. This option lets you place the dimension text to conform to **JIS** (Japanese Industrial Standards) representation. In this setting, **DIMTAD=3.**

Note

The Horizontal and Vertical Placement options selected are reflected in the dimensions shown in the Preview window.

Horizontal. Horizontal Placement controls the Horizontal justification of the dimension text. To display the available options, choose the down arrow button to display the **Horizontal** drop-down list (Figure 8-26). The selected setting is stored in the **DIMJUST** system variable. The default option is **Centered**. The following is the list of the available options with **DIMJUST** values (Figure 8-27).

Option	**Description**	**DIMJUST**
Centered	Places text between extension lines	0
1st Extension Line	Places text next to first extension line	1
2nd Extension Line	Places text next to second extension line	2
Over 1st Extension	Places text aligned with and above the first extension line	3
Over 2nd Extension	Places text aligned with and above the second extension line	4

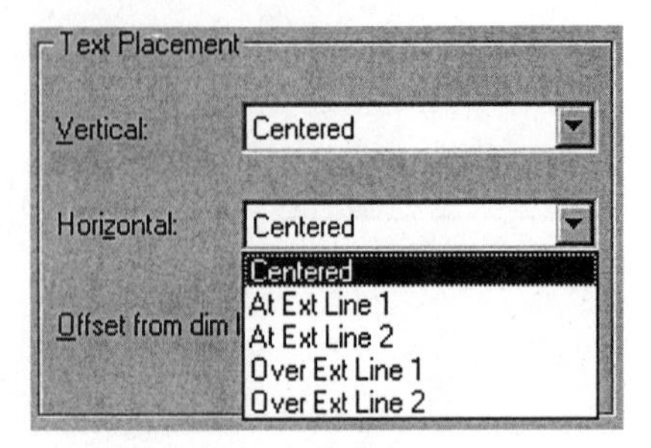

Figure 8-26 Horizontal Placement options of the ***Text*** *tab of the* ***New Dimension Style*** *dialog box*

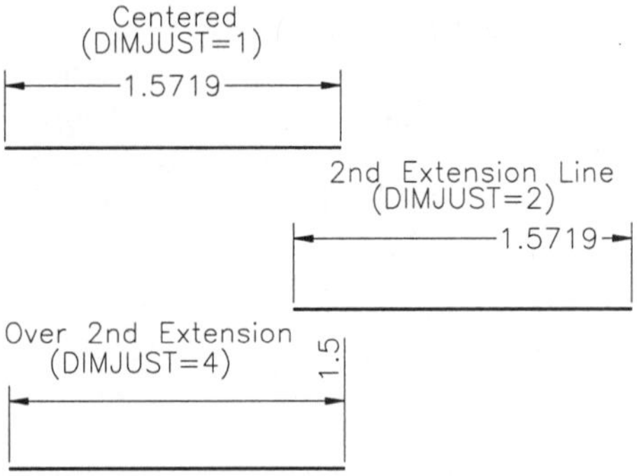

Figure 8-27 Using Horizontal Placement options to position text

Offset from dim line. The **Offset from dim line:** edit box is used to specify the distance between the dimension line and the dimension text (Figure 8-28). You can enter the text gap you need in this edit box. The Text Gap value is also used as the measure of minimum length for the segments of the dimension line and in basic tolerance. The default value specified in this box is 0.09 units. The value of this setting is stored in the **DIMGAP** system variable.

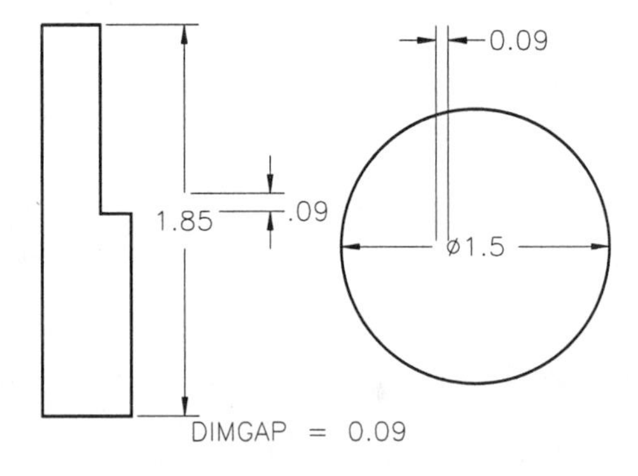

Figure 8-28 Dimension text gap, ***DIMGAP***

Text Alignment

You can use the **Text Alignment** area options (Figure 8-29) to specify the alignment of the dimension text with the dimension line. These options can be used to control the alignment of the dimension text for linear, radius, and diameter dimensions and the values are stored in the **DIMTIH** and **DIMTOH** system variables. By default, the Horizontal option is selected and both the inside and outside dimension text is drawn horizontally.

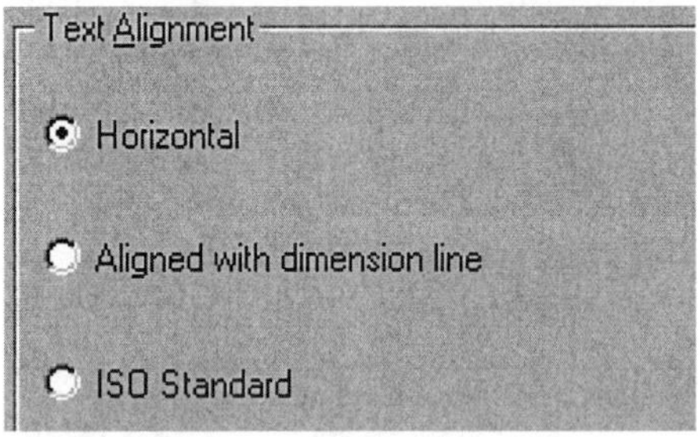

Figure 8-29 Text Alignment area of the ***Text*** *tab of the* ***New Dimension Style*** *dialog box*

Horizontal. By default, the dimension text is drawn horizontally with respect to the UCS (user coordinate system). Therefore, the **Horizontal** radio button is selected. The alignment of the dimension line does not affect text alignment. Selecting this option turns both the **DIMTIH** and **DIMTOH** system variables **on** (Figure 8-30 and Figure 8-31). The text is drawn horizontally even if the dimension line is at an angle.

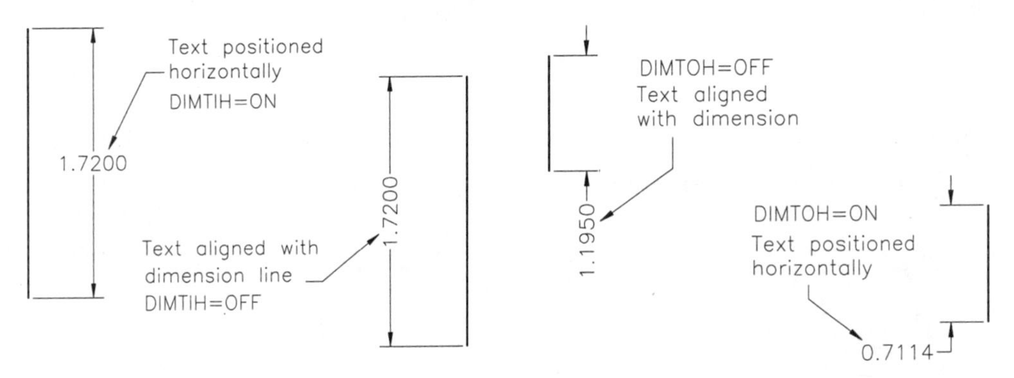

*Figure 8-30 Dimension text inside extension lines horizontal, **DIMTIH***

*Figure 8-31 Dimension text outside extension lines horizontal, **DIMTOH***

Aligned with dimension line. Selecting this radio button aligns the text with the dimension line and both the system variables **DIMTIH** and **DIMTOH** are off.

ISO Standard. By default, the dimension text is drawn horizontally with respect to the UCS (user coordinate system). Therefore, the **Horizontal** radio button is selected. The alignment of the dimension line does not affect the text alignment. If you select the **ISO Standard** radio button, the dimension text is aligned with the dimension line only when the dimension text is inside the extension lines (Figure 8-31). Selecting this option turns the system variable **DIMTOH** on; that is, the dimension text outside the extension line is horizontal regardless of the angle of the dimension line.

Exercise 2 *Mechanical*

Draw Figure 8-32. Then set the values in the **Lines and Arrows** and **Text** tab of the **New Dimension Style** dialog box to dimension the drawing as shown in the figure. (Baseline Spacing = 0.25, Extension beyond dimension lines = 0.10, Offset from Origin = 0.05, Arrow size = 0.09.) Assume the missing dimensions and set **DIMTXT**= 0.09.

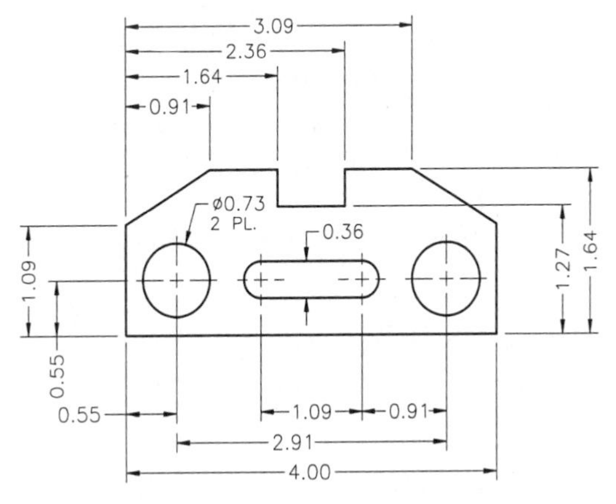

Figure 8-32 Drawing for Exercise 2

FITTING DIMENSION TEXT AND ARROWHEADS

Fit Tab

The **Fit** tab of the **New Dimension Style** dialog box gives you options that control the placement of dimension lines, arrowheads, leaderlines, text, and the overall dimension scale (Figure 8-33).

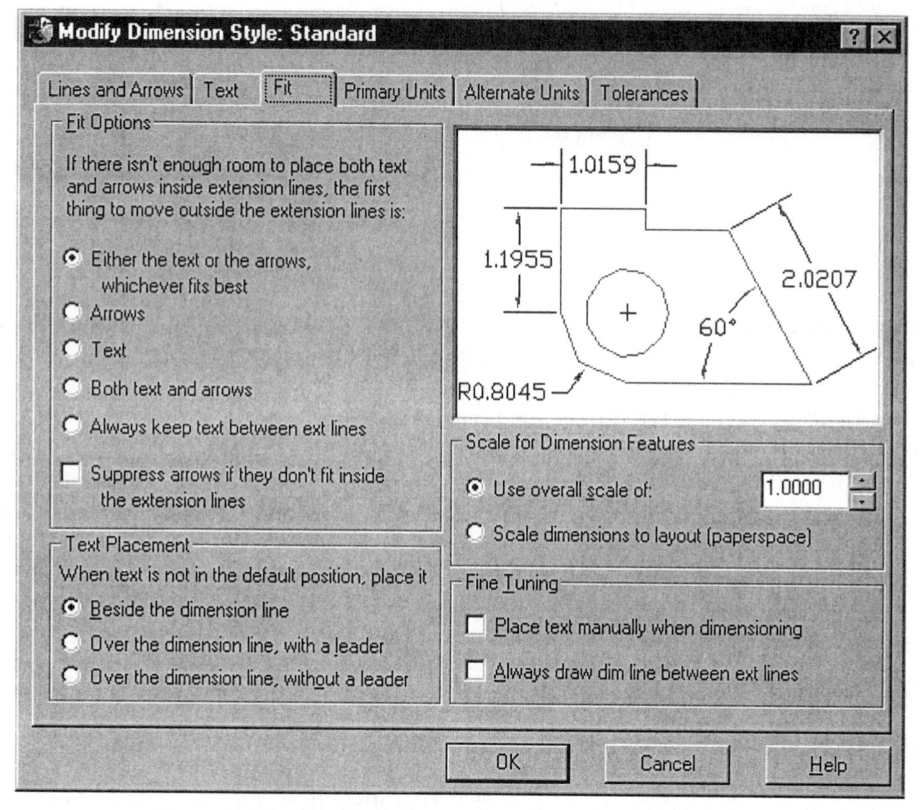

Figure 8-33 ***Fit*** *tab of the* ***New Dimension Style*** *dialog box*

Fit Options

In the **Fit Options** area of the **Fit** tab, AutoCAD LT displays the available options for fitting the arrows and dimension text between the extension lines. The value of the **Fit Options** is stored in the **DIMATFIT**, **DIMTMOVE**, **DIMTIX**, and **DIMSOXD** system variables. The **Fit** options are described next. Selecting any of these radio buttons, sets priorities for moving the text and arrowheads outside the extension lines, if space between the extension lines is not enough to fit both of them.

Either the text or the arrows whichever fits best. This is the default option. In this option, AutoCAD LT places the dimension where it fits best between the extension lines. In this setting, **DIMTMOVE=0**, **DIMATFIT=3**.

Arrows. When you select this option, AutoCAD LT places the text and arrowheads inside the extension lines if there is enough space to fit both. If space is not available for both arrows and text, the text is placed inside the extension lines and the arrows are placed outside the extension lines. If there is not enough space for text, both text and arrowheads are placed outside the extension lines. In this setting, **DIMTMOVE=0, DIMATFIT=1** (Figure 8-34).

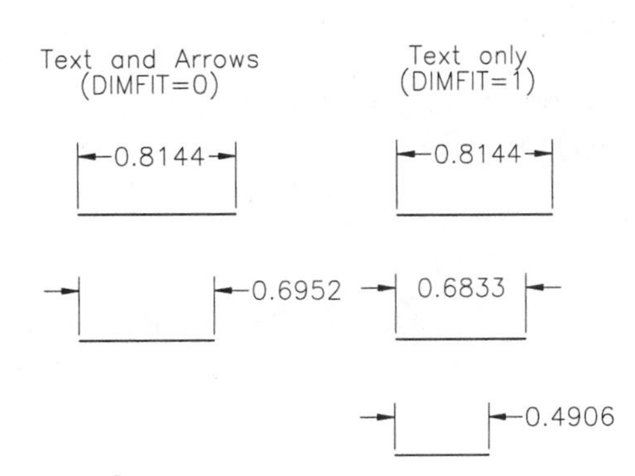

Figure 8-34 Both Text and Arrows whichever fits best and Arrows options

Text. When you select this option, AutoCAD LT places the text and arrowheads inside the extension lines if there is enough space to fit both. If there is enough space to fit the arrows, the arrows will be placed inside the extension lines and the dimension text moves outside the extension lines. If there is not enough space for either text or arrowheads, both are placed outside the extension lines. In this setting, **DIMTMOVE=0, DIMATFIT=2**.

Both Text and Arrows. If you select this option, AutoCAD LT will place the arrows and dimension text between the extension lines if there is enough space available to fit both. Otherwise, both text and arrowheads are placed outside the extension lines. In this setting, **DIMTMOVE=0, DIMATFIT=0** (Figure 8-34).

Always keep text between extension lines. This option always keeps the text between extension lines even in cases where AutoCAD LT would not do so. Selecting this radio button does not affect radius and diameter dimensions. The value is stored in the **DIMTIX** variable and the default value is **off**.

Suppress arrows if they don't fit inside the extension lines. If you select this check box, it suppresses the display of arrowheads and dimension line outside the extension line if there is insufficient space between the extension lines. The value is stored in the **DIMSOXD** variable and the default value is **off**.

Text Placement Options

The **Text Placement** area of the **Fit** tab of the **New Dimension Style** dialog box gives you options regarding positioning of dimension text when it is moved from the default position. The value is stored in the **DIMTMOVE** variable. The options are discussed as follows:

Beside the dimension line. This option places the dimension text beside the dimension line and the value for the **DIMTMOVE** variable is 0.

Over the dimension line, with a leader. Selecting this option places the dimension text away from the dimension line and a leader line is created, which connects the text to the dimension line. But, if the dimension line is too close to the text, a leader is not drawn. The Horizontal placement decides whether the text is placed to the right or left of the leader. In this setting, **DIMTMOVE=1**.

Over the dimension line, without a leader. In this option, AutoCAD LT does not create a leader line if there is insufficient space to fit the dimension text between the extension lines. The dimension text can be moved freely, independent of the dimension line. In this setting, **DIMTMOVE=2**.

Scale for Dimension Features

Sets the value for overall Dimension scale or scaling to paper space (Figure 8-35).

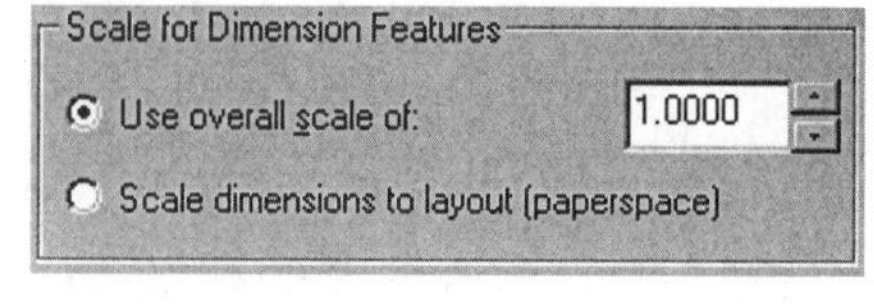

Figure 8-35 *Scale for Dimension Features area of the **Fit** tab*

Use Overall Scale Of. The current general scaling factor that pertains to all of the size-related dimension variables, like text size, center mark size, and arrowhead size, is displayed in the **Use overall scale of** edit box. You can alter the scaling factor to your requirement by entering the scaling factor of your choice in the **Use overall scale of:** edit box. Altering the contents of this box alters the value of the **DIMSCALE** variable, since the current scaling factor is stored in it. The overall scale (**DIMSCALE**) is not applied to the measured lengths, coordinates, angles, or tolerance. The default value for this variable is 1.0 [Figure 8-36(a)]; and, in this case, the dimensioning variables assume their preset values and the drawing is plotted at full scale. The scale factor is the reciprocal of the drawing size so the drawing is to be plotted at half size. The overall scale factor (**DIMSCALE**) will be the reciprocal of 1/2, which is 2 [Figure 8-36(b)].

Note

*If you are in the middle of the dimensioning process and you change the **DIMSCALE** value and save the changed setting in a dimension style file, the dimensions with that style will be updated.*

Scale dimensions to layout (paper space). If you select the **Scale dimensions to layout (Paper Space)** radio button, the scale factor between the current model space viewport and floating viewport (paper space) is computed automatically. Also, by selecting this radio button, you disable the **Use overall scale of:** edit box (it is disabled in the dialog box) and **DIMSCALE** is set to 0. When **DIMSCALE** is assigned a value of 0, AutoCAD LT calculates an acceptable default value based on the scaling between the current model space viewport and paper space. If you are in paper space (**TILEMODE=0**), or are not using the **Scale dimensions to layout (paper Space)** feature, AutoCAD LT sets **DIMSCALE** to 1; otherwise, AutoCAD LT calculates a scale factor that makes it possible to plot text sizes, arrow sizes, and other scaled distances at the values in which they have been previously set. (For further details, see "Model Space and Paper Space Dimensioning" in Chapter 7.)

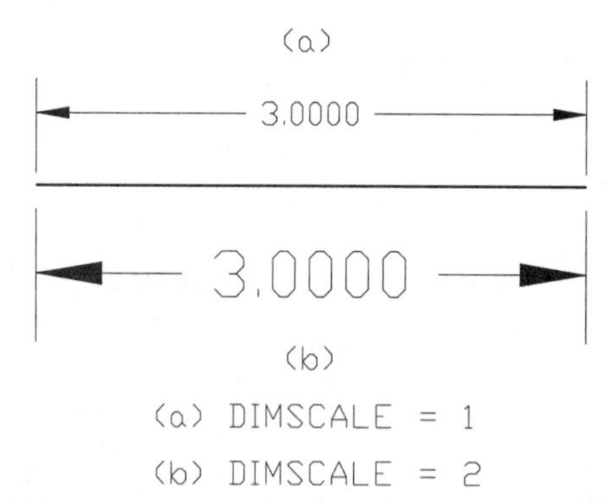

Figure 8-36 *Using overall scaling to scale dimensions, **DIMSCALE***

Fine Tuning Options

The Fine Tuning area of the **Fit** tab of the **New Dimension Style** dialog box, provides additional options governing placement of dimension text. The options are as follows:

Place text manually when dimensioning. When you dimension, AutoCAD LT places the dimension text in the middle of the dimension line (if there is enough space). If you select the **Place text manually when dimensioning** check box, you can position the dimension text anywhere along the dimension line (Figure 8-37). You will also notice that when you select this check box, the Horizontal Justification is ignored. This setting is saved in the **DIMUPT** system variable. The default value of this variable is **off**. If you set the **DIMUPT** variable on, AutoCAD LT will place the dimension text at the point you have specified as the dimension line location. This enables you to position dimension text anywhere along the dimension line.

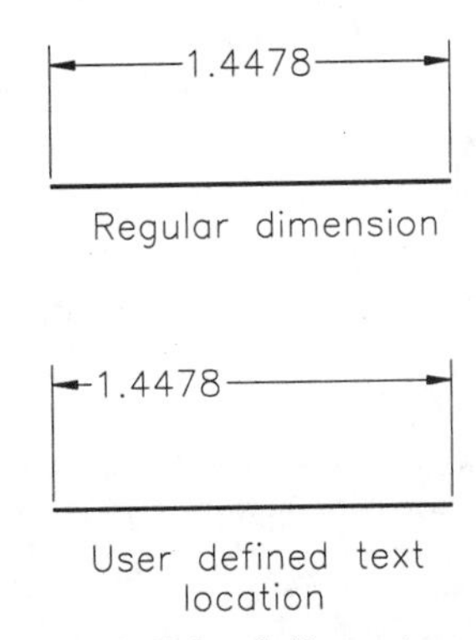

Figure 8-37 *User-defined dimension text position*

Always draw dimension line between extension lines. If the dimension text and the dimension lines are outside the extension lines and you want the dimension lines to be placed between the extension lines, select the **Always draw dimension lines between extension lines** check box. When you select this option in the radius and diameter dimensions (when default text placement is horizontal), the dimension line and arrows are drawn inside the circle or arc, and the text and leader are drawn outside (see Figure 8-38). When you select the **Always draw dimension line between extension lines** check box, the **DIMTOFL** variable is set to on by AutoCAD LT. The default setting is off, resulting in generation of the dimension line outside the extension lines when the dimension text is located outside the extension lines.

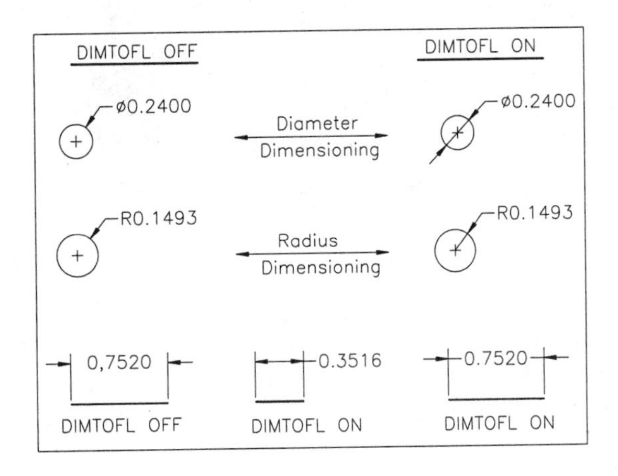

Figure 8-38 *Always draw dimension line between extension line option*

FORMATTING PRIMARY DIMENSIONS UNITS

Primary Units Tab

You can use the **Primary Units** tab of the **New Dimension Style** dialog box to control the dimension text format and precision values (Figure 8-39). You can use this tab to control Units, Dimension Precision, and Zero Suppression for dimension measurements. AutoCAD LT lets you attach a user-defined prefix or suffix to the dimension text. For example, you can define the diameter symbol as a prefix by entering %%C in the **Prefix:** edit box; AutoCAD LT will automatically attach the diameter symbol in front of the dimension text. Similarly, you can define a unit type, such as **mm**, as a suffix; AutoCAD LT will then attach **mm** at the end of every dimension text. This tab also enables you to define zero suppression, precision, and dimension text format.

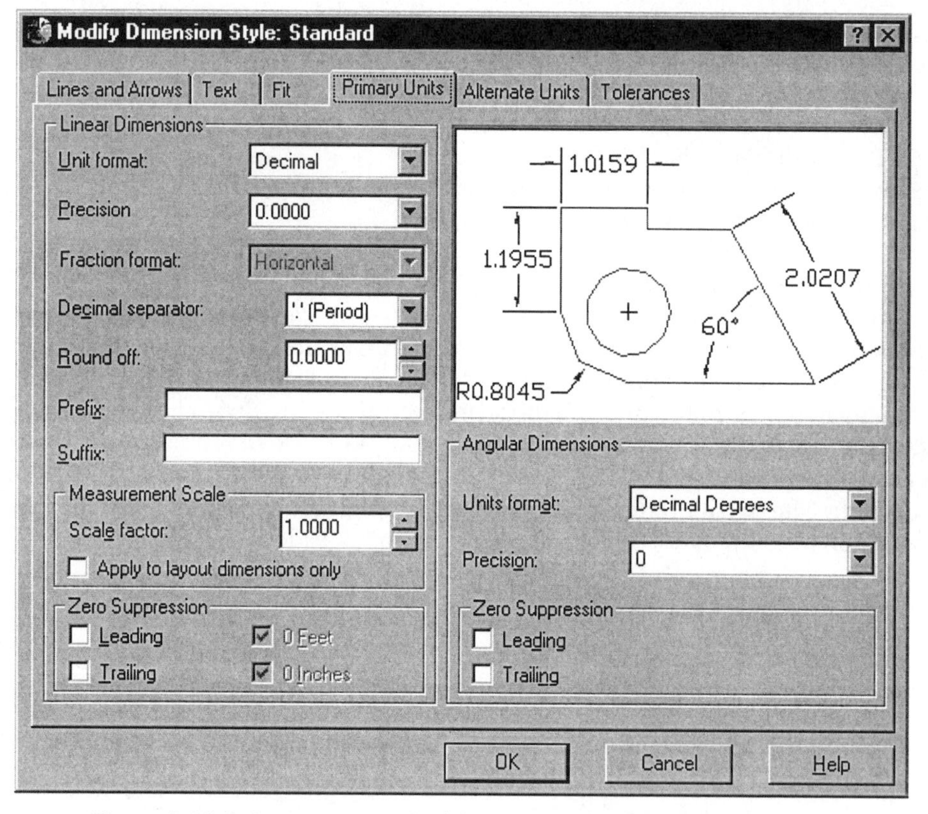

Figure 8-39 ***Primary Units*** *tab of the* ***New Dimension Style*** *dialog box*

Linear Dimensions

The **Linear Dimensions** area of the **Primary Units** tab of the **New Dimension Style** dialog box consists of options like the Unit format, Prefix, and Suffix, and so on, which are described next.

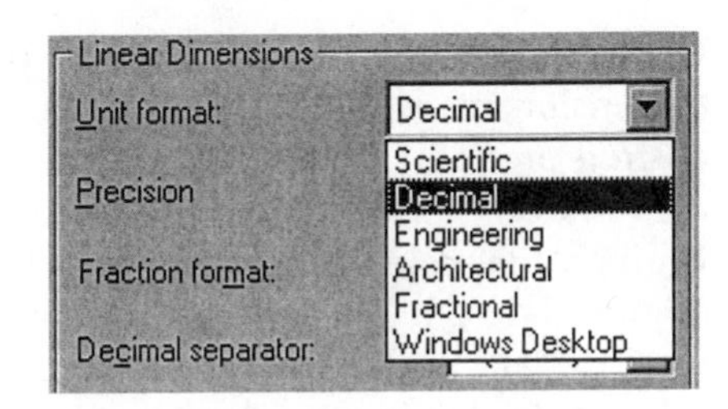

Figure 8-40 Unit Format: drop-down list

Unit Format. When you choose the down arrow button, AutoCAD LT displays a drop-down list of Unit formats, such as Decimal, Scientific, Architectural to select from (Figure 8-40). You can select one of the listed units from the **Unit format** drop-down list to use when dimensioning. Notice that by selecting a dimension unit format, the drawing units (which you might have selected by using the **UNITS** command) are not affected. The unit setting for linear dimensions is stored in the **DIMUNIT** system variable.

Precision. You can also control unit precision by using the **Precision:** drop-down list. The setting for precision (number of decimal places) is saved in **DIMDEC**.

Fraction Format. Sets the format for fractions in the dimension text. You can select a format from the **Fraction format:** drop-down list. The options are Diagonal, Horizontal, and not stacked. The value is stored in the **DIMFRAC** variable.

Decimal Separator. The **Decimal separator:** drop-down list gives you options which you can select to set as decimal separators. For example, Period [.], Comma [,] or Space [ ]. If you have selected Windows desktop units in the **Unit Format:** drop-down list, AutoCAD LT uses the Decimal symbol settings. The value is stored in the **DIMDSEP** variable.

Figure 8-41 Round off area

Round off. The **Round off:** edit box sets the value for rounding off dimension values (see Figure 8-41). For example, if the **Round off:** edit box has a value of .05, a value of .06 will round off to .10. The number of decimal places of the round off value you enter in the edit box should be less than or equal to the value in the **Precision:** edit box. The value is stored in the **DIMRND** variable and the default value in the **Round Off:** edit box is 0. Also see Figure 8-42.

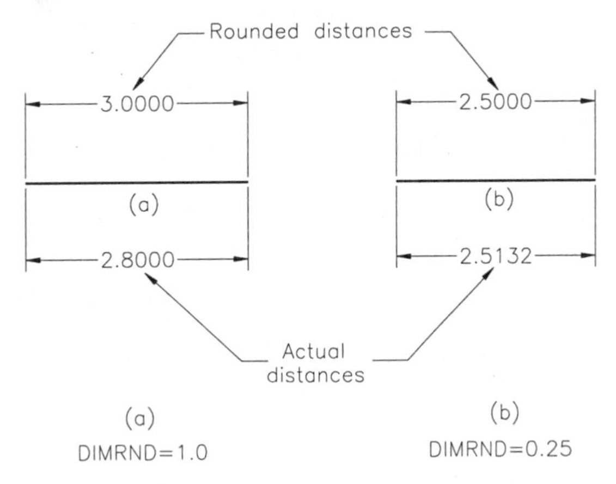

Figure 8-42 *Rounding dimension measurements,* ***DIMRND***

Prefix. You can append a prefix to the dimension measurement by entering the desired prefix in the **Prefix:** edit box. The dimension text is converted into the **Prefix<dimension measurement>** format [Figure 8-42(a)]. For example, if you enter the text "Ht" in the Prefix: edit box, "Ht" will be placed in front of the dimension text. The prefix string is saved in the **DIMPOST** system variable.

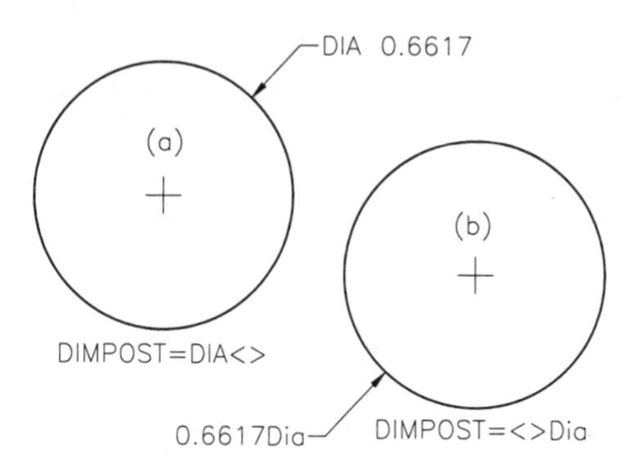

Figure 8-42 *Using text prefix (a) and text suffix (b) in dimensioning,* ***DIMPOST***

Note

Once you specify a prefix, default prefixes such as ***R*** *in radius dimensioning and ø in diameter dimensioning are cancelled.*

Suffix. Just like appending a prefix, you can append a suffix to the dimension measurement by entering the desired suffix in the **Suffix**: edit box in the **Linear Dimensions** area of the **Primary Units** tab of the **New Dimension Style** dialog box [Figure 8-42 (b)]. For example, if you enter the text cm in the **Suffix:** edit box, the dimension text will have <dimension measurement>cm format. AutoCAD LT stores the suffix string in the **DIMPOST** variable.

The **DIMPOST** variable is used to define a prefix or suffix to the dimension text. **DIMPOST** takes a string value as its argument. For example, if you want to have a suffix for centimeters, set **DIMPOST** to cm. To establish a prefix to a dimension text, type the prefix text string and then "<>".

When you define a prefix with the **DIMPOST** variable, this prefix nullifies any default prefixes like R in radius dimensions. For angular dimensions, the <> mechanism is used. To define a prefix or suffix or to prefix the dimension text in **DIMPOST**, the <> mechanism is used. In this way, AutoCAD LT treats the **DIMPOST** values like text entered at the **Enter dimension text:** prompt. For example, if you want to have the prefix **Radius** and the suffix **cm** with the default text, you can set the **DIMPOST** value in the following manner:

Command: **DIMPOST** Enter
Enter new value for DIMPOST, or . for none <" ">: **Radius <> cm**

If you want to add the user-defined text above or below the dimension line, use the separator symbol **\X**. The text that precedes the \X is aligned with the dimension line and positioned above the line. The text that follows the \X is positioned below the dimension line and aligned with the line. If you want to add more lines, use the symbol **\P**. (For a detailed description of these symbols, see the **MTEXT** Command in Chapter 5.) The following example illustrates the use of these symbols:

Command: **DIMPOST** Enter
Enter new value for DIMPOST, or . for none <" ">: **Radius <>\X[In centimeters]**

DIMPOST value text	You enter	Value	Type	Dimension
Default (nil)	*null response*	*1*	*Linear*	*3.00*
Len <>	*null response*	*1*	*Linear*	*Len 3.00*
Default (nil)	*null response*	*3*	*Radial*	*R3.00*
<> R	*null response*	*3*	*Radial*	*3.00 R*
<> R	*<>Radius in cm*	*3*	*Radial*	*3.00 Radius in cm*
nil	*Radius <>*	*3*	*Radial*	*Radius 3.00*
mm	*null response*	*1*	*Linear*	*3.00mm*

You can also use control codes and special characters to enter special prefix and suffix symbols (See **TEXT** Command in Chapter 5).

Measurement scale. This area allows you options to set linear scale factors as follows:

Scale factor. You can specify a global scale factor for linear dimension measurements by entering the desired scale factor in the **Scale factor:** edit box. All the linear distances measured by dimensions, which includes radii, diameters, and coordinates, are multiplied by the existing value in the **Scale factor:** edit box. For example, if the value of the **Scale factor:** edit box is set to 2, two unit segments will be dimensioned as 4 units (2 x 2). The angular dimensions are not affected. In this manner, the value of the linear scaling factor affects the contents of the default (original) dimension text (Figure 8-43). Default value for linear scaling is 1. With the default value, the dimension text generated is the actual measurement of the object being dimensioned. The linear scaling value is saved in the **DIMLFAC** variable.

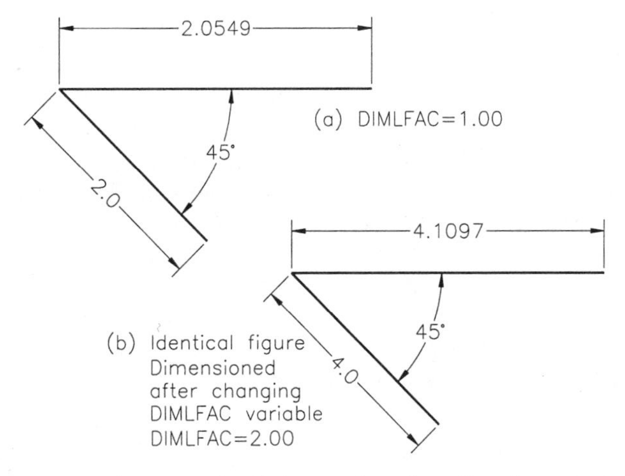

Figure 8-43** Changing the dimension length scaling factor,* ***DIMLFAC

Note

The linear scaling value is not exercised on rounding a value or on plus or minus tolerance values. Therefore, changing the linear scaling factor will not affect the tolerance values.

Apply to layout dimensions only. When you select the **Apply to layout dimensions only** check box you apply the measurement scale factor value only to dimensions in the layout. The value is stored as a negative value in the **DIMLFAC** variable. If you change the **DIMLFAC** from the **Dim:** prompt, AutoCAD LT displays the viewport option to calculate the **DIMLFAC**. First, the **TILEMODE** should be set to 0 (paper space), and then invoke the **MVIEW** command to get the Viewport option.

Command: **DIM** Enter
Dim: **DIMLFAC** Enter
Enter new value for dimension variable, or Viewport <current>: **V** Enter
Select viewport to set scale: *Select the desired viewport.*

The **DIMLFAC** variable is automatically adjusted to the zoom scale factor of the model space viewport and assigned a negative value. In model space, AutoCAD LT ignores the negative value of **DIMLFAC** and assigns it a value of 1.

Note

*Selecting the Viewport option can be accessed only from paper space (**TILEMODE** = 0).*

Zero Suppression. When Engineering or Architectural units are being used, the **Leading** and **Trailing** check boxes are disabled. If you want to suppress the feet portion of a feet-and-inches dimension when the distance is less than 1 foot (when there is a 0 in the feet portion of the text), select the **0 Feet** check box. For example, if you select the **0 Feet** check box, the dimension text 0'-8 3/4" becomes 8 3/4". By default, the 0 Feet and 0 Inches value is suppressed. If you want to suppress the inches part of a feet-and-inches dimension when the distance in the feet portion is an integer value and the inches portion is zero, select the **0 Inches** check box. For example, if you select the **0 Inches** check box, the dimension text 3'-0" becomes 3'.

If you want to suppress the leading zeros in all of the distances measured in decimals, select the **Leading** check box. For example, by selecting this box, 0.0750 becomes .0750. If you want to suppress the trailing zeros in all the distances measured in decimals, select the **Trailing** check box. For example, by selecting this box, 0.0750 becomes 0.075. AutoCAD LT stores zero suppression as an integer value in the **DIMZIN** variable in the following manner:

If you select the **0 Feet** check box, an integer value 3 is stored in **DIMZIN** by AutoCAD LT. If you select the **0 Inches** check box, an integer value 2 is stored in **DIMZIN**. If you select the **Leading** check box, an integer value 4 is stored in **DIMZIN** by AutoCAD LT. If you select the **Trailing** check box, an integer value 8 is stored in **DIMZIN**. A combination of not selecting the **0 Feet** and **0 Inches** check boxes results in display of the 0 feet as well as the 0 inches of a measurement. For this combination, the value stored in **DIMZIN** is 1. Selecting both the **Trailing** and **Leading** check boxes has a **DIMZIN** value of 12.

Remember that, by default, **the 0 Feet** and **0 Inches** check boxes are selected. This is displayed by a mark in these two check boxes. The following table shows the result of selecting each of the **DIMZIN** values.

DIMZIN Value	Meaning
0 (default)	Suppress zero feet and zero inches
1	Include zero feet and exactly zero inches
2	Include zero feet, suppress zero inches
3	Include zero inches, suppress zero feet

The following table illustrates the preceding results with examples.

DIMZIN Value	Examples			
0	2'-0 2/3"	8"	2'	4/5"
1	2'-0 2/3"	0'-8"	2'-0"	0'-0 4/5"
2	2'-0 2/3"	0'-8"	2'	0'-0.4/5"
3	2'-0 2/3"	8"	2'-0"	4/5"

If the dimension has feet and a fractional inch part, the number of inches is included even if it is 0. This is independent of the **DIMZIN** setting. For example, a dimension such as 1'-2/3" never exists. It will be in the form 1'-0 2/3".

The integer values 0-3 of the **DIMZIN** variable control the feet-and-inch dimension only. If you set **DIMZIN** to 4 (0 + 4 = 4), the leading zeros will be omitted in all decimal dimensions; for example, 0.2600 becomes .2600. If you set **DIMZIN** to 8 (4 + 4 = 8), the trailing zeros are omitted; for example, 4.9600 becomes 4.96. If you set **DIMZIN** to 12 (4 + 8 = 12), the leading and the trailing zeroes are omitted; for example, 0.2300 becomes .23. The same thing applies to zero suppression for tolerance values (**Tolerance** tab). The setting of Zero Suppression for tolerance is stored in the **DIMTZIN** system variable.

Angular Dimensions

This area gives you options to control the units format, precision, and zero suppression for Angular units.

Units Format. The **Units format:** drop-down list displays a list of unit formats for you to select from. The value governing the unit setting for angular dimensions is stored in **DIMAUNIT**.

Precision. You can select the number of decimal place display from the **Precision:** drop-down list. The value is stored in the **DIMADEC** variable.

Zero Suppression. Similar to linear dimensioning, you can suppress the **leading**, **trailing**, neither or both by selecting the respective check boxes in the Angular dimensions area of the **Primary units** tab. The value is stored in the **DIMAZIN** variable.

FORMATTING ALTERNATE UNITS

Alternate Units Tab

By default, the **Alternate Units** tab of the **New Dimension Style** dialog box is disabled and the value of the **DIMALT** variable is turned off. If you want to perform alternate units dimensioning, select the **Display Alternate Units** check box. By doing so, AutoCAD LT activates the various edit boxes (Figure 8-44). This tab sets the format, precision, angles, placement, scale, and so on, for the alternate units in use. In this tab you can specify the values which will be applied to alternate dimensions.

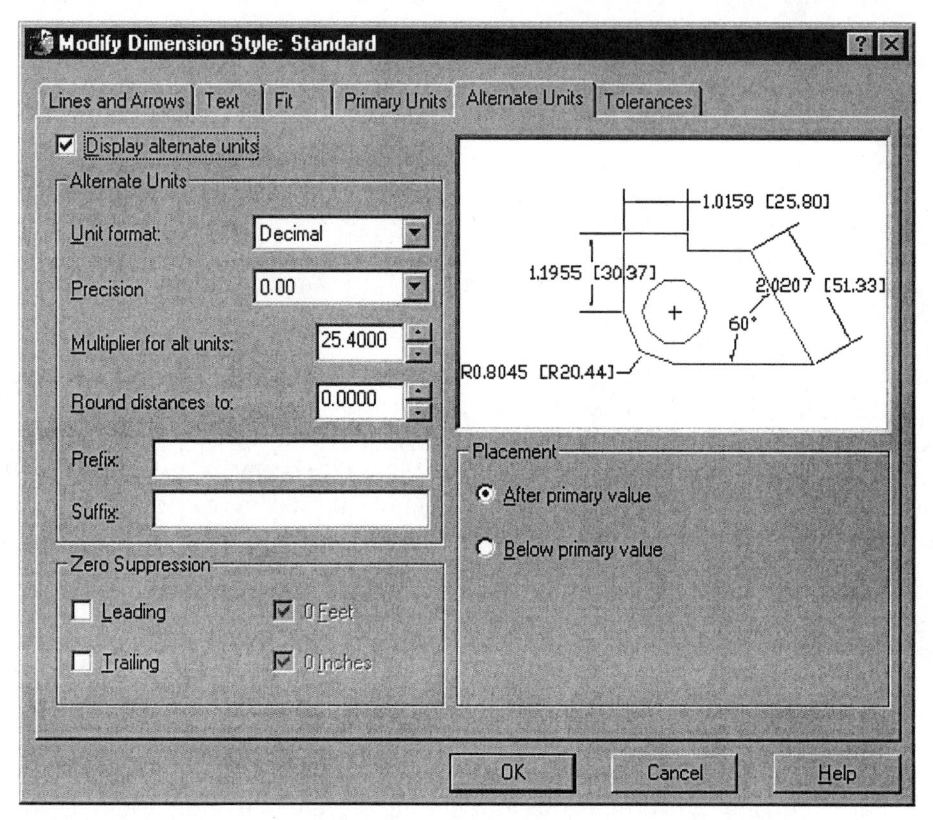

Figure 8-44 ***Alternate Units*** *tab of the* ***New Dimension Style*** *dialog box*

Alternate Units

This area is identical to the Linear Units area of the **Primary Units** tab and gives options to set the format for all dimension types except Angular.

Unit format. You can select a unit format to apply to the alternate dimensions from the **Unit format:** drop-down list. It has the following options which include Scientific, Decimal, Engineering, Architectural stacked, Fractional stacked, Architectural, Fractional, and Windows desktop. The value is stored in the **DIMALTU** variable. The relative size of fractions is governed by the **DIMTFAC** variable.

Precision. You can select the precision value from the **Precision:** drop-down list. The value is stored in the **DIMALTD** variable.

Multiplier for Alternate units. To generate a value in the alternate system of measurement, you need a factor with which all the linear dimensions will be multiplied. The value for this factor can be entered in the **Multiplier for alt units:** edit box located in the Alternate units area of the **Alternate Units** tab of the **New Dimension Style** dialog box. The default value of 25.4 is for dimensioning in inches with alternate units in millimeters. This scaling value (contents of the **Multiplier for alt units:** edit box) gets stored in the **DIMALTF** variable.

Round distances to. In this edit box you can enter a value to which you want all your measurements (made in alternate units) rounded off to. This value is stored in the **DIMALTRND** system variable. For example if you enter a value of .25 in the **Round distances to:** edit box, all the alternate units get rounded off to the nearest .25 unit.

Prefix and Suffix. The **Prefix:** and **Suffix:** edit boxes are similar to the edit boxes in the Linear units area of the **Primary units** tab. You can enter the text or symbols that you want to precede or follow the alternate dimension text. The value is stored in the **DIMAPOST** variable. You can also use control codes and special characters to display special symbols.

Zero Suppression

This area allows you to suppress the leading or trailing zeros in decimal unit dimensions by selecting either, both, or none of the **Trailing** and **Leading** check boxes. Similarly, selecting the **0 Feet** check box, suppresses the zeros in the feet area of the dimension, when the dimension value is less than a foot. Selecting the **0 inches** check box suppresses the zeros in the inches area of the dimension. For example, 1'-0" becomes 1'. The **DIMALTZ** variable controls the suppression of zeros for alternate unit dimension values. The values that are between 0 to 3 affect feet-and-inch dimensions only.

Placement

This area gives options that control the positioning of the Alternate units. The value is stored in the **DIMAPOST** variable.

After Primary Value. Selecting the **After primary value** radio button, places the alternate units dimension text after the primary units. This is the default option.

Below Primary Value. Selecting the **Below primary value** radio button, places the alternate units dimension text below the primary units.

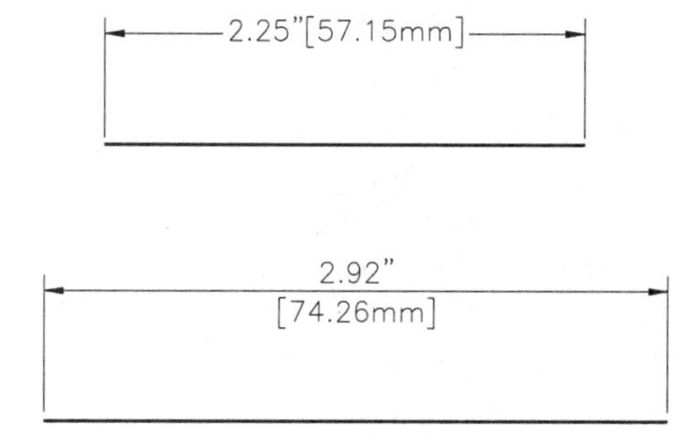

Figure 8-45 *Dimensioning with alternate units*

Figure 8-45 illustrates the result of entering information in the **Alternate Units** tab. The decimal places get saved in the **DIMALTD** variable, the scaling value (contents of the **Multiplier for alt units:** edit box) in the **DIMALTF** variable, and the suffix string (contents of the **Suffix:** edit box) in the **DIMAPOST** variable. Similarly, the units format for alternate units in **DIMALTU**, and suppression of zeros for alternate units decimal values in **DIMALTZ**.

Tolerances Tab

The **Tolerances** tab (Figure 8-46) allows you to set values for options that control the format and display of tolerance dimension text including the alternate unit tolerance dimension text.

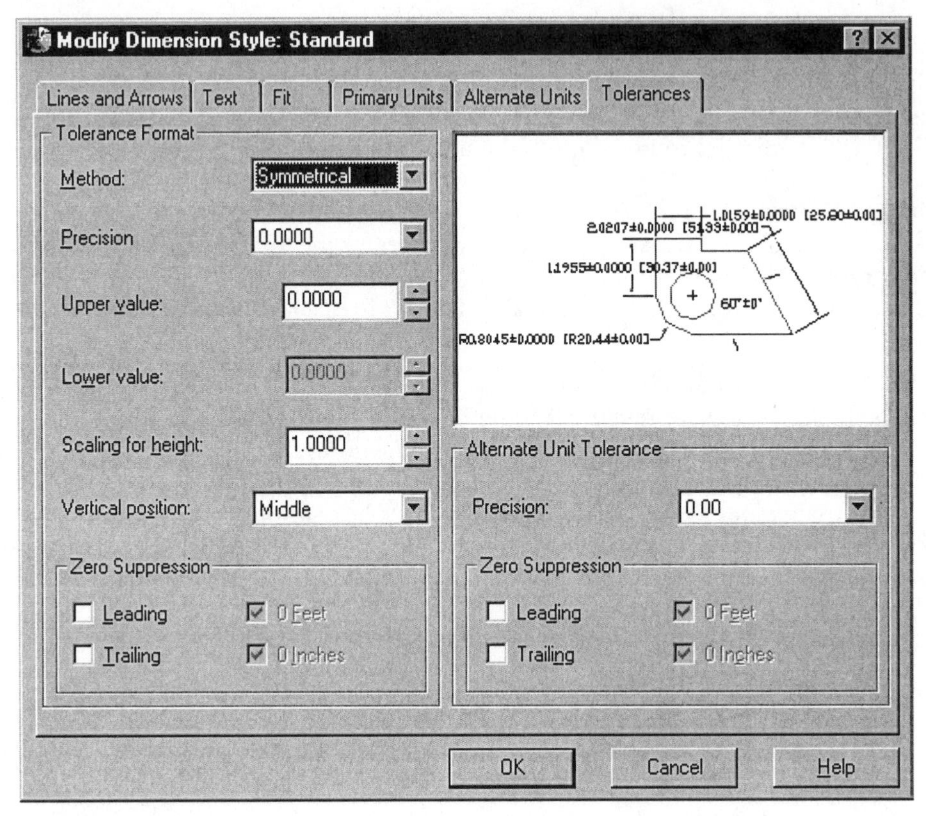

Figure 8-46 ***Tolerances*** *tab of the* ***New Dimension Style*** *dialog box*

Tolerance Format

The **Tolerance format** area of the **Tolerances** tab (Figure 8-46) lets you specify the tolerance method, tolerance value, position of tolerance text, precision and the height of the tolerance text. For example, if you do not want a dimension to deviate more than plus 0.01 and minus 0.02, you can specify this by selecting **Deviation** from the **Method:** drop-down list and then specifying the plus and minus deviation in the **Upper Value:** and the **Lower Value:** edit boxes. When you dimension, AutoCAD LT will automatically append the tolerance to the dimension. The **DIMTP** variable sets the maximum (or upper) tolerance limit for the dimension text and **DIMTM** variable sets the minimum (or lower) tolerance limit for dimension text. Different settings and their effects on relevant dimension variables are explained in following sections:

Method. The **Method:** drop-down list lets you select the tolerance method. The tolerance methods supported by AutoCAD LT are **Symmetrical**, **Deviation**, **Limits**, and **Basic**. These tolerance methods are described next.

None. Selecting the **None** option sets the **DIMTOL** variable to 0 and does not add tolerance values to the dimension text, that is, the **Tolerances** tab is disabled.

Symmetrical. If you select **Symmetrical**, the **Lower Value:** edit box is disabled and the value specified in the **Upper Value:** edit box is applied to both plus and minus tolerance. For example, if the value specified in the **Upper Value:** edit box is 0.05, the tolerance appended to the dimension text is ±0.05. The value of **DIMTOL** is set to 1 and the value of **DIMLIM** is set to 0.

Deviation. If you select the **Deviation** tolerance method, the values in the **Upper Value:** and **Lower Value:** edit boxes will be displayed as plus and minus dimension tolerances. If you enter values for the plus and minus tolerances, AutoCAD LT appends a plus sign (+) to the positive values of the tolerance and a negative sign (-) to the negative values of the tolerance. For example, if the upper value of the tolerance is 0.005 and the lower value of the tolerance is 0.002, the resulting dimension text generated will have a positive tolerance of 0.005 and a negative tolerance of 0.002 (Figure 8-47).

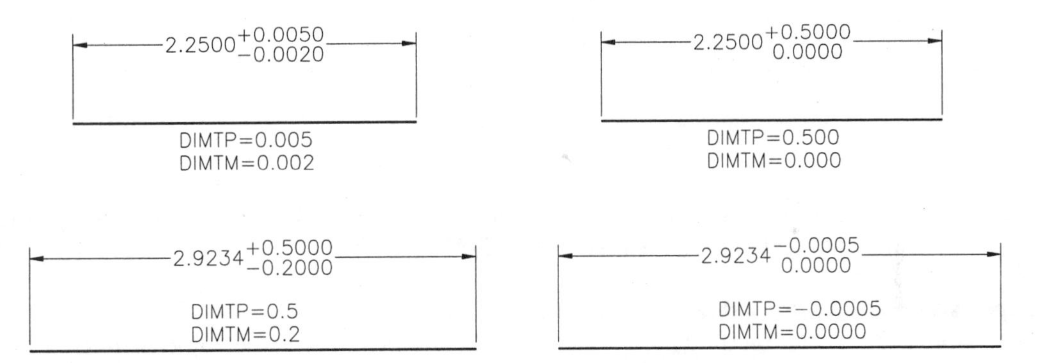

Figure 8-47 Dimensioning with deviation tolerance

Figure 8-48 Dimensioning with deviation tolerance (when one tolerance value equals 0)

Even if one of the tolerance values is 0, a sign is appended to it (Figure 8-48). On specifying the deviation tolerance, AutoCAD LT sets the **DIMTOL** variable value to 1 and the **DIMLIM** variable value to 0. The values in the **Upper Value:** and **Lower Value:** edit boxes are saved in the **DIMTP** and **DIMTM** system variables, respectively.

Limits. If you select the **Limits** tolerance method from the **Method** drop-down list, AutoCAD LT adds the upper value (contents of the **Upper Value** edit box) to the dimension text (actual measurement) and subtracts the lower value (contents of the **Lower Value** edit box) from the dimension text. The resulting values are drawn with the dimension text (Figure 8-49). Selecting the **Limits** tolerance method results in setting the **DIMLIM** variable value to 1 and the **DIMTOL** variable value to 0. The numeral values in the **Upper Value**

and **Lower Value** edit boxes are saved in the **DIMTP** and **DIMTM** system variables, respectively.

Basic. A basic dimension text is dimension text with a box drawn around it (Figure 8-50). The basic dimension is also called a reference dimension. Reference dimensions are used primarily in geometric dimensioning and tolerances. The basic dimension can be realized by selecting the basic tolerance method. The distance provided around the dimension text (distance between dimension text and the rectangular box) is stored as a negative value in the **DIMGAP** variable. The negative value signifies basic dimension. The default setting is off, resulting in the generation of dimensions without the box around the dimension text.

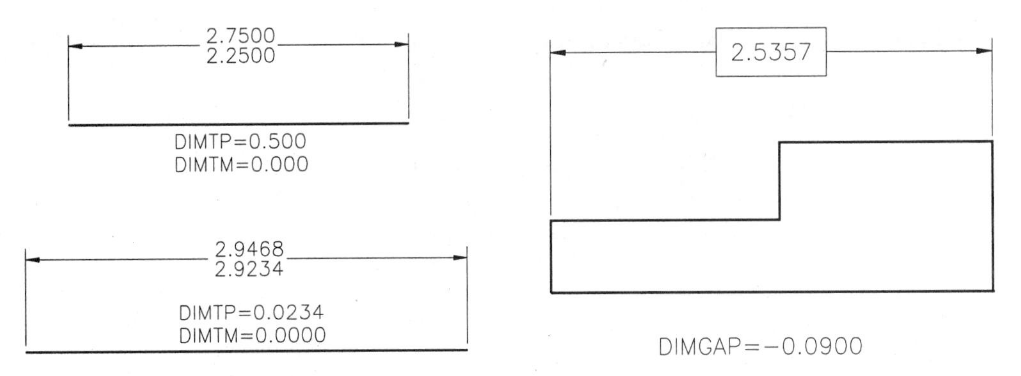

Figure 8-49 Dimensioning with limits tolerance

*Figure 8-50 Basic dimension (**DIMGAP** assigned a negative value)*

Precision. The **Precision:** drop-down list lets you select the number of decimal places for the tolerance dimension text. The value is stored in **DIMTDEC** variable.

Upper Value and Lower Value. In the **Upper value:** edit box, the positive upper or maximum value is entered. If the method of tolerances is symmetrical, the same value is used as the lower value also. The value is stored in **DIMTP** variable. In the **Lower:** edit box the lower or minimum value is entered. The value is stored in **DIMTM**.

Vertical Position. It lets you justify the dimension tolerance text for deviation and symmetrical methods only. Three alignments are possible: with the **Bottom**, **Middle**, or **Top** of the main dimension text. If you select the Limits tolerance method, the **Vertical position:** drop-down list is automatically disabled. The settings are saved in the **DIMTOLJ** system variable (Bottom=0, Middle=1, and Top=2).

Scaling for Height. The **Scaling for height** edit box lets you specify the height of the dimension tolerance text relative to the dimension text height (see Figure 8-51). The default value is 1; the height of the tolerance text is the same as the dimension text height. If you want the tolerance text to be 75 percent of the dimension height text, enter 0.75 in the **Scaling for height** edit box. The ratio of the tolerance height to the dimension text height is calculated by AutoCAD LT and then stored in the **DIMTFAC** variable. **DIMTFAC = Tolerance Height/Text Height.**

Zero Suppression. This area controls the zero suppression in the dimension tolerance text depending on which of the check boxes are selected. Selecting the **Leading** check box suppresses the leading zeros in all decimal dimension text. For example, 0.2000 becomes .2000. Trailing zeros are of significance to the tolerance. Selecting the **Trailing** check box suppresses the trailing zeros in all decimal dimensions. For example, 0.5000 becomes 0.5. Similarly, selecting both the boxes suppresses both the trailing and leading zeros and selecting none, suppresses none. If you select **0 Feet** check box, the zeros in the feet portion of the tolerance dimension text are suppressed if the dimension value is less than a foot. Similarly, selecting the **0 Inches** check box, suppresses the zeros in the inches portion of the dimension text. The value is stored in the **DIMTZIN** variable.

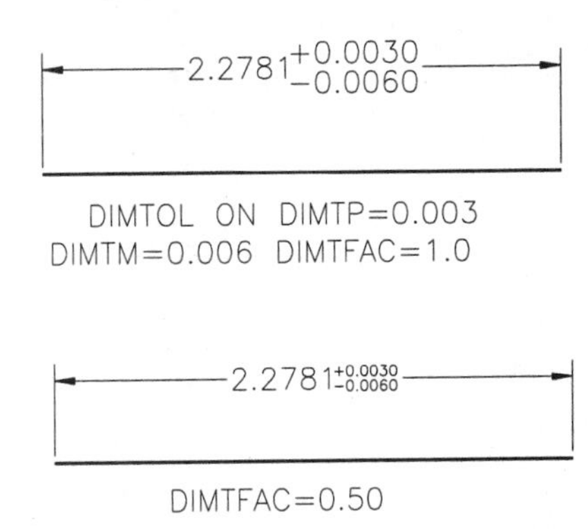

Figure 8-51 *Tolerance height,* ***DIMTFAC***

Alternate Unit Tolerance

Defines the precision and zero suppression settings for Alternate unit tolerance values.

Precision. From the **Precision** drop-down list you can select the precision value for alternate unit tolerance dimensions. It sets the number of decimal places to be displayed in the dimension text. The value is stored in the **DIMALTTD** variable.

Zero Suppression. Selecting the respective check boxes controls the suppression of the **Leading** and **Trailing** zeros in decimal values and the suppression of zeros in the Feet and Inches portions for dimensions in the feet and inches format. The value is stored in the **DIMALTTZ** variable.

Exercise 3 *Mechanical*

Draw Figure 8-52. Then set the values in the various tabs of the **New Dimension Style** dialog box to dimension it as shown. (Baseline Spacing = 0.25, Extension beyond dim lines = 0.10, Offset from origin = 0.05, Arrowhead size = 0.09, Text height = 0.09). Assume the missing dimensions.

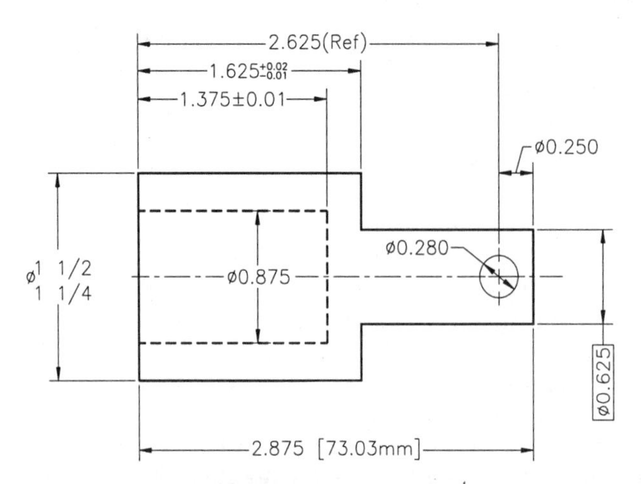

Figure 8-52 Drawing for Exercise 3

OTHER DIMENSIONING VARIABLES

Positioning Dimension Text (DIMTVP)

You can position the dimension text with respect to the dimension line by using the **DIMTVP** system variable (Figure 8-53). In certain cases, **DIMTVP** is used with **DIMTAD** to control the vertical position of the dimension text. **DIMTVP** value applies only when the **DIMTAD** is off. To select the vertical position of the dimension text to meet your requirement (over or under the dimension line), you must first calculate the numerical value by which you want to offset the text from the dimension line. The vertical placing of the text is done by offsetting the dimension text. The magnitude of the offset of dimension text is a product of text height and **DIMTVP** value. If the value of **DIMTVP** is 1, **DIMTVP** acts as **DIMTAD**. For example, if you want to position the text 0.25 units from the dimension line, the value of **DIMTVP** is calculated as follows:

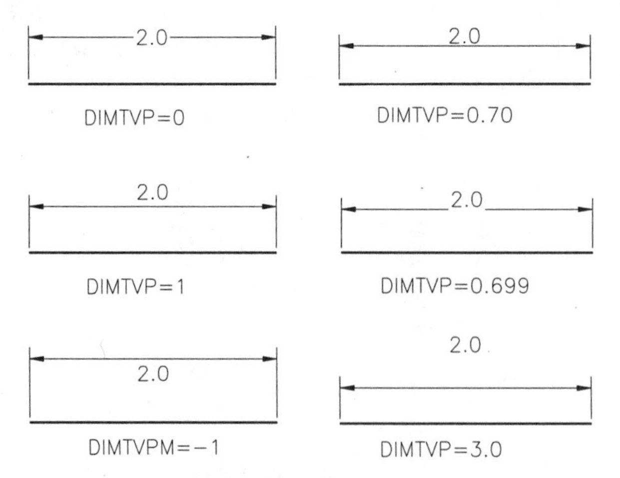

***Figure 8-53** DIMTVP, dimension text vertical position*

DIMTVP=Relative Position value/Text Height value
DIMTVP=0.25/ 0.09=2.7778

The value 2.7778 is stored in the dimension variable **DIMTVP**. If the absolute value is less than 0.70, the dimension line is broken to accommodate the dimension text. Relative positioning is not effective on angular dimensions.

Horizontal Placement of Text (DIMTIX and DIMSOXD)

DIMTIX System Variable. With the help of the **DIMTIX** (dimension text inside extension lines) variable, the dimension text can be placed between the extension lines even if it would normally be placed outside the lines (Figure 8-54). This can be achieved when **DIMTIX** is on (1). If DIMTIX is off (the default setting), the placement of the dimension text depends on the type of dimension. For example, if the dimensions are linear or angular, the text will be placed inside the extension lines by AutoCAD LT, if there is enough space available. For the radius and diameter dimensions, the text is placed outside the object being dimensioned.

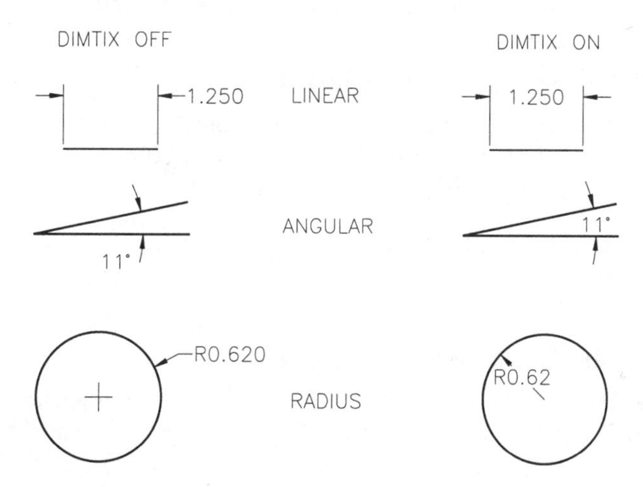

Figure 8-54 DIMTIX, dimension text inside extension lines

DIMSOXD System Variable. If you want to place the dimension text inside the extension lines, you will have to turn the **DIMTIX** variable on. If you want to suppress the dimension lines and the arrowheads, you will have to turn the **DIMDSOXD** (dimension suppress outside extension dimension lines) variable on. **DIMSOXD** suppresses the drawing of dimension lines and the arrowheads when they are placed outside the extension lines. If **DIMTIX** is on and **DIMSOXD** is off, and there is not enough space inside the extension lines to draw the dimension lines, the lines will be drawn outside the extension lines. In such a situation, if both **DIMTIX** and **DIMSOXD** are on, the dimension line will be totally suppressed. **DIMSOXD** works only when **DIMTIX** is on. The default value for **DIMSOXD** and **DIMTIX** is off (Figure 8-55).

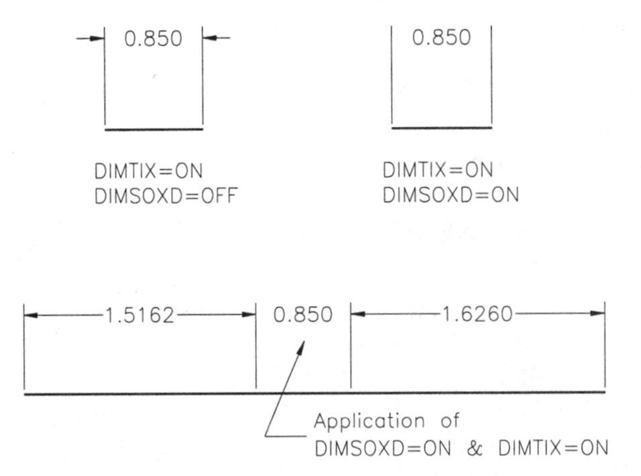

***Figure 8-55 DIMSOXD** and **DIMTIX** suppress outside extension lines and dimension text inside extension lines*

Redisplay of Dimension (DIMSHO)

The associative dimension computes and redisplays dynamically as the dimension is dragged while you are editing it. This feature is controlled by the **DIMSHO** system variable. By default, it is on, which means that the dimension will be redisplayed as it is dragged. Although it is a good feature, dynamic dragging sometimes can be very slow. In that case you can turn it off by setting **DIMSHO** to off. When you dimension a circle or an arc, the **DIMSHO** settings are ignored. The value of **DIMSHO** is not saved with the dimension style.

DIMENSION STYLE FAMILIES

The dimension style feature of AutoCAD LT lets the user define a dimension style with values that are common to all dimensions. For example, the arrow size, dimension text height, or color of the dimension line are generally the same in all types of dimensioning, such as linear, radial, diameter, and angular. These dimensioning types belong to the same family because they have some characteristics in common. In AutoCAD LT, this is called a **dimension style family,** and the values assigned to the family are called **dimension style family values.**

After you have defined the dimension style family values, you can specify variations on it for other types of dimensions, such as radial and diameter. For example, if you want to limit the number of decimal places to two in radial dimensioning, you can specify that value for radial dimensioning. The other values will stay the same as the family values to which this dimension type belongs. When you use the radial dimension, AutoCAD LT automatically uses the style that was defined for radial dimensioning; otherwise, it creates a radial dimension with the values as defined for the family. After you have created a dimension style family, any changes in the parent style are applied to family members if the particular property is the same. Special suffix codes are appended to the dimension style family name that correspond to different dimension types. For example, if the dimension style family name is MYSTYLE and you define a diameter type of dimension, AutoCAD LT will append $4 at the end of the dimension style family name. The name of the diameter type of dimension will be MYSTYLE$4. The following are the suffix codes for different types of dimensioning:

Suffix Code	Dimension Type	Suffix Code	Dimension Type
0	Linear	2	Angular
3	Radius	4	Diameter
6	Ordinate	7	Leader

Example 1

The following example illustrates the concepts of dimension style families, Figure 8-56.

1. Specify the values for the dimension style family.
2. Specify the values for linear dimension.
3. Specify the values for diameter and radius dimension.
4. After creating the dimension style, use it to dimension the given drawing.

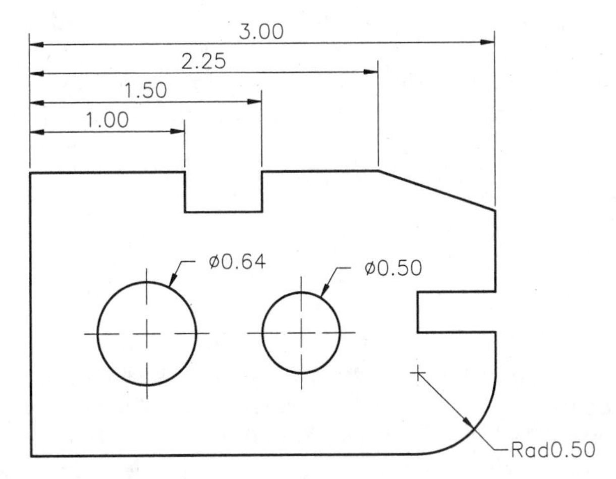

Figure 8-56 Dimensioning using dimension style families

Chapter 8

Step 1

Invoke the **Dimension Style Manager** dialog box by entering the **DDIM** or **DIMSTYLE** command at the AutoCAD LT Command prompt. You can also invoke it from the **Dimension** toolbar, or the **Dimension** or **Format** menus. If you have not defined any dimension style, AutoCAD LT will display STANDARD in the **Styles:** list box. Select the STANDARD style from the **Styles:** list box.

Step 2

Choose the **New** button, to display the **New Dimension Style** dialog box. In this dialog box, enter MYSTYLE in the **New Style Name:** edit box. Select STANDARD from the **Start With:** drop-down list. Also select **All dimensions** from the **Use for:** drop-down list. Now, choose the **Continue** button to display the **New Dimension Style: MYSTYLE** dialog box. In this dialog box, choose the **Lines and Arrows** tab and enter the following values:

Baseline Spacing: 0.15 **Extension beyond dim line: 0.07** **Offset from origin: 0.03**
Arrow size: 0.09 **Center Mark for circle, Size: 0.05**

In the **Text** tab, change the following values:

Text Height: 0.09 **Offset from dimension line: 0.03**

After entering the values, choose the **OK** button to return to the **Dimension Style Manager** dialog box. This dimension style contains the values that are common to all dimension types.

Step 3
Now, choose the **New...** button again in the **Dimension Style Manager** dialog box to display the **Create New Dimension Style:** dialog box. AutoCAD LT displays **Copy of MYSTYLE** in the **New Style name:** edit box. Select MYSTYLE from the **Start with:** drop down list if it is not already selected. From the **Use for:** drop-down list select **Linear dimensions**. Choose the **Continue** button to display the **New Dimension Style** dialog box and set the following values:

1. Select the **Aligned with dimension line** radio button in the Text alignment area.
2. In the Text placement area, from the **Vertical:** drop-down list, select the **Above** option.

In the **Primary units** tab, change the following:

1. **Unit format:** to Decimal units.
2. Dimension precision to two decimal places.

Step 4
Choose the **OK** button to return to the **Dimension Style Manager** dialog box. Here choose the **New** button again to display the **New Dimension Style** dialog box. Select MYSTYLE from the **Start with:** drop-down list. Also select **Diameter dimension** type from the **Use for:** drop-down list. Choose the **Continue** button to display the **New dimension Style** dialog box once again. Choose the **Primary units** tab and set the **Precision:** to two decimal places. In the **Lines and Arrows** tab, select **Line** from the Center mark for circle **Type:** drop-down list. Choose the **OK** button to return to the **Dimension Style Manager** dialog box.

In this dialog box, choose the **New** button to display the **New Dimension Style** dialog box once again. Select MYSTYLE from the **Start with:** drop-down list and **Radius dimensions** from the **Use for:** drop-down list. Choose the **Continue** button to display the **New Dimension Style** dialog box. Choose the **Primary Units** tab and set the precision to two decimal places and enter **Rad** in the **Prefix:** edit box. In the **Fit** tab, select the **Text** radio button in the **Fit Options** area. Choose the **OK** button to return to the **Dimension Style Manager** dialog box. Here, select MYSTYLE from the **Styles:** list box and choose the **Set current** button. Choose the **Close** button to exit the dialog box.

Step 5
Use the linear and baseline (**DIMLIN** and **DIMBASE**) commands to draw the linear dimensions as shown in Figure 8-55. Notice that when you enter any linear dimensioning command, AutoCAD LT automatically uses the values that were defined for the linear type of dimensioning.

Step 6

Use the diameter dimensioning (**DIMDIA**) command to dimension the circles as shown in Figure 8-56. Again notice that the dimensions are drawn according to the values specified for the diameter type of dimensioning.

Step 7

Now, similarly use the radius dimensioning (**DIMRADIUS**) command and dimension the arc as shown in Figure 8-56.

USING DIMENSION STYLE OVERRIDES

Most of the dimension characteristics are common in a production drawing. The values that are common to different dimensioning types can be defined in the dimension style family. However, at times you might have different dimensions. For example, you may need two types of linear dimensioning: one with tolerance and one without. One way to draw these dimensions is to create two dimensioning styles. You can also use the dimension variable overrides to override the existing values. For example, you can define a dimension style (MYSTYLE) that draws dimensions without tolerance. Now, to draw a dimension with tolerance or update an existing dimension, you can override the previously defined value. You can override the values through the **Dimension Style Manager** dialog box or by setting the variable values at the Command prompt. The following example illustrates how to use the dimension style overrides.

Example 2

In this example, you will update the overall dimension (3.00) so that the tolerance is displayed with the dimension. You will also add two linear dimensions, as shown in Figure 8-57.

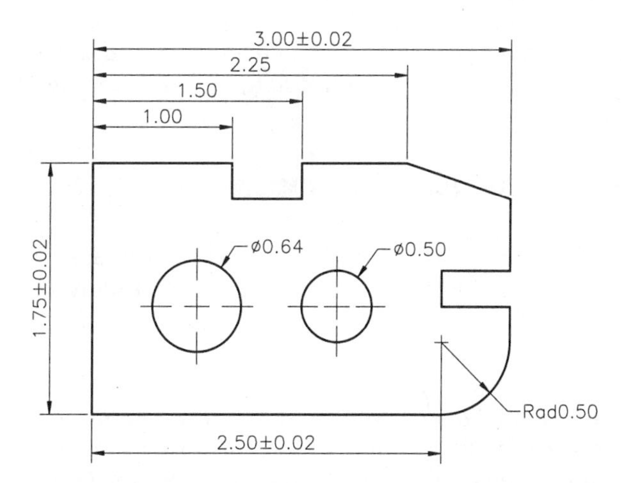

Figure 8-57 Overriding the dimension style values

Using Dimension Styles/ Overrides

Step 1

Invoke the **Dimension Style Manager** dialog box. Select MYSTYLE from the **Styles:** list box and choose the **Override** button to display the **Override Current Style** dialog box. The options in this dialog box are identical to the **New Dimension Style** dialog box discussed earlier in the chapter. Choose the **Tolerance** tab and specify the tolerance, symmetrical with upper value 0.02 and precision of two decimal places. Choose the **OK** button to exit the dialog box (this does not save the style). Notice that the **<style overrides>** is displayed under MYSTYLE, indicating that the style overrides the MYSTYLE dimension style.

This **<style overrides>** is displayed till you save it under a new name or under the Style it is displayed under or until you delete it. The Style overrides can be modified and compared also by choosing the **Modify** and **Compare** buttons respectively. Selecting the **style overrides** and right- clicking, displays a shortcut menu. Choosing the **Save to current Style** option from the shortcut menu saves the overrides to the current style. Choosing the **Rename** option, allows you to rename the style override and save it as a new style.

Step 2

Use the **-DIMSTYLE** command to apply the change to the existing dimensions.

Command: **-DIMSTYLE**
Current dimension style: MYSTYLE
Current dimension overrides:
DIMTOL On
Enter a dimension style option
[Save/Restore/STatus/Variables/Apply/?] <Restore>: **A**
Select objects: *Select the dimension that you want to update.*

After you select the dimension, AutoCAD LT will update the dimension and the tolerance will be appended to the selected dimension. If you create a new dimension, the tolerance value will be displayed automatically with the dimension, unless you make MYSTYLE (dimension style) current.

Using the PROPERTIES Command*

Step 1

You can also use the **PROPERTIES** command to modify a dimension. Select the dimension you want to modify and enter **PROPERTIES** at the Command prompt. You can also choose **Properties** from the **Modify** menu.

Step 2

AutoCAD LT will display the **Object Properties Window**. In this window, select Tolerances and specify the tolerance for linear dimensions, Symmetrical with Upper Value 0.02 and Precision of two decimal places. Choose the **X** button to exit the **Object Properties Window**. The dimension will be updated to new specifications (See Chapter 7, Editing Dimensions, for details). If you want to save the modified settings to a new style, select the modified dimension and right-click to display a shortcut menu. Choose the **Dim Style > Save as New style** option. Enter a name in the **Style Name:** edit box of the **Save As New Dimension Style** dialog box.

Using the DIMOVERRIDE Command

You can also use the **DIMOVERRIDE** command to override a dimension value. If you want to have tolerance displayed with the dimension, make sure the tolerances are specified. You can also choose **Override** from the **Dimension** menu.

Command: **DIMOVERRIDE** [Enter]
Enter dimension variable name to override or [Clear overrides]: **DIMTOL** [Enter]
Enter new value for dimension variable <Off> New value: **ON** [Enter]
Enter dimension variable name to override: **DIMTP** [Enter]
Enter new value for dimension variable <0.0000> New value: **0.02** [Enter]
Enter dimension variable name to override: **DIMTM** [Enter]
Enter new value for dimension variable <0.0000> New value: **0.02** [Enter]
Enter dimension variable name to override: [Enter]
Select objects: *Select the object that you want to update.*

You can also update a dimension by entering **Update** at the **Dim:** prompt or choosing the **Dimension Update** button from the **Dimension** toolbar. For details, see Chapter 7, Editing Dimensions.

COMPARING AND LISTING DIMENSION STYLES

Choosing the **Compare** button in the **Dimension Style Manager** dialog box displays the **Compare Dimension Styles** dialog box where you can compare the settings two dimensions styles or list all the settings of one of them (Figure 8-58).

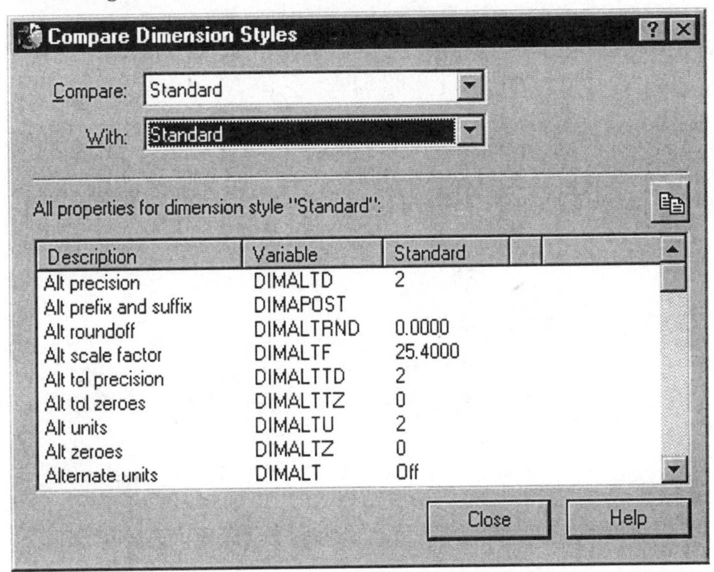

***Figure 8-58 Compare Dimension Styles** dialog box*

The **Compare:** and the **With:** drop-down lists display the dimension styles in the current drawing and selecting dimension styles from the respective lists compare the two styles. In the **With:** drop-down list, if we select **None** or the same style as selected from the **Compare:** drop-down list, all the properties of the selected style are displayed. The Comparison results are displayed under three headings: **Description** of the Dimension Style property, the **System Variable** controlling a particular settings, and the **values of the variable** which differ in the two styles in comparison. The print to clipboard button prints the comparison results to the Windows clipboard from where they can be pasted to other Windows applications.

The Save, Restore, and Apply options of the **-DIMSTYLE** command were discussed earlier in this chapter (see the section "Creating and Restoring Dimension Styles"). You can also use this command to ascertain the status of a dimension style or compare a dimension style with the current style.

Comparing Dimension Styles

You can compare the current dimension style with another style by using the Restore option and appending the tilde (~) symbol in front of the dimension style name.

Command: **-DIMSTYLE** [Enter]
Current dimension style:
Enter a dimension style option
[Save/Restore/STatus/Variables/Apply/?] <Restore>: **R** [Enter]
Enter dimension style name, [?] or <select dimension>: **~Standard** [Enter]

AutoCAD LT will display a listing of dimension variable names and their values for the standard dimension style and the current dimension style. Only those variables that have different values in the current and the named styles are listed.

Differences between STANDARD and current settings:

STANDARD		*Current Setting*
DIMSOXD	*Off*	*On*
DIMTAD	*0*	*1*
DIMTIX	*Off*	*On*
DIMTOL	*Off*	*On*
DIMTP	*0.0000*	*0.5000*
DIMTVP	*0.0000*	*2.0000*

Listing Dimension Styles

The **STatus** option of the **-DIMSTYLE** command displays the settings of the current dimensioning status. You can use the question mark **(?)** to display the named dimension styles in the current drawing.

Command: **-DIMSTYLE** [Enter]
Current dimension style:
Enter a dimension style option
[Save/Restore/STatus/Variables/Apply/?] <Restore>: **?** [Enter]
Enter dimension style(s) to list<*>: *Press ENTER.*

If you select the **Variables** option, AutoCAD LT will display the dimension status of the named dimension style or the dimension style that is associated with the selected dimension.

Command: **-DIMSTYLE** Enter
Current dimension style:
Enter a dimension style option
[Save/Restore/STatus/Variables/Apply/?] <Restore>: **V** Enter
Enter dimension style name, [?] or <select dimension>: **MYSTYLE** Enter

You can also use the **STATUS** command from the **Dim:** mode to list the dimension styles ina drawing.

USING EXTERNALLY REFERENCED DIMENSION STYLES

The externally referenced dimensions cannot be used directly in the current drawing. When you Xref a drawing, the drawing name is appended to the style name and the two are separated by the vertical bar (|) symbol. It uses the same syntax as other externally dependent symbols. For example, if the drawing (FLOOR) has a dimension style called DECIMAL and you Xref this drawing in the current drawing, AutoCAD LT will rename the dimension style to FLOOR|DECIMAL. You cannot make this dimension style current, nor can you modify or override it. However, you can use it as a template to create a new style. To accomplish this, invoke the **Dimension Style Manager** dialog box. If the **Don't list styles in Xrefs** check box is selected, the styles in the Xref are not displayed. Clear this check box to display the Xref dimension styles and choose the **New...** button. In the **New Style Name:** edit box of the **New Dimension Style** dialog box, enter the name of the dimension style. AutoCAD LT will create a new dimension style with the same values as those of the externally referenced dimension style (FLOOR|DECIMAL).

Self-Evaluation Test

Answer the following questions, and then compare your answers to the correct answers given at the end of this chapter.

1. What is the function of dimension styles (**DIMSTYLE** command)? ________________
____.

2. How can you create a dimension style of your requirement? ________________
____.

3. Fill in the command and entries required to scale the dimensioning variables that reflect sizes, distances, or offsets (such as arrow size, text size) by a value of 2.0.
Command: __________
DIM: __________

Enter new value for dimension variable <current>: __________

4. The default dimension style file name is __________.

5. Dimension arrows have the same color as the __________ because arrows constitute a part of __________.

6. The size of the arrow block is determined by the value stored in the Arrow size: edit box. (T/F)

7. The center marks are created by using the __________ , __________ , and **Radius** dimensioning commands.

8. When you select Arrows option, AutoCAD LT places the text and arrowheads __________.

9. If you want to add the user-defined text above or below the dimension line, use the separator symbol __________.

10. A basic dimension text is dimension text with a __________ drawn around it.

11. A predefined text height (specified in the **STYLE** command) __________ any other setting for the dimension text height.

12. The **Text** tab of the **New Dimension Style** dialog box lets you specify the text style, text height, text gap, and __________ of the dimension text.

13. The **Suppress:** check boxes control the drawing of the __________ and __________ dimension lines.

14. The overall scale (**DIMSCALE**) is not applied to the measured lengths, coordinates, __________ , or __________.

15. You can restore an existing dimension style as the current style with the **RESTORE** Command. (T/F)

Review Questions

1. You can use the __________ dialog box to control dimension styles and dimension variables.

2. You can also create dimension styles from the command line by entering __________ at the Command prompt.

3. The __________ tab of the **New Dimension Style** dialog box can be used to specify the dimensioning attributes (variables) that affect the geometry of the dimensions.

4. The first and second dimension lines are determined by __________ extension line origins.

5. The **Suppress**: check boxes control the display of the __________.

6. Extension is the distance the __________ should extend past the dimension line.

7. AutoCAD LT has provided __________ standard termination symbols you can apply at each end of the dimension line.

8. The size of the __________ is determined by the value stored in the **Arrow size:** edit box.

9. You cannot replace the default arrowheads at the end of the dimension lines. (T/F)

10. When **DIMSCALE** is assigned a value of __________, AutoCAD LT calculates an acceptable default value based on the scaling between the current model space viewport and paper space.

11. If you use the **DIMCEN** command, a positive value will create a center mark, and a negative value will create a __________.

12. You can control the dimension format through the __________.

13. If you select the __________ check box, you can position the dimension text anywhere along the dimension line.

14. When you select the **Arrows** option, AutoCAD LT places the text and arrowheads __________ the extension lines if there is enough space to fit both.

15. The linear scaling value is not exercised on tolerance values. (T/F)

16. You can append a prefix to the dimension measurement by entering the desired prefix in the **Prefix**: edit box of the __________ dialog box.

17. If you want to add the user-defined text above or below the dimension line, use the separator symbol __________.

18. If you want to add more lines, use the symbol __________.

19. If you select the **Limits** tolerance method from the **Method:** drop-down list, AutoCAD LT __________ the upper value to the dimension and __________ the lower value from the dimension text.

20. The **Text** tab of the __________ dialog box lets you specify the text style, text height, text gap, and __________ of the dimension text.

21. After you have defined the dimension style family values, you can specify variations on it

for other types of dimensions, such as __________, and __________.

22. The values that are common to different dimensioning types can be defined in the dimension __________.
23. Use the __________ command to apply the change to the existing dimensions.

24. You can also use the __________ command to modify a dimension.

25. You can also use the __________ command to override a dimension value.

26. You can compare the current dimension style with another style by using the Restore option and appending the __________ symbol in front of the dimension style name.

27. What is the dimension style family, and how does it help in dimensioning? __________
__________.

28. What are the different ways to access the **Dimension Style Manager** dialog box? __________
__________.

29. Explain various methods that can be used to change the value of a dimension variable.
__________.

30. How can you use the dimension style that has been defined in the Xref drawing? Explain.
__________.

31. Explain dimension overrides.__________
__________.

32. What are the different ways to override an existing dimension? __________
__________.

33. How can you compare a named dimension style with the current style? __________
__________.

34. Explain the use of the **DIMOVERRIDE** command and how it works. __________
__________.

35. How can you compare the settings of the dimension variables of the current dimension style with those of another dimension style? __________
__________.

36. Which dimensioning command is used to display the list of all dimension variables and their current settings? __________
__________.

37. In oblique dimensioning, the extension lines are always drawn perpendicular to the dimension line. (T/F)

38. A group of dimension variables with some setting is termed a dimension style. (T/F)

39. Dimension style cannot have a name. (T/F)

40. The dimension style used to create a particular dimension can be revoked later by selecting that dimension. (T/F)

41. When the **DIMTVP** variable has a negative value, the dimension text is placed below the dimension line. (T/F)

42. The named dimension style associated with the dimension being updated with the **OVERRIDE** command is not updated. (T/F)

Exercises

Exercise 4 thru' 9 *Mechanical*

Make the drawings as shown in Figure 8-59 through Figure 8-64. You must create dimension style files and specify values for different dimension types, such as linear, radial, diameter, and ordinate. Assume the missing dimensions.

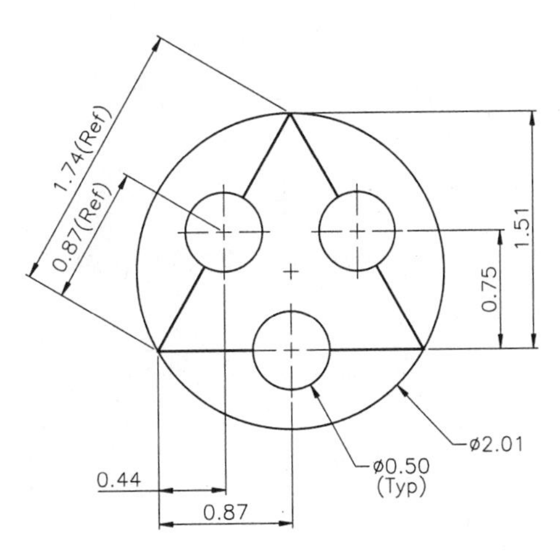

Figure 8-59 *Drawing for Exercise 4*

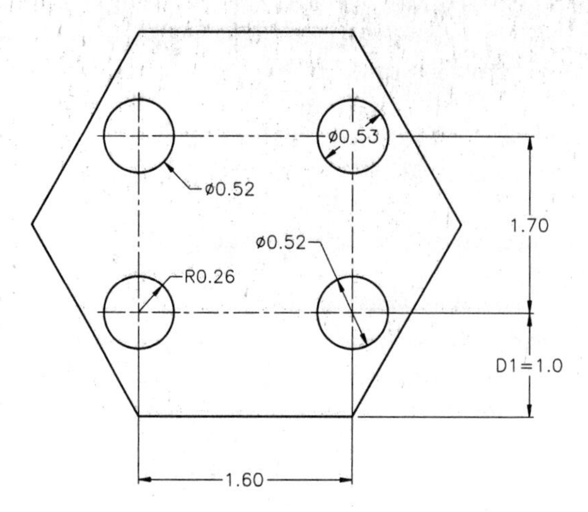

Figure 8-60 Drawing for Exercise 5

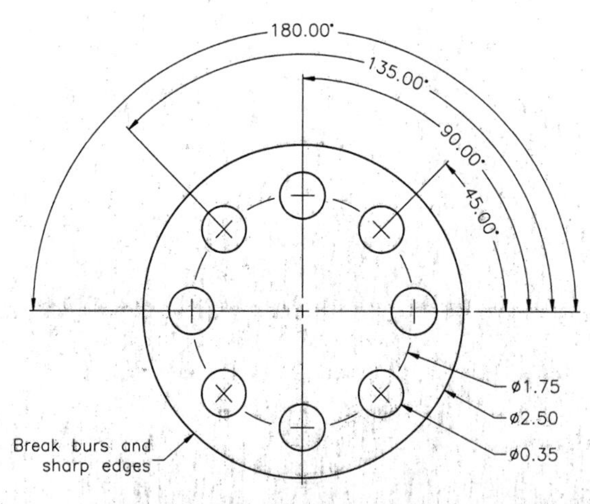

Figure 8-61 Drawing for Exercise 6

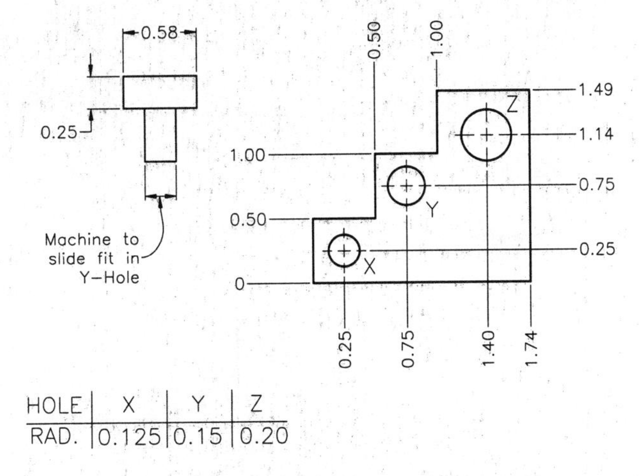

Figure 8-62 *Drawing for Exercise 7*

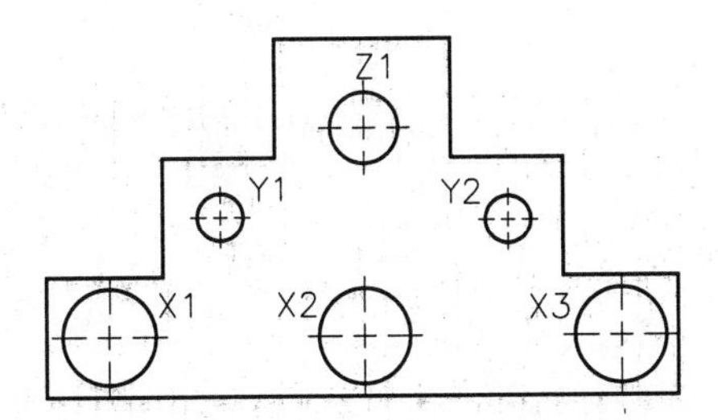

HOLE	X1	X2	X3	Y1	Y2	Z1
DIM.	R0.2	R0.2	R0.2	R0.1	R0.1	R0.15
QTY.	1	1	1	1	1	1
X	0.25	1.375	2.50	0.75	2.0	1.375
Y	0.25	0.25	0.25	0.75	0.75	1.125
Z	THRU	THRU	THRU	1.0	1.0	THRU

Figure 8-63 *Drawing for Exercise 8*

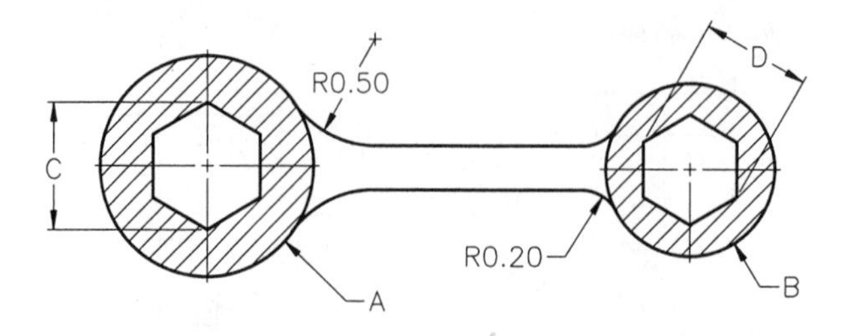

SPANNER NO.	A	B	C	D
S1	0.85	0.65	0.50	0.38
S2	1	0.75	0.59	0.44
S3	1.15	0.88	0.67	0.52
S4	1.25	0.95	0.74	0.56

Figure 8-64 *Drawing for Exercise 9*

Exercise 10

Mechanical

Make the drawing as shown in Figure 8-65. You must create dimension style files and specify values for different dimension types. Assume the missing dimensions.

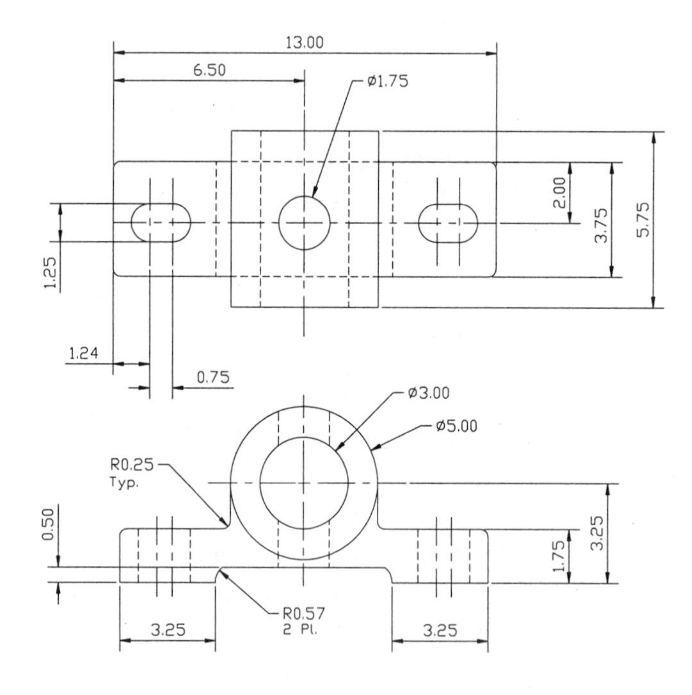

Figure 8-65 *Drawing for Exercise 10*

Problem Solving-Exercise 1 *Architecture*

Draw and dimension the floor plan as shown in Figure 8-66. Assume the missing dimensions or measure the relative dimensions from the given drawing (The drawing is not to scale).

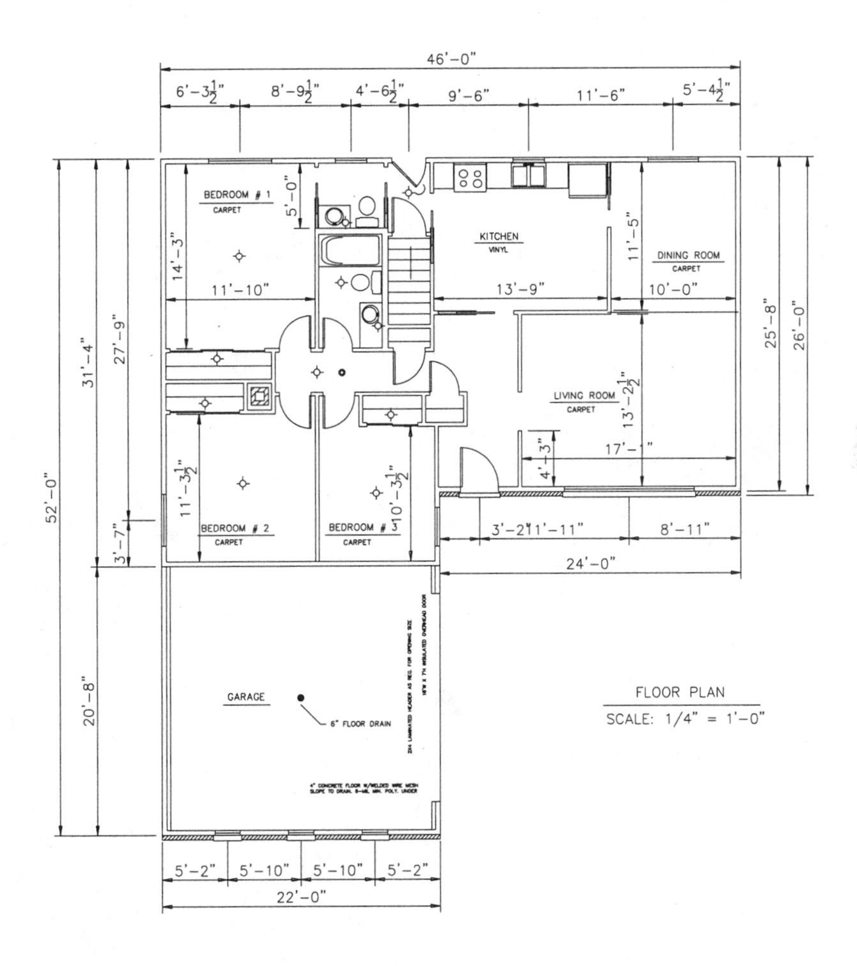

Figure 8-66 Drawing for Problem-Solving Exercise 1

Answers to Self-Evaluation Test

1 -Invokes **Dimension Style Manager** dialog box **2** -By using the **Dimension Style Manager** dialog box or by assigning values to dimensioning variables, **3** -**DIM/ DIMSCALE/** 2, **4** -Standard, **5** -dimension line/ dimension line, **6** -T, **7** -Center/ Diameter, **8** -inside, **9** -\X, **10** -box, **11** -Overrides, **12** -color, **13** -first/ second, **14** -angles/ tolerance, **15** -T

Chapter 9

Geometric Dimensioning and Tolerancing

Learning Objectives

After completing this chapter, you will be able to:

- *Use geometric tolerance components to specify tolerances.*
- *Use feature control frames and geometric characteristics symbols.*
- *Use tolerance values and the material condition modifier.*
- *Use complex feature control frames.*
- *Combine geometric characteristics and create composite position tolerancing.*
- *Use projected tolerance zones.*
- *Use feature control frames with leaders.*

IMPORTANCE OF GEOMETRIC DIMENSIONING AND TOLERANCING

One of the most important parts of the design process is giving the dimensions and tolerances, since every part is manufactured from the dimensions given in the drawing. Therefore, every designer must understand and have a thorough knowledge of the standard practices used in industry to make sure that the information given on the drawing is correct and can be understood by other people. Tolerancing is equally important, especially in the assembled parts. Tolerances and fits determine how the parts will fit. Incorrect tolerances could result in a product that is not usable.

In addition to dimensioning and tolerancing, the function and the relationship that exists between the mating parts is important if the part is to perform the way it was designed. This aspect of the design process is addressed by **geometric dimensioning and tolerancing**, generally known as **GDT**. Geometric dimensioning and tolerancing is a means to design and manufacture parts with respect to actual function and the relationship that exists between different features of the same part or the features of the mating parts. Therefore, a good design is not achieved by just giving dimensions and tolerances. The designer has to go beyond dimensioning and think of the intended function of the part and how the features of the part are going to affect its function. For example, in Figure 9-1, which shows a part with the required dimensions and tolerances, there is no mention of the relationship between the pin and the plate. Is the pin perpendicular to the plate? If it is, to what degree should it be perpendicular? Also, it does not mention on which surface the perpendicularity of the pin is to be measured. A design like this is open to individual interpretation based on intuition and experience. This is where geometric dimensioning and tolerancing play an important part in the product design process.

Figure 9-2 has been dimensioned using geometric dimensioning and tolerancing. The feature symbols define the datum (reference plane) and the permissible deviation in the perpendicularity of the pin with respect to the bottom surface. From a drawing like this, the chances of making a mistake are minimized. Before discussing the application of AutoCAD LT commands in geometric dimensioning and tolerancing, you need to understand the following feature symbols and tolerancing components.

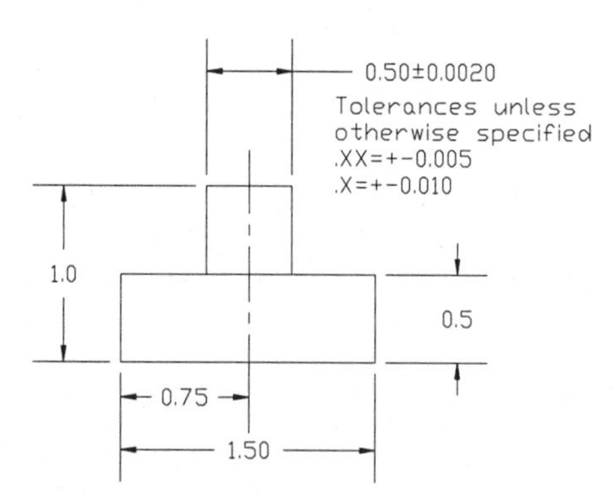

Figure 9-1 *Using traditional dimensioning and tolerancing technique*

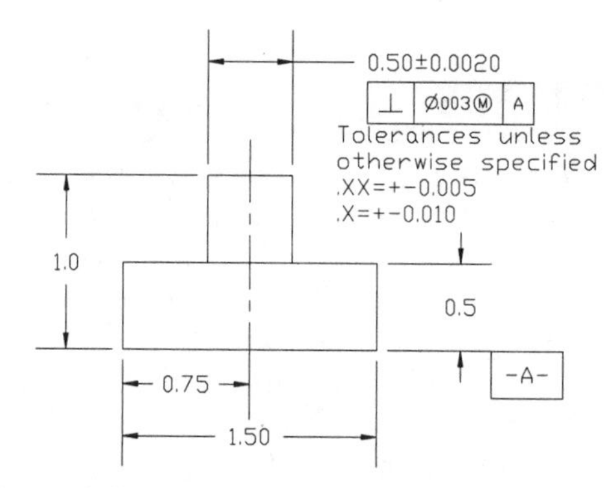

Figure 9-2 *Using geometric dimensioning and tolerancing*

GEOMETRIC CHARACTERISTICS AND SYMBOLS

Figure 9-3 shows the geometric characteristics and symbols used in geometric dimensioning and tolerancing. These symbols are the building blocks of geometric dimensioning and tolerancing.

KIND OF FEATURE	TYPE OF FEATURE	CHARACTERISTICS	
RELATED	LOCATION	Position	⌖
		Concentricity or coaxiality	◎
		Symmetry	⌯
	ORIENTATION	Parallelism	//
		Perpendicularity	⊥
		Angularity	∠
INDIVIDUAL	FORM	Cylindricity	⌭
		Flatness	▱
		Circularity or roundness	○
		Straightness	—
INDIVIDUAL or RELATED	PROFILE	Profile of a surface	⌓
		Profile of a line	⌒
RELATED	RUNOUT	Circular runout	↗
		Total runout	⌰

Figure 9-3 Geometric characteristics and symbols used in geometric dimensioning and tolerancing

GEOMETRIC TOLERANCING

Toolbar:	Dimension > Tolerance
Menu:	Dimension > Tolerance
Command:	TOLERANCE

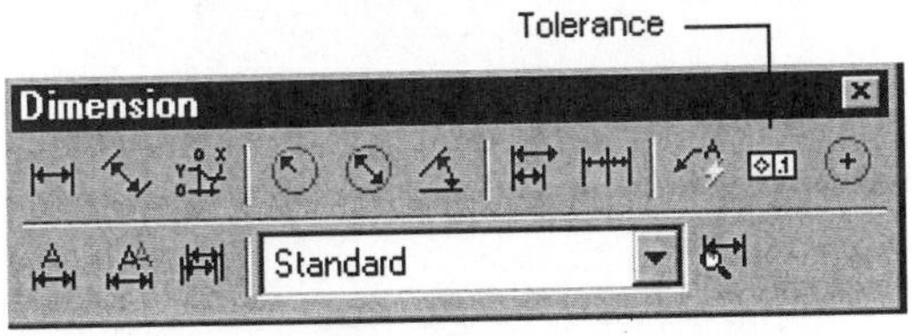

Tolerance

Dimension

Standard

*Figure 9-4 Choosing **Tolerance** from the **Dimension** toolbar*

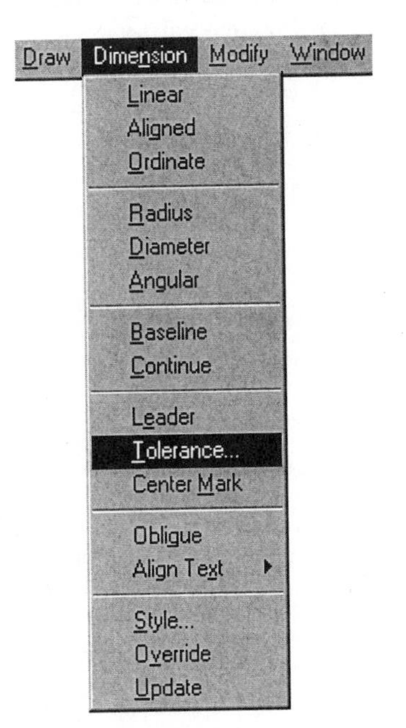

*Figure 9-5 Choosing **Tolerance** from the **Dimension** menu*

Geometric Tolerancing displays the deviations of profile, orientation, form, location and runout of a feature.In AutoCAD LT, geometrical tolerancing is displayed by feature control frames. The frames contain all the information about tolerances for a single dimension.To display feature control frames with the various tolerancing parameters, the specifications are entered in the **Geometric Tolerancing** dialog box (Figure 9-6). This dialog box can be invoked by entering the **TOLERANCE** command at the AutoCAD LT 2000 Command prompt. The dialog box can also be invoked by choosing **Tolerance** from the **Dimension** menu or toolbar (Figures 9-4 and 9-5).

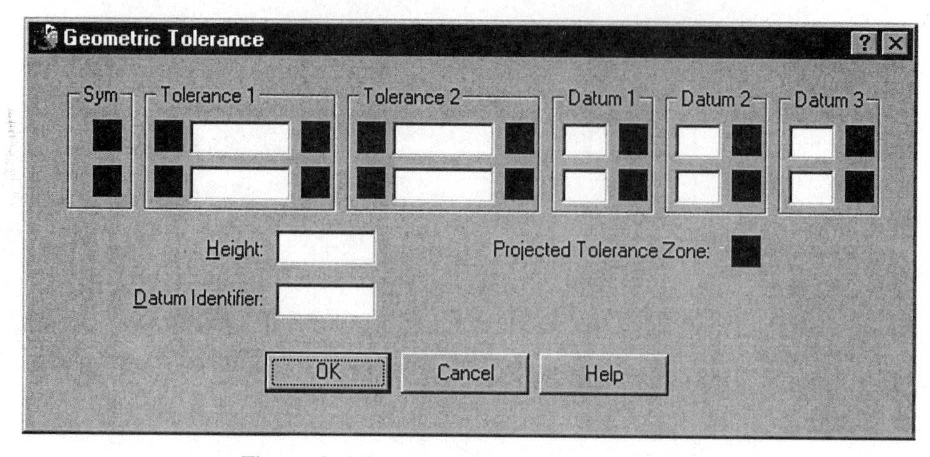

Figure 9-6 ***Geometric Tolerance*** *dialog box*

The following list presents the geometric tolerancing components; Figure 9-7 shows their placement in the tolerance frame:

Feature control frame
Tolerance value
Material condition modifier
Geometric characteristics symbol
Tolerance zone descriptor
Datum reference

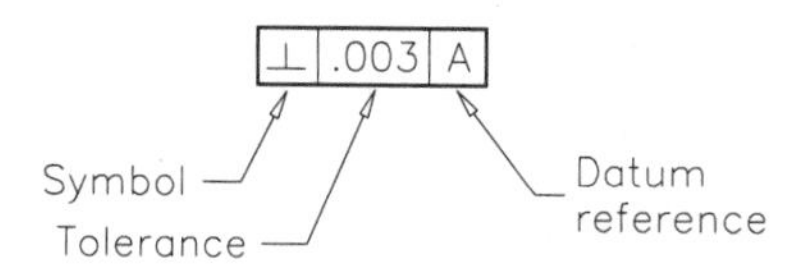

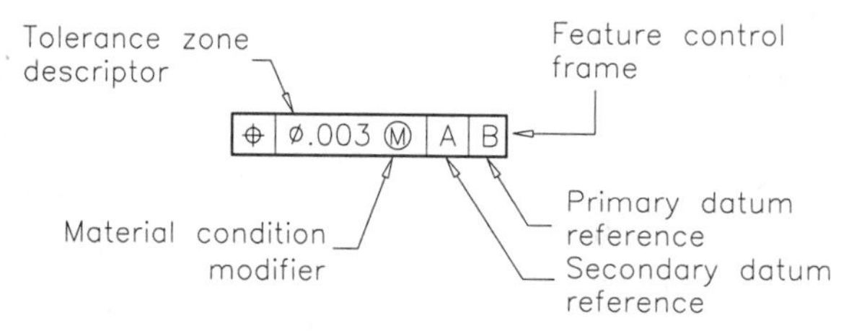

Figure 9-7 *Components of geometric tolerance*

Feature Control Frame

The **feature control frame** is a rectangular box that contains the geometric characteristics symbols and tolerance definition. The box is automatically drawn to standard specifications; you do not need to specify its size. You can copy, move, erase, rotate, and scale feature control frame. You can also snap to them using various Object snap modes. You can edit feature control frames using the **DDEDIT** command or you can also edit them using **GRIPS**. System variable **DIMCLRD** controls the color of the feature control frame. System variable **DIMGAP** controls the gap between the feature control frame and the text.

Geometric Characteristics Symbol

The geometric characteristics symbols indicate the characteristics of a feature, for example, straightness, flatness, and perpendicularity, etc. You can access the **Symbol** dialog box (Figure 9-8). by selecting the **Sym** edit box in the **Geometric Tolerance** dialog box, which is displayed when you enter the **TOLERANCE** command or choose it from the **Dimension** toolbar or menu. The different symbols can be selected from the **Symbol** dialog box.

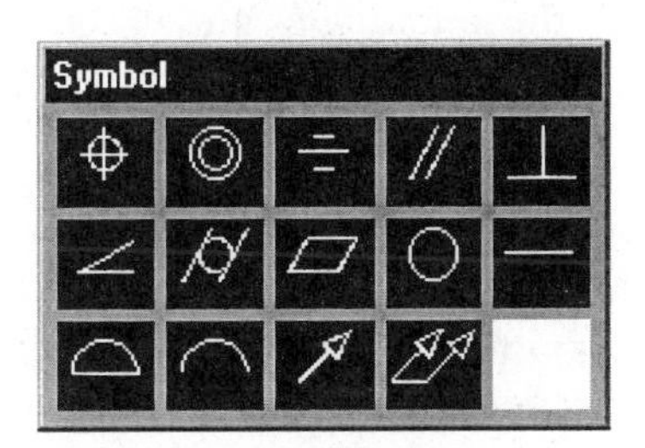

Figure 9-8 ***Symbol*** *dialog box*

Tolerance Value and Tolerance Zone Descriptor

The tolerance value specifies the tolerance on the feature as indicated by the tolerance zone descriptor. For example, a value of .003 indicates that the feature must be within a 0.003 tolerance zone. Similarly, ϕ.003 indicates that this feature must be located at true position within 0.003 diameter. The tolerance value can be entered in the **Value** edit box of the **Geometric Tolerance** dialog box. The tolerance zone descriptor can be invoked by selecting the box labeled **Dia**, located to the left of the **tolerance** edit box. The **Geometric Tolerance** dialog box can be invoked by the **TOLERANCE** command or by choosing **TOLERANCE** from the **Dimension** menu or toolbar. System variable **DIMCLRT** controls the color of the tolerance text, variable **DIMTXT** controls the tolerance text size, and variable **DIMTXSTY** controls the style of the tolerance text. Using the **Projected Tolerance Zone**, inserts a projected tolerance zone symbol, which is an encircled P, after the projected tolerance zone value.

Material Condition Modifier

The **Material Condition** modifier specifies the material condition when the tolerance value takes effect. For example, ϕ.003(M) indicates that this feature must be located at true position within 0.003 diameter at maximum material condition (MMC). The material condition modifier symbol can be selected from the **Material Condition** dialog box (Figure 9-9). This dialog box can be invoked by selecting the **MC** button located just after **Tolerance Value** edit box in the **Geometric Tolerance** dialog box.

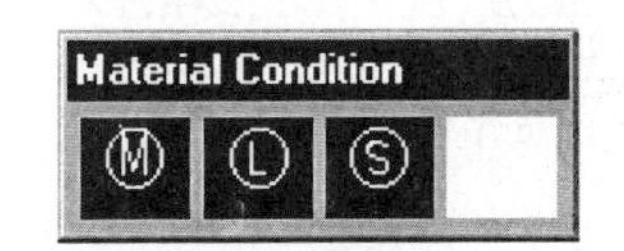

Figure 9-9 ***Material condition*** *dialog box*

Datum

The datum is the origin, surface, or feature from which the measurements are made. The datum is also used to establish the geometric characteristics of a feature. The datum feature symbol consists of a reference character enclosed in a feature control frame. You can create the datum feature symbol by entering characters (like -A-) in the **Datum Identifier** edit box in the **Geometric Tolerance** dialog box and then selecting a point where you want to establish that datum.

You can also combine datum references with geometric characteristics. AutoCAD LT automatically positions the datum references on the right end of the feature control frame.

Example 1

In the following example, you will create a feature control frame to require a perpendicularity specification (see Figure 9-10).

Step 1. Use the **TOLERANCE** command to display the **Geometric Tolerance** dialog box. Select the **Sym** edit box to display the **Symbol** dialog box. Choose the **perpendicularity** symbol in the dialog box and then choose the **OK** button.

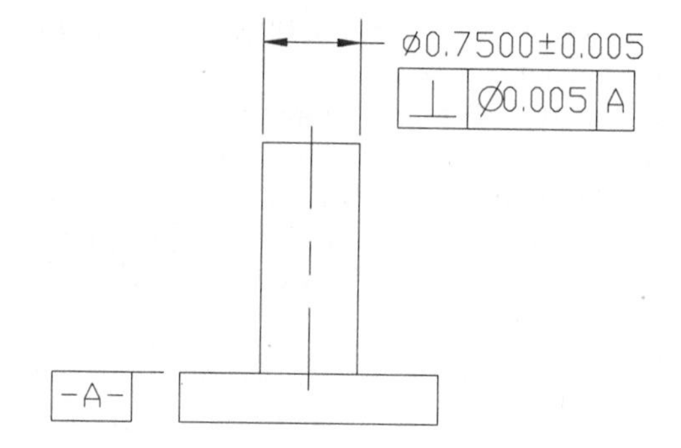

Figure 9-10 Drawing for Example 1

Step 2. The perpendicularity symbol will be displayed in the **Sym** edit box in the first row of the **Geometric Tolerance** dialog box. Select the **Dia** edit box, located just before the Tolerance Value edit box, in the **Tolerance 1** area on the first row. A diameter symbol will appear, to denote a cylindrical tolerance zone.

Step 3. Select the **Tolerance Value** edit box in the **Tolerance 1** area in the first row and enter **0.005**.

Step 4. Select the **Datum** edit box in the **Datum 1** area in the first row and enter **A**.

Step 5. Choose the **OK** button to accept the changes made in the **Geometric Tolerance** dialog box. The **Enter tolerance location:** prompt is displayed in the Command line area and the **Feature Control Frame** is attached to the cursor at its middle left point. Select a point to insert the frame.

Step 6. To place the datum symbol, use the **TOLERANCE** command to display the **Geometric Tolerance** dialog box. Select the **Datum Identifier** edit box, and enter **-A-**.

Step 7. Choose the **OK** button to accept the changes to the **Geometric Tolerance** dialog box, and then select a point to insert the frame.

COMPLEX FEATURE CONTROL FRAMES

Combining Geometric Characteristics

Sometimes it is not possible to specify all geometric characteristics in one frame. For example, Figure 9-11 shows the drawing of a plate with a hole in the center.

In this part, it is determined that surface C must be perpendicular to surfaces A and B within 0.002 and 0.004, respectively. Therefore, we need two frames to specify the geometric characteristics of surface C. The first frame specifies the allowable deviation in perpendicularity of surface C with respect to surface A. The second frame specifies the allowable deviation in perpendicularity of surface C with respect to surface B. In addition to these two frames, we need a third frame that identifies datum surface C.

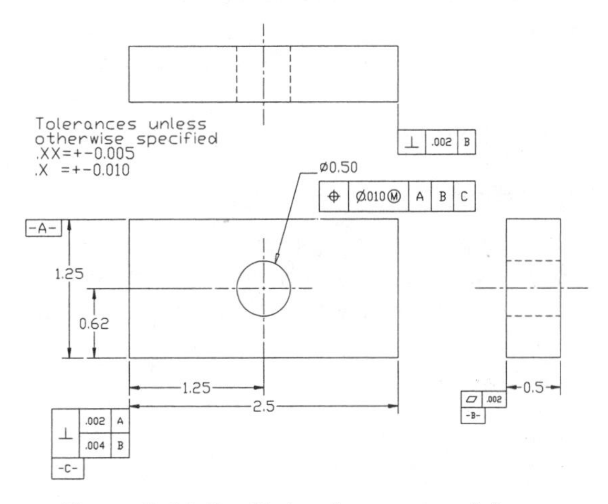

Figure 9-11 Combining feature control frames

All the three feature control frames can be defined in one instance of the **TOLERANCE** command.

1. Enter the **TOLERANCE** command to invoke the **Geometric Tolerance** dialog box (see Figure 9-12). Select **Sym** edit box to display **Symbol** dialog box. Choose the **perpendicular** symbol, and then choose the **OK** button. AutoCAD LT will display the selected symbol in the first row **Sym** edit box.

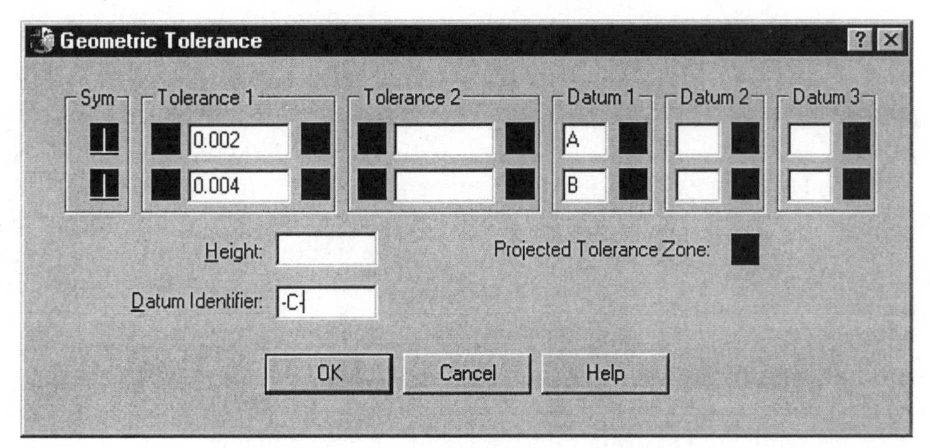

***Figure 9-12 Geometric Tolerance** dialog box*

2. Enter **.002** in the first row **Tolerance Value** edit box, and enter **A** in the first row of **Datum 1** edit box.
3. Select the second row **Sym** edit box to display the **Symbol** dialog box. Choose the **perpendicular** symbol, and then choose the **OK** button. AutoCAD LT will display the selected symbol in the second row **Sym** edit box in the **Geometric Tolerance** dialog box.

Chapter 9

4. Enter **.004** in the second row **Tolerance Value** edit box, and enter **B** in the second row **Datum 1** edit box.
5. In the **Datum Identifier** edit box enter **-C-**, and then choose the **OK** button to exit the dialog box.
6. In the graphics screen, select the position to place the frame.

Composite Position Tolerancing

Sometimes the accuracy required within a pattern is more important than the location of the pattern with respect to the datum surfaces. To specify such a condition, composite position tolerancing may be used.For example, Figure 9-13 shows four holes (pattern) of diameter

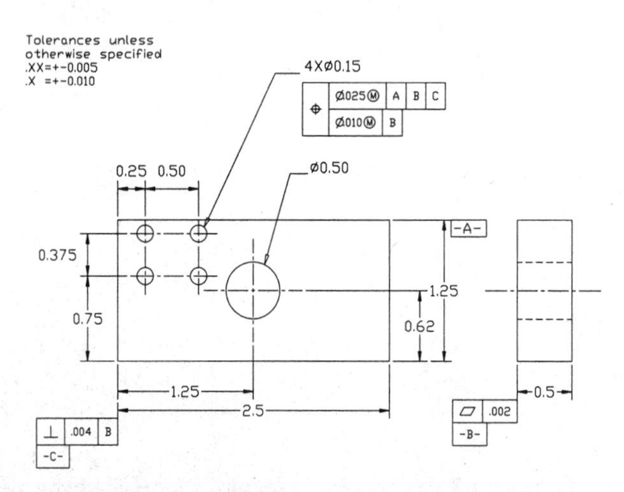

Figure 9-13 Using composite position tolerancing

0.15. The design allows a maximum tolerance of 0.025 with respect to datums A, B, and C at the maximum material condition (holes are smallest). The designer wants to maintain a closer positional tolerance (0.010 at MMC) between the holes within the pattern. To specify this requirement, the designer must insert the second frame. This is generally known as composite position tolerancing. AutoCAD LT provides the facility to create the two composite position tolerance frames by means of the **Geometric Tolerance** dialog box. The composite tolerance frames can be created as follows:

1. Enter the **TOLERANCE** command to invoke the **Geometric Tolerance** dialog box. Select the **Sym** edit box to display the **Symbol** dialog box. Choose the position symbol,and then Choose the **OK** button to display the symbol in the **Sym** edit box.
2. In the first row of the **Geometric Tolerance** dialog box, enter the geometric characteristics and the datum references required for the first position tolerance frame.
3. In the second row of the **Geometric Tolerance** dialog box,enter the geometric characteristics and the datum references required for the second position tolerance frame.
4. When you have finished entering the values, choose the **OK** button in the **Geometric Tolerance** dialog box, and then select the point where you want to insert the frames. AutoCAD LT will create the two frames and automatically align them with common position symbol, as shown in Figure 9-13.

Projected Tolerance Zone

Figure 9-14 shows two parts joined with a bolt. The lower part is threaded, and the top part has a drilled hole. When these two parts are joined, the bolt that is threaded in the lower part will have the orientation error that exists in the threaded hole. In other words, the error in the threaded hole will extend beyond the part thickness, which might cause interference, and the parts may not assemble. To avoid this problem, projected tolerance is used. The projected tolerance establishes a tolerance zone that extends above the surface. In Figure 9-14, the position tolerance for the threaded hole is 0.010, which extends 0.25 above the surface (datum A). By using the projected tolerance, you can ensure that the bolt is within the tolerance zone up to the specified distance.

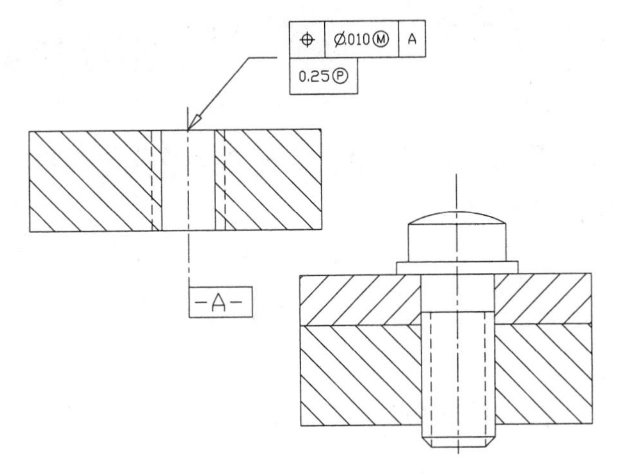

Figure 9-14 *Using composite position tolerancing*

You can use the AutoCAD LT GDT feature to create feature control frames for the projected tolerance zone as follows:

1. Enter the **TOLERANCE** command to invoke the **Geometric Tolerance** dialog box. Choose the position symbol from the **Symbol** dialog box by selecting the **Sym** edit box, and then choose the **OK** button to exit from this dialog box.

2. In the first row of the **Geometric Tolerance** dialog box, enter the geometric characteristics and the datum references required for the first position tolerance frame (see Figure 9-15).

3. In the **Height** edit box, enter the height of the tolerance zone (0.25 for the given drawing) and select the edit box to the right of Projected Tolerance Zone. The projected tolerance zone symbol will be displayed in the box.

4. Once you are done entering the values, choose the **OK** button in the **Geometric Tolerance** dialog box, and then select the point where you want to insert the frames. AutoCAD LT will create the two frames and automatically align them, as shown in Figure 9-15.

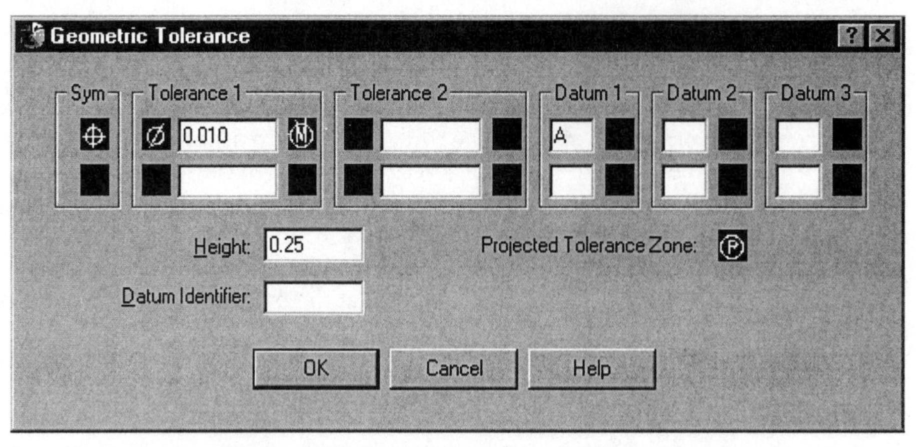

***Figure 9-15 Geometric Tolerance** dialog box*

USING FEATURE CONTROL FRAMES WITH THE LEADER COMMAND

The **LEADER** command has the Tolerance option, which allows you to create the feature control frame and attach it to the end of the leader extension line. The following is the prompt sequence for using the **LEADER** command with the Tolerance option:

Command: **LEADER** [Enter]
Specify leader start point: *Select a point where you want the arrow (P1).*
Specify next point: *Select a point (P2).*
Specify next point or [Annotation/Format/Undo]<Annotation>: *Select a point (P3).*
Specify next point or [Annotation/Format/Undo]<Annotation>: *Press ENTER.*
Enter first line of annotation text or <options>: *Press ENTER.*
Enter an annotation option [Tolerance/Copy/Block/None/Mtext]<Mtext>: **T** [Enter]

When you select the Tolerance (**T**) option, AutoCAD LT will display the **Geometric Tolerance** dialog box. Select the **Sym** edit box and then from the **Symbol** dialog box choose the desired symbol, to return to the **Geometric Tolerance** dialog box. Enter the required values, and choose the **OK** button to exit the dialog box. The feature control frame with the defined geometric characteristics will be inserted at the end of the extension line, as shown in Figure 9-16.

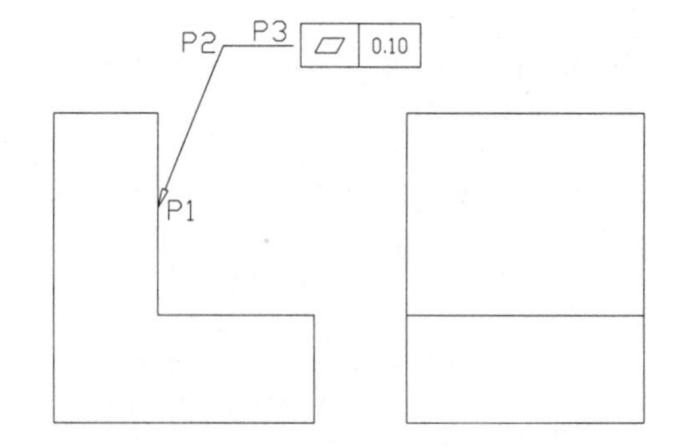

***Figure 9-16** Using feature control frame with the leader*

Note

*When you choose **Leader** from the **Dimension** menu, press ENTER to choose the **Settings** option to display the **Leader Settings** dialog box. You can select the **Tolerance** radio button in the*

Annotation Type area to set the leader for Tolerance symbol. The default option is ***Mtext.***

Example 2

In the following example, you will create a leader with a combination feature control frame to control runout and cylindricity (Figure 9-17).

Step 1. Use the **LEADER** command to draw the leader.

Command: **LEADER** [Enter]
Specify leader start point: *Select a point where you want the arrow.*
Specify next point: *Select the point for the first bend of the leader.*
Specify next point or [Annotation/Format/Undo]<Annotation>: *Select the endpoint.*
Specify next point or [Annotation/Format/Undo]<Annotation>: *Press ENTER.*
Enter first line of annotation text or <options>: *Press ENTER.*
Enter an annotation option [Tolerance/Copy/Block/None/Mtext]<Mtext>: **T** [Enter]

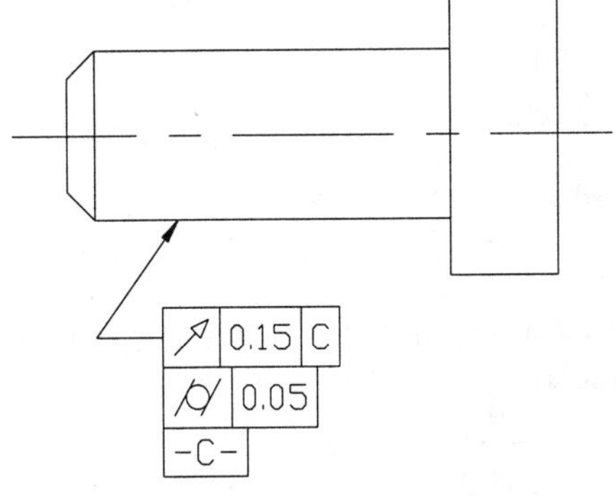

Figure 9-17 *Drawing for Example 2*

Step 2. The **Geometric Tolerance** dialog box is displayed. Select the **Sym** edit box and from the **Symbol** dialog box, choose the **runout** symbol in the dialog box, and then return to the **Geometric Tolerance** dialog box.

Step 3. The **runout** symbol will be displayed in the **Sym** edit box on the first row of the **Geometric Tolerance** dialog box. Select the Value edit box in the Tolerance 1 area on the first row, and enter **0.15**.

Step 4. Select the **Datum** edit box in the Datum 1 area on the first row, and enter **C**.

Step 5. Select the **Sym** edit box on the second row of the **Geometric Tolerance** dialog box to display the **Symbol** dialog box. Choose the **cylindricity** symbol in the dialog box to return to the **Geometric Tolerance** dialog box.

Step 6. The **cylindricity** symbol will be displayed in the **Sym** edit box on the second row of the **Geometric Tolerance** dialog box. Select the **Tolerance Value** edit box in the Tolerance 1 area on the second row, and enter **0.05**.

Step 7. Select the **Datum** Identifier edit box, and enter **-C-**.

Step 8. Choose the **OK** button to accept the changes to the **Geometric Tolerance** dialog box, and the control frames will be drawn at the end of the leader.

Self-Evaluation Test

Answer the following questions, and then compare your answers to the correct answers given at the end of the chapter.

1. What is the advantage of showing tolerances and fits in a drawing? ____.

2. Geometric dimensioning and tolerancing is generally known as ________.

3. List the components of geometric tolerance. ____.

4. Give three examples of geometric characteristics that indicate the characteristics of a feature. ____.

5. Give an example of a material condition modifier that specifies the material condition when the tolerance value takes effect. ____.

6. What is the projected tolerance zone? Explain. ____.

7. The **LEADER** command has the **Tolerance** option, which allows you to create the feature control frame and attach it to the end of the leader extension line. (T/F)

8. Name two characteristics of the location feature. ______________________.

9. Explain maximum material condition. ____.

Review Questions

1. One of the most important parts of the design process is to give the dimensions and tolerances. Why? ____.

2. Geometric dimensioning and tolerancing is a means to design and manufacture parts with respect to actual function and the relationship that exists between different features. (T/F)

3. ________ symbols are the building blocks of geometric dimensioning and tolerancing.

4. The feature control frame is a circular shape that contains geometric characteristics symbolsand tolerance definition. (T/F)

5. Give an example of tolerance value that specifies the tolerance on the feature as indicated by the tolerance zone descriptor.

6. The datum is the ________, ________, or feature from which the measurements are made.

7. Sometimes the accuracy required within a pattern is more important than the location of the pattern with respect to the datum surfaces. To specify such a condition, composite position tolerancing may be used. (T/F)

8. You cannot use the AutoCAD LT GDT feature to create feature control frames for the projected tolerance zone. (T/F)

9. Draw a picture of a feature control frame and explain its features.

10. Name three characteristics of the form feature.________________

Exercises

Exercise 1 — *Mechanical*

Draw Figure 9-18. Then use the **TOLERANCE** and **LEADER** commands to draw the geometric tolerances as shown.

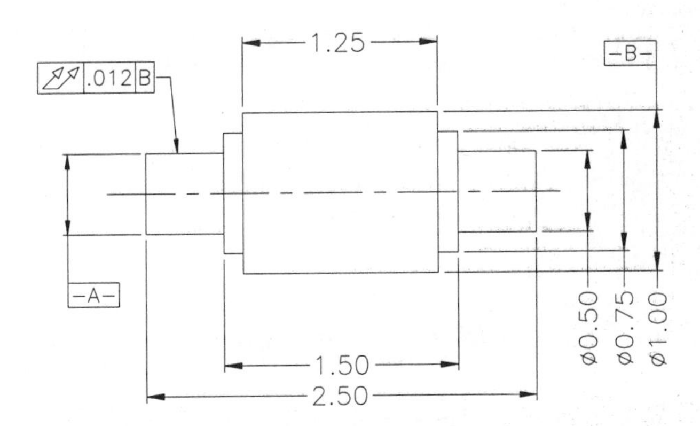

Figure 9-18 Drawing for Exercise 1

Project Exercise 2-1 — *Mechanical*

Generate the drawing in Figure 9-19 according to the dimensions shown in the figure. Save the drawing as TOOLORGA.

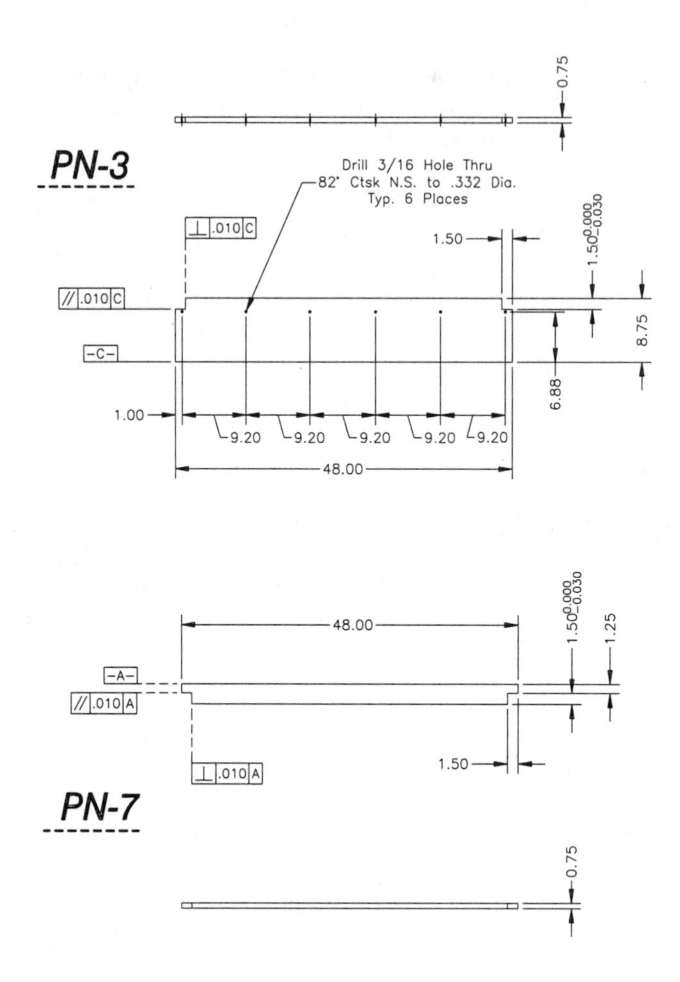

Figure 9-19 Drawing for Project Exercise 2-1

Project Exercise 2-2

Mechanical

Generate the drawing in Figure 9-20 according to the dimensions shown in the figure. Save the drawing as TOOLORGB.

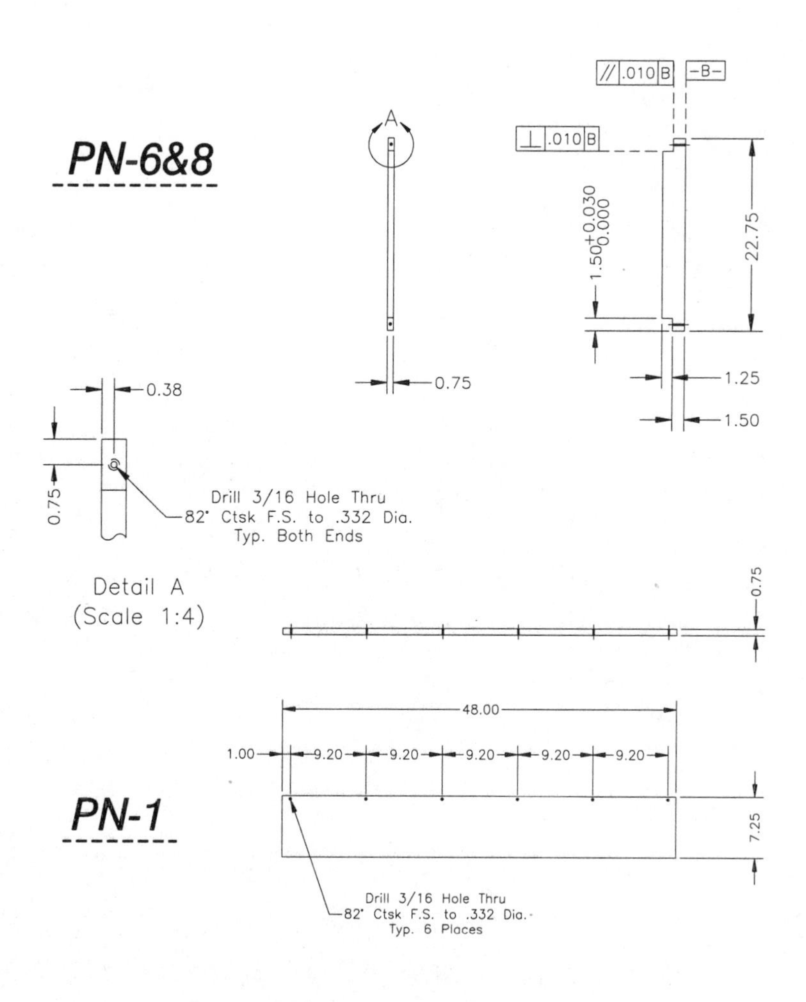

Figure 9-20 *Drawing for Project Exercise 2-2*

Chapter 9

Answers to Self-Evaluation Test

1 - Tolerances and fits determine how the parts will fit. Incorrect tolerances could result in a product that is not functional. **2** - GDT, **3** - Feature control frame, geometric characteristics symbol, tolerance value, tolerance zone descriptor, material condition modifier, datums. **4** - Straightness, flatness, perpendicularity. **5** - Example: f.003(M) indicates that this feature must be located at true position within 0.003 diameter at maximum material condition (MMC). **6** - The projected tolerance establishes a tolerance zone that extends above the surface. **7** - T, **8** - Position, concentricity, symmetry. **9** - When the part has maximum material; for example, if the part has a hole, the material will be maximum when the hole diameter is minimum.

Chapter 10

Editing with GRIPS

Learning Objectives

After completing this chapter, you will be able to:

- *Adjust grip settings.*
- *Select objects with grips.*
- *Stretch, move, rotate, scale, and mirror objects with grips.*
- *Use grips system variables.*

EDITING WITH GRIPS

Grips provide a convenient and quick means of editing objects. With grips you can stretch, move, rotate, scale, mirror objects, change properties, and load the Web browser. Grips are small squares that are displayed on an object at its definition points when the object is selected. The number of grips depends on the selected object. For example, a line has three grip points, a polyline has two, and an arc has three. Similarly, a circle has five grip points and a dimension (vertical) has five. When you select the **Enable grips** and the **Noun/Verb Selection** check boxes in the **Selection** tab of the **Options** dialog box, a small square (aperture box) at the intersection of the crosshairs is displayed (Figure 10-1).

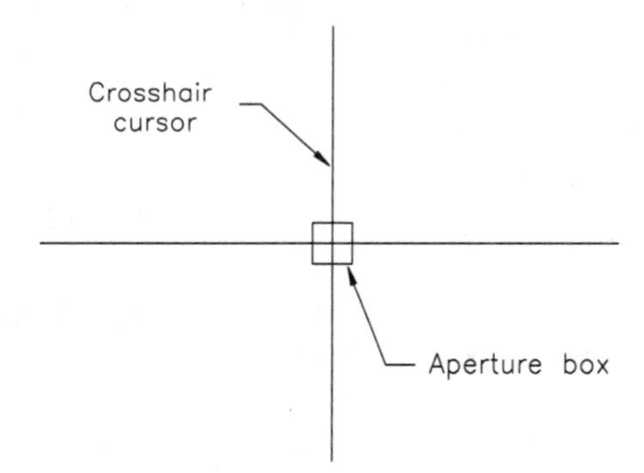

***Figure 10-1** Aperture box at the intersection of crosshair cursor*

Note

*AutoCAD LT also displays a small square (aperture box) at the intersection of crosshairs when the **PICKFIRST** (Noun/Verb Selection) system variable is set to 1 (On).*

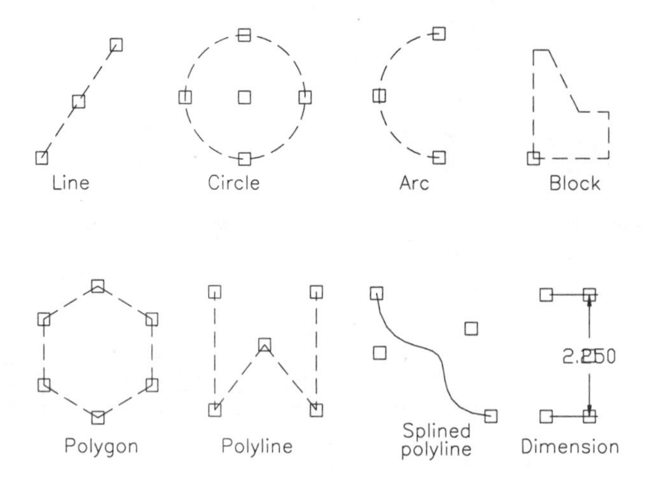

Figure 10-2 Grip location of various objects

ADJUSTING GRIP SETTINGS

Menu:	Tools > Options
Command:	OPTIONS

Grip settings can be adjusted through the **Selection** tab of the **Options** dialog box (Figure 10-4).

Command: **OPTIONS** [Enter]

The **Selection** tab of the **Options** dialog box has the following areas:

1. Grips
2. Grip Size

Grips

The **Grips** area has two check boxes: **Enable Grips** and **Enable Grips Within Blocks**. The grips can be enabled by selecting the **Enable Grips** check box. They can also be enabled by setting the **GRIPS** system variable to 1.

Command: **GRIPS** [Enter]
Enter new value for GRIPS <0>: 1

The second check box, **Enable Grips Within Blocks**, enables the grips within a block. If you select this box, AutoCAD LT will display grips for every object in the block. If you disable the display of grips within a block, the block will have only one grip at its insertion point.

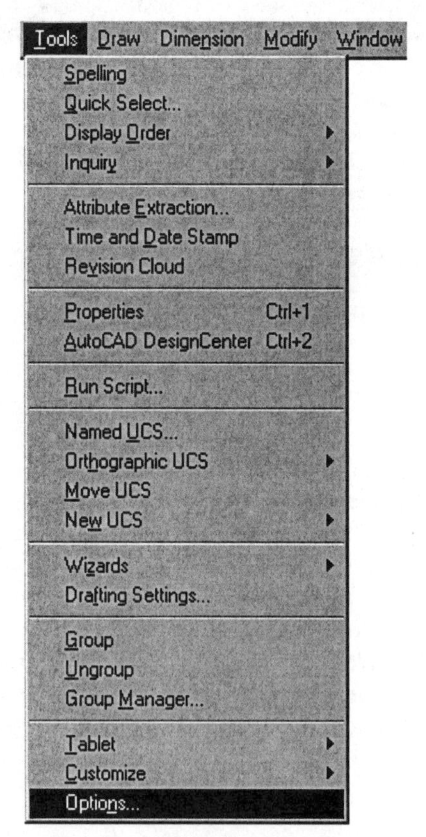

***Figure 10-3** Invoking the **Options** dialog box from the **Tools** menu*

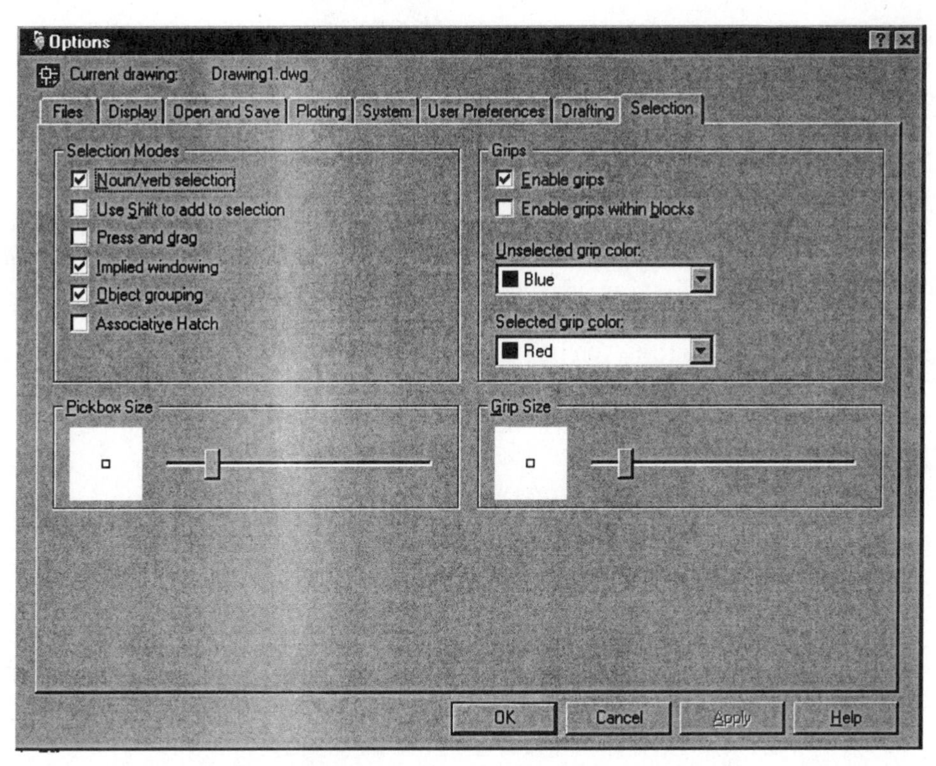

Figure 10-4 ***Selection*** *tab of the* ***Options*** *dialog box*

You can also enable the grips within a block by setting the system variable **GRIPBLOCK** to 1 (On). If **GRIPBLOCK** is set to 0 (Off), AutoCAD LT will display only one grip for a block at its insertion point (Figure 10-5).

Note

If the block has a large number of objects, and if ***GRIPBLOCK*** *is set to 1 (On), AutoCAD LT will display grips for every object in the block. Therefore, it is recommended that you set the system variable* ***GRIPBLOCK*** *to 0 or clear the* ***Enable Grips Within Blocks*** *check box in the* ***Selection*** *tab of the* ***Options*** *dialog box.*

Grip Colors

The **Grips** area of the **Selection** tab of the **Options** dialog box has two drop-down lists, **Unselected grip color:** and **Selected grip color:**. These two drop-down lists give you options of colors to select from, for unselected and selected grips respectively. When you select the **More** option in either of these drop-down lists, AutoCAD LT displays the standard **Select Color** dialog box from which you can select a desired color. By default, the unselected grips have blue color

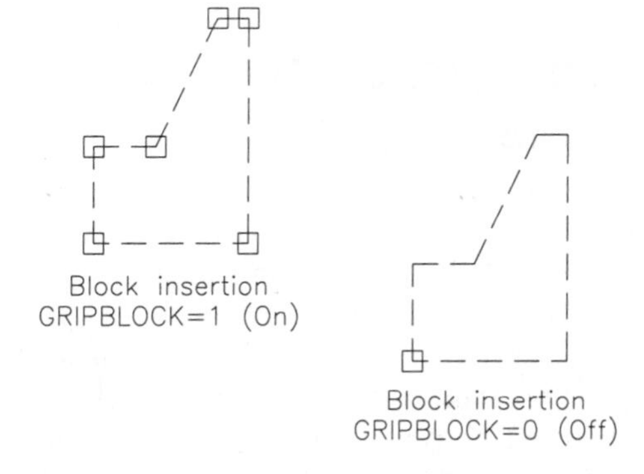

Figure 10-5 Block insertion with ***GRIPBLOCK*** *set to 1 and to 0*

(frame of the box) and the selected grips have red color (filled red square). The color of the unselected grips can also be changed by using the **GRIPCOLOR** system variable. Similarly, the color of the selected grips can also be changed by using the **GRIPHOT** system variable.

Grip Types. Grips can be classified into three types: hot grips, warm grips, and cold grips. When you select an object, the grips are displayed at the definition points of the object, and the object is highlighted by displaying it as a dashed line. These grips are called warm grips (blue). Now, if you select a grip on this object, the grip becomes a hot grip (filled red square). Once the grip is hot, the object can be edited. To cancel the grip, press ESC. If you press ESC once more, the hot grip changes to cold grip. When the grip is cold, the object is not highlighted, as shown in Figure 10-6. You can also snap to a cold grip.

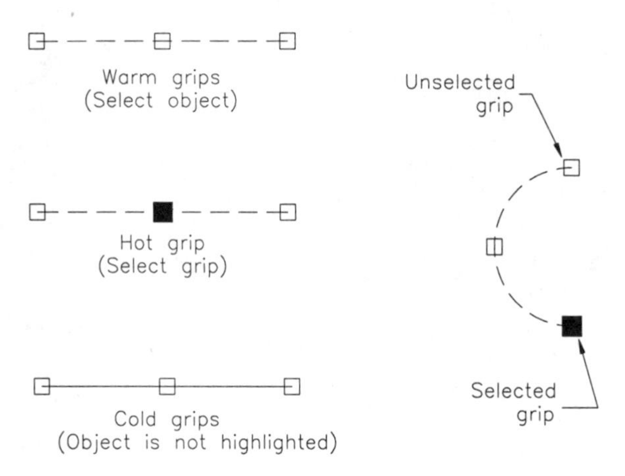

Figure 10-6 *Hot, warm, cold, selected, and unselected grips*

Grip Size

The **Grip Size** area of the **Selection** tab of the **Options** dialog box consists of a slider bar and a rectangular box that displays the size of the grip. To adjust the size of the grip, move the slider box left or right. The size of the grip can also be adjusted by using the system variable **GRIPSIZE**. **GRIPSIZE** is defined in pixels, and its value can range from 1 to 255 pixels.

STRETCHING OBJECTS WITH GRIPS (Stretch Mode)

If you select an object, AutoCAD LT displays grips (warm grips) at the definition points of the object. When you select a grip for editing, you are automatically in the **Stretch** mode. The Stretch mode has a function similar to the **STRETCH** command. When you select a grip, it acts as a base point and is called a base grip. You can also select several grips by holding the SHIFT key down and then select the warm grips. Now, release the SHIFT key and select one of the hot grips to stretch them simultaneously. The geometry between the selected base grips is not altered. You can also make copies of the selected objects or define a new base point. When selecting grips on text objects, blocks, midpoints of lines, centers of circles and ellipses and point objects in the stretch mode, the selected objects are moved to a new location. The following example illustrates the use of the Stretch mode.

1. Use the **PLINE** command to draw a W-shaped figure as shown in Figure 10-7(a).

2. Select the object that you want to stretch [Figure 10-7(a)]. When you select the object, grips will be displayed at the endpoints of each object. A polyline has two grip points. If you use the **LINE** command to draw the object, AutoCAD LT will display three grips for each object.

3. Hold the SHIFT key down, and select the grips that you want to stretch [Figure 10-7(b)]. The selected grips will become hot grips, and the color of the grip will change from blue to red.

Note
By holding down the SHIFT key, you can select several grips. If you do not hold down the SHIFT key, only one grip can be selected.

4. Choose one of the selected (hot grip) grips, and specify a point to which you want to stretch the line [Figure 10-7(c)]. When you select a grip, the following prompt is displayed in the Command prompt area:

   STRETCH
   Specify stretch point or [Base point/Copy/Undo/eXit]:

   The Stretch mode has several options: **Base point**, **Copy**, **Undo**, and **eXit**. You can use the **Base point** option to define the base point and the **Copy** option to make copies.

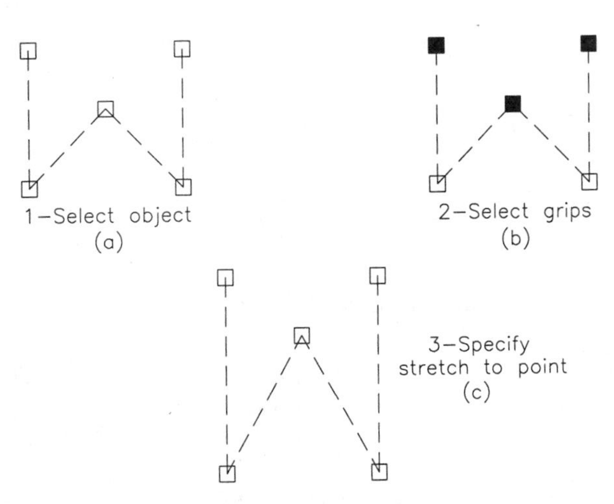

Figure 10-7 *Using the* ***Stretch*** *mode to stretch the lines*

5. Select the grip where the two lines intersect, as shown in Figure 10-9(a). Select the **Copy** option from the shortcut menu (Figure 10-8) by right-clicking your pointing device, or

enter C for copy at the command line, then select the points as shown in Figure 10-9(b). Each time you select a point, AutoCAD LT will make a copy.

If you press the SHIFT key when specifying the point to which the object is to be stretched, without selecting the **Copy** option, then also AutoCAD LT allows you to make multiple copies of the selected object. Also if you press the SHIFT key again when specifying the next point, the cursor snaps to a point whose location is based on the distance between the first two points, that is the distance between the selected object and the location of the copy of the selected object.

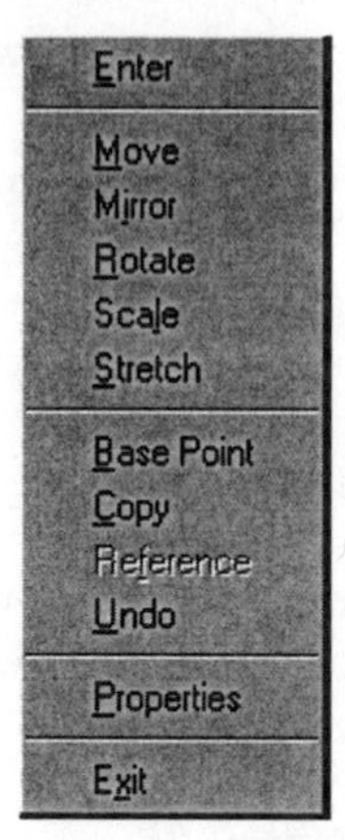

Figure 10-8 *Selecting different Grip options from the shortcut menu*

6. Make a copy of the drawing as shown in Figure 10-9(c). Select the object, and then select the grip where the two lines intersect. When AutoCAD LT displays the ****STRETCH**** prompt, select the **Base Point** option from the shortcut menu or enter B at the Command prompt. Select the bottom left grip as the base point, and then give the displacement point as shown in Figure 10-9(d).

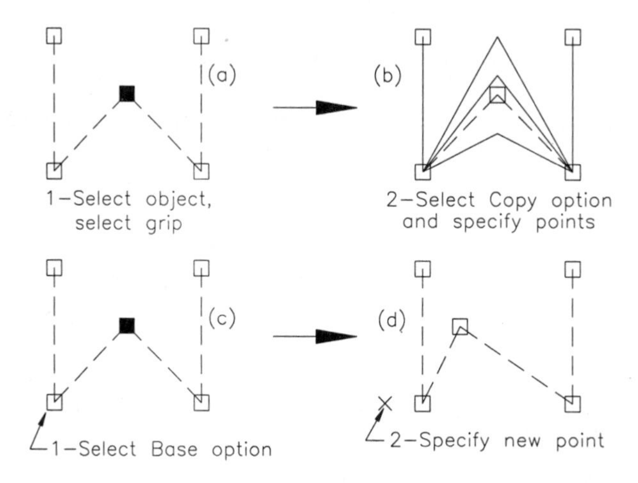

Figure 10-9 *Using the Stretch mode's **Copy** and **Base point** options*

7. To exit the grip mode, right-click to display the shortcut menu and then select **eXit**. You can also enter X at the Command: prompt or press ESC to exit.

Note

*You can select an option (**Copy** or **Base** **Point**) from the shortcut menu that can be invoked by right-clicking your pointing device after selecting a grip. The different modes can also be selected from the shortcut menu. You can also cycle through all the different modes by selecting a grip and pressing the ENTER key or the SPACEBAR. You can also enter keyboard shorcuts like **ST** for **STRETCH**.*

MOVING OBJECTS WITH GRIPS
(Move Mode)

The Move mode lets you move the selected objects to a new location. When you move objects, the size of the objects and their angle do not change. You can also use this mode to make copies of the selected objects or to redefine the base point. The following example illustrates the use of Move mode.

1. Use the **LINE** command to draw the shape as shown in Figure 10-10(a). When you select the objects, grips will be displayed at the definition points and the object will be highlighted.

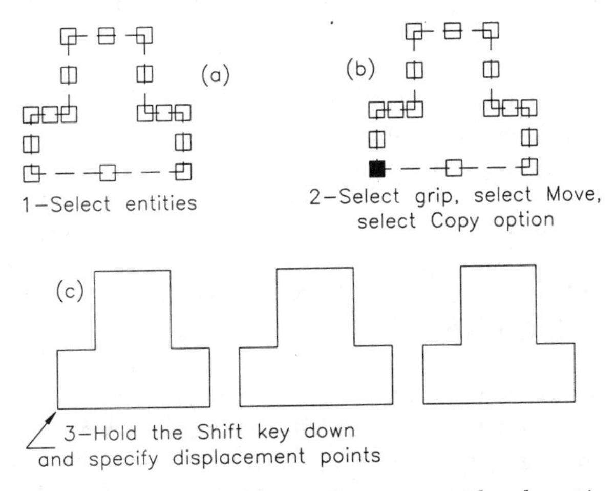

***Figure 10-10** Using the **Move** mode to move and make copies of the selected objects*

2. Select the grip located at the lower left corner, and then select **Move** from the shortcut menu. You can also invoke the Move mode by entering **MOVE** or **MO** at the keyboard or giving a null response by pressing the SPACEBAR or ENTER key. AutoCAD LT will display the following prompt in the Command prompt area:

   MOVE
   Specify move point or [Base point/Copy/Undo/eXit]:

3. Hold down the SHIFT key, and then enter the first displacement point. The distance between the first and the second object defines the snap offset for subsequent copies. While holding down the SHIFT key, move the screen crosshair to the next snap point and select the point. AutoCAD LT will make a copy of the object at that location. If you release the SHIFT key, you can specify any point where you want to place a copy of the object. You can also enter coordinates to specify the displacement.

ROTATING OBJECTS WITH GRIPS (Rotate Mode)

The **Rotate** mode allows you to rotate objects around the base point without changing their size. The options of Rotate mode can be used to redefine the base point, specify a reference angle, or make multiple copies that are rotated about the specified base point. You can access the Rotate mode by selecting the grip and then selecting **Rotate** from the shortcut menu, or by entering **ROTATE** or **RO** at the keyboard or giving a null response twice by pressing the SPACEBAR or the ENTER key. The following example illustrates the use of Rotate mode.

1. Use the **LINE** command to draw the shape as shown in Figure 10-11(a). When you select the objects, grips will be displayed at the definition points and the shape will be highlighted.

2. Select the grip located at the lower left corner and then invoke the Rotate mode. AutoCAD LT will display the following prompt:

   ROTATE
   Specify rotation angle or [Base point/Copy/Undo/Reference/eXit]:

3. At this prompt, enter the rotation angle. AutoCAD LT will rotate the selected objects by the specified angle [Figure 10-11(b)].

4. Make a copy of the original drawing as shown in Figure 10-11(c). Select the objects, and then select the grip located at the lower left corner of the object. Invoke the Rotate mode and then select the **Copy** option from the shortcut menu or enter C (Copy) at the prompt. Enter the rotation angle. AutoCAD LT will rotate a copy of the object through the specified angle [Figure 10-11(d)].

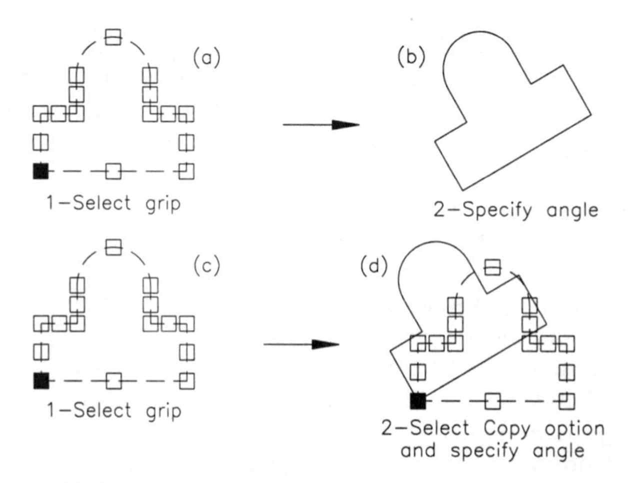

***Figure 10-11** Using the **ROTATE** mode to rotate and make copies of the selected objects*

5. Make another copy of the object as shown in Figure 10-12(a). Select the object, and then select the grip at point (P0). Access the Rotate mode and **Copy** option as described earlier. Select the **Reference** option from the shortcut menu or enter R (Reference) at the following prompt:

 **ROTATE (multiple) **
 Specify rotation angle or [Base point/Copy/Undo/Reference/eXit]: R Enter
 Specify reference angle <0>: Select the grip at (P1).
 Specify second point: Select the grip at (P2).
 Specify new angle or [Base point/Copy/Undo/Reference/eXit]: 45 Enter

In response to the **Specify reference angle <0>:** prompt, select the grips at points (P1) and (P2) to define the reference angle. When you enter the new angle, AutoCAD LT will rotate and insert a copy at the specified angle [Figure 10-12(c)]. For example, if the new angle is 45 degrees, the selected objects will be rotated about the base point (P0) so that the line P1P2 makes a 45-degree angle with respect to the positive X axis.

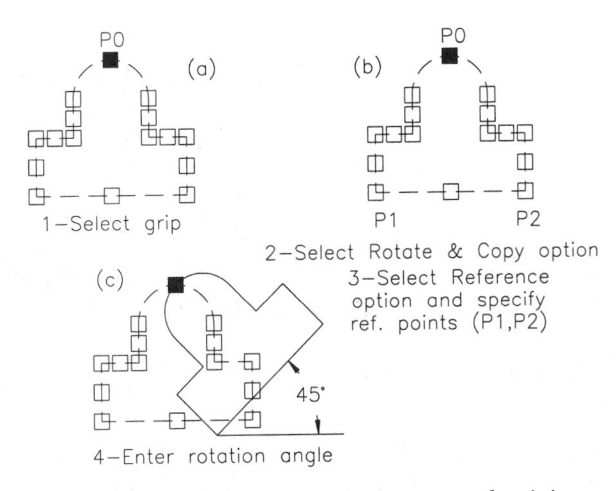

***Figure 10-12** Using the **ROTATE** mode to rotate by giving a reference angle*

SCALING OBJECTS WITH GRIPS
(Scale Mode)

The **Scale** mode allows you to scale objects with respect to the base point without changing their orientation. The options of Scale mode can be used to redefine the base point, specify a reference length, or make multiple copies that are scaled with respect to the specified base point. You can access the **Scale** mode by selecting the grip and then selecting **Scale** from the shortcut menu, or entering **SCALE** or **SC** on the keyboard, or giving a null response three times by pressing the SPACEBAR or the ENTER key. The following example illustrates use of the **Scale** mode.

1. Use the **PLINE** command to draw the shape as shown in Figure 10-13(a). When you select the objects, grips will be displayed at the definition points, and the object will be highlighted.

2. Select the grip located at the lower left corner as the base grip, then invoke the **SCALE** mode. AutoCAD LT will display the following prompt in the Command prompt area.

   SCALE
   Specify scale factor or [Base point/Copy/Undo/Reference/eXit]:

3. At this prompt enter the scale factor or move the cursor and select a point to specify a new size. AutoCAD LT will scale the selected objects by the specified scale factor [Figure 10-13(b)]. If the scale factor is less than 1 (<1), the objects will be scaled down by the specified factor. If the scale factor is greater than 1, the objects will be scaled up.

4. Make a copy of the original drawing as shown in Figure 10-13(c). Select the objects, and then select the grip located at the lower left corner of the object. Invoke the Scale mode. At the following prompt, enter C (Copy), and then enter B for base point:

   **SCALE (multiple) **
   Specify scale factor or [Base point/Copy/Undo/Reference/eXit]: **B** Enter

5. At the **Specify base point prompt:**, select the point (P0) as the new base point, and then enter R at the following prompt.

   **SCALE (multiple) **
   Specify scale factor or [Base point/Copy/Undo/Reference/eXit]: **R** Enter
   Specify reference length <1.000>: *Select grips at (P1) and (P2).*

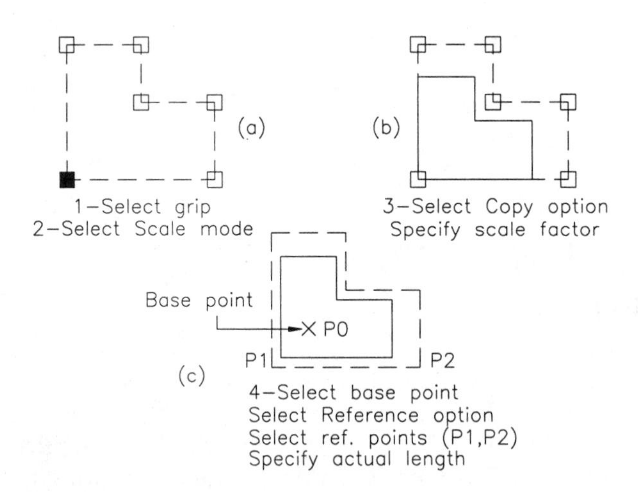

Figure 10-13 *Using the SCALE mode to scale and make copies of selected objects*

After specifying the reference length at the **Specify new length or [Base point/Copy/Reference/eXit]** prompt, enter the actual length of the line. AutoCAD LT will scale the objects so that the length of the bottom edge is equal to the specified value [Figure 10-13(c)].

MIRRORING OBJECTS WITH GRIPS (Mirror Mode)

The **Mirror** mode allows you to mirror the objects across the mirror axis without changing the size of the objects. The mirror axis is defined by specifying two points. The first point is the base point, and the second point is the point that you select when AutoCAD LT prompts for the second point. The options of the Mirror mode can be used to redefine the base point and make a mirror copy of the objects. You can access the Mirror mode by selecting a grip and then choosing **Mirror** from the shortcut menu, or by entering **MIRROR** or **MI** at the keyboard, or giving a null response four times by pressing the SPACEBAR or the ENTER key four times. The following is the example for the Mirror mode.

1. Use the **PLINE** command to draw the shape as shown in Figure 10-14(a). When you select the object, grips will be displayed at the definition points and the object will be highlighted.

2. Select the grip located at the lower right corner (P1), and then invoke the Mirror mode. The following prompt is displayed:

   MIRROR
   Specify second point or [Base point/Copy/Undo/eXit]:

3. At this prompt, enter the second point (P2). AutoCAD LT will mirror the selected objects with line P1P2 as the mirror axis, Figure 10-14(b).

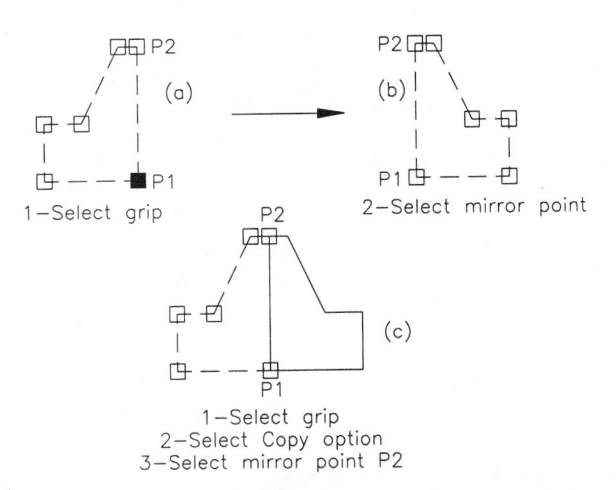

Figure 10-14 *Using the Mirror mode to create a mirror image of selected objects*

4. Make a copy of the original figure as shown in Figure 10-14(c). Select the object, and then select the grip located at the lower right corner (P1) of the object. Invoke the Mirror mode and then select the **Copy** option to make mirror image while retaining the original object. Alternatively, you can also hold down the SHIFT key and make several mirror copies by specifying the second point.

5. Select point (P2) in response to the prompt **Specify second point or [Base point/Copy/ Undo/eXit]:**. AutoCAD LT will create a mirror image, and the original object will be retained.

Note

*You can use some editing commands, such as **ERASE, MOVE, ROTATE, SCALE, MIRROR,** and **COPY,** on an object with warm grips (Selected object), provided the system variable **PICKFIRST** is set to 1 (On).*

You cannot select an object once you select a grip (when the grip is hot).

If you want to remove an object from the selection set displaying grips, press the SHIFT key and then select the particular object. This object, which is removed from the selection set, will now not be highlighted although the grips are still displayed (Cold grips).

CHANGING PROPERTIES USING GRIPS

You can also use the grips to change the properties of a single or multiple object. To change the properties of an object, select the object to display the grips and then right-click to display the shortcut menu. In the shortcut menu, select the **Properties** option to display the **Object Properties window**. If you select a circle, AutoCAD LT will display Circle in the Selection drop-down list in the **Object Properties window**. Similarly, if you select text, Text is displayed in the drop-down list. If you select several objects, AutoCAD LT will display all the objects in the selection drop-down list of the **Object Properties window**. You can use this window to change the properties (color, layer, linetype, linetypes scale, lineweight, plot style and thickness and so on) of the gripped objects. (The **Object Properties window** has been discussed in detail in Chapter 4.) Similarly, you can use grips to edit hatch objects.

LOADING HYPERLINKS

You can also use the grips to open a file associated with the hyperlink. For example, the hyperlink could start a word processor, or activate the Web browser and load a Web page that is embedded in the selected object. If you want to launch the Web browser that provides hyperlinks to other Web pages, select the URL embedded object and then right-click to display the shortcut menu. In the shortcut menu, select the **Hyperlink** option and AutoCAD LT will automatically load the Web browser. When you move the cursor over or near the object that contains a hyperlink, AutoCAD LT displays the hyperlink information with the cursor.

EDITING GRIPPED OBJECTS

You can also edit the properties of the gripped objects by using the **Object Properties** toolbar (Figure 10-15). The gripped objects are created when you select objects without invoking a

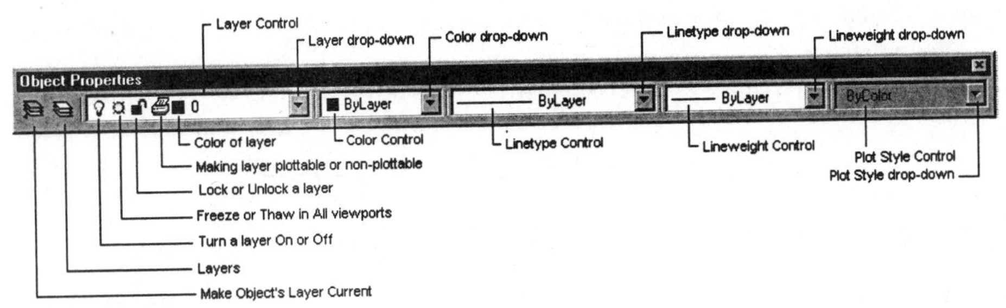

*Figure 10-15 Using the **Object Properties** toolbar to change properties of the gripped objects*

command. The gripped objects are highlighted and will display grips (rectangular boxes) at their grip points. For example, if you want to change the color of the gripped objects, select the Color drop-down list in the **Object Properties** toolbar and then select a color. The color of the gripped objects will change to the selected color. Similarly, if you want to change the layer, lineweight or linetype of the gripped objects, select the linetype, lineweight or layer from the corresponding drop-down lists. If the gripped objects have been drawn in different layers or have different colors, linetypes or lineweights, the Layer Control, Color Control, Linetype Control, Lineweight Control boxes will appear blank. You can also change the Plot Style of the selected objects.

GRIP SYSTEM VARIABLES

System variable	Default	Setting	Function
GRIPS	1	1=On, 0=Off	Enables or disables Grip mode
GRIPBLOCK	0	1=On, 0=Off	Controls the display of grips in a block
GRIPCOLOR	5	1-255	Specifies the color of unselected grips
GRIPHOT	1	1-255	Specifies the color of selected grips
GRIPSIZE	3	1-255	Specifies the size of the grip box in pixels

Self-Evaluation Test

Answer the following questions, and then compare your answers to the correct answers given at the end of this chapter.

1. What editing functions can you perform with grips?________________________
____.

2. A grip is a small square that is displayed on an object at its __________ points.
3. The number of grips depends on the selected object. (T/F)
4. A line has ________ grip points and a polyline has __________ .
5. The grip settings can be adjusted with the **GRIPS** command. (T/F)
6. The **Options** dialog box is activated by entering __________ at the Command prompt.
7. You can enable grips within a block by setting the system variable __________ to 1 (On).
8. The color of the unselected grips can also be changed by using the __________system variable.
9. You can access the Mirror mode by selecting a grip and then entering __________ or __________ from the keyboard or giving a null response four times by pressing the SPACEBAR four times.

Review Questions

1. If you select a grip of an object, the grip becomes a cold grip. (T/F)
2. To cancel the grip, press the SHIFT and C keys. (T/F)
3. The **GRIPSIZE** is defined in pixels, and its value can range from __________ to __________ pixels.
4. The size of the grip can also be adjusted by using the system variable __________.
5. When you select a grip for editing, you are automatically in the __________ mode.
6. The __________ mode lets you move the selected objects to a new location.
7. What happens if you hold down the SHIFT key and then enter the first displacement point? ____.
8. The Rotate mode allows you to rotate objects around the base point without changing their size. (T/F)

9. The __________ mode allows you to scale the objects with respect to the base point without changing their orientation.

10. The Mirror mode allows you to mirror the objects across the __________ without changing the size of the objects.

11. The **Go to URL** option allows you to ______________________________.

12. The **Properties** option allows you to ______________________________.

Exercises

Exercise 1 *Mechanical*

1. Use the **PLINE** or **LINE** command to make the drawing as shown in Figure 10-16(a).
2. Dimension the drawing, and place the dimensions as shown in Figure 10-16(a).
3. Use the grips to correct the dimensions and place them as shown in Figure 10-16(b).

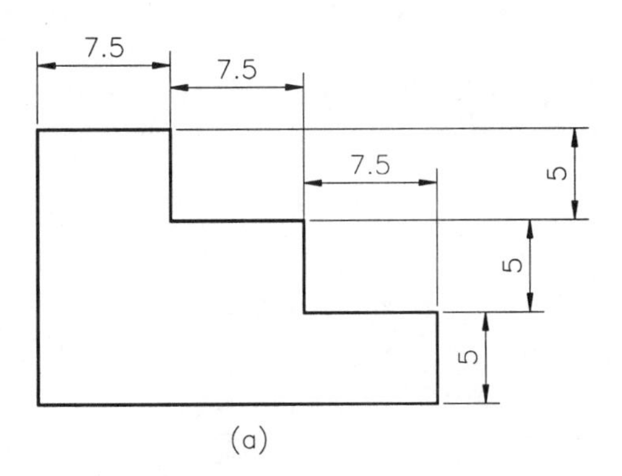

Figure 10-16(a) *Drawing for Exercise 1*

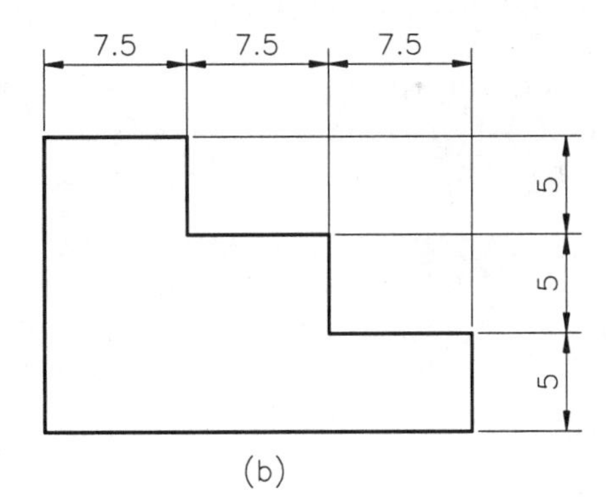

Figure 10-16(b) *Drawing for Exercise 1*

Exercise 2 *Genral*

1. Use the **PLINE** command to draw the shape as shown in Figure 10-17(a).
2. Use grips (Stretch mode) to get the shape as shown in Figure 10-17(b).
3. Use the Rotate and Stretch modes to get the copies as shown in Figure 10-17(c)

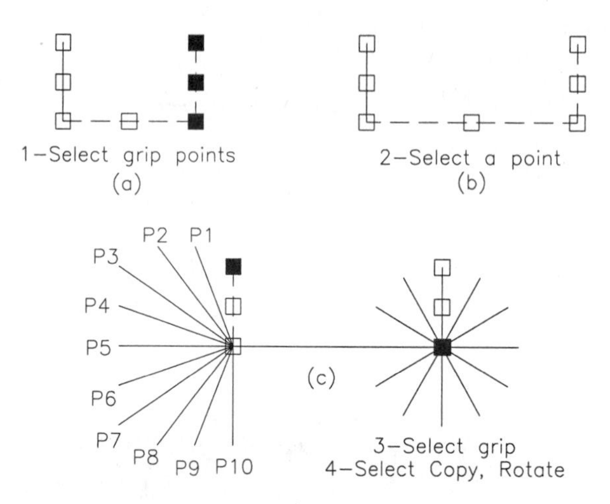

Figure 10-17 Drawing for Exercise 2

Exercise 3 *Genral*

1. Use the **PLINE** or **LINE** command to make the drawing as shown in Figures 10-18(a) and 10-19(a).
2. Use grips to get the shapes as shown in Figures 10-18(b) and 10-19(b). (Do not use any AutoCAD LT command except GRIPS. Points (P1), (P2), and (P3) are midpoints. Also, X=5 units)

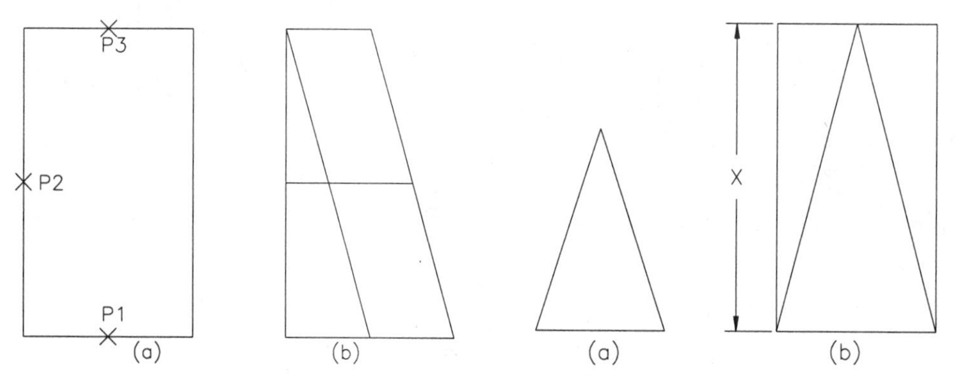

Figure 10-18 Drawing for Exercise 3

Figure 10-19 Drawing for Exercise 3

Exercise 4 Mechanical

1. Make the drawing as shown in Figure 10-20(a). Assume the dimensions where necessary.
2. Dimension the drawing as shown in Figure 10-20(b).
3. Edit the dimensions as shown in Figure 10-20(b), using GRIPS.

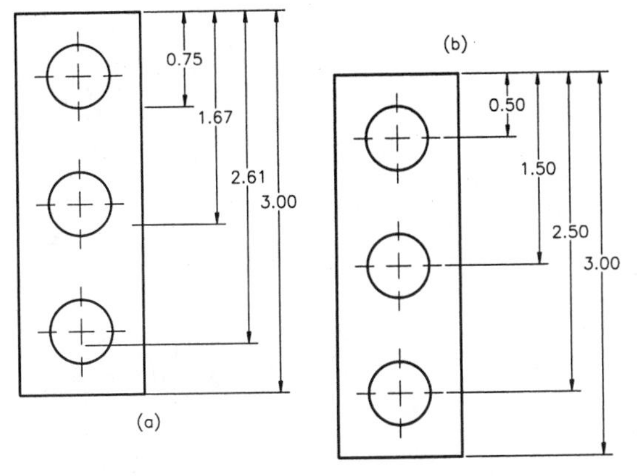

***Figure 10-20** Drawing for Exercise 4, showing dimensions (a) before and (b) after editing*

Answers to Self-Evaluation Test.
1 -With GRIPS you can stretch, move, rotate, scale, and mirror objects, **2** -definition, **3** -T, **4** -three, two, **5** -F, (**Options** dialog box), **6** -**OPTIONS**, **7** -**GRIPBLOCK**, **8** -**GRIPCOLOR**, **9** -MIRROR, MI.

Chapter 11

Hatching

Learning Objectives

After completing this chapter, you will be able to:

- *Use the **BHATCH** command to hatch an area.*
- *Use boundary hatch options and predefined, as well as user-defined, hatch patterns.*
- *Specify pattern properties.*
- *Preview and apply hatching.*
- *Use advanced hatching options.*
- *Edit associative hatch and hatch boundary.*
- *Hatch inserted blocks.*
- *Align hatch lines in adjacent hatch areas.*
- *Hatch by using the **HATCH** command at the Command prompt.*

HATCHING

In many drawings (such as sections of solids or sections of objects), the area must be filled with some pattern. Different filling patterns make it possible to distinguish between different parts or components of an object. Also, the material the object is made of can be indicated by the filling pattern. Filling the objects with a pattern is known as hatching (Figure 11-1). This hatching process can be accomplished by using the **BHATCH** or **HATCH** command.

Before using the **BHATCH** and **HATCH** commands, you need to understand some terms that are used when hatching. The following subsection describes some of the terms.

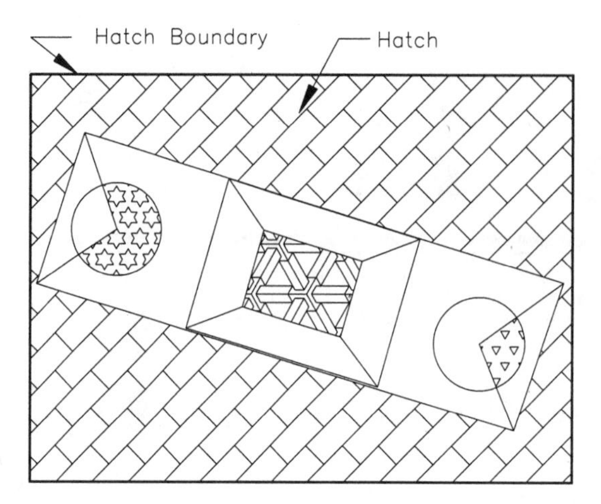

Figure 11-1 Illustration of hatching

Hatch Patterns

AutoCAD LT 2000 supports a variety of hatch patterns (Figure 11-2).

Figure 11-2 Some hatch patterns

Every hatch pattern is comprised of one or more hatch lines. These lines are placed at specified angles and spacing. You can change the angle and the spacing between the hatch lines. These lines may be broken into dots and dashes, or may be continuous, as required. The hatch pattern is trimmed or repeated, as required, to fill exactly the specified area. The lines comprising the hatch are drawn in the current drawing plane. The basic mechanism behind hatching is that the line objects of the pattern you have specified are generated and incorporated in the desired area in the drawing. Although a hatch can contain many lines, AutoCAD LT normally groups them together into an internally generated object and treats them as such for all practical purposes. For example, if you want to perform an editing operation, such as erasing the hatch, all you need to do is select any point on the hatch. The object created for the hatch lines gets deleted automatically when all references to it are deleted (only after the drawing is saved and reopened). If you want to break a pattern into individual lines to edit an individual line, you can use AutoCAD's **EXPLODE** command.

Hatch Boundary

Hatching can be used on parts of a drawing enclosed by a boundary. This boundary may be lines, circles, arcs, plines, 3D faces, or other objects, and at least part of each bounding object must be displayed within the active viewport. The **BHATCH** command automatically defines the boundary, whereas in the case of the **HATCH** command you have to define the boundary by selecting the objects that form the boundary of the hatch area.

THE BHATCH COMMAND

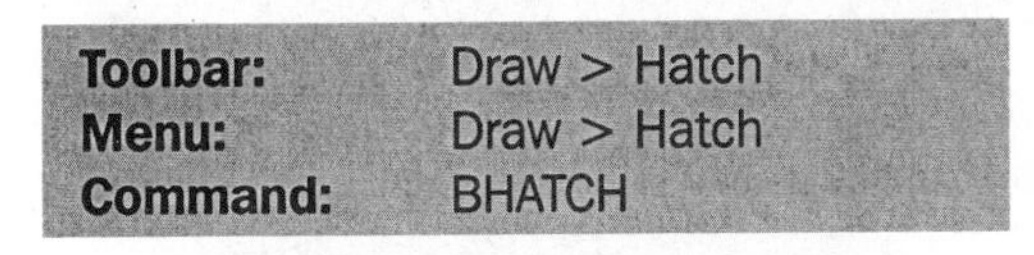

Toolbar: Draw > Hatch
Menu: Draw > Hatch
Command: BHATCH

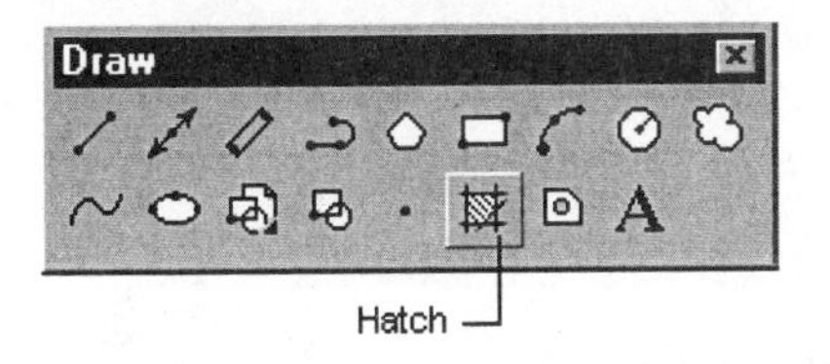

***Figure 11-3** Choosing **Hatch** from the **Draw** toolbar*

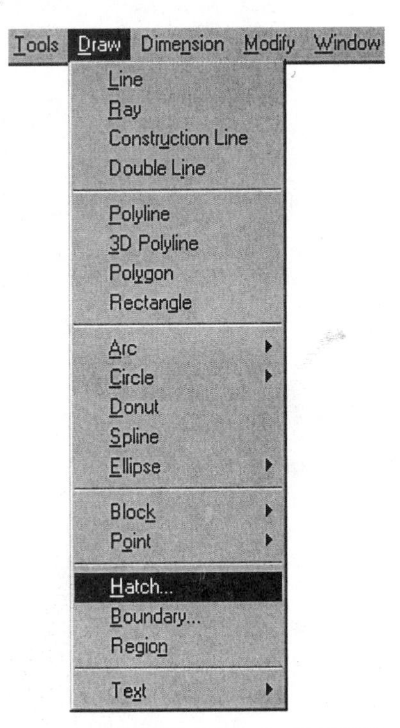

***Figure 11-4** Invoking the **HATCH** command from the **Draw** menu*

The **BHATCH** (boundary hatch) command allows you to hatch a region enclosed within a boundary (closed area) by selecting a point inside the boundary or by selecting the objects to be hatched. This command automatically designates a boundary and ignores any other objects (whole or partial) that may not be a part of this boundary. One of the advantages of this command is that you do not have to select each object comprising the boundary of the area you want to hatch as in the case of the **HATCH** command (discussed later). This is because this command defines a boundary by creating a polyline from all the objects comprising the boundary. By default, this polyline gets deleted; however, if you want to retain it, you can specify that. This command also allows you to preview the hatch before actually applying it.

When you invoke the **BHATCH** command, the **Boundary Hatch** dialog box is displayed through which you can perform the hatching operation (Figure 11-5).

Command: **BHATCH** Enter

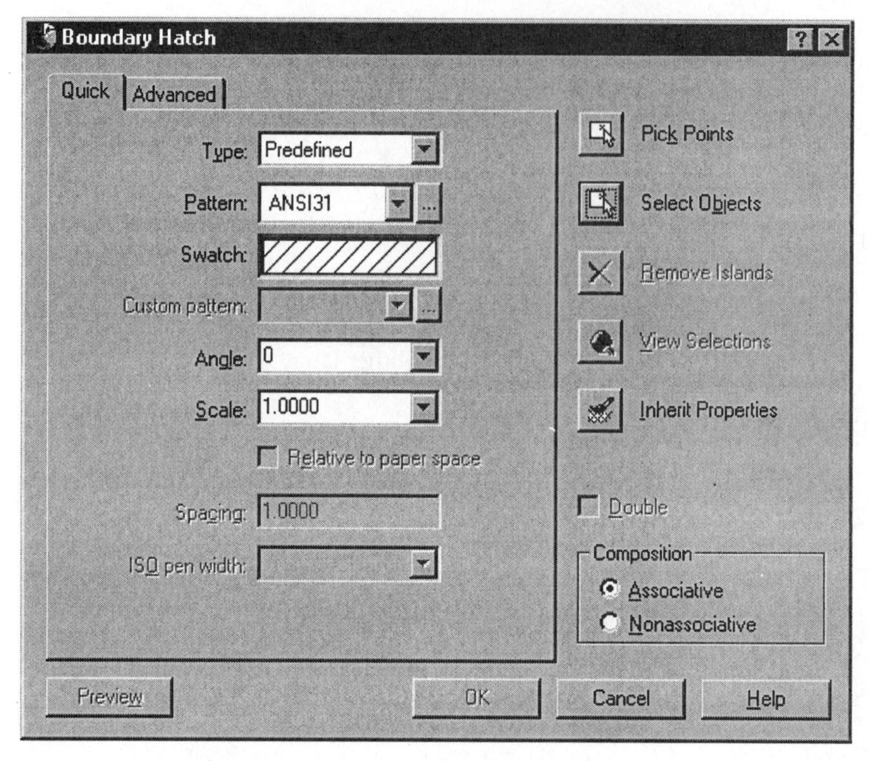

Figure 11-5 ***Boundary Hatch*** *dialog box*

Example 1

In this example, you will hatch a circle using the default hatch settings. Later in the chapter you will learn how to change the settings to get a desired hatch pattern.

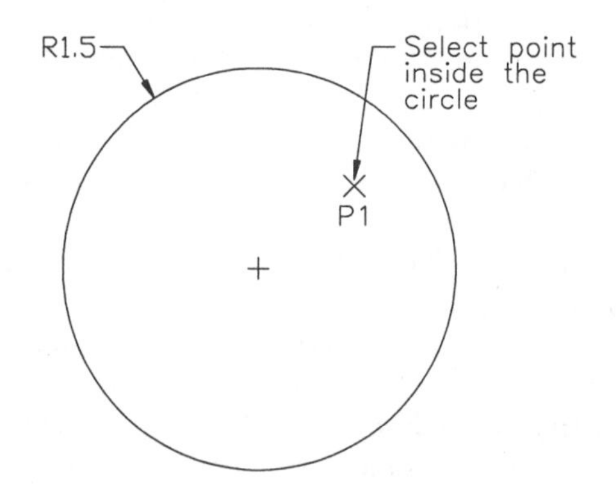

Figure 11-6 Using the ***BHATCH*** *command to hatch an object*

1. Invoke the **Boundary Hatch** dialog box (Figure 11-5) by entering **BHATCH** at the Command prompt.

2. Choose the **Pick Points** button from the **Boundary Hatch** dialog box (in the **Quick** tab). The dialog box temporarily closes.

3. Select a point inside the circle (P1) (Figure 11-6). After choosing an internal point, you can preview the hatching by choosing the **Preview** button. You can also right-click to get the shortcut menu, where you can preview the hatching by choosing **Preview**. You can also change the Island detection modes and undo selection from the shortcut menu.

4. Press ENTER to complete the point specification, and the dialog box reopens.

5. Choose the **OK** button to apply the hatch to the selected object.

BOUNDARY HATCH OPTIONS

The **Boundary Hatch** dialog box has several options that let you control various aspects of hatching, such as pattern type, scale, angle, boundary parameters, and associativity. The following is a description of these options. The options have been grouped by the tab name in which they are found in the **Boundary Hatch** dialog box. The **Boundary Hatch** dialog box has two tabs. The description of the tabs are as follows:

Quick Tab

The **Quick** Tab (**Boundary Hatch** dialog box) defines the appearance of the hatching patterns that can be applied (Figure 11-5). For quick hatching of a drawing object this dialog box can be used easily.

Type. This displays the type of the pattern that can be used for hatching drawing objects. This area lets you choose the type of hatch pattern you want. You can choose a predefined hatch pattern that comes with AutoCAD LT. The default hatch pattern is the predefined pattern named ANSI31. You can also choose the custom or a user-defined hatch pattern by selecting the pattern type from the drop-down list.

Pattern. The **Pattern** drop-down list display all available predefined hatch patterns. The selected pattern is stored in the **HPNAME** system variable. ANSI31 is the default pattern in the **HPNAME** system variable. The pattern option is available only if the selected pattern is predefined. See Figures 11-7(a), 11-7(b), and 11-8.

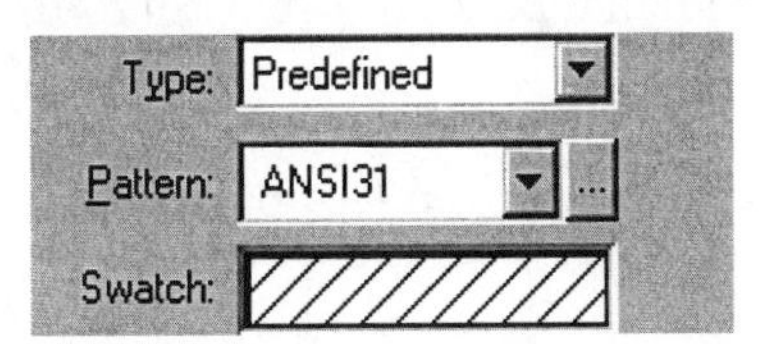

***Figure 11-7(a)** Selecting pattern type from the **Boundary Hatch** dialog box*

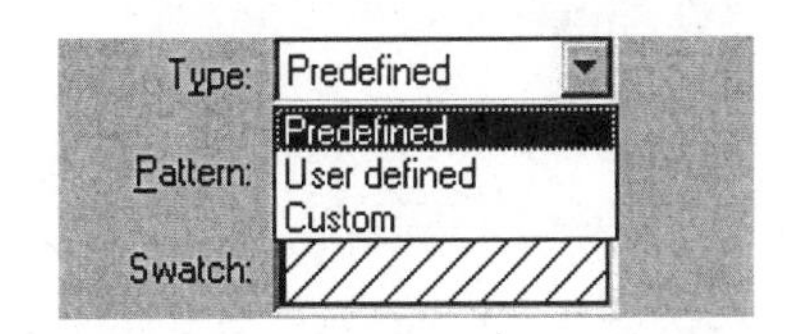

***Figure 11-7(b)** Available hatch options*

When you choose the [...] button, the **Hatch Pattern Palette** dialog box (Figure 11-9) is displayed that shows the images and names of the available hatch patterns. There are four tabs in the dialog box.

ANSI. This tab displays all the ANSI defined hatch patterns shipped with AutoCAD LT. ANSI31 pattern is default.

ISO This tab displays all the ISO defined hatch patterns shipped with AutoCAD .

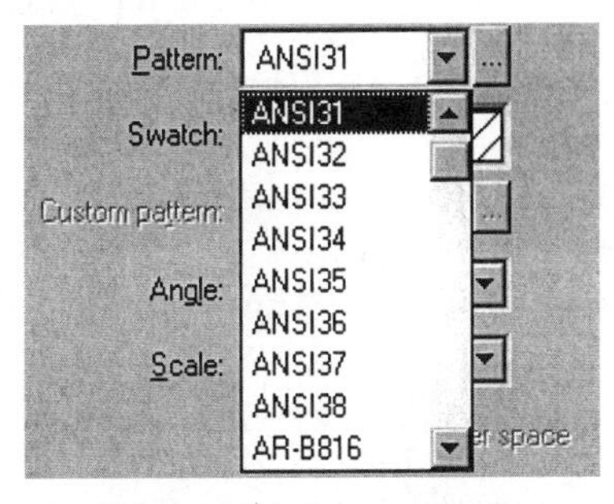

***Figure 11-8** Pattern list box*

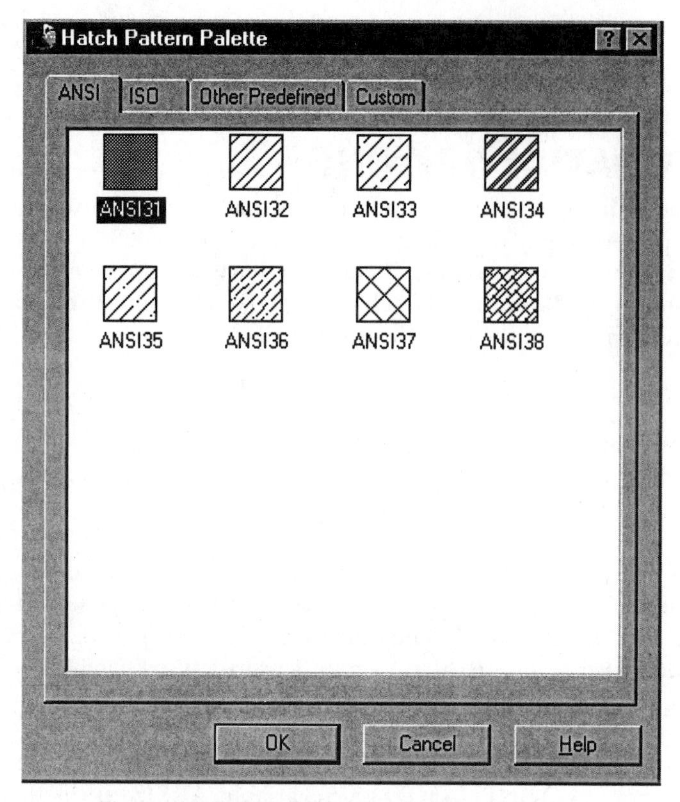

*Figure 11-9 **Hatch Pattern Palette** dialog box*

Other Predefined. This displays all the predefined hatch patterns other than ANSI or ISO patterns.

Custom. This displays all custom patterns described in any custom PAT files that is added in the search path of AutoCAD LT 2000.

Swatch. This box displays the preview of the selected pattern.

Custom Pattern. This drop-down list displays all the available custom hatch patterns, if any.The six most recent custom hatch patterns used appears in the list. The selected pattern is stored in **HPNAME** system variable. If you choose the Pattern [...] button, the **Hatch Pattern Palette** dialog box (**Custom** tab) appears. This tab displays all the available custom patterns, if any.

Angle. This Pattern Properties option lets you rotate the hatch pattern with respect to the X axis of the current UCS. You can enter the angle of rotation value of your choice in the edit box next to the **Angle**: edit-box. This value is stored in the **HPANG** system variable. The angle of hatch lines of a hatch pattern is governed by the values specified in the hatch definition.For example, in the ANSI31 hatch pattern definition, the specified angle of hatch lines is 45 degrees. If you select an angle of 0, the angle of hatch lines will be at 45. If you enter an angle of 45 degrees, the angle of the hatch lines will be at 90 degrees (Figure 11-10).

Scale. This Pattern Properties option lets you expand or contract the pattern, that is, scale the hatch pattern. You can enter the scale factor of your choice in the edit box next to the **Scale:** option. This value is stored in the **HPSCALE** system variable. Entering a value of 1 does not mean that the distance between the hatch lines is 1 unit. The distance between the hatch lines and other parameters of a hatch pattern is governed by the values specified in the hatch definition. For example, in the ANSI31 hatch pattern definition, the specified distance between the hatch lines is 0.125. If you select a scale factor of 1, the distance between the lines will be 0.125. If you enter a scale factor of 0.5, the distance between the hatch lines will be 0.5 x 0.125 = 0.0625.

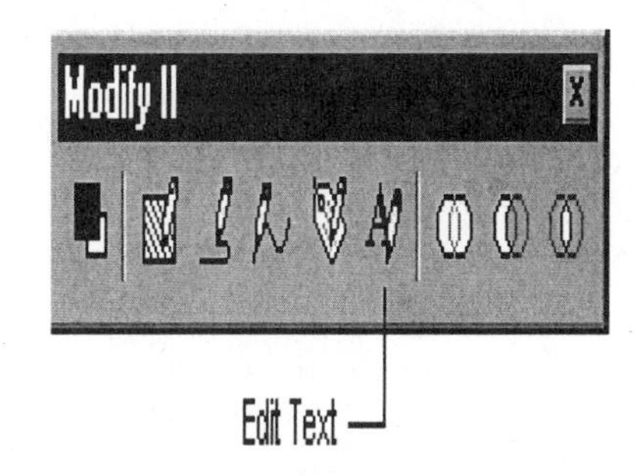

Figure 11-10 *Selecting pattern properties (Angle and Scale) from the **Boundary Hatch** dialog box*

Relative to Paper Space. If this check box is selected on then AutoCAD LT 2000 will automatically scale the hatched pattern to the paper space units. This option can be used to display the hatch pattern at a scale that is appropriate to your layout. This option is available only in a layout.

Spacing. This displays the spacing between the lines in a predefined hatch pattern. The spacing is stored in the **HPSPACE** system variable. The default value of spacing is 1.0000.

ISO Pen Width. The **ISO pen width** option is only valid for ISO hatch patterns and linetypes. When you select any ISO pattern from the **Pattern** drop-down list, this option is enabled. The pen width can be selected by choosing the desired value from the ISO pen width drop-down list (Figure 11-11). The value selected specifies the ISO-related pattern scaling.

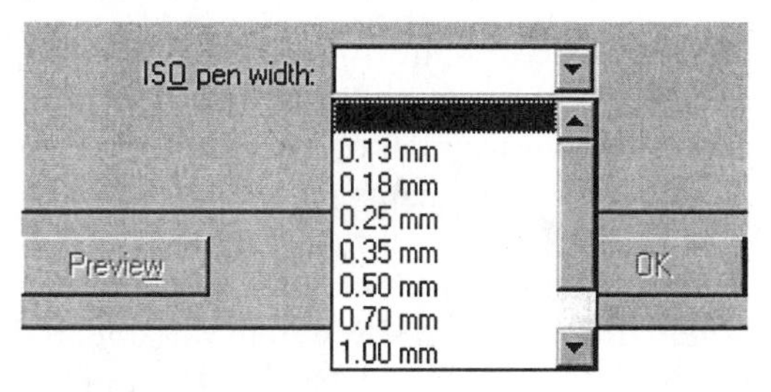

Figure 11-11 *Selecting ISO Pen Width*

User-Defined Hatch Pattern

If you want to define a simple pattern, you can select **User-Defined** pattern type in the **Type** drop-down list. You will notice that the **Angle**: and **Spacing**: edit boxes, and the **Double** check box are enabled in the dialog box. See Figure 11-12.

Angle. You can specify an angle by entering a value in the **Angle:** edit box. This value is considered with respect to the X axis of the current UCS. This value is stored in the **HPANG** system variable.

Spacing. This edit box lets you specify the space between the hatch lines. The entered spacing value is stored in the **HPSPACE** system variable.

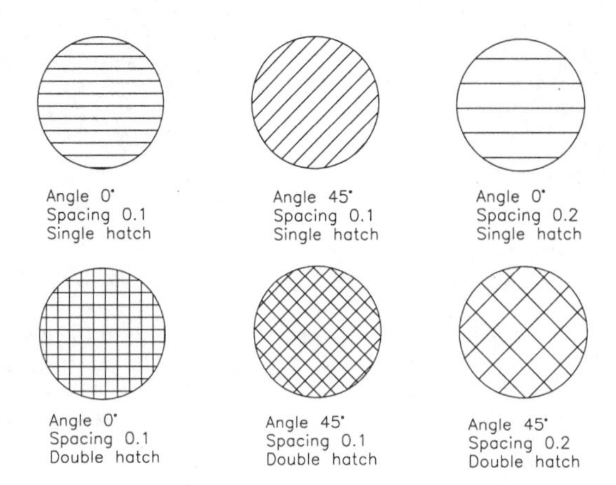

Figure 11-12 *Specifying angle and spacing for User-defined hatch*

Double. This option makes AutoCAD LT draw a second set of lines at right angles to the original lines. If the **Double** check box is selected, AutoCAD LT sets the **HPDOUBLE** system variable to 1.

Custom Hatch Pattern

When you select the **Custom** option in the **Type** drop-down list, you will notice that the **Custom Pattern**, **Scale** and **Angle** edit boxes are enabled in the dialog box.

Custom Pattern. In AutoCAD LT, hatch patterns are normally stored in the file named **ACLT.PAT**. You can select a hatch pattern from an individual file by choosing the [...] button to display the custom tab of the **Hatch Pattern Palette** dialog box type and then entering the name of the stored hatch pattern in the **Custom pattern:** edit box. The Custom pattern type can also be selected from the **Custom** drop-down list . You can do this if you know the name of the hatch pattern. For example, you can enter BRASS or STEEL as a hatch pattern. The name is held in the **HPNAME** system variable. If AutoCAD LT does not locate the entered pattern in the **ACLT.PAT** file, it searches for it in a file with the same name as the pattern.

Exercise 1 *Mechanical*

In this exercise, you will hatch the given drawing using the hatch pattern named STEEL. Set the hatch scale and the angle to match the drawing shown in Figure 11-13.

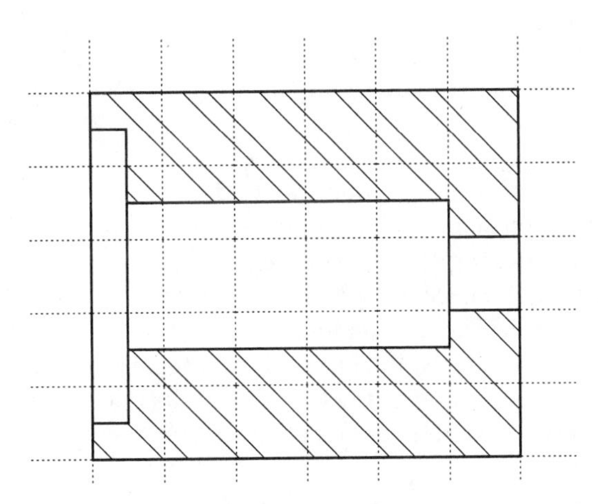

Figure 11-13 Drawing for Exercise 1

Selecting Hatch Boundary

The boundary selection options available in the **Boundary Hatch** dialog box allow you to define the hatch boundary by selecting a point inside the area or by selecting the objects. The other options are for removing islands, viewing the selection, and setting advanced options.

Pick Points. This option makes AutoCAD LT automatically construct a boundary. When you select a point inside an object, the object and every other object inside it is selected as the boundary. To select this option, choose the **Pick Points** button in the **Boundary Hatch** dialog box. The following prompts appear:

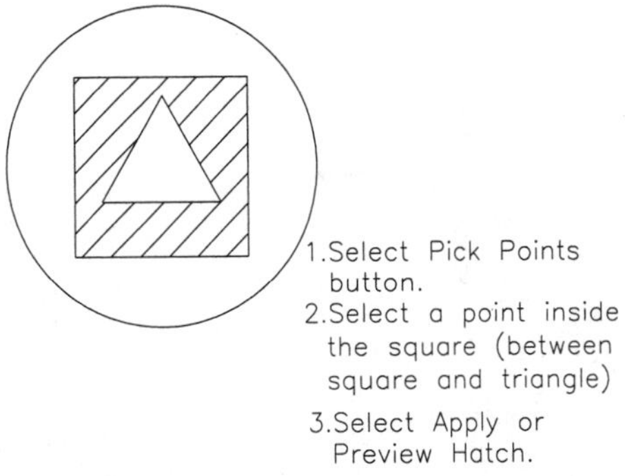

1.Select Pick Points button.
2.Select a point inside the square (between square and triangle)
3.Select Apply or Preview Hatch.

***Figure 11-14** Defining multiple hatch boundaries by selecting a point*

Select internal point: *Select a point inside the object to hatch.*
Selecting everything ...
Selecting everything visible...
Analyzing the selected data...
Analyzing internal islands...
Select internal point: *Select a point or press ENTER to end selection.*

In case you want to hatch an object and leave another object contained in it alone, select a point inside the object you want to hatch. By default, in the **BHATCH** command, selecting a point creates multiple boundaries, as all the objects inside the selected object are also selected. In Figure 11-14, the **Pick Points** button has been used to hatch the square but not the triangle simply by selecting the point inside the square but outside the triangle (using the default settings of the dialog box).

Boundary Definition Error. This dialog box (Figure 11-15) is displayed if AutoCAD LT finds that the selected point is not inside the boundary or that the boundary is not closed.

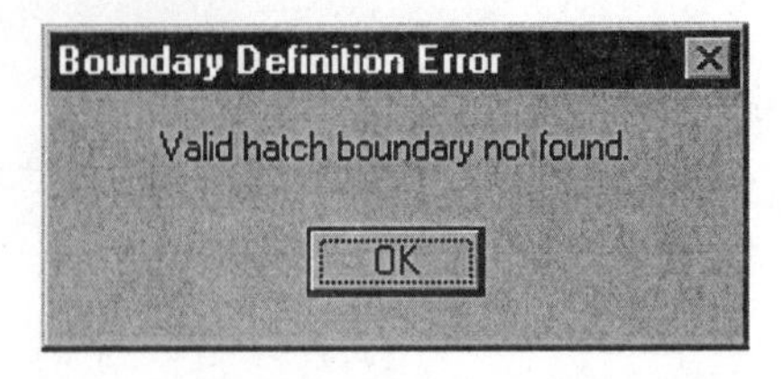

Figure 11-15 ***Boundary Definition Error*** *dialog box*

Select Objects. This option in the **Boundary Hatch** dialog box lets you select objects that form the boundary for hatching. When you select this option, AutoCAD LT will prompt you to select objects. You can select the objects individually or use other object selection methods. In Figure 11-16, the **Select Objects** button is used to select the triangle. This uses the triangle as the hatch boundary, and everything inside it (text) will also be hatched. So, at the next **Select objects**: prompt, select the text so that it is also selected and hence, not hatched.

Remove Islands. This option is used to remove any islands from the hatching area. For example, if you have a rectangle and circles as shown in Figure 11-17 and you use the **Pick Points** option to select a point inside the rectangle, AutoCAD LT will select both the circle and the rectangle. To remove the circle (island), you can use the **Remove Islands** option. When you select this option, AutoCAD LT will prompt you to select the islands to remove.

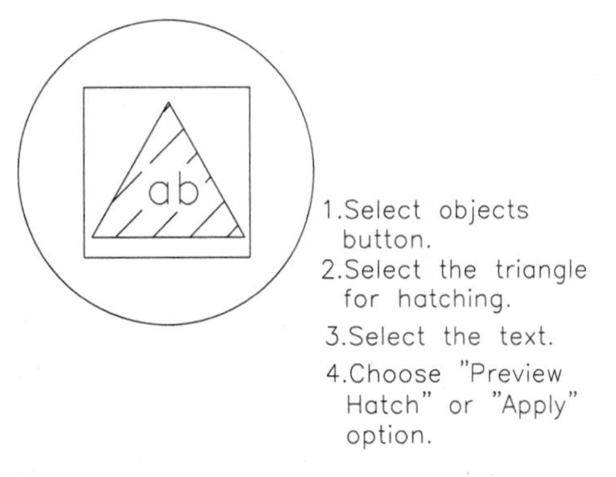

ab
1.Select objects button.
2.Select the triangle for hatching.
3.Select the text.
4.Choose "Preview Hatch" or "Apply" option.

Figure 11-16 Using the ***Select Objects*** *option to specify the hatching boundary*

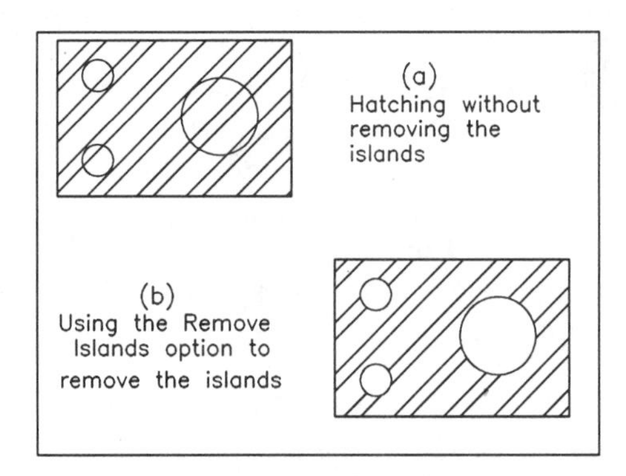

Figure 11-17 Using the ***Remove Islands*** *option to remove islands from the hatch area*

View Selections. This option lets you view the selected boundaries and closes the dialog box temporarily. This option is unavailable when you have not selected any points or boundaries.

Inherit Properties. This option hatches the specified boundaries using the hatch properties of one object. If you select the associative hatch object whose properties you want the hatch to inherit, you can right-click in the drawing area to use the shortcut menu to toggle between the **Select Objects** and **Pick Internal Point** options to specify boundaries. When you choose the **Inherit Properties** button, the following command sequence appears:

Select associative hatch object: *Select a hatch pattern.*
Inherited Properties: Name <current>, Scale <current>, Angle <current>
Select Internal point: *Select a point inside the object to be hatched.*

When you press ENTER, the **Boundary Hatch** dialog box reappears with the name of the copied hatch pattern in the **Pattern**: edit box. The inherited hatch pattern is also displayed in the **Swatch** box. The selected pattern now becomes the current hatch pattern. If you then want to adjust the properties, such as angle or scale, you can do so.

Composition

Composition controls whether the hatch is associative or nonassociative with the hatch boundary. By default, the hatch associativity is on. One of the major advantages with the associative hatch feature is that you can edit the hatch pattern or edit the geometry that is hatched. After editing, AutoCAD LT will automatically regenerate the hatch and the hatch geometry to reflect the changes. The hatch pattern can be edited by using the **HATCHEDIT** command, and the hatch geometry can be edited by using grips or some AutoCAD LT editing commands.

Preview

You can use this option after you have selected the area to be hatched to see the hatching before it is actually applied. When you choose the **Preview** button, the dialog box is cleared and the object selected for hatching is temporarily filled with the specified hatch pattern. Previewing can also be done by right-clicking in the drawing area to get the shortcut menu; choose **Preview**. Pressing ENTER redisplays the **Boundary Hatch** dialog box so that the hatching operation can be completed and modifications made, if required. This option is available only after you have selected the area to be hatched.

OK

By choosing the **OK** button, you can apply hatching in the area specified by the hatch boundary. This option works whether or not you have previewed your hatch. This button is disabled if the hatch boundary is not defined.

Exercise 2 *Mechanical*

In this exercise, you will hatch the front view section of the drawing in Figure 11-18 using the hatch pattern for BRASS. Two views, top and front, are shown. In the top view, the cutting plane line indicates how the section is cut and the front view shows the full section view of the object. The section lines must be drawn only where the material is actually cut.

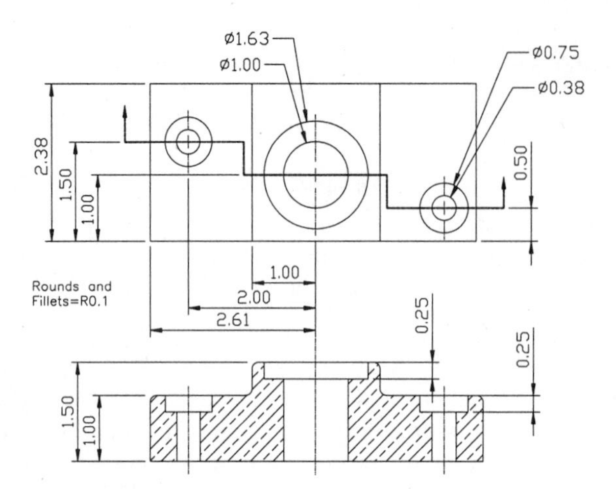

Figure 11-18 Drawing for Exercise 2

Advanced Tab (Boundary Hatch Dialog Box)

Defining a boundary by specifying an internal point is quite simple in the case of small and less complicated drawings. It may take more time in the case of large, complicated drawings because AutoCAD LT examines everything that is visible in the current viewport. Thus, the larger and more complicated the drawing, the more time it takes to locate the boundary edges. In such cases, you can improve the hatching speed by setting parameters in the **Advanced** Tab of the **Boundary Hatch** dialog box (Figure 11-19).

Island Detection Style

The island detection style can be specified by choosing style from **Island detection style** area (Figure 11-20). The styles are displayed with the effects in the image box. There are three styles from which you can choose: **Normal**, **Outer**, and **Ignore**. These styles are discussed next (Figure 11-21).

Normal. This style hatches inward starting at the outermost boundary. If it encounters an internal boundary, it turns off the hatching. An internal boundary causes the hatching to turn off until another boundary is encountered. In this manner, alternate areas of the selected object are hatched, starting with the outermost area. Thus, areas separated from the outside of the hatched area by an odd number of boundaries are hatched, while those separated by an even number of boundaries are not. AutoCAD LT 2000 stores Normal style code by adding**, N** to the pattern name under the **HPNAME** system variable.

Outer. This particular option also lets you hatch inward from an outermost boundary, but the hatching is turned off if an internal boundary is encountered. Unlike the previous case, it does not turn the hatching on again. The hatching process in this case starts from both ends of each hatch line; only the outermost level of the structure is hatched, hence the name **Outer**. AutoCAD LT stores Outer style code by adding, **O** to the pattern name under the **HPNAME** system variable.

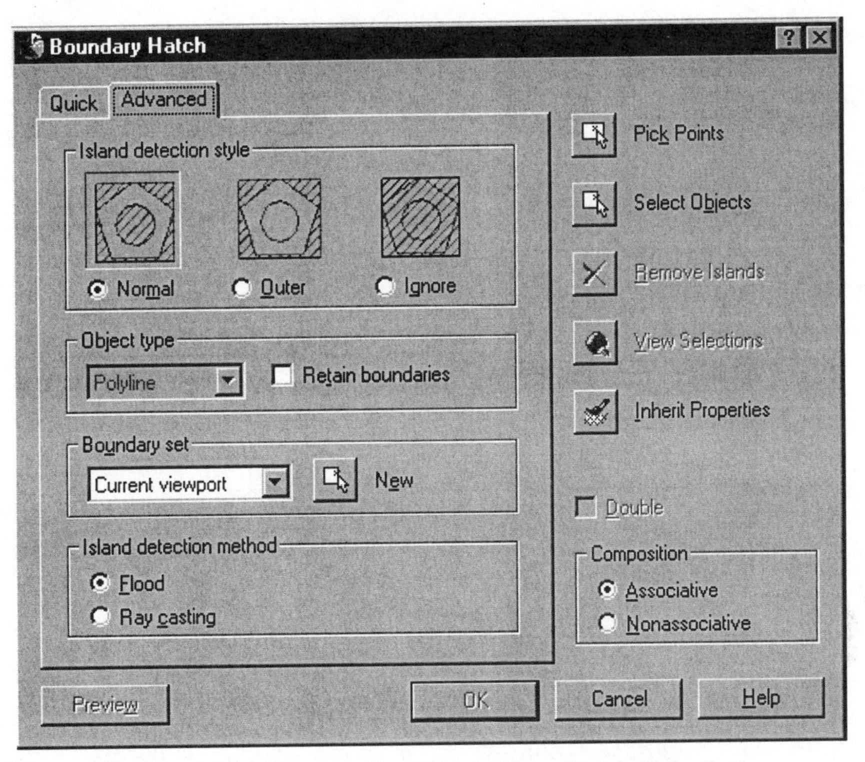

***Figure 11-19 Advanced** Tab (**Boundary Hatch** dialog box)*

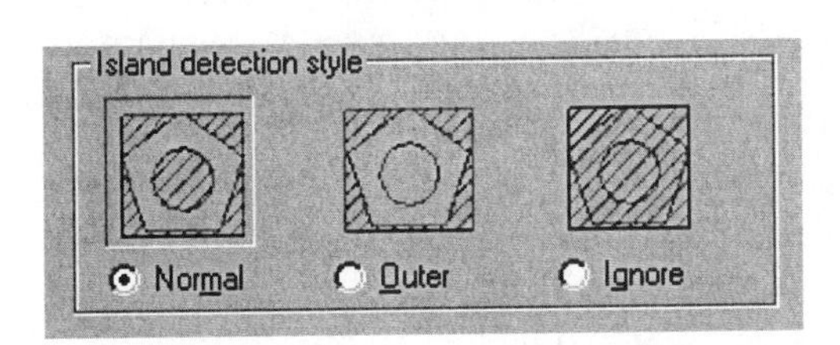

***Figure 11-20** Island detection styles*

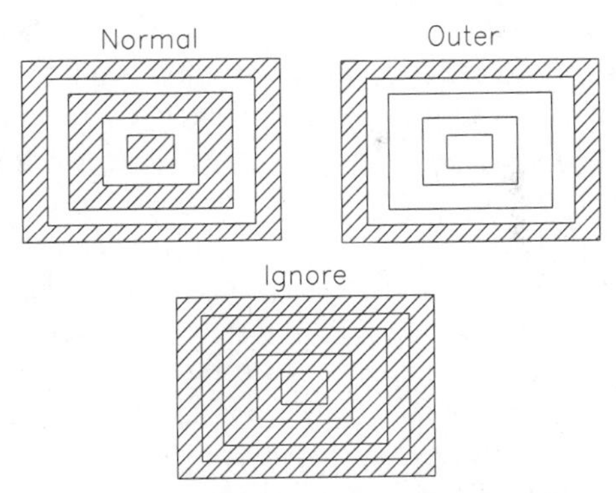

***Figure 11-21** Using hatching styles*

Ignore. In this option, all areas bounded by the outermost boundary are hatched, ignoring any hatch boundaries that are within the outer boundary. AutoCAD LT stores Ignore style code by adding**, I** to the pattern name under the **HPNAME** system variable.

The **Normal**, **Outer**, and **Ignore** options are also available in the shortcut menu when you select an internal point in the drawing object and right-click in the drawing area.

Object Type

Object type specifies whether to keep the boundaries as objects or not, and AutoCAD LT applies object types to those objects.

Retain Boundaries. It can be used to retain the defined boundary. If the hatching was successful and you want to leave the boundary as a polyline or region to use it again, you can select the **Retain Boundaries** check box.

Object Type. The **Object type** option controls the type of object AutoCAD LT will create when you define a boundary. It has two options: Polyline and Region. Regions are two-dimensional areas which can be created from closed shapes or loops. You can choose the desired option from the drop-down list. **Object type** option is available only when **Retain Boundaries** is selected.

Boundary Set

You can specify the boundary set by using the **Advanced** Tab of the **Boundary Hatch** dialog box. The boundary set comprises the objects that the **BHATCH** command uses when constructing the boundary. The default boundary set comprises everything that is visible in the current viewport. A boundary can be produced faster by specifying a boundary set because, in this case, AutoCAD LT does not have to examine everything on screen.

New. This option is used to create a new boundary set. All the dialog boxes are cleared from the screen to allow selection of objects to be included in the new boundary set. While constructing the boundary set, AutoCAD LT uses only those objects that you select and that are hatchable. If a boundary set already exists, it is eliminated for the new one. If you do not select any hatchable objects, no new boundary set is created; AutoCAD LT retains the current set if there is one. Once you have formed a boundary set, you will notice that the **Existing Set** option gets added in the drop-down list. When you invoke the **BHATCH** command and you have not formed a boundary set, there is only one option, **Current Viewport** available in the drop-down list. The **Advanced** Tab of the **Boundary Hatch** dialog box reappears after you have selected the **New** button and selected objects to create a selection set. The benefit of creating a selection set is that when you select a point or select the objects to define the hatch boundary, AutoCAD LT will search only for the objects that are in the selection set. By confining the search to the objects in the selection set, the hatching process is faster. If you select an object that is not a part of the selection set, AutoCAD LT ignores objects that does not exist in the boundary set defined by using **Pick Points**. When a boundary set is formed, it becomes the default for hatching until you exit the **BHATCH** command or select the **Current Viewport** option from the drop-down list.

Note

The selection of hatch style carries meaning only if the objects to be hatched are nested (that is, one or more selected boundaries is within another boundary).

Island Detection Method

Island Detection Method specifies whether to include objects within the outermost boundary as boundary objects. These internal objects are known as **islands**.

Flood. This option includes islands as boundary objects.

Ray Casting. This option runs a line from the point specified to the nearest object and then traces the boundary in a counterclockwise direction, thus excluding islands as boundary objects.

RAY CASTING OPTIONS

You can use the **ray casting** technique to define a hatch boundary. The ray casting options are available only through the **-BHATCH** command. In ray casting, AutoCAD LT casts an imaginary ray in all directions or a particular specified direction and locates the object that is nearest to the point you have selected. After locating the object, it takes a left turn to trace the boundary. If the first ray is intercepted by an internal area or internal text, it results in a boundary definition error. To avoid boundary definition errors, or to force the ray casting in a particular direction, you can use the Ray Casting options. These options control the direction in which AutoCAD LT casts the ray to form a hatch boundary. The available options are **Nearest**, **+X**, **-X**, **+Y**, and **-Y** . The prompt sequence for the Ray Casting options is:

Command: **-BHATCH**
Current hatch Pattern: ANSI31
Specify internal point or [Properties/Select/Remove islands/Advanced]: **A**
Enter an option [Boundary set/Retain boundary/Island detection/Style/Associativity]: **I**
Do you want island detection? [Yes/No] <Y>: **N**
Enter type of ray casting [Nearest/+X/-X/+Y/-Y/Angle] <Nearest>: *Specify an option.*

Nearest. The **Nearest** option is selected by default. When this option is used, AutoCAD LT sends an imaginary line to the nearest hatchable object, then takes a turn to the left and continues the tracing process in an effort to form a boundary around the internal point. To make this process work properly, the point you select inside the boundary should be closer to the boundary than any other object that is visible on the screen. This may be difficult when the space between two boundaries is very narrow. In such cases, it is better to use one of the other four options. Figure 11-22 shows you some of the conditions in which these options can be used.

+X, -X, +Y, and -Y options. These options can be explained better by studying Figure 11-22.

Figure 11-22(a) shows that the points that can be selected are those that are nearest to the right-hand edge of the circumference of the circle because the ray casting takes place in the direction of the positive X.

Figure 11-22(b) shows that the points that can be used for selection are those that are nearest to the left-hand edge of the circle because the ray casting takes place in the negative X direction.

In Figure 11-22(c), the ray casting takes place in the positive Y direction, so the points that can be used for selection are the ones closest to the upper edge of the circle.

In Figure 11-22(d), the ray casting takes place in the negative Y direction, so the internal points that can be selected are the ones closest to the lower edge of the circle.

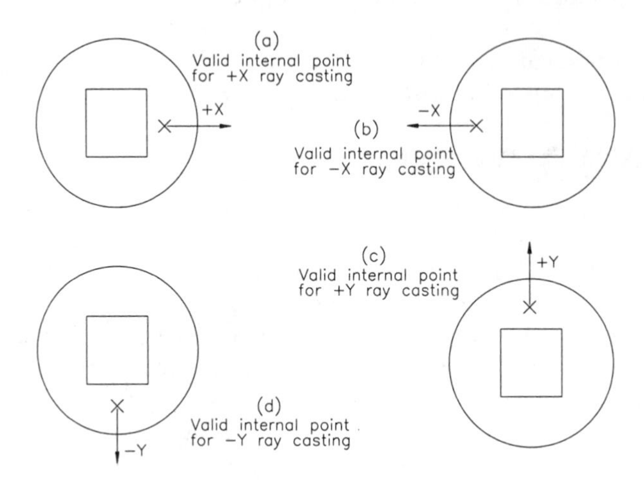

Figure 11-22 *Valid points for casting a ray in different directions to select the circle as boundary*

The effect of hatching for these ray casting options is shown Figures 11-23(a) and (b).

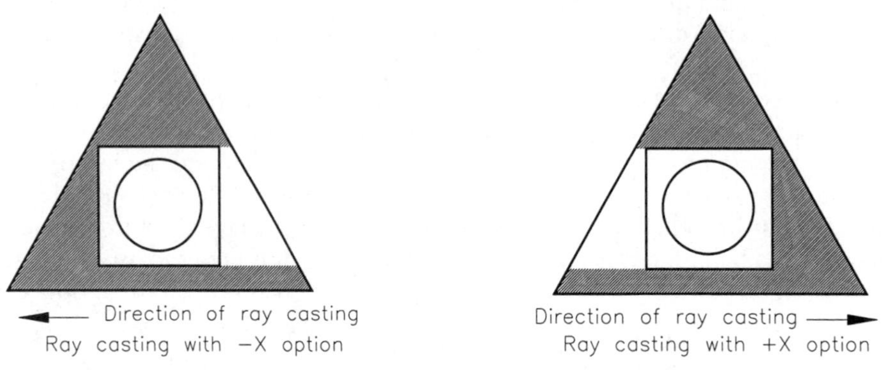

Figure 11-23(a) *Effect of hatching for ray casting*

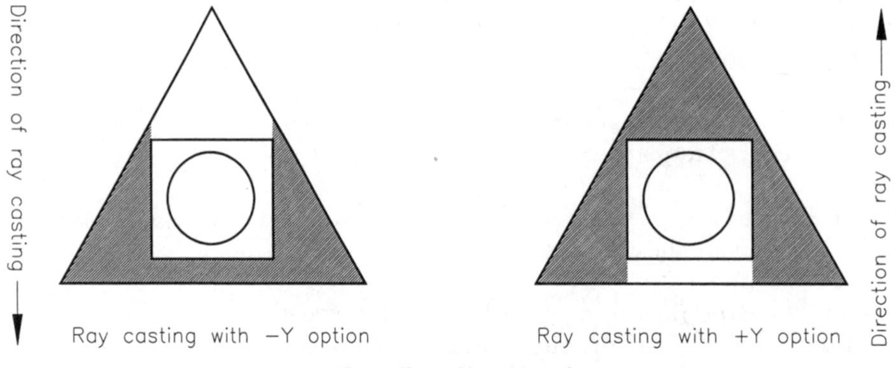

Figure 11-23(b) *Effect of hatching for ray casting*

HATCHING AROUND TEXT, ATTRIBUTES, SHAPES, AND SOLIDS

The hatch lines do not pass through the text, attributes, or shapes present in an object being hatched because AutoCAD LT places an imaginary box around these objects that does not allow the hatch lines to pass through it. You must select the text/attribute/shape along with the object in which it is placed when defining the hatch boundary. If multiple line text is to be written, use the **MTEXT** command. Figure 11-24 shows you the selection of boundaries for hatching using the window selection method.

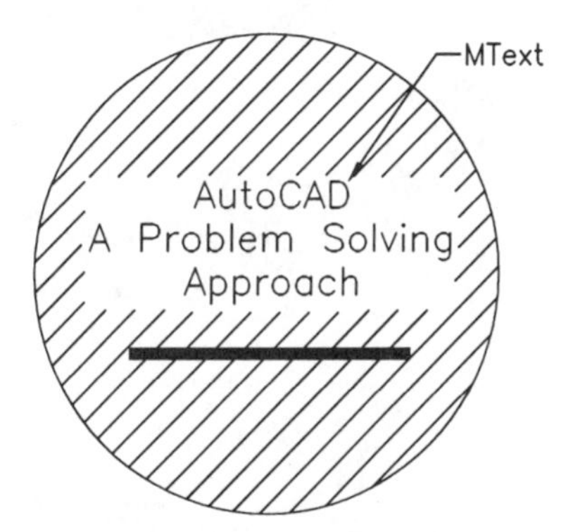

Figure 11-24 *Hatching around text*

EDITING ASSOCIATIVE HATCH PATTERNS

Using the HATCHEDIT Command

Toolbar:	Modify II > Edit Hatch
Menu:	Modify > Object > Hatch
Command:	HATCHEDIT

Figure 11-25 *Choosing* ***Edit Hatch*** *from the* ***Modify II*** *toolbar*

The **HATCHEDIT** command can be used to edit the hatch pattern. When you enter this command and select the hatch for editing, the **Hatch Edit** dialog box (Figure 11-26) is displayed on the screen. The **Hatch Edit** dialog box can also be invoked from the shortcut menu after selecting a hatched object, right-clicking in the drawing area, and choosing **Hatch Edit**. The command sequence is as follows:

Command: **HATCHEDIT**
Select associative hatch objects: *Select a hatch object.*

AutoCAD LT will display the **Hatch Edit** dialog box, where the hatch pattern can be changed or modified accordingly.

You can redefine the hatch pattern by entering the new hatch pattern name in the **Pattern**: edit box. You can also change the scale or angle by entering the new value in the **Scale**: or **Angle**: edit box. You can also define the hatch style by choosing the **Advanced** Tab and then selecting **Normal**, **Outer**, or **Ignore** styles. If you want to copy the properties from an existing hatch pattern, choose the **Inherit Properties** button, and then select the hatch. In Figure 11-27, the object is hatched using the ANSI31 hatch pattern. Using the **HATCHEDIT** command, you can select the hatch and then edit it using the dialog box to get the hatch as shown in Figure 11-28. You can also edit an existing hatch through the command line by entering **-HATCHEDIT** at the Command prompt.

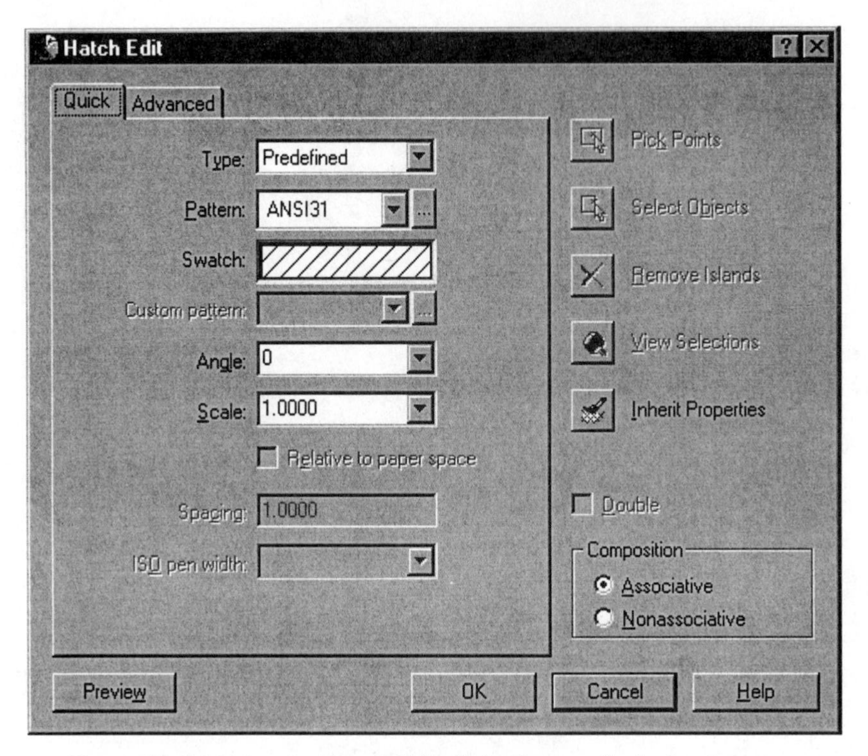

***Figure 11-26** Using the **Hatch Edit** dialog box to edit the hatch pattern*

Command: **-HATCHEDIT**
Select associative hatch object: *Select the hatch pattern.*
Enter hatch option [Disassociate/Style/Properties] <Properties>: *Press ENTER.*
Enter a pattern name or [?/Solid/User defined] <ANSI31>: *Enter the pattern name.*
Specify a scale for the pattern <1.0000>: *Enter scale.*
Specify an angle for the pattern <0>: *Enter angle.*

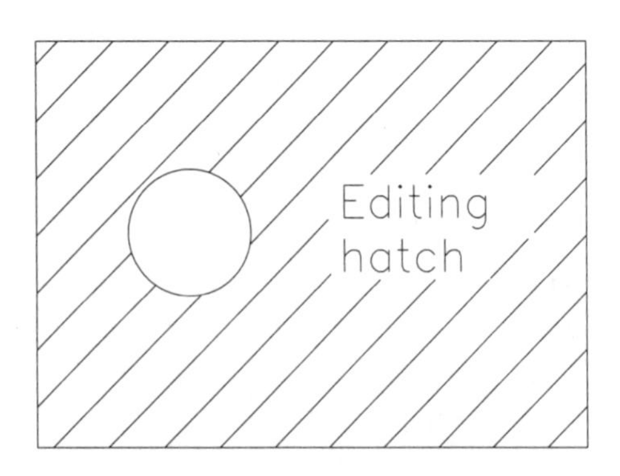

***Figure 11-27 ANSI31** hatch pattern*

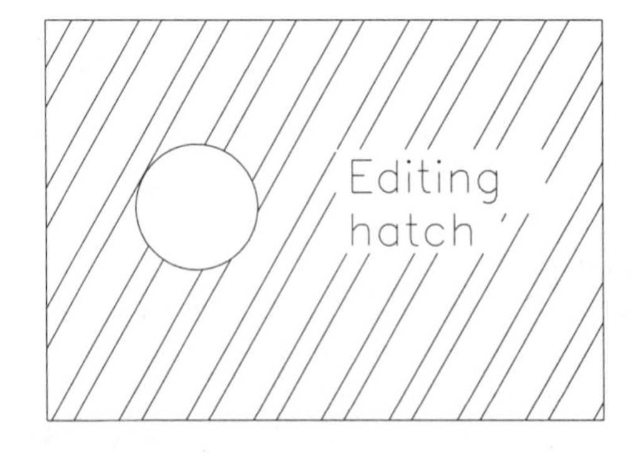

***Figure 11-28** Using the **HATCHEDIT** command to edit the hatch pattern*

USING PROPERTIES COMMAND*

Toolbar:	Standard > Properties
Menu:	Modify, Tools > Properties
Command:	PROPERTIES

You can also use the **PROPERTIES** command to edit the hatch pattern. When you select a hatch pattern for editing, AutoCAD LT displays the object **Properties** Window. The categories in the object **Properties** Window (Figure 11-29) are as follows:

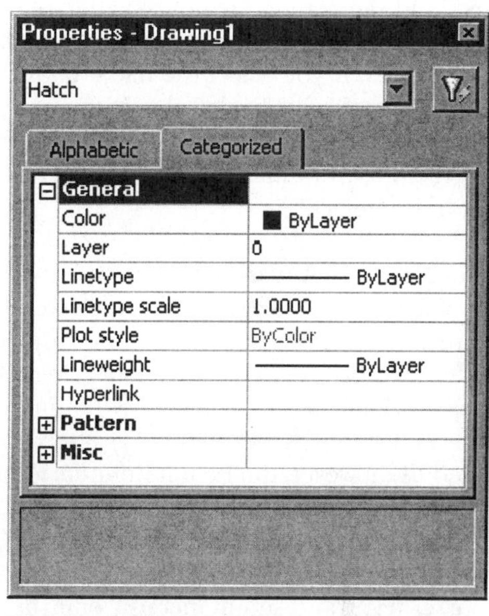

*Figure 11-29 Object **Properties** window (Hatch)*

General. You can change the hatch pattern properties like Color, Layer, Linetype, Linetype scale, Lineweight, and so on, by selecting the **General** category and changing the respective properties from the available drop-down list. The drop-down arrow is displayed when you select an item.

Pattern. In this category you can change the Pattern Type, Pattern name, Angle, Scale, Spacing, ISO pen width and double properties. When you choose the [...] button in **Type** field, AutoCAD LT 2000 displays the **Hatch Pattern Type** dialog box (Figure 11-30). When you choose the [...] button in the **Pattern name** field, AutoCAD LT displays the **Hatch Pattern Palette** dialog box with the default pattern ANSI31 selected. The other properties can be changed manually.

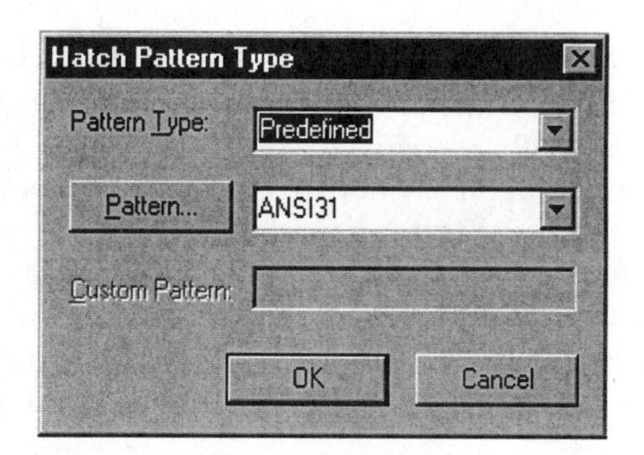

*Figure 11-30 **Hatch Pattern Type** dialog box*

Misc. In this category, Elevation, Associative, and Island detection style properties of hatching can be changed. Elevation can be changed manually while the Associative and Island detection styles can be changed from the available drop-down list.

EDITING HATCH BOUNDARY

Using Grips

One of the ways you can edit the hatch boundary is by using grips. You can select the hatch pattern or the hatch boundaries. If you select the hatch pattern, the hatch highlights and a grip is displayed. However, if you select an object that defines the hatch boundary, the object grips are displayed at the defining points of the selected object. Once you change the boundary definition, AutoCAD LT will reevaluate the hatch boundary and then hatch the area. When you edit the hatch boundary, make sure that there are no open spaces in the hatch boundary. AutoCAD LT will not create a hatch if the outer boundary is not closed. Figure 11-31 shows the hatch after moving the circle and text, and shortening the bottom edge of the hatch boundary. The objects were edited by using grips. If you select the hatch pattern, AutoCAD LT will display the hatch object grip at the centroid of the hatch. You can select the grip to make it hot and then edit the hatch object.

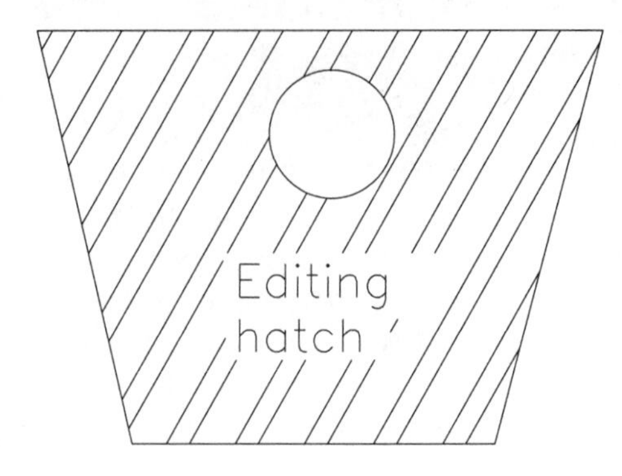

Figure 11-31 *Using Object Grips to edit the hatch boundary*

If the **Retain Boundaries** option was off when the hatch was created, all objects selected or found by the selection or point selection processes during the hatch creation are associated with the hatch as boundary objects. Editing any of them may affect the hatch. If, however, the Retain Boundaries option was on when the hatch was created, only the retained polyline boundary created by **BHATCH** is associated with the hatch as its boundary object. Editing it will affect the hatch, and editing the objects selected or found by selection or point selection processes during the hatch created will have no effect on the hatch. You can, of course, select and simultaneously edit the objects selected or found by the selection or point selection processes, as well as the retained polyline boundary created by the **BHATCH** command.

Using AutoCAD LT Editing Commands

When you use the editing commands, such as **MOVE**, **COPY**, **SCALE**, **STRETCH**, and **MIRROR**, associativity is maintained, provided all objects that define the boundary are selected for editing. If any object is missing, the associativity will be lost and AutoCAD LT will display the message **Hatch boundary associativity removed**. When you rotate or scale an associative hatch, the new rotation angle and the scale factor are saved with the hatch object data. This data is then used to update the hatch. If you explode an associative hatch pattern, the associativity between the hatch pattern and the defining boundary is removed. Also, when the hatch object is exploded, each line in the hatch pattern becomes a separate object.

HATCHING BLOCKS AND XREF DRAWINGS

The hatching procedure in AutoCAD LT works on inserted blocks and xref drawings (see Figure 11-32). The internal structure of a block is treated by the **HATCH** and **BHATCH** commands as if the block were composed of independent objects. If you are using the **BHATCH** command, you can use the Pick Points button to generate the desired hatch in a block. If you want to hatch a block using the **BHATCH** command, the objects inside the blocks should be parallel to the current UCS. If the block is comprised of objects such as arcs, circles, or pline arc segments, they **need not be** uniformly scaled for hatching.

When you xref a drawing, you can hatch any part of the drawing that is visible. When you detach the xref drawing, the hatch pattern and its boundaries are not detached.

You can also create a hatch pattern from the Command prompt line by preceding the **BHATCH** command with a dash (-). The following is the command prompt sequence for the **-BHATCH** command:

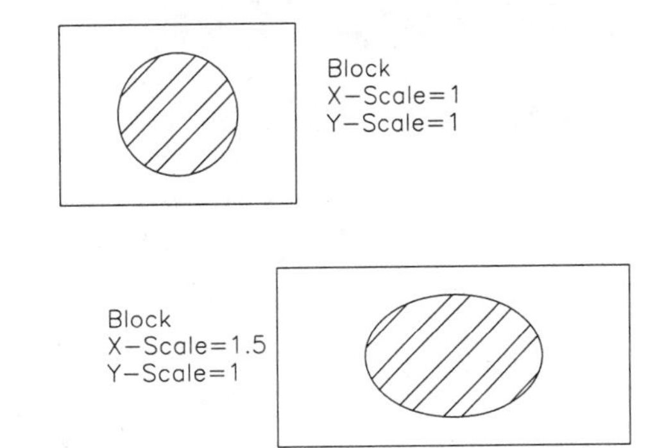

Figure 11-32 *Hatching inserted blocks*

Command: **-BHATCH**
Specify internal point or [Properties/Select/Remove islands/Advanced]: **P**
Enter a pattern name or [?/Solid/User defined] <ANSI31>: *Enter the pattern name.*
Specify a scale for the pattern <1.0000>: *Enter scale.*
Specify an angle for the pattern <0>: *Enter angle.*
Specify internal point or [Properties/Select/Remove islands/Advanced]: *Select internal point.*

PATTERN ALIGNMENT DURING HATCHING (SNAPBASE Variable)

Pattern alignment is an important feature of hatching, since on many occasions you need to hatch adjacent areas with similar or sometimes identical hatch patterns while keeping the adjacent hatch patterns properly aligned. Proper alignment of hatch patterns is taken care of automatically by generating all lines of every hatch pattern from the same reference point. The reference point is normally at the origin point (0,0). Figure 11-33 shows two adjacent hatch areas. If you hatch these areas, the hatch lines may not be aligned, as shown in Figure 11-33(a). To align them, you can use the **SNAPBASE** variable so that the hatch lines are aligned, as shown in Figure 11-33(b). The reference point for hatching can be changed using the system variable **SNAPBASE**. The command prompt sequence is:

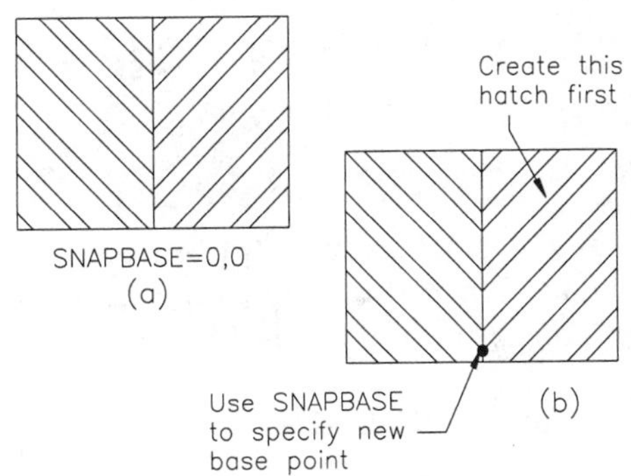

Figure 11-33 *Aligning hatch patterns using the* ***SNAPBASE*** *variable*

Command: **SNAPBASE**
Enter new value for SNAPBASE <0.0000, 0.0000>: *Enter the new reference point.*

Note

You should set the base point back to (0,0) when you are done hatching the object.

Exercise 3 — *Mechanical*

In this exercise, you will hatch the given drawing using the hatch pattern ANSI31. Use the **SNAPBASE** variable to align the hatch lines shown in the drawing (Figure 11-34).

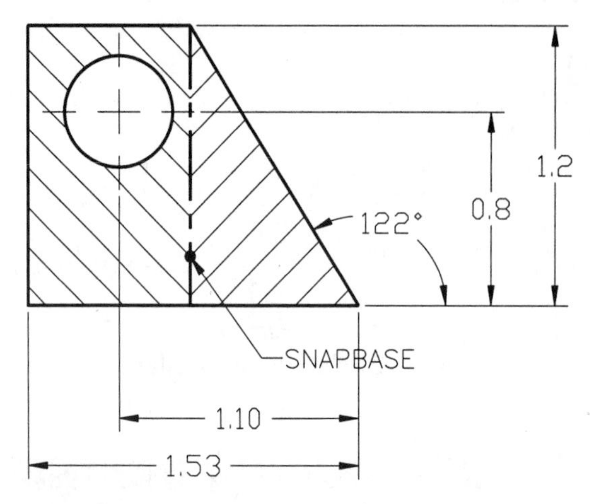

Figure 11-34 *Drawing for Exercise 3*

THE BOUNDARY COMMAND

Menu:	Draw > Boundary
Command:	BOUNDARY

The **BOUNDARY** command is used to create a polyline or a region by defining a hatch boundary. When this command is entered, AutoCAD LT displays the **Boundary Creation** dialog box, shown in Figure 11-35.

Command: **BOUNDARY**

The options shown in the **Boundary Creation** dialog box are identical to the options in the **Advanced** Tab (**Boundary Hatch** dialog box) discussed earlier, except for the **Pick Points** option. The **Pick Points** option in the **Boundary Creation** dialog box is used to create a boundary by selecting a point inside the objects.

When you select the **Pick Points** option, the dialog box is cleared from the screen and the following prompt is displayed:

Select internal point:

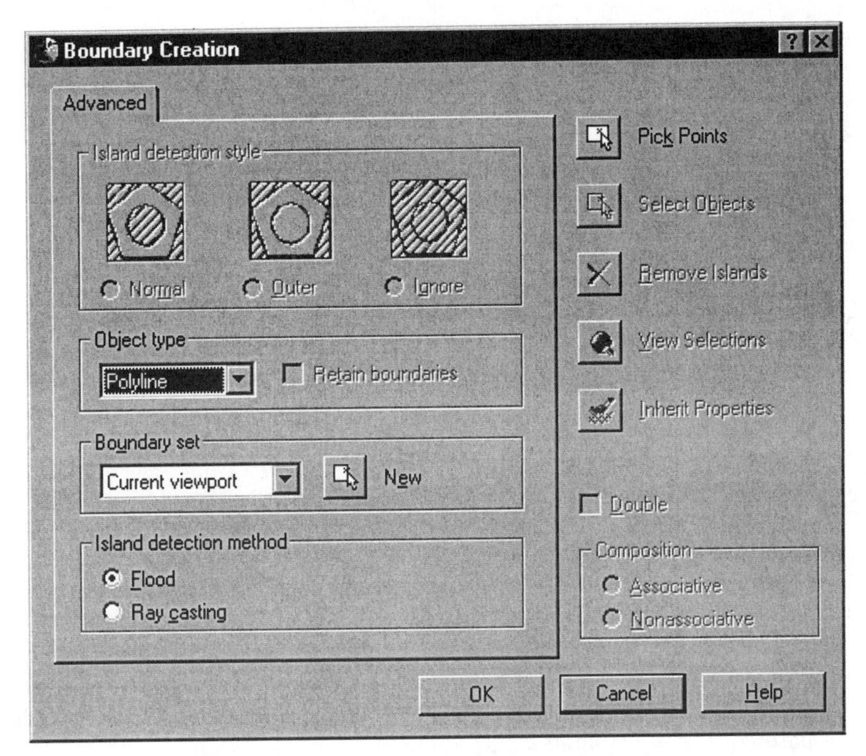

Figure 11-35 ***Boundary Creation*** *dialog box*

Once you select an internal point, a polyline or a region is formed around the boundary (Figure 11-36). To end this process, press ENTER at the **Select internal point:** prompt. The polyline or region is determined by the **Object Type** you have specified in the **Boundary Creation** dialog box. If you want to edit the boundary, you can select the boundary by using the last object selection option.

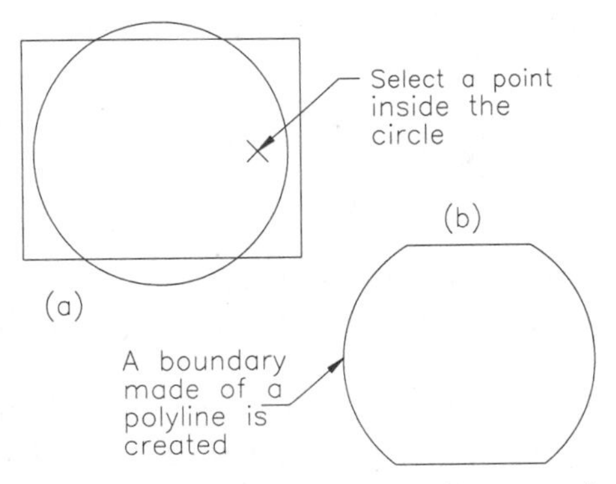

Figure 11-36 *Using* ***BOUNDARY*** *command to create a hatch boundary*

Exercise 4 — *Mechanical*

First see Figure 11-37. Then make the drawing shown in Figure 11-38 using the **BOUNDARY** command to create a hatch boundary. Copy the boundary from the drawing shown in Figure 11-37, and then hatch.

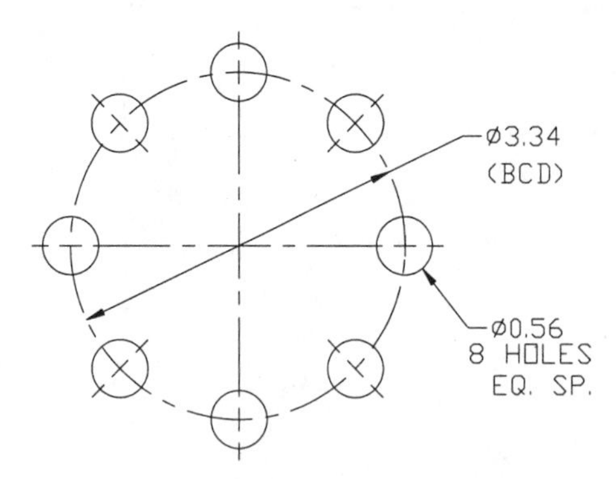

Figure 11-37 Drawing for Exercise 4

Figure 11-38 Final drawing for Exercise 4

OTHER FEATURES OF HATCHING

1. In AutoCAD LT the hatch patterns are separate objects. The objects store the information about the hatch pattern boundary with reference to the geometry that defines the pattern for each hatch object.

2. When you save a drawing, the hatch patterns are stored with the drawing. Therefore, the hatch patterns can be updated even if the hatch pattern file that contains the definitions of hatch patterns is not available.

3. If the system variable **FILLMODE** is 0 (Off), the hatch patterns are not displayed. After changing the value, you must use the **REGEN** command to regenerate the drawing to see the changes.

4. You can edit the boundary of a hatch pattern even when the hatch pattern is not visible (**FILLMODE**=0).

5. The hatch patterns created with earlier releases are automatically converted to AutoCAD LT 2000 hatch objects when the hatch pattern is edited.

6. When you save an AutoCAD LT 2000 drawing in LT 97 format, the hatch objects are automatically converted to LT 97 hatch blocks.

7. You can use the **CONVERT** command to change the pre-Release 14 hatch patterns to AutoCAD LT 2000 objects. You can also use this command to change the pre-LT 97 polylines to AutoCAD LT 2000 optimized format to save memory and disk space.

HATCHING BY USING THE HATCH COMMAND

Hatches can also be created with the **HATCH** command. However, such hatches are not associated with the boundary. Also, all selected objects must meet endpoint to endpoint to form a continuous series for a closed boundary because the **HATCH** command does not create its own boundary. You can invoke this command by entering **HATCH** at the AutoCAD LT Command prompt. When you enter the **HATCH** command, it does not display a dialog box as in the case of the **BHATCH** command. The prompt sequence is:

Command: **HATCH**
Enter a pattern name or [?/Solid/User defined] <ANSI31>: *Enter a pattern name.*

If you know the name of the pattern, enter it. You can also use the following options in response to this prompt:

? Option

If you enter ? As the response to the **Enter a pattern name or [?/Solid/User defined] <ANSI31>:** prompt, the following prompt is displayed:

Enter pattern(s) to list <*>:

If you give a null response to this last prompt, the names and descriptions of all defined hatch patterns are displayed in the AutoCAD LT Text window. If you are not sure about the name of the pattern you want to use, you can select this option before entering a name. You can also list a particular pattern by specifying it at the prompt.

Solid Option

This option is used to fill the selected boundary with solid. If you enter **S** as the response to the **Enter a pattern name or [?/Solid/User defined] <ANSI31>:** prompt, the following prompt is displayed:

Select objects to define hatch boundary or <direct hatch>
Select objects: *Select a boundary for filling or press ENTER for direct hatch.*

If you select a boundary, the object is highlighted; by pressing ENTER at the next **Select objects**: prompt, you can complete the command and the selected area will be solid-filled.

If you press ENTER at the initial **Select objects:** prompt, you will be allowed to create a boundary for filling. The prompt sequence for the direct hatch option is:

Select objects to define hatch boundary or <direct hatch>
Select objects: *Press ENTER for direct hatch.*
Retain polyline boundary [Yes/No] <N>: *Enter* **Y** *to retain the polyline,* **N** *to lose it.*
Specify start point: *Select first point of the boundary.*
Specify next point or [Arc/ Close/ Length/ Undo]: *Select the next point of the boundary.*
Specify next point or [Arc/ Close/ Length/ Undo]: *Select the next point of the boundary.*
Specify next point or [Arc/ Close/ Length/ Undo]: **C** (*Enter* **C** *to close the boundary.*)
Specify start point for new boundary or <apply hatch>: *Press ENTER to complete the command.*

The created boundary will be solid-filled.

User-defined Option

This option is used for creating a user-defined array or double-hatched grid of straight lines. To invoke this option, enter **U** at the **Enter a pattern name or [?/Solid/User defined] <ANSI31>**: prompt. After you enter **U**, the following prompts are displayed:

Specify angle for crosshatch lines <0>:
Specify spacing between the lines <1.0000>:
Double hatch area? [Yes/No] <N>

At the **Specify angle for crosshatch lines <0>**: prompt, you need to specify the angle at which you want the hatch to be drawn. This angle is measured with respect to the X axis of the current UCS. You can also specify the angle by selecting two points on the screen. This value is stored in the **HPANG** system variable.

At the second prompt, **Specify spacing between the lines <1.0000>:**, you can specify the space between the hatch lines. The default value for spacing is 1.0000. The specified spacing value is stored in the **HPSPACE** system variable.

At the third prompt, **Double hatch area? [Yes/No] <N>:**, you can specify whether you want double hatching. The default is no (**N**), and in this case, double hatching is disabled. You can enter **Y** (YES) if you want double hatching, and in this case, AutoCAD LT draws a second set of lines at right angles to the original lines. If the Double hatch is active, AutoCAD LT sets the **HPDOUBLE** system variable to 1.

Hatching by Specifying the Pattern Name

When you enter the **HATCH** command, you can also specify the hatch name of the hatch pattern at the pattern prompt. With the hatch name you can also specify the hatch style. The format is as follows:

Command: **HATCH**
Enter a pattern name or [?/Solid/User defined] <ANSI31>: **Stars, I** *(Enter a pattern name and style.)*
Specify a scale for the pattern <1.0000>:
Specify an angle for the pattern <0>:

By specifying **Stars, I** you are specifying the hatch pattern as Stars and the hatch style as Ignore. There are three hatch styles: **Normal**, **Outer**, and **Ignore**. They have been discussed earlier in this chapter in the "**Advanced** Tab of **Boundary Hatch** dialog box" subsection.

All the stored hatch patterns are assigned an initial scale (size) and a rotation of 0 degrees with respect to the positive X axis. You can expand or contract according to your needs. You can scale the pattern by specifying a scale factor. You can also rotate the pattern with respect to the positive direction of the current UCS by entering a desired angle in response to the Angle prompt. You can even specify the angle manually by selecting or entering two points. For example, Figure 11-39 illustrates the hatch pattern ANSI31 with different scales and angles. The scale and angle are shown with the figures. Once you are finished specifying the pattern,

AutoCAD LT prompts you to select objects for hatching or to create a direct hatch option (discussed earlier).

Select objects to define hatch boundary or <direct hatch>
Select objects: *Select a boundary for hatching or press ENTER for direct hatch.*

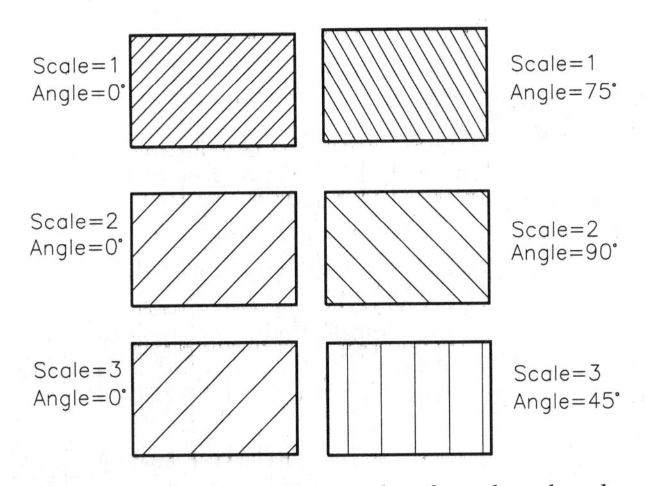

Figure 11-39 *Using different values for scale and angle*

Defining the boundaries in the **HATCH** command is a bit tedious compared with defining them with the **BHATCH** command. In the **BHATCH** command, the boundaries are automatically defined, whereas with the **HATCH** command you need to define the boundaries by selecting the objects. The objects that form the boundary must intersect at their endpoints, and the objects you select must completely define the boundary. You can use any object selection method to select the objects that define the boundary of the area to be hatched and the objects that you want to exclude from hatching. For example, in the first illustration in Figure 11-40, no hatching takes place because only two lines lie completely within the window and the boundary is not closed. In such a case, AutoCAD LT displays the following message:

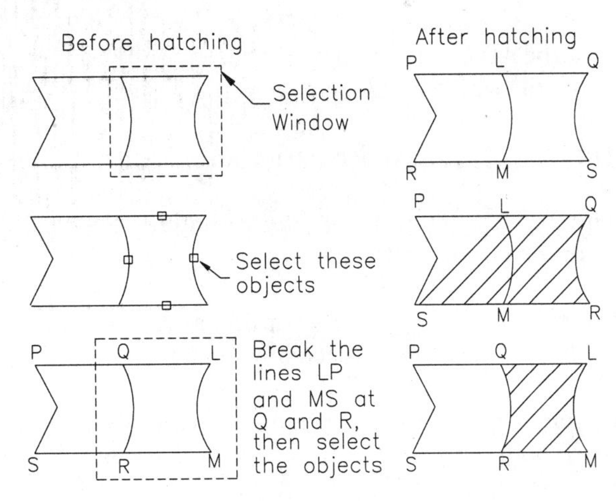

Figure 11-40 *Selecting objects using the* ***HATCH*** *command*

Unable to hatch the boundary

In the second illustration, the boundary is selected by selecting objects. In this illustration, the lines LQ and MR extend beyond the hatch boundary, which makes the hatching outside the intended hatch area. You must break the top and bottom line at points Q and R so that lines LQ, QR, RM, and ML define the hatch boundary. Now, if you select the objects, AutoCAD LT will hatch the area bound by the four selected objects, as shown in the third illustration.

Self-Evaluation Test

Answer the following questions, and then compare your answers to the correct answers given at the end of this chapter.

1. The material the object is made of can be indicated by the hatch pattern. (T/F)

2. The __________ command automatically defines the boundary, whereas in the case of the __________ command you have to define the boundary by selecting the objects that form the boundary of the hatch area.

3. The **Boundary Hatch** dialog box can be invoked by entering __________ at the Command prompt.

4. You can define a simple pattern by selecting the **User defined** pattern type in the **Type** drop-down list . (T/F)

5. One of the ways to specify a hatch pattern from the group of stored hatch patterns is by selecting a pattern from the down arrow in the **Pattern:** drop-down list. (T/F)

6. If AutoCAD LT does not locate the entered pattern in the __________ file, it searches for it in a file with the same name as the pattern.

7. The __________ option lets you rotate the hatch pattern with respect to the X axis of current UCS.

8. The __________ option in the **Boundary Hatch** dialog box lets you select objects that form the boundary for hatching.

9. You can use the __________ option after you have selected the area to be hatched to see the hatching before it is actually applied.

10. By selecting the __________ button you can apply hatching in the area specified by the hatch boundary.

11. When a boundary set is formed, it does not become the default for hatching. (T/F)

12. The selection of **Normal** hatch style carries meaning only if the objects to be hatched are __________.

13. In Nearest ray casting, AutoCAD LT casts an imaginary ray in all directions and locates the object that is __________ to the point that you have selected.

14. You can edit the hatch boundary by using grips. (T/F)

15. Different filling patterns make it possible to distinguish between different parts or components of an object. (T/F)

Review Questions

1. The hatching process can be accomplished by using the __________ or __________ command.

2. The hatch boundary may be comprised of Line, Circle, Arc, Plines, 3D face, or other objects, and each must be at least partially displayed within the active viewport. (T/F)

3. The **BHATCH** command does not allow you to hatch a region enclosed within a boundary (closed area) by selecting a point inside the boundary. (T/F)

4. One of the advantages of the __________ command is that you don't have to select each object comprising the boundary of the area you want to hatch, as in the case of the __________ command.

5. You can cycle through the predefined hatch patterns by selecting the image box in the Hatch Pattern Palette dialog box. (T/F)

6. If the **Double** check box is selected, AutoCAD LT sets the __________ system variable to 1.

7. The **Iso Pen Width** option is valid only for __________ .

8. One way to select a predefined hatch pattern is to select the **Custom** pattern type and then enter the name of the stored hatch pattern in the **Custom Pattern:** edit box. (T/F)

9. If AutoCAD LT does not locate the entered pattern in the ACLT.PAT file, it does not load any hatch pattern file. (T/F)

10. The __________ option lets you scale the hatch pattern.

11. A **Boundary Definition Error** dialog box is displayed if AutoCAD LT finds that the selected point is inside the boundary or that the boundary is not closed. (T/F)

12. The __________ option is used to remove the islands from the hatching area.

13. The __________ option lets you view the selected boundary.

14. If you want to have the same hatching pattern, style, and properties as that of an existing hatch on the screen, select the __________ button.

15. The **Associative** radio button controls the __________ of hatch with the hatch boundary.

16. You can improve the speed of hatching by setting parameters in the **Advanced Options** dialog box. (T/F)

17. The **Object Type** option in the **Advanced** tab of the **Boundary Hatch** dialog box controls the type of object AutoCAD LT will create when you define a boundary. It has two options: __________ and __________.

18. The hatching style cannot be specified by selecting a style in the Island detection style area. (T/F)

19. There are three hatching styles from which you can choose: __________, __________, and __________.

20. If you select the **Ignore** style, all areas bounded by the outermost boundary are hatched, ignoring any hatch boundaries that are within the outer boundary. (T/F)

21. Ray casting is enabled only when Island detection is turned on. (T/F)

22. You can use ray casting options to control the way AutoCAD LT casts the ray to form a hatch boundary. The available options are __________, __________, __________, __________, and __________.

23. The **HATCHEDIT** command can be used to edit the hatch pattern. (T/F)

24. You can also edit the hatch pattern from the Command prompt line by preceding the **BHATCH** command with a dash (-). (T/F)

25. When you use editing commands such as **MOVE**, **COPY**, **SCALE**, **STRETCH**, and **MIRROR**, associativity is lost. (T/F)

26. If you explode an associative hatch pattern, the associativity between the hatch pattern and the defining boundary is not affected. (T/F)

27. The hatching procedure in AutoCAD LT does not work on inserted blocks. (T/F).

28. If the block comprises objects such as arcs, circles or pline arc segments, it must be exploded before hatching. (T/F)

29. To align the hatches in adjacent hatch areas, you can use the __________ variable.

30. The specified hatch spacing value is stored in the __________ system variable.

Exercises

Exercise 5 *Mechanical*

Hatch the drawings in Figures 11-41 and 11-42 using the hatch pattern to match.

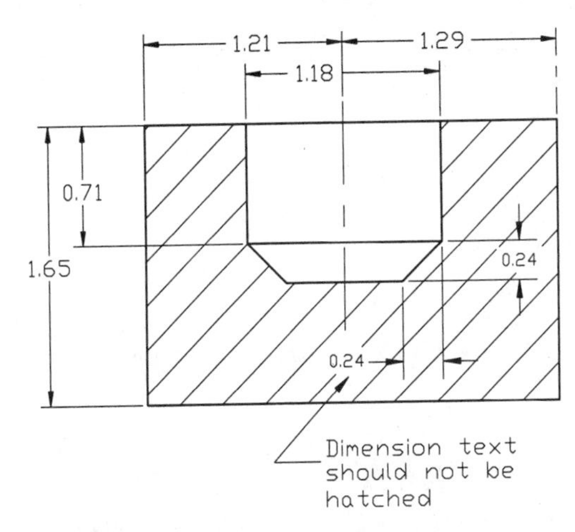

Figure 11-41 *Drawing for Exercise 5*

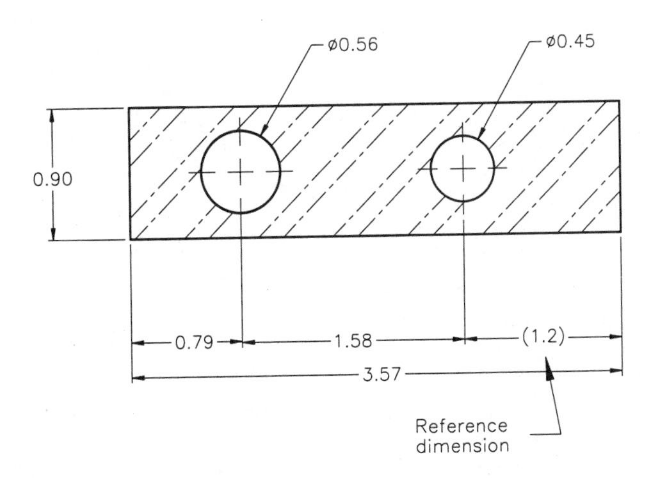

Figure 11-42 *Drawing for Exercise 5*

Exercise 6 *General*

Hatch the drawing in Figure 11-43 using the hatch pattern ANSI31. Use the **SNAPBASE** variable to align the hatch lines as shown in the drawing.

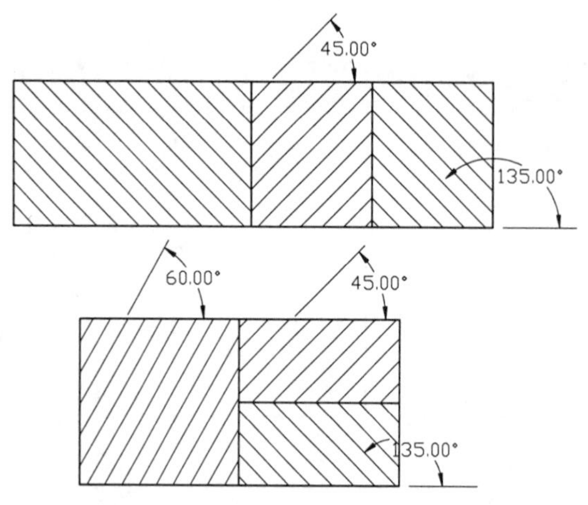

Figure 11-43 *Drawing for Exercise 6*

Exercise 7 *Mechanical*

Figure 11-44 shows the top and front views of an object. It also shows the cutting plane line. Based on the cutting plane line, hatch the front views in section. Use the hatch pattern of your choice.

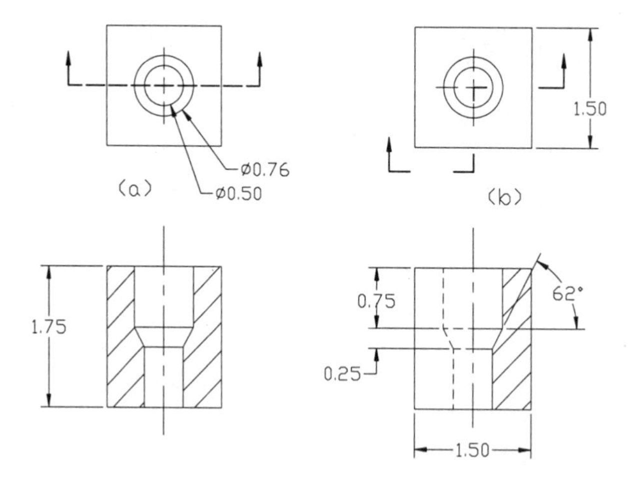

Figure 11-44 *Drawing for Exercise 7*

Exercise 8 *Mechanical*

Figure 11-45 shows the top and front views of an object. Hatch the front view in full section. Use the hatch pattern of your choice.

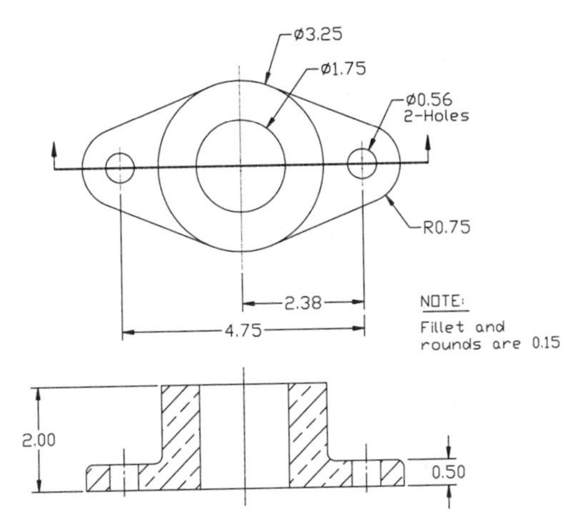

Figure 11-45 Drawing for Exercise 8

Exercise 9 — *Mechanical*

Figure 11-46 shows the front and side views of an object. Hatch the side view in half-section. Use the hatch pattern of your choice.

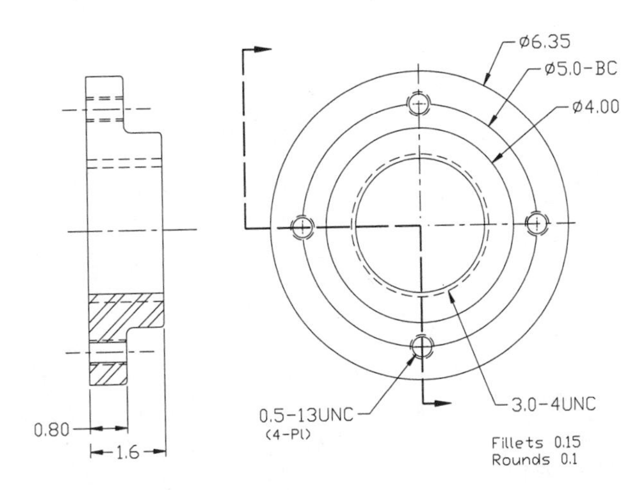

Figure 11-46 Drawing for Exercise 9

Problem-Solving Exercise 1 — *Mechanical*

Figure 11-47 shows the front and side views of an object. Hatch the aligned section view. Use the hatch pattern of your choice.

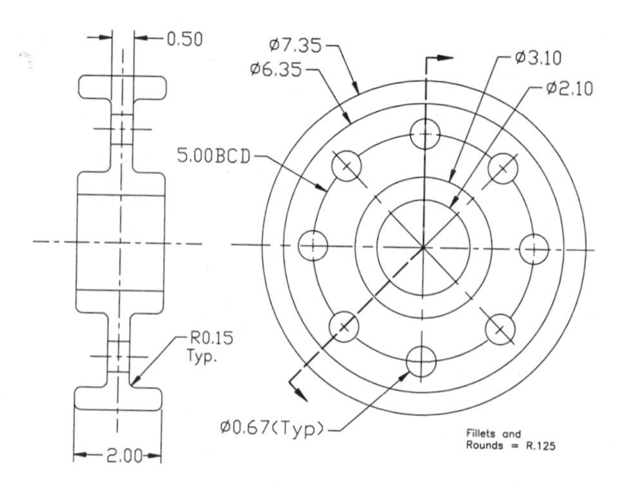

Figure 11-47 Drawing for Problem-Solving Exercise 1

Problem Solving Exercise 2 — *Mechanical*

Figure 11-48 shows the front view, side view, and the detail "A" of an object. Hatch the side view and draw the detail drawing as shown.

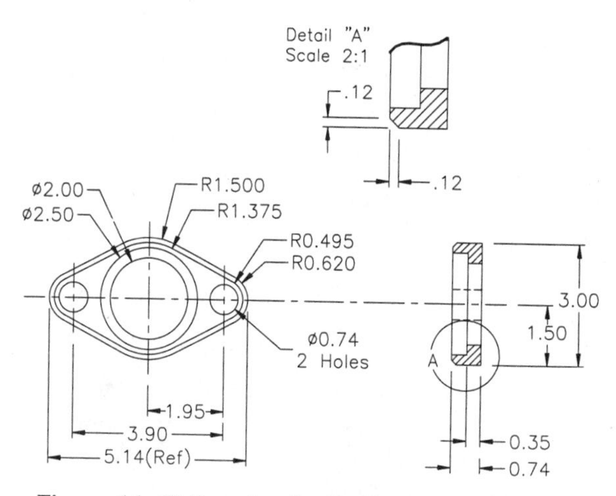

Figure 11-48 Drawing for Problem-Solving Exercise 2

Answers to Self-Evaluation Test

1 - T, **2** - **BHATCH, HATCH**, **3** - **BHATCH**, **4** - T, **5** - T, **6** - ACLT.PAT, **7** - Angle, **8** - Select Objects, **9** - Preview, **10** - Apply, **11** - F, **12** - Nested, **13** - Nearest, **14** - T, **15** - T

Chapter 12

Blocks

Learning Objectives

After completing this chapter, you will be able to:
- *Create blocks with the* ***BLOCK*** *command.*
- *Insert blocks with the* ***INSERT*** *command.*
- *Perform editing operations on blocks.*
- *Create drawing files using the* ***WBLOCK*** *command.*
- *Split a block into individual objects using the* ***EXPLODE*** *and* ***XPLODE*** *commands.*
- *Define the insertion basepoint for a drawing.*
- *Rename blocks and delete unused blocks.*

THE CONCEPT OF BLOCKS

The ability to store parts of a drawing, or the entire drawing, so that they need not be redrawn when needed again in the same drawing or another drawing is extremely beneficial to the user. These parts of a drawing, entire drawings, or symbols (also known as blocks) can be placed (inserted) in a drawing at the location of your choice, in the desired orientation, with the desired scale factor. The block is given a name (block name) and the block is referenced (inserted) by its name. All the objects within a block are treated as a single object. You can **MOVE**, **ERASE**, or **LIST** the block as a single object; that is, you can select the entire block simply by selecting a point on it. As far as the edit and inquiry commands are concerned, the internal structure of a block is immaterial, since a block is treated as a primitive object, like a polygon. If a block definition is changed, all references to the block in the drawing are updated to incorporate the changes made to the block.

A block can be created using the **BLOCK** command. You can also create a drawing file using the **WBLOCK** command. The main difference between the two is that a wblock can be inserted in any other drawing, but a block can be inserted only in the drawing file in which it was created.Another feature of AutoCAD LT is that instead of inserting a symbol as a block (which results in adding the content of the referenced drawing to the drawing in which it is inserted), you can reference drawings (Xref). This means that the contents of the referenced drawing are not added to the current drawing file, although they become part of that drawing on the screen. This is explained in detail in Chapter 14, External References.

Advantages of Using Blocks

Blocks offer many advantages; here are some of them.

1. Drawings often have some repetitive feature. Instead of drawing the same feature again and again, you can create a block of that feature and insert it wherever required with the desiredscale factor and orientation. This style of working helps you to reduce drawing time and better organize your work.

2. Another advantage of using blocks is that they can be drawn and stored for future use. You can thus create a custom library of objects required for different applications. For example, if your drawings are concerned with gears, you could create blocks of gears and then integrate these blocks with custom menus (see the customizing section). In this manner, you could create an application environment of your own in AutoCAD LT.

3. The size of a drawing file increases as you add objects to it. AutoCAD LT keeps track of information about the size and position of each object in the drawing; for example, the points, scale factors, radii; and so on. If you combine several objects into a single object by forming a block out of them by using **BLOCK** command, there will be a single scale factor, rotation angle, position, and so on, for all objects in the block, thereby saving storage space. Each object repeated in multiple block insertions needs to be defined only once in the block definition. Ten insertions of a gear made of 43 lines and 41 arcs require only 94 objects (10 + 43 + 41), while ten individually drawn gears require 840 objects [10 x (43+41)].In other words whenever the block is inserted, it is not the actual block that is placed but just the raster image of the block (Block Reference) that is placed which takes very less space in the memory and thus saves the storage space.

4. If the specification for an object changes, the drawing needs to be modified. This is a very tedious task if you need to detect each location where the change is to be made and edit it individually. But if this object has been defined as a block, you can make the necessary changes and then redefine it with same insertion base point and same name, and where ever the object appears, it will get revised automatically (Figure 12-1).

5. Different attributes (textual information) can be included in blocks.

6. You can create symbols and then store them as blocks with the **BLOCK** command. Later on, with the **INSERT** command, you can insert the blocks in the drawing in which they were defined. There is no limit to the number of times you can insert a block in a drawing.

7. You can store symbols as drawing files using the **WBLOCK** command and later insert those symbols into any other drawing using the **INSERT** command.

8. Blocks can have varying X, Y, and Z scales and rotation angles from one insertion to another.

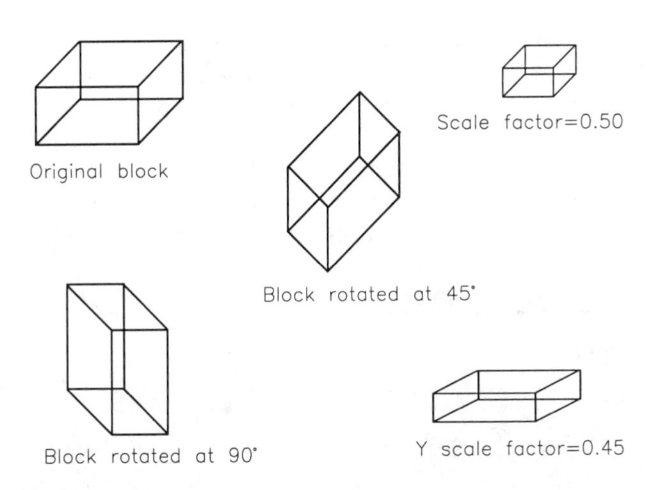

Figure 12-1 Block with different specifications

FORMATION OF BLOCKS

Drawing Objects for Blocks

The first step in creating blocks is to draw the object(s) to be converted into a block. You can consider any symbol, shape, or view that may be used more than once for conversion into a block (Figure 12-2). Even a drawing that is to be used more than once can be inserted as a block. If you do not have the object to be converted into a block, use the relevant AutoCAD LT commands to draw it.

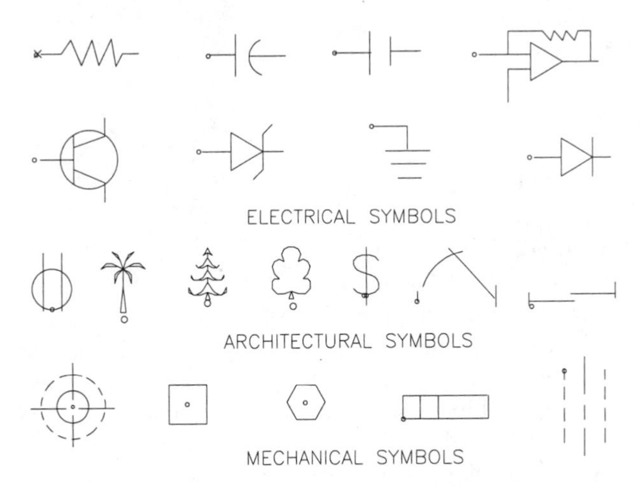

Figure 12-2 Symbols created to store in a library

Checking Layers

You should be particularly careful about the layers on which the objects to be converted into a block are drawn. The objects must be drawn on layer 0 if you want the block to inherit the linetype and color of the layer on which it is inserted. If you want the block to retain the linetype and color of the layer on which it was drawn, draw the objects defining the block on that layer. For example, if you want the block to have the linetype and color of layer OBJ, draw the objects comprising the block on layer OBJ. Now, even if you insert the block on any layer, the linetype and color of the block will be those of layer OBJ. If the objects in a block are drawn with the color, linetype, and lineweight as BYBLOCK, this block when inserted assumes the properties of the current layer. To change the layer associated with an object, use the **Properties Window**, **LAyer** option of the **CHPROP** command or any other relevant command.

CONVERTING OBJECTS INTO A BLOCK

BLOCK Command

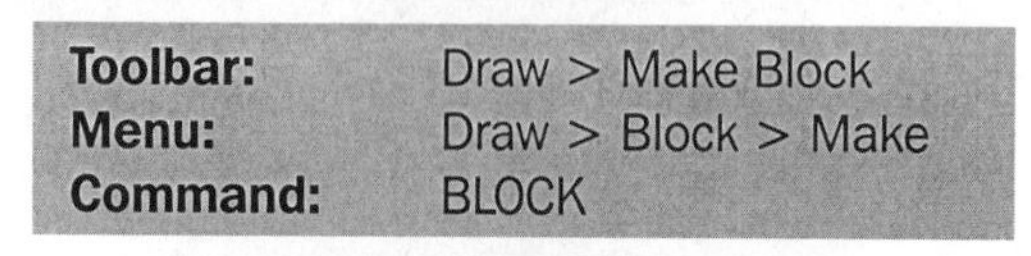

Toolbar: Draw > Make Block
Menu: Draw > Block > Make
Command: BLOCK

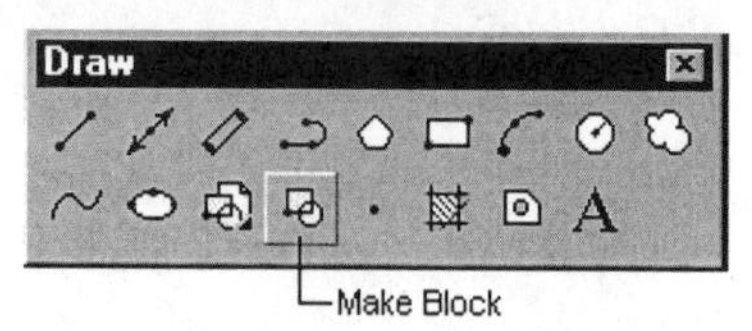

***Figure 12-3** Make Block button in the **Draw** toolbar*

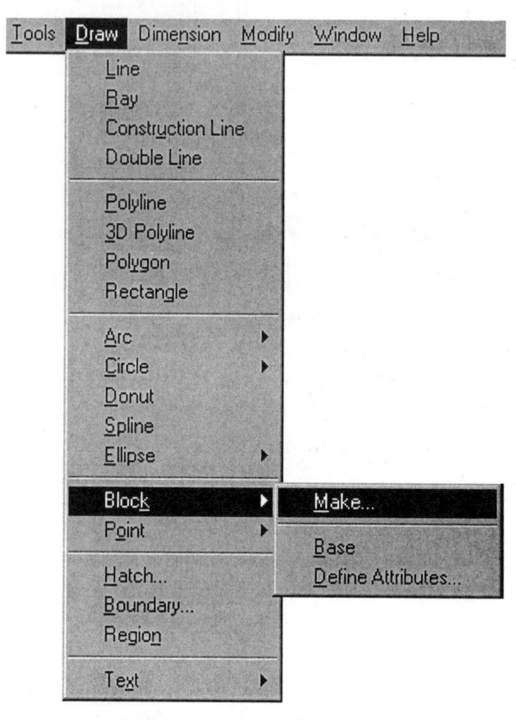

***Figure 12-4** Invoking the **BLOCK** command from the **Draw** menu*

When you invoke the **BMAKE** or **BLOCK** command, the **Block Definition** dialog box (Figure 12-5) is displayed. You can use the **Block Definition** dialog box to save any object as a block.

Command: **BLOCK** Enter

In the **Block Definition** dialog box, you can enter the name of the block you want to create or redefine, in the **Name:** edit box. The block name can have upto 255 characters. The name can contain letters, digits, blank spaces as well as special characters like the $ (dollar sign), - (hyphen), and _ (underscore) if they are not being used for any other purpose by AutoCAD LT. The block name is controlled by the **EXTNAMES** system variable and the default value is 1. If the **EXTNAMES** is set to 0, the block name can be only 31 characters long.

If a block already exists with the block name that you have specified in the **Name:** edit box, an AutoCAD LT dialog box, warning that the block name already exists, is displayed at the time of choosing **OK** button in the **Block Definition** dialog box. In this dialog box, you can either redefine the existing block by choosing the **Yes** button or you can exit it by choosing the **No** button.

After you have specified a block name, you are required to specify the insertion base point.

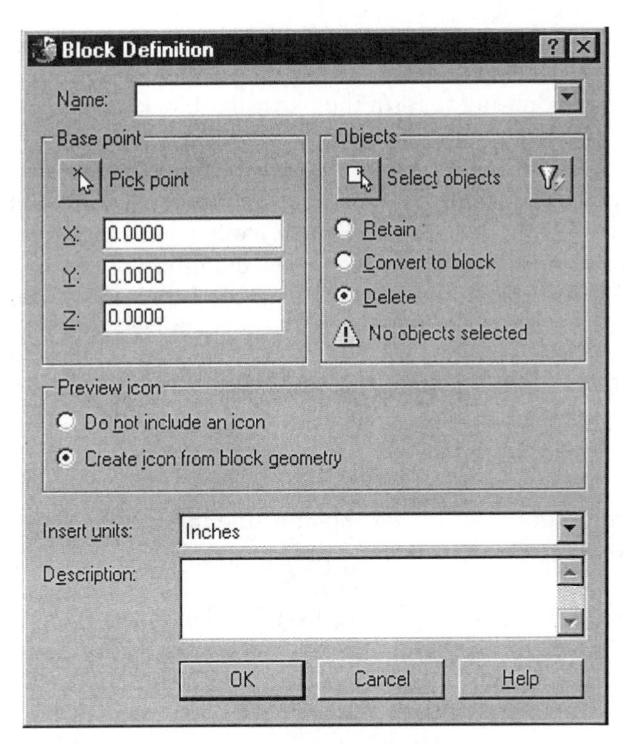

*Figure 12-5 **Block Definition** dialog box*

This point is used as a reference point to insert block. Usually, either the center of the block or the lower left corner of the block is defined as the insertion base point. Later on, when you insert the block, you will notice that the block appears at the insertion point, and you can insert the block with reference to this point. The point you specify as the insertion base point is taken as the origin point of the block's coordinate system. You can specify the insertion point by choosing the **Pick point** button in the **Base Point** area of the dialog box. The dialog box is temporarily removed, and you can select a base point on the screen. You can also enter the coordinates in the **X**, **Y**, and **Z** edit boxes instead of selecting a point on screen.

After specifying the Insertion Base point, you are required to select the objects that will constitute the block. Until the objects are not selected, AutoCAD LT displays a warning: **No objects selected**, at the bottom of the **Objects** area. Choose the **Select objects** button. The dialog box is temporarily removed from the screen. You can select the objects on the screen using any selection method. After completing the selection process, right-click or press the ENTER key to return to the dialog box.

The **Object** area of the **Block Definition** dialog box also has a **Quick Select** button. If you choose this button, the **Quick Select** dialog box is displayed which allows you to define a selection set based on properties of objects. **Quick select** is used in cases where the drawings are very large and complicated. In this area, selecting the **Retain** radio button, retains the selected objects that form the block as individual objects and does not convert them into a block after the block has been defined. Selecting the **Convert to Block** radio button (default option), converts the selected objects into a block after you have defined the block and selecting the **Delete** radio button, deletes the selected objects from the drawing after the block has been defined. After objects have been selected, the bottom of the **Object** area displays the number of objects selected in the drawing to form the particular block.

The **Preview icon** area, gives you the following options: Selecting the **Do not include an icon** radio button, does not create a preview icon and the **Preview Image box** is not displayed in this area of the dialog box. Selecting the **Create icons from block geometry** radio button creates a **Preview icon** which is saved with the block definition in the drawing. Whenever you select a block name in the **Name** drop-down list of the **Block Definition** dialog box or the **AutoCAD Design Center**, a Preview image is displayed. Similarly, in the **Description** text box, if you have entered a description of the block, it is stored with the block definition and displayed on selecting the block name in the **Name** drop-down list of the **Block Definition** dialog box or the **AutoCAD DesignCenter**. Adding a preview image and description text in the block definition, makes it easier to identify a block.

The **Insert Units** drop-down list displays the units the inserted block in the drawing will be scaled to on insertion. The default value is Unitless.

The **Name** drop-down list displays all the blocks defined in that drawing. This drop-down list contains the list of all block names created in the drawing. This way you can verify whether the block you have defined has been saved. Choose the **OK** button to complete defining block.

-BLOCK Command

You can use the **-BLOCK** command to create blocks from the command line. The following prompt sequence saves a symbol as a block named SYMBOLX.

> Command: **-BLOCK** Enter
> Enter block name or [?]: SYMBOLX Enter
> Specify insertion base point: *Select a point.*
> Select objects: *Select all the objects comprising the block.*
> Select objects: *Press ENTER when selection is complete.*

AutoCAD LT acknowledges the creation of the block by removing the objects defining the block from the screen. However, you can get the objects back at the same position by entering **OOPS**.

You can use the **?** option to list all the blocks created in the drawing. The prompt sequence is:

> Command: **-BLOCK** Enter
> Enter block name or [?]: ? Enter
> Enter block(s) to list <*>: Enter

Information similar to the following is displayed in the AutoCAD LT Text Window:

> *Defined blocks*
> *"BLK1"*
> *"BLK2"*
> *"SYMBOLX"*

User Blocks	*External References*	*Dependent Blocks*	*Unnamed Blocks*
n	*n*	*n*	*n*

Here **n** denotes a numeric value.

Defined blocks. The names of all the blocks defined in the drawing.

User blocks. The number of blocks created by the user.

External references. The number of external references. These are the drawings that are referenced with the **XREF** command.

Dependent blocks. The number of externally dependent blocks. These are the blocks that are present in a referenced drawing.

Unnamed blocks. The number of unnamed blocks in the drawing. These are objects, such as associative dimensions and hatch patterns.

Exercise 1 *General*

Draw a unit radiuscircle, and use the **BLOCK** command to create a block. Name the block CIRCLE. Select the center as the insertion point. Use the **?** option of the **-BLOCK** command to see if the block CIRCLE was saved. Figure 12-6 shows the steps involved in creating a block.

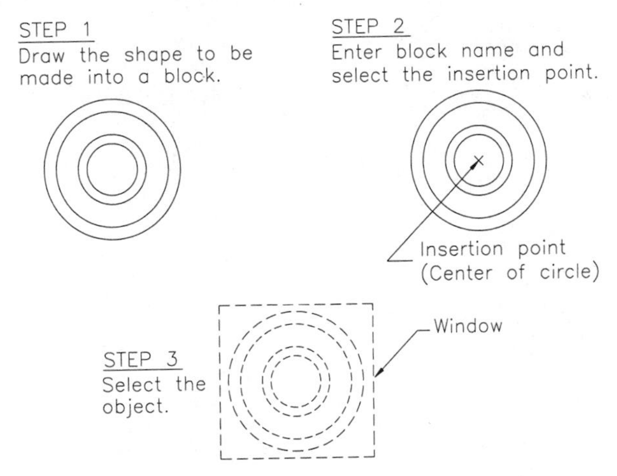
STEP 1
Draw the shape to be made into a block.
STEP 2
Enter block name and select the insertion point.
Insertion point (Center of circle)
Window
STEP 3
Select the object.

***Figure 12-6** Using the **BLOCK** command*

INSERTING BLOCKS

Insertion of a predefined block is possible with the **INSERT** command. With these commands, insertion is carried out using the **Insert** dialog box. In the case of the **-INSERT** command, you can insert a block by entering this command at the command line. You should determine the layer on which you want to insert the block, the location where you want to insert it, its size, and the angle by which you want the block to be rotated. If the layer on which you want to insert the block is not current, you can use the **Current** button in the **Layer Properties Manager** dialog box to make it current.

INSERT COMMAND

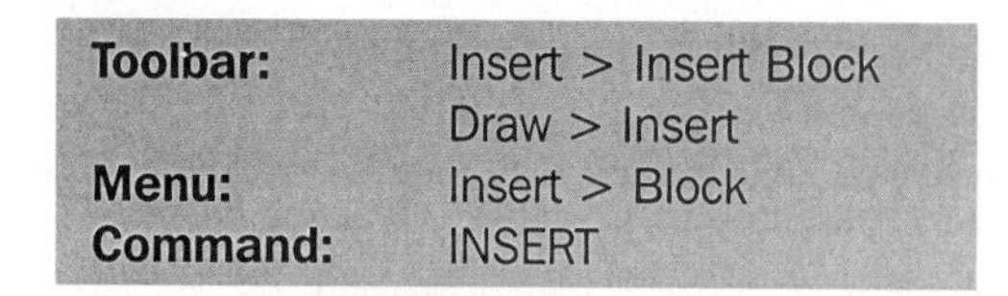

Toolbar: Insert > Insert Block
Draw > Insert
Menu: Insert > Block
Command: INSERT

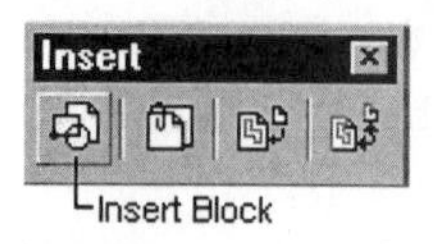

Figure 12-7 *Invoking the* ***INSERT*** *command from the* ***Insert*** *toolbar*

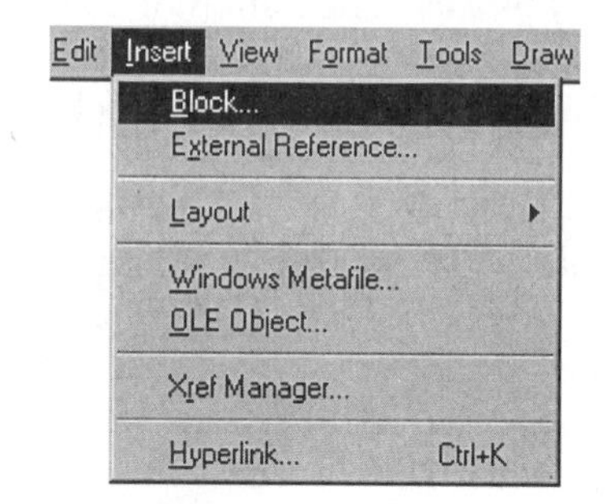

Figure 12-8 *Invoking the* ***INSERT*** *command from the* ***Insert*** *menu*

When you invoke the **INSERT** command, the **Insert** dialog box (Figure 12-9) is displayed on the screen. You can specify the different parameters of the block to be inserted in the **Insert** dialog box.

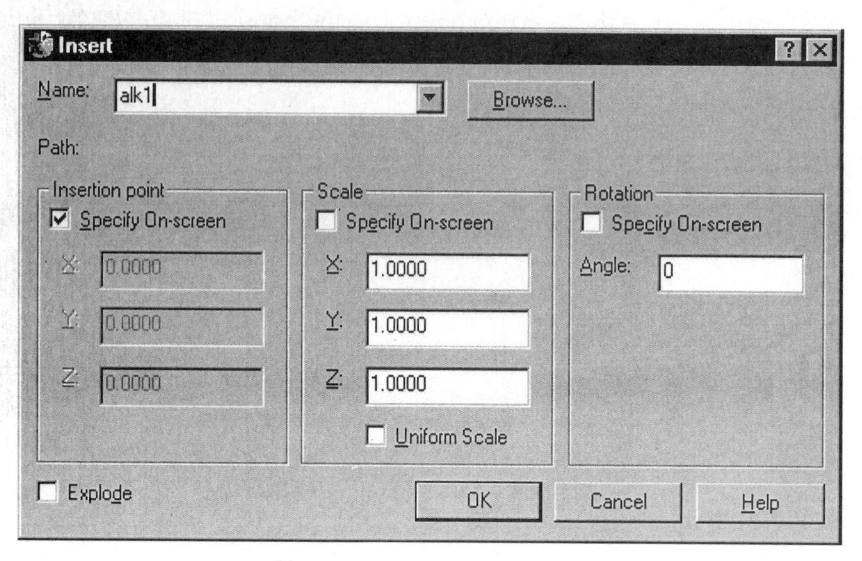

Figure 12-9 ***Insert*** *dialog box*

Name: Edit Box

This edit box is used to specify the name of the block to be inserted. Enter the name of the block to be inserted. You can also select the Block name from the **Name** drop-down list. The **Browse** button is used to insert external files. Choosing this button displays the **Select Drawing File** dialog box from where you can select a file to be inserted as a block. In the **Insert** dialog box, whichever block name you select from the **Name** drop-down list, is displayed in the

Name edit box. For example, to insert a block named BLK1, select this name from the **Name** drop-down list or enter it in the **Name** edit box. **Path** displays the path of the block or file to be inserted. If you want to insert a drawing file as a block, choose the **Browse** button in the **Insert** dialog box. When you choose the **Browse** button, the **Select Drawing File** dialog box (Figure 12-10) is displayed. You can select the drawing file from the listed drawing files existing in the current directory. You can change the directory by selecting the desired directory in the **Look in** drop-down list. Once you select the drawing file in this dialog box and choose the **Open** button, the drawing file name is displayed in the **Name** drop-down list of the **Insert** dialog box. If you want to change the block name, just change the name in the **Name** edit box. In this manner, the drawing can be inserted with a different block name. Changing the original drawing does not affect the inserted drawing.

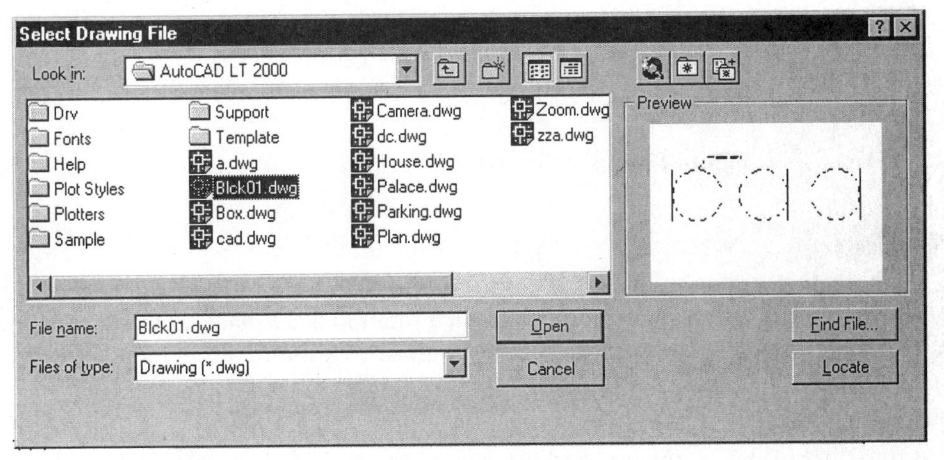

Figure 12-10 ***Select Drawing File*** *dialog box*

Insertion Point

When a block is inserted, the coordinate system of the block is aligned parallel to the current UCS. In this area, you can specify the X, Y, and Z coordinate locations of the block insertion point. If you select the **Specify On-screen** check box, you can specify the insertion point on the screen. By default, the **Specify On-screen** check box is selected; hence, you can specify the insertion point with a pointing device.

Scale

In this area, you can specify the X, Y, and Z scale factors of the block to be inserted. By selecting the **Specify On-screen** check box, you can specify the scale of the block at which it has to be inserted, on the screen. If the **Uniform scale** check box is selected, the X scale factor value is assumed by the Y and Z scale factors. All the dimensions in the block are multiplied by the X, Y, and Z scale factors you specify. By specifying negative scale factors for X and Y, you can obtain a mirror image of the block. These scale factors allow you to stretch or compress a block along X and Y axes, respectively, according to your needs. You can also insert 3D objects into a drawing by specifying the third scale factor (since 3D objects have three dimensions): Z scale factor. Figure 12-11 shows variations of X and Y scale factors for the block, CIRCLE, during insertion.

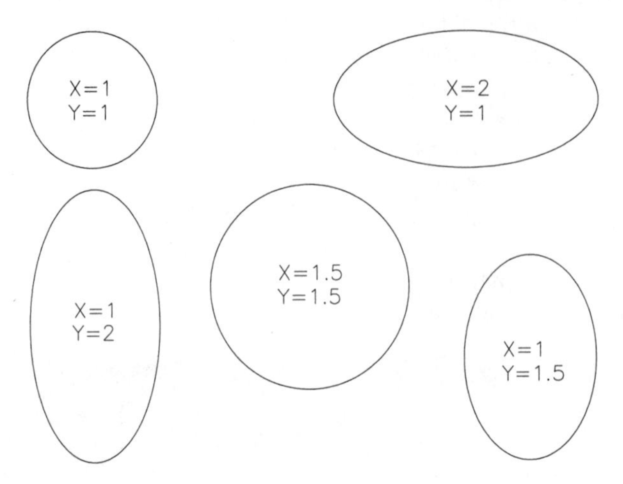

Figure 12-11 *Block inserted with different scale factors*

Rotation

You can enter the angle of rotation for the block to be inserted in the **Angle** edit box. The insertion point is taken as the location about which rotation takes place. Selecting the **Specify On-screen** check box allows you to specify the angle of rotation on the screen.

Explode Check Box

By selecting this check box, the inserted block is inserted as a collection of individual objects. The function of the **Explode** check box is identical to that of the **EXPLODE** command. Once a block is exploded, the X, Y, and Z scale factors are identical. Hence, you are provided access to only one scale factor edit box (X scale factor). This scale factor is assigned to the Y and Z scale factors. You must enter a positive scale factor value.

After entering the relevant information in the dialog box, choose the **OK** button. Once you do so, the **Insert** dialog box is removed from the screen and if you have selected the **Specify On-Screen** check boxes, you can specify the insertion point, scale, and angle of rotation with a pointing device or at the command line. The following prompts appear in the command line:

Specify insertion point or [Scale/X/Y/Z/Rotate/PScale/PX/PY/PZ/PRotate]:
Enter X scale factor or specify opposite corner or [Corner/XYZ] <1>:
Enter Y scale factor <Use X scale factor>:
Specify rotation angle <0>:

You can also specify these features in the **Insert** dialog box. When you clear the **Specify On-screen** check box, it activates the **Insertion Point**, **Scale**, and **Rotation** areas in the dialog box. Now you can enter these parameters in the respective edit boxes.

-INSERT COMMAND

Command: -INSERT

As mentioned before, predefined blocks can also be inserted in a drawing with the **-INSERT** command. The following is the prompt sequence for the **-INSERT** command:

Command: **-INSERT** Enter
Enter block name or [?] <Current>: *Enter the name of the block to be inserted.*

To display the list of blocks in the drawing, enter ? at the **Enter block name or [?]:** prompt. The next prompt issued is:

Enter block(s) to list <*>: Enter

If you press ENTER at this prompt, AutoCAD LT displays the list of blocks available in the current drawing. If you want to insert a drawing file as a block, enter the file name at the **Enter block name or [?]:** prompt. To create a block with a different name from the drawing file, enter **blockname= file name** at the **Enter block name or[?]:** prompt. If you enter ~ (tilde character) at the **Enter block name or [?]:** prompt, the **Select Drawing File** dialog box is displayed on the screen. If you enter an * (asterisk) before entering the block name, the block is automatically exploded upon insertion.

The next prompt asks you to specify the insertion point. The insertion point can be selected accurately by selecting or by using Osnaps. You can also enter the coordinate values of the insertion point. If you make changes to a block and want to update the block in the existing drawing, enter **blockname=** at the **Specify insertion point:** prompt. AutoCAD LT will display the prompt:

Block "current" already exists. Redefine it? [Yes/No] <No>: *Enter y to redefine it.*

The command prompt for specifying the insertion point is as follows:

Specify insertion point or [Scale/X/Y/Z/Rotate/PScale/PX/PY/PZ/PRotate]: *Select the point where you want the insertion base point for the block.*

The next two prompts ask you to specify X and Y scale factors.

Enter X scale factor, specify opposite corner or [Corner/XYZ]: *Select a point, enter a number, or press ENTER.*
Enter Y scale factor <Use X scale factor>: *Select a point, enter a number, or press ENTER.*

You can use the **XYZ** option to insert the 3D object. For this option, an additional prompt for specifying the Z scale factor is provided.

At the next prompt, you can specify the angle of rotation for the block to be inserted.

Specify rotation angle <0>: *Select a point or enter an angle.*

You can enter the numeral value of the rotation angle at the **Specify rotation angle <0>** prompt. You can also select a point to specify the angle. In this case, AutoCAD LT measures the angle of an imaginary line made by joining the insertion point and the point you have selected. The block is rotated by this angle.

Inserting the Mirror Image of a Block

Scaling of the block being inserted takes place by multiplying the X and Y dimensions of the block by the X and Y scale factors (specified in the **Enter X scale factor, specify opposite corner or [Corner/XYZ]** and **Enter Y scale factor <Use X scale factor>** prompts, respectively). If a negative value is specified for the X or Y scale factors, you can get a mirror image of a block. For instance, a scale factor of -1 for X and -1 for Y will mirror the block in the opposite quadrant of the coordinate system. There will be no change in the block size since the magnitude of the scale factor for X and Y variables is 1. For example, we have a block named ARC. The effect of negative scale factor on the block can be marked by a change in position of insertion point marked by a (X) (Figure 12-12) for the following prompt sequence:

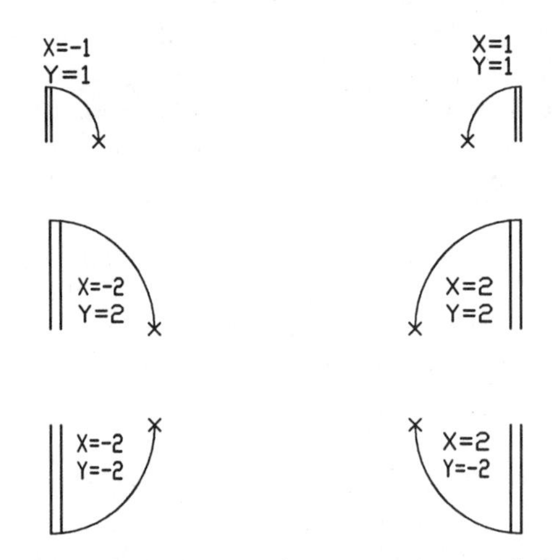

***Figure 12-12** Block inserted using negative scale factors*

Command: **-INSERT** Enter
Enter block name or [?] <current>: **ARC** Enter
Specify insertion point or [Scale/X/Y/Z/Rotate/PScale/PX/PY/PZ/PRotate]: *Select a point.*
Enter X scale factor, specify opposite corner or [Corner/XYZ] <1>: **-1** Enter
Enter Y scale factor <Use X scale factor>: **-1** Enter
Specify rotation angle <0>: Enter

Using the Corner Option

With the **Corner** option, you can specify both X and Y scale factors at the same time. When you invoke this option, and **DRAGMODE** is turned off, you are prompted to specify the other corner (the first corner of the box is the insertion point). You can also enter a coordinate value instead of moving the cursor. The length and breadth of the box are taken as the X and Y scale factors for the block. For example, if the X and Y dimensions (length and width) of the box are the same as that of the block, the block will be drawn without any change. Points selected should be above and to the right of the insertion block. Mirror images will be produced if points selected are below or to the left of the insertion point. The prompt sequence is:

Command:**-INSERT** Enter
Enter block name or [?]: **SQUARE** Enter
Specify insertion point or [Scale/X/Y/Z/Rotate/PScale/PX/PY/PZ/PRotate]: *Select a point.*
Enter X scale factor, specify opposite corner or [Corner/XYZ]: **C** Enter
Specify opposite corner: *Select a point as the corner.*

The **Corner** option also allows you to use a dynamic scaling technique when **DRAGMODE** is turned on (default). You can also move the cursor at the **Enter X scale factor** prompt to change the block size dynamically and, when the size meets your requirements, select a point.

Note

*You should try to avoid using the **Corner** option if you want to have identical X and Y scale factors because it is difficult to select a point whose X distance = Y distance, unless the **SNAP** mode is on. If you use the **Corner** option, it is better to specify the X and Y scale factors explicitly or select the corner by entering coordinates.*

Example 1

Using the dialog box, insert a block SQUARE into a drawing. The block insertion point is (1,2). The X scale factor is 2 units, the Y scale factor is 2 units, and the angle of rotation is 35 degrees. It is assumed that the block SQUARE is already defined in the current drawing.

Command: **INSERT** Enter

The **Insert** dialog box is displayed on the screen. In the **Name** edit box, enter SQUARE or select the Block SQUARE from the **Name** drop-down list. Clear the **Specify On-screen** check box. Now, under the **Insertion point** area, enter 1 in the **X** edit box and 2 in the **Y** edit box. Under the **Scale** area, enter 2 in the **X** edit box and 2 in the **Y** edit box. Under **Rotation**, enter 35 in the **Angle** edit box. Choose the **OK** button to insert the block on the screen.

Exercise 2 *General*

a. Insert the block CIRCLE created in Exercise 1. Use different X and Y scale factors to get different shapes after inserting this block.
b. Insert the block CIRCLE created in Exercise 1. Use the **Corner** option to specify the scale factor.

PRESETTING THE ROTATION, ANGLE, AND SCALE FACTORS

The scale factor and the rotation of a block can be preset before specifying the block's position of insertion. This option is ideal if you want to see the scale and/or angle of the block while the block is being dragged before actually being inserted into the drawing. Once you preset any or both scale factors and angles of rotation, the effect of these values is reflected in the image of the block. For the options with P as a prefix, the preset values can be negated, if desired, when inserting the block. To realize the presetting function, enter one of the following options at the **Specify insertion point** prompt.

Scale	**Xscale**	**Yscale**	**Zscale**	**Rotate**
PScale	**PXscale**	**PYscale**	**PZscale**	**PRotate**

Scale

If you enter Scale at the **Specify Insertion point** prompt, you are asked to enter the scale factor. After entering the scale factor, the block assumes the specified scale factor, and AutoCAD LT lets you drag the block until you locate the insertion point on the screen. The X, Y, and Z axes are uniformly scaled by the specified scale factor. Once the insertion point is specified, you are not prompted for the scale factors, so you cannot change the scale factor, and the block is drawn with the previously specified (preview) scale factor. When you use the **Scale** option the prompt sequence is:

Command: **-INSERT** Enter
Enter block name or [?]<current>: *Enter the block name.*
Specify insertion point or [Scale/X/Y/Z/Rotate/PScale/PX/PY/PZ/PRotate]: **SCALE** Enter
Specify scale factor for XYZ axes: *Enter a value to preset general scale factor.*
Specify insertion point: *Select the insertion point.*
Specify rotation angle <0>: *Specify angle of rotation.*

X

With **X** as the response to the **Specify insertion point** prompt, you can specify the X scale factor before specifying the insertion point. This option works similarly to the **Scale** option. The prompt sequence when you use the **X** option is:

Command: **-INSERT** Enter
Enter block name or [?]: *Enter the block name.*
Specify insertion point or [Scale/X/Y/Z/Rotate/PScale/PX/PY/PZ/PRotate]: **X** Enter
Specify X scale factor: *Enter a value to preset X scale factor.*
Specify insertion point: *Select the insertion point.*
Specify rotation angle <0>: *Specify angle of rotation.*

Y

With **Y** as the response to the **Specify insertion point** prompt, you can specify the Y scale factor before specifying the insertion point. This option works similarly to the **Scale** option.

Z

If you enter **Z** at the **Specify insertion point** prompt, you are asked to specify the Z scale factor before specifying the insertion point.

Rotate

If you enter **Rotate** at the **Specify insertion point** prompt, you are asked to specify the rotation angle. You can specify the angle by specifying two points. Dragging is resumed only when you specify the angle, and the block assumes the specified rotation angle. Once the insertion point is specified, you cannot change the rotation angle, and the block is drawn with the previously specified rotation angle. The prompt sequence for this option is:

Command: **-INSERT** Enter
Enter block name or [?]: *Enter the block name.*
Specify insertion point or [Scale/X/Y/Z/Rotate/PScale/PX/PY/PZ/PRotate]: ROTATE Enter
Specify rotation angle: *Enter a value to preview the rotation angle.*
Specify insertion point: *Select the insertion point.*
Enter X scale factor, specify opposite corner, or [Corner/XYZ] <1>: *Specify the X scale factor.*
Enter Y scale factor <Use X scale factor>: *Specify the Y scale factor.*

PScale

The **PScale** option is similar to the **Scale** option. The difference is that in the case of the **PScale** option, the scale factor specified reflects only in the display of the block while it is dragged to a desired position. After you have specified the insertion point for the block, you are again prompted for a scale factor. If you do not enter the scale factor, the block is drawn exactly as it was when created. The prompt sequence for this option is:

Command: **-INSERT** Enter
Enter block name or [?]: *Enter the block name.*
Specify insertion point or [Scale/X/Y/Z/Rotate/PScale/PX/PY/PZ/PRotate]: **PSCALE** Enter
Specify preview scale factor for XYZ axes: *Enter a value to preview the general scale factor.*
Specify insertion point: *Specify the insertion point for the block.*
Enter X scale factor, specify opposite corner, or [Corner/XYZ] <1>: *Specify the X scale factor.*
Enter Y scale factor <Use X scale factor>: *Specify the Y scale factor.*
Specify rotation angle <0>: *Specify angle of rotation.*

PX

This option is similar to the **PScale** option. The only difference is that the X scale factor specified is reflected only in the display of the block while it is dragged to a desired position. After you have specified the insertion point for the block, you are again prompted for the X scale factor and the Y scale factor. If you do not enter the scale factor, the block is drawn exactly as it was when created.

Command: **-INSERT** Enter
Enter block name or [?]: *Enter the block name.*
Specify insertion point or [Scale/X/Y/Z/Rotate/PScale/PX/PY/PZ/PRotate]: **PX** Enter
Specify preview X scale factor: *Enter a value to preview the X scale factor.*
Specify insertion point: *Specify the insertion point for the block.*
Enter X scale factor, specify opposite corner, or [Corner/XYZ] <1>: *Specify the X scale factor.*
Enter Y scale factor <Use X scale factor>: *Specify the Y scale factor.*
Specify rotation angle <0>: *Specify angle of rotation.*

PY

This option is similar to the **PX** option. The difference is that in the **PY** option, you can preview the Y scale factor.

PZ

This option is similar to the **PX** option. The difference is that in the **PZ** option, you can preview the Z scale factor.

PRotate

The **PRotate** option is similar to the **Rotate** option. The difference is that the rotation angle specified reflects only in the display of the block while it is dragged to a desired position. After you have specified the insertion point for the block, you are again prompted for an angle of rotation. If you do not enter the rotation angle, the block is drawn exactly as it was when created (with the same angle of rotation as specified on the creation of the block). The prompt sequence is:

Command: **-INSERT** [Enter]
Enter block name or [?] <current>: *Enter the block name.*
Specify insertion point or [Scale/X/Y/Z/Rotate/PScale/PX/PY/PZPRotate]: **PROTATE** [Enter]
Specify preview rotation angle: *Specify the preview value for rotation.*
Specify insertion point: Specify the insertion point for the block.
Enter X scale factor, specify opposite corner, or [Corner/XYZ]: Specify the X scale factor.
Enter Y scale factor < Use X scale factor>: *Specify the Y scale factor.*
Specify rotation angle <0>: *Specify angle of rotation.*

Exercise 3 *General*

a. Construct a triangle and form a block of it. Name the block TRIANGLE. Now, preset the Y scale factor so that on insertion, the Y scale factor of the inserted block is 2.
b. Insert the block TRIANGLE with preset rotation of 45 degrees. After defining the insertion point, enter an X and a Y scale factor of 2.

USING AUTOCAD DESIGNCENTER TO INSERT BLOCKS*

You can use the **AutoCAD DesignCenter** to copy or insert blocks from existing drawings to current drawings. On the **Standard** toolbar, choose the **AutoCAD DesignCenter** button to display the **Design Center** window. In this window, choose the **Tree View Toggle** and the **Desktop** buttons to display the **tree pane** on the left side. Expand **My Computer** to display the C:/Program/ACAD LT 2000/Sample/Designcenter folder. Click on this folder to display its contents. Select a drawing file you wish to use to insert blocks from and click on the plus sign adjacent to the drawing. Select **Blocks** by clicking on it; a list of blocks in the drawing are displayed. Select the block you wish to insert and drag and drop the block into the current drawing. Now, you can move it to the desired location. You can also right-click on the block name to display a shortcut menu. Select the **Insert as Block** option from the shortcut menu. The **Insert** dialog box is displayed. Specify the **Insertion point**, **scale**, **rotation angle** in the respective edit boxes. If you select the **Specify On-Screen** check boxes you are allowed to specify these parameters on the screen.

LAYERS, COLORS, LINETYPES, AND LINEWEIGHTS* FOR BLOCKS

A block possesses the properties of the layer on which it is drawn. The block may comprise objects drawn on several different layers, with different colors, linetypes and lineweights. All this information is preserved in block. At the time of insertion, each object in the block is drawn on its original layer with the original linetype, lineweight, and color, irrespective of the current drawing layer, object color, object linetype and object lineweights. You may want all instances of a block to have identical layer, linetype properties, lineweight, and color. This can be achieved by allocating all the properties explicitly to objects forming the block. On the other hand, if you want linetype and color of each instance of a block to be set according to the linetype and color of the layer on which it is inserted, draw all the objects forming the block on layer 0 and set the color, lineweight and linetype to BYLAYER. Objects with a BYLAYER color, linetype and lineweight can have their colors, linetypes, and lineweights changed after insertion by changing the layer settings. If you want the linetype, lineweight, and color of each instance of a block to be set according to the current explicit linetype, lineweight and color at the time of insertion, set the color, lineweight, and linetype of its objects to BYBLOCK. You can use the **PROPERTIES**, **CHPROP** or **CHANGE** command to change some of the characteristics associated with a block (such as layer). For a further description, refer to the **CHPROP** command or the **Properties** option of the **CHANGE** command [Chapter 16, Changing Properties and Points of an Object (**CHANGE** Command)].

Note

The block is inserted on the layer that is current, but the objects comprising the block are drawn on the layers on which they were drawn when the block was being defined.

For example, assume block B1 includes a square and a triangle that were originally drawn on layer X and layer Y, respectively. Let the color assigned to layer X be red and to layer Y be green. Also, let the linetype assigned to layer X be continuous and for layer Y be hidden. Now, if we insert B1 on layer L1 with color yellow and linetype dot, block B1 will be on layer L1, but the square will be drawn on layer X with color red and linetype continuous, and the triangle will be drawn on layer Y with color green and linetype hidden.

The **BYLAYER** option instructs AutoCAD LT to give objects within the block the color and linetype of the layers on which they were created. There are three exceptions:

1. If objects are drawn on a special layer (layer 0), they are inserted on the current layer. These objects assume the characteristics of the current layer (the layer on which the block is inserted) at the time of insertion, and can be modified after insertion by changing that layer's settings.

2. Objects created with the special color BYBLOCK are generated with the color that is current at the time of insertion of the block. This color may be explicit or BYLAYER. You are thus allowed to construct blocks that assume the current object color.

3. Objects created with the special linetype BYBLOCK are generated with the linetype that is prevalent at the time the block is inserted. Blocks are thus constructed with the current object linetype, which may be BYLAYER or explicit.

Note

If a block is inserted on a frozen layer, the block is not shown on the screen.

If you are providing drawing files to others for their use, using only BYLAYER settings provide the greatest compatibility with varying office standards for layer/ color/ linetype, and so on, because they can more easily be changed after insertion.

NESTING OF BLOCKS

The concept of having one block within another block is known as the **nesting of blocks**. For example, you can insert several blocks and then select them with the **BLOCK** command to create another block. Similarly, if you use **INSERT** command to insert a drawing that contains several blocks into current drawing, it creates a block containing nested blocks in the current drawing. There is no limit to the degree of nesting. The only limitation in nesting of blocks is that blocks that reference themselves cannot be inserted. The nested blocks must have different block names. Nesting of blocks affects layers, colors, and linetypes. The general rule is:

If an inner block has objects on layer 0, or objects with linetype or color BYBLOCK, these objects may be said to behave like fluids. They "float up" through the nested block structure until they find an outer block with fixed color, layer, or linetype. These objects then assume the characteristics of the fixed layer. If a fixed layer is not found in the outer blocks, then the objects with color or linetype BYBLOCK are formed; that is, they assume the color white and the linetype CONTINUOUS.

To make the concept of nested blocks clear, consider the following example:

Example 2

1. Draw a rectangle on layer 0, and form a block of it named X.
2. Change the current layer to OBJ, and set its color to red and linetype to hidden.
3. Draw a circle on OBJ layer.
4. Insert the block X in the OBJ layer.
5. Combine the circle with the block X (rectangle) to form a block Y.
6. Now, insert block Y in any layer (say, layer CEN) with color green and linetype continuous.

You will notice that block Y is generated in color red and linetype hidden. Normally, block X, which is nested in block Y and created on layer 0, should have been generated in the color (green) and linetype (continuous) of the layer CEN. The reason for this is that the object (rectangle) on layer 0 floated up through the nested block structure and assumed the color and linetype of the first outer block (Y) with a fixed color (red), layer (OBJ), and linetype (hidden). If both the blocks (X and Y) were on layer 0, the objects in block Y would assume color and linetype of the layer on which the block was inserted.

Exercise 4 *General*

1. Change the color of layer 0 to red.
2. Draw a circle and form a block of it, B1, with color BYBLOCK. It appears white because color is set to BYBLOCK.

3. Set the color to BYLAYER and draw a rectangle. The color of the rectangle is red.
4. Insert block B1. Notice that the block B1 (circle) assumes red color.
5. Create another block B2 consisting of Block B1 (circle) and rectangle.
6. Create a layer L1 with green color and hidden linetype. Insert block B2 in layer L1.
7. Explode block B2. Notice the change.
8. Explode block B1, circle. You will notice that the circle changes to white because it was drawn with color set to BYBLOCK.

Exercise 5 *General*

Part A

1. Draw a unit square on layer 0 and make it a block named B1.
2. Make a circle with radius 0.5 and change it into block B2 (Figure 12-13).
3. Insert block B1 into a drawing with an X scale factor of 3 and a Y scale factor of 4.
4. Now, insert the block B2 in the drawing and position it at the top of B1.
5. Make a block of the entire drawing and name it PLATE.
6. Insert the block Plate in the current layer.
7. Create a new layer with different colors and linetypes and insert blocks B1, B2, and Plate.

Keep in mind the layers on which the individual blocks and the inserted block were made.

Part B

Try nesting the blocks drawn on different layers and with different linetypes.

Part C

Change the layers and colors of the different blocks you have drawn so far.

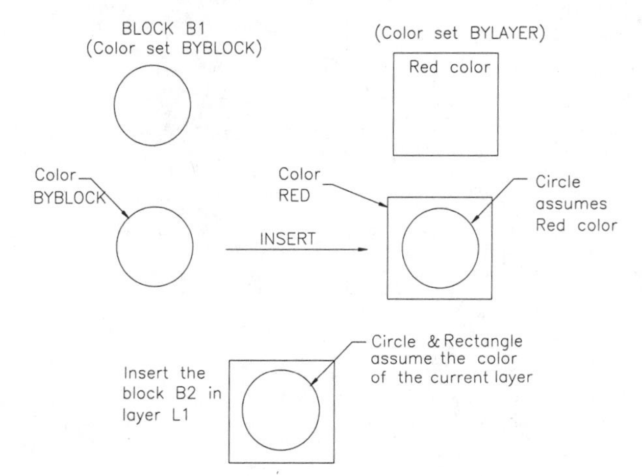

Figure 12-13 *Blocks versus layers and colors*

CREATING DRAWING FILES (WBLOCK COMMAND)

Command: WBLOCK

The blocks or symbols created by the **BLOCK** command can be used only in the drawing in which they were created. This is a shortcoming because you may need to use a particular block in different drawings. The **WBLOCK** command is used to export symbols by writing them to new drawing files that can then be inserted in any drawing. With the **WBLOCK** command, you can create a drawing file (.DWG extension) of specified blocks or selected objects in the current drawing. All the used named objects (linetypes, layers, styles, system variables) of the current drawing are taken by the new drawing created with the **WBLOCK** command. This block can then be inserted in any drawing.

When the **WBLOCK** command is invoked, the **Write Block** dialog box is displayed (Figure 12-14). This dialog box converts blocks into drawing files and also saves objects as drawing files. It has two main areas: the **Source** area and the **Destination** area.

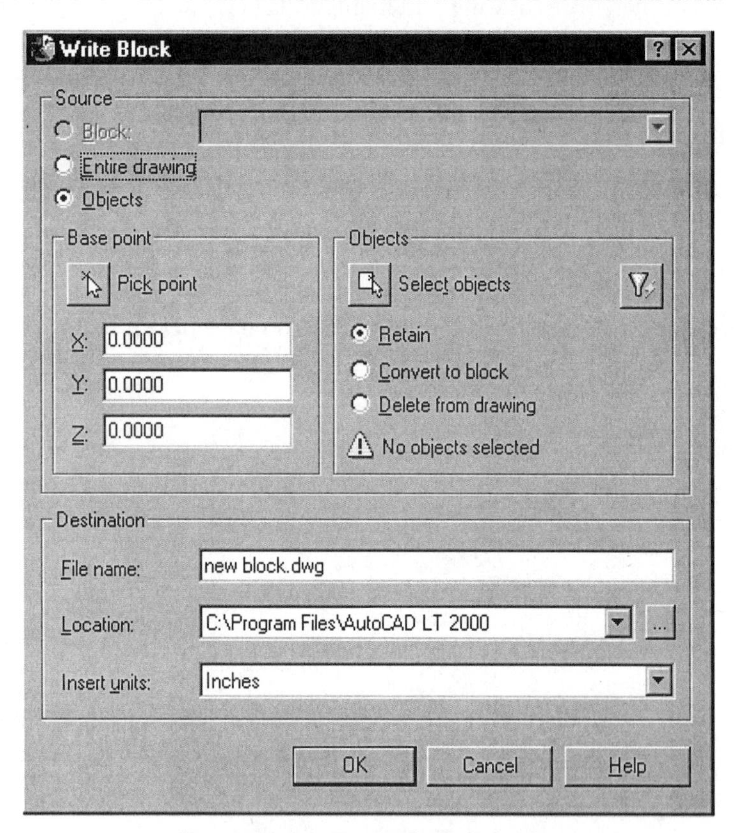

*Figure 12-14 **Write Block** dialog box*

The **Source** area allows you to select objects and blocks, specify insertion base points and convert them into drawing files. In this area of the dialog box, depending on the selection made by you, different default settings are displayed. If nothing is selected, the **Objects** radio button is shown as selected and in the **Destination** area, the **File Name** edit box displays **new**

block.dwg as the new file name. The **Block** drop-down list is also not available. If a single block is selected, the **Block** radio button is shown as selected and the name of the selected block is displayed in the **Block** edit box. **Base point** and **Objects** areas are not available. In the **Destination** area, **File Name** edit box displays the name of the selected block. The **Block** drop-down list displays all the block names in the current drawing. Selecting the **Entire drawing** radio button, selects the current drawing as a block and saves it as a file. Using this option, makes the **Base point** and **Objects** areas unavailable and is equivalent to entering **-WBLOCK/** * at the command line.

The **Base point** area allows you to specify the base point of a block which it uses as an insertion point. You can either enter values in the **X**, **Y**, and **Z** edit boxes or choose the **Pick insertion point** button to select it on the screen. The default value is 0,0,0. The **Objects** area allows you to select objects to save as a file. You can use the **Select objects** button to select objects or use the **Quick select** button to set parameters in the **Quick Select** dialog box to select objects in the current drawing. If the **Retain** radio button is selected in the **Objects** area, the selected objects in the current drawing are kept as such after they have been saved as a new file. If the **Convert to block** radio button is selected, the selected objects in the current drawing are converted into a block with the same name as the new file after being saved as a new file. Selecting the **Delete** radio button deletes the selected objects from the current drawing after they have been saved as a file. The number of objects selected is displayed at the bottom of the **Objects** area.

The **Destination** area sets the File name, location and units of the new file in which the selected objects are saved. In the **File name** edit box you can specify the file name of the block or the selected objects. From the **Location** drop-down list you can select a path for the new file. You can also choose the **[...]** button to display the **Browse for folder** window, where you can specify a path for where the new file will be saved. From the **Insert Units** drop-down list, you can select the units the new file will use when inserted as a block. The settings for units are stored in the **INSUNITS** system variable and the default option Unitless has a value of 0. On specifying the required information in the dialog box, choose **OK**. The objects or the block are saved as a new file in path specified by you. A **WBLOCK Preview** window with new file contents is displayed.

USING THE -WBLOCK COMMAND

When the **-WBLOCK** command is invoked, AutoCAD LT displays the **Create Drawing File** dialog box (Figure 12-15), except when the system variable **FILEDIA** is set to 0. This dialog box displays a list of all the drawing files in the current directory. Enter the name of the output drawing file in the **File name** edit box and then choose the **Save** button. If the file name you specify is identical to some other file in the same directory path, AutoCAD LT displays a message whether you want to replace the existing drawing with the new drawing. Right-clicking on a file name in the list box, displays a shortcut menu that allows you to Select, open, print, delete, rename, move, cut, copy, and list properties of the particular drawing file.

After choosing the **Save** button, the dialog box is cleared from the screen, and AutoCAD LT issues the following prompt asking you for the name of an existing block that you want to convert into a permanent symbol:

Enter name of existing block or
[=(block=output file)/*(whole drawing)] <define new drawing>: *Enter the name of a predefined block.*

The drawing file with the name specified in the dialog box is created from the predefined block.

The = Sign

If you want to assign the same name to the output drawing file as that of the existing block, enter an = (equal) sign as a response to the **Enter name of existing block** prompt and enter same file name in the **File name** edit box of the **Create Drawing File** dialog box.

If a block by the name you have already assigned to the output file does not exist in the current drawing, AutoCAD LT gives an appropriate message that no block by that name exists in the current drawing, and the **Enter name of existing block or** prompt is repeated.

Creating a New WBlock

If you have not yet created a block, but want to create an output drawing file of the objects in the current drawing, enter the **-WBLOCK** command and press ENTER instead of specifying a block name at the **Enter name of existing block** prompt, you are then required to select an insertion point and select the objects to be incorporated in the drawing file. The prompt sequence in this case is:

Command: **-WBLOCK** Enter

Enter a file name in the **File name** edit box of the **Create Drawing File** dialog box.

Enter name of existing block or
[=(block=output file)/*(whole drawing)] <define new drawing>: Enter
Specify insertion base point: *Select the insertion point.*
Select objects: *Select the objects to be incorporated in the drawing file.*
Select objects: *Press ENTER when finished selecting.*

Storing an Entire Drawing as a WBlock

You can also store an entire drawing as a WBlock. In other words, the current file is copied into a new one specified in the **File name** edit box. To do so, respond to the prompt **Enter name of existing block** with an asterisk (*). The prompt sequence is:

Command: **-WBLOCK** Enter

Enter a file name in the **File name** edit box of the **Create Drawing File** dialog box.

Enter name of existing block or
[=(block=output file)/*(whole drawing)] <define new drawing>:* Enter

Here, the entire drawing is saved in the file in the directory you have specified with the insertion base point as 0,0,0. However you can change the insertion base point using the **BASE** command

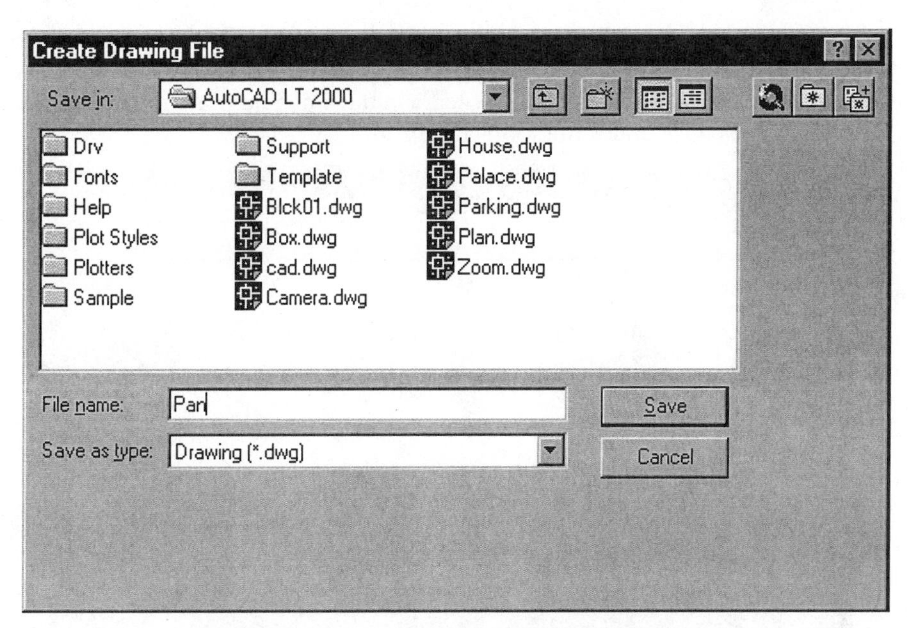

***Figure 12-15 Create Drawing File** dialog box*

which is discussed in detail later on in this chapter. The effect is identical to the **SAVE** command. The lone difference is that if you use the * response in the **-WBLOCK** command, all unused named objects are automatically purged. In this manner, the size of a drawing containing a number of unused blocks is reduced by a considerable amount. When the **FILEDIA** is set to 0, entering **-WBLOCK** at the command prompt, does not display the **Create Drawing File** dialog box and displays prompts on the command line as follows:

Command: **-WBLOCK** Enter
Enter name of output file: *Enter a name for the output file.*
Enter name of existing block or
[=(block=output file)/*(whole drawing)] <define new drawing>: *Enter a name of an existing block, enter=, enter* or press ENTER.*

The existing block whose name you have entered will be saved as a drawing file. If you enter =, the existing block and the output file have the same name. Entering * at the **Enter name of existing block or** prompt, saves the entire drawing to a new output file. Pressing ENTER allows you to define a new drawing. Once the file is defined, the selected objects are deleted from the drawing. They can be restored by using the **OOPS** command.

This sequence is similar to that of the **-BLOCK** command. The drawing is saved in the current drawing symbol table and written to the disk as a drawing file.

Export Data Dialog Box

You can also use the **Export Data** dialog box (Figure 12-16) to create drawing files. The **Export Data** dialog box can be invoked from the **File** menu (**File > Export**), or by entering **EXPORT** at the Command prompt.

In this dialog box, you can specify information about the drawing file to be created. Select the Block (*.dwg) format from the **Save as type** drop-down list, enter the name of the drawing file in the **File name** edit box, and then choose the **Save** button. The dialog box is cleared from the screen and AutoCAD LT issues the following prompt:

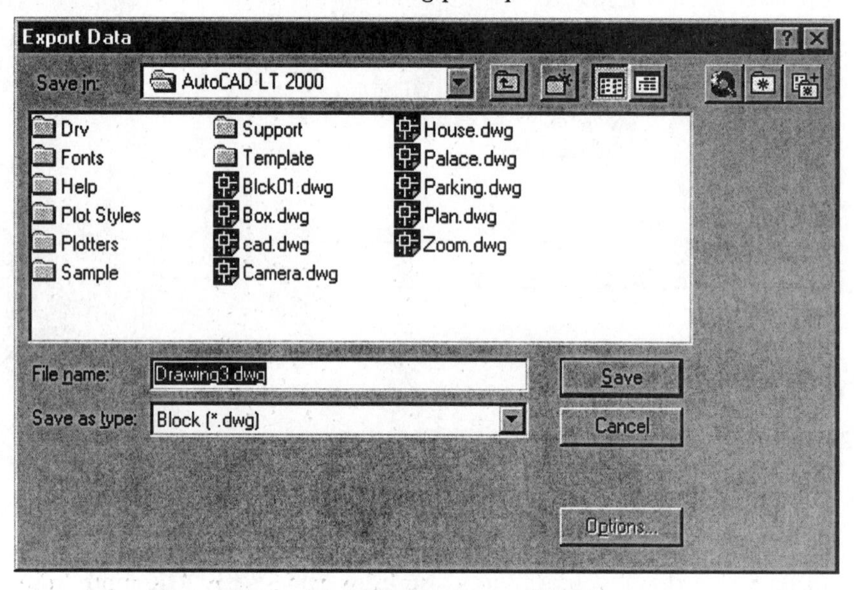

***Figure 12-16 Export Data** dialog box*

Enter name of existing block or
[=(block=output file)/*(whole drawing)] <define new drawing>: *Enter the name of a predefined block.*

The drawing file with the name specified in the dialog box is created from the predefined block.

DEFINING THE INSERTION BASE POINT (BASE COMMAND)

Menu:	Draw > Block > Base
Command:	BASE

The **BASE** command lets you set the insertion base point for a drawing just as you set the base insertion point in the prompt sequence of the **-BLOCK** command. This base point is defined so that when you insert the drawing into some other drawing, the specified base point is placed on the insertion point. By default, the base point is at the origin (0,0,0). The prompt sequence is:

Command: **BASE** Enter
Enter base point <current>: *Specify the base point or press ENTER.*

Exercise 6 — General

a. Create a drawing file named CHAIR using the **WBLOCK** command. Get a listing of your .DWG files and make sure that CHAIR.DWG is listed. Quit the drawing editor.

b. Begin a new drawing and insert the drawing file into the drawing. Save the drawing.

EDITING BLOCKS

Breaking Up a Block

A block may be comprised of different basic objects, such as lines, arcs, polylines, and circles. All these objects are grouped together in the block and are treated as a single object. To edit any particular object of a block, the block needs to be "exploded," or split into independent parts. This is especially useful when an entire view or drawing has been inserted and a small detail needs to be corrected. This can be done in the following ways.

Entering * as the Prefix of Block Name. As we discussed, individual objects within a block cannot be edited unless the block is broken up (exploded). To insert a block as a collection of individual objects, you need to enter an asterisk before its name. The prompt sequence for exploding the block BLOCKPLAN while inserting is:

Command: **INSERT** [Enter]
Enter block name or [?] <current>: ***BLOCKPLAN** [Enter]
Specify insertion point or [Scale/X/Y/Z/Rotate/PScale/PX/PY/PZ/PRotate]: *Select the insertion point.*
Scale factor <1>: [Enter]
Specify rotation angle <0>: [Enter]

The inserted object will no longer be treated as a block. The figure consists of various objects that can be edited separately.

Using the EXPLODE Command

Toolbar:	Modify > Explode
Menu:	Modify > Explode
Command:	EXPLODE

As we mentioned before, the other method to break a block into individual objects is by using the **EXPLODE** command (Figure 12-17). The prompt sequence is:

Command: **EXPLODE** [Enter]
Select objects: *Select the block to be exploded.*

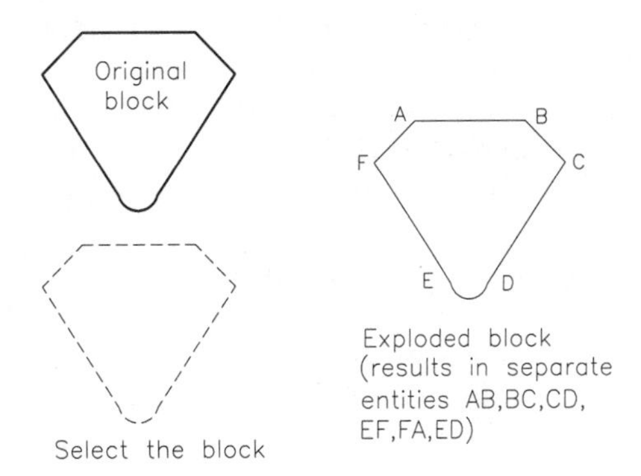

Figure 12-17 *Using the* ***EXPLODE*** *command*

The **EXPLODE** command explodes a block into its component objects regardless of the scale factors. It also does not have control over the properties like, layer, linetype, lineweight, and color of the component objects. When a block is exploded, there may be no visible change in the drawing. The drawing is identical except that the color and linetype may have changed because of floating layers, colors, or linetypes. The exploded block is now a group of objects that can be edited individually. To check whether the breaking of the block has taken place, select any object that was formerly a part of the block; only that particular object should be highlighted. After a block is exploded, the block definition continues to be in the block symbol table. After exploding a block you can modify it and then redefine it by using the **BLOCK** command. The **Block Definition** dialog box is displayed where you can choose the name of the block you wish to redefine from the **Name:** drop-down list and then, use the dialog box options and redefine it. You can also use the **-BLOCK** or the **INSERT** commands to redefine a block. Once you redefine a block, it automatically gets updated in the current drawing and for future use. Blocks inserted External references cannot be exploded.

Using the XPLODE command*

Command: XPLODE

With the **XPLODE** command, you can explode a block or blocks into component objects and simultaneously control their properties like Layer, linetype, color, and lineweight. The scale factor of the object to be exploded should be equal and absolute, that is 1,-1,1. The command prompts are as follows:

Command: **XPLODE** Enter
Select objects: Use *any object selection method and select objects and then press ENTER.*

On pressing ENTER, AutoCAD LT reports the total number of objects selected and also the number of objects that cannot be exploded. It also further prompts for an option that decides whether changes in the properties of the component objects will be made individually or globally. The prompt is:

Enter an option [Individually/Globally] <Globally>: *Enter i, g, or press ENTER to accept the default option.*

If you enter **i** at the **Enter an option [Individually/Globally] <Globally>** prompt, AutoCAD LT will modify each object individually, one at a time. The command prompt is:

Enter an option [All/Color/LAyer/LType/Inherit from parent block/Explode] <Explode>: *Select an option.*

The options available are discussed below:

All. Sets all the properties like color, layer, linetype, and lineweight of selected objects after exploding them. AutoCAD LT prompts you to enter new color, linetype, lineweight, and layer name for the exploded component objects.

Color. Sets the color of the exploded objects. The prompt is:

Enter new color for exploded objects [Red/Yellow/Cyan/Blue/Magenta/White/BYLayer/BYBlock]<BYLAYER>: *Enter a color, option or press ENTER.*

On entering BYLAYER, the component objects take on the color of the exploded object's layer and on entering BYBLOCK, they take on the color of the exploded object.

Layer. Sets the layer of the exploded objects. The default option is inheriting the current layer. The command prompt is as follows:

Enter new layer name for exploded objects <current>: *Enter an existing layer name or press ENTER*

LType. Sets the linetype of the components of the exploded object. The command prompt is:

Enter new linetype name for exploded objects <BYLAYER>: *Enter a linetype name or press ENTER.*

Inherit from Parent Block. Sets the properties of the component objects to that of the exploded parent object, provided the component objects are drawn on layer 0 and the color, lineweight, and linetype are BYBLOCK.

Explode. Explodes the selected object exactly as in the **EXPLODE** command.

Selecting the **Globally** option, applies changes to all the selected objects at the same time and the options are similar to the ones discussed in the **Individually** option.

RENAMING BLOCKS

Menu:	Format > Rename
Command:	RENAME

Blocks can be renamed with the **RENAME** command. AutoCAD LT displays the **Rename** dialog box (Figure 12-18) which allows you to modify the name of an existing block.

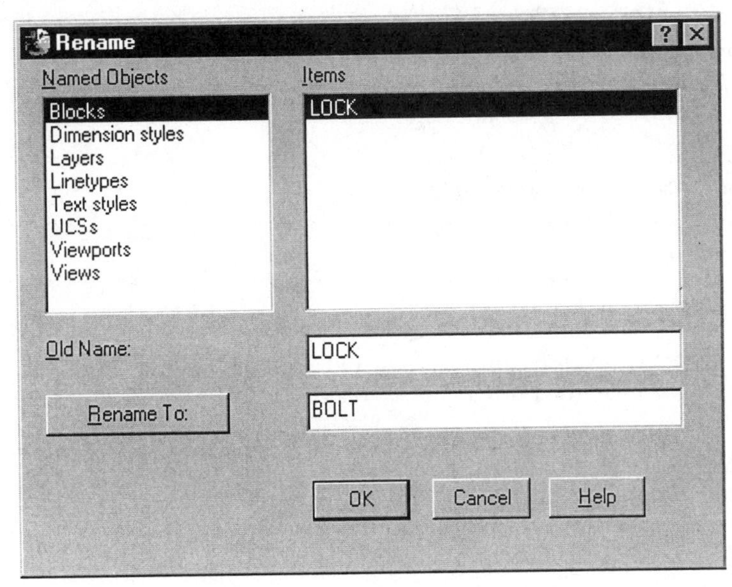

***Figure 12-18 Rename** dialog box*

The **Named Objects** list box displays the categories of object types that can be renamed, such as: blocks, layers, dimension styles, linetypes, plot and text styles, UCSs, views, and tiled viewports. You can rename all of these except layer 0 and continuous linetype. On selecting **Blocks** from the **Named Objects** list, the **Items** list box displays all the block names in the current drawing. When you select a block name you wish to rename from the **Items** list, the block name is displayed in the **Old Name** edit box. You can enter a block name you wish to modify in the **Old Name** edit box or select it from the **Items** list box. Enter the new name to be assigned to the block in the **Rename To** edit box. Choosing the **Rename To** button, applies the change in name to the old name. Choose **OK** to exit the dialog box.

If you enter **-RENAME** at the Command prompt, command prompts are displayed at the Command line. You can rename a block using the command line. For example, if you want to rename a block named FIRST to SECOND, the prompt sequence is:

Command: **-RENAME** Enter
Enter object type to rename
[Block/Dimstyle/LAyer/LType/Style/Ucs/VIew/VPort]: **B** Enter
Enter old block name: **FIRST** Enter
Enter new block name: **SECOND** Enter

DELETING UNUSED BLOCKS

Menu:	File > Drawing Utilities > Purge
Command:	PURGE

The **PURGE** command is used to delete the blocks that are not used in the drawing. Entering **PURGE** at the Comand prompt displays the **Purge** dialog box shown in Figure 12-19.

Dialog Box Options

View items you can purge

When you choose this radio button, all the unused items in the drawing that can be deleted are displayed.

View items you cannot purge

As you choose this radio button, all the items in the drawing that are used and cannot be purged are displayed.

Confirm each items to be purged

If this check box is not cleared, the **Confirm Purge** dialog box is displayed every time you purge an item. To confirm the purge, choose **Yes** button.

Purge nested items

Nested items are generally those that are within the blocks and can be purged from within the blocks. This check box is used to remove all the nested items when the **Purge All** button is chosen or the purge is performed on all the items.

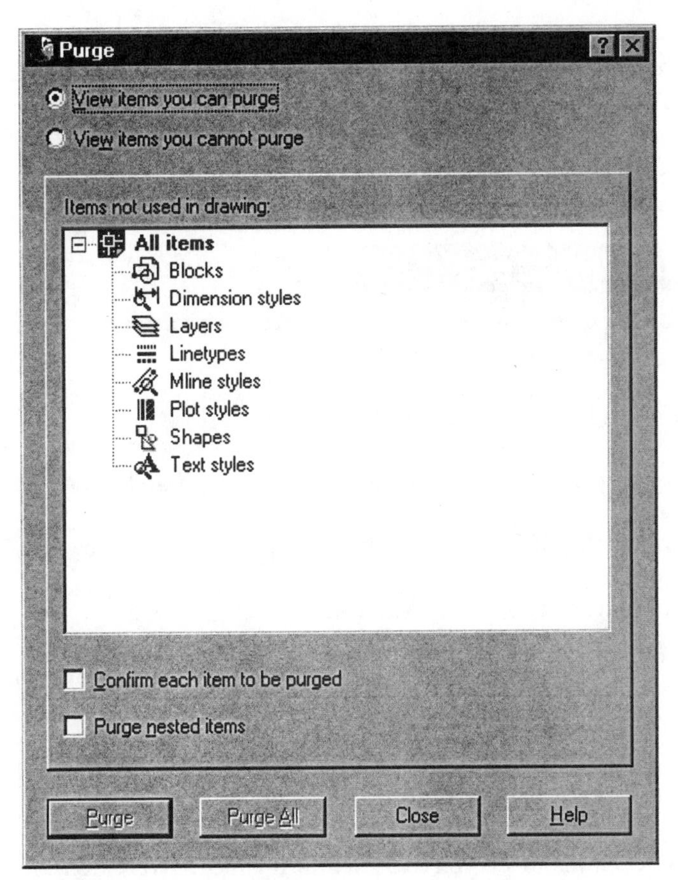

*Figure 12-19 **Purge** dialog box*

Purge

This button is chosen to delete the unused items you have selected in the tree view.

Purge All

As you choose this button, all the unused items in the drawing are deleted. However before deleting each items, AutoCAD LT displays **Confirm Purge** dialog box for each items even if the **Confirm each item to be purged** check box is cleared.

Deleting unused blocks using command line

You can also delete unused blocks using the command line with the help of the **-PURGE** command. For example, if you want to delete an unused block named SECOND, the prompt sequence is:

Command: **-PURGE** [Enter]
Enter type of unused objects to purge
[Blocks/Dimstyles/LAyers/LTypes/Plotstyles/SHapes/textSTyles/Mlinestyles/ All]: **B** [Enter]
Enter names to purge <*>: **SECOND** [Enter]

Verify each name to be purged? [Yes/No] <Y> Enter
Purge block SECOND? <N>: **Y** Enter

If there are no objects to remove, AutoCAD LT displays a message that there are no unreferenced objects to purge.

Self-Evaluation Test

Answer the following questions, and then compare your answers with the correct answers given at the end of this chapter.

1. Individual objects forming a block can be erased. (T/F)

2. A block can be mirrored by providing a scale factor of -1 for X and -1 for Y during insertion. (T/F)

3. Blocks created by the **BLOCK** command can be used in any drawing. (T/F)

4. An existing block cannot be redefined. (T/F)

5. At the time of insertion, each object that makes up a block is drawn on the current layer with the current linetype and color. (T/F)

6. If a block that is inserted is on a frozen layer, the block is generated only if portions of the block lie on nonfrozen (thawed) layers. (T/F)

7. The **WBLOCK** command lets you create a drawing file (.DWG extension) of a block defined in the current drawing. (T/F)

8. The **RENAME** command can be used to change the name of a drawing file. (T/F)

9. The __________ command is used to place a previously created block in a drawing.

10. You can delete unreferenced blocks with the __________ command.

Review Questions

1. Only blocks can be scaled. (T/F)

2. Blocks can be scaled or rotated upon insertion. (T/F)

3. Drawing files created by the **WBLOCK** command can be used in any drawing. (T/F)

4. An entire drawing can be converted into a block. (T/F)

5. Invoking the **INSERT** command causes generation of the **Insert** dialog box. (T/F)

6. The objects in a block possess the properties of the layer on which they are drawn, such as color and linetype. (T/F)

7. If the objects forming a block were drawn on layer 0 with color and linetype BYLAYER, then at the time of the insertion, each object that makes up a block is drawn on the current layer with the current linetype and color. (T/F)

8. Objects created with the special color BYBLOCK are generated with the color that is current at the time the block was inserted. (T/F)

9. Objects created with the special linetype BYBLOCK are generated with the linetype that is current at the time the block was inserted. (T/F)

10. The color, linetype, and layer on which a block is drawn can be changed with the Properties option of the **CHANGE** command. (T/F)

11. Suppose objects in Block1 have color BYBLOCK. If you incorporate Block1 into Block2 and insert with the current color set to green, Block1 also assumes the green color. (T/F)

12. Suppose objects in Block1 have color BYLAYER. If you incorporate Block1 into Block2 and insert it with the current color set to red, Block1 will assume the color of Block2, not the current color of the object's layer(s). (T/F)

13. The **WBLOCK** command allows you to convert an existing block into a drawing file. (T/F)

14. If a block was inserted with nonuniform scales (unequal X and Y), it can be exploded. (T/F)

15. After exploding an object, the object remains identical except that the color and linetype may change because of floating layers, colors, or linetypes. (T/F)

16. **WBLOCK**-asterisk method has the same effect as the **PURGE** command. The only difference is that with the **PURGE** command, deletion takes place automatically. (T/F)

17. The **RENAME** command can be used to change the name of a block. (T/F)

18. The __________ command is used to place a previously created block in a drawing.

19. The Xscale factor of a block can be preset before specifying its position of insertion by entering __________ at the **Specify insertion point** prompt.

20. Each X, Y, and Z scale factor of a block can be preset to a uniform value before specifying its position of insertion by entering __________ at the **Specify insertion point** prompt.

21. The __________ option lets you redefine the preset Y scale factor even after you have specified the insertion point for the block.

22. The concept of blocks within blocks is known as the __________.

Exercises

Exercise 7 — *Mechanical*

Draw part (a) of Figure 12-20 and define it as a block named A. Then, using the block insert command, insert the block in the plate as shown.

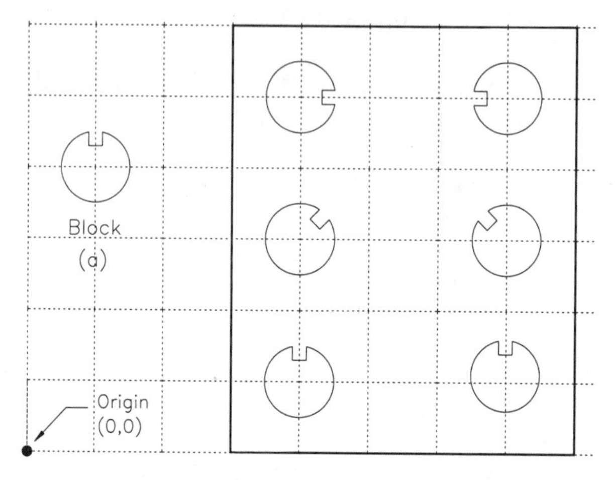

***Figure 12-20** Drawing for Exercise 7*

Exercise 8 — *Piping*

Draw the diagrams in Figure 12-21 using blocks.
a. Create a block for the valve Figure 12-21(a).
b. Use a thick polyline for the flow lines.

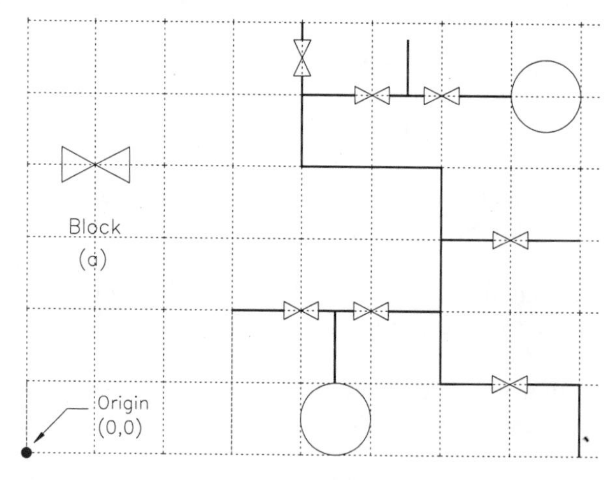

***Figure 12-21** Drawing for Exercise 8*

Exercise 9 *Mechanical*

Draw part (a) of Figure 12-22 and define it as a block named B. Then, using the relevant insertion method, generate the pattern as shown. Note that the pattern is rotated at 30 degrees.

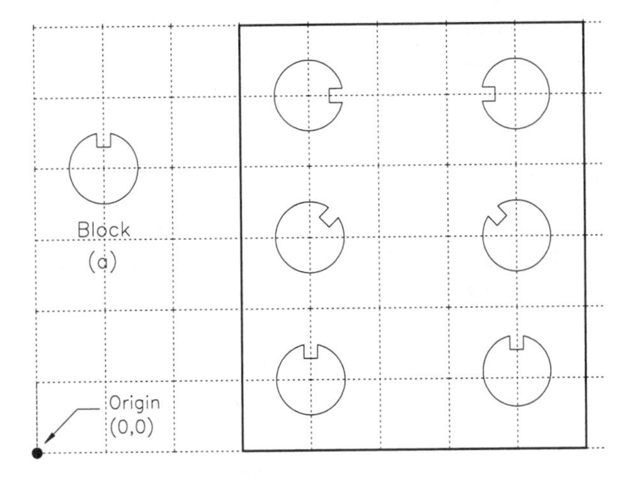

Figure 12-22 *Drawing for Exercise 9*

Exercise 10 *Electronics*

Draw the circuit diagram as shown in Figure 12-23 using blocks. Create the blocks first, and then insert the blocks to complete the circuit diagram.

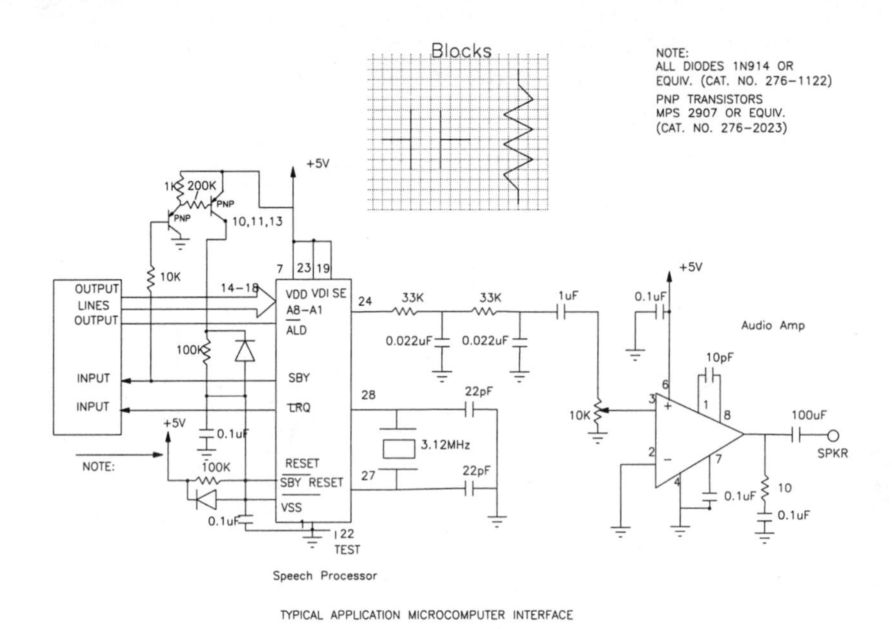

***Figure 12-23** Drawing for Exercise 10*

Answers to Self-Evaluation Test

1 - F, **2** - T, **3** - F, **4** - F, **5** - F, **6** - F, **7** - T, **8** - F, **9** - **INSERT**, **10** - **PURGE**

Chapter 13

Defining Block Attributes

Learning Objectives

After completing this chapter, you will be able to:
- *Understand what attributes are and how to define attributes with a block.*
- *Edit attribute tag names.*
- *Insert blocks with attributes and assign values to attributes.*
- *Extract attribute values from the inserted blocks.*
- *Control attribute visibility.*
- *Perform global and individual editing of attributes.*
- *Insert a text file in a drawing to create bill of material.*

ATTRIBUTES

AutoCAD has provided a facility that allows the user to attach information to blocks. This information can then be retrieved and processed by other programs for various purposes. For example, you can use this information to create a bill of material, find the total number of computers in a building, or determine the location of each block in a drawing. Attributes can also be used to create blocks (such as title blocks) with prompted or preformatted text, to control text placement. The information associated with a block is known as **attribute value** or simply **attribute**. AutoCAD references the attributes with a block through tag names.

Before you can assign attributes to a block, you must create an attribute definition by using the **ATTDEF** command. The attribute definition describes the characteristics of the attribute. You can define several attribute definitions (tags) and include them in the block definition. Each time you insert the block, AutoCAD will prompt you to enter the value of the attribute. The attribute value automatically replaces the attribute tag name. The information (attribute values) assigned to a block can be extracted and written to a file by using AutoCAD's **ATTEXT** command. This file can then be inserted in the drawing as a table or processed by other programs to analyze the data. The attribute values can be edited by using the **ATTEDIT** command. The display of attributes can be controlled with **ATTDISP** command.

DEFINING ATTRIBUTES

ATTDEF Command

Menu:	Draw > Block > Define Attributes
Command:	ATTDEF

When you invoke the **ATTDEF** command, the **Attribute Definition** dialog box (Figure 13-1) is displayed. The block attributes can be defined through this dialog box. When you create an attribute definition, you must define the mode, attributes, insertion point, and text information for each attribute. All this information can be entered in the dialog box. Following is a description of each area of the **Attribute Definition** dialog box.

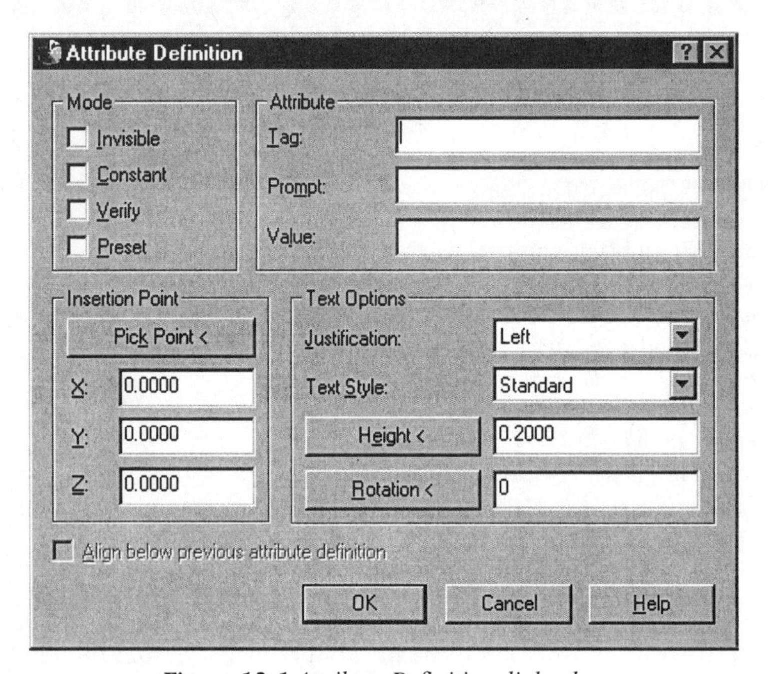

Figure 13-1 Attribute Definition dialog box

Mode. The **Mode** area of the **Attribute Definition** dialog box (Figure 13-2) has four options: **Invisible**, **Constant**, **Verify**, and **Preset**. These options determine the display and edit features of the block attributes. For example, if an attribute is invisible, the attribute is not displayed on the screen. Similarly, if an attribute is constant, its value is predefined and cannot be changed. These options are described below:

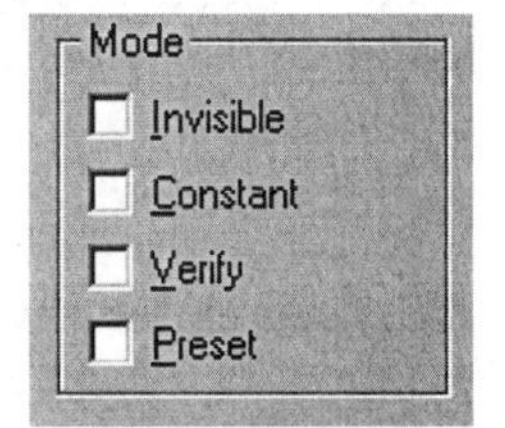

*Figure 13-2 **Mode** area of the **Attribute Definition** dialog box*

Invisible. This option lets you create an attribute that is not visible on the screen. This mode is useful when you do not want the attribute values to be displayed on the screen to avoid cluttering the drawing. Also, if the attributes are invisible, it takes less time to regenerate the drawing. If you want to make the invisible attribute visible, use the **ATTDISP** command discussed later in this chapter [section "Controlling Attribute Visibility (**ATTDISP** COMMAND)"].

Constant. This option lets you create an attribute that has a fixed value and cannot be changed after block insertion. When you select this mode, the **Prompt** edit box and the **Verify** and **Preset** check boxes are disabled.

Verify. This option allows you to verify the attribute value you have entered when inserting a block, by asking you twice for the data. If the value is incorrect, you can correct it by entering the new value.

Preset. This option allows you to create an attribute that is automatically set to default value. The prompt is not displayed and an attribute value is not requested when you insert a block with attributes using this option to define a block attribute. Unlike a constant attribute, the preset attribute value can later be edited.

Attribute. The **Attribute** area (Figure 13-3) of the **Attribute Definition** dialog box has three edit boxes: **Tag**, **Prompt**, and **Value**. To enter a value, you must first select the corresponding edit box and then enter the value. You can enter up to 256 characters in these edit boxes.

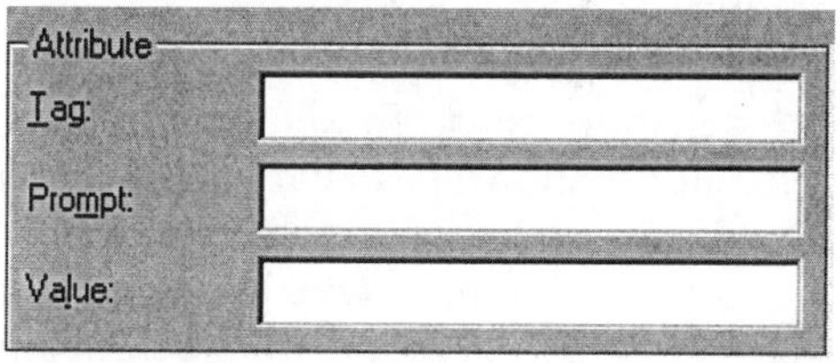

Figure 13-3 ***Attribute*** *area of the* ***Attribute Definition*** *dialog box*

Tag. This is like a label that is used to identify an attribute. For example, the tag name COMPUTER can be used to identify an item. The tag names can be uppercase, lowercase, or both. Any lowercase letters are automatically converted into uppercase. The tag name cannot be null. Also, the tag name must not contain any blank spaces. You should select a tag name that reflects the contents of the item being tagged. For example, the tag name COMP or COMPUTER is an appropriate name for labeling computers.

Prompt. The text that you enter in the **Prompt** edit box is used as a prompt when you insert a block that contains the defined attribute. If you have selected the **Constant** option in the **Mode** area, the **Prompt** edit box is disabled because no prompt is required if the attribute is constant. If you do enter nothing in the **Prompt** edit box, the entry made in the **Tag** edit box is used as the prompt.

Value. The entry in the **Value** edit box defines the default value of the specified attribute; that is, if you do not enter a value, it is used as the value for the attribute. The entry of a value is optional.

Insertion Point. The **Insertion Point** area of the **Attribute Definition** dialog box (Figure 13-4) lets you define the insertion point of block attribute text. You can define the insertion point by entering the values in the **X**, **Y**, and **Z** edit boxes or by selecting the **Pick Point** button. If you select this button, the dialog box clears, and you can enter the X, Y, and Z values of the insertion point at the command line or specify the point by selecting a point on the screen. When you are done specifying the insertion point, the **Attribute Definition** dialog box reappears.

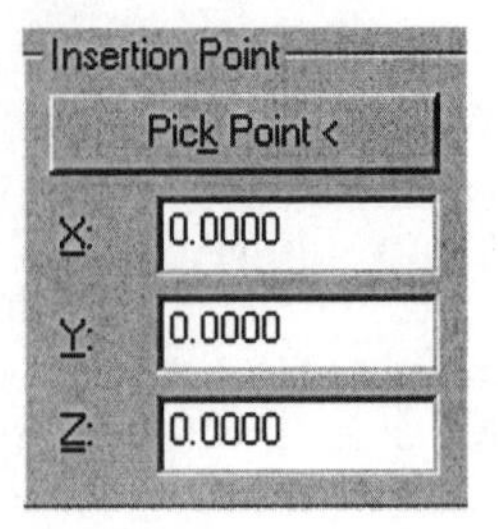

*Figure 13-4 Insertion Point area of the **Attribute Definition** dialog box*

Just under the **Insertion Point** area of the dialog box is a check box labeled **Align below previous attribute definition**. You can use this box to place the subsequent attribute text just below the previously defined attribute automatically. This check box is disabled if no attribute has been defined. When you select this check box, the **Insertion Point** area and the **Text Options** areas are disabled because AutoCAD assumes previously defined values for text such as text height, text style, text justification, and text rotation. Also, the text is automatically placed on the following line. After insertion, the attribute text is responsive to the setting of the **MIRRTEXT** system variable.

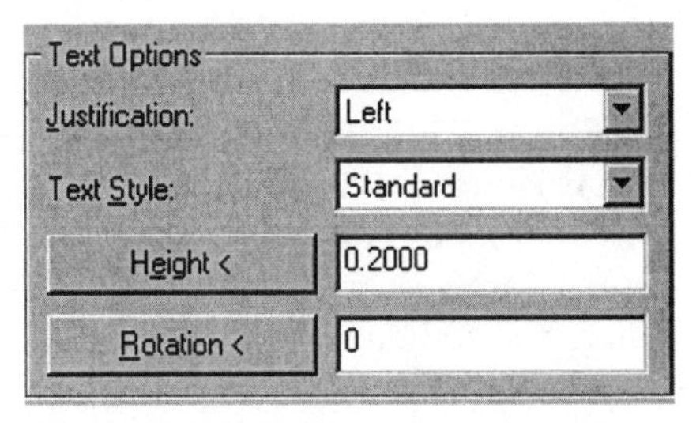

*Figure 13-5 **Text Options** area of **Attribute Definition** dialog box*

Text Options. The **Text Options** area of the **Attribute Definition** dialog box (Figure 13-5) lets you define the justification, text style, height, and rotation of the attribute text. To set the text justification, select justification type in the **Justification** drop-down list. Similarly, you can use the **Text Style** drop-down list to define the text style. You can specify the text height and text rotation in the **Height** and **Rotation** edit boxes. You can also define the text height by selecting the **Height** button. If you select this button, AutoCAD temporarily exits the dialog box and lets you enter the value from the command line. Similarly, you can define the text rotation by selecting the **Rotation** button and then entering the rotation angle at the command line.

Note

The text style must be defined before it can be used to specify the text style.

*If you select a style that has the height predefined, AutoCAD automatically disables the **Height** edit box.*

*If you have selected the **Align** option for the text justification, the **Height** and **Rotation** edit box are disabled.*

*If you have selected the **Fit** option for the text justification, the **Rotation** edit box is disabled.*

After you complete the settings in the **Attribute Definition** dialog box and choose **OK**, the attribute tag text is inserted in the drawing at the specified insertion point.

Using Command Line

You can define block attributes through the command line by entering the **-ATTDEF** command.

Command: **-ATTDEF**
Current attribute modes: Invisible=N Constant=N Verify=N Preset=N
Enter an option to change [Invisible/Constant/Verify/Preset] <done>:
Enter attribute tag name:

The default value of all attribute modes is **N** (No). To reverse the default mode, enter **I**, **C**, **V**, or **P**. For example, if you enter **I**, AutoCAD will change the Invisible mode from **N** to **Y** (Yes). This will make the attribute visible. After setting the modes, press ENTER to go to the next prompts where you can enter the attribute tag, attribute prompt, and the attribute values.

Enter attribute tag name:
Enter attribute prompt:
Enter default attribute value:

The entry at this prompt defines the default value of the specified attribute; that is, if you do not enter a value, it is used as the value for the attribute. The entry of a value is optional. If you have selected the **Constant** mode, AutoCAD displays the following prompt:

Enter attribute tag name:
Enter attribute value:

Next, AutoCAD displays the following text prompts:

Current text style: "Standard" Text height: *current*
Specify start point of text or [Justify/Style]:
Specify height <*current*>:
Specify rotation angle of text <*current*>:

After you respond to these prompts, the attribute tag text will be placed at the specified location. If you press ENTER at **Specify start point of text or [Justify/Style]:**, AutoCAD will automatically place the subsequent attribute text just below the previously defined attribute, and it assumes previously defined text values such as text height, text style, text justification, and text rotation. Also, the text is automatically placed on the following line.

Example 1

In this example, you will define the following attributes for a computer and then create a block using the **BLOCK** command. The name of the block is COMP.

Mode	Tag name	Prompt	Default value
Constant	ITEM		Computer
Preset, Verify	MAKE	Enter make:	CAD-CIM
Verify	PROCESSOR	Enter processor type:	Unknown
Verify	HD	Enter Hard-Drive size:	100MB
Invisible,Verify	RAM	Enter RAM:	4MB

1. Draw the computer as shown in (Figure 13-6). Assume the dimensions, or measure the

dimensions of the computer you are using for AutoCAD.

2. Invoke the **ATTDEF** command. The **Attribute Definition** dialog box is displayed (Figure 13-7).

3. Define the first attribute as shown in the preceding table. Select **Constant** in the **Mode** area because the mode of the first attribute is constant. In the **Tag**: edit box, enter the tag name, **ITEM**. Similarly, enter Computer in the **Value** edit box. The **Prompt** edit box is disabled because the variable is constant.

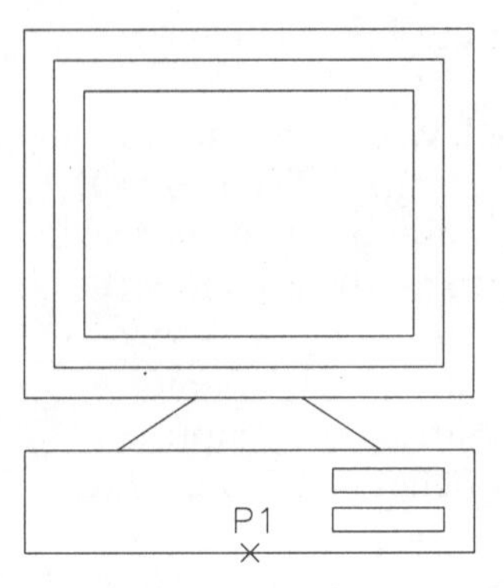

Figure 13-6 *Drawing for Example 1*

4. In the **Insertion Point** area, choose the **Pick Point** button to define the text insertion point. Select a point below the drawing of the computer.

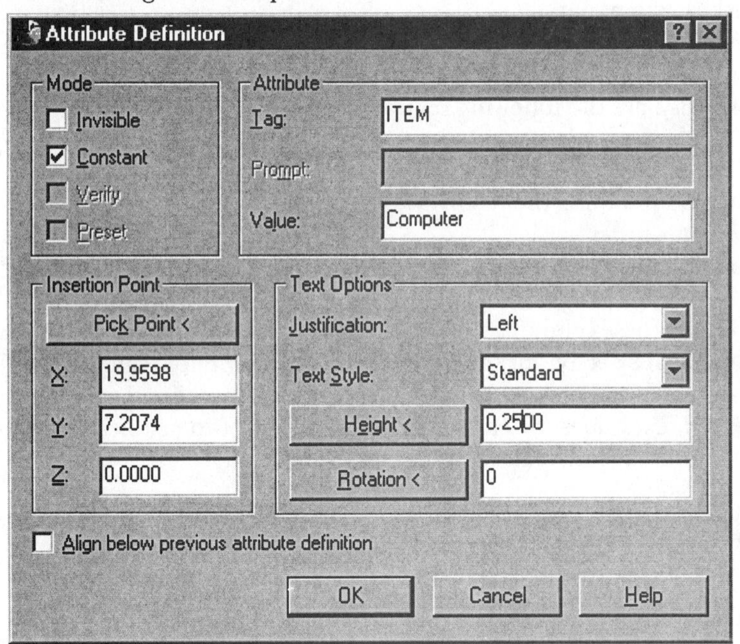

Figure 13-7 *Enter information in the* ***Attribute Definition*** *dialog*

5. In the **Text Options** area, select the justification, style, height, and rotation of the text.

6. Choose the **OK** button when you are done entering information in the **Attribute Definition** dialog box.

7. Enter **ATTDEF** at the Command prompt to invoke the **Attribute Definition** dialog box. Enter the Mode and Attribute information for the second attribute shown in the table at

the beginning of Example 1. You need not define the **Insertion Point** and **Text Options**. Select the **Align below previous attribute definition** that is located just below the Insertion Point area. When you check this box, the Insertion Point and Text Justification areas are disabled. AutoCAD places the attribute text just below the previous attribute text.

Figure 13-8 Define attributes below the computer drawing

8. Define the remaining attributes (Figure 13-8).

9. Use the **BLOCK** command to create a block. The name of the block is COMP, and the insertion point of the block is P1, midpoint of the base. When you select the objects for the block, make sure you also select the attributes. The order of attribute selection controls the order of prompts.

EDITING ATTRIBUTE TAGS

Using the DDEDIT Command

Toolbar:	Modify II > Edit Text
Menu:	Modify > Object >Text
Command:	DDEDIT

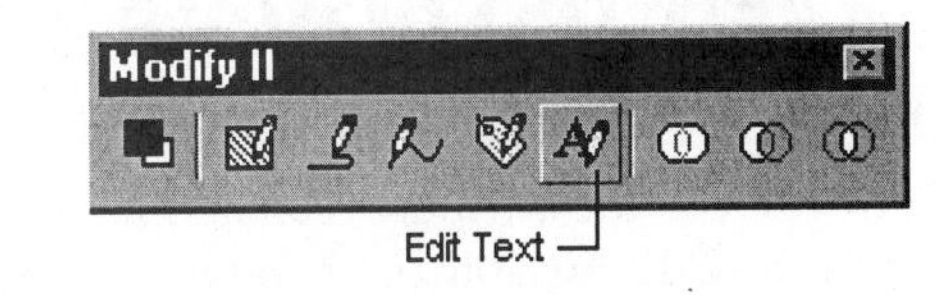

Figure 13-9 ***Edit Text*** *button in the* ***ModifyII*** *toolbar*

The **DDEDIT** command lets you edit text and attribute definitions. After invoking this command, AutoCAD will prompt to **select an annotation object or [Undo]:**. If you select an attribute definition created by attribute definition, the **Edit Attribute Definition** dialog box is displayed and lists the tag name, prompt, and default value of the attribute (Figure 13-10).

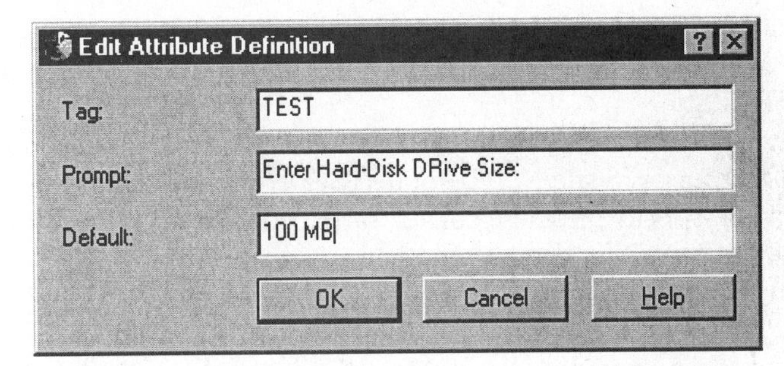

Figure 13-10 ***Edit Attribute Definition*** *dialog box*

You can select the edit boxes and enter the changes. Once you are done making the required changes, choose the **OK** button in the dialog box. After you exit the dialog box, AutoCAD will prompt you to select text or attribute object (Attribute tag). If you are done editing, press ENTER to return to the command line.

Command: **DDEDIT**
Select an annotation object or [Undo]: *Select the attribute tag.*

Using the CHANGE Command

You can also use the **CHANGE** command to edit text or attribute objects. The following is the Command prompt sequence for the **CHANGE** command:

Command: **CHANGE**
Select objects: *Select attribute objects.*
Select objects: «
Specify change point or [Properties]: «
Specify new text insertion point <no change>: «
Enter new text style <current>: «
Specify new height <current>: «
Specify new rotation angle <0>: «
Enter new tag <current>: *Enter new tag name or* «
Enter new prompt <current>: *Enter new prompt or* «
Enter new default value <current>: *Enter new default or* «

INSERTING BLOCKS WITH ATTRIBUTES

Using the Dialog Box

The value of the attributes can be specified during block insertion, either at the command line or in the **Edit Attributes** dialog box. The **Edit Attributes** dialog box is invoked by setting the system variable **ATTDIA** to **1** and then using the **INSERT** command to invoke **Insert** dialog box (discussed earlier in Chapter 12). The default value for **ATTDIA** is **0** which disables the dialog box. When you insert blocks with attributes ,check that the **Mode** is not **Constant** while defining it using the **ATTDEF** command because a **Constant Mode** block cannot be edited. The command sequence for **INSERT** command is as follows:

Command: **ATTDIA** Enter
Enter new value for ATTDIA <0>: **1**

Command: **-INSERT** Enter
Enter block name or [?] <current>: *Enter block name that has attributes in it.*
Specify insertion point or [Scale/X/Y/Z/Rotate/PScale/PX/PY/PZ/PRotate]: *Specify a point.*
Enter X scale factor, specify opposite corner, or [Corner/XYZ] <1>: *Enter X scale factor.*
Enter Y scale factor <use X scale factor>: *Enter Y scale factor.*
Specify rotation angle <0>: *Enter rotation angle.*

After you respond to these prompts, AutoCAD will display the **Edit Attributes** dialog box (Figure 13-11). This displays the prompts and their default values that have been entered at the time of attribute definition. If there are more attributes, they can be accessed by using the **Next** or **Previous** buttons. You can enter the attribute values in the edit box located next to the attribute prompt. The block name is displayed at the top of the dialog box. After entering the new attribute values, choose **OK** button; AutoCAD will place these attribute values at the specified location.

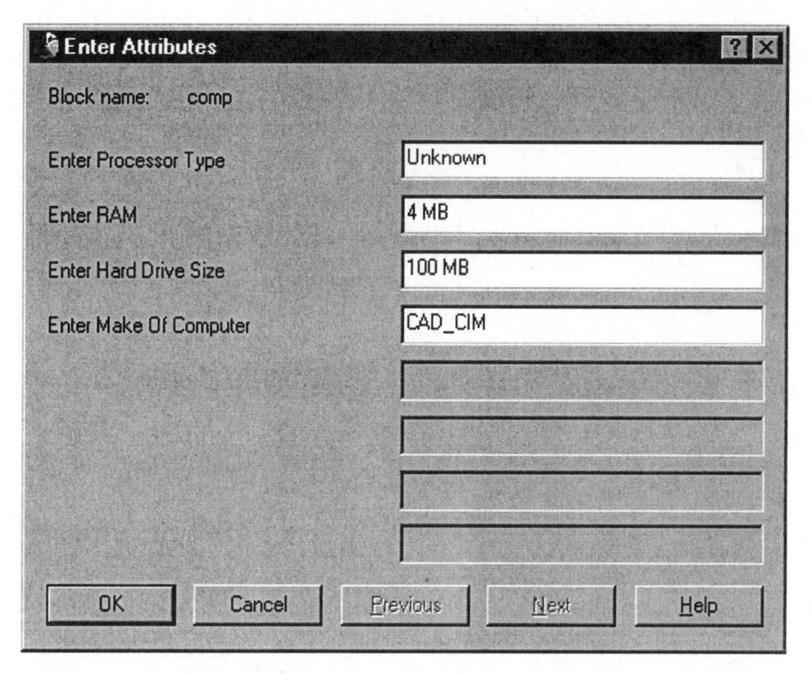

Figure 13-11 *Enter attribute values in the* ***Enter Attributes*** *dialog box*

Note

If you use the dialog box to define the attribute values, the Verify mode is ignored because the ***Enter Attributes*** *dialog box allows the user to examine and edit the attribute values.*

Using the Command Line

You can also define attributes from the command line by setting the system variable **ATTDIA** to **0** (default value). When you use the **-INSERT** command with **ATTDIA** set to **0**, AutoCAD does not display the **Enter Attributes** dialog box. Instead, AutoCAD will prompt you to enter the attribute values for various attributes that have been defined in the block. To define the attributes from the command line, enter the **-INSERT** command at the Command: prompt. After you define the insertion point, scale, and rotation, AutoCAD will display the following prompt:

Enter attribute values

It will be followed by the prompts that have been defined with the block using the ATTDEF command. For example:

Enter processor type <Unknown>:
Enter RAM <4MB>:
Enter Hard-Drive size <100MB>:

Example 2

In this example, you will use the **-INSERT** command to insert the block (COMP) that was defined in Example 1. The following is the list of the attribute values for computers.

ITEM	MAKE	PROCESSOR	HD	RAM
Computer	Gateway	486-60	150MB	16MB
Computer	Zenith	486-30	100MB	32MB
Computer	IBM	386-30	80MB	8MB
Computer	Del	586-60	450MB	64MB
Computer	CAD-CIM	Pentium-90	100 Min	32MB
Computer	CAD-CIM	Unknown	600MB	Standard

1. Make the floor plan drawing as shown in Figure 13-12 (assume the dimensions).

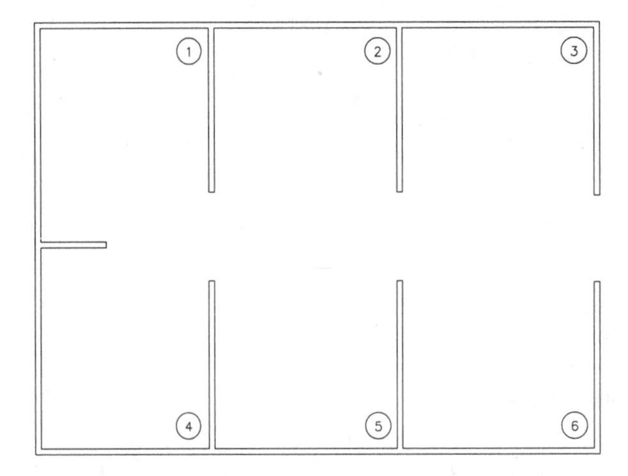

***Figure 13-12** Floor plan drawing for Example 2*

2. Set the system variable **ATTDIA** to 1. Use the **-INSERT** command to insert the blocks, and define the attribute values in the **Enter Attributes** dialog box (Figure 13-13).

    Command: **-INSERT**
    Enter block name or [?] <current>: **COMP**
    Specify insertion point or [Scale/X/Y/Z/Rotate/PScale/PX/PY/PZ/PRotate]: *Specify a point.*
    Enter X scale factor, specify opposite corner, or [Corner/XYZ] <1>: Enter
    Enter Y scale factor <use X scale factor>: Enter
    Specify rotation angle <0>: Enter

3. Repeat the **-INSERT** command to insert other blocks, and define their attribute values as shown in Figure 13-14.

4. Save the drawing for further use.

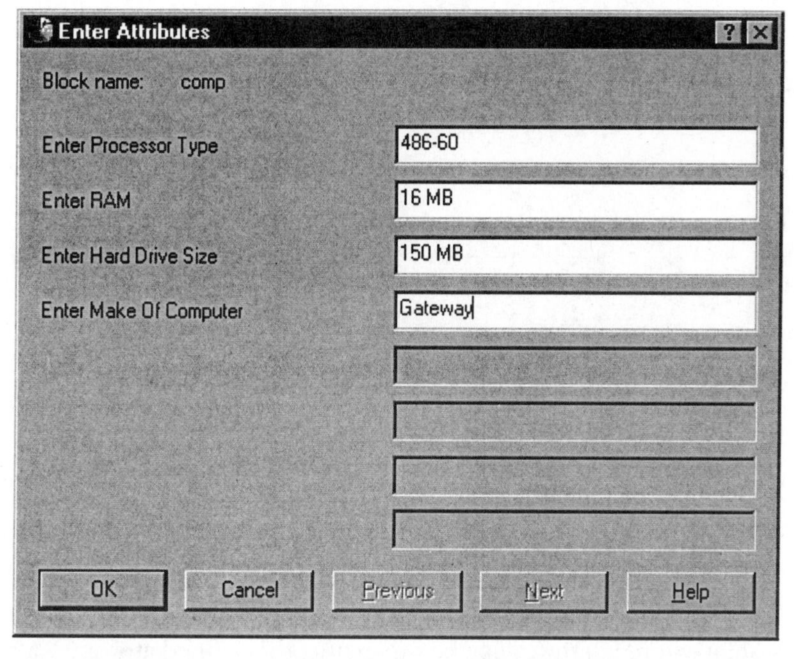

Figure 13-13 *Enter attribute values in the* ***Enter Attributes*** *dialog box*

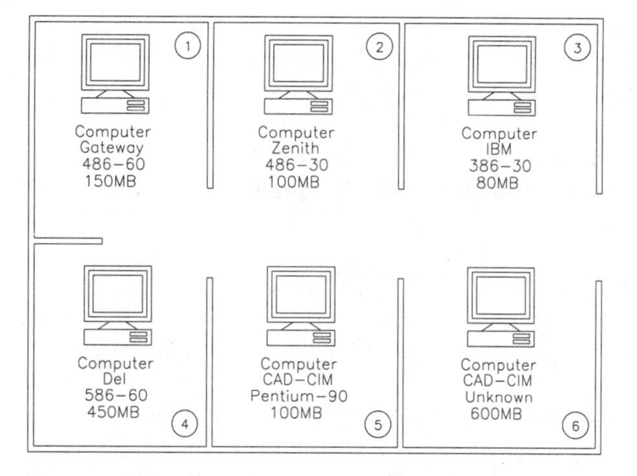

Figure 13-14 *The floor plan after inserting blocks and defining their attributes*

EXTRACTING ATTRIBUTES

Using the Dialog Box (ATTEXT)

Menu :	Tools > Attribute Extraction
Command:	ATTEXT

To use the **Attribute Extraction** dialog box (Figure 13-15) for extracting the attributes, enter

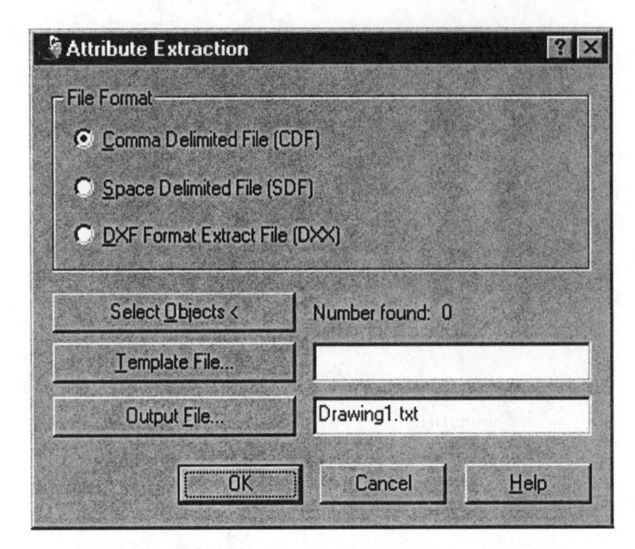

Figure 13-15 ***Attribute Extraction*** *dialog box*

ATTEXT at the Command prompt. The information about the file format, template file, and output file must be entered in the dialog box to extract the defined attribute. Also, you must select the blocks whose attribute values you want to extract.

File Format. This area of the dialog box lets you select the file format of the extracted data. You can select Comma Delimited File, Space Delimited File, or DXF Format Extract File (DXX). The format selection is determined by the application that you plan to use to process the data.

Comma Delimited File (CDF). In CDF format, each character field is enclosed in single quotes, and the records are separated by a delimiter (comma by default). CDF file is a text file with the extension **.TXT.**

Space Delimited File (SDF). In SDF format, the records are of fixed width as specified in the template file. The records are not separated by a comma, and the character fields are not enclosed in single quotes. The SDF file is a text file with the extension **.TXT**.

DXF Format Extract File (DXX). If you select the Drawing Interchange File format, the template file name and the Template File edit box in the **Attribute Extraction** dialog box are automatically disabled. DXF™ -format extraction do not requires any template. The file created by this option contains only block references, attribute values, and end-of-sequence objects. The extension of these files is **.DXX**.

Select Objects<. This button closes the dialog box so that you can use the pointing device to select blocks with attributes. When the **Attribute Extraction** dialog box reopens, Number Found displays the number of objects you have selected.

Template File. The **template file** allows you to specify the attribute values you want to extract and the information you want to retrieve about the block. It also lets you format the display of the extracted data. The file can be created by using any text editor, such as Notepad, Windows

Write, or WordPad. You can also use a word processor or a database program to write the file. The template file must be saved as an **ASCII** file and the extension of the file must be **.TXT**. The following are the fields that you can specify in a template file (the comments given on the right are for explanation only; they must not be entered with the field description):

BL:LEVEL	Nwww000	(Block nesting level)
BL:NAME	Cwww000	(Block name)
BL:X	Nwwwddd	(X coordinate of block insertion point)
BL:Y	Nwwwddd	(Y coordinate of block insertion point)
BL:Z	Nwwwddd	(Z coordinate of block insertion point)
BL:NUMBER	Nwww000	(Block counter)
BL:HANDLE	Cwww000	(Block's handle)
BL:LAYER	Cwww000	(Block insertion layer name)
BL:ORIENT	Nwwwddd	(Block rotation angle)
BL:XSCALE	Nwwwddd	(X scale factor of block)
BL:YSCALE	Nwwwddd	(Y scale factor of block)
BL:ZSCALE	Nwwwddd	(Z scale factor of block)
BL:XEXTRUDE	Nwwwddd	(X component of block's extrusion direction)
BL:YEXTRUDE	Nwwwddd	(Y component of block's extrusion direction)
BL:ZEXTRUDE	Nwwwddd	(Z component of block's extrusion direction)
Attribute tag		(The tag name of the block attribute)

The extract file may contain several fields. For example, the first field might be the item name and the second field might be the price of the item. Each line in the template file specifies one field in the extract file. Any line in a template file consists of the name of the field, the width of the field in characters, and its numerical precision (if applicable). For example:

```
ITEM       N015002
BL:NAME    C015000
```

Where **BL:NAME** ------ Field name
Blankspaces --- Blank spaces (must not include the tab character)
C ------------------ Designates a character field
N ------------------ Designates a numerical field
015 --------------- Width of field in characters
002 --------------- Numerical precision

BL:NAME or **ITEM** Indicates the field names; can be of any length.

C Designates a character field; that is, the field contains characters or it starts with characters. If the file contains numbers or starts with numbers, then C will be replaced by N. For example, **N015002**.

015 Designates a field that is 15 characters long.

002 Designates the numerical precision. In this example, the numerical precision is 2, two places following the decimal. The decimal point and the two digits following decimal are **included in the field width.**

In the next example, (000), the numerical precision, is not applicable because the field does not have any numerical value (the field contains letters only).

After creating a template file, in AutoCAD 2000, if you select the **Template File** button, the **Output File** dialog box (Figure 13-16) is displayed, where you can browse and select the required template file.

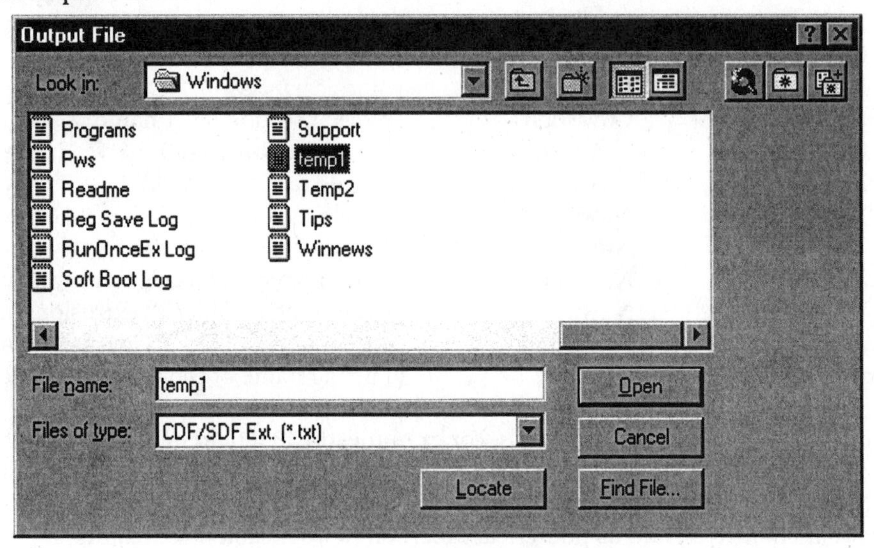

***Figure 13-16 Output File** dialog box*

Note

You can put any number of spaces between the field name and the character C or N (ITEM N015002). However, you must not use the tab characters. Any alignment in the fields must be done by inserting spaces after the field name.

In the template file, a field name must not appear more than once. The template file name and the output file name must be different.

The template file must contain at least one field with an attribute tag name because the tag names determine which attribute values are to be extracted and from which blocks. If several blocks have different block names but the same attribute tag, AutoCAD will extract attribute values from all selected blocks. For example, if there are two blocks in the drawing with the attribute tag PRICE, then when you extract the attribute values, AutoCAD will extract the value from both blocks (if both blocks were selected). To extract the value of an attribute, the tag name must match the field name specified in the template file. AutoCAD automatically converts the tag names and the field names to uppercase letters before making a comparison.

Output File. This button specifies the name of the output file. Enter the file name in the box, or choose the **Output File** button to search for existing template files from **Output File** dialog box (Figure 13-17). AutoCAD appends the **.TXT** file extension for CDF or SDF files and the **.DXF** file extension for DXF files.

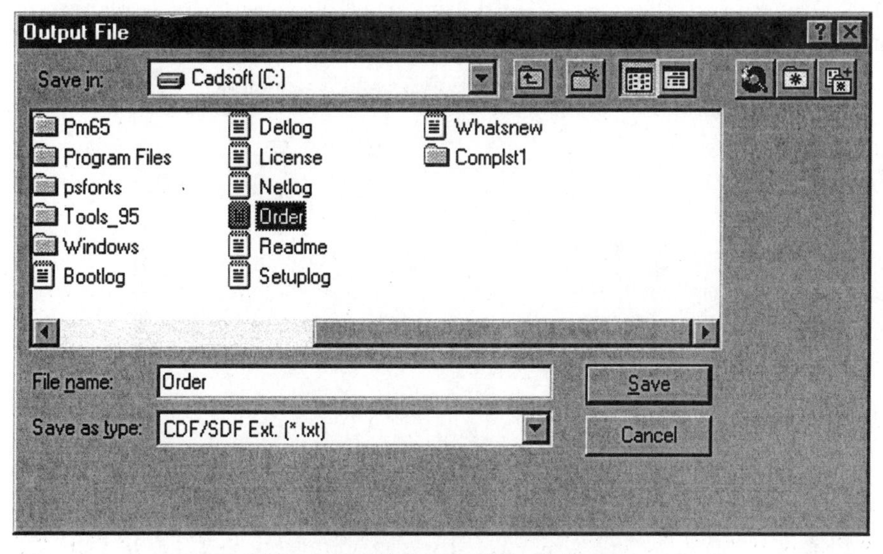

***Figure 13-17 Output File** dialog box*

Example 3

In this example, you will write a template file for extracting the attribute values as defined in Example 2. These attribute values must be written to a file **COMPLST1.TXT** and the values arranged as shown in the following table:

Field width in characters

< 10 >	< 12 >	< 10 >	< 12 >	< 10 >	< 10 >
COMP	Computer	Gateway	486-60	150MB	**16MB**
COMP	Computer	Zenith	486-30	100MB	**32MB**
COMP	Computer	IBM	386-30	80MB	**8MB**
COMP	Computer	Del	586-60	450MB	**64MB**
COMP	Computer	**CAD-CIM**	Pentium-90	100 Min	**32MB**
COMP	Computer	**CAD-CIM**	**Unknown**	600MB	Standard

1. Load the drawing you saved in Example 2.

2. Use the Windows Notepad to write the following template file. You can use any text editor or word processor to write the file. After writing the file, save it as an ASCII file under the file name **TEMP1.TXT**. Exit the Notepad and access AutoCAD.

```
BL:NAME      C010000      (Block name, 10 spaces)
Item         C012000      (Item, 12 spaces)
```

Make	C010000	(Computer make, 10 spaces)
Processor	C012000	(Processor type, 12 spaces)
HD	C010000	(Hard drive size, 10 spaces)
RAM	C010000	(RAM size, 10 spaces)

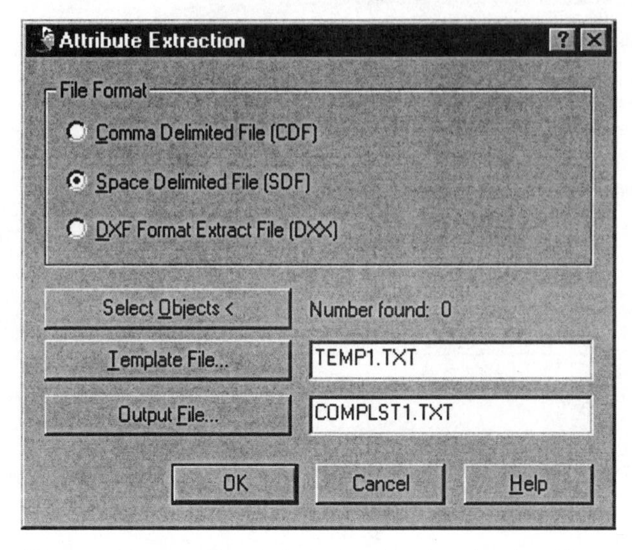

***Figure 13-18** Enter information in the **Attribute Extraction** dialog box*

3. Use the **ATTEXT** command to invoke the **Attribute Extraction** dialog box (Figure 13-18), and select the Space Delimited File (SDF) radio button.

4. Choose the **Select Objects <** button to select the objects (blocks) present on the screen. You can select the objects by using the Window or Crossing option. After selection is complete, right-click your pointing device to display the dialog box again.

5. In the **Template File** edit box, enter the name of the template file, **TEMP1.TXT.**

6. In the **Output File** edit box, enter the name of the output file, **COMPLST1.TXT.**

7. Choose the **OK** button in the **Attribute Extraction** dialog box.

8. Use the Notepad again to list the output file, **COMPLST1.TXT.** The output file will be similar to the file shown at the beginning of Example 3.

Using the Command Line (-ATTEXT)

You can also extract the attributes from the command line by entering the **-ATTEXT** command at the Command prompt. When you enter this command, AutoCAD will first prompt you to enter the type of output file. You could select CDF, SDF, DXF, or Objects. If you select the **Objects** option, AutoCAD will prompt you to select the objects whose attributes you want to extract. After you select the objects, AutoCAD will return the prompt, this time without the **Objects** option. The following is the command prompt sequence for this command.

Command: **-ATTEXT**
Enter extraction type or enable object selection [Cdf/Sdf/Dxf/Objects] <C>: **O**
Select objects: *Select objects.*
Select objects: Enter
Enter attribute extraction type [Cdf/Sdf/Dxf] <C>: **C** or **S**

If you enter **C** or **S** (CDF or SDF), AutoCAD 2000 will display the **Select Template File** dialog box. You can browse for the required template file. (Template files were discussed earlier in this chapter in the subsection "Template File.")

After you select the template file name, AutoCAD 2000 will display the **Create extract file** dialog box (Figure 13-19). The extract file is the output file where you want to write the attribute values.

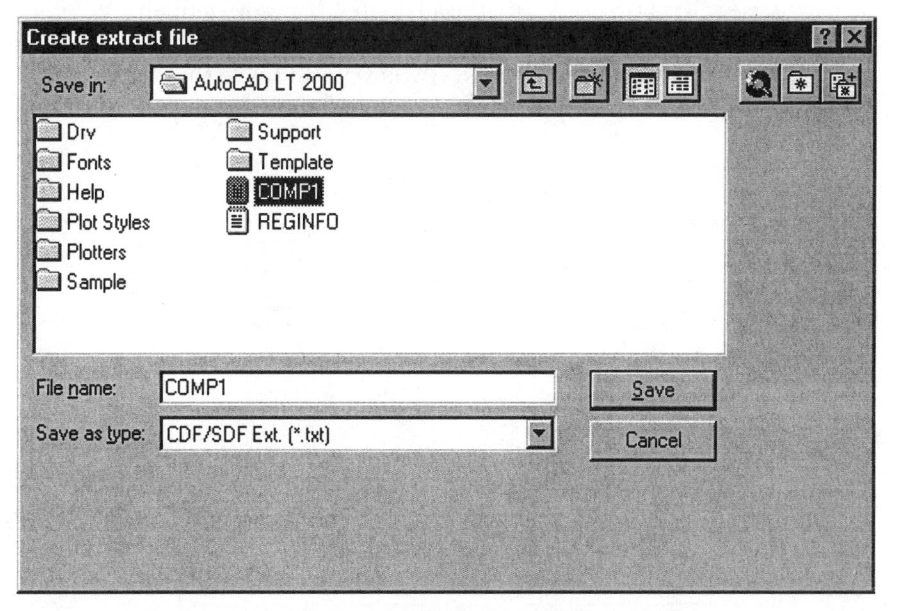

***Figure 13-19 Create extract file** dialog box*

If you do not enter a file name, AutoCAD assumes that the name of the extract file is the same as the drawing file name with the file extension **.TXT** or **.DXF**, depending on the attribute extract format. If you enter the file name and have selected the CDF or SDF file extraction format, the file extension must be **.TXT**; if you have selected the .DXF format, the file extension must be **.DXF**. After you enter the file name, AutoCAD will extract the attribute values from the blocks and write the data to the output file.

CONTROLLING ATTRIBUTE VISIBILITY (ATTDISP COMMAND)

Menu:	View > Display > Attribute Display
Command:	ATTDISP

The **ATTDISP** command allows you to change the visibility of all attribute values. Normally, the attributes are visible unless they are defined invisible by using the Invisible mode. The invisible attributes are not displayed, but they are a part of the block definition. The prompt sequence is:

Command: **ATTDISP**
Enter attribute visibility setting [Normal/ON/OFF] <current>:

If you enter ON and press ENTER, all attribute values will be displayed, including the attributes that are defined with the Invisible mode.

If you select OFF, all attribute values will become invisible. Similarly, if you enter N (Normal), AutoCAD will display the attribute values the way they are defined; that is, the attributes that were defined invisible will stay invisible and the attributes that were defined visible will become visible. In Example 2, the RAM attribute was defined with the Invisible mode; therefore, the RAM values are not displayed with the block. If you want to make the RAM attribute values visible (Figure 13-20), enter the **ATTDISP** command and then turn it ON.

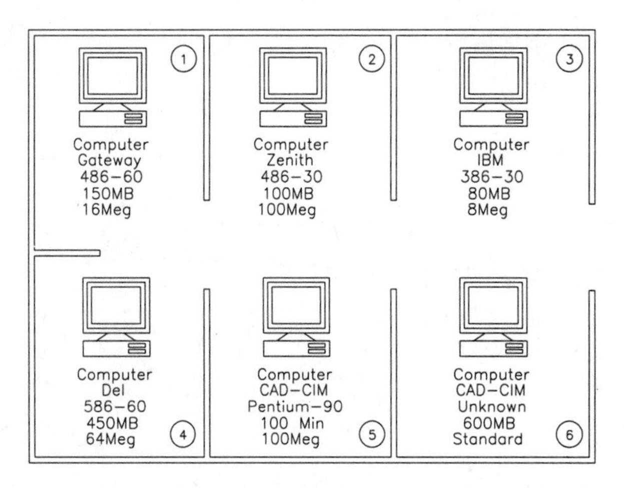

Figure 13-20 *Using the* ***ATTDISP*** *command to make the RAM attribute values visible*

EDITING ATTRIBUTES (ATTEDIT COMMAND)

Single Block

Toolbar:	Modify II > Edit Attribute
Menu:	Modify > Attribute > Single
Command:	ATTEDIT

The **ATTEDIT** command allows you to edit the block attribute values through the **Edit Attributes** dialog box. When you enter this command, AutoCAD prompts you to select the block whose values you want to edit. After selecting the block, the **Edit Attribute** dialog box is displayed. The dialog box shows the prompts and the attribute values of the selected block. If an attribute has been defined with **Constant** mode, it is not displayed in the dialog box because a constant attribute value cannot be edited. To make any changes, choose the respective edit box and enter the new value. After you choose the **OK** button, the attribute values are updated in the selected block.

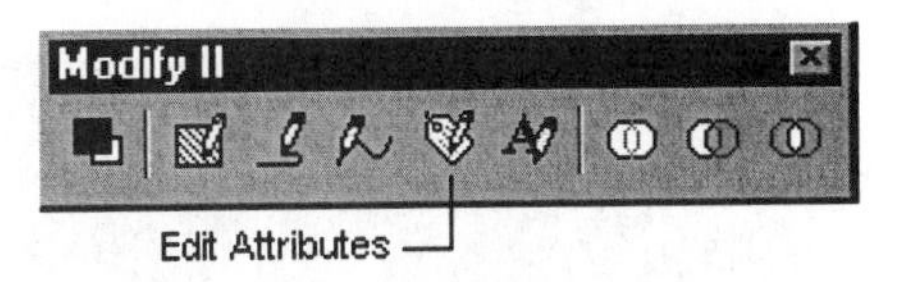

Figure 13-21 ***Modify II*** *toolbar*

Command: **ATTEDIT**
Select block reference: *Select a block with attributes.*

If the selected block has no attributes, AutoCAD will display the alert message **That block has no editable attributes**. Similarly, if the selected object is not a block, AutoCAD again displays the alert message **That object is not a block**.

Note

*You cannot use the **ATTEDIT** command to do global editing of attribute values.*

*You cannot use the **ATTEDIT** command to modify position, height, or style of the attribute value.*

Example 4

In this example you will use the **ATTEDIT** command to change the attribute of the first computer (16MB to 16 Meg), which is located in Room-1.

1. Load the drawing that was created in Example 2. The drawing has six blocks with attributes. The name of the block is COMP, and it has six defined attributes, one of them invisible. Zoom in so that the first computer is displayed on the screen (Figure 13-22).

2. At the AutoCAD Command prompt, enter the **ATTEDIT** command. AutoCAD will prompt you to select a block. Select the block, first computer located in Room-1.

   Command: **ATTEDIT**
   Select block reference: *Select a block.*

   AutoCAD will display the **Edit Attributes** dialog box (Figure 13-23), which shows the attribute prompts and the attribute values.

3. Edit the values, and choose the **OK** button in the dialog box. When you exit the dialog box, the attribute values are updated.

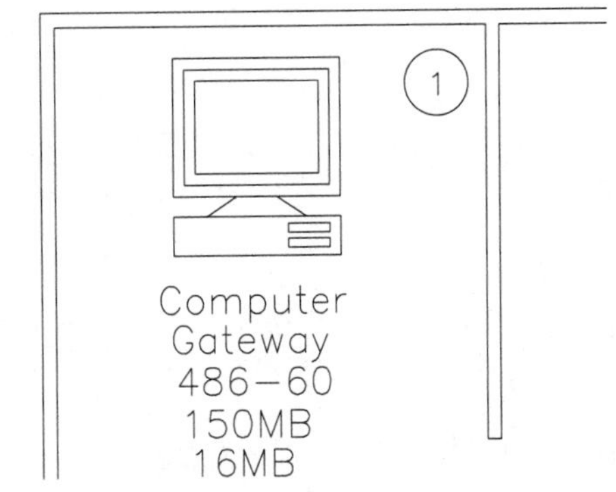

Figure 13-22 *Zoomed view of the first computer*

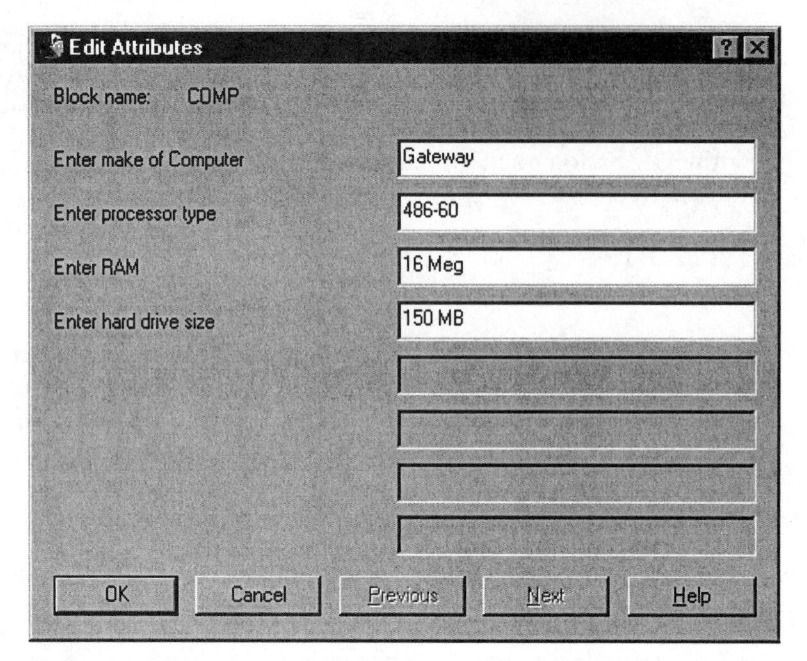

Figure 13-23 *Editing attribute values using the* ***Edit Attributes*** *dialog box*

Global Editing of Attributes

Menu:	Modify > Attribute > Global
Command:	-ATTEDIT

The **-ATTEDIT** command allows you to edit the attribute values independently of the blocks that contain the attribute reference. For example, if there are two blocks, COMPUTER and TABLE, with the attribute value PRICE, you can globally edit this value (PRICE) independently of the block that references these values. You can also edit the attribute values one at a time. For example, you can edit the attribute value (PRICE) of the block TABLE without affecting the value of the other block, COMPUTER. When you enter the **-ATTEDIT** command, AutoCAD LT displays the following prompt:

Command: **-ATTEDIT**
Edit attributes one at a time? [Yes/No] <Y>: **N**
Performing global editing of attribute values

If you enter **N** at this prompt, it means that you want to do the global editing of the attributes. However, you can restrict the editing of attributes by block names, tag names, attribute values, and visibility of attributes on the screen.

Editing Visible Attributes Only

After you select global editing, AutoCAD LT will display the following prompt:

Edit only attributes visible on screen? [Yes/No] <Y>: **Y**

If you enter **Y** at this prompt, AutoCAD will edit only those attributes that are visible and displayed on the screen. The attributes might have been defined with the Visible mode, but if they are not displayed on the screen they are not visible for editing. For example, if you zoom in, some of the attributes may not be displayed on the screen. Since the attributes are not displayed on the screen, they are invisible and cannot be selected for editing.

Editing All Attribute

If you enter **N** at the earlier-mentioned prompt, AutoCAD flips from graphics to text screen and displays the following message on the screen:

Drawing must be regenerated afterwards.

Now, AutoCAD will edit all attributes even if they are not visible or displayed on the screen. Also, changes that you make in the attribute values are not reflected immediately. Instead, the attribute values are updated and the drawing regenerated after you are done with the command.

Editing Specific Blocks

Although you have selected global editing, you can confine the editing of attributes to specific blocks by entering the block name at the prompt. For example:

Enter block name specification <*>: **COMP**

When you enter the name of the block, AutoCAD will edit the attributes that have the given block (COMP) reference. You can also use the wild-card characters to specify the block names. If you want to edit attributes in all blocks that have attributes defined, press ENTER.

Editing Attributes with Specific Attribute Tag Names

Like blocks, you can confine attribute editing to those attribute values that have the specified tag name. For example, if you want to edit the attribute values that have the tag name MAKE, enter the tag name at the following AutoCAD prompt:

Enter attribute tag specification <*>: **MAKE**

When you specify the tag name, AutoCAD will not edit attributes that have a different tag name, even if the values being edited are the same. You can also use the wild-card characters to specify the tag names. If you want to edit attributes with any tag name, press ENTER.

Editing Attributes with a Specific Attribute Value

Like blocks and attribute tag names, you can confine attribute editing to a specified attribute value. For example, if you want to edit the attribute values that have the value 100 MB, enter the value at the following AutoCAD prompt:

Enter attribute value specification <*>: **100MB**

When you specify the attribute value, AutoCAD will not edit attributes that have a different value, even if the tag name and block specification are the same. You can also use the wild-card characters to specify the attribute value. If you want to edit attributes with any value, press ENTER.

Sometimes the value of an attribute is null, and these values are not visible. If you want to select the null values for editing, make sure you have not restricted the global editing to visible attributes. To edit the null attributes, enter \ at the following prompt:

Enter attribute value specification <*>: \

After you enter this information, AutoCAD will prompt you to select the attributes. You can select the attributes by selecting individual attributes or by using one of the object selection options (Window, Crossing, and so on).

Select Attributes: *Select the attribute values parallel to the current UCS only.*

After you select the attributes, AutoCAD will prompt you to enter the string you want to change and the new string. AutoCAD will retrieve the attribute information, edit it, and then update the attribute values.

Enter string to change:
Enter new string:

The following is the complete command prompt sequence of the **-ATTEDIT** command. It is assumed that the editing is global and for visible attributes only.

Command: **-ATTEDIT**
Edit attributes one at a time? [Yes/No] <Y>: **N**
Performing global editing of attribute values.
Edit only attributes visible on screen? [Yes/No] <Y>: **N**
Drawing must be regenerated afterwards.
Enter block name specification <*>:
Enter attribute tag specification <*>:
Enter attribute value specification <*>:
Enter string to change:
Enter new string:

Note

*AutoCAD regenerates the drawing at the end of the command automatically unless system variable **REGENAUTO** is off, which controls automatic regeneration of the drawing.*

Example 5

In this example, you will use the drawing from Example 2 to edit the attribute values that are highlighted in the following table. The tag names are given at the top of the table (ITEM, MAKE, PROCESSOR, HD, RAM). The RAM values are invisible in the drawing.

	ITEM	MAKE	PROCESSOR	HD	RAM
COMP	Computer	Gateway	486-60	150MB	**16MB**
COMP	Computer	Zenith	486-30	100MB	**32MB**
COMP	Computer	IBM	386-30	80MB	**8MB**
COMP	Computer	Del	586-60	450MB	**64MB**
COMP	Computer	**CAD-CIM**	Pentium-90	100 Min	**32MB**
COMP	Computer	**CAD-CIM**	**Unknown**	600MB	Standard

Make the following changes in the **highlighted** attribute values (Figure 13-24).

1. Change Unknown to Pentium.
2. Change CAD-CIM to Compaq.
3. Change MB to Meg for all attribute values that have the tag name RAM. (No changes should be made to the values that have the tag name HD.)

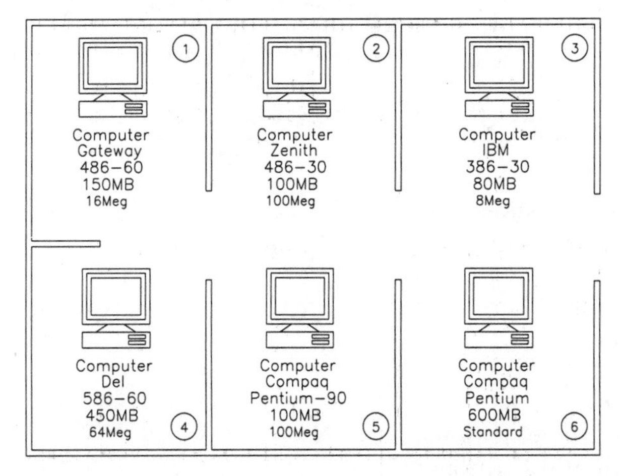

Figure 13-24 *Using **-ATTEDIT** to change the attribute values*

The following is the prompt sequence to change the attribute value **Unknown to Pentium**.

1. Enter the **-ATTEDIT** command at the Command prompt. At the next prompt, enter **N** (No).

   Command: **-ATTEDIT**
   Edit attributes one at a time? [Yes/No] <Y>: **N**
   Performing global editing of attribute values.

2. We want to edit only those attributes that are visible on the screen, so press ENTER at the following prompt:

   Edit only attributes visible on screen? [Yes/No] <Y>: *Press ENTER.*

3. As shown in the table, the attributes belong to a single block, COMP. In a drawing, there could be more blocks. To confine the attribute editing to the COMP block only, enter the

name of the block (COMP) at the next prompt.

Enter block name specification <*>: **COMP**

4. At the next two prompts, enter the attribute tag name and the attribute value specification. When you enter these two values, only those attributes that have the specified tag name and attribute value will be edited.

   Enter attribute tag specification<*>: **Processor**
   Enter attribute value specification<*>: **Unknown**

5. Next, AutoCAD will prompt you to select attributes. Use the Crossing option to select all blocks. AutoCAD will search for the attributes that satisfy the given criteria (attributes belong to the block COMP, the attributes have the tag name Processor, and the attribute value is Unknown). Once AutoCAD locates such attributes, they will be highlighted.

   Select Attributes:

6. At the next two prompts, enter the string you want to change, and then enter the new string.

   Enter string to change: **Unknown**
   Enter new string: **Pentium**

7. The following is the command prompt sequence to change the make of the computers from **CAD-CIM** to **Compaq**.

   Command: **-ATTEDIT**
   Edit attributes one at a time? [Yes/No] <Y>: **N**
   Performing global editing of attribute values.
   Edit only attributes visible on screen? [Yes/No] <Y>: *Press ENTER.*
   Drawing must be regenerated afterwards.
   Enter block name specification <*>: **COMP**
   Enter attribute tag specification <*>: **MAKE**
   Enter attribute value specification <*>:
   Select Attributes:
   Enter string to change: **CAD-CIM**
   Enter new string: **Compaq**

8. The following is the command prompt sequence to change **MB** to **Meg**.

   Command: **-ATTEDIT**
   Edit attributes one at a time? [Yes/No] <Y>: **N**
   Performing global editing of attribute values.

At the next prompt, you must enter **N** because the attributes you want to edit (tag name, RAM) are not visible on the screen.

Edit only attributes visible on screen? [Yes/No] <Y>: **N**
Drawing must be regenerated afterwards.
Enter block name specification <*>: **COMP**

At the next prompt, about the tag specification, you must specify the tag name because the text string MB also appears in the hard drive size (tag name, HD). If you do not enter the tag name, AutoCAD will change all MB attribute values to Meg.

Enter attribute tag specification <*>: **RAM**
Enter attribute value specification <*>:
Select Attributes:
Enter string to change: **MB**
Enter new string: **Meg**

9. Use the **ATTDISP** command to display the invisible attributes on the screen.

Command: **ATTDISP**
Enter attribute visibility setting [Normal/ON/OFF] <ON>: **ON**

Individual Editing of Attributes

The **-ATTEDIT** command can also be used to edit the attribute values individually. When you enter this command, AutoCAD will prompt **Edit attributes one at a time? [Yes/No] <Y>:**. At this prompt, press ENTER to accept the default or enter **Y.** The next three prompts are about block specification, attribute tag specification, and attribute value specification, which were discussed in the previous section, in this Chapter. These options let you limit the attributes for editing. For example, if you specify a block name, AutoCAD will limit the editing to those attributes that belong to the specified block. Similarly, if you also specify the tag name, AutoCAD will limit the editing to the attributes in the specified block and with the specified tag name.

Command: **-ATTEDIT**
Edit attributes one at a time? [Yes/No] <Y>: Enter
Enter block name specification<*>: Enter
Enter attribute tag specification<*>: Enter
Enter attribute value specification<*>: Enter
Select Attributes:

At the **Select Attributes:** prompt, select the objects by choosing the objects or by using an object selection option such as Window, Crossing, WPolygon, CPolygon, or Box. By using these options you can further limit the attribute values selected for editing. After you select the objects, AutoCAD will mark the first attribute it can find with an **X**. The next prompt is:

Enter an option [Value/Position/Height/Angle/Style/Layer/Color/Next] <N>:

Value. The **Value** option lets you change the value of an attribute. To change the value, enter **V** at this prompt. AutoCAD will display the following prompt:

Enter type of value modification [Change/Replace] <R>:

The **Change** option allows you to change a few characters in the attribute value. To select the **Change** option, enter Change or **C** at the prompt. AutoCAD will display the next prompt:

Enter string to change:
Enter new string:

At the **Enter string to change**: prompt, enter the characters you want to change and press ENTER. At the next prompt, **Enter new string:**, enter the new string.

Note

*You can use ? and * in the string value. When these characters are used in string values, AutoCAD does not interpret them as wild-card characters.*

To use the **Replace** option, enter **R** or press ENTER at the **Enter type of value modification [Change/Replace] <R>:** prompt. AutoCAD will display the following prompt:

Enter new Attribute value:

At this prompt, enter the new attribute value. AutoCAD will replace the string bearing the **X** mark with the new string. If the new attribute is null, the attribute will be assigned a null value.

Position, Height, Angle. You can change the position, height, or angle of an attribute value by entering, respectively, **P**, **H**, or **A** at the following prompt:

Enter an option [Value/Position/Height/Angle/Style/Layer/Color/Next] <N>:

The **Position** option lets you define the new position of the attribute value. AutoCAD will prompt you to enter the new starting point, center point, or endpoint of the string. If the string is aligned, AutoCAD will prompt for two points. You can also define the new height or angle of the text string by entering, respectively, **H** or **A** at the prompt.

Layer and Color. The **Layer** and **Color** options allow you to change the layer and color of the attribute. For a color change, you can enter the new color by entering a color number (1 through 255), a color name (red, green, and so on), BYLAYER, or BYBLOCK.

Example 6

In this example, you will use the drawing in Example 2 to edit the attributes individually (Figure 13-25). Make the following changes in the attribute values.

a. Change the attribute value 100 Min to 100MB.
b. Change the height of all attributes with the tag name RAM to 0.075 units.

1. Load the drawing that you had saved in Example 2.

2. At the AutoCAD Command prompt, enter the **-ATTEDIT** command. The following is the command prompt sequence to change the value of 100 Min to 100 MB.

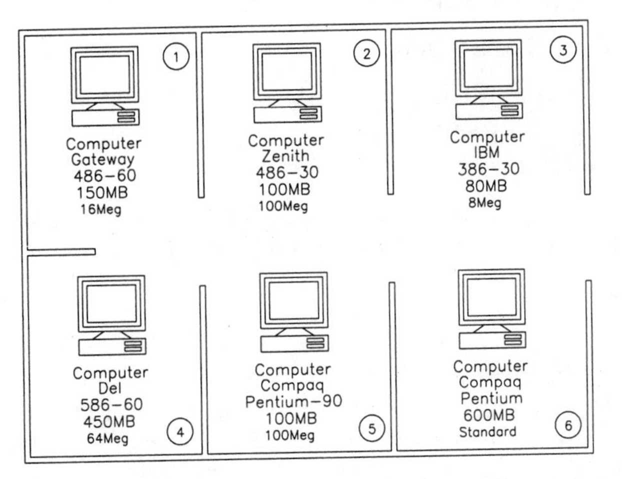

***Figure 13-25** Using **-ATTEDIT** to change the attribute values individually*

Command: **-ATTEDIT**
Edit attributes one at a time? [Yes/No] <Y>: Enter
Enter block name specification <*>: **COMP**
Enter attribute tag specification <*>: Enter
Enter attribute value specification <*>: Enter
Select Attributes: *Select the attribute.*
Enter an option [Value/Position/Height/Angle/Style/Layer/Color/Next] <N>: **V**
Enter type of value modification [Change/Replace] <R>: **C**
Enter string to change: \ **Min**
Enter new string: **100MB**

When AutoCAD prompts **Enter string to change:**, enter the characters you want to change. In this example, the characters **Min** are preceded by a space. If you enter a space, AutoCAD displays the next prompt, **Enter new string:**. If you need a leading blank space, the character string must start with a backslash (\), followed by the desired number of blank spaces.

3. To change the height of the attribute text, enter the **-ATTEDIT** command as just shown. When AutoCAD displays the following prompt, enter **H** for height.

Enter an option [Value/Position/Height/Angle/Style/Layer/Color/Next] <N>: **H**
Specify new height <current>: **0.075**

After you enter the new height and press ENTER, AutoCAD will change the height of the text string that has the **X** mark. AutoCAD will then repeat the last prompt. Use the **Next** option to move the **X** mark to the next attribute. To change the height of other attribute values, repeat these steps.

INSERTING TEXT FILES IN THE DRAWING

Using MTEXT Command

Toolbar:	Draw > Multiline Text
Menu:	Draw > Text > Multiline Text
Command:	MTEXT

You can insert a text file by selecting the **Import Text** option in the **Multiline Text Editor** dialog box. You can invoke this dialog box by using the **MTEXT** command.

Command: **MTEXT**

Next, AutoCAD prompts you to enter the insertion point and other corner of the paragraph text box. After you enter these points, the **Multiline Text Editor** dialog box appears on screen. To insert the text file **COMPLST1.TXT** (created in Example 3), choose the **Import Text** button. AutoCAD displays the **Open** dialog box (Figure 13-26).

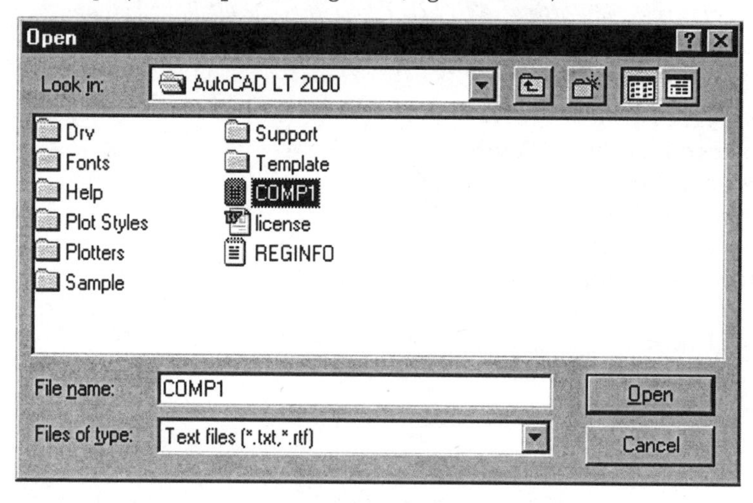

Figure 13-26 ***Open*** *dialog box*

In this dialog box, you can select the text file COMPLST1 and then choose the **Open** button. The imported text is displayed in the text area of the **Multiline Text Editor** dialog box (Figure 13-27). Note that only ASCII files are properly interpreted.

Now choose the **OK** button to get the imported text in the selected area on the screen (Figure 13-28).You can also use the **Multiline Text Editor** dialog box to change the text style, height, direction, width, rotation, line spacing, and attachment.

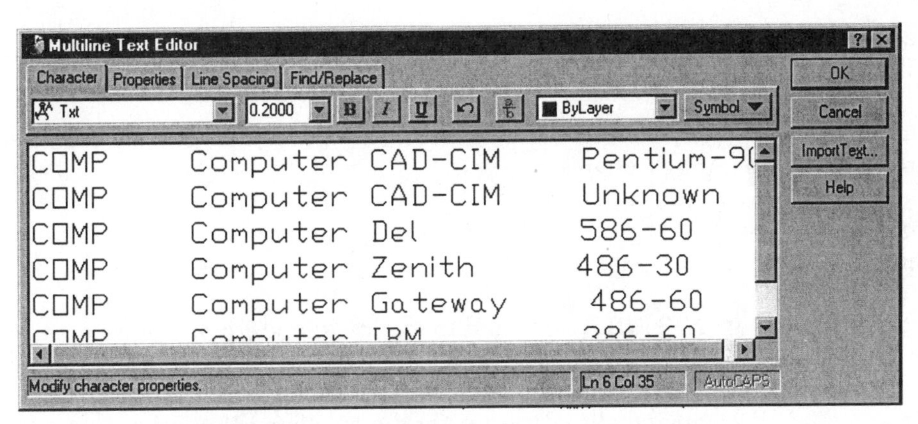

Figure 13-27 ***Multiline Text Editor*** *dialog box displaying the imported text*

COMP	Computer	CAD-CIM	Pentium-90	100 Min
COMP	Computer	CAD-CIM	Unknown	600MB
COMP	Computer	Del	586-60	450MB
COMP	Computer	Zenith	486-30	100MB
COMP	Computer	Gateway	486-60	150MB
COMP	Computer	IBM	386-30	80MB

Figure 13-28 Imported text file on the screen

Self-Evaluation Test

Answer the following questions, and then compare your answers to the correct answers given at the end of this chapter.

1. The **Verify** option allows you to verify the attribute value you have entered when inserting a block. (T/F)

2. Unlike a constant attribute, the **Preset** attribute cannot be edited. (T/F)

3. For tag names, any lowercase letters are automatically converted to uppercase. (T/F)

4. The entry in the **Value:** edit box of the **Attribute Definition** dialog box defines the __________ of the specified attribute.

5. If you have selected the **Align** option for the text justification, the **Height** < and **Rotation** < edit boxes are __________.

6. You can use the __________ command to edit text or attribute definitions.

7. The default value of the **ATTDIA** variable is __________, which disables the dialog box.

8. In the Space Delimited File, the records are of fixed width as specified in the _________ File.

9. In the _________ File, the records are not separated by a comma and the character fields are not enclosed in single quotes.

10. You must not use the _________character in template files. Any alignment in the fields must be done by inserting spaces after the field name.

11. You cannot use the **ATTEDIT** command to modify the position, height, or style of the attribute value. (T/F)

12. The _________ command allows you to edit the attribute values independently of the blocks that contain the attribute reference.

13. The **-ATTEDIT** command can also be used to edit the attribute values individually. (T/F)

14. You can use ? and * in the string value. When these characters are used in string values, AutoCAD does not interpret them as wild-card characters. (T/F)

15. You can insert a text file in the drawing by entering the _________ command at the AutoCAD Command prompt.

Review Questions

1. Give two major uses of defining block attributes _________________________.

2. The information associated with a block is known as _________ or _________.

3. You can define the block attributes by entering _________ at the Command prompt.

4. The most convenient way to define the block attributes is by using the **Attribute Definition** dialog box, which can be invoked by entering _________.

5. What are the options in the **Mode** area of the **Attribute Definition** dialog box?

6. The **Constant** option lets you create an attribute that has a fixed value and cannot be changed later. (T/F)

7. What is the function of the **Preset** option? .

8. The attribute value is requested when you use the **Preset** option to define a block attribute. (T/F)

9. Name the three edit boxes in the **Attribute area** of the **Attribute Definition** dialog box.

10. The **tag** is like a label that is used to identify an attribute. (T/F)

11. The tag names can only be uppercase. (T/F)

12. The tag name cannot be null. (T/F)

13. The tag name can contain a blank space. (T/F)

14. If you select the **Constant** option in the **Mode** area of the **Attribute Definition** dialog box, the **Prompt:** edit box is __________ because no prompt is required if the attribute is __________.

15. If you do not enter anything in the **Prompt:** edit box, the entry made in the **Tag:** edit box is used as the prompt. (T/F)

16. What option, button, or check box should you select in the **Attribute Definition** dialog box to automatically place the subsequent attribute text just below the previously defined attribute?

17. The text style must be defined before it can be used to specify the text style. (T/F)

18. If you select a style that has the height predefined, AutoCAD automatically disables the **Height** < edit box in the **Attribute Definition** dialog box. (T/F)

19. If you have selected the **Fit** option for the text justification, the __________ edit box is disabled.

20. What is the difference between the **ATTDEF** and the **-ATTDEF** commands?

21. The __________ command lets you edit both text and attribute definitions.

22. The value of the block attributes can be specified in the **Edit Attribute Definition** dialog box. (T/F)

23. The **Edit Attribute Definition** dialog box is invoked by using the __________ or __________ command with the system variable __________ set to 1.

24. If you use the **Enter Attributes** dialog box to define the attribute values, the **Verify** mode is __________ because the **Enter Attributes** dialog box allows the user to examine and edit the attribute values.

25. You can also define attributes from the command line by setting the system variable __________ to 0.

26. To use the **Attribute Extraction** dialog box for extracting the attributes, enter __________

at the Command prompt.

27. You can select the Comma Delimited File, Space Delimited File, or Drawing Interchange File. The format selection is determined by the text editor you use. (T/F)

28. In the Comma Delimited File, each character field is enclosed in _________ and each record is separated by a _________.

Exercises

Exercise 1 *Electronics*

In this exercise, you will define the following attributes for a resistor and then create a block using the **BLOCK** command. The name of the block is RESIS.

Mode	Tag name	Prompt	Default value
Verify	RNAME	Enter name	RX
Verify	RVALUE	Enter resistance	XX
Verify, Invisible	RPRICE	Enter price	00

1. Draw the resistor as shown in Figure 13-29.

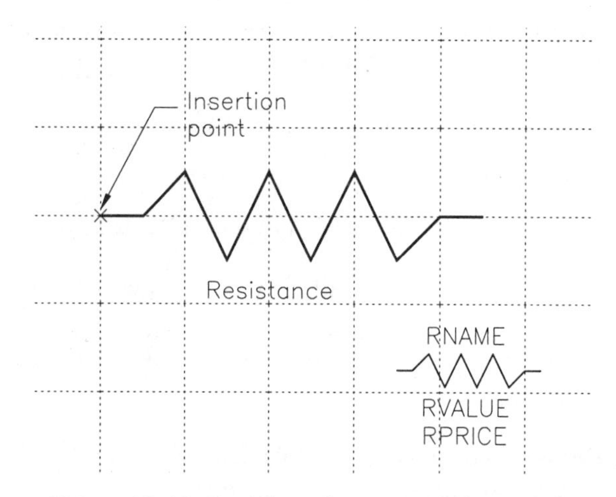

***Figure 13-29** Drawing of a resistor for Exercise 1*

2. Enter **ATTDEF** at the AutoCAD Command prompt to invoke the **Attribute Definition** dialog box.

3. Define the attributes as shown in the preceding table, and position the attribute text as shown in Figure 13-29.

4. Use the **BLOCK** command to create a block. The name of the block is RESIS, and the insertion point of the block is at the left end of the resistor. When you select the objects for the block, make sure you also select the attributes.

Exercise 2 — *Electronics*

In this exercise, you will use the **INSERT** command to insert the block that was defined in Exercise 1 (RESIS). The following is the list of the attribute values for the resistances in the electric circuit.

RNAME	RVALUE	RPRICE
R1	35	0.32
R2	27	0.25
R3	52	0.40
R4	8	0.21
RX	10	0.21

1. Draw the electric circuit diagram as shown in Figure 13-30 (assume the dimensions).

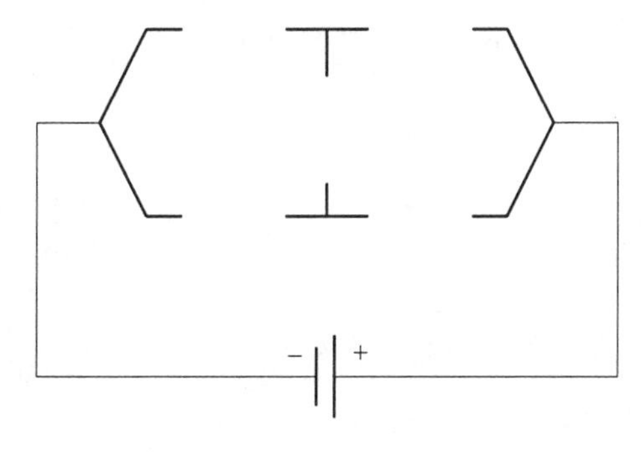

Figure 13-30 Drawing of the electric circuit diagram without resistors for Exercise 2

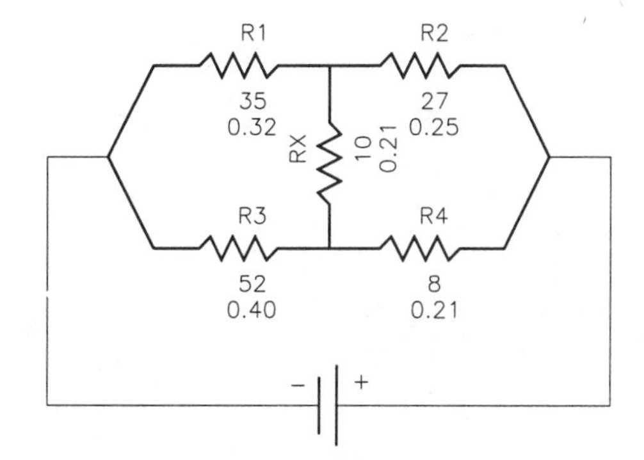

Figure 13-31 Drawing of the electric circuit diagram with resistors for Exercise 2

2. Set the system variable **ATTDIA** to 1. Use the **INSERT** command to insert the blocks, and define the attribute values in the **Enter Attributes** dialog box.

3. Repeat the **INSERT** command to insert other blocks, and define their attribute values as given in the table. Save the drawing as **ATTEXR2.DWG** (Figure 13-31).

Exercise 3 — *Electronics*

In this exercise, you will write a template file for extracting the attribute values as defined in Exercise 2. These attribute values must be written to a file **RESISLST.TXT** and arranged as shown in the following table.

	Field width in characters		
< 10 >	< 10 >	< 10 >	< 10 >
RESIS	R1	35	0.32
RESIS	R2	27	0.25
RESIS	R3	52	0.40

Chapter 13

RESIS	R4	8	0.21
RESIS	RX	10	0.21

1. Load the drawing **ATTEXR2** that you saved in Exercise 2.

2. Use the Windows Notepad to write the template file. After writing the file, save it as an ASCII file under the file name **TEMP2.TXT.**

3. In the AutoCAD screen, use the **ATTEXT** command to invoke the **Attribute Extraction** dialog box, and select the Space Delimited File (SDF) radio button.

4. Select the objects (blocks). You can also select the objects by using the Window or Crossing option.

5. In the Template File edit box, enter the name of the template file, **TEMP2.TXT.**

6. In the Output File edit box, enter the name of the output file, **RESISLST.TXT.**

7. Choose the **OK** button in the **Attribute Extraction** dialog box.

8. Use the Windows Notepad to list the output file, **RESISLST.TXT**. The output file should be similar to the file shown in the beginning of Exercise 3.

Exercise 4 *Electronics*

In this exercise, you will use the **ATTEDIT** or **-ATTEDIT** command to change the attributes of the resistances that are highlighted in the following table. You will also extract the attribute values and insert the text file in the drawing.

1. Load the drawing **ATTEXR2** that was created in Exercise 2. The drawing has five resistances with attributes. The name of the block is RESIS, and it has three defined attributes, one of them invisible.

2. Use the AutoCAD **ATTEDIT** or **-ATTEDIT** command to edit the values that are **highlighted** in the following table.

RESIS	R1	40	0.32
RESIS	R2	29	0.25
RESIS	R3	52	0.45
RESIS	R4	8	0.25
RESIS	R5	10	0.21

3. Extract the attribute values, and write the values to a text file.

4. Use the **MTEXT** command to insert the text file in the drawing.

Exercise 5 *Electronics*

Use the information given in Exercise 3 to extract the attribute values, and write the data to the output file. The data in the output file should be Comma Delimited CDF. Use the **ATTEXT** and **-ATTEXT** commands to extract the attribute values.

Exercise 6 *Electronics*

In this exercise, you will draw the circuit diagram as shown in Figure 13-32, define the attributes, and then extract the attributes to create a bill of materials.

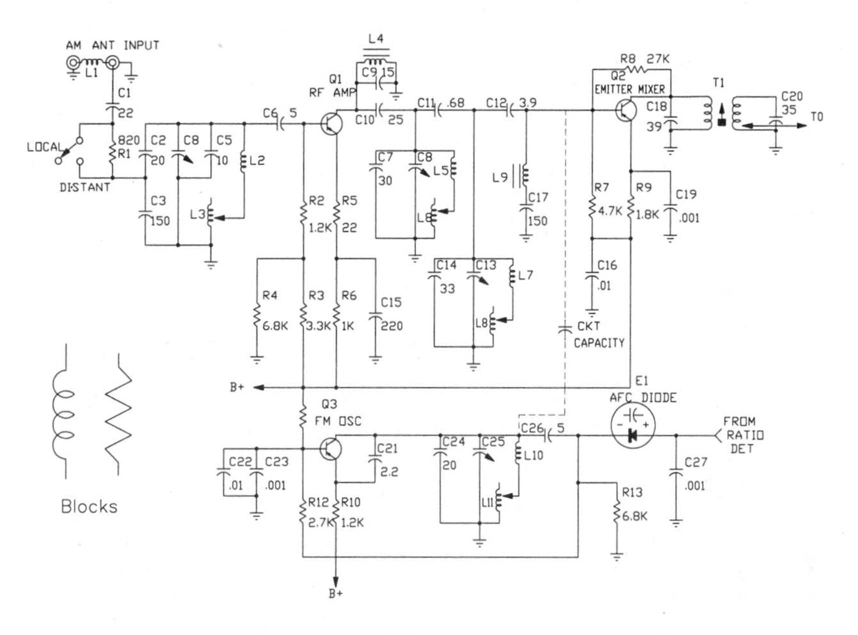

Figure 13-32 Drawing of the circuit diagram for Exercise 6

Answers to Self-Evaluation Test

1 - T, **2** - F, **3** - T, **4** - default value, **5** - disabled, **6** - **CHANGE**, **7** - 0, **8** - template, **9** - Space Delimited, **10** - tab, **11** - F, **12** - **ATTEDIT**, **13** - T, **14** - T, **15** - **MTEXT**

Chapter 14

External References

Learning Objectives

After completing this chapter, you will be able to:

- *Understand external references and their applications.*
- *Understand dependent symbols.*
- *Use the **XREF** command and its options.*
- *Use the **Attach**, **Unload**, **Reload**, **Detach**, and **Bind** options.*
- *Change the path of a drawing.*
- *Use the **XBIND** command to add dependent symbols.*
- *Understand demand loading.*
- *Use **XCLIPFRAME** and **XLOADCTL** system variables, and **XATTACH** command.*

EXTERNAL REFERENCES

The external reference feature lets you reference an external drawing without making that drawing a permanent part of the existing drawing. For example, assume that we have an assembly drawing ASSEM1 that consists of two parts, SHAFT and BEARING. The SHAFT and BEARING are separate drawings created by two CAD operators or provided by two different vendors. We want to create an assembly drawing from these two parts. One way to create an assembly drawing is to insert these two drawings as blocks by using the **INSERT** command. Now assume that the design of BEARING has changed due to customer or product requirements. To update the assembly drawing we have to make sure that we insert the BEARING drawing after the changes have been made. If we forget to update the assembly drawing, then the assembly drawing will not reflect the changes made in the piece part drawing. In a production environment, this could have serious consequences.

You can solve this problem by using the **external reference** facility, which lets you link the piece part drawings with the assembly drawing. If the xref drawings (piece part) get updated, the changes are automatically reflected in the assembly drawing. This way, the assembly drawing stays updated no matter when the changes were made in the piece part drawings. There is no limit to the number of drawings that you can reference. You can also have **nested references**. For example, the piece part drawing BEARING could be referenced in the SHAFT drawing, then the SHAFT drawing could be referenced in the assembly drawing ASSEM1. When you open or plot the assembly drawing, AutoCAD LT automatically loads the referenced drawing SHAFT and the nested drawing BEARING.

If you use the **INSERT** command to insert the piece parts, the piece parts become a permanent part of the drawing and, therefore, the drawing has a certain size. However, if you use the external reference feature to link the drawings, the piece part drawings are not saved with the assembly drawing. AutoCAD LT only saves the reference information with the assembly drawing; therefore, the size of the drawing is minimized. Like blocks, the xref drawings can be scaled, rotated, or positioned at any desired location, but they cannot be exploded. You can also use only a part of the Xref by making clipped boundary of Xrefs.

DEPENDENT SYMBOLS

If you use the **INSERT** command to insert a drawing, the information about the named objects is lost if the names are duplicated. If they are unique, it is imported. The **named objects** are entries such as blocks, layers, text styles, and layers. For example, if the assembly drawing has a layer HIDDEN with green color and HIDDEN linetype, and the piece part BEARING has a layer HIDDEN with blue color and HIDDEN2 linetype, then when you insert the BEARING drawing in the assembly drawing, the values set in the assembly drawing will override the values of the inserted drawing (Figure 14-1). Therefore, in the assembly drawing, the layer HIDDEN will retain green color and HIDDEN linetype, ignoring the layer settings of the inserted drawing. Only those layers that have the same names are affected. Remaining layers that have different layer names are added to the current drawing.

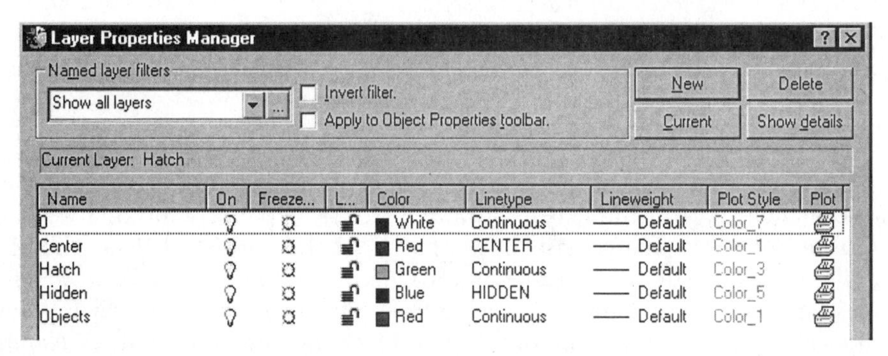

Figure 14-1 *Layer settings of the current drawing override the layers of the inserted drawing*

In the xref drawings, the information about named objects is not lost because AutoCAD LT will create additional named objects such as the specified layer settings as shown in Figure 14-2. For xref drawings, these named objects become dependent symbols (features such as layers, linetypes, object color, text style, and so on).

The layer HIDDEN of the xref drawing (BEARING) is appended with the name of the xref drawing BEARING, and the two are separated by the vertical bar symbol (|). Also these layers appear faded. The layer name HIDDEN changes to BEARING|HIDDEN. Similarly, CENTER is renamed BEARING|CENTER and OBJECT is renamed BEARING|OBJECT (Figure 14-2). The information added to the current drawing is not permanent. It is added only when the xref drawing is loaded. If you detach the xref drawing, the dependent symbols are automatically erased from the current drawing.

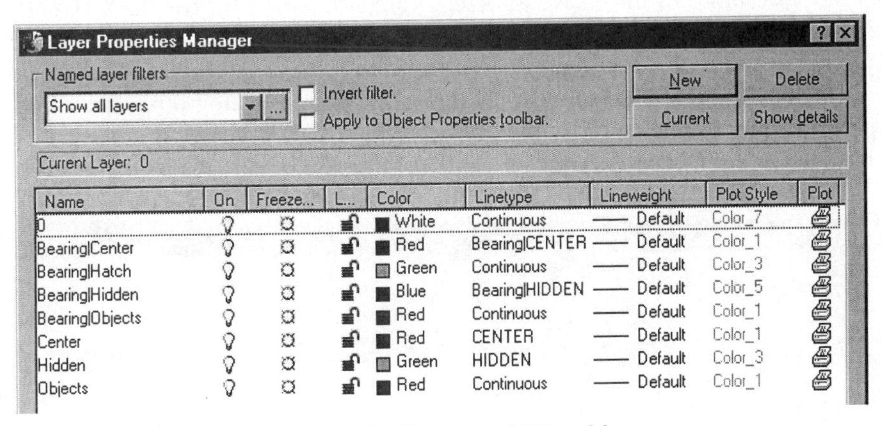

Figure 14-2 Xref creates additional layers

When you Xref a drawing, AutoCAD LT does not let you reference the symbols directly. For example, you cannot make the dependent layer, BEARING|HIDDEN, current. Therefore, you cannot add any objects to that layer. However, you can change the color, linetype, lineweight, or visibility (on/off, freeze/thaw) of the layer in the current drawing. If the system variable **VISRETAIN** is set to 0 (default), the settings are retained for the current drawing session. When you save and exit the drawing, the changes are discarded and the layer settings return to their default status. If the **VISRETAIN** variable is set to 1, layer settings such as color, linetype, on/off, and freeze/thaw are retained; the settings are saved with the drawing and are used when you open the drawing the next time. Whenever, you open a drawing or plot it, AutoCAD LT reloads each Xref in the drawing and as a result, the latest updated version of the drawing is loaded automatically.

XREF COMMAND

Managing Xrefs in a Drawing*

Toolbar:	Insert > External Reference
	Reference > External Reference
Menu:	Insert > Xref Manager
Command:	XREF

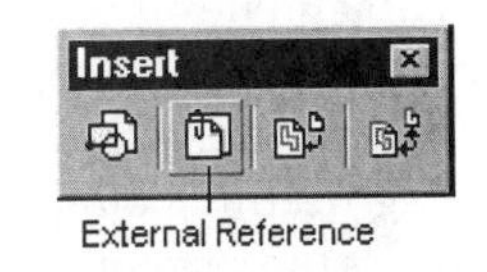

*Figure 14-3 **Insert** toolbar*

When you invoke the **XREF** command, AutoCAD LT displays the **Xref Manager** dialog box (Figure 14-4). The **Xref Manager** dialog box displays the status of each Xref in the current drawing and the relation between the various Xrefs. It allows you to attach a new xref, detach, unload, load an existing one, change an attachment to an overlay, or an overlay to an attachment. It also allows you to edit an xref's path and bind the xref definition to the drawing. You can also select an Xref in the current drawing and then right-click drawing area; a shortcut menu is displayed. Choosing the **Xref**

Manager option from the shortcut menu displays the **Xref Manager** dialog box. The Upper-left corner of the dialog box has two buttons: **List View** and **Tree View**. Choosing the **List View** button displays the Xrefs in the drawing in an alphabetical list; this is the default view. It has the following headings: **Reference name**, **Status** (whether the xref is loaded, unloaded, unreferenced, not found or unresolved), **Size**, **Type** (attachment or overlay), **Date** (last saved on), and **Saved Path**. Choosing any of these headings sorts and lists the Xrefs in the current drawing according to that particular title. For example, choosing **Reference Name** sorts and lists the xrefs as per name. You can also rename a xref in this dialog box. Choosing the **Tree View** button displays the xrefs in the drawing in a hierarchical tree view. It displays information on nested xrefs and their relationship with one another.

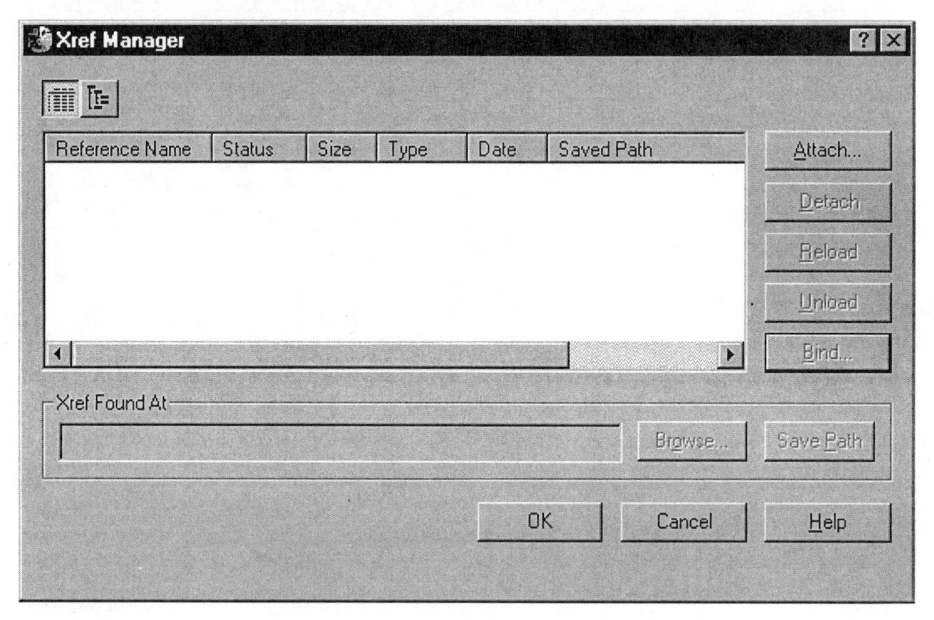

***Figure 14-4 Xref Manager** dialog box*

Attaching an Xref Drawing (Attach Option)

The **Attach** button of the **Xref Manager** dialog box is used to attach an xref drawing to the current drawing. This option can be invoked by choosing the **Attach** button in the **Xref Manager** dialog box. The following examples illustrate the process of attaching an xref to the current drawing. In this example, it is assumed that there are two drawings, SHAFT and BEARING. SHAFT is the current drawing that is loaded on the screen (Figure 14-5); the BEARING drawing is saved on the disk. We want to xref the BEARING drawing in the SHAFT drawing.

1. The first step is to make sure that the SHAFT drawing is on the screen (draw the shaft drawing with assumed dimensions). (One of the drawings does not need to be on the screen. You could attach both drawings, BEARING and SHAFT, to an existing drawing, even if it is a blank drawing.)

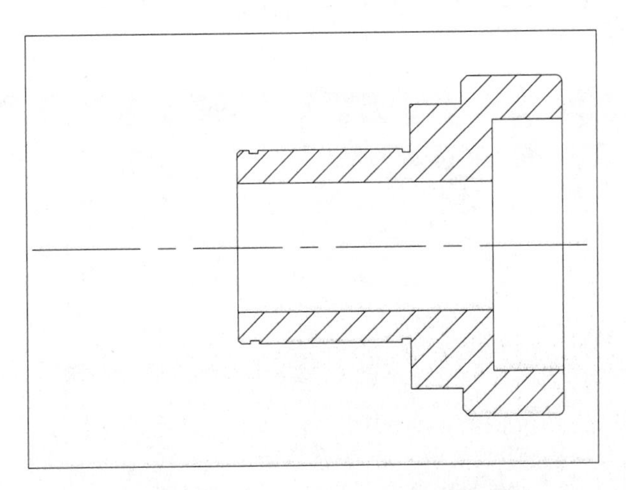

Figure 14-5 Current drawing, SHAFT

2. Invoke the **XREF** command to display the **Xref Manager** dialog box. In this dialog box, choose the **Attach** button. If you choose **External Reference** from the **Insert** menu, the **Select Reference File** dialog box (Figure 14-6) is displayed. Select the drawing that you want to attach (BEARING), and then choose the **Open** button; the **External reference** dialog box is displayed on the screen (Figure 14-7).

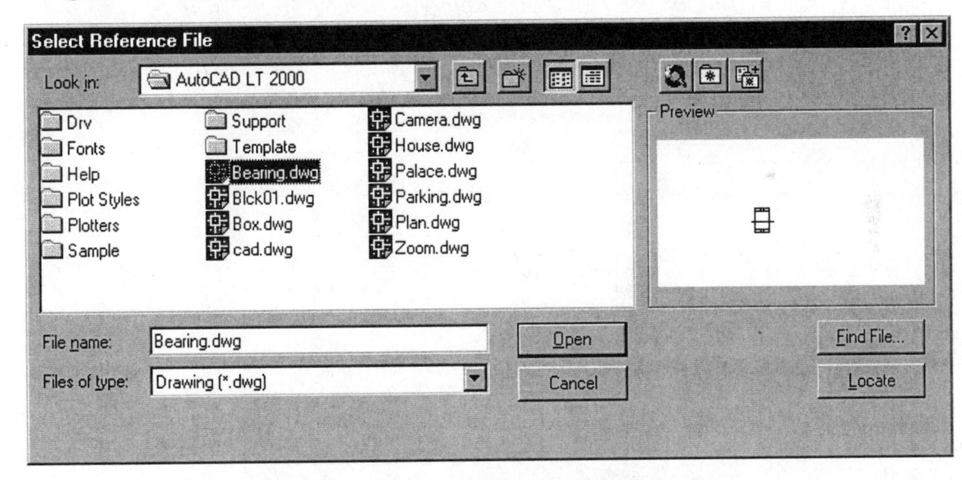

***Figure 14-6 Select Reference File** dialog box*

In the **External Reference** dialog box (Figure 14-7) you can select the name of the file to attach from the **Name** drop-down list. The selected file name is displayed in the **Name** text box and the **Path:** of the file is also displayed below the **Name** text box. Under the **Reference Type** area, select the **Attachment** radio button if it is not already selected (default option). You can either specify the insertion point, scale, and rotation angle in the respective **X**, **Y**, **Z** and **Angle** edit boxes or select the **Specify On-screen** check boxes to use the pointing device to specify them on the screen. Choose the **OK** button in the **External Reference** dialog box and then specify the Insertion, Scale and Angle parameters on the screen. After attaching the BEARING drawing, save the current drawing with the file name SHAFT (Figure 14-8).

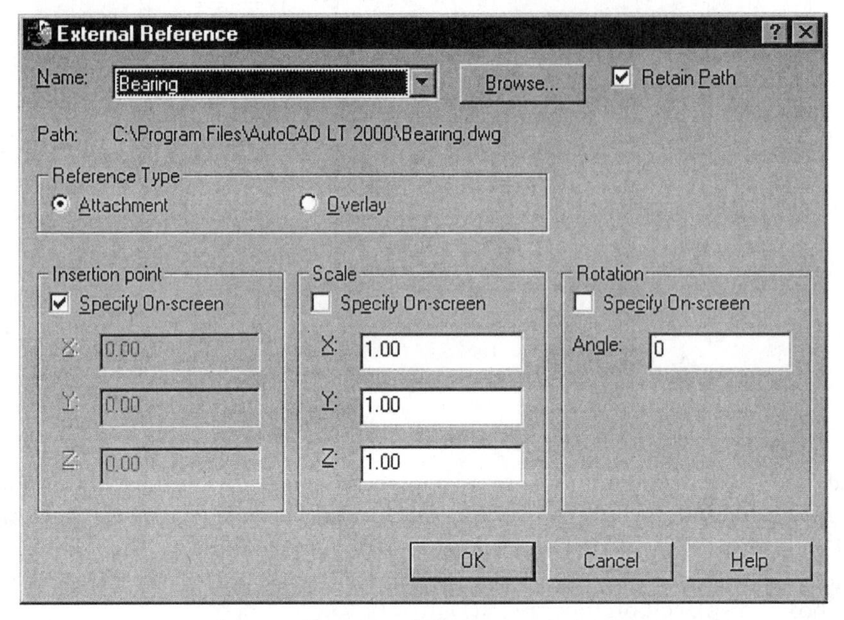

Figure 14-7 External Reference *dialog box*

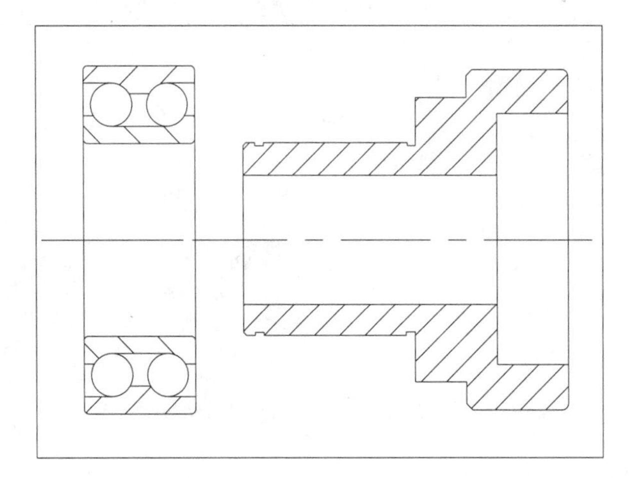

Figure 14-8 *Attaching xref drawing, BEARING*

You can also use the **-XREF** command to attach a drawing from the command line. If the **FILEDIA** system variable is set to 0, AutoCAD LT will not display the **Select Reference File** dialog box. Entering the ~ (tilde) symbol at the **Enter name of file to attach** prompt displays the **Select Reference File** dialog box. If the file name has more than 31 characters, or has spaces or invalid characters, AutoCAD LT will prompt you to enter a name different from the drawing name at the **Enter name of file to attach** prompt. Enter NEW NAME=OLD NAME at the **Enter name of file to attach** prompt. After specifying the name of the file to attach, press ENTER. AutoCAD LT will now prompt you to specify the insertion point, scale factor, and the rotation angle. The command prompts are as follows:

Command: **-XREF** Enter
Enter an option [?/Bind/Detach/Path/Unload/Reload/Overlay/Attach] <Attach>: Enter
Enter name of file to attach: BEARING *(If **FILEDIA**= 0.)*
Specify insertion point or
[Scale/X/Y/Z/Rotate/PScale/PX/PY/PZ/PRotate]: *Select a point.*

3. Load the drawing BEARING, and make the changes shown in Figure 14-9 (draw polylines on the sides). Now, save the drawing with the file name BEARING.

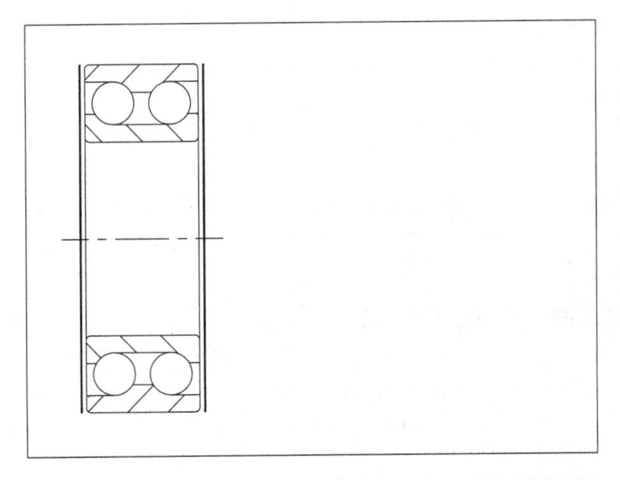

Figure 14-9 *Modifying the xref drawing, BEARING*

4. Load the drawing SHAFT on the screen. You will notice (Figure 14-10) that the xref drawing BEARING is automatically updated. This is the most useful feature of the **XREF** command. You could also have inserted the BEARING drawing as a block, but if you had updated the BEARING drawing, the drawing in which it was inserted would not have been updated automatically.

When you attach an xref drawing, AutoCAD LT remembers name of the attached drawing. If you xref the drawing again, AutoCAD LT displays a message to that effect as shown:

Xref "BEARING" has already been defined.
Using existing definition.
Specify insertion point or
[Scale/X/Y/Z/Rotate/PScale/PX/PY/PZ/PRotate]: *Indicate where to place another copy of the xref.*

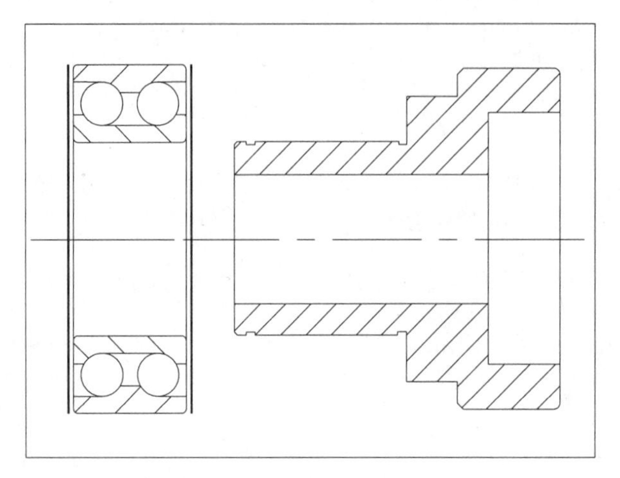

Figure 14-10 *After loading the drawing SHAFT, BEARING is automatically updated*

Note

*If the xref drawing you want to attach is currently being edited, AutoCAD LT will attach the drawing that has been saved earlier through the **SAVE**, **WBLOCK**, or **QUIT** command.*

Points to Remember About Xref

1. When you enter the name of the xref drawing, AutoCAD LT checks for block names and xref names. If a block exists with the same name as the name of the xref drawing in the current drawing, the **XREF** command is terminated and an error message is displayed.

2. If you are using **-XREF** command with **FILEDIA** set to 0, you can enter a ~ (tilde) at the **Enter name of file to attach** prompt. The tilde sign displays the **Select Reference File** dialog box. Entering **?** at the **Enter an option [?/Bind/Detach/Path/Unload/Reload/Overlay/Attach] <Attach>** prompt, displays a list of xref names, path, and type present in the current drawing.

3. When you xref a drawing, the objects that are in the model space are attached. Any objects that are in the paper space are not attached to the current drawing.

4. The layer 0, DEFPOINTS, and the linetype CONTINUOUS are treated differently. The current drawing layers 0, DEFPOINTS, and the linetype CONTINUOUS will override the layers and linetypes of the xref drawing. For example, if the layer 0 of the current drawing is white and the layer 0 of the xref drawing is red, the white color will override the red.

5. The xref drawings can be nested. For example, if the BEARING drawing contains the reference INRACE and you xref the BEARING drawing to the current drawing, the INRACE drawing is automatically attached to the current drawing. If you detach the BEARING drawing, the INRACE drawing gets detached automatically.

6. When you attach an xref drawing using the **-XREF** command with **FILEDIA** set to 0, you can assign the xref drawing a different name. For example, if the name of the xref drawing is BEARING and you want to assign it the name OBEARING, the prompt sequence is as follows:

   Command: **-XREF** Enter
   Enter an option [?/Bind/Detach/Path/Unload/Reload/Overlay/Attach] <Attach>: Enter
   Enter name of file to attach: OBEARING= BEARING
   Specify insertion point or
   [Scale/X/Y/Z/Rotate/PScale/PX/PY/PZ/PRotate]: *Select a point.*

7. When you xref a drawing, AutoCAD LT stores the name and path of the drawing. If the name of the xref drawing or the path where the drawing was originally stored has changed or you cannot find it in the path specified in the **Options** dialog box, AutoCAD LT cannot load the drawing, plot it, or use the **Reload** option of the **XREF** command.

Detaching an Xref Drawing (Detach Option)

The **Detach** option can be used to detach the xref drawings. If there are any nested Xref drawings defined with the xref drawings, they are also detached. Once a drawing is detached, it is erased from the screen. To detach the xref drawings, select the file names in the **Xref Manager** dialog box and then choose the **Detach** button. You can also use the **-XREF** command to detach the xref drawings.

Command: **-XREF** Enter
Enter an option [?/Bind/Detach/Path/Unload/Reload/Overlay/Attach] <Attach>: **D** Enter
Enter Xref name(s) to detach: *Enter names of xref drawings.*

When AutoCAD LT prompts for xref to detach, you can enter the name of one xref drawing or the name of several drawings separated by commas. You can also enter * (asterisk), in which case all referenced drawings, including the nested drawings, will be detached.

Updating an Xref Drawing (Reload Option)

When you load a drawing, AutoCAD LT automatically loads the referenced drawings. The **Reload** option of the **XREF** command lets you update the xref drawings and nested xref drawings any time. You do not need to exit the drawing editor and then reload the drawing. To reload the xref drawings, invoke the **XREF** command and select the drawings in the **Xref Manager** dialog box, then choose the **Reload** button. AutoCAD LT will scan for the referenced drawings and the nested xref drawings.

The **Reload** option is generally used when the xref drawings are currently being edited and you want to load the updated drawings. The xref drawings are updated based on what is saved on the disk. Therefore, before reloading an xref drawing, you should make sure that the xref drawings that are being edited have been saved. If AutoCAD LT encounters an error while loading the referenced drawings, the **XREF** command is terminated, and the entire reload operation is cancelled. You can also reload the xref drawings by using **-XREF** command. When you enter the **-XREF** command, AutoCAD LT will prompt you to enter the name of the xref drawing. You could enter the name of one xref drawing or the name of several drawings separated by commas. If you enter * (asterisk), AutoCAD LT will reload all xref and nested xref drawings.

Command: **-XREF** Enter
Enter an option [?/Bind/Detach/Path/Unload/Reload/Overlay/Attach] <Attach>: **R** Enter
Enter Xref name(s) to reload: *Enter names of xref drawings.*

Unloading an Xref Drawing (Unload Option)

The **Unload** option allows you to unload the definition of xref drawings. However, AutoCAD LT retains the pointer to the xref drawings. When you unload the xref drawings, the drawings are not displayed on the screen. You can reload the xref drawings by using the **Reload** option. It is recommended that you unload the referenced drawings if they are not being used. Also, after unloading the xref drawings, the drawings load much faster and need less memory.

Adding an Xref Drawing (Bind Option)

The **Bind** option lets you convert the xref drawings to blocks in the current drawing. The bound drawings, including the nested xref drawings (that are no longer xrefs), become a permanent part of the current drawing. The bound drawing cannot be detached or reloaded. To bind the xref drawings, select the file names in the **Xref Manager** dialog box, and then choose the **Bind** button to display **Bind Xrefs** dialog box (Figure 14-11). AutoCAD LT provides two methods to bind xref drawing.

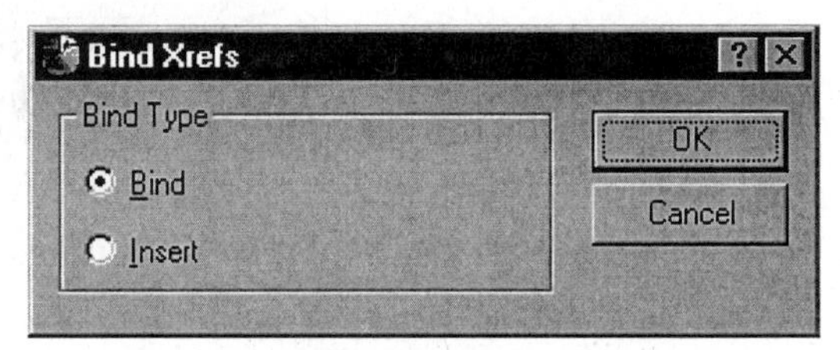

***Figure 14-11 Bind Xrefs** dialog box*

Bind. When you use the **Bind** option, AutoCAD LT binds the selected xref definition to the current drawing. For example, if you xref the drawing BEARING that has a layer named OBJECT, a new layer BEARING|OBJECT is created in the current drawing. When you bind this drawing, the xref dependent layer BEARING|OBJECT will become a locally defined layer BEARING0OBJECT (Figure14-12). If the BEARING0OBJECT layer already exits, AutoCAD LT will automatically increment the number, and the layer name becomes BEARING1OBJECT.

Insert. When you use the **Insert** option, AutoCAD LT inserts the xref drawing. For example, if you xref the drawing BEARING that has a layer named OBJECT, a new layer BEARING|OBJECT is created in the current drawing. If you use the **Insert** option to bind the xref drawing, the layer name BEARING|OBJECT is renamed as OBJECT (Figure 14-12). If the object layer already exists, then the values set in the current drawing override the values of the inserted drawing.

You can use this option when you want to send a copy of your drawing to a customer for review. Since all the xref drawings are a part of the existing drawing, you do not need to include the xref drawings or the path information. You can also use this option to safeguard the master drawing from accidental editing of the piece parts. You can also use the **-XREF** command to bind the xref drawings.

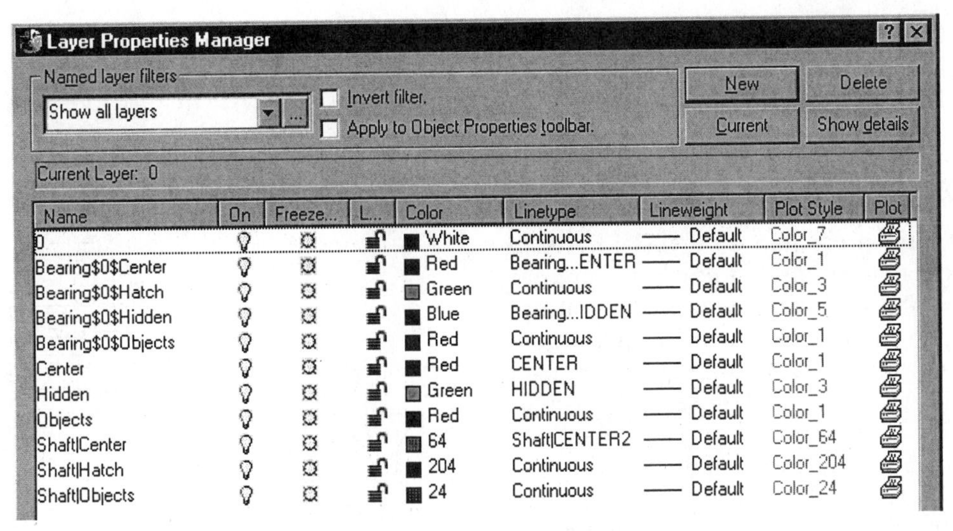

Figure 14-12 ***Layer Properties Manager*** *dialog box*

Command: **-XREF** Enter
Enter an option [?/Bind/Detach/Path/Unload/Reload/Overlay/Attach] <Attach>: **B** Enter
Enter Xref name(s) to bind: *Enter names of xref drawings.*

Editing an XREF's Path (Path Option)

You can edit the path where the xref drawing is located by editing the path found in **Xref Found At** edit box in the **Xref Manager** dialog box. You can also use the **Browse** button to display the **Select New path** dialog box where you can locate the drawing to be used as xref. Choosing the **Open** button in this dialog box, returns you to the **Xref manager** dialog box. The drawing path and name selected is displayed in the **Xref found at** and **Name** edit boxes respectively. To save the path, use the **Save Path** button. For example, if the drawing was originally in the C:\CAD\Proj1 subdirectory and the drawing has been moved to A:\Parts directory, the path must be edited so that AutoCAD LT can load the xref drawings. When you enter the path name, AutoCAD LT will try to locate the file in the specified directory. If the referenced drawing is not found in the specified path, AutoCAD LT will automatically search the directories specified in the **Options** dialog box. If AutoCAD LT cannot locate the specified file, it will display an error message, prompt you to enter a new path, and display the new file name in the drawing as marker text. When using the **Browse** button in the **Xref Manager** dialog box to locate the drawing or editing the path in the **Xref found at** edit box and choosing the **Save path** button, the drawing is loaded and replaces the marker text in the drawing. You can also use the **-XREF** command to change the path.

Command: **-XREF** Enter
Enter an option [?/Bind/Detach/Path/Unload/Reload/Overlay/Attach] <Attach>: **P** Enter
Edit Xref name(s) to edit path: **BEARING** Enter
Old path: **C:\CAD\Proj1\Bearing** Enter
Enter new path:

When AutoCAD LT prompts **Edit xref name(s) to edit path:**, you can enter the name of one xref drawing or the names of several drawings separated by commas. You can also enter * (asterisk), in which case AutoCAD LT will prompt you for the path name of each xref drawing. The path name stays unchanged if you press ENTER when AutoCAD LT prompts for a new path name.

OVERLAY OPTION

One of the problems with the **XREF Attach** option is that you cannot have circular reference. For example, assume you are designing the plant layout of a manufacturing unit. One person is working on the floor plan (see Figure 14-13), and the second person is working on the furniture layout in the offices. The names of the drawings are FLOORPLN and OFFICES, respectively. The person working on the office layout uses the **XREF Attach** option to insert the FLOORPLN drawing so that he or she has the latest floor plan drawing. The person who is working on the floor plan wants to XREF the OFFICES drawing. Now, if the **XREF Attach** option is used to reference the drawing, AutoCAD LT displays an error message because by inserting the OFFICES drawing a circular reference is created [Figure 14-13(a)]. To overcome this problem, the **Overlay** option can be used to overlay the OFFICES drawing. This is a very useful option because the **Overlay** option lets different operators avoid circular reference and share the drawing data without affecting the drawing. This option can be invoked by selecting the **Overlay** radio button in the **External Reference** dialog box or by entering O at the **Enter an option [?/Bind/Detach/Path/Unload/Reload/Overlay/Attach] <Attach>** prompt. Selecting the **Overlay** option, displays the **Enter Name of file to Overlay** dialog box, if the **FILEDIA** system variable is 1. Otherwise, the command prompt is:

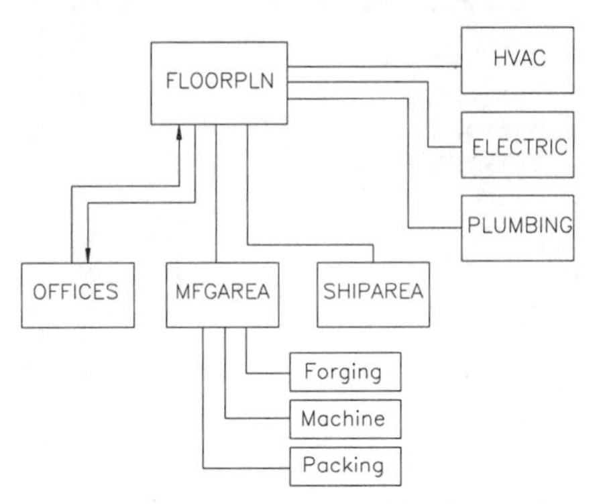

Figure 14-13 *Drawing files hierarchy*

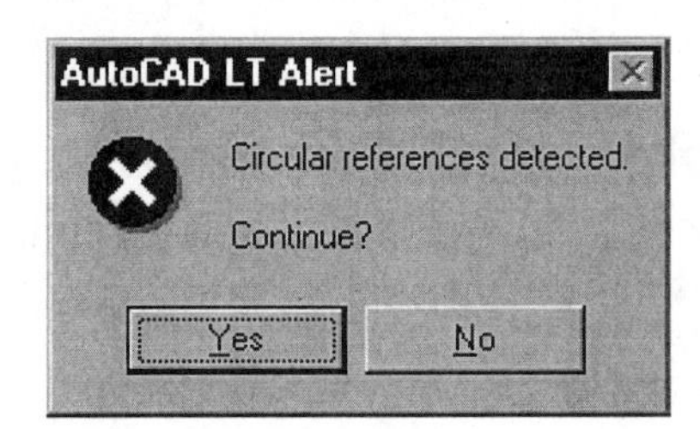

Figure 14-13(a) *AutoCAD LT Alert message box*

Enter name of file to overlay: Enter *the name of the file*

When a drawing that has a nested overlay is overlaid, the nested overlay is not visible in the current drawing.

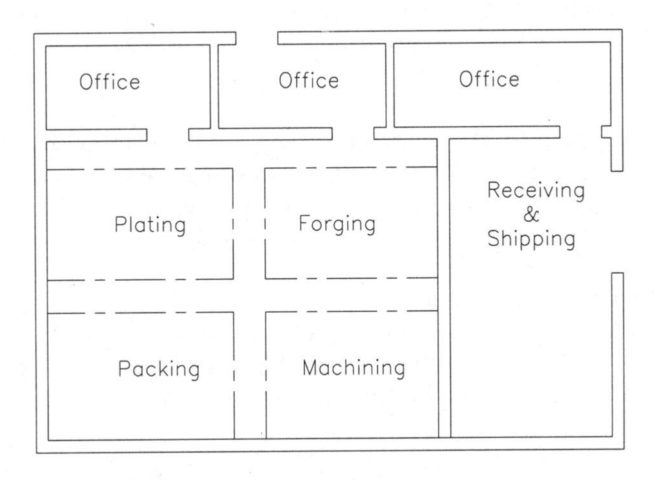

Figure 14-14 Sample plant layout drawing

Example 1

In this example you will use the **Attach** and **Overlay** options to attach and reference the drawings. Two drawings, PLAN and PLANFORG, are given. The PLAN drawing (Figure 14-15) consists of the floor plan layout, and the PLANFORG drawing (Figure 14-16) has the details of the forging section only. The CAD operator who is working on the PLANFORG drawing wants to xref the PLAN drawing for reference. Also, the CAD operator working on the PLAN drawing should be able to xref the PLANFORG drawing to complete the project. The following steps illustrate how to accomplish the defined task without creating a circular reference.

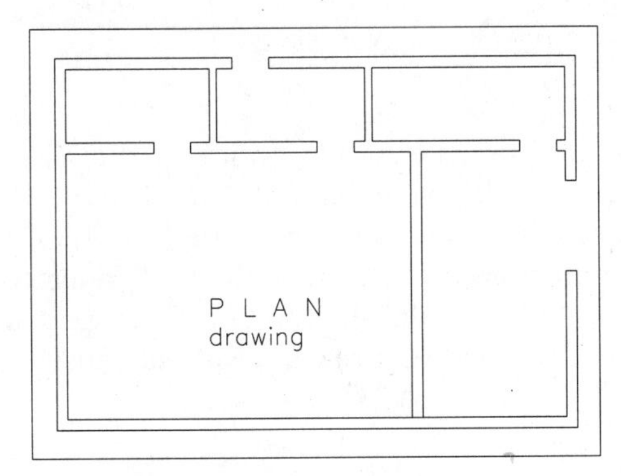

Figure 14-15 PLAN drawing

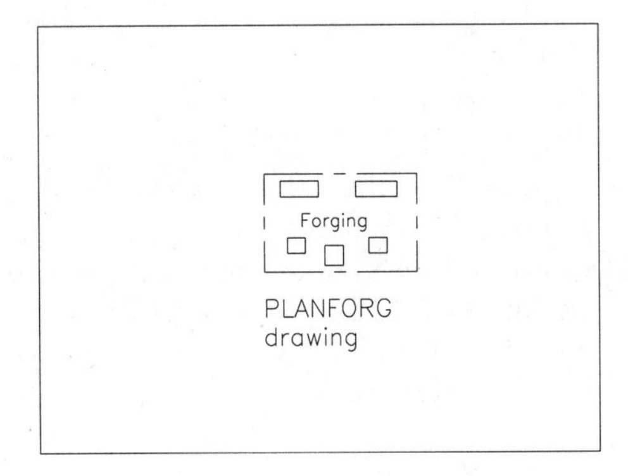

Figure 14-16 PLANFORG drawing

How circular reference is caused:

1. Load the drawing PLANFORG, and invoke the **XREF** command and choose the **Attach** option to attach the PLAN drawing. Now the drawing consists of PLANFORG and PLAN. Save the drawing.

2. Open the drawing file PLAN, and invoke the **XREF** command and attach the PLAN drawing. AutoCAD LT will prevent you from attaching the drawing because it causes circular reference.

   One possible solution is for the operator working on the PLANFORG drawing to detach the PLAN drawing. This way, the PLANFORG drawing does not contain any reference to the PLAN drawing and would not cause any circular reference. The other solution is to use the **-XREF** command's **Overlay** option, as follows.

How to prevent circular reference:

3. Open the drawing PLANFORG (Figure 14-17), and use the **Overlay** option of the **-XREF** command (or Select the **Overlay** radio button in the **External reference** dialog box) to overlay the PLAN drawing. The PLAN drawing is overlaid on the PLANFORG drawing (Figure 14-18).

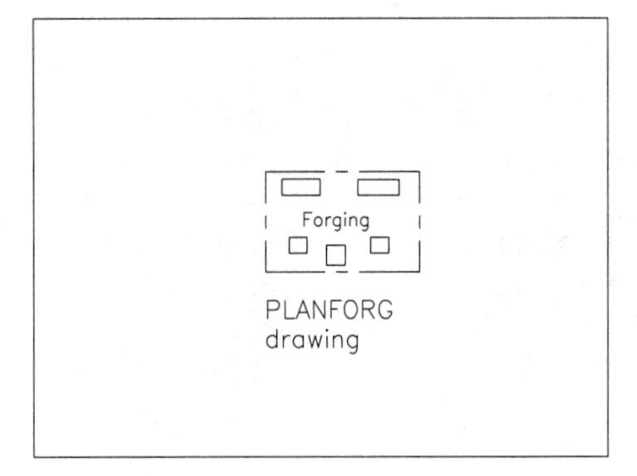

Figure 14-17 *PLANFORG drawing*

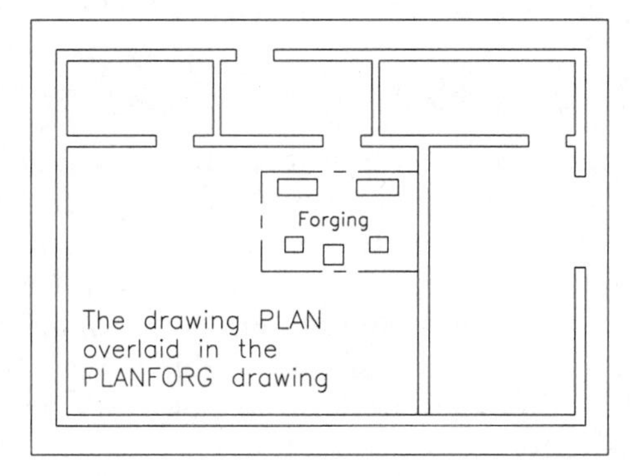

Figure 14-18 *PLANFORG drawing after overlaying the PLAN drawing*

4. Open the drawing file PLAN (Figure 14-19), and use the **Attach** option of the **XREF** command to attach the PLANFORG drawing. You will notice that only the PLANFORG drawing is attached (Figure 14-20). The drawing that was overlaid in the PLANFORG drawing (PLAN) does not appear in the current drawing. This way, the CAD operator working on the PLANFORG drawing can overlay the PLAN drawing, and the CAD operator working on the PLAN drawing can attach the PLANFORG drawing, without causing a circular reference.

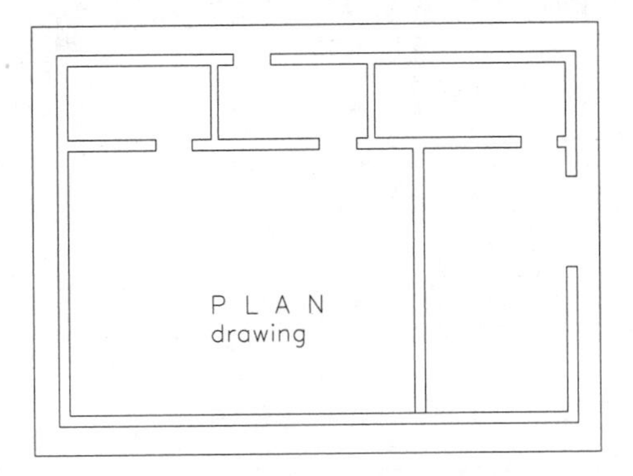

Figure 14-19 PLAN drawing

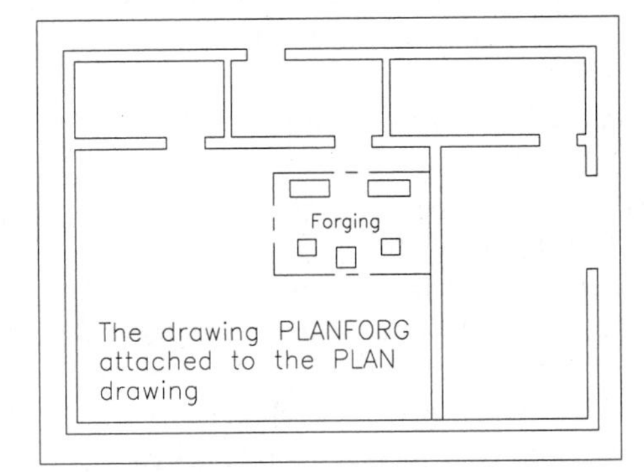

Figure 14-20 PLAN drawing after attaching the PLANFORG drawing

XATTACH COMMAND

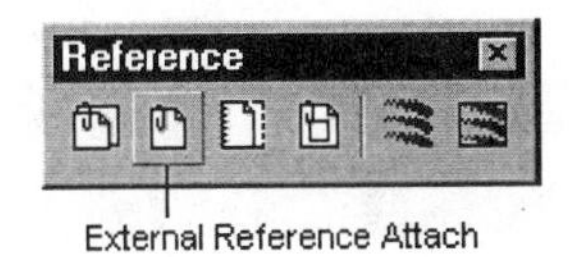

*Figure 14-21 **Reference** toolbar*

Toolbar:	Reference > External Reference Attach
Menu:	Insert > External Reference
Command:	XATTACH

If you want to attach a drawing without invoking the **Xref Manager** dialog box, you can use the **XATTACH** command. When you invoke this command, AutoCAD LT displays the **Select Reference file** dialog box. This command makes it easier to attach a drawing, since most of the xref operations involve simply attaching a drawing file. After you have selected the drawing file to attach, the **External Reference** dialog box is displayed. Select the **Attachment** radio button under the **Reference Type** area. Specify the insertion point, scale, and rotation angle on screen or in the respective edit boxes.

Note

*When you attach or reference a drawing that has a drawing order created by using the **DRAWORDER** command, the drawing order is not maintained in the xref. To correct the drawing order, first open the xref drawing and specify the drawing order in it. Now, using the **WBLOCK** command convert it into a drawing file. Use the **XATTACH** command to now attach the newly created drawing file to the current drawing. This way the drawing order will be maintained. AutoCAD LT maintains a log file for xref drawings if the **XREFCTL** system variable is on. This file lists information about the date and time of loading and other operations being carried out.*

Using AutoCAD DesignCenter to Attach a Drawing As Xref*

Choose the **AutoCAD DesignCenter** button in the **Standard** toolbar to display the AutoCAD LT **DesignCenter** window (Figure 14-22). In the **AutoCAD DesignCenter** toolbar, choose the **Tree toggle** button to display the **tree pane**. Expand the **Tree view** and double-click the folder whose contents you want to view. From the list of drawings diplayed in the palette, right-click the drawing you wish to attach as an xref. A shortcut menu is displayed. Choose the **Attach as Xref** option, the **External reference** dialog box is displayed. The **Name** edit box displays the name of the selected file to be inserted as an xref and **Path** displays the path of the file. Select the **Attachment** radio button in the **Reference type** area, if not already selected. Specify the **Insertion point**, **Scale**, and **Rotation** in the respective edit boxes or select the **Specify On-screen** check boxes to specify this information on the screen. Choose **OK** to exit the dialog box.

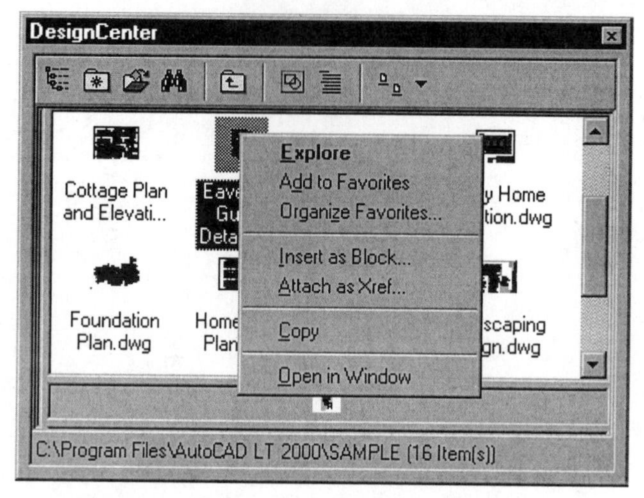

*Figure 14-22 AutoCAD **Design Center** window*

ADDING DEPENDENT SYMBOLS TO A DRAWING (XBIND COMMAND)

Toolbar:	Reference > External Reference Bind
Menu:	Modify > Object > External Reference Bind
Command:	XBIND

You can use the **XBIND** command to add the selected dependent symbols of the xref drawing to the current drawing. The following example describes how to use the **XBIND** command:

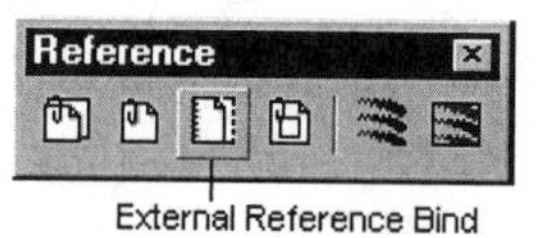

*Figure 14-23 **Reference** toolbar*

1. Load the drawing BEARING that was created earlier when discussing the **Attach** option of the **XREF** command. Make sure the drawing has the following layer setup; otherwise, create the following layers.

Layer Name	**Color**	**Linetype**
0	White	Continuous
Object	Red	Continuous
Hidden	Blue	Hidden2
Center	White	Center2
Hatch	Green	Continuous

2. Draw a circle and use the **BLOCK** command to create a block. The name of the block is SIDE. Save the drawing as BEARING.

3. Start a new drawing with the following layer setup.

Layer Name	**Color**	**Linetype**
0	White	Continuous

Object	Red	Continuous
Hidden	Green	Hidden

4. Use the **XATTACH** command and attach the BEARING drawing to the current drawing. When you xref the drawing, the layers will be added to the current drawing, as discussed earlier in this chapter.

5. Invoke the **XBIND** command, and the **Xbind** dialog box (Figure 14-24) is displayed on the screen. If you want to bind the blocks defined in the xref drawing BEARING, double-click on BEARING and then double-click on the Block icon. AutoCAD LT lists the blocks defined in the xref drawing (Bearing). Select the block BEARING|SIDE and then select the **Add** button; the block name will be added to the **Definitions to Bind** list box. AutoCAD LT will bind the block with the current drawing, and the name of the block will change to BEARING0SIDE. If you want to insert the block, you must enter the new block name (BEARING0SIDE). You can also rename the block to another name that is easier to use. If the block contains reference to another xref drawing, AutoCAD LT binds that xref drawing and all its dependent symbols to the current drawing. Once you **XBIND** the dependent symbols, AutoCAD LT does not delete them. For example, the block BEARING|SIDE will not be deleted when you detach the xref drawing or end the drawing session.

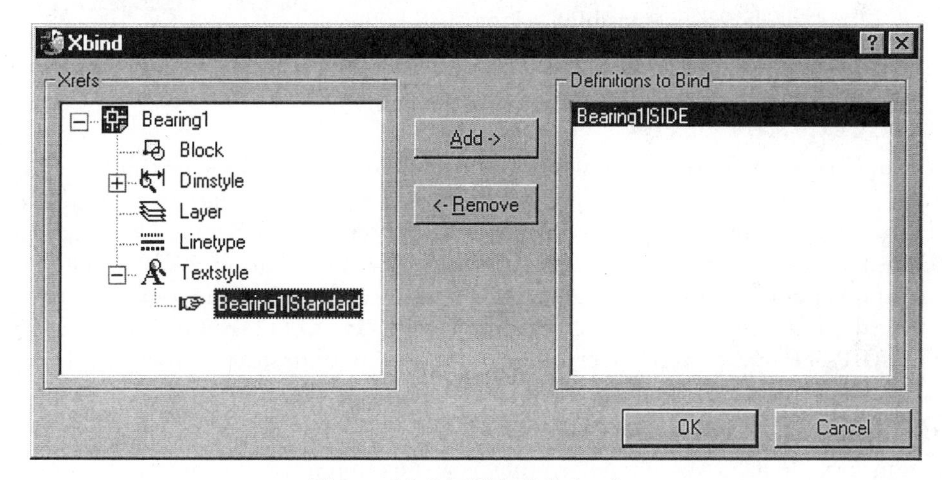

***Figure 14-24 Xbind** dialog box*

You can also use the **-XBIND** command to bind the selected dependent symbols of the xref drawing, using the command line.

Command: **-XBIND** [Enter]
Enter symbol type to bind [Block/Dimstyle/LAyer/LType/Style]: **B** [Enter]
Enter dependent Block name(s): **BEARING|SIDE** [Enter]
1 Block(s) bound.

6. Similarly, you can bind the dependent symbols, BEARING|STANDARD (text style), BEARING|HIDDEN, and BEARING|OBJECT layers of the xref drawing. You can also use the **-XBIND** command as follows:

Command: **-XBIND** [Enter]
Enter symbol type to bind [Block/Dimstyle/LAyer/LType/Style]: **B** [Enter]
Enter dependent Layer name(s): **BEARING|HIDDEN, BEARING|OBJECT** [Enter]
2 Layer(s) bound.

The layer names will change to BEARING0HIDDEN and BEARING0OBJECT. If the layer name BEARING0HIDDEN was already there, the layer will be named BEARING1HIDDEN. These two layers become a permanent part of the current drawing. If the xref drawing is detached or the current drawing is saved, the layers are not discarded. The other way of adding the dependent symbols to the current drawing is by using the **Bind** option of the **XREF** command. The **Bind** option makes the xref drawing and the dependent symbols a permanent part of the drawing.

Displaying Clipping Frame (XCLIPFRAME)

You can use the **XCLIPFRAME** system variable to turn the clipping boundary on or off. This system variable can be set by choosing the **External Reference Clip Frame** button from the **Reference** toolbar, or by entering **XCLIPFRAME** at the Command prompt. When the value is 0 (default) the clipping boundary is not displayed (not visible). When it is 1, the clipping boundary is displayed (visible).

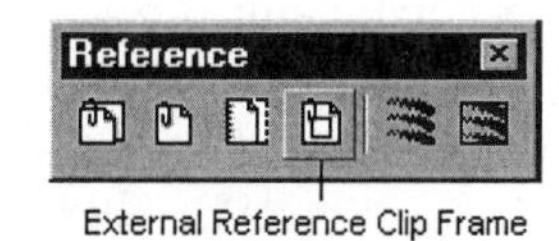

Figure 14-25 ***Reference*** *toolbar*

DEMAND LOADING

The demand loading feature loads only that part of the referenced drawing that is required in the existing drawing. For example, if you have attached a drawing and then clipped it, only a part of the drawing (clipped portion) is displayed on the screen. The demand loading provides a mechanism by which only the clipped portion of the referenced drawing will be loaded. This makes the xref operation more efficient, especially when the drawing is reopened. This feature works with the following system variables: **XLOADPATH**, **XLOADCTL**, and **INDEXCTL**. The **XLOADPATH** system variable creates a temporary path to store demand loaded Xrefs.

XLOADCTL System Variable

The system variable **XLOADCTL** determines how the referenced drawing file is loaded. The following are the three possible settings of **XLOADCTL** system variable:

Setting	Features
0	1. Turns off demand loading. 2. Loads entire xref drawing file. 3. The File is locked on the server and other users cannot edit the xref drawing.
1	1. Turns on demand loading. 2. The referenced file is kept open. 3. Makes the referenced file read only for other users.
2	1. Turns on demand loading with copy option. 2. A copy of referenced drawing is opened.

3. Other users can access and edit the original referenced drawing file.

You can set the value of **XLOADCTL** at the command line or in the **External Reference** area of the **Open and Save** tab of the **Options** dialog box (Figure 14-26).

Command: **XLOADCTL** [Enter]
Enter new value for XLOADCTL <1>: *Enter new value.*

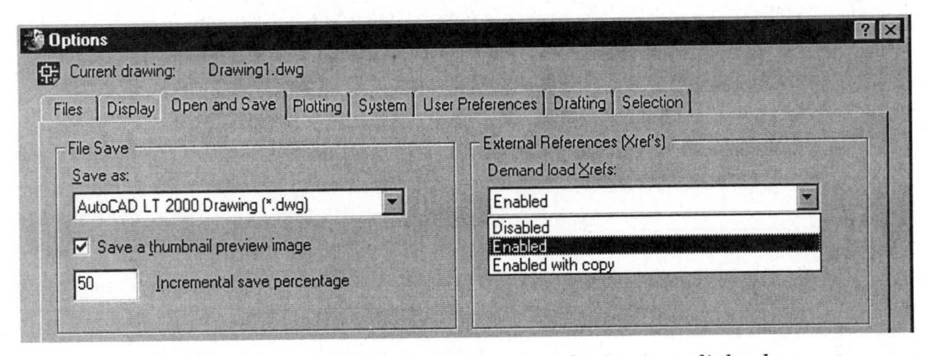

Figure 14-26 *Setting* ***XLOADCTL****, using the* ***Options*** *dialog box*

Spatial and Layer Indexes

As mentioned previously, the demand loading improves performance when the drawing contains referenced files. To make it work effectively and to take full advantage of demand loading, the **INDEXCTL** system variable must be set to 3; for example, you must store a drawing with Layer and Spatial indexes. The **INDEXCTL** variable controls the creation of layer index and spatial index. The layer index maintains a list of objects in different layers and the spatial index contains lists of objects based on their location in 3D space. The following are the settings of the **INDEXCTL** system variable:

Setting	**Features**
0	No index created.
1	Layer index created.
2	Spatial index created.
3	Layer and spatial index created.

Self-Evaluation Test

Answer the following questions, and then compare your answers to the correct answers given at the end of this chapter.

1. If the assembly drawing has been created by inserting a drawing, the drawing will be updated automatically if a change is made in the drawing that was inserted. (T/F)

2. The external reference facility helps you to keep the drawing updated no matter when the changes were made in the piece part drawings. (T/F)

3. AutoCAD LT saves only the reference information with the assembly drawing, and therefore, the size of the drawing is minimized. (T/F)

4. The ____________________ are entries such as blocks, layers, text styles, and layers.

5. What are dependent symbols? ____
____.

6. Can you add objects to a dependent layer? (Y/N)

7. If the **VISRETAIN** variable is set to __________, layer settings such as color, linetype, on/off, and freeze/thaw are retained. The settings are saved with the drawing and are used when you xref the drawing the next time.

8. If a block exists with the same name as the name of the xref drawing, the **XREF** command is terminated and an error message is displayed on the screen. (T/F)

9. When you enter ~ (tilde) in response to the **Enter name of file to attach:** prompt, AutoCAD LT displays the **Select reference file** dialog box, even if the **FILEDIA** system variable is turned off. (T/F)

10. If you insert a drawing using the **INSERT** command, the information about the named objects is not lost. (T/F)

Review Questions

1. The external reference feature lets you reference an external drawing without making that drawing a permanent part of the existing drawing. (T/F)

2. If the assembly drawing has been created by inserting a drawing, what needs to be done to update the drawing if a change has been made in the drawing that was inserted? Explain.
____
____.

3. The external reference facility lets you link the piece part drawings with the assembly drawing. (T/F)

4. If the xref drawings get updated, the changes are not automatically reflected in the assembly drawing when you open the assembly drawing. (T/F)

5. There is a limit to the number of drawings you can reference. (T/F)

6. It is not possible to have nested references. (T/F)

7. If you use an external reference feature to link the drawings, the piece parts are saved with the assembly drawing. (T/F)

8. Like blocks, the xref drawings can be scaled, rotated, or positioned at any desired location. (T/F)

9. What are the named objects? Explain. ____
____
____.

10. What happens to the information related to named objects when you insert a drawing?
____
____.

11. In the ____________________ drawings the information regarding dependent symbols is not lost.

12. For xref drawings, the ____________________ are features such as layers, linetypes, object color, and text style.

13. The information added to the current drawing is not permanent. It is added only when the xref drawing is loaded. (T/F)

14. If you detach the xref drawing, the dependent symbols are not automatically erased from the current drawing. (T/F)

15. When you xref a drawing, AutoCAD LT does not let you reference the symbols directly. For example, you cannot make the dependent layer BEARING|HIDDEN current. (T/F)

16. You can change the color, linetype, or visibility (on/off, freeze/thaw) of the dependent layer. (T/F)

17. If the system variable ____________________ is set to 0 (default), the settings are retained for the current drawing session. When you save the drawing, the changes are discarded, and the layer settings return to their default status.

18. The _________________ option of the **XREF** command can be used to attach an xref drawing to the current drawing.

19. When you attach an xref drawing, AutoCAD LT remembers the name of the attached drawing. At the time you attach the xref drawing, the previous xref drawing name becomes the default name. (T/F)

20. If the xref drawing you want to attach is currently being edited, AutoCAD LT will not attach the drawing. (T/F)

21. What is demand loading? Explain. ____
____
____.

22. What is the function of the **XLOADCTL** system variable? Explain. ______________
____
____.

23.What is the function of the **XCLIPFRAME** system variable? Explain. ____________
____
____.

Exercises

Exercise 1 *Mechanical*

In this exercise, you will start a new drawing and xref the drawings Part-1 and Part-2. You will also edit one of the piece parts to correct the size and use the **XBIND** command to bind some of the dependent symbols to the current drawing. Following are detailed instructions for completing this exercise.

1. Start a new drawing, Part-1, and set up the following layers.

Layer Name	**Color**	**Linetype**
0	White	Continuous
Object	Red	Continuous
Hidden	Blue	Hidden2
Center	White	Center2
Dim-Part1	Green	Continuous

2. Draw Part-1 with dimensions as shown in Figure 14-27. Save the drawing as Part-1.

3. Start a new drawing, Part-2, and set up the following layers.

Layer Name	**Color**	**Linetype**
0	White	Continuous
Object	Red	Continuous
Hidden	Blue	Hidden
Center	White	Center
Dim-Part2	Green	Continuous
Hatch	Magenta	Continuous

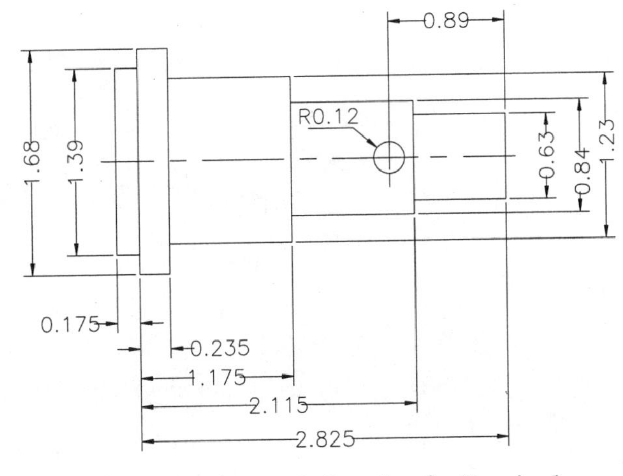

Figure 14-27 *Part-1, Drawing for Exercise 1*

4. Draw Part-2 with dimensions as shown in the Figure 14-28. Save the drawing as Part-2.

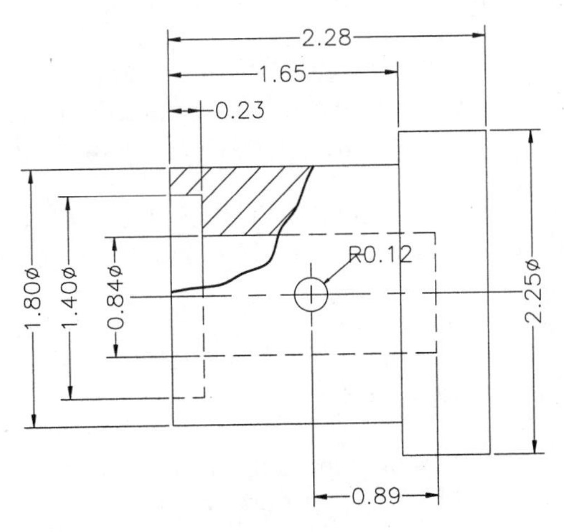

Figure 14-28 *Part-2, Drawing for Exercise 1*

5. Start a new drawing, ASSEM1, and set up the following layers.

Layer Name	Color	Linetype
0	White	Continuous
Object	Blue	Continuous
Hidden	Yellow	Hidden

6. Xref the two drawings Part-1 and Part-2 so that the centers of the two drilled holes coincide. Notice the overlap as shown in Figure 14-29. Save the assembly drawing as ASSEM1.

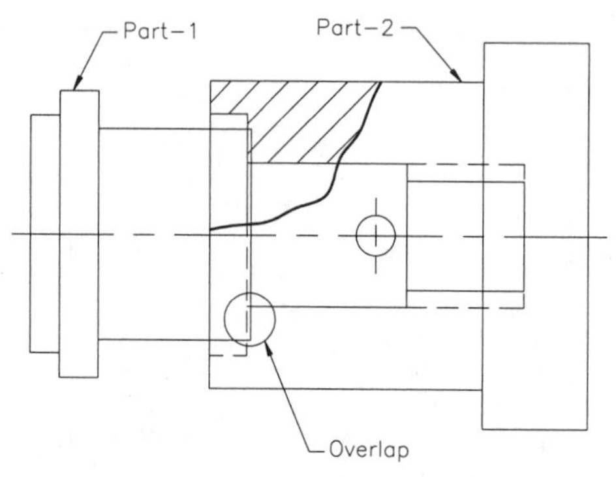

Figure 14-29 *Assembly drawing after attaching Part-1 and Part-2*

7. Open the drawing Part-1 and correct the mistake so that there is no overlap. You can do it by editing the line (1.175 dimension) so that the dimension is 1.160.

8. Open the assembly drawing ASSEM1 and notice the change in the overlap. The assembly drawing gets updated automatically.

9. Study the layers and notice how AutoCAD LT renames the layers and linetypes assigned to each layer. Check to see if you can make the layers belonging to Part-1 or Part-2 current.

10. Use the **XBIND** command to bind the OBJECT and HIDDEN layers that belong to drawing Part-1. Check again to see if you can make one of these layers current.

11. Use the **Detach** option to detach the xref drawing Part-1. Study the layer again, and notice that the layers that were edited with the **XBIND** command have not been erased. Other layers belonging to Part-1 are erased.

12. Use the **Bind** option of the **XREF** command to bind the xref drawing Part-2 with the assembly drawing ASSEM1. Open the xref drawing Part-1 and add a border or make any changes in the drawing. Now, open the assembly drawing ASSEM1 and check to see if the drawing is updated.

Project Exercise 2-3 *Mechanical*

Generate the drawing in Figure 14-30 according to the dimensions shown in the figure. Save the drawing as TOOLORGC. (See Chapters 9 and 21 for other parts of Project Exercise 2)

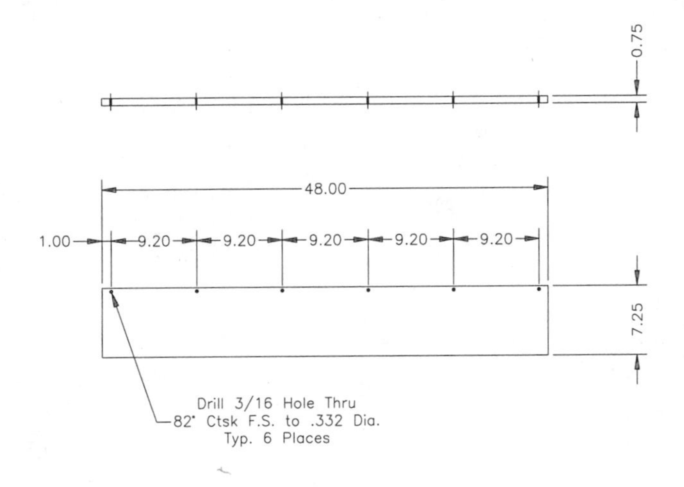

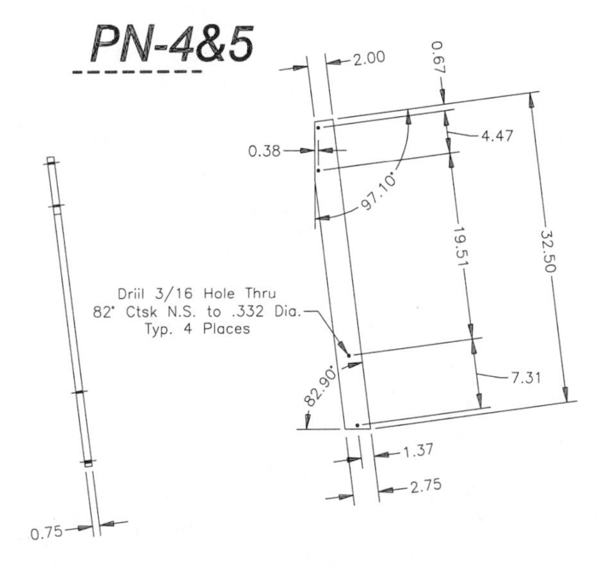

Figure 14-30 Drawing for Project Exercise 2-3

Project Exercise 2-4 *Mechanical*

Where possible, use the **XREF** command, to generate the assembly drawing as shown in Figure 14-31. Also, give the bill of materials and the part numbers as shown. (See Chapters 9 and 23 for other parts of Project Exercise 2)

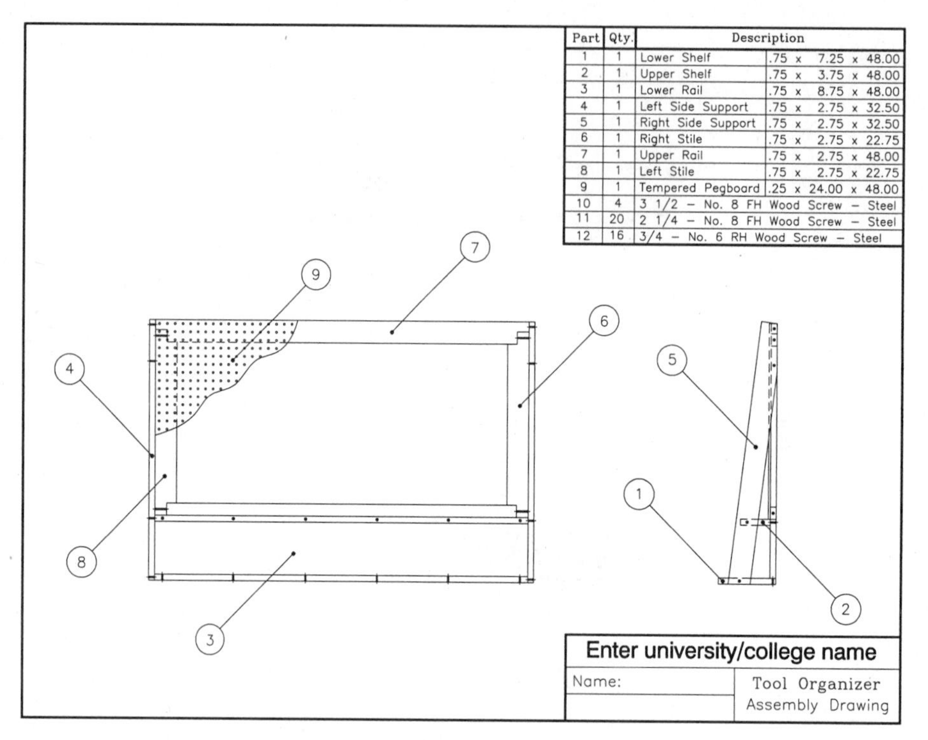

Part	Qty.	Description	
1	1	Lower Shelf	.75 x 7.25 x 48.00
2	1	Upper Shelf	.75 x 3.75 x 48.00
3	1	Lower Rail	.75 x 8.75 x 48.00
4	1	Left Side Support	.75 x 2.75 x 32.50
5	1	Right Side Support	.75 x 2.75 x 32.50
6	1	Right Stile	.75 x 2.75 x 22.75
7	1	Upper Rail	.75 x 2.75 x 48.00
8	1	Left Stile	.75 x 2.75 x 22.75
9	1	Tempered Pegboard	.25 x 24.00 x 48.00
10	4	3 1/2 – No. 8 FH Wood Screw – Steel	
11	20	2 1/4 – No. 8 FH Wood Screw – Steel	
12	16	3/4 – No. 6 RH Wood Screw – Steel	

Figure 14-31 *Drawing for Project Exercise 2-4*

Answers to Self-Evaluation Test

1 - F, **2** - T, **3** - T, **4** - named objects, **5** - For xref drawings, the dependent symbols are features such as layers, linetypes, object color, and text style, **6** - N, **7** - 1, **8** - T, **9** - T, **10** - F

Chapter 15

Plotting Drawings and Draw Commands

Learning Objectives

After completing this chapter, you will be able to:

- *Set plotter specifications and plot drawings.*
- *Use Plot Styles.*
- *Create* ***DOUBLE LINES*** *using* ***DLINE*** *command.*
- *Create polylines using* ***REVCLOUD*** *command.*
- *Draw construction lines using the* ***XLINE*** *and* ***RAY*** *commands.*
- *Draw* ***NURBS*** *splines using the* ***SPLINE*** *command.*
- *Edit* ***NURBS*** *splines using the* ***SPLINEDIT*** *command.*
- *Digitize drawings.*

PLOTTING DRAWINGS IN AUTOCAD LT

When you are done with a drawing, you can store it on the computer storage device, such as the hard drive or diskettes. However, to get a hard copy of the drawing you should plot the drawing on a sheet of paper using a plotter or printer. With the help of pen plotters you can obtain a high-resolution drawing.

PLOT Command

Toolbar:	Standard > Plot
Menu:	File > Plot
Command:	PLOT

The **PLOT** command is used to plot a drawing. After invoking the **PLOT** command, the **Plot** dialog box (Figure 15-1) is displayed. You can also invoke the **Plot** dialog box from the shortcut menu by right-clicking in the **Model** tab or **Layout** tab and choosing **Plot**.

Command: **PLOT**

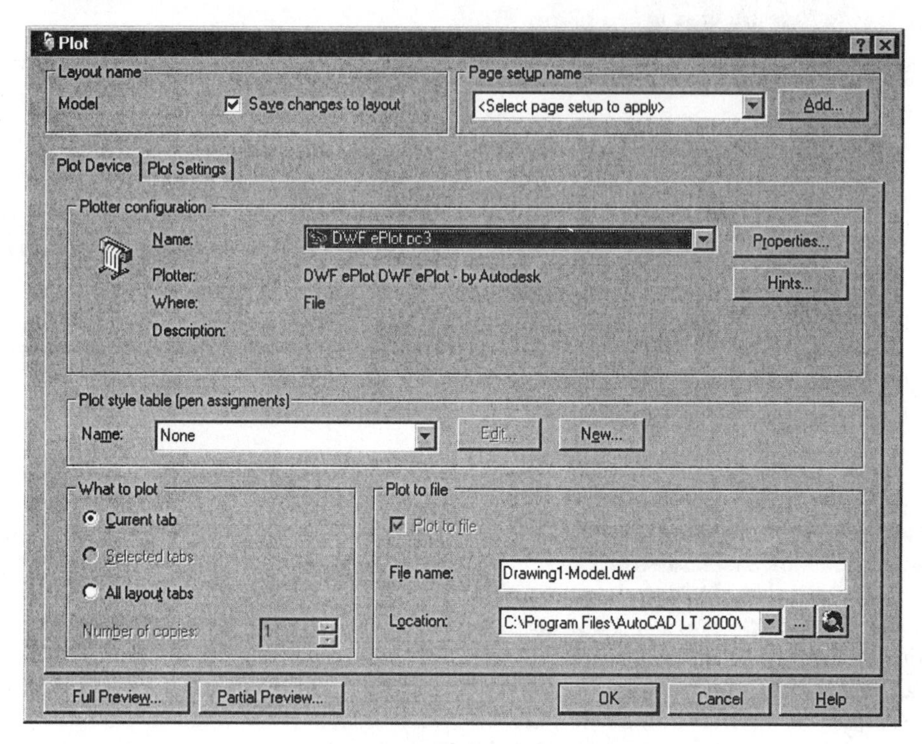

Figure 15-1 ***Plot*** *dialog box*

Some values in this dialog box were set when AutoCAD LT was first configured. You can examine these values; if they conform to your requirements, you can start plotting directly. If you want to alter the plot specifications, you can do so through the options provided in the **Plot** dialog box. The available plot options are described next.

Layout Name Area

This area displays the current layout name or displays "Selected layouts" if multiple tabs are selected. For example, if the **Model** tab is currently selected and you choose the **PLOT** command, the Layout Name shows "Model."

Save Changes to layout. This check box saves the changes made in the **Plot** dialog box in the layout (Figure 15-2). This option is not available when multiple layouts are selected.

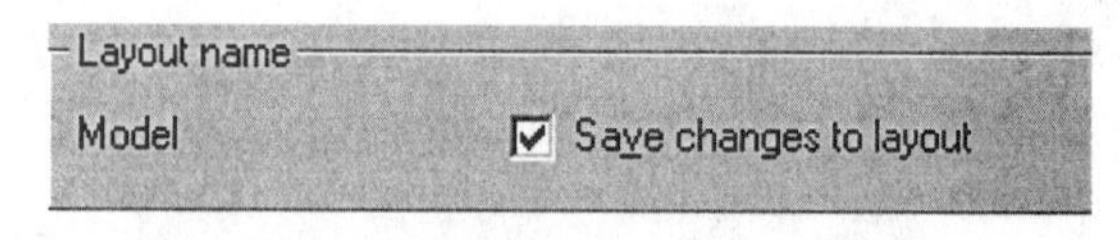

Figure 15-2 Layout name area of ***Plot*** *dialog box*

Page Setup Name Area*

The **Page setup name** drop-down list displays all the saved and named page setups. You can choose the base for the current page setup on a named page setup or you can add a new named page setup by choosing the **Add** button.

Add. You can add a new page setup for the current layout by choosing the **Add** button. When you choose this button, AutoCAD LT 2000 displays the **User Defined Page Setups** dialog box (Figure 15-3). You can create, delete, rename or import page setups. All the settings applied to the current layout are saved in the user defined page setup file. The description of the **User Defined Page Setups** dialog box is as follows:

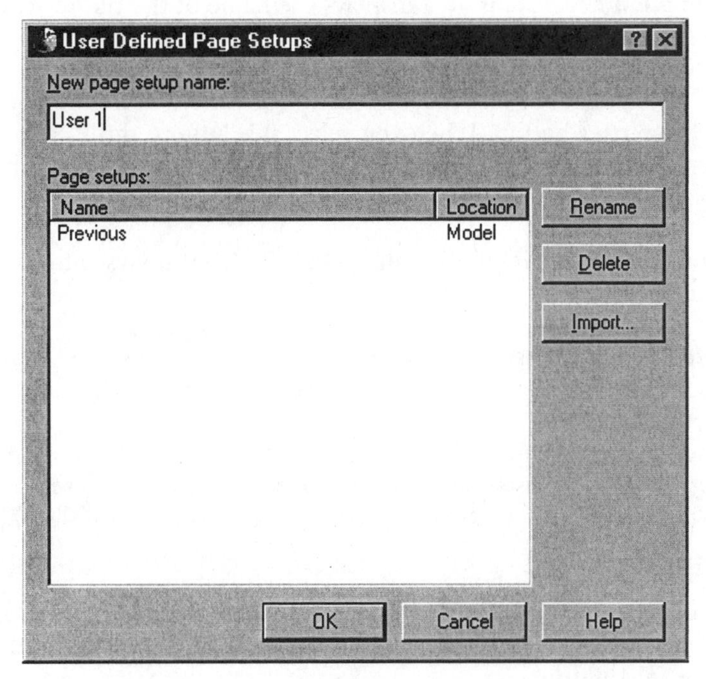

*Figure 15-3 **User Defined Page Setups** dialog box*

New page setup name. You can save the Name of the new user-defined page setup in this box. Then the current page setup can be based upon the named page setup.

Page setups. This box lists all the named and saved user-defined page setups. You can select a named page setup from the list to apply to the current layout.

Rename. You can select any user-defined page setup from the list and rename the page setup.

Delete. You can select a user-defined page setup from the list and delete it by choosing the **Delete** button.

Import. You can import a user-defined page setup from another drawing and apply it to a layout in the current drawing. If you choose the **Import** button, AutoCAD LT displays the standard file selection dialog box, where you can browse through the files for the page setup selection.

Plot Device Tab (Plot Dialog Box)

In this tab, information about the current configured plotter, plot style table, the layout or layouts to be plotted and information about plotting to a file is displayed. The tab contains the following parameters:

Plotter configuration area. This area displays the name of the plotter currently selected from the **Name** drop-down list with its location and description.

Properties. If you want to check the information about the configured printer or plotter, choose the **Properties** button. When you select this button, the **Plotter Configuration Editor** (discussed later in the "Editing Plotter Configuration" subsection of this chapter) is displayed.

Hints. If you choose the **Hints** button, AutoCAD LT displays information about the specific plotting device.

Plot Style Table (pen assignments) area. This area in the **Plot** dialog box sets the plot style table, edits the current plot style table, or creates a new plot style table.

Name. **Name** drop-down list displays the plot style table assigned to the current Model tab or layout tab and also lists the available plot style tables. If you select more than one layout tab with different plot style tables assigned to them, the list displays "**Varies**."

Edit. If you choose the **Edit** button, AutoCAD LT displays the **Plot Style Table Editor** (discussed later in the "Using Plot Styles" subsection of this chapter), where you can edit the selected plot style table. There are three tabs: **General**, **Table View**, and **Form View** in the **Plot Style Table Editor**.

New. If you choose the **New** button, AutoCAD LT displays the **Add Color-Dependent Plot Style Table-Wizard** dialog box, which can be used to create a new plot style table.

What to Plot area. This area in the **Plot** dialog box defines what you want to plot whether you select the **Model** tab, a **layout** tab, or multiple layout tabs. You can also plot a multiple number of copies of the plots.

Current Tab. If this radio button is selected, AutoCAD LT plots the current **Model** or **layout** tab. If you select multiple tabs, the tab that shows its viewing area is plotted.

Selected Tabs. If you select this radio button, AutoCAD LT plots the multiple Model or layout tabs that were selected previously by holding down CTRL during the selection of tabs. If you select only one tab, this option is unavailable.

All Layout Tabs. This option in the **Plot** dialog box plots all layout tabs, independent of your selection of layout tabs.

Number of Copies. This box shows the number of copies that will be plotted. If multiple layouts and copies are selected, any layouts that are set for plotting to a file or AutoSpool produces a single plot. You can plot the desired number of plots by changing the value in this box.

Plot to File area. This is used for plotting a drawing to a file instead of plotter. AutoCAD LT plots output to a file rather than to the plotter.

Plot to File. If you select this check box, AutoCAD LT plots the drawing to a file.

File Name. You can specify the file name of the plot. The default plot file name is the drawing name with the tab name, separated by a hyphen, with a .plt file extension.

Location. This box displays the location of the directory where the plot file will be stored. The default location of the plot file is the directory where the drawing file resides.

[...]. If you choose this button, AutoCAD LT displays a standard **Browse for Folder** dialog box, where you can choose a location of the directory to store a plot file.

Browse the Web. By choosing this button, you can browse the Web and send the plot file to other locations via email.

Plot Settings Tab (Plot Dialog Box)

The **Plot Settings** tab in the **Plot** dialog box controls the paper size, orientation, scale of plotting, plot offset, and other plotting options. The description of this tab is as follows:

Paper Size and Paper Units area. The **Paper Size and Paper Units** area displays available standard paper sizes for the selected plotting device. Actual paper size is indicated by the width (in X-axis direction) and the height (in Y-axis direction). The paper size selected is saved with a layout and overrides the PC3 file settings created by **Add-A-Plotter-Wizard** (discussed in the "Plotter Manager command" subsection of this chapter). If there is no plotter selected, AutoCAD displays the full standard paper size list and is available for selection. If a raster image is plotted, such as a **BMP** or **TIFF** file, the size of the plot is specified in terms of pixels, not in inches or millimeters.

Plot device. The name of the currently selected plot device is displayed here.

Paper size. All the available paper sizes are displayed in the drop-down list. You can select any size from the list to make it current.

Printable area. The actual dimension of the area on the paper that is used for plotting based on the current paper size selected.

Inches or mm. The radio button specifies the units to be used for plotting.

Drawing orientation area. This area specifies the orientation of the drawing on the paper for the plotters that support landscape or portrait orientation. You can change the drawing orientation by selecting **Portrait**, or **Landscape**, or **Plot upside-down** check box. The paper icon shows the media orientation of the selected paper and the letter icon shows the orientation of the drawing on the page.

Plot area. In this area you can specify the portion of the drawing to be plotted. You can also control the way the plotting will be carried out. The various options in this area are described next.

Layout. If you select this radio button, AutoCAD LT plots everything within the margins of the specified paper size and the origin is calculated from 0,0 in the layout. This option is available only when a layout is selected.

Limits. If you select the **Limits** radio button, the whole area defined by the drawing limits is plotted. The exception is when the current view is not the top view. In this case, the **Limits** option works exactly as the **Extents** option by scaling all the objects in the drawing to fit into the plotting area.

Extents. If you select this option, the section of drawing that currently holds objects is plotted. In this way, this option resembles the **ZOOM Extents** option. If you add objects to the drawing, they are also included in the plot because the extents of the drawing may also be altered. If you reduce the drawing extents by erasing, moving, or scaling objects, you must use the **ZOOM Extents** or **ZOOM All** option. Only after this does the **Extents** option understand the extents of the drawing to be plotted. If you are in model space, the plot is created in relation to the model space extents; if you are in paper space, the plot is created in relation to the paper space extents. If you invoke the Extents option when the perspective view is on and the position of the camera is not outside the drawing extents, the following message is displayed: **PLOT Extents incalculable, using display**. In this case, the plot is created as it would be with the **Display** option. This option is disabled if the current drawing does not have extents.

Display. If you select this radio button while you are in model space, the whole current viewport is plotted. If you are in paper space, the whole current view is plotted.

View. Selecting the **View** radio button enables you to plot a view that was created with the **VIEW** command. The view must be defined in the current drawing. If no view has been created, the **View** radio button and the drop-down are disabled. These two options can be activated by creating a view in the current drawing.
To select a view for plotting, select a named view you want to plot from the drop-down list (Figure 15-4) and then choose **OK**. In this case, the specifications of the plot depend on the specifications of the named view.

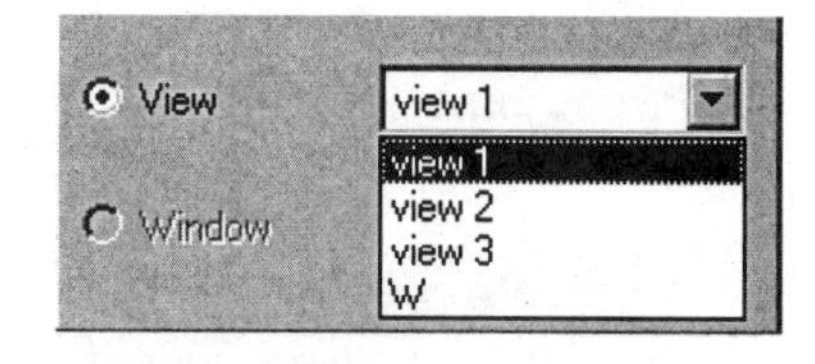

Figure 15-4 Selecting a named view for plotting

Window. With this option, you can specify the section of the drawing to be plotted. The section of the drawing to be plotted is defined by a lower left corner and an upper right corner of the section. A window can be defined by choosing the **Window <** button. The following command sequence is displayed:

Specify window for printing
Specify first corner: *Specify a point.*
Specify other corner: *Specify a point.*

When you are defining the window, the dialog box disappears temporarily and when the selection is completed, the box reappears.

Plot Scale area. This area controls the drawing scale of the plot area. The default scale setting is **1:1** when you are plotting a layout. But if you are plotting in the **Model** tab, the default setting is **Scaled to Fit**. When you are selecting a standard scale from the **Scale** drop-down list, the scale is displayed in the **Custom** edit box. You can change the scale factor manually in the **Custom** edit box.You can also scale lineweights in proportion to the plot scale. The settings are controlled by the lineweight system variables. Lineweights generally specifies the linewidth of the printable objects and are plotted with original linewidth size regardless of the plot scale.

Plot Offset area. This area of the **Plot Settings** tab (**Plot** dialog box) specifies an offset of the plotting area from the lower-left corner of the paper. In the layout tab, the lower-left corner of a specified plot area is positioned at the lower-left margin of the paper. If you select the Center the plot check box, AutoCAD LT automatically centers the plot on the paper by calculating the X and Y offset values. You can also specify an offset from the origin by entering a positive or negative value in the **X** and **Y** edit boxes. The plotter unit values are either in inches or millimeters on the paper.

Plot Options area. This area specifies the various options like lineweights, plot styles, and the current plot style table. You can select whether the lineweights are to be plotted.

Plot object lineweights. If you select this check box, AutoCAD LT plots the drawing with the lineweights. Plotting with lineweights plots the drawing with the specified line widths.

Plot with plot styles. By selecting **Plot with plot style,** AutoCAD LT plots using the plot styles applied to objects in the drawing and defined in the plot style table. The different property characteristics associated with the different style definitions are stored in the plot style tables and can be easily attached to the geometry. This setting replaces pen mapping in earlier versions of AutoCAD LT.

Plot paperspace last. By selecting **Plot paperspace last**, you get an option of plotting model space geometry first. Usually paper space geometry is plotted before model space geometry. This option is available only when multiple layout is selected.

Hide objects. If this is selected, AutoCAD LT plots layouts with hidden lines removed for objects in the layout environment (paper space). But in model space, hidden line removal of objects in viewports is controlled by **Viewports Hide** property in the **Properties** window. The hidden lines are displayed in the plot preview, but not in layout.

Plot Preview

You can view the plot on the specified paper size before actually plotting it by selecting the **Full Preview** or **Partial Preview** buttons in the Plot dialog box. AutoCAD LT provides two types of Plot Previews: partial and full.

Full Preview. When you choose the **Full Preview** button AutoCAD LT displays the drawing on the screen just as it would be plotted on the paper. **Full Preview** takes more time than the Partial preview because regeneration of the drawing takes place. Once regeneration is performed, the dialog boxes on the screen are removed temporarily, and an outline of the paper size is shown. The **Plot Preview** (Figure 15-5) with the **Zoom Realtime** icon is displayed. This Realtime icon can be used to zoom in and out interactively by holding the pick button down and then moving the pointing device. You can right-click to display the shortcut menu and then select **Exit** to exit the preview. You can also use the other options from the shortcut menu.

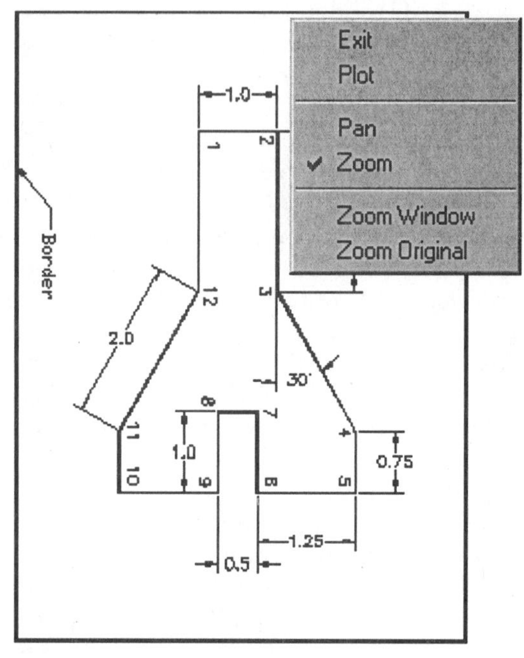

Figure 15-5 Full Plot Preview with shortcut menu

Partial. To generate a **Partial Preview** of a plot, choose the **Partial Preview** button in the **Plot** dialog box. The **Partial Plot Preview** dialog box (Figure 15-6) is displayed. The paper size is graphically represented by the white paper icon. The paper size is also given numerically. The dashed rectangle shows the printable area. The dimensions of the printable area is also shown numerically. The blue rectangle is the section in the paper that is used by the image. This area is also known as the **effective area**. The size of the effective area is also given numerically. If the paper boundary and the effective area boundary overlap (are the same), this is graphically represented by a red triangle. If you define the origin of the plot so that the effective area is not

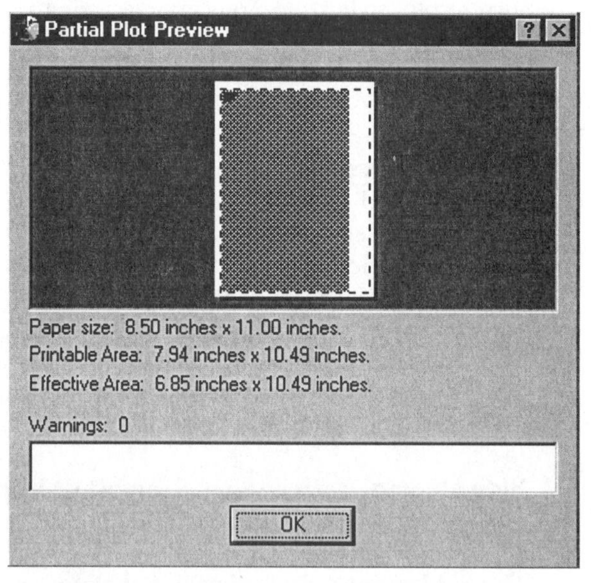

*Figure 15-6 **Partial Plot Preview** dialog box*

accommodated in the graphic area of the dialog box, AutoCAD displays a warning. Hence, with the help of a partial preview, you can accurately see how the plot will fit on the paper. If there is something wrong with the specifications of the plot, AutoCAD provides a warning message so that corrections can be made before actual plotting takes place. After you have finished with all the settings and other parameters, if you choose the **OK** button, AutoCAD starts plotting the drawing in the file or plotters as specified. AutoCAD displays the **Plot Progress** dialog box (Figure 15-7), where you can view the actual progress in plotting.

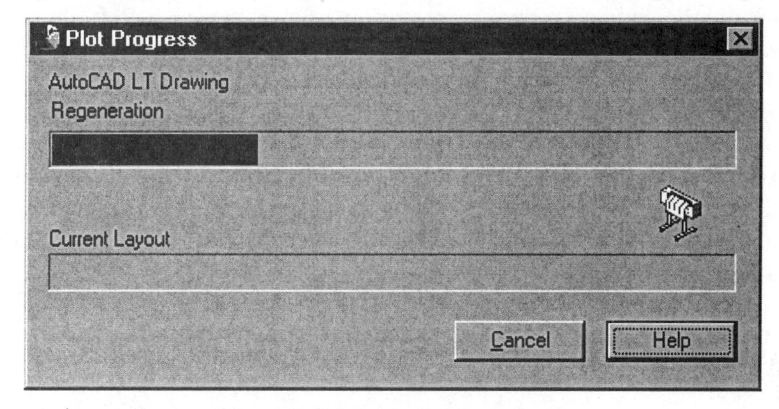

*Figure 15-7 **Plot progress** dialog box*

PLOTTERMANAGER COMMAND

Menu:	File > Plotter Manager
Command:	PLOTTERMANAGER

This is a new feature of AutoCAD LT 2000. If you enter **PLOTTERMANAGER** at the Command prompt, AutoCAD will display **Plotter Manager** (Figure 15-8). You can also invoke Plotter Manager by choosing **Plotter Manager** from the **File** menu. The **Plotter Manager** is basically a Windows Explorer window. **Plotter Manager** displays all the configured Plotters with a option of **Add-a-Plotter-Wizard.**

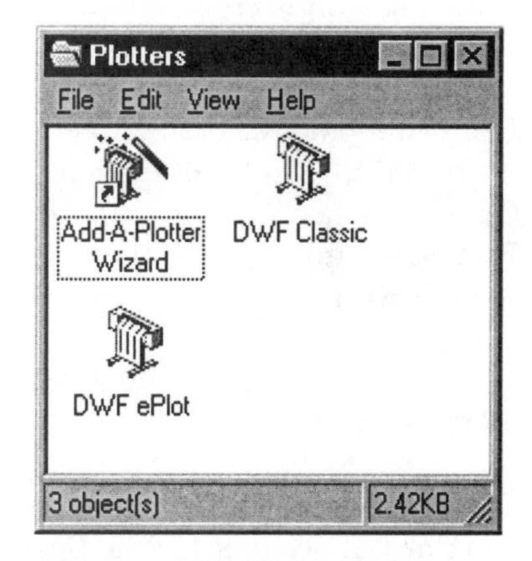

Figure 15-8 Plotter Manager

Add-a-Plotter-Wizard. If you double-click your pointing device on **Add-a-Plotter-wizard** in **Plotter Manager,** AutoCAD guides you to configure a nonsystem plotter for plotting your drawing files. AutoCAD LT 2000 stores all the information of a configured plotter in configured plot (PC3) files. The **PC3** file are stored in **AutoCAD LT 2000\Plotters** folder by default. The steps for configuring a new plotter using **Add-a-Plotter-Wizard** is as follows:

1. Open the **Plotter Manager** by choosing from the **File** menu.
2. In the **Plotter Manager**, double-click the **Add-a-Plotter-Wizard.**
3. In the **Add-a-Plotter-Wizard**, carefully read the Introduction page, and then choose the **Next** button to advance to the **Add Plotter -Introduction Begin** page.

4. On the **Add Plotter - Begin** page, choose the **My Computer** radio button (selected by default). Choose the **Next** button.
5. On the **Add Plotter - Plotter Model** page, select a manufacturer and model of your nonsystem plotter from the list and choose the **Next** button. If your plotter is not present in the list of available plotters, and you have a driver disk for your plotter, choose the **Have Disk** button to locate the **HIF** file from the driver disk, and install the driver supplied with your plotter.
6. If you want to import configuring information from a **Pcp** or a **Pc2** file created with a previous version of AutoCAD LT, you can choose the **Import File** button in the **Add Plotter - Import Pcp or Pc2** screen (Optional) and select the file.
7. On the **Add Plotter- Ports** page, select the port from the list to use when plotting. Choose **Next.**
8. On the **Add Plotter - Plotter Name** page, you can specify the name of the currently configured plotter or the default name will be entered automatically. Choose **Next.**
9. When you reach the **Add Plotter - Finish** page, you can choose the **Finish** button to exit the **Add-a-Plotter-wizard.**

A **PC3** file for the newly configured plotter will be displayed in the **Plotters** window and you can use the plotter for plotting in the list of devices.

Editing Plotter Configuration*

You can also change the default settings for the plotter by choosing the **Edit Plotter Configuration** button on the **Add Plotter - Finish** page. AutoCAD LT 2000 will display the **Plotter Configuration Editor** (Figure 15-9). The other methods of invoking the **Plotter Configuration Editor** are as follows:

From Windows Explorer, select the **PC3** file for editing (by default, **PC3** files are stored in the **AutoCAD LT 2000\Plotters** folder) and double-click the file or right-click the file and choose **Open** from the shortcut menu. You can also invoke the **Plotter Configuration Editor** by choosing **Page Setup** from the **File** menu and then choose the **Properties** button. Or, choose **Plot** from the **File** menu and then choose the **Properties** button.

The **Plotter Configuration Editor** has three tabs. The description of the three tabs are as follows:

General tab. This tab contains basic information about the configured plotter or the PC3 file. You can make changes in the **Description** area. The rest of the information in the tab is read only. This tab contains information of the configured plotter file name, plotter driver type, HDI driver file version number, name of the system printer (if any), and the location and name of the PMP file (if any calibration file is attached to the PC3 file).

Ports tab. The **Ports** tab contains information about the communication between the plotting device and your computer. You can choose between a serial (local), parallel (local), or network port. The default settings for parallel ports is **LPT1** and **COM1** for serial ports. You can also change the port name if your device is connected to a different port.

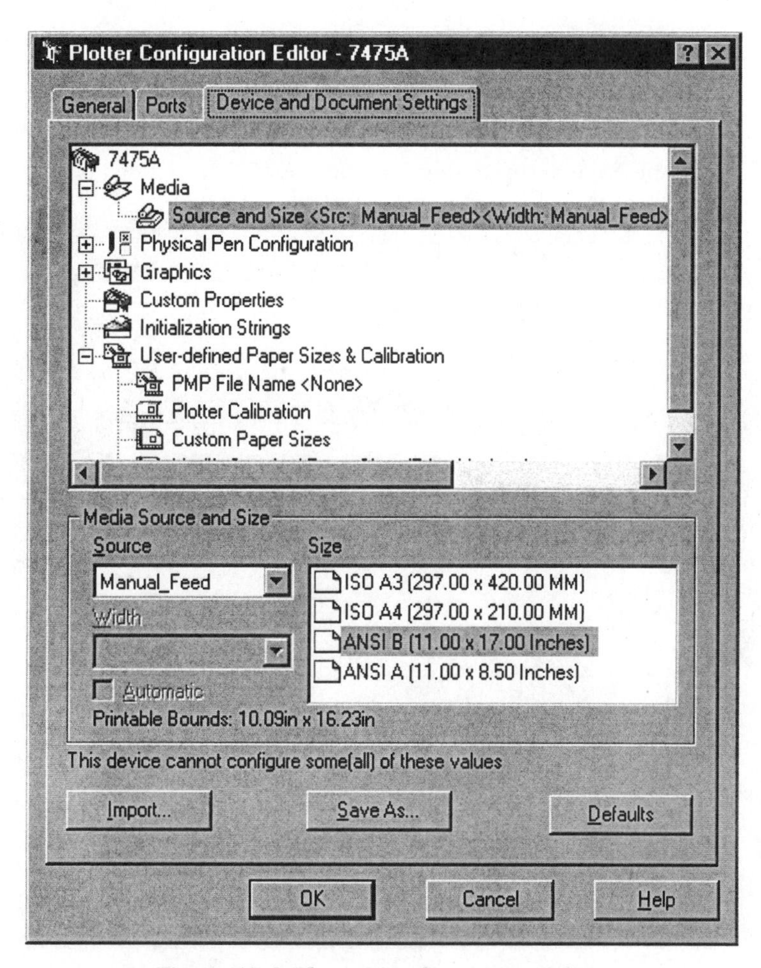

Figure 15-9 Plotter Configuration Editor

Device and Document Settings tab. The **Device and Document Settings** tab contains plotting options. Additional options are available on the **Device and Document Settings** tab depending on your configured plotting device. For example, if you configure a nonsystem plotter you have the option to modify the pen characteristics. You can select any plotter properties from the tree displayed in the tab and change the values as required. You can also import information and save the settings as a new PC3 file.

PAGESETUP COMMAND*

Toolbar:	Layouts > Page Setup
Menu:	File > Page Setup
Command:	PAGESETUP

This is a new feature in AutoCAD LT 2000. The **PAGESETUP** command can be used for specifying the layout page, plotting device, paper size, and settings for all the new layouts. The **PAGESETUP** command can also be invoked from the shortcut menu by right clicking in the **Model** or **Layout** tab and choosing **Page Setup**.

The command sequence for the **PAGESETUP** command is as follows:

Command: **PAGESETUP**

AutoCAD LT displays the **Page Setup** dialog box (Figure 15-10).

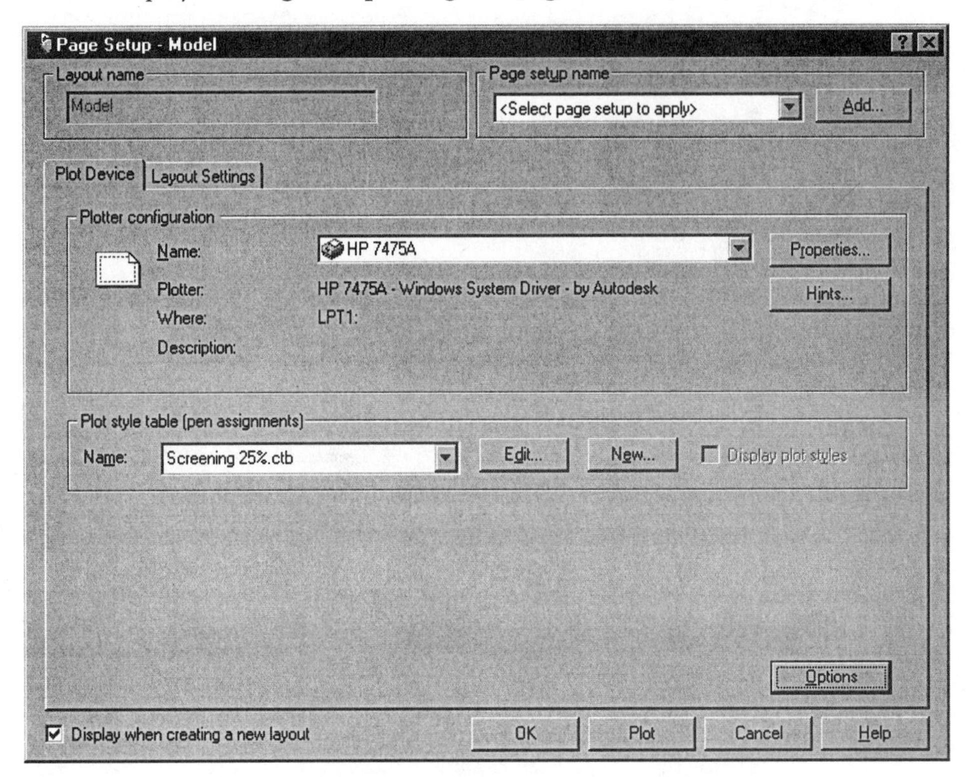

Figure 15-10 ***Page Setup*** *dialog box*

The **Page Setup** dialog box specifies the page layout and current plotting device settings, and is displayed every time you choose a layout tab in a drawing session. The layout settings are stored with the layout. The current Layout name is displayed in the **Layout name** box, and in the **Page setup name** area, a list of all the named and saved page setups are displayed. You can choose to base the current page setup on a named page setup or you can add a new named page setup by choosing the **Add** button, AutoCAD LT displays the **User Defined Page Setups** dialog box (discussed earlier in this chapter in "**PLOT** Command"). The **Page Setup** dialog box has two tabs; the **Plot Device** tab and the **Layout Settings** tab.

Plot Device tab. This tab specifies the current plotting device for plotting the layout. This tab is similar to the **Plot Device** tab of the **Plot** dialog box except for the **What to Plot** and **Plot to file** areas. The additional feature in this tab is the **Options** button. If you choose the **Options** button, AutoCAD LT displays the **Options** dialog box with **Output** tab selected (The **Options** dialog box is described in Chapter 3). You can make necessary changes in this dialog box.

Layout Settings tab. This tab specifies the layout settings of the drawing such as paper size, drawing orientation, plot area, plot scale, plot offset, and other plotting options. This tab is similar to the **Plot Settings** tab of the **Plot** dialog box (discussed earlier in the chapter).

The settings specified in the **Page Setup** dialog box are stored with the layout. You can also Plot the current layout from the **Page Setup** dialog box by choosing the **Plot** button after specifying all the necessary parameters in the dialog box. You can control whether the plotting should include lineweights and whether the lineweights are in proportion to the plot scale.

PSETUPIN COMMAND*

Command:	PSETUPIN

This is a new feature added in AutoCAD LT 2000. You can import a User-defined Page setup into a new drawing layout by using this command. The **PSETUPIN** command facilitates importing a saved and named page setup from one drawing into a new drawing. The settings of the named page setup can be applied to layouts in the new drawing. If you enter **PSETUPIN** at the Command prompt, a standard file selection dialog box is displayed, where you can locate a drawing (.dwg) file whose pagesetups have to be imported. After you select a drawing file, AutoCAD LT displays the **Import user defined page setups** dialog box (Figure 15-11). To import a User-defined Page setup file using the AutoCAD command line, the prompt sequence is as follows:

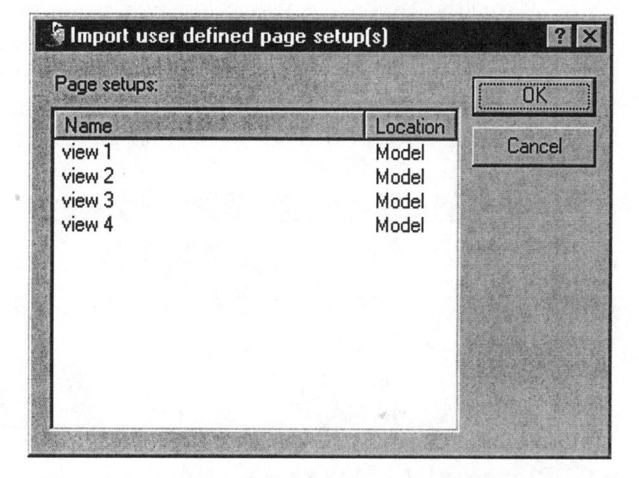

Figure 15-11 Import User Defined Page Setup(s) dialog box

Command: **-PSETUPIN**

If the system variable **FILEDIA** is set to 1, AutoCAD LT displays a standard file selection dialog box, but if **FILEDIA** is set to 0, AutoCAD LT displays the following prompt sequence:

Enter file name: *Specify the file name to be imported*
Enter user defined page setup to import or [?]: *Specify the named page setup or ? for listing of available setups.*

PCINWIZARD COMMAND*

Menu:	Tools > Wizards > Import AutoCAD R14/LT97/LT98 Plot Settings
Command:	PCINWIZARD

If you want to import a **PCP** or **PC2** configuration file, or plot settings created by previous releases of AutoCAD into the **Model** tab or current layout for the drawing, you can use **PCINWIZARD** command to display the **Import PCP or PC2 Plot Settings Wizard**. All the information from a **PCP** or **PC2** file regarding plot area, rotation, plot offset, plot optimization, plot to file, paper size, plot scale, and pen mapping can be imported. After you specify the name of the file for importing, you can modify the plot settings for the current layout.

USING PLOT STYLES*

Plot Styles is a new object property feature of AutoCAD LT 2000. This feature can change the complete look of a plotted drawing. You can use this feature to override a drawing object's color, linetype, and lineweight. You can change the end, join and fill styles of the drawing, and also change the output effects like dithering, gray scales, pen assignments, and screening. Basically, you can use **Plot Styles** effectively to plot the same drawing in various ways.

Every object and layer in the drawing has a plot style property. The plot style characteristics are defined in the plot style tables attached to the Model tab, layouts, and viewports within the layouts. You can attach and detach different plot style tables to get different looks for your plots. Basically, there are two plot style modes: Color-Dependent and Named plot style. **Color-dependent** plot styles (**.CTB** files) are based on object color and there are **255** color-dependent plot styles. **Named** plot (**.STB** files) styles are independent of object color and you can assign any plot style to any object regardless of that object's color. Every drawing in AutoCAD LT 2000 is in either of the modes.

STYLESMANAGER COMMAND*

Menu:	File > Plot Style Manager
Command:	STYLESMANAGER

The **STYLESMANAGER** command is a new feature of AutoCAD LT 2000. If you enter **STYLESMANAGER** at the Command prompt, AutoCAD LT displays the **Plot Styles Manager** window (Figure 15-12).

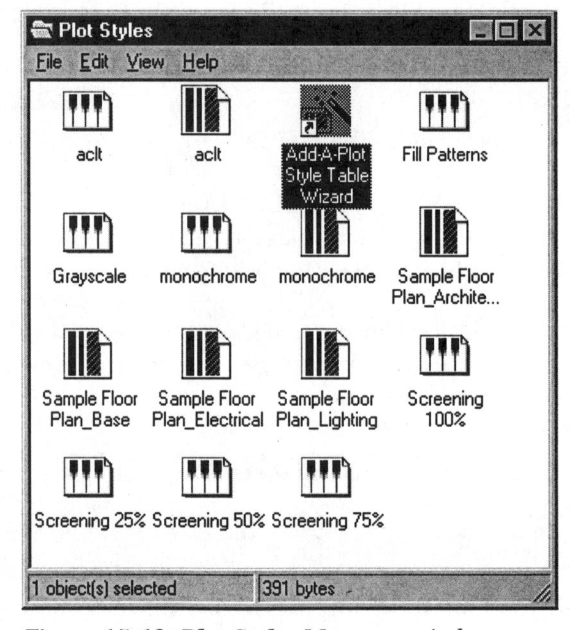

Figure 15-12 Plot Styles Manager window

Add-A-Plot Style Table Wizard. If you want to add a new Plot Style Table to your drawing, you can double-click on **Add-a-Plot-Style-Table Wizard** in the **Plot Styles** window. AutoCAD LT will display the wizard. Following are the steps for creating a new plot style table using the wizard:

1. Read the Introduction page of the **Add Plot Style Table** dialog box and choose the **Next** button.
2. In the **Begin** page, select the **Start from scratch** radio button and choose **Next**.
3. In the **Pick Plot Style Table** page, select the **Named Plot Style Table** or **Color-**

Dependent Plot Style Table radio buttons according to the requirement and choose **Next.**

4. In the **File name** page, enter a file name for the new plot style table and choose **Next.**
5. A new Plot Style Table will be created with default settings. Choose **Finish** and the new Plot Style Table will be added to the **Plot Styles Manager**, which can be used for plotting.

Plot Style Table Editor. If you want to change the setting of the new Plot Style Table created by **Add-a-Plot Style Table Wizard**, choose **Plot Style Table Editor** button in the **Finish** page and AutoCAD LT will display the **Plot Style Table Editor** (Figure 15-13). To edit a previously created Plot Style table, double-click the file in the **Plot Styles Manager** window and the **Plot Style Table Editor** is displayed.

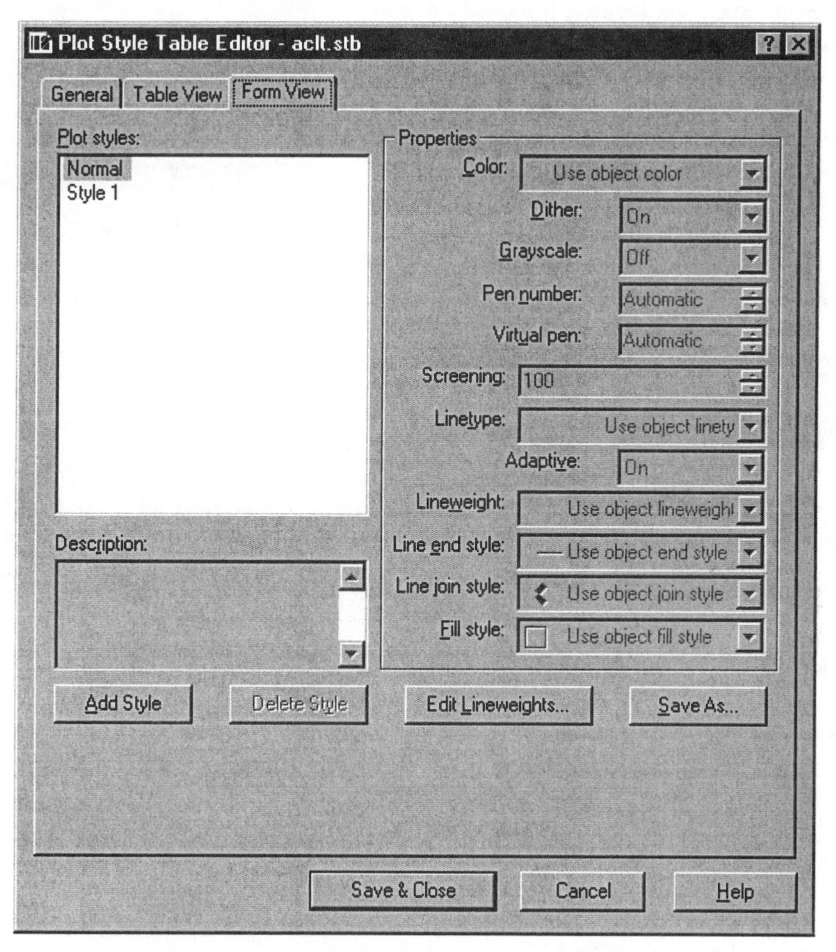

*Figure 15-13 **Plot Style Table Editor** dialog box*

The **Plot Style Table Editor** has three tabs: **General**, **Table View**, and **Form View**. You can edit all the properties of an existing plot style table using the different tabs. The description of the tabs are as follows:

General tab. This tab provides the information about the file name, location of the file, version, and scale factor. All the information except the description are read only.

Table View tab. This tab displays all the plot styles available with their properties in tabular form. You can edit the existing styles or add new styles by choosing the **Add Style** button. A new column with default style name **Style1** will be added in the table. You can change the style name and the properties in the table.

Form View tab. It displays all the properties in a form. All the available plot styles are displayed in the **Plot Styles** list box. Select any style and edit its properties in the **Properties** area. If you want to add a new style,choose **Add Style** button, AutoCAD LT will display the **Add Plot Style** window (Figure 15-14) with default name **Style1**. When you choose the **OK** button, the new style will be added in the **Plot Styles** list box and you can select it for editing.If you want to delete a style, select the style and choose the **Delete Style** button.

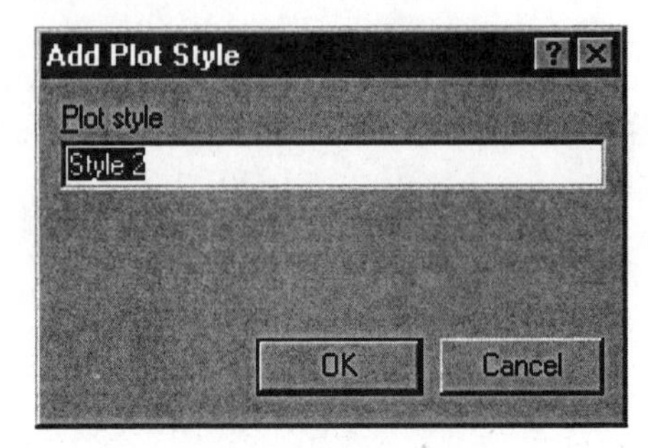

Figure 15-14 Add Plot Style window

After editing, choose the **Save & Close** button to save and return to the **Plot Styles** dialog box.

PLOTSTYLE COMMAND*

Command: PLOTSTYLE

The **PLOTSTYLE** command can be used to set the current plot style for new objects. When you enter **PLOTSTYLE** at the Command prompt, if there is no object selected then AutoCAD LT displays the **Current Plot Style** dialog box (Figure 15-15). But if any object selection is there in the drawing AutoCAD LT, displays **Select Plot Style** dialog box (Figure 15-16). The parameters in the **Current Plot Style** dialog box are as follows:

Current Plot Style. This displays the name of the current plot style.

Plot Style List. Lists all the available plot styles that can be assigned to an object, including the default plot style, **NORMAL**.

Active Plot Style Table. This dropdown list displays the name of the plot style table attached to the current layout or viewport.

Editor. If you choose the **Editor** button, AutoCAD LT displays the **Plot Style Table Editor**.

Attached To. This displays to whichever tab, **Model** tab or a **layout**, the plot style table is attached

The **Select Plot Style** dialog box displays the name of the Original and New plot styles to be assigned to the selected object, in addition to the above features.

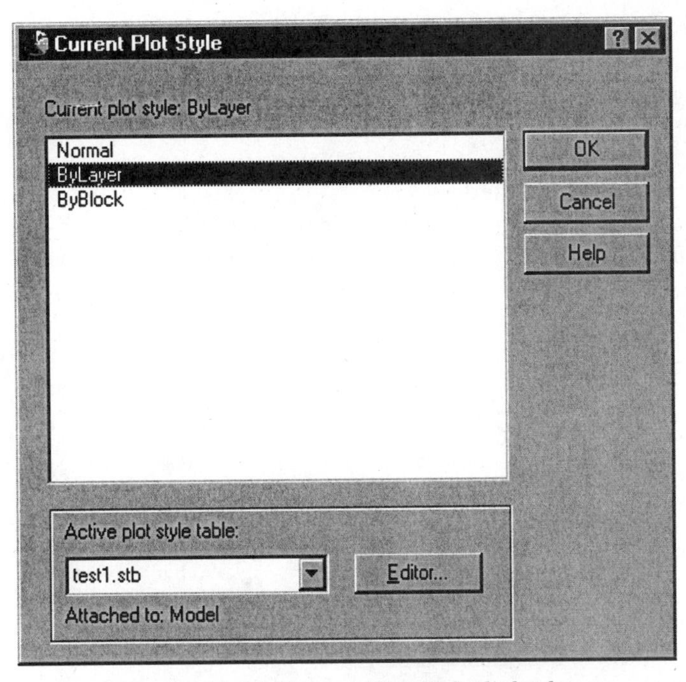

***Figure 15-15 Current Plot Style** dialog box*

Note

*This command is avaliable only for drawings that use Named Plot Styles.To create a drawing using Named Plot Styles ,select the **Use Named Plot Styles** option in the **Plotting** tab of the **Options** dialog box.*

The **Use Named Plot Styles** option affects New Drawings only.

You can also select Plotstyle using the command line. The command prompt sequence is as follows:

Command: **-PLOTSTYLE**
Current plot style is "*current style*"
Enter an option [?/Current] :

At the last prompt, enter the name of a style or type **?** and AutoCAD LT will display a list of all the available styles in the AutoCAD LT Text Window.

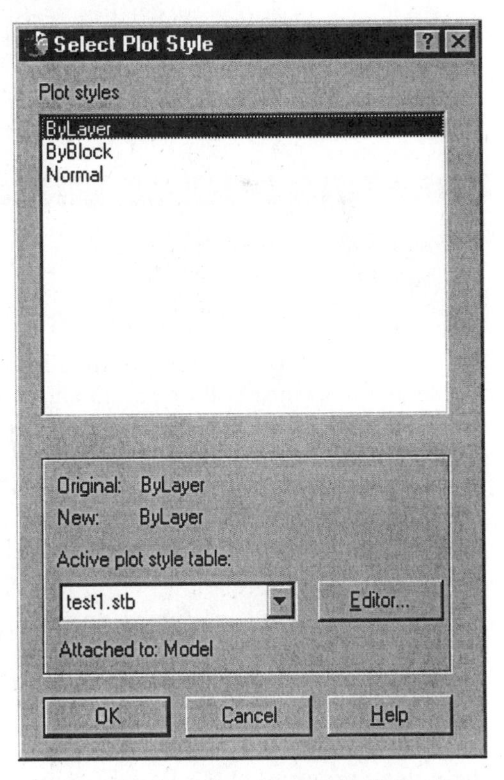

***Figure 15-16 Select Plot Style** dialog box*

Exercise 1 *General*

Make the drawing given in Figure 15-17, and plot it according to the following specifications. It is assumed that the plotter supports six pens.

1. The object lines must be plotted with Pen-1.
2. The dimension lines and centerlines must be plotted with Pen-2.
3. The border and title block must be plotted with Pen-6.

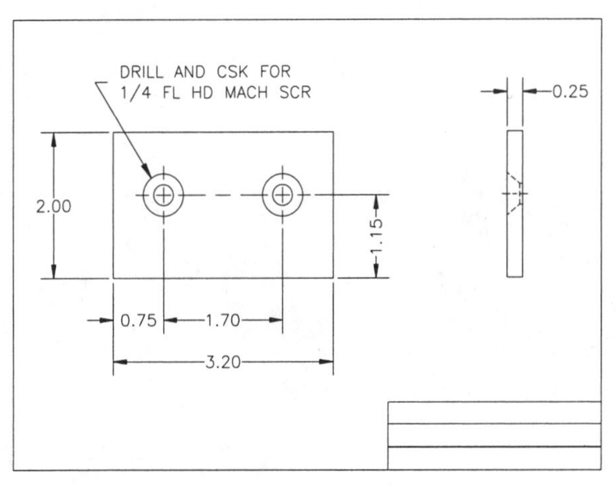

Figure 15-17 *Drawing for Exercise 1*

CREATING DOUBLE LINES (USING DLINE COMMAND)

Toolbar: Draw > Doubleline
Menu: Draw > Doubleline
Command: DLINE

The AutoCAD LT Doubleline feature allows you to create doublelines that consist of two parallel lines or arcs. Double lines are drawn like lines. You can also draw double connected lines. AutoCAD LT treats each line segment forming a double line as an indiviual object. The prompt sequence to create a simple doubleline is:

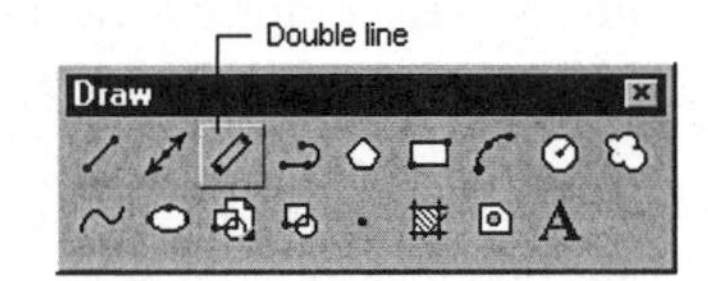

Figure 15-18 *Selecting the* ***DLINE*** *command from the toolbar*

Command: **DLINE** «
Specify start point or [Break/Caps/Dragline/Offset/Snap/Undo/Width]: *Select a start point*
Specify next point or [Arc/Break/CAps/CLose/Dragline/Snap/Undo/Width]: *Select the ending point*

The different options of the **DLINE** command are described below:

Start Point

This is the default option and requires you to specify a point either by entering its coordinates or by picking the point on the screen with the pointing device (example mouse).

Break

If you require that a line should be broken when it intersects a double line,as shown in Figure 15-19, set **Break** to ON. The default option is ON. Set **Break** OFF if you do not want to have breaks at doubleline intersections. The prompt sequence to set **Break** to ON is :

Command : **DLINE** [Enter]
Specify start point or [Break/Caps/Dragline/Offset/Snap/Undo/Width]: **B** [Enter]
Break dlines at start and end points [OFf/ON]<ON>: [Enter]

Caps

With this option you can close the ends of a double line and obtain clean corners as shown in the Figure 15-20.

Command: **DLINE** [Enter]
Specify start point or [Break/Caps/Dragline/Offset/Snap/Undo/Width]: **C** [Enter]
Enter option for drawing endcaps[Both/End/None/Start/Auto]<Auto>: [Enter]

At this prompt you can enter any of these options. If you press ENTER, that is if you select **Auto**, AutoCAD LT caps all the ends of the double line except those that are snapped to an object. With the rest of the options, AutoCAD LT caps the ends specified even if they are snapped to an object. Select **Both** if you want both ends of a double line to be capped. To specify the end of the double line which you want capped, select **End** or **Start**. **Start** caps the double line at its beginning while **End** caps the end you pick last. If you do not want any ends to be capped, select **None**.

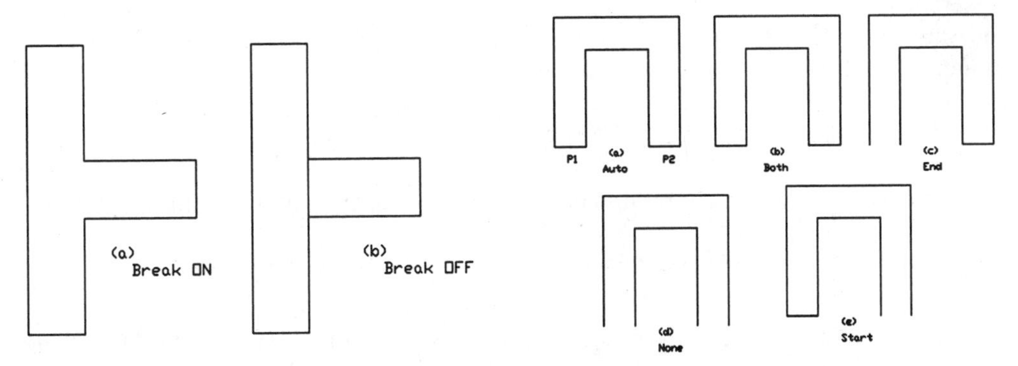

Figure 15-19 *Using the Break Option*

Figure 15-20 *Using the Caps option*

Dragline

This option is used to set the location of the pick point with respect to the double line. You can set the pick point to be located at the center, or on the left or right leg of the double line. To ascertain the left and right sides of the double line, imagine looking towards the end of the double line from its start point. The default option is set to center of the double line. The prompt sequence to set the pick point to be on the right leg of the double line is:

Command: **DLINE** «
Specify start point or [Break/Caps/Dragline/Offset/Snap/Undo/Width]: **D** «
Enter offset from center or dragline position option [Left/Center/Right] <0.0000>: **R** «

Another way of setting the location of the pick point is by setting the offset distance from the center. You can enter a negative value to set the pick point to the left of the center of the double line. A positive value sets the pick point to the right of the center of the double line. See Figure 15-21.

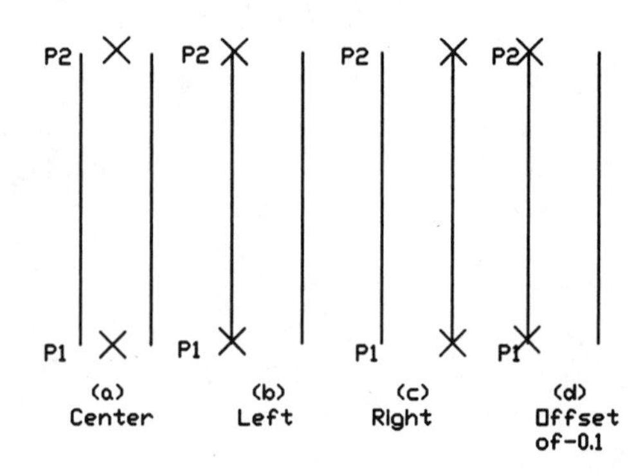

Figure 15-21 *Using Dragline and Offset option*

Offset

With this option (Figure 15-21), you can start a double line at a specified distance and direction from a base point. Enter O at the prompt you receive after entering the **DLINE** command. The subsequent prompts are as follows:

Specify base point to offset from: *Specify the point from which you want to offset*
Specify point to indicate offset direction: *Specify point*
Specify distance of offset<Default>Enter

At this prompt you can also enter a distance or pick point .

Snap

If you want a double line to start or end at an existing entity, set the **snap** option of the **DLINE** command. AutoCAD LT prompts you as follows:

Enter Snap option[Size/ON/OFF]<ON>:

You can set the snap on or off or set a snap size.To set a snap size enter S at the above prompt.

Enter size of snap area (1-10) <10 <Default>: *Select a number between 1 and 10*

This size is in pixels and refers to the search area set up by AutoCAD LT to find an object to snap to. The larger the number you can enter, the farther the cursor can be made from the object, that is, AutoCAD LT sets up a larger area to search for the object. AutoCAD LT breaks up the object if the **Break** option is ON.

Undo

If you draw one or more double lines by mistake, you can undo them in the **DLINE** command itself. With the **Undo** option, you can erase the last picked vertex of the double line. Thus, you can undo a series of double line segments in the order reverse to that you have drawn them while you are still in the **DLINE** command.

Width

You can assign any width to the double line using the Width option of the **DLINE** command. The prompt sequence to set the width of 0.5 for the double line is:

Command: **DLINE** «
Specify start point or [Break/Caps/Dragline/Offset/Snap/Undo/Width]: **W** «
Specify width of dline<Default>: 0.5 «

After you enter the DLINE command and enter the start point of the DLINE, the next prompt displayed is:

Specify next point or [Arc/Break/CAps/Dragline/Snap/Undo/Width]:

These options are discussed below :

Arc

You can draw double line arcs with the **DLINE** command. The prompt sequence is :

Command: **DLINE** «
Specify start point or [Break/Caps/Dragline/Offset/Snap/Undo/Width]: *Select a start point.*
Specify next point or [Arc/Break/CAps/Dragline/Snap/Undo/Width]: **A** «
Break/CAps/CEnter/CLose/Dragline/Endpoint/Line/Snap/Undo/Width/<secondpoint>:

Here, you pick the next point of the double line arc. The prompt sequence specifying its third point is:

Endpoint: *Select endpoint of the arc.*

The prompt sequence to draw an arc by specifying its center point is:

Break/CAps/CEnter/CLose/Dragline/Endpoint/Line/Snap/Undo/Width/<secondpoint>: **CE** «
Center point: *Specify the center point.*
Angle/<Endpoint>: «
Endpoint: *Specify the endpoint.*

You can also specify the included angle by entering **A** at the **Angle/<Endpoint>**: prompt.

Angle/<Endpoint>: **A** «

Included Angle: *Enter the included angle, for example, 45.*

Another method of drawing a double line arc is by specifying the Endpoint first. The prompt sequence to draw an arc by this method is:

Command: **DLINE** «
Specify start point or [Break/Caps/Dragline/Offset/Snap/Undo/Width]: *Select a start point.*
Specify next point or [Arc/Break/CAps/Dragline/Snap/Undo/Width]: **A** «
Break/CAps/CEnter/CLose/Dragline/Endpoint/Line/Snap/Undo/Width/ <secondpoint>:
E «
Endpoint: *Specify the endpoint.*
Included angle: *Specify the angle* «

Line

Line option allows you to change from double line arcs to drawing double lines. AutoCAD LT continues to make double line arcs once you select the **Arc** option. To resume drawing double lines type L at the **Break/CAps/CEnter/CLose/Dragline/Endpoint/Line/Snap/Undo/Width/ <second point>:** prompt.

Close

You can close a double line with this option of the **DLINE** command. When you select this option, AutoCAD LT closes the double line with an arc or a line depending on the mode which isactive.**Close** option works only if the double line consists of two or more line segments or one or more arcs. After closing the double line , AutoCAD LT exits the **DLINE** command. The prompt sequence for closing an arc segment is:

Command: **DLINE** «
Specify start point or [Break/Caps/Dragline/Offset/Snap/Undo/Width]: *Select a start point.*
Specify next point or [Arc/Break/CAps/Dragline/Snap/Undo/Width]: **A** «
Break/CAps/CEnter/CLose/Dragline/Endpoint/Line/Snap/Undo/Width/<second point>:
E «
Endpoint: *Specify the endpoint.*
Included angle: *Specify the angle* «
Break/CAps/CEnter/CLose/Dragline/Endpoint/Line/Snap/Undo/Width/<second point>
CL«

CREATING A POLYLINE (USING REVCLOUD COMMAND)*

Menu:	Tools > Revision Cloud
Toolbar:	Draw > Revcloud
Command:	REVCLOUD

This command creates a polyline of sequential arcs to form a cloud shape. The prompt sequence for using this command is as follows:

Command : **REVCLOUD** «
Current Arc Length : **0.500**
Specify Start point or [Arc length]: *Specify a point and drag or enter **a** to set arc length* «
Guide crosshairs along cloud path....

As you drag and draw, different arcs of the cloud with varied lengths are drawn. When the start point and endpoints meet, the revision cloud is completed and you get a message.

Revision Cloud Finished.

If you choose the **Arc Length** option, you can define the length of the arcs to be drawn and all the arcs drawn are of constant length.

The resulting object is a polyline. Figure 15-22 shows the use of the **REVCLOUD** command.

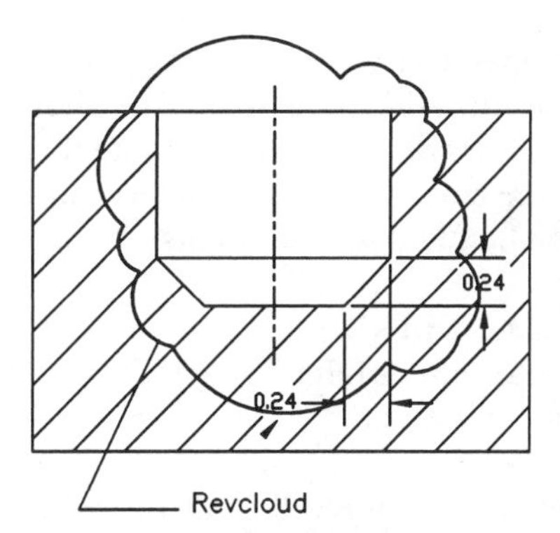

Figure 15-22 *Using **REVCLOUD** command*

Note
***REVCLOUD** stores the last used arc length in the system registry. The value is multiplied by **DIMSCALE** to provide consistency when the program is used with drawings that have different scale factors.*

DRAWING CONSTRUCTION LINES (XLINE AND RAY COMMANDS)

Toolbar:	Draw > Construction Line
Menu:	Draw > Construction Line
Command:	XLINE

The **XLINE** and **RAY** commands can be used to draw construction or projection lines. An **xline** is a 3D line that extends to infinity on both ends. Since the line is infinite in length, it does not have any endpoints. A **ray** is a 3D line that extends to infinity on only one end. The other end of the ray has a finite endpoint. The xlines and rays have zero extents. This means that the extents of the drawing will not change if you use the commands that change the drawing extents, such as the **ZOOM** command with the **All** option. Most of the object snap modes work with both xlines and rays, with some limitations: You cannot use the Endpoint object snap with the xline because by definition an xline does not have any endpoints. However, for rays you can use the Endpoint snap on one end only. Also, xlines and rays take the properties

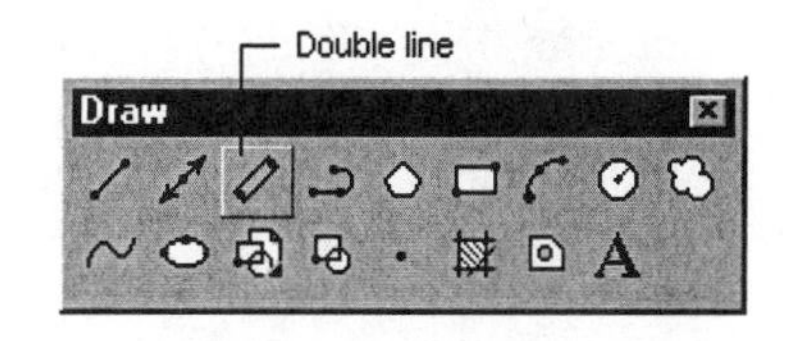

Figure 15-23 Construction Line *from the **Draw** toolbar*

of the layer in which they are drawn. The linetype will be continuous even if the linetype assigned to the layer is not continuous (Figure 15-24). The **RAY** command can be invoked from the **Draw** menu (Select **Draw > Ray**), or by entering **RAY** at the Command prompt.

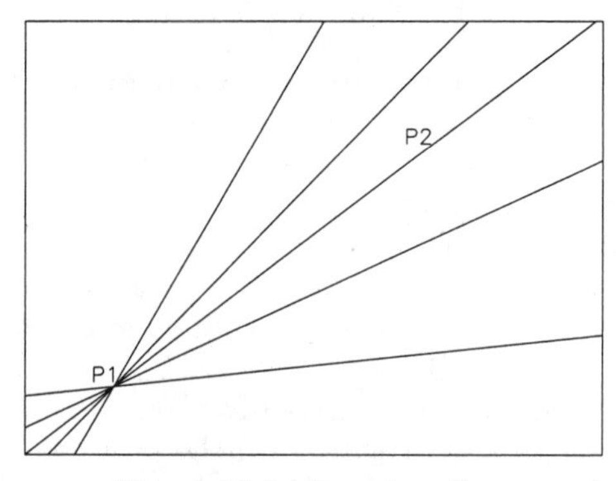

Figure 15-24 *Drawing xlines*

XLINE Options

Point. If you select the default option, AutoCAD will prompt you to select two points, the **Specify a point:** and the **Specify through point:**. After you select the first point, AutoCAD will dynamically rotate the xline with the cursor. When you select the second point, an xline will be created that passes through the first and second points.

Command: **XLINE**
Specify a point or [Hor/Ver/Ang/Bisect/Offset]: *Specify a point.*
Specify through point: *Specify the second point.*

Horizontal. This option will create horizontal xlines of infinite length that pass through the selected points. The xlines will be parallel to the X axis of the current UCS (Figure 15-25).

Vertical. This option will create vertical xlines of infinite length that pass through the selected points. The xlines will be parallel to the Y axis of the current UCS (Figure 15-25).

Angular. This option will create xlines of infinite length that pass through the selected point at a specified angle. The angle can be specified by entering a value at the keyboard. You can also use the reference option by selecting a line object and then specifying an angle relative to the selected line (Figure 15-26). The following is the prompt sequence for the **Angular** option:

Command: **XLINE**
Specify a point or [Hor/Ver/Ang/Bisect/Offset]: **Ang**
Enter angle of xline (0) or [Reference]: **R**
Select a line object: *Select a line, or polyline, or xline, or ray.*
Enter angle of xline <0>: *Enter angle (The angle will be measured counterclockwise with respect to the selected line.)*
Specify through point: *Specify the second point (from where you want the xline to pass).*

Bisect. This option will create an xline that passes through the angle vertex and bisects the angle. Specify the angle by selecting two points. The xline created using this option will lie in the plane defined by the selected points. You can use the object snaps to select the points on the existing objects. The following is the prompt sequence for this option (Figure 15-27):

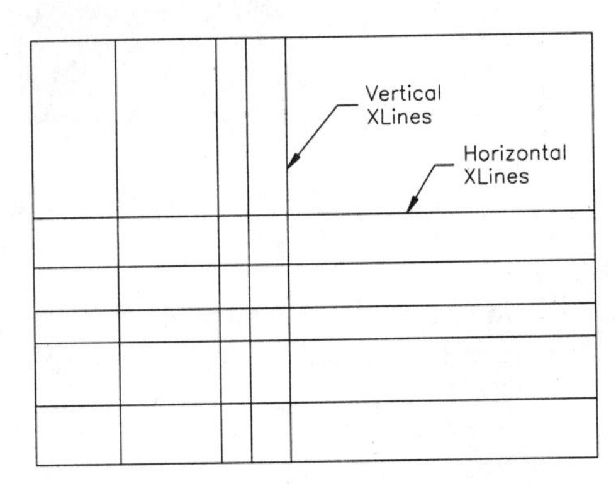

Figure 15-25 Using the Horizontal and Vertical options to draw xlines

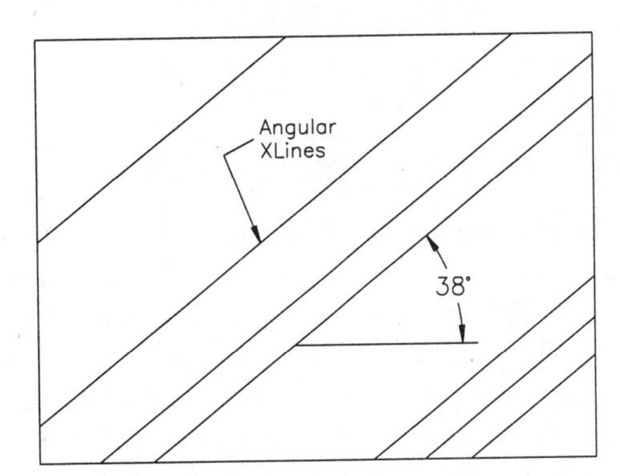

Figure 15-26 Using the Angular option to draw xlines

Command: **XLINE**
Specify a point or [Hor/Ver/Ang/Bisect/Offset]: **B**
Specify angle vertex point: *Enter a point (P1).*
Specify angle start point: *Enter a point (P2).*
Specify angle end point: *Enter a point (P3).*
(You can use object snaps to select points).
Specify angle end point: *Press ENTER.*

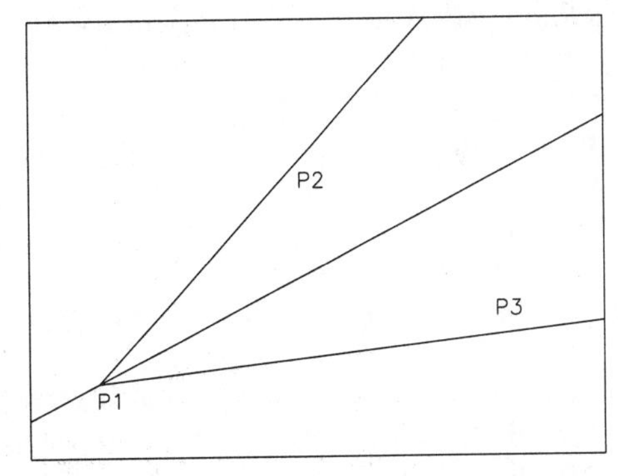

Figure 15-27 Using the Bisect option to draw xlines

Offset. The **Offset** option creates xlines that are parallel to the selected line/xline at the specified offset distance. You can specify the offset distance by entering a numerical value or by selecting two points on the screen. If you select the **Through** option, the offset line will pass through the selected point. (This option works like the **OFFSET** editing command.)

RAY Command

A ray is a 3D line similar to the xline construction line with the difference being that it extends to infinity only in one direction. It starts from a point you specify and extends to infinity through the specified point. To exit the **RAY** command, press ESC or ENTER. The prompt sequence is:

Command: **RAY**
Specify start point: *Select the starting point for the ray.*
Specify through point: *Specify the second point.*

CREATING NURBS

Toolbar:	Draw > Spline
Menu:	Draw > Spline
Command:	SPLINE

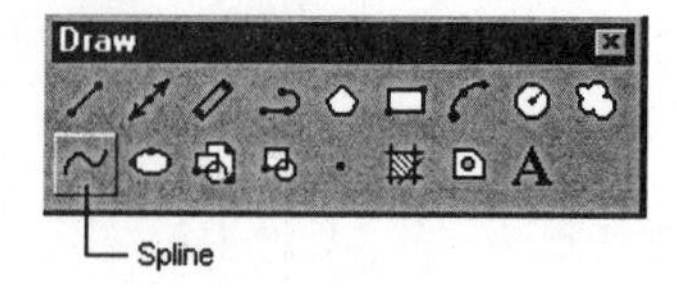

***Figure 15-28** Invoking **SPLINE** from the **Draw** toolbar*

NURBS can be created using the **SPLINE** command. NURBS is an acronym for **NonUniform Rational Bezier-Spline**. The spline created with the **SPLINE** command is different from the spline created using the **PLINE** command. The nonuniform aspect of the spline enables the spline to have sharp corners because the spacing between the spline elements that constitute a spline can be irregular. Rational means that irregular geometry, such as arcs, circles, and ellipses, can be combined with free-form curves. The Bezier-spline (B-spline) is the core that enables accurate fitting of curves to input data with Bezier's curve-fitting interface. The following is the command prompt sequence for creating the spline shown in Figure 15-29:

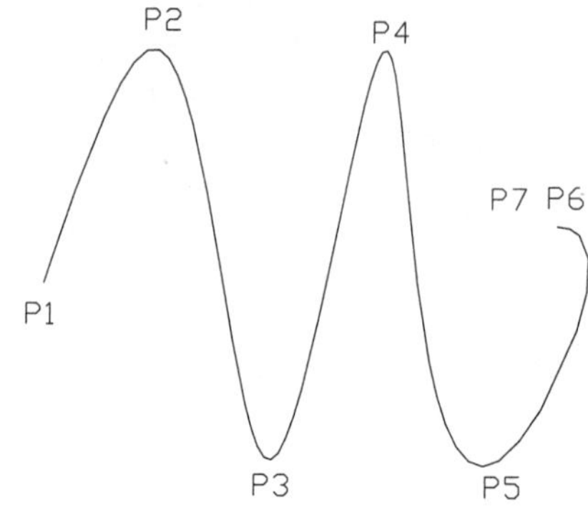

***Figure 15-29** Using the **SPLINE** command to draw splines*

Command: **SPLINE**
Specify first point or [Object]: *Select point (P1).*
Specify next point: *Select the second point (P2).*
Specify next point or [Close/Fit tolerance] <start tangent>: *Select point (P3).*
Specify next point or [Close/Fit tolerance] <start tangent>: *Select point (P4).*
Specify next point or [Close/Fit tolerance] <start tangent>: *Select point (P5).*
Specify next point or [Close/Fit tolerance] <start tangent>: *Select point (P6).*
Specify next point or [Close/Fit tolerance] <start tangent>: *Press ENTER.*
Specify start tangent: *Press ENTER for default.*
Specify end tangent: *Select point (P7).*

SPLINE Command Options

Object. This option allows you to change quadratic or cubic spline-fit polylines to equivalent splines and also, deletes the polylines. The original splined polyline is deleted if the system variable **DELOBJ** is set to 1. To change a polyline into a splined polyline, use the **PEDIT** command.

Command: **SPLINE**
Specify first point or [Object]: **O**
Select objects to convert to splines ...
Select objects: *Select a 2D splined polyline.*

Close. This option allows you to close the NURBS. When you use this option, AutoCAD LT will automatically join the endpoint with the start point of the spline, and you will be prompted to define the start tangent only.

Fit Tolerance. This option allows you to control the fit of the spline between specified points. If you enter a smaller value, the spline will pass through the defined points as closely as possible (Figure 15-30).If the fit tolerance value is 0, the spline passes through the fit points.

Figure 15-30 *Creating a spline with a Fit Tolerance of 2*

Command: **SPLINE**
Specify first point or [Object]: *Select point1.*
Specify next point: *Select second point.*
Specify next point or [Close/Fit tolerance] <start tangent>: **F**
Specify fit tolerance<current>: *Enter a value.*
Specify next point or [Close/Fit tolerance] <start tangent>: *Select third point.*

Start and End Tangents. This allows you to control the tangency of the spline at the start point and endpoint of the spline. If you press ENTER at these prompts, AutoCAD LT will use the default value. By default, the tangency is determined by the slope of the spline at the specified point.

EDITING SPLINES (SPLINEDIT COMMAND)

Toolbar:	Modify II > Edit Spline
Menu:	Modify > Spline
Command:	SPLINEDIT

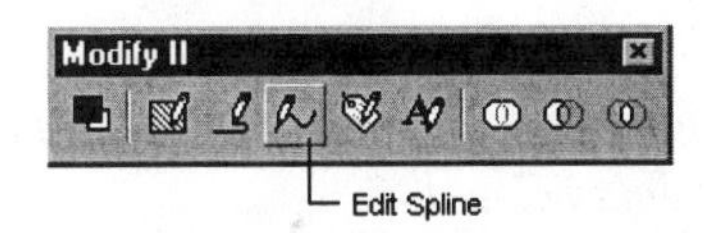

Figure 15-31 *Selecting the* ***Edit Spline*** *button from the* ***Modify II*** *toolbar*

NURBS can be edited using the **SPLINEDIT** command. With this command, you can fit data in the selected spline, close or open the spline, move vertex points, and refine or reverse a spline. You can also invoke **SPLINEDIT** command from the shortcut menu by selecting a spline and right-clicking in the drawing area and choosing **Spline Edit**. The following is the prompt sequence of **SPLINEDIT** command:

Command: **SPLINEDIT**
Select spline: *Select the spline to be selected.*
Enter an option [Fit data/Close/Move vertex/Refine/rEverse/Undo]:

SPLINEDIT Command Options

Fit Data. When you draw a spline, the spline is fit to the specified points (data points). The **Fit Data** option allows you to edit these points (fit data points) (Figure 15-32). For example, if you want to redefine the start and end tangents of a spline, select the **Fit Data** option; then select the **Tangents** option, as follows:

Command: **SPLINEDIT**
Select spline:
Enter an option [Fit data/Close/Move vertex/Refine/rEverse/Undo]: **F**
Enter a fit data option
[Add/Close/Delete/Move/Purge/Tangents/toLerance/eXit] <eXit>: **T**
Specify start tangent or [System default]: *Select a point (P0).*
Specify end tangent or [System default]: *Select a point (P7).*

Close. This option allows you to close or open a spline. When you select the Close option, AutoCAD lets you open, move the vertex, or refine or reverse the spline. The following is the command prompt sequence for closing a polyline (Figure 15-33):

Enter an option [Fit data/Close/Move vertex/Refine/rEverse/Undo]: **C**
Enter an option [Open/Move vertex/Refine/rEverse/Undo/eXit] <eXit>: **X**

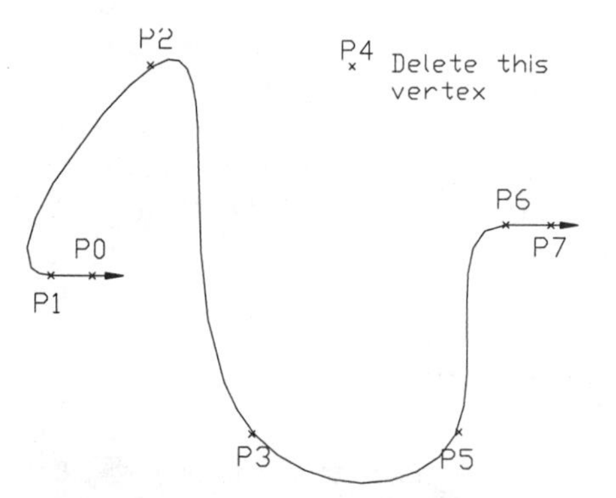

Figure 15-32 Using the ***SPLINEDIT*** *command to fit data points*

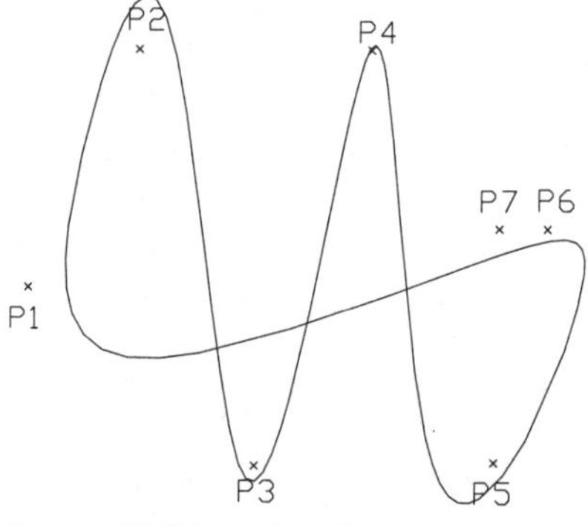

Figure 15-33 Using the ***SPLINEDIT*** *command to close a spline*

Move Vertex. When you draw a spline, it is associated with the Bezier control frame. The Move Vertex option allows you to move the vertices of the control frame. To display the frame with the spline, set the value of **SPLFRAME** system variable to 1. The following is the command prompt sequence for moving one of the vertex points (Figure 15-34):

Enter an option [Fit data/Close/Move vertex/Refine/rEverse/Undo]: **M**
Specify new location or [Next/Previous/Select point/eXit] <N>: **S**
Specify control point <exit>: *Select a point (P1).*
Specify new location or [Next/Previous/Select point/eXit] <N>: *Enter new location (P0).*
Specify new location or [Next/Previous/Select point/eXit] <N>: **X**

Refine. This option allows you to refine a spline by adding more control points in the spline, elevating the order, or adding weight to vertex points. For example, if you want to add more control points to a spline, the command prompt sequence is as follows (Figure 15-35):

Enter an option [Fit data/Close/Move vertex/Refine/rEverse/Undo]: **R**
Enter a refine option [Add control point/Elevate order/Weight/eXit] <eXit>: **A**
Specify a point on the spline <exit>: *Select a point.*

Specify a point on the spline <exit>: *Press ENTER.*
Enter a refine option [Add control point/Elevate order/Weight/eXit] <eXit>: **X**

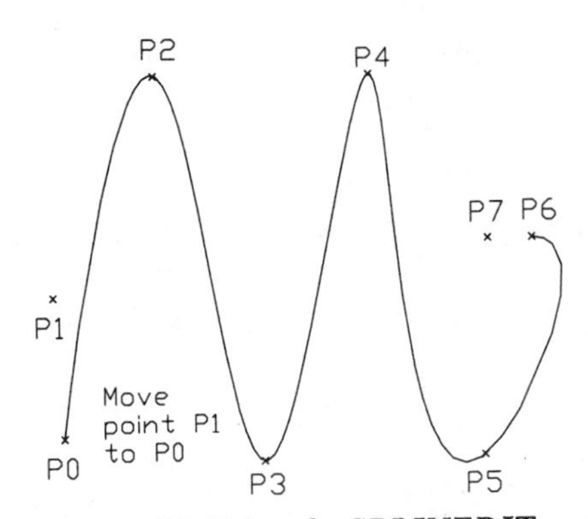

Figure 15-34 *Using the* ***SPLINEDIT*** *command to move a vertex point*

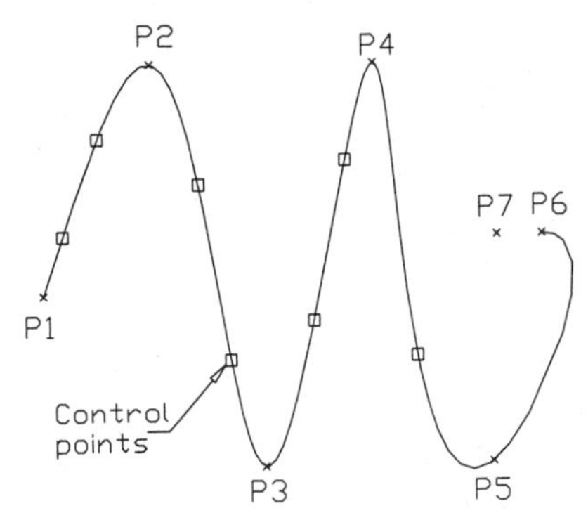

Figure 15-35 *Using the* ***SPLINEDIT*** *command to refine a spline*

Reverse. This option allows you to reverse the spline direction.

Undo. This option will undo the previous option.

Exit. This option exits the command line.

Exercise 2 — *General*

Draw the illustration shown at the top of Figure 15-36. Then use the **SPLINE** and **SPLINEDIT** commands to obtain the illustration shown at the bottom. (Assume the missing dimensions.)

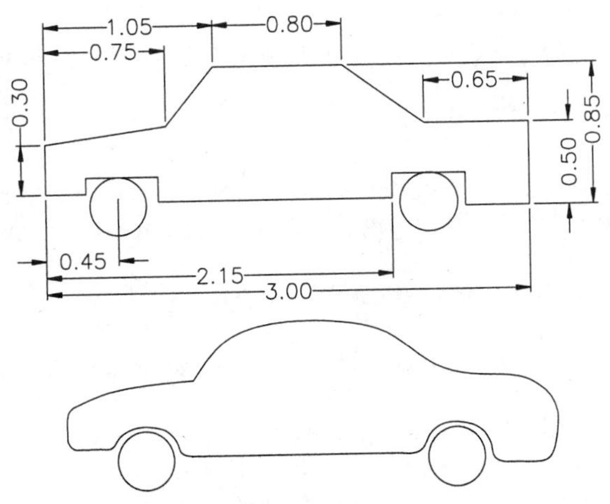

Figure 15-36 *Drawing for Exercise 2*

DIGITIZING DRAWINGS

In digitizing, the information in the drawing is transferred to a CAD system by locating the points on the drawing that is placed on the top surface of the digitizing tablet. Sometimes it is easier to digitize an existing drawing than to create the drawing again on a CAD system. The digitizing of drawings is an important application in industry. The advantages of a drawing based on a CAD system have prompted companies to digitize drawings.

If you want to digitize a drawing, you should have a digitizer as large as the size of the largest drawing to be digitized so that you do not have to realign the drawing sheet. But situations often arise in which the digitizer is not as large as the drawing to be digitized. In such cases, after digitizing a part of the drawing (as large as the digitizer can accommodate), digitize the other parts until the entire drawing is digitized. Digitizing can be carried out in the Tablet mode. In the Tablet mode, AutoCAD LT uses the digitizing tablet as a digitizer, not as a screen pointing device. In this mode, the coordinate system of the paper drawings can be mapped directly into AutoCAD LT. The Tablet mode can be turned on or off with the **TABLET** command or by pressing function key F4.

Command: **TABLET**
Enter an option [ON/OFF/CAL/CFG]: **ON** (or **OFF**)

TABLET COMMAND

Menu:	Tools > Tablet
Command:	TABLET

The first step in digitizing a drawing is to configure the tablet so that the maximum possible area on the tablet can be used. You can do this with the **TABLET** command.

Command: **TABLET**

At the next prompt, enter CFG (Configuration option) to configure the tablet.

Enter an option [ON/OFF/CAL/CFG]: **CFG**

Then, you are prompted to specify number of tablet menus you want. When digitizing drawings, the whole tablet area is used for digitizing the drawing; hence, there are no tablet menus.

Enter the number of tablet menus desired (0-4)<>: **1**

At the next prompt, enter **Y** so that reconfiguration of the tablet is possible:

Do you want to respecify the menu area 1? [Yes/No] <N>: **Y**
Digitize upper left corner of area 1: *Specify the upper left corner of the pointing area.*
Digitize lower left corner of menu area 1: *Specify the lower left corner of the pointing area.*
Digitize lower right corner of menu area 1:*Specify the lower right corner of the pointing area*
Do you want to specify the menu area 1? [Yes/No]<N>: **N**

The next step in digitizing a drawing is to calibrate the tablet. For this, position the drawing to be digitized to the tablet with the help of adhesive tape. Calibration is possible in model space and paper space.

Command: **TABLET**
Enter an option [ON/OFF/CAL/CFG]: **CAL**

After you select or enter **CAL**, the Tablet mode is turned on and the screen crosshairs cursor disappears. The process of calibration involves digitizing two or more points on the drawing and then entering their coordinate values. The points to be selected can be anywhere on the drawing. Once you have specified the position of the points and the coordinate values of these positions, AutoCAD automatically calibrates the digitizing area.

Digitize point 1: *Select the first point.*
Enter coordinates for point 1: *Enter the coordinates of the first point selected.*
Digitize point 2: *Select the second point.*
Enter coordinates for point 2: *Enter the coordinates of the second point selected.*
Digitize point 3 (or ENTER to end): *Select the third point.*
Enter coordinates for point 3: *Enter the coordinates of the second point selected.*
Digitize point 4 (or ENTER to end): *Press ENTER to end.*

AutoCAD will display the Text Window displaying the transformation information (Figure 15-37).

Enter transformation type [Orthogonal/Affine/Projective/Repeat table] <Repeat>: **A**

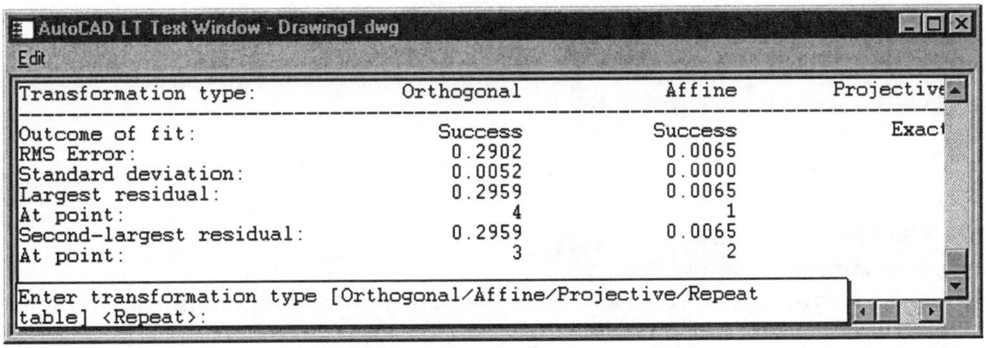

AutoCAD LT Text Window - Drawing1.dwg
Edit

Transformation type:	Orthogonal	Affine	Projective
Outcome of fit:	Success	Success	Exact
RMS Error:	0.2902	0.0065	
Standard deviation:	0.0052	0.0000	
Largest residual:	0.2959	0.0065	
At point:	4	1	
Second-largest residual:	0.2959	0.0065	
At point:	3	2	

Enter transformation type [Orthogonal/Affine/Projective/Repeat table] <Repeat>:

Figure 15-37 *Tranformation Table*

You can enter any number of points. The more points you enter, the more accurate is the digitizing. If you have entered only two points, AutoCAD LT will automatically compute orthogonal transformation. If you have entered three or more points, AutoCAD LT will compute orthogonal, affine, and projective transformations and determine which best fits the configuration of the selected points. After configuring and calibrating the tablet, the crosshairs cursor is redisplayed on the screen. Now, you can use AutoCAD LT commands to digitize the drawing.

Floating Screen Pointing Area

You can configure the tablet for two pointing areas: the fixed pointing area and the floating pointing area. Before you configure the tablet, the entire tablet surface is the fixed screen pointing area. At this point the digitizer tablet performs like a pointing device. You can use the CFG option of the **TABLET** command to specify the new fixed screen pointing area. When you enter this command, AutoCAD will prompt you to specify the two points: lower left corner and upper right corner of the new screen pointing area. Now, if you move the digitizer puck within the fixed pointing area, the cursor moves on entire screen because there is one-to-one correspondence between the fixed pointing area on the digitizer and the screen. You can also define a floating screen pointing area that is the same as the screen pointing area or you can specify a different area. The following is the prompt sequence to specify a floating screen pointing area:

Command: **TABLET**
Enter an option [ON/OFF/CAL/CFG]: **CFG**
Enter the number of tablet menus desired (0-4)<>: 0
Do you want to respecify the Fixed Screen pointing area?[Yes/No]<N>: **N**
Do you want to specify the Floating Screen pointing area? [Yes/No]<N>: **Y**
Do you want the Floating Screen area to be the same size as the Fixed Screen pointing area? [Yes/No] <Y>: **N**
Digitize lower left corner of the Floating Screen pointing area: *Specify the lower left corner of the pointing area.*
Digitize upper right corner of the Floating Screen pointing area: *Specify the upper right corner of the pointing area.*

Next, AutoCAD will prompt you to specify the toggle key. By default, you can use the function key **F12** to toggle between the floating and fixed screen pointing area or you can define a button of the digitizer puck as the toggle key. You can still use the function key **F4** to turn the Tablet mode on or off to digitize a drawing.

Self-Evaluation Test

Answer the following questions, and then compare your answers to the correct answers given at the end of this chapter.

1. The settings for the plot parameters are saved in the __________.

2. If you want to store the plot in a file and not have it printed directly on a plotter, select the __________ check box in the **Plot** dialog box.

3. The scale for the plot can be specified in the __________ edit boxes in the **Plot** dialog box.

4. In partial preview, the __________ is graphically represented by the white rectangle. The blue rectangle is the section in the paper that is used by the __________ .

5. The size of a plot can be specified by selecting any paper size from the __________ drop-down list in the **Plot** dialog box.

6. Different objects in the same drawing can be plotted in different colors, with different linetypes and line widths. (T/F)

7. Full preview takes __________ time than Partial preview.

8. If you select the __________ radio button in the **Plot** dialog box, the section of drawing that currently holds objects is plotted.

Review Questions

Plot

1. The __________ command is used to plot a drawing.

2. By selecting the View radio button in the **Plot** dialog box, you can plot a __________ view that was created with the __________ command in the current drawing.

3. If you do not want the hidden lines of the objects created in paper space to be plotted, select the __________ of Plot Settings tab in the **Plot** dialog box.

4. You can view the plot on the specified paper size before actually plotting it by selecting the __________ in the **Plot** dialog box.

5. In a Partial preview, if the __________ boundary and the __________ boundary overlap, this is graphically represented by a red and blue dashed line.

6. By selecting the ______ button in the **Plot** dialog box, the drawing is displayed on the screen just as it would be plotted on the paper.

Line, Ray, and Spline

7. The **XLINE** and **RAY** commands can be used to draw __________ or __________ lines.

8. An xline is a 3D line that extends to __________ on both ends.

9. NURBS can be created using the __________ command. NURBS is an acronym for __________.

10. NURBS splines can be edited using the __________ command.

Digitizing

11. The Tablet mode can be turned on and off with the help of the __________ command.

12. The process of calibration involves digitizing __________ two points on the drawing and then entering their __________ values.

13. The digitizing process can be used to convert a drawing on paper to a CAD system. (T/F)

Exercises

Exercise 3 *General*

Make the drawing given in Figure 15-38, and plot it according to the following specifications. It is assumed that the plotter supports six pens.

1. The drawing is to be plotted on 10 x 8 inch paper.
2. The border lines (polylines) must be 0.01 inches wide when plotted.
3. Change the **DIMSCALE** and text size so that the dimension text height and text height are each 0.125 inches when the drawing is plotted.
4. The object lines must be plotted with Pen-1.
5. The dimension lines and centerlines must be plotted with Pen-2.
6. The border and title block must be plotted with Pen-3.

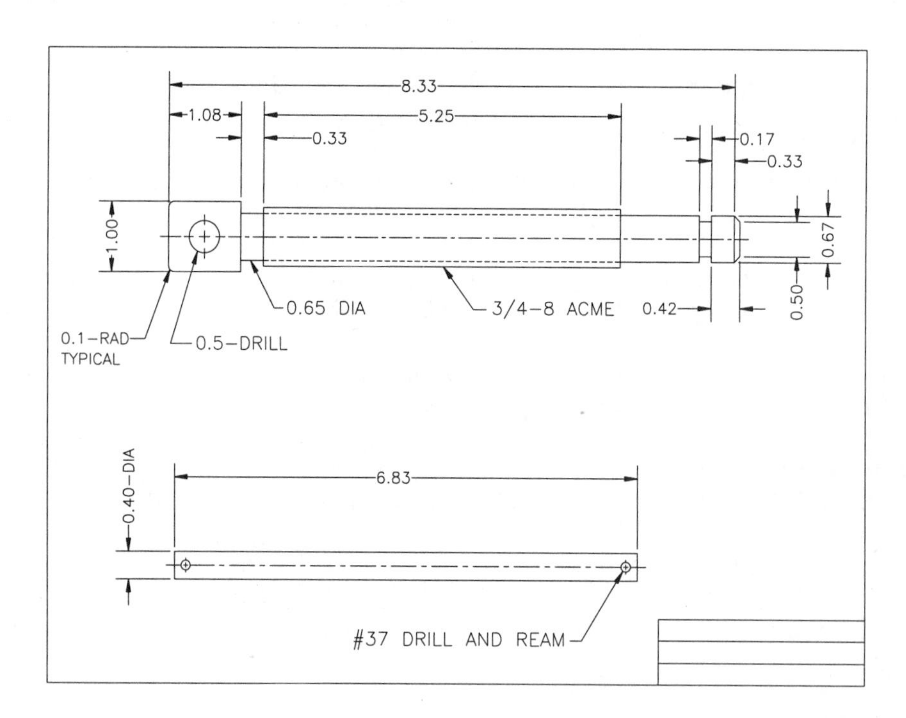

Figure 15-38 *Drawing for Exercise 3*

Answers to Self-Evaluation Test

1 - PC3, **2** - **Plot to File**, **3** - Plotted Inches = Drawing Units, **4** - paper, image, **5**- Paper Size, **6**- T, **7**- more, **8**- **Extents**

Chapter 16

Object Grouping and Editing Commands

Learning Objectives

After completing this chapter, you will be able to:
- *Use the **GROUP** command to group objects.*
- *Select and cycle through defined groups.*
- *Change properties and points of objects using the **CHANGE** command.*
- *Perform editing operations on polylines using the **PEDIT** command.*
- *Explode compound objects using the **EXPLODE** command.*
- *Undo previous commands using the **UNDO** command.*
- *Rename named objects using the **RENAME** command.*
- *Remove unused named objects using the **PURGE** command.*
- *Change the display order of objects using **DRAWORDER** command.*
- *Use the **AutoCAD DesignCenter** window to manage content*

GROUP MANAGER (GROUP COMMAND)*

Toolbar:	Group > Group Manager
Menu:	Tools > Group Manager
Command:	GROUP

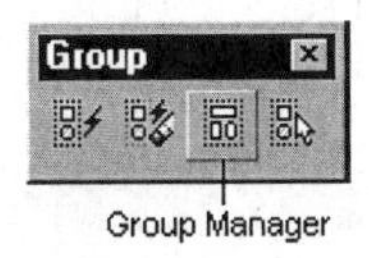

***Figure 16-1** Invoking Group Manager from the Group toolbar.*

You can use the **Group Manager** dialog box (Figure 16-2) to group AutoCAD LT objects and assign a name to the group. Once you have created groups, you can select the objects by group name. The individual characteristics of an object are not affected by forming groups. Groups are simply a mechanism that enables you to form groups and edit objects by groups. It makes the object selection process easier and faster. Objects can be members of several groups.

GROUP MANAGER DIALOG BOX OPTIONS*

The **Group Manager** dialog box (Figure 16-2) provides the following options.

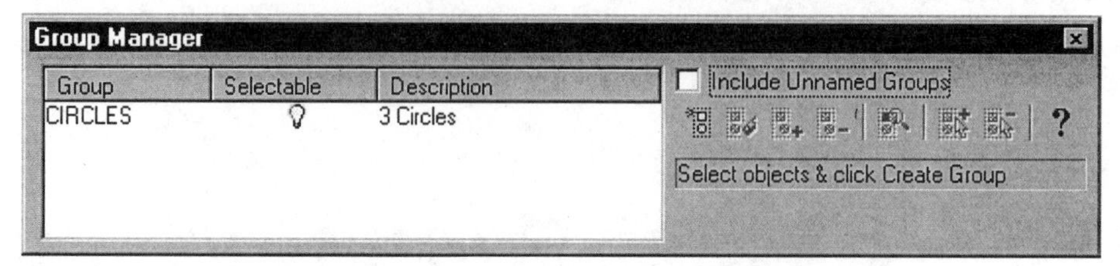

*Figure 16-2 **Group Manager** dialog box*

Group

The **Group** list in the **Group Manager** dialog box displays the names of the existing groups. The list box also displays whether a group is **selectable** or not. A **selectable** group is one, in which all members are selected on selecting a single member. It also displays description regarding the group.

Note

*If the value of the **PICKSTYLE** System Variable is set to 1 or 3, entire Group is selected upon selecting any one object from the group. If the value is set to 0, entire group is not selected upon selecting any one object from the group.*

Create Group

Choosing this button allows you to create a new group consisting of the objects you have selected. Names assigned to the groups are displayed under **Group** heading. You have to first select the objects and then choose this button to create a new group.

Ungroup

This button of the **Group Manager** dialog box deletes the selected group from the **Group** list and also removes the association of the objects in the group. This option is used when you want do remove a specified group from the list.

Add to Group

Choosing this button allows you to add the selected objects to the group already created and selected under **Group** list. This option is used when you want to add some more objects to the group already created.

Remove from Group

Choosing this button removes the selected objects from the group which is selected under **Group** list. You can remove a few selected objects from the group of number of objects by choosing this option.

Note

*Objects can be added or removed from a selection set by the **PICKADD** System Variable. If the value of this System Variable is set to 1, you can add or remove the objects from the group. If the value is set to 0, the current selection set replaces the previous selection set.*

Details

Choosing this button displays the **Group Manager - Details** dialog box (Figure 16-3). This dialog box gives all the details regarding the group.

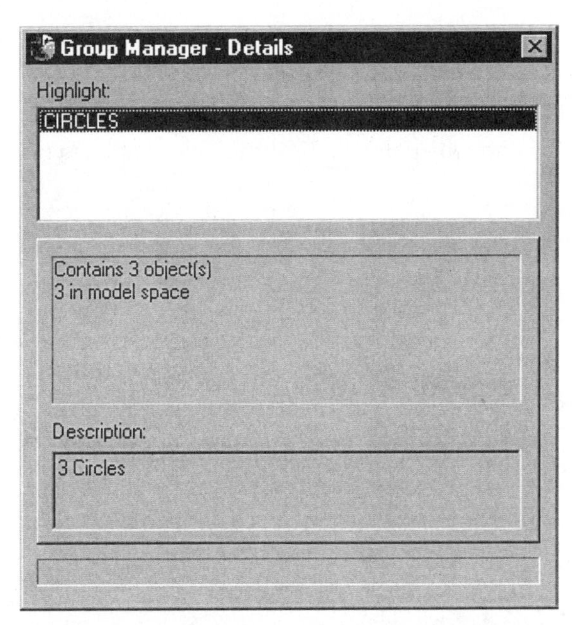

Figure 16-3 *Group Manager - Details dialog box.*

Select Group

As you choose this button, all of the objects in the specified group are highlighted. You have to first select the group name under the **Group** list and then choose the **Select Group** button to highlight the objects in the selected group.

Deselect Group

As you choose this button a specified group is removed from the selection set. This option is used when more than one group is selected at the same time. In this case you can remove a particular group from the selection set by selecting the group under the **Group** list and then choosing this button.

Include Unnamed

The **Include Unnamed** check box is used to display the names of the unnamed groups in the **Group Manager** dialog box. The unnamed groups are created when you copy all objects in a named object. AutoCAD LT automatically groups and assigns a name to the copied objects.

The format of the name is AX (for example, *A1, *A2, *A3). If you select the **Include Unnamed** check box, the unnamed object names (*A1, *A2, *A3, ...) will be displayed in the dialog box. The unnamed groups can also be created by not assigning a name to the group.

USING COMMAND LINE

You can also use the command line from of the **GROUP** command by entering **-GROUP**.

Command: **-GROUP** [Enter]
Enter a group option
[?/Add/Remove/Ungroup/REName/Selectable/Create] <Create>:

SELECTING GROUPS

You can select a group by name by entering G at the **Select objects** prompt.

Command: **MOVE** [Enter]
Select objects: **G** [Enter]
Enter group name: *Enter group name.*
4 found
Select objects: [Enter]

If you select any member of a selectable group at the **Select objects** prompt, then all the group members are also selected. The **PICKSTYLE** system variable should be set to 1 or 3. Also, the group selection is turned on or off by pressing CTRL+A.

Command: **ERASE** [Enter]
Select objects: CTRL+A *(Press the CTRL and A keys together.)*
<Groups on> *Select an object that belongs to a group.* *(If the group has been defined as selectable, all objects belonging to that group will be selected.)*

If the group has not been defined as selectable, you cannot select all objects in the group, even if you turn the group selection on by pressing the CTRL+A keys.

Note

The combination of CTRL and A keys is used as a toggle to turn the group selection on or off.

CYCLING THROUGH GROUPS

When you use the **GROUP** command to form object groups, AutoCAD LT lets you sequentially highlight the groups of which the selected object is a member. For example, let us assume that an object belongs to two different groups, and you want to highlight the objects in those groups. To accomplish this, press CTRL at the **Select objects** prompt and select the object that belongs to different groups. AutoCAD LT will highlight the objects in one of the groups; press the pick button of your pointing device to cycle through the groups (Figure 16-4). The prompt sequence is:

Command: **ERASE** [Enter]
Select objects: *Hold down CTRL.*
<Cycle on> *Select the object that belongs to different groups. (Press the pick button on your pointingdevice tocycle through the groups.)*
Select objects: [Enter]

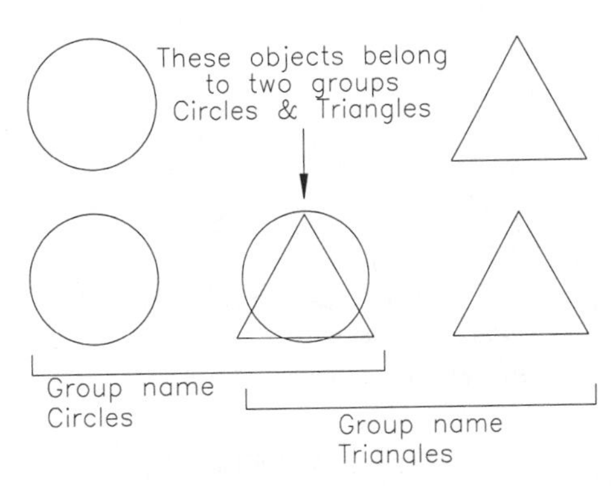

Figure 16-4 *Object selection cycling*

Also, if objects are very close or directly on top of one another, such that they lie within the same selection pickbox, you can press CTRL at the **Select objects:** prompt and repeatedly click your pointing device until the object you want is highlighted. Press ENTER to select it.

CHANGING PROPERTIES AND POINTS OF AN OBJECT (CHANGE COMMAND)

Command: CHANGE

With the help of the **CHANGE** command, you can change some of the characteristics associated with an object, such as location, color, layer, lineweight, and linetype. The **CHANGE** command has two options, **Properties** and **Change point**.

Change Point Option

You can change various features and the location of an object with the **Change point** option of the **CHANGE** command. For example, to change the endpoint of a line or a group of lines, the prompt sequence is:

Command: **CHANGE** [Enter]
Select objects: *Select the line.*
Select objects: [Enter]
Select change point or [Properties]: *Specify a point to be used as a new endpoint.*

When the **ORTHO** mode is on, using the **CHANGE** command to change the endpoint of a line or group of lines makes them parallel to either the X axis or the Y axis.

To change various features associated with the Text, the prompt sequence is:

Command: **CHANGE** [Enter]
Select objects: *Select the text.*
Select objects: [Enter]
Specify new change point or [Properties]: [Enter]
Specify new text insertion point <no change>: *Specify the new text insertion point (location of text) or press ENTER for no change.*

Enter new text style <current>: *Enter the name of the new text style or press ENTER to accept the current style.*

The next prompt is available only for text styles that do not have a fixed height.

Specify new text height <current>: *Specify the new text height or press ENTER to accept the current value.*
Specify new rotation angle <current>: *Specify the new rotation angle or press ENTER to accept the current value.*
Enter new text <current>: *Enter the new text or press ENTER to accept the current text.*

The properties of an attribute definition text can be changed just as you change the text. The prompt sequence for changing the properties of an attribute definition text is as follows:

Command: **CHANGE** [Enter]
Select objects: *Select the attribute definition text.*
Select objects: [Enter]
Specify new change point or [Properties]: [Enter]
Specify new text insertion point <no change>: *Specify the new attribute definition text insertion point (location).*
Enter new text style <current>: *Enter the name of the new text style or press ENTER to accept the current style.*

The next prompt is available only for text styles that do not have a fixed height.

Specify new height <current>: *Specify the new attribute definition text height.*
Specify new rotation angle <current>: *Specify the new rotation angle.*
Enter new tag <current>: *Enter the tag.*
Enter new prompt <current>: *Enter the new prompt.*
Enter new default value <current>: *Enter the new default value.*

You can also change the position of an existing block and specify a new rotation angle. The prompt sequence is as follows:

Command: **CHANGE** [Enter]
Select objects: *Select the block.*
Select objects: [Enter]
Specify change point or [Properties]: [Enter]
Specify new block insertion point <no change>: *Specify the new block insertion point.*
Specify new block rotation angle <current>: *Specify the new rotation angle for the block.*

The radius of a circle can be changed with the **Change point** option of the **CHANGE** command by specifying the new radius in the case of the circle. To change the radius of a circle, the prompt sequence is:

Command: **CHANGE** [Enter]
Select objects: *Select the circle.*
Select objects: [Enter]
Specify change point or [Properties]: *Select a point to specify the radius of the circle.*

If more than one circle is selected, after the first circle is modified, the same prompts are repeated for the next circle.

The Properties Option

The **Properties** option can be used to change the characteristics associated with an object.

Changing the Layer of an Object. If you want to change the layer on which an object exists and other characteristics associated with layers, you can use the **LAyer** option of the **Properties** option. The prompt sequence is:

Command: **CHANGE** [Enter]
Select objects: *Select the object whose layer you want to change.*
Select objects: *If you have finished selection, press ENTER.*
Specify change point or [Properties]: **P** [Enter]
Enter property to change [Color/Elev/LAyer/LType/ltScale/LWeight/Thickness]: **LA** [Enter]
Enter new layer name <0>: *Enter a new layer name.*
Enter property to change [Color/Elev/LAyer/LType/ltScale/LWeight/Thickness]: [Enter]

The selected object is now placed on the desired layer.

Changing the Color of an Object. You can change the color of the selected object with the **Color** option of the **CHANGE** command. If you want to change the color and linetype to match the layer, you can use the **Properties** option of the **CHANGE** command; then, change the color or linetype by entering BYLAYER as follows:

Command: **CHANGE** [Enter]
Select objects: *Select the block to change.*
Select objects: *If you have finished the selection, press ENTER.*
Specify change point or [Properties]: **P** [Enter]
Enter property to change [Color/Elev/LAyer/LType/ltScale/LWeight/Thickness]: **C** [Enter]
Enter new color <current>: BYLAYER *(Or enter the color name.)*
Specify property to change [Color/Elev/LAyer/LType/ltScale/LWeight/Thickness]: [Enter]

Blocks originally created on layer 0 assume the color of the new layer; otherwise (if it was not created on layer 0), it will retain the color of the layer on which it was created.

Changing the Thickness of an Object. You can change the thickness of the selected object in the following manner:

Command: **CHANGE** [Enter]
Select objects: *Select the object whose thickness you want to change.*
Select objects: [Enter]

Specify change point or [Properties]: **P** [Enter]
Enter property to change [Color/Elev/LAyer/LType/ltScale/LWeight/Thickness]: **T** [Enter]
Enter new thickness <current thickness>: *Specify the new thickness.*

Changing the Lineweight of an Object. You can change the lineweight of the selected object by using the **LWeight** option of the **CHANGE** command. Since the lineweight values are predefined, if you enter a value that is not predefined, a predefined value of lineweight which comes closest to the specified value is assigned to the selected object. The prompt sequence is:

Command: **CHANGE** [Enter]
Select objects: *Select objects whose lineweight you wish to change.*
Select objects: *If you have finished selecting objects, press ENTER.*
Specify change point or [Properties]: **P** [Enter]
Enter property to change [Color/Elev/LAyer/LType/ltScale/LWeight/Thickness]: **LW** [Enter]
Enter new lineweight <current>: *Specify new lineweight.*

You can see the effect of changing the thickness of an object in 3D view [viewpoint (1,-1,1)]. Similarly, you can change the elevation, linetype, and linetype scale of the selected objects. You can change the Z-axis elevation of a selected object provided all the points in the particular object have the same Z value.

PROPERTIES COMMAND*

Toolbar:	Standard > Properties
Menu:	Tools > Properties
Command:	PROPERTIES

Instead of using the **CHANGE** command and then the **Properties** option, you can use the **PROPERTIES** command to change the properties of an object. The **PROPERTIES** command can also be invoked by selecting an object and right-clicking to display a shortcut menu; choose **Properties** from the shortcut menu.

Command: **PROPERTIES** [Enter]

The **Object Properties Window** is displayed (Figure 16-5), from where you can change different properties of the selected object. Select objects whose properties you wish to change; the **Properties window** displays the properties of the selected object. Right-clicking in the **Properties** window displays a shortcut menu where you can choose to **Allow Docking** or **Hide** the window.

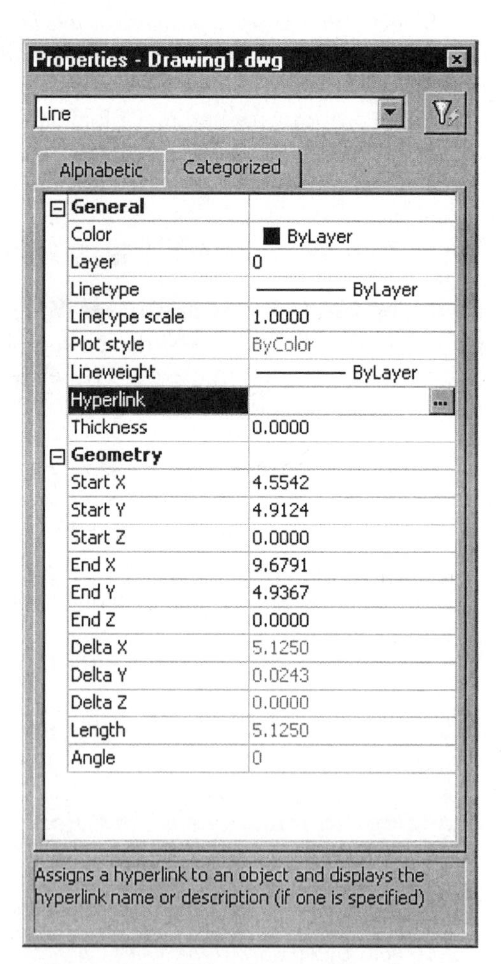

***Figure 16-5 Object Properties** windoow*

Color

When you select **Color** from the **General** list, the color of the selected object is displayed. From the **Color** drop-down list you can select any other color you wish to assign to the selected object. Choosing the **Other** option displays the **Select Color** dialog box (Figure 16-6). You can select the new color you want to assign to the selected object by selecting the desired color in this dialog box, and then choosing the **OK** button.

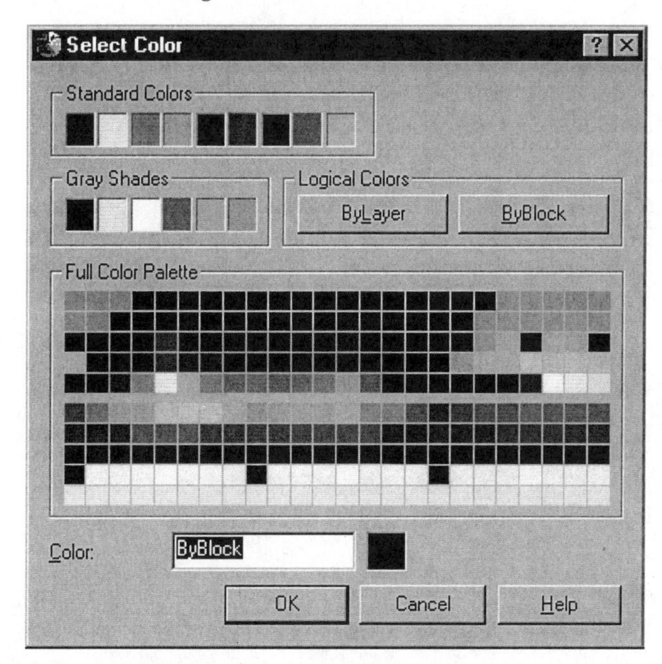

*Figure 16-6 **Select Color** dialog box*

Layer

Selecting **Layer** in the **General** list of the **properties** window, displays the name of the layer of the object. You can select the new layer you want to assign to the selected object from the **Layer** drop-down list.

Linetype

From the **General** list in the **Properties** window, select **Linetype**. From the Linetype drop-down list, you can select the new linetype you want to assign to the selected object from the list of linetypes.

Linetype Scale

You can enter the new linetype scale factor of the selected object in the **Linetype Scale** edit box.

Plot Style

The current Plot style is displayed in the **Plot Style** edit box under the **General** list of the **Properties** window. You can select a different plot style for the selected object from the **Plot**

Style drop-down list of Plot styles available in the current drawing.

Lineweight

Similarly, you can select a different lineweight value from the **Lineweight** drop-down list to assign to the selected object.

Hyperlink

Selecting this option under the **General** list, assigns a hyperlink to an object. The **Hyperlink** name and description assigned to the object is displayed. Choosing the **[...]** button, displays the **Insert Hyperlink** dialog box where you can enter the path of the URL or file you want to link the selected object to (Figure 16-7).

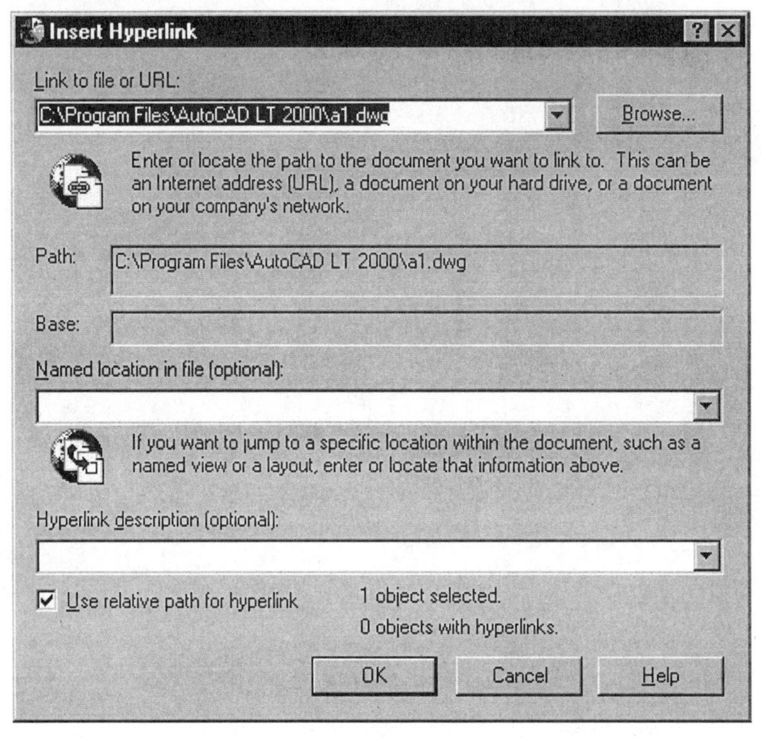

Figure 16-7 Insert Hyperlink *dialog box*

Thickness

You can enter the new 3D solid thickness of the selected object in the **Thickness** edit box. The current value of the selected object is displayed in the **Thickness** edit box.

You can use the **CHPROP** command to change the properties of an object using the command line.

Command: **CHPROP** Enter
Select objects: *Select the object whose properties you want to change.*
Select objects: Enter

Enter property to change [Color/LAyer/LType/ltScale/LWeight/Plotstyle/Thickness]: *Specify the property.*

You can also use the **Object Properties** toolbar to change properties of the selected objects. This option is discussed in Chapter 10, Editing with Grips.

Exercise 1 — *General*

Draw a hexagon on layer OBJ in red color. Let the linetype be hidden. Now, use the **PROPERTIES** command to change the layer to some other existing layer, the color to yellow, and the linetype to continuous. Use the **LIST** command to verify that the changes have taken place.

EXPLODE COMMAND

Toolbar:	Modify > Explode
Menu:	Modify > Explode
Command:	EXPLODE

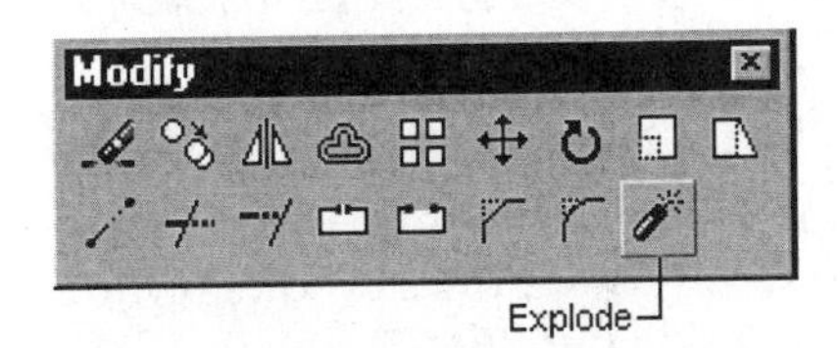

Figure 16-8 Choosing the ***Explode*** *button from the* ***Modify*** *toolbar*

The **EXPLODE** command is used to split compound objects such as blocks, polylines, regions, polyface meshes, polygon meshes, multilines, 3D solids, 3D meshes, bodies, or dimensions into the objects that make them up (Figure 16-9). For example, if you explode a polyline or a 3D polyline, the result will be ordinary lines or arcs (tangent specification and width are not considered). When a 3D polygon mesh is exploded, the result is 3D faces. Polyface meshes are turned into 3D faces, points, and lines. Upon exploding 3D solids, the planar surfaces of the 3D solid turn into regions, and nonplanar surfaces turn into bodies. Multilines are changed to lines. Regions turn into lines, ellipses, splines, or arcs. Two-dimensional polylines lose their width and tangent specifications; 3D polylines explode into lines. When a body is exploded, it changes into single-surface bodies, curves, or regions. When a leader is exploded, the components are lines, splines, solids, block inserts, text or Mtext, tolerance objects and so on. Mtext explodes into single line text. This command is especially useful when you have inserted an entire drawing and you need to alter a small detail.

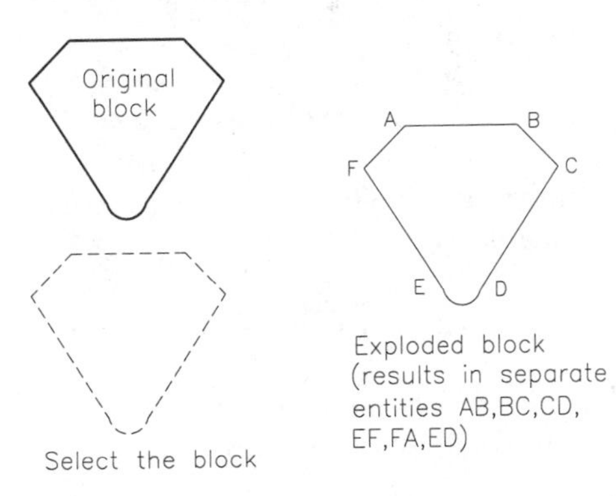

Figure 16-9 Using the ***EXPLODE*** *command*

Command: **EXPLODE** [Enter]
Select objects: *Select the block.*

When a block or dimension is exploded, there is no visible change in the drawing. The drawing remains the same except that the color and linetype may have changed because of floating layers, colors, or linetypes. The exploded block is turned into a group of objects that can be modified separately. To check whether the explosion of the block has taken place, select any object that was a part of the block. If the block has been exploded, only that particular object will be highlighted. With the **EXPLODE** command, only one nesting level is exploded at a time. Hence, if there is a nested block or a polyline in a block and you explode it, the inner block or the polyline will not be exploded. Attribute values are deleted when a block is exploded, and the attribute definitions are redisplayed. If a block is non-uniformly scaled, the circles in it are exploded into ellipses and the arcs are exploded into ellipses.

*While inserting a block using the **Insert** dialog box, you can insert the block as separate objects by selecting the **Explode** check box in the dialog box or using * in front of the block name when using the **-INSERT** command.*

POLYLINE EDITING (USING PEDIT COMMAND)

A polyline can assume various characteristics such as width, linetype, joined polyline, and closed polyline. You can edit polylines, polygons, or rectangles to attain the desired characteristics using the **PEDIT** command. In this section, we will be discussing how to edit simple 2D polylines. The following are the editing operations that can be performed on an existing polyline using the **PEDIT** command. They are discussed in detail later in this chapter.

1. A polyline of varying widths can be converted to a polyline of uniform width.
2. An open polyline can be closed and a closed one can be opened.
3. All bends and curved segments between two vertices can be removed to make a straight polyline.
4. A polyline can be split up into two.
5. Individual polylines or polyarcs connected to one another can be joined into a single polyline.
6. The appearance of a polyline can be changed by moving and adding vertices.
7. Curves of arcs and B-spline curves can be fit to all vertices in a polyline, with the specification of the tangent of each vertex being optional.
8. The linetype generation at the vertices of a polyline can be controlled.

PEDIT Command

Toolbar:	Modify II > Edit Polyline
Menu:	Modify > Polyline
Command:	PEDIT

Figure 16-10 Edit Polyline in ModifyII toolbar

You can also invoke the **PEDIT** command by selecting a polyline and right-clicking to display a shortcut menu, where you can choose **Polyline Edit**.

You can use the **PEDIT** command to edit any type of polyline. The prompt sequence is:

Command: **PEDIT** [Enter]
Select polyline: *Select the polyline to be edited using any selection method.*

If the selected line is an arc or a line (not a polyline), AutoCAD LT issues the following prompt:

Object selected is not a polyline.
Do you want to turn it into one? <Y>

If you want to turn the line or arc into a polyline, respond by entering a Y and pressing ENTER or by just pressing ENTER (null response). Otherwise, you can enter N and exit to the Command prompt. The subsequent prompts and editing options depend on which type of polyline has been selected. In the case of a simple 2D polyline, the next prompt displayed is:

Enter an option [Close/Join/Width/Edit vertex/Fit/Spline/Decurve/Ltype gen/Undo]: *Enter an option or press ENTER to end command.*

C (Close) Option. This option is used to close an open polyline. **Close** creates the segment that connects the last segment of the polyline to the first. You will get this option only if the polyline is not closed. Figure 16-11 illustrates this option.

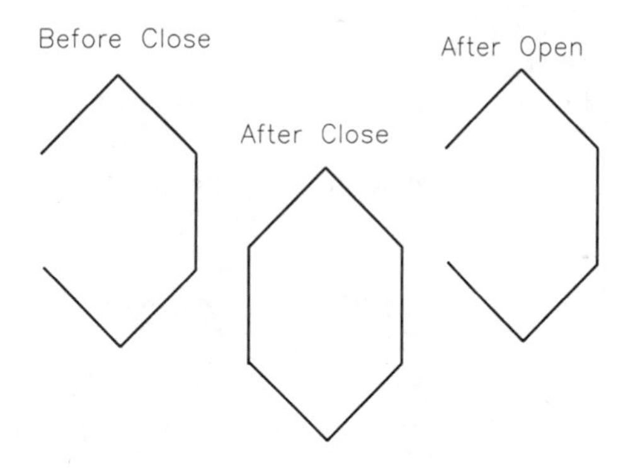

*Figure 16-11 The **Close** option*

O (Open) Option. If the selected polyline is closed, the **Close** option is replaced by the **Open** option. Entering O, for open, removes the closing segment.

J (Join) Option. This option appends lines, polylines, or arcs whose endpoints meet a selected polyline at any of its endpoints and adds (joins) them to it (Figure 16-12). This option can be used only if a polyline is open. After this option has been selected, AutoCAD LT asks you to select objects. You may also select the polyline itself. Once you have chosen the objects to be joined to the original polyline, AutoCAD LT examines them to determine whether any of them has an endpoint in common with the current polyline, and joins such an object with the original polyline. The search is then repeated using new endpoints. AutoCAD LT will not join if the endpoint of the object does not exactly meet the polyline. The line touching a polyline at its endpoint to form a T will not be joined. If two lines meet a polyline in a Y shape, only one of them will be selected, and this selection is unpredictable. To verify which lines have been added to the polyline, use the **LIST** command or select a part of the object. All the segments that are joined to polyline will be highlighted. Figure 16-13 illustrates conditions that are required to join polylines.

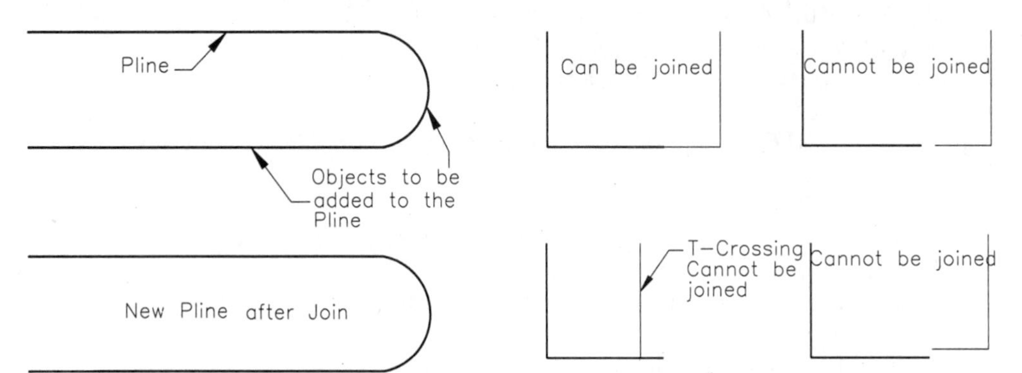

*Figure 16-12 Using the **Join** option*

*Figure 16-13 Conditions for using the **Join** option*

Width (W) Option. The **W** option allows you to define a new, unvarying width for all segments of a polyline (Figure 16-14). It changes the width of a polyline with a constant or a varying width. The desired new width can be specified either by entering the width at the keyboard or by specifying the width as the distance between two specified points. Once the width has been specified, the polyline assumes it. The following prompt sequence will change the width of the given figure from 0.02 to 0.05 (Figure 16-15):

Command: **PEDIT** [Enter]
Select polyline: *Select the polyline.*
Enter an option [Close/Join/Width/Edit vertex/Fit/Spline/Decurve/Ltype gen/Undo]: **W** [Enter]
Specify new width for all segments: **0.05** [Enter]

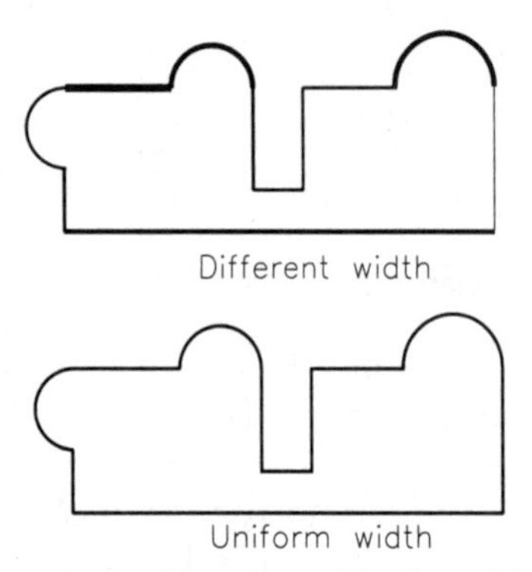

Figure 16-14 Making the width of a polyline uniform

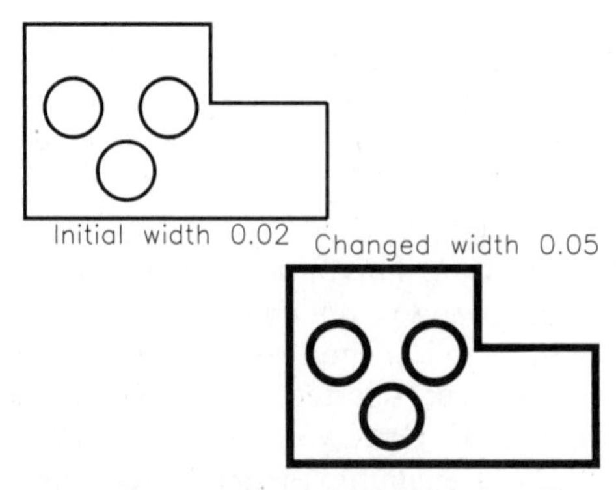

Figure 16-15 Entering a new width for all segments

Note

*Circles drawn using the **CIRCLE** command cannot be changed to polylines. Polycircles can be drawn using the **Pline Arc** option (by drawing two semicircular polyarcs) or by using the **DONUT** command.*

Edit Vertex (E) Option. The **Edit vertex** option lets you select a vertex of a polyline and perform different editing operations on the vertex and the segments following it. A polyline segment has two vertices. The first one is at the start point of the polyline segment; the other one is at the endpoint of the segment. When you invoke this option, an X marker appears on the screen at the first vertex of the selected polyline. If a tangent direction has been specified for this particular vertex, an arrow is generated in that direction. After this option has been selected, the next prompt appears with a list of options for this prompt. The prompt sequence is:

Command: **PEDIT** [Enter]
Select polyline: *Select the polyline to be edited.*
Enter an option [Close/Join/Width/Edit vertex/Fit/Spline/Decurve/Ltype gen/Undo]: [Enter]
Enter a vertex editing option
[Next/Previous/Break/Insert/Move/Regen/Straighten/Tangent/Width/eXit]<N>: *Enter an option or press ENTER to accept default.*

The options available for the **Edit vertex** option are discussed next.

Next and Previous Options. These options move the X marker to the next or the previous vertex of a polyline. The default value in the **Edit vertex** is one of these two options. The option that is selected as default is the one you chose last. In this manner, the **Next** and **Previous** options help you to move the X marker to any vertex of the polyline by selecting one of these two options, and then pressing ENTER to reach the desired vertex. These options cycle back and forth between first and last vertices, but cannot move past the first or last vertices, even if the polyline is closed (Figure 16-16).

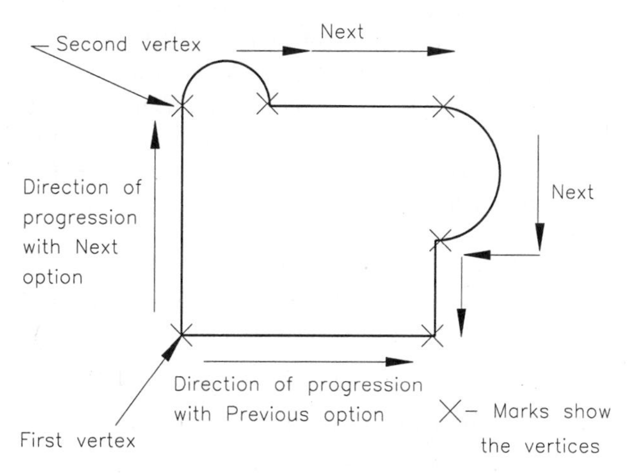

***Figure 16-16** The **Next** and **Previous** options*

The prompt sequence is:

Command: **PEDIT** Enter
Select polyline: *Select the polyline to be edited.*
Enter an option [Close/Join/Width/Edit vertex/Fit/Spline/Decurve/Ltype gen/Undo]: Enter
Enter a vertex editing option
[Next/Previous/Break/Insert/Move/Regen/Straighten/Tangent/Width/eXit] <N>:*Enter N or P.*

Break Option. With the **Break** option, you can divide a polyline into two parts (Figure 16-17). The division of the polyline can be specified at one vertex or at two different vertices. By specifying two different vertices, all the polyline segments and vertices between the specified vertices are erased. If one of the selected vertices is at the endpoint of the polyline, the **Break** option will erase all the segments between the first vertex and the endpoint of the polyline. The exception to this is that AutoCAD LT does not erase the entire polyline if you specify the first vertex at the start point (first vertex) of the polyline and the second vertex at the endpoint (last vertex) of the polyline. If both vertices are at the endpoint of the polyline, or only one vertex is specified and its location is at the endpoint of the polyline, no change is made to the polyline. The last two selections of vertices are treated as invalid by AutoCAD LT, which acknowledges this by displaying the message ***Invalid***.

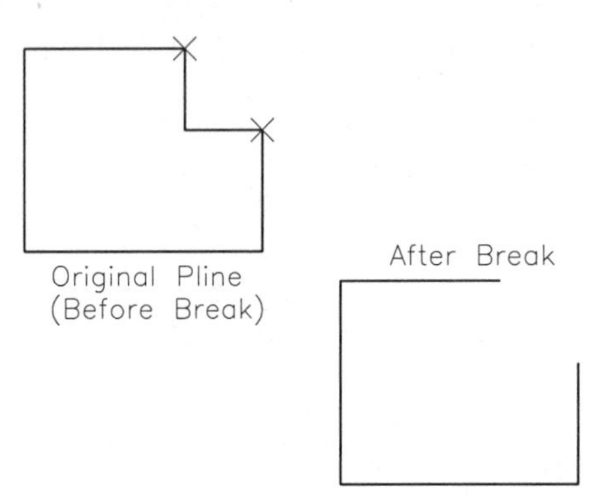

Figure 16-17 *Using the* ***Break*** *option*

To use the **Break** option, first you need to move the marker to the first vertex where you want the split to start. Placement of the marker can be achieved with the help of the **Next** and **Previous** options. Once you have selected the first vertex to be used in the **Break** operation, invoke the **Break** option by entering B. AutoCAD LT takes the vertex where the marker (X) is placed as the first point of breakup. The next prompt asks you to specify the position of the next vertex for breakup. You can enter GO if you want to split the polyline at one vertex only, or use the **Next** or **Previous** option to specify the position of next vertex. The prompt sequence is:

Command: **PEDIT** Enter
Select polyline: *Select the polyline to be edited.*
Enter an option [Close/Join/Width/Edit vertex/Fit/Spline/Decurve/Ltype gen/Undo]: Enter
Enter a vertex editing option
[Next/Previous/Break/Insert/Move/Regen/Straighten/Tangent/Width/eXit]<N>: *Enter N or P to locate the first vertex for the* ***Break*** *option.*

Enter a vertex editing option
[Next/Previous/Break/Insert/Move/Regen/Straighten/Tangent/Width/eXit]<N>: **B** [Enter]

Once you invoke the **Break** option, AutoCAD LT treats the vertex where the marker (X) is displayed as the first point for splitting the polyline. The next prompt issued is:

Enter an option [Next/Previous/Go/eXit] <N>: *Move the X marker to specify the position of next vertex for breakup.*

Enter GO if you want to split the polyline at one vertex only, or use the **Next** or **Previous** option to specify the position of the next vertex for breakup. Entering the **Go** option deletes the polyline segment between the two markers specified using the **Next** and **Previous** options. Choosing the **eXit** option, exits the **break** option.

Insert Option. The **Insert** option allows you to define a new vertex to the polyline (Figure 16-18). You can invoke this option by entering I for Insert. You should invoke it only after moving the marker (X) to the vertex that is located immediately before the new vertex. This is because the new vertex is inserted immediately after the vertex with the X mark.

Command: **PEDIT** [Enter]
Select polyline: *Select the polyline to be edited.*
Enter an option [Close/Join/Width/Edit vertex/Fit/Spline/Decurve/Ltype gen/Undo]: **E** [Enter]
Enter a vertex editing option
[Next/Previous/Break/Insert/Move/Regen/Straighten/Tangent/Width/eXit]<N>: *Move the marker to the vertex prior to which the new vertex is to be inserted.*
Enter a vertex editing option
[Next/Previous/Break/Insert/Move/Regen/Straighten/Tangent/Width/eXit]<N>: **I** [Enter]
Specify location for new vertex: *Move the cursor and select to specify the location of the new vertex.*

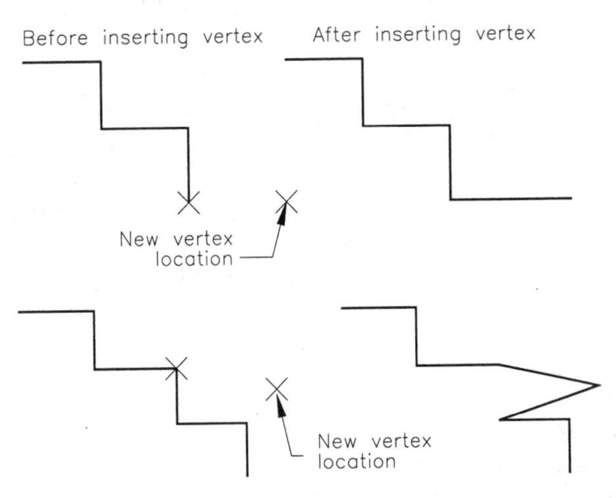

***Figure 16-18** Using the **Insert** option to define new vertex points*

Move Option. This option lets you move the X-marked vertex to a new position (Figure 16-19). Before invoking the **Move** option, you must move the X marker to the vertex you want to relocate by selecting the **Next** or **Previous** option. The prompt sequence is:

Command: **PEDIT** [Enter]
Select polyline: *Select the polyline to be edited.*
Enter an option [Close/Join/Width/Edit vertex/Fit/Spline/Decurve/Ltype gen/Undo]: **E** [Enter]
Enter a vertex editing option
[Next/Previous/Break/Insert/Move/Regen/Straighten/Tangent/Width/eXit]<N>: *Enter N or P to move the X marker to the vertex you want to relocate.*
Enter a vertex editing option
[Next/Previous/Break/Insert/Move/Regen/Straighten/Tangent/Width/eXit]<N>: **M** [Enter]
Specify new location for marked vertex: *Specify the new location for the selected vertex by selecting the new location or entering its coordinate values.*

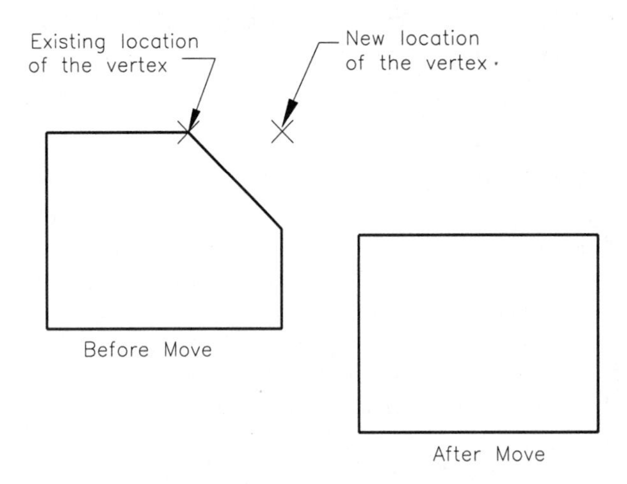

Figure 16-19 ***Move*** *option*

Regen Option. The **Regen** option regenerates the polyline to display the effects of edits you have made. It is used most often with the **Width** option.

Straighten Option. The **Straighten** option can be used to straighten polyline segments or arcs between specified vertices (Figure 16-20). It deletes the arcs, line segments, or vertices between the two specified vertices and substitutes them with one polyline segment.

The prompt sequence is:

Command: **PEDIT** [Enter]
Select polyline: *Select the polyline to be edited.*
Enter an option [Close/Join/Width/Edit vertex/Fit/Spline/Decurve/Ltype gen/Undo]: **E** [Enter]
Enter a vertex editing option

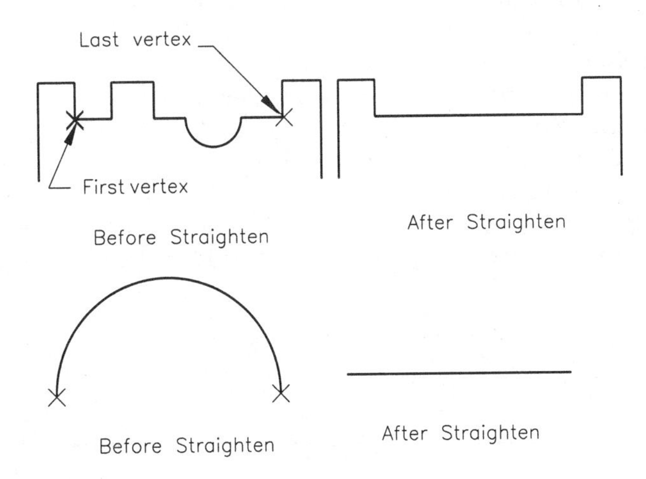

***Figure 16-20** Using the **Straighten** option to straighten polylines*

[Next/Previous/Break/Insert/Move/Regen/Straighten/Tangent/Width/eXit]<N>: *Move the marker to the desired vertex with the **Next** or **Previous** option.*
Enter a vertex editing option
[Next/Previous/Break/Insert/Move/Regen/Straighten/Tangent/Width/eXit]<N>: **S** Enter
Enter a vertex editing option
[Next/Previous/Break/Insert/Move/Regen/Straighten/Tangent/Width/eXit]<N>: *Move the marker to the next desired vertex.*
Enter an option [Next/Previous/Go/eXit] <N>: **G** Enter

The polyline segments between two marker locations are replaced by a single straight line segment. If you specify a single vertex, the segment following the specified vertex is straightened, if it is curved.

Tangent Option. The **Tangent** option is used to associate a tangent direction to the current vertex (marked by X). The tangent direction is used in curve fitting. This option is discussed in detail in the subsequent section on curve fitting. The prompt is:

Specify direction of vertex tangent: *Specify a point or enter an angle.*

You can specify the direction by entering an angle at the **Specify direction of vertex tangent:** prompt or by selecting a point to express the direction with respect to the current vertex.

Width Option. The **Width** option lets you change the starting and the ending widths of a polyline segment that follows the current vertex (Figure 16-21). By default, the ending width is equal to the starting width; hence, you can get a polyline segment of uniform width by accepting the default value at the **Specify ending width for next segment <starting width>** prompt. You can specify different starting and ending widths to get a varying-width polyline. The prompt sequence is:

Enter an option [Close/Join/Width/Edit vertex/Fit/Spline/Decurve/Ltype gen/Undo]: Enter
Enter a vertex editing option
[Next/Previous/Break/Insert/Move/Regen/Straighten/Tangent/Width/eXit] <N>: *Move the marker to the starting vertex of the segment whose width is to be altered.*
Enter a vertex editing option
[Next/Previous/Break/Insert/Move/Regen/Straighten/Tangent/Width/eXit] <N>: **W** Enter
Specify starting width for next segment <current>: *Enter the revised starting width.*
Specify ending width for next segment <starting width>: *Enter the revised ending width.*

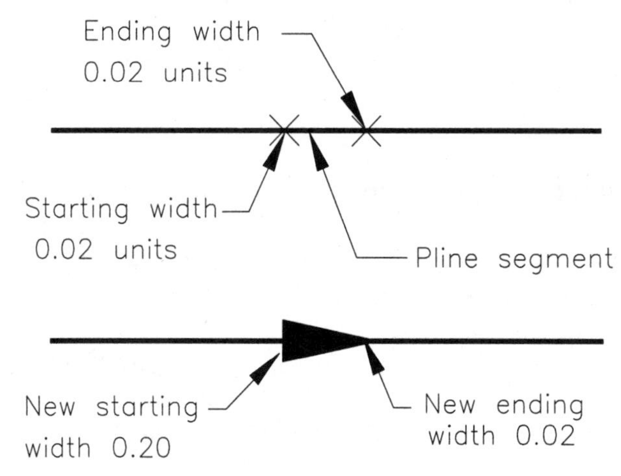

Figure 16-21 *Using the* ***Width*** *option to change the width of a polyline*

The segment with the revised widths is redrawn only after invoking the **Regen** option or when you exit the vertex mode editing.

eXit Option. This option lets you exit the Vertex mode editing and return to the main **PEDIT** prompt.

Fit Option. The **Fit** or **Fit curve** option generates a curve that passes through all the corners (vertices) of the polyline, using the tangent directions of the vertices (Figure 16-22). The curve is composed of a series of arcs passing through the corners (vertices) of the polyline. This option is used when you draw a polyline with sharp corners and need to convert it into a series of smooth curves. An example of this is a graph. In a graph, we need to show a curve by joining a series of plotted points. The process involved is called curve fitting; hence, the name of this option. These vertices of the polyline are also known as the control points. The closer together these control points are, the smoother the curve. Hence, if the **Fit** option does not give optimum results, insert more vertices into the polyline or edit the tangent directions of vertices; then, use the **Fit** option on the polyline. Before using this option you may give each vertex a tangent direction. The curve is then constructed, keeping in mind the tangent directions you have specified. The following prompt sequence illustrates the **Fit** option:

Command: **PEDIT** [Enter]
Select polyline: *Select the polyline to be edited.*
Enter an option [Close/Join/Width/Edit vertex/Fit/Spline/Decurve/Ltype gen/Undo]: **F** [Enter]

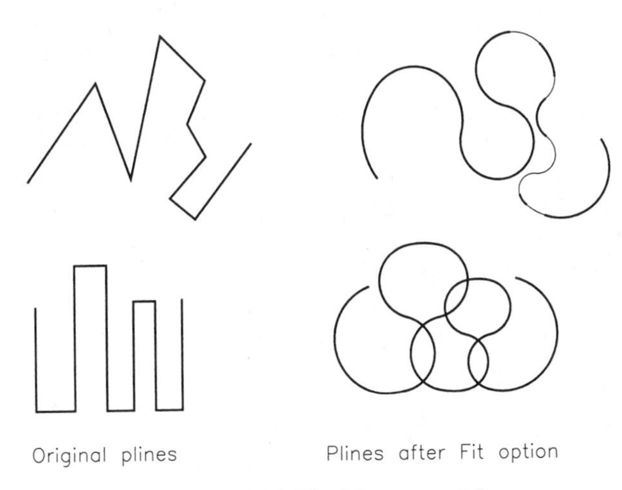

Figure 16-22 *The* ***Fit curve*** *option*

If the tangent directions need to be edited, use the **Edit vertex** option of the **PEDIT** command. Move the X marker to each of the vertices that need to be changed. Now, you can invoke the **tangent** option, and either enter the tangent direction in degrees or select points. The chosen direction is expressed by an arrow placed at the vertex. The prompt sequence is:

Command: **PEDIT** [Enter]
Select polyline: *Select the polyline to be edited.*
Enter an option [Close/Join/Width/Edit vertex/Fit/Spline/Decurve/Ltype gen/Undo]: **E** [Enter]
Enter a vertex editing option
[Next/Previous/Break/Insert/Move/Regen/Straighten/Tangent/Width/eXit]<N>: **T** [Enter]
Specify direction of vertex tangent: *Specify a direction in + or - degrees, or select a point in the desired direction. Press ENTER.*

Once the tangent directions are specified, use the **eXit** option to return to the previous prompt and use its **Fit curve** option.

Spline Option. The **Spline** option (Figure 16-23) also smoothens the corners of a straight segment polyline, as does the **Fit** option, but the curve passes through only the first and the last control points (vertices), except in the case of a closed polyline. The spline curve is stretched toward the other control points (vertices) but does not pass through them, as in the case of the **Fit** option. The greater the number of control points, the greater the force with which the curve is stretched toward them. The prompt sequence is as follows:

Command: **PEDIT** Enter
Select polyline: *Select the polyline.*
Enter an option [Close/Join/Width/Edit vertex/Fit/Spline/Decurve/Ltype gen/Undo]: S Enter

The generated curve is a B-spline curve. The **frame** is the original polyline without any curves in it. If the original polyline has arc segments, these segments are straightened when the spline's frame is formed. A frame that has width produces a spline curve that tapers smoothly from the width of the first vertex to that of the last. Any other width specification between the first width specification and the last is neglected. When a spline is formed from a polyline, the frame is displayed as a polyline with zero width and continuous linetype. Also, AutoCAD LT saves its frame so that it may be restored to its original form. Tangent specifications on control point vertices do not affect spline construction. By default, the spline frames are not shown on screen, but you may want them displayed for reference. In such a case, system variable **SPLFRAME** needs to be manipulated. The default value for this variable is zero. If you want to see the spline frame as well, set it to 1. The prompt sequence is:

Command: **SPLFRAME** Enter
New value for SPLFRAME <0>: **1** Enter

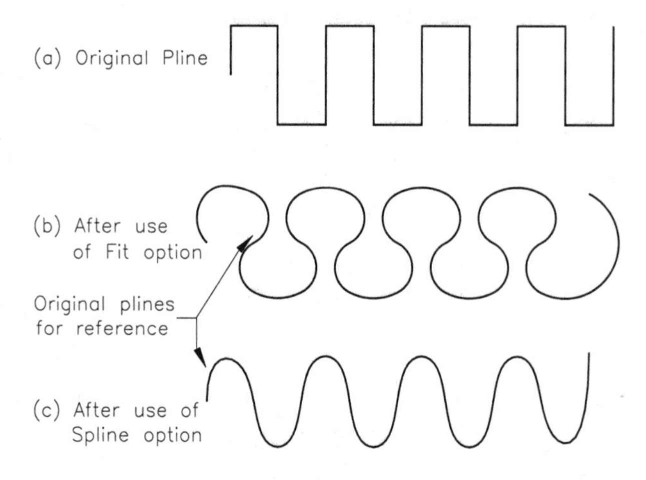

***Figure 16-23** The **Spline** option*

Now, whenever the **Spline** option is used on a polyline, the frame will also be displayed. Most editing commands such as, **MOVE**, **ERASE**, **COPY**, **MIRROR**, **ROTATE**, and **SCALE** work similarly for both the Fit curves and Spline curves. They work on both the curve and its frame, whether the frame is visible or not. The **EXTEND** command changes the frame by adding a vertex at the point the last segment intersects with the boundary. If you use any of the above mentioned commands and then use the **Decurve** option to decurve the Spline curve and later use the **Spline** option again, the same Spline curve is generated. The **BREAK**, **TRIM**, and **EXPLODE** commands delete the frame. The **DIVIDE**, **MEASURE**, **FILLET**, **CHAMFER**, **AREA**, and **HATCH** commands recognize only the Spline curve and do not consider the frame. The **STRETCH** command first stretches the frame and then fits the Spline curve to it.

When you use the **Join** option of the **PEDIT** command, the Spline curve is decurved and the original spline information is lost. The **Next** and **Previous** options of the **Edit vertex** option of the **PEDIT** command moves the marker only to points on the frame whether visible or not. The **Break** option discards the spline curve. **Insert, Move, Straighten** and **Width** options refit the Spline curve. Object Snaps consider the Spline curve and not the frame, therefore, if you wish to snap to the frame control points, restore the original frame.

There are two types of spline curves:

Quadratic B-spline
Cubic B-spline

Both of them pass through the first and the last control points, which is characteristic of the spline curve. The cubic curves are very smooth. The cubic curve passes through the first and last control points, and the curve is closer to the other control points. The quadratic curves are not as smooth as the cubic ones, but they are smoother than the curves produced by the **Fit curve** option. The quadratic curve passes through the first and last control points, and the rest of the curve is tangent to the polyline segments between the remaining control points (Figure 16-24).

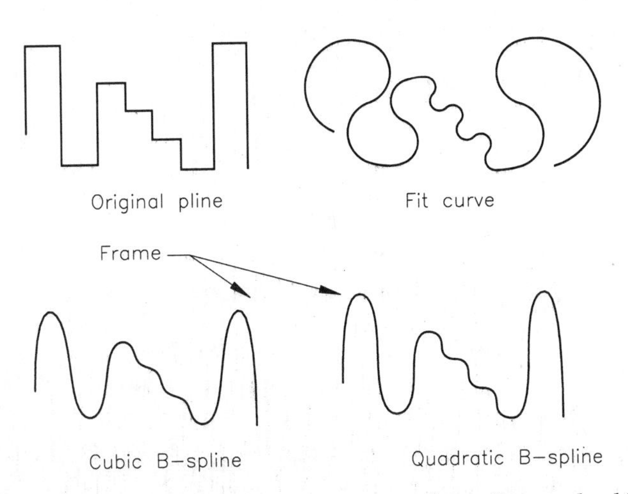

***Figure 16-24** Comparison of fit curve, quad B-spline, and cubic B-spline*

Generation of Different Types of Spline Curves. If you want to edit a polyline into a B-spline curve, you are first required to enter a relevant value in the **SPLINETYPE** system variable. A value of 5 produces the quadratic curve, whereas 6 produces a cubic curve.

Command: **SPLINETYPE** [Enter]
Enter new value for SPLINETYPE <6>: **5** [Enter]

SPLINESEGS. The system variable **SPLINESEGS** governs the number of line segments used to construct the spline curves, so you can use this variable to control the smoothness of the curve. The default value for this variable is 8. With this value, a reasonably smooth curve that does not need much regeneration time is generated. The greater the value of this variable, the smoother the curve, the greater the regeneration time, and the more space occupied by the drawing file.

Figure 16-25 shows cubic curves with different values for the **SPLINESEGS** parameter.

The prompt sequence for setting **SPLINESEGS** to a value of 500 is:

Command: **SPLINESEGS** [Enter]
Enter new value for SPLINESEGS <8>: **500** [Enter]

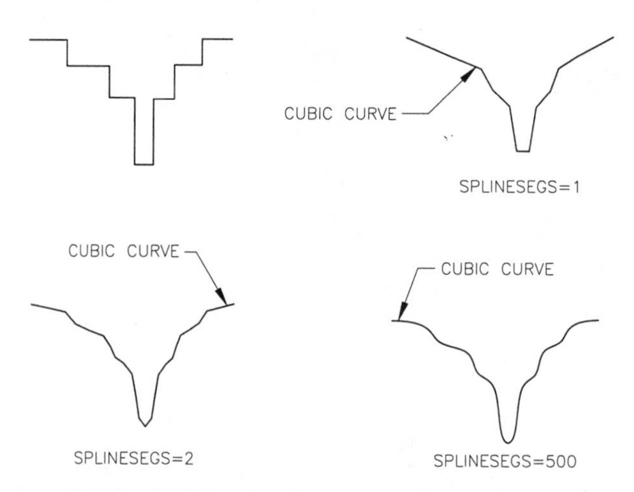

Figure 16-25 *Using the* ***SPLINESEGS*** *variable*

Decurve Option. The **Decurve** option straightens the curves generated after using the **Fit** or **Spline** option on a polyline; they return to their original shape (Figure 16-26). Polyline segments are straightened using the **Decurve** option. The vertices inserted after using the **Fit** or **Spline** option are also removed. Information entered for tangent reference is retained for use in future Fit curve operations. You can also use this command to straighten out any curve drawn with the help of the **Arc** option of the **PLINE** command.

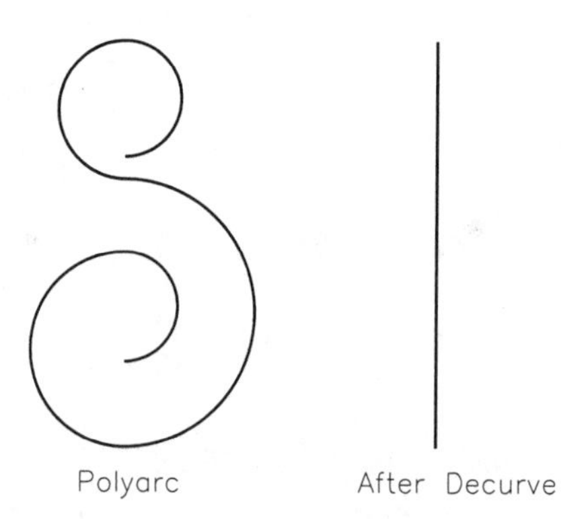

Figure 16-26 *The* ***Decurve*** *option*

The prompt sequence is:

Command: **PEDIT** [Enter]
Select polyline: *Select the polyline to be edited.*
Enter an option [Close/Join/Width/Edit vertex/Fit/Spline/Decurve/Ltype gen/Undo]: [Enter]

If you use edit commands like the **BREAK** or **TRIM** commands on spline curves, the **Decurve** option cannot be used.

Ltype gen Option. You can use this option to control the linetype pattern generation for linetypes other than Continuous with respect to the vertices of the polyline (Figure 16-27). This option has two modes: ON and OFF. If turned on, this option generates the linetype in a continuous pattern with respect to the vertices of the polyline. If you turn this option off, it generates the linetype with a dash at each vertex. This option is not applicable to polylines with tapered segments. The linetype generation for new polylines can be controlled with the help of the **PLINEGEN** system variable, which acts as a toggle. The command prompt is as follows:

Enter polyline linetype generation option [ON/OFF] <Off>:

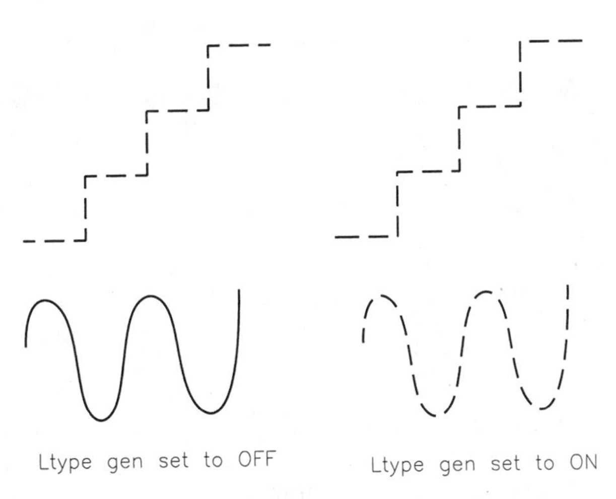

Figure 16-27 *Using the **Ltype gen** option*

U (UNDO) Option. The **Undo** option negates the effect of the most recent **PEDIT** operation. You can go back as far as you need to in the current **PEDIT** session by using the **Undo** option repeatedly until you get the desired screen. If you started editing by converting an object into a polyline, and you want to change the polyline back to the object from which it was created, the **Undo** option of the **PEDIT** command will not work. In this case, you will have to exit to the Command prompt and use the **UNDO** command to undo the operation.

Exercise 2 — General

a. Draw a line from point (0,0) to point (6,6). Convert the line into a polyline with starting width 0.30 and ending width 0.00. Then, convert the polyline back into the original line.

b. Draw a polyline of varying width segments and use the **Join** option to join all the segments into one polyline. Before joining the different segments, make the widths of all the segments uniform.

Exercise 3 — General

a. Draw a staircase-shaped polyline, then use different options of the **PEDIT** command to generate a fit curve, a quadratic B-spline, and a cubic B-spline. Then, convert the curves back into the original polyline.

b. Draw square-wave shaped polylines, and use different options of the **Edit vertex** option to navigate around the polyline, split the polyline into two at the third vertex, insert more vertices at different locations in the original polyline, and convert the square-wave shaped polyline into a straight-line polyline.

UNDOING COMMANDS (UNDO COMMAND)

Toolbar:	Standard > Undo
Menu:	Edit > Undo
Command:	UNDO

The **LINE** and **PLINE** commands have the **Undo** option, which can be used to undo (nullify) changes made within these commands. The **UNDO** command can be used to undo a previous command or to undo more than one command at one time. This command can be selected by entering **UNDO** at the Command prompt. **Undo** is also available in the **Edit** menu and the **Standard** toolbar, but this can only undo the previous command and only one command at a time. If you right-click in the drawing area a shortcut menu is displayed. Choose **Undo**. It is equivalent to **UNDO 1** or using entering **U** at the Command prompt. The **U** command can undo only one operation at a time. You can use **U** as many times as you want until AutoCAD LT displays the message: **Nothing to undo**. When an operation can not be undone, the message name is displayed but no action takes place. External commands like **PLOT** can not be undone. The prompt sequence for the **UNDO** command is:

Command: **UNDO** Enter
Enter the number of operations to UNDO or [Auto/Control/BEgin/End/Mark/Back]: *Enter a positive number, an option or press ENTER to undo a single operation.*

The various options of this command are discussed next.

Number (N) Option

This is the default option. This number represents the number of previous command sequences to be deleted. For example, if the number entered is 1, the previous command is undone; if the number entered is 4, the previous four commands are undone, and so on. This is identical to invoking the **U** command four times, except that only one regeneration takes place. AutoCAD LT lets you know which commands were undone by displaying a message after you press ENTER. The prompt sequence is:

Command: **UNDO** [Enter]
Enter the number of operations to UNDO or [Auto/Control/BEgin/End/Mark/Back]: [Enter]
PLINE LINE CIRCLE

Auto (A) Option

Enter A to invoke this option. The following prompt is displayed:

Enter UNDO Auto mode [ON/OFF] <current>: *Select ON or OFF or press ENTER to accept the default.*

If Auto is On, any group of commands that was issued by one menu item is undone together. All of them are considered to be a single group and are undone as a single command. If the **Auto** option is Off, each command in the group is treated separately.

Control (C) Option

This option lets you determine how many of the options are active in the **UNDO** command. You can disable the options you do not need. With this option you can even disable the **UNDO** command. To access this option, type C. You will get the following prompt:

Enter an UNDO control option [ALL/None/One] <All>:

The options of the **Control** option are discussed next.

All (A). The **All** option activates all the features (options) of the **UNDO** command.

None (N). This option turns off **UNDO** and the **U** command. If you have used the **BEgin** and **End** options or **Mark** and **Back** options to create **UNDO** information, all of that information is undone.

Command: **UNDO** [Enter]
Enter the number of operations to UNDO [Auto/Control/BEgin/End/Mark/Back]: **C** [Enter]
Enter an UNDO control option [ALL/None/One] <All>: **NONE** [Enter]

The prompt sequence for the **UNDO** command after issuing the **None** option is:

Command: **UNDO** [Enter]
Enter an UNDO control option [All/None/One] <All>:

To enable the **UNDO** options again, you must enter the **All** or **One** (one mode) option. If you try to use the **U** command while the **UNDO** command has been disabled, AutoCAD LT gives you the following message:

Command: **U** Enter
U command disabled. Use **UNDO** command to turn it on.

One (O). This option restrains the **UNDO** command to a single operation. All **UNDO** information saved earlier during editing is scrapped.

Enter the number of operations to UNDO or [Auto/Control/BEgin/End/Mark/Back]: **C** Enter
Enter an UNDO control option [All/None/One] <All>: **O** Enter

If you then enter the **UNDO** command, you will get the following prompt:

Command: **UNDO** Enter
Control/<1>:

In response to this last prompt, you can now either press ENTER to undo only the previous command, or go into the **Control** options by entering C. AutoCAD LT acknowledges undoing the previous command with messages like:

Command: **UNDO** Enter
Control/<1>: Enter
CIRCLE
Everything has been undone

BEgin (BE) and End (E) Options

A group of commands is treated as one command for the **U** and **UNDO** commands by embedding the commands between the **BEgin** and **End** options of the **UNDO** command. If you expect to remove a group of successive commands, you can use this option. Since all of the commands after the **BEgin** option and before the **End** option are treated as a single command by the **U** command, they can be undone by a single **U** command. For example, the following sequence illustrates the possibility of removal of two commands:

Command: **UNDO** Enter
Enter the number of operations to UNDO or [Auto/Control/BEgin/End/Mark/Back]: **BE** Enter
Command: **CIRCLE** Enter
Specify center point for Circle or [3P/2P/Ttr (tan tan radius)]: *Specify the center.*
Specify radius of circle or [Diameter] <current>: *Specify the radius of the circle.*
Command: **PLINE** Enter
Specify start point or [Arc/Close/Halfwidth/Length/Undo/Width]: *Select first point.*
Specify next point: *Select the next point.*
Specify start point or [Arc/Close/Halfwidth/Length/Undo/Width]: *Press ENTER.*
Command: **UNDO** Enter
Enter the number of operations to UNDO or [Auto/Control/BEgin/End/Mark/Back]: **E** Enter
Command: **U** Enter

To start the next group once you are finished specifying the current group, use the **End** option to end this group. Another method is to enter the **BEgin** option to start the next group while the current group is active. This is equivalent to issuing the **End** option followed by the **BEgin** option. The group is complete only when the **End** option is invoked to match a **BEgin** option. If **U** or the **Undo** command is issued after the **BEgin** option has been invoked and before the **End** option has been issued, only one command is undone at a time until it reaches the juncture where the **BEgin** option has been entered. If you want to undo the commands issued before the **BEgin** option was invoked, you must enter the **End** option so that the group is complete. This is demonstrated in Example 2.

Example 2

Enter the following commands in the same sequence as given, and notice the changes that take place on screen.

CIRCLE
POLYGON
UNDO BEgin
PLINE
TRACE
U
DONUT
UNDO End
DTEXT
U
U
U
U

The first **U** command will undo the **TRACE** command. If you repeat the **U** command, the **PLINE** command will be undone. Any further invoking of the **U** command will not undo any previously drawn object (**POLYGON** and **CIRCLE**, in this case), because after the **PLINE** is undone you have an **UNDO BEgin**. Only after you enter **UNDO End** can you undo the **POLYGON** and the **CIRCLE**. In the example, the second **U** command will undo the **TEXT** command, the third **U** command will undo the **DONUT** and **PLINE** commands (these are enclosed in the group), the fourth **U** command will undo the **POLYGON** command, and the fifth **U** command will undo the **CIRCLE** command. When the commands in a group are undone, the name of each command or operation is not displayed as it is undone; only the name, **GROUP**, is displayed.

Note

*You can use the **BEgin** option only when the **UNDO Control** is set to **All**.*

Mark (M) and Back Options

The **Mark** option installs a marker in the Undo file. The **Back** option lets you undo all the operations until the mark. In other words, the **Back** option returns the drawing to the point where the previous mark was inserted. For example, if you have completed a portion of your drawing and do not want anything up to this point to be deleted, you insert a marker and then proceed. Then, even if you use the **UNDO Back** option, it will work only until the marker. You can insert multiple markers, and with the help of the **Back** option you can return to the successive mark points. The following prompt sequence illustrates this:

Command: **UNDO** Enter
Enter the number of operations to UNDO or [Auto/Control/BEgin/End/Mark/Back]: **M** Enter

After proceeding further with your work, if you want to use the **UNDO Back** option:

Command: **UNDO** Enter
Enter the number of operations to UNDO or [Auto/Control/BEgin/End/Mark/Back]: **B** Enter

Once all the marks have been exhausted with the successive **Back** options, any further invoking of the **Back** option displays the message: **This will undo everything. OK? <Y>**.

If you enter Y (Yes) at this prompt, all the operations carried out since you entered the current drawing session will be undone. If you enter N (No) at this prompt, the Back option will be disregarded. You cannot undo certain commands and system variables, for example, **DIST**, **LIST**, **DELAY**, **NEW**, **OPEN**, **QUIT**, **AREA**, **HELP**, **PLOT**, **QSAVE** and **SAVE**, among many more. Actually, these commands have no effect that can be undone. Commands that change operating modes (**GRID**, **UNITS**, **SNAP**, **ORTHO**) can be undone, though the effect may not be apparent at first. This is the reason why AutoCAD LT displays the command names as they are undone.

REDO COMMAND

Toolbar:	Standard > Redo
Menu:	Edit > Redo
Command:	REDO

If you right-click in the drawing area, a shortcut menu is displayed. Choose **Redo** to invoke the **REDO** command. The **REDO** command brings back the objects you removed previously using the **U** and **UNDO** commands. This command undoes the **UNDO** command, but it must be entered immediately after the **UNDO** command.

Command: **REDO** Enter

The objects previously undone reappear on the screen.

RENAMING NAMED OBJECTS (RENAME COMMAND)

Menu:	Format > Rename
Command:	RENAME

You can edit the names of named objects such as blocks, dimension styles, layers, linetypes, styles, UCS, views, and viewports using the **Rename** dialog box (Figure 16-28). You can select the type of named object from the list provided in the **Named Objects** area of the dialog box.

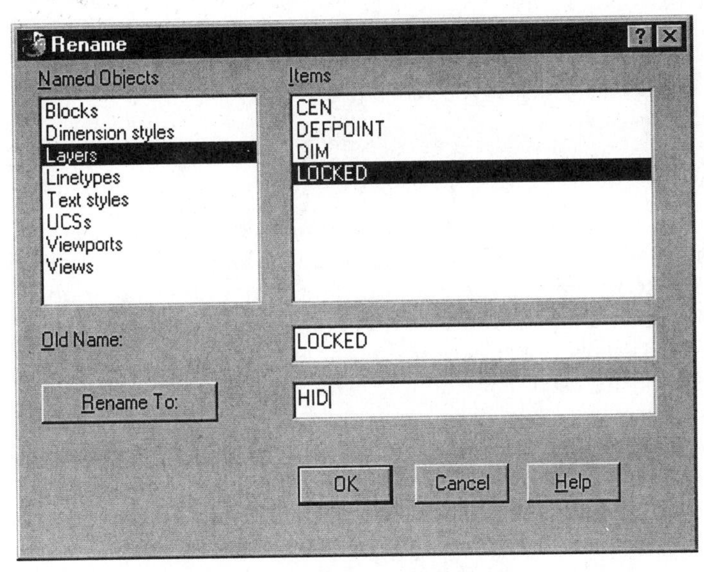

*Figure 16-28 **Rename** dialog box*

Corresponding names of all the objects of the specified type that can be renamed are displayed in the **Items** area. For example, if you want to rename the layer named LOCKED to HID, the process will be:

1. Select the **Layer** option from the list in the **Named Objects** area. All the layer names that can be renamed are displayed in the **Items** area.

2. Select LOCKED from the list in the Items area, such that LOCKED is displayed in the **Old Name:** edit box.

3. Enter HID in the **Rename To:** edit box, and choose the **OK** button.

Now the layer named LOCKED is renamed to HID. You can rename blocks, dimension styles, linetypes, styles, UCS, views, and viewports in the same way. You can enter **-RENAME** at the Command prompt to use the command line. The prompt sequence for changing the name of the block SQUARE to PLATE1 is:

Command: **-RENAME** [Enter]
Enter object type to rename
[Block/Dimstyle/LAyer/LType/Style/Ucs/VIew/VPort]: **B** [Enter]
Enter old block name: **SQUARE** [Enter]
Enter new block name: **PLATE1** [Enter]

Note

*The **RENAME** command cannot be used to rename drawing files created using the **WBLOCK** command. You can change the name of a drawing file by using the **Rename** option of your Operating System.*

REMOVING UNUSED NAMED OBJECTS (PURGE COMMAND)

Menu:	File > Drawing Utilities > Purge
Command:	PURGE

The **PURGE** command is another editing operation used for deletion. You can delete unused named objects such as blocks, layers, dimension styles, linetypes, text styles, and shapes with the help of the **PURGE** command. When you create a new drawing or open an existing one, AutoCAD LT records the named objects in that drawing and notes other drawings referencing the named objects. Usually only a few of the named objects in the drawing (such as layers, linetypes, and blocks) are used. For example, when you create a new drawing, the prototype drawing settings may contain various text styles, blocks, and layers that you do not want to use. Also, you may want to delete particular unused named objects, such as unused blocks, in an existing drawing. Deleting inactive named objects is important and useful because doing so reduces the space occupied by the drawing. As you enter **PURGE** in the command line, the **Purge** dialog box is displayed (Figure 16-29). You can select the named objects you want to delete from the tree view. You can use this command any time in the drawing session.

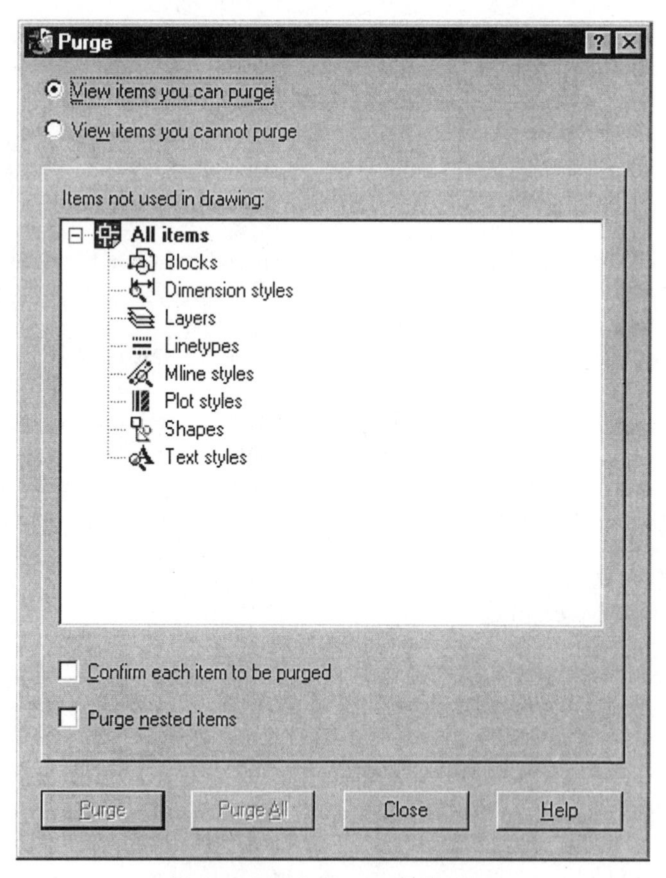

Figure 16-29 ***Purge*** *dialog box*

Dialog Box Options

View items you can purge

When you choose this radio button, all the unused items in the drawing that can deleted are displayed in the tree view window.

View items you cannot purge

As you choose this radio button, all of the items in the drawing that are used and cannot be purged are displayed in the tree view window.

Confirm each items to be purged

If this check box is selected, **Confirm Purge** dialog box is displayed every time you delete an item. To confirm the purge, choose **Yes** button.

Purge nested items

Nested items are generally those that are within the blocks and can be purged from within the blocks. This check box is used to remove all the nested items when the **Purge All** button is chosenor the purge is performed on all the items.

Purge

This button is chosen to delete the unused items you have selected in the tree view.

Purge All

As you choose this button, all the unused items in the drawing are deleted. However before deleting each item, AutoCAD LT displays **Confirm Purge** dialog box for each item even if the **Confirm each items to be purged** check box is cleared.

The **-WBLOCK asterisk** option has the same effect as the **PURGE** command. The only difference is that in the case of the **-WBLOCK asterisk** option, unused named objects are removed automatically.

If there are no objects to remove, AutoCAD LT displays a message that there are no unreferenced objects to purge.

DELETING UNUSED BLOCKS USING COMMAND LINE

You can also delete unused blocks using command line with the help of the **-PURGE** command. For example, if you want to delete an unused block named SQUARE, the prompt sequence is:

Command: **-PURGE** Enter
Enter type of unused objects to purge
[Blocks/Dimstyles/LAyers/LTypes/Plotstyles/textSHapes/textSTyles/Mlinestyles/ All]:
B Enter
Enter names to purge <*>: **SQUARE** Enter
Verify each name to be purged? [Yes/No] <Y> Enter
Purge block SECOND? <N>: **Y** Enter

Note

Standard objects created by AutoCAD LT (such as layer 0, STANDARD text style, and linetype CONTINUOUS) cannot be removed by the ***PURGE*** *command, even if these objects are not used.*

OBJECT SELECTION MODES

Menu:	Tools > Options
Command:	OPTIONS

When you select a number of objects, the selected objects form a **selection set**. Selection of the objects is controlled in the **Options** dialog box (Figure 16-30) which is invoked by the **OPTIONS** command. Five selection modes are provided in the **Selection** tab of this dialog box. You can select any one of these modes or a combination of various modes.

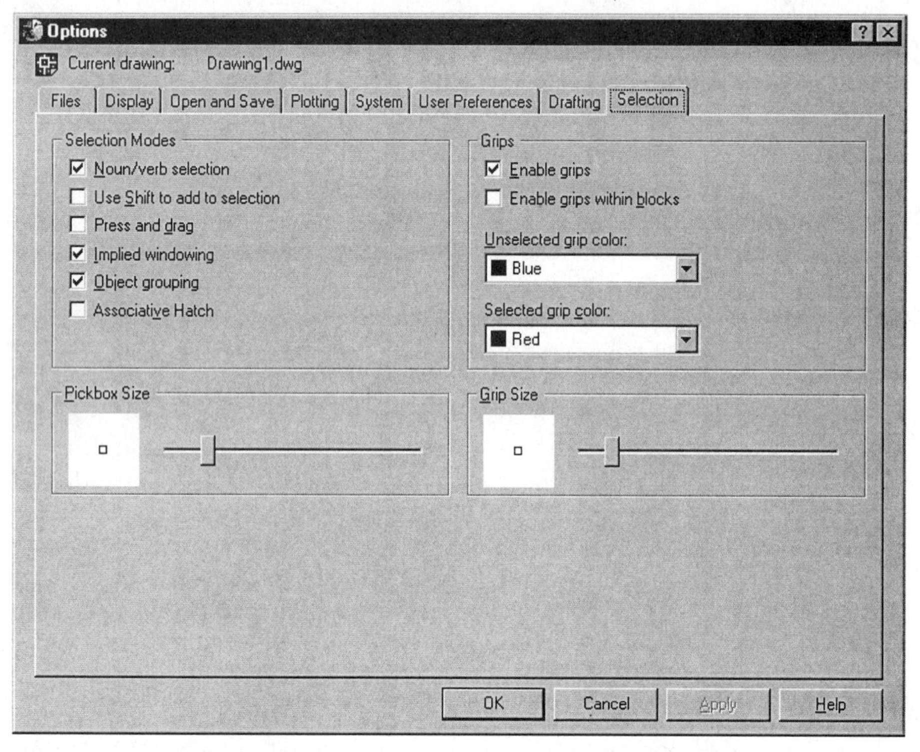

Figure 16-30 Selection *tab of the* ***Options*** *dialog box*

Noun/Verb Selection

By selecting the **Noun/Verb Selection** check box, you can select the objects (noun) first and then specify the operation (verb) (command) to be performed on the selection set. If this selection mode is chosen, a small pick box is displayed at the intersection of the graphic cursor (crosshairs). This mode is active by default. For example, if you want to move some objects when **Noun/Verb Selection** is enabled, first select the objects to be moved, and then invoke the **MOVE** command. The objects selected are highlighted automatically when the **MOVE** command is invoked, and AutoCAD LT does not issue any **Select objects** prompt. The following commands can be used on the selected objects when the **Noun/Verb Selection** mode is active:

ARRAY	**BLOCK**	**CHANGE**	**CHPROP**	**COPY**
DVIEW	**EXPLODE**	**ERASE**	**LIST**	**MIRROR**
MOVE	**PROPERTIES**	**ROTATE**	**SCALE**	
STRETCH	**WBLOCK**			

The following are some of the commands that are not affected by the **Noun/Verb Selection** mode. You are required to specify the objects (noun) on which an operation (command/verb) is to be performed, after specifying the command (verb):

BREAK	**CHAMFER**	**DIVIDE**	**EXTEND**	**FILLET**
MEASURE	**OFFSET**	**TRIM**		

When the **Noun/Verb Selection** mode is active, the **PICKFIRST** system variable is set to 1 (On). In other words, you can activate the **Noun/Verb Selection** mode by setting **PICKFIRST** to 1 (On).

Command: **PICKFIRST**
Enter new value for PICKFIRST <0>: **1** Enter

Use Shift to Add to Selection

The next option in the **Selection Modes** area of the **Selection** tab of the **Options** dialog box is **Use Shift to add to selection**. Selecting this option establishes additive selection mode which is the normal method of most Windows programs. In this mode, you have to hold down the SHIFT key when you want to add objects to the selection set. For example, suppose X, Y, and Z are three objects on the screen. Select object X. It is highlighted and put in the selection set. After selecting X, and while selecting object Y, if you do not hold down the SHIFT key, object Y only is highlighted and it replaces object X in the selection set. On the other hand, if you hold down the SHIFT key while selecting Y, it is added to the selection set (which contains X), and the resulting selection set contains both X and Y. Also, both X and Y are highlighted. To summarize the concept, objects are added to the selection set only when the SHIFT key is held down while objects are selected. Objects can be discarded from the selection set by reselecting these objects while the SHIFT key is held down. If you want to clear an entire selection set quickly, draw a blank selection window anywhere in a blank drawing area. You can also right-click to display a shortcut menu. Choose **Deselect All**. All selected objects in the selection set are discarded from it.

When the **Use Shift to add to selection** mode is active, the **PICKADD** system variable is set to 1 (On). In other words, you can activate the **Use Shift to add to selection** mode by setting **PICKADD** to On.

Command: **PICKADD** Enter
Enter new value for PICKADD <0>: **1** Enter

Press and Drag

This selection mode is used to govern the way you can define a Selection window or Crossing window. When this option is selected, the window can be created by pressing the pick button to select one corner of the window and continuing to hold down the pick button and dragging the cursor to define the other diagonal point of the window. When you have the window you want, release the pick button. If the **Press and Drag** mode is not active, you have to select twice to specify the two diagonal corners of the window to be defined.

When the **Press and Drag** mode is active, the **PICKDRAG** system variable is set to 1 (On). In other words, you can activate the **Press and Drag** mode by setting **PICKDRAG** to On.

Command: **PICKDRAG** [Enter]
Enter new value for PICKDRAG <0>: **1** [Enter]

Implied Windowing

By selecting this option, you can automatically create a Window or Crossing selection when the **Select objects:** prompt is issued. The selection window or crossing window in this case is created in the following manner:

At the **Select objects:** prompt, select a point in empty space on the screen. This becomes the first corner point of the selection window. After this, AutoCAD LT asks you to specify the other corner point of the selection window. If the first corner point is to the right of the second corner point, a Crossing selection is defined; if the first corner point is to the left of the second corner point, a Window selection is defined. If this option is not active, or if you need to select the first corner in a crowded area where selecting would select an object, you need to specify Window or Crossing at the **Select objects:** prompt, depending on your requirement.

When the **Implied Windowing** mode is active, the **PICKAUTO** system variable is set to 1 (On). In other words, you can activate **Implied Windowing** mode by setting **PICKAUTO** to On.

Command: **PICKAUTO** [Enter]
Enter new value for PICKAUTO <0>: **1** [Enter]

Object Grouping

This turns the automatic group selection on and off. When this option is on and you select a member of a group, the whole group is selected. You can also activate this option by setting the value of the **PICKSTYLE** system variable to 1. (Groups were discussed earlier in this chapter.)

Associative Hatch

If the **Associative Hatch** check box is selected in the **Selection modes** area of the **Selection** tab of the **Options** dialog box, the boundary object is also selected when an associative hatch is selected. You can also select this option by setting the value of the **PICKSTYLE** system variable to 2.

Pickbox Size. The **Pick box** slider bar controls the size of the pickbox. The size ranges from 0 to 20. The default size is 3. You can also use the **PICKBOX** system variable.

CHANGING DISPLAY ORDER
Using DRAWORDER Command

Toolbar:	Modify II > Draworder
Menu:	Tools > Display Order
Command:	DRAWORDER

Figure 16-31 Invoking the ***DRAWORDER*** *command from the* ***Modify II*** *toolbar*

When you create objects, they are drawn in the order they were created. For example, if you insert a raster image and then draw a wide polyline over the image, the polyline will be displayed at the top of the raster image. At the time of plotting the raster image will be plotted first and then the polyline. You can use the **DRAWORDER** command to change the order of display and also the order of plotting.

Command: **DRAWORDER** [Enter]
Enter object ordering option [Above object/Under object/Front/Back] <Back>: *Select an option or press ENTER.*

Above Object. With the **Above Object** option the selected object is moved above the specified object. When several objects are selected for reordering, the relative order of the selected objects is maintained. For example, in Figure 16-32 the **Above Object** option moves the objects 2, 3, and 5 above the specified object (4). When you select the **Above** or **Under object** options, the next command prompt is:

Select reference object: *Select the object for reference.*

Figure 16-32 Using the ***DRAWORDER*** *command to change the display order of objects*

Under Object. The **Under Object** option moves the selected object below the specified object. For example, if a wide polyline segment is overlapping an image (Figure 16-33). You can bring the image to the front (make the overlapping segment of the polyline invisible) by changing the display order of the image or polyline. Similarly, you can also change the display order two overlapping images.

Front. The **Front** option moves the selected objects to the top of the display order. For example, in Figure 16-33, the **Front** option moves the object to the top of the display order.

Back. The **Back** option moves the selected objects to the bottom of the display order.

When you change the display order of the objects several times, the display order might get disordered. You can use the **REGEN** command to restore the display order you set. AutoCAD LT cannot control the display order for nested objects in blocks and external references. The

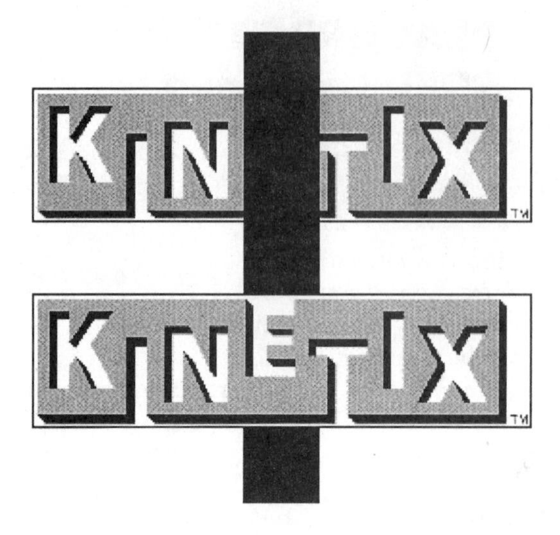

Figure 16-33 *Effect of changing the display order*

display order of the nested objects is determined by the order in which the objects were created. Similarly, the **PSOUT** command ignores the display order information and displays the objects in the same order in which they were created. By changing the display order the file size increases by 16 bytes for each object affected by the display order. This increases the file size by 1 to 5 percent.

Using the SORTENTS System Variable

The **DRAWORDER** command turns on all **Object Sorting methods** options on the **User Preferences** tab of the **Options** dialog box (Figure 16-34), which results in slower regeneration of drawings.

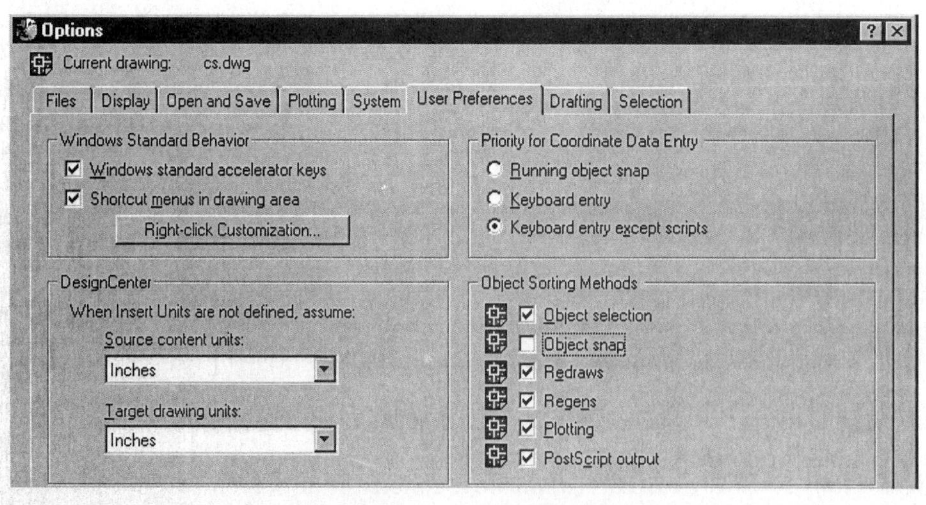

Figure 16-34 Object Sorting methods *area in the* ***User Preferences*** *tab of the* ***Options*** *dialog box*

The **SORTENTS** variable governs the display of object sort order operations using following values:

0 - **SORTENTS** is disabled, 1 - Sorts for object selection, 2 - Sorts for object snap, 4 - Sorts for redraw, 8 - Sorts for **MSLIDE** slide creation, 16 - Sorts for **REGEN** command, 32 - Sorts for plotting, 64 - Sorts for PostScript output. More than one options can be selected by specifying the sum of the values of these options. This variable is saved in AutoCAD LT configuration file, not the drawing file. The initial value of **SORTENTS** is 96. This value specifies sort operations for plotting and PostScript output. When **SORTENTS** variable is set to 0, AutoCAD LT displays the objects in the most efficient way.

When you use the **DRAWORDER** command the values changes to 127. If you change the value of **SORTENTS** to a value other than 127, all display order information is ignored.

Command: **SORTENTS** Enter
Enter new value for SORTENTS <current>:

MANAGING CONTENT*
Using AutoCAD DesignCenter*

Toolbar:	Standard > AutoCAD DesignCenter
Menu:	Tools > AutoCAD DesignCenter
Command:	ADCENTER

When you choose the **AutoCAD DesignCenter** button in the **Standard** toolbar displays the **AutoCAD DesignCenter** window (Figure 16-35). In its default position it is docked to the left of the screen. It can be moved to any other location by clicking and dragging it with the grab bars located on the top of the window to any other position on the screen. You can also resize it by clicking the borders and dragging them to the right or left. Right-clicking on the title bar of the window displays a shortcut menu which gives options to **Allow docking**, **Hide**, **Move**, and **Resize** the **AutoCAD DesignCenter** window. The CTRL+2 key acts as toggle between hiding the window and displaying it. Also, double-clicking on the title bar of the window, docks the **AutoCAD DesignCenter** window if it is undocked and vice versa.

The **AutoCAD DesignCenter** window is used for locating and organizing drawing data, for inserting blocks, layers, external references and other customized drawing content from either your own files, local drives, network, or the internet and for viewing contents of a file. You can even access and use contents between files or from the internet.

When you choose the **Tree View Toggle** button on the **AutoCAD DesignCenter** toolbar, it displays the **Tree Pane** with a Tree view of the contents of the drives. You can also right-click in the window and choose **Tree** from the shortcut menu which is displayed. Now, the window is divided into two parts, the **Tree pane** and the **palette**. The palette displays folders, files, objects in a drawing, images, web based content and custom content. You can also resize both the tree pane and the palette by clicking and dragging the bar between them to the right or the left. In the Tree pane you can browse the contents of any folder by clicking on the plus sign (+) adjacent to it. Further expanding the contents of a file displays the following categories:

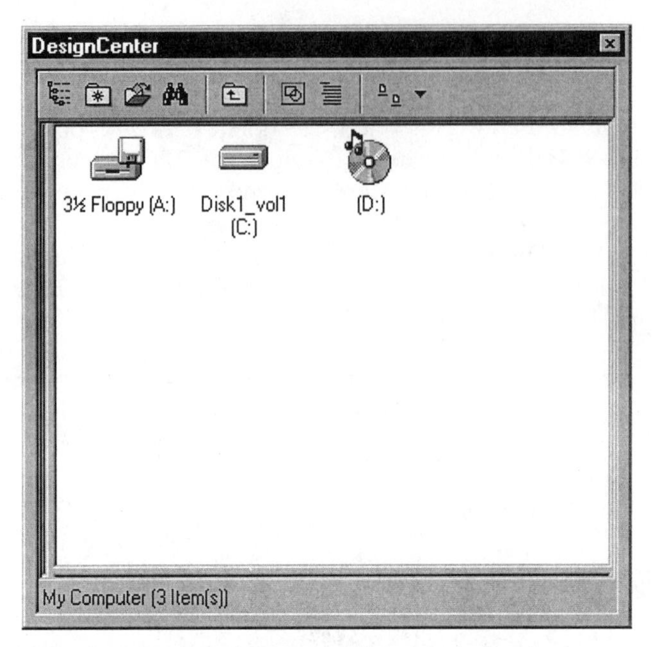

***Figure 16-35 AutoCAD DesignCenter window** in the default position.*

Blocks, **Dimstyles**, **Layers**, **Linetypes**, **Textstyles**, **Layout**, and **External References**. Clicking on any one of these categories in the **Tree pane** displays the listing under the selected category in the palette (Figure 16-36). Alternately, right-clicking a particular folder, file, or category of the file contents, displays a shortcut menu. The **Explore** option in this shortcut menu also further expands the selected folder, file, or category of contents to display the listing of contents respectively. Choosing the **Preview** button displays an image of the selected object or file in a **Preview pane** in the Palette. Choosing the **Description** button displays a brief text description of the selected item if it has one in the **Description** box.

As discussed earlier in Chapters 12 and 14, you can drag and drop any of the contents into the current drawing or add them by double-clicking on the contents. These are then reused as part of the current drawing. When you double-click on specific X-refs and Blocks, AutoCAD LT displays the **External Reference** dialog box and the **Insert** dialog box to help in attaching the External Reference and inserting the block, respectively. Similarly, when you right-click a specific linetype, layer, textstyle, layout or dimstyle, in the palette, a shortcut menu is displayed which gives you an option to **Add** or **Copy**. Right-clicking a block displays the options of **Insert Block** or **Copy** and right-clicking an Xref displays the options of **Attach Xref** or **Copy** in the shortcut menus.

The other options in the shortcut menu are **Add to Favorites**, **Organize Favorites**, and **Open in window**. Choosing **Add to Favorites** adds the selected file or folder to the Favorites folder, which contains the most often accessed files and folders. **Organize Favorites** allows you to reorganize the contents of the Favorites folder, and **Open in window** opens the selected drawing file as a current drawing.

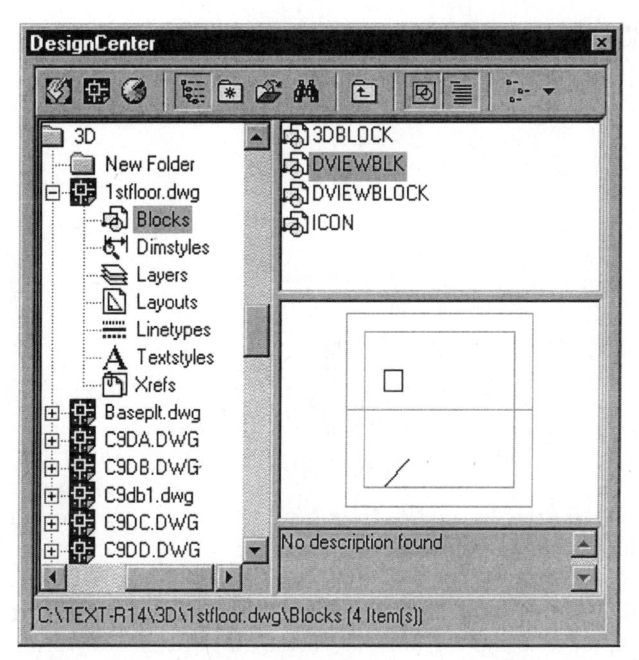

***Figure 16-36** The **AutoCAD LT DesignCenter window** showing the Tree pane, the palette, the Preview pane and the Description box*

Choosing the **Desktop** button in the **AutoCAD DesignCenter** toolbar, lists all the local and network drives. The **Open Drawings** button lists the drawings that are open, including the current drawing which is being worked on. The **History** button lists the last twenty locations accessed through the **AutoCAD DesignCenter**. The **Tree View Toggle** button displays or hides the tree pane with the tree view of the contents in a hierarchial form. This option is disabled when you choose the **History** button. Choosing the **Favorites** button displays shortcuts to files and folders that are accessed frequently by you and are stored in the **Favorites** folder. This reduces the time you take to access these files or folders from their normal location. Choosing the **Load** button displays the **Load DesignCenter Palette** dialog box (Figure 16-38) whose options are similar to the standard **Select file** dialog box. When you select a file here and choose the **Open** button, AutoCAD LT displays the selected file and its contents in the **AutoCAD DesignCenter** window.

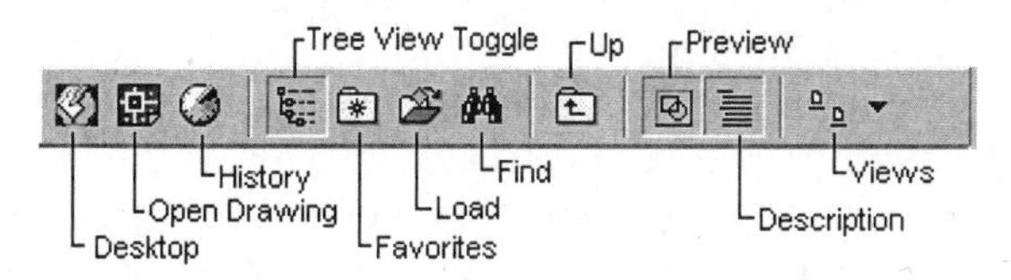

***Figure 16-37** The **AutoCAD DesignCenter** toolbar*

Choosing the **Find** button displays the **Find** dialog box (Figure 16-39). By using this dialog box, you can search for drawings that contain specific Blocks, textstyles, dimstyles, layers, layouts, external references, or linetypes, which you can select from the **Look for** drop-down list, or you can conduct a search based on the following criteria:

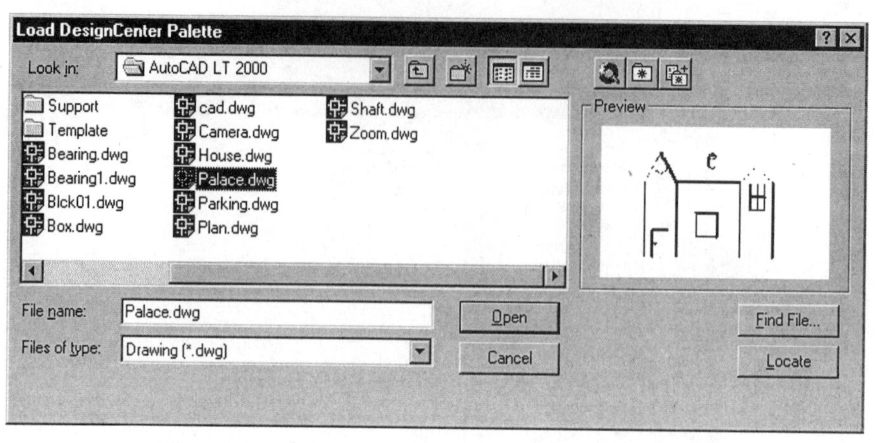

Figure 16-38 Load DesignCenter Palette dialog box

1. **Date last modified on.** Choose the **Date Modified** tab in the **Find** dialog box and select the appropriate radio buttons there.

2. **Text in the block name, block and drawing description, attribute tag and value.** You can select the category to look for from the **Containing:** drop-down list in the **Advanced** tab of the **Find** dialog box. The text it contains can be entered in the **Contains text** edit box. You can also specify the **At least** or **At most** sizes in the **Size is:** edit boxes to narrow down the search.

3. **Words which are part of the File name, Title, Subject, Author's name, keyword text stored in the summary tab of the Drawing Properties.** Enter the words in the **Search for word(s)** edit box and the field can be selected from the **In the field(s)** drop-down list of the **Drawings** tab of the **Find** dialog box.

After entering the apt criteria for searching the specific file, choose the **Find Now** button. The Name, Location, File Size, Type, and Date last modified on is displayed at the bottom of the dialog box.

The **Up** button moves one level up in the tree structure from the current location. The **Views** button gives four display format options for the contents of the palette: **Large icons**, **Small icons**, **List**, and **Details**. The **List** option lists the contents in the palette while the **Details** option, gives a detailed list of the contents in the palette with the Name, File Size, and Type.

Right-clicking in the palette displays a shortcut menu with all the options provided in the AutoCAD LT DesignCenter in addition to **Add to Favorites**, **Organize favorites**, and **Refresh** options. The **Refresh** option refreshes the palette display if you have made any changes to it.

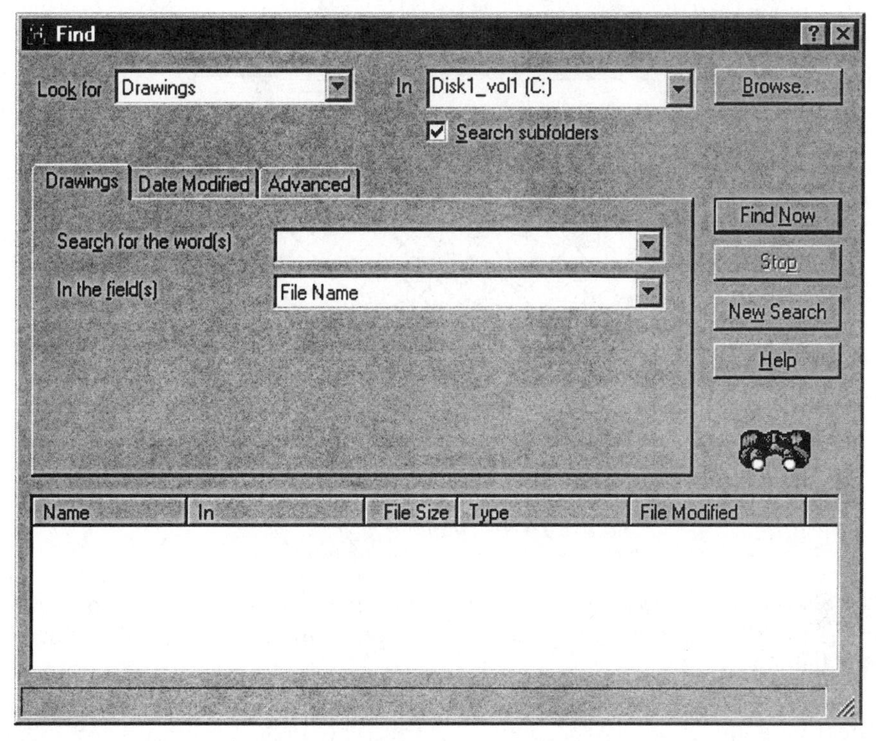

Figure 16-39 ***Find*** *dialog box*

Example 3

Use the **AutoCAD DesignCenter** window to locate and view contents of the drawing KITCHENS. Also, use the **AutoCAD DesignCenter** to insert a block from this drawing and also import a layer and Textstyle from the 1st floor architectural.dwg located in the Sample folder. Also import a dimstyle from Wilhome.dwg. Use these to make a drawing of a Kitchen plan (MYKITCHEN.dwg) and add text and dimensions to it as shown in Figure 16-40.

1. Open a new drawing using the **Start from Scratch** button. Make sure to select the **English (feet and inches)** radio button in the **Create New Drawing** or the **Startup** dialog box.

2. Draw a rectangle of 8" x 7".

3. Choose the **AutoCAD DesignCenter** button in the **Standard** toolbar. The **AutoCAD DesignCenter** window is displayed in its default location, docked to the left of the drawing screen.

4. In the **AutoCAD DesignCenter** toolbar, choose the **Tree View Toggle** button to display the **Tree pane** and the **Palette** (if not already displayed).

5. Now, choose the **Find** button. The **Find** dialog box is displayed. Select **Drawings** from the **Look For** drop-down list. Also select **C:** from the **In** drop-down list. Select the **Search subfolders** check box if it is not selected already. In the **Drawings** tab, type KITCHENS

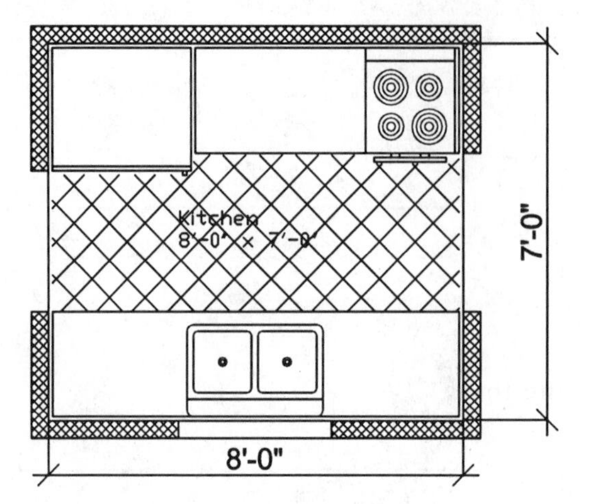

Figure 16-40 Drawing for Example 3

in the **Search for the word(s)** edit box. Choose the **Find Now** button to commence the search. After the drawing has been located, its details and path are displayed at the bottom of the dialog box.

6. In the Tree pane, expand My Computer to display the C:\Program Files\Auto CAD LT 2000\SAMPLE\DesignCenter folder. Now, click on it to display its contents in the Palette. You can also right-click to display the shortcut menu and choose **Explore** from it.

7. Right-click on KITCHEN.dwg in the palette to display the shortcut menu. Choose **Open in Window**. KITCHENS.dwg is now the current drawing. You can attach any drawing you want to use as an Xref to the current drawing by choosing the **Attach as Xref** option in the shortcut menu. After you have viewed the contents of this drawing after using the various **ZOOM** commands, close this drawing.

8. Double-click on KITCHENS.dwg in the Palette to display its contents or expand the contents by clicking on the plus sign adjacent to the File name, in the Tree pane.

9. Click on Blocks, in the Tree Pane to display the list of blocks in the drawing, in the Palette. Right-click on the block KITCHEN LAYOUT-7 x 8 ft. to display a shortcut menu. Choose **Insert Block** to display the **Insert** dialog box. Alternately, you can double-click on the block to display the **Insert** dialog box. Select the **Specify On-Screen** check boxes and the Explode check box. Choose **OK** to exit the dialog box. Now, insert this block in the rectangle you had drawn such that the rectangle overlaps the floor area of the kitchen plan.

10. Now, double-click on 1ST FLOOR ARCHITECTURAL.DWG located in the SAMPLE folder in the same directory to display its contents in the Palette.

11. Click on Layers in the Tree Pane to display the layers in the drawing. Drag and drop or double-click the layer DIMENSIONS (COTAS) from this drawing to the current drawing.

Now, you can use this layer for dimensioning in the current drawing after making it the current layer.

12. Now, click on Textstyles to display the list of Textstyles in the drawing. Select ROMAN SIMPLEX and drag and drop it in the current drawing. You can use this textstyle for adding text to the current drawing.

13. Similarly, click on the WILHOME.DWG in the same folder. Select Dimstyles to display the Dimension styles being used in the drawing. Select ARCH and drag and drop it into the current drawing.

14. Use the imported data to add dimensions and text to the current drawing and complete it as shown in the Figure 16-41.

15. Save the current drawing as MYKITCHEN.DWG.

Self-Evaluation Test

Answer the following questions, and then compare your answers to the answers given at the end of this chapter.

1. Only the **CHANGE** command can be used to change the properties associated with a object. (T/F)

2. The **PURGE** command can be used only when you enter the drawing editor and before the drawing database has been changed by operations involving addition or deletion of objects in the drawing file. (T/F)

3. The **One** option of the **UNDO** command restrains the **U** and **UNDO** commands to a single operation. (T/F)

4. If the last (closing) segment of the polyline was drawn as a polyline and not by using the **Close** option, the effect of the **Open** option will not be visible. (T/F)

5. The **Control** option lets you determine how many of the **UNDO** options you want active. (T/F)

6. The **Width** option of the **PEDIT** command can be used to change the width of a polyline with a constant and unvarying width. (T/F)

7. You can move past the first and last vertices in a closed polyline by using either the **Next** option or the **Previous** option. (T/F)

8. The system variable __________ controls the number of line segments used to construct the spline curves.

9. If you want to edit a polyline into a B-spline curve, you have to set the system variable __________. A value equal to __________ produces the __________ curve, whereas a value equal to __________ produces a __________ curve.

10. The group names can be up to __________ characters long and can include special characters __________ .

11. The **Include Unnamed** check box is used to display the names of the unnamed objects in the **Group Manager** dialog box. (T/F)

12. The **Selectable** option allows you to define a group that is selectable. (T/F)

13. The **Join** option of the **PEDIT** command can be used only if a polyline is open. (T/F)

14. The **Width** option of the **PEDIT** command's main prompt can be used to change the starting and ending widths of a polyline separately to a desired value. (T/F)

Review Questions

1. You can use the __________ command to group AutoCAD LT objects and assign a name to the group.

2. You can also use the **GROUP** command from the command line by entering __________ at the Command prompt.

3. You can select the entire group by just selecting one member of that perticular group if the group is __________ .

4. A group can be selected by entering __________ at the AutoCAD LT **Select objects** prompt.

5. When you use the **GROUP** command to form object groups, AutoCAD LT lets you sequentially highlight the groups of which the selected object is a member. (T/F)

6. The color, linetype, lineweight, and layer on which a block is drawn can be changed with the help of the **CHANGE** command. (T/F)

7. After exploding an object, the object remains identical except that the color and linetype may change because of floating layers, colors, or linetypes. (T/F)

8. The **PURGE** command has the same effect as the **WBLOCK-asterisk** method. The only difference is that with the **PURGE** command, deletion takes place automatically. (T/F)

9. The **RENAME** command can be used to change the name of a drawing file created by the **WBLOCK** command. (T/F)

10. The **RENAME** command can be used to change the name of a block. (T/F)

11. If the **Auto** option of the **UNDO** command is off, any group of commands used to insert an object is undone together. (T/F)

12. If the selected polyline is closed, the **Close** option is replaced by the **Open** option. (T/F)

13. Circles drawn using the **CIRCLE** command can be changed to polylines. (T/F)

14. When you enter the **Edit vertex** option of the **PEDIT** command, an X marker appears on the screen at the second vertex of the polyline. (T/F)

15. If a tangent direction has been specified for the vertex selected with the **Edit vertex** option, an arrow is generated in that direction. (T/F)

16. With the **Break** option of the **PEDIT** command, if there are any polyline segments or vertices between the vertices you have specified, they will be erased. (T/F)

17. The **Break** operation is valid if both the specified vertices are located at the endpoint of the polyline, or only one vertex is specified and its location is at the endpoint of the polyline. (T/F)

18. The **Move** option lets you move the X-marked vertex to a new position. (T/F)

19. The **Straighten** option of the **PEDIT** command deletes the arcs, line segments, or vertices between the desired two vertices and substitutes them with a single line segment. (T/F)

20. The **Width** option of the **Edit vertex** option of the **PEDIT** command allows you to change the starting and ending widths of an existing polyline segment. (T/F)

21. The spline curve passes through only the first and last vertices and pulls toward the other vertices but does not pass through them, unlike the **Fit** option. (T/F)

22. The fit curve passes only through the first and last vertices of the polyline. (T/F)

23. The **Decurve** option removes the vertices inserted after using the **Fit** or the **Spline** options. (T/F)

24. The **Decurve** option cannot be used to straighten out any curve drawn with the help of the **Arc** option of the **PLINE** command. (T/F)

25. Quadratic curves are extremely smooth. The quadratic curve passes through the first and last control points, and the curve is closer to the other control points. (T/F)

26. The __________ command can be used to undo any previous command or to undo several commands at once.

27. If you enter 6 at the **Enter the number of operations to undo or [Auto/Control/BEgin/End/Mark/Back]:** prompt, __________ previous commands will be undone.

28. The __________ option of the **UNDO** command disables the **UNDO** and **U** commands entirely.

29. To activate the **UNDO** options if they are disabled, you must enter either __________ or __________ options.

30. A group of commands is caused to be treated as a single command for the **U** and **UNDO** commands by the combined work of the __________ and __________ **UNDO** options.

Exercises

Exercise 4 *Graphics*

Draw part (a) in Figure 16-41; then, using the **CHANGE** and relevant **PEDIT** options, convert it into parts (b), (c), and (d). The linetype used in part (d) is HIDDEN.

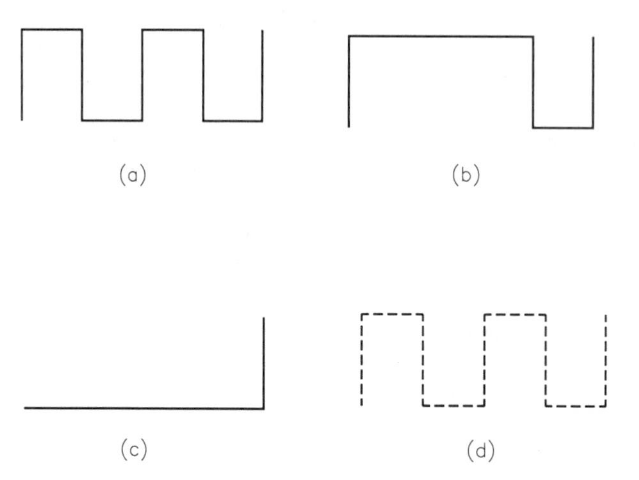

Figure 16-41 *Drawing for Exercise 4*

Exercise 5 *Mechanical*

Draw the object in Figure 16-42 using the **LINE** command. Change the object to a polyline with a width of 0.01.

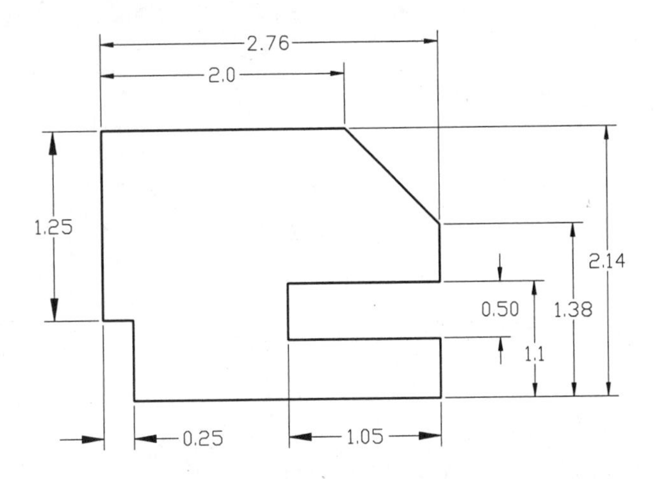

Figure 16-42 Drawing for Exercise 5

Exercise 6 *Graphics*

Draw part (a) in Figure 16-43; then, using the relevant **PEDIT** options, convert it into drawing (b).

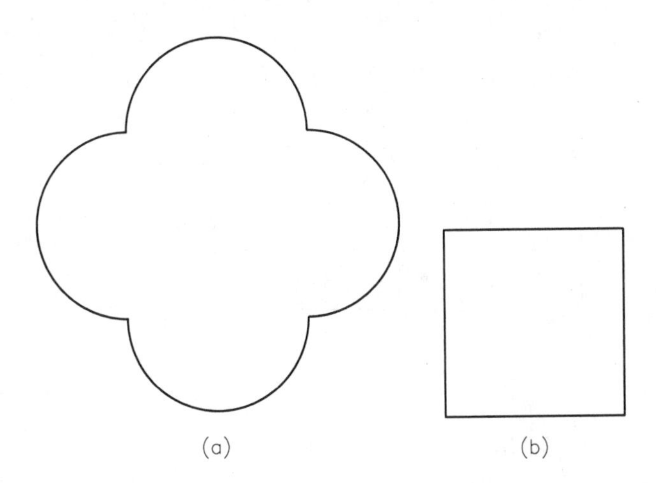

Figure 16-43 Drawing for Exercise 6

Exercise 7 — *Graphics*

Draw part (a) in Figure 16-44; then, using the relevant **PEDIT** options, convert it into drawings (b), (c), and (d). Identify the types of curves.

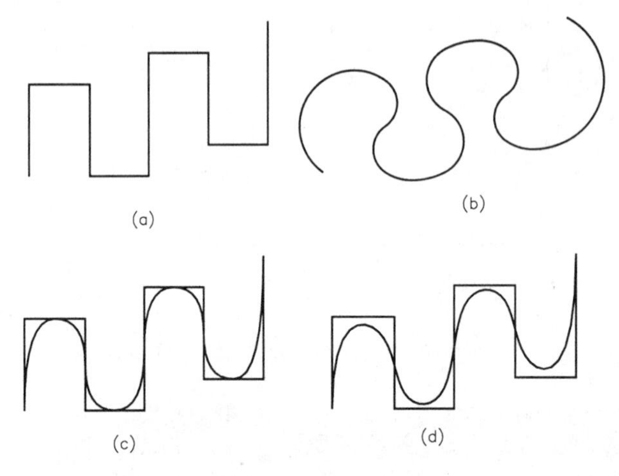

Figure 16-44 *Drawing for Exercise 7*

Exercise 8 — *Graphics*

Draw part (a) in Figure 16-45; then, using the relevant **PEDIT** options, convert it into drawing (b).

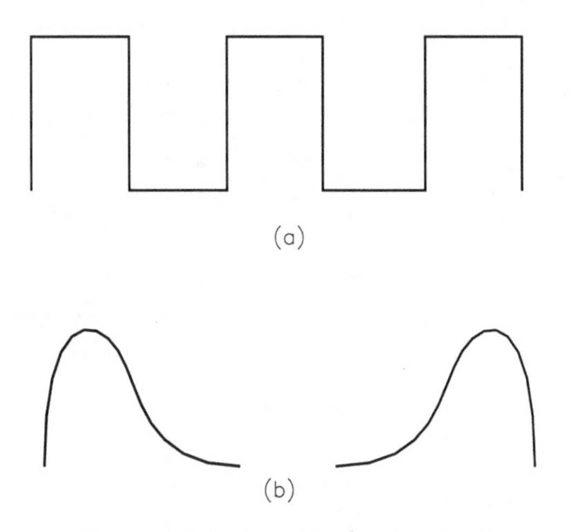

Figure 16-45 *Drawing for Exercise 8*

Exercise 9 *Mechanical*

Draw part (a) in Figure 16-46; then, using the relevent **PEDIT** options, convert it into drawing (b).

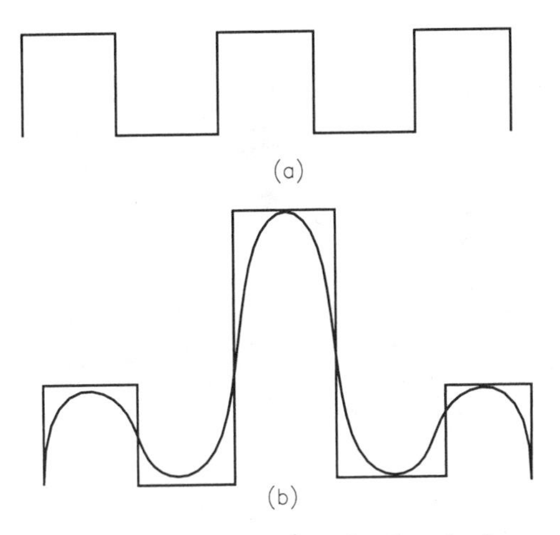

Figure 16-46 *Drawing for Exercise 9*

Answers to Self-Evaluation Test
1 - F, **2** - F, **3** - T, **4** - F, **5** - T, **6** - T, **7** - F, **8** - **SPLINESEGS**, **9** - **SPLINETYPE**, 5, quadratic, 6, cubic, **10** - 255, * , and ?, **11** - T, **12** - T, **13** - T, **14** - F

Chapter 17

Inquiry Commands, Data Exchange, and Object Linking and Embedding

Learning Objectives

After completing this chapter, you will be able to:
- *Make inquiries about drawings and objects.*
- *Calculate area using the **AREA** command.*
- *Calculate distances using the **DIST** command.*
- *Identify a position on the screen using the **ID** command.*
- *List information about drawings and objects using the **LIST** command.*
- *Import and export .dxf files using the **SAVEAS** and **OPEN** commands.*
- *Understand the embedding and the linking functions of the **OLE** feature of Windows.*

MAKING INQUIRIES ABOUT A DRAWING (INQUIRY COMMANDS)

When you create a drawing or examine an existing one, you often need some information about the drawing. In manual drafting, you inquire about the drawing by performing measurements and calculations manually. Similarly, when drawing in an AutoCAD LT environment, you will need to make inquiries about data pertaining to your drawing. The inquiries can be about the distance from one location on the drawing to another, the area of an object like a polygon or circle, coordinates of a location on the drawing, and so on. AutoCAD LT keeps track of all the details pertaining to a drawing. Since inquiry commands are used to obtain information about the selected objects, these commands do not affect the drawings in any way. The following is the list of Inquiry commands:

AREA	DIST	ID	LIST	MASSPROP
TIME	DWGPROPS			

For most of the Inquiry commands, you are prompted to select objects; once the selection is complete, AutoCAD LT switches from graphics mode to text mode, and all the relevant information about the selected objects is displayed. For some commands information is displayed in the AutoCAD LT Text Window. The display of the text screen can be tailored to your requirements with the help of a pointing device; hence, by moving the text screen to one side, you can view the drawing screen and the text screen simultaneously. If you select the **minimize** button or select the close button, you will return to the graphics screen. You can also return to the graphics screen by entering **GRAPHSCR** command at the Command prompt. Similarly, you can return to the AutoCAD LT Text Window by entering **TEXTSCR** at the Command prompt.

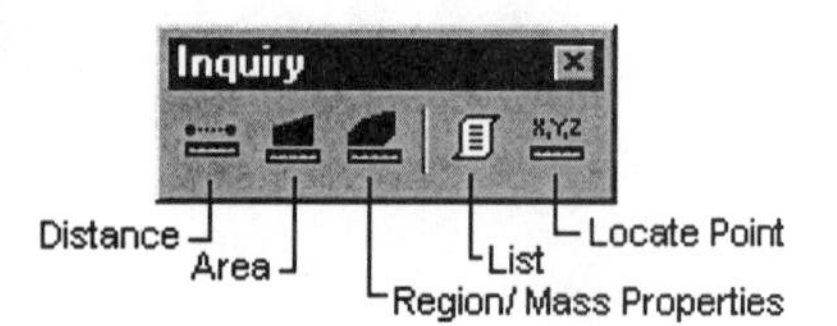

*Figure 17-1 Selecting Inquiry commands from the **Inquiry** toolbar*

AREA COMMAND

Toolbar:	Inquiry > Area
Menu:	Tools > Inquiry > Area
Command:	AREA

Finding the area of a shape or an object manually is time-consuming. In AutoCAD LT, the **AREA** command is used to automatically calculate the area of an object in square units. This command saves time when calculating the area of shapes, especially when the shapes are complicated.

You can use the default option of the **AREA** command to calculate the area and perimeter or circumference of the space enclosed by the sequence of specified points. For example, to find the area of an object (one which is not formed of a single object) you have created with the help of the **LINE** command (Figure 17-2), you need to select all the vertices of that object. By selecting the points, you define the shape of the object whose area is to be found. This is the default method for determining the area of an object. The only restriction is that all the points you specify be in a plane parallel to the XY plane of the current UCS. You can make the best possible use of object snaps such as ENDpoint, INTersect, and TANgent, or even use running Osnaps, to help you select the vertices quickly and accurately. For AutoCAD LT to find the area of a shape, the shapes need not have been drawn with polylines; nor do the lines need to be closed. However, curves must be approximated with

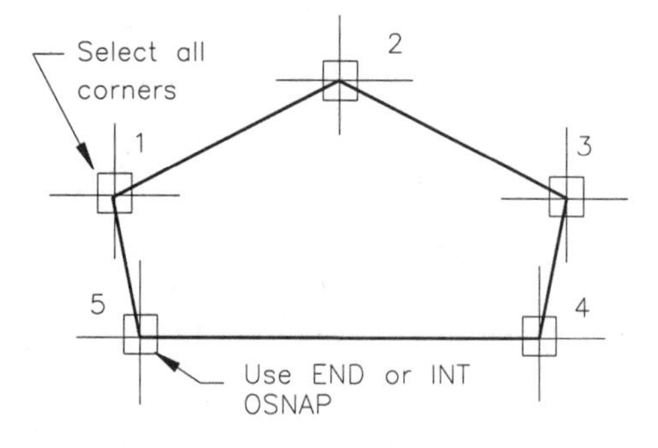

*Figure 17-2 Using the **AREA** command*

short straight segments. In such cases, AutoCAD LT computes the area by assuming that the first point and the last point are joined. The prompt sequence in this case is:

Command: **AREA** Enter
Specify first corner point or [Object/Add/Subtract]: *Specify first point.*
Specify next corner point or press ENTER for total: *Specify the second point.*
Specify next corner point or press ENTER for total: *Continue selecting until all the points enclosing the area have been selected.*
Specify next corner point or press ENTER for total: Enter
Area = X, Perimeter = Y

Here, X represents the numerical value of the area and Y represents the circumference/perimeter. It is not possible to accurately determine the area of a curved object, such as an arc, with the default (Point) option. However, the approximate area under an arc can be calculated by specifying several points on the given arc. If the object whose area you want to find is not closed (formed of independent segments) and has curved lines, you should use the following steps to determine the accurate area of such an object:

1. Convert all the segments in that object into polylines using the **PEDIT** command.
2. Join all the individual polylines into a single polyline. Once you have performed these operations, the object becomes closed and you can then use the **Object** option of the **AREA** command to determine the area.

If you specify two points on the screen, the **AREA** command will display the value of the area as 0.00; the perimeter value is the distance between the two points.

Object option. You can use the **Object** option of the **AREA** command to find the area of objects such as polygons, circles, polylines, regions, solids, and splines. If the selected object is a polyline or polygon, AutoCAD LT displays the area and perimeter of the polyline. In case of open polylines, the area is calculated assuming that the last point is joined to the first point but the length of this segment is not added to the polyline length unlike the default option. If the selected object is a circle, ellipse, or planar closed spline curve, AutoCAD LT will provide information about its area and circumference. For a 3D polyline, all vertices must lie in a plane parallel to the XY plane of the current UCS. The extrusion direction of a 2D polyline whose area you want to determine should be parallel to the Z axis of the current UCS. In case of polylines which have a width, the area and length of the polyline is calculated using the centerline. If any of these conditions is violated, an error message is displayed on the screen. The prompt sequence is:

Command: **AREA** Enter
Specify first corner point or [Object/Add/Subtract]: **O** Enter
Select objects : *Select an object* Enter
Area = (X), Circumference = (Y)

The X represents the numerical value of the area, and Y represents the circumference/perimeter.

Note

*In many cases, the easiest and most accurate way to find the area of an area enclosed by multiple objects is to use the **BOUNDARY** command to create a polyline, then use the **AREA Object** option.*

Add option. Sometimes you want to add areas of different objects to determine a total area. For example, in the plan of a house, you need to add the areas of all rooms to get the total floor area. In such cases, you can use the **Add** option. Once you invoke this option, AutoCAD LT activates the **Add** mode. By using the **First corner point** option at the **Specify first corner point or [Object/Subtract]:** prompt, you can calculate the area and perimeter by selecting points on the screen. Pressing ENTER after you have selected points defining the area which is to be added, calculates the total area, since the **Add** mode is on. The command prompt is:

Specify next corner point or press ENTER for total (ADD mode):

If the polygon whose area is to be added, is not closed, the area and perimeter are calculated assuming a line is added which connects the first point to the last point to close the polygon. The length of this area is added in the perimeter. The **Object** option adds the areas and perimeters of selected objects. While using this option, if you select an open polyline, the area is calculated considering the last point is joined to the first point but the perimeter does not consider the length of this assumed segment, unlike the **First corner point** option. When you select an object, the area of the selected object is displayed on the screen. At this time the total area is equal to the area of the selected object. When you select another object, AutoCAD LT displays the area of the selected object as well as the combined area (total area) of the previous object and the currently selected object. In this manner you can add areas of different objects. Until the **Add** mode is active, the string ADD mode is displayed along with all subsequent object selection prompts to remind you that the **Add** mode is active. When the **AREA** command is invoked, the total area is initialized to zero.

Subtract option. The action of the **Subtract** option is the reverse of that of the **Add** option. Once you invoke this option, AutoCAD LT activates the **Subtract** mode. The **First corner point** and **Object** options work similar to the way they work in the ADD mode. When you select an object, the area of the selected object is displayed on the screen. At this time, the total area is equal to the area of the selected object. When you select another object, AutoCAD LT displays the area of the selected object as well as the area obtained by subtracting the area of the currently selected object from the area of the previous object. In this manner, you can subtract areas of objects from the total area. Until the **Subtract** mode is active, the string SUBTRACT mode is displayed along with all subsequent object selection prompts, to remind you that the SUBTRACT mode is active. To exit the **AREA** command, press ENTER (null response) at the **Specify first corner point or [Object/Add/Subtract]:** prompt. The prompt sequence for these two modes for Figure 17-3 is:

Command: **AREA** [Enter]
Specify first corner point or [Object/Add/Subtract]: **A** [Enter]
Specify first corner point or [Object/Subtract]: **O** [Enter]
(ADD mode) Select objects: *Select the polyline.*
Area = 2.4438, Perimeter = 6.4999

Total area = 2.4438
(ADD mode) Select objects: [Enter]
Specify first corner point or [Object/Subtract]: **S** [Enter]
Specify first corner point or [Object/Add]: **O** [Enter]
(SUBTRACT mode) Select object: *Select the circle.*
Area = 0.0495, Circumference = 0.7890
Total area = 2.3943
(SUBTRACT mode) Select objects: *Select the second circle.*
Area = 0.0495, Circumference = 0.7890
Total area = 2.3448
(SUBTRACT mode) Select object: [Enter]
Specify first corner point or [Object/Add]: [Enter]

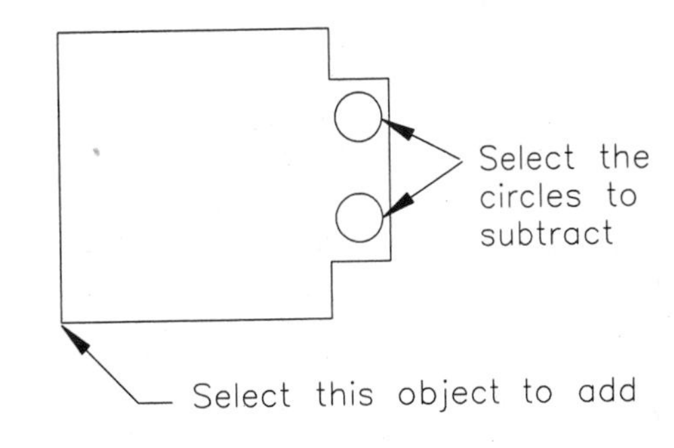

Figure 17-3 Using the ***Add*** *and* ***Subtract*** *options*

The **AREA** and **PERIMETER** system variables hold the area and perimeter (or circumference in the case of circles) of the previously selected polyline (or circle). Whenever you use the **AREA** command, the **AREA** variable is reset to zero.

CALCULATING DISTANCE BETWEEN TWO POINTS (DIST COMMAND)

Toolbar:	Inquiry > Distance
Menu:	Tools > Inquiry > Distance
Command:	DIST

The **DIST** command is used to measure the distance between two selected points (Figure 17-4). The angles that the selected points make with the X axis and the XY plane are also displayed. The measurements are displayed in current units. Delta X (horizontal displacement), delta Y (vertical displacement), and delta Z are also displayed. The distance computed by the **DIST** command is saved in the **DISTANCE** variable. The prompt sequence is:

Command: **DIST** [Enter]
Specify first point: *Specify a point.*
Specify second point: *Specify a point.*

AutoCAD LT returns the following information:

Distance = *Calculated distance between the two points.*
Angle in XY plane = *Angle between the two points in the XY plane.*
Angle from XY plane = *Angle the specified points make with the XY plane.*
Delta X = *Change in X*, Delta Y = *Change in Y*, Delta Z = *Change in Z.*

If you enter a single number or fraction at the **Specify first point:** prompt, AutoCAD LT will display that number in the current unit of measurement.

Command: **DIST**
First point: 3-3/4 *(Enter a number or a fraction.)*
Distance = 3.7500

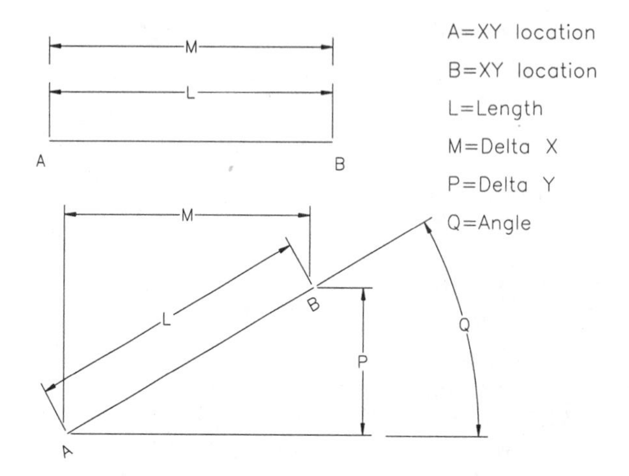

Figure 17-4 *Using the **DIST** command*

Note

The Z coordinate is used in 3D distances. If you do not specify the Z coordinates of the two points between which you want to know the distance, AutoCAD LT takes the current elevation as the Z coordinate value.

IDENTIFYING A POSITION ON THE SCREEN (ID COMMAND)

Toolbar:	Inquiry > Locate Point
Menu:	Tools > Inquiry > ID Point
Command:	ID

The **ID** command identifies the position of a point you specify and tells you its coordinates.

Command: **ID** [Enter]
Specify point: *Specify the point to be identified.*
X = X coordinate Y = Y coordinate Z = Z coordinate

AutoCAD LT takes the current elevation as the Z coordinate value. If an **Osnap** mode is used to snap to a 3D object in response to the **Specify point:** prompt, the Z coordinate displayed will be that of the selected feature of the 3D object. You can also use the **ID** command to identify the location on the screen. This can be realized by entering the coordinate values you want to locate on the screen. AutoCAD LT identifies the point by drawing a blip mark at that location (**BLIPMODE** should be On). For example, say you want to find the position on the screen where X = 2.345, Y = 3.674, and Z = 1.0000 is located. The prompt sequence is:

Command: **ID** Enter
Specify point: **2.345,3.674,1.00** Enter
X = 2.345 Y = 3.674 Z = 1.0000

The coordinates of the point specified in the **ID** command are saved in the **LASTPOINT** system variable. You can locate a point with respect to the **ID** point by using the relative or polar coordinate system. You can also snap to this point by typing @.

Command: **ID**
Specify point: *Select a point.*
Command: **LINE**
Specify first point: @ *(The line will snap to the ID point.)*

You can also use the **X/Y/Z point filters** to specify points on the screen. You can respond to the **Specify point:** prompt with any desired combination of .X, .Y, .Z, .XY, .XZ, .YZ. The 2D or 3D point is specified by specifying individual (intermediate) points and forming the desired point from the selected X, Y, and Z coordinates of the intermediate points. More simply, you can specify a 2D or 3D point by supplying separate information for the X, Y, and Z coordinates. For example, say you want to identify a point whose X, Y, and Z coordinates are marked on the screen separately. This can be achieved in the following manner:

Command: **ID** Enter
Specify point: **.X** Enter
of *Specify the location whose X coordinate is the X coordinate of the final desired point to be identified.*
of (need YZ): *Specify the location whose YZ coordinate is the YZ coordinate of the final desired point to be identified.*
X = X coordinate Y = Y coordinate Z = Z coordinate

A blip mark is formed at the point of intersection of the specified X, Y, and Z coordinates. This blip mark identifies the desired point.

LISTING INFORMATION (LIST COMMAND)

Toolbar:	Inquiry > List
Menu:	Tools > Inquiry > List
Command:	LIST

The **LIST** command displays all the data pertaining to the selected objects. The prompt sequence is:

Command: **LIST** Enter
Select objects: *Select objects whose data you want to list.*
Select objects: Enter

Chapter 17

Once you select the objects to be listed, AutoCAD LT shifts you from the graphics screen to the AutoCAD LT Text Window. The information displayed (listed) varies from object to object. Information on an object's type, its coordinate position with respect to the current UCS (user coordinate system), the name of the layer on which it is drawn, and whether the object is in model space or paper space is listed for all types of objects. If the color, lineweight, and the linetype are not BYLAYER, they are also listed. Also, if the thickness of the object is greater than 0, that is also displayed. The elevation value is displayed in the form of a Z coordinate. If an object has an extrusion direction different from the Z axis of the current UCS, the object's extrusion direction is also provided.

More information based on the objects in the drawing is also provided. For example, for a line the following information is be displayed:

1. The coordinates of the endpoints of the line.
2. Its length (in 3D).
3. The angle made by the line with respect to the X axis of the current UCS.
4. The angle made by the line with respect to the XY plane of the current UCS.
5. Delta X, delta Y, delta Z: this is the change in each of the three coordinates from the start point to the endpoint.
6. The name of the layer in which the line was created.
7. Whether the line is drawn in Paper space or Model space.

The center point, radius, true area, and circumference of circles is displayed. For polylines, this command displays the coordinates. In addition, for a closed polyline, its true area and perimeter are also given. If the polyline is open, AutoCAD LT lists its length and also calculates the area by assuming a segment connecting the start point and endpoint of the polyline. In the case of wide polylines, all computation is done based on the centerlines of the wide segments. For a selected viewport, the **LIST** command displays whether the viewport is on and active, on and inactive, or off. Information is also displayed about the status of Hideplot and the scale relative to paper space. If you use the **LIST** command on a polygon mesh, the size of the mesh (in terms of M, X, N), the coordinate values of all the vertices in the mesh, and whether the mesh is closed or open in M and N directions are all displayed. As mentioned before, if all the information does not fit on a single screen, AutoCAD LT pauses to allow you to press ENTER to continue the listing.

Exercise 1 *Mechanical*

Draw Figure 17-5 and save it as EX1. Using INQUIRY commands, determine the following values:

a. Area of the hexagon.
b. Perimeter of the hexagon.
c. Perimeter of the inner rectangle.
d. Circumference of each circle.
e. Area of the hexagon minus the inner rectangle.
f. Use the **LIST** command to get the database listing.

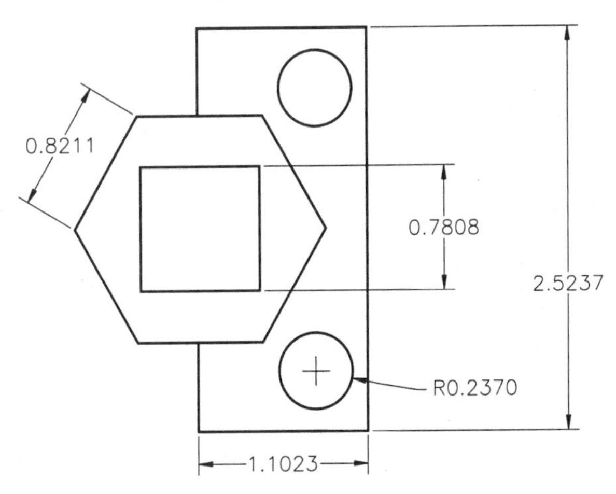

Figure 17-5 *Drawing for Exercise 1*

LISTING INFORMATION ABOUT REGIONS (MASSPROP COMMAND)

Toolbar:	Inquiry > Mass properties
Menu:	Tools > Inquiry > Mass Properties
Command:	MASSPROP

With the **MASSPROP** command, you can determine volumetric information, such as principle axes, center of gravity, and moment of inertia of 2D and 3D objects.

Command: **MASSPROP** Enter
Select objects: *Select an object (region).*

For detailed explanation of the **MASSPROP** command, refer to Chapter 21, Drawing and Viewing 3D Objects.

CHECKING TIME-RELATED INFORMATION (TIME COMMAND)

Menu:	Tools > Inquiry > Time
Command:	TIME

The time and date maintained by your system are used by AutoCAD LT to provide information about several time factors related to the drawings. Hence, you should be careful about setting the current date and time in your computer. The **TIME** command can be used to display information pertaining to time related to a drawing and the drawing session. The display obtained by invoking the **TIME** command is similar to the following:

Command: **TIME**

Current time:	Monday, April 26 1999 at 11:22:35.069 AM
Times for this drawing:	
Created:	Thursday, April 15 1999 at 09:34:42.157 AM
Last updated:	Friday, April 16 1999 at 02:54:25.700 PM
Total editing time:	0 days 07:52:16.205
Elapsed timer (on)	0 days 00:43:07.304
Next automatic save in:	0 days 00:54:27:153

Enter an option [Display/ON/OFF/Reset]:

The foregoing display gives you information on the following.

Current Time

Provides today's date and the current time.

Drawing Creation Time

Provides the date and time that the current drawing was created. The creation time for a drawing is set to the system time when the **NEW**, **WBLOCK**, or **SAVE** command is used to create that drawing file.

Last Updated Time

Provides the most recent date and time you saved the current drawing. In the beginning, it is set to the drawing creation time, and it is modified every time you use the **QUIT** or **SAVE** command to save the drawing.

Total Editing Time

This tells you the total time spent on editing the current drawing since it was created. If you terminate the editing session without saving the drawing, the time you have spent on that editing session is not added to the total time spent on editing the drawing. Also, the last update time is not revised.

Elapsed Timer

This timer operates while you are in AutoCAD LT. You can stop this timer by entering OFF at the **Enter an option [Display/ON/OFF/Reset]:** prompt. To activate the timer, enter ON. If you want to know how much time you have spent on the current drawing or part of the drawing in the current editing session, use the **Reset** option as soon as you start working on the drawing or part of the drawing. This resets the user-elapsed timer to zero. By default, this timer is ON. If you turn this timer OFF, the time accumulated in this timer up to the time you turned it OFF will be displayed.

Next Automatic Save In Time

This tells you when the next automatic save will be performed. The automatic save time interval can be set in the **Options** dialog box (**Open and Save** tab) or with the **SAVETIME** system variable. If the time interval has been set to zero, the **TIME** command displays the following message:

Next automatic save in: <disabled>

If the time interval is not set to zero, and no editing has taken place since the previous save, the **TIME** command displays the following message:

Next automatic time save in: <no modification yet>

If the time interval is not set to zero, and editing has taken place since the previous save, the **TIME** command displays the following message:

Next automatic time save in: 0 days hh:mm:ss.msec

hh stands for hours
mm stands for minutes
ss stands for seconds
msec stands for milliseconds

At the end of the display of the **TIME** command, AutoCAD LT prompts:

Enter an option [Display/ON/OFF/Reset]:

The information displayed by the **TIME** command is static; that is, the information is not updated dynamically on the screen. If you respond to the last prompt with DISPLAY (or D), the display obtained by invoking the **TIME** command is repeated. This display contains updated time values.

With the ON response, the user-elapsed timer is started, if it was off. As mentioned earlier, when you enter the drawing editor, by default the timer is on. The OFF response is just the opposite of the ON response and stops the user-elapsed time, if it is on. With the **Reset** option, you can set the user-elapsed time to zero.

Displaying Drawing Properties (DWGPROPS Command*)

Menu:	File > Drawing Properties
Command:	DWGPROPS

On choosing **Drawing Properties** from the **File** Menu, the **Drawing Properties** dialog box is displayed (Figure 17-6). This dialog box has four tabs under which information about the drawing is displayed. The information displayed in this dialog box helps you look for the drawing more easily. The tabs are as follows:

General. This tab displays general properties about the drawing like the **Type**, **Size,** and **Location**.

Summary. The **Summary** tab displays predefined properties like the author, title, and subject.

Custom. This tab displays custom file properties including values assigned by you.

Statistics. This tab stores and displays data such as the file size and data such as the dates when the drawing was last saved on or modified on.

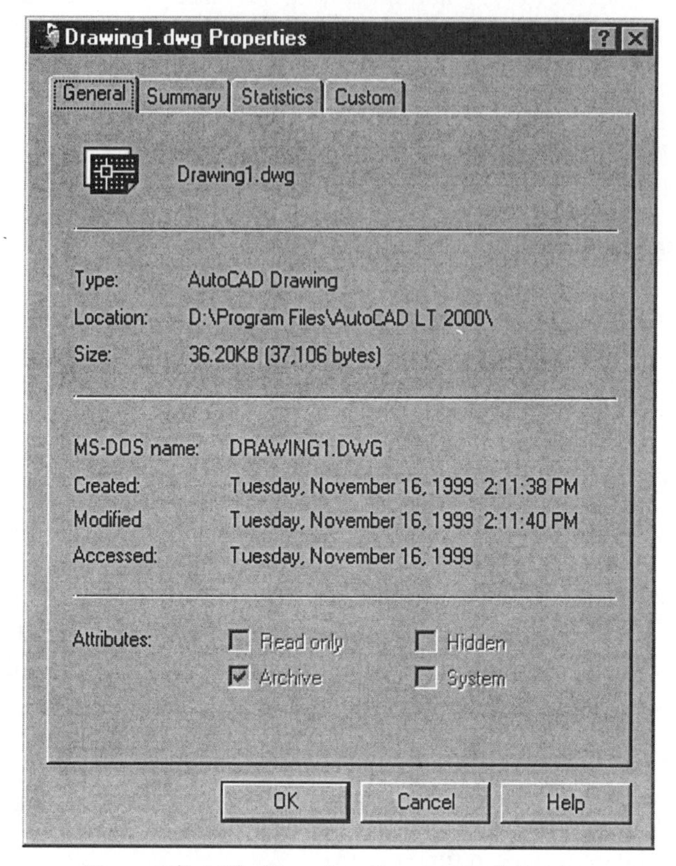

*Figure 17-6 The **Drawing Properties** dialog box*

WHOHAS COMMAND*

This command displays the **Select File to Query** dialog box for you to select the opened drawing whose ownership information you want to view. The command displays information about the user who has the particular drawing opened at that time.

DATA EXCHANGE IN AUTOCAD LT

Different companies have developed different software for applications such as CAD, desktop publishing, and rendering. This nonstandardization of software has led to the development of various data exchange formats that enable transfer (translation) of data from one data processing software to another. We will now discuss various data exchange formats provided in AutoCAD LT. AutoCAD LT uses the .DWG format to store drawing files. This format is not recognized by most other CAD software, such as Intergraph, CADKEY, and MicroStation. To solve this problem so that files created in AutoCAD LT can be transferred to other CAD software for further use, AutoCAD LT provides various data exchange formats, such as DXF (data interchange file) and DXB (binary drawing interchange).

DXF FILE FORMAT (DATA INTERCHANGE FILE)

The DXF file format generates a text file in ASCII code from the original drawing. This allows any computer system to manipulate (read/write) data in a DXF file. Usually, DXF format is used for CAD packages based on microcomputers. For example, packages like SmartCAM use DXF files. Some desktop publishing packages, such as Pagemaker and Ventura Publisher, also use DXF files.

Creating a Data Interchange File

The **SAVE** or the **SAVEAS** command is used to create an ASCII file with a .DXF extension from an AutoCAD LT drawing file. Once you invoke any of these commands, the **Save Drawing As** dialog box (Figure 17-7) is displayed. By default, the DXF file to be created assumes the name of the drawing file from which it will be created. However, you can specify a file name of your choice for the DXF file by typing the desired file name in the **File Name:** edit box. Select the extension as **DXF [*dxf]**. This can be observed in the **Save as type:** drop-down list where you can select the output file format. You can also use the **-WBLOCK** command. The **Create Drawing** dialog box is displayed, where you can enter the name of the file in the **File Name:** edit box and select **DXF [*dxf]** from the **Save as type:** drop-down list.

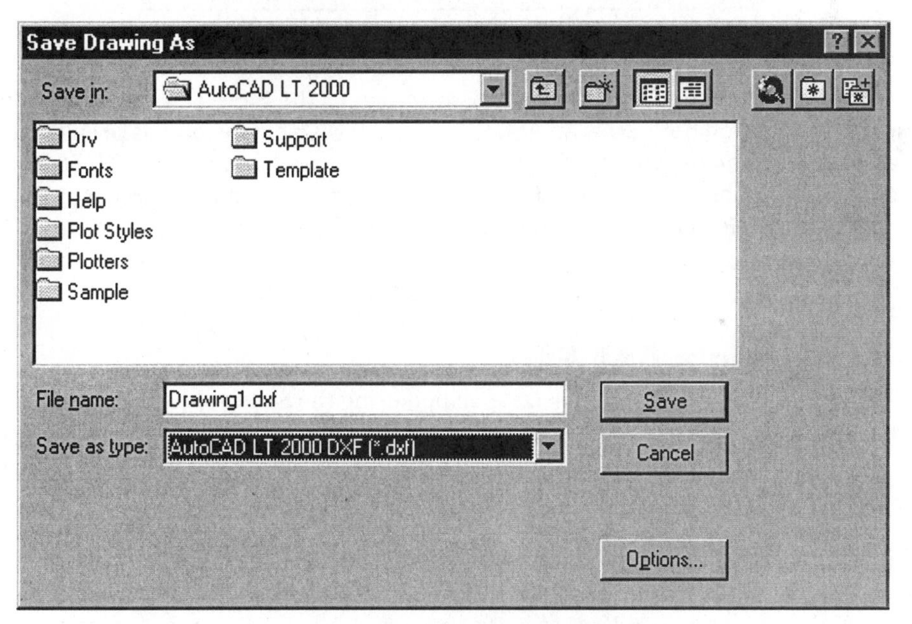

Figure 17-7 Save Drawing As dialog box

Choose the **Options** button to display the **Saveas Options** dialog box (Figure 17-8). In this dialog box, choose the **DXF options** tab and enter the degree of accuracy for the numeric values. The default value for the degree of accuracy is six decimal places, however, this results in less accuracy than the original drawing, which is accurate to 16 places. You can enter a value between 0 and 16 decimal places.

In this dialog box, you can also select the **Select Objects** check box, which allows you to specify objects you want to include in the DXF file. In this case, the definitions of named objects such as block definitions, text styles, and so on, are not exported. Selecting the **Save thumbnail preview image** check box, saves a preview image with the file that can be previewed in the **Preview** window of the **Select File** dialog box.

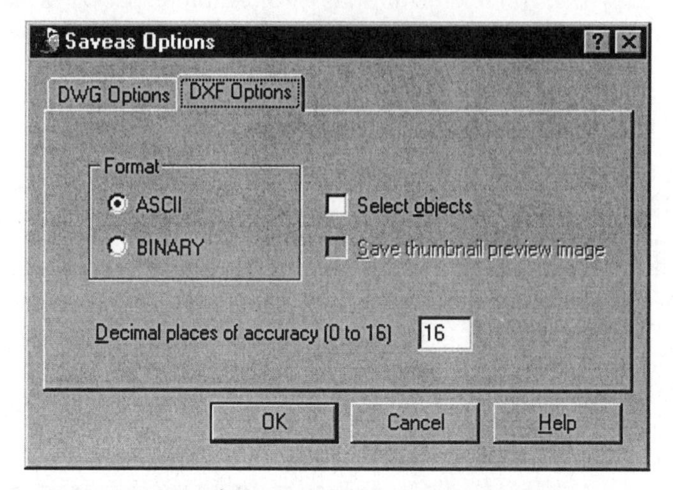

*Figure 17-8 **Saveas Options** dialog box*

Select the **ASCII** radio button. Choose **OK** to return to the **Save Drawing As** dialog box. Choose the **Save** button here. Now an ASCII file with a .DXF extension has been created, and this file can be accessed by other CAD systems. This file contains data on the objects specified. By default, DXF files are created in ASCII format. However, you can also create binary format files by selecting the **Binary** radio button in the **Saveas Options** dialog box. Binary DXF files are more efficient and occupy only 75 percent of the ASCII DXF file. You can access a file in binary format more quickly than the same file in ASCII format.

Information in a DXF File

The DXF file contains data on the objects specified using the **Select objects** option in the **Save Drawing As** dialog box. You can change the data in this file to your requirement. To examine the data in this file, load the ASCII file in word processing software. A DXF file is composed of the following parts.

Header. In this part of the drawing database, all the variables in the drawing and their values are displayed.

Classes. This section deals with the internal database.

Tables. All the named objects, such as linetypes, layers, blocks, text styles, dimension styles, and views, are listed in this part.

Blocks. The objects that define blocks and their respective values are displayed in this part.

Entities. All the entities in the drawing, like Circle, and so forth, are listed in this part.

Objects. Objects in the drawing are listed in this part.

Converting DXF Files into a Drawing File (OPEN Command)

You can import a DXF file into a new AutoCAD LT drawing file with the **OPEN** command. From the **File** menu choose **Open**.

Command: **OPEN**

After you invoke the **OPEN** command, the **Select File** dialog box (Figure 17-9) is displayed. From the **Files of Type:** drop-down list, select **DXF [*dxf]**. In the **File Name:** edit box enter the name of the file you want to import into AutoCAD LT or select the file from the list. Choose the **Open** button. Once this is done, the specified DXF file is converted into a standard DWG file, regeneration is carried out, and the file is inserted into the new drawing. Now you can perform different operations on this file just as with other drawing files. You can also use the **INSERT** command to insert a DXF file into the current drawing.

Chapter 17

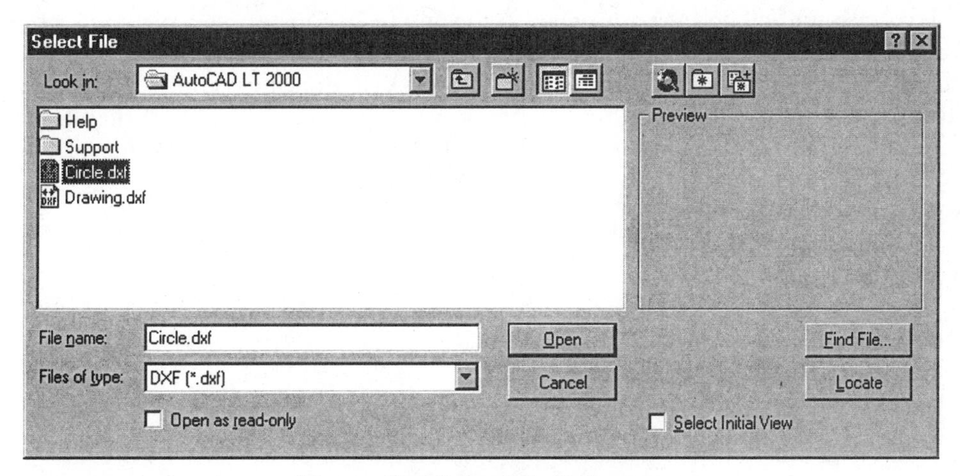

*Figure 17-9 **Select File** dialog box*

Creating and Using a Windows WMF File

Menu:	File > Export
Command:	EXPORT, WMFOUT

The Windows Metafile File format file contains screen vector and raster graphics format. In the **Export Data** dialog box or the **Create WMF file** dialog box, enter the file name. Select the objects you want to save in this file format. Select **Metafile [*wmf]** from the **Save as type:** drop-down list. The extension .wmf is appended to the file name.

The **WMFIN** command displays the **Import WMF** dialog box. Window metafiles are imported as blocks in AutoCAD LT. Select the wmf file you want to import and choose the **Open** button. Specify an insertion point, rotation angle and scale factor. Specify scaling by entering a **scale factor**, using the **corner** option to specify an imaginary box whose dimensions correspond to the scale factor or entering **xyz** to specify 3D scale factors. You can also invoke the **WMFIN** command by choosing **Windows Metafile** from the **Insert** menu.

Creating a BMP File

Command: BMPOUT

This is used to create bitmap images of the objects in your drawing. Entering **BMPOUT** displays the **Create BMP File** dialog box. Enter the file name and choose **Save**. Select the objects to save as bitmap. The file extension .bmp is appended to the file name.

DATA INTERCHANGE THROUGH RASTER FILES

Until now we have discussed importing and exporting files in the DXF file format. To uphold the accuracy of the drawing, the DXF file includes almost all the information about the original drawing file. The accuracy is maintained at the expense of DXF file size and degree of complexity of these files. There are many applications in which accuracy is not very important, like desktop publishing. In such applications, you are concerned primarily with image presentation. A very simple and effective way of storing an image for import/export is in the form of **raster files**. In a raster file, information is stored in the form of a dot pattern on the screen. This bit pattern is also known as a **bit map**. For example, in a raster file a picture is stored in the form of information about the position and color of the screen pixels. AutoCAD LT allows you to add raster images to the vector base AutoCAD LT drawings, and view and plot the resulting file. We will be discussing three types of raster files: TIFF files, TGA files, and BMP files. These formats make it possible to transfer a file from AutoCAD LT to other software.

TIFF (tagged image file format)

TGA (targa format)

BMP (bitmap format)

RASTER IMAGES

A raster image consists of small square-shaped dots known as pixels. In a colored image, the color is determined by the color of pixels. The raster images can be moved, copied, or clipped,and used as a cutting edge with the **TRIM** command. They can also be modified by using grips. You can also control the image contrast, transparency, and quality of the image. AutoCAD LT stores images in a special temporary image swap file whose default location is the Windows Temp directory. You can change the location of this file modifying it under **Temporary File Location** in the **Files** tab of the **Options** dialog box.

The images can be 8-bit gray, 8-bit color, 24-bit color, or bitonal. When image transparency is set to On, the image file formats with transparent pixels is recognized by AutoCAD LT and transparency is allowed. The transparent images can be in color or gray scale. AutoCAD LT supports the following file formats.

Image Type	File Extension	Description
BMP	.bmp, .dib, .rle	Windows and OS/2 Bitmap Format
CALS-I	.gp4, .mil, .rst	Mil-R-Raster I
FLIC	.flc, .fli	Flic Autodesk Animator Animation
GEOSPOT	.bil	GeoSPOT (BIL files must be accompanied

		with HDR and PAL files with connection data in the same directory.)
IG4	.ig4	Image Systems group 4
IGS	.igs	Image Systems Grayscale
JFIF or JPEG	.jpg, jpeg	Joint Photographics Expert group
PCX	.pcx	Picture PC Paintbrush Picture
PICT	.pct	Picture Macintosh Picture
PNG	.png	Portable Network Graphic
RLC	.rlc	Run-length Compressed
TARGA	.tga	True Vision Raster based Data format
TIFF/LZW	.tif	Taffed image file format

When you store images as Tiled Images, that is, in the Tagged Image File Format [TIFF], you can edit or modify any portion of the image; only the modified portion is regenerated thus saving time. Tiled images load much faster compared to nontiled images.

Managing Raster Images (Image Manager Dialog Box*)

Toolbar:	Reference > Image
Command:	IMAGE

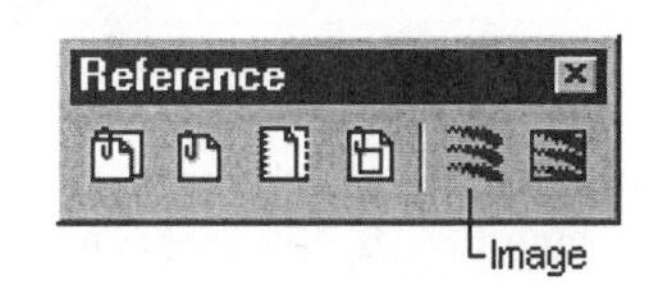

***Figure 17-10** Invoking the **IMAGE** command from the **Reference** toolbar*

When you invoke the **IMAGE** command, AutoCAD LT displays the **Image Manager** dialog box (Figure 17-11). You can also invoke the **Image Manager** dialog box by selecting an image and right-clicking to display a shortcut menu. Choose **Image > Image Manager**. You can view image information either as a list view or as a tree view by choosing the respective buttons located in the upper-left corner of the **Image Manager** dialog box. The **List View** displays the names of all the images in the drawing, its loading status, size, date last modified on, and its search path. The **Tree View** displays the images in a hierarchy which shows its nesting levels within blocks and Xrefs. The **Tree View** does not display the Status, size, or any other information about the image file. The F3 key displays the List View and the F4 key displays the Tree View. You can rename an image file in this dialog box. To insert an image in an AutoCAD LT file, you must use AutoCAD 2000 for it. In AutoCAD 2000, you can attach and detach the image files into AutoCAD LT. The different options in the **Image Manager** dialog box are:

Reload. Reloads an image. The changes made to the image since the last insert will be loaded on the screen. You can change the status of the image file in the **Image Manager** dialog box by double-clicking on the current status. It changes from **Unload** to **Reload** and vice versa.

Unload. Unloads an image. An image is unloaded when it is not needed in the current drawing session. When you unload an image, AutoCAD LT retains information about the location and size of the image and the image boundary is displayed. If you reload the image, the image will appear at the same point and in the same size as the image was before unloading. Unloading the raster images enhances AutoCAD LT performance. Also, the unloaded images are not plotted, but unloading does not unlink the file from the drawing. If multiple images are to be loaded and the memory is insufficient, AutoCAD LT automatically unloads them.

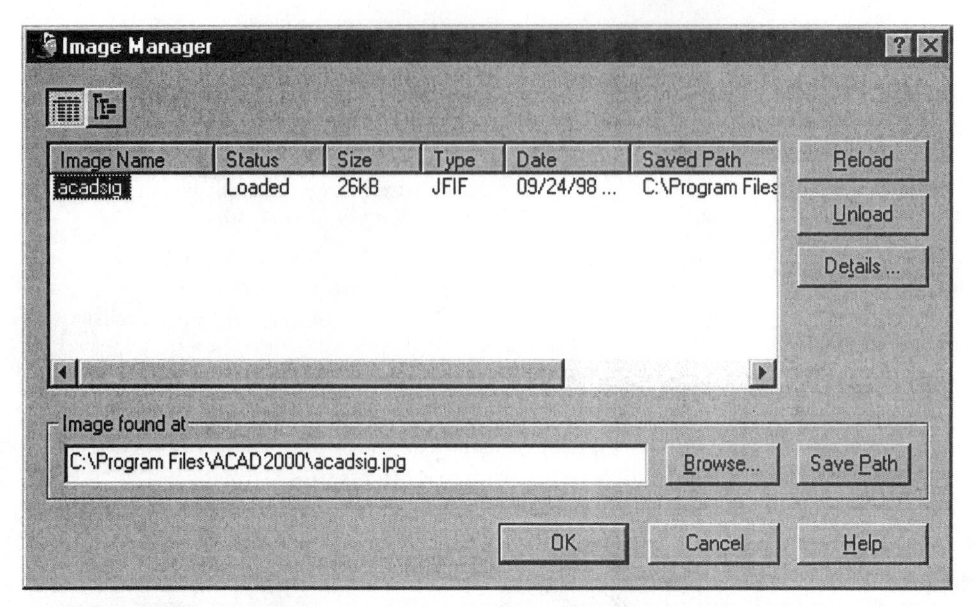

Figure 17-11 ***Image Manager*** *dialog box*

Details. Displays the **Image File Details** box, which lists information about the image like file name, saved path, file creation date, file size, file type, color and color path, pixel width and height, resolution, and default size. It also displays the image in the preview box.

Image found at. This edit box displays the path of the selected image. You can edit this path and choose the **Save path** button to save the new path. If you have changed the path of an image, choose the **Browse** button to display the **Select Image File** dialog box. In this dialog box, locate the image file and then choose the **Open** button. The new path is displayed in the **Image found at** edit box. Choose the **Save Path** button to save this new path. This path is also displayed in the **Saved path** column of the dialog box.

Save Path. Saves the current path of the image.

Browse. Displays the **Select Image File** dialog box.

Using the -IMAGE Command

The options in the **Image Manager** dialog box can also be used from the command line when you use the **-IMAGE** command. The Command prompts are as follows:

Command: **-IMAGE**
Enter image option [?/Path/Reload/Unload] <Unload>: *Enter an option.*

Entering ? at the above prompt gives the next prompt as follows:

Images to list<*>: *Press ENTER.*

The AutoCAD LT Text window lists information about the image.

EDITING RASTER IMAGE FILES

Image Frame

Toolbar:	Reference > Image Frame
Menu:	Modify > Object > Image Frame
Command:	IMAGEFRAME

The **IMAGEFRAME** command is used to turn the image boundary on or off. If the image boundary is off, the image cannot be selected with the pointing device and hence, cannot be accidently moved or modified. The command prompt is as follows:

Command: **IMAGEFRAME**
Enter image frame setting [ON/OFF] <OFF>:

Other Editing Commands

You can use other editing commands like **COPY**, **MOVE**, and **STRETCH** to edit the raster image. You can also use the image as the trimming edge for trimming objects. However, you cannot trim an image. You can insert the raster image several times or make multiple copies of it. Each copy could have a different clipping boundary. You can also edit the image using grips. You can also use the **Object Properties** window to change the image layer, boundary linetype, and color.

Scaling Raster Images

The scale of the inserted image is determined by the actual size of the image and the unit of measurement (inches, feet, and so on). For example, if the image is 1" x 1.26" and you insert this image with a scale factor of 1, the size of the image on the screen will be 1 AutoCAD LT unit by 1.26 AutoCAD LT units. If the scale factor is 5, the image will be five times larger. The image that you want to insert must contain the resolution information (DPI). If the image does not contain this information, AutoCAD LT treats the width of the image as one unit.

POSTSCRIPT FILES

PostScript is a page description language developed by Adobe Systems. It is used mostly in DTP (desktop publishing) applications. AutoCAD LT allows you to work with PostScript files. You can create and export PostScript files as well as convert PostScript files into regular AutoCAD LT drawing files (import). PostScript images have higher resolution than raster images. The extension for these files is .EPS (Encapsulated PostScript).

PSOUT Command

Command:	PSOUT

As just mentioned, any AutoCAD LT drawing file can be converted into a PostScript file. This can be accomplished with the **PSOUT** command. Once the **PSOUT** command is invoked, AutoCAD LT displays the **Create PostScript File** dialog box (Figure 17-12).

In the **File Name:** edit box, enter the name of the PostScript (EPS) file you want to create. Then you can choose the **Save** button to accept the default setting and create the PostScript file. You can also choose the **Options** button to change the settings through the **PostScript Out Options** dialog box (Figure 17-13) and then save the file. The **PostScript Out Options** dialog box has the following options:

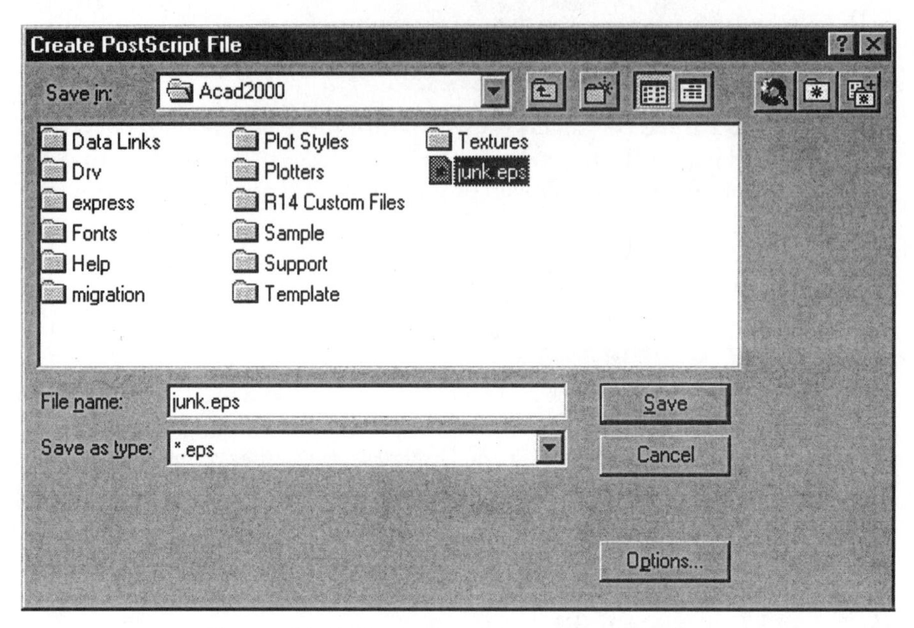

Figure 17-12 ***Create PostScript File*** *dialog box*

Prolog Section Name

In this edit box, you can assign a name for a prolog section to be read from the acad.psf file.

What to plot

The **What to plot** area of the dialog box has the following options:

Display. If you specify this option when you are in model space, the image in the current viewport is saved in the specified EPS file. Similarly, if you are in paper space, the current view is saved in the specified EPS file.

Extents. If you use this option, the PostScript file created will contain the section of the AutoCAD LT drawing that currently holds objects. In this way, this option resembles the **ZOOM Extents** option. If you add objects to the drawing, they are also included in the PostScript file to be created because the extents of the drawing are also altered. If you reduce the drawing extents by erasing, moving, or scaling objects, then you must use the **ZOOM Extents** or **ZOOM All** option. Only then does the **Extents** option of the **PSOUT** command understand the extents of the drawing to be exported. If you are in model space, the PostScript file is created in relation to the model space extents; if you are in paper space, the PostScript file is created in relation to the paper space extents. If you invoke the **PSOUT** command **Extents** option when perspective view is on and the position of camera is not out of the drawing extents, the following message is displayed:

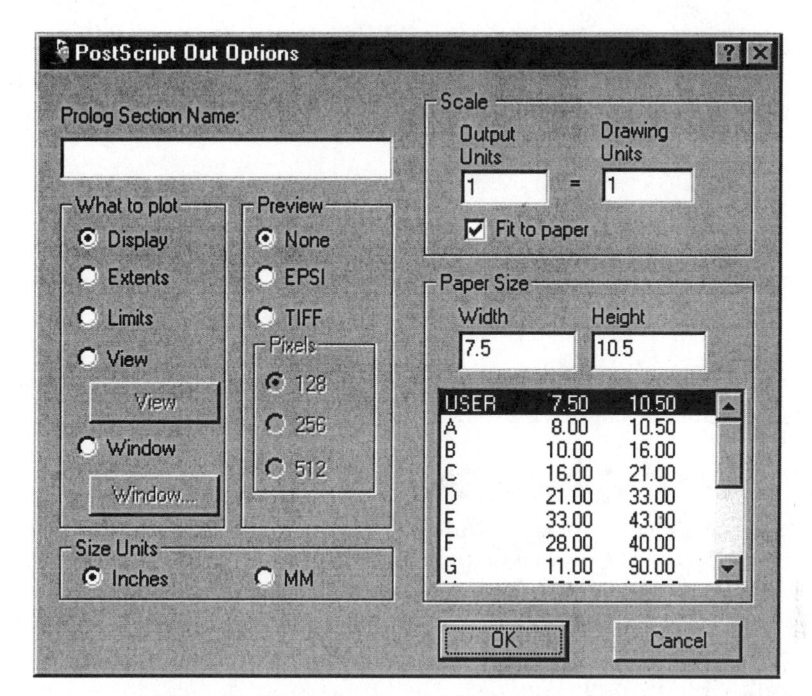

***Figure 17-13 PostScript Out Options** dialog box*

Chapter 17

PLOT Extents incalculable, using display

In such cases, the EPS file is created as it would be created with the **Display** option.

Limits. With this option, you can export the whole area specified by the drawing limits. If the current view is not the plan view [viewpoint (0,0,1)], the **Limits** option exports the area just as the **Extents** option would.

View. Any view created with the **VIEW** command can be exported with this option. When this radio button is activated, the **View** button is available. Choose the **View** button to display the **View Name** dialog box from where you can select the view.

Window. In this option, you need to specify the area to be exported with the help of a window. When this radio button is selected, the **Window** button is also available. Choose the **Window** button to display the **Window Selection** dialog box where you can select the **Pick** button and then specify the two corners of the window on the screen. You can also enter the coordinates of the two corners in the **Window Selection** dialog box.

Preview

The **Preview** area of the dialog box has two types of formats for preview images: **EPSI** and **TIFF**. If you want a preview image with no format, select the **None** radio button. If you select **TIFF** or **EPSI**, you are required to enter the pixel resolution of the screen preview in the **Pixels** area. You can select a preview image size of 128 x 128, 256 x 256, or 512 x 512.

Size Units

In this area, you can set the paper size units to **Inches** or **Millimeters** by selecting their corresponding radio buttons.

Scale

In this area, you can set an explicit scale by specifying how many drawing units are to be output per unit. You can select the **Fit to paper** check box so that the view to be exported is made as large as possible for the specified paper size.

Paper Size

You can select a size from the list or enter a new size in the **Width** and **Height** edit boxes to specify a paper size for the exported PostScript image.

OBJECT LINKING AND EMBEDDING

With Windows, it is possible to work with different Windows-based applications by transferring information between them. You can edit and modify the information in the original Windows application, and then update this information in other applications. This is made possible by creating links between the different applications and then updating those links, which in turn updates or modifies the information in the corresponding applications. This linking is a function of the OLE feature of Microsoft Windows. The OLE feature can also join together separate pieces of information from different applications into a single document. AutoCAD LT and other Windows-based applications, such as Microsoft Word, Notepad, and Windows WordPad support the Windows OLE feature.

For the OLE feature, you should have a source document where the actual object is created in the form of a drawing or a document. This document is created in an application called a **server** application. AutoCAD LT for Windows and Paintbrush can be used as server applications. Now this source document is to be linked to (or embedded in) the **compound** (destination) document, which is created in a different application, known as the **container** application. AutoCAD LT for Windows, Microsoft Word, and Windows WordPad can be used as container applications.

Clipboard

The transfer of a drawing from one Windows application to another is performed by copying the drawing or the document from the server application to the Clipboard. The drawing or document is then pasted in the container application from the Clipboard; hence, a Clipboard is used as a medium for storing the documents while transferring them from one Windows application to another. The drawing or the document on the Clipboard stays there until you copy a new drawing, which overwrites the previous one, or until you exit Windows. You can save the information present on the Clipboard with the .CLP extension.

Object Embedding

You can use the embedding function of the OLE feature when you want to ensure that there is no effect on the source document even if the destination document has been changed through the server application. Once a document is embedded, it has no connection with the source.

Although editing is always done in the server application, the source document remains unchanged. Embedding can be accomplished by means of the following steps. In this example, AutoCAD LT for Windows is the server application and Windows WordPad is the container application.

1. Create a drawing in the server application (AutoCAD LT).

2. Open Windows WordPad (container application) from the Accessories group in the Program.

3. It is preferable to arrange both the container and the server windows so that both are visible (Figure 17-14).

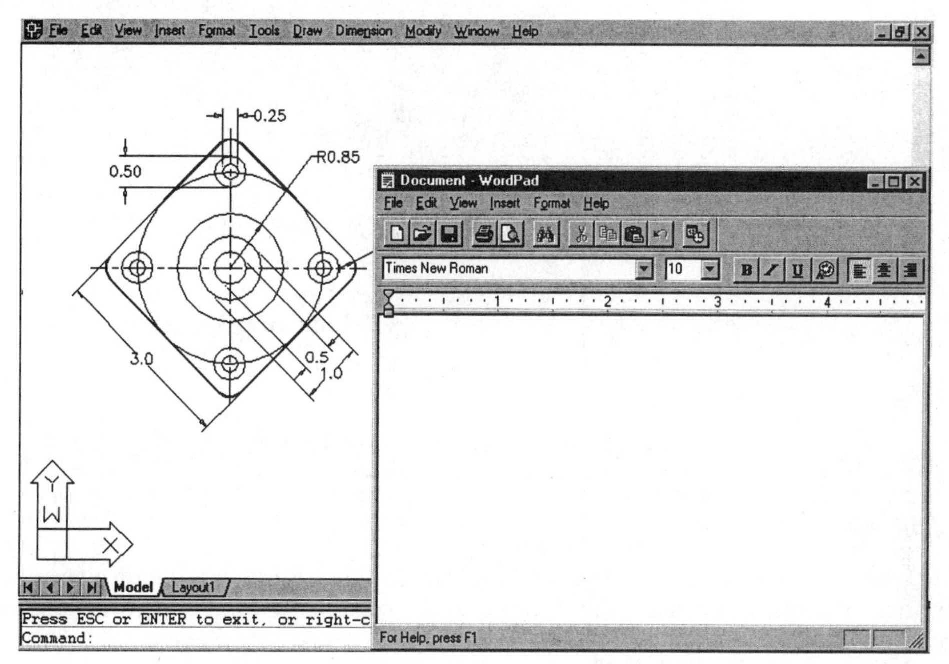

Figure 17-14 AutoCAD graphics screen with the WordPad window

4. In the AutoCAD LT graphics screen, use the **COPYCLIP** command. This command can be used in AutoCAD LT for embedding the drawings. This command can be invoked from the **Standard** toolbar by choosing the **Copy to Clipboard** button, from the **Edit** menu (Choose **Copy**), or by entering **COPYCLIP** at the command line.

   Command: **COPYCLIP**

   The next prompt, **Select objects:**, allows you to select the entities you want to transfer. You can either select the full drawing by entering ALL or select some of the entities by selecting them. You can use any of the selection set options for selecting the objects. With this command the selected objects are automatically copied to the Windows Clipboard.

5. After the objects are copied to the Clipboard, make the WordPad window active. To get the drawing from the Clipboard to the WordPad application (client), select the **Paste** in the WordPad application. Choose **Paste** from the **Edit** menu in Windows WordPad (Figure 17-15). You can also use **Paste Special** from the **Edit** menu, which will display the **Paste Special** dialog box (Figure 17-16). In this dialog box, select the **Paste** radio button (default) for embedding, and then choose **OK**. The drawing is now embedded in the WordPad window.

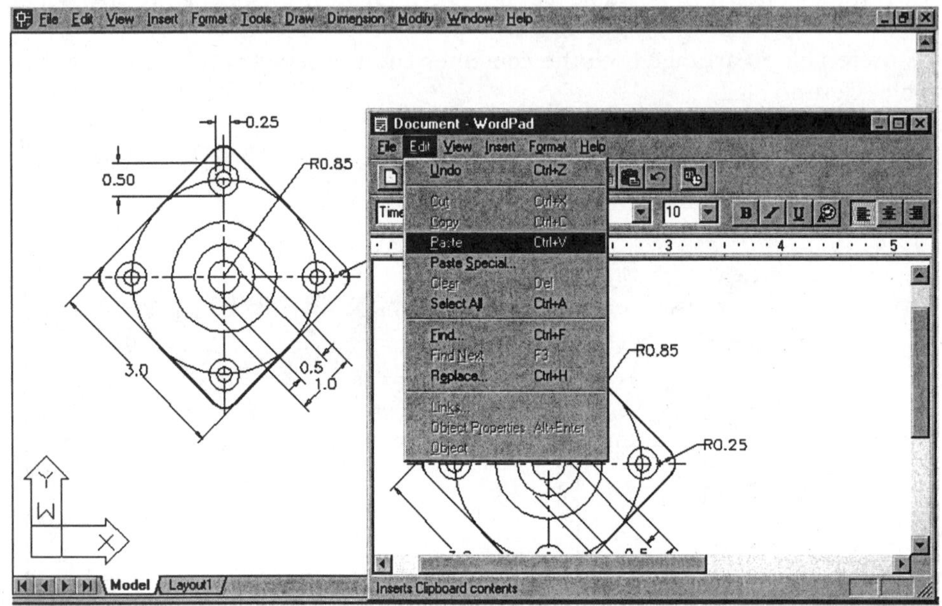

Figure 17-15 *Pasting a drawing to the WordPad application by selecting **Paste** from the **Edit** menu*

6. Your drawing is now displayed in the Write window, but it may not be displayed at the proper position. You can get the drawing in the current viewport by moving the scroll button up or down in the WordPad window. You can also save your embedded drawing by choosing **Save** from the **File** menu. It displays a **Save As** dialog box where you can enter a file name. You can now exit AutoCAD LT.

7. You can now edit your embedded drawing. Editing is performed in the server application, which in this case is AutoCAD LT for Windows. You can get the embedded drawing into the server application (AutoCAD LT) directly from the container application (WordPad) by double-clicking on the drawing in WordPad. The other method is by choosing **Edit Drawing Object** in the **Edit** menu. (This menu item has replaced **Object**, which was present before pasting the drawing.)

8. Now you are in AutoCAD LT, with your embedded drawing displayed on the screen, but as a temporary file with a file name, such as [Drawing in Document]. Here you can edit the drawing by changing the color and linetype or by adding and deleting text, entities, and so on. In Figure 17-17 the two upper circles and their dimensions (diameter 25 and 50) have been erased, and a hexagon (diameter 50) has been drawn in its place.

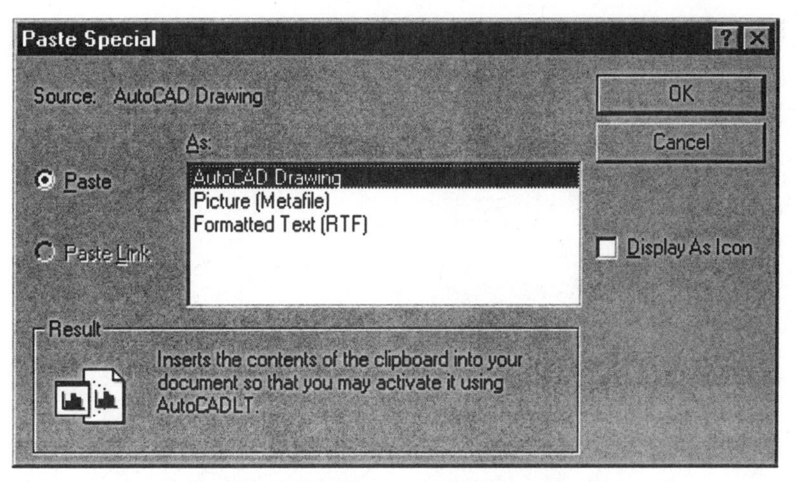

Figure 17-16 Paste Special *dialog box*

9. After you have finished modifying your drawing, choose **Update WordPad** from the **File** menu in the server (AutoCAD LT). This menu item has replaced the previous **Save** menu item. When you choose **Update**, AutoCAD LT automatically updates the drawing in Wordpad (container application). Now you can exit this temporary file in AutoCAD LT.

10. This completes the embedding function so you can exit the container application. While exiting, a dialog box that asks whether or not to save changes in WordPad is displayed.

Linking Objects

The linking function of OLE is similar to the embedding function. The only difference is that here a link is created between the source document and the destination document. If you edit the source, you can simply update the link, which automatically updates the client. This allows you to place the same document in a number of applications, and if you make a change in the source document, the clients will also change by simply updating the corresponding links. Consider AutoCAD LT for Windows to be the server application and Windows WordPad to be the container application. Linking can be performed by means of the following.

1. Open a drawing in the server application (AutoCAD LT). If you have created a new drawing, then you must save the drawing before you can link it with the container application.

2. Open Windows WordPad (the container application) from **Accessories** in the **Program** directory.

3. It is preferable to arrange both the container and the server windows so that both are visible.

4. In the AutoCAD LT graphics screen, use the **COPYLINK** command. This command can be used in AutoCAD LT for linking the drawing. This command can be invoked from the **Edit** menu (Choose **Copy Link**) or by entering **COPYLINK** at the Command prompt. The prompt sequence is:

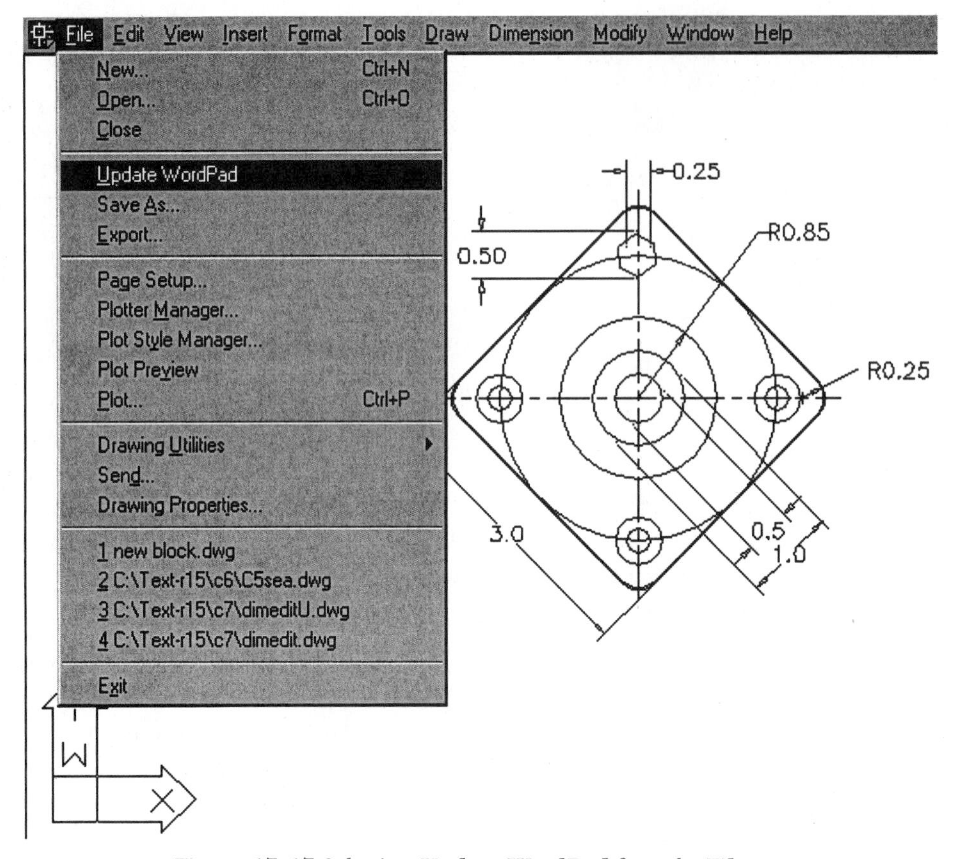

Figure 17-17 *Selecting* ***Update WordPad*** *from the* ***File*** *menu*

Command: **COPYLINK**

The **COPYLINK** copies the whole drawing in the current viewport directly to the Clipboard. Here you cannot select the objects for linking. If you want only a portion of the drawing to be linked, you can zoom into that view so that it is displayed in the current viewport prior to invoking the **COPYLINK** command. This command also creates a new view of the drawing having a name OLE1. Now you can exit AutoCAD LT.

5. Make the WordPad window active. To get the drawing from the Clipboard to the Write (container) application, choose **Paste Special** from the **Edit** menu, which will display the **Paste Special** dialog box. In this dialog box, select the **Paste Link** radio button for linking. Choose **OK**. The drawing is now linked to the WordPad window.

6. Your drawing is now displayed in the WordPad window. You can also save your linked drawing by choosing **Save** from the **File** menu. It displays a **Save As** dialog box where you can enter a file name.

7. You can now edit your linked drawing. Editing can be performed in the server application, which in this case is AutoCAD LT for Windows. You can get the linked drawing in the

server (AutoCAD LT) directly from the client (WordPad) by double-clicking on the drawing in WordPad. The other method is by choosing **Edit Linked Drawing Object** in the **Edit** menu. (This menu item has replaced **Object**, which was present before pasting the drawing.)

8. Now you are in AutoCAD LT, with the original drawing displayed on the screen. You can edit the drawing by changing the color and linetype or by adding and deleting text, entities, etc. Then save your drawing in AutoCAD LT by using the **SAVE** command. You can now exit AutoCAD LT.

9. You will notice that the drawing is automatically updated, and the changes made in the source drawing are present in the destination drawing also. This automatic updating is dependent on the selection of the **Automatic** radio button (default) in the **Links** dialog box (Figure 17-18). The **Links** dialog box can be invoked by choosing **Links** from the **Edit** menu. For updating manually, you can select the **Manual** radio button in the dialog box. In the manual case, after making changes in the source document and saving it, you need to invoke the **Links** dialog box and then choose the **Update Now** button; then, choose **Cancel**. This will update the drawing in the container application and display the updated drawing on the WordPad.

10. Exit the container application after saving the updated file.

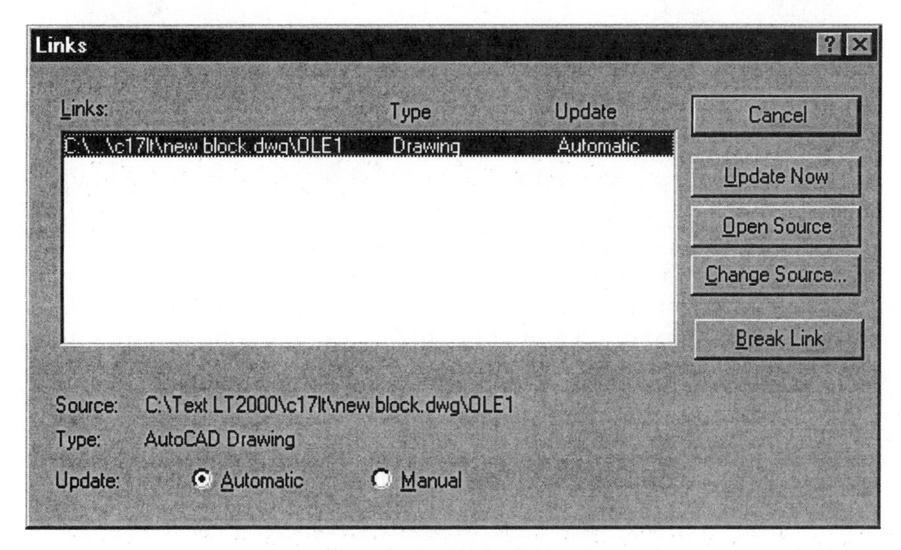

Figure 17-18 ***Links*** *dialog box*

Linking Information into AutoCAD LT*

Similarly, you can also embed and link information from a server application into an AutoCAD LT drawing. You can also drag selected OLE objects from another application into AutoCAD LT, provided this application supports Microsoft Activex and the application is running and visible on the screen. Dragging and dropping is like cutting and pasting. If you press the CTRL key while you drag the object it is copied to AutoCAD LT. Dragging and dropping an OLE object into AutoCAD LT embeds it into AutoCAD LT.

Linking Objects into AutoCAD LT*

Start any server application like the Windows Wordpad and open a document in it. Select the information you wish to use in AutoCAD LT with your pointing device and choose **Copy** from the **Edit** menu or choose the **Copy** button in the toolbar to copy this data to the Clipboard. Open the AutoCAD LT drawing you wish to link this data to. Choose **Paste Special** from the **Edit** menu or use the **PASTESPEC** command. The **Paste Special** dialog box is displayed (Figure 17-19). In the **As:** list box, select the data format you wish to use. For example, for a word pad document, select **WordPad document**. Picture format uses a Metafile format. Select the **Paste Link** radio button to paste the contents of the Clipboard to the current drawing. If you select the **Paste** radio button, the data is embedded and not linked. Choose **OK** to exit the dialog box. The data is displayed in the drawing and can be positioned as needed.

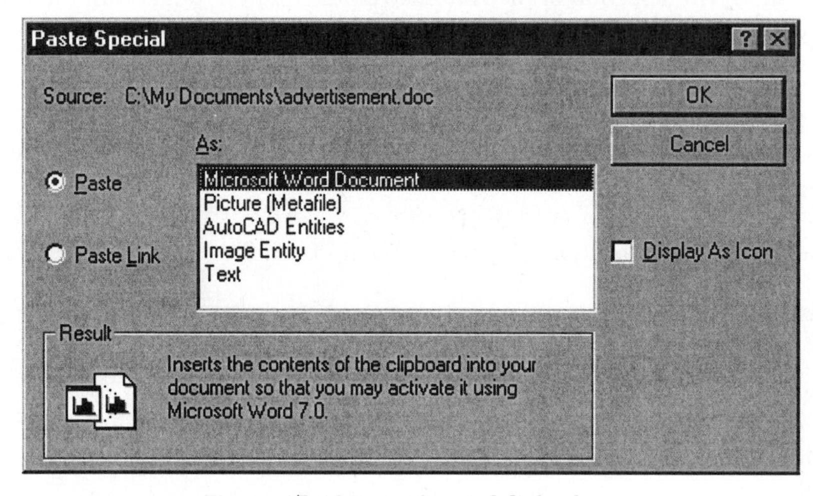

*Figure 17-19 **Paste Special** dialog box*

You can also use the **INSERTOBJ** command by entering INSERTOBJ at the Command prompt, by choosing **Ole Object** from the **Insert** menu, or by choosing the **OLE Object** button in the **Insert** toolbar. This command links an entire file to a drawing from within AutoCAD LT. Using this command displays the **Insert Object** dialog box (Figure 17-20).

Select the **Create from File** radio button. Also select the **Link** check box. Choosing the **Browse** button, displays the **Browse** dialog box. Select a file you want to link from the list box or enter a name in the **File name:** edit box and choose the **Insert** button. The path of the file is displayed in the **File:** edit box. If you select the **Display as icon** check box, an icon is also displayed in the dialog box. Choose **OK** to exit the dialog box, the selected file is linked to the AutoCAD LT drawing.

AutoCAD LT updates the links automatically by default, whenever the server document changes, but you can use the **OLELINKS** command to display the **Links** dialog box (Figure 17-21) where you can change these settings. This dialog box can also be displayed by choosing **OLE Links** from the **Edit** menu. In the **Links** dialog box, select the link you want to update and then choose the **Update Now** button. Then choose the **Close** button. If the server file location changes or if it is renamed, you can choose the **Change Source** button in the **Links** dialog box to display the **Change Source** dialog box. In this dialog box, locate the server filename and

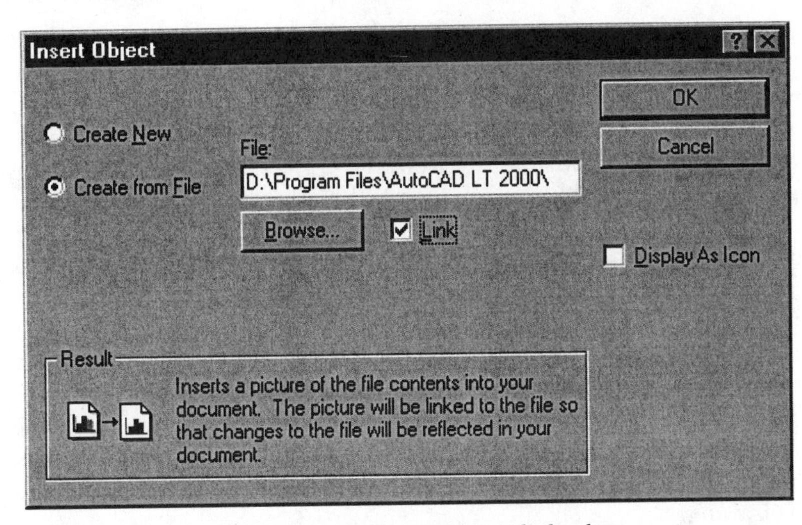

*Figure 17-20 **Insert Object** dialog box*

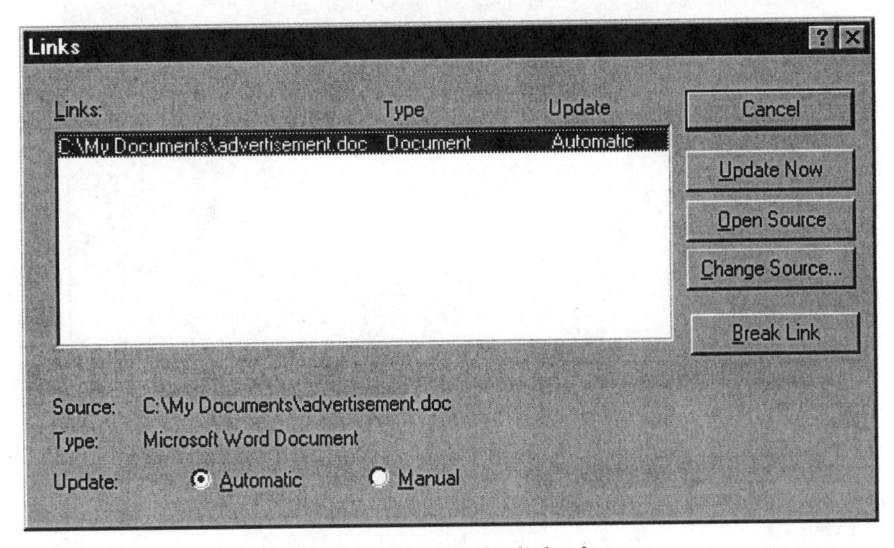

*Figure 17-21 **Links** dialog box*

location and choose the **Open** button. You can also choose the **Break Link** button in the **Links** dialog box to disconnect the inserted information from the server application. This is done when the linking between the two is not required anymore.

Embedding Objects into AutoCAD LT*

Open the server application and select the data you want to embed into the AutoCAD LT drawing. Copy this data to the Clipboard by choosing the **Copy** button in the toolbar or choosing **Copy** from the **Edit** menu. Open the AutoCAD LT drawing and choose **Paste** from the AutoCAD LT **Edit** menu. You can also use the command **PASTECLIP**. The selected information is embedded into the AutoCAD LT drawing.

You can also create and embed an object into an AutoCAD LT drawing starting from AutoCAD LT itself using the **INSERTOBJ** command. You can also choose **OLE Object** from the **Insert** menu or choose the **OLE Object** button from the **Insert** toolbar. The **Insert Object** dialog box is displayed (Figure 17-22). In this dialog box, select the **Create New** radio button and select the application you wish to use from the **Object Type:** list box. Choose **OK**. The selected application opens, now, you can create the information you wish to insert in the AutoCAD LT drawing and save it before closing the application. The **OLE Properties** dialog box is displayed (Figure 17-23) where you can resize and rescale the inserted objects. This dialog is displayed by default. If you do not want to display this dialog box, clear the **Display dialog when pasting OLE objects** check box in the **OLE Properties** dialog box. You can also clear the **Display OLE Properties Dialog** check box in the **Systems** tab of **Options** dialog box. You can edit information embedded in the AutoCAD LT drawing by opening the server application by double-clicking on the inserted OLE object. You can also select the object and right-click to display a shortcut menu. Choose **Object** > **Edit**. After editing the data in the server application, choose **Update** from **File** menu to reflect the modifications in the AutoCAD LT drawing.

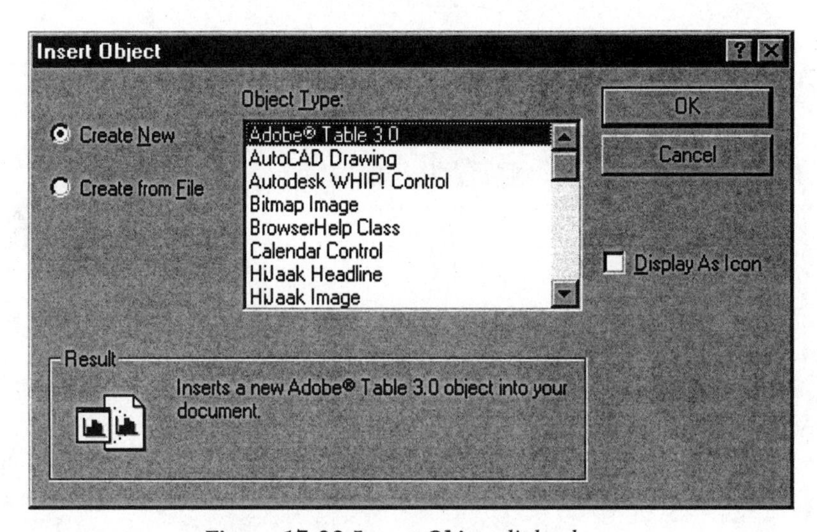

***Figure 17-22 Insert Object** dialog box*

Working with OLE Objects*

Select an OLE Object and right-click to display a shortcut menu; choose **Properties**. The **OLE Properties** dialog box is displayed (Figure 17-23). You can also select the OLE Object and use the **OLESCALE*** command. Specify a new height and a new width in the **Height:** and **Width:** edit boxes in the **Size** area or Under **Scale**, enter a value in percentage of the current values in the **Height:** and **Width:** edit boxes. Here, if you select the **Lock aspect Ratio** check box, whenever you change either the height or the width under **Scale** or **Size**, the respective width or height changes automatically to maintain the aspect ratio. If you want to change only the height or only the Width, clear this check box. Choose **OK** to apply changes.

Choosing the **Reset** button restores the selected OLE objects to their original size, that is, the size they were when inserted. If the AutoCAD LT drawing contains an OLE object with text with different fonts and you wish to select and modify specific text. You can select a particular

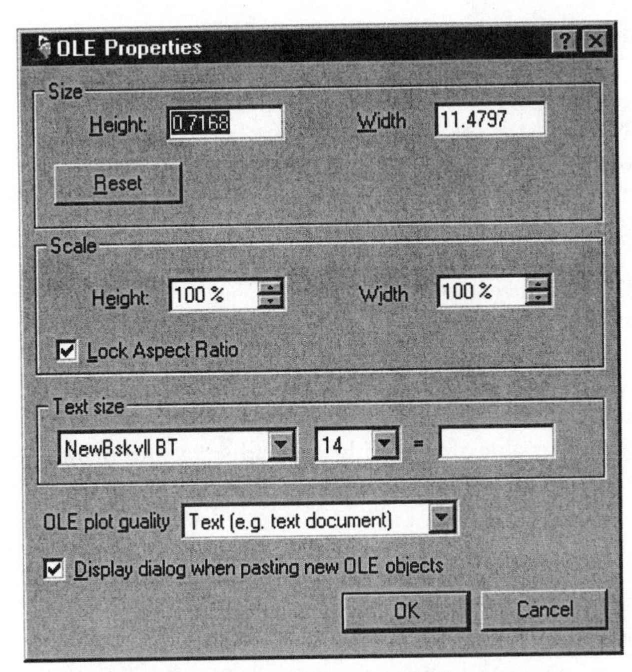

*Figure 17-23 **OLE Properties** dialog box*

font and point size from the drop-down lists under the **Text size** area and enter in the box after the = sign the value in drawing units. For example if you wish to select text in the Times Roman font, of point size 10 and modify it to size 0.5 drawing units, select Times Roman and 10 point size from the drop-down lists and in the text box after the = sign enter .5. All the text which is in Times Roman and is of point size 10 will change to 0.5 drawing units in height. The pointing device can also be used to modify and scale an OLE Object in the AutoCAD LT drawing. Selecting the object displays the object frame and the move cursor. The move cursor allows you to select and drag the object to a new location. The middle handle allows you to select the frame and stretch it. It does not scale objects proportionately. The corner handle scales the object proportionately.

Select an OLE object and right-click to display the shortcut menu. Choosing **Cut** removes the object from the drawing and pastes it on the Clipboard, **Copy** places a copy of the selected object on the clipboard; and **Clear** removes the object from the drawing and does not place it on the clipboard. Choosing **Object** displays the **Convert, Open**, and **Edit** options. Choosing **Convert**, displays the **Convert** dialog box where you can convert objects from one type to another and **Edit** opens the object in the Server application where you can edit it and update it in the current drawing. **Undo** cancels the last action. **Bring to Front** and **Send to Back** options, place the OLE objects in the front of or back of the AutoCAD LT objects. The **Selectable** option, turns the selection of the OLE object on or off. If the **Selectable** option is on, the object frame is visible and the object is selected.

If you want to change the layer of an OLE object, select the object and right-click to display the shortcut menu, choose **Cut**. The selected object is placed on the clipboard. Change the current layer to the one you want to change the OLE object's layer to using the **Layer**

Properties Manager dialog box. Now, choose **Paste** from the **Edit** menu to paste the contents of the clipboard in the AutoCAD LT drawing. The OLE object is pasted in the New layer, in its original size.

The **OLEHIDE** system variable controls the display of OLE objects in AutoCAD LT. The default value is 0, that is, all the OLE objects are visible. The different values and their effects are as follows:

0 All OLE objects are visible
1 OLE objects are visible in paper space only
2 OLE objects are visible in model space only
3 No OLE objects are visible

The **OLEHIDE** system variable affects both screen display and printing.

Self-Evaluation Test

Answer the following questions, and then compare your answers to the answers given at the end of this chapter.

1. The object whose area you want to find with the help of the **AREA** command must be a closed object. (T/F)

2. If you quit the editing session without saving the drawing, the time you spent on that editing session is added to the total time spent on editing the drawing. Also, the last update time is revised. (T/F)

3. The angle between two points can be measured with the help of the __________ command.

4. The __________ command displays all the information pertaining to the selected objects.

5. __________ time provides the most recent date and time you edited the current drawing.

6. You can set the automatic save time interval with the **Options** dialog box or with the __________ system variable.

7. You can import a DXF file into an AutoCAD LT drawing file with the __________ command.

8. The __________ system variable holds the value of distance computed.

9. The __________ command is used to identify the coordinate values of a point on the screen.

10. The __________ command displays all the information pertaining to the selected objects

in the drawing.

11. The _________ is used as a medium for storing the documents while transferring them from one Windows application to another.

12. The _________ command can be used in AutoCAD LT for embedding drawings.

13. You can edit your embedded drawing in the _________ application.

14. You can get an embedded drawing into the server application directly from the container application by _________ on the drawing.

15. The _________ command can be used in AutoCAD LT for linking a drawing.

16. The **COPYLINK** command copies a drawing in the _________ to the Clipboard.

Review Questions

Inquiry

1. Inquiry commands are used to obtain information about the drawn figures. (T/F)

2. The default method of specifying the object whose area you want to find is by specifying all the vertices of the object. (T/F)

3. The **AREA** and **PERIMETER** system variables are reset to zero whenever you invoke the **AREA** command. (T/F)

4. To find the circumference or perimeter of an object, you can use the _________ command.

5. You can use the _________ option to add subsequently measured areas to the running total.

6. You can use the _________ option to subtract subsequently measured areas from the running total.

7. The distance between two points can be measured with the _________ command.

8. The _________ command can be used to display the data pertaining to time related to a drawing and the drawing session.

9. Drawing _________ provides the date and time the current drawing was created.

10. The drawing creation time for a drawing is set to the system time when either the _________

command, the _________ command, or the _________ command was used to create that drawing file.

Data Exchange

11. The _________ command is used to create an ASCII format file with the .DXF extension from AutoCAD LT drawing files.

12. With the **Binary** option of the **Save As Options** dialog box, you can also create binary format files. Binary DXF files are _________ efficient and occupy only 75 percent of the ASCII DXF file. File access for files in binary format is _________ than for the same file in ASCII format.

13. In a _________ file, information is stored in the form of a dot pattern on the screen. This bit pattern is also known as _________ .

Exercises

Exercise 2 — *Mechanical*

Draw Figure 17-24 without dimensions, and determine the indicated parameters using the required commands. Select the **LIST** command. Select **TIME**, and note the time in the drawing editor. Save the drawing.

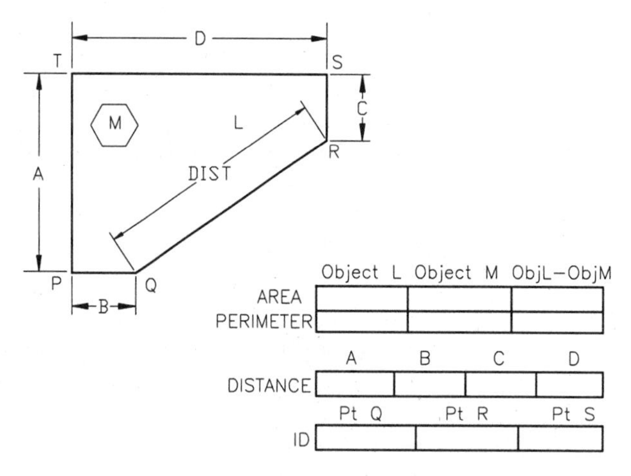

Figure 17-24 *Drawing for Exercise 2*

Answers to Self-Evaluation Test

1 - F, **2** - F, **3** - **DIST**, **4** - **LIST**, **5** - Last Update, **6** - **SAVETIME**, **7** - **OPEN**, **8** - **DISTANCE**, **9** - **ID**, **10** - **LIST**, **11** - clipboard, **12** - **COPYCLIP**, **13** - server, **14** - double-clicking, **15** - **COPYLINK**, **16** - current viewport

Chapter 18

Technical Drawing with AutoCAD LT

Learning Objectives

After completing this chapter, you will be able to:

- *Understand the concepts of multiview drawings.*
- *Understand X, Y, Z axes; XY, YZ, XZ planes; and parallel planes.*
- *Draw orthographic projections and position the views.*
- *Dimension a drawing.*
- *Understand the basic dimensioning rules.*
- *Draw sectional views using different types of sections.*
- *Hatch sectioned surfaces.*
- *Understand how to use auxiliary views and how to draw them.*
- *Create assembly and detail drawings.*

MULTIVIEW DRAWINGS

When designers design a product, they visualize the shape of the product in their minds. To represent that shape on paper or to communicate the idea to other people, they must draw a picture of the product or its orthographic views. Pictorial drawings, such as isometric drawings, convey the shape of the object, but it is difficult to show all of its features and dimensions in an isometric drawing. Therefore, in industry, multiview drawings are the accepted standard for representing products. Multiview drawings are also known as **orthographic projection drawings**. To draw different views of an object, it is very important to visualize the shape of the product. The same is true when you are looking at different views of an object to determine its shape. To facilitate visualizing the shapes, you must picture the object in 3D space with reference to the X, Y, and Z axes. These reference axes can then be used to project the image into different planes. This process of visualizing objects with reference to different axes is, to some extent, natural in human beings. You might have noticed that sometimes when looking at objects that are at an angle, people tilt their heads. This is a natural reaction, an effort to position the object with respect to an imaginary reference frame (X, Y, Z axes).

UNDERSTANDING X, Y, Z AXES

To understand the X, Y, and Z axes, imagine a flat sheet of paper on the table. The horizontal edge represents the positive X axis and the other edge, the edge along the width of the sheet, represents the positive Y axis. The point where these two axes intersect is the origin. Now, if you draw a line perpendicular to the sheet passing through the origin, the line defines the positive Z axis (Figure 18-1). If you project the X, Y, and Z axes in the opposite direction beyond the origin, you will get the negative X, Y, and Z axes (Figure 18-2).

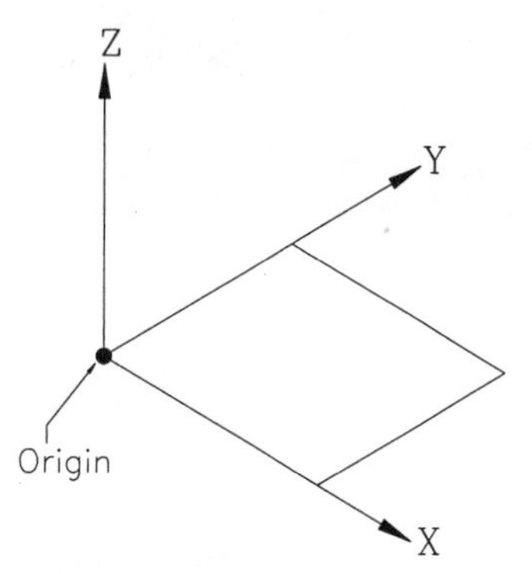

Figure 18-1 *X, Y, Z axes*

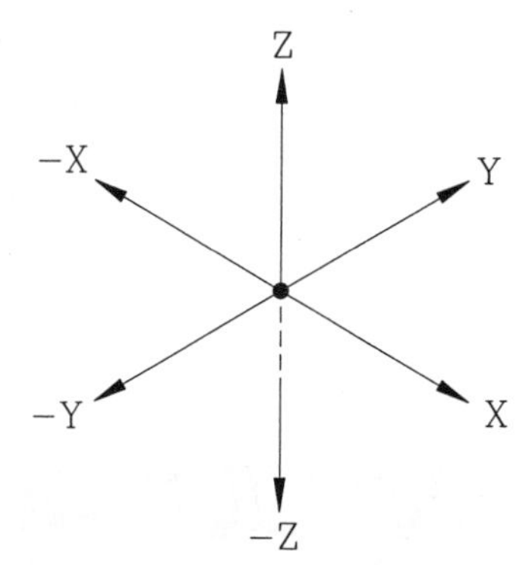

Figure 18-2 *Positive and negative axes*

The space between the X and Y axes is called the XY plane. Similarly, the space between the Y and Z axes is called the YZ plane, and the space between the X and Z axes is called the XZ plane (Figure 18-3). A plane that is parallel to these planes is called a parallel plane (Figure 18-4).

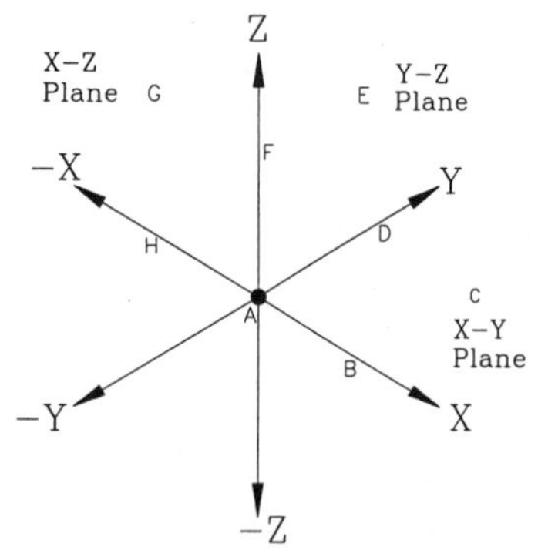

Figure 18-3 *XY, YZ, and XZ planes*

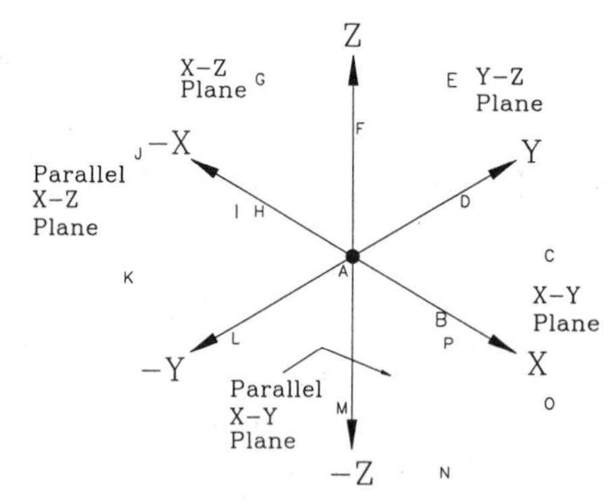

Figure 18-4 *Parallel planes*

ORTHOGRAPHIC PROJECTIONS

The first step in drawing an orthographic projection is to position the object along the imaginary X, Y, and Z axes. For example, if you want to draw orthographic projections of the step block shown in Figure 18-5, position the block so that the far left corner coincides with the origin, and then align the block with the X, Y, and Z axes.

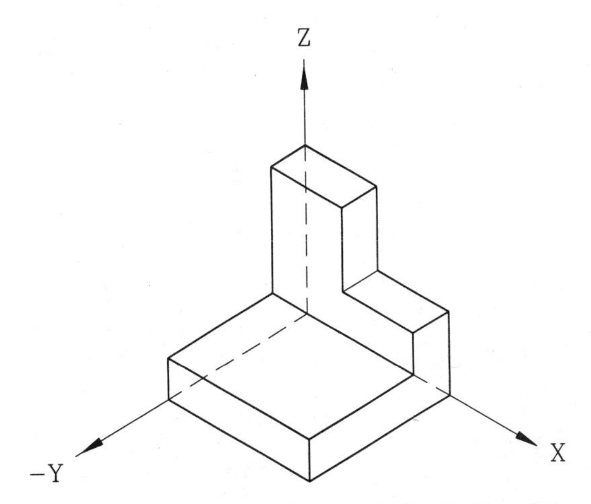

Figure 18-5 *Aligning the object with the X, Y, and Z axes*

Now you can look at the object from different directions. Looking at the object along the negative Y axis and toward the origin is called the front view. Similarly, looking at the object from the positive X direction is called the right-hand side view. To get the top view, you look at the object from the positive Z axis. See Figure 18-6.

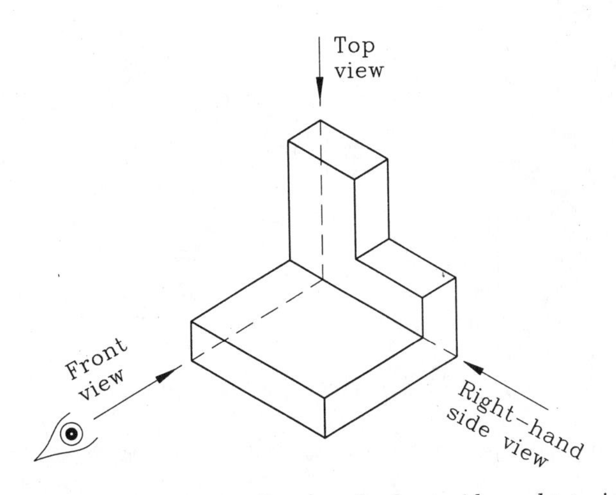

Figure 18-6 *Viewing directions for front, side, and top views*

To draw the front, side, and top views, project the points onto the parallel planes. For example, if you want to draw the front view of the step block, imagine a plane parallel to the XZ plane located at a certain distance in front of the object. Now, project the points from the object onto the parallel plane (Figure 18-7), and join the points to complete the front view.

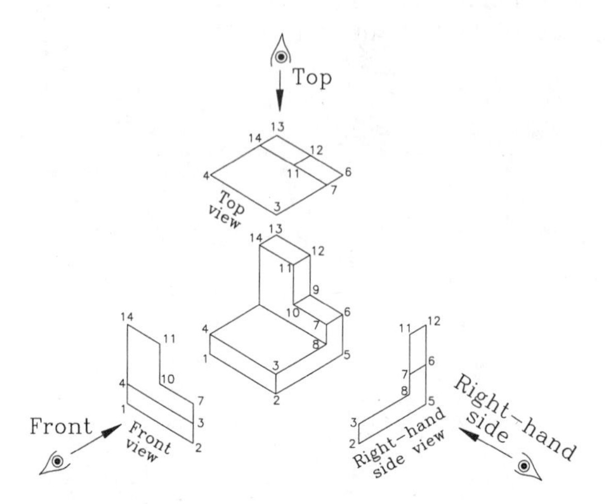

Figure 18-7 *Projecting points onto parallel planes*

Repeat the same process for the side and top views. To represent these views on paper, position the views as shown in Figure 18-8.

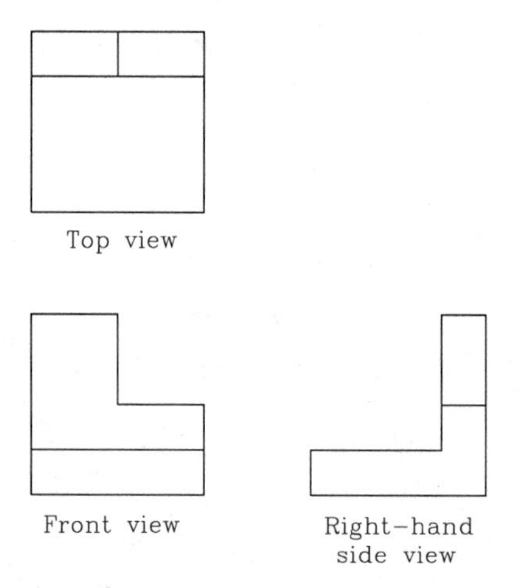

Figure 18-8 *Representing views on paper*

Another way of visualizing different views is to imagine the object enclosed in a glass box (Figure 18-9).

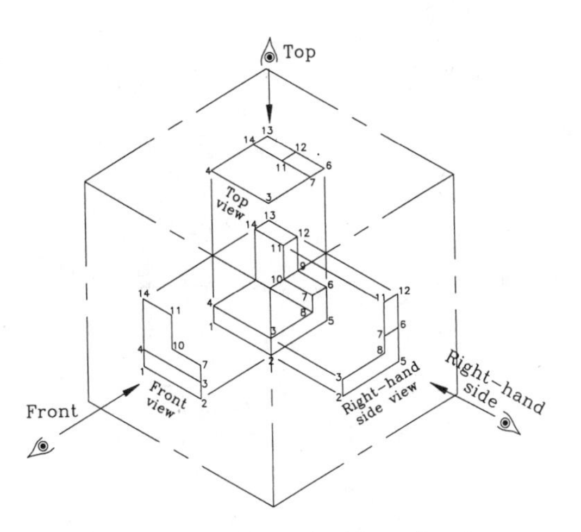

Figure 18-9 *Objects inside a glass box*

Now, look at the object along the negative Y axis and draw the front view on the front glass panel. Repeat the process by looking along the positive X and Z axes, and draw the views on the right-hand side and the top panel of the box (Figure 18-10).

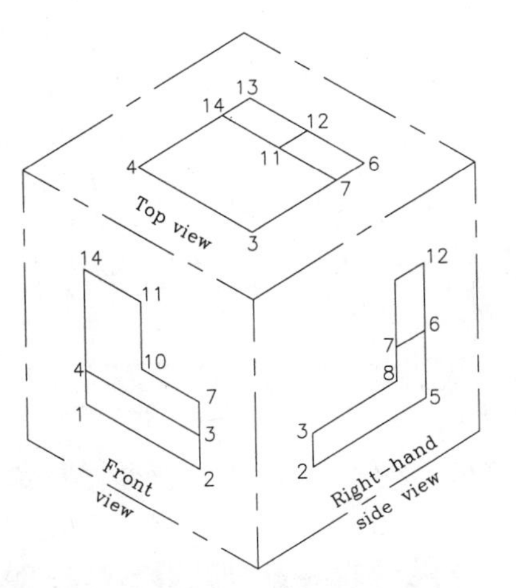

Figure 18-10 *Front, top, and side views*

To represent the front, side, and top views on paper, open the side and the top panel of the glass box (Figure 18-11). The front panel is assumed to be stationary.

After opening the panels through 90 degrees, the orthographic views will appear as shown in Figure 18-12.

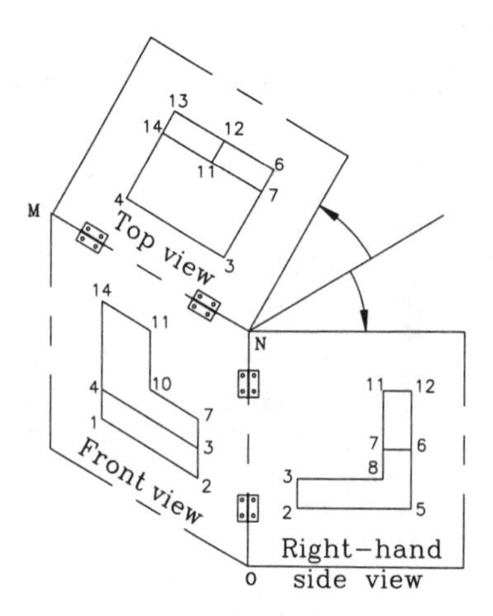

Figure 18-11 Open the side and the top panel

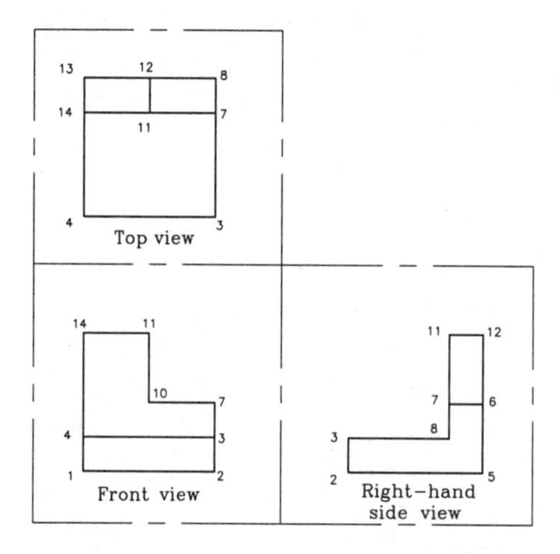

Figure 18-12 Views after opening the box

POSITIONING ORTHOGRAPHIC VIEWS

Orthographic views must be positioned as shown in Figure 18-13.

The right-hand side view must be positioned directly on the right side of the front view. Similarly, the top view must be directly above the front view. If the object requires additional views, they must be positioned as shown in Figure 18-14.

The different views of the step block are shown in Figure 18-15.

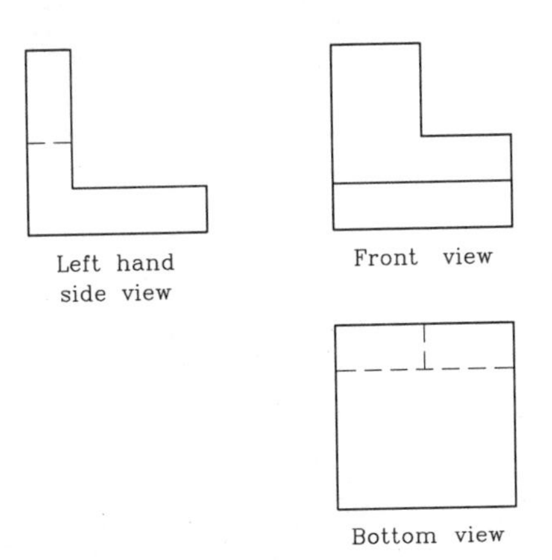

Figure 18-13 Positioning orthographic views

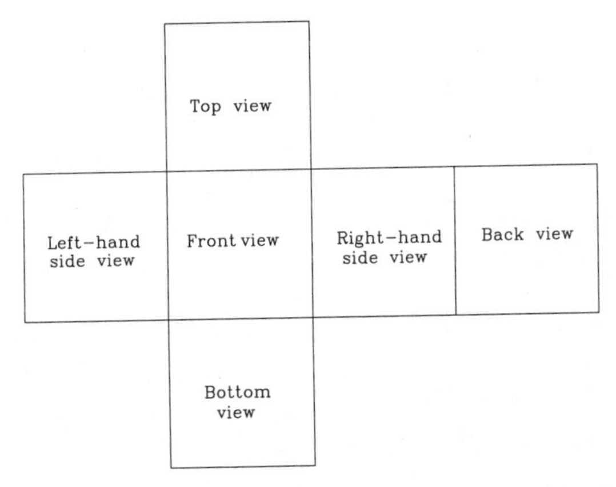

Figure 18-14 Standard placement of orthographic views

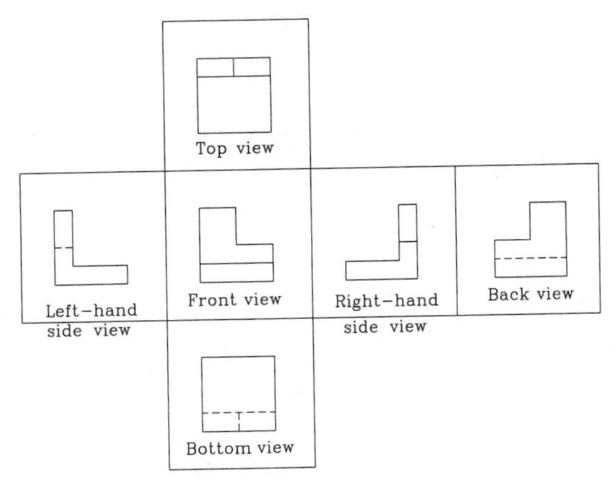

Figure 18-15 Different views of the step block

Example 1

In this example, you will draw the required orthographic views of the object in Figure 18-16.

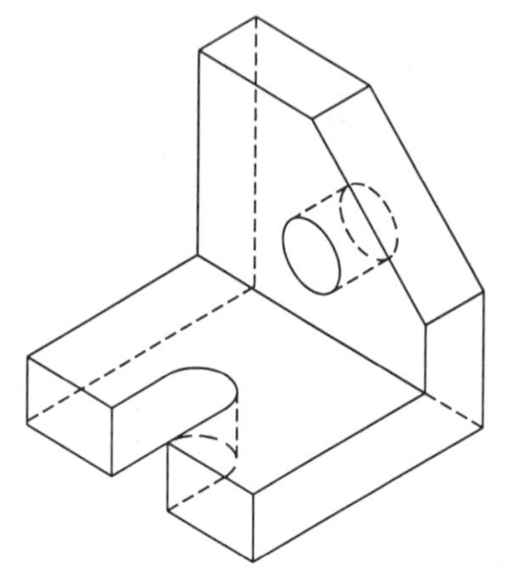

Figure 18-16 *Step block with hole and slot*

Drawing the orthographic views of an object involves the following steps.

Step 1. Look at the object, and determine the number of views required to show all of its features. For example, the object in Figure 18-16 will require three views only (front, side, and top).

Step 2. Based on the shape of the object, select the side you want to show as the front view. Generally the front view is the one that shows the maximum number of features or that gives a better idea about the shape of the object. Sometimes the front view is determined by how the part will be assembled in an assembly.

Step 3. Picture the object in your mind, and align it along the imaginary X, Y, Z axes (Figure 18-17).

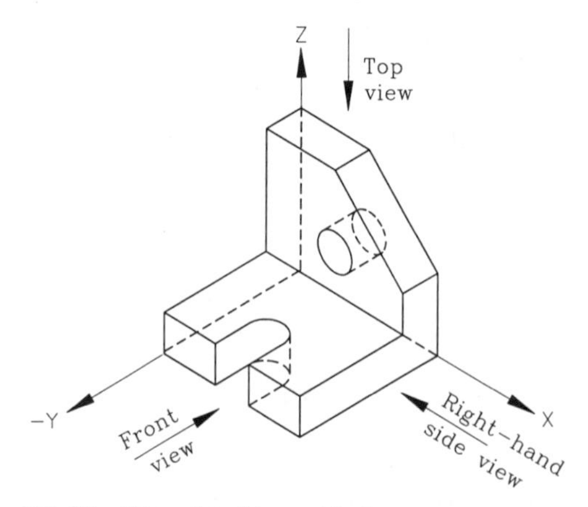

Figure 18-17 *Align the object with the imaginary X, Y, Z axes*

Step 4. Look at the object along the negative Y axis, and project the image on the imaginary XZ parallel plane, Figure 18-18.

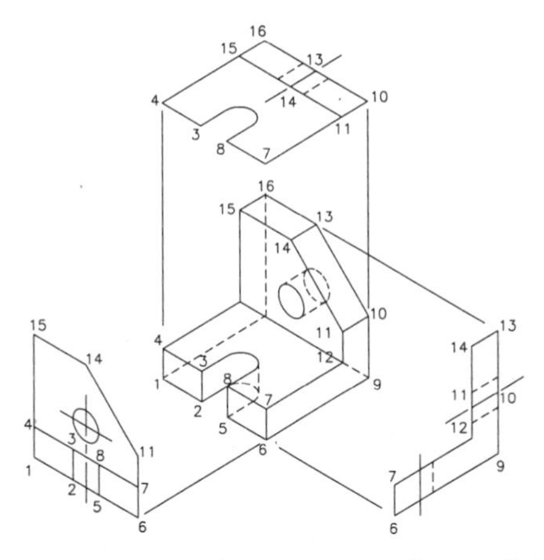

Figure 18-18 Project the image onto the parallel planes

Step 5. Draw the front view of the object according to the given dimensions. If there are any hidden features, they must be drawn with hidden lines. The holes and slots must be shown with the centerlines.

Step 6. To draw the right-hand side view, look at the object along the positive X axis and project the image onto the imaginary YZ parallel plane.

Step 7. Draw the right-hand side view of the object according to the given dimensions. If there are any hidden features, they must be drawn with hidden lines. The holes and slots, when shown in side view, must have one centerline.

Step 8. Similarly, draw the top view to complete the drawing. Figure 18-19 shows different views of the given object.

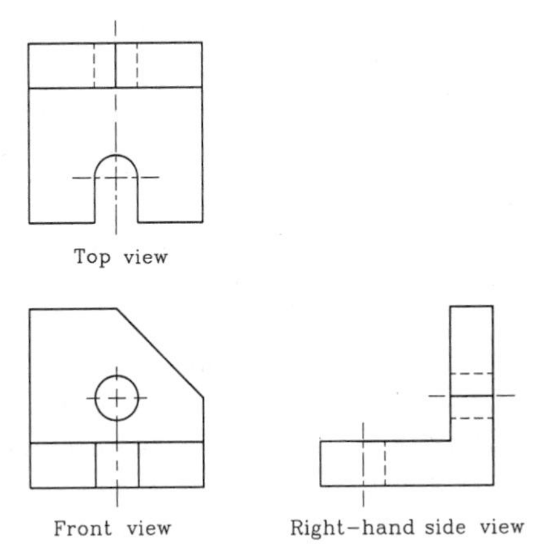

Figure 18-19 Front, side, and top views

Exercises 1 through 4 — Mechanical

Draw the required orthographic views of the following objects in Figures 18-20 through 18-23. The distance between the dotted lines is 0.5 units.

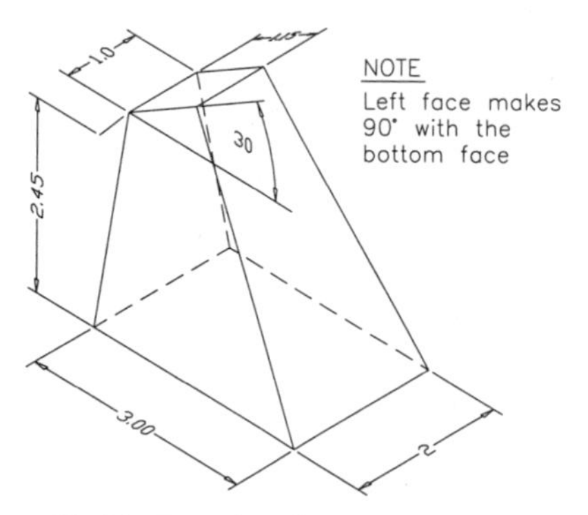

Figure 18-20 *Drawing for Exercise 1 (the object is shown as a surfaced wireframe model)*

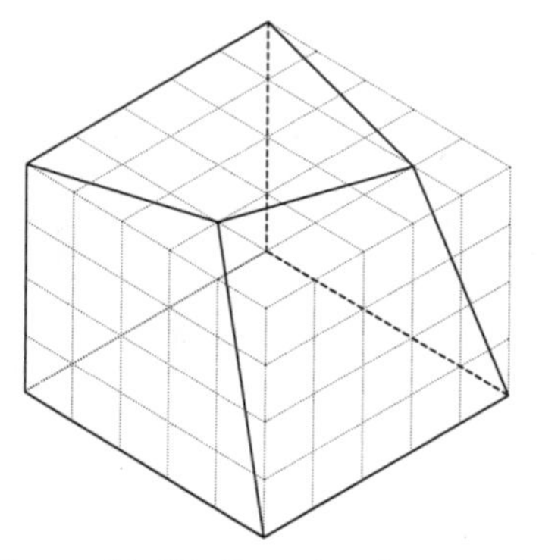

Figure 18-21 *Drawing for Exercise 2*

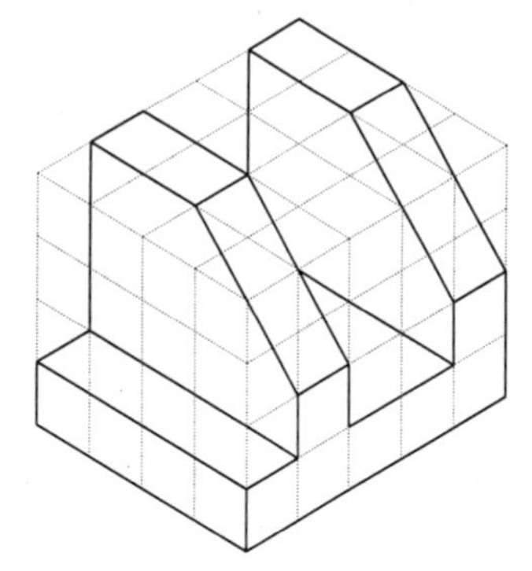

Figure 18-22 *Drawing for Exercise 3*

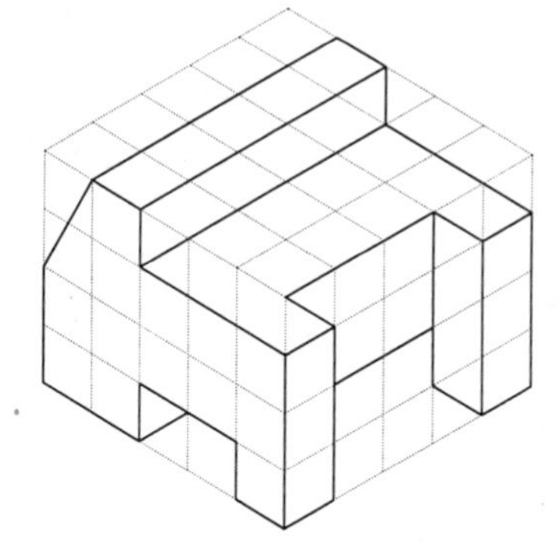

Figure 18-23 *Drawing for Exercise 4*

DIMENSIONING

Dimensioning is one of the most important things in a drawing. When you dimension, you not only give the size of a part, you give a series of instructions to a machinist, an engineer, or an architect. The way the part is positioned in a machine, the sequence of machining operations, and the location of different features of the part depend on how you dimension it. For example, the number of decimal places in a dimension (2.000) determines the type of machine that will be used to do that machining operation. The machining cost of such an operation is significantly higher than a dimension that has only one digit after the decimal (2.0). If you are using a computer numerical control (CNC) machine, locating a feature may not be a problem, but the number of pieces you can machine without changing the tool depends on the tolerance assigned to a dimension. A closer tolerance (+.0001 -.0005) will definitely increase the tooling cost and ultimately the cost of the product. Similarly, if a part is to be forged or cast, the radius of the edges and the tolerance you provide to these dimensions determine the cost of the product, the number of defective parts, and the number of parts you get from the die.

When dimensioning, you must consider the manufacturing process involved in making a part and the relationships that exist among different parts in an assembly. If you are not familiar with any operation, get help. You must not assume things when dimensioning or making a piece part drawing. The success of a product, to a large extent, depends on the way you dimension a part. Therefore, never underestimate the importance of dimensioning in a drawing.

Dimensioning Components

A dimension consists of the following components (Figure 18-24):

Extension line
Arrows or tick marks
Dimension line
Leader lines
Dimension text

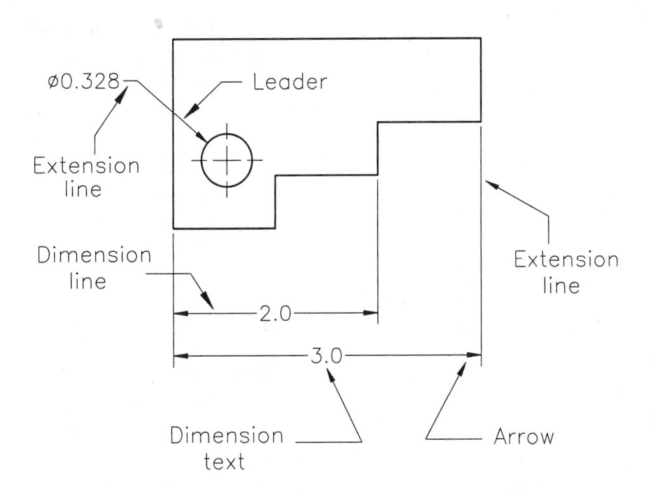

Figure 18-24 *Dimensioning components*

The **extension lines** are drawn to extend the points that are dimensioned. The length of the extension lines is determined by the number of dimensions and the placement of the dimension lines. These lines are generally drawn perpendicular to the surface. The **dimension lines** are drawn between the extension lines, at a specified distance from the object lines. The **dimension text** is a numerical value that represents the distance between the points. The dimension text can also consist of a variable (A, B, X, Y, Z12,...), in which case the value assigned to the variable is defined in a separate table. The dimension text can be centered around the dimension line or at the top of the dimension line. **Arrows or tick marks** are drawn at the end of the dimension line to indicate the start and end of the dimension. **Leader lines** are used when dimensioning a circle, arc, or any nonlinear element of a drawing. They are also used to attach a note to a feature or to give the part numbers in an assembly drawing.

Basic Dimensioning Rules

1. You should make the dimensions in a separate layer/layers. This makes it easy to edit or control the display of dimensions (freeze, thaw, lock, unlock). Also, the dimension layer/layers should be assigned a unique color so that at the time of plotting you can assign the desired pen to plot the dimensions. This helps to control the line width and contrast of dimensions at the time of plotting.

2. The distance of the first dimension line should at least 0.375 units (10 units for metric drawing) from the object line. In CAD drawing, this distance may be 0.75 to 1.0 units (19 to 25 units for metric drawings). Once you decide on the spacing, it should be maintained throughout the drawing.

3. The distance between the first dimension line and the second dimension line must be at least 0.25 units. In CAD drawings, this distance may be 0.25 to 0.5 units (6 to 12 units for metric drawings). If there are more dimension lines (parallel dimensions), the distances between them must be same (0.25 to 0.5 units). Once you decide on the spacing (0.25 to 0.5), the same spacing should be maintained throughout the drawing. An appropriate snap setting is useful for maintaining this spacing. If you are using baseline dimensioning, you can use AutoCAD LT's **DIMDLI** variable to set the spacing. You must present the dimensions so that they are not crowded, especially when there is not much space (Figure 18-25).

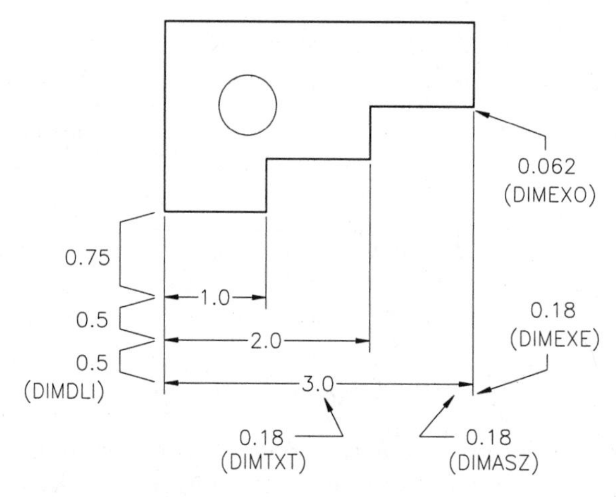

Figure 18-25 *Arrow size, text height, and spacing between dimension lines*

4. For parallel dimension lines, the dimension text can be staggered if there is not enough room between the dimension lines to place the dimension text. You can use the AutoCAD LT Object Grips feature or the **DIMTEDIT** command to stagger the dimension text (Figure 18-26).

Note

You can change grid, snap or UCS origin, and snap increment to make it easier to place the dimensions. You can also add the following lines to the AutoCAD LT menu file or toolbar buttons:

SNAP;R;\0;SNAP;0.25;GRID;0.25
SNAP;R;0,0;0;SNAP;0.25;GRID;0.5

The first line sets the snap to 0.25 units and allows the user to define the new origin of snap and grid display. The second line sets the grid to 0.5 and snap to 0.25 units. It also sets the origin for grid and snap to (0,0).

5. All dimensions should be given outside the view. However, the dimensions can be shown inside the view if they can be easily understood there and cause no confusion with other dimensions (Figure 18-27).

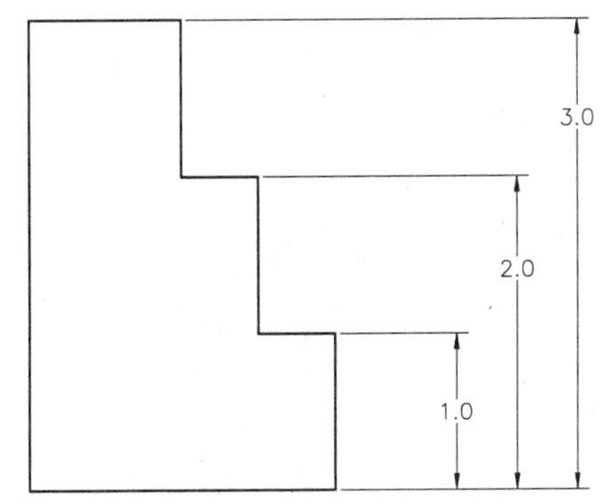

Figure 18-26 *Staggered dimensions*

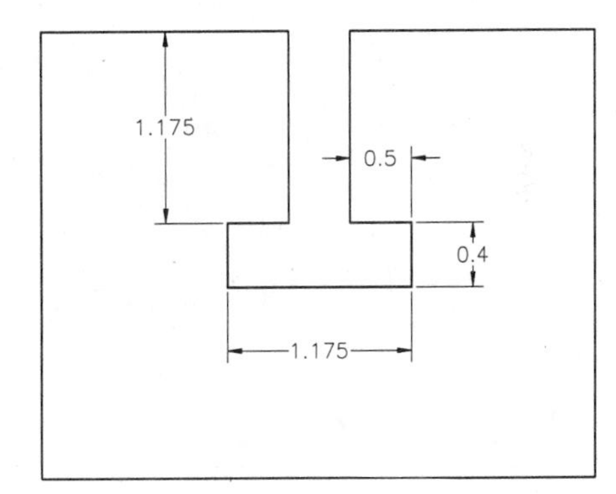

Figure 18-27 *Dimensions inside the view*

6. Dimension lines should not cross extension lines (Figure 18-28). You can accomplish this by giving the smallest dimension first and then the next largest dimension (Figure 18-29).

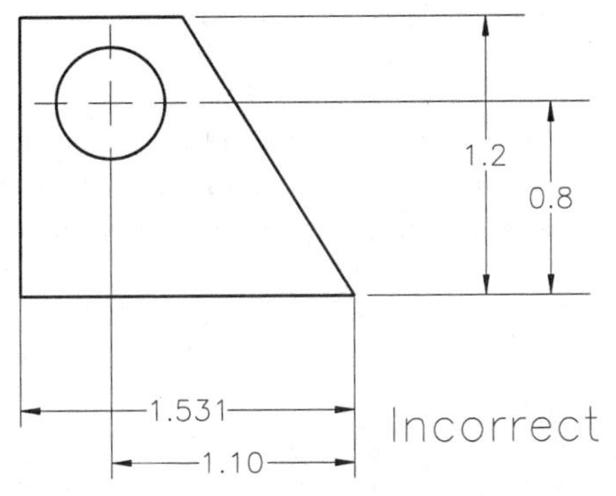

Figure 18-28 *Dimension lines should not cross extension lines*

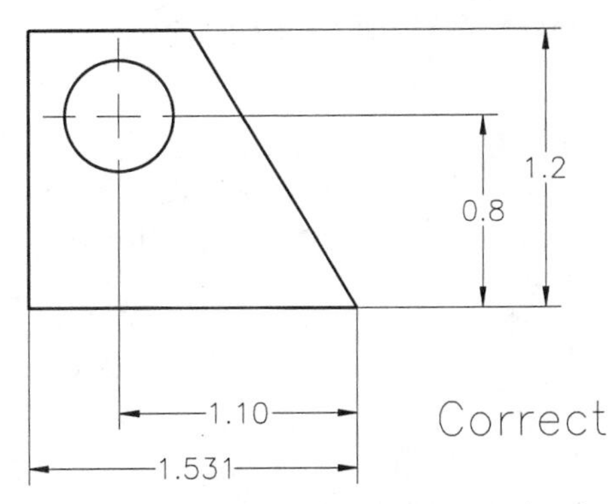

Figure 18-29 *Smallest dimension must be given first*

7. If you decide to have the dimension text aligned with the dimension line, then all dimension text in the drawing must be aligned (Figure 18-30). Similarly, if you decide to have the dimension text horizontal or above the dimension line, then to maintain uniformity in the drawing, all dimension text must be horizontal (Figure 18-31) or above the dimension line (Figure 18-32).

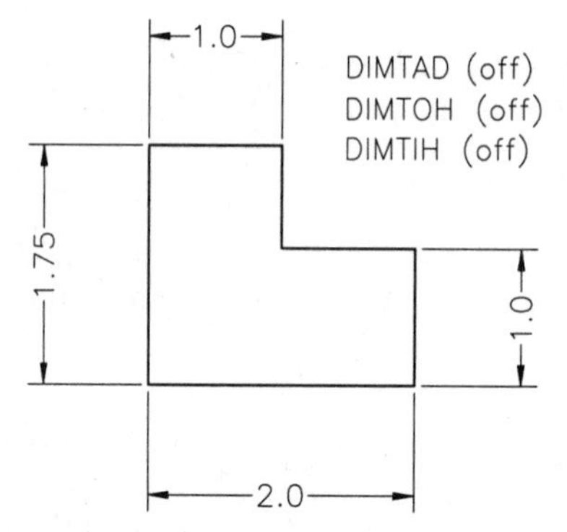

***Figure 18-30** Dimension text aligned*

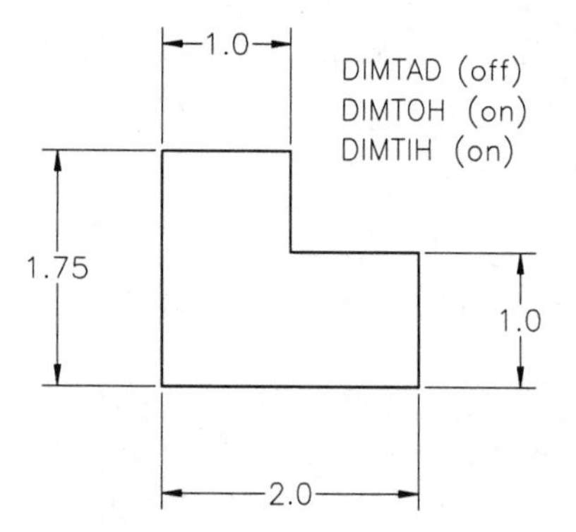

***Figure 18-31** Dimension text horizontal*

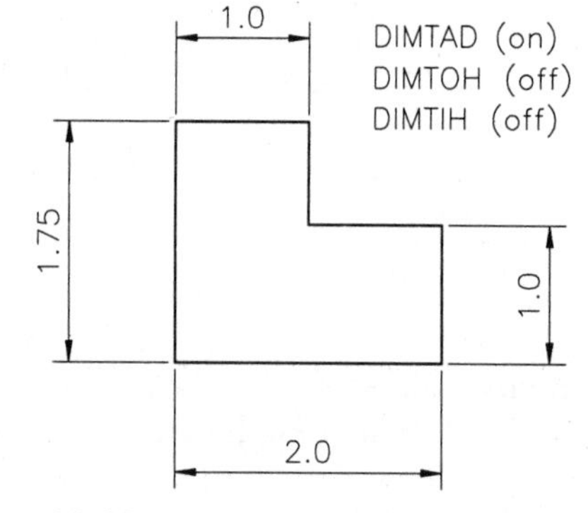

***Figure 18-32** Dimension text above dimension line*

8. If you have a series of continuous dimensions, they should be placed in a continuous line (Figure 18-33). Sometimes you may not be able to give the dimensions in a continuous line even after adjusting the dimension variables. In that case, give dimensions that are parallel (Figure 18-34).

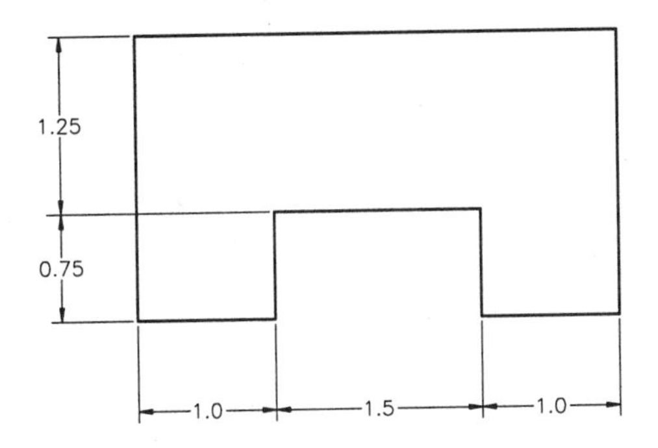

Figure 18-33 *Dimensions should be continuous*

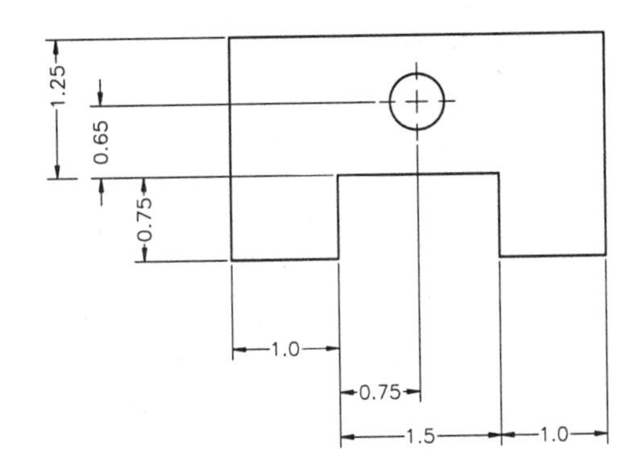

Figure 18-34 *Parallel dimensions*

9. You should not dimension with hidden lines. The dimension must be given where the feature is visible (Figure 18-35). However, in some complicated drawings you might be justified to dimension a detail with a hidden line.

10. The dimensions must be given where the feature that you are dimensioning is obvious and shows the contour of the feature (Figure 18-36).

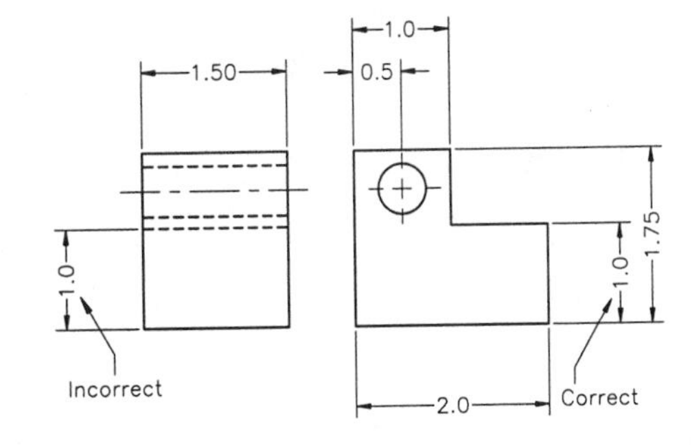

Figure 18-35 *Do not dimension with hidden lines*

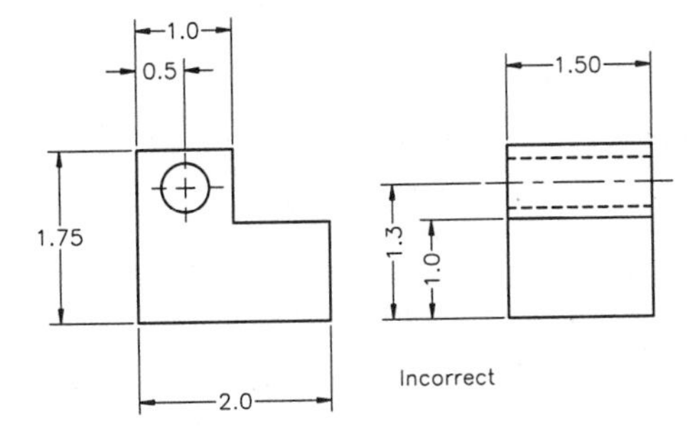

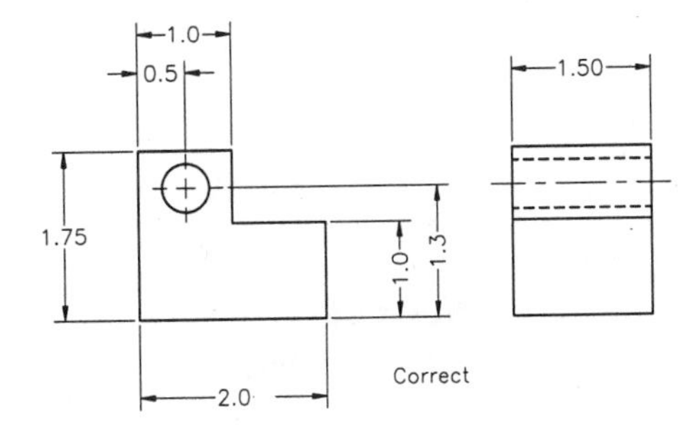

Figure 18-36 *Dimensions should be given where they are obvious*

11. The dimensions must not be repeated; this makes it difficult to update a dimension, and sometimes the dimensions might get confusing (Figure 18-37).

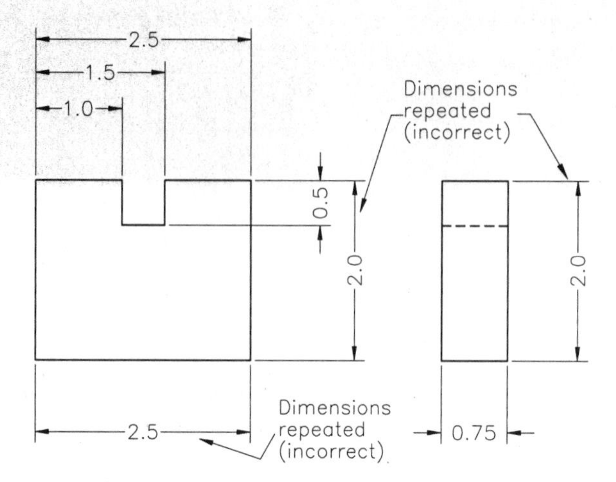

Figure 18-37 *Dimension must not be repeated*

12. The dimensions must be given depending on how the part will be machined and the relationship that exists between different features of the part (Figure 18-38).

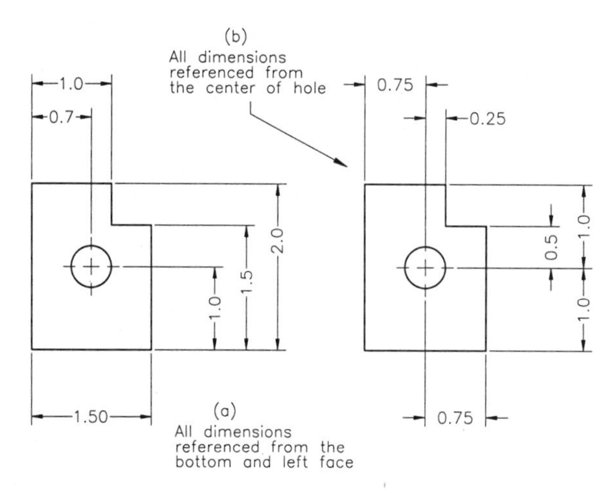

Figure 18-38 *When dimensioning, consider the machining processes involved*

13. When a dimension is not required but you want to give it for reference, it must be a reference dimension. The reference dimension must be enclosed in parentheses (Figure 18-39).

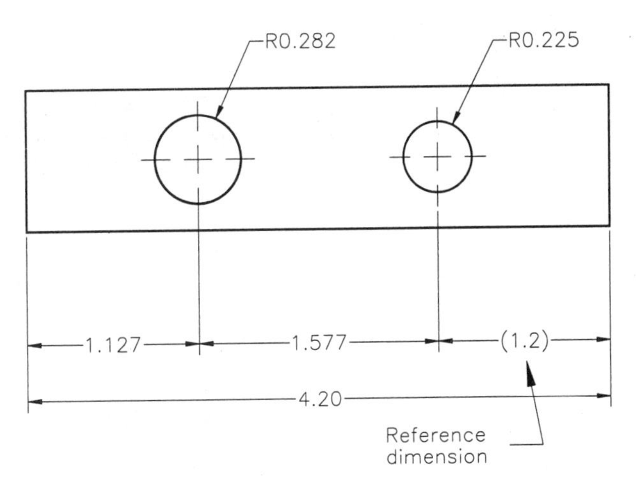

Figure 18-39 Reference dimensions

14. If you give continuous (incremental) dimensions for dimensioning various features of a part, the overall dimension must be omitted or given as a reference dimension (Figure 18-40). Similarly, if you give the overall dimension, one of the continuous (incremental) dimensions must be omitted or given as a reference dimension. Otherwise, there will be a conflict in tolerances. For example, the total positive tolerance on the three incremental dimensions shown in Figure 18-40 is 0.06. Therefore, the maximum size based on the incremental dimensions is (1 + 0.02) + (1 + 0.02) + (1 + 0.02) = 3.06. Also, the positive tolerance on the overall 3.0 dimension is 0.02. Based on this dimension, the overall size of the part must not exceed 3.02. This causes a conflict in tolerances: with incremental dimensions, the total tolerance is 0.06, whereas with the overall dimension the total tolerance is only 0.02.

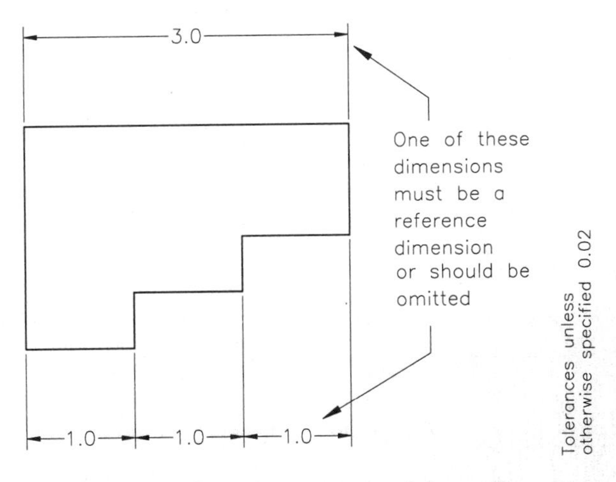

Figure 18-40 Referencing or omitting a dimension

15. If the dimension of a feature appears in a section view, you must not hatch the dimension text (Figure 18-41). You can accomplish it by selecting the dimension object when defining the hatch boundary. You can also accomplish this by drawing a rectangle around the dimension text and then hatching the area after excluding the rectangle from the hatch boundary. (You can also use the **EXPLODE** command to explode the dimension and then exclude the dimension text from hatching. This is not recommended because by exploding a dimension the associativity of the selected dimension is lost.)

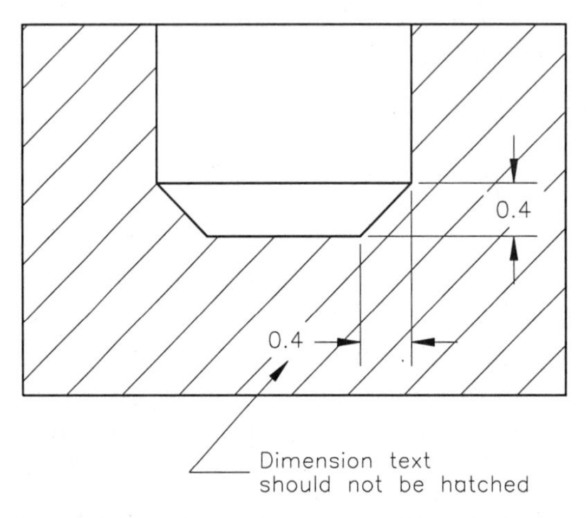

Figure 18-41 *Dimension text should not be hatched*

16. When dimensioning a circle, the diameter should be preceded by the diameter symbol (Figure 18-42). AutoCAD LT automatically puts the diameter symbol in front of the diameter value. However, if you override the default diameter value, you can use %%c followed by the value of the diameter (%%c1.25) to put the diameter symbol in front of the diameter dimension.

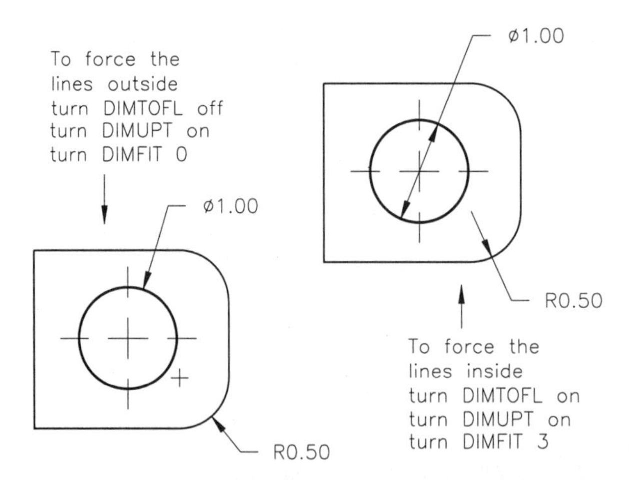

Figure 18-42 *Diameter should be preceded by the diameter symbol*

17. The circle must be dimensioned as a diameter, never as a radius. The dimension of an arc must be preceded by the abbreviation R (R1.25), and the center of the arc should be indicated by drawing a small cross. You can use the AutoCAD LT **DIMCEN** variable to control the size of the cross. If the value of this variable is 0, AutoCAD LT does not draw the cross in the center when you dimension an arc or a circle. You can also use the **DIMCENTER** command or the **CENTER** option of the **DIM** command to draw a cross at the center of the arc or circle.

18. When dimensioning an arc or a circle, the dimension line (leader) must be radial. Also, you should place the dimension text horizontally (Figure 18-43).

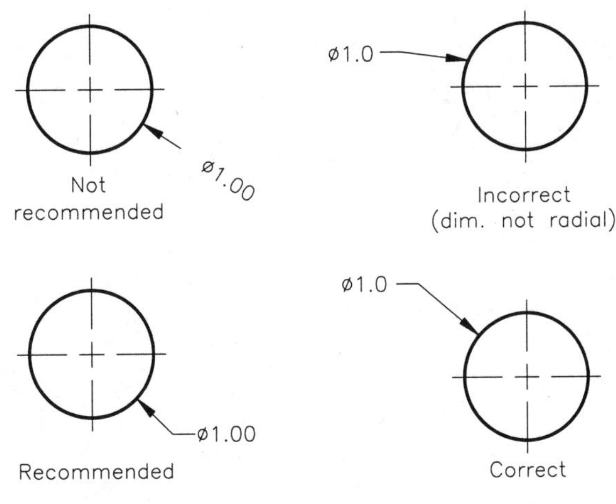

Figure 18-43 *Dimensioning a circle*

19. A chamfer can be dimensioned by specifying the chamfer angle and the distance of the chamfer from the edge. A chamfer can also be dimensioned by specifying the distances as shown in Figure 18-44.

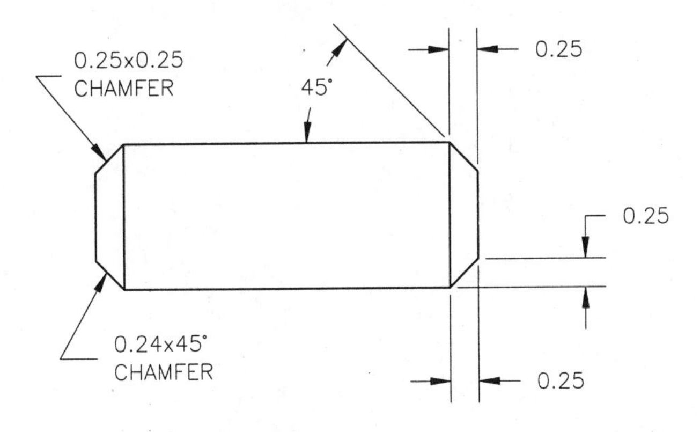

Figure 18-44 *Different ways of specifying chamfer*

20. A dimension that is not to scale should be indicated by drawing a straight line under the dimension text (Figure 18-45). You can draw this line by using the **DDEDIT** command. When you invoke this command, AutoCAD LT will prompt you to select an annotation object. When you select the object, the **Multiline Text Editor** dialog box is displayed. Select the text (< >), and then choose the **U** (underline) button.

21. A bolt circle should be dimensioned by specifying the diameter of the bolt circle, the diameter of the holes, and the number of holes in the bolt circle (Figure 18-46).

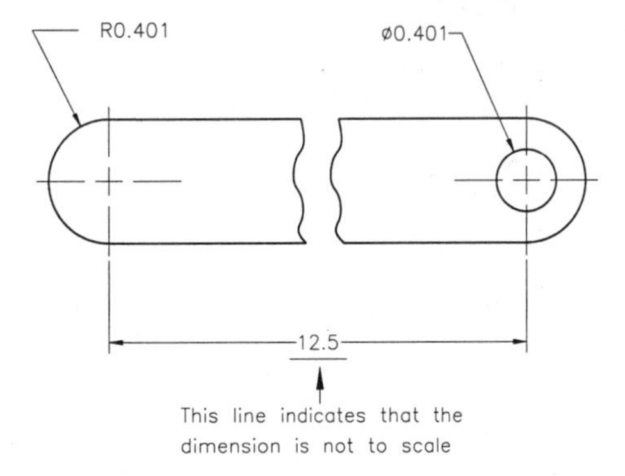

Figure 18-45 Specifying dimensions that are not to scale

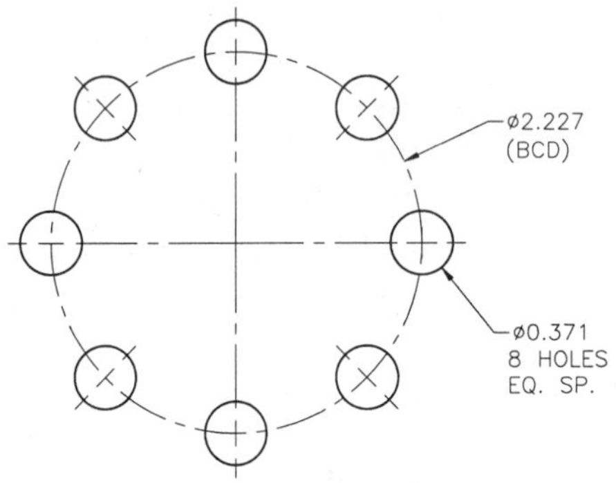

Figure 18-46 Dimensioning a bolt circle

Exercises 5 through 10 — *Mechanical*

Draw the required orthographic views of the following objects, and then give the dimensions (refer to Figures 18-47 through 18-52). The distance between the grid lines is 0.5 units.

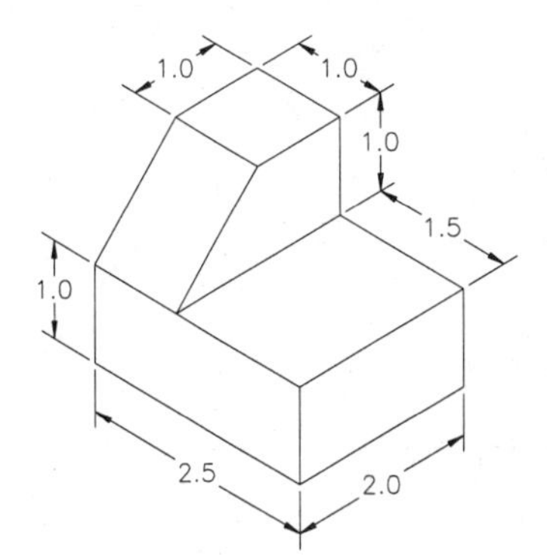

Figure 18-47 Drawing for Exercise 5

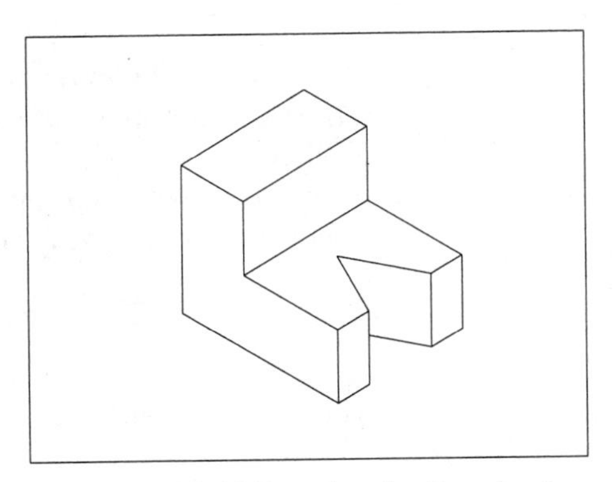

Figure 18-48 Drawing for Exercise 6

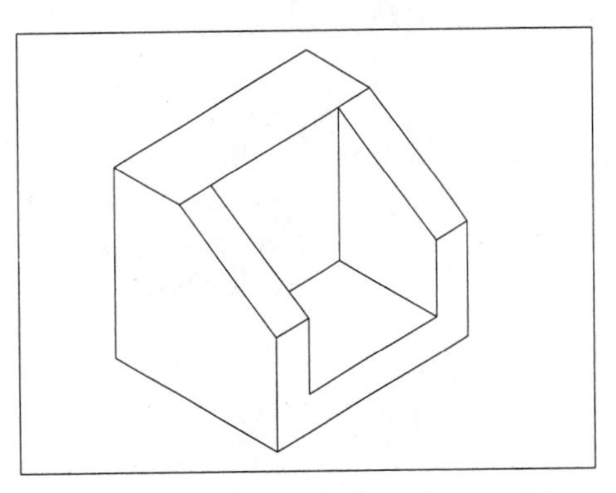

Figure 18-49 Drawing for Exercise 7

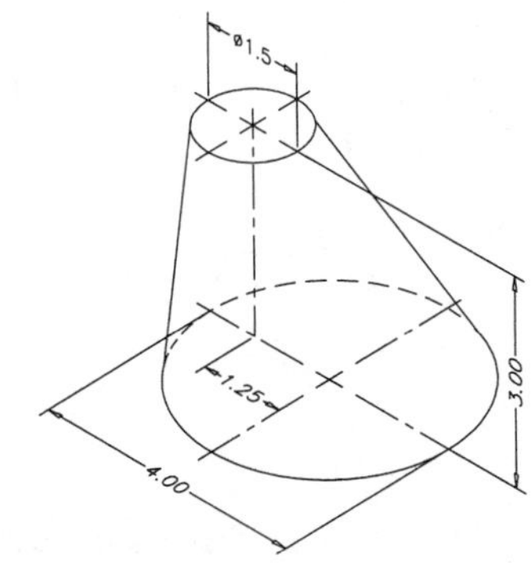

Figure 18-50 Drawing for Exercise 8

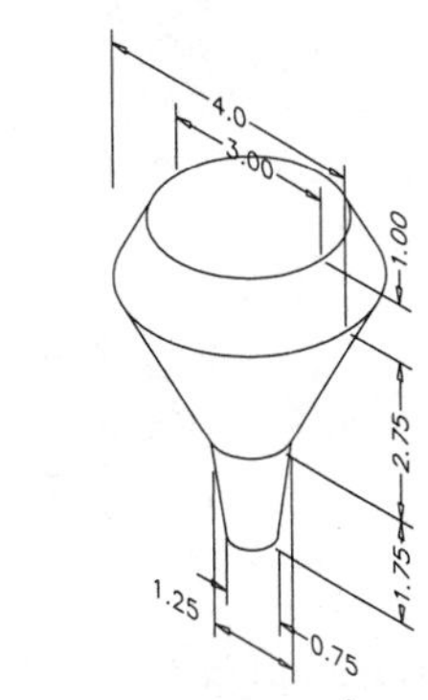

Figure 18-51 Drawing for Exercise 9

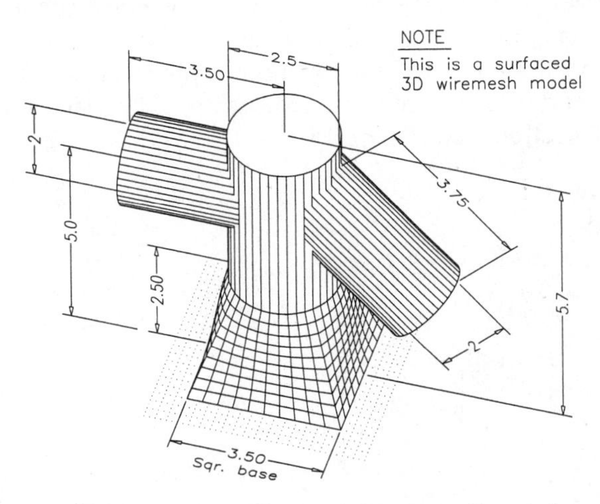

Figure 18-52 Drawing for Exercise 10 (assume the missing dimensions)

SECTIONAL VIEWS

In the principal orthographic views, the hidden features are generally shown by hidden lines. In some objects, the hidden lines may not be sufficient to represent the actual shape of the hidden feature. In such situations, sectional views can be used to show the features of the object that are not visible from outside. The location of the section and the direction of sight depend on the shape of the object and the features that need to be shown. Several ways to cut a section in the object are discussed next.

Full Section

Figure 18-53 shows an object that has a drilled hole, a counterbore, and a taper. In the orthographic views, these features will be shown by hidden lines (Figure 18-54).

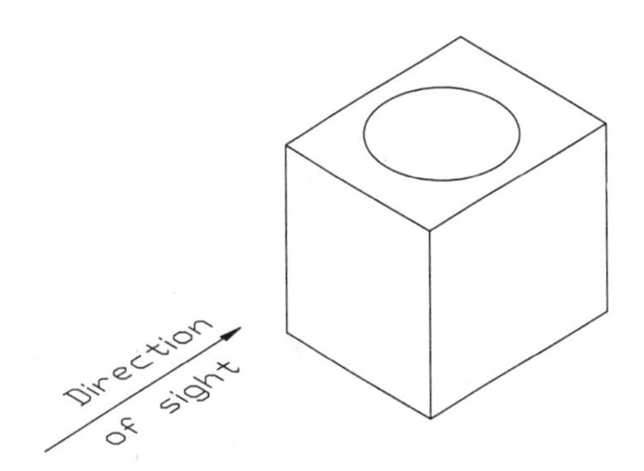

Figure 18-53 *Rectangular object with hole*

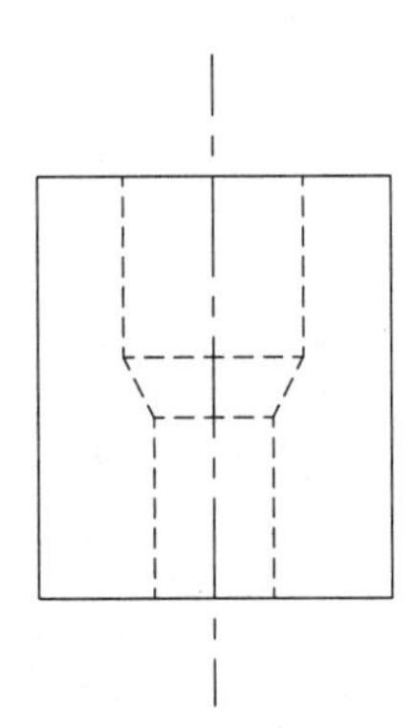

Figure 18-54 *Front view without section*

To better represent the hidden features, the object must be cut so the hidden features are visible. In the **full section**, the object is cut along the entire length of the object. To get a better idea of a full section, imagine that the object is cut into two halves along the centerline, as shown in Figure 18-55. Now remove the left half of the object and look at the right half in the direction that is perpendicular to the sectioned surface. The view you get after cutting the section is called a **full section** view (Figure 18-56).

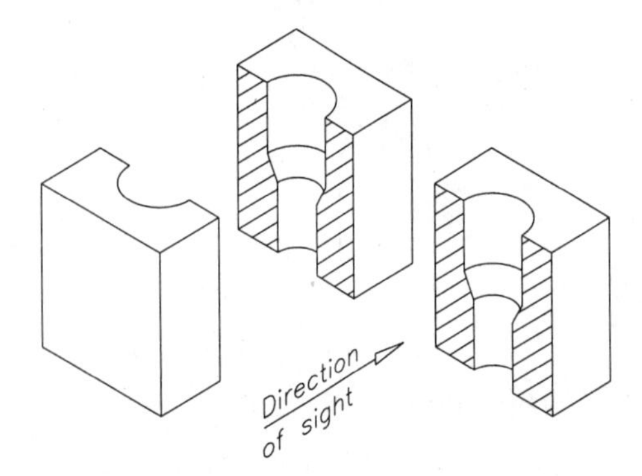

Figure 18-55 *One half of the object removed*

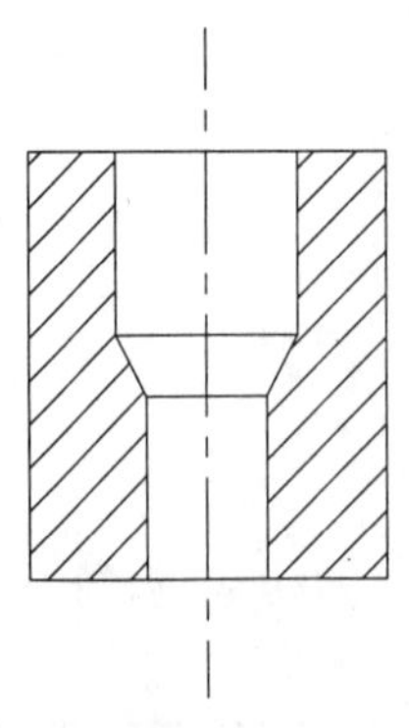

Figure 18-56 *Front view in full section*

In this section view, the features that would be hidden in a normal orthographic view are visible. Also, the part of the object where the material is actually cut is indicated by section lines. If the material is not cut, the section lines are not drawn. For example, if there is a hole, no material is cut when the part is sectioned, and so the section lines must not be drawn through that area of the section view.

Half Section

If the object is symmetrical, it is not necessary to draw a full section view. For example, in Figure 18-58 the object is symmetrical with respect to the centerline of the hole, so a full section is not required. Also, in some objects it may help to understand and visualize the shape of the hidden details better to draw the view in half section. In half section, one quarter of the object is cut, as shown in Figure 18-57. To draw the view in half section, imagine one quarter of the object removed, and then look in the direction that is perpendicular to the sectioned surface.

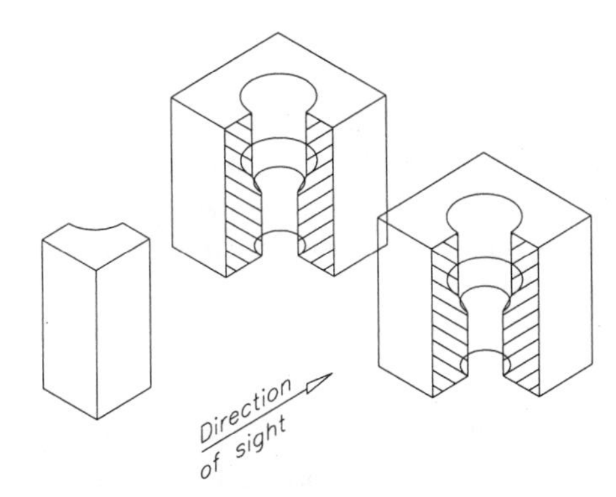

Figure 18-57 *One quarter of the object removed*

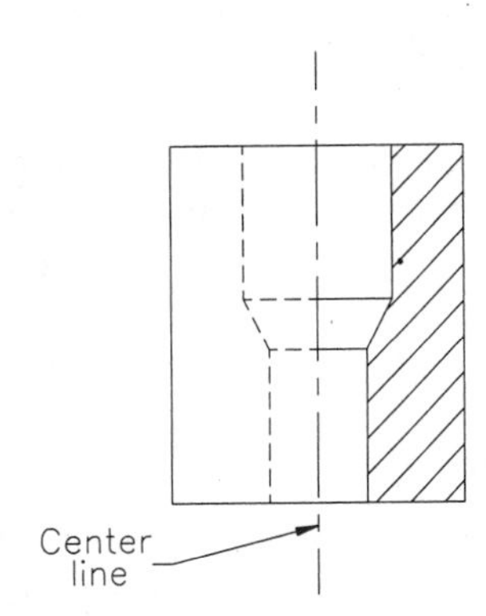

Figure 18-58 *Front view in half section*

You can also show the front view with a solid line in the middle, as in Figure 18-59. Sometimes the hidden lines representing the remaining part of the hidden feature are not drawn, as in Figure 18-60.

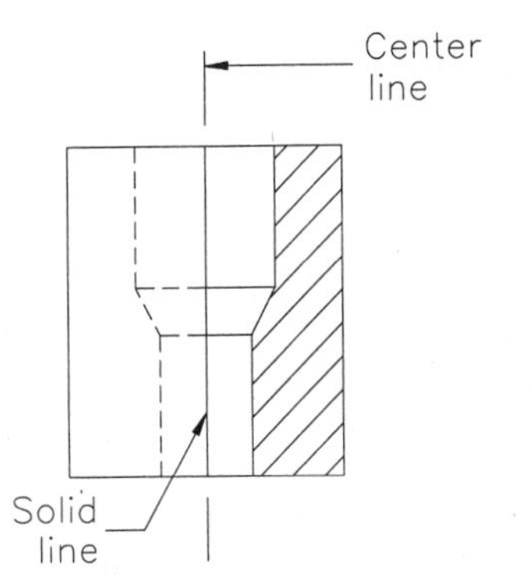

Figure 18-59 *Front view in half section*

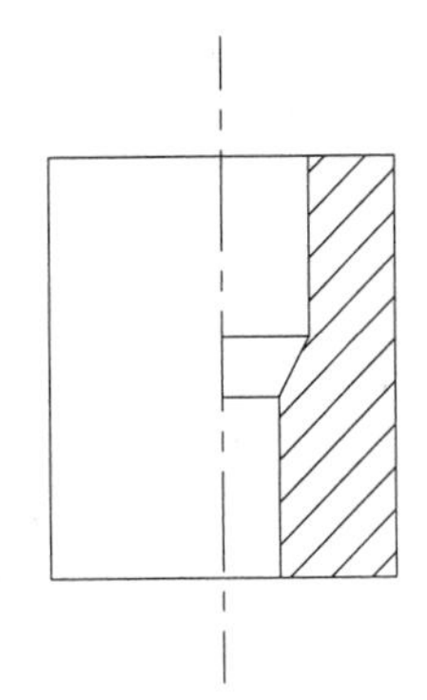

Figure 18-60 *Front view in half section*

Broken Section

In the **broken section**, only a small portion of the object is cut to expose the features that need to be drawn in section. The broken section is designated by drawing a thick zigzag line in the section view (Figure 18-61).

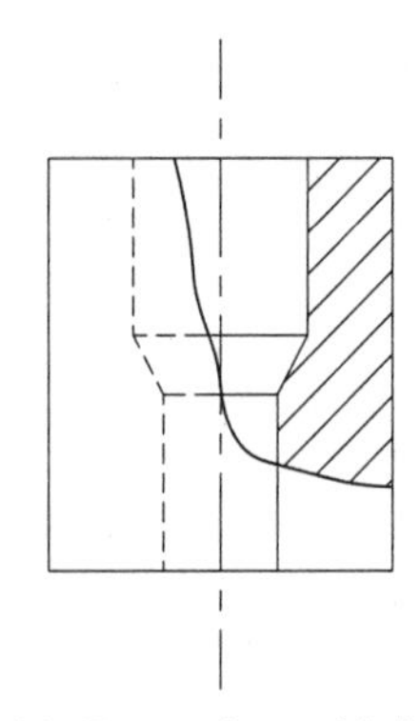

Figure 18-61 *Front view with broken section*

Revolved Section

The **revolved section** is used to show the true shape of the object at the point where the section is cut. The revolved section is used when it is not possible to show the features clearly in any principal view. For example, for the object in Figure 18-62, it is not possible to show the actual shape of the middle section in the front, side, or top view. Therefore, a revolved section is required to show the shape of the middle section.

The revolved section involves cutting an imaginary section through the object and then looking at the sectioned surface in a direction that is perpendicular to it. To represent the shape, the view is revolved 90 degrees and drawn in the plane of the paper, as shown Figure 18-62. Depending on the shape of the object, and for clarity, it is recommended to provide a break in the object so that the object lines do not interfere with the revolved section.

Removed Section

The **removed section** is similar to the revolved section, except that it is shown outside the object. The removed section is recommended when there is not enough space in the view to show the revolved section or if the scale of the section is different from the parent object. The removed section can be shown by drawing a line through the object at the point where the revolved section is desired and then drawing the shape of the section, as in Figure 18-63.

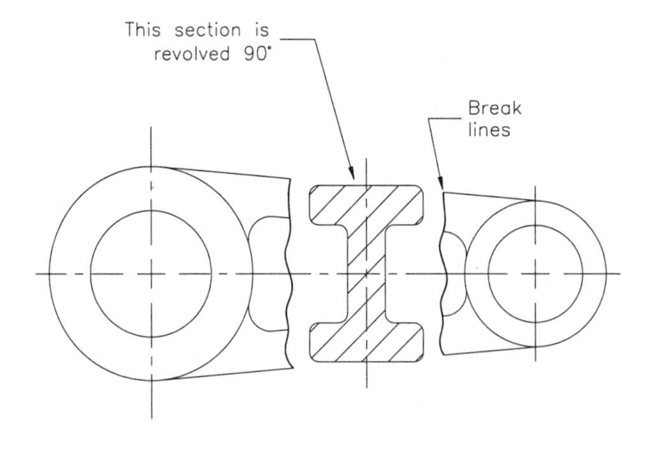

Figure 18-62 *Front view with revolved section*

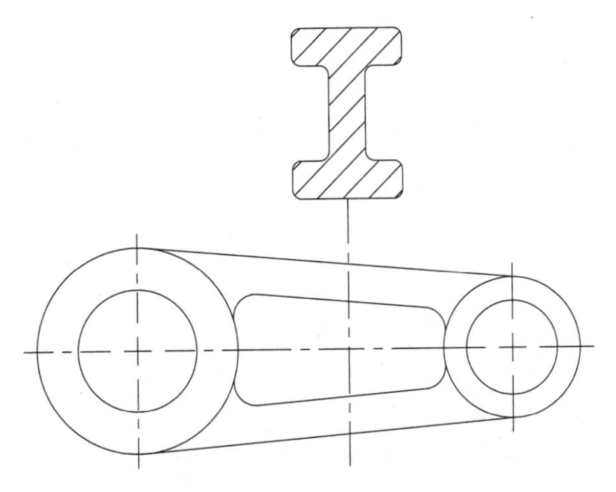

Figure 18-63 Front view with removed section

The other way of showing a removed section is to draw a cutting plane line through the object where you want to cut the section. The arrows should point in the direction in which you are looking at the sectioned surface. The section can then be drawn at a convenient place in the drawing. The removed section must be labeled as shown in Figure 18-64. If the scale has been changed, it must be mentioned with the view description.

Offset Section

The **offset section** is used when the features of the object that you want to section are not in one plane. The offset section is designated by drawing a cutting plane line that is offset through the center of the features that need to be shown in section (Figure 18-65). The arrows indicate the direction in which the section is viewed.

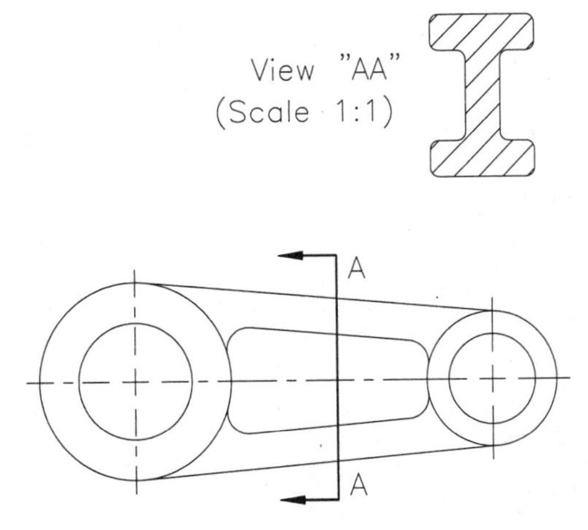

Figure 18-64 Front view with removed section

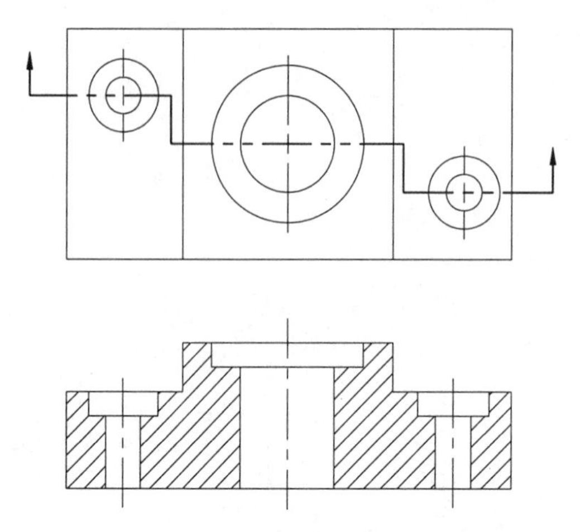

Figure 18-65 Front view with offset section

Aligned Section

In some objects, cutting a straight section might cause confusion in visualizing the shape of the section. Therefore, the aligned section is used to represent the shape along the cutting plane (Figure 18-66). Such sections are widely used in circular objects that have spokes, ribs, or holes.

Cutting Plane Lines

Cutting plane lines are thicker than object lines (Figure 18-67). You can use the **PLINE** command to draw the polylines of desired width, generally 0.005 to 0.01. However, for drawings that need to be plotted, you should assign a unique color to the cutting plane lines and then assign that color to the slot of the plotter that carries a pen of the required tip width. (For details, see Chapter 15, Plotting Drawings and Draw Commands.)

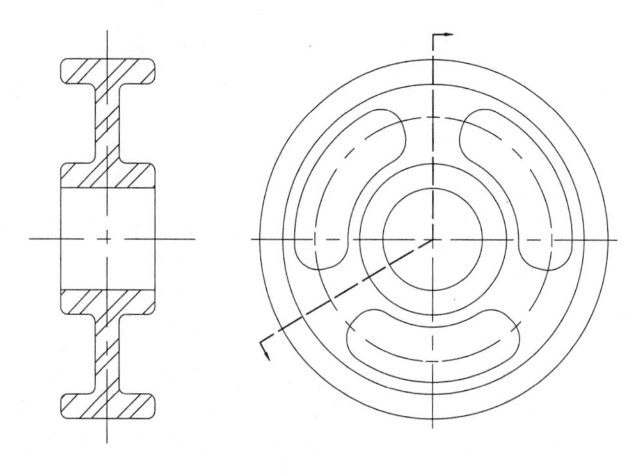

Figure 18-66 Side view in section (aligned section)

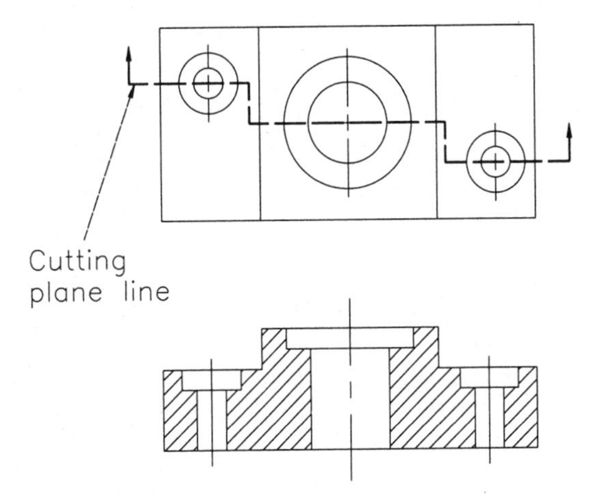

Figure 18-67 *Cutting plane line*

In industry, generally three types of lines are used to show the cutting plane for sectioning. The first line consists of a series of dashes 0.25 units long. The second type consists of a series of long dashes separated by two short dashes (Figure 18-68). The length of the long dash can vary from 0.75 to 1.5 units, and the short dashes are about 0.12 units long. The space between the dashes should be about 0.03 units.

Sometimes the cutting plane lines might clutter the drawing or cause confusion with other lines in the drawing. To avoid this problem, you can show the cutting plane by drawing a short line at the end of the section (Figures 18-68 and 18-69). The line should be about 0.5 units long.

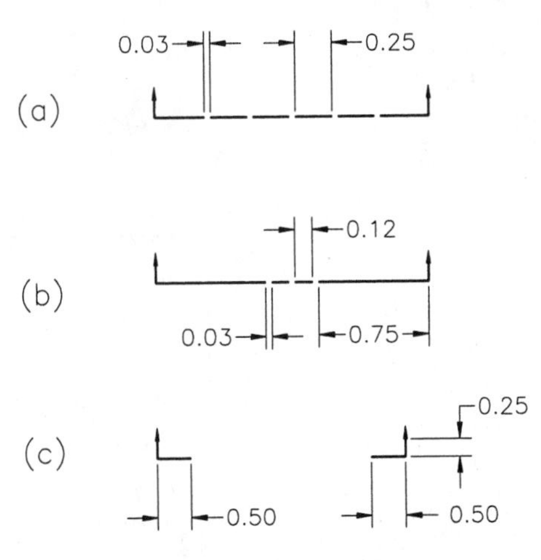

Figure 18-68 *Cutting plane lines*

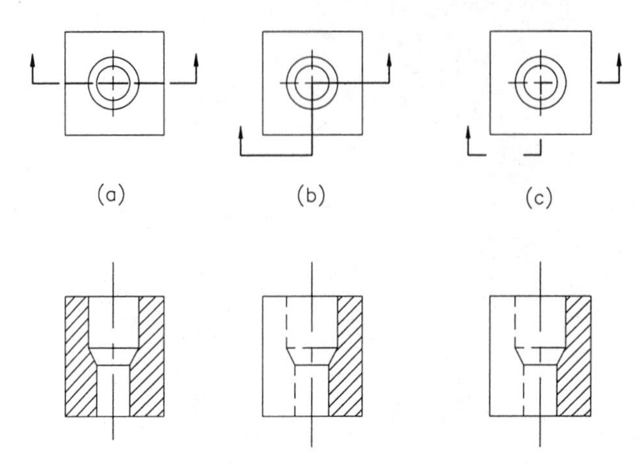

Figure 18-69 *Application of cutting plane lines*

Note

In AutoCAD LT, you can define a new linetype that you can use to draw the cutting plane lines. See Chapter 25, Creating Linetypes and Hatching Patterns, for more information on defining linetypes. Add the following lines to the ***ACLT.LIN*** *file, and then load the linetypes before assigning it to an object or a layer:*

**CPLANE1,___ ___ ___ ___*
A,0.25,-0.03
**CPLANE2,___ ___ ___ ___*
A,1.0,-0.03,0.12,-0.03,0.12,-0.03

Spacing for Hatch Lines

The spacing between the hatch (section) lines is determined by the space that is being hatched (Figure 18-70). If the hatch area is small, the spacing between the hatch lines should be smaller compared with a large hatch area.

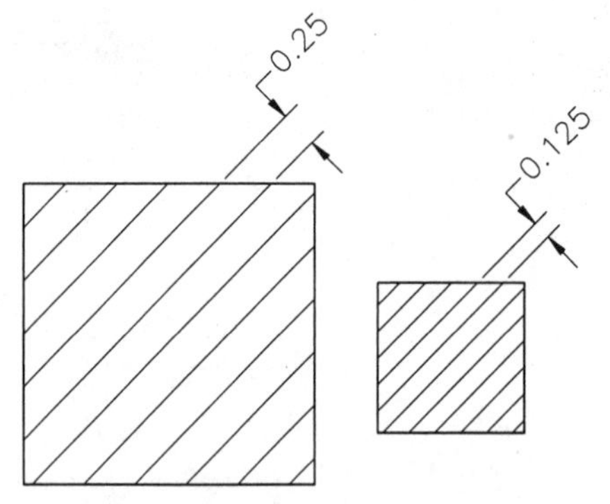

Figure 18-70 *Hatch line spacing*

In AutoCAD LT, you can control the spacing between the hatch lines by specifying the **scale factor** at the time of hatching. If the scale factor is 1, the spacing between the hatch lines is the same as defined in the hatch pattern file for that particular hatch. For example, in the following hatch pattern definition, the distance between the lines is 0.125:

***ANSI31, ANSI Iron, Brick, Stone masonry**
45, 0, 0, 0, .125

When the hatch scale factor is 1, the line spacing will be 0.125; if the scale factor is 2, the spacing between the lines will be 0.125 x 2 = 0.25.

Direction of Hatch Lines

The angle for the hatch lines should be 45 degrees. However, if there are two or more hatch areas next to one another representing different parts, the hatch angle must be changed so that the hatched areas look different (Figure 18-71).

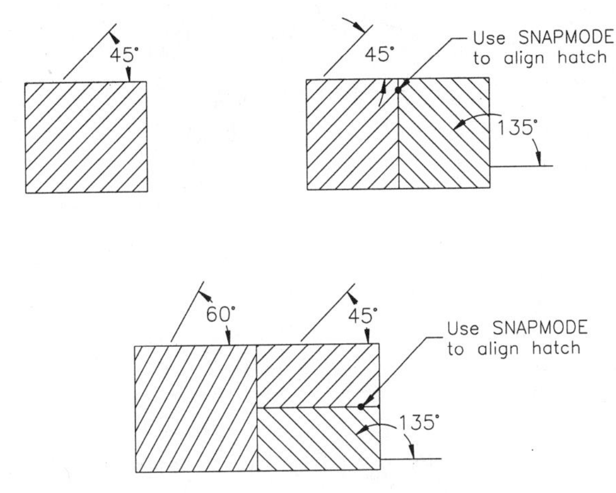

Figure 18-71 Hatch angle for adjacent parts

Also, if the hatch lines fall parallel to any edge of the hatch area, the hatch angle should be changed so that the lines are not parallel to any object line (Figure 18-72).

Points to Remember

1. Some parts, such as bolts, nuts, shafts, ball bearings, fasteners, ribs, spokes, keys, and other similar items that do not show any important feature, if sectioned, should not be shown in section.

2. Hidden details should not be shown in the section view unless the hidden lines represent an important detail or help the viewer to understand the shape of the object.

3. The section lines (hatch lines) must be thinner than the object lines. You can

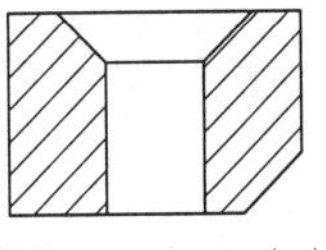

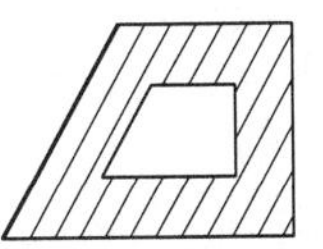

Figure 18-72 Hatch angle

accomplish this by assigning a unique color to hatch lines and then assigning the color to that slot on the plotter that carries a pen with a thinner tip.

4. The section lines must be drawn on a separate layer for display and editing purposes.

Exercises 11 and 12 *Mechanical*

In the following drawings (Figures 18-73 and 18-74), the views have been drawn without a section. Draw these views in section as indicated by the cutting plane lines in each object.

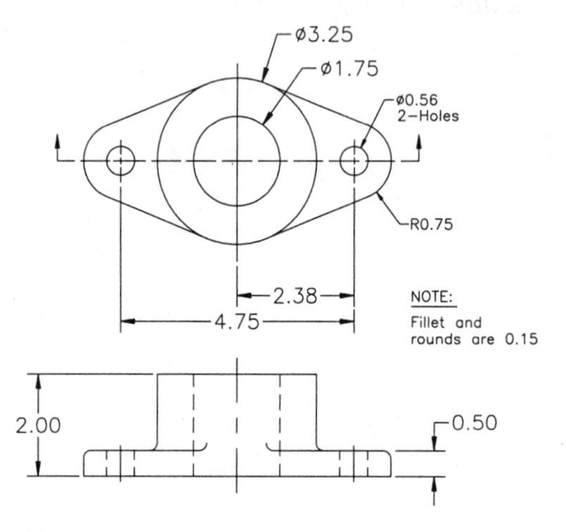

Figure 18-73 Drawing for Exercise 11: draw the front view in section

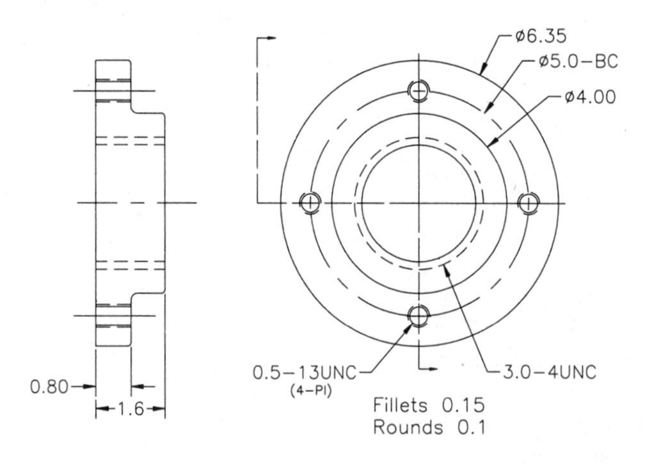

Figure 18-74 Drawing for Exercise 12: draw the left-hand side view in section

Exercises 13 and 14

Mechanical

Draw the required orthographic views for the following objects in Figures 18-75 and 18-76. Show the front view in section when the object is cut so that the cutting plane passes through the holes. Also, draw the cutting plane lines in the top view.

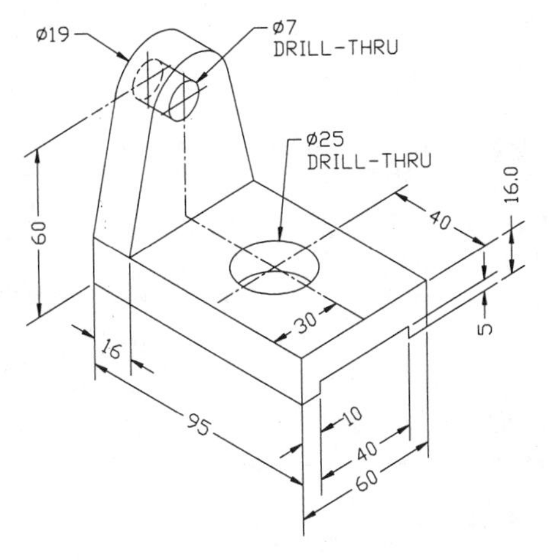

Figure 18-75 *Drawing for Exercise 13: draw the front view in full section*

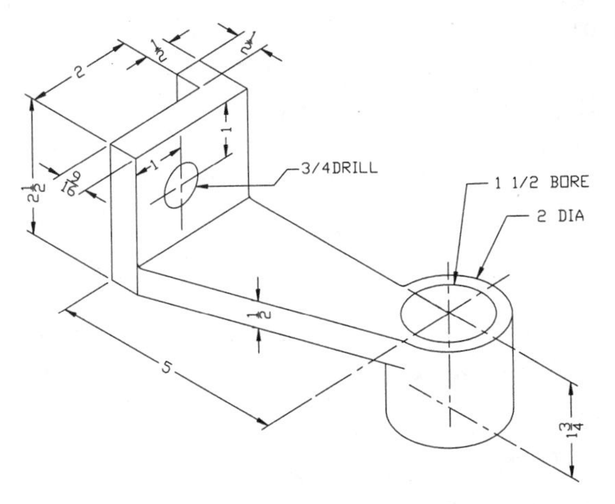

Figure 18-76 *Drawing for Exercise 14: draw the front view with offset section*

Exercise 15 — Mechanical

Draw the required orthographic views for the objects in Figures 18-77 and 18-78 with the front view in section. Also, draw the cutting plane lines in the top view to show the cutting plane. The material thickness is 0.25 units. (The object has been drawn as a surfaced 3D wiremesh model.)

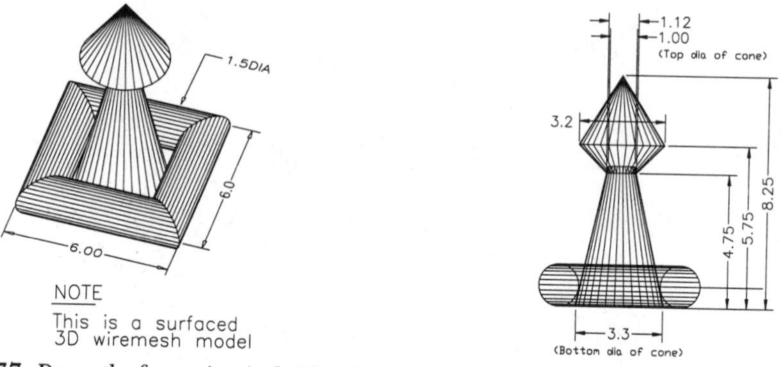

Figure 18-77 Draw the front view in half section *Figure 18-78 Draw the front view in half section*

AUXILIARY VIEWS

As discussed earlier, most objects generally require three principal views (front view, side view, and top view) to show all features of the object. Round objects may require just two views. Some objects have inclined surfaces. It may not be possible to show the actual shape of the inclined surface in one of the principal views. To get the true view of the inclined surface, you must look at the surface in a direction that is perpendicular to the inclined surface. Then you can project the points onto the imaginary auxiliary plane that is parallel to the inclined surface. The view you get after projecting the points is called the **auxiliary view**, as shown in Figures 18-79 and 18-80.

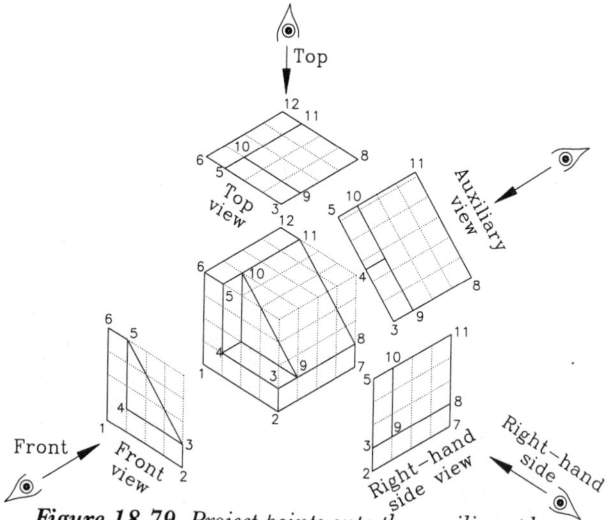

Figure 18-79 Project points onto the auxiliary plane

The auxiliary view in Figure 18-80 shows all features of the object as seen from the auxiliary view direction. For example, the bottom left edge is shown as a hidden line. Similarly, the lower right and upper left edges are shown as continuous lines. Although these lines are technically correct, the purpose of the auxiliary view is to show the features of the inclined surface. Therefore, in the auxiliary plane you should draw only those features that are on the inclined face, as shown in Figure 18-81. Other details that will help to understand the shape of the object may also be included in the auxiliary view. The remaining lines should be ignored because they tend to cause confusion in visualizing the shape of the object.

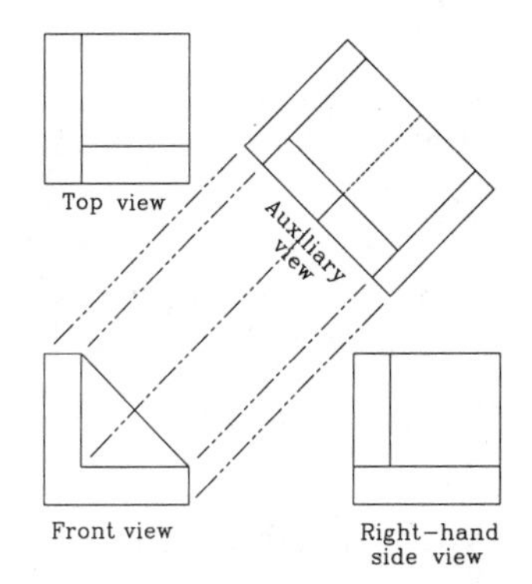

Figure 18-80 *Auxiliary, front, side, and top views*

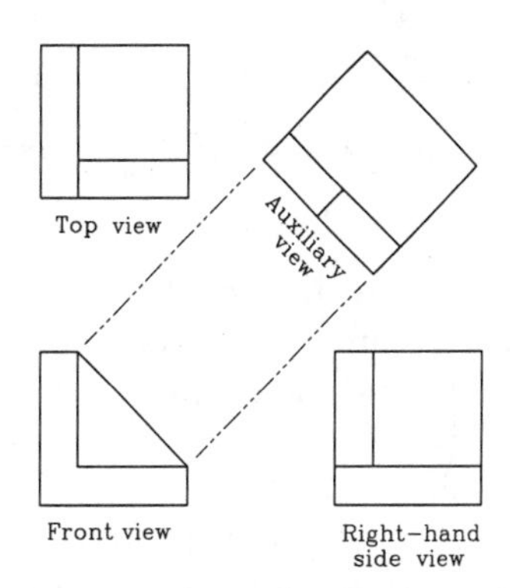

Figure 18-81 *Auxiliary, front, side, and top views*

How to Draw Auxiliary Views

The following example illustrates how to use AutoCAD LT to generate an auxiliary view.

Example 2

Draw the required views of the hollow triangular block with a hole in the inclined face as shown in Figure 18-82. (The block has been drawn as a solid model.)

The following steps are involved in drawing different views of this object.

Step 1. Draw the required orthographic views: the front view, side view, and the top view as shown in Figure 18-83(a). The circles on the inclined surface appear like ellipses in the front and top views. These ellipses may not be shown in the orthographic views because they tend to clutter the views [Figure 18-83(b)].

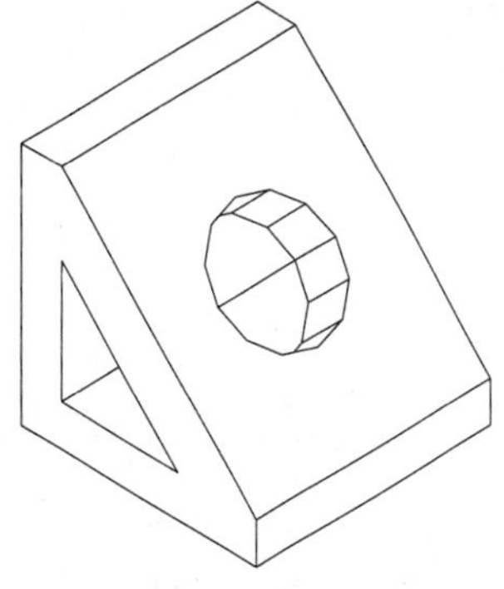

Figure 18-82 *Hollow triangular block*

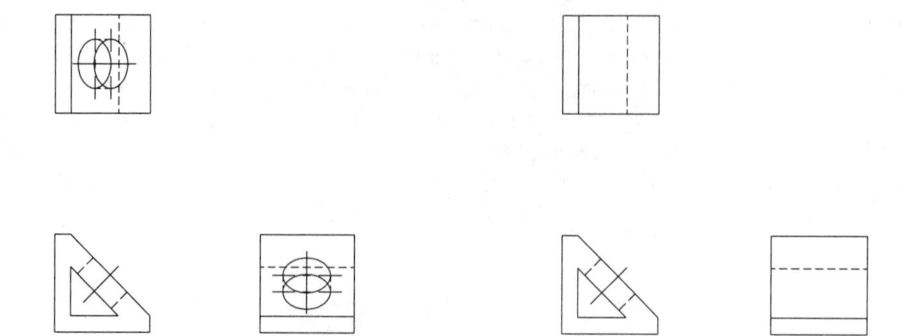

Figure 18-83(a) *Front, side, and top views*

Figure 18-83(b) *The ellipses may not be shown in the orthographic views*

Step 2. Determine the angle of the inclined surface. In this example, the angle is 45 degrees. Use the **SNAP** command to rotate the snap by 45 degrees (Figure 18-84) or rotate the UCS by 45 degrees around the Z axis.

Command: **SNAP** [Enter]
Specify snap spacing or [ON/OFF/Aspect/Rotate/Style/Type] <0.5000>: **R** [Enter]
Specify base point <0.0000,0.0000>: *Select P1.*
Specify rotation angle <0>: **45** [Enter]

Using the rotate option, the snap will be rotated by 45 degrees and the grid lines will also be displayed at 45 degrees (if the **GRID** is on). Also, one of the grid lines will pass through point P1 because it was defined as the base point.

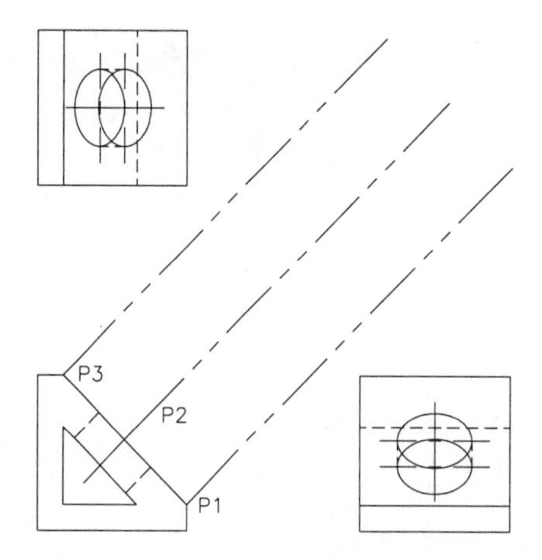

Figure 18-84 *Grid lines at 45 degrees*

Step 3. Turn **ORTHO** on, and project points P1, P2, and P3 from the front view onto the auxiliary plane. Now you can complete the auxiliary view and give dimensions. The projection lines can be erased after the auxiliary view is drawn.

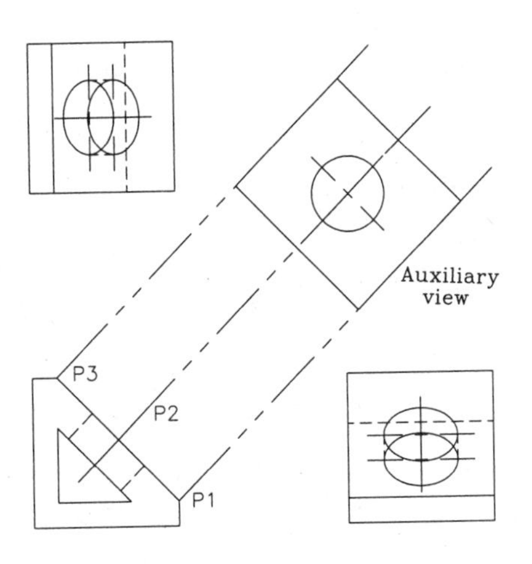

Figure 18-85 Auxiliary, front, side, and top views

Exercise 16

General

Draw the required orthographic and auxiliary views for the object in Figure 18-86. The object is drawn as a surfaced 3D wiremesh model.

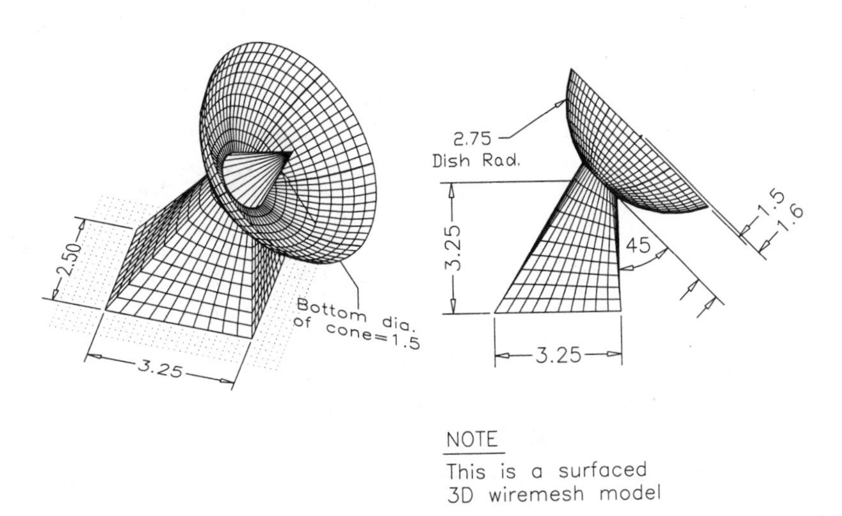

Figure 18-86 Drawing for Exercise 16: draw the required orthographic and auxiliary views

Exercises 17 and 18

General

Draw the required orthographic and auxiliary views for the following objects in Figures 18-87 and 18-88. The objects are drawn as 3D solid models and are shown at different angles (viewpoints are different).

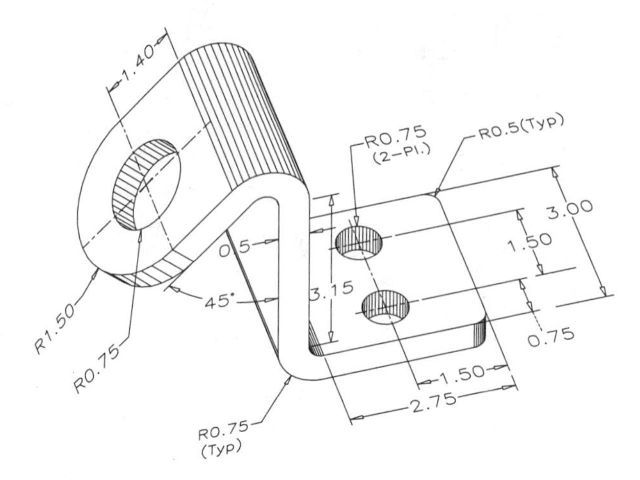

Figure 18-87 *Drawing for Exercise 17: draw the required orthographic and auxiliary views*

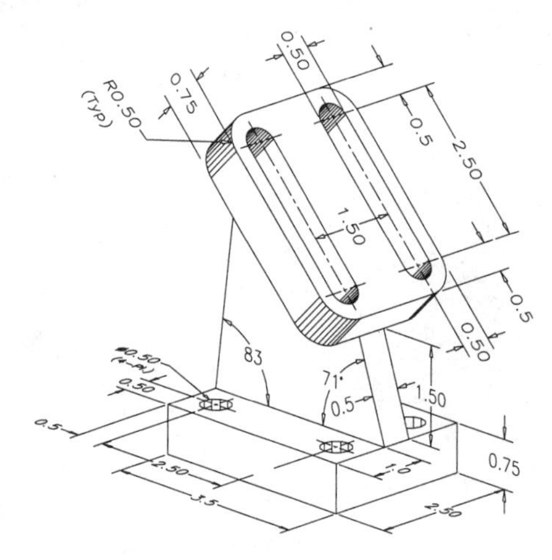

Figure 18-88 *Drawing for Exercise 18: draw the required orthographic and auxiliary views*

DETAIL DRAWING, ASSEMBLY DRAWING, AND BILL OF MATERIALS

Detail Drawing

A detail drawing is a drawing of an individual component that is a part of the assembled product. Detail drawings are also called **piece part drawings**. Each detail drawing must be drawn and dimensioned to completely describe the size and shape of the part. It should also contain information that might be needed in manufacturing the part. The finished surfaces should be indicated by using symbols or notes and all necessary operations on the drawing. The material of which the part is made and the number of parts that are required for production of the assembled product must be given in the title block. Detail drawings should also contain part numbers. This information is used in the bill of materials and the assembly drawing. The part numbers make it easier to locate the drawing of a part. You should make a detail drawing of each part, regardless of the part's size, on a separate drawing. When required, these detail drawings can be inserted in the assembly drawing by using the **XREF** command.

Assembly Drawing

The **assembly drawing** is used to show the parts and their relative positions in an assembled product or a machine unit. The assembly drawing should be drawn so that all parts can be shown in one drawing. This is generally called the main view. The main view may be drawn in full section so that the assembly drawing shows nearly all the individual parts and their locations. Additional views should be drawn only when some of the parts cannot be seen in the main view. The hidden lines, as far as possible, should be omitted from the assembly drawing because they clutter the drawing and might cause confusion. However, a hidden line may be drawn if it helps to understand the product. Only assembly dimensions should be shown in the assembly drawing. Each part should be identified on the assembly drawing by the number used in the detail drawing and in the bill of materials. The part numbers should be given as shown in Figure 18-89. It consists of a text string for the detail number, a circle (balloon), a leader line, and an arrow or dot. The text should be made at least 0.2 inches (5 mm) high and enclosed in a 0.4 inch (10 mm) circle (balloon). The center of the circle must be located not less than 0.8 inches (20 mm) from the nearest line on the drawing. Also, the leader line should be radial with respect to the circle (balloon). The assembly drawing may also contain an exploded isometric or isometric view of the assembled unit.

Bill of Materials

A **bill of materials** is a list of parts placed on an assembly drawing just above the title block (Figure 18-89). The bill of materials contains the part number, part description, material, quantity required, and drawing numbers of the detail drawings (Figure 18-90). If the bill of materials is placed above the title block, the parts should be listed in ascending order so that the first part is at the bottom of the table (Figure 18-91). The bill of materials may also be placed at the top of the drawing. In that case, the parts must be listed in descending order with the first part at the top of the table. This structure allows room for any additional items that may be added to the list.

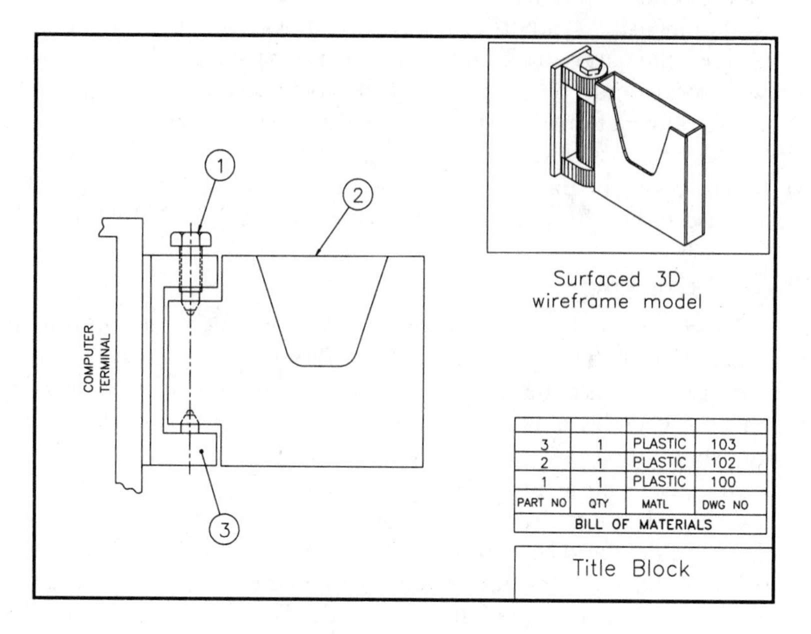

Figure 18-89 *Assembly drawing with title block, bill of materials, and surfaced 3D wireframe model*

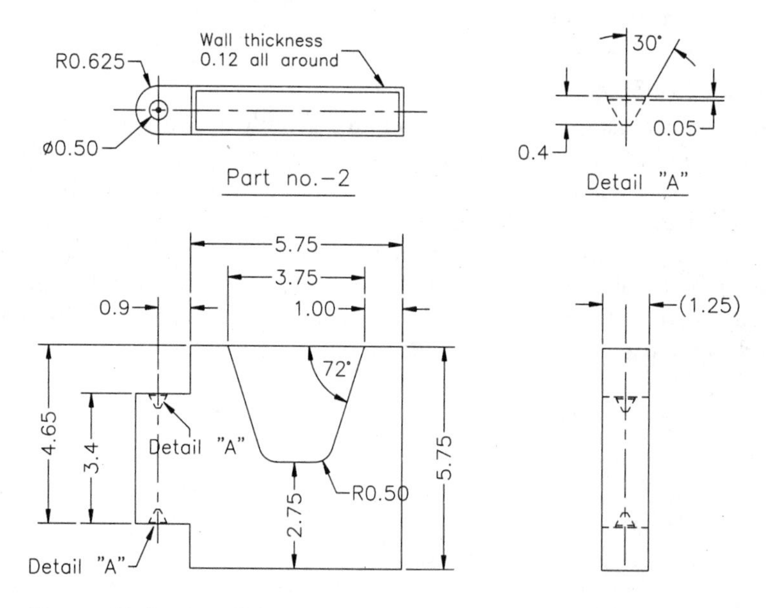

Figure 18-90 *Detail drawing (piece part drawing) of Part Number 2*

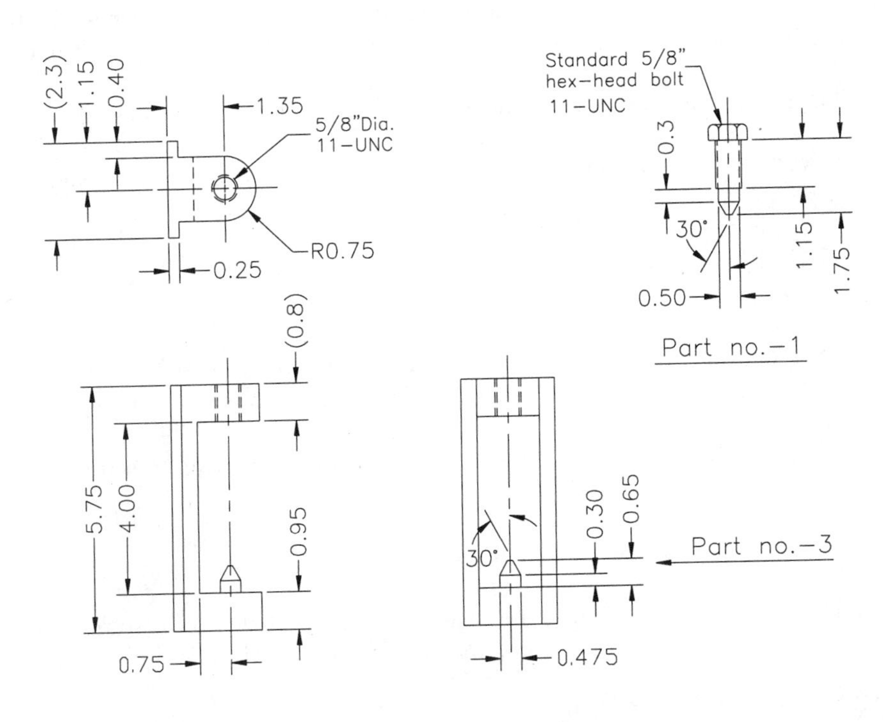

Figure 18-91 Detail drawings of Part Numbers 1 and 3

Self-Evaluation Test

Answer the following questions, and then compare your answers to the correct answers given at the end of this chapter.

1. If you draw a line perpendicular to the X and Y axes, this line defines the __________ axis.

2. The front view shows the maximum number of features or gives a better idea about the shape of the object. (T/F)

3. The number of decimal places in a dimension (for example, 2.000) determines the type of machine that will be used to do that machining operation. (T/F)

4. The dimension layer/layers should be assigned a unique color so that at the time of plotting you can assign a desired pen to plot the dimensions. (T/F)

5. All dimensions should be given inside the view. (T/F)

6. The dimensions can be shown inside the view if the dimensions can be easily understood there and cause no confusion with other object lines. (T/F)

7. The circle must be dimensioned as a(n) __________, never as a radius.

8. If you give continuous (incremental) dimensioning for dimensioning various features of a part, the overall dimension must be omitted or given as a(n) __________ dimension.

9. If the object is symmetrical, it is not necessary to draw a full section view. (T/F)

10 The removed section is similar to the __________section, except that it is shown outside the object.

11. In AutoCAD LT you can control the spacing between hatch lines by specifying the __________ at the time of hatching.

12. When dimensioning, you must consider the manufacturing process involved in making a part and the relationship that exists between different parts in an assembly. (T/F)

13. The distance between the first dimension line and the second dimension line must be __________ to __________units.

14. The reference dimension must be enclosed in __________.

15. A dimension must be given where the feature is visible. However, in some complicated drawings you might be justified to dimension a detail with a hidden line. (T/F)

Review Questions

1. Multiview drawings are also known as __________ drawings.

2. The space between the X and Y axes is called the XY plane. (T/F)

3. A plane that is parallel to the XY plane is called a parallel plane. (T/F)

4. If you look at the object along the negative Y axis and toward the origin, you will get the side view. (T/F)

5. The top view must be directly below the front view. (T/F)

6. Before drawing orthographic views, you must look at the object and determine the number of views required to show all features of the object. (T/F)

7. By dimensioning, you are not only giving the size of a part, you are giving a series of instructions to a machinist, an engineer, or an architect. (T/F)

8. What are the components of a dimension? ______________________________
____.

9. Why should you make the dimensions in a separate layer/layers? ______________________
____.

10. The distance of the first dimension line should be __________ to __________ units from the object line.

11. You can change grid, snap or UCS origin, and snap increments to make it easier to place the dimensions. (T/F)

12. For parallel dimension lines, the dimension text can be __________ if there is not enough room between the dimension lines to place the dimension text.

13. You should not dimension with hidden lines. (T/F)

14. A dimension that is not to scale should be indicated by drawing a straight line under the dimension text. (T/F)

15. The dimensions must be given where the feature that you are dimensioning is obvious and shows the contour of the feature. (T/F)

16. When dimensioning a circle, the diameter should be preceded by the __________.

17. When dimensioning an arc or a circle, the dimension line (leader) must be __________.

18. In radial dimensioning, you should place the dimension text vertically. (T/F)

19. If the dimension of a feature appears in a section view, you must hatch the dimension text. (T/F)

20. The dimensions must not be repeated; this makes it difficult to update a dimension, and the dimensions might get confusing. (T/F)

21. A bolt circle should be dimensioned by specifying the __________ of the bolt circle, the __________ of the holes, and the __________ of the holes in the bolt circle.

22. In the __________ section, the object is cut along the entire length of the object.

23. The part of the object where the material is actually cut is indicated by drawing __________ lines.

24. The __________ section is used to show the true shape of the object at the point where the section is cut.

25. The __________ section is used when the features of the object that you want to section are not in one plane.

26. Cutting plane lines are thinner than object lines. You can use the AutoCAD LT PLINE command to draw the polylines of desired width. (T/F)

27. The spacing between the hatch (section) lines is determined by the space being hatched. (T/F)

28. The angle for the hatch lines should be 45 degrees. However, if there are two or more hatch areas next to one another, representing different parts, the hatch angle must be changed so that the hatched areas look different. (T/F)

29. Some parts, such as bolts, nuts, shafts, ball bearings, fasteners, ribs, spokes, keys, and other similar items that do not show any important feature, if sectioned, must be shown in section. (T/F)

30. Section lines (hatch lines) must be thicker than object lines. You can accomplish this by assigning a unique color to hatch lines and then assigning the color to that slot on the plotter carrying a pen with a thicker tip. (T/F)

31. The section lines must be drawn on a separate layer for __________ and __________ purposes.

32. In a broken section, only a __________ of the object is cut to __________ the features that need to be drawn in section.

33. The assembly drawing is used to show the __________ and their __________ in an assembled product or a machine unit.

34. In an assembly drawing, additional views should be drawn only when some of the parts cannot be seen in the main view. (T/F)

35. Why shouldn't hidden lines be shown in an assembly drawing? ____________________
____.

Exercises

Exercises 19 through 24 *Mechanical*

Draw the required orthographic views of the following objects (the isometric view of each object is given in Figures 18-92 through 18-97). The dimensions can be determined by counting the number of grid lines. The distance between the isometric grid lines is assumed to be 0.5 units. Also dimension the drawings.

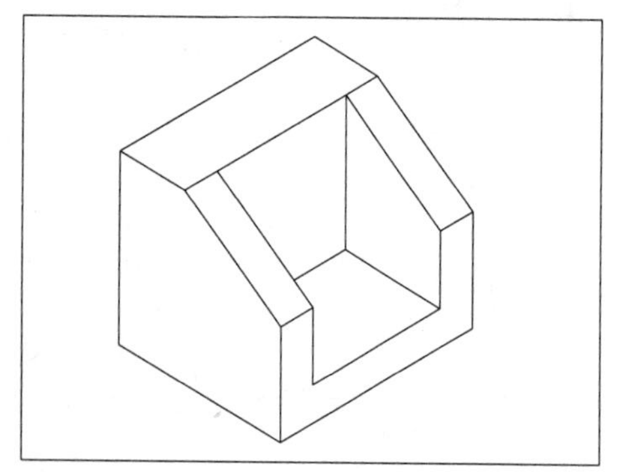

Figure 18-92 *Drawing for Exercise 19*

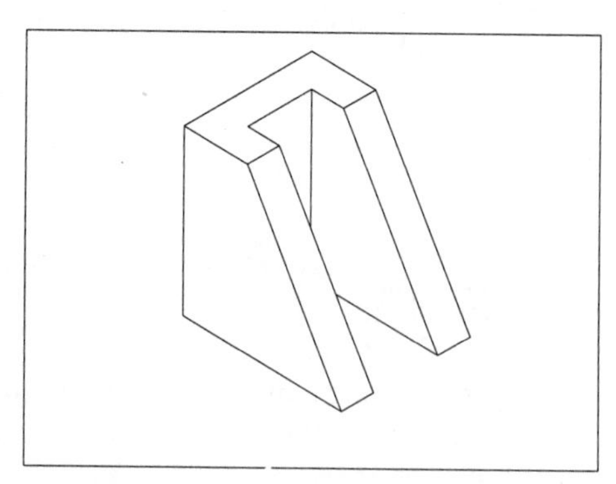

Figure 18-93 *Drawing for Exercise 20*

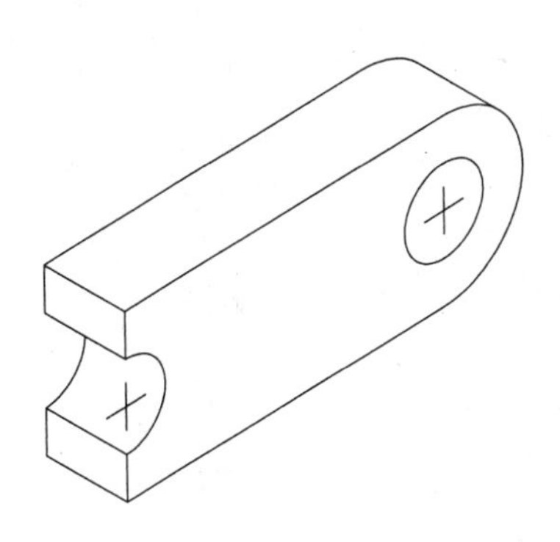

Figure 18-94 Drawing for Exercise 21

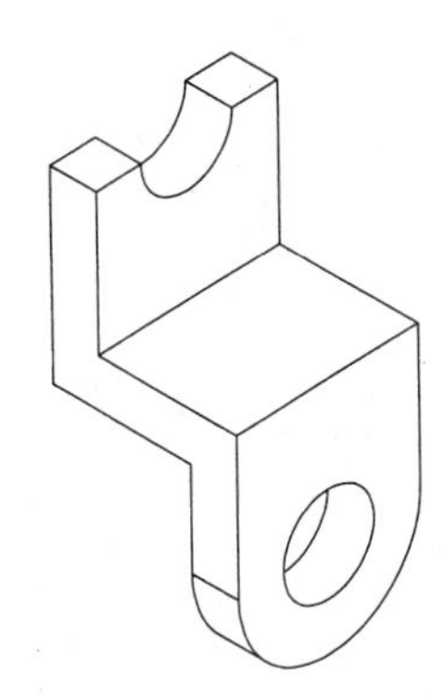

Figure 18-95 Drawing for Exercise 22

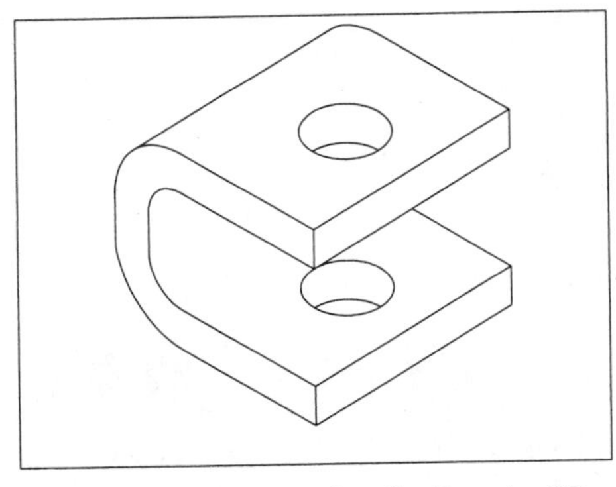

Figure 18-96 Drawing for Exercise 23

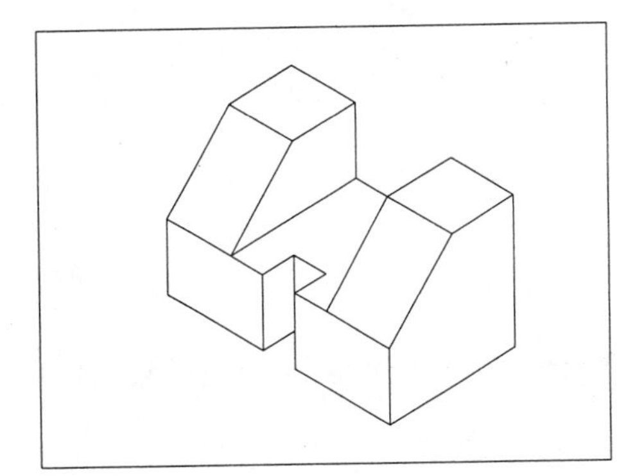

Figure 18-97 Drawing for Exercise 24

Exercise 25 — *Electronics*

Make the drawing as shown in Figure 18-98. Use layers and dimension the drawings where shown.

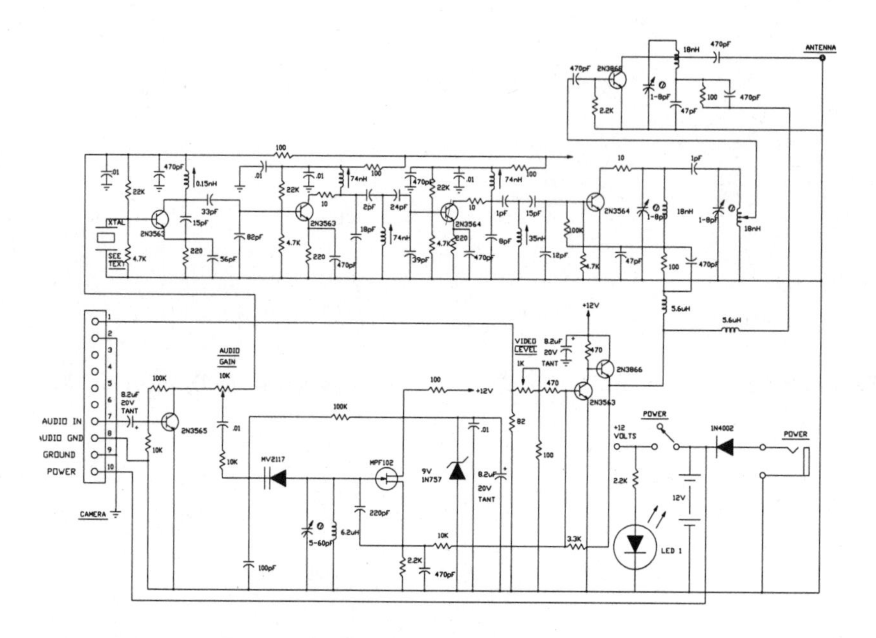

Figure 18-98 Drawing for Exercise 25

Exercises 26 through 30 *Mechanical*

Make the drawings as shown in Figures 18-99 through Figure 18-103. Use layers and dimension the drawings where shown.

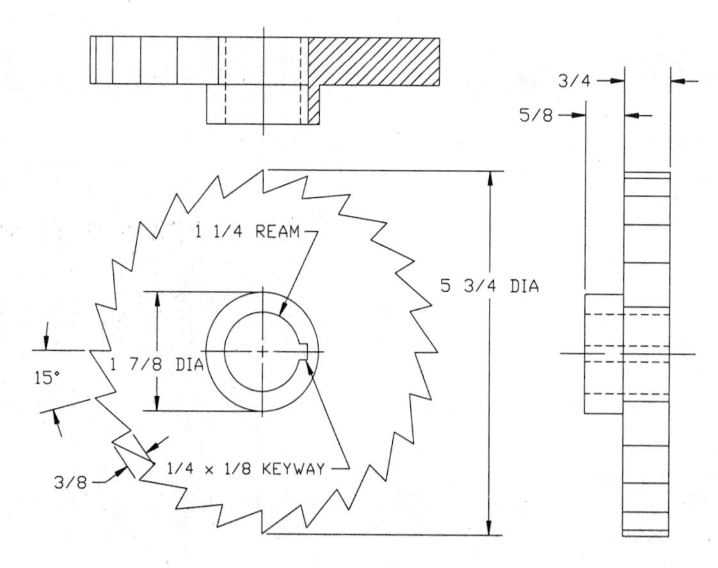

Figure 18-99 Drawing for Exercise 26

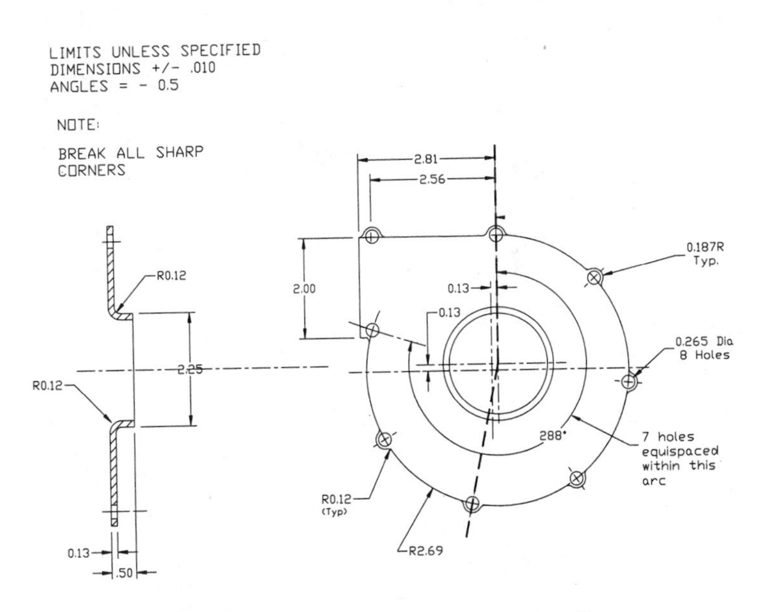

Figure 18-100 Drawing for Exercise 27

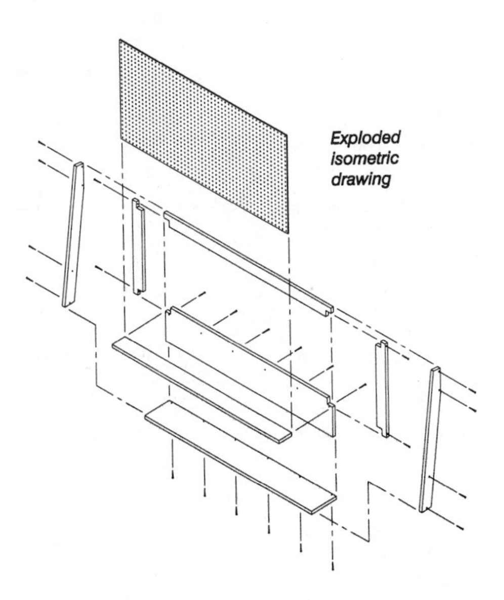

Figure 18-101 Drawing for Exercise 28, exploded isometric drawing of Tool Organizer (For dimensions, see Project Exercise 2 in Chapters 9, 14, and 23)

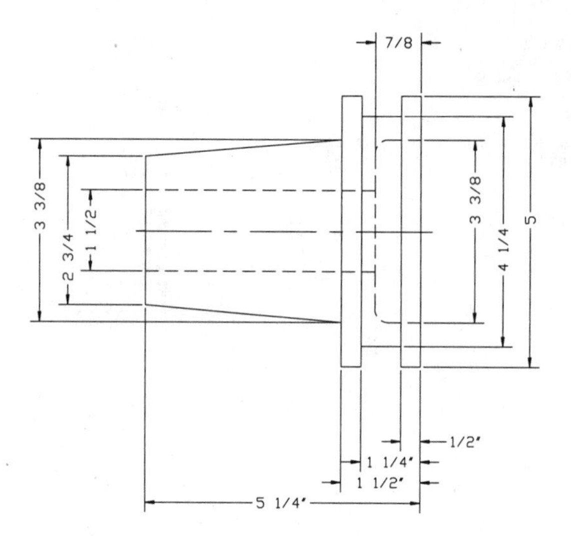

Figure 18-102 *Drawing for Exercise 29*

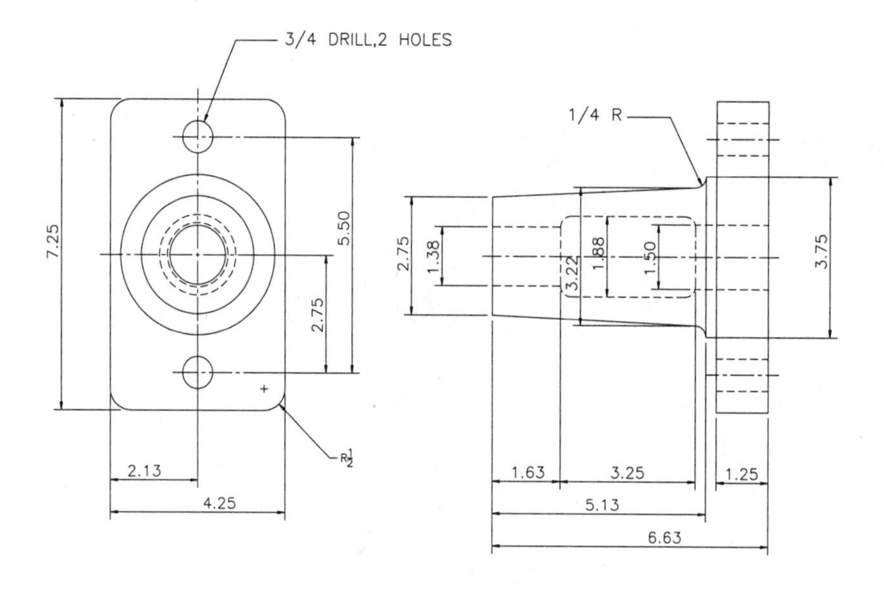

Figure 18-103 *Drawing for Exercise 30*

Answers to Self-Evaluation Test

1 - Z, **2** - T, **3** - T, **4** - T, **5** - F, **6** - T, **7** - diameter, **8** - reference, **9** - T, **10** - revolved, **11** - scale factor, **12** - T, **13** - 0.25 to 0.5, **14** - parentheses, **15** - T

Chapter 19

Isometric Drawing

Learning Objectives

After completing this chapter, you will be able to:

- *Understand isometric drawings, isometric axes, and isometric planes.*
- *Set isometric grid and snap.*
- *Draw isometric circles in different isoplanes.*
- *Dimension isometric objects.*
- *Place text in an isometric drawing.*

ISOMETRIC DRAWINGS

Isometric drawings are generally used to help visualize the shape of an object. For example, if you are given the orthographic views of an object (Figure 19-1), it takes time to put information together to visualize the shape. However, if an isometric drawing is given (Figure 19-2), it is

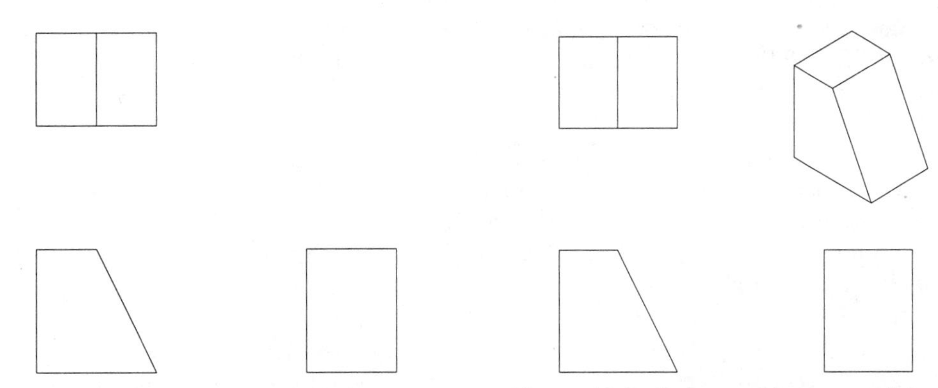

Figure 19-1 *Orthographic views of an object*

Figure 19-2 *Orthographic views with isometric drawing*

much easier to conceive the shape of the object. Thus, isometric drawings are widely used in industry to aid in understanding products and their features.

An isometric drawing should not be confused with a three-dimensional (3D) drawing. An isometric drawing is just a two-dimensional representation of a three-dimensional drawing in a 2D plane. A 3D drawing is a 3D model of an object on the X,Y and Z axes. An isometric drawing is a two-dimensional (2D) drawing in a 2D plane. A 3D drawing is a true 3D model of the object. The model can be rotated and viewed from any direction. A 3D model can be a wireframe model, surface model, or solid model.

ISOMETRIC PROJECTIONS

The word isometric means "**equal measure**" because the three angles between the three principal axes of an isometric drawing are each 120 degrees (Figure 19-3). An isometric view is obtained by rotating the object 45 degrees around the imaginary vertical axis, and then tilting the object forward through a 35°16' angle. If you project the points and the edges on the frontal plane, the projected length of the edges will be approximately 81 percent (isometric length/actual length = 9/11) shorter than the actual length of the edges. However, isometric drawings are always drawn to full scale because their purpose is to help the user visualize the shape of the object. Isometric drawings are not meant to describe the actual size of the object. The actual dimensions, tolerances, and feature symbols must be shown in the orthographic views. Also, you should avoid showing any hidden lines in the isometric drawings, unless they show an important feature of the object or help in understanding the shape of the object.

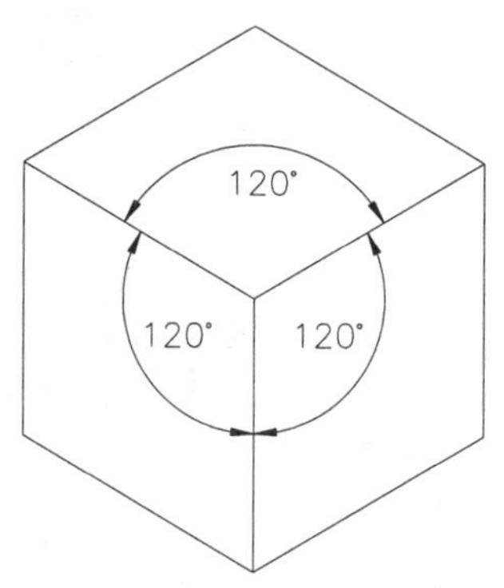

***Figure 19-3** Principal axes of an isometric drawing*

ISOMETRIC AXES AND PLANES

Isometric drawings have three axes: **right horizontal axis** (P0,P1), **vertical axis** (P0,P2), and **left horizontal axis** (P0,P3). The two horizontal axes are inclined at 30 degrees to the horizontal, or X, axis (X1,X2). The vertical axis is at 90 degrees (Figure 19-4).

When you draw an isometric drawing, the horizontal object lines are drawn along or parallel to the horizontal axis. Similarly, the vertical lines are drawn along or parallel to the vertical axis. For example, if you want to make an isometric drawing of a rectangular block, the vertical edges of the block are drawn parallel to the vertical axis. The horizontal edges on the right-hand side of the block are drawn parallel to the right horizontal axis (P0,P1), and the horizontal edges on the left side of the block are drawn parallel to the left horizontal axis (P0,P3). It is important to remember that the **angles do not appear true** in isometric drawings. Therefore, the edges or surfaces that are at an angle are drawn by locating the endpoints of the edges. The lines that are parallel to the isometric axes are called **isometric lines**. The lines that are not parallel to the isometric axes are called **nonisometric lines**.

Similarly, the planes can be **isometric planes** or **nonisometric planes**.

Isometric drawings have three principal planes, **isoplane right**, **isoplane top**, and **isoplane left**, as shown in (Figure 19-5). The isoplane right (P0,P4,P10,P6) is the plane as defined by the vertical axis and the right horizontal axis. The isoplane top (P6,P10,P9,P7) is the plane as defined by the right and left horizontal axes. Similarly, the isoplane left (P0,P6,P7,P8) is defined by the vertical axis and the left horizontal axis.

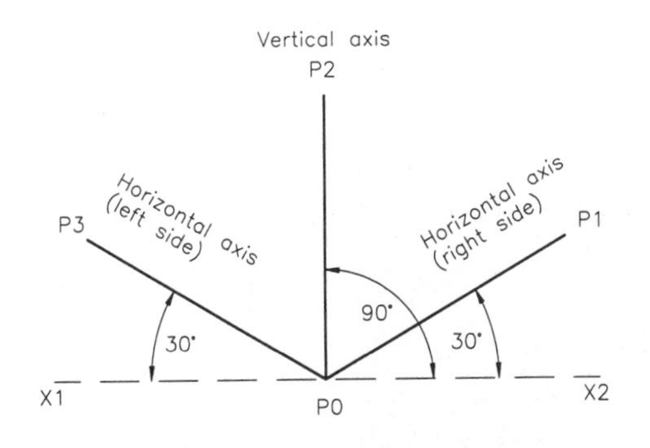

Figure 19-4 Isometric axes

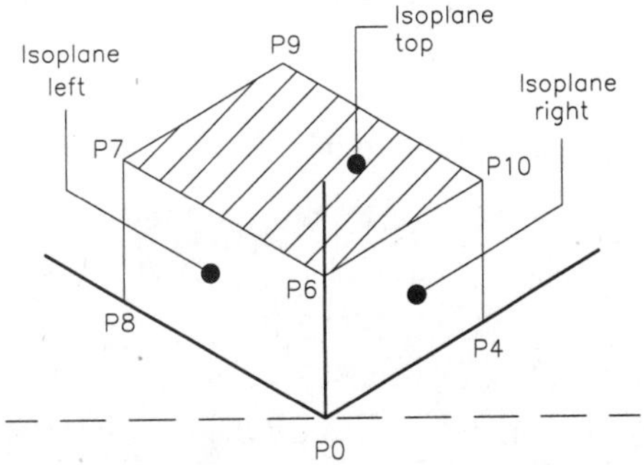

Figure 19-5 Isometric planes

SETTING THE ISOMETRIC GRID AND SNAP

You can use the **SNAP** command to set the isometric grid and snap. The isometric grid lines are displayed at 30 degrees to the horizontal axis. Also, the distance between the grid lines is determined by the vertical spacing, which can be specified by using the **GRID** or **SNAP** command. The grid lines coincide with the three isometric axes, which makes it easier to create isometric drawings. The following command sequence illustrates the use of the **SNAP** command to set the isometric grid and snap of 0.5 units (Figure 19-6).

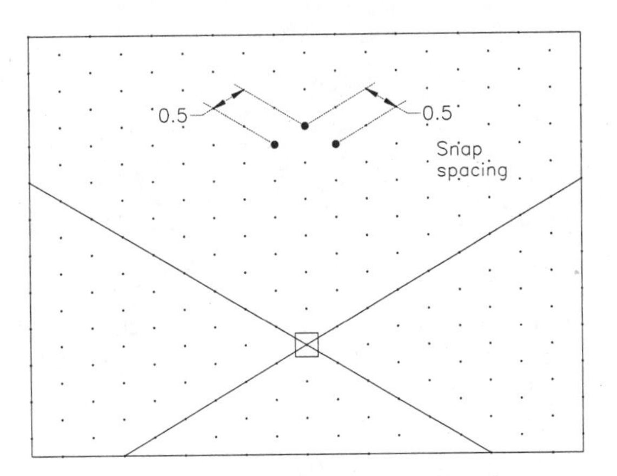

Figure 19-6 Setting isometric grid and snap

Command: **SNAP**
Specify snap spacing or [ON/OFF/Aspect/Rotate/Style/Type] <0.5000>: **S**

Enter snap grid style [Standard/Isometric] <S>: **I**
Specify vertical spacing <0.5000>: *Enter new snap distance.*

Note

*When you use the **SNAP** command to set the isometric grid, the grid lines may not be displayed. To display the grid lines, turn the grid on using the **GRID** command or press function key F7.*

You cannot set the aspect ratio for the isometric grid. Therefore, the spacing between the isometric grid lines will be the same.

You can also set the isometric grid and snap by using the **Drafting Settings** dialog box, Figure 19-7, that can be invoked by entering **DSETTINGS** at the Command prompt.

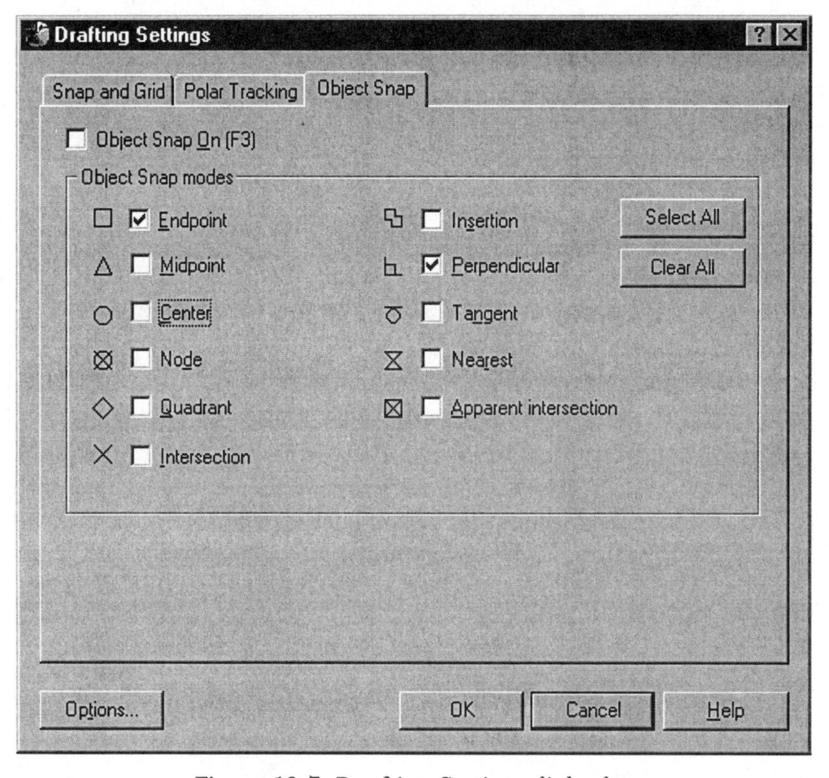

***Figure 19-7 Drafting Settings** dialog box*

You can also access this dialog box by selecting **Drafting Settings** from the **Tools** menu. You can also invoke the dialog box by right-clicking **Snap**, **Grid**, **Polar**,or **Osnap**, on the status bar and choosing **Settings** from the shortcut menu.

The isometric snap and grid can be turned on/off by choosing the **On** box located in the **Snap and Grid/Object Snap** tabs of the **Drafting Settings** dialog box. The **Snap and Grid** tab also contains the radio buttons to set the snap type and style. To display the grid on the screen, make sure the grid is turned on.

When you set the isometric grid, the display of the crosshairs also changes. The crosshairs are displayed at an isometric angle, and their orientation depends on the current isoplane. You can toggle among isoplane right, isoplane left, and isoplane top by pressing the CTRL and E keys simultaneously (Ctrl+E) or using the function key F5. You can also toggle among different isoplanes by using the **Drafting Settings** dialog box or by entering the **ISOPLANE** command at the Command prompt:

Command line: **ISOPLANE**
Enter isometric plane setting [Left/Top/Right] <Top>: **T**
Current Isoplane: **Top**

The Ortho mode is often useful when drawing in Isometric mode. In Isometric mode, Ortho aligns with the axes of the current isoplane.

Example 1

In this example, you will create the isometric drawing in Figure 19-8.

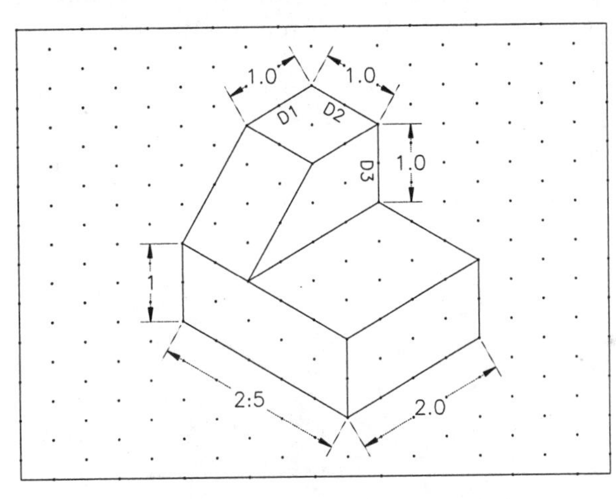

Figure 19-8 *Drawing for Example 1*

1. Use the **SNAP** command to set the isometric grid and snap. The snap value is 0.5 units.

   Command: **SNAP**
   Specify snap spacing or [ON/OFF/Aspect/Rotate/Style/Type] <0.5000>: **S**
   Enter snap grid style [Standard/Isometric] <S>: **I**
   Specify vertical spacing <0.5000>: **0.5** (*or press ENTER.*)

2. Change the isoplane to isoplane left by pressing the CTRL and E keys. Enter the **LINE** command and draw lines between points P1, P2, P3, P4, and P1 as shown in Figure 19-9.

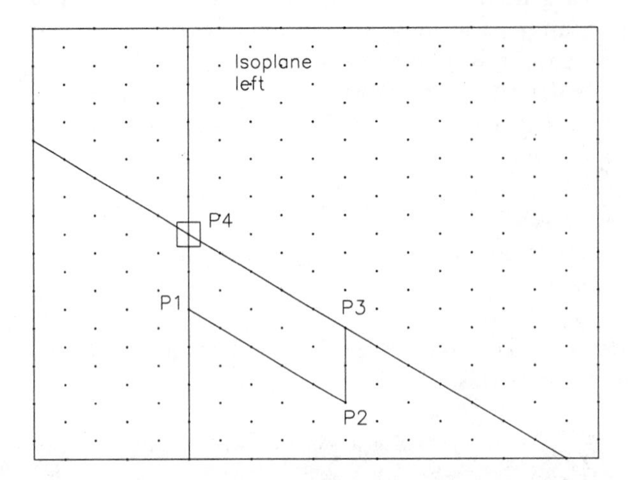

Figure 19-9 *Draw the bottom left face*

3. Change the isoplane to isoplane right by pressing the CTRL and E keys. Enter the **LINE** command and draw the lines as shown in Figure 19-10.

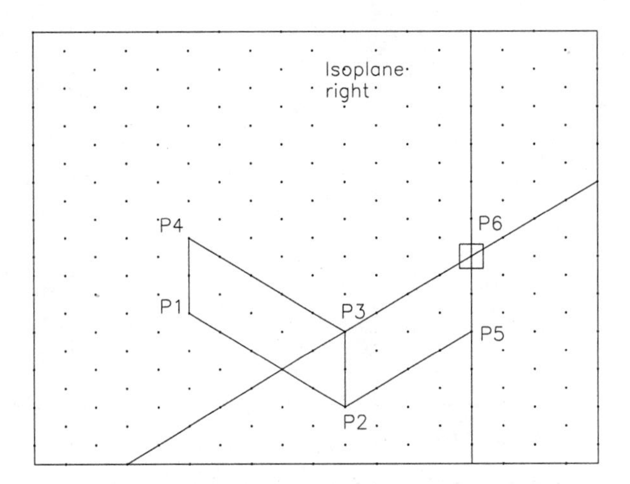

Figure 19-10 *Draw the bottom right face*

4. Change the isoplane to isoplane top by pressing the **CTRL** and **E** keys. Enter the **LINE** command and draw the lines as shown in Figure 19-11.

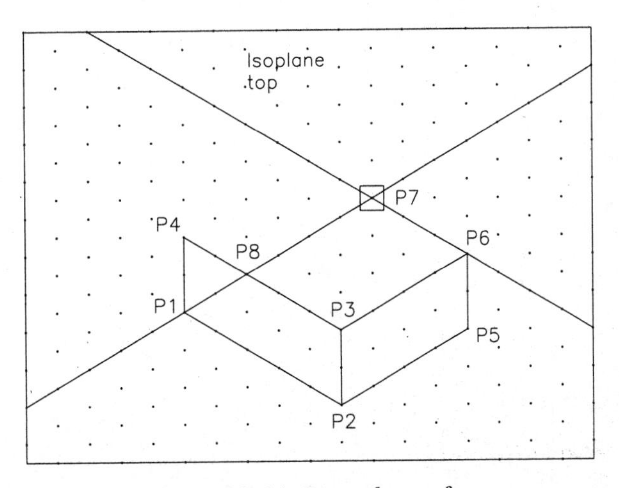

Figure 19-11 *Draw the top face*

5. Draw the remaining lines (Figure 19-12) according to the dimensions given in Figure 19-8.

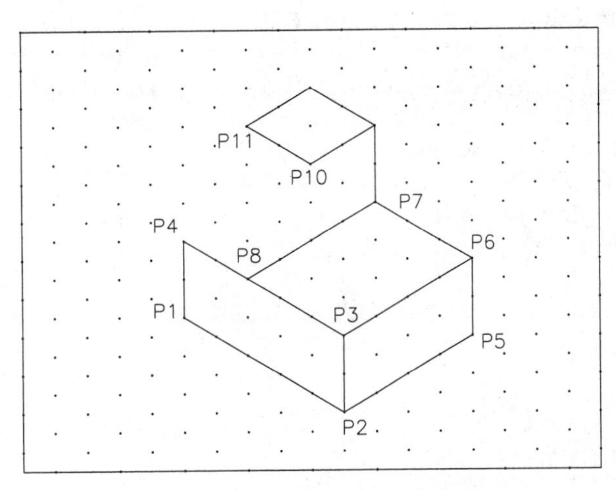

Figure 19-12 *Draw the remaining lines*

6. The front left end of the object is tapered at an angle. In isometric drawings, oblique surfaces (surfaces at an angle to the isometric axis) cannot be drawn like other lines. You must first locate the endpoints of the lines that define the oblique surface and then draw lines between those points. To complete the drawing of Figure 19-8, draw a line from P10 to P8 and from P11 to P4 (Figure 19-13).

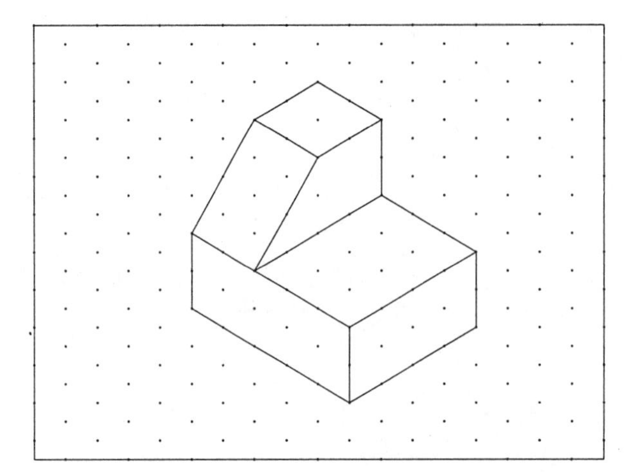

Figure 19-13 Drawing the tapered face

DRAWING ISOMETRIC CIRCLES

Isometric circles are drawn by using the **ELLIPSE** command and then selecting the **Isocircle** option. Before you enter the radius or diameter of the isometric circle, you must make sure you are in the required isoplane. For example, if you want to draw a circle in the right isoplane, you must toggle through the isoplanes until the required isoplane (right isoplane) is displayed. You can also set the required isoplane current before entering the **ELLIPSE** command. The crosshairs and the shape of the isometric circle will automatically change as you toggle through different isoplanes. After you enter the radius or diameter of the circle, AutoCAD will draw the isometric circle in the correct plane (Figure 19-14).

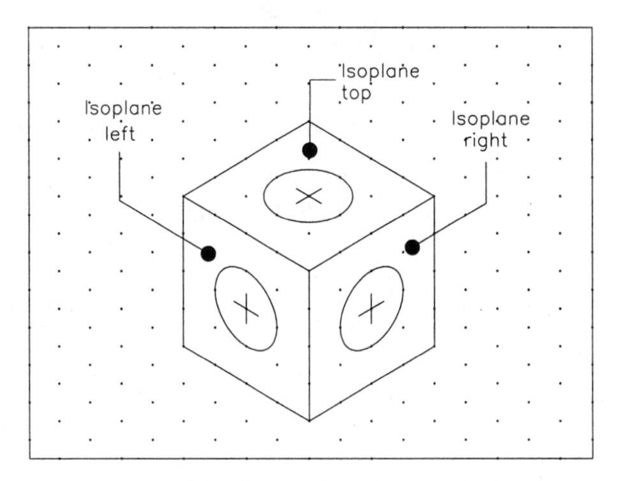

Figure 19-14 Drawing isometric circles

Command: **ELLIPSE**
Specify axis endpoint of ellipse or [Arc/Center/Isocircle]: **I**

Specify center of isocircle: *Select a point.*
Specify radius of isocircle or [Diameter]: *Enter circle radius.*

Note

*You must have the isometric snap on for the **ELLIPSE** command to display the **Isocircle** option with the **ELLIPSE** command. If the isometric snap is not on, you cannot draw an isometric circle.*

DIMENSIONING ISOMETRIC OBJECTS

Isometric dimensioning involves two steps: (1) dimensioning the drawing using the standard dimensioning commands; (2) editing the dimensions to change them to oblique dimensions.

Example 2

The following example illustrates the process involved in dimensioning an isometric drawing.

In this example, you will dimension the isometric drawing you created in Example 1.

1. Dimension the drawing as shown in Figure 19-15.

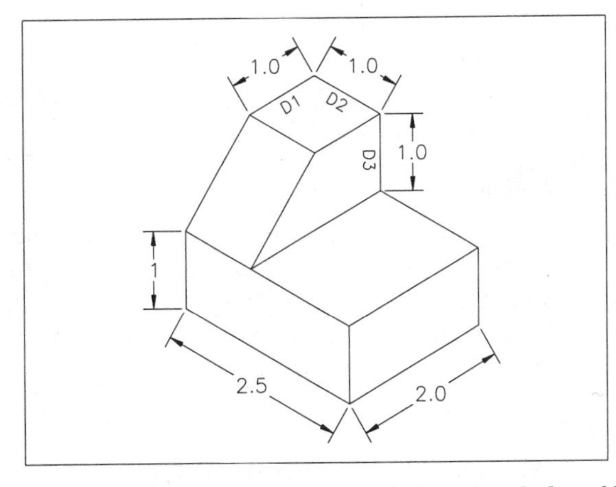

Figure 19-15 *Dimensioning an isometric drawing, before obliquing*

You can use the aligned or vertical dimensioning command to give the dimensions. When you select the points, you must use the Intersection or Endpoint object snap to grab the endpoints of the object you are dimensioning. AutoCAD automatically leaves a gap between the object line and the extension line as specified by the **DIMGAP** variable.

2. The next step is to edit the dimensions. First, enter the **DIM** command, and then enter **OBLIQUE** at the **Dim:** prompt or use the **Oblique** option of the **DIMEDIT** command. After you select the dimension you want to edit, AutoCAD will prompt you to enter the oblique angle. The oblique angle is determined by the angle the extension line of the isometric dimension makes with the positive X axis.

Command: **DIM**
Dim: **OBLIQUE**
Select object: Select the dimension (D1).
Enter oblique angle (Press ENTER for none): **30**

For example, the extension line of the dimension labeled D1 makes a 150-degree angle with the positive X axis [Figure 19-16(a)]; therefore, the oblique angle is 150 degrees. Similarly, the extension lines of the dimension labeled D2 makes a 30-degree angle with the positive X axis [Figure 19-16(b)]; therefore, the oblique angle is 30 degrees. After you edit all dimensions, the drawing should appear as shown in Figure 19-17.

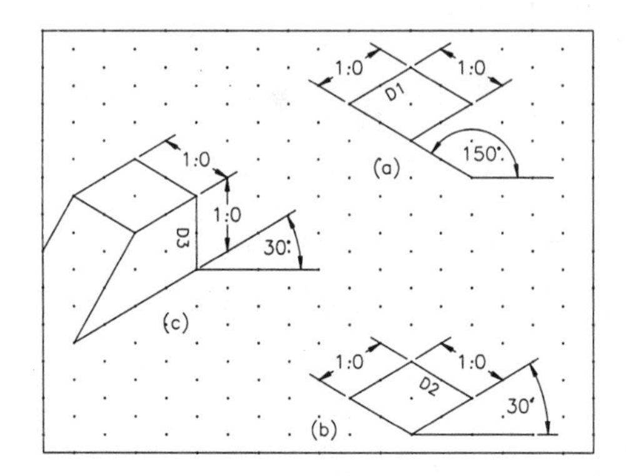

Figure 19-16 *Determining the oblique angle*

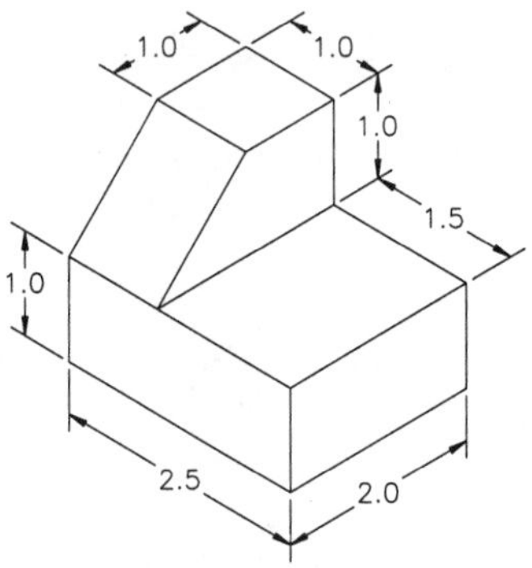

Figure 19-17 *Object with isometric dimensions*

ISOMETRIC TEXT

You cannot use regular text when placing text in an isometric drawing because the text in an isometric drawing is obliqued at positive or negative 30 degrees. Therefore, you must create two text styles with oblique angles of 30 degrees and negative 30 degrees. You can use the **STYLE** command and the **Text Style** dialog box to create a new text style as described here. (For more details, refer to "**STYLE** Command" in Chapter 5.)

Command: **STYLE**
Enter name of text style or [?] <Standard>: **ISOTEXT1**
Specify full font name or font filename (TTF or SHX) <txt>: **ROMANS**
Specify height of text <0.0000>: **0.075**
Specify width factor <1.0000>: *Press ENTER.*
Specify obliquing angle <0>: **30**
Display text backwards? [Yes/No] <N>: **N**
Display text upside-down? [Yes/No] <N>: **N**
Vertical? <N> **N**

Similarly, you can create another text style, **ISOTEXT2**, with a negative 30-degree oblique angle. When you place the text in an isometric drawing, you must also specify the rotation angle for the text. The text style you use and the text rotation angle depend on the placement of the text in the isometric drawing, as shown in Figure 19-18.

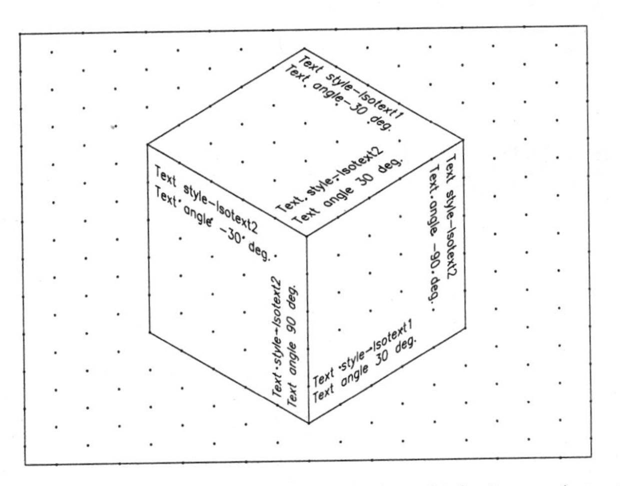

Figure 19-18 Text style and rotation angle for isometric text

Self-Evaluation Test

Answer the following questions, and then compare your answers with the correct answers given at the end of this chapter.

1. The word isometric means "__________" because the three angles between the three principal axes of an isometric drawing are each __________ degrees.

2. The ratio of isometric length to actual length in an isometric drawing is approximately __________.

3. The angle between the right isometric horizontal axis and the X axis is __________ degrees.

4. Isometric drawings have three principal planes: isoplane right, isoplane top, and __________.

5. What key combination or function key can you use to toggle among isoplane right, isoplane left, and isoplane top? __________.

6. You can only use the aligned dimension option to dimension an isometric drawing. (T/F)

7. Must the isometric snap be on to display the **Isocircle** option with the **ELLIPSE** command? (Y/N)

8. Do you need to specify the rotation angle when placing text in an isometric drawing? (Y/N). If yes, what are the possible angles? __________.

9. The lines that are not parallel to the isometric axes are called _________.

10. You should avoid showing any hidden lines in isometric drawings. (T/F)

Review Questions

1. Isometric drawings are generally used to help in _________ the shape of an object.

2. An isometric view is obtained by rotating the object _________ degrees around the imaginary vertical axis, and then tilting the object forward through a _________ angle.

3. If you project the points and the edges onto the frontal plane, the projected length of the edges will be approximately _________ percent shorter than the actual length of the edges.

4. When should hidden lines be shown in an isometric drawing?

5. Isometric drawings have three axes: right horizontal axis, vertical axis, and _________.

6. The angles do not appear true in isometric drawings. (T/F)

7. What are the lines called that are parallel to the isometric axis? _________.

8. What commands can you use to set the isometric grid and snap? _________.

9. Isometric grid lines are displayed at _________ degrees to the horizontal axis.

10. It is possible to set the aspect ratio for the isometric grid. (T/F)

11. You can also set the isometric grid and snap by using the **Drafting Settings** dialog box, which can be invoked by entering _________ at the Command prompt.

12. Isometric circles are drawn by using the **ELLIPSE** command and then selecting the _________ option.

13. Can you draw an isometric circle without turning the isometric snap on? (Yes/No)

14. Only aligned dimensions can be edited to change them to oblique dimensions. (T/F)

15. To place the text in an isometric drawing, you must create two text styles with oblique angles of _________ degrees and negative _________ degrees.

Exercises

Exercises 1 through 6 — General

Draw the following isometric drawings (Figures 19-19 through 19-24). The dimensions can be determined by counting the number of grid lines. The distance between the isometric grid lines is assumed to be 0.5 units. Dimension the odd-numbered drawings.

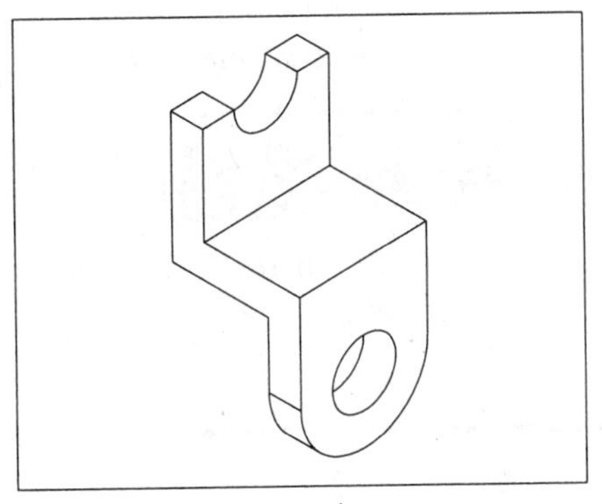

***Figure 19-19** Drawing for Exercise 1*

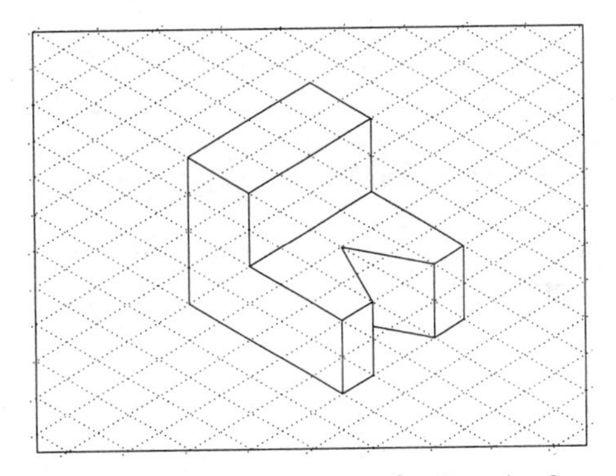

***Figure 19-20** Drawing for Exercise 2*

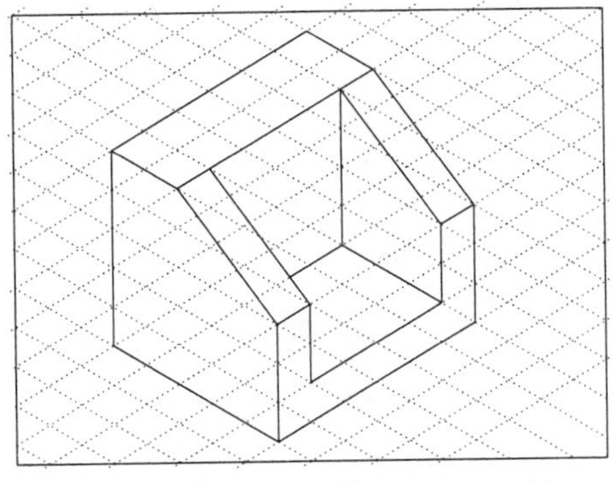

***Figure 19-21** Drawing for Exercise 3*

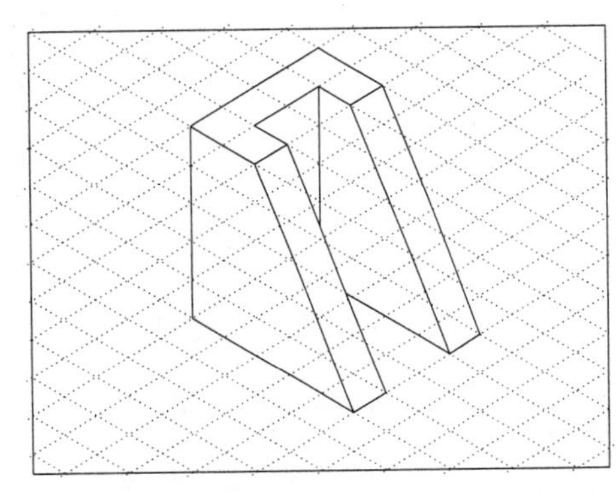

***Figure 19-22** Drawing for Exercise 4*

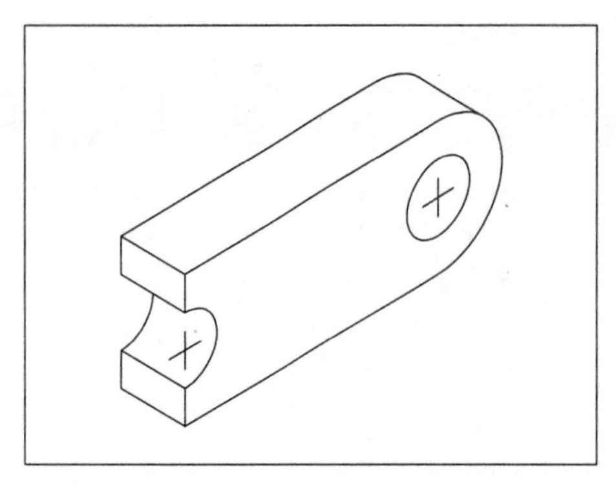

Figure 19-23 Drawing for Exercise 5

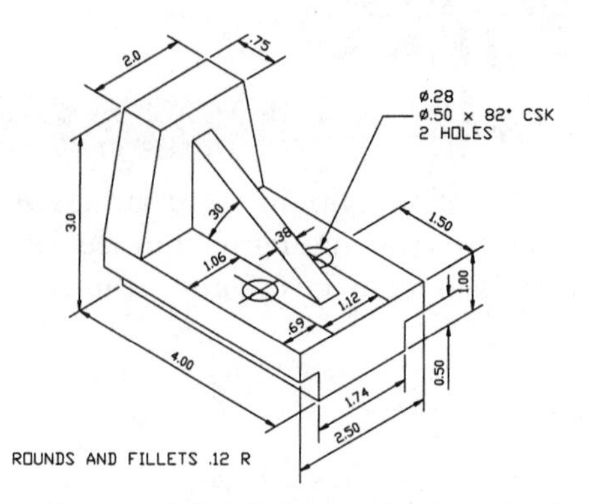

Figure 19-24 Drawing for Exercise 6

Project Exercise 1-7 *Mechanical*

Draw the exploded isometric drawing of the work bench as shown in Figure 19-25. (See Chapters 1, 2, 6, 7, and 23 for other drawings of Project Exercise 1.)

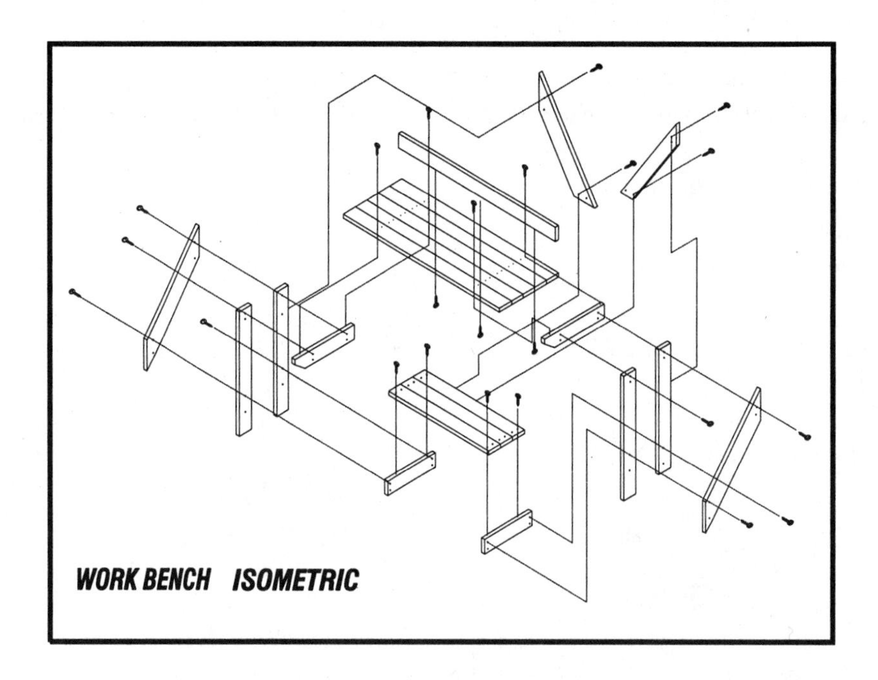

Figure 19-25 Drawing for Project Exercise 1-7

Answers to Self-Evaluation Test

1 - equal measure, 120, **2** - 9/11, **3** - 30, **4** - isoplane left, **5** - CTRL and E, or F5, **6** - F, **7** - Yes, **8** - Yes, 30 degrees, -30 degrees, 90 degrees, etc., **9** - nonisometric lines, **10** - T

Chapter 20

User Coordinate System

Learning Objectives

After completing this chapter, you will be able to:

- *Use and understand the user coordinate system (UCS) and the world coordinate system (WCS).*
- *Position the UCS icon using different options of the* ***UCSICON*** *command.*
- *Use the UCS to create a 3D Wireframe drawing.*
- *Use different options to redefine the UCS using the* ***UCS*** *command.*
- *Using the various tabs of the* ***UCS*** *dialog box.*

THE WORLD COORDINATE SYSTEM (WCS)

When you get into the AutoCAD LT drawing editor, by default, the WCS (world coordinate system) is established. In the WCS, the X, Y, and Z coordinates of any point are measured with reference to the fixed origin (0,0,0). This origin is located at the lower left corner of the screen by default. The objects you have drawn up to now use the WCS. This coordinate system is fixed and cannot be moved. The WCS is used mostly in 2D drawings. The world coordinate system and the user coordinate system (UCS), enable you to relocate and reorient the origin and X, Y, and Z axes, and were introduced in AutoCAD Release 10. Earlier releases used only the fixed origin (0,0,0).

UCSICON COMMAND

Menu:	View > Display > UCS Icon
Command:	UCSICON

The orientation of the origin of the current UCS is represented by the UCS icon (Figure 20-1). This icon is a graphic reminder of the direction of the UCS axes and the location of the origin. This icon also tells you what the viewing direction is, relative to the UCS XY plane. The default location of this symbol is a little to the right and a little above the lower left corner of the viewport. The plus (+) sign on the icon indicates that the UCS icon is placed at the origin of the current UCS. The absence of the W on the Y axis indicates that the current coordinate system is not the WCS. If there is a box at the icon's base, it means you are viewing the UCS from above; the absence of the box indicates that you are viewing the UCS from below. Choosing **Display > UCS Icon > On**, from the **View** menu, when a check mark is displayed next to the **On** option, turns off the display of the UCS Icon. Choose the **On** option again to turn on the display of the UCS Icon. A check mark besides **On** indicates that the display of the icon is on. The **UCSICON** command also controls display of this symbol. The prompt sequence is:

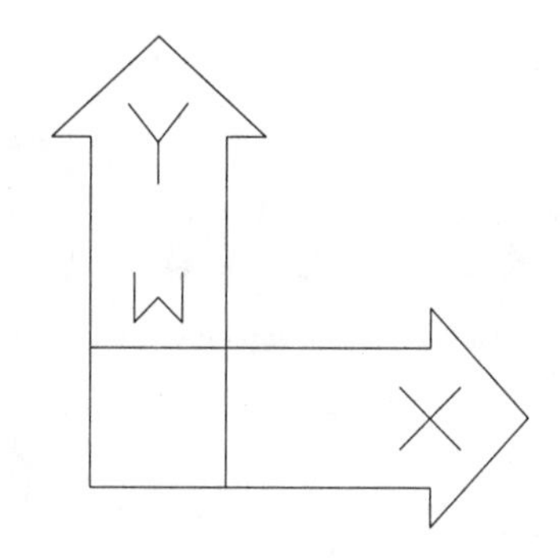

Figure 20-1 *World coordinate system (WCS) icon*

Command: **UCSICON** [Enter]
Enter an option [ON/OFF/All/Noorigin/ORigin] <ON>:

Also see Figure 20-2 for model space and paper space icons.

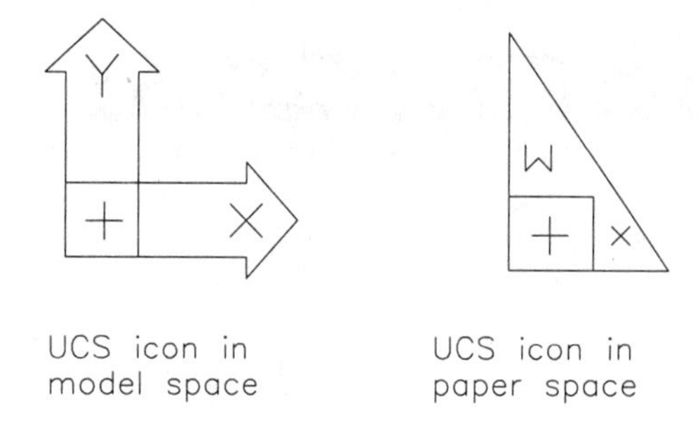

Figure 20-2 *Model space and paper space icons*

The different options of the **UCSICON** command are as follows:

ON Option. This option displays the UCS icon in the current viewport.

Command: **UCSICON** [Enter]
Enter an option [ON/OFF/All/Noorigin/ORigin]<ON>: **ON** [Enter]

OFF Option. This option disables the UCS icon in the current viewport, such that it is no more displayed in the current viewport.

Enter an option [ON/OFF/All/Noorigin/ORigin]<ON>: **OFF** [Enter]

All Option. This option is used to change the UCS icons in all viewports. Normally, changes are made only to the icon in the current viewport. For example, to turn off the UCS icons in all the viewports, the prompt sequence will be:

Enter an option [ON/OFF/All/Noorigin/ORigin]<default>: **ALL** [Enter]
Enter an option [ON/OFF/Noorigin/ORigin] <ON>: **OFF** [Enter]

NO (Noorigin) Option. Makes AutoCAD LT display the icon, if enabled, at lower left corner of the screen, whatever the position of the current UCS origin may be. This is the default setting.

Enter an option [ON/OFF/All/Noorigin/ORigin]<default>: **N** Enter

OR (ORigin) Option. This option makes AutoCAD LT display the UCS icon (if enabled) at the origin (0,0,0) of the current coordinate system. If the current origin is off screen or it is not possible to position the icon at the origin without being clipped at the viewport edges, it is displayed at the lower left corner of the screen. Sometimes your view direction may be along the edge of the current UCS (or within one degree of the edge direction). In such cases, the UCS icon is replaced by a broken-pencil icon, to indicate that selecting locations with the pointing device is meaningless.

Enter an option [ON/OFF/All/Noorigin/ORigin]<default>: **OR** Enter

You can also choose **Display** > **UCS Icon** > **Origin** from the **View** menu to display the UCS Icon at the origin. A check mark next to Origin, indicates that the UCS Icon is displayed at the Origin of the coordinate system. The value of the current UCS origin is stored in the **UCSORG** system variable. The UCS can be different for different viewports if the **UCSVP** system variable is set to 1. When we change the UCS of the current viewport, it does not reflect in the UCSs of other active viewports if the value of **UCSVP** is set to 1; that is, the value of each UCS is stored in the specific viewport.

THE USER COORDINATE SYSTEM (UCS)

The user coordinate system (UCS) helps you to establish your own coordinate system. Changing the placement of the base point of your drawing is accomplished by changing the position/orientation of the coordinate system. As already mentioned, the WCS is fixed, whereas the UCS can be moved and rotated to any desired position to conform to the shape of the object being drawn. Alterations in the UCS are manifested in the position and orientation of the UCS icon symbol, originally located in the lower left corner of the screen. The UCS is used mostly in 3D drawings, where you may need to specify points that vary from each other along the X, Y, and Z axes. It is also useful for relocating the origin or rotating the XY axes in 2D work, such as ordinate dimensioning, drawing auxiliary views, or controlling hatch alignment.

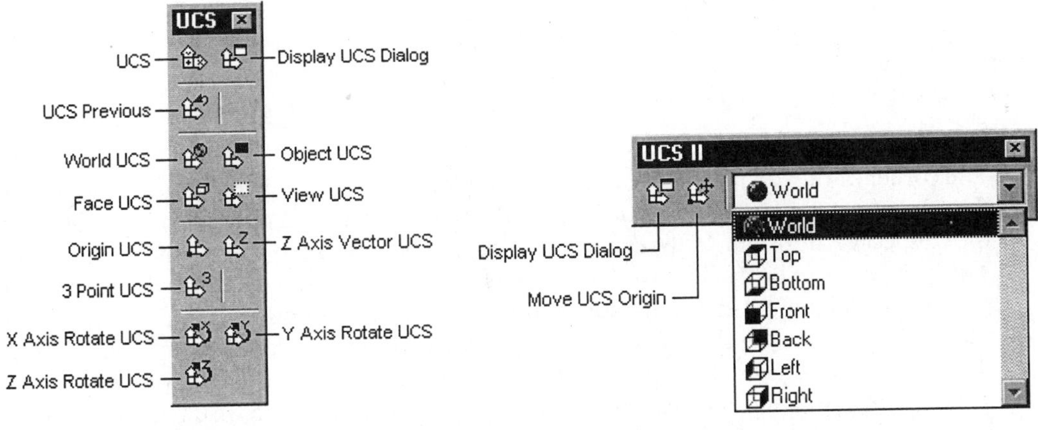

Figure 20-3(a) UCS toolbar

Figure 20-3(b) UCS II toolbar

Example 1

To illustrate the use of the UCS, draw a tapered rectangular block with a circle on the inclined side of it. The center of the circle is located at a given distance from the lower left corner of the tapered side. This problem can be broken into the following steps:

1. Draw the bottom side of the tapered block.
2. Draw the top side.
3. Draw the other sides.
4. Draw the circle on the inclined side of the block.

To make it easier to visualize and draw the object, the 3D object is shown in Figure 20-4. Follow the steps just given to draw an identical figure.

1. Start by drawing the bottom rectangle (Figure 20-5). Enter the **LINE** command, and select a point on the screen to draw the first line. Continue drawing lines to complete the rectangle shown in Figure 20-4. Following is the command sequence:
   Command: **LINE** Enter
   Specify first point: **1,1** Enter
   Specify next point or [Undo]: **@2.20,0** Enter
   Specify next point or [Undo]: **@0,1.30** Enter
   Specify next point or [Close/Undo]: **@-2.20,0** Enter
   Specify next point or [Close/Undo]: **C** Enter

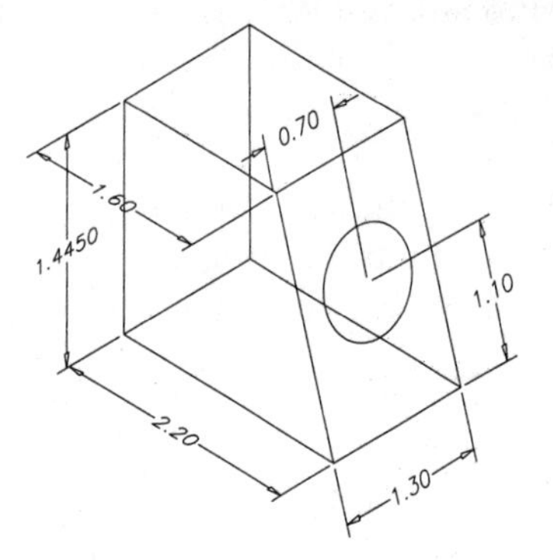

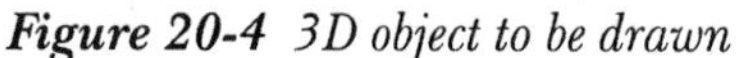
Figure 20-4 *3D object to be drawn*

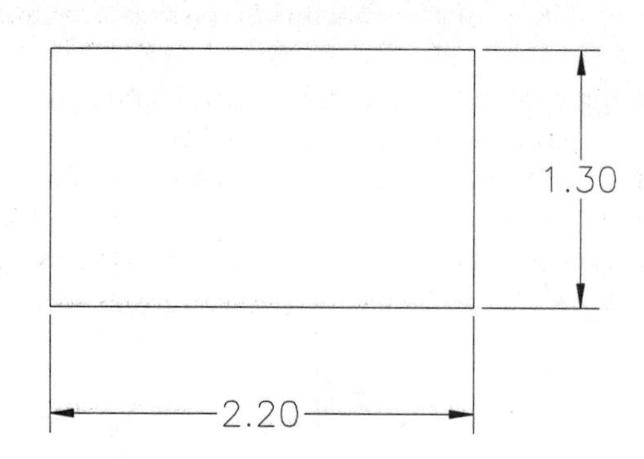

Figure 20-5 *Step 1: draw the bottom rectangle*

2. The corner of the top rectangle of the tapered block can be established with the WCS, but this could cause some confusion in locating the points since each point must be referenced with respect to the origin. With relative coordinates, you must keep track of the Z coordinate. This problem can be easily solved with the UCS by moving the origin of the coordinate system to a plane 1.4450 units above and parallel to the bottom plane. This can be accomplished by using the **UCS** command (described in detail later). Once you establish the new origin, it is much easier to draw the top side, since the drawing plane is 1.4450 units above the plane on which the bottom side was drawn (the distance between the bottom side and the top side is 1.4450 units) (Figure 20-6). The prompt sequence is:

Command: **UCS** [Enter]
Current ucs name: *WORLD*
Enter an option [New/Move/orthoGraphic/Prev/Restore/Save/Del/?/World]<World>: **N** [Enter]
Specify origin of new UCS or [ZAxis/3point/OBject/Face/View/X/Y/Z] <0,0,0>: **0,0,1.445** [Enter]

After you change the UCS, the prompt sequence to draw the top side is:

Command: **LINE** [Enter]
Specify first point: **1,1** [Enter]
Specify next point or [Undo]: **@1.6,0** [Enter]
Specify next point or [Undo]: **@0,1.3** [Enter]
Specify next point or [Close/Undo]: **@-1.6,0** [Enter]
Specify next point or [Close/Undo]: **C** [Enter]

3. Since you are looking at the object from the top side, the three edges of the top side will overlap the corresponding bottom edges; hence, you will not be able to see these three sides. Once the top and bottom are drawn, you can see the 3D shape of the object using the **VPOINT** command (discussed in detail in the next chapter) (Figure 20-7):

Command: **VPOINT** [Enter]
*** Switching to the WCS ***
Current view direction: VIEWDIR= 0.0000,0.0000,1.0000
Specify a view point or [Rotate] <display compass and tripod>: **1,-1,1** [Enter]
Join the corners of the top rectangle with the corresponding corners of the bottom rectangle (Figure 20-8). You can use the **LINE** command to draw the edges. Use the ENDpoint or INTersect snap mode to grab the corners.

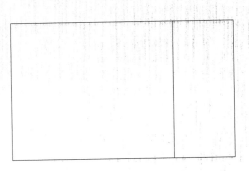

Figure 20-6 *Step 2: draw the top rectangle*

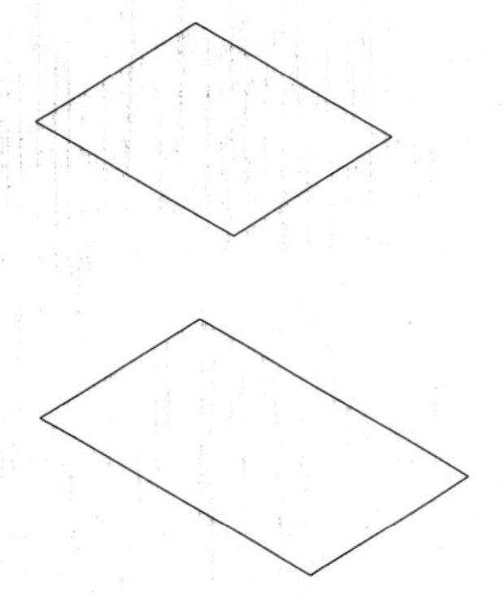

Figure 20-7 *Step 3: change Vpoint to (1,-1,1)*

4. The last step is to draw the circle on the inclined surface of the object (Figure 20-9). To do this, you need to make the inclined side (surface) the current drawing plane. This can be accomplished with the **3point** option of the **UCS** command (discussed later in this chapter). The prompt sequence is:

Command: **UCS** [Enter]
Current ucs name: *CURRENT*
Enter an option [New/Move/orthoGraphic/Prev/Restore/Save/Del/?/World]<World>: **N** [Enter]
Specify origin of new UCS or [ZAxis/3point/OBject/Face/View/X/Y/Z] <0,0,0>: **3** [Enter]
Specify new origin point <0,0,0>: *Specify the lower left corner (INT osnap) of the inclined side as the origin point (P1).*
Specify point on positive portion of X-axis <4.2000,1.0000,-1.4450>: *Select the endpoint of the bottom edge of the inclined surface (P2), using the INT osnap.*
Specify point on positive-Y portion of the UCS XY plane <2.2000,1.0000,-1.4450>: *Select the endpoint of the left edge of the inclined surface (P3), using the INT osnap.*

Now that you have aligned the current drawing plane with the inclined surface, you can draw the circle. We know the X and Y axes displacements are between the lower corner of the inclined surface and the center of the circle. Use these X and Y displacements to specify the center of the circle, and then draw the circle. The command sequence is:

Command: **CIRCLE** [Enter]
Specify center point for circle or [3P/2P/Ttr (tan tan radius)]: **0.7,1.10** [Enter]
Specify radius of circle or [Diameter]: *Specify the desired radius.*

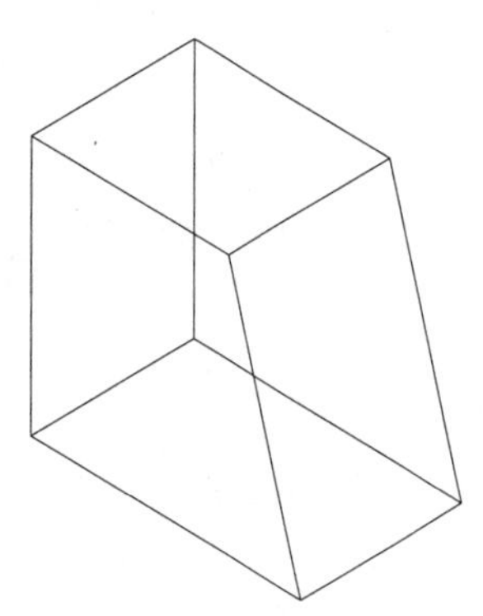

Figure 20-8 *Join the corresponding corners*

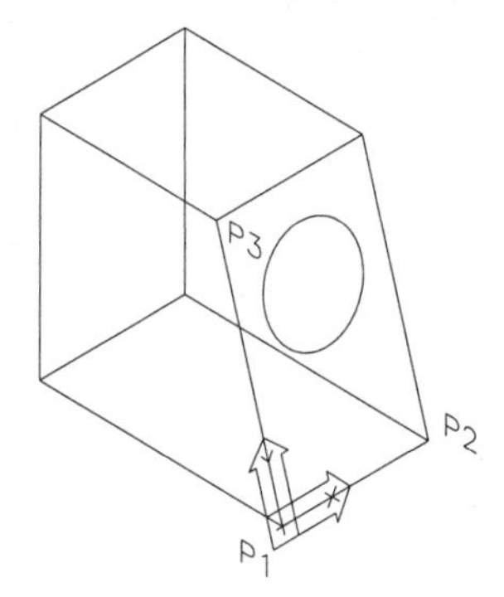

Figure 20-9 *Step 4: position the UCS icon and draw a circle*

Using the user coordinate system, the circle can be drawn in the plane on which the inclined surface lies, and the center of the circle can easily be specified just by entering the displacements about the X and Y axes of the current UCS.

UCS COMMAND

Toolbar:	UCS > UCS
Menu:	Tools > New UCS
Command:	UCS

As already discussed, the WCS is fixed, whereas the UCS enables you to set your own coordinate system. For certain views of the drawing, it is better to have the origin of measurements at some other point on or relative to your drawing objects. This makes locating the features and dimensioning the objects easier. The origin and orientation of a coordinate system can be redefined using the **UCS** command. The prompt sequence is:

Command: **UCS** Enter
Current ucs name: *WORLD*
Enter an option [New/Move/orthoGraphic/Prev/Restore/Save/Del/?/World] <World>: *Select an option*.

If the **UCSFOLLOW** system variable is set to 0, any change in the UCS, does not affect the drawing view.

W (World) Option

With this option, you can set the current UCS back to the world coordinate system. To invoke this option, choose the **World UCS** button from the **UCS** toolbar or from the **UCS** flyout in the **Standard** toolbar. You can also choose **New UCS > World** from the **Tools** menu or press ENTER or enter WORLD (or W) at the following prompt:

Command: **UCS** Enter
Enter an option [New/Move/orthoGraphic/Prev/Restore/Save/Del/?/World] <World>: Enter

New Option*

The **New** option defines the origin of a new UCS by using any one of the options described below. The command prompt is :

Command: **UCS** Enter
Enter an option [New/Move/orthoGraphic/Prev/Restore/Save/Del/?/World] <World>: N Enter
Specify origin of new UCS or [ZAxis/3point/OBject/Face/View/X/Y/Z] <0,0,0>

O (Origin) Option

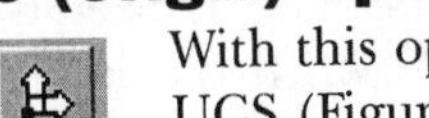

With this option, you can define a new UCS by changing the origin of the current UCS (Figure 20-10). The directions of the X, Y, and Z axes remain unaltered. To invoke this option, choose the **Origin UCS** button from the **UCS** toolbar or from the **UCS** flyout in the **Standard** toolbar. You can also choose **New UCS > Origin** from the **Tools** menu or enter ORIGIN (or O) at the following prompt:

Command: **UCS** Enter
Enter an option [New/Move/orthoGraphic/Prev/Restore/Save/Del/?/World] <World>: N Enter
Specify origin of new UCS or [ZAxis/3point/OBject/Face/View/X/Y/Z] <0,0,0>: *Select point (P4) using ENDpoint snap (Figure 20-10).*

At the **Specify origin of new UCS or [ZAxis/3point/OBject/Face/View/X/Y/Z] <0,0,0>** prompt you need to specify any point relative to the current UCS. This point acts as the new origin point. For example, if you want to specify the new origin point at (2,1,0) with respect to the current UCS, enter 2,1,0 at the **Specify origin of new UCS or [ZAxis/3point/OBject/Face/View/X/Y/Z] <0,0,0>** prompt. You can also select a point with a pointing device to specify the location of the new origin. Repeat the command and select point (P3) as new the origin.

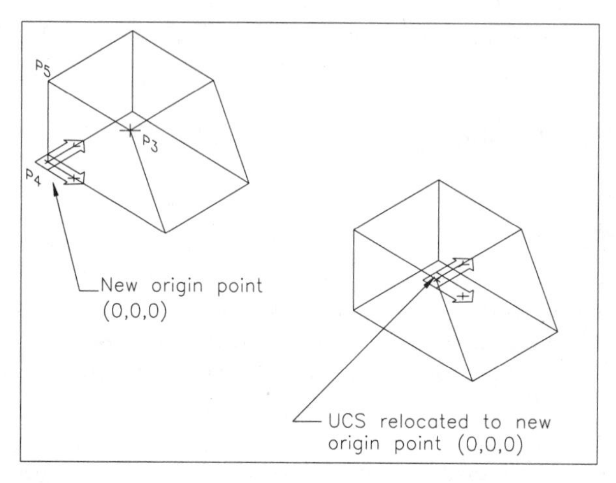

***Figure 20-10** Relocating the origin with the **Origin** option*

Note

If you do not provide a Z coordinate for the origin, the Z coordinate is assigned the current elevation value.

ZA (ZAxis) Option

With this option of the UCS command, you can change the coordinate system by selecting the origin point of the coordinate system and a point on the positive Z axis (Figure 20-11). After you specify a point on the Z axis, AutoCAD LT determines the X and Y axes of the new coordinate system accordingly. This option can be invoked by choosing the **Z Axis Vector UCS** button from the **UCS** toolbar or the **UCS** flyout in the **Standard** toolbar. You can also choose **New UCS > Z Axis Vector** from the **Tools** menu or by entering ZAXIS (or ZA) at the following prompt:

Command: **UCS** Enter
Current ucs name: *WORLD*
Enter an option [New/Move/orthoGraphic/Prev/Restore/Save/Del/?/World]<World>: **N** Enter
Specify origin of new UCS or [ZAxis/3point/OBject/Face/View/X/Y/Z] <0,0,0>: **ZA** Enter
Specify new origin point <0,0,0>: *Specify the origin point (P4).*
Specify point on positive portion of Z-axis <current>: *Specify a point on the positive Z axis (P6)*

If you give a null response for the **Specify point on the positive portion of Z axis <current>:**

prompt, the Z axis of the new coordinate system will be parallel to (in the same direction as) the Z axis of the previous coordinate system. Null responses to the origin point and the point on the positive Z axis establish a new coordinate system in which the direction of the Z axis is identical to that of the previous coordinate system; however, the X and Y axes may be rotated around the Z axis. The positive Z axis direction is also known as the extrusion direction. In Figure 20-11, the UCS has been relocated by specifying the origin point and a point on positive the portion of the Z axis.

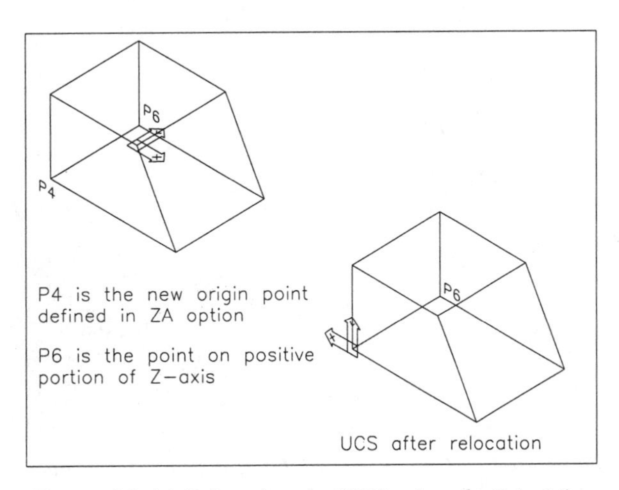

***Figure 20-11** Relocating the UCS using the **ZA** option*

3 (3point) Option

With this option, you can establish a new coordinate system by specifying the new origin point, a point on the positive X axis, and a point on the positive Y axis (Figure 20-12). The direction of the Z axis is determined by applying the right-hand rule. Align your thumb with the positive X axis and your index finger with the positive Y axis. Now bend your middle finger so it is perpendicular to the thumb and the index finger. The direction of the middle finger is the direction of the positive Z axis. The **3point** option of the **UCS** command changes the orientation of the UCS to any angled surface. This option can be invoked by selecting the **3 Point UCS** button from the **UCS** toolbar or the **UCS** flyout in the **Standard** toolbar. You can also choose **New UCS > 3Point** in the **Tools** menu or enter 3POINT (or 3) at the following prompt:

Command: **UCS** Enter
Current ucs name: *WORLD*
Enter an option [New/Move/orthoGraphic/Prev/Restore/Save/Del/?/World] <World>: Enter
Specify origin of new UCS or [ZAxis/3point/OBject/Face/View/X/Y/Z] <0,0,0>: **3** Enter
Specify new origin point <0,0,0>: *Specify the origin point.*
Specify point on positive portion of X-axis <current>: *Specify a point on the positive X axis.*
Specify point on positive-Y portion of the UCS XY plane <current>: Specify *a point on the positive Y axis.*

A null response to the **Specify new origin point <0,0,0>:** prompt will lead to a coordinate system in which the origin of the new UCS is identical to that of the previous UCS. Similarly, null responses to the Point on X or Y axis prompt will lead to a coordinate system in which the X or Y axis of the new UCS is parallel to that of the previous UCS. In Figure 20-12, the UCS has been relocated by specifying three points (the origin point, a point on the positive portion of the X axis, and a point on the positive portion of the Y axis).

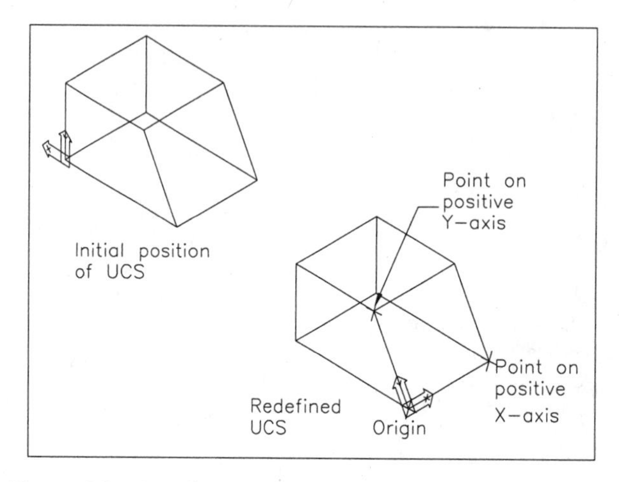

*Figure 20-12 Relocating the UCS using the **3point** option*

OB (OBject) Option

With the **OB** (**OBject**) option of the **UCS** command, you can establish a new coordinate system by pointing to any object in an AutoCAD LT drawing except a 3D polyline, 3D solid, 3D mesh, viewport object, mtext, mline, xline, ray, leader, ellipse, region, or spline. The positive Z axis of the new UCS is in the same direction as the positive Z axis of the object selected. If the X and Z axes are given, the new Y axis is determined by the right-hand rule. This option can be invoked by selecting the **Object UCS** button from the **UCS** toolbar or the **UCS** flyout in the **Standard** toolbar. You can also choose **New UCS > Object** from the **Tools** menu or enter **OBject** (or **OB**) at the following prompt:

Command: **UCS** Enter
Current ucs name: *WORLD*
Enter an option [New/Move/orthoGraphic/Prev/Restore/Save/Del/?/World]<World>: **N** Enter
Specify origin of new UCS or [ZAxis/3point/OBject/Face/View/X/Y/Z] <0,0,0>: **OB** Enter
Select object to align UCS: *Select the circle.*

In Figure 20-13, the UCS is relocated using the **OBject** option and is aligned to the circle. The origin and the X axis of the new UCS are determined by the following rules.

Arc. When you select an arc, the center of the arc becomes the origin for the new UCS. The X axis passes through the endpoint of the arc that is closest to the point selected on the object.

Circle. The center of the circle becomes the origin for the new UCS, and the X axis passes through the point selected on the object (Figure 20-13).

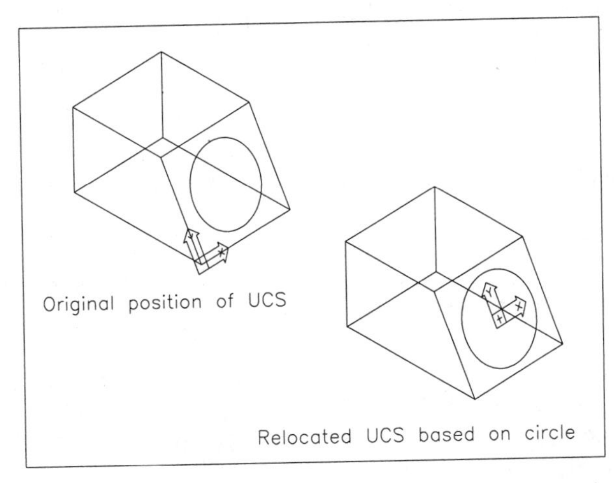

*Figure 20-13 Relocating the UCS using the **OBject** option*

Line. The new UCS origin is the endpoint of the line nearest the point selected on the line. The X axis is defined so that the line lies in the ZX plane of the new UCS. Therefore, in the new UCS the Y coordinate of the second endpoint of the line is 0.

Trace. The origin of the new UCS is the "from" point of the Trace. The new X axis lies along the centerline of the Trace.

Dimension. The middle point of the dimension text becomes the new origin. The X axis direction is identical to the direction of the X axis of the UCS that existed when the dimension was drawn.

Point. The position of the point is the new UCS origin.

Solid. The origin of the new UCS is the first point of the solid. The X axis of the new UCS lies along the line between the first and second points of the solid.

2D Polyline. The start point of the polyline or polyarc is treated as the new UCS origin. The X axis extends from the start point to the next vertex.

Text/ Insert/ Attribute definition. The insertion point of the object becomes the new UCS origin. The new X axis is defined by the rotation of the object around its positive Z axis. Hence, the object you select will have a rotation angle of zero in the new UCS.

Note

The XY plane of the new UCS will be parallel to the XY plane existing when the object was drawn; however, X and Y axes may be rotated.

Face (F) Option*

This option aligns the new UCS with the selected face of the solid object. You can invoke this option by choosing the **Face UCS** button from the **UCS** toolbar or the **UCS** flyout in the **Standard** toolbar. You can also choose **New UCS > Face** from the **Tools** menu or enter FACE (or F) at the following prompt:

Command: **UCS** [Enter]
Current ucs name: *WORLD*
Enter an option [New/Move/orthoGraphic/Prev/Restore/Save/Del/?/World]<World>: **N** [Enter]
Specify origin of new UCS or [ZAxis/3point/OBject/Face/View/X/Y/Z] <0,0,0>: **F** [Enter]
Select face of solid object: *Select within the boundary or the edge of the face to select.*
Enter an option [Next/Xflip/Yflip] <accept>:

The **Next** option, locates the new UCS on the next, adjacent face or the back face of the selected edge. **Xflip** rotates the new UCS by 180 degrees around the X axis and **Yflip** rotates it around the Y axis. Pressing ENTER at the **Enter an option [Next/Xflip/Yflip] <accept>:** accepts the location of the new UCS as specified.

Note

To use this option a 3D object is needed which has to be created in some similar compatible software.

V (View) Option*

The **V (View)** option of the **UCS** command lets you define a new UCS whose XY plane is perpendicular to your current viewing direction. In other words, the new XY plane is parallel to the screen. The origin of the UCS defined in this option remains unaltered. This option is used mostly to view a drawing from an oblique viewing direction. This option can be invoked by selecting the **View UCS** button from the **UCS** toolbar or the **UCS** flyout in the **Standard** toolbar. You can also choose **New UCS > View** from the **Tools** menu or enter VIEW (or V) at the following prompt:

Command: **UCS** [Enter]
Current ucs name: *WORLD*
Enter an option [New/Move/orthoGraphic/Prev/Restore/Save/Del/?/World] <World>: N
Specify origin of new UCS or [ZAxis/3point/OBject/Face/View/X/Y/Z] <0,0,0>: V [Enter]

Note

*To achieve the desired UCS, it is often easier to use the **OBject** or **View** option to establish the UCS orientation, and then use the **Origin** option to locate it, rather than using other options in a single **UCS** command.*

X/Y/Z Options*

With these options, you can rotate the current UCS around a desired axis.

Command: **UCS** [Enter]
Current ucs name: *WORLD*

Enter an option [New/Move/orthoGraphic/Prev/Restore/Save/Del/?/World] <World>: **N** Enter
Specify origin of new UCS or [ZAxis/3point/OBject/Face/View/X/Y/Z] <0,0,0>: *Specify the axis around which you want to rotate the UCS.*
Specify rotation angle about Z axis <90>: Specify the angle of rotation

You can specify the angle by entering the angle value at the **Specify rotation angle about Z axis <90>:** prompt or by selecting two points on the screen with the help of a pointing device. You can specify a positive or a negative angle. The new angle is taken relative to the X axis of the existing UCS. The **UCSAXISANG** system variable stores the default angle by which the UCS is rotated around the specified axis, using the **X/ Y/ Z** options of the **New** option of the **UCS** command. The right-hand rule is used to determine the positive direction of rotation around an axis. As already described, the right-hand rule states that if you visualize gripping with your right hand the axis about which you want the rotation, with your thumb pointing in the positive direction of that axis, the direction in which your fingers close indicates the positive rotation angle.

In Figure 20-14, the UCS is relocated using the X option by specifying an angle about the X axis. The upper left part of the figure shows the UCS setting before the UCS was relocated with the X option. Enter the following prompt:

Command: **UCS** Enter
Current ucs name: *WORLD*
Enter an option [New/Move/orthoGraphic/Prev/Restore/Save/Del/?/World] <World>:

Specify origin of new UCS or [ZAxis/3point/OBject/Face/View/X/Y/Z] <0,0,0>: **X** Enter
Specify rotation angle about Z axis <current>: **90** Enter

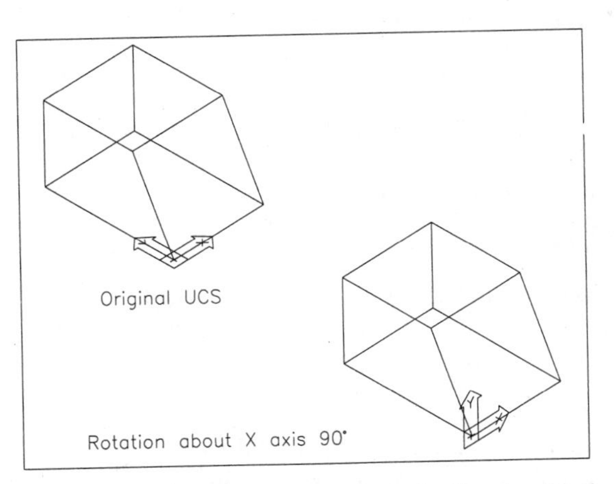

Figure 20-14 Rotating the UCS about the X axis with the ***X*** *option*

In Figure 20-15, the UCS is relocated using the **Y** option by specifying an angle about the Y axis. The upper left part of the figure shows the UCS setting before the UCS was relocated with the **Y** option. Enter the following prompt:

Command: **UCS** Enter
Current ucs name: *WORLD*
Enter an option [New/Move/orthoGraphic/Prev/Restore/Save/Del/?/World] <World>: **N** Enter
Specify origin of new UCS or [ZAxis/3point/OBject/Face/View/X/Y/Z] <0,0,0>: **Y** Enter
Specify rotation angle about Z axis <current>: 90

Original UCS
Rotation about Y axis 90°

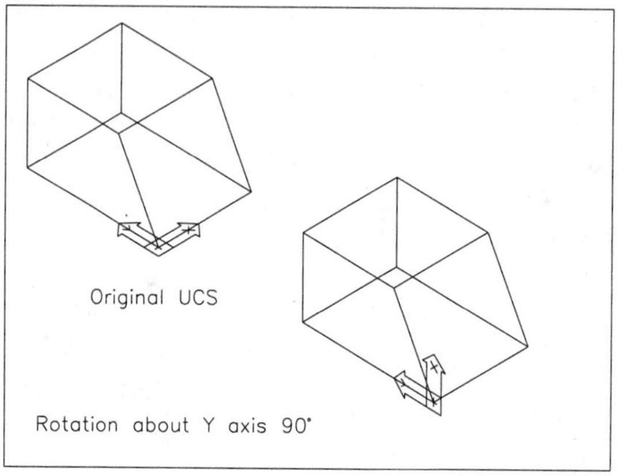

Figure 20-15 *Rotating the UCS about the Y axis using the* ***Y*** *option*

In Figure 20-16, the UCS is relocated using the **Z** option by specifying an angle about the Z axis. The upper left part of the figure shows the UCS setting before the UCS was relocated with the **Z** option. Enter the following prompt:

Command: **UCS** Enter
Current ucs name: *WORLD*
Enter an option [New/Move/orthoGraphic/Prev/Restore/Save/Del/?/World] <World>: **N** Enter
Specify origin of new UCS or [ZAxis/3point/OBject/Face/View/X/Y/Z] <0,0,0>: **Z** Enter
Specify rotation angle about Z axis <current>: **90** Enter

Rotation about the Z axis is commonly used as an alternative to using the **SNAP** command to rotate the snap and grid.

Move Option*

This option redefines the UCS by moving the UCS Origin in the positive or negative direction along the Z axis with respect to the current UCS Origin. The orientation of the XY plane remains the same. You can invoke this command by choosing the **Move UCS Origin** button in the **UCS II** toolbar. You can also choose **Move UCS** from the **Tools** menu or enter MOVE or M at the following prompt:

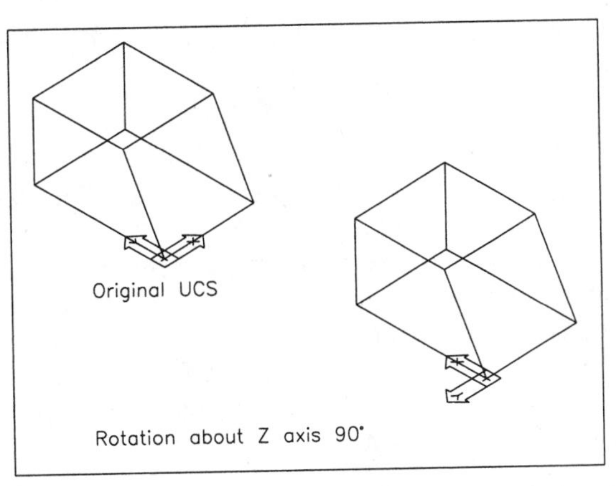

Figure 20-16 Rotating the UCS about the Z axis using the ***Z*** *option*

Command: **UCS** [Enter]
Current ucs name: *WORLD*
Enter an option [New/Move/orthoGraphic/Prev/Restore/Save/Del/?/World] <World>: **M** [Enter]
Specify new origin point or [Zdepth] <0,0,0>: *Specify a point or enter value by which the current UCS origin will move along the Z axis.*

The **New Origin point** option, moves the current UCS origin to the one specified. The **Zdepth** option specifies the distance by which the UCS origin moves along the Z axis. When we use the **Move UCS** option, in case, there are multiple viewports in the drawing, the change in the UCS is applied only to the current viewport. The **UCS Move** option is not added to the Previous list.

Orthographic Option

This option allows you to set current any one of the six Orthographic UCSs provided with AutoCAD LT. They are: **Top, Bottom, Front, Back, Left,** and **Right**. You can specify any one of these UCSs by choosing from the **UCS II** toolbar drop-down list of predefined UCSs. The Orthographic UCSs are used normally when one is viewing and editing models. You can also select any one of the Orthographic UCSs from the List box in the **Orthographic UCSs** tab of the **UCS** dialog box and then choose the **Set current** button to apply it. You can also choose **Orthographic UCS** from the **Tools** menu or enter ORTHGRAPHIC (or G) at the command prompt as follows:

Command: **UCS** [Enter]
Current ucs name: *WORLD*
Enter an option [New/Move/orthoGraphic/Prev/Restore/Save/Del/?/World] <World>: **G** [Enter]
Enter an option [Top/Bottom/Front/BAck/Left/Right] <Top>: *Specify an option.*

The Orthographic UCSs are set with respect to the WCS. The **UCSBASE** system variable controls the UCS upon which the orthographic settings are based; the initial value is the World Coordinate System.

P (Previous) Option

The **P** (**Previous**) option restores the previous UCS. The last ten user coordinate system settings are saved by AutoCAD LT. You can go back to the previous ten UCS settings in the current space using the **Previous** option. If **TILEMODE** is off, the last 10 coordinate systems in paper space and in model space are saved. This option can be invoked by selecting the **Previous UCS** button from the **UCS** toolbar or the **UCS** flyout in the **Standard** toolbar. You can select the **Previous** option from the **UCS II** toolbar drop-down list or from the list box of Preset UCSs in the **Names UCSs** tab of the **UCS** dialog box. You can also enter PREVIOUS (or P) at the following prompt:

Command: **UCS** Enter
Enter an option [New/Move/orthoGraphic/Prev/Restore/Save/Del/?/World] <World>: **P** Enter

R (Restore) Option

With this option of the UCS command, you can restore a previously saved named UCS. Once a saved UCS is restored, it becomes the current UCS. The viewing direction of the saved UCS is not restored. Choose **Named UCSs...** from the **Tools** menu to displays the **UCS** dialog box. The **Named UCSs** tab of this dialog box displays a list of the Named UCSs in the list box, select a UCS you want to set current and choose the **Set Current** button. You can also restore a named UCS by selecting it from the **UCS II** toolbar drop-down list. This option can also be invoked by entering RESTORE (or R) at the following prompt:

Command: **UCS** Enter
Current ucs name: *WORLD*
Enter an option [New/Move/orthoGraphic/Prev/Restore/Save/Del/?/World]<World>: **R** Enter

After this, AutoCAD LT issues the following prompt:

Enter name of UCS to restore or [?]: *Enter the name of the UCS you want to restore.*

You can list the UCS names by entering ? at this prompt. Next AutoCAD LT prompts:

Enter UCS name(s) to list <*>:

You can use wild-cards, or if you want to list all the UCS names, give a null response.

S (Save) Option

With this option, you can name and save the current UCS. When you are naming the UCS, the following should be kept in mind:

1. The name can be up to 255 characters long.
2. The name can contain letters, digits, blank spaces and the special characters $ (dollar), - (hyphen), and _ (underscore).

The following is the prompt sequence:

Command: **UCS** Enter
Current ucs name: *WORLD*
Enter an option [New/Move/orthoGraphic/Prev/Restore/Save/Del/?/World]<World>: **S** Enter

At the next prompt, enter a valid name for the UCS. AutoCAD LT saves it as a UCS.

Enter name to save current UCS or [?]:

Just as in the case of the **Restore** option, you can list the UCS names by entering ? at the prompt. Next, AutoCAD LT prompts:

Enter UCS name(s) to list <*>:

If you want to list all the UCS names, press ENTER.

D (Delete) Option

The **D** (**Delete**) option is used to delete the selected UCS from the list of saved coordinate systems. To invoke this option, enter DELETE (or D) at the following prompt:

Command: **UCS** Enter
Current ucs name: *WORLD*
Enter an option [New/Move/orthoGraphic/Prev/Restore/Save/Del/?/World]<World>: **D** Enter
Enter UCS name(s) to delete <none>:

The UCS name you enter at this prompt is deleted. You can delete more than one UCS by separating the UCS names with commas or by using wild cards. If you delete a UCS which is current, it is renamed **UCS Unnamed**. Right-clicking a named UCS in the List box of the **Named UCSs** tab of the **UCS** dialog box, displays a shortcut menu. Choose **Delete** here to delete the selected UCS.

? Option

By invoking this option, you can list the name of the specified UCS. This option gives you the name, origin, and X, Y, and Z axes of all of the coordinate systems relative to the existing UCS. If the current UCS has no name, it is listed as WORLD or UNNAMED. The choice between these two names depends on whether the current UCS is the same as the WCS. This option can be invoked by entering ? at the following prompt:

Command: **UCS** Enter
Current ucs name: *WORLD*
Enter an option [New/Move/orthoGraphic/Prev/Restore/Save/Del/?/World]<World>: **?** Enter
Enter UCS name(s) to list<*>: Enter

MANAGING UCS THROUGH DIALOG BOX (UCSMAN Command)

Toolbar:	UCS > Display UCS Dialog
Menu:	Tools > Named UCS
Command:	UCSMAN

The **UCSMAN** command displays the **UCS** dialog box (Figure 20-17). This dialog box can be used to restore the previously saved UCS configuration. You can also change the UCS to any of the orthographic UCSs, specify UCS icon and settings, and name and rename UCSs. It has three tabs: **Named UCSs** tab, **Orthographic UCSs** tab and **Settings** tab.

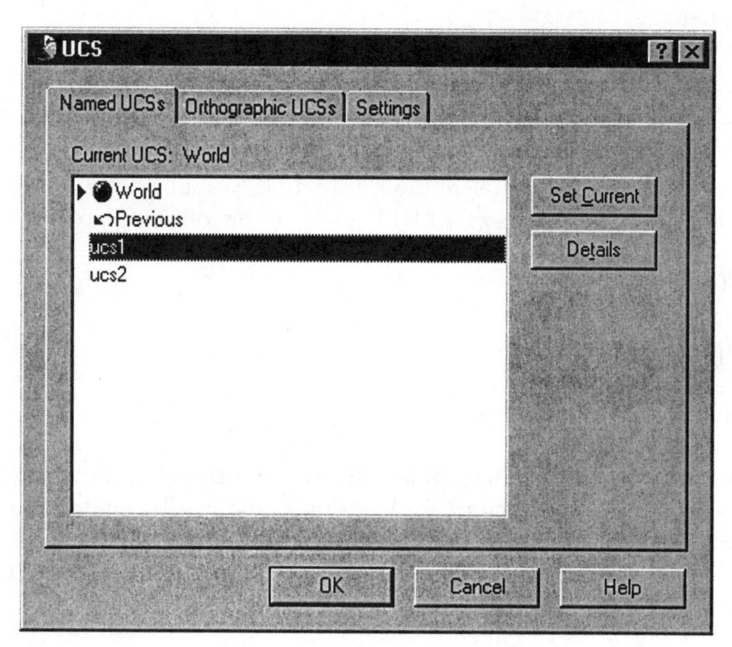

Figure 20-17 UCS dialog box

Named UCSs Tab

The list of all the coordinate systems defined (saved) on your system is displayed in the UCS Names list of the dialog box. The first entry in this list is always the **World coordinate system**. The next entry is **Previous**, if you have defined any other coordinate systems in the current editing session. Selecting the **Previous** entry and then choosing the **OK** button repeatedly enables you to go backward through the coordinate systems defined in the current editing session. **Unnamed** is the next entry in the list if you have not named the current coordinate system. If there are a number of viewports and unnamed settings only the current viewport UCS name is displayed in the list. The current coordinate system is indicated by a small pointer icon to the right of the coordinate system name. The current UCS name is also displayed next to **Current UCS:**. If you want to make some other coordinate system current, select that coordinate system name in the UCS Names list, and then choose the **Set Current** button. To delete a coordinate system, select that coordinate system name, and then right-click to display a shortcut menu. The options are **Set Current**, **Delete**, **Rename** and **Delete**. Choose the **Delete** button to delete the selected UCS name. To rename a coordinate system, select that coordinate system name, and then right-click to display the shortcut menu and choose **Rename**. Now, enter the desired new name. You can also double-click the name to be modified and then change the name.

Note

All the changes and updating of the UCS information in the drawing are carried out only after you select the ***OK*** *button.*

If you want to check the current coordinate system's origin and X, Y, and Z axis values, select a UCS from the list and then choose the **Details** button. The **UCS Details** dialog box (Figure 20-18) containing that information is then displayed. You can also choose **Details** from the shortcut menu displayed on right-clicking a specific UCS name in the list box.

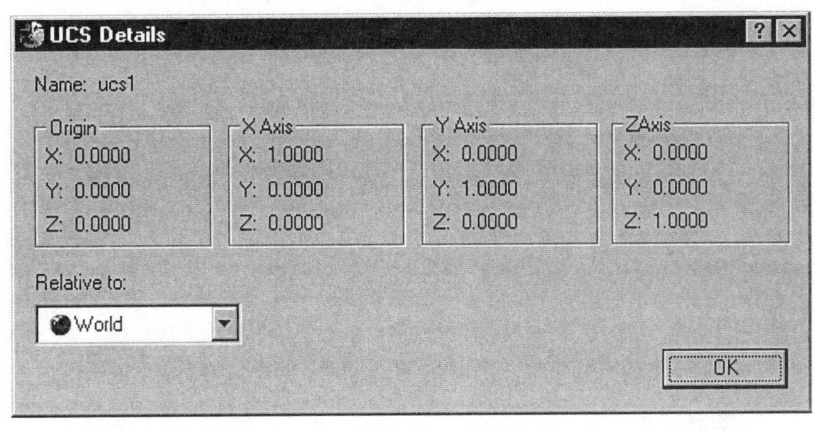

Figure 20-18 UCS Details *dialog box*

Setting UCS to Preset Orthographic UCSs Using the Orthographic UCSs Tab

The **UCS** dialog box with the **Orthographic UCSs** tab is displayed also on choosing **Orthographic UCS > Preset...** from the **Tools** menu. You can select any one of the preset orthographic UCSs from the list in the **Orthographic UCSs** tab of the **UCS** dialog box (Figure 20-19). Selecting **Top** and choosing the **Set Current** button, results in creation of the UCS icon in the top view, also known as the plan view. You can also double-click the UCS name you want to set as current to make it current. You can also right-click the specific orthographic UCS name in the list and choose **Set Current** from the shortcut menu displayed. The **Current UCS:** displays the name of the current UCS name. If the current settings have not been saved

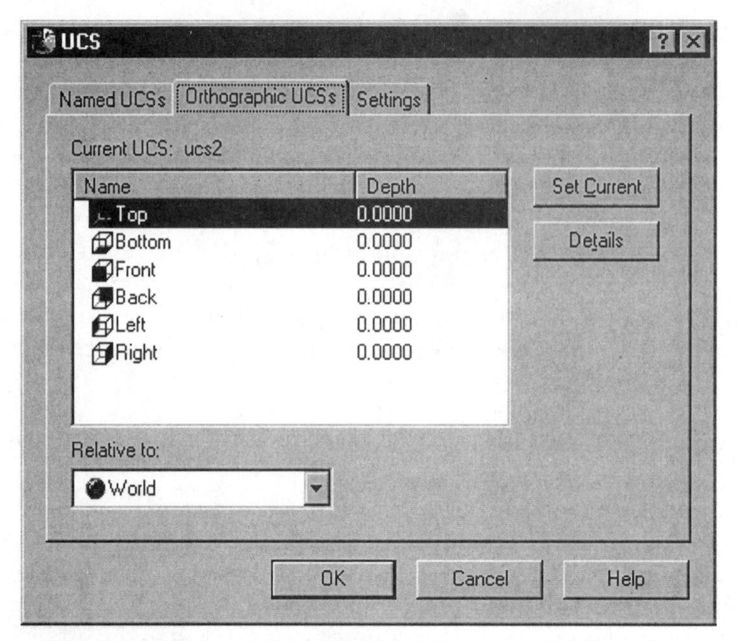

Figure 20-19 ***Orthographic UCSs*** *Tab of the* ***UCS*** *dialog box*

and named, the **Current UCS** name displayed is **Unnamed**.

The **Depth** field in the list box of this dialog box, displays the distance between the XY plane of the selected orthographic UCS setting and a parallel plane passing through the origin of the UCS base setting. The system variable **UCSBASE** stores the name of the UCS which is considered the base settings; that is, defines the origin and orientation. You can enter or modify values of **Depth** by double-clicking the Depth value of the selected UCS in the list box to display the **Orthographic UCS depth** dialog box (Figure 20-20), where you can enter new depth values. You can also right-click a specific Orthographic UCS in the list box to display a shortcut menu. Choose **Depth** to display the **Orthographic UCS depth** dialog box. Enter a value in the **Front Depth** edit box or choose the **Select new origin** button to specify a new origin or depth on the screen.

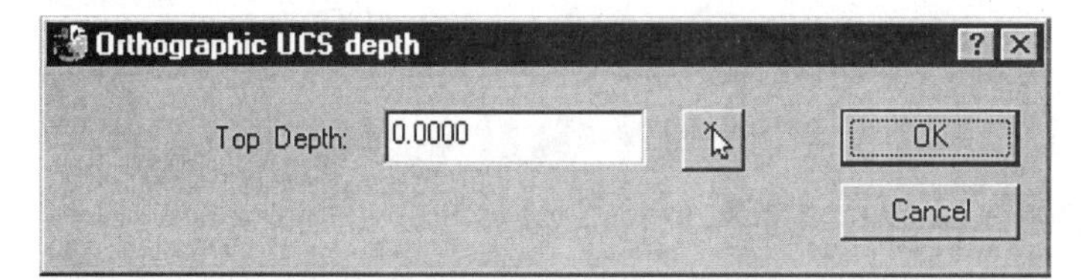

Figure 20-20 Orthographic UCS Depth *dialog box*

In the **Orthographic UCSs** tab of the **UCS** dialog box, you are given the choice of establishing a new UCS relative to a named UCS or to the WCS. This choice is offered from the **Relative to:** drop-down list. This list lists all the named UCSs in the current drawing. By default **World** is the base coordinate system. If you have set current any one of the preset orthographic UCSs and the you select a UCS other than World in the **Relative to** drop-down list, the Current UCS changes to **Unnamed**. Choosing the **Details** button, displays the **UCS Details** dialog box with the origin and X, Y, Z coordinate values of the selected UCS. You can also choose **Details** from the shortcut menu that is displayed on right-clicking a UCS in the list box. The shortcut menu also has a **Reset** option. Choosing **Reset** from the shortcut menu restores the origin of the selected orthographic UCS to it's default location (0,0,0).

Settings Tab

The **Settings** tab of the **UCS** dialog box (Figure 20-21), displays and modifies UCS and UCS icon settings of a specified viewport.

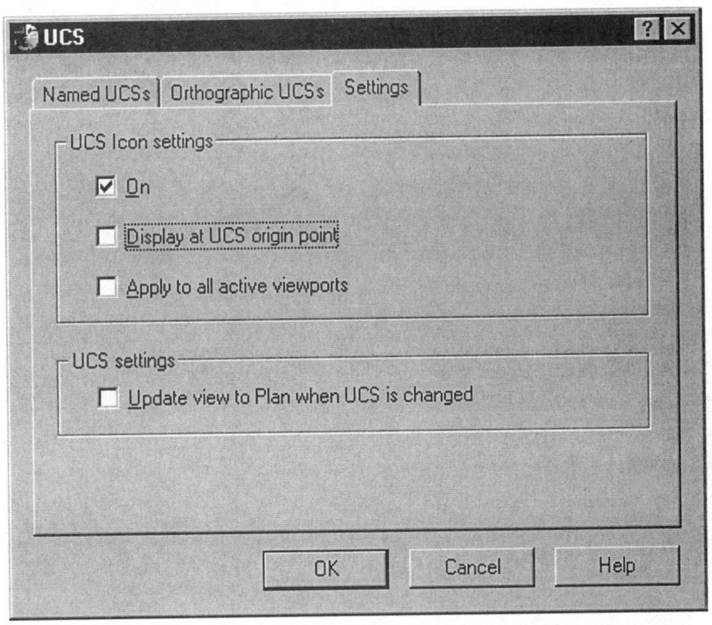

Figure 20-21 ***Settings*** *tab of the* ***UCS*** *dialog box*

UCS Icon Settings

Selecting the **On** check box, displays the UCS icon in the current viewport, it is similar to using the **UCSICON** command to set the display of the UCS icon to on or off. If you select the **Display at UCS origin point** check box, the UCS icon is displayed at the origin point of the coordinate system in use in the current viewport. If the origin point is not visible in the current viewport or if the check box is cleared, the UCS icon is displayed in the lower left corner of the viewport. Selecting the **Apply to all active viewports** check box, applies the current UCS icon settings to all active viewports in the current drawing.

UCS Setting

This area of the **UCS** dialog box **Settings** tab, specifies UCS settings for the current viewport. When you select the **Update view to plan** check box, the plan view is restored when the UCS in the viewport is changed. Also, when the selected UCS is restored, the plan view is restored. The value is stored in the **UCSFOLLOW** system variable.

SYSTEM VARIABLES

The coordinate value of the origin of the current UCS is held in the **UCSORG** system variable. The X and Y axis directions of the current UCS are held the in **UCSXDIR** and **UCSYDIR** system variables, respectively. The name of the current UCS is held in the **UCSNAME** variable. All these variables are read-only. If the current UCS is identical to the WCS, the **WORLDUCS** system variable is set to 1; otherwise, it holds the value 0. The current UCS icon setting can be examined and manipulated with the help of the **UCSICON** system variable. This variable holds the UCS icon setting of the current viewport. If more than one viewport is active, each one can have a different value for the **UCSICON** variable. If you are in paper space, the **UCSICON** variable will contain the setting for the UCS icon of the paper space. The **UCSFOLLOW** system variable controls the automatic display of a plan view when you switch from one UCS to another. If **UCSFOLLOW** is set to 1, a plan view is automatically displayed when you switch from one UCS to another. The **UCSAXISANG** variable stores the default angle value for the X, Y, or Z axis around which the UCS is rotated using the **X**, **Y**, **Z** options of the **New** option of the UCS command. The **UCSBASE** variable stores the name of the UCS that acts as the base; that is, defines the origin and orientation of the orthographic UCS setting.

Self-Evaluation Test

Answer the following questions, and then compare your answers to the correct answers given at the end of this chapter.

1. When you get into the AutoCAD LT environment (that is, the AutoCAD LT drawing editor), by default the _________ is established as the coordinate system.

2. The _________ coordinate system can be moved and rotated to any desired position.

3. The _________ command controls display of the icon symbol.

4. The UCS icon is a graphic reminder only of the direction of the UCS origin. (T/F)

5. The _________ option of the **UCSICON** command enables (displays) the UCS icon in the current viewport.

6. With the _________ option of the **UCSICON** command, changes can be made to the UCS icons in all active viewports.

7. The **ORigin** option of the **UCSICON** command makes AutoCAD LT display the icon, if enabled, at the lower left corner of the screen, whatever the position of the current UCS origin may be. (T/F)

8. When using the **ORigin** option of the **UCSICON** command, if the current origin is off screen or it is not possible to position the icon at the origin without being clipped at the viewport edges, it is displayed at the lower left corner of the screen. (T/F)

9. The origin and orientation of a coordinate system can be redefined using the _________ command.

10. When changing the UCS with the **Origin** option of the **New** option of the **UCS** command, the directions of the **X**, **Y**, and **Z** axes can be altered. (T/F)

11. If you give a null response to the **Specify point on positive portion of Z-axis:** prompt of the **ZAxis** option of the **New** option of the **UCS** command, the Z axis of the new coordinate system will be _________to the Z axis of the previous coordinate system.

12. With the **OBject** option of the **New** option of the UCS command, you can establish a new coordinate system by pointing to any object in an AutoCAD LT drawing, including a 3D polyline, 3D solid, 3D mesh, viewport object, mtext, xline, ray, leader, ellipse, region, or spline. (T/F)

13. The **View** option of the **New** option of the **UCS** command lets you define a new UCS whose XY plane is parallel to your current viewing direction. (T/F)

14. If **TILEMODE** is off, the last ten coordinate systems in both paper space and model space are saved. (T/F)

15. Once a saved UCS is restored, it becomes the _________ UCS. The viewing direction of the saved UCS is not restored.

Review Questions

1. The user coordinate system (UCS) helps you to establish your own coordinate system. (T/F)

2. To generate the UCS icon at the new location specified in the **UCS** command, you can use the **UCSICON** command. (T/F)

3. The orientation and the origin of the current UCS are graphically represented by the _________.

4. The + sign on the icon reflects that the UCS icon is placed at the _________ of the current UCS.

5. Absence of the _________ on the Y axis indicates that the current coordinate system is not the WCS.

6. The _________ option of the **UCSICON** command disables the UCS icon in the current viewport.

7. The **NOorigin** option makes AutoCAD LT display the UCS icon (if enabled) at the origin (0,0,0) of the current coordinate system. (T/F)

8. Sometimes the user's view direction may be edge-on to the current UCS (or within 1 degree of edge-on). In such cases, the UCS icon is replaced by a _________ icon as an indication that the pointing locations are meaningless.

9. The **Origin** option of the **New** option of the **UCS** command lets you define a new UCS by changing the _________ of the current UCS.

10. Using the _________ option of the **New** option of the **UCS** command allows you to change the coordinate system by establishing the origin point and a point on the positive Z axis.

11. The positive X axis direction is also known as the extrusion direction. (T/F)

12. With the _________ option of the **New** option of the **UCS** command, you can establish a new coordinate system by specifying the new origin point, a point on the positive X axis, and a point on the positive Y axis.

13. If you have changed the coordinate system with the **OBject** option of the **New** option of the **UCS** command, the positive Z axis of the new UCS is in the same direction as the positive Z axis of the object selected. (T/F)

14. With the _________ option of the **New** option of the **UCS** command, you can rotate the current the UCS around the Z axis by specifying the angle value at the **Specify rotation angle about Z axis <0.0>:** prompt.

15. The _________ option restores the previous UCS.

16. With the **Restore** option of the **UCS** command, you can restore a previously saved named UCS. (T/F)

17. The name of a UCS can be only up to 21 characters long. (T/F)

18. The **D** (**Delete**) option of the **UCS** command is used to delete the selected UCS from the list of _________ coordinate systems.

19. By invoking the _________option of the **UCS** command, you can list the name of the specified UCS.

20. The _________ option of the **UCS** command gives you the name, origin, and X, Y, and Z axes of all coordinate systems relative to the existing UCS.

21. In the UCS **Orthographic UCSs** tab of the **UCS** dialog box, you are given the choice of establishing a new UCS relative to any named UCS or the _________.

22. If the current UCS is identical to the WCS, the **WORLDUCS** system variable is set to 0; otherwise, it holds the value 1. (T/F)

23. The __________system variable holds the UCS icon setting of the current viewport.

24. The __________system variable controls the automatic display of a plan view when you switch from one UCS to another.

25. If **UCSFOLLOW** is set to 0, a plan view is automatically displayed when you switch from one UCS to another. (T/F)

26. With the __________ option of the **UCS** command, you can name and save the current UCS.

27. With the __________ option of the **UCS** command, you can set the current UCS to be identical to the world coordinate system.

28. You can set the orientation of the UCS to any one of the orthographic settings in the UCS dialog box with the help of the __________ command.

29. The coordinate value of the origin of the current UCS is held in the __________ system variable.

30. The __________ dialog box lists all the coordinate systems.

Exercises

Exercises 1 through 4 *Mechanical*

Draw the given Figures 20-22 through 20-25 using the **UCS** command and its options to position the UCS icon so that you can draw the lines on different faces. The dimensions can be determined by counting the number of grid lines. The distance between the isometric grid lines can be assumed to be 0.5 or 1.0 units.

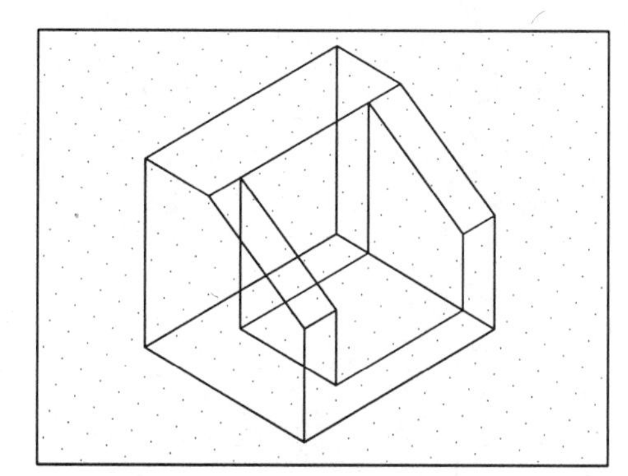

***Figure 20-22** Drawing for Exercise 1*

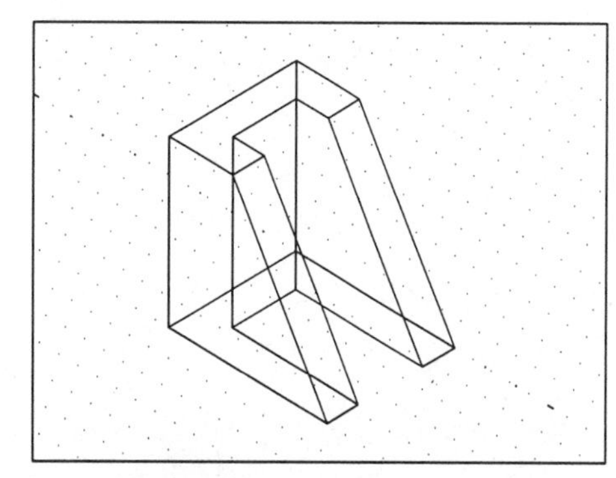

***Figure 20-23** Drawing for Exercise 2*

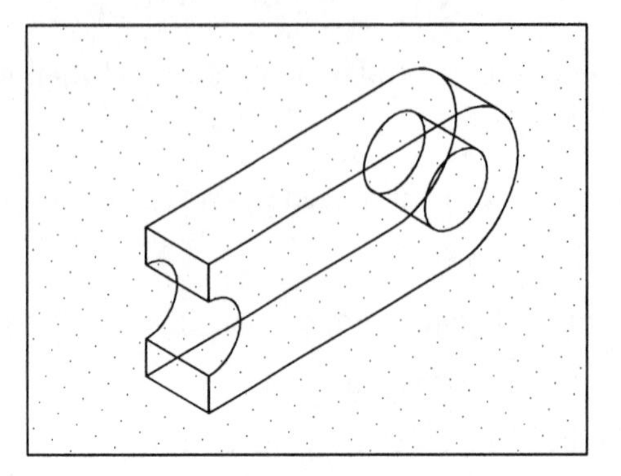

Figure 20-24 Drawing for Exercise 3

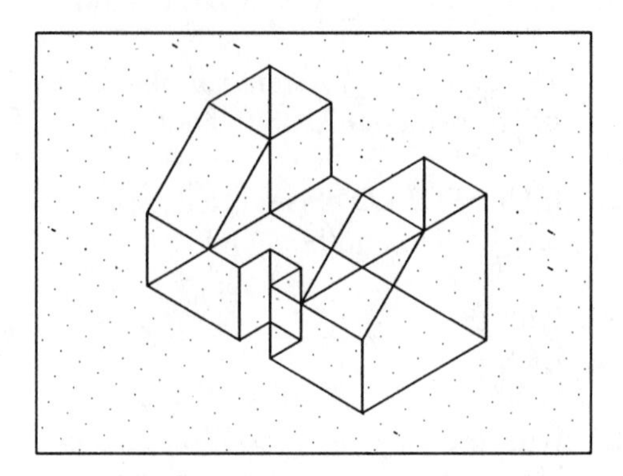

Figure 20-25 Drawing for Exercise 4

Answers to Self-Evaluation Test

1 - world coordinate system, **2** - user, **3** - **UCSICON**, **4** - F, **5** - **On**, **6** - **All**, **7** - F, **8** - T, **9** - **UCS**, **10** - F, **11** - parallel, **12** - F, **13** - F, **14** - T, **15** - current

Chapter 21

Drawing and Viewing 3D objects

Learning Objectives

After completing this chapter, you will be able to:
- *Use the **ELEV** command.*
- *Set up viewpoints with the **VPOINT** and **PLAN** commands.*
- *Use the **HIDE** command.*
- *Understand regions and how to create them.*
- *Create 3D objects.*
- *Dynamically view 3D objects using the **DVIEW** command.*
- *Create Shaded models.*
- *Analyze regions using the **MASSPROP** command.*

3D DRAWINGS

Until the preceding chapter, all the drawings you have seen or drawn have been two-dimensional. These drawings were generated in the XY plane which can be considered as a working plane. You also viewed these drawings from the Z axis, and the picture you saw in this perspective is also known as the plan view. Once you start working with 3D objects, you will often need to view the drawing from different angles. For example, you may require the 3D view, the top view, or the front view of the object in space. In AutoCAD LT, it is possible to view an object from any position in model space. The point from which you view the drawing is known as viewpoint. A viewpoint for the current viewport can be defined using the **VPOINT** command. To draw 3D objects you have to understand the concept of shifting the working plane above or below the world XY plane. This can be done using the **ELEV** command.

ELEV Command

Command:	ELEV

The **ELEV** command is used to shift the working plane (can be considered as XY plane) for drawing objects above or below the world XY plane. In other words using this command you can set the default Z value for all points prompt. Defining positive value shifts the work plane above the world XY plane, and defining the negative value shifts the work plane below the world XY plane see Figure 21-1. Another option provided by this command is that of setting thickness for the objects see Figure 21-2. Thickness can be defined as the distance by which a 2D object is extruded or extended in Z direction. Defining positive value for thickness extrudes the objects in positive Z direction, and defining negative thickness value extrudes the objects in negative Z direction. The prompt sequence is as follows:

Specify new default elevation <current>: *Specify a value or press ENTER.*
Specify new default thickness <current>: *Specify a value or press ENTER.*

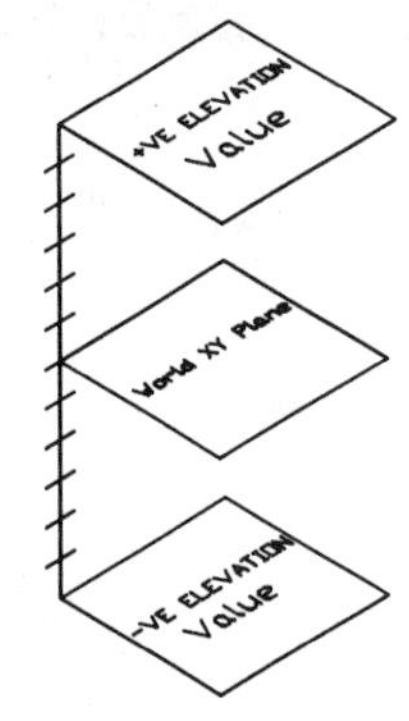

Figure 21-1 *Defining positive, zero, and negative value for elevation*

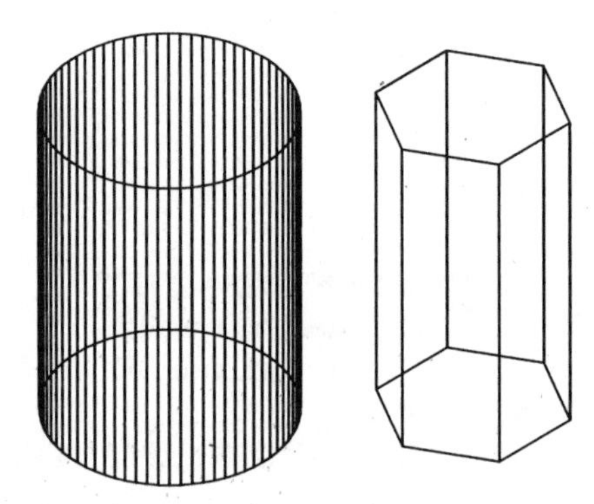

Figure 21-2 *Circle and Polygon with thickness.*

Note

Thickness value entered using tthe ***ELEV*** *command is not valid for rectangle drawn using* ***RECTANG*** *command. This command itself gives the option for thickness.*

To draw an ellipse with thickness, the value for the system variable ***PELLIPSE*** *should be set to 1.*

VIEWING OBJECTS IN 3D SPACE

AutoCAD LT has provided the **VPOINT,** and **DVIEW** commands to view objects in 3D space. (The **DVIEW** command is discussed later in this chapter.)

VPOINT Command

Menu:	Views > 3DViews > Vpoint
Command:	VPOINT

As mentioned earlier, the **VPOINT** command lets you define a viewpoint. Once you set a viewpoint, all the objects in the current viewport are regenerated and the objects are displayed as if you were viewing them from the newly defined line of sight.

Command: **VPOINT** [Enter]
Current view direction: VIEWDIR= current
Specify a view point or [Rotate] <display compass and tripod>:

At this prompt, you are supposed to enter the coordinates of the viewpoint. Regardless of the actual method or point used, the resulting view is looking towards (0,0,0) through the entire drawing, along the line of sight that the vector from that point to (0,0,0) defines or that the angle used in the **Rotate** option defines. You can enter X, Y, and Z coordinate values of the viewpoint of your choice at the **Specify a view point or [Rotate] <display compass and tripod>** prompt. The following are some of the viewpoint settings and the sort of views obtained from each. This table will give you a fair idea of the viewing directions and viewpoints in 3D space.

Vpoint Value	View Displayed
0,0,1	Top
0,0,-1	Bottom
0,-1,0	Front
0,1,0	Rear
1,0,0	Right side
-1,0,0	Left side
1,-1,-1	Bottom, front, right side
-1,-1,-1	Bottom, front, left side
1,1,-1	Bottom, rear, right side
-1,1,-1	Bottom, rear, left side
1,-1,1	Top, front, right side
-1,-1,1	Top, front, left side
1,1,1	Top, rear, right side
-1,1,1	Top, rear, left side

Note
*The **VPOINT**, and **DVIEW** commands cannot be used in paper space. You can use these commands in the model space.*

Using Axes Tripod and Compass to Set the Viewpoint

If you are not well-acquainted with visualizing objects in 3D space and find it difficult to imagine points in 3D space, give a null response at the **Specify a view point or [Rotate] <display compass and tripod>** prompt instead of entering coordinates or choose **3D Views > VPOINT** from **View** menu. This makes AutoCAD LT display a compass and an axis tripod (Figure 21-3). Now you can specify the viewpoint by

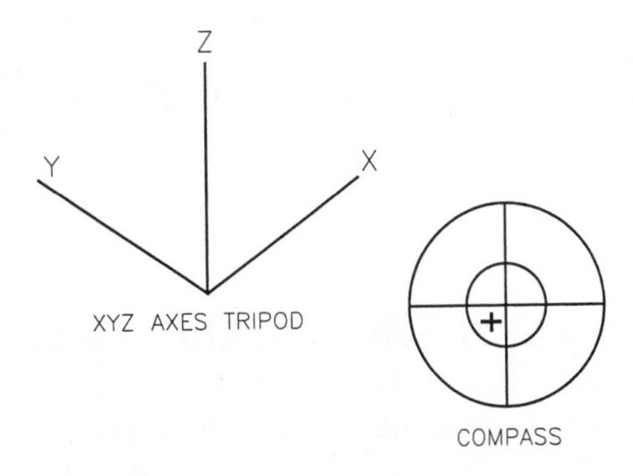

Figure 21-3 *Axes tripod and compass*

moving the pointing device. You will notice that as you move the pointing device, the X, Y, Z axes tripod also moves, and so do the small crosshairs in the compass. You can now select the viewpoint with respect to the position of the X, Y, Z axes in the tripod. The compass displayed is a 2D symbol of a globe. The center point of the circle is the north pole (0,0,1), and by placing the crosshairs on the north pole you can see the top view of the drawing on the screen. The inner circle in the globe depicts the equator (n,n,0). The outer circle represents the south pole (0,0,-1). By selecting crosshair locations on the screen in this manner, you can get any viewpoint (Figures 21-4 and 21-5).

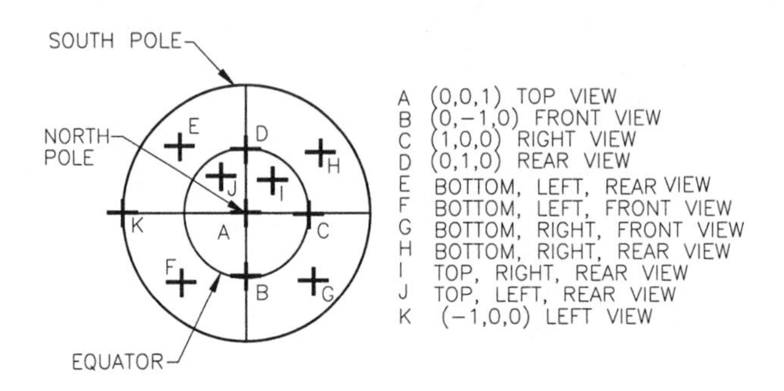

Figure 21-4 *Various locations on the compass and their respective views*

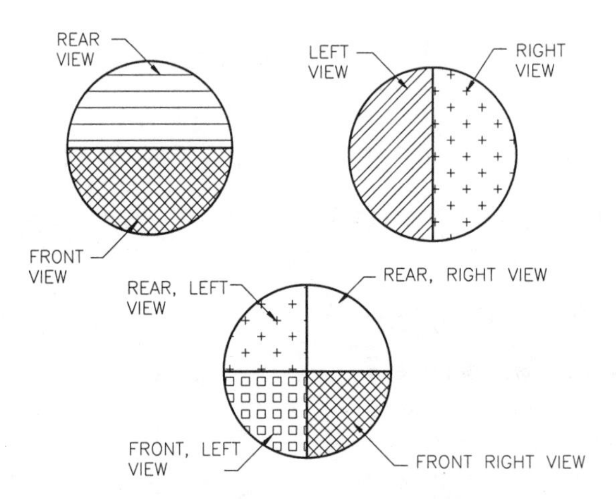

Figure 21-5 *Various sections of the compass*

Using the Rotate Option to Set the Viewpoint

With the **Rotate** option of the **VPOINT** command, you can set a new viewpoint by specifying values for two angles. The first angle you specify establishes the rotation (clockwise or counterclockwise) in the XY plane from the X axis. By default, the X axis is drawn at a 0-degree angle. The second angle lets you specify the angle at which the XY plane is up (or down). On selecting the **Rotate** option, you are provided with the following prompts:

Enter angle in XY plane from X axis <current>:
Enter angle from XY plane <current>:

For example, if you specify an angle value of 0 degrees for the angle in the XY plane from the X axis, and 90 degrees for the angle from the XY plane, the resulting viewpoint will be located along the Z axis; you will get the top view of the object, Figure 21-6.

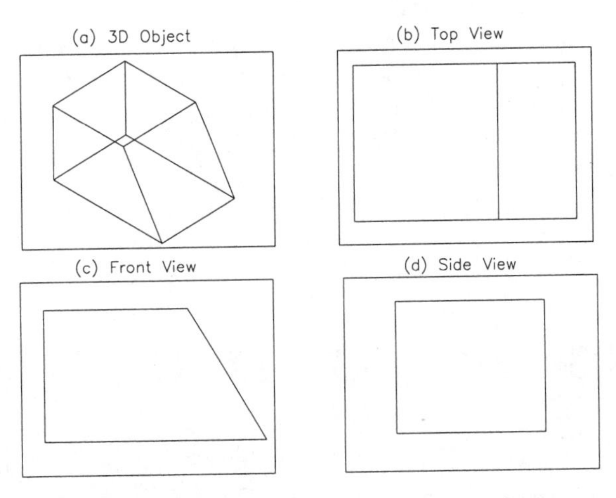

***Figure 21-6** Different views of a 3D object*

Changing the Viewpoint Using Predefined Views

Toolbar:	View Standard > View flyout
Menu:	View > Named Views
Command:	VIEW

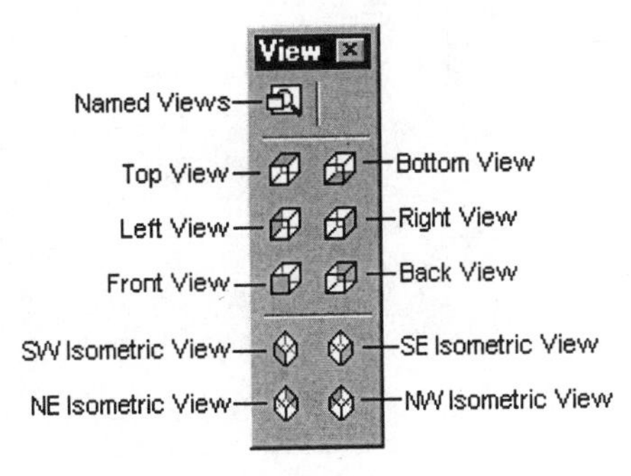

***Figure 21-7** View toolbar*

You can also change the viewpoint by selecting one of the ten options: **Iso SW**, **Iso SE**, **Iso NE**, **Iso NW**, **Top**, **Bottom**, **Left**, **Right**, **Front**, and **Back**, **Figure 21-7**. The effects on the viewpoint are as follows:

Viewpoint Option	Corresponding Viewpoint
Top,	(0,0,1)
Bottom	(0,0,-1)
Left	(-1,0,0)
Right	(1,0,0)
Front	(0,-1,0)
Back	(0,1,0)
Iso SW	(-1,-1,1)
Iso SE	(1,-1,1)

Iso NE (1,1,1)
Iso NW (-1,1,1)

As discussed earlier in Chapter 5, invoking the **VIEW** command displays the **View** dialog box, here you can save and restore views in the **Named Views** tab of the dialog box. The **Orthographic and Isometric** tab (Figure 21-8) allows you to restore predefined Isometric and Orthographic views. You can select a specific orthographic and isometric view from the list box. The **Current View** displays the name of the current view and a small icon is placed next to the view you have set current. After you have selected a specific view, choose the **Set Current** button and choose **OK** to set the selected view as current. You can also double-click the specific view or right-click it to display a shortcut menu and choose **Set Current**. When an orthographic view is set as current, the drawing is zoomed to its extents. The **Relative to:** drop-down list displays a list of all the named UCSs relative to which the orthographic views can be set current; the default is **World**. The **UCSBASE** variable stores the value for the name of the UCS that is selected as base for the selected view. Selecting the **Restore orthographic UCS with View** check box restores a UCS associated with the particular view when it is set current. The **UCSORTHO** variable stores the value for this option. To find out detailed information about the UCS associated with a selected named or orthographic view, choose the **Details** button in the **Named Views** tab of this dialog box.

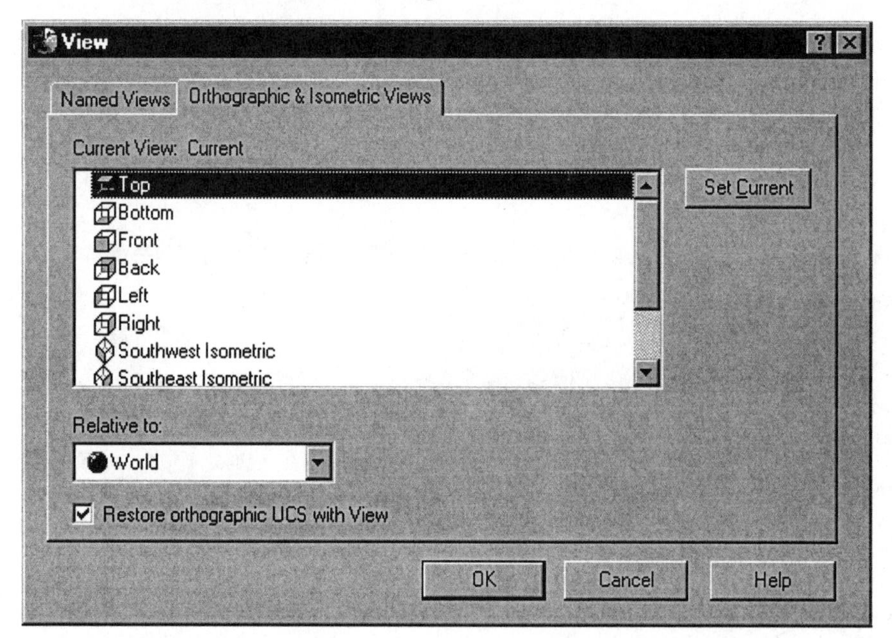

Figure 21-8 *The* ***Orthographic and Isometric Views*** *tab of the* ***View*** *dialog box*

SETTING PLAN VIEWS (PLAN COMMAND)

Menu:	View > 3D Views > Plan View
Command:	PLAN

The **PLAN** command can be used to generate the plan view of an object. The viewpoint used to obtain the plan view is always (0,0,1) relative to the current UCS, the WCS, or a previously saved UCS. In other words, a plan view is defined as the view obtained when you observe the object along the Z axis. Remember that the plan view is not necessarily the view obtained by looking at the object from the top. If you are using the WCS, the plan view is obtained by looking at the object from top. If you want the plan view relative to current UCS, the plan view depends totally on the direction of the Z axis. Since you can change the orientation of the X, Y, and Z axes with the UCS command, you can have a different plan view for same object depending on the orientation of the X, Y, and Z axes.

Command: **PLAN** [Enter]
Enter an option [Current ucs/Ucs/World] <Current >:

If you have multiple viewports, the **PLAN** command generates the plan view in the current viewport only. If Perspective is on, the use of the **PLAN** command turns it off. Clipping is also turned off. The **PLAN** command cannot be used in Paper space. The prompt sequence for the **PLAN** command has the following three options.

Current UCS. This is the default option. If you invoke this option, the plan view relative to the current UCS will be generated. The display is regenerated to fit the drawing extents in the current viewport.

UCS. With the help of this option, you can generate the plan view relative to a previously defined UCS. You are prompted for the name of the UCS. You can list the names of all the saved coordinate systems by entering ? as the response to the prompt.

World. This option generates the plan view relative to the WCS. The display is regenerated to fit the drawing extents on the screen.

If the **UCSFOLLOW** system variable is set to 1, any change in the UCS changes the Plan view.

HIDE COMMAND

Menu:	View > Hide
Command:	HIDE

In 3D objects drawn by giving thickness, some of the lines lie behind other lines or objects. Sometimes you may not want these hidden lines to show up on screen, to enhance the clarity of object. In order to do so, you can use the **HIDE** command. The prompt sequence for using the **HIDE** command is:

Command: **HIDE** [Enter]

Once you enter this command, the screen is regenerated, and all the hidden lines are suppressed in the drawing. The hidden lines are again included in the drawing when the next regeneration takes place. Circles, solids, wide polyline segments, and polygons with thicknesses are treated as opaque surfaces; hence, any object behind them is suppressed once the **HIDE** command is used, Figure 21-9. Circles, solids, and wide polyline segments are considered to be solid objects if they are extruded. In this case, the bottom and top faces are also considered, Figure 21-10. Objects on frozen layers or layers that have been turned off are not taken into consideration by the **HIDE** command.

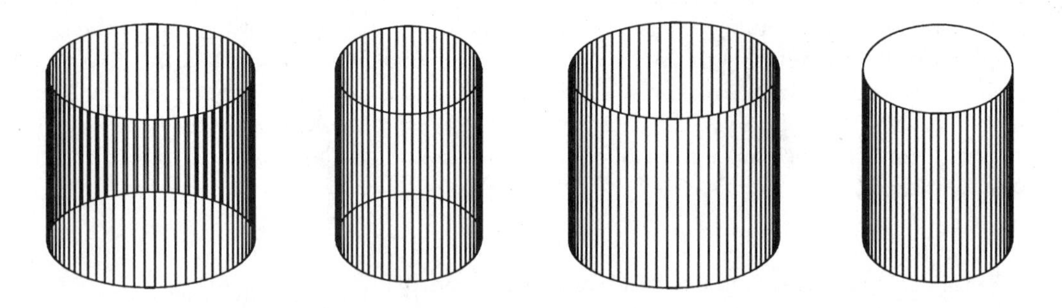

Figure 21-9 Ellipse and Circle with thickness and before using the HIDE command

Figure 21-10 Ellipse and Circle with thickness before and after using the HIDE command.

Exercise 1 *General*

Draw the objects shown in Figures 21-11 and 21-12 using the **Elevation** and **Thickness** option of the **ELEV** command. Then view it from various viewpoints using the **VPOINT** command.

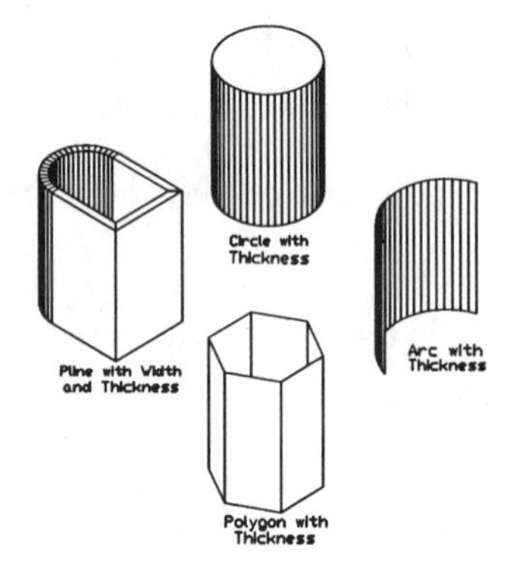

Figure 21-11 Drawing for Exercise 1

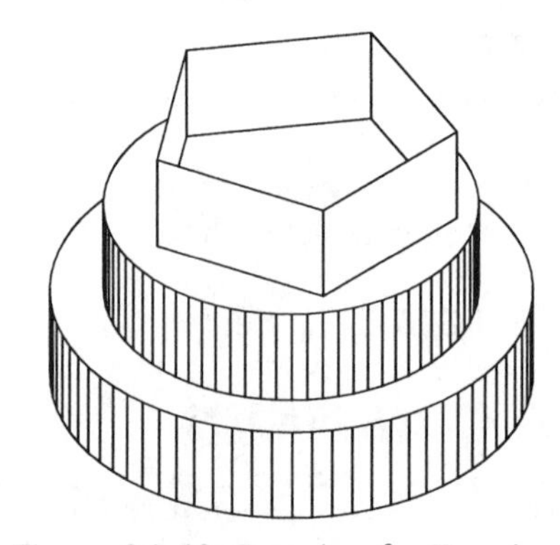

Figure 21-12 Drawing for Exercise 1

Creating a 2D Region (REGION Command)

Toolbar: Draw > Region
Menu: Draw > Region
Command: REGION

You can use the **REGION** command to impart the properties of 3D objects on a 2D object (Figure 21-13). AutoCAD LT will create a region from the selected closed polyline, line, circle, arc, ellipse, and splines and then remove the polyline from the drawing unless the **DELOBJ** system variable is set to 0.

1. Use the **REGION** command to create a region.

Command: **REGION** [Enter]
Select objects: *Select the polyline.*
Select objects: [Enter]
1 loop extracted
1 region created

2. Use the **UCS** command to move the UCS icon to the center point of the arc (5,4). (When AutoCAD LT prompts for the origin point, you can also use the CENter object snap to grab the center point of the arc.)

Command: **UCS** [Enter]
Current UCS name: *WORLD*
Enter an option [New/Move/orthoGraphic/Prev/Restore/Save/Del/?/World]<World>: **N** [Enter]
Specify new origin of new UCS or [ZAxis/3point/OBject/Face/View/X/Y/Z]<0,0,0>: **5,4** [Enter]

3. Draw 0.5-diameter circles located on the 4.0-diameter-bolt circle (Figure 21-14). (You can also create circles by drawing one circle and then using the **ARRAY** command to create the remaining circles.)

Command: **CIRCLE** [Enter]
Specify center point or [3P/2P/TtR(tan tan radius)]: **@2<60** [Enter]
Specify radius of circle or [Diameter]: **D** [Enter]
Specify diameter of circle: **0.5** [Enter]

Command: **ARRAY** [Enter]
Select objects: *Select 0.5 diameter circle.*
Enter type of array [Rectangular/Polar] <R>: **P** [Enter]
Specify center point of array: **0,0** [Enter]
Enter number of items in the array: **3** [Enter]
Specify the angle to fill (+=CCW, -=CW)<360>: **240** [Enter]
Rotate arrayed objects? [Yes/ No]<Y>: [Enter]

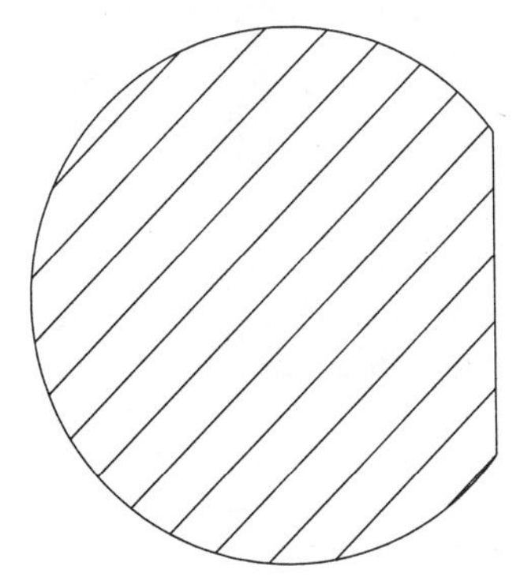

Figure 21-13 *Creating a 2D Region*

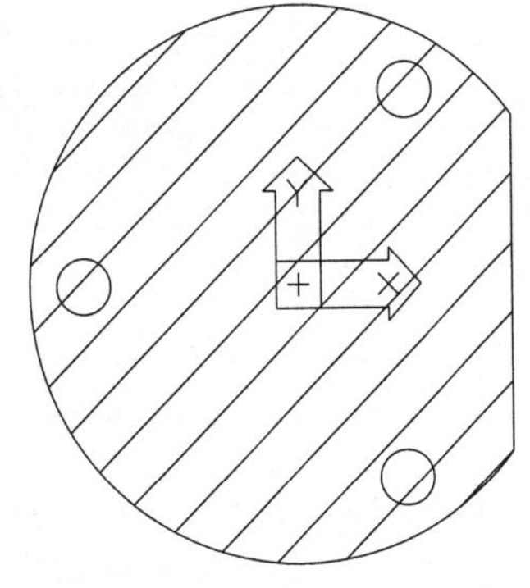

Figure 21-14 *Creating a circular array*

4. Use the **REGION** command to create a region from the circles.

Subtracting Regions (SUBTRACT)

Toolbar:	Modify II > Subtract
Menu:	Modify > Region > Subtract
Command:	SUBTRACT

You can use the **SUBTRACT** command to subtract regions. When you use this command, AutoCAD LT will prompt you to select two selection sets. Once you have selected the objects, the objects of the second selection set will be subtracted from the first selection set (Figure 21-15).

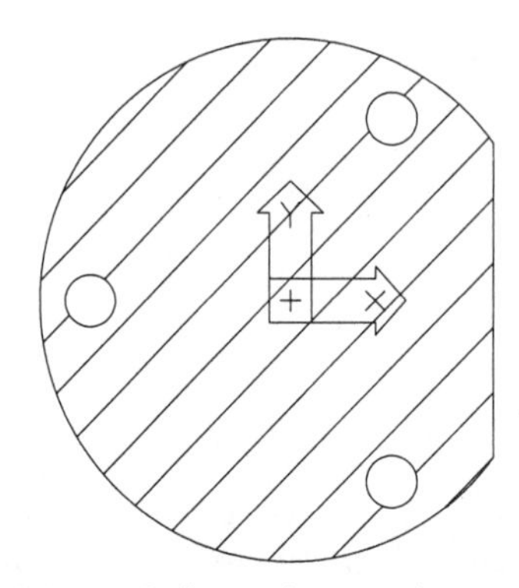

***Figure 21-15** Subtracting regions using the SUBTRACT command*

Command: **SUBTRACT** Enter
Select regions to subtract from...
Select objects: *Select the region.*
Select objects: Enter
Select regions to subtract...
Select objects: *Select three 0.5-diameter circles.*
Select objects: Enter

The hatch lines may not get removed from the small circles until you use the **MOVE** command and move the outer profile.

Hatching a Region (Using BHATCH Command). When you create a region, the mesh lines are not visible. To hatch a region, you can use the **BHATCH** command. The hatch pattern, the hatch angle, and the hatch size can be set in the **Boundary Hatch** dialog box. You can use any hatch pattern defined in the AutoCAD LT ACLT.PAT file. To get hatch as shown in Figure 21-15 use the **ANSI31** hatch pattern and a scale factor of 3.5. You can also create regions with the **BOUNDARY** and **BHATCH** commands.

Uniting Regions (UNION)

Toolbar:	Modify II > Union
Menu:	Modify > Region > Union
Command:	UNION

The **UNION** command can be used to create composite regions, Figure 21-16 and 21-17. Several regions can be combined in the same command. However, the solids cannot be combined with the regions.

Command: **UNION** Enter
Select objects: *Select the regions.*
Select objects: Enter

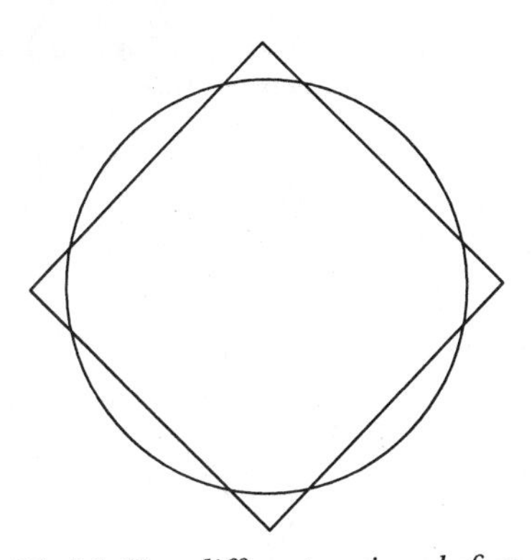

Figure 21-16 *Two different regions before union*

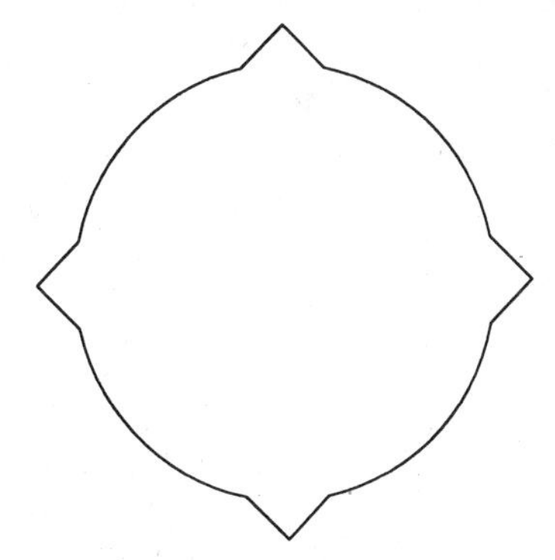

Figure 21-17 *Single region created by union of two different regions*

Creating Composite Regions (INTERSECT)

Toolbar:	Modify II > Intersect
Menu:	Modify > Region > Intersect
Command:	INTERSECT

The **INTERSECT** command creates a composite region from objects that intersect (Figure 21-18). This command can be invoked from the **Modify II** toolbar, from the **Modify** menu (choose **Region > Intersect**), or by entering **INTERSECT** at the Command prompt. If the objects do not intersect, they do not have any common overlapping area or volume; therefore, **INTERSECT** can be used with them, but it will delete them and prompt "Null region created - deleted", Figure 21-19.

Command: **INTERSECT** [Enter]
Select objects: *Select the regions.*
Select objects: [Enter]

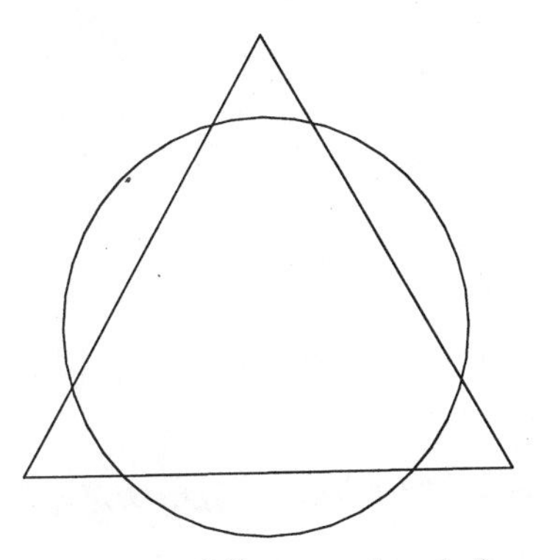

Figure 21-18 *Two different regions before intersecting*

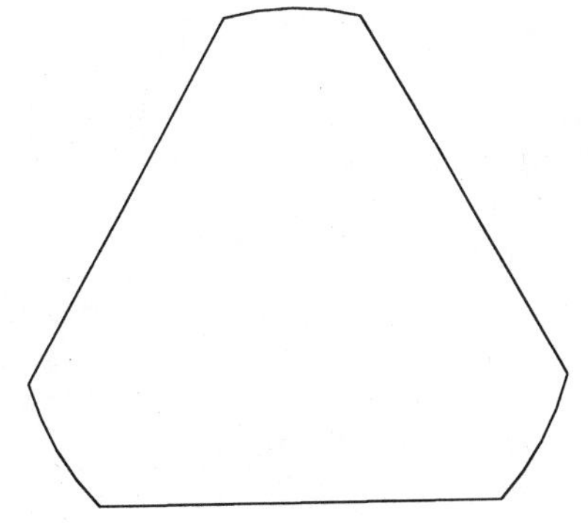

Figure 21-19 *Single region created by intersecting two regions*

Exercise 2 — *General*

Draw the objects shown in Figure 21-20 by creating different regions and then bringing them together. Assume the dimensions.

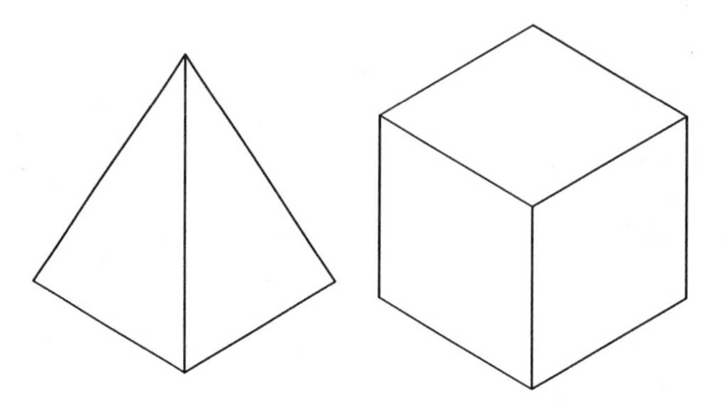

Figure 21-20 *Drawing for Exercise 2*

DRAWING 3D POLYLINES (3DPOLY COMMAND)

Menu:	Draw > 3D Polyline
Command:	3DPOLY

You can draw 3D polylines using the **3DPOLY** command (Figure 21-21). The working of the **3DPOLY** and **PLINE** commands is similar. The exceptions are that a third dimension (Z) is added to the polyline, and with the **3DPOLY** you can only draw straight line segments without variable widths.

Command: **3DPOLY** [Enter]
Specify start point of polyline: *Specify a point where you want to start the 3D polyline.*
Specify endpoint of line or [Undo]: *Specify the endpoint on the screen.*
Specify endpoint of line or [Undo]: *Specify the endpoint on the screen.*
Specify endpoint of line or [Close/Undo]: *Specify the endpoint on the screen or enter an option or press ENTER when finished drawing the 3D polyline.*

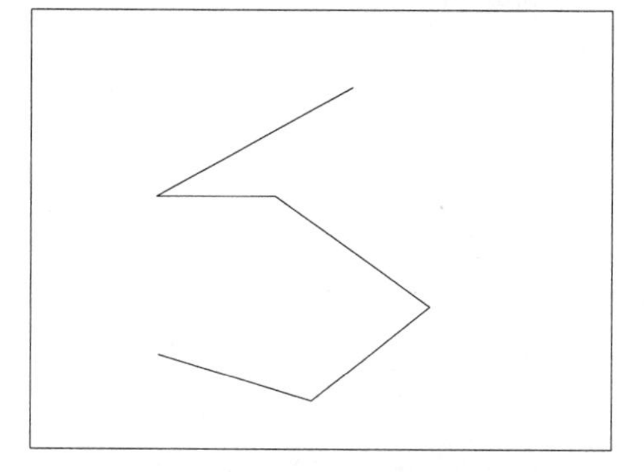

Figure 21-21 *Using the* ***3DPOLY*** *command*

The **3DPOLY** command provides the **Close** and **Undo** options.

Close Option

This option draws the connecting segment between the start point of the first polyline segment and the endpoint of the last polyline segment. Hence, you need to have at least two lines.

Undo Option

The **Undo** option reverses the effect of the previous operation. If you use the **Undo** option, the last polyline segment is erased, and a rubber-band line is fixed to the previous point while you are still in the **3DPOLY** command.

Certain editing operations cannot be performed on 3D polylines: joining, curve fitting with arc segments, and giving tangent specifications to the 3D polyline. You can use the **PEDIT** command to edit 3D polylines.

CYLINDRICAL COORDINATE SYSTEM

The **cylindrical coordinate system** (Figure 21-22) is also a modification of the polar coordinate system. The location of any point is described by three things:

1. Distance from the present UCS origin.
2. Angle in the XY plane.
3. Z dimension.

This coordinate system is used mostly to locate points on a cylindrical shape. For example, a point 4 units from the UCS origin, at an angle of 25 degrees from the X axis, and having a Z dimension of 6 units would be represented in the following format:

<Distance from the UCS origin><Angle from X axis><Z dimension>
@4<25,6.0

SPHERICAL COORDINATE SYSTEM

The spherical coordinate system (Figure 21-23) is a modification of the polar coordinate system. The location of any point is described by three things:

1. Distance from the present UCS origin.
2. Angle in the XY plane.
3. Angle from the XY plane.

This coordinate system is somewhat similar to using longitude, latitude, and the distance from the center of earth to find the location of a point on the earth. The distance from the present UCS origin is analogous to the distance from the center of the earth. The angle in the XY plane is analogous to the longitudinal measurement. The angle from the XY plane is analogous to the latitudinal measurement. The spherical coordinate system is used mostly to locate points on a spherical surface, Figure 21-23. For example, a point 7 units from the UCS origin, at an angle of 60 degrees from the X axis, and at an angle of 50 degrees from XY plane would be represented in the following format:

<Distance from the UCS origin><Angle from X axis><Angle from XY plane>
@7<60<50

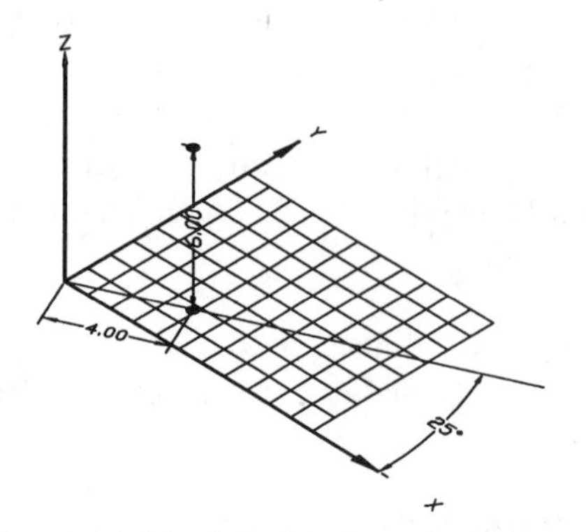

Figure 21-22 *Cylindrical coordinate system*

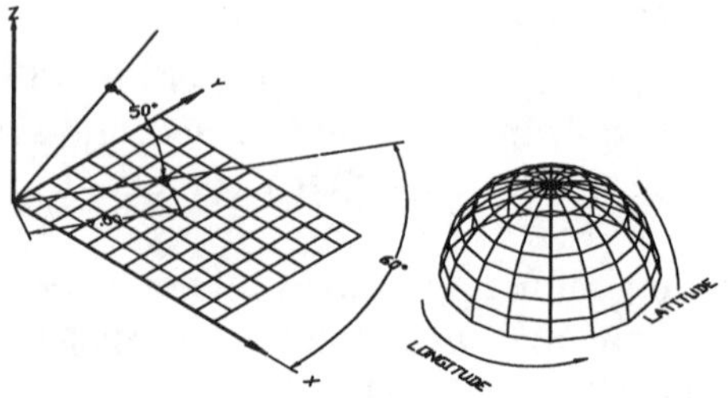

Figure 21-23 *Spherical coordinates system*

DRAWING 3D OBJECTS

AutoCAD LT has provided two methods to create a 3D model. You can create a simple 3D model for visualization by specifying thickness values for the 2D objects, thus extruding or extending them in the Z direction. You can also use wireframe modeling to create an accurate representation of the 3D objects. To view 3D objects properly, change the viewpoint using the **VPOINT** command and then use the **HIDE** command. Two methods that can be used to create 3D models are:

Adding Thickness to a 2D Objects

Thickness is the property that creates a 3D model from a 2D model. It is defined as the distance by which the 2D object is extruded or extended in the Z direction above or below its original location in the space. If the value of thickness is positive, the object is extended in positive Z direction and if the value of thickness is negative, the object gets extended in negative Z direction, Figures 21-24, and 21-25.

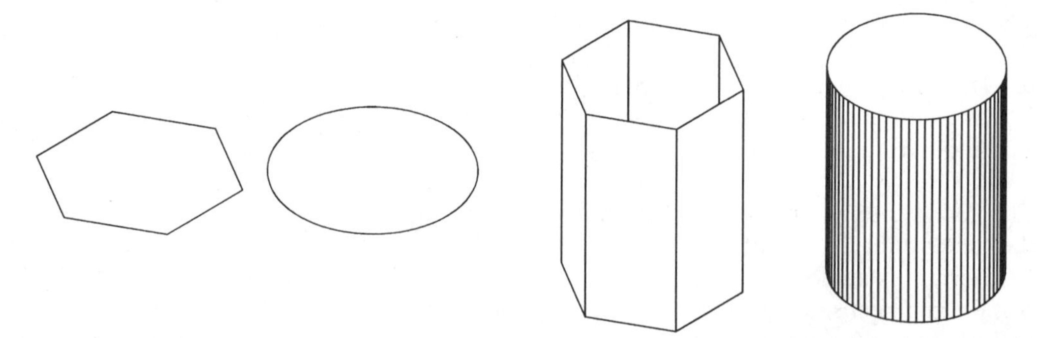

Figure 21-24 *Polygon and Circle without thickness*

Figure 21-25 *Polygon and Circle with thickness*

Creating 3D Wireframe Models

Wireframe models are just an edge or a skeletal representation of 3D objects using lines and curves. These models are just like a 3D model made by joining a number of matchsticks together. A wireframe model can be created by entering 3D coordinates for lines, using 3D Polylines, shifting the work plane using the **ELEV** command, or by moving and copying the 2D objects after creating them. Figure 21-26 shows various wireframe models. Since these models are just a skeletal representation of the 3D objects, so you can see through them.

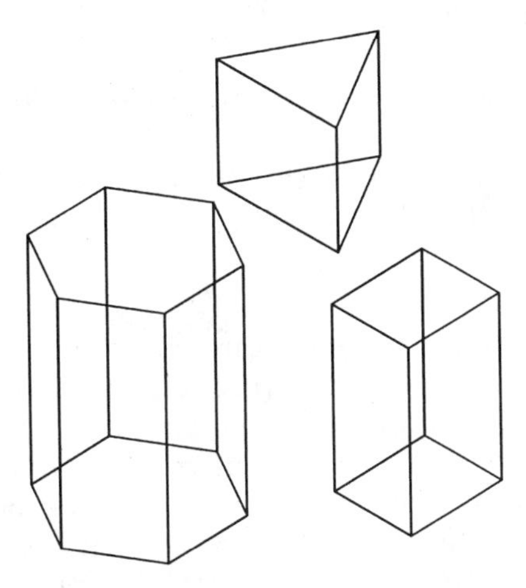

Figure 21-26 *Wireframe models*

DYNAMIC VIEWING OF 3D OBJECTS (DVIEW COMMAND)

With the **DVIEW** command, you can create a parallel projection or a perspective view of objects on the screen. Basically, **DVIEW** is an improvement over the **VPOINT** command in that the **DVIEW** command allows you to visually maneuvre around 3D objects to obtain different views. As just mentioned, the **DVIEW** command offers you a choice between parallel viewing and perspective viewing. The difference between the two is that in parallel view, parallel lines in the 3D object remain parallel, whereas in perspective view, parallel lines meet at a vanishing point. These definitions suggest that with the **VPOINT** command, only parallel viewing is possible. When the Perspective view is current, **ZOOM**, **PAN**, **'ZOOM**, **'PAN**, **DSVIEWER** commands and scroll bars are not available. The **DVIEW** command uses the camera and target concept to visualize an object from any desired position in 3D space. The position from which you want to view the object (the viewer's eye) is the camera, and the focus point is the target. The line formed between these two points is the line of sight, also known as the viewing direction. To get different viewing directions, you can move the camera or target or both. Once you have the required viewing direction, you can change the distance between the camera and the target by moving the camera along the line of sight. The field of view can be changed by attaching a wide-angle lens or a telephoto lens to the camera. You can pan or twist (rotate) the image. The hidden lines in the object being viewed can be suppressed. You can clip those portions of model that you do not want to see. All these options of the **DVIEW** command demonstrate how useful this command is. The **'ZOOM**, **DSVIEWER**, **PAN** commands and Scroll bars are not available in the **DVIEW** command.

Command: **DVIEW** [Enter]
Select objects or <use DVIEWBLOCK>: *Select the objects you want to view dynamically or press ENTER to use the DVIEWBLOCK.*
Enter option
[CAmera/TArget/Distance/POints/PAn/Zoom/TWist/CLip/Hide/Off/Undo]: *Specify a point or enter an option.*

The point you specify is the start point for dragging. As you move the cursor, the viewing direction changes. The command prompt is:

Enter direction and magnitude angles: *Enter angles between 0 to 360 degrees or specify on screen.*

Direction angle determines the front of the view and the magnitude angle determines the how far the view is. The different **DVIEW** command options are discussed next.

CAmera Option

With the **CAmera** option, you can rotate the camera about the target point. To invoke the CAmera option, enter CA at the **Enter option [CAmera/TArget/Distance/POints/PAn/Zoom/TWist/CLip/Hide/Off/Undo]:** prompt. Once you have invoked this option, the drawing is still, but you can maneuver the camera up and down (above or below the target) or you can move it left or right (clockwise or counterclockwise around the target). Remember that when you are moving the camera, the target is stationary, and that when you are moving the target, the camera is stationary. The next prompt AutoCAD LT displays is:

Specify camera location, or enter angle from XY plane, or [Toggle (angle in)] <current>:

By default, this prompt asks you for the angle from the XY plane of the camera to the current UCS. This is nothing but the vertical (up or down) movement of the camera about the target. There are two ways to specify the angle.

1. You can specify the angle of rotation by moving the graphics cursor in the graphics area until you attain the desired rotation and then choosing the pick button of your pointing device. You will notice that as you move the cursor in the graphics area, the angle of rotation is continuously displayed in the status bar. Also, as you move the cursor, the object also moves dynamically. If you give a horizontal movement to the camera at the **Specify camera location, or enter angle from XY plane, or [Toggle (angle in)] <current>** prompt, notice that the angle value displayed in the status bar does not change. This is because at the **Specify camera location, or enter angle from XY plane, or [Toggle (angle in)] <current>** prompt, you are required to specify the vertical movement.

2. The other way to specify the angle of rotation is by entering the required angle of rotation value at the prompt. An angle of 90 degrees (default value) from the XY plane makes the camera look at the object straight down from the top side of the object, providing you with the top view (plan view). In this case, the line of sight is perpendicular to the XY plane of the current UCS. An angle of negative 90 degrees makes the camera look at the object straight up from the bottom side of the object. As you move the camera toward the top side of the object, the angle value increases. If you do not want to specify a vertical

movement, press ENTER.

Once you have specified the angle from the XY plane, the next prompt displayed is:

Specify camera location, or enter angle in XY plane from X axis, or [Toggle (angle from)] <current>:

This prompt asks you for the angle of the camera in the XY plane from the X axis. This is nothing but the horizontal (right or left) movement of the camera. You can enter the angle value in the range of -180 to 180 degrees. Moving the camera toward the right side of the object (counterclockwise) can be achieved by increasing the angle value; moving the camera toward the left side of the object (clockwise) can be achieved by decreasing the angle value. You can also toggle between these two prompts using the **Toggle (angle from)** or **Toggle (angle in)** option in the two prompts. The use of the **toggle** option can be illustrated as follows.

Let's say you want to enter only the angle in the XY plane from the X axis. There are two ways to do this. The first is to give a null response to the **Specify camera location, or enter angle from XY plane, or [Toggle (angle in)] <default>:** prompt. The other is by using the **Toggle (angle in)** option. Hence, to switch from the **Specify camera location, or enter angle from XY plane, or [Toggle (angle in)] <default>:** prompt to the **Specify camera location, or enter angle in XY plane from X axis, or [Toggle (angle from)] <current>:** prompt, you will have the following prompt sequence:

Specify camera location, or enter angle from XY plane, or [Toggle (angle in)] <default>: Enter
Specify camera location, or enter angle in XY plane from X axis, or [Toggle (angle from)] <current>:

In the same manner, if you want to switch from the **Specify camera location, or enter angle in XY plane from X axis, or [Toggle (angle from)] <current>:** prompt to the **Specify camera location, or enter angle from XY plane, or [Toggle (angle in)] <default>:** prompt, the following will be the prompt sequence:

Specify camera location, or enter angle in XY plane from X axis, or [Toggle (angle from)] <current>: **T** Enter
Specify camera location, or enter angle from XY plane, or [Toggle (angle in)] <default>:

You have gone through the basic concept of the **CAmera** option of the **DVIEW** command. Now, you will work out some examples to apply the camera concept.

Note

The ***DVIEWBLOCK*** *is automatically displayed when you use* ***DVIEW*** *without selecting objects. You can define or redefine your own substitute* ***DVIEWBLOCK****, just like any other symbol block. Just be sure to define it to a 1x1x1 unit size so that it will be properly scaled in the* ***DVIEW*** *command.*

Example 1

In Figure 21-27, you have four sections. The house shown can be obtained by pressing ENTER at the **Select objects or <use DVIEWBLOCK>** prompt. The drawing of the house with a window, an open door, and a chimney obtained on the screen is a block named **DVIEWBLOCK**.

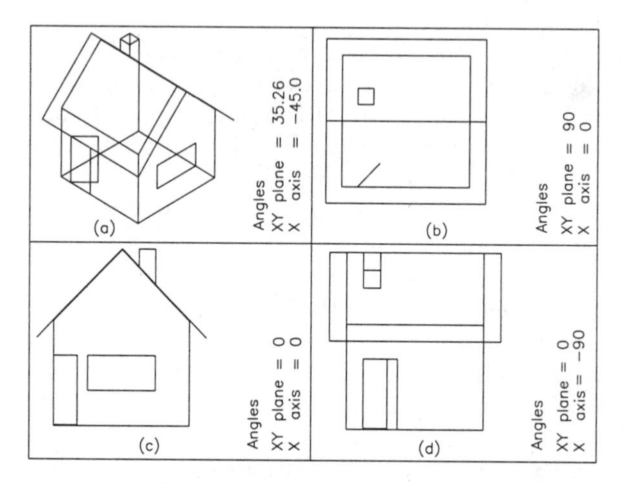

***Figure 21-27 CAmera** option of the **DVIEW** command*

As you use various options of the **DVIEW** command, the block is updated to reflect the changes. Once you come out of the **DVIEW** command, the entire drawing is regenerated, and the view obtained depends on the view you have selected using the various options of the **DVIEW** command. In this chapter, you will use this block to demonstrate the effect of various options of the **DVIEW** command. However, you can make a custom block of your own. The block should be of unit size. Set the lower left corner of the block as the origin point. The different views of the house block we see in Figure 21-27 can be obtained with the **CAmera** option of the **DVIEW** command. First, use the **DVIEW** command to get the image of the house on the screen.

Command: **DVIEW** Enter
Select objects or <use DVIEWBLOCK>: Enter

Now you have the image of the house on the screen. For Figure 21-27(a) (3D view), the following is the prompt sequence:

Enter option
[CAmera/TArget/Distance/POints/PAn/Zoom/TWist/CLip/Hide/Off/Undo]: **CA** Enter
Specify camera location, or enter angle from XY plane, or [Toggle (angle in)] <current>: **35.26** Enter
Specify camera location, or enter angle in XY plane from X axis, or [Toggle (angle from)] <current>: **-45** Enter
Enter option
[CAmera/TArget/Distance/POints/PAn/Zoom/TWist/CLip/Hide/Off/Undo]: *Press ENTER.*
Regenerating model.

Once you exit the **DVIEW** command, the figure of the house is removed from the screen. For Figure 21-27(b) (top view), the following is the prompt sequence:

Command: **DVIEW** [Enter]
Select objects or <use DVIEWBLOCK>: [Enter]
Enter option
[CAmera/TArget/Distance/POints/PAn/Zoom/TWist/CLip/Hide/Off/Undo]: **CA** [Enter]
Specify camera location, or enter angle from XY plane, or [Toggle (angle in)] <current>:
90 [Enter]
Specify camera location, or enter angle in XY plane from X axis, or [Toggle (angle from)]
<current>: **0** [Enter]
Enter option
[CAmera/TArget/Distance/POints/PAn/Zoom/TWist/CLip/Hide/Off/Undo]: *Press ENTER.*
Regenerating model.

For Figure 21-27(c) (right side view), the following is the prompt sequence:

Command: **DVIEW** [Enter]
Select objects or <use DVIEWBLOCK>: [Enter]
Enter option
[CAmera/TArget/Distance/POints/PAn/Zoom/TWist/CLip/Hide/Off/Undo]: **CA**
Specify camera location, or enter angle from XY plane, or [Toggle (angle in)] <current>:
0.00 [Enter]
Specify camera location, or enter angle in XY plane from X axis, or [Toggle (angle from)]
<current>: **0.00** [Enter]
Enter option
[CAmera/TArget/Distance/POints/PAn/Zoom/TWist/CLip/Hide/Off/Undo]: *Press ENTER.*
Regenerating model.

For Figure 21-27(d) (front view), the following is the prompt sequence:

Command: **DVIEW** [Enter]
Select objects or <use DVIEWBLOCK>: [Enter]
Enter option
[CAmera/TArget/Distance/POints/PAn/Zoom/TWist/CLip/Hide/Off/Undo]: **CA** [Enter]
Specify camera location, or enter angle from XY plane, or [Toggle (angle in)] <current>:
0.00 [Enter]
Specify camera location, or enter angle in XY plane from X axis, or [Toggle (angle from)]
<current>: **-90.00** [Enter]
Enter option
[CAmera/TArget/Distance/POints/PAn/Zoom/TWist/CLip/Hide/Off/Undo]: *Press ENTER.*
Regenerating model.

TArget Option

With the **TArget** option you can rotate the target point with respect to the camera. To invoke the **TArget** option, enter TA at the **Enter option [CAmera/ TArget/ Distance/ POints/ PAn/**

Zoom/ TWist/ CLip/ Hide/ Off/ Undo]: prompt. Once you have invoked this option, the drawing is still, but you can maneuver the target point up or down or left or right about the camera. When you move the target, the camera is stationary. The prompt sequence for the **TArget** option is:

Command: **DVIEW** Enter
Select objects or <use DVIEWBLOCK>: Enter
Enter option
[CAmera/TArget/Distance/POints/PAn/Zoom/TWist/CLip/Hide/Off/Undo]: **TA** Enter
Specify camera location, or enter angle from XY plane, or [Toggle (angle in)] <current>:
Specify the angle about which the target will lie above or below the camera.
Specify camera location, or enter angle in XY plane from X axis, or [Toggle (angle from)] <current>: *Specify the angle about which the target will lie left or right of the camera.*
Enter option
[CAmera/TArget/Distance/POints/PAn/Zoom/TWist/CLip/Hide/Off/Undo]: Enter

The prompt sequence does not reveal the difference between the **CAmera** option and the **TArget** option. The difference lies in the actual angle of view. For example, if you specify an angle of 90 degrees in the **Specify camera location, or enter angle from XY plane, or [Toggle (angle in)] <current>:** prompt of the **CAmera** option, your viewing direction is from the top of the object toward the bottom; if you specify the same angle in the same prompt for the **TArget** option, your viewing direction is from the bottom of the object toward the top.

Example 2

The different views of the house block in Figure 21-28 can be obtained with the **TArget** option of the **DVIEW** command.

First, use the **DVIEW** command to display the image of the house on the screen.

Command: **DVIEW** Enter
Select objects or <use DVIEWBLOCK>: Enter

For Figure 21-28(a) (3D view), the following is the prompt sequence:

Enter option
[CAmera/TArget/Distance/POints/PAn/Zoom/TWist/CLip/Hide/Off/Undo]: **TA** Enter
Specify camera location, or enter angle from XY plane, or [Toggle (angle in)] <current>:
-35.26 Enter
Specify camera location, or enter angle in XY plane from X axis, or [Toggle (angle from)] <current>: **135** Enter
Enter option
[CAmera/TArget/Distance/POints/PAn/Zoom/TWist/CLip/Hide/Off/Undo]:
Regenerating model.

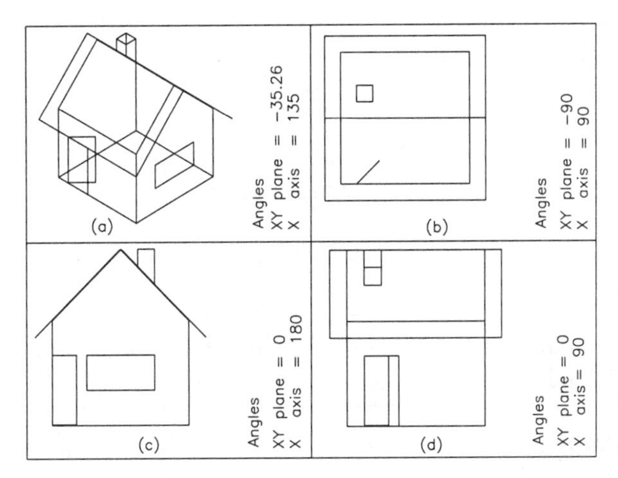

***Figure 21-28 TArget** option of the **DVIEW** command*

For Figure 21-28(b) (top view), the following is the prompt sequence:

Command: **DVIEW** Enter
Select objects or <use DVIEWBLOCK>: Enter
Enter option
[CAmera/TArget/Distance/POints/PAn/Zoom/TWist/CLip/Hide/Off/Undo]: **TA** Enter
Specify camera location, or enter angle from XY plane, or [Toggle (angle in)] <current>: **-90.00** Enter
Specify camera location, or enter angle in XY plane from X axis, or [Toggle (angle from)] <current>: **90.00** Enter
Enter option
[CAmera/TArget/Distance/POints/PAn/Zoom/TWist/CLip/Hide/Off/Undo]: *Press ENTER.*
Regenerating model

For Figure 21-28(c) (right side view), the following is the prompt sequence:

Command: **DVIEW** Enter
Select objects or <use DVIEWBLOCK>: Enter
Enter option
[CAmera/TArget/Distance/POints/PAn/Zoom/TWist/CLip/Hide/Off/Undo]: **TA** Enter
Specify camera location, or enter angle from XY plane, or [Toggle (angle in)] <current>: **0.00** Enter
Specify camera location, or enter angle in XY plane from X axis, or [Toggle (angle from)] <current>: **180.00** Enter
Enter option
[CAmera/TArget/Distance/POints/PAn/Zoom/TWist/CLip/Hide/Off/Undo]: *Press ENTER.*
Regenerating model

For Figure 21-28(d) (front view), the following is the prompt sequence:

> Command: **DVIEW** [Enter]
> Select objects or <use DVIEWBLOCK>: [Enter]
> Enter option
> [CAmera/TArget/Distance/POints/PAn/Zoom/TWist/CLip/Hide/Off/Undo]: **TA** [Enter]
> Specify camera location, or enter angle in XY plane from X axis, or [Toggle (angle from)] <default>: **0.00** [Enter]
> Specify camera location, or enter angle in XY plane from X axis, or [Toggle (angle from)] <default>: **90.00** [Enter]
> Enter option
> [CAmera/TArget/Distance/POints/PAn/Zoom/TWist/CLip/Hide/Off/Undo]: *Press ENTER.*
> Regenerating model.

Distance Option

As mentioned before, the line obtained on joining the camera position and the target position is known as the line of sight. The **Distance** option can be used to move the camera toward or away from the target along the line of sight. Invoking the **Distance** option enables perspective viewing. Since we have not used the **Distance** option until now, all the previous views were in parallel projection. In perspective display, the objects nearer to the camera appear bigger than objects that are farther away from the camera position. In other words, in perspective views the parallel lines meet at a vanishing point. Another noticeable difference is that the regular coordinate system icon is replaced by the perspective icon. This icon acts as a reminder that perspective viewing is enabled. The **Distance** option can be invoked by entering **D** at the **Enter option [CAmera/TArget/Distance/POints/PAn/Zoom/TWist/CLip/Hide/Off/Undo]:** prompt. Once you invoke the **Distance** option, AutoCAD LT prompts you to specify the new distance between the camera and the target.

> Specify new camera-target distance <1.0000>: *Enter the desired distance between the camera and the target.*

On the top side of the screen, a slider bar appears. It is marked from 0x to 16x. The current distance is represented by the 1x mark. This is verified by the fact that the slider bar moves right or left with respect to the 1x mark on the slider bar. As you move the slider bar toward the right, the distance between the camera and the target increases; as you move the slider bar toward the left, the distance between the camera and the target decreases. For example, when you move the slider bar to the 16x mark, the distance between the camera and the target increases 16 times, or you can say that the camera moved away from the target on the line of sight 16 times the previous distance. The distance between the camera and the target is dynamically displayed in the status line. If you cannot display the entire object on the screen by moving the slider bar to the 16x mark, enter a larger distance at the keyboard. To revert to parallel viewing, invoke the **Off** option. If you want to magnify the drawing without turning the Perspective viewing on, you can use the **Zoom** option.

Example 3

Let Figure 21-29(a) be the default figure, in which the distance between camera and target is 4 units, which corresponds to the 1x mark on the slider bar. You can change the distance between camera and target and get different views as follows:

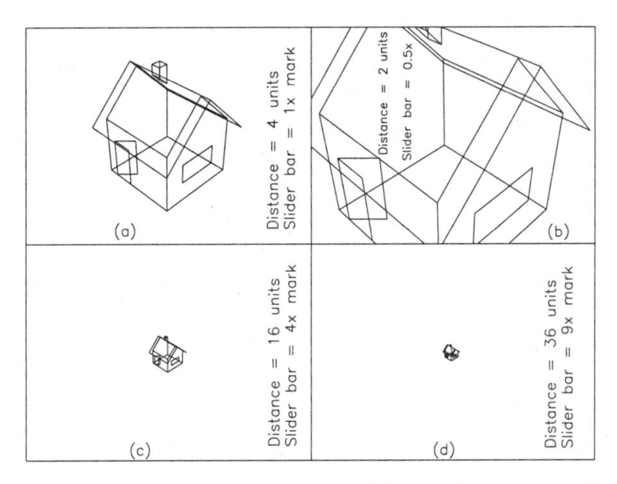

Figure 21-29 Distance *option of the* ***DVIEW*** *command*

For Figure 21-29(b):

Command: **DVIEW** Enter

Select objects or <use DVIEWBLOCK>: Enter
Enter option
[CAmera/TArget/Distance/POints/PAn/Zoom/TWist/CLip/Hide/Off/Undo]: **D** Enter
Specify new camera-target distance <4.0000>: **2.0** Enter

For Figure 21-29(c):

Command: **DVIEW** Enter
Select objects or <use DVIEWBLOCK>: Enter
Enter option
[CAmera/TArget/Distance/POints/PAn/Zoom/TWist/CLip/Hide/Off/Undo]: **D** Enter
Specify new camera-target distance <4.0000>: **16.0** Enter *or move the pointer to the 4x mark on slider bar.*

For Figure 21-29(d):

Command: **DVIEW** Enter
Select objects or <use DVIEWBLOCK>: Enter
Enter option
[CAmera/TArget/Distance/POints/PAn/Zoom/TWist/CLip/Hide/Off/Undo]: **D** Enter

Specify new camera-target distance <4.0000>: **36.0** [Enter] *or move the pointer to the 9x mark on slider bar.*

POints Option

With the **POints** option, you can specify the camera and target positions (points) in X, Y, Z coordinates. You can specify the X, Y, Z coordinates of the point by any method used to specify points, including object snap and .X, .Y, .Z point filters. The X, Y, Z coordinate values are with respect to the current UCS. If you use the object snap to specify the points, you must type the name of the object snap. This option can be invoked by entering PO at the **Enter option [CAmera/TArget/Distance/POints/PAn/Zoom/TWist/CLip/Hide/Off/Undo]:** prompt.

Command: **DVIEW** [Enter]
Select objects or <use DVIEWBLOCK>: [Enter]
Enter option
[CAmera/TArget/Distance/POints/PAn/Zoom/TWist/CLip/Hide/Off/Undo]: **PO** [Enter]

The target point needs to be specified first. A rubber-band line is drawn from the current target position to the drawing crosshairs. This is the line of sight.

Specify target point <current>: *Specify the location of the target or press ENTER.*

Once you have specified the target point, you are prompted to specify the camera point. A rubber-band line is drawn between the target point and the drawing crosshairs. This helps you to place the camera relative to the target.

Enter camera point <current>: *Specify the location of the camera press ENTER.*

Establishment of the new target point and camera point should be carried in parallel projection. If you specify these two points while the perspective projection is active, the perspective projection is temporarily turned off, until you specify the camera and target points. Once this is done, the object is again displayed in perspective. If the viewing direction is changed by the new target location and camera location, the preview image is regenerated to show the change.

Example 4

In Figure 21-30, the target is located at the lower corner of the house, and the camera is located at the corner of the chimney.

In Figure 21-31, the camera is located at the lower corner of the house, and the target is located at the corner of the chimney. Both these points are marked by cross marks.

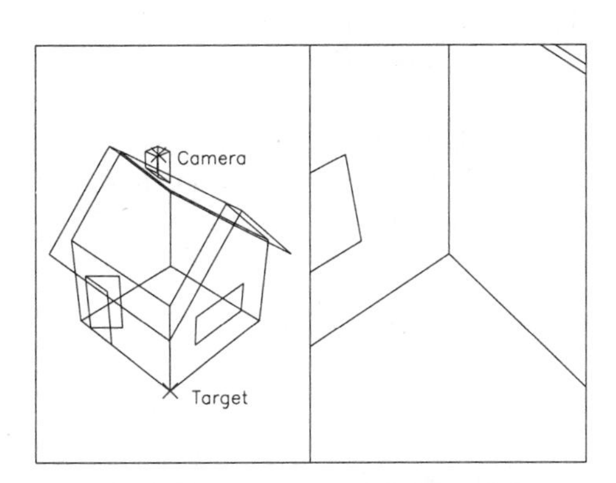

Figure 21-30 ***Point*** *option of the* ***DVIEW*** *command*

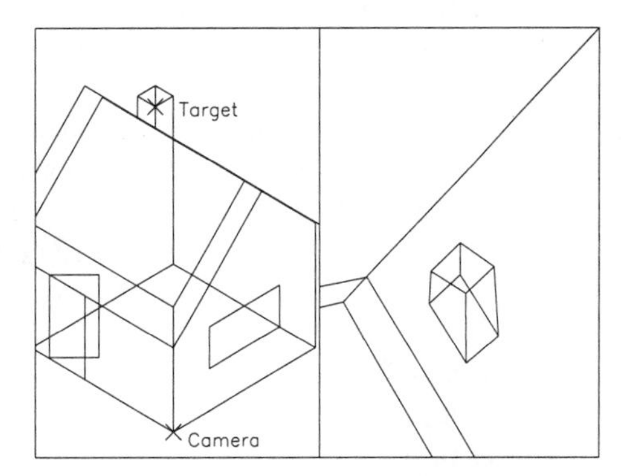

Figure 21-31 ***Point*** *option of the* ***DVIEW*** *command*

PAn Option

The **PAn** option of the **DVIEW** command resembles the **PAN** command. This option lets you shift the entire drawing with respect to the graphics display area. Just as with the **PAN** command, you have to specify the pan distance and direction by specifying two points. You must use a pointing device to specify the two points if perspective viewing is active. The prompt sequence for this option is:

Command: **DVIEW** [Enter]
Select objects or <use DVIEWBLOCK>: [Enter]
Enter option
[CAmera/TArget/Distance/POints/PAn/Zoom/TWist/CLip/Hide/Off/Undo]: **PA** [Enter]
Specify displacement base point: *Specify the first point.*
Specify second point: *Specify the second point.*

Zoom Option

The **Zoom** option of the **DVIEW** command resembles the **ZOOM** command. With the help of this option, you can enlarge or reduce the drawing. This option can be invoked by entering Z at the following prompt:

Enter option
[CAmera/TArget/Distance/POints/PAn/Zoom/TWist/CLip/Hide/Off/Undo]: **Z** [Enter]
Specify lens length <50.000mm>: *Specify the new lens length.*

Just as with the **Distance** option, in the **Zoom** option a slider bar marked from 0x to 16x is displayed on the top side of the screen. The default position on the slider bar is 1x. Two ways to zoom can now be specified. The first is when perspective is enabled. In this case, zooming is defined in terms of lens length. The 1x mark (default position) corresponds to a 50.000 mm lens length. As you move the slider bar toward the right, the lens length increases, and as you move the slider bar toward the left, the lens length decreases. For example, when you move the slider bar to the 16x mark, the lens length increases 16 times, which is 16 x 50.000 mm = 800.000 mm. You can simulate the telephoto effect by increasing the lens length; by reducing the lens length, you can simulate the wide-angle effect. The lens length is dynamically displayed in the status bar. If perspective is not enabled, zooming is defined in terms of the zoom scale factor. In this case, the **Zoom** option resembles the **ZOOM Center** command, and the center point lies at the center of the current viewport. The 1x mark (default position) corresponds to a scale factor of 1. As you move the slider bar toward the right, the scale factor increases; as you move the slider bar toward the left, the scale factor decreases. For example, when you move the slider bar to the 16x mark, the scale factor increases 16 x 1 = 16 times. The scale factor is dynamically displayed on the status line.

Command: **DVIEW** Enter
Select objects or <use DVIEWBLOCK>: *Select the objects.*
Enter option
[CAmera/TArget/Distance/POints/PAn/Zoom/TWist/CLip/Hide/Off/Undo]: **Z** Enter
Specify zoom scale factor <1>: *Specify the scale factor.*

Example 5

Let's see the effect of the **Zoom** option in perspective projection. In Figure 21-32(a), the lens length is set to 25mm. This can be realized in the following manner:

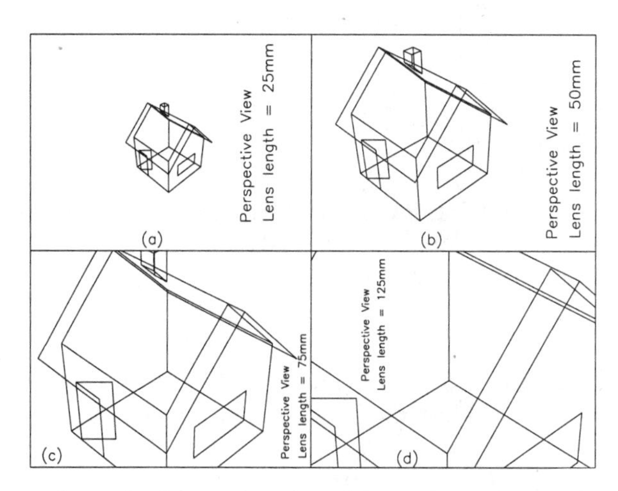

***Figure 21-32 Zoom** option of the **DVIEW** command*

Command: **DVIEW** Enter
Select objects or <use DVIEWBLOCK>: Enter
Enter option
[CAmera/TArget/Distance/POints/PAn/Zoom/TWist/CLip/Hide/Off/Undo]: **Z** Enter
Specify lens length <50.000mm>: **25** Enter

Similarly, for Figure 21-32(b), (c), and (d), you can set lens lengths to 50 mm, 75 mm, and 125 mm, respectively.

TWist Option

The **TWist** option allows you to rotate (twist) the view around the line of sight. You can also say that the object on the screen is rotated around the center point of the screen because the display is always adjusted so that the target point is at the center of the screen. If you use a pointing device to specify the angle, the angle value is dynamically displayed on the status line. A rubber-band line is drawn from the center (target point) to the drawing crosshairs, and as you move the crosshairs with a pointing device, the object on the screen is rotated around the line of sight. You can also enter the angle of twist from the keyboard. The angle of twist is measured in a counterclockwise direction starting from the right side.

Command: **DVIEW** [Enter]
Select objects or <use DVIEWBLOCK>: [Enter]
Enter option
[CAmera/TArget/Distance/POints/PAn/Zoom/TWist/CLip/Hide/Off/Undo]: **TW** [Enter]
Specify view twist angle <0.00>: *Specify the angle of rotation (twist).*

Example 6

In Figure 21-33, different twist angles have been specified. For (a) the twist angle is 0, for (b) it is 338 degrees, for (c) it is 37 degrees, and for (d) it is 360 degrees.

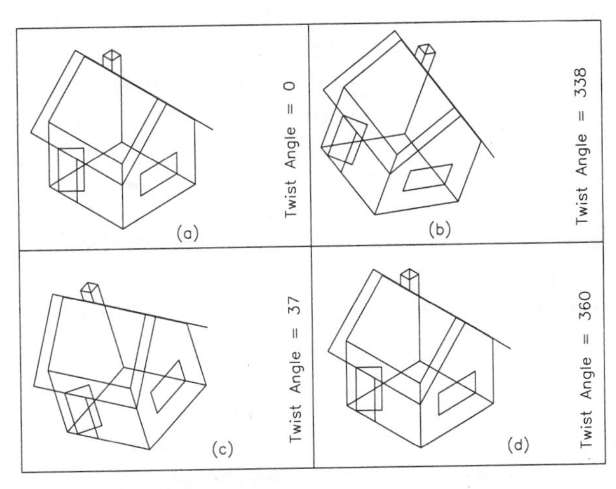

***Figure 21-33 TWist** option of the **DVIEW** command*

CLip Option

The **CLip** option can be used to clip sections of the drawing (Figure 21-34). AutoCAD LT uses two invisible clipping planes to realize clipping. These clipping walls can be positioned anywhere on the screen and are perpendicular to the line of sight (line between target and camera). Once you position the clipping planes, AutoCAD LT conceals all the lines that are in front of the front clipping plane or behind the back clipping plane. The **CLip** option can be used in both parallel and perspective projections. When perspective is enabled, the front clipping plane is automatically enabled.

Note

If you have specified a positive distance, the clipping plane is placed between the target and the camera. If the distance you have specified is negative, the clipping plane is placed beyond the target.

The prompt sequence for the **CLip** option is:

Enter option
[CAmera/TArget/Distance/POints/PAn/Zoom/TWist/CLip/Hide/Off/Undo]: **CL** [Enter]
Enter clipping option [Back/Front/Off]<Off>: **B** [Enter]
Specify distance from target or [ON/OFF] <current>:

Once you have specified which clipping plane you want to set (the back clipping plane in our case), a slider bar appears on the screen. As you move the pointer toward the right side of the slider bar, the negative distance between the target and the clipping plane increases. As you move toward the left side of the slider bar, the positive distance between the target and the clipping plane increases; hence, a greater portion of the drawing is clipped. The rightmost mark on the slider bar corresponds to a distance equal to the distance between the target and the farthermost point on the object you want to clip. After specifying the distance between one of the clipping planes (back clipping plane in this prompt sequence), you need to invoke the **CLip** option again and to specify the position of the front clipping plane in terms of the distance between the front clipping plane and the target. As you move the slider bar toward the right, the negative distance between the target and the front clipping plane increases. As the negative distance increases, a greater portion of the front side of the drawing is clipped. The rightmost mark on the slider bar corresponds to a distance equal to the distance between the target and the back clipping plane.

Enter option
[CAmera/TArget/Distance/POints/PAn/Zoom/TWist/CLip/Hide/Off/Undo]: **CL** [Enter]
Enter clipping option [Back/Front/Off] <Off>: **F** [Enter]
Specify distance from target or [set to Eye(camera)/ON/OFF] <1.0000>: *Specify the distance from target.*

To illustrate the concepts behind clipping, let us display two objects (a sphere and a cone). Let the distance between the two be 10 units. Draw a line between the center of the sphere and the center of the cone. Use the **POints** option to position the target at the center of the sphere and the camera at the midpoint of the line joining the two center points. In this way, you have defined the line between the two center points as the line of sight.

Now that you have defined the target and the camera, you will see the sphere and the cone overlapping because both of them are aligned along the line of sight. You may wonder how you are able to see the cone, since it lies behind the camera position that is analogous to the eye. This is because in parallel projection, the camera is not analogous to the eye, but the line of sight is; hence, you see everything that lies in the field of vision (determined by the **Zoom** option) about the line of sight. Now, when you use the **CLip** option and the **Front** suboption to positioned at the camera point. Now you cannot see the objects (cone in our case) that are behind the camera point because the default position of the front clipping plane is at the camera position; hence, any object in front of the front clipping plane is clipped. Also, if you have placed the front clipping plane behind the camera position and the perspective projection is on, the camera position is used as the front clipping plane position.

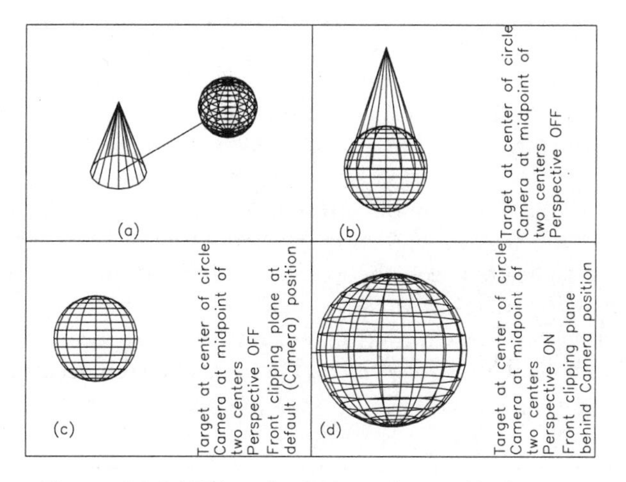

***Figure 21-34** Using the **CLip** option to clip the view*

If you change the front clipping plane position, and then at some stage you want to go back to the default front clipping plane position, you can use the **Eye** option. In perspective, without invoking the **CLip** option, the front clipping plane is placed at the camera position; hence, you are able to see only what lies in front of the camera.

Example 7

In Figure 21-35, the front clipping plane has been established at a distance of 0.50 units. The distance between the camera and the target is 3.50 units. After invoking the **CLip** option, the prompt sequence is:

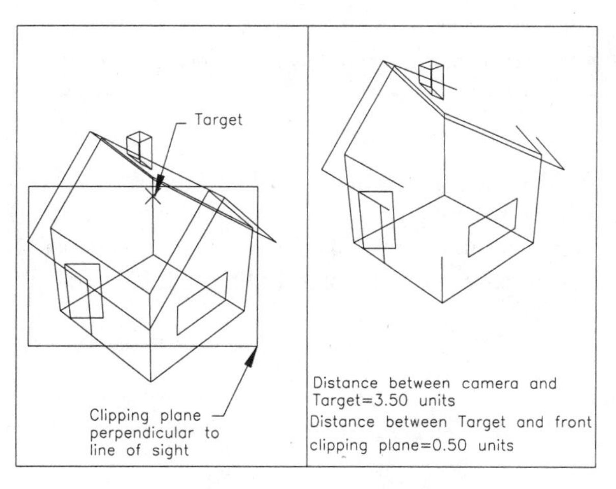

***Figure 21-35 Front** clipping with the **DVIEW** command*

Enter clipping option [Back/Front/Off]<Off>: **F** Enter
Specify distance from target or [set to Eye(Camera)/On/OFF] <3.5000>: **0.50** Enter

In Figure 21-36, the back clipping plane has been established at a distance of 0.50 units. The distance between the camera and the target is 3.50 units. After invoking the **CLip** option, the prompt sequence is:

Enter clipping option [Back/Front/Off]<Off>: **B** Enter
Specify distance from target or [ON/OFF] <1.0000>: **0.50** Enter

You might have noticed that if you combine the clipped shape in Figure 21-36 and the clipped shape in Figure 21-36, you get the original shape of the house. The reason is that in the first clipped figure, whatever lies between the camera and the front clipping plane is clipped, whereas in the second clipped figure, whatever lies behind the back clipping plane is clipped. Since the distances between the clipping planes and the target, and between camera and target are identical in both figures, then if you combine the clipped figures, you will get the original shape.

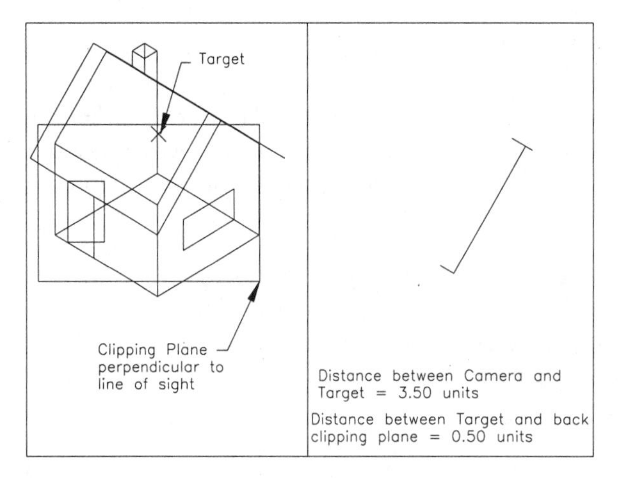

Figure 21-36 Back clipping with the DVIEW command

Hide Option

In most 3D drawings, some of the lines lie behind other lines or objects. Sometimes, you may not want these hidden lines to show up on the screen, so you can use the **Hide** option (Figure 21-37). This option is similar to the **HIDE** command. To invoke **Hide** option, enter H at the **Enter option [CAmera/TArget/Distance/POints/PAn/Zoom/TWist/CLip/Hide/Off/Undo]** prompt.

Off Option

The **Off** option turns the perspective projection off. The prompt sequence is:

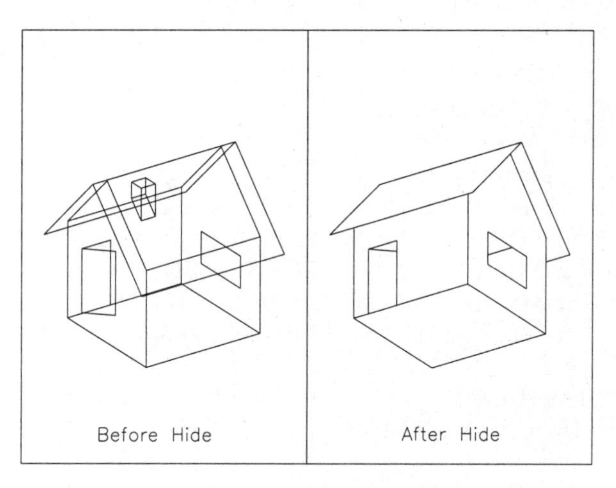

Figure 21-37 Hide *option of the* ***DVIEW*** *command*

Enter option
[CAmera/TArget/Distance/POints/PAn/Zoom/TWist/CLip/Hide/Off/Undo]: **O** Enter

When the perspective projection is turned off, you will notice that the perspective icon is replaced by the regular UCS icon.

Undo Option

The **Undo** option is similar to the **UNDO** command. The **Undo** option nullifies the result of the previous **DVIEW** operation. Just as in the case of the **UNDO** command, you can use this option a number of times to undo the results of multiple **DVIEW** operations. To invoke this option, enter U at the following prompt:

Enter option
[CAmera/TArget/Distance/POints/PAn/Zoom/TWist/CLip/Hide/Off/ Undo]: **U** Enter

CREATING SHADED IMAGES

In a 3D object the **HIDE** command can be used to hide the hidden line in the object. By hiding the lines, you can get a better idea about the shape of the object. To get a more realistic image of the object, you can use the **SHADE** and **SHADEMODE** commands.

SHADE Command

Command: SHADE

As you use the **SHADE** command, the hidden lines of the object are removed and the shaded image of the object is displayed. The color of the shaded image depends upon the color of the object. The single light source used here to shade the object is assumed to be placed just over the user's shoulder. Any subsequent changes made in the drawing are not displayed until the **SHADE** command is used again. You have to use the **REGEN** or **ZOOM** command to regenerate the drawing and display the 3D wireframe model.

SHADEMODE Command*

Command: SHADEMODE

The **SHADEMODE** command performs a hide and then creates a flat shaded picture in the current viewport. The shading uses a single light source that is assumed to be located just behind the user over the shoulder. The shaded image can be edited and the **UNDO** command cannot be used to undo shading. Use the **REGEN** or **ZOOM** command to regenerate the drawing and display the 3D wire frame of the object. This command can also be invoked by choosing **Shade** from the **View** menu. The command prompts are:

Command: **SHADEMODE** Enter
Enter option [2D wireframe/Hidden]<current>: *Enter an option.*

Setting the Shading Method

You can use the **SHADEDGE** system variable to set the shading method. If **SHADEDGE** is 0, AutoCAD LT creates a shaded image with no edges highlighted. If **SHADEDGE** is 1, AutoCAD LT creates a shaded image with edges highlighted in the background color. To see the effect of these two shading methods you need 256 color display.

If **SHADEDGE** is 2, AutoCAD LT paints the surfaces of the object in the background color and displays the visible edges in the object's color (Figure 21-38). If **SHADEDGE** is 3 (default setting), AutoCAD LT paints the faces in the object's color and displays the edges in the background color (Figure 21-39).

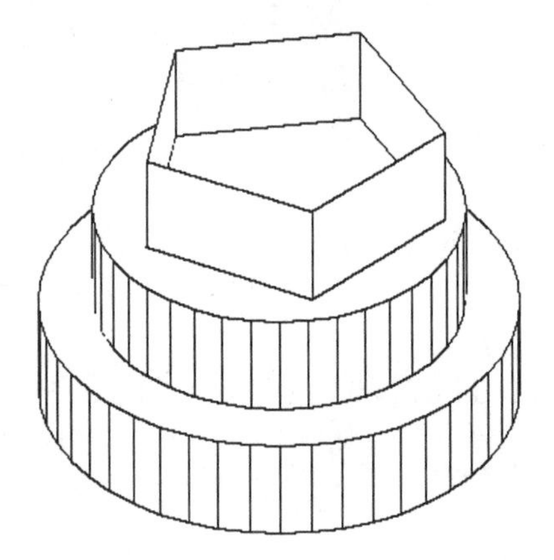

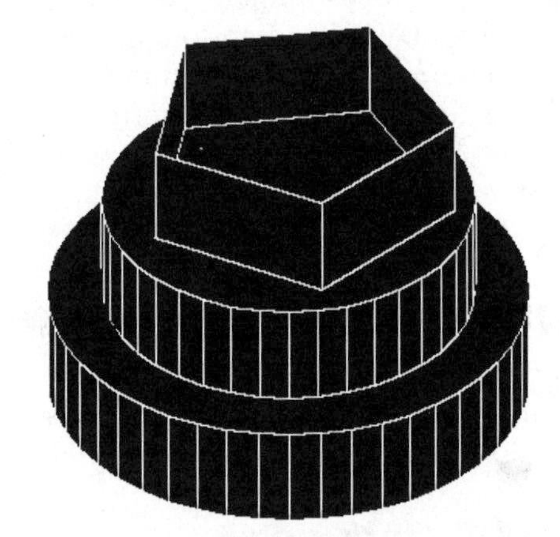

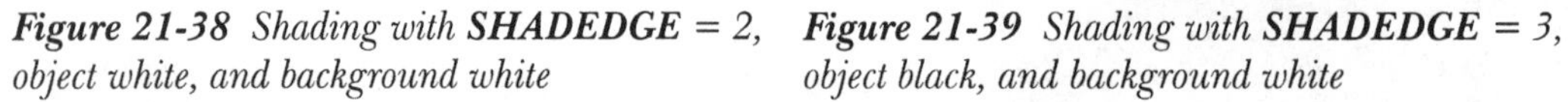

Figure 21-38 *Shading with* ***SHADEDGE*** *= 2, object white, and background white*

Figure 21-39 *Shading with* ***SHADEDGE*** *= 3, object black, and background white*

Setting Diffuse Reflection

The system variable **SHADEDIF** is used to control the amount of light diffused that AutoCAD LT uses to calculate the shade for each surface. The default value of **SHADEDIF** is 70. It can range from 0 to 100.

Analyzing Regions (MASSPROP)

The **MASSPROP** command can be used to analyze a region. This command will automatically calculate the mass properties of the region. When you enter this command, AutoCAD LT will list the properties of the region.

Command: **MASSPROP** Enter
Select objects: *Select the region.*

For Coplanar and Noncoplanar Regions. The information displayed for coplanar regions is similar to the following:

```
———————————— REGIONS ————————————

Area:          13.6480
Perimeter:     21.2475
Boundary box:          X: 3.3610 — 8.5063
               Y: 4.1330 — 6.4829
Centroid:      X: 5.9337
               Y: 5.3127
Moments of inertia: X: 390.8867
               Y: 512.2000
Product of inertia:    XY: 430.2535
Radii of gyration:     X: 5.3516
               Y: 6.1260
Principal moments and X-Y directions about centroid
               I: 5.6626 along [1.0000 0.0000]
               J: 31.6536 along [0.0000 1.0000]

Write analysis to a file ?<N>:
```

If you enter Y (Yes) at this last prompt, the **Create Mass and Area Properties File** dialog box (Figure 21-40) is displayed. All the file names of the .MPR type are listed. You can enter the name of the file in the **File Name:** edit box. The file is automatically given the .MPR extension.

The various terms displayed on the screen as a result of invoking the **MASSPROP** command are explained next.

Area. It is the enclosed area of the region.

Perimeter. Total length of inside and outside loops of region.

Bounding Box. For regions that are coplanar with the XY plane of the current UCS, the bounding box is defined by the diagonally opposite corners of a rectangle that encloses the region. For regions that are not coplanar with the XY plane of the current UCS, the bounding box is defined by the diagonally opposite corners of a 3D box that encloses the region.

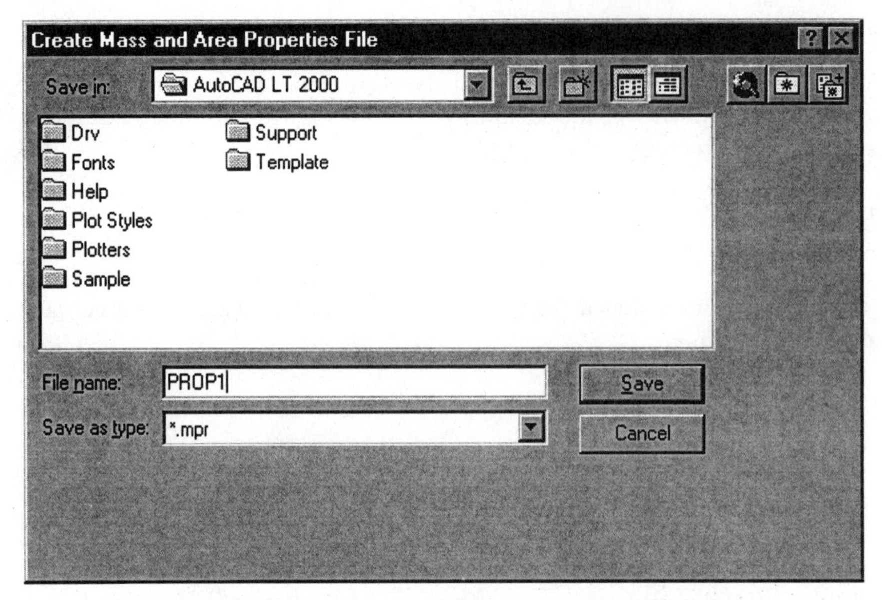

Figure 21-40 ***Create Mass and Area Properties File*** *dialog box.*

Centroid. This provides the coordinates of the center of area for the selected region.

Moments of Inertia. This property provides the mass moments of inertia of a region about the two axes. The equation used to calculate this value is:

area_moments_of_inertia = area_of_interest * (square of radius)

Products of Inertia. The value obtained with this property helps to determine the force resulting in the motion of the object. The equation used to calculate this value is:

product _of_inertia YX,XZ = mass * dist centroid_to_YZ * dist centroid_to_XZ

Radii of Gyration. The equation used to calculate this value is:

gyration_radii = (moments_of_inertia/body_mass)1/2

Principal Moments and X-Y-Z Directions about Centroid. This property provides you with the highest, lowest, and middle value for the moment of inertia about an axis passing through the centroid of the object.

Self-Evaluation Test

Answer the following questions, and then compare your answers to the correct answers given at the end of this chapter.

1. The two options given by the **ELEV** command are _________ and _________.
2. The view displayed by entering 0,0,-1 at **Specify a view point or [Rotate] <display compass and tripod>:** prompt is called _________.
3. The default viewpoint is _________ and the view obtained with this viewpoint is called the plan view.
4. You can define the viewpoint by defining the values of two angles using _________ option of **VPOINT** command.
5. To display **View** dialog box, the command is _________.
6. If you have multiple viewports, the **PLAN** command generates the plan view in only the _________ viewport.
7. _________ command imparts the properties of 3D objects on 2D objects.
8. Command used to join two different regions is _________.
9. To draw polyline in space, command is _________.
10. The line formed between the target and the camera is known as _________, or _________.
11. The _________ option allows you to rotate the camera about the target point.
12. If perspective is disabled, the zooming is defined in terms of the _________.
13. The _________ option allows you to shift from perspective viewing to parallel viewing.
14. The 1x mark (default position) corresponds to the scale factor of _________.
15. When you move the slider bar to 16x mark, the scale factor is _________.

Review Questions

1. **ELEV** command shifts the working plane about _________.
2. Negative value for the thickness extrudes the object in _________ direction.
3. Negative value for the elevation shifts the workplane in _________ direction.

4. The ________ command lets you define a viewpoint.

5. Once you set a viewpoint, ________ in the current viewport are regenerated and the objects are displayed as if you were viewing them from the direction of the newly defined viewpoint.

6. The default viewpoint is ________ and the view obtained with this viewpoint is called the plan view.

7. If you give a null response at the **Specify a view point or [Rotate] <display compass and tripod>:** prompt, AutoCAD LT displays a(n) ________ and ________ on the screen.

8. You can specify the viewpoint in the axis and tripod method by moving the ________ in the ________ with a pointing device.

9. As you move the small crosshairs in the compass, the ________ also moves to reflect the changes in the viewpoint.

10. You can set the viewpoint by specifying its coordinate location or by specifying the rotation in the XY plane from the X axis and the ________ .

11. The ________ command can be used to generate the plan view of an object relative to the ________ , ________ , or ________ .

12. If you have multiple viewports, the **PLAN** command generates the plan view of only the ________ viewport.

13. Perspective projection and clipping are turned ________ when you use the **PLAN** command.

14. You can use the ________ command to suppress the display of lines that lie behind other lines or objects.

15. 3D polylines can be drawn using the ________ command.

16. The difference between the **3DPOLY** and **PLINE** commands is that a ________ is added to the polyline, and with the **3DPOLY** you can only draw ________ segments without variable ________ .

17. To find the location of a point using the cylindrical coordinate system, you need to know:
    1. ____
    2. ____
    3. ____

18. To find the location of a point using the spherical coordinate system, you need to know:
    1. ____________________
    2. ____________________
    3. ____________________

19. Use the spherical coordinate system to find the location of a point 5 units from the UCS origin, making an angle of 45 degrees in the XY plane, and at an angle of 70 degrees from the XY plane.

20. In the XZ option, the mirroring plane is aligned with the __________ plane and passes through the specified point.

Exercises

Exercise 3 *General*

Draw the base circle (Figure 21-41) with elevation 0 and thickness 2, and then draw the polygon and smaller circle with elevation 2 and thickness 4. The diameters of the circles are 6 and 3, and the edge of polygon is 4.

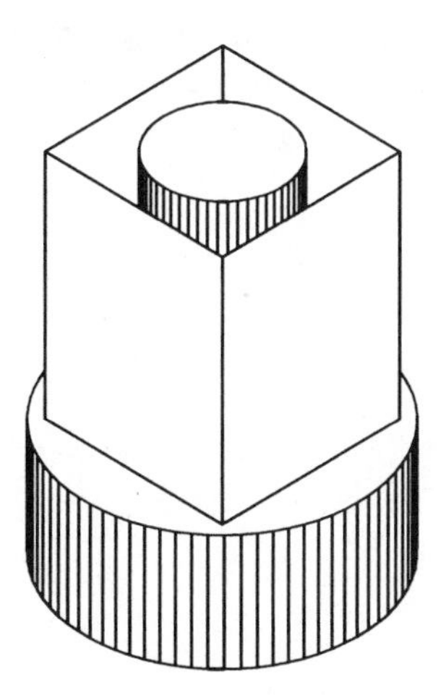

Figure 21-41 *Drawing for Exercise 3*

Exercise 4 *General*

Draw the objects in Figure 21-42. Assume the dimensions.

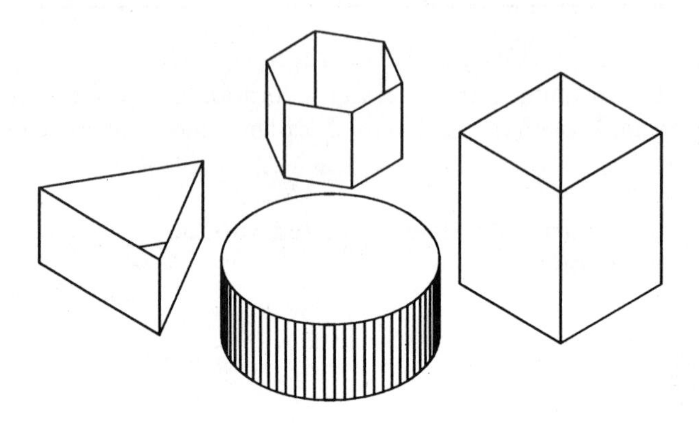

Figure 21-42 *Drawing for Exercise 4*

Answers to Self-Evaluation Test

1 - Elevation, Thickness, 2 - Bottom **3** - 0,0,1 **4** - Rotate, **5 - VIEW, 6** - Active, **7 - REGION, 8 - UNION, 9 - 3DPOLY, 10** - line of sight or view direction, **11 - CAmera, 12** zoom scale factor, **13 - Off, 14** - 1, **15** - 16 times.

Chapter 22

Model Space Viewports, Paper Space Viewports, and Layouts

Learning Objectives

After completing this chapter, you will be able to:

- *Create viewports in model space and paper space using the* ***VPORTS*** *command.*
- *Create viewports in paper space using the* ***-VPORTS*** *and* ***MVIEW*** *commands.*
- *Use paper space and model space in paper space.*
- *Shift from paper space to model space using the* ***MSPACE*** *command.*
- *Shift from model space to paper space using the* ***PSPACE*** *command.*
- *Control the visibility of viewport layers with the* ***VPLAYER*** *command.*
- *Set linetype scaling in paper space using the* ***PSLTSCALE*** *system variable.*

MODEL SPACE VIEWPORTS

Normally, in AutoCAD LT, a single viewport is displayed on the screen. AutoCAD LT allows you to create multiple viewports that can be used to display different views of the same object (Figure 22-1). A viewport is a rectangular part of the graphics area of the screen. Viewports can be used in different ways and for different purposes. The first reason to create viewports and divide the display screen into a number of parts is usually to create a model or layout. This is done using the **VPORTS** command. You are in the **tiled viewport** mode when you use this command. This is the default mode in AutoCAD LT. When you want to use the **VPORTS** command to create multiple tiled viewports, the **TILEMODE** system variable is set to 1 (default value). An arrangement of tiled viewports is a display function and cannot be plotted.

The second purpose is to have different parts or different views of your drawing in different viewports. Each viewport can contain a different 2D or 3D view of your drawing, and you can use the **PAN** or **ZOOM** command to display different portions or different levels of detail of the drawing in each viewport. The Snap, Grid, and UCS icon modes can be set separately for each viewport.

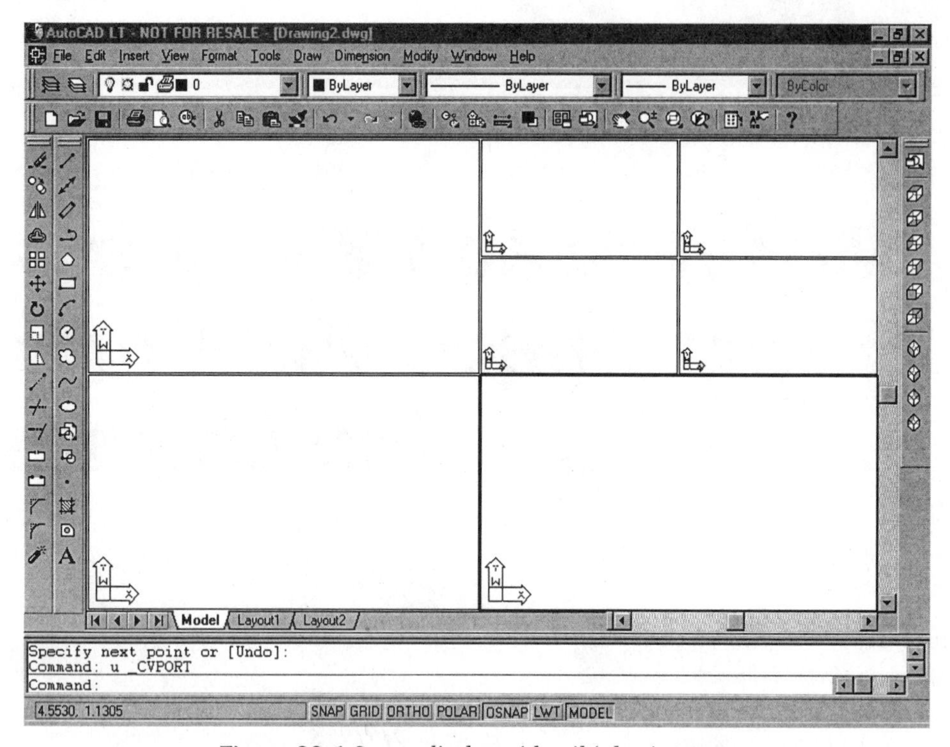

***Figure 22-1** Screen display with miltiple viewports*

DISPLAYING VIEWPORTS AS TILED AREAS (VPORTS COMMAND)

Toolbar:	Layout > Display Viewports Dialog
Menu:	View > Viewports > New Viewports
Command:	VPORTS

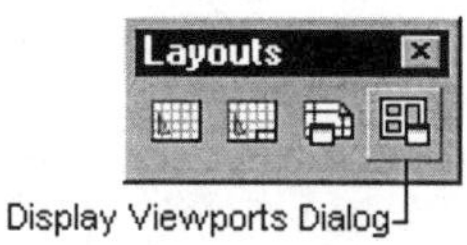

***Figure 22-2** Invoking the **VPORTS** command from the **Layouts** toolbar*

As mentioned earlier, the display screen can be divided into multiple nonoverlapping tiled viewports whose number depends on the equipment and the operating system on which AutoCAD LT is running. Each tiled viewport contains a view of the drawing. The tiled viewports must touch at the edges without overlapping one another. While using tiled viewports, you are not allowed to edit, rearrange, or turn individual viewports on or off. These viewports are created using the **VPORTS** command when the system variable **TILEMODE** is set to 1; the **Model** tab is active. The prompt sequence is:

Command: **TILEMODE** Enter
Enter new value for TILEMODE <0>: **1** Enter

As required, we have set the system variable **TILEMODE** to 1. You can also set **TILEMODE** to 1 by choosing the PAPER button located in the status bar or choosing the **Model** tab. Now we will use the **VPORTS** command.

Command: **VPORTS** Enter

The **Viewports** dialog box is displayed. You can use this dialog box to create new viewport configurations and save them. The options in the dialog box vary depending upon whether you choose the **Model** tab or the **Layout** tab. The **Viewports** dialog box has the following two tabs: **New Viewports** tab (Figure 22-3) and **Named Viewports** tab.

New Viewports Tab

You can enter a name for the viewport configuration you wish to create in the **New name** edit box. If you do not enter a name in this edit box, the viewport configuration created by you is not saved and hence, cannot be used in layout. A list of standard viewport configurations is listed in the **Standard viewports** list box. This list also contains the ***Active Model Configuration*** which is the current viewport configuration. From the **Standard viewports** list, you can select any one of the listed standard viewport configurations to apply and a

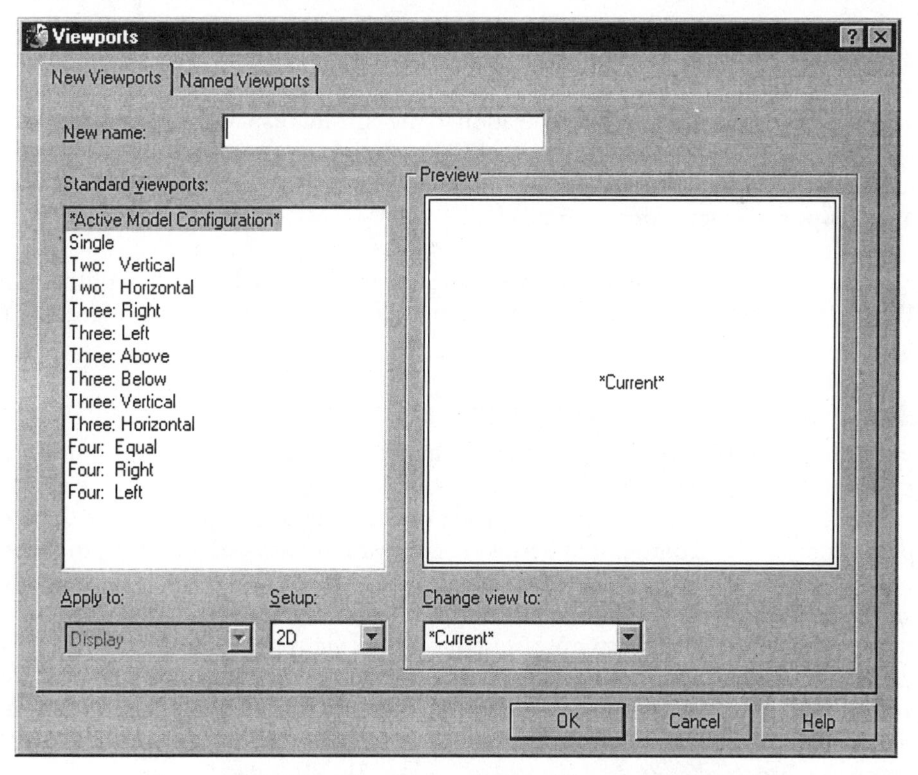

***Figure 22-3** New **Viewports** tab of the **Viewports** dialog box*

preview image of the selected configuration is displayed in the **Preview** window. The **Apply to** drop-down list has two options: **Display** and **Current viewport**. Selecting the **Display** option applies the selected viewport configuration to the entire display and selecting the **Current viewport** option applies the selected viewport configuration to only the current viewport. From the **Setup** drop-down list, you can select **2D** or **3D**. Selecting the **2D** option, creates the new viewport configuration with the current view of the drawing in all the viewports initially. Using the **3D** option, applies the standard orthogonal and isometric views to the viewports. For example if your configuration has 4 viewports, they are assigned the **Top**, **Front**, **Left** and **South East Isometric** views respectively. You can also modify these standard orthogonal and isometric views by selecting from the **Change view to** drop-down list and replacing the existing view in the selected viewport. For example you can select the viewport which is assigned the **Top view** and then choose **Bottom view** from the **Change view to** drop-down list to replace it. The Preview image in the **Preview** window reflects the changes you make. If you are using the **2D** option, you can select a named viewport configuration to replace the selected one. Choose **OK** to exit the dialog box and apply the created or selected configuration to the current display in the drawing. When you name and save a viewport configuration, it saves information about the number and position of viewports, the viewing direction and zoom factor and the grid, snap, coordinate system and UCS icon settings of the viewports.

Named Viewports Tab

The **Named Viewports** tab of the **Viewports** dialog box (Figure 22-4), displays the name of the current viewport next to **Current name**. The names of all the saved viewport configurations in a drawing are displayed in the **Named viewports** list box. You can select any one of the named viewport configurations and apply it to the current display. A preview image of the selected configuration is displayed in the **Preview** window. Choose **OK** to exit the dialog box and apply the selected viewport configuration to the current display. In the **Named viewports** list box, you can select a name and right-click to display a shortcut menu. Choosing **Delete** deletes the selected viewport configuration and choosing **Rename** allows you to rename the selected viewport configuration.

After we have discussed how to make a viewport current, we will describe all of the options of the **-VPORTS** command. You should know how to make a viewport current before using any of the options.

Making a Viewport Current

The viewport you are working in is called the current viewport. You can display several model space viewports on the screen, but you can work in only one of them at a time. You can switch from one viewport to another even when you are in the middle of a command. The current viewport is indicated by a border that is heavy compared with the borders of the other viewports. Also, the graphics cursor appears as a drawing cursor (screen crosshairs) only when it is within the current viewport. Outside the current viewport this cursor appears as an arrow cursor. You can enter points and select objects only from the current viewport. To make a viewport current, you can select it with the pointing device. Another way of making a viewport current is by assigning its identification number to the **CVPORT** system variable. The identification numbers of the named viewport configurations are not listed in the display. The options of the **-VPORTS** command are discussed next.

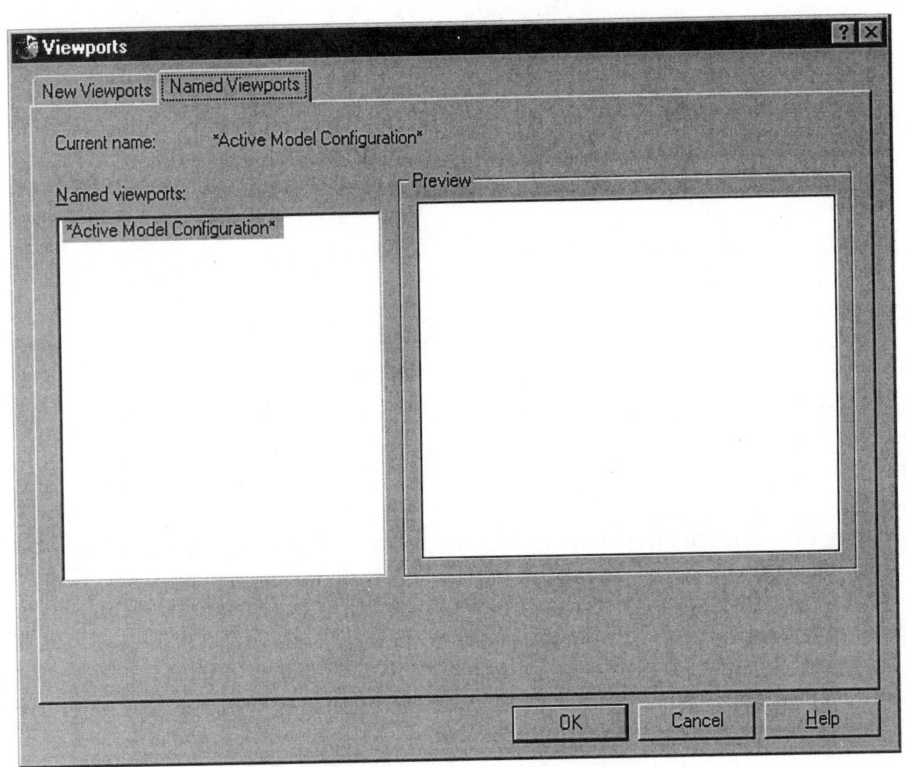

*Figure 22-4 **Named Viewports** tab of the **Viewports** dialog box*

-VPORTS Command Options

-VPORTS command gives you number of options depending upon wether you are working with tiled viewports or floating viewports. The options you get while working with floating viewports are discussed later in this chapter.

Save option. This option is used to name and save the current viewport configuration. The prompt sequence is:

> Command: **-VPORTS** [Enter]
> Enter an option [Save/Restore/Delete/Join/SIngle/?/2/3/4]<3>: **S** [Enter]
> Enter name for new viewport configuration or [?]: *Enter the name.*

The naming conventions are the same as those used earlier. You can use up to 31 characters and wild cards. Entering the **?** symbol causes the following prompt to be displayed:

> Enter name(s) of viewport configuration(s) to list <*>:

As a response to this prompt, you can press ENTER to get a listing of all the saved viewport configurations.

Restore option. This option restores any saved viewport configuration. You can enter the name of the viewport configuration to restore, or you can enter **?** to display a list of all saved viewport configurations. The prompt sequence is:

Command: **-VPORTS** Enter
Enter an option [Save/Restore/Delete/Join/SIngle/?/2/3/4]<3>: **R** Enter
Enter name of viewport configuration to restore or [?]: *Enter the name.*

You can use the **?** symbol in the same way as for the **Save** option.

Delete option. You can delete a previously saved viewport configuration using this option. The prompt sequence is:

Command: **-VPORTS** Enter
Enter an option [Save/Restore/Delete/Join/SIngle/?/2/3/4]<3>: **D** Enter
Enter name(s) of viewport configurations to delete <none>: *Enter the name.*

If you enter a wrong name, the system gives the following message: **No matching viewport configurations found.**

Join option. With this option, you can join two adjacent viewports into a single viewport. The view in the resulting viewport depends on which one of the two viewports you specified as the dominant viewport when they were joined. You can also select this option from the **View** menu, choose **View > Viewports > Join**.

Command: **-VPORTS** Enter
Enter an option [Save/Restore/Delete/Join/SIngle/?/2/3/4]<3>: **J** Enter
Select dominant viewport <current viewport>: *Select the dominant viewport.*
Select viewport to join: *Select other viewport.*

SIngle option. With this option, you can have only one viewport on the screen, and the view in this viewport depends on which one of the viewports was current when you selected the **SIngle** option. This is similar to choosing **Viewports > 1 Viewport** from the **View** menu.

? option. Invoking this option displays the identification number and screen positions of all the active viewports and the names and screen positions of the saved viewport configurations. The position of the viewports is defined by the lower left and upper right corners. The value for these corners is between (0.0,0.0) for the lower left corner of the graphics area and (1.0,1.0) for the upper right corner of the graphics area. The current viewport is the first one listed. You can also use wild-card characters to filter names. For example, entering sp* will list all configurations whose names start with sp.

2 option. This option is used to divide the current viewport into two equal parts. You can also choose **Viewports > 2 Viewports** from the **View** menu.

Enter an option [Save/Restore/Delete/Join/SIngle/?/2/3/4]<3>: **2** Enter
Enter a configuration option [Horizontal/Vertical] <Vertical>:

If you press ENTER at this prompt, then the current viewport is divided vertically. If you enter **Horizontal**, the current viewport is divided horizontally.

3 option. You invoke the **3 option** to divide the current viewport into three viewports in the following manner:

Enter an option [Save/Restore/Delete/Join/SIngle/?/2/3/4]<3>: **3** [Enter]

The next prompt is:

Enter a configuration option [Horizontal/Vertical/Above/Below/Left/Right] <Right>:

When you enter **Horizontal** at this prompt, the current viewport is divided horizontally into three equal parts. The **Vertical** option divides the current viewport vertically into three equal parts. The **Above**, **Below**, **Left**, and **Right** options divide the current viewport into one large and two smaller viewports. The placement of the larger viewport depends on which one of these options you specify. For example, if you select the **Above** option, the current viewport is divided horizontally into two viewports and the lower viewport is further divided vertically into two viewports. You can also invoke this option by choosing **Viewports > 3 Viewports** from the **View** menu.

4 option. Using this option, you can divide the current viewport into four equal viewports. You can also choose **Viewports > 4 Viewports** from the **View** menu.

Note

By dividing, subdividing, and joining viewports, you can create nearly any viewport configuration you want, so long as all viewports are rectangular.

OBTAINING PLAN VIEWS AUTOMATICALLY (UCSFOLLOW SYSTEM VARIABLE)

As mentioned before, you can obtain the plan view of an object by using the **PLAN** command. If you have different viewports, you can obtain the plan view relative to the current UCS by activating a viewport, setting the UCS to your requirement, and using the default option of the **PLAN** command. This operation can be automated by using the **UCSFOLLOW** system variable. When **UCSFOLLOW** is assigned a value of 1, a plan view is automatically generated if you change the UCS. By default, this variable is assigned the value 0. All of the viewports have the **UCSFOLLOW** facility; hence, you need to specify the **UCSFOLLOW** setting separately for each viewport. The prompt sequence is:

Command: **UCSFOLLOW** [Enter]
Enter new value for UCSFOLLOW <0>: **1** [Enter]

WORKING WITH FLOATING VIEWPORTS (VPORTS COMMAND)

Invoking the **VPORTS** command in the **Layout** tab also displays the **Viewports** dialog box. This dialog box is used to create a floating viewports configuration. Floating viewports can be overlapping and can be plotted at the same time. They are used for arranging the final layout of a drawing. The **Viewports** dialog box has the same two tabs as the **Viewports** dialog box displayed in the **Model** tab (Tilemode =1), but the options vary.

New Viewports Tab

In this tab of the **Viewports** dialog box (Figure 22-5), the current viewport configuration name is displayed next to **Current name**. The **Standard viewports** list box displays all the standard viewport configurations. Selecting a viewport configuration from this list, displays a preview image in the **Preview** window with all the default views assigned to individual viewports. In the **Viewport Spacing** edit box, you can enter a value for the distance you want to keep between individual viewports. Selecting **2D** from the **Setup** drop-down list, displays the current view in all the viewports of the current viewport configuration initially. Selecting the **3D** option, a set of standard orthogonal views are applied to the viewports in the configuration. You can replace a selected viewport configuration with one selected from the **Change view to** drop-down list. In case of a **3D** setup, you can replace the default orthogonal view applied to individual viewports by selecting one from the **Change view to** drop-down list.

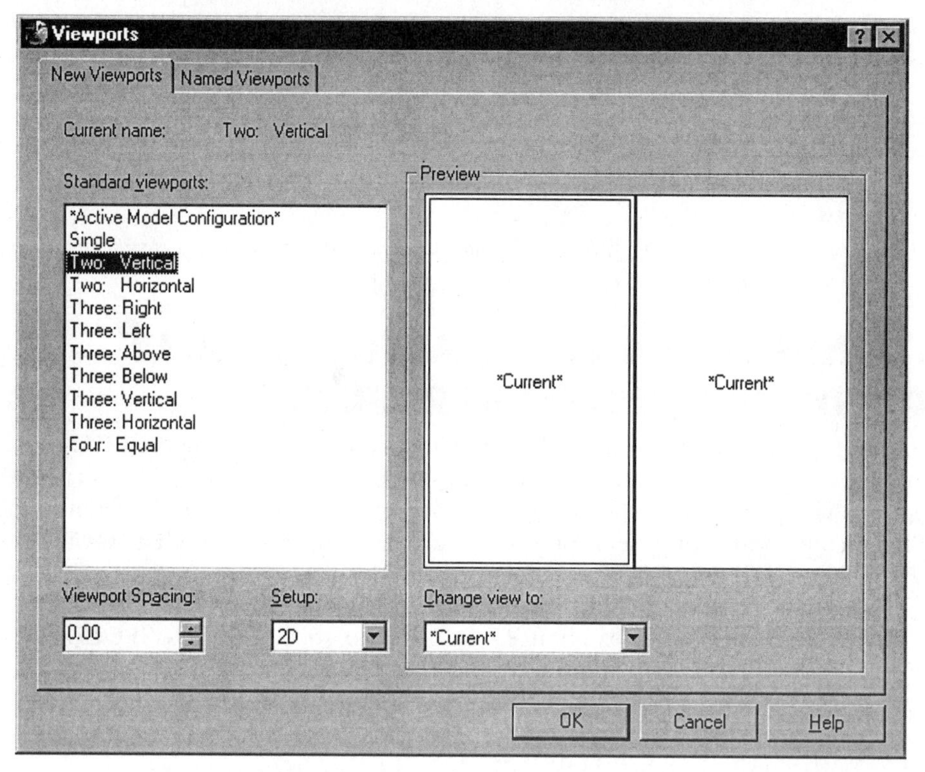

*Figure 22-5 **New Viewports** tab of the **Viewports** dialog box*

Named Viewports Tab

The **Named Viewports** tab displays any saved and named tiled configurations for you to use in the current layout. You cannot save and name a floating viewport configuration.

-VPORTS COMMAND

In AutoCAD LT, you can create tiled viewports or floating viewports (untiled viewports). Tiled viewports are created by using the **VPORTS** command when the **TILEMODE** system variable is set to 1 (model space). The same command is used to create floating viewports in paper space (Figure 22-6) when you are in the layout tab. The **-VPORTS** command can be used when in the **Layout** tab to create and use floating viewports. If you invoke this command in the floating model space (TILEMODE set to 0), AutoCAD LT shifts you from model space to paper space, until the **-VPORTS** command is active. Once the **-VPORTS** command is over, AutoCAD LT returns you to model space. The command prompts are as follows:

Command: **-VPORTS** Enter
Specify corner of viewport or
[ON/OFF/Fit/Hideplot/Lock/Restore/2/3/4]<Fit>:

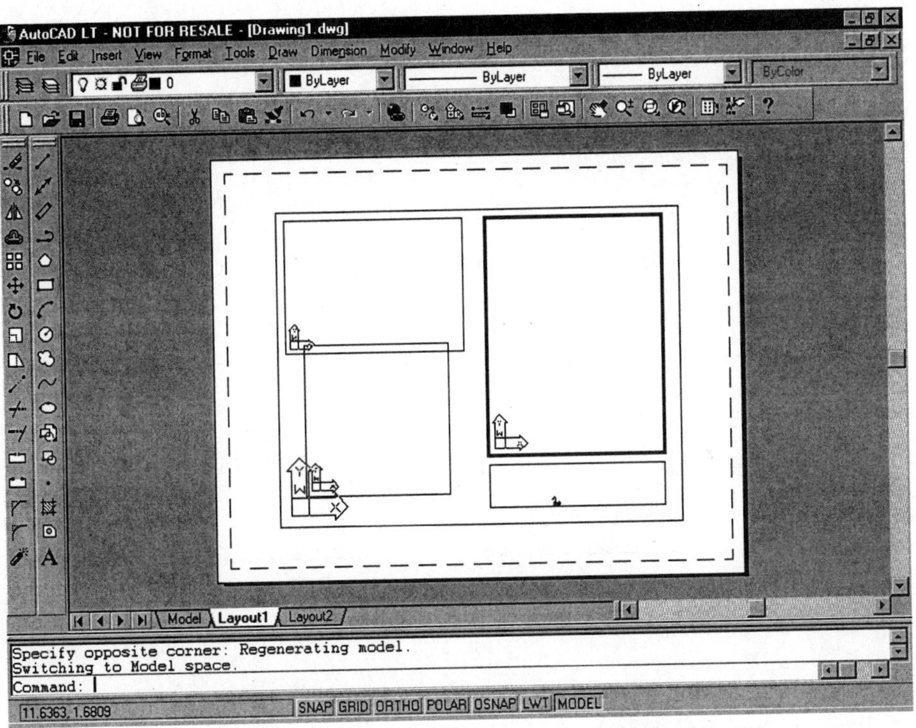

***Figure 22-6** Floating viewports using the **-VPORTS** command*

The purpose of creating untiled viewports in paper space by using the **-VPORTS** command is that viewports are created in paper space, and these viewports can be edited as objects like circles, polygons, or text. Various settings of the first viewport created with the **-VPORTS** command, such as grid, current view, and snap, are taken from the previous viewport that was current in paper space. You can also alter the size of the floating viewports and adjust them on the screen. Since viewports in paper space are just like objects, it is not possible to edit the model in paper space. For this, you need to shift from paper space to floating model space with the help of the **MSPACE** command or you can choose PAPER on the status bar.

When you enter the **-VPORTS** command, AutoCAD LT prompts you either to select one of the options listed in the prompt line or to select the first corner point of the paper space viewport window.

-VPORTS Command Options

Corner of viewport option. The **Corner of viewport** option (Figure 22-7) is the default option. With this option, you can create a single viewport. The size of the viewport is determined by the two diagonal points you specify. Once you specify the two points, a viewport is created.

Hideplot option. With the **Hideplot** option (Figure 22-8), you can choose the viewports from which you want to remove the hidden lines when you plot in paper space. If you set **Hideplot** to ON, hidden lines are not plotted. If you set **Hideplot** to OFF, hidden lines are plotted.

Command: **-VPORTS** [Enter]
Specify corner of viewport or
[ON/OFF/Fit/Hideplot/Lock/Object/Polygonal/Restore/2/3/4]<Fit>: **H** [Enter]
Hidden line removal for plotting [ON/OFF]: **ON** [Enter]
Select objects: *Select the viewports.*

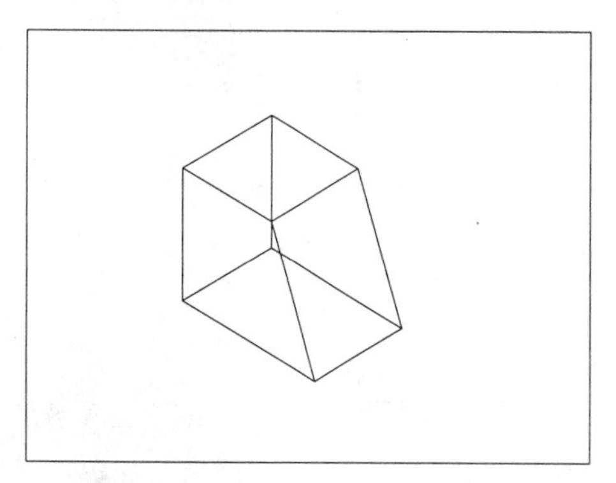

Figure 22-7 -VPORTS *command,* ***Corner of viewport*** *option*

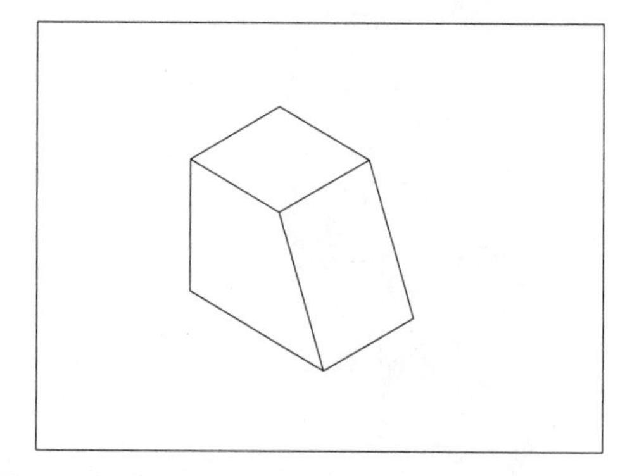

Figure 22-8 -VPORTS *command,* ***Hideplot*** *option*

Note

When you turn ***Hideplot*** *On, the hidden line removal takes place only at the time of plotting. It does not affect the display of the lines on screen (hidden line will be displayed). You can use the* ***HIDE*** *command to see the effect of hidden line removal.*

ON/OFF option. These options are used to turn the display in the selected viewports on or off. By default, the display inside the current viewport is on. You should turn off the viewports that are not required so that regeneration is not carried out on viewports that are turned off. Objects are not displayed in the viewports that are turned off.

Note

The system variable ***MAXACTVP*** *controls the number of viewports that display the objects. For example, if* ***MAXACTVP*** *is set to 16, only 16 viewports can simultaneously display the objects. The remaining viewports will not display the objects unless one or more viewports that display the objects are turned off or the* ***MAXACTVP*** *system variable is set to a higher value.*

If you zoom or use ***REGEN*** *in the viewport that does not display the object, AutoCAD LT will automatically turn off one of the other viewports to force display of the objects in the viewport.*

You will not be able to work in the floating model space if all viewports are turned off; you must create a new viewport or turn on one of the viewports.

Fit option. This option generates the viewport that fits the current screen display. For example, if you want to create a single viewport, use the **Fit** option. The size of the viewport can be controlled by setting the limits or by using the **ZOOM** command to display the desired area.

Lock option*. You can turn the viewport locking on or off for a specific viewport. This option restricts any accidental modifications to scaling in a viewport. The command prompts are:

Viewport View Locking [ON/OFF]: **ON** [Enter]
Select Objects: *Select the viewports you want to lock.*

Restore option. If you have used the **VPORTS** command to create and save a viewport configuration, the **Restore** option of the **-VPORTS** option can be used to obtain that configuration by specifying the name of the configuration (Figure 22-9). Once you specify the configuration name, you can specify the location and size of the viewports by using the **Fit** or **First Corner** option.

Enter window configuration name or [?] <default>: *Specify the name of the configuration you want to restore.*
Specify first corner or [Fit]<Fit>:
Select first corner.
Specify opposite corner: *Select second corner.*

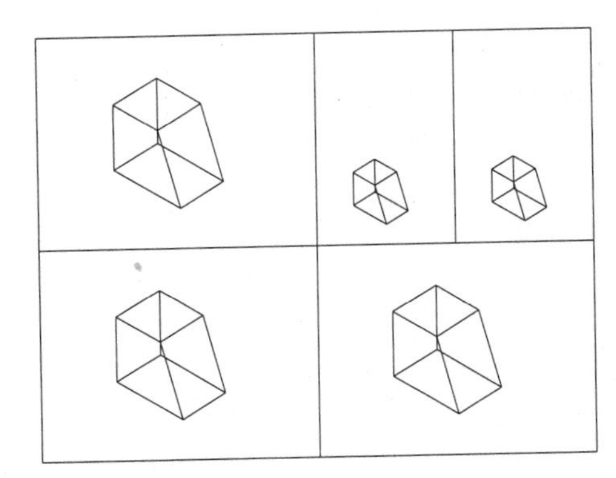

Figure 22-9 -VPORTS *command,* ***Restore*** *option*

2 option. You can invoke the **2** option to create two viewports in the area you specify. This option is similar to the **2** option of the **-VPORTS** command. Once you enter 2 at the **Specify corner of viewport or [ON/OFF/Fit/Hideplot/Lock/Restore/2/3/4]<Fit>:** prompt, AutoCAD LT prompts you for the location and arrangement of the viewports. The prompt is:

Enter viewport arrangement [Horizontal/Vertical] <Vertical>:

At the prompt, you can choose whether the rectangular area you have specified is to be horizontally or vertically divided. The default arrangement is vertical (Figure 22-10), but you can change it to horizontal (Figure 22-11) by entering HORIZONTAL at the prompt. Once you have specified the arrangement of the viewport, the next two prompts are:

Specify first corner or [Fit]<Fit>:
Specify opposite corner:

These prompts are identical to the **Fit** and **First corner** options of the **-VPORTS** command. For example, if you enter the first corner, the next prompt asks you to specify the opposite corner of the rectangular area in which the two viewports will be created. Once these two points are defined, AutoCAD LT will fit the two viewports within the specified area. The **Fit** option will fit the viewports in the current screen display.

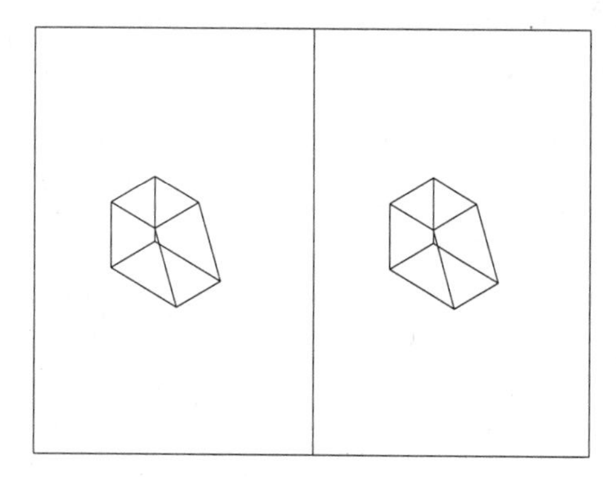

Figure 22-10 -VPORTS command, 2 option with vertical viewports

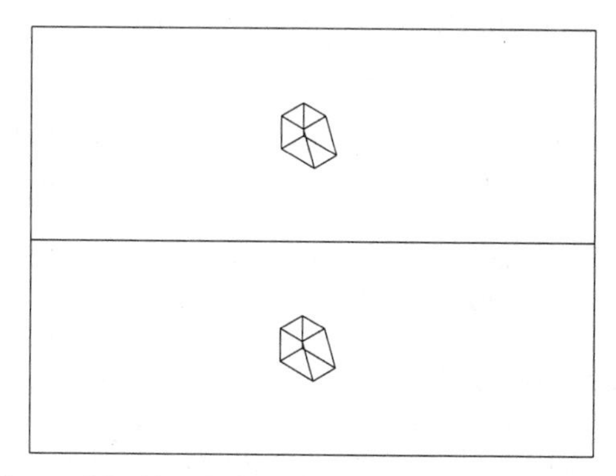

Figure 22-11 -VPORTS command, 2 option with horizontal viewports

3 option. You can invoke the **3** option to create three viewports in the area you specify. This option is similar to the **3** option of the **-VPORTS** command in the **Model** tab. Once you enter **3** at the **Specify corner of viewport or [ON/ OFF/ Fit/Hideplot/Fit/Lock/Restore/2/3/4] <Fit>** prompt, AutoCAD LT prompts for the location and arrangement of the viewports. The prompt is:

Enter viewport arrangement [Horizontal/Vertical/Above/Below/Left/Right] <Right>:

At the prompt, you can choose whether the rectangular area you have specified is to be divided into three viewports horizontally or vertically, and whether the third viewport is to be placed above, below, to the left of, or to the right of the other two viewports (Figures 22-12 through 22-17). The default option is **Right**, with the third viewport placed on the right side of the other two viewports. If you invoke the **Horizontal** option, the space you have specified for the viewports is divided into three equal horizontal viewports.

If you invoke the **Vertical** option, the space you have specified for the viewports is divided into three equal vertical viewports. The **Above** option places the third viewport on top of the other two viewports. The **Below** option places the third viewport under the other two viewports. If you use the **Left** option, the third viewport is placed on the left side of the other two viewports.

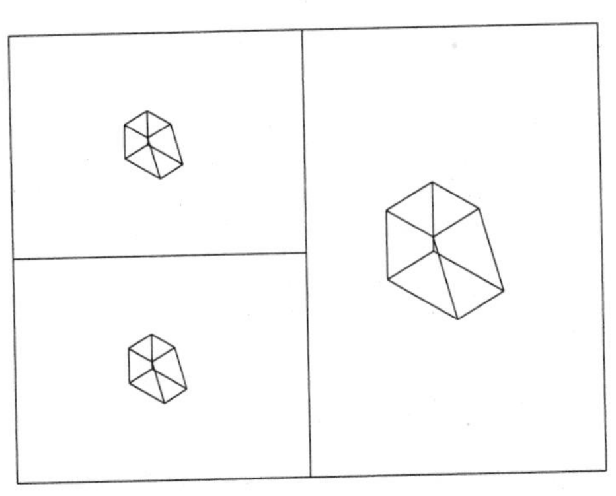

***Figure 22-12 -VPORTS** command, **3** option with the third viewport on right*

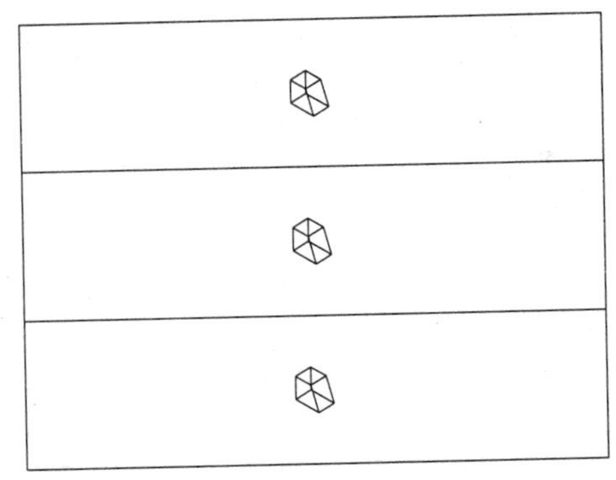

***Figure 22-13 -VPORTS** command, **3** option with horizontal viewports*

Once you have specified the arrangement of the viewport, the next two prompts are:

Specify first corner or [Fit]<Fit>:
Specify opposite corner:

These prompts are identical to the **Fit** and **First Corner** options of the **-VPORTS** command. For example, if you specify the first corner, the next prompt asks you to specify the opposite corner of the rectangular area in which the three viewports will be created. Once these two points are defined, AutoCAD LT will fit the three viewports within the specified area. The **Fit** option will fit the viewports in the current screen display.

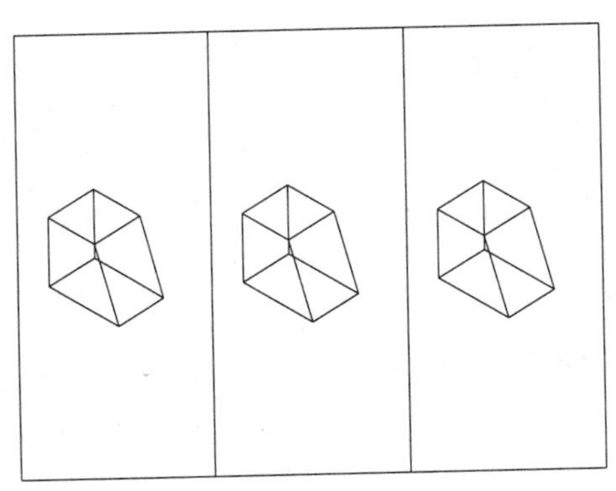

***Figure 22-14 -VPORTS** command, **3** option with the vertical viewports*

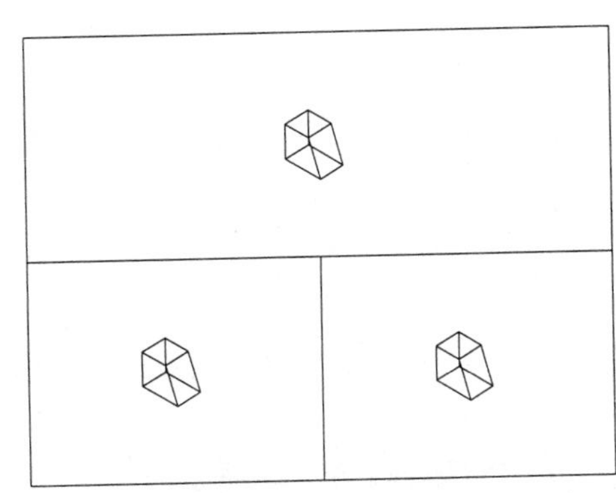

***Figure 22-15 -VPORTS** command, **3** option with one viewport above*

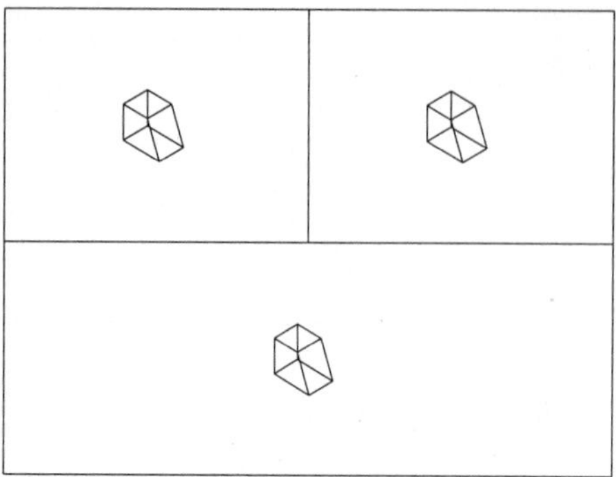

Figure 22-16 -VPORTS command, 3 option with one viewport below

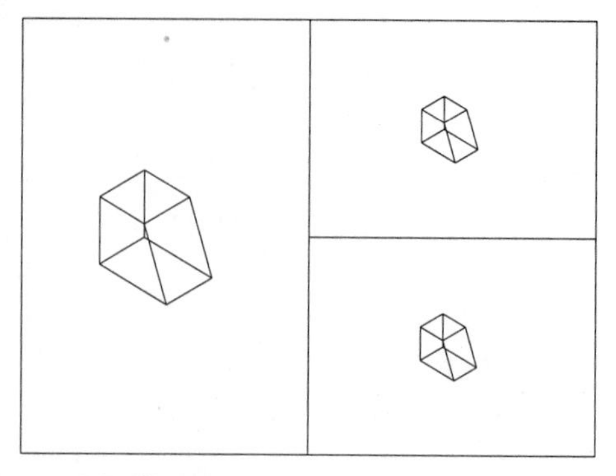

Figure 21-17 -VPORTS command, 3 option with one viewport on left

4 option. Using the **4** option, you can divide the area you have specified for the viewports into four equal viewports. If you use the **Fit** option, the graphics area on the screen is divided into four equal viewports (Figure 22-18).

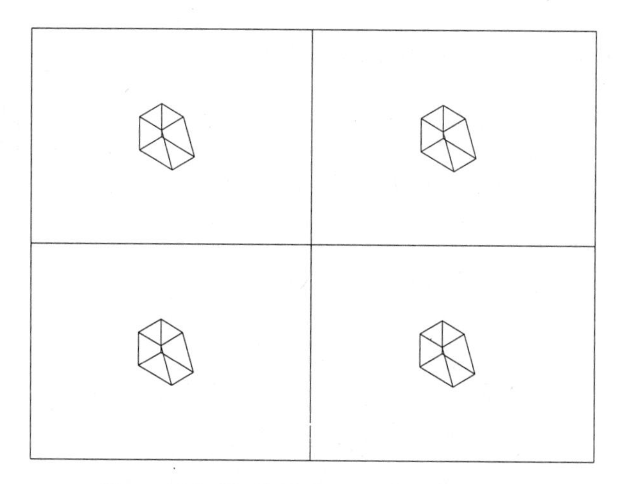

Figure 22-18 -VPORT command, 4 option

The Viewports Toolbar*

The **Display Viewports Dialog** button invokes the **Viewports** dialog box depending on whether you are in **Model** tab or **Layout** tab. The **Single Viewport** button creates a single viewport depending on whether you are in **Model** tab or in the **Layout** tab. The **Viewport Scale Control** drop-down list displays a list of scale factors to choose from that can be applied to a selected viewport.

When you have selected a viewport and you right-click, a shortcut menu appears. **Display Viewport Objects** gives you the option of either displaying the objects in the selected viewport or not. **Display Locked** gives you an option to either Lock or unlock the selected viewport. **Hide Plot** allows you to either turn hideplot on or off. Choosing **Properties** from the shortcut menu, displays the **Object Properties** window. The **Misc** area lists all the properties of the

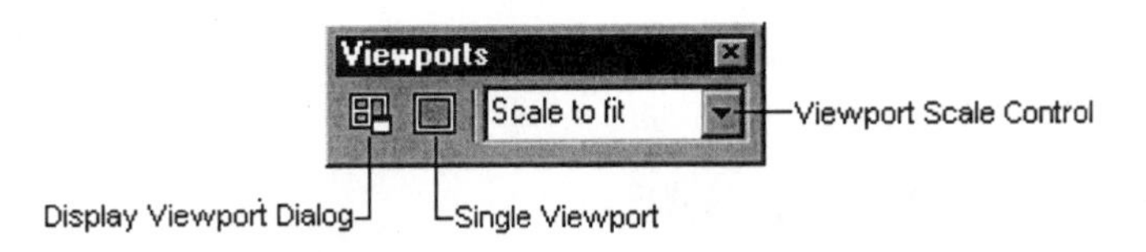

Figure 22-19 ***Viewports*** *toolbar*

selected viewport. You can turn the viewport on or off by choosing **Yes** or **No** from the **On** drop-down list. You can similarly choose options to turn on or off properties like **Display locked**, **Hideplot** and specifying the **Custom Scale** for the viewwport.

MVIEW COMMAND

Menu:	View > Viewports > 1 Viewports, 2 Viewports, 3 Viewports or 4 Viewports
Command:	MVIEW

The **MVIEW** command creates floating viewports and allows various options to work with them. The **TILEMODE** system variable should be set to 0. If you invoke this command in the floating Model space while you are in the **Layout** tab, AutoCAD LT shifts you to Paper space from Model space until the **MVIEW** command is over. Once the **MVIEW** command is over, you are returned to model space. The command prompt is:

Command: **MVIEW**
Specify corner of viewport
or[ON/OFF/Fit/Hideplot/Lock/Restore/2/3/4]<Fit>: *Enter an option or specify the viewport on screen by specifying points by using your pointing device.*

The options displayed by the **MVIEW** command are the same as the **-VPORTS** command in the **Layout** tab and have been discussed earlier in this Chapter.

MODEL SPACE AND PAPER SPACE

Model Space (MSPACE Command)

Status bar:	PAPER
Command:	MSPACE

The **MSPACE** command is used to shift from paper space to floating model space (Figure 22-20). Before using this command, you must set **TILEMODE** to 0 (Off); you should be in the **Layout** tab. To shift from paper space to model space, at least one of the floating viewports must be on.

Command: **MSPACE** [Enter]

AutoCAD LT acknowledges switching into Mspace by issuing the **MSPACE** message at the command line. Also, in the status bar PAPER is changed to MODEL. You can also switch to model space by double-clicking a viewport in Paper space.

Paper Space (PSPACE Command)

Status bar:	MODEL
Command:	PSPACE

This command is used to shift from model space to paper space (Figure 22-21). Before using this command, you must set **TILEMODE** to 0 (Off); you should be in the Layout tab.

Command: **PSPACE** [Enter]

You can also switch to Paper space by double-clicking an area of Paper space.

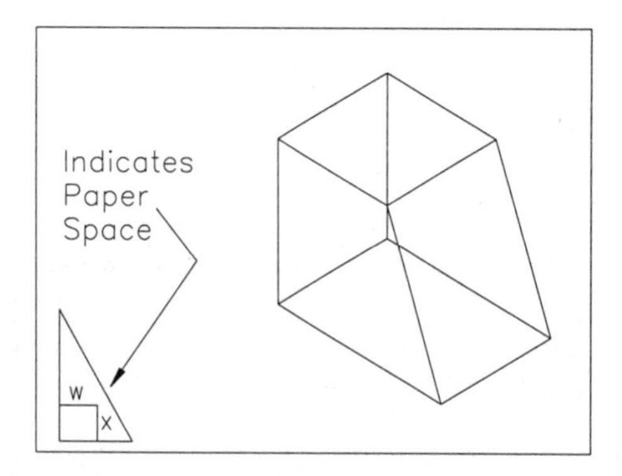

***Figure 22-20** Using the **MSPACE** command to switch to model space*

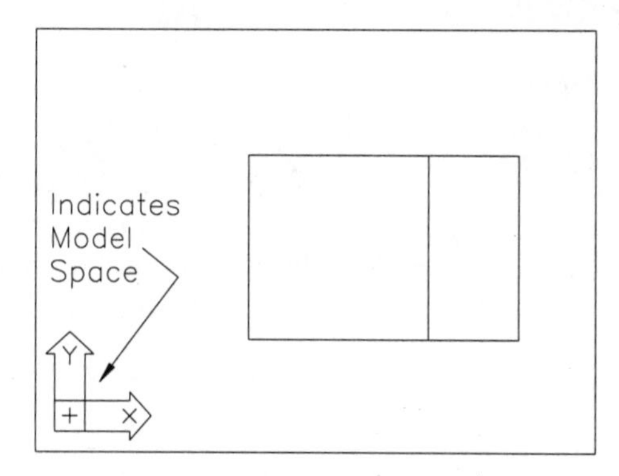

***Figure 22-21** Using the **PSPACE** command to switch to paper space*

MANIPULATING THE VISIBILITY OF VIEWPORT LAYERS (VPLAYER COMMAND)

Command:	VPLAYER

You can control the visibility of layers inside a floating viewport with the **VPLAYER** or **LAYER** command. The **On/Off** or **Freeze/Thaw** option of the **LAYER** command controls the visibility of layers globally, including the viewports. However, with the **VPLAYER** command you can control the visibility of layers in individual floating viewports. For example, you can use the **VPLAYER** command to freeze a layer in the selected viewport. The contents of this layer will not be displayed in the selected viewports, although in the other viewports, the contents are displayed. This command can be used from either model space or paper space. The only restriction is that **TILEMODE** be set to 0 (Off); that is, you can use this command only in the **Layout** tab.

Command: **VPLAYER** [Enter]
Enter an option [?/Freeze/Thaw/Reset/Newfrz/Vpvisdflt]:

VPLAYER Command Options

When you enter the **VPLAYER** command at the Command prompt, AutoCAD LT returns a prompt line that displays the available options:

Command: **VPLAYER** Enter
Enter an option [?/Freeze/Thaw/Reset/Newfrz/Vpvisdflt]:

? option. You can use this option to obtain a listing of the frozen layers in the selected viewport. When you enter ?, AutoCAD LT displays the following prompt:

Select a viewport:

At this prompt, select the viewport for which you want a listing of the frozen layers. If you are in model space, AutoCAD LT will temporarily shift you to paper space to let you select the viewport. The complete prompt sequence of this option is:

Command: **VPLAYER** Enter
Enter an option [?/Freeze/Thaw/Reset/Newfrz/Vpvisdflt]: **?** Enter
Switching to Paper space
Select a viewport: *Select a viewport.*
Layers currently frozen in viewport 1:
DIM2
DIM3
Switching to Model space
Enter an option [?/Freeze/Thaw/Reset/Newfrz/Vpvisdflt]:

Freeze option. The **Freeze** option is used to freeze a layer (or layers) in one or more viewports (Figure 22-22). When you select this option, AutoCAD LT displays the following prompt:

Enter layer name(s) to freeze: *Enter the layer name(s).*

In response to the prompt, specify the name of the layer you want to freeze. If you want to specify more than one layer, the layer names must be separated by commas. You can also use

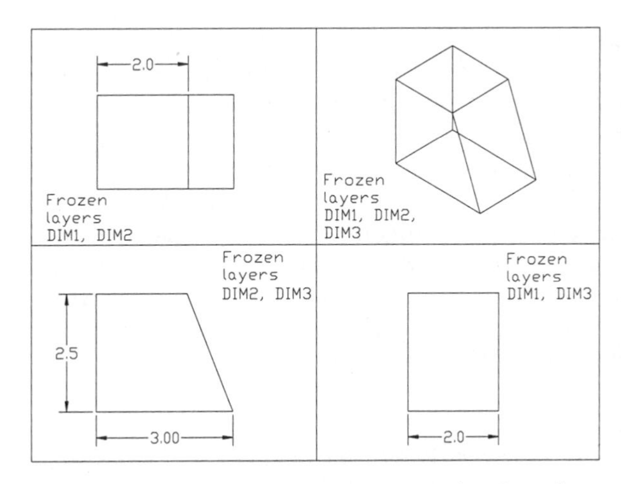

***Figure 22-22** Using the **VPLAYER** command to freeze layers in viewports*

wild cards to specify the names of the layers you want to freeze. Once you have specified the name of the layer(s), AutoCAD LT prompts you to select the viewport(s) in which you want to freeze the specified layer(s). The prompt is:

Enter an option [All/Select/Current]<current>: *Enter an option or press ENTER to accept default.*

The **All** option applies the changes to all viewports. **Select** allows you to specify a particular viewport to which you wish to apply changes. The **Current** option applies changes to only the current viewport.

Thaw option. With this option, you can thaw the layers that have been frozen in viewports using **VPLAYER Freeze** or the **LAYER** command. Layers that have been frozen, thawed, turned on, or turned off globally are not affected by **VPLAYER Thaw**. For example, if a layer has been frozen, the objects on the frozen layer are not regenerated on any viewport even if **VPLAYER Thaw** is used to thaw that layer in any viewport. The prompt sequence for the **VPLAYER** command with the **Thaw** option is:

Command: **VPLAYER** Enter
Enter an option [?/Freeze/Thaw/Reset/Newfrz/Vpvisdflt]: **THAW** Enter
Enter layer name(s) to thaw: *Specify the layer(s) to be thawed.*

If you want to specify more than one layer, separate the layer names with commas. The next prompt lets you specify the viewport(s) in which you want to thaw the specified frozen layers.

Enter an option [All/Select/Current]<Current>: *Specify the viewports.*

Reset option. With the **Reset** option, you can set the visibility of layer(s) in the specified viewports to their current default setting. The visibility defaults of a layer can be set by using the **Vpvisdflt** option of the **VPLAYER** command. The following is the prompt sequence of the **VPLAYER** command with the **Reset** option:

Command: **VPLAYER** Enter
Enter an option [?/Freeze/Thaw/Reset/Newfrz/Vpvisdflt]: **RESET** Enter
Enter layer name(s) to reset: *Specify the names of the layer(s) you want to reset.*
Enter an option [All/Select/Current]<current>: *Select the viewports in which you want to reset the specified layer to its default setting.*

Newfrz (New freeze) option. With this option, you can create new layers that are frozen in all viewports. This option is used mainly where you need a layer that is visible only in one viewport. This can be accomplished by creating the layer with **VPLAYER Newfrz**, and then thawing that particular layer in the viewport where you want to make the layer visible. The following is the prompt sequence of the **VPLAYER** command with the **Newfrz** option:

Command: **VPLAYER** Enter
Enter an option [?/Freeze/Thaw/Reset/Newfrz/Vpvisdflt]: **NEWFRZ** Enter
Enter name(s) of new layers frozen in all viewports: *Specify the name of the frozen layer(s) you want to create.*

If you want to specify more than one layer, separate the layer names with commas. After you specify the name(s) of the layer(s), AutoCAD LT creates frozen layers in all viewports. Also, the default visibility setting of the new layer(s) is set to Frozen; hence, if you create any new viewports, the layers created with **VPLAYER Newfrz** are also frozen in the new viewports.

Vpvisdflt (Viewport visibility default) option. With this option, you can set a default for the visibility of layer(s) in the subsequent new viewports. When a new viewport is created, the frozen/thawed status of any layer depends on the **Vpvisdflt** setting for that particular layer.

Command: **VPLAYER** Enter
Enter an option [?/Freeze/Thaw/Reset/Newfrz/Vpvisdflt]: **VPVISDFLT** Enter
Enter layer name(s) to change viewport visibility: *Specify the name(s) of the layer(s) whose default viewport visibility you want to set.*

Once you have specified the layer name(s), AutoCAD LT prompts:

Enter a viewport visibility option [Frozen/Thawed]<Thawed>:

At this prompt, enter FROZEN or F if you want to set the default visibility to frozen. To set the default visibility to thawed, press ENTER or enter T for Thawed.

Controlling Viewport through the Layer Properties Manager Dialog Box

You can use the **Layer Properties Manager** dialog box (Figure 22-23) to perform certain functions of the **VPLAYER** command, such as freezing/thawing layers in viewports.

Active VP Freeze. When the **TILEMODE** is turned off, you can freeze or thaw the selected layers in the current floating viewport by selecting the **Active viewport Freeze** toggle icon. The frozen layers will still be visible in other viewports.

New VP Freeze. If you want to freeze some layers in the new floating viewports, then select the **New VP Freeze** toggle icon. AutoCAD LT will freeze the layers in subsequently created new viewports without affecting the viewports that already exist (see Chapter 3, Drawing Aids, for more information).

If you start drawing on the frozen layer, objects drawn in the new viewport will not be displayed in the new viewport; however, in other viewports, the objects drawn in the new viewport will appear.

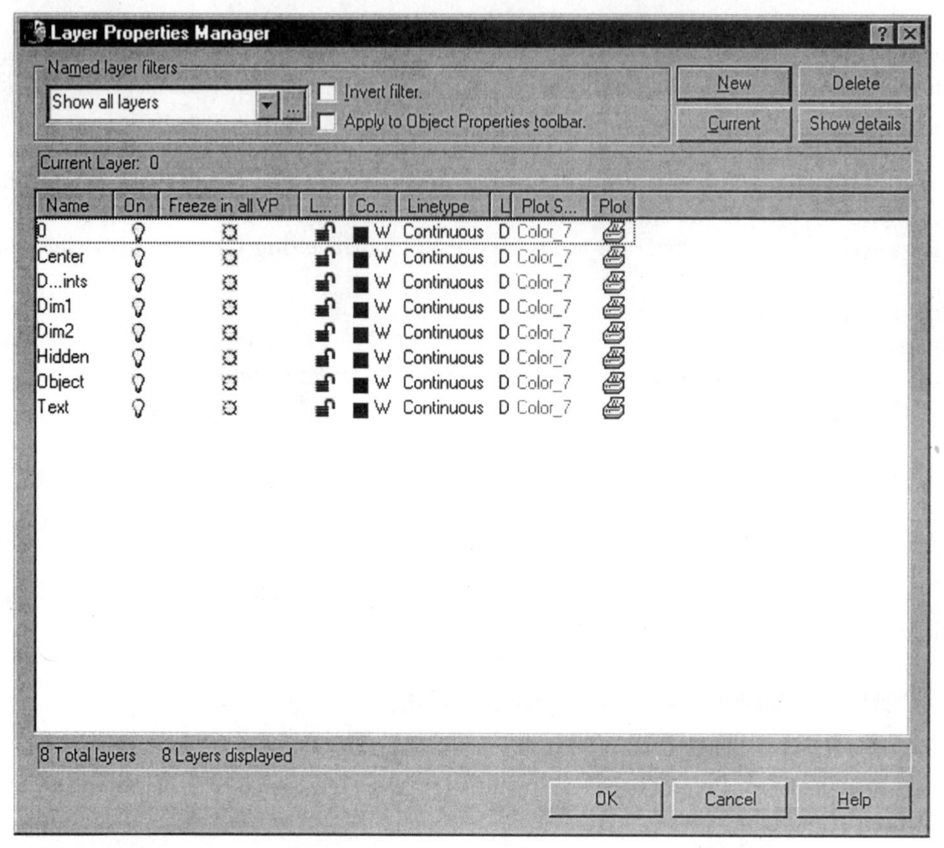

Figure 22-23 *Controlling viewport display through the* ***Layer Properties Manager*** *dialog box*

PAPER SPACE LINETYPE SCALING (PSLTSCALE SYSTEM VARIABLE)

By default, the linetype scaling is controlled by the **LTSCALE** system variable. Therefore, the display size of the dashes depends on the **LTSCALE** factor, on the drawing limits, or on the drawing units. If you have different viewports with different zoom (XP) factors, the size of the dashes will be different for these viewports. Figure 22-24 shows three viewports with different sizes and different zoom (XP) factors. You will notice that the dash length is different in each of these three viewports.

Generally, it is desirable to have identical line spacing in all viewports. This can be achieved with the **PSLTSCALE** system variable. By default, **PSLTSCALE** is set to 0. In this case, the size of the dashes depends on the **LTSCALE** system variable and on the zoom (XP) factor of the viewport where the objects have been drawn. If you set **PSLTSCALE** to 1 and **TILEMODE** to 0, the size of the dashes for objects in the model space are scaled to match the **LTSCALE** of objects in the paper space viewport, regardless of the zoom scale of the viewports. In other words, if **PSLTSCALE** is set to 1, even if the viewports are zoomed to different sizes, the length of the dashes will be identical in all viewports. Figure 22-25 shows three viewports with different limits. Notice that the dash length is identical in all three viewports.

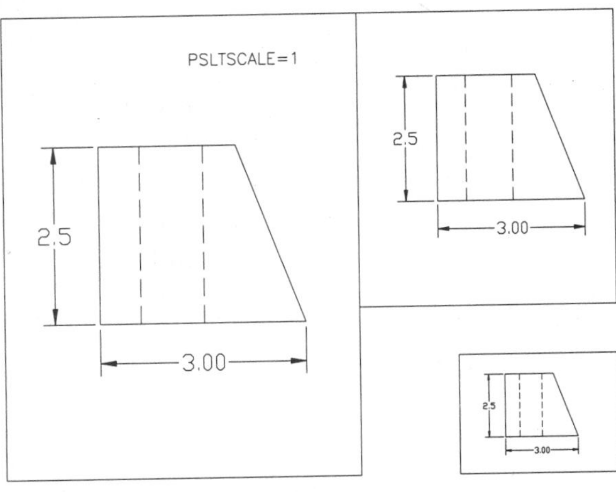

*Figure 22-24 When **PSLTSCALE** is 0 and **TILEMODE** is 0, the dash length depends on the size of the viewport*

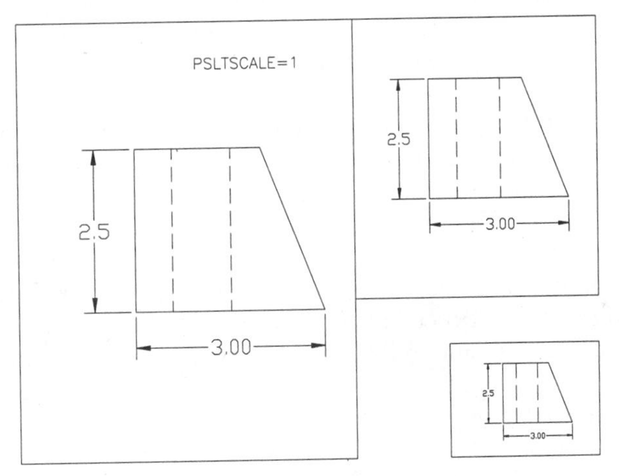

*Figure 22-25 When **PSLTSCALE** is 1 and **TILEMODE** is 0, the dash length is the same in all viewports*

CREATING AND WORKING WITH LAYOUTS (LAYOUT COMMAND)*

Toolbar:	Layouts > New Layout
Menu:	Insert > Layout > New Layout
Command:	LAYOUT

The **LAYOUT** command is used to create a new layout. It also allows you to rename, copy, save and delete existing layouts. A drawing which has been designed in the **Model** tab can be composed for plotting in the **Layout** tab. You can also add title blocks, more than one viewport and other annotations. The command prompts are:

Enter layout option [Copy/Delete/New/Template /Rename/SAveas/Set/?]<Set>:

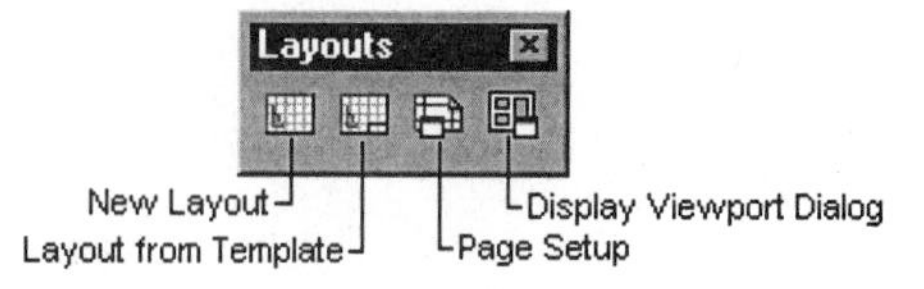

*Figure 22-26 The **Layouts** toolbar*

New option. Creates a new layout. The command prompt is :

Enter new layout name <Layout 3>: *Specify a new name for the new layout or press ENTER to accept the default name.*

A new tab with the new layout name appears in the drawing.

Copy option. This option copies a layout. The command prompts are:

Enter name of layout to copy <current>: *Enter the name of the layout you wish to copy.*
Enter layout name for copy <default>:

If you do not enter a name for the copy of the layout, the name of the copied layout is assumed with an incremental number in brackets next to it. For example, Layout 1 is copied as Layout1(2). The name of the new layout appears as a new tab next to the copied **Layout** tab.

Delete option. This option deletes an exiting layout. The most recently created layout is the default layout. The **Model** tab cannot be deleted. The command prompt is:

Enter name of layout to delete <current>:

Template option. This option creates a new template based on an existing layout template in either .dwg or .dwt files. This option invokes the **Select File** dialog box if the **FILEDIA** system variable is 1. The layout and geometry from the specified template or drawing file is inserted into the current drawing. After the DWT or DWG file is selected, you can select one or more layouts from the **Insert** dialog box.

Choosing the **Layout from Template** button from the **Layouts** toolbar creates a layout using an existing template or drawing file. You can also choose **Layout > Layout from Template** from the **Insert** menu.

Rename option. This option allows you to rename a layout. The command prompts are:

Enter layout to rename <current>:
Enter new layout name:

Layout names have to be unique and can contain upto 255 characters, out of which only 32 are displayed in the tab. The characters in the name are not case sensitive.

Save option. Saves a layout in the drawing template file. If the **FILEDIA** system variable is set to 1, the **Create Drawing File** dialog box appears where you can enter a file name in the

File name: edit box. The command line prompts are:

Enter layout to save to template <Default>:

As you enter the name of the layout you want to save as template file, AutoCAD LT displays **Create Drawing File** dialog box, Figure 22-27, where you can enter the name of the template file.

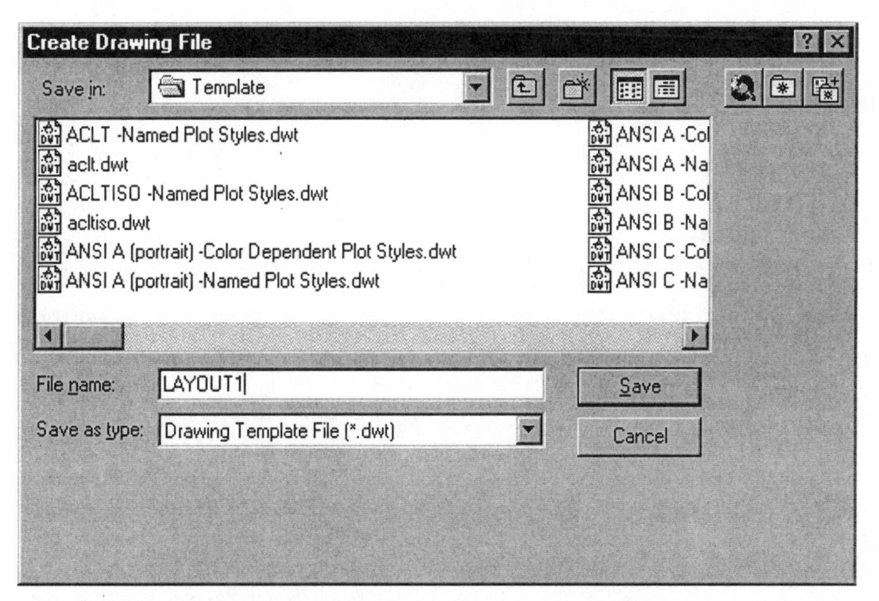

***Figure 22-26 Create Drawing File** dialog box for entering the name of tempelate file*

Set option. Sets a layout as the current layout. The command prompt is:

Enter layout to make current<last>:

? option. This option lists all the layouts in the current drawing.

You can choose any of the above **LAYOUT** command options from the shortcut menu that appear on right-clicking any of the **Layout** tabs.

PAGESETUP COMMAND*

Toolbar:	Layouts > Page Setup
Menu:	File > Page Setup
Command:	PAGESETUP

You can also right-click the **Model** or any of the **Layout** tabs and choose **Page Setup** from the shortcut menu that is displayed. This command allows you to specify the Layout and Plot device settings for each new layout. On invoking this command, the **Page Setup** dialog box is displayed. The details of that have been discussed in Chapter 15, Plotting Drawings and Draw Commands.

Every time you select a new **Layout** tab, the **Page Setup** dialog box is displayed automatically. These default settings are controlled in the **Layout elements** area of the **Display** tab of the **Options** dialog box.

LAYOUTWIZARD COMMAND*

Menu:	Insert > Layouts > Layout Wizard
	Tools > Wizards > Create Layout
Command:	LAYOUTWIZARD

This command displays the **Layout Wizard** which guides you step by step through the process of creating a new layout.

Example 1

In this example, you will learn how to create a drawing in the model space and then use the paper to plot the drawing.

1. Based on the drawing size, calculate the limits. Since the drawing is 50' x 48', the limits should be about 75' x 60'.

2. Set the limits and layers, then draw the floor plan as shown in Figure 22-28. The overall dimensions are given in the drawing in Figure 22-29; assume other dimensions.

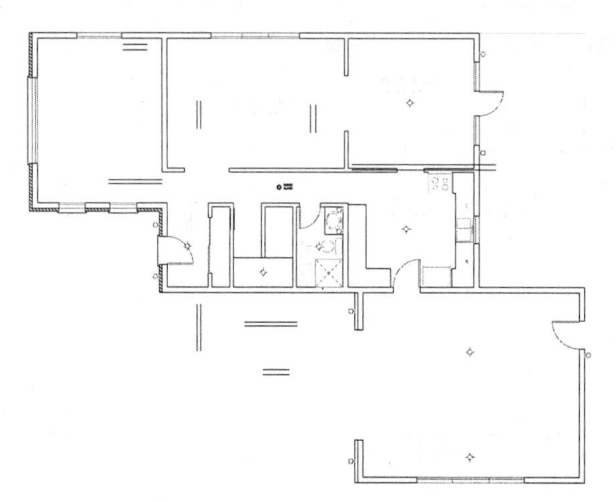

Figure 22-28 *Floor plan drawing for Example 1*

3. Use the **LAYOUT** command to create a new layout. You can also create a new layout by right-clicking on the **Model** or one of the **layout** tabs to display the shortcut menu. In the shortcut menu select **New layout**.
   Command: **LAYOUT** [Enter]
   Enter layout option [Copy/Delete/New/Template/Rename/SAveas/Set/?] <set>: **N** [Enter]
   Enter new layout name <Layout3>: *Specify new name or press ENTER to accept default name.*

A new tab with the new layout name is displayed next to the last **layout** tab.

4. Next, choose the new layout (**Layout3**) tab, AutoCAD LT displays the **Page Setup-Layout3** dialog box. Select the **Plot Device** tab and then the select the printer or plotter that you want to use, from the **Name:** drop-down list. In this example, **HP Lasejet4000** is used. Choose the **Layout Settings** tab and then select the paper size that is supported by your plotting device from the **Paper size:** drop-down list. In this example, the paper size is 8.5 x 11. Choose the **OK** button to accept the settings and exit the dialog box. AutoCAD LT displays the new layout (**Layout3**) on the screen with the default viewport. Use the **ERASE** command to erase the default viewport.

5. Use the **MVIEW** command to create a viewport of 11" x 8.5" size. Once you create the viewport, the drawing will automatically appear in the viewport. Draw the border and title block according to your requirements.

6. Use the **MSPACE** command to switch to model space or double-click the viewport to switch to Model space and use the **ZOOM** command to zoom the drawing to 1/96XP. In this example, it is assumed that the scale factor is 1:96 (1/8"=1'); therefore, the zoom factor is 1/96XP.

7. Create a dimension style with the text height of 1/8" and the text above the dimension line. Also, based on the given drawing, change other dimension settings and select the **Scale dimensions to layout (Paper Space)** radio button in the **Fit** tab of the **New Dimension Style** dialog box.

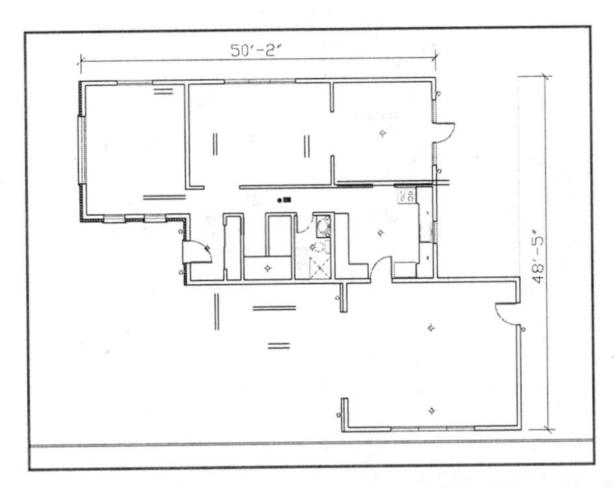

Figure 22-29 *Drawing plotted in paper space with 1:1 scale*

8. Now, give the two overall dimensions as shown in Figure 22-29. When you dimension, AutoCAD LT automatically adjusts the dimensions. For example, the dimension text size will increase so that when you plot the drawing the dimension text will be 1/8".

Note

For more information about dimensioning in paper space, see "Model Space and Paper Space Dimensioning" in Chapter 7.

9. Use the **PSPACE** command or double-click on the paper area to switch to paper space and then plot the drawing at 1:1 scale (Figure 22-30).

10. Use the **MSPACE** command to switch to model space or double-click the viewport and then use the **ZOOM** command to zoom to 1/144XP. The drawing display will reduce.

11. To adjust the dimensions to the correct size, use the **UPDATE** command at the **DIM** prompt and select the dimensions to be updated. AutoCAD LT will automatically change the dimensions so that when the drawing is plotted, the dimension are of the right size.

12. Plot the drawing and compare the two drawings. You will notice that the dimension text height is the same in both drawings.

This way, you can use the model space to draw and dimension your drawings, and then use the paper space to draw the border, title block, and other annotations. When you are done, you can plot the drawings 1:1 size from the paper space.

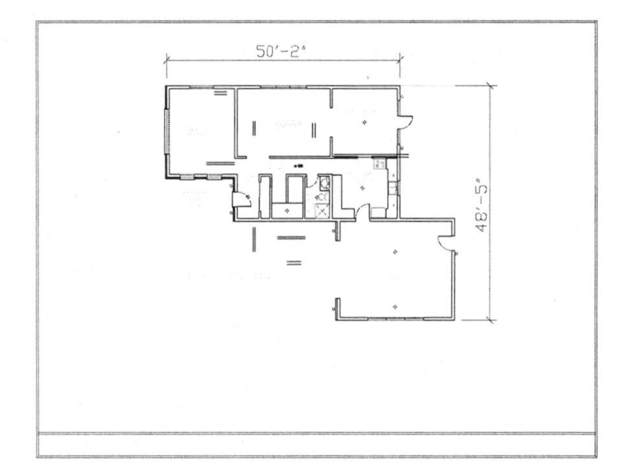

Figure 22-30 *Drawing plotted in paper space with 1:1 scale*

Self-Evaluation Test

Answer the following questions, and then compare your answers to the correct answers given at the end of this chapter.

1. What are the two types of viewports in AutoCAD LT? ______________________________

____

2. Tiled viewports are created by using the __________ command when the system variable __________ is set to 1 (model space).

3. With the __________ option, you can choose the floating viewports for which you want to remove the hidden lines when you plot in paper space.

4. If you set **Hideplot** to **Off**, hidden lines are not plotted. (T/F)

5. To shift from paper space to model space, at least two viewports must be active and on. (T/F)

6. The **VPLAYER** command can be used from either model space or paper space. (T/F)

7. To use the **VPLAYER** command, the only restriction is that **TILEMODE** be set to 1 (On). (T/F)

8. The **Freeze** option of the **VPLAYER** command is used to freeze a layer (or layers) in only one viewport. (T/F)

9. Layers that have been frozen, thawed, switched on, or switched off globally are not affected by the __________ command.

Review Questions

1. The __________ command can be used to create viewports and divide the display screen into a number of parts (viewports).

2. An arrangement of viewports created with the **VPORTS** command is a display function and __________ be plotted.

3. When the __________ variable is assigned a value of 1, a plan view is automatically generated if you change the UCS in a viewport.

4. The **MVIEW** command is used to create ________________ in paper space.

5. Before using the **-VPORTS** command, you must change the system variable _______ to 0.

6. If you invoke the **MVIEW** command in the floating model space (**TILEMODE** set to 0), AutoCAD LT shifts you from __________ to __________ , until the **MVIEW** command is not over.

7. It is possible to alter the size of the floating viewports and adjust them on the screen. (T/F)

8. Name five options of the **VPLAYER** command: __________, __________, __________, __________, __________.

9. You should turn off the viewports that are not required because ____________________
____.

10. The system variable __________ controls the number of viewports that will display the objects.

11. You will not be able to work in floating model space if any viewport is turned off. (T/F)

12. What happens if you turn **Hideplot** on? ________________________________
____.

13. The __________ command is used to shift from paper space to model space.

14. Before using the **MSPACE** command, you must set ________ to 0 (Off).

15. The __________ command is used to shift from model space to paper space.

16. You can control the visibility of layers inside the viewport with the __________ command.

17. With the _________ command, you can control the visibility of layers in individual viewports.

18. With the _________ option, you can set the visibility of layer(s) in the specified viewports to their current default setting.

19. The visibility defaults of subsequent new layers can be set by using the ________ option of the **VPLAYER** command.

20. With the __________ option, you can create new layers that are frozen in all viewports.

21. With the __________ option, you can set the default for the visibility of subsequent layer(s).

22. The **New VP Freeze** toggle icon governs the thawing and freezing of selected layers in ______ viewport.

Exercises

Exercise 1 — *General*

a. Draw the computer as shown in Figure 22-31 using regions. Take the dimensions from the computer you are using. For your convenience, two images of the same computer are shown. The first one has the lines hidden, and the second one (Figure 22-32) shows all lines, including the lines that are behind the visible surfaces.

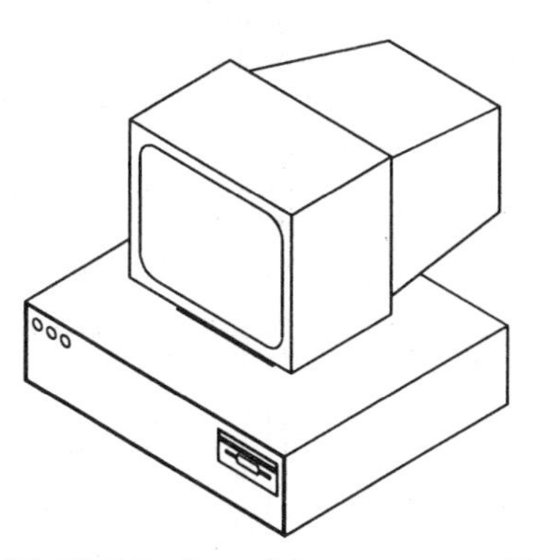

Figure 22-31 *3D view of the computer without hidden lines*

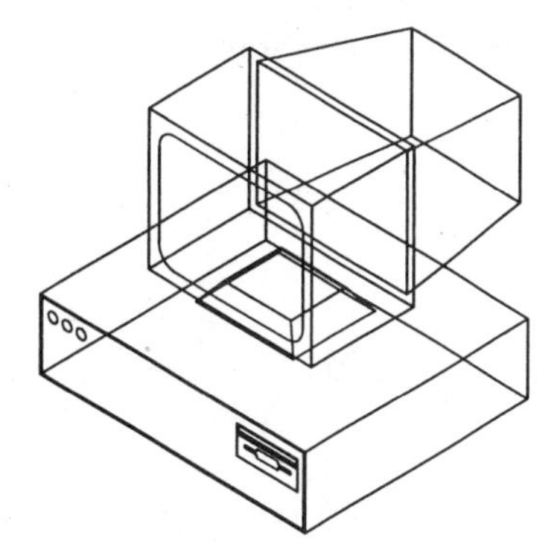

Figure 22-32 *3D view of the computer with hidden lines*

Construction of the given figure can be divided into the following steps:

1. Drawing the CPU box (central processing unit).
2. Drawing the base of the monitor.
3. Drawing the monitor.

b. Once you have drawn the computer, create four viewports in tiled model space and obtain different views of the computer as shown in the screen capture (Figure 22-33). You can use the **VPOINT** command to obtain the different views.

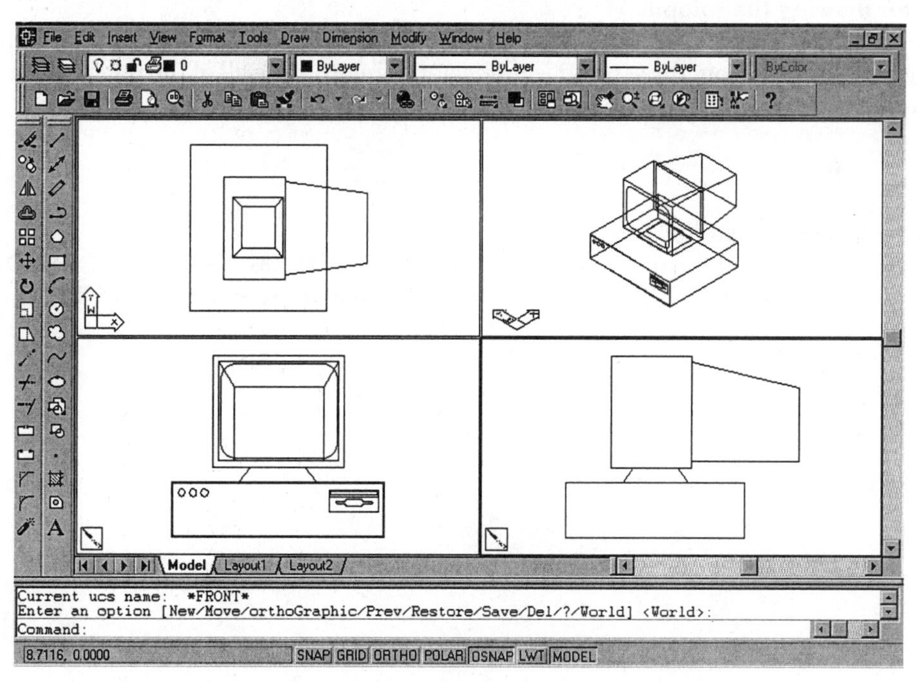

Figure 22-33 *Four viewports in the model space*

c. Next, use the **MVIEW** command to create four viewports as shown in Figure 22-34. Center the figure of the computer in 3D view in all four of the viewports.

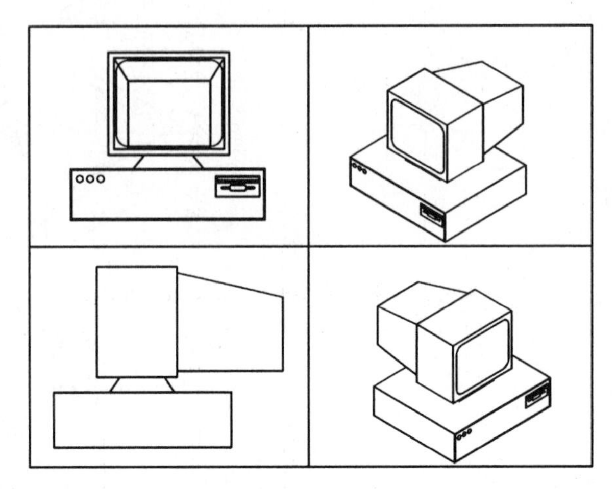

Figure 22-34 *Using **MVIEW** options to obtain different displays of the computer*

Hint for drawing the computer:

Draw the figure first in wireframe using **LINE** and **3DPOLY** command then make regions taking the reference from the wireframe models and finally bring all of the regions together as shown in the figure.

Answers to Self-Evaluation Test

1 - tiled viewports and floating viewports, **2** - **VPORTS**, **TILEMODE**, **3** - **HIDEPLOT**,**4** -F, **5** - F (One), **6** - F, **7** - F (**TILEMODE** set to 0), **8** - F (One or more), **9** - **VPLAYER**.

Chapter 23

Template Drawings

Learning Objectives

After completing this chapter, you will be able to:
- *Create template drawings.*
- *Load template drawings using dialog boxes and the command line.*
- *Do initial drawing setup.*
- *Customize drawings with layers and dimensioning specifications.*
- *Customize drawings with viewports and paper space.*

CREATING TEMPLATE DRAWINGS

One way to customize AutoCAD LT is to create template drawings that contain initial drawing setup information and, if desired, visible objects and text. When the user starts a new drawing, the settings associated with the template drawing are automatically loaded. If you start a new drawing from scratch, AutoCAD LT loads default setup values. For example, the default limits are (0.0,0.0), (12.0,9.0) and the default layer is 0 with white color and continuous linetype. Generally, these default parameters need to be reset before generating a drawing on the computer using AutoCAD LT. A considerable amount of time is required to set up the layers, colors, linetypes, limits, snaps, units, text height, dimensioning variables, and other parameters. Sometimes, border lines and a title block may also be needed.

In production drawings, most of the drawing setup values remain the same. For example, the company title block, border, layers, linetypes, dimension variables, text height, LTSCALE, and other drawing setup values do not change. You will save considerable time if you save these values and reload them when starting a new drawing. You can do this by making template drawings, which can contain the initial drawing setup information, set according to company specifications. They can also contain border, title block, tolerance table, block definitions, floating viewports in paper space, and perhaps some notes and instructions that are common to all drawings.

STANDARD TEMPLATE DRAWINGS

The AutoCAD LT software package comes with standard template drawings like aclt.dwt, acltiso.dwt, ansi_a.dwt, din_a.dwt, iso_a4.dwt, jis_a0.dwt. The ansi, din, and iso template drawings are based on the drawing standards developed by ANSI (American National Standards Institute), DIN (German), and ISO (International Organization for Standardization). When you start a new drawing, AutoCAD LT displays the **Create New Drawing** dialog box on the screen. To load the template drawing, select the **Use a Template** button and AutoCAD LT will display the list of standard template drawings. From this list you can select any template drawing according to your requirements. If you want to start a drawing with default setting, select the **Start from Scratch** button in the **Create New Drawing** dialog box. The following are some of the system variables, with the default values that are assigned to a new drawing:

System Variable Name	Default Value
AREA	0.0000
BLIPMODE	0
CHAMFERA	0.5000
CHAMFERB	0.5000
CECOLOR	Bylayer
DIMALT	Off
DIMALTD	2
DIMALTF	25.4000
DIMPOST	None
DIMASO	On
DIMASZ	0.1800
DRAGMODE	2
ELEVATION	0.0000
FILLMODE	1
FILLETRAD	0.5000
GRIDMODE	0
GRIPBLOCK	0
LIMMIN	0.0000,0.0000
LIMMAX	12.0000,9.0000
LTSCALE	1.0000
MIRRTEXT	1
ORTHOMODE	0
TILEMODE	1
TRIMMODE	1
UNITMODE	0

Example 1

Create a template drawing with the following specifications. The name of the template drawing is **PROTO1**.

Limits	18.0,12.0
Snap	0.25
Grid	0.50

Text height	0.125
Units	Decimal
	2-digits to the right of decimal point
	Decimal degrees
	2-digits to the right of decimal point
	0 angle along positive X axis (east)
	Angle positive if measured counterclockwise

Start AutoCAD LT and display the **Startup** dialog box. Select the **Start from Scratch** button in the **Startup** dialog box (Figure 23-1). You can also invoke the **Create New Drawing** dialog box by selecting **New** in the **File** menu or by entering **NEW** at AutoCAD LT Command prompt and then selecting the **Start from Scratch** button.

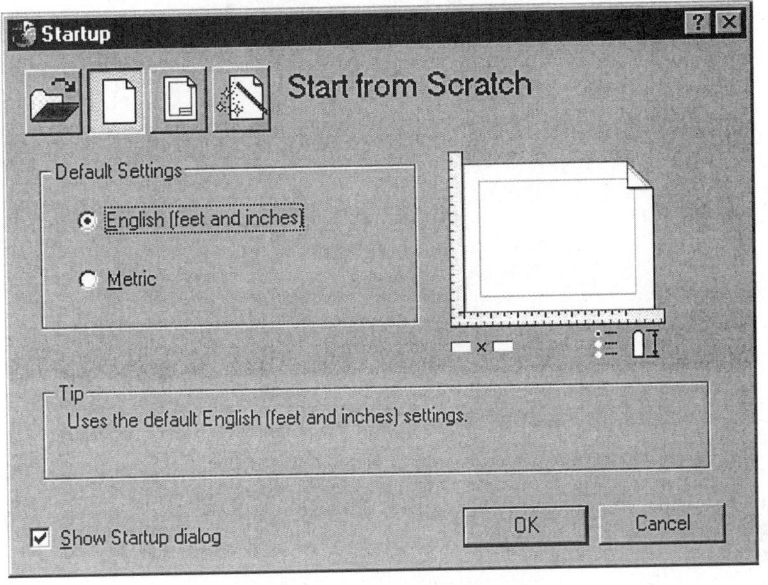

***Figure 23-1 Startup** dialog box*

Once you are in the Drawing Window, use the following AutoCAD LT commands to set up the values as given in Example 1.

Command: **LIMITS**
Specify lower left corner or [ON/OFF] <0.0000,0.0000>: **0,0**
Specify upper right corner <12.0000,9.0000>: **18.0,12.0**

Command: **SNAP**
Specify snap spacing or [ON/OFF/Aspect/Rotate/Style/Type] <0.5000>: **0.25**

Command: **GRID**
Specify grid spacing(X) or [ON/OFF/Snap/Aspect] <0.5000>: **0.50**

Command: **SETVAR**
Enter variable name or [?]: **TEXTSIZE**
Enter new value for TEXTSIZE <0.2000>: **0.125**

Using the Dialog Box. You can use the **Drawing Units** dialog box (Figure 23-2) to set the units. To invoke the **Drawing Units** dialog box, enter **UNITS** at AutoCAD LT Command prompt or select **Units** in the **Format** menu.

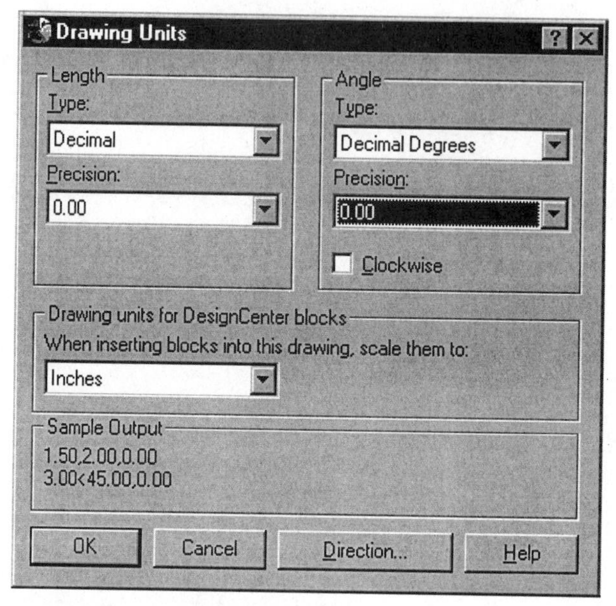

*Figure 23-2 **Drawing Units** dialog box*

Using UNITS Command. You can also use the **-UNITS** command to set the units.

Command: **-UNITS**

Report formats:	**(Examples)**
1. Scientific	1.55E+01
2. Decimal	15.50
3. Engineering	1'-3.50"
4. Architectural	1'-3 1/2"
5. Fractional	15 1/2

With the exception of the Engineering and Architectural formats, these formats can be used with any basic units of measurement. For example, Decimal mode is perfect for metric units as well as decimal English units.

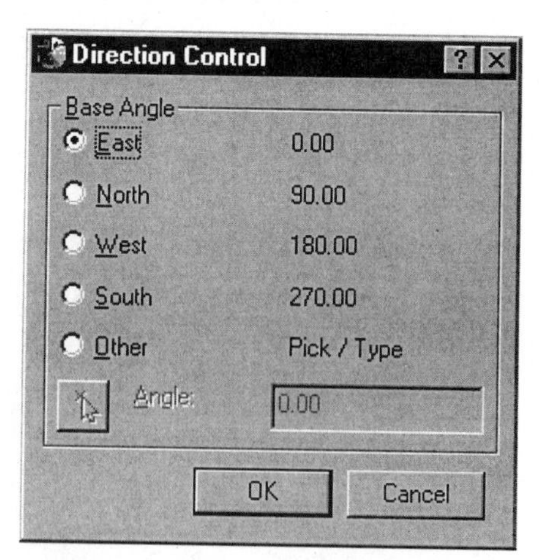

*Figure 23-3 **Direction Control** dialog box*

Enter choice, 1 to 5<2>: **2**
Enter the number of digits to right of decimal point (0 to 8)<4>: **2**

Systems of angle measure:	**(Examples)**
1. Decimal degrees	45.0000
2. Degrees/minutes/seconds	45d0'0"
3. Grads	50.0000g
4. Radians	0.7854r
5 Surveyor's units	N 45d0'0"

Enter choice, 1 to 5<1>: **1**
Enter number of fractional places for display of angles (0 to 8)<0>: **2**

Direction for angle 0.00:

East	3 o'clock	= 0.00
North	12 o'clock	= 90.00
West	9 o'clock	= 180.00
South	6 o'clock	= 270.00

Enter direction for angle 0.00<0.00>: **0**
Measure angles clockwise? [Yes/No]<N>: **N**

Now, save the drawing as **PROTO1.DWT** using AutoCAD LT's **SAVE** or **SAVEAS** command. You must select template (Drawing Template File [*.dwt]) from the **Save as type:** list box in the dialog box. This drawing is now saved as PROTO1.DWT on the default drive. You can also save this drawing on a floppy diskette in drives A or B.

LOADING A TEMPLATE DRAWING

Using the Dialog Box

You can use the template drawing any time you want to start a new drawing. To use the preset values of the template drawing, start AutoCAD LT or enter the **NEW** command (Command: **NEW**) to start a new drawing. AutoCAD LT displays the **Create New Drawing** dialog box (Figure 23-4).

You can also start a new drawing by selecting the **File** menu and then selecting the **New** option from this menu. To load the template drawing, select the **Use a Template** button; AutoCAD LT will display the list of template drawings. Select **PROTO1** template drawing and then select the **OK** button to exit the dialog box. AutoCAD LT will start a new drawing that will have the same setup as that of template drawing, **PROTO1**.

You can have several template drawings, each with a different setup. For example, **PROTOB** for a 18" by 12" drawing, **PROTOC** for a 24" by 18" drawing, **PROTOD** for a 36" by 24" drawing, and **PROTOE** for a 48" by 36" drawing. Each template drawing can be created according to user-defined specifications. You can then load any of these template drawings as discussed previously.

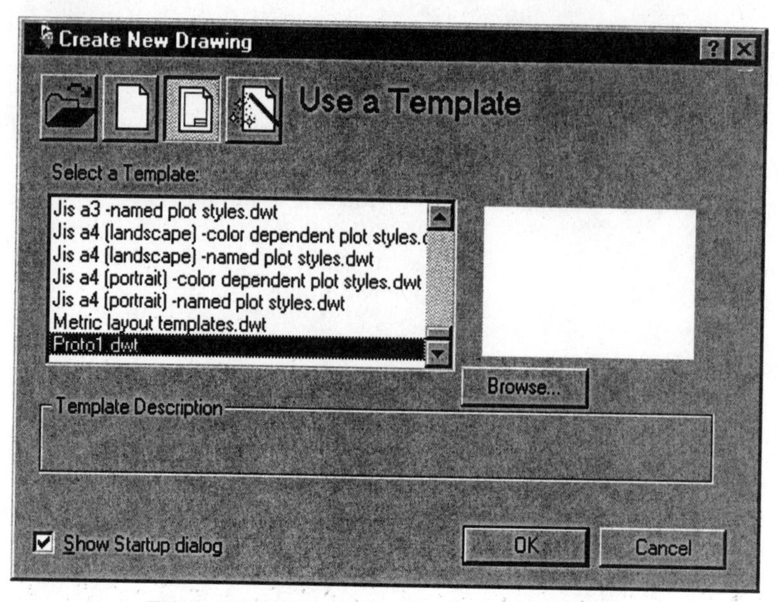

Figure 23-4 ***Create New Drawing*** *dialog box*

Using the NEW Command

You can create a new drawing without using the dialog boxes by assigning a value of 0 to the AutoCAD LT system variable **FILEDIA**. To start a new drawing, enter **NEW** at the Command prompt (Command: **NEW**) and AutoCAD LT will prompt you to enter the name of the drawing.

Command: **NEW**
Enter template file name or [. (for none)] <current>: *Enter template drawing name.*

CUSTOMIZING DRAWINGS WITH LAYERS AND DIMENSIONING SPECIFICATIONS

Most production drawings need multiple layers for different groups of objects. In addition to layers, it is a good practice to assign different colors to different layers to control the line width at the time of plotting. You can generate a template drawing that contains the desired number of layers with linetypes and colors according to your company specifications. You can then use this template drawing to make a new drawing. The next example illustrates the procedure used for customizing a drawing with layers, linetypes, and colors.

Example 2

You want to create a template drawing (**PROTO2**) that has a border and the company's title block, as shown in Figure 23-5. In addition, you want the following initial drawing setup:

Limits	48.0,36.0
Snap	1.0
Grid	4.00

Text height	0.25
PLINE width	0.02
Ltscale	4.0

DIMENSIONS

DIMSCALE	4.0
DIMTAD	1
DIMTIX	ON
DIMTOH	OFF
DIMTIH	OFF
DIMSCALE	25

(Use the **DIMSTYLE** command to save these values in the MYDIM1 dimension style file.)

LAYERS

Layer Names	Line Type	Color
0	Continuous	White
OBJ	Continuous	Red
CEN	Center	Yellow
HID	Hidden	Blue
DIM	Continuous	Green
BOR	Continuous	Magenta

Chapter 23

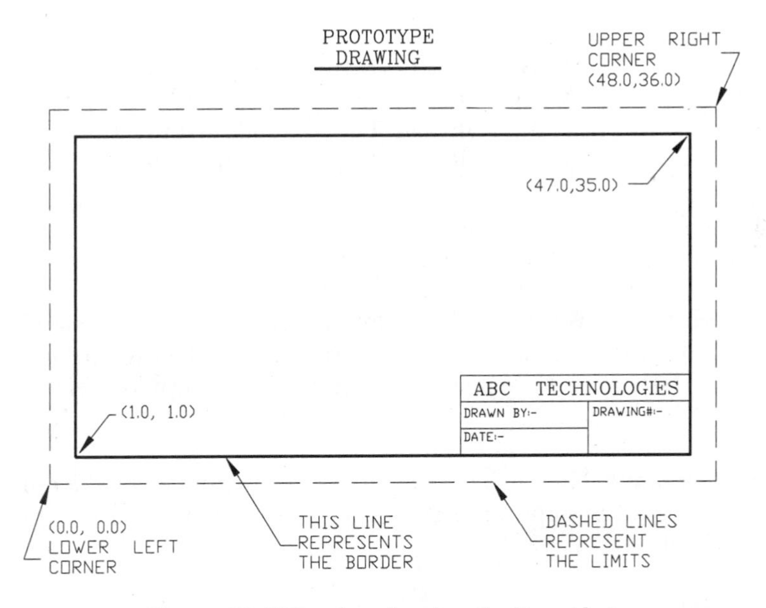

***Figure 23-5** Template drawing for Example 2*

Start a new drawing with default parameters. You can do this by selecting the **Start from Scratch** button in the **Create New Drawing** dialog box. Once you are in the drawing editor, use the AutoCAD LT commands to set up the values as given for this example. Also, draw a border and a title block as shown in Figure 23-5. In this figure the hidden lines indicate the drawing limits. The border lines are 0.5 units inside the drawing limits. For the border lines, use a polyline of width of 0.02 units. Use the following command sequence to produce the template drawing for Example 2:

Command: **LIMITS**
Specify lower left corner or [ON/OFF] <0.00,0.00>: **0,0**
Specify upper right corner <18.00,12.00>: **48.0,36.0**

Command: **SNAP**
Specify snap spacing or [ON/OFF/Aspect/Rotate/Style/Type] <0.25>: **1.0**

Command: **GRID**
Specify grid spacing(X) or [ON/OFF/Snap/Aspect] <0.50>: **4.00**

Command: **SETVAR**
Enter variable name or [?]: **TEXTSIZE**
Enter new value for TEXTSIZE <0.13>: **0.25**

Command: **PLINEWID**
Enter new value for PLINEWID <0.0000>: **0.02**

Command: **PLINE**
Specify start point: **1.0,1.0**
Current line-width is **0.02**
Specify next point or [Arc/Close/Halfwidth/Length/Undo/Width]: **47,1**
Specify next point or [Arc/Close/Halfwidth/Length/Undo/Width]: **47,35**
Specify next point or [Arc/Close/Halfwidth/Length/Undo/Width]: **1,35**
Specify next point or [Arc/Close/Halfwidth/Length/Undo/Width]: **C**

Command: **LTSCALE**
Enter new linetype scale factor<Current>: **1.0**

Using the Dimension Style Manager Dialog Box. You can use the **Dimension Style Manager** dialog box (Figure 23-6) to set the dimension variables. To invoke the **Dimension Style Manager** dialog box, enter **DDIM** at AutoCAD LT Command prompt or select **Style** in the **Dimension** menu.

Using the DIM Command. You can also use the **DIM** command to set the dimensions. The following is the command prompt sequence for setting the dimension variables and saving the dimension style file.

Command: **DIM**
Dim: **DIMSCALE**
Enter new value for dimension variable <Current>: **1.0**

Dim: **DIMTIX**
Enter new value for dimension variable <Off>: **ON**

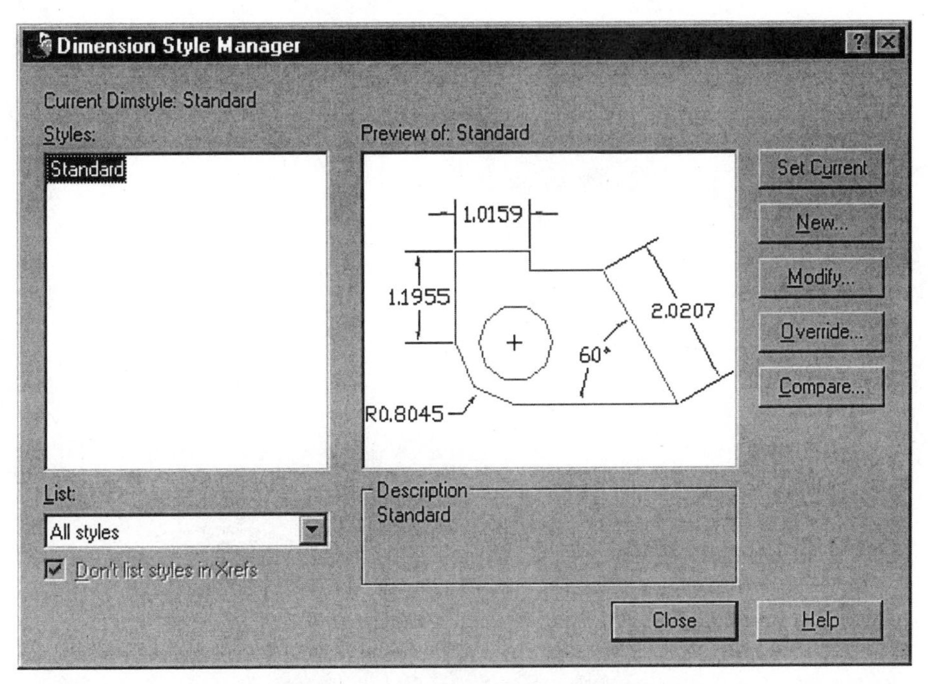

*Figure 23-6 **Dimension Style Manager** dialog box*

Dim: **DIMTAD**
Enter new value for dimension variable <0>: **1**
Dim: **DIMTOH**
Enter new value for dimension variable <On>: **OFF**
Dim: **DIMTIH**
Enter new value for dimension variable <On>: **OFF**
Dim: **STYLE**
New text style <Standard>: **MYDIM1**
Dim: **ESC** *(Press the ESCAPE key.)*

Using the Layer Properties Manager Dialog Box. You can use the **Layer Properties Manager** dialog box (Figure 23-8) to set the layers and linetypes in a drawing. To invoke this dialog box, enter **LAYER** at AutoCAD LT Command prompt or choose **Layer** from the **Format** menu.

Using the LAYER command. You can also use the **-LAYER** command to set the layers and linetypes from the command prompt. The following is the prompt sequence for setting the layers and linetypes.

Command: **-LAYER**
Enter an option
[?/Make/Set/New/ON/OFF/Color/Ltype/LWeight/Plot/Freeze/Thaw/LOck/Unlock]: **N**
Enter name list for new layer(s): **OBJ,CEN,HID,DIM,BOR**

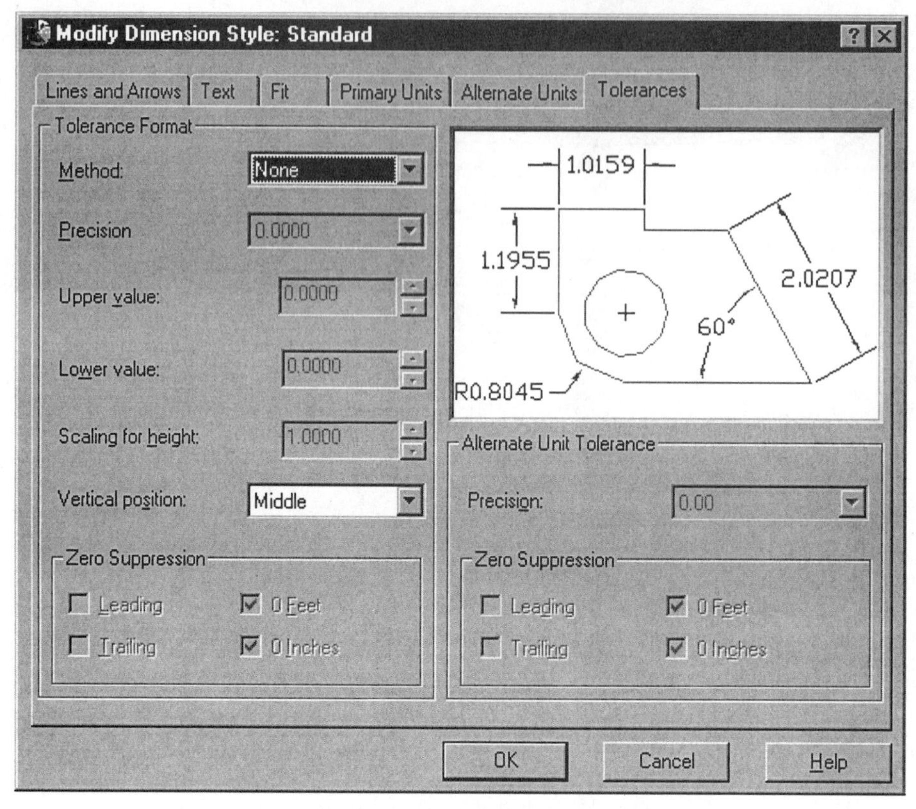

*Figure 23-7 **Modify Dimension Style** dialog box*

Enter an option
[?/Make/Set/New/ON/OFF/Color/Ltype/LWeight/Plot/Freeze/Thaw/LOck/Unlock]: **L**
Enter loaded linetype name or [?]<Continuous>: **HIDDEN**
Enter name list of layer(s) for linetype "HIDDEN" <0>: **HID**

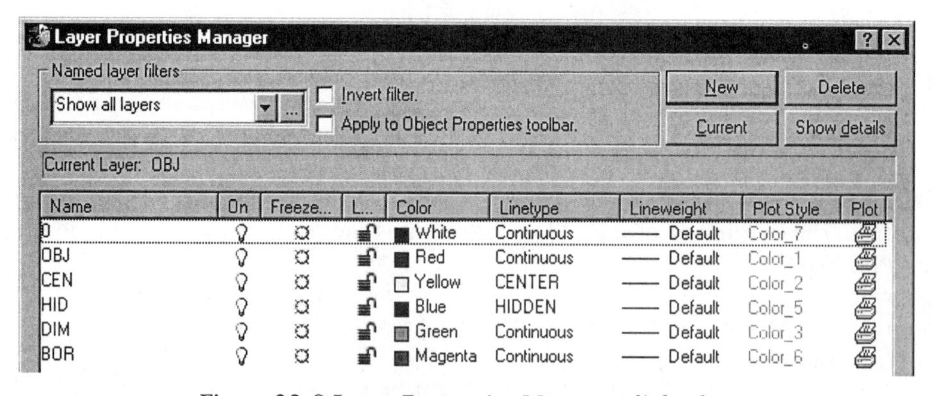

*Figure 23-8 **Layer Properties Manager** dialog box*

Enter an option
[?/Make/Set/New/ON/OFF/Color/Ltype/LWeight/Plot/Freeze/Thaw/LOck/Unlock]: **L**
Enter loaded linetype name or [?]<Continuous>: **CENTER**
Enter name list of layer(s) for linetype "HIDDEN" <0>: **CEN**

Enter an option
[?/Make/Set/New/ON/OFF/Color/Ltype/LWeight/Plot/Freeze/Thaw/LOck/Unlock: **C**
Enter color name or number (1-255): **RED**
Enter name list of layer(s) for color 1 (red) <0>: **OBJ**

Enter an option
[?/Make/Set/New/ON/OFF/Color/Ltype/LWeight/Plot/Freeze/Thaw/LOck/Unlock: **C**
Enter color name or number (1-255): **Yellow**
Enter name list of layer(s) for color 2 (yellow) <0>: **CEN**

Enter an option
[?/Make/Set/New/ON/OFF/Color/Ltype/LWeight/Plot/Freeze/Thaw/LOck/Unlock: **C**
Enter color name or number (1-255): **Blue**
Enter name list of layer(s) for color 5 (blue) <0>: **HID**

Enter an option
[?/Make/Set/New/ON/OFF/Color/Ltype/LWeight/Plot/Freeze/Thaw/LOck/Unlock: **C**
Enter color name or number (1-255): **GREEN**
Enter name list of layer(s) for color 3 (green) <0>: **DIM**

Enter an option
[?/Make/Set/New/ON/OFF/Color/Ltype/LWeight/Plot/Freeze/Thaw/LOck/Unlock: **C**
Enter color name or number (1-255): **MAGENTA**
Enter name list of layer(s) for color 6 (magenta) <0>: **BOR**
Enter an option
[?/Make/Set/New/ON/OFF/Color/Ltype/LWeight/Plot/Freeze/Thaw/LOck/Unlock: *Press ENTER.*

Next, add the title block and the text as shown in Figure 23-5. After completing the drawing, save it as PROTO2.DWT. You have created a template drawing (PROTO2) that contains all of the information given in Example 2.

CUSTOMIZING DRAWINGS ACCORDING TO PLOT SIZE AND DRAWING SCALE

You can generate a template drawing according to plot size and scale. For example, if the scale is 1/16" = 1' and the drawing is to be plotted on a 36" by 24" area, you can calculate drawing parameters like limits, **DIMSCALE**, and **LTSCALE** and save them in a template drawing. This will save considerable time in the initial drawing setup and provide uniformity in the drawings. The next example explains the procedure involved in customizing a drawing according to a certain plot size and scale. (Note, you can also use the paper space to specify the paper size and scale.)

Example 3

Generate a template drawing with the following specifications (Figure 23-9). The name of the template drawing is **PROTO3**.

Plotted sheet size	36" by 24"
Scale	1/8" = 1.0'
Snap	3'
Grid	6'
Text height	1/4" on plotted drawing
Ltscale	Calculate
Dimscale	Calculate
Units	Architectural
	16-denominator of smallest fraction
	Angle in degrees/minutes/seconds
	4-number of fractional places for display of angles
	0 angle along positive X axis
	Angle positive if measured counterclockwise
Border	Border should be 1" inside the edges of the plotted drawing sheet, using PLINE 1/32" wide when plotted (Figure 23-9)

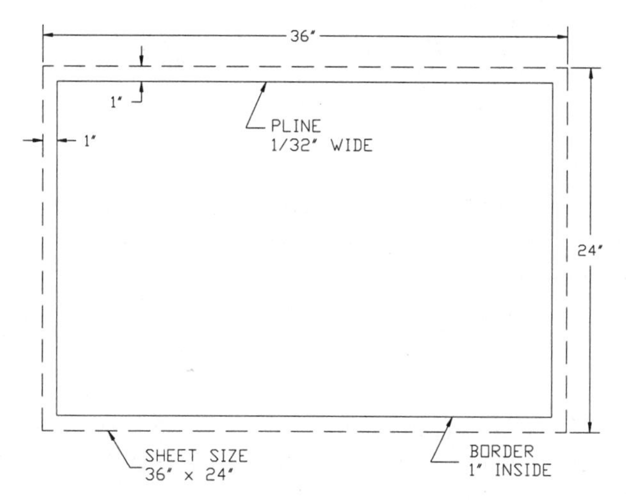

Figure 23-9 *Border of template drawing*

In this example, you need to calculate some values before you set the parameters for the template drawing. For example, the limits of the drawing depend on the plotted size of the drawing and the scale of the drawing. Similarly, **LTSCALE** and **DIMSCALE** depend on the limits of the drawing. The following calculations explain the procedure for finding the values of limits, ltscale, dimscale, width for pline and text height. Use the calculated values for the template drawing.

Limits

Given:
Sheet size 36" x 24"
Scale 1/8" = 1'
or 1" = 8'

Calculate:
XLimit
YLimit
Since sheet size is 36" x 24" and scale is 1/8"=1'
Therefore, XLimit = 36 x 8' = 288'
YLimit = 24 x 8' = 192'

Text height

Given:
Text height when plotted = 1/4"
Sheet size 36" x 24"
Scale 1/8" = 1'

Calculate:
Text height
Since scale is 1/8" = 1'
or 1/8" = 12"
or 1" = 96"
Therefore, scale factor = 96
Text height = 1/4" x 96
= 24" = 2'

Ltscale and Dimscale

Known:
Since scale is 1/8" = 1'
or 1/8" = 12"
or 1" = 96"

Calculate:
Ltscale and Dimscale
Since scale factor = 96
Therefore, LTSCALE = Scale factor = 96
Similarly, DIMSCALE = 96
(All dimension variables, like DIMTXT and DIMASZ, will be multiplied by 96.)

Pline Width

Given:
Scale is 1/8" = 1'

Calculate:
PLINE width
Since scale is 1/8" = 1'
or 1" = 8'
or 1" = 96"

Therefore,
PLINE width = 1/32 x 96 = 3"

After calculating the parameters, use the following AutoCAD LT commands to set up the drawing, then save the drawing as **PROTO3.DWT.**

Command: **-UNITS**

Report formats:	**(Examples)**
1. Scientific	1.55E+01
2. Decimal	15.50
3. Engineering	1'-3.50"
4. Architectural	1'-3 1/2"
5. Fractional	15 1/2

With the exception of Engineering and Architectural formats, these formats can be used with any basic unit of measurement. For example, decimal mode is perfect for metric units as well as decimal English units.

Enter choice, 1 to 5<2>: **4**
Enter denominator of smallest fraction to display
(1, 2, 4, 8, 16, 32, 64, 128 or 256) <16>: **16**

Systems of angle measure:	**(Examples)**
1. Decimal degrees	45.0000
2. Degrees/minutes/seconds	45d0'0"
3. Grads	50.0000g
4. Radians	0.7854r
5. Surveyor's units	N 45d0'0" E

Enter choice, 1 to 5<1>: **2**
Number of fractional places for display of angles (0 to 8)<0>: **4**

Direction for angle 0.00:

East	3 o'clock	= 0d0'0"
North	12 o'clock	= 90d0'0"
West	9 o'clock	= 180d0'0"
South	6 o'clock	= 270d0'0"

Enter direction for angle 0d0'0"<0d0'0">: *Press ENTER.*
Measure angles clockwise? [Yes/No]<N>: **N**

Command: **LIMITS**
Specify lower left corner or [ON/OFF] <0'-0",0'-0">: **0,0**
Specify upper right corner <4'-0",3'-0">: **288',192'**

Command: **SNAP**
Specify snap spacing or [ON/OFF/Aspect/Rotate/Style/Type] <0'-1">: **3'**

Command: **GRID**
Specify grid spacing(X) or [ON/OFF/Snap/Aspect] <0'-4">: **6'**

Command: **SETVAR**
Enter variable name or [?] <current>: **TEXTSIZE**
Enter new value for TEXTSIZE <current>: **2'**

Command: **LTSCALE**
Enter new linetype scale factor <1.0000>: **96**

Command: **DIM**
Dim: **Dimscale**
Enter new value for dimension variable <1.0000> New value: **96**
Dim: **Style**
New text style <STANDARD>: **MYDIM2**

Command: **PLINE**
Specify start point: **8',8'**
Current line-width is **0.0000**
Specify next point or [Arc/Close/Halfwidth/Length/Undo/Width]: **W**
Specify starting width<0.00>: **3**
Specify ending width<0'-3">: *Press ENTER.*
Specify next point or [Arc/Close/Halfwidth/Length/Undo/Width]: **280',8'**
Specify next point or [Arc/Close/Halfwidth/Length/Undo/Width]: **280',184'**
Specify next point or [Arc/Close/Halfwidth/Length/Undo/Width]: **8',184'**
Specify next point or [Arc/Close/Halfwidth/Length/Undo/Width]: **C**

Review Questions

1. The three standard template drawings that come with AutoCAD LT software are _________, __________ , and _________.
2. To use a template file, select the __________ button in the **Create New Drawing** dialog box.
3. To start a drawing with default setup, select the __________ button in the **Create New Drawing** dialog box.

4. To save the drawing as a template file, select the ________ option from the **Save as type** list box in the **Save Drawing As** dialog box.

5. To create a new drawing without using a dialog box, assign a value of 0 to the ________ system variable.

6. The default value of DIMSCALE is ________.

7. The default value for DIMTXT is ________.

8. The default value for SNAP is ________.

9. You can use the ________ dialog box to set the layers and linetypes.

10. You can use the ________ dialog box to set the dimension variables.

11. Architectural units can be selected by using AutoCAD LT's ________ or ________ commands.

12. If plot size is 36" x 24", and the scale is 1/2" = 1', then XLimit = ________ and YLimit = ________.

13. If the plot size is 24" x 18", and the scale is 1 = 20, the XLimit = ________ and YLimit = ________.

14. If the plot size is 200 x 150 and limits are (0.00,0.00) and (600.00,450.00), the **LTSCALE** factor = ________. (S.Factor=600/200=3 LTSCALE=3)

Exercises

Exercise 1

General

Generate a template drawing (**PROTOE2**) with the following specifications:

Limits	48.0,36.0
Snap	0.5
Grid	2.0
Text height	0.25
PLINE width	0.03
Ltscale	Calculate
Dimscale	Calculate
Plot size	10.5" x 8"

LAYERS

Layer Names	Line Type	Color
0	Continuous	White
OBJECT	Continuous	Green
CENTER	Center	Magenta
HIDDEN	Hidden	Blue
DIM	Continuous	Red
BORDER	Continuous	Cyan
GRID	Dotted	Green

Exercise 2 *General*

Generate a template drawing with the following specifications and save it with the name **PROTOE3**:

Plotted sheet size	36" x 24" (Figure 23-10)
Scale	1/2" = 1.0'
Text height	1/4" on plotted drawing
Ltscale	Calculate
Dimscale	Calculate
Units	Architectural 32-denominator of smallest fraction to display Angle in degrees/minutes/seconds 4-number of fractional places for display of angles 0d0'0"-direction for angle Angle positive if measured counterclockwise
Border	Border is 1-1/2" inside the edges of the plotted drawing sheet, using PLINE 1/32" wide when plotted (Figure 23-10)

LAYERS

Layer Names	Line Type	Color
0	Continuous	White
HID	Hidden	Green
CEN	Center	Cyan
DIM	Continuous	Blue
BOR	Continuous	Red
GRID	Dotted	Magenta

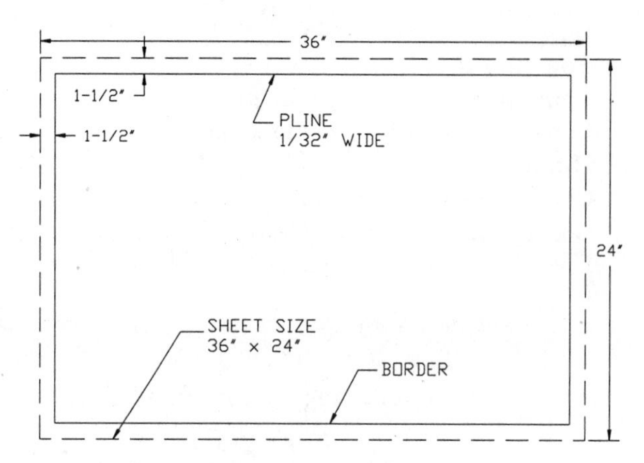

Figure 23-10 *Drawing for Exercise 2*

Chapter 24

Script Files and Slide Shows

Learning Objectives

After completing this chapter, you will be able to:

- *Write script files and use the* **SCRIPT** *command to run script files.*
- *Use the* **RSCRIPT** *and* **DELAY** *commands in script files.*
- *Invoke script files when loading AutoCAD LT.*
- *Create a slide show.*
- *Preload slides when running a slide show.*

WHAT ARE SCRIPT FILES?

AutoCAD LT has provided a facility called **script files** that allows you to combine different AutoCAD LT commands and execute them in a predetermined sequence. The commands can be written as a text file using any text editor like Notepad. These files, generally known as script files, have extension **.SCR** (example: **PLOT1.SCR**). A script file is executed with the AutoCAD LT **SCRIPT** command.

Script files can be used to generate a slide show, do the initial drawing setup, or plot a drawing to a predefined specification. They can also be used to automate certain command sequences that are used frequently in generating, editing, or viewing a drawing. Scripts cannot access dialog boxes or menus. When commands that open the file and plot dialog boxes are issued from a script file, AutoCAD LT runs the command line version of the command instead of opening the dialog box.

Example 1

Write a script file that will perform the following initial setup for a drawing (**SCRIPT1.SCR**). It is assumed that the drawing will be plotted on 12x9 size paper (Scale factor for plotting = 4).

Ortho	On	Zoom	All
Grid	2.0	Text height	0.125
Grid	Off	Ltscale	4.0
Snap	0.5	Dimscale	4.0
Limits	0,0 48.0,36.0		

Before writing a script file, you need to know the AutoCAD LT commands and the entries required in response to the command prompts. To find out the sequence of the prompt entries, you can type the command at the keyboard and then respond to different prompts. The following is a list of AutoCAD LT commands and prompt entries for Example 1:

Command: **ORTHO**
Enter mode [ON/OFF] <OFF>: **ON**

Command: **GRID**
Specify grid spacing(X) or [ON/OFF/Snap/Aspect] <0.5000>: **2.0**

Command: **GRID**
Specify grid spacing(X) or [ON/OFF/Snap/Aspect] <1.0>: **OFF**

Command: **SNAP**
Specify snap spacing or [ON/OFF/Aspect/Rotate/Style/Type] <0.5000>: **0.5**

Command: **LIMITS**
Reset Model space limits:
Specify lower left corner or [ON/OFF] <0.0000,0.0000>: **0,0**
Specify upper right corner <12.0000,9.0000>: **48.0,36.0**

Command: **ZOOM**
Specify corner of window, enter a scale factor (nX or nXP), or
[All/Center/Dynamic/Extents/Previous/Scale/Window] <real time>: **A**

Command: **SETVAR**
Enter variable name or [?]: **TEXTSIZE**
Enter new value for TEXTSIZE <0.2000>: **0.125**

Command: **LTSCALE**
Enter new linetype scale factor <1.0000>: **4.0**

Command: **SETVAR**
Enter variable name or [?] <TEXTSIZE>: **DIMSCALE**
Enter new value for DIMSCALE <1.0000>: **4.0**

Once you know the AutoCAD LT commands and the required prompt entries, you can write the script file using any text editor. The following file is a listing of the script file for Example 1:

```
ORTHO
ON
GRID
2.0
GRID
OFF
SNAP
0.5
LIMITS
0,0
48.0,36.0
ZOOM
ALL
SETVAR
TEXTSIZE
0.125
LTSCALE
4.0
SETVAR
DIMSCALE
4.0
```

Notice that the commands and the prompt entries in this file are in the same sequence as mentioned before. You can also combine several statements in one line, as shown in the following list:

```
ORTHO ON
GRID 2.0 GRID OFF
SNAP 0.5 SNAP ON
LIMITS 0,0 48.0,36.0 ZOOM ALL
SETVAR TEXTSIZE 0.125
LTSCALE 4.0
SETVAR DIMSCALE 4.0
```

Note

In the script file, a space is used to terminate a command or a prompt entry. Therefore, spaces are very important in these files. Make sure there are no extra spaces, unless they are required to press ENTER more than once.

*After you change the limits, it is a good practice to use the **ZOOM** command with the **All** option to display the new limits on the screen.*

AutoCAD LT ignores and does not process any lines that begin with a semicolon (;). This allows you to put comments in the file.

SCRIPT COMMAND

The AutoCAD LT **SCRIPT** command allows you to run a script file while you are in the drawing editor. To execute the script file, type the **SCRIPT** command and press ENTER. AutoCAD LT will prompt you to enter the name of the script file. You can accept the default file name or enter a new file name. The default script file name is the same as the drawing name. If you want to enter a new file name, type the name of the script file without the file extension (**.SCR**). (The file extension is assumed and need not be included with the file name.)

To run the script file of Example 1, type the **SCRIPT** command and press ENTER; AutoCAD LT will display the **Select Script File** dialog box (Figure 24-1). In the dialog box, select the file SCRIPT1 and then choose the **Open** button. You will see the changes taking place on the screen as the script file commands are executed.

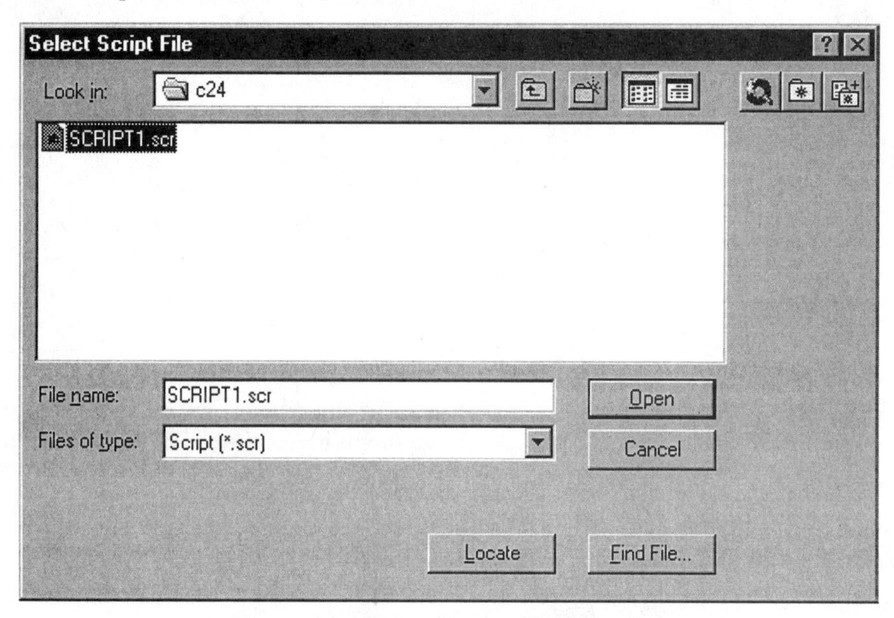

Figure 24-1 Select Script File *dialog box*

You can also enter the name of the script file at the Command prompt by setting **FILEDIA**=0. The format of the **SCRIPT** command is:

Command: **FILEDIA**
Enter new value for FILEDIA <1>: **0**
Command: **SCRIPT**
Enter script file name <default>: *Script file name.*

For example:
Command: **SCRIPT**
Script file <CUSTOM>: **SCRIPT1**
Where **<CUSTOM>** ------------- Default drawing file name
SCRIPT1 ----------------- Name of the script file

Example 2

Write a script file that will set up the following layers with the given colors and linetypes (filename **SCRIPT2.SCR**).

Layer Names	Color	Linetype	Line Weight
Object	Red	Continuous	default
Center	Yellow	Center	default
Hidden	Blue	Hidden	default
Dimension	Green	Continuous	default
Border	Magenta	Continuous	default
Hatch	Cyan	Continuous	0.05

As mentioned earlier, you need to know the AutoCAD LT commands and the required prompt entries before writing a script file. For Example 2, you need the following commands to create the layers with the given colors and linetypes:

Command: **LAYER**
Enter an option
[?/Make/Set/New/ON/OFF/Color/Ltype/LWeight/Plot/Freeze/Thaw/LOck/Unlock]: **N**
Enter name list for new layer(s): **OBJECT,CENTER,HIDDEN,DIM,BORDER,HATCH**

Enter an option
[?/Make/Set/New/ON/OFF/Color/Ltype/LWeight/Plot/Freeze/Thaw/LOck/Unlock]: **L**
Enter loaded linetype name or [?] <Continuous>: **CENTER**
Enter name list of layer(s) for linetype "CENTER" <0>: **CENTER**

Enter an option
[?/Make/Set/New/ON/OFF/Color/Ltype/LWeight/Plot/Freeze/Thaw/LOck/Unlock]: **L**
Enter loaded linetype name or [?] <Continuous>: **HIDDEN**
Enter name list of layer(s) for linetype "HIDDEN" <0>: **HIDDEN**

Enter an option
[?/Make/Set/New/ON/OFF/Color/Ltype/LWeight/Plot/Freeze/Thaw/LOck/Unlock]: **C**
Enter color name or number (1-255): **RED**
Enter name list of layer(s) for color 1 (red) <0>:**OBJECT**

Enter an option
[?/Make/Set/New/ON/OFF/Color/Ltype/LWeight/Plot/Freeze/Thaw/LOck/Unlock]: **C**
Enter color name or number (1-255): **YELLOW**
Enter name list of layer(s) for color 2 (yellow) <0>: **CENTER**

Enter an option
[?/Make/Set/New/ON/OFF/Color/Ltype/LWeight/Plot/Freeze/Thaw/LOck/Unlock]: **C**
Enter color name or number (1-255): **BLUE**
Enter name list of layer(s) for color 5 (blue)<0>: **HIDDEN**

Enter an option
[?/Make/Set/New/ON/OFF/Color/Ltype/LWeight/Plot/Freeze/Thaw/LOck/Unlock]: **C**
Enter color name or number (1-255): **GREEN**
Enter name list of layer(s) for color 3 (green)<0>: **DIM**

Enter an option
[?/Make/Set/New/ON/OFF/Color/Ltype/LWeight/Plot/Freeze/Thaw/LOck/Unlock]: **C**
Enter color name or number (1-255): **MAGENTA**
Enter name list of layer(s) for color 6 (magenta)<0>: **BORDER**

Enter an option
[?/Make/Set/New/ON/OFF/Color/Ltype/LWeight/Plot/Freeze/Thaw/LOck/Unlock]: **C**
Enter color name or number (1-255): **CYAN**
Enter name list of layer(s) for color 4 (cyan)<0>: **HATCH**

Enter an option
[?/Make/Set/New/ON/OFF/Color/Ltype/LWeight/Plot/Freeze/Thaw/LOck/Unlock]: **LW**
Enter lineweight (0.0mm - 2.11mm): **0.05**
Enter name list of layers(s) for lineweight 0.05mm <0>: **HATCH**
[?/Make/Set/New/ON/OFF/Color/Ltype/LWeight/Plot/Freeze/Thaw/LOck/Unlock]:
Press ENTER.

The following file is a listing of the script file that creates different layers and assigns the given colors and linetypes to these layers:

```
;This script file will create new layers and
;assign different colors and linetypes to layers
LAYER
NEW
OBJECT,CENTER,HIDDEN,DIM,BORDER,HATCH
L
CENTER
CENTER
L
HIDDEN
HIDDEN
C
RED
OBJECT
C
YELLOW
CENTER
C
BLUE
HIDDEN
C
GREEN
```

DIM
C
MAGENTA
BORDER
C
CYAN
HATCH
(This is a blank line to terminate the **LAYER** command. End of script file.)

Example 3

Write a script file that will rotate the circle and the line, as shown in Figure 24-2, around the lower endpoint of the line through 45-degree increments. The script file should be able to produce a continuous rotation of the given objects with a delay of two seconds after every 45-degree rotation (file name **SCRIPT3.SCR**).

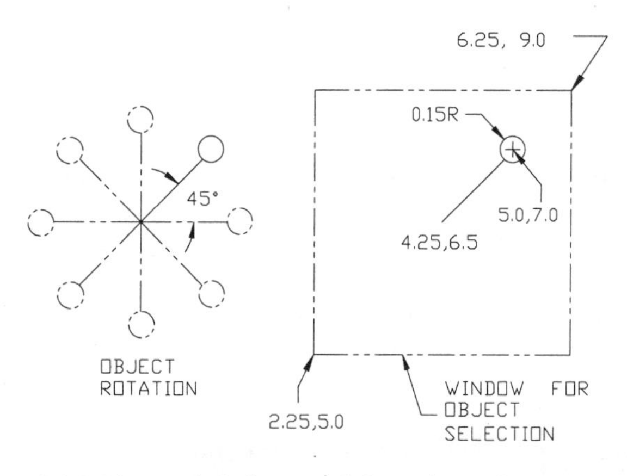

Figure 24-2 *Line and circle rotated through 45-degree increments*

Before writing the script file, enter the required commands and the prompt entries at the keyboard. Write down the exact sequence of the entries in which they have been entered to perform the given operations. The following is a listing of the AutoCAD LT command sequence needed to rotate the circle and the line around the lower endpoint of the line:

Command: **ROTATE** *(Enter **ROTATE** command.)*
Current positive angle in UCS: ANGDIR=counterclockwise ANGBASE=0
Select objects: **W** *(Window option to select object.)*
Specify first corner: **2.25, 5.0**
Specify opposite corner: **6.25, 9.0**
Select objects: *Press ENTER.*
Specify base point: **4.25,6.5**
Specify rotation angle or [Reference]: **45**

Once the AutoCAD LT commands, command options, and their sequences are known, you can write a script file. As mentioned earlier, you can use any text editor to write a script file.

The following file is a listing of the script file that will create the required rotation of the circle and line. The line and the circle are already drawn on the screen at the given location.

```
ROTATE          1
W               2
2.25,5.0        3
6.25,9.0        4
                (Blank line for Return.)    5
4.25,6.5        6
45              7
```

Line 1
ROTATE
In this line, **ROTATE** is an AutoCAD LT command that rotates the objects.

Line 2
W
In this line, W is the Window option for selecting the objects that need to be edited.

Line 3
2.25,5.0
In this line, 2.25 defines the X coordinate and 5.0 defines the Y coordinate of the lower left corner of the object selection window.

Line 4
6.25,9.0
In this line, 6.25 defines the X coordinate and 9.0 defines the Y coordinate of the upper right corner of the object selection window.

Line 5
Line 5 is a blank line that terminates the object selection process.

Line 6
4.25,6.5
In this line, 4.25 defines the X coordinate and 6.5 defines the Y coordinate of the base point for rotation.

Line 7
45
In this line, 45 is the incremental angle for rotation.

Note

One of the limitations of the script files is that all the information has to be contained within the file. These files do not let you enter information. For instance, in Example 3, if you want to use the Window option to select the objects, the Window option (W) and the two points that define this window must be contained within the script file. The same is true for the base point and all other information that goes in a script file.

RSCRIPT COMMAND

The AutoCAD LT **RSCRIPT** command allows the user to execute the script file indefinitely until canceled. It is a very desirable feature when the user wants to run the same file continuously. For example, in the case of a slide show for a product demonstration, the **RSCRIPT** command can be used to run the script file again and again until it is terminated by pressing the ESC (ESCAPE) key from the keyboard. Similarly, in Example 3, the rotation command needs to be repeated indefinitely to create a continuous rotation of the objects. This can be accomplished by adding **RSCRIPT** at the end of the file, as in the following file:

```
ROTATE
W
2.25,5.0
6.25,9.0
                (Blank line for Return.)
4.25,6.5
45
RSCRIPT
```

The **RSCRIPT** command on line 8 will repeat the commands from line 1 to line 7, and thus set the script file in an indefinite loop. The script file can be stopped by pressing the ESC or the BACKSPACE key.

Note

You cannot provide conditional statements in a script file to terminate the file when a particular condition is satisfied.

DELAY COMMAND

In the script files, some of the operations happen very quickly and make it difficult to see the operations taking place on the screen. It might be necessary to intentionally introduce a pause between certain operations in a script file. For example, in a slide show for a product demonstration, there must be a time delay between different slides so that the audience has enough time to see them. This is accomplished by using the AutoCAD LT **DELAY** command, which introduces a delay before the next command is executed. The general format of the **DELAY** command is:

Command: DELAY Time

Where **Command** ------ AutoCAD LT command prompt
DELAY ---------- **DELAY** command
Time ------------ Time in milliseconds

The **DELAY** command is to be followed by the delay time in milliseconds. For example, a delay of 2,000 milliseconds means that AutoCAD LT will pause for approximately two seconds before executing the next command. It is approximately two seconds because computer processing speeds vary. The maximum time delay you can enter is 32,767 milliseconds (about 33 seconds). In Example 3, a two-second delay can be introduced by inserting a **DELAY** command line between line 7 and line 8, as in the following file listing:

ROTATE
W
2.25,5.0
6.25,9.0
(Blank line for Return.)
4.25,6.5
45
DELAY 2000
RSCRIPT

The first seven lines of this file rotate the objects through a 45-degree angle. Before the **RSCRIPT** command on line 8 is executed, there is a delay of 2,000 milliseconds (about two seconds). The **RSCRIPT** command will repeat the script file that rotates the objects through another 45-degree angle. Thus, a slide show is created with a time delay of two seconds after every 45-degree increment.

RESUME COMMAND

If you cancel a script file and then want to continue it, you can do so by using the AutoCAD LT **RESUME** command.

Command: **RESUME**

The **RESUME** command can also be used if the script file has encountered an error that causes it to be suspended. The **RESUME** command will skip the command that caused the error and continue with the rest of the script file. If the error occurred when the command was in progress, use a leading apostrophe with the **RESUME** command (**'RESUME**). The prompt sequence to invoke the **RESUME** command in transparent mode is:

Command: **'RESUME**

COMMAND LINE SWITCHES

The command line switches can be used as arguments to the aclt.exe file that launches AutoCAD LT. You can also use the **Options** dialog box to set the environment or by adding a set of environment variables in the autoexec.bat file. The command line switches and environment variables override the values set in the **Options** dialog box for the current session only. These command line switches do not alter the system registry. The following is the list of the command line switches:

Switch	Function
/c	Controls where AutoCAD LT stores and searches for the hardware configuration file
/b	Designates a script to run after AutoCAD LT starts
/t	Specifies a template to use when creating a new drawing
/v	Designates a particular view of the drawing to be displayed upon start-up of AutoCAD LT

INVOKING A SCRIPT FILE WHEN LOADING AUTOCAD LT

The script files can also be run when loading AutoCAD LT, without getting into the drawing editor. The format of the command for running a script file when loading AutoCAD LT is:

Drive>AutoCAD LT 2000 [existing-drawing] [/t template] [/v view] /b script-file

In the following example, AutoCAD LT will open the existing drawing (Mydwg1) and then run the script file (Setup), Figure 24-3.

Example

C:\AutoCAD LT 2000>aclt.exe Mydwg1 /b Setup

Where **AutoCAD LT 2000** ------ AutoCAD LT 2000 subdirectory containing AutoCAD LT system files
aclt.exe ---------- aclt command to start AutoCAD LT
MyDwg1 -------- Existing drawing file name
Setup ------------ Name of the script file

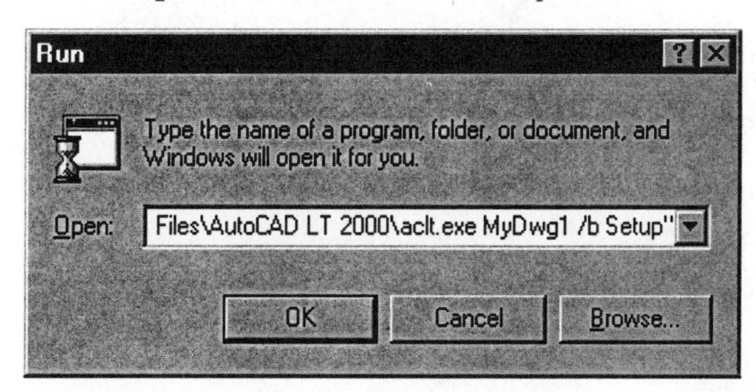

Figure 24-3 *Invoking script file when loading AutoCAD LT using the* ***Run*** *dialog box*

In the following example, AutoCAD LT will start a new drawing with the default name (Drawing), using the template file temp1, and then run the script file (Setup).

Example

C:\Program Files\AutoCAD LT 2000\aclt.exe /t temp1 /b Setup

Where **temp1** ------------ Existing template file name
Setup ------------ Name of the script file

or

C:\AutoCAD LT 2000\aclt.exe /t c:\AutoCAD LT 2000\mytemp\temp1 /b Setup
Where **c:\AutoCAD LT 2000\mytemp** - Path name

In the following example, AutoCAD LT will start a new drawing with the default name (Drawing), and then run the script file (Setup).

Example
C:\AutoCAD LT 2000\aclt.exe /b Setup
Where **Setup** ------------ Name of the script file

Here, it is assumed that the AutoCAD LT system files are loaded in the AutoCAD LT 2000 directory.

Note
For invoking a script file when loading AutoCAD LT, the drawing file or the template file specified in the command must exist in the search path. You cannot start a new drawing with a given name.

You should avoid abbreviations to prevent any confusion. For example, a C can be used as a close option when you are drawing lines. It can also be used as a command alias for drawing a circle. If you use both of these in a script file, it might be confusing.

Example 4

Write a script file that can be invoked when loading AutoCAD LT and create a drawing with the following setup (filename SCRIPT4.SCR):

Grid	3.0
Snap	0.5
Limits	0,0
	36.0,24.0
Zoom	All
Text height	0.25
Ltscale	3.0
Dimscale	3.0

Layers

Name	Color	Linetype
Obj	Red	Continuous
Cen	Yellow	Center
Hid	Blue	Hidden
Dim	Green	Continuous

First, write a script file and save the file under the name SCRIPT4.SCR. The following file is a listing of this script file that does the initial setup for a drawing:

```
GRID 2.0
SNAP 0.5
LIMITS 0,0 36.0,24.0 ZOOM ALL
SETVAR TEXTSIZE 0.25
LTSCALE 3
SETVAR DIMSCALE 3.0
LAYER NEW
OBJ,CEN,HID,DIM
L CENTER CEN
L HIDDEN HID
C RED OBJ
C YELLOW CEN
C BLUE HID
C GREEN DIM
```

(Blank line for ENTER.)

After you have written and saved the file, quit the drawing editor. To run the script file, SCRIPT4, select Start, Run, and then enter the following command line:

C:\AutoCAD LT2000\aclt.exe /t C:\EX4 D:\SCRIPT4

Where **aclt.exe** ---------- aclt.exe to load AutoCAD LT
EX4 -------------- Drawing filename
SCRIPT4 ------- Name of the script file

Here it is assumed that the template file EX4 is on C drive and script file SCRIPT4 is on D drive, Figure 24-4.

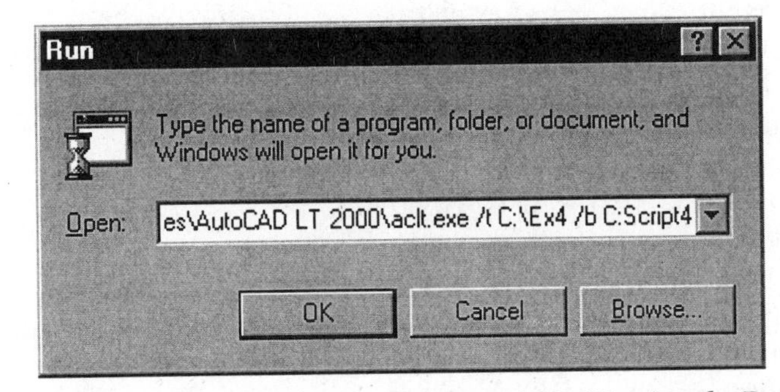

***Figure 24-4** Invoking script file when loading AutoCAD LT using the **Run** dialog box*

When you enter this line, AutoCAD LT is loaded and the file EX4.DWT is opened. The script file, SCRIPT4, is then automatically loaded and the commands defined in the file are executed.

In the following example, AutoCAD LT will start a new drawing with the default name (Drawing), and then run the script file (SCRIPT4). Here, it is assumed that the AutoCAD LT system files are loaded in the AutoCAD LT 2000 directory.

C:\AutoCAD LT 2000\aclt.exe /b SCRIPT4

Where **SCRIPT4** ------- Name of the script file

Example 5

Write a script file that will plot a 36" by 24" to maximum plot size, using your system printer/plotter. Use the Window option to select the drawing to be plotted.

Before writing a script file to plot a drawing, find out the plotter specifications that must be entered in the script file to obtain the desired output. To determine the prompt entries and their sequences to set up the plotter specifications, enter the AutoCAD LT **-PLOT** command at the keyboard. Note the entries you make and their sequence (the entries for your printer or plotter will probably be different). The following is a listing of the plotter specification:

Command: **-PLOT**
Detailed plot configuration? [Yes/No] <No>: **Yes**
Enter a layout name or [?] <Model>: **Model**
Enter an output device name or [?] <HP LaserJet 4000 Series PCL 6>: *Press ENTER.*
Enter paper size or [?] <Letter (8 1/2 x 11 in)>: *Press ENTER.*
Enter paper units [Inches/Millimeters] <Inches>: **I**
Enter drawing orientation [Portrait/Landscape] <Landscape>: *Press ENTER.*
Plot upside down? [Yes/No] <No>: **No**
Enter plot area [Display/Extents/Limits/View/Window] <Display>: **W**
Enter lower left corner of window <0.000000,0.000000>: **0,0**
Enter upper right corner of window <0.000000,0.000000>: **36,24**
Enter plot scale (Plotted Inches=Drawing Units) or [Fit] <Fit>: **F**
Enter plot offset (x,y) or [Center] <0.00,0.00>: **0,0**
Plot with plot styles? [Yes/No] <Yes>: **Yes**
Enter plot style table name or [?] (enter . for none) <>: .
Plot with lineweights? [Yes/No] <Yes>: **Yes**
Remove hidden lines? [Yes/No] <No>: **No**
Write the plot to a file [Yes/No] <N>: **No**
Save changes to model tab [Yes/No]? <N> **No**
Proceed with plot [Yes/No] <Y>: **Y**
Effective plotting area: 7.07 wide by 10.60 high
Plotting viewport 2.

Now you can write the script file by entering the responses to these prompts in the file. The following file is a listing of the script file that will plot a 36" by 24" drawing on 9" by 6" paper after making the necessary changes in the plot specifications. The comments on the right are not a part of the file.

```
Plot
y
                    (Blank line for ENTER, selects default layout.)
                    (Blank line for ENTER, selects default printer.)
                    (Blank line for ENTER, selects the default paper size.)
I
L
N
```

```
w
0,0
36,24
F
0,0
Y
.                    (Enter . for none)
Y
N
N
N
Y
```

Note

You can use a blank line to accept the default value for a prompt. A blank line in the script file will cause a Return. However, you must not accept the default plot specifications because the file might have been altered by another user or by another script file. Therefore, always enter the actual values in the file so that when you run a script file, it does not take the default values.

Exercise 1

Write a script file that will plot a 288' by 192' drawing on a 36" x 24" sheet of paper. The drawing scale is 1/8" = 1'. (The filename is SCRIPT6.SCR. In this example assume that AutoCAD LT is configured for the HPGL plotter and the plotter description is HPGL-Plotter.)

WHAT IS A SLIDE SHOW?

AutoCAD LT provides a facility using script files to combine the slides in a text file and display them in a predetermined sequence. In this way, you can generate a slide show for a slide presentation. You can also introduce a time delay in the display so that the viewer has enough time to view a slide.

A drawing or parts of a drawing can also be displayed by using the AutoCAD LT display commands. For example, you can use **ZOOM**, **PAN**, or other commands to display the details you want to show. If the drawing is very complicated, it takes quite some time to display the desired information, and it may not be possible to get the desired views in the right sequence. However, with slide shows you can arrange the slides in any order and present them in a definite sequence. In addition to saving time, this will also help to minimize the distraction that might be caused by constantly changing the drawing display. Also, some drawings are confidential in nature and you may not want to display some portions or views of them. By making slides, you can restrict the information that is presented through them. You can send a slide show to a client without losing control of the drawings and the information that is contained in them.

WHAT ARE SLIDES?

A **slide** is the snapshot of a screen display; it is like taking a picture of a display with a camera. The slides do not contain any vector information like AutoCAD LT drawings, which means that the entities do not have any information associated with them. For example, the slides do not retain any information about the layers, colors, linetypes, start point, or endpoint of a line or viewpoint. Therefore, slides cannot be edited like drawings. If you want to make any changes in the slide, you need to edit the drawing and then make a new slide from the edited drawing.

MSLIDE COMMAND

Slides are created by using the AutoCAD LT **MSLIDE** command. If **FILEDIA** is set to 0, the command will prompt you to enter the slide file name.

Command: **MSLIDE**
Enter name of slide file to create <Default>: *Slide file name.*

Example
Command: **MSLIDE**
Slide File: <Drawing1> SLIDE1
Where **Drawing1** ------- Default slide file name
SLIDE1 --------- Slide file name

In the preceding example, AutoCAD LT will save the slide file as **SLIDE1.SLD**. If **FILEDIA** is set to 1, the **MSLIDE** command displays the **Create Slide File** dialog box (Figure 24-5) on the screen. You can enter the slide file name in this dialog box.

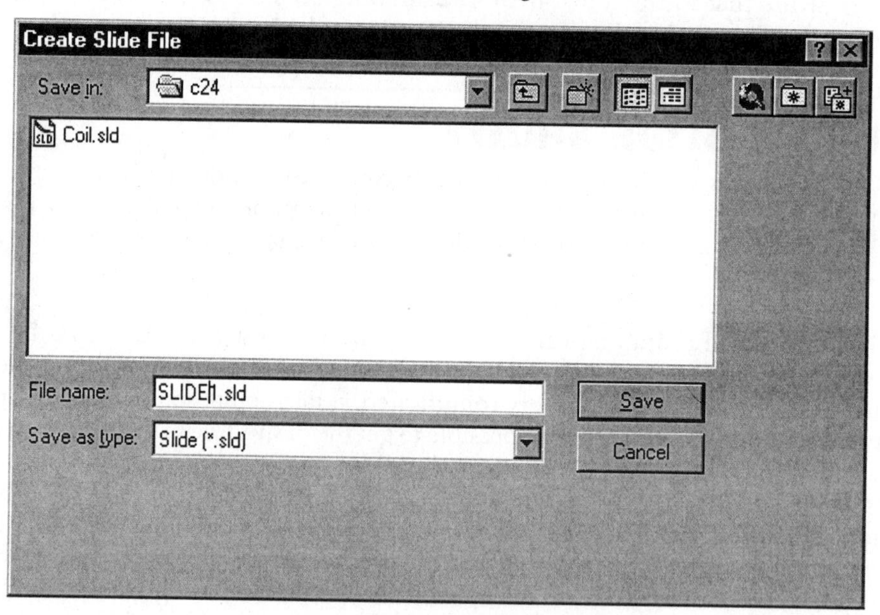

***Figure 24-5 Create Slide File** dialog box*

Note
*In model space, you can use the **MSLIDE** command to make a slide of the existing display in the current viewport.*

If you are in the paper space viewport, you can make a slide of the display in the paper space that includes any floating viewports.

When the viewports are not active, the ***MSLIDE*** *command will make a slide of the current screen display.*

VSLIDE COMMAND

To view a slide, use the **VSLIDE** command. If **FILEDIA** is 0, AutoCAD LT will prompt you to enter the slide filename. Enter the name of the slide to view and press ENTER. Do not enter the extension after the slide filename. AutoCAD LT automatically assumes the extension **.SLD**.

Command: **VSLIDE**
Enter name of slide file to create<Default>: *Enter name.*

Example
Command: **VSLIDE**
Slide file <Drawing1>: SLIDE1
Where **Drawing1** ------- Default slide file name
SLIDE1 --------- Name of slide file

If FILEDIA is set to 1, the VSLIDE command displays the **Select Slide File** dialog box (Figure 24-6). You can select the slide to view it.

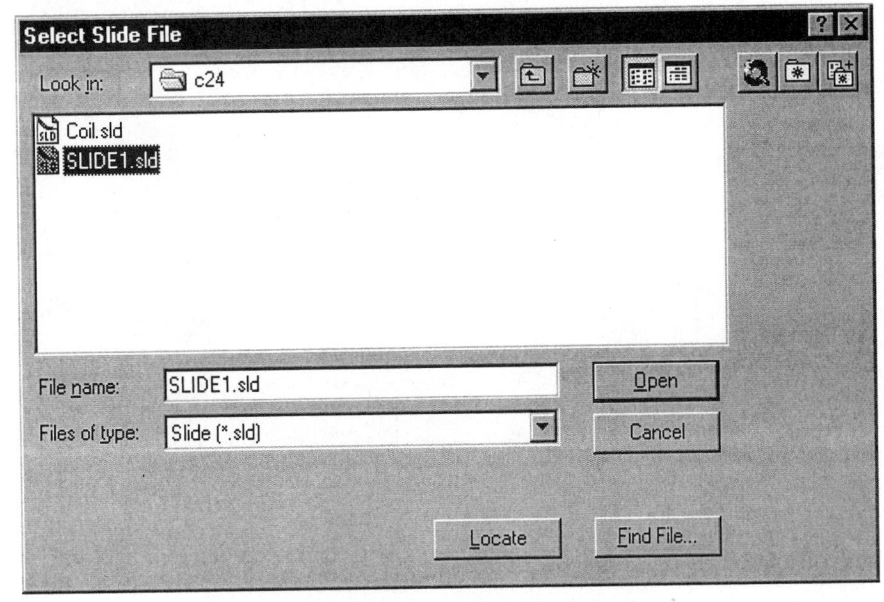

Figure 24-6 Select Slide File *dialog box*

Note

After viewing a slide, you can use the AutoCAD LT ***REDRAW*** *command to remove the slide display and return to the existing drawing on the screen.*

Any command that is automatically followed by a redraw will also display the existing drawing. For example, AutoCAD LT ***GRID, ZOOM ALL,*** *and* ***REGEN*** *commands will automatically return to the existing drawing on the screen.*

You can view the slides on high-resolution or low-resolution monitors. Depending on the resolution of the monitor, AutoCAD LT automatically adjusts the image. However, if you are using a high-resolution monitor, it is better to make the slides on the same monitor to take full advantage of that monitor.

Example 6

Write a script file that will generate a slide show of the following slide files, with a time delay of 15 seconds after every slide:

SLIDE1, SLIDE2, SLIDE3, SLIDE4

The first step in a slide show is to create the slides. Figure 24-7 shows the drawings that have been saved as slide files **SLIDE1, SLIDE2, SLIDE3,** and **SLIDE4**. The second step is to find out the sequence in which you want these slides to be displayed, with the necessary time delay, if any, between slides. Then you can use any text editor to write the script file with the extension **.SCR**.

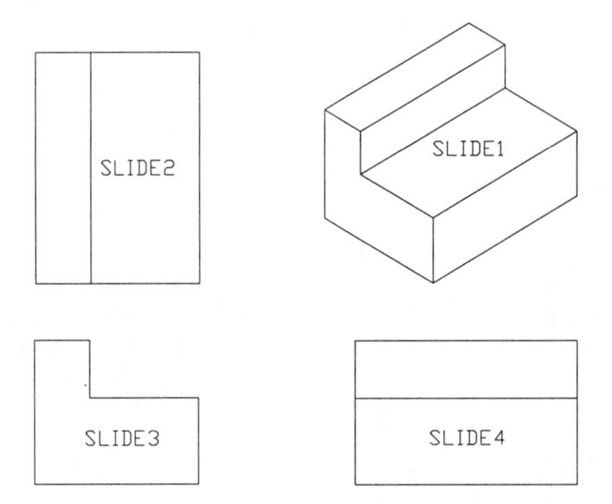

Figure 24-7 *Slides for slide show*

The following file is a listing of the script file that will create a slide show of the slides in Figure 24-7. The name of the script file is **SLDSHOW1**.

```
VSLIDE SLIDE1
DELAY 15000
```

```
VSLIDE SLIDE2
DELAY 15000
VSLIDE SLIDE3
DELAY 15000
VSLIDE SLIDE4
DELAY 15000
```

To run this slide show, type SCRIPT in response to the AutoCAD LT Command prompt. Now, select the name of the script file **(SLDSHOW1)** in the **Select Script File** dialog box. The slides will be displayed on the screen, with an approximate time delay of 15 seconds between them.

PRELOADING SLIDES

In the script file of Example 6, VSLIDE SLIDE1 in line 1 loads the slide file, **SLIDE1**, and displays it on screen. After a pause of 15,000 milliseconds, it starts loading the second slide file, **SLIDE2**. Depending on the computer and the disk access time, you will notice that it takes some time to load the second slide file; the same is true for other slides. To avoid the delay in loading slide files, AutoCAD LT has provided a facility to preload a slide while viewing the previous slide. This is accomplished by placing an asterisk (*) in front of the slide filename.

```
VSLIDE SLIDE1          (View slide, SLIDE1.)
VSLIDE *SLIDE2         (Preload slide, SLIDE2.)
DELAY 15000            (Delay of 15 seconds.)
VSLIDE                 (Display slide, SLIDE2.)
VSLIDE *SLIDE3         (Preload slide, SLIDE3.)
DELAY 15000            (Delay of 15 seconds.)
VSLIDE                 (Display slide, SLIDE3.)
VSLIDE *SLIDE4
DELAY 15000
VSLIDE
DELAY 15000
RSCRIPT                (Restart the script file.)
```

Example 7

Write a script file to generate a continuous slide show of the slide files SLD1, SLD2, SLD3, with a time delay of two seconds between slides. The slide files are located in different subdirectories, as shown in Figure 24-8. The subdirectory **SUBDIR1** is the current subdirectory.

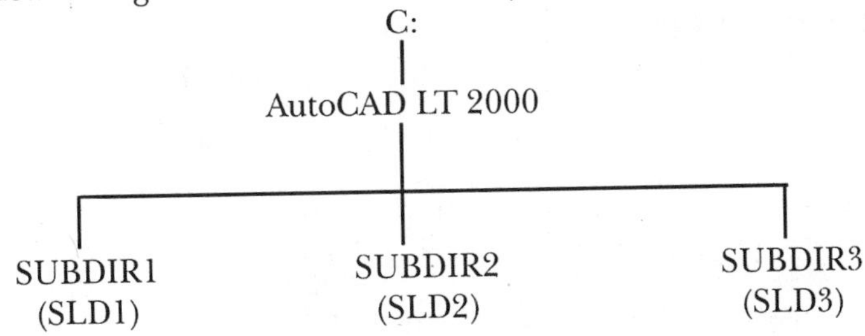

Figure 24-8 *Subdirectories of the C drive*

Where **C:** -------------------- *Root directory.*
AutoCAD LT 2000 ------ *Subdirectory where the AutoCAD LT files are loaded.*
SUBDIR1 ----------------- *Drawing subdirectory.*
SUBDIR2 ----------------- *Drawing subdirectory.*
SUBDIR3 ----------------- *Drawing subdirectory.*
SLD1 ------------------- *Slide file in SUBDIR1 subdirectory.*
SLD2 ------------------- *Slide file in SUBDIR2 subdirectory.*
SLD3 ------------------- *Slide file in SUBDIR3 subdirectory.*

The following file is the listing of the script files that will generate a slide show for the slides in Example 7:

```
VSLIDE SLD1
DELAY 2000
VSLIDE C:\AutoCAD LT 2000\SUBDIR2\SLD2
DELAY 2000
VSLIDE C:\AutoCAD LT 2000\SUBDIR3\SLD3
DELAY 2000
RSCRIPT
```

Line 1
VSLIDE SLD1
In this line, the AutoCAD LT command **VSLIDE** loads the slide file **SLD1**. Since, in this example, we are assuming you are in the subdirectory **SUBDIR1** and the first slide file, **SLD1**, is located in the same subdirectory, it does not require any path definition.

Line 2
DELAY 2000
This line uses the AutoCAD LT **DELAY** command to create a pause of approximately two seconds before the next slide is loaded.

Line 3
VSLIDE C:\AutoCAD LT 2000\SUBDIR2\SLD2
In this line, the AutoCAD LT command **VSLIDE** loads the slide file **SLD2**, located in the subdirectory **SUBDIR2**. If the slide file is located in a different subdirectory, you need to define the path with the slide file.

Line 5
VSLIDE C:\AutoCAD LT2000\SUBDIR3\SLD3
In this line, the **VSLIDE** command loads the slide file SLD3, located in the subdirectory SUBDIR3.

Line 7
RSCRIPT
In this line, the **RSCRIPT** command executes the script file again and displays the slides on the screen. This process continues indefinitely until the script file is canceled by pressing the ESC key or the BACKSPACE key.

Review Questions

SCRIPT FILES

1. AutoCAD LT has provided a facility of _________ that allows you to combine different AutoCAD LT commands and execute them in a predetermined sequence.

2. The _________ files can be used to generate a slide show, do the initial drawing setup, or plot a drawing to a predefined specification.

3. Before writing a script file, you need to know the AutoCAD LT _________ and the _________ required in response to the command prompts.

4. In a script file, you can _________ several statements in one line.

5. In a script file, the _________ are used to terminate a command or a prompt entry.

6. The AutoCAD LT _________ command is used to run a script file.

7. When you run a script file, the default script file name is the same as the _________ name.

8. When you run a script file, type the name of the script file without the file _________.

9. The limitation of script files is that all the information has to be contained ______ the file.

10. The AutoCAD LT _________ command allows you to re-execute a script file indefinitely until the command is canceled.

11. You cannot provide a _________ statement in a script file to terminate the file when a particular condition is satisfied.

12. The _________ command introduces a delay before the next command is executed.

13. The **DELAY** command is to be followed by _________ in milliseconds.

14. If the script file was canceled and you want to continue the script file, you can do so by using the AutoCAD LT ____________________ command.

SLIDE SHOWS

15. AutoCAD LT provides a facility through _________ files to combine the slides in a text file and display them in a predetermined sequence.

16. A _________ can also be introduced in the script file so that the viewer has enough time to view a slide.

17. Slides are the __________ of a screen display.

18. Slides do not contain any __________ information, which means that the entities do not have any information associated with them.

19. Slides __________ edited like a drawing.

20. Slides can be created using the AutoCAD LT __________ command.

21. Slide file names can be up to __________ characters long.

22. In model space, you can use the **MSLIDE** command to make a slide of the __________ display in the __________ viewport.

23. To view a slide, use the AutoCAD LT __________ command.

24. If you want to make any change in the slide, you need to ______________ the drawing, then make a new slide from the edited drawing.

Exercises

SCRIPT FILES

Exercise 2 *General*

Write a script file that will do the following initial setup for a new drawing:

Limits	0,0 24,18
Grid	1.0
Snap	0.25
Ortho	On
Snap	On
Zoom	All
Pline width	0.02
PLine	0,0 24,0 24,18 0,18 0,0
Units	Decimal units
	Number of decimal digits (2)
	Decimal degrees
	Number of decimal digits (2)
	Direction of 0 angle (3 o'clock)
	Angle measured counterclockwise
Ltscale	1.5

Layers

Name	Color	Linetype
Obj	Red	Continuous
Cen	Yellow	Center
Hid	Blue	Hidden
Dim	Green	Continuous

Exercise 3 *General*

Write a script file that will **PLOT** a given drawing according to the following specifications. (Use the plotter for which your system is configured and adjust the values accordingly.)

Plot, using the Window option
Window size (0,0 24,18)
Do not write the plot to file
Size in Inch units
Plot origin (0.0,0.0)
Maximum plot size (8.5,11 or the smallest size available on your printer/plotter)
90 degree plot rotation
No removal of hidden lines
Plotting scale (Fit)

Exercise 4 *General*

Write a script file that will continuously rotate a line in 10-degree increments around its midpoint (Figure 24-9). The time delay between increments is one second.

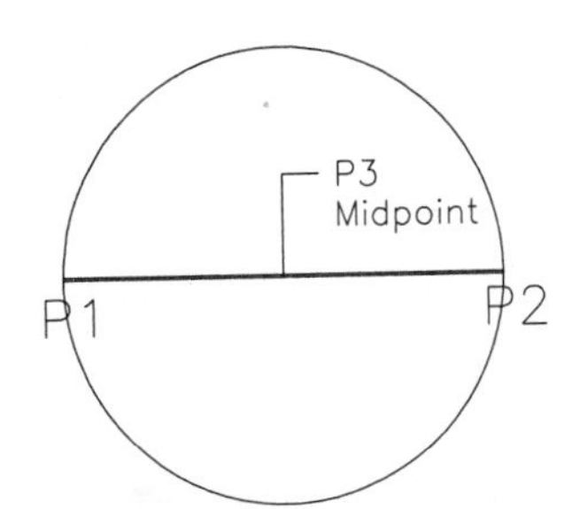

***Figure 24-9** Drawing for Exercise 4*

SLIDE SHOWS

Exercise 5 — General

Make the slides shown in Figure 24-10 and write a script file for a continuous slide show. Provide a time delay of 3 seconds after every slide. (You do not have to use the slides shown in Figure 24-10; you can use any slides of your choice.)

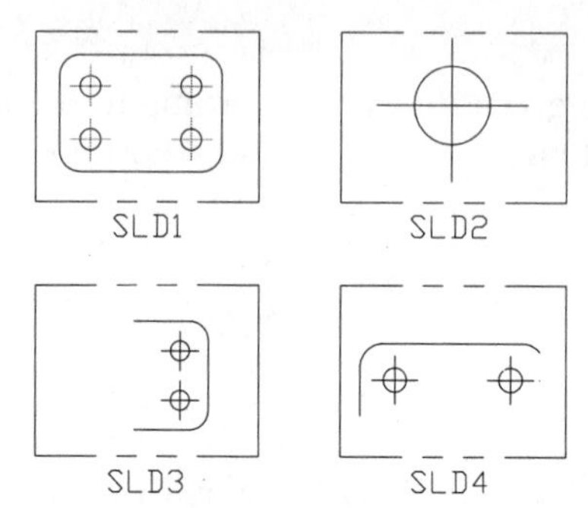

Figure 24-10 Slides for slide show

Chapter 25

Creating Linetypes and Hatch Patterns

Learning Objectives

After completing this chapter, you will be able to:

Create Linetypes:
- *Write linetype definitions.*
- *Create different linetypes.*
- *Create linetype files.*
- *Determine* ***LTSCALE*** *for plotting the drawing to given specifications.*
- *Define alternate linetypes and modify existing linetypes.*
- *Create string complex linetypes.*

Create Hatch Patterns:
- *Understand hatch pattern definition.*
- *Create new hatch patterns.*
- *Determine the effect of angle and scale factor on hatch.*
- *Create hatch patterns with multiple descriptors.*
- *Define custom hatch pattern file.*

STANDARD LINETYPES

The AutoCAD LT software package comes with a library of standard linetypes that has several linetypes, including ISO linetypes. These linetypes are saved in the **ACLT.LIN** file. You can modify existing linetypes or create new ones.

LINETYPE DEFINITION

All linetype definitions consist of two parts: **header line and pattern line.**

Header Line

The **header line** consists of an asterisk (*) followed by the name of the linetype and the linetype description. The name and linetype description should be separated by a comma. If there is no description, the comma that separates the linetype name and the description is not required.

The format of the header line is:

*** Linetype Name, Description**

Example

***HIDDENS,__ __ __ __ __ __**

Where * ------------------- Asterisk sign
HIDDENS ------ Linetype name
, -------------------- Comma
__ __ __ __ ------ Linetype description

All linetype definitions require a linetype name. When you want to load a linetype or assign a linetype to an object, AutoCAD LT recognizes the linetype by the name you have assigned to the linetype definition. The names of the linetype definition should be selected to help the user recognize the linetype by its name. For example, a linetype name LINEFCX does not give the user any idea about the type of line. However, a linetype name like DASHDOT gives a better idea about the type of line that a user can expect.

The linetype description is a textual representation of the line. This representation can be generated by using dashes, dots, and spaces at the keyboard. The graphic is used by AutoCAD LT when you want to display the linetypes on the screen by using the AutoCAD LT **LINETYPE** command with the ? option or by using the dialog box. The linetype description cannot exceed 47 characters.

Pattern Line

The **pattern line** contains the definition of the line pattern. The definition of the line pattern consists of the alignment field specification and the linetype specification. The alignment field specification and the linetype specification are separated by a comma.

The format of the pattern line is:

Alignment Field Specification, Linetype Specification

Example

A,.75,-.25,.75

Where **A** ------------------ Alignment field specification
, -------------------- Comma
.75,-.25,.75 ----- Linetype specification

The letter used for alignment field specification is A. This is the only alignment field supported by AutoCAD LT; therefore, the pattern line will always start with the letter A. The linetype specification defines the configuration of the dash-dot pattern to generate a line. The maximum number for dash length specification in the linetype is 12, provided the linetype pattern definition fits on one 80-character line.

ELEMENTS OF LINETYPE SPECIFICATION

All linetypes are created by combining the basic elements in a desired configuration. There are three basic elements that can be used to define a linetype specification.

Dash	(Pen down)
Dot	(Pen down, 0 length)
Space	(Pen up)

Example

______ . ______ . ______ . ______

Where . -------------------- Dot (pen down with 0 length)
Blank space --------------- Space (pen up)
______ -------------------- Dash (pen down with specified length)

The dashes are generated by defining a positive number. For example, .5 will generate a dash 0.5 units long. Similarly, spaces are generated by defining a negative number. For example, -.2 will generate a space 0.2 units long. The dot is generated by defining a 0 length.

Example

A,.5,-.2,0,-.2,.5

Where **0** ------------------- Dot (zero length)
-.2 ----------------- Length of space (pen up)
.5 ------------------ Length of dash (pen down)

CREATING LINETYPES

Before creating a linetype, you need to decide the type of line you want to generate. Draw the line on a piece of paper and measure the length of each element that constitutes the line. You need to define only one segment of the line, because the pattern is repeated when you draw a line. Linetypes can be created or modified by one of the following methods:

Using the AutoCAD LT LINETYPE command
Using a text editor (such as Notepad).

Consider the following example, which creates a new linetype, first using the AutoCAD LT **LINETYPE** command and then using a text editor.

Example 1

Using the AutoCAD LT **LINETYPE** command, create linetype DASH3DOT (Figure 25-1) with the following specifications:

Length of the first dash 0.5
Blank space 0.125
Dot
Blank space 0.125
Dot
Blank space 0.125
Dot
Blank space 0.125

Using the AutoCAD LT Linetype Command

To create a linetype using the AutoCAD LT **LINETYPE** command, first make sure that you are in the drawing editor. Then enter the **-LINETYPE** command and select the **Create** option to create a linetype, Figure 25-1.

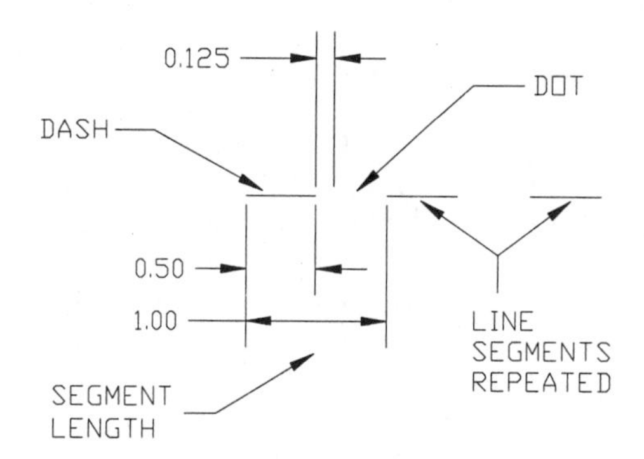

Figure 25-1 Linetype specifications of DASH3DOT

Command: **-LINETYPE**
Enter an option [?/Create/Load/Set]: **C**

Enter the name of the linetype and the name of the library file in which you want to store the definition of the new linetype.

Enter name of linetype to create: **DASH3DOT**

If **FILEDIA**=1, the **Create or Append Linetype File** dialog box (Figure 25-2) will appear on the screen. If **FILEDIA**=0, AutoCAD LT will prompt you to enter the name of the file.

Enter linetype file name for new linetype definition <default>: **ACLT**

If the linetype already exists, the following message will be displayed on the screen:

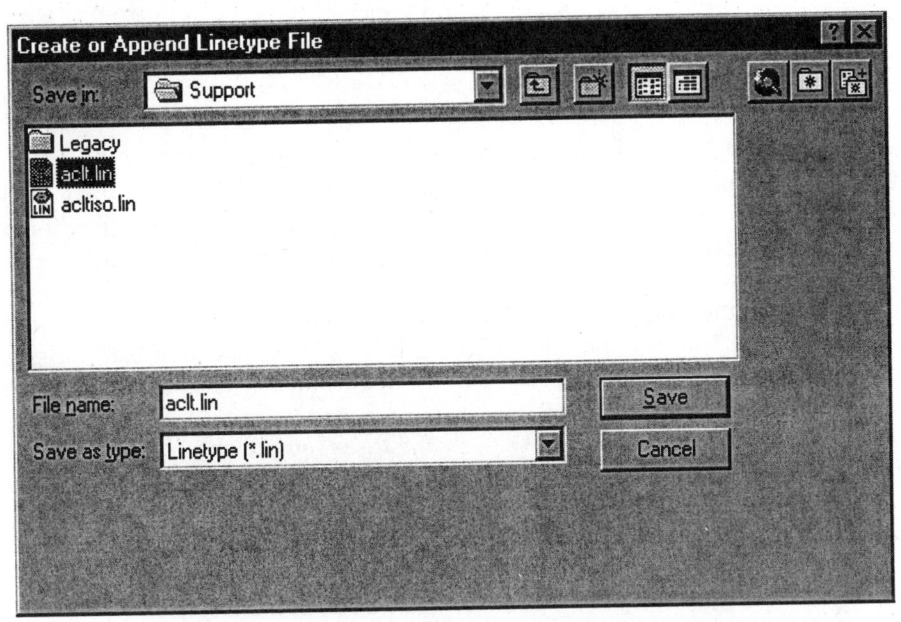

Figure 25-2 Create or Append Linetype File *dialog box*

Wait, checking if linetype already defined...
"Linetype" already exists in this file. Current definition is:
alighment, dash-1, dash-2, ____.
Overwrite?<N>

If you want to redefine the existing line style, enter Y; otherwise, type N or press RETURN to choose the default value of N. You can then repeat the process with a different name of the linetype.

After entering the name of the linetype and the library file name, AutoCAD LT will prompt you to enter the descriptive text and the pattern of the line.

Descriptive text: ***DASH3DOT,____ . . . ____ . . . ____**
Enter linetype pattern (on next line):
A,.5,-.125,0,-.125,0,-.125,0,-.125

Descriptive Text

***DASH3DOT,____ . . . ____ . . . ____**

For the descriptive text, you have to type an asterisk (*) followed by the name of the linetype. For Example 1, the name of the linetype is DASH3DOT. The name *DASH3DOT can be followed by the description of the linetype; the length of this description cannot exceed 47 characters. In this example, the description is dashes and dots ____ . . . ____. It could be any text or alphanumeric string. The description is displayed on the screen when you list the linetypes.

Pattern

A,.5,-.125,0,-.125,0,-.125,0,-.125

The line pattern should start with an alignment definition. Currently, AutoCAD LT supports only one type of alignment—A. Therefore, it is automatically displayed on the screen when you select the **LINETYPE** command with the **Create** option. After entering **A** for pattern alignment, you must define the pen position. A positive number (.5 or 0.5) indicates a "pen-down" position, and a negative number (-.25 or -0.25) indicates a "pen-up" position. The length of the dash or the space is designated by the magnitude of the number. For example, 0.5 will draw a dash 0.5 units long, and -0.25 will leave a blank space of 0.25 units. A dash length of 0 will draw a dot (.). Here are the pattern definition elements for Example 1:

.5	**pen down**	**0.5 units long dash**
-.125	**pen up**	**.125 units blank space**
0	**pen down**	**dot**
-.125	**pen up**	**.125 units blank space**
0	**pen down**	**dot**
-.125	**pen up**	**.125 units blank space**
0	**pen down**	**dot**
-.125	**pen up**	**.125 units blank space**

After you enter the pattern definition, the linetype (DASH3DOT) is automatically saved in the **ACLT.LIN** file. The linetype (DASH3DOT) can be loaded using the AutoCAD LT **LINETYPE** command and selecting the **Load** option.

Note

The name and the description must be separated by a comma (,). The description is optional. If you decide not to give one, omit the comma after the linetype name DASH3DOT.

Using a Text Editor

You can also use a text editor (like Notepad) to create a new linetype. Using the text editor, load the file and insert the lines that define the new linetype. The following file is a partial listing of the ACLT.LIN file after adding a new linetype to the file:

```
*BORDER, Border__ __ . __ __ . __ __ . __ __ . __ __ . __ __ .
A,.5,-.25,.5,-.25,0,-.25
*BORDER2, Border(.5x)__.__.__.__.__.__.__.__.__.__.__.__.__.
A,.25,-.125,.25,-.125,0,-.125
*BORDERX2, Border(2x)____ ____ . ____ ____ . ____ ____ .
A,1.0,-.5,1.0,-.5,0,-.5

*CENTER, Center____ _ ____ _ ____ _ ____ _ ____ _ ____
A,1.25,-.25,.25,-.25
*CENTER2, Center(.5x)___ _ ___ _ ___ _ ___ _ ___ _ ___ _ ___
A,.75,-.125,.125,-.125
```

```
*CENTERX2,________ __ ________ __ ________ __ _
A,2.5,-.5,.5,-.5
*DASHDOT,__ . __ . __ . __ . __ . __ . __ . __ .
A,.5,-.25,0,-.25
|
|
|
*HIDDEN2,Hidden(.5x)_ _ _ _ _ _ _ _ _ _ _ _ _ _ _ _ _ _ _ _ _ _ _ _
A,.125,-.0625
*HIDDENX2,Hidden(2x)____ ____ ____ ____ ____ ____ ____ ____ ____ ____
A,.5,-.25
*PHANTOM,Phantom______ __ __ ______ __ __ ______ __ __
A,1.25,-.25,.25,-.25,.25,-.25
|
|
|
*HOT_WATER_SUPPLY,Hot water supply ---- HW ---- HW ---- HW ----
A,.5,-.2,["HW",STANDARD,S=.1,R=0.0,X=-0.1,Y=-.05],-.2
*GAS_LINE,Gas line ----GAS----GAS----GAS----GAS----GAS----GAS--
A,.5,-.2,["GAS",STANDARD,S=.1,R=0.0,X=-0.1,Y=-.05],-.25
*ZIGZAG,Zig zag /\/\/\/\/\/\/\/\/\/\/\/\/\/\/\/\/\/\/\/\/\/\/\/\/\/\/\/\/\/\/\/\/\/
A,.0001,-.2,[ZIG,ltypeshp.shx,x=-.2,s=.2],-.4,[ZIG,ltypeshp.shx,r=180,x=.2,s=.2],-.2
*DASH3DOT,____ . . . ____ . . . ____
A,.5,-.125,0,-.125,0,-.125,0,-.125
```

The last two lines of this file define the new linetype, DASH3DOT. The first line contains the name DASH3DOT and the description (___ . . .___). The second line contains the alignment and the pattern definition. Save the file and then load the linetype using the AutoCAD LT **LINETYPE** command with the **Load** option. Make the linetype current and draw some lines. The lines and polylines that this linetype will generate are shown in Figure 25-3.

Figure 25-3 *Lines created by linetype DASH3DOT*

Note

If you change the LTSCALE factor, all lines in the drawing are affected by the new ratio.

CREATING LINETYPE FILES

You can start a new linetype file and then add the line definitions to this file. Use any text editor like Notepad to start a new file (NEWLT.LIN) and then add the following two lines to the file to define the DASH3DOT linetype.

***DASH3DOT,___ . . . ___ . . . ___**
A,.5,-.125,0,-.125,0,-.125,0,-.125

You can load the DASH3DOT linetype from this file using AutoCAD LT's **LINETYPE** command and then selecting the LOAD option.

Command: **-LINETYPE**
Enter an option [?/Create/Load/Set]: **L**
Enter linetype(s) to load: **DASHDOT**
Enter name of linetype file to search <default>: **NEWLT**

You can also load the linetype by using the **Linetype Manager** dialog box that can be invoked by entering **LINETYPE** at the Command prompt or by choosing **Linetype** from the **Format** menu and choosing the Load button. The **Load and Reload Linetypes** dialog box is displayed from where you can choose the **File** button to display the **Select Linetype File** dialog box (Figure 25-4). Select NEWLT.LIN file from the specified directory.

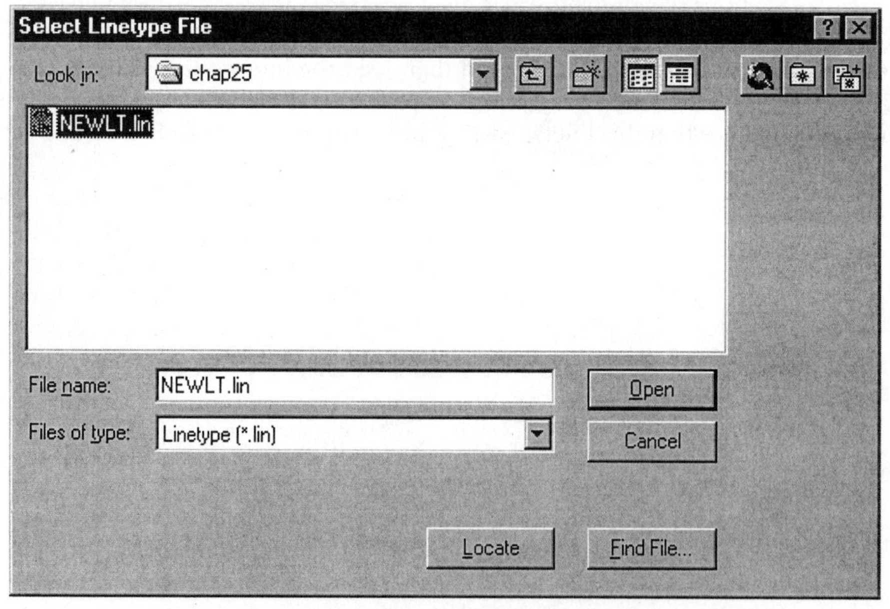

***Figure 25-4** Loading linetype using **Select Linetype File** dialog box*

ALIGNMENT SPECIFICATION

The alignment specifies the pattern alignment at the start and the end of the line, circle, or arc. In other words, the line would always start and end with the dash (___). The alignment definition "A" requires that the first element be a dash or dot (pen down), followed by a negative (pen up) segment. The minimum number of dash segments for alignment A is two. If there is not enough space for the line, AutoCAD LT will draw a continuous line.

For example, in the linetype DASH3DOT of Example 1, the length of each line segment is 1.0 (.5 + .125 + .125 + .125 + .125 = 1.0). If the length of the line drawn is less than 1.00, the line will be drawn as a continuous line (Figure 30-5). If the length of the line is 1.00 or greater, the line will be drawn according to DASH3DOT linetype. AutoCAD LT automatically adjusts the length of the dashes and the line will always start and end with a dash. The length of the starting and ending dashes will be at least half the length of the dash as specified in the file. If the length of the dash as specified in the file is 0.5, the length of the starting and ending dashes will be at least 0.25. To fit a line that starts and ends with a dash, the length of these dashes can also increase as shown in Figure 25-5.

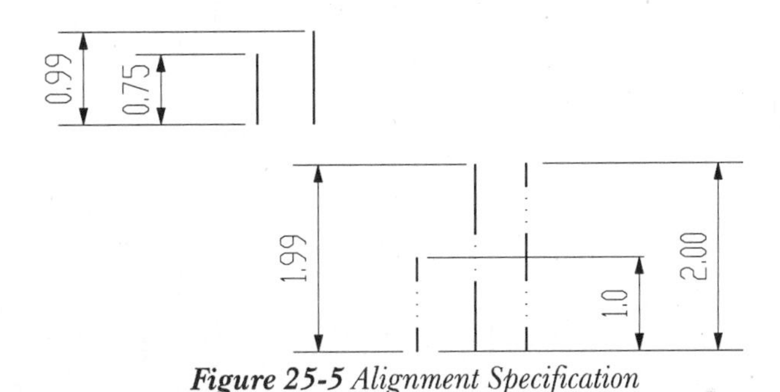

Figure 25-5 Alignment Specification

LTSCALE COMMAND

As we mentioned previously, the length of each line segment in the DASH3DOT linetype is 1.0 (.5 + .125 + .125 + .125 + .125 = 1.0). If you draw a line that is less than 1.0 units long, AutoCAD LT will draw a single dash that looks like a continuous line (Figure 25-6). This problem can be rectified by changing the linetype scale factor variable **LTSCALE** to a smaller value. This can be accomplished by using AutoCAD LT's **LTSCALE** command:

Command: **LTSCALE**
Enter new linetype scale factor <default>: *New value.*

The default value of **LTSCALE** variable is 1.0. If it is changed to 0.75, the length of each segment is reduced by 0.75 (1.0 x 0.75 = 0.75). If you draw a line 0.75 units or longer, it will be drawn according to definition of DASH3DOT (___ . . . ___) (Figures 25-6, 25-7, and 25-8).

The appearance of the lines is also affected by the limits of the drawing. Most of the AutoCAD LT linetypes work fine for a drawing that has the limits 12,9. Figure 25-9 shows a line of linetype DASH3DOT that is four units long and the limits of the drawing are 12,9. If you

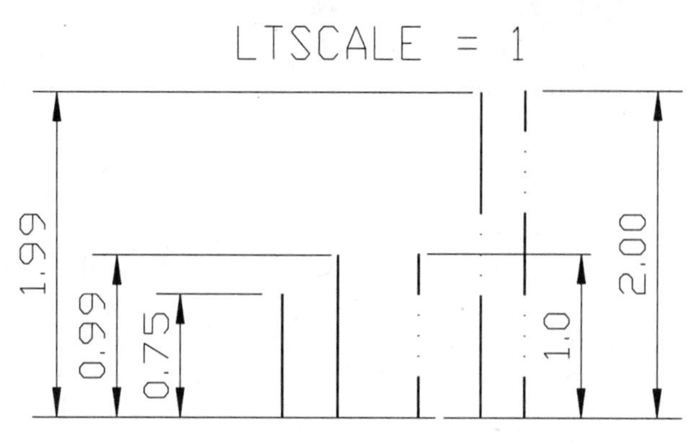

Figure 25-6 Alignment when Ltscale = 1

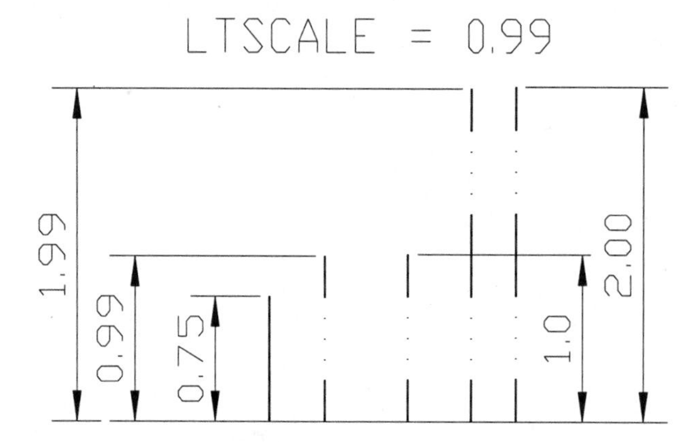

Figure 25-7 Alignment when Ltscale = 0.99

increase the limits to 48,36 the lines will appear as continuous lines. If you want the line to appear same as before **on the screen**, the LTSCALE should be changed. Since the limits of the drawing have increased four times, the LTSCALE should also increase by the same amount. If you change the scale factor to four, the line segments will also increase by a factor of four. As shown in Figure 25-9, the length of starting and the ending dash has increased to one unit.

In general, the approximate LTSCALE factor for **screen display** can be obtained by dividing the X-limit of the drawing by the default X-limit (12.00).

LTSCALE factor for SCREEN DISPLAY = X-limits of the drawing/12.00

Example

Drawing limits are 48,36
LTSCALE factor for screen display= 48/12 = 4

Drawing sheet size is 36,24 and scale is 1/4" = 1'
LTSCALE factor for screen display = 12 x 4 x (36 / 12) = 144

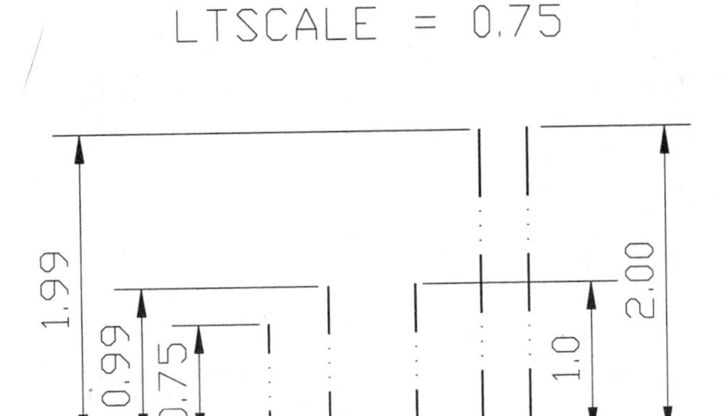

Figure 25-8 Alignment when Ltscale = 0.75

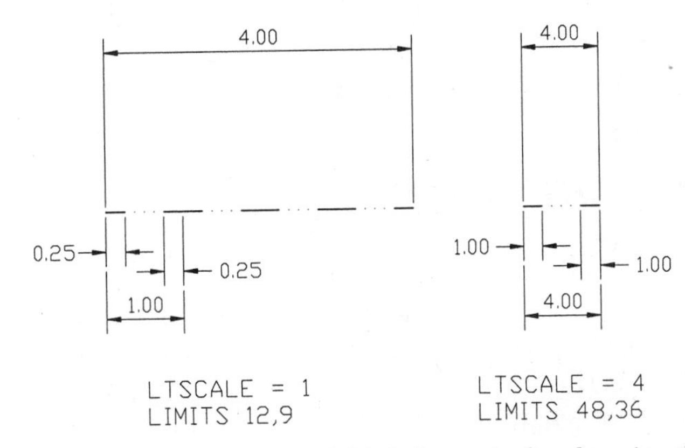

Figure 25-9 *Linetype DASH3DOT before and after changing the LTSCALE factor*

LTSCALE FACTOR FOR PLOTTING

The LTSCALE factor for plotting depends on the size of the sheet you are using to plot the drawing. For example, if the limits are 48 by 36, the drawing scale is 1:1, and you want to plot the drawing on a 48" by 36" size sheet, the LTSCALE factor is 1. If you check the specification of a hidden line in the ACLT.LIN file, the length of each dash is 0.25. Therefore, when you plot a drawing with 1:1 scale, the length of each dash in a hidden line is 0.25.

However, if the drawing scale is 1/8" = 1' and you want to plot the drawing on a 48" by 36" paper, the LTSCALE factor must be 96 (8 x 12 = 96). The length of each dash in the hidden line will increase by a factor of 96 because the LTSCALE factor is 96. Therefore, the length of each dash will be 24 units (0.25 x 96 = 24). At the time of plotting, the scale factor for plotting must be 1:96 to plot the 384' by 288' drawing on a 48" by 36" size paper. Each dash of the hidden line that was 24" long on the drawing will be 0.25 (24/96 = 0.25) inch long when plotted. Similarly, if the desired text size on the paper is 1/8", the text height in the drawing must be 12" (1/8 x 96 = 12").

Ltscale Factor for PLOTTING = Drawing Scale

Sometimes your plotter may not be able to plot a 48" by 36" drawing or you might like to decrease the size of the plot so that the drawing fits within a specified area. To get the correct dash lengths for hidden, center, or other lines, you must adjust the LTSCALE factor. For example, if you want to plot the previously mentioned drawing in a 45" by 34" area, the correction factor is:

Correction factor	= 48/45
	= 1.0666
New LTSCALE factor	= LTSCALE factor x Correction factor
	= 96 x 1.0666
	= 102.4

New Ltscale Factor for PLOTTING = Drawing Scale x Correction Factor

Note

If you change the LTSCALE factor, all lines in the drawing are affected by the new ratio.

ALTERNATE LINETYPES

One of the problems with the LTSCALE factor is that it affects all the lines in the drawing. As shown in Figure 25-10(a), the length of each segment in all DASH3DOT type lines is approximately equal, no matter how long the lines. You might want to have a small segment length if the lines are small and a longer segment length if the lines are long. You can accomplish this by using CELTSCALE (discussed later in this chapter) or by defining an alternate linetype with a different segment length. For example, you can define a linetype DASH3DOT and DASH3DOTX with different line pattern specifications.

```
*DASH3DOT,____ . . . ____ . . . ____ . . . ____
A,0.5,-.125,0,-.125,0,-.125,0,-.125
*DASH3DOTX,_______ . . . _________
A,1.0,-.25,0,-.25,0,-.25,0,-.25
```

In DASH3DOT linetype the segment length is one unit, whereas in DASH3DOTX linetype the segment length is two units. You can have several alternate linetypes to produce the lines with different segment lengths. Figure 25-10(b) shows the lines generated by DASH3DOT and DASH3DOTX.

Note

Although you might have used different linetypes with different segment lengths, the lines will be affected equally when you change the LTSCALE factor. For example, if the LTSCALE factor is 0.5, the segment length of DASH3DOT line will be 0.5 and the segment length of DASH3DOTX will be 1.0 units.

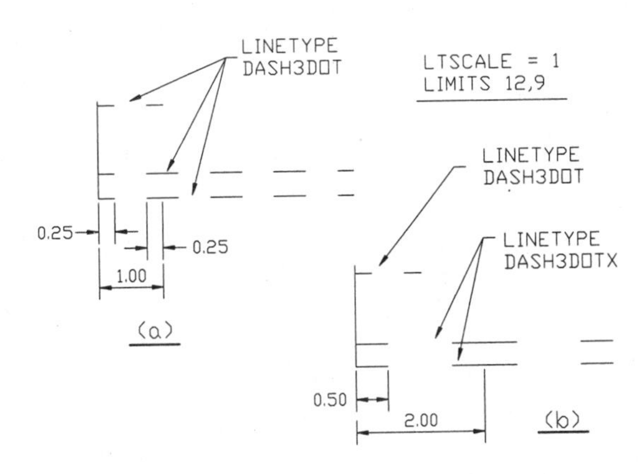

Figure 25-10 Linetypes generated by DASH3DOT and DASH3DOTX

MODIFYING LINETYPES

You can also modify the linetypes that are defined in the ACLT.LIN file. You need a text editor, such as Notepad, to modify the linetype. For example, if you want to change the dash length of the border linetype from 0.5 to 0.75, load the file, then edit the pattern line of the border linetype. The following file is a partial listing of the ACLT.LIN file after changing the Border and Centerx2 linetypes.

```
;; AutoCAD LT Linetype Definition file
;; Version 2.0
;; Copyright 1991, 1992, 1993, 1994, 1996 by Autodesk, Inc.
;;
*BORDER,Border__ __ . __ __ . __ __ . __ __ . __ __ . __ __ . __ __ . __ __ .
A,.75,-.25,.75,-.25,0,-.25
*BORDER2,Border (.5x) __.__.__.__.__.__.__.__.__.__.
A,.25,-.125,.25,-.125,0,-.125
*BORDERX2,Border (2x) ____ ____ . ____ ____ . ___
A,1.0,-.5,1.0,-.5,0,-.5

*CENTER,Center ____ _ ____ _ ____ _ ____ _ ____
A,1.25,-.25,.25,-.25
*CENTER2,Center (.5x) ___ _ ___ _ ___ _ ___ _ ___
A,.75,-.125,.125,-.125
*CENTERX2,Center (2x) ________ __ ________ __ _____
A,3.5,-.5,.5,-.5

*DASHDOT,Dash dot __ . __ . __ . __ . __ . __ . __ . __
A,.5,-.25,0,-.25
*DASHDOT2,Dash dot (.5x) _._._._._._._._._._._._._._.
A,.25,-.125,0,-.125
*DASHDOTX2,Dash dot (2x) ____ . ____ . ____ . ___
A,1.0,-.5,0,-.5
```

```
*DASHED,Dashed __ __ __ __ __ __ __ __ __ __ __ __ __ _
A,.5,-.25
*DASHED2,Dashed (.5x) _ _ _ _ _ _ _ _ _ _ _ _ _ _ _ _ _ _
A,.25,-.125
*DASHEDX2,Dashed (2x) ____  ____  ____  ____  ____  ___
A,1.0,-.5

*DIVIDE,Divide ____ . . ____ . . ____ . . ____ . . ____
A,.5,-.25,0,-.25,0,-.25
*DIVIDE2,Divide (.5x) __..__..__..__..__..__..__..__.._
A,.25,-.125,0,-.125,0,-.125
*DIVIDEX2,Divide (2x) ________  . .  ________  . . _
A,1.0,-.5,0,-.5,0,-.5

*DOT,Dot . . . . . . . . . . . . . . . . . . . . . . . .
A,0,-.25
*DOT2,Dot (.5x) ........................................
A,0,-.125
*DOTX2,Dot (2x) .  .  .  .  .  .  .  .  .  .  .  .  .  .
A,0,-.5

*HIDDEN,Hidden __ __ __ __ __ __ __ __ __ __ __ __ __ __
A,.25,-.125
*HIDDEN2,Hidden (.5x) _ _ _ _ _ _ _ _ _ _ _ _ _ _ _ _ _ _ _
A,.125,-.0625
*HIDDENX2,Hidden (2x) ____ ____ ____ ____ ____ ____ ____
A,.5,-.25

*PHANTOM,Phantom ______ __ __ ______ __ __ ______
A,1.25,-.25,.25,-.25,.25,-.25
*PHANTOM2,Phantom (.5x) ___ _ _ ___ _ _ ___ _ _ ___ _ _
A,.625,-.125,.125,-.125,.125,-.125
*PHANTOMX2,Phantom (2x) ____________  ____  ____  _
A,2.5,-.5,.5,-.5,.5,-.5
;;  ISO 128 (ISO/DIS 12011) linetypes
;;  The size of the line segments for each defined ISO line, is
;;  defined for an usage with a pen width of 1 mm. To use them with
;;  the other ISO predefined pen widths, the line has to be scaled
;;  with the appropriate value (e.g. pen width 0,5 mm -> ltscale 0.5).
;;
*ACAD_ISO02W100,ISO dash __ __ __ __ __ __ __ __ __ __ __ __ __
A,12,-3
*ACAD_ISO03W100,ISO dash space __    __    __    __    __    __
A,12,-18
*ACAD_ISO04W100,ISO long-dash dot ____ . ____ . ____ . ____ . _
A,24,-3,.5,-3
*ACAD_ISO05W100,ISO long-dash double-dot ____ .. ____ .. ____ .
```

```
A,24,-3,.5,-3,.5,-3
*ACAD_ISO06W100,ISO long-dash triple-dot ____ ... ____ ... ____
A,24,-3,.5,-3,.5,-3,.5,-3
*ACAD_ISO07W100,ISO dot . . . . . . . . . . . . . . . . . . . .
A,.5,-3
*ACAD_ISO08W100,ISO long-dash short-dash ____ __ ____ __ ____ _
A,24,-3,6,-3
*ACAD_ISO09W100,ISO long-dash double-short-dash ____ __ __ ____
A,24,-3,6,-3,6,-3
*ACAD_ISO10W100,ISO dash dot __ . __ . __ . __ . __ . __ . __ .
A,12,-3,.5,-3
*ACAD_ISO11W100,ISO double-dash dot __ __ . __ __ . __ __ . __ _
A,12,-3,12,-3,.5,-3
*ACAD_ISO12W100,ISO dash double-dot __ . . __ . . __ . . __ . .
A,12,-3,.5,-3,.5,-3
*ACAD_ISO13W100,ISO double-dash double-dot __ __ . . __ __ . . _
A,12,-3,12,-3,.5,-3,.5,-3
*ACAD_ISO14W100,ISO dash triple-dot __ . . . __ . . . __ . . . _
A,12,-3,.5,-3,.5,-3,.5,-3
*ACAD_ISO15W100,ISO double-dash triple-dot __ __ . . . __ __ . .
A,12,-3,12,-3,.5,-3,.5,-3,.5,-3
;;  Complex linetypes
;;
;;  Complex linetypes have been added to this file.
;;  These linetypes were defined in LTYPESHP.LIN in
;;  Release 13, and are incorporated in ACAD.LIN in
;;  Release 14.
;;
;;  These linetype definitions use LTYPESHP.SHX.
;;
*FENCELINE1,Fenceline circle ----0-----0----0-----0----0-----0--
A,.25,-.1,[CIRC1,ltypeshp.shx,x=-.1,s=.1],-.1,1
*FENCELINE2,Fenceline square ----[]-----[]----[]-----[]----[]---
A,.25,-.1,[BOX,ltypeshp.shx,x=-.1,s=.1],-.1,1
*TRACKS,Tracks -|-|-|-|-|-|-|-|-|-|-|-|-|-|-|-|-|-|-|-|-|-|-|-|-
A,.15,[TRACK1,ltypeshp.shx,s=.25],.15
*BATTING,Batting SSSSSSSSSSSSSSSSSSSSSSSSSSSSSSSSSSSSSSSSSSSSSSS
A,.0001,-.1,[BAT,ltypeshp.shx,x=-.1,s=.1],-.2,[BAT,ltypeshp.shx,r=180,x=.1,s=.1],-.1
*HOT_WATER_SUPPLY,Hot water supply ---- HW ---- HW ---- HW ----
A,.5,-.2,["HW",STANDARD,S=.1,R=0.0,X=-0.1,Y=-.05],-.2
*GAS_LINE,Gas line ----GAS----GAS----GAS----GAS----GAS----GAS--
A,.5,-.2,["GAS",STANDARD,S=.1,R=0.0,X=-0.1,Y=-.05],-.25
*ZIGZAG,Zig zag /\/\/\/\/\/\/\/\/\/\/\/\/\/\/\/\/\/\/\/\/\/\/
A,.0001,-.2,[ZIG,ltypeshp.shx,x=-.2,s=.2],-.4,[ZIG,ltypeshp.shx,r=180,x=.2,s=.2],-.2
```

Chapter 25

Example 2

Create a new file, **NEWLINET.LIN**, and define a linetype, VARDASH, with the following specifications:

Length of first dash 1.0
Blank space 0.25
Length of second dash 0.75
Blank space 0.25
Length of third dash 0.5
Blank space 0.25
Dot
Blank space 0.25
Length of next dash 0.5
Blank space 0.25
Length of next dash 0.75

Use a text editor and insert the following lines that define the new linetype VARDASH. Save the file as NEWLINET.LIN.

```
*VARDASH,———— ——— —— . —— ——— ————
A,1,-.25,.75,-.25,.5,-.25,0,-.25,.5,-.25,.75,-.25
```

The type of lines that this linetype will generate are shown in Figure 25-11.

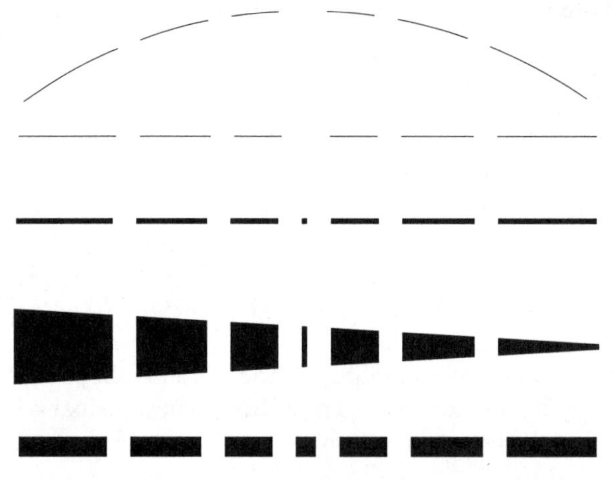

Figure 25-11 Lines generated by linetype VARDASH

CURRENT LINETYPE SCALING (CELTSCALE)

Like **LTSCALE**, the **CELTSCALE** system variable controls the linetype scaling. The difference is that **CELTSCALE** determines the current linetype scaling. For example, if you set the **CELTSCALE** to 0.5, all lines drawn after setting the new value for **CELTSCALE** will have the linetype scaling factor of 0.5. The value is retained in the **CELTSCALE** system variable. The first line (a) in Figure 25-12 is drawn with the **CELTSCALE** factor of 1 and the second line (b) is drawn with the **CELTSCALE** factor of 0.5. The length of the dashed is reduced by a factor of 0.5 when the **CELTSCALE** is 0.5.

The **LTSCALE** system variable controls the global scale factor. For example, if **LTSCALE** is set to 2, all lines in the drawing will be affected by a factor of 2. The net scale factor is equal to the product of **CELTSCALE** and **LTSCALE**. Figure 25-12(c) shows a line that is drawn with **LTSCALE** of 2 and **CELTSCALE** of 0.25. The net scale factor is = **LTSCALE** x **CELTSCALE** = 2 x 0.25 = 0.5.

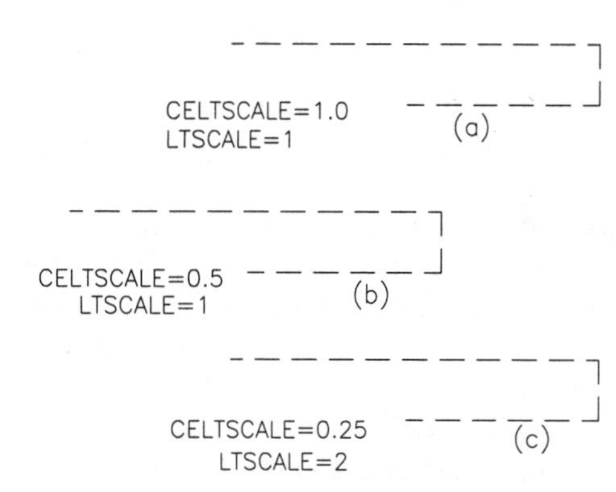

***Figure 25-12** Using **CELTSCALE** to control current linetype scaling*

COMPLEX LINETYPES

AutoCAD LT has provided a facility to create complex linetypes. The string complex linetype has a text string inserted in the line. The facility of creating complex linetypes increases the functionality of lines. For example, if you want to draw a line around a building that indicates the fence line, you can do it by defining a complex linetype that will automatically give you the desired line with the text string (Fence).

Creating a String Complex Linetype

When writing the definition of a string complex linetype, the actual text and its attributes must be included in the linetype definition. The format of the string complex linetype is:

["String", Text Style, Text Height, Rotation, X-Offset, Y-Offset]

String. It is the actual text that you want to insert along the line. The text string must be enclosed in quotation marks (" ").

Text Style. This is the name of the text style file that you want to use for generating the text string. The text style must be predefined.

Text Height. This is the actual height of the text, if the text height defined in the text style is 0. Otherwise, it acts as a scale factor for the text height specified in the text style. In Figure 25-13, the height of the text is 0.1 units.

Rotation. This is the rotation of the text string with respect to the positive X axis. The angle is always measured with respect to the positive X axis, no matter what AutoCAD LT's direction setting. The angle can be specified in radians (r), grads (g), or degrees (d). The default is degrees.

X-Offset. This is the distance of the lower left corner of the text string from the endpoint of the line segment measured along the line. If the line is horizontal, then the X-Offset distance is measured along the X axis. In Figure 25-13, the X-Offset distance is 0.05.

Y-Offset. This is the distance of the lower left corner of the text string from the endpoint of the line segment measured perpendicular to the line. If the line is horizontal, then the Y-Offset distance is measured along the Y axis. In Figure 25-13, the Y-Offset distance is -0.05. The distance is negative because the start point of the text string is 0.05 units below the endpoint of the first line segment.

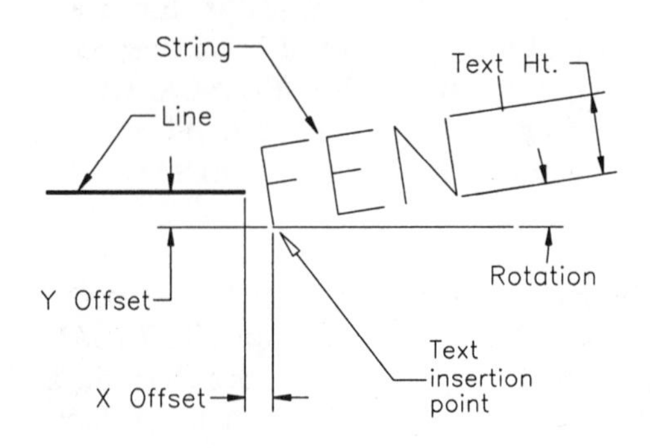

Figure 25-13 The attributes of a string complex linetype

Example 3

In the following example, you will write the definition of a string complex linetype that consists of the text string "Fence" and line segments. The length of each line segment is 0.75. The height of the text string is 0.1 units, and the space between the end of the text string and the following line segment is 0.05 (Figure 25-14).

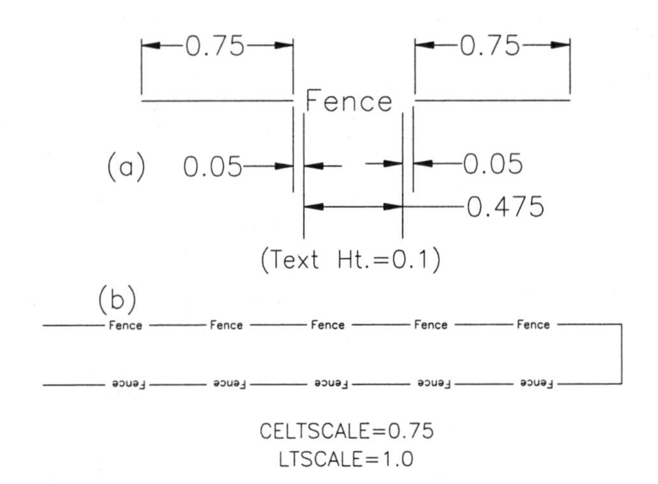

Figure 25-14 The attributes of a string complex linetype and line specifications for Example 3

Step 1

Before writing the definition of a new linetype, it is important to determine the line specification. One of the ways this can be done is to actually draw the lines and the text the way you want them to appear in the drawing. Once you have drawn the line and the text to your satisfaction, measure the distances needed to define the string complex linetype. In this example, the values are given as follows:

Text string =	Fence
Text style =	Standard
Text height =	0.1
Text rotation =	0
X-Offset =	0.05
Y-Offset =	-0.05
Length of the first line segment =	0.75
Distance between the line segments =	0.575

Step 2

Use a text editor to write the definition of the string complex linetype. You can add the definition to the AutoCAD LT **ACLT.LIN** file or create a separate file. The extension of the file must be **.LIN**. The following file is the listing of the **FENCE.LIN** file for Example 3. The name of the linetype is NEWFence.

```
*NEWFence1,New fence boundary line
A,0.75,["Fence",Standard,S=0.1,A=0,X=0.05,Y=-0.05],-0.575
or
A,0.75,-0.05,["Fence",Standard,S=0.1,A=0,X=0,Y=-0.05],-0.525
```

Step 3

To test the linetype, load the linetype using the **LINETYPE** command with the **Load** option, and assign it to a layer. Draw a line or any object to check if the line is drawn to the given specifications. Notice that the text is always drawn along X axis. Also, when you draw a line at an angle, polyline, circle, or spline, the text string does not align with the object (Figure 25-15).

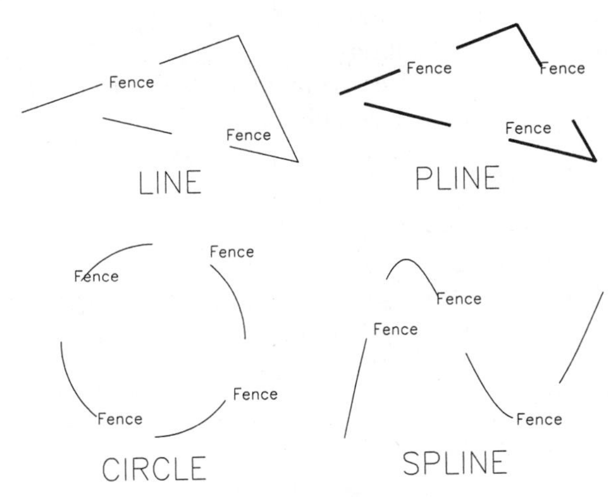

***Figure 25-15** Using string complex linetype with angle A = 0*

Step 4

In the NEWFence linetype definition, the specified angle is 0 degrees (Absolute angle A = 0). Therefore, when you use the NEWFence linetype to draw a line, circle, polyline, or spline, the text string (Fence) will be at zero degrees. If you want the text string (Fence) to align with the polyline (Figure 25-16), spline, or circle, specify the angle as relative angle (R=0) in the NEWFence linetype definition. The linetype definition with relative angle R = 0:

```
*NEWFence2,New fence boundary line
A,0.75,["Fence",Standard,S=0.1,R=0,X=0.05,Y=-0.05],-0.575
```

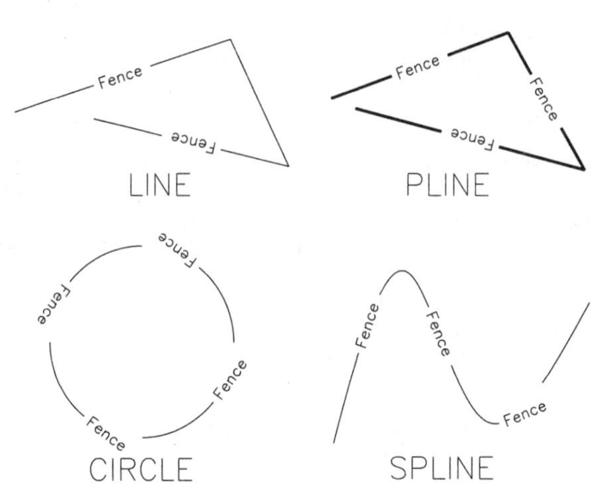

Figure 25-16 Using a string complex linetype with angle R = 0

Step 5

In Figure 25-16, you might notice that the text string is not properly aligned with the circumference of the circle. This is because AutoCAD LT draws the text string in a direction that is tangent to the circle at the text insertion point. To resolve this problem, you must define the middle point of the text string as the insertion point. Also, the line specifications should be measured accordingly. Figure 25-17 gives the measurements of the NEWFence linetype with the middle point of the text as the insertion point.

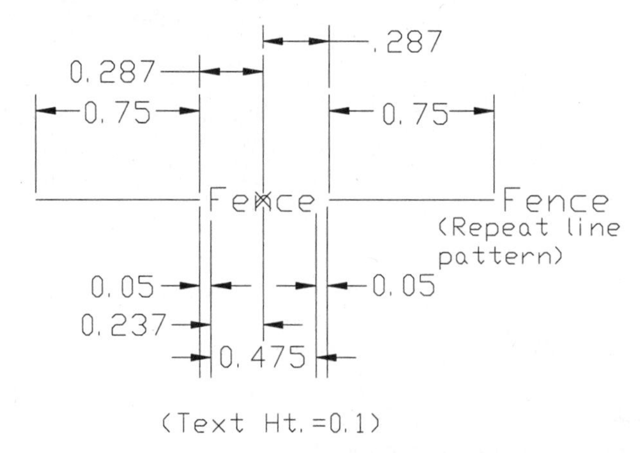

Figure 25-17 Specifications of a string complex linetype with the middle point of the text string as the text insertion point

The following is the linetype definition for NEWFence linetype:

```
*NEWFence3,New fence boundary line
A,0.75,-0.287,["FENCE",Standard,S=0.1,X=-0.237,Y=-0.05],-0.287
```

Note

If no angle is defined in the line definition, it defaults to angle R = 0.

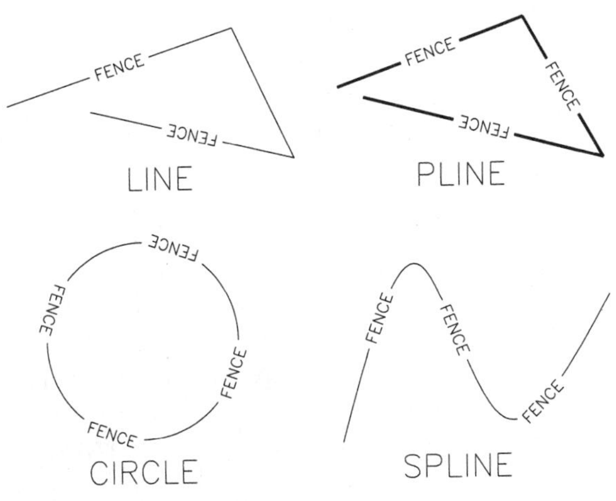

Figure 25-18 *Using a string complex linetype with the middle point of the text string as the text insertion point*

HATCH PATTERN DEFINITION

The AutoCAD LT software comes with a hatch pattern library file, **ACLT.PAT**, that contains 67 hatch patterns. These hatch patterns are sufficient for general drafting work. However, if you need a different hatch pattern, AutoCAD LT lets you create your own. There is no limit to the number of hatch patterns you can define.

The hatch patterns you define can be added to the hatch pattern library file, **ACLT.PAT**. You can also create a new hatch pattern library file, provided the file contains only one hatch pattern definition, and the name of the hatch is the same as the name of the file. The hatch pattern definition consists of the following two parts: **header line and hatch descriptors**.

Header Line

The **header line** consists of an asterisk (*) followed by the name of the hatch pattern. The hatch name is the name used in the hatch command to hatch an area. After the name, you can give the hatch description, which is separated from the hatch name by a comma (,). The general format of the header line is:

***HATCH Name [, Hatch Description]**

Where ***** ------------------------------ Asterisk
HATCH Name ----------- Name of hatch pattern

Hatch Description ------ Description of hatch pattern

The description can be any text that describes the hatch pattern. It can also be omitted, in which case, a comma should not follow the hatch pattern name.

Example
***DASH45, Dashed lines at 45 degrees**
Where **DASH45** --------- Hatch name
Dashed lines at 45 degrees ------ Hatch description

Hatch Descriptors

The **hatch descriptors** consist of one or more lines that contain the definition of the hatch lines. The general format of the hatch descriptor is:

Angle, X-origin, Y-origin, D1, D2 [,Dash Length.....]
Where **Angle** ------------ Angle of hatch lines
X-origin --------- X coordinate of hatch line
Y-origin --------- Y coordinate of hatch line
D1 ---------------- Displacement of second line (Delta-X)
D2 ---------------- Distance between hatch lines (Delta-Y)
Length ----------- Length of dashes and spaces (Pattern line definition)

Example
45,0,0,0,0.5,0.5,-0.125,0,-0.125
Where **45** ---------------- Angle of hatch line
0 ------------------ X-Origin
0 ------------------ Y-Origin
0 ------------------ Delta-X
0.5 ---------------- Delta-Y
0.5 ---------------- Dash (pen down)
-0.125 ------------ Space (pen up)
0 ------------------ Dot (pen down)
-0.125 ------------ Space (pen up)
0.5,-0.125,0,-0.125 Pattern line definition

Hatch Angle

X-origin and Y-origin. The hatch angle is the angle that the hatch lines make with the positive X axis. The angle is positive if measured counterclockwise (Figure 25-19), and negative if the angle is measured clockwise. When you draw a hatch pattern, the first hatch line starts from the point defined by X-origin and Y-origin. The remaining lines are generated by offsetting the first hatch line by a distance specified by delta-X and delta-Y. In Figure 25-20(a), the

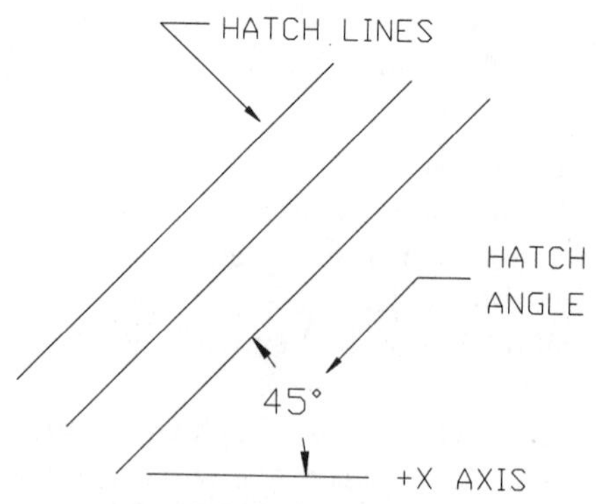

Figure 25-19 Hatch angle

first hatch line starts from the point with the coordinates X = 0 and Y = 0. In Figure 25-20(b) the first line of hatch starts from a point with the coordinates X = 0 and Y = 0.25.

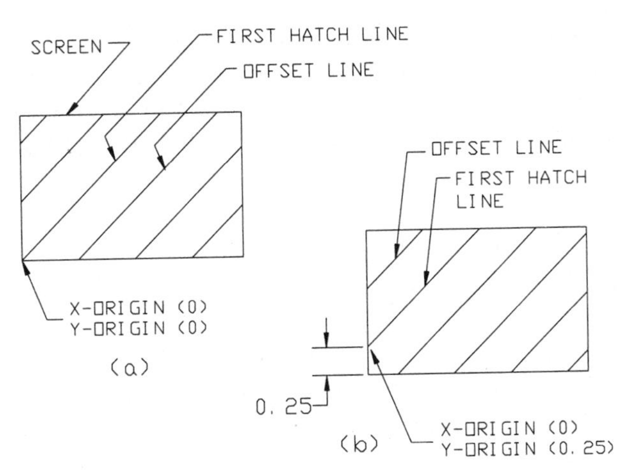

Figure 25-20 X-origin and Y-origin of hatch lines

Delta-X and Delta-Y. Delta-X is the displacement of the offset line in the direction in which the hatch lines are generated. For example, if the lines are drawn at a 0-degree angle and delta-X = 0.5, the offset line will be displaced by a distance delta-X (0.5) along the 0-angle direction. Similarly, if the hatch lines are drawn at a 45-degree angle, the offset line will be displaced by a distance delta-X (0.5) along a 45-degree direction (Figure 25-21). Delta-Y is the displacement of the offset lines measured perpendicular to the hatch lines. For example, if delta-Y = 1.0, the space between any two hatch lines will be 1.0 (Figure 25-21).

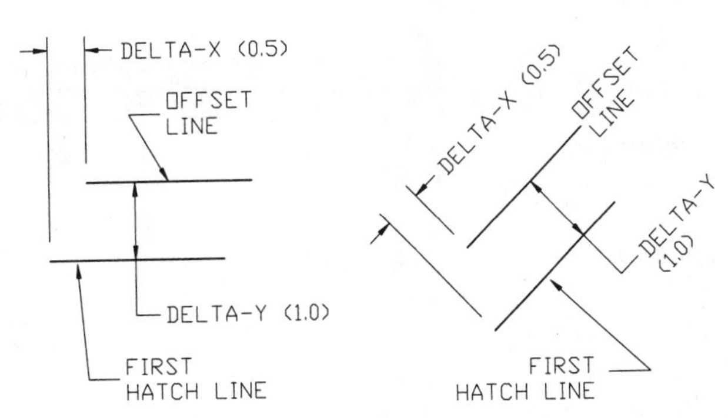

Figure 25-21 Delta-X and delta-Y of hatch lines

HOW HATCH WORKS

When you hatch an area, AutoCAD LT generates an infinite number of hatch lines of infinite length. The first hatch line always passes through the point specified by the X-origin and Y-origin. The remaining lines are generated by offsetting the first hatch line in both directions. The offset distance is determined by delta-X and delta-Y. All selected entities that form the boundary of the hatch area are then checked for intersection with these lines. Any hatch lines found within the defined hatch boundaries are turned on, and the hatch lines outside the hatch boundary are turned off, as shown in Figure 25-22. Since the hatch lines are generated by offsetting, the hatch lines in different areas of the drawing are automatically aligned relative to the drawing's snap origin. Figure 25-22(a) shows the hatch lines as computed by AutoCAD LT. These lines are not drawn on the screen; they are shown here for illustration only. Figure 25-22(b) shows the hatch lines generated in the circle that was defined as the hatch boundary.

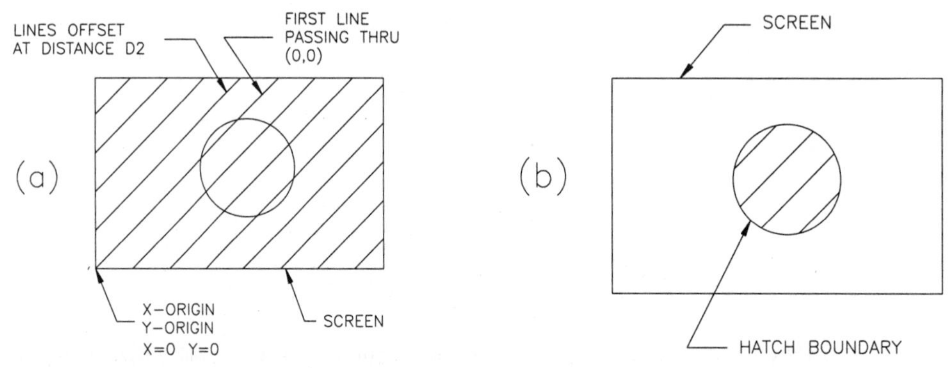

Figure 25-22 *Hatch lines outside the hatch boundary are turned off*

SIMPLE HATCH PATTERN

It is good practice to develop the hatch pattern specification before writing a hatch pattern definition. For simple hatch patterns it may not be that important, but for more complicated hatch patterns you should know the detailed specifications. Example 4 illustrates the procedure for developing a simple hatch pattern.

Example 4

Write a hatch pattern definition for the hatch pattern shown in Figure 25-23, with the following specifications:

Name of the hatch pattern =	HATCH1
X-Origin =	0
Y-Origin =	0
Distance between hatch lines =	0.5
Displacement of hatch lines =	0
Hatch line pattern =	Continuous

This hatch pattern definition can be added to the existing **ACLT.PAT** hatch file. You can use any text editor (like Notepad) to write the file. Load the **ACLT.PAT** file that is located in **AutoCAD LT 2000\SUPPORT** directory and insert the following two lines at the end of the file.

```
*HATCH1,Hatch Pattern for Example 4
45,0,0,0,.5
        Where 45 ----------------- Hatch angle
               0 ----------------- X-origin
               0 ----------------- Y-origin
               0 ----------------- Displacement of second hatch line
              .5 ----------------- Distance between hatch lines
```

The first field of hatch descriptors contains the angle of the hatch lines. That angle is 45 degrees with respect to the positive X axis. The second and third fields describe the X and Y coordinates of the first hatch line origin. The first line of the hatch pattern will pass through this point. If the values of the X-origin and Y-origin were 0.5 and 1.0, respectively, then the first line would pass through the point with the X coordinate of 0.5 and the Y coordinate of 1.0, with respect to the drawing origin 0,0. The remaining lines are generated by offsetting the first line, as shown in Figure 25-23.

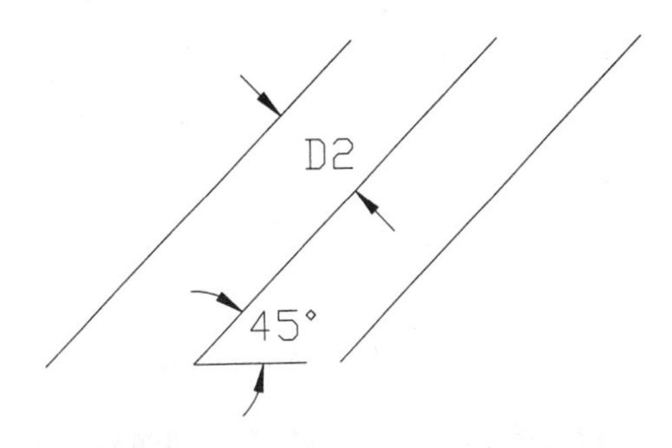

Figure 25-23 *Hatch pattern angle and offset distance*

EFFECT OF ANGLE AND SCALE FACTOR ON HATCH

When you hatch an area, you can alter the angle and displacement of hatch lines you have specified in the hatch pattern definition to get a desired hatch spacing. You can do this by entering an appropriate value for angle and scale factor in AutoCAD LT **HATCH** command.

Command: **HATCH**
Enter a pattern name or [?/Solid/User defined] <default>: **Hatch1**
Specify a scale for the pattern <default>: **1**
Specify an angle for the pattern<default>: **0**

To understand how the angle and the displacement can be changed, hatch an area with the hatch pattern HATCH1 in Example 4. You will notice that the hatch lines have been generated according to the definition of hatch pattern HATCH1. Notice the effect of hatch angle and scale factor on the hatch. Figure 25-24(a) shows a hatch that is generated by the AutoCAD LT **HATCH** command with a 0-degree angle and a scale factor of 1.0. If the angle is 0, the hatch will be generated with the same angle as defined in the hatch pattern definition (45 degrees in Example 4). Similarly, if the scale factor is 1.0, the distance between the hatch lines will be the same as defined in the hatch pattern definition (0.5 in Example 4). Figure 25-24(b) shows a hatch that is generated when the hatch scale factor is 0.5. If you measure the distance between the successive hatch lines, it will be 0.5 x 0.5 = 0.25. Figures 25-24(c) and (d) show the hatch when the angle is 45 degrees and the scale factors are 1.0 and 0.5, respectively. You can enter any value in response to the **HATCH** command prompts to generate hatch lines at any angle and with any line spacing.

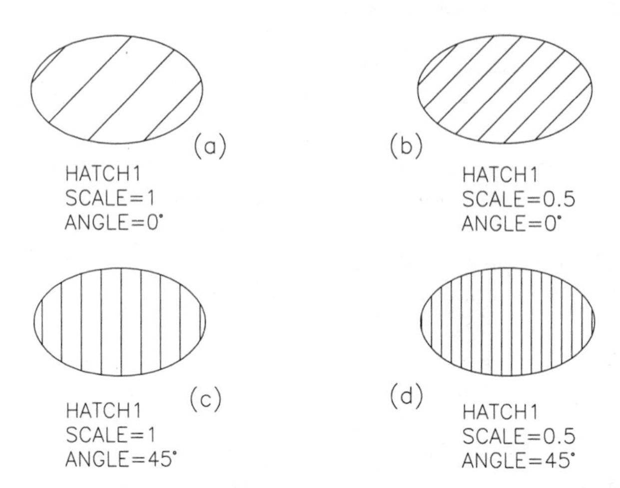

Figure 25-24 *Effect of angle and scale factor on hatch*

HATCH PATTERN WITH DASHES AND DOTS

The lines you can use in a hatch pattern definition are not restricted to continuous lines. You can define any line pattern to generate a hatch pattern. The lines can be a combination of dashes, dots, and spaces in any configuration. However, the maximum number of dashes you can specify in the line pattern definition of a hatch pattern is six. Example 5 uses a dash-dot line to create a hatch pattern.

Example 5

Write a hatch pattern definition for the hatch pattern shown in Figure 25-25, with the following specifications:

Name of the hatch pattern	HATCH2
Hatch angle =	0
X-origin =	0
Y-origin =	0
Displacement of lines (D1) =	0.25
Distance between lines (D2) =	0.25
Length of each dash =	0.5
Space between dashes and dots =	0.125
Space between dots =	0.125

You can use any text editor (Notepad) to edit the **ACLT.PAT** file. The general format is:

***HATCH NAME, Hatch Description**
Angle, X-Origin, Y-Origin, D1, D2 [,Dash Length.....]

Substitute the values from Example 5 in the header line and field descriptor:

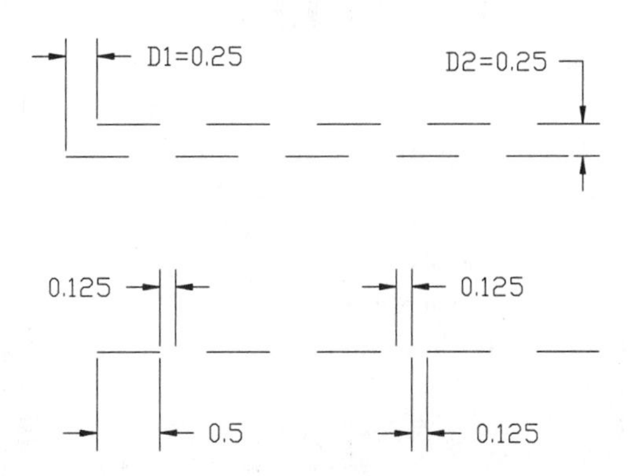

Figure 25-25 *Hatch lines made of dashes and dots*

*HATCH2,Hatch with dashes and dots
0,0,0,0.25,0.25,0.5,-0.125,0,-0.125,0,-0.125

Where **0** ------------------- Angle
0 ------------------- X-origin
0 ------------------- Y-origin
0.25 ------------------- Delta-X
0.25 ------------------- Delta-Y
0.5 ------------------- Length of dash
-0.125 ------------------- Space (pen up)
0 ------------------- Dot (pen down)
-0.125 ------------------- Space (pen up)
0 ------------------- Dot
-0.125 ------------------- Space

The hatch pattern this hatch definition will generate is shown in Figure 25-26. Figure 25-26(a) shows the hatch with 0-degree angle and a scale factor of 1.0. Figure 25-26(b) shows the hatch with a 45-degree angle and a scale factor of 0.5.

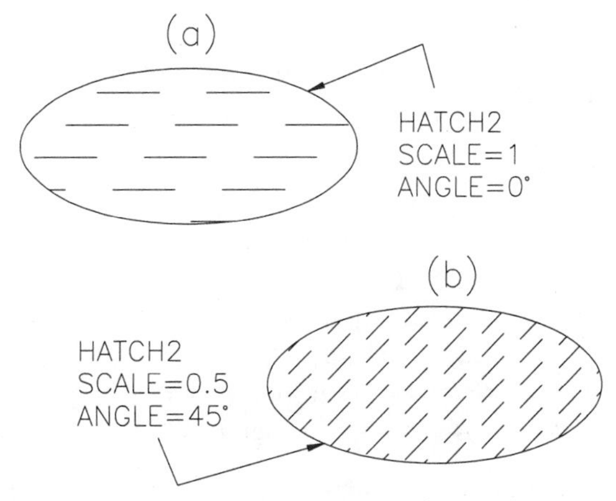

Figure 25-26 *Hatch pattern at different angles and scales*

HATCH WITH MULTIPLE DESCRIPTORS

Some hatch patterns require multiple lines to generate a shape. For example, if you want to create a hatch pattern of a brick wall, you need a hatch pattern that has four hatch descriptors to generate a rectangular shape. You can have any number of hatch descriptor lines in a hatch pattern definition. It is up to the user to combine them in any conceivable order. However, there are some shapes you cannot generate. A shape that has a nonlinear element, like an arc, cannot be generated by hatch pattern definition. However, you can simulate an arc by defining short line segments because you can use only straight lines to generate a hatch pattern. Example 6 uses three lines to define a triangular hatch pattern.

Example 6

Write a hatch pattern definition for the hatch pattern shown in Figure 25-27, with the following specifications:

Name of the hatch pattern = HATCH3
Vertical height of the triangle = 0.5
Horizontal length of the triangle = 0.5
Vertical distance between the triangles = 0.5
Horizontal distance between the triangles = 0.5

Each triangle in this hatch pattern consists of the following three elements: a vertical line, a horizontal line, and a line inclined at 45 degrees.

Vertical Line. For the vertical line, the specifications are (Figure 25-28):

Hatch angle = 90 degrees
X-origin = 0
Y-origin = 0
Delta-X (D1) = 0
Delta-Y (D2) = 1.0
Dash length = 0.5
Space = 0.5

Substitute the values from the vertical line specification in various fields of the hatch descriptor to get the following line:

90,0,0,0,1,.5,-.5

Where **90** ----------------- Hatch angle
0 ----------------- X-origin
0 ----------------- Y-origin
0 ----------------- Delta-X
1 ----------------- Delta-Y
.5 ----------------- Dash (pen down)
-.5 ----------------- Space (pen up)

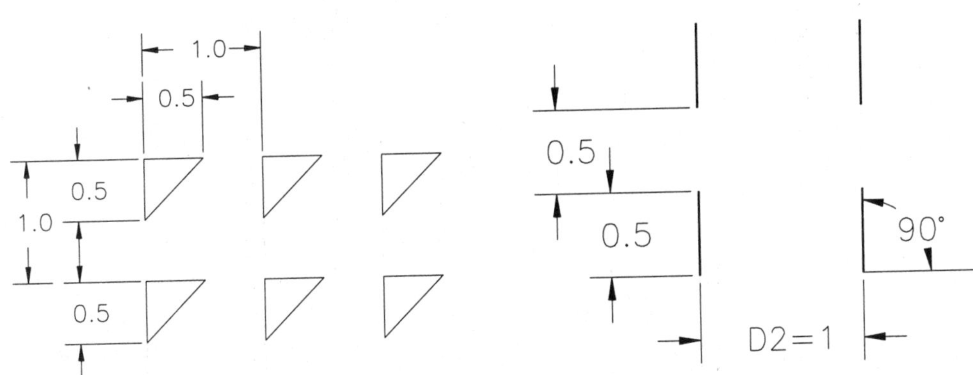

Figure 25-27 Triangle hatch pattern

Figure 25-28 Vertical line

Horizontal Line. For the horizontal line (Figure 25-29), the specifications are:

Hatch angle =	0 degrees
X-origin =	0
Y-origin =	0.5
Delta-X (D1) =	0
Delta-Y (D2) =	1.0
Dash length =	0.5
Space =	0.5

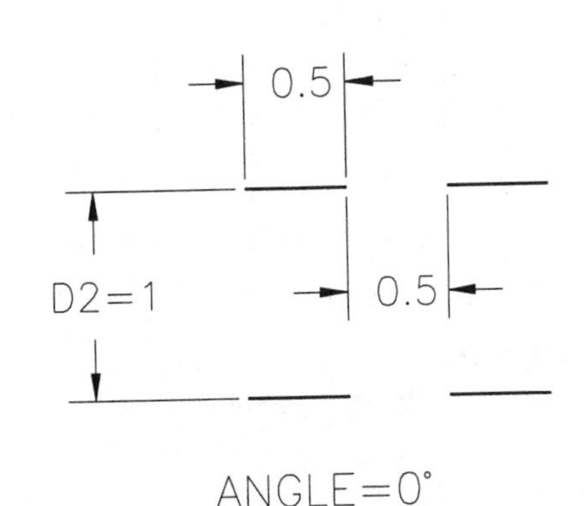

Figure 25-29 Horizontal line

The only difference between the vertical line and the horizontal line is the angle. For the horizontal line, the angle is 0 degrees, whereas for the vertical line, the angle is 90 degrees. Substitute the values from the vertical line specification to obtain the following line:

0,0,0.5,0,1,.5,-.5

Where **0** ------------------- Hatch angle
0 ------------------- X-origin
0.5 ------------------- Y-origin
0 ------------------- Delta-X
1 ------------------- Delta-Y
.5 ------------------- Dash (pen down)
-.5 ------------------- Space (pen up)

Line Inclined at 45 Degrees. This line is at an angle; therefore, you need to calculate the distances delta-X (D1) and delta-Y (D2), the length of the dashed line, and the length of space. Figure 25-30 shows the calculations to find these values.

Hatch angle =	45 degrees
X-Origin =	0

Y-Origin =	0
Delta-X (D1) =	0.7071
Delta-Y (D2) =	0.7071
Dash length =	0.7071
Space =	0.7071

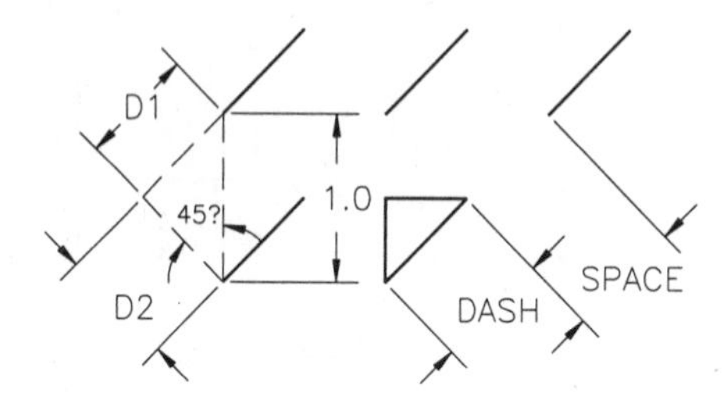

D1 = 1.0 x COS 45 D2 = 1.0 x SIN 45

D1 = 0.7071 D2 = 0.7071

DASH = SQRT(0.5^2 +0.5^2)
= .7071
SPACE = DASH = .7071

Figure 25-30 *Line inclined at 45 degrees*

After substituting the values in the general format of the hatch descriptor, you will obtain the following line:

45,0,0,.7071,.7071,.7071,-.7071

Where **45**	----------------	Hatch angle
0	----------------	X-origin
0	----------------	Y-origin
.7071	----------------	Delta-X
.7071	----------------	Delta-Y
.7071	----------------	Dash (pen down)
-.7071	----------------	Space (pen up)

Now you can combine these three lines and insert them at the end of the **ACLT.PAT** file.

Figure 25-31 shows the hatch pattern that will be generated by this hatch pattern (HATCH3). In Figure 25-31(a) the hatch pattern is at a 0-degree angle and the scale factor is 0.5. In Figure 25-31(b) the hatch pattern is at a -45-degree angle and the scale factor is 0.5.

The following file is a partial listing of the **ACLT.PAT** file, after adding the hatch pattern definitions from Examples 5, 6, and 7.

```
*SOLID, Solid fill
45, 0,0, 0,.125
*angle,Angle steel
0, 0,0, 0,.275, .2,-.075
```

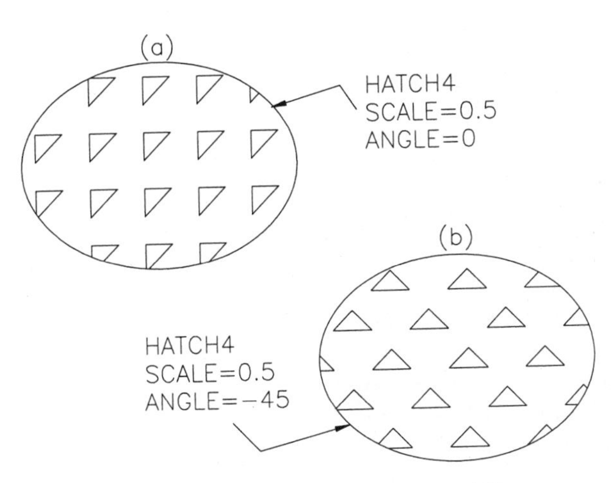

Figure 25-31 *Hatch generated by HATCH3 pattern*

```
90, 0,0, 0,.275, .2,-.075
*ansi31,ANSI Iron, Brick, Stone masonry
45, 0,0, 0,.125
*ansi32,ANSI Steel
45, 0,0, 0,.375
45, .176776695,0, 0,.375
*ansi33,ANSI Bronze, Brass, Copper
45, 0,0, 0,.25
45, .176776695,0, 0,.25, .125,-.0625
*ansi34,ANSI Plastic, Rubber
45, 0,0, 0,.75
45, .176776695,0, 0,.75
45, .353553391,0, 0,.75
45, .530330086,0, 0,.75
*ansi35,ANSI Fire brick, Refractory material
45, 0,0, 0,.25
45, .176776695,0, 0,.25, .3125,-.0625,0,-.0625
*ansi36,ANSI Marble, Slate, Glass
45, 0,0, .21875,.125, .3125,-.0625,0,-.0625
|
|
|
*swamp,Swampy area
0, 0,0, .5,.866025403, .125,-.875
90, .0625,0, .866025403,.5, .0625,-1.669550806
90, .078125,0, .866025403,.5, .05,-1.682050806
90, .046875,0, .866025403,.5, .05,-1.682050806
60, .09375,0, .5,.866025403, .04,-.96
120, .03125,0, .5,.866025403, .04,-.96
```

```
*trans,Heat transfer material
0, 0,0, 0,.25
0, 0,.125, 0,.25, .125,-.125
*triang,Equilateral triangles
60, 0,0, .1875,.324759526, .1875,-.1875
120, 0,0, .1875,.324759526, .1875,-.1875
0, -.09375,.162379763, .1875,.324759526, .1875,-.1875
*zigzag,Staircase effect
0, 0,0, .125,.125, .125,-.125
90, .125,0, .125,.125, .125,-.125
*HATCH1,Hatch at 45 Degree Angle
45,0,0,0,.5
*HATCH2,Hatch with Dashes & Dots:
0,0,0,.25,.25,0.5,-.125,0,-.125,0,-.125
*HATCH3,Triangle Hatch:
90,0,0,0,1,.5,-.5
0,0,0.5,0,1,.5,-.5
45,0,0,.7071,.7071,.7071,-.7071
```

CUSTOM HATCH PATTERN FILE

As mentioned earlier, you can add the new hatch pattern definitions to the **ACLT.PAT** file. There is no limit to the number of hatch pattern definitions you can add to this file. However, if you have only one hatch pattern definition, you can define a separate file. It has the following three requirements:

1. The name of the file has to be the same as the hatch pattern name.
2. The file can contain only one hatch pattern definition.
3. The hatch pattern name—and, therefore, the hatch file name—should be unique.
4. If you want to save the hatch pattern on the A drive, then the drive letter (A:) should precede the hatch name. For example, if the hatch name is HATCH3, the header line will be ***A:HATCH3, Triangle Hatch:** and the file name **HATCH3.PAT**.

```
*HATCH3,Triangle Hatch:
90,0,0,0,1,.5,-.5
0,0,0.5,0,1,.5,-.5
45,0,0,.7071,.7071,.7071,-.7071
```

Note

The hatch lines can be edited after exploding the hatch with the AutoCAD LT ***EXPLODE*** *command. After exploding, each hatch line becomes a separate object.*

It is good practice not to explode a hatch because it increases the size of the drawing database. For example, if a hatch consists of 100 lines, save it as a single object. However, after you explode the hatch, every line becomes a separate object and you have 99 additional objects in the drawing.

Keep the hatch lines in a separate layer to facilitate editing of the hatch lines.

Assign a unique color to hatch lines so that you can control the width of the hatch lines at the time of plotting.

Review Questions

CREATING LINETYPES

1. The AutoCAD LT __________ command can be used to create a new linetype.
2. The AutoCAD LT __________ command can be used to load a linetype.
3. The AutoCAD LT __________ command can be used to change the linetype scale factor.
4. In AutoCAD LT, the linetypes are saved in the __________ file.
5. The linetype description should not be more than __________ characters long.
6. A positive number denotes a pen __________ segment.
7. The segment length __________ generates a dot.
8. AutoCAD LT supports only __________ alignment field specification.
9. A line pattern definition always starts with __________.
10. A header line definition always starts with __________.

CREATING HATCH PATTERNS

11. The **ACLT.PAT** file contains __________ number of hatch pattern definitions.
12. The header line consists of an asterisk, the pattern name, and __________.
13. The first hatch line passes through a point whose coordinates are specified by __________ and __________.
14. The perpendicular distance between the hatch lines in a hatch pattern definition is specified by __________.

15. The displacement of the second hatch line in a hatch pattern definition is specified by __________.

16. The maximum number of dash lengths that can be specified in the line pattern definition of a hatch pattern is __________.

17. The hatch lines in different areas of the drawing will automatically __________ since the hatch lines are generated by offsetting.

18. The hatch angle as defined in the hatch pattern definition can be changed further when you use the AutoCAD LT __________ command.

19. When you load a hatch pattern, AutoCAD LT looks for the hatch pattern in ________ file.

20. The hatch lines can be edited after __________ the hatch by using the AutoCAD LT __________ command.

Exercises

Creating Linetypes

Exercise 1 — *General*

Using the AutoCAD LT **LINETYPE** command, create a new linetype "DASH3DASH" with the following specifications:

Length of the first dash 0.75
Blank space 0.125
Dash length 0.25
Blank space 0.125
Dash length 0.25
Blank space 0.125
Dash length 0.25
Blank space 0.125

Exercise 2 — *General*

Use a text editor to create a new file, **NEWLT2.LIN**, and a new linetype, DASH2DASH, with the following specifications:

Length of the first dash 0.5
Blank space 0.1
Dash length 0.2
Blank space 0.1
Dash length 0.2
Blank space 0.1

Exercise 3 General

Using AutoCAD LT's **LINETYPE** command, create a linetype DASH3DOT with the following specifications:

Length of the first dash 0.75
Blank space 0.25
Dot
Blank space 0.25
Dot
Blank space 0.25
Dot
Blank space 0.25

Exercise 4 General

a. Write the definition of a string complex linetype (Hot water line) as shown in Figure 25-32(a).

b. Write the definition of a string complex linetype (Gas line) as shown in Figure 25-32(b).

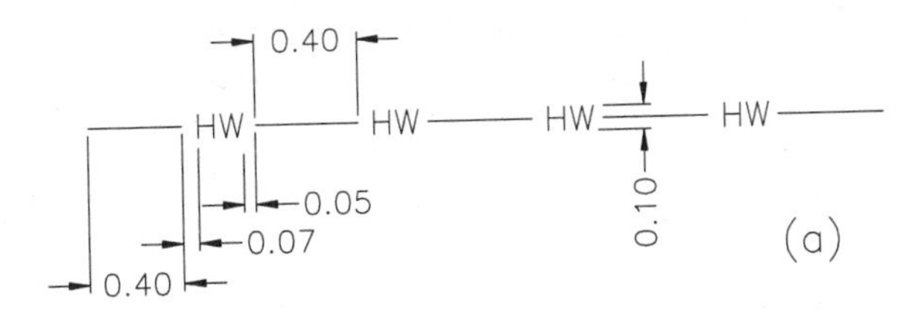

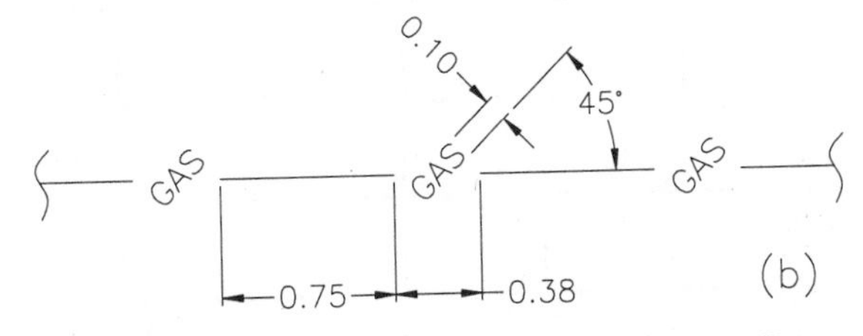

Figure 25-32 Specifications for string a complex linetype

Creating Hatch Patterns

Exercise 5 General

Determine the hatch specifications and write a hatch pattern definition for the hatch pattern in Figure 25-33.

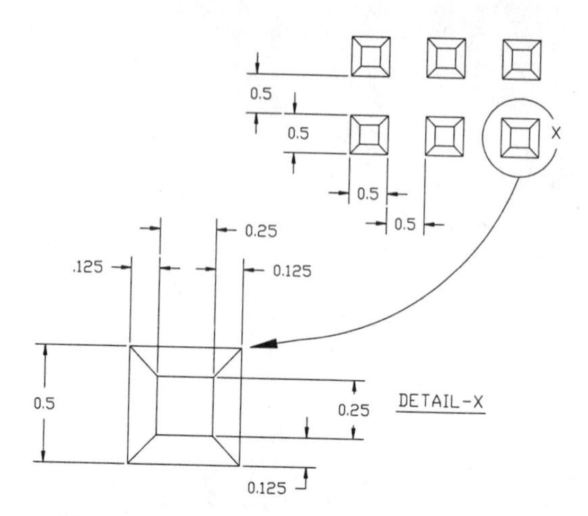

Figure 25-33 Hatch pattern for Exercise 5

Exercise 6

General

Determine the hatch specifications and write a hatch pattern definition for the hatch pattern as shown in Figure 25-34. Use this hatch to hatch a circle or rectangle.

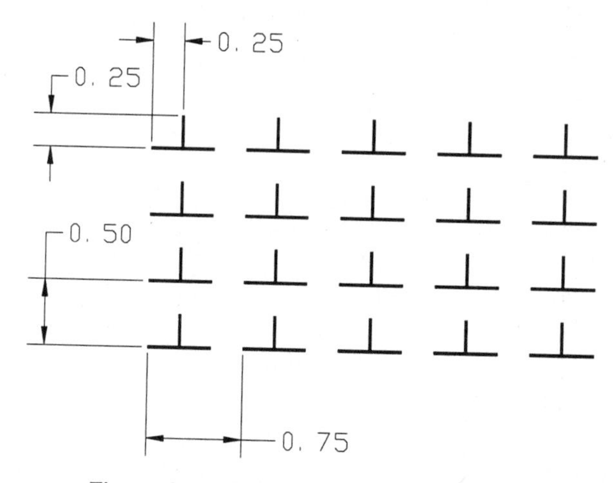

Figure 25-34 Hatch pattern for Exercise 6

Chapter 26

Customizing the ACLT.PGP File, DIESEL: A String Expression Language

Learning Objectives

After completing this chapter, you will be able to:

- *Customize the **ACLT.PGP** file.*
- *Abbreviate commands by defining command aliases.*
- *Use the **REINIT** command to reinitialize the PGP file.*
- *Use DIESEL to customize the status bar.*
- *Use the **MODEMACRO** system variable.*
- *Write macro expressions using DIESEL.*

WHAT IS THE ACLT.PGP FILE?

AutoCAD LT software comes with the program parameters file **ACLT.PGP**, which defines aliases for the AutoCAD LT commands. When you install AutoCAD LT, this file is automatically copied on the **AutoCAD LT 2000\ SUPPORT** subdirectory of the hard drive.

The file contains command aliases of some frequently used AutoCAD LT commands. For example, the command alias for the **LINE** command is L. If you enter L at the Command prompt **(Command: L)**, AutoCAD LT will treat it as the **LINE** command. The **ACLT.PGP** file also contains comment lines that give you some information about different sections of the file.

The following file is a partial listing of the standard **ACLT.PGP** file. Some of the lines have been deleted to make the file shorter.

```
; ACLT.PGP - Command Alias Definitions
; Copyright (C) 1997-1999 by Autodesk, Inc.

; Each time you open a new or existing drawing, AutoCAD LT searches
; the support path and reads the first ACLT.pgp file that it finds.

; You can abbreviate any AutoCAD LT command by defining an
; alias for it in aclt.pgp.
; Recommendation: back up this file before editing it.

; Command alias format:
;  <Alias>,*<Full command name>

; The following are guidelines for creating new command aliases.
; 1. An alias should reduce a command by at least two characters.
;    Commands with a control key equivalent, status bar button,
;    or function key do not require a command alias.
; 2. Try the first character of the command, then try the first two,
;    then the first three.
; 3. Once an alias is defined, add suffixes for related aliases:
;    Examples: L for Line, LT for Linetype.
; 4. Use a hyphen to differentiate between command line and dialog
;    box commands.
;    Example: B for Block, -B for -Block.
;
; Exceptions to the rules include AA for Area, T for Mtext, X for Explode.

3P,      *3DPOLY
A,       *ARC
AA,      *AREA
AD,      *ATTDISP
AE,      *ATTEDIT
AR,      *ARRAY
-AR,     *-ARRAY
-AT,     *-ATTDEF
ATE,     *ATTEDIT
-ATE,    *-ATTEDIT
ATTE,    *-ATTEDIT
AV       *DSVIEWER
AX,      *ATTEXT
B,       *BLOCK
-B,      *-BLOCK
BA,      *BASE
BH,      *BHATCH
```

```
BM,     *BLIPMODE
BO,     *BOUNDARY
-BO,    *-BOUNDARY
BR,     *BREAK
C,      *CIRCLE
CE,     *ADCENTER
CH,     *PROPERTIES
-CH,    *CHANGE
CHA,    *CHAMFER
CL,     *COPYLINK
CO,     *COPY
COL,    *COLOR
COLOUR, *COLOR
CP,     *COPY
D,      *DIMSTYLE
D1      *DIM1
DAD     *ATTDEF
DAL,    *DIMALIGNED
DAN,    *DIMANGULAR
DBA,    *DIMBASELINE
DCE,    *DIMCENTER
DCO,    *DIMCONTINUE
DDI,    *DIMDIAMETER
DED,    *DIMEDIT
DI,     *DIST
DIV,    *DIVIDE
|
|
L,      *LINE
LA,     *LAYER
-LA,    *-LAYER
LE,     *QLEADER
LEN,    *LENGTHEN
LI,     *LIST
LINEWEIGHT, *LWEIGHT
LO,     *-LAYOUT
LS,     *LIST
LT,     *LINETYPE
-LT,    *-LINETYPE
LTYPE,  *LINETYPE
-LTYPE, *-LINETYPE
LTS,    *LTSCALE
LW,     *LWEIGHT
M,      *MOVE
MA,     *MATCHPROP
ME,     *MEASURE
```

```
MI,     *MIRROR
MO,     *PROPERTIES
MR,     *MREDO
MS,     *MSPACE
MT,     *MTEXT
MV,     *MVIEW
N,      *NEW
O,      *OFFSET
OO,     *OOPS
OP,     *OPTIONS
OR,     *ORTHO
OS,     *OSNAP
-OS,    *-OSNAP
P,      *PAN
-P,     *-PAN
PA,     *PASTESPEC
PE,     *PEDIT
PL,     *PLINE
PO,     *POINT
POL,    *POLYGON
PR,     *OPTIONS
PRCLOSE, *PROPERTIESCLOSE
PROPS,  *PROPERTIES
PRE,    *PREVIEW
PRINT,  *PLOT
PS,     *PSPACE
PU,     *PURGE
-PU,    *-PURGE
QT,     *QTEXT
R,      *REDRAW
RC,     *REVCLOUD
RD,     *REVDATE
RE,     *REGEN
REA,    *REGENALL
REC,    *RECTANGLE
REG,    *REGION
|
|
SP,     *SPELL
SPL,    *SPLINE
SPE,    *SPLINEDIT
ST,     *STYLE
SU,     *SUBTRACT
T,      *MTEXT
-T,     *-MTEXT
TA,     *TABLET
```

```
TH,      *THICKNESS
TI,      *TILEMODE
TO,      *TOOLBAR
TOL,     *TOLERANCE
TR,      *TRIM
TX,      *TEXT
UC,      *DDUCS
UCP,     *DDUCSP

; Aliases for Hyperlink/URL AutoCAD LT97/LT98 compatibility
SAVEURL,    *SAVEAS
OPENURL,    *OPEN
INSERTURL, *INSERT

; Aliases for commands discontinued in AutoCAD LT:
BUPDATE,     *BLOCKICON
DDATTDEF,    *ATTDEF
DDATTEXT,    *ATTEXT
DDCHPROP,    *PROPERTIES
DDCOLOR,     *COLOR
DDLMODES,    *LAYER
DDLTYPE,     *LINETYPE
DDMODIFY,    *PROPERTIES
DDOSNAP,     *OSNAP
DDRENAME,    *RENAME
POLAR,       *DSETTINGS
```

SECTIONS OF THE ACLT.PGP FILE

The contents of the AutoCAD LT program parameters file **(ACLT.PGP)** can be categorized into three sections. These sections merely classify the information that is defined in the **ACLT.PGP** file. They do not have to appear in any definite order in the file, and they have no section headings. For example, the comment lines can be entered anywhere in the file; the same is true with AutoCAD LT command aliases. The **ACLT.PGP** file can be divided into these three sections: **comments,** and **command aliases**.

Comments

The comments of ACLT.PGP file can contain any number of comment lines and can occur anywhere in the file. Every comment line must start with a semicolon (;) (This is a comment line). Any line that is preceded by a semicolon is ignored by AutoCAD LT. You should use the comment line to give some relevant information about the file that will help other AutoCAD LT users to understand, edit, or update the file.

Command Aliases

It is time-consuming to enter AutoCAD LT commands at the keyboard because it requires typing the complete command name before pressing ENTER. AutoCAD LT provides a facility that can be used to abbreviate the commands by defining aliases for the AutoCAD LT commands.

This is made possible by the AutoCAD LT program parameters file (**ACLT.PGP** file). Each command alias line consists of two fields **(L, *LINE)**. The first field **(L)** defines the alias of the command; the second field **(*LINE)** consists of the AutoCAD LT command. The AutoCAD LT command must be preceded by an asterisk for AutoCAD LT to recognize the command line as a command alias. The two fields must be separated by a comma. The blank lines and the spaces between the two fields are ignored. In addition to AutoCAD LT commands, you can also use aliases for AutoLISP command names, provided the programs that contain the definition of these commands are loaded.

Example 1

Add the following AutoCAD LT command aliases to the AutoCAD LT program parameters file (**ACLT.PGP**).

Command Aliases

Abbreviation	Command	Abbreviation	Command
ELL	Ellipse	TM	Trim
CY	Copy	CHA	Chamfer
OFF	Offset	SH	Stretch
SL	Scale	MIR	Mirror

The **ACLT.PGP** file is an ASCII text file. To edit this file you can use any text editor (Notepad or Wordpad). The following is a partial listing of the **ACLT.PGP** file after insertion of the lines for the command aliases of Example 1. **The line numbers are not a part of the file; they are shown here for reference only.** The lines that have been added to the file are highlighted in bold.

```
WO,      *WMFOUT      1
X,       *EXPLODE     2
XA,      *XATTACH     3
XB,      *XBIND       4
-XB,     *-XBIND      5
XL,      *XLINE       6
XR,      *XREF        7
-XR,     *-XREF       8
Z,       *ZOOM        9
ELL,     *ELLIPSE     10
CY,      *COPY        11
OFF,     *OFFSET      12
SL,      *SCALE       13
TM,      *TRIM        14
CHA,     *CHAMFER     15
SH,      *STRETCH     17
MIR,     *MIRROR      16
```

Explanation

Lines 10 and 11
ELL, *ELLIPSE
CY, *COPY
Line 10 defines the alias **(ELL)** for the AutoCAD LT command **ELLIPSE**, and the next line defines the alias **(CY)** for the **COPY** command. The AutoCAD LT commands must be preceded by an asterisk. You can put any number of spaces between the alias abbreviation and the AutoCAD LT command.

REINIT COMMAND

When you make any changes in the **ACLT.PGP** file, there are two ways to reinitialize the **ACLT.PGP** file. One is to quit AutoCAD LT and then reenter it. When you start AutoCAD LT, the **ACLT.PGP** file is automatically loaded.

You can also reinitialize the **ACLT.PGP** file by using the AutoCAD LT **REINIT** command. The **REINIT** command lets you reinitialize the I/O ports, digitizer, display, and AutoCAD LT program parameters file, **ACLT.PGP**. When you enter the **REINIT** command, AutoCAD LT will display the **Re-initialization** dialog box (Figure 26-1). To reinitialize the **ACLT.PGP** file, select the corresponding toggle box, and then choose **OK**. AutoCAD LT will reinitialize the program parameters file **(ACLT.PGP)**, and then you can use the command aliases defined in the file.

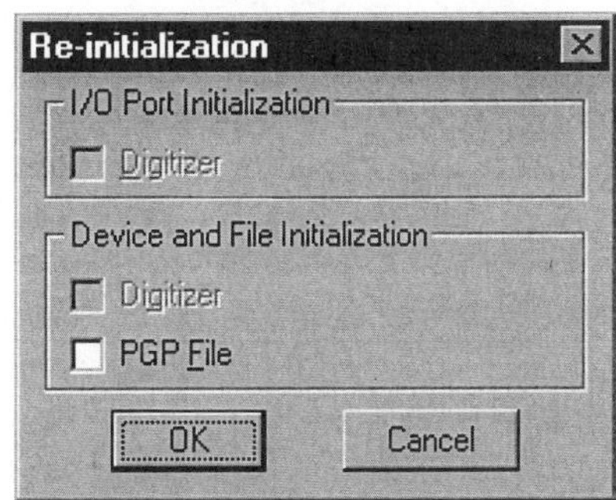

Figure 26-1 ***Re-initialization*** *dialog box*

DIESEL

DIESEL (Direct Interpretively Evaluated String Expression Language) is a string expression language. It can be used to display a user-defined text string (macro expression) in the status bar by altering the value of the AutoCAD LT system variable **MODEMACRO**. The value assigned to **MODEMACRO** must be a string, and the output it generates will be a string. It is fairly easy to write a macro expression in DIESEL, and it is an important tool for customizing AutoCAD LT. You can write the definition of the **MODEMACRO** expression in the menu files. A detailed explanation of the DIESEL functions and the use of DIESEL in writing a macro expression is given later in this chapter.

STATUS BAR

When you are in AutoCAD LT, a status bar is displayed at the bottom of the graphics screen (Figure 26-2). This line contains some useful information and tools that will make it easy to change the status of some AutoCAD LT functions. To change the status, you must click on the buttons. For example, if you want to display grid lines on the screen, click on the **GRID** button. Similarly, if you want to switch to paper space, click on **MODEL**. The status bar contains the following information:

Coordinate Display. The coordinate information displayed in the status bar can be static or

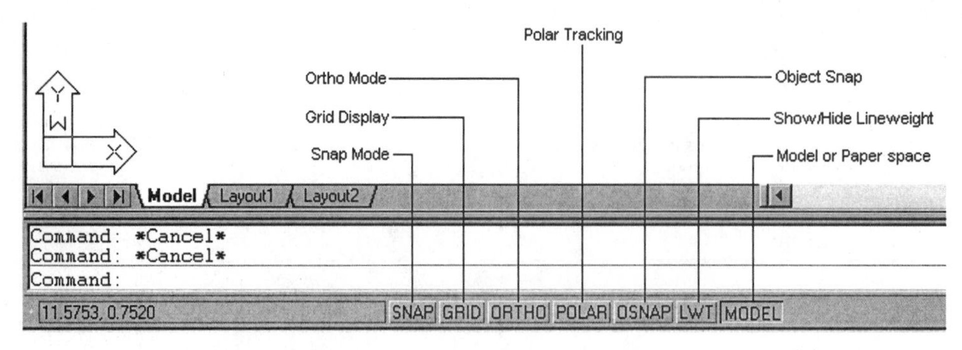

Figure 26-2 *Default status bar display*

dynamic. If the coordinate display is static, the coordinate values displayed in the status bar change only when you specify a point. However, if the coordinate display is dynamic (default setting), AutoCAD LT constantly displays the absolute coordinates of the graphics cursor with respect to the UCS origin. AutoCAD LT can also display the polar coordinates (length<angle) if you are in an AutoCAD LT command.

SNAP. If SNAP is on, the cursor snaps to the snap point.

GRID. If GRID is on, grid lines are displayed on the screen.

ORTHO. If ORTHO is on, a line can be drawn in vertical or horizontal direction.

POLAR. When the POLAR snap is on, the the cursor snaps to polar angles as set in the Polar Tracking tab of the **Drafting Settings** dialog box.

OSNAP. If OSNAP is on, you can use the running object snaps. If OSNAP is off, the running object snaps are temporarily disabled. The status of OSNAP (Off or On) does not prevent you from using regular object snaps.

LWT. When Lineweight button is on, the objects are displayed with the assigned width.

MODEL and PAPER Space. AutoCAD LT displays MODEL in the status bar when you are working in the model space. If you are working in the paper space, AutoCAD LT will display PAPER in place of MODEL.

MODEMACRO SYSTEM VARIABLE

The AutoCAD LT system variable **MODEMACRO** can be used to display a new text string in the status bar. You can also display the value returned by a macro expression using the DIESEL language, which is discussed in a later section of this chapter. MODEMACRO is a system variable and you can assign a value to this variable by entering MODEMACRO at the Command prompt or by using the SETVAR command. For example, if you want to display **Customizing AutoCAD LT** in the status bar, enter **SETVAR** at the Command: prompt and then press ENTER. AutoCAD LT will prompt you to enter the name of the system variable. Enter **MODEMACRO** and then press ENTER again. Now you can enter the text you want to display in the status bar. After you enter **Customizing AutoCAD LT** and press ENTER, the status bar will display the new text.

Command: **MODEMACRO**
or
Command: **SETVAR**
Enter variable name or ?: **MODEMACRO**
Enter new value for MODEMACRO, or . for none<“”>: **Customizing AutoCAD LT**

You can also enter MODEMACRO at the Command prompt and then enter the text that you want to display in the status bar.

Command: **MODEMACRO**
Enter new value for MODEMACRO, or . for none<“”>: **Customizing AutoCAD LT**

Once the value of the **MODEMACRO** variable is changed, it retains that value until you enter a new value, start a new drawing, or open an existing drawing file. If you want to display the standard text in the status line, enter a period (.) at the prompt **Enter new value for MODEMACRO, or . for none <“”>:**. The value assigned to the **MODEMACRO** system variable is not saved with the drawing, in any configuration file, or anywhere in the system.

Command: **MODEMACRO**
Enter new value for MODEMACRO, or . for none<“”>: **.**

CUSTOMIZING THE STATUS BAR

The information contained in the status bar can be divided into two parts: toggle functions and coordinate display. The toggle functions part consists of the status of Snap, Grid, Ortho, Polar Tracking, Object Snap, Lineweight, and Model Space (Figure 26-2). The coordinate display displays the X, Y, and Z coordinates of the cursor. The status bar can be customized to your requirements by assigning a value to the AutoCAD LT system variable **MODEMACRO**. The value assigned to this variable is displayed left-justified in the status bar at the bottom of the AutoCAD LT window. The number of characters that can be displayed in the status line depends on the system display and the size of the AutoCAD LT window. The coordinate display field cannot be changed or edited.

The information displayed in the status bar is a valuable resource. Therefore, you must be careful when selecting the information to be displayed in the status bar. For example, when working on a project, you may like to display the name of the project in the status bar. If you are using several dimensioning styles, you could display the name of the current dimensioning style (DIMSTYLE) in the status bar. Similarly, if you have several text files with different fonts, the name of the current text file (TEXTSTYLE) and the text height (TEXTSIZE) can be displayed in the status bar. Therefore, the information that should be displayed in the status bar depends on you and the drawing requirements. AutoCAD LT lets you customize this line and have any information displayed in the status bar that you think is appropriate for your application.

MACRO EXPRESSIONS USING DIESEL

You can also write a macro expression using DIESEL to assign a value to the **MODEMACRO** system variable. The macro expressions are similar to AutoLISP functions, with some differences. For example, in DIESEL, the drawing name can be obtained by using the macro expression **$(getvar,dwgname)**. The DIESEL macro expressions return only string values. The format of a macro expression is:

$(function-name,argument1,argument2, ...)

Example

$(getvar,dwgname)

Here, **getvar** is the name of the DIESEL string function and **dwgname** is the argument of the function. There must not be any spaces between different elements of a macro expression. For example, spaces between the $ sign and the open parentheses are not permitted. Similarly, there must not be any spaces between the comma and the argument, **dwgname**. All macro expressions must start with a $ sign.

The following example illustrates the use of a macro expression using DIESEL to define and then assign a value to the **MODEMACRO** system variable.

Example 2

Using the AutoCAD LT **MODEMACRO** command, redefine the status bar to display the following information:

Project name (Cust-ACLT)
Name of the drawing (DEMO)
Name of the current layer (OBJ)

Note that in this example the project name is Cust-ACLT, the drawing name is DEMO, and the current layer name is OBJ.

Before entering the **MODEMACRO** command, you need to determine how to retrieve the required information from the drawing database. For example, here the project name **(Cust-ACLT)** is a user-defined name that lets you know the name of the current project. This project name is not saved in the drawing database. The name of the drawing can be obtained using the DIESEL string function **GETVAR $(getvar,dwgname)**. Similarly, the **GETVAR** function can also be used to obtain the name of the current layer, **$(getvar,clayer)**. Once you determine how to retrieve the information from the system, you can use the **MODEMACRO** system variable to obtain the new status line. For Example 2, the following DIESEL expression will define the required status bar.

Command: **MODEMACRO**
Enter new value for MODEMACRO, or . for none<“”>:**Cust-ACLT N:$**
(GETVAR,dwgname) L:$(GETVAR,clayer)

Explanation

Cust-ACLT

Cust-ACLT is assumed to be the project name you want to display in the status bar.

N:$(GETVAR,dwgname)

Here, N: is used as an abbreviation for the drawing name. The GETVAR function retrieves the name of the drawing from the system variable **dwgname** and displays it in the status bar, next to N:.

L:$(GETVAR,clayer)

Here L: is used as an abbreviation for the layer name. The GETVAR function retrieves the name of the current layer from the system variable **clayer** and displays it in the status bar.

The new status bar is shown in Figure 26-3.

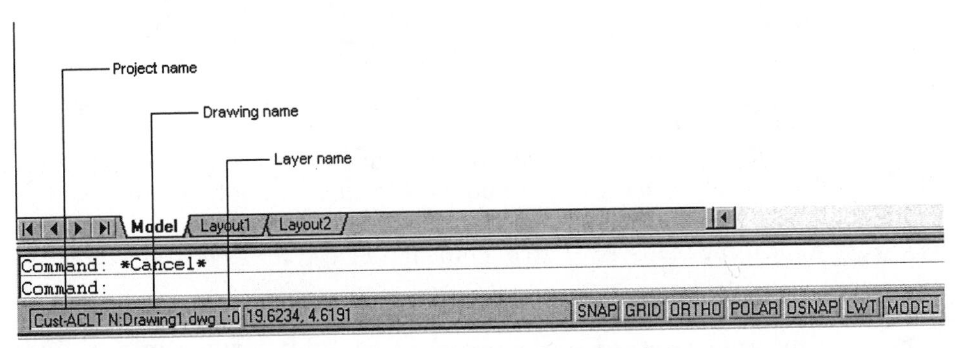

***Figure 26-3** Status bar for Example 2*

Example 3

Using the AutoCAD LT **MODEMACRO** command, redefine the status bar to display the following information, Figure 26-4:

Name of the current textstyle
Size of text
User-elapsed time in minutes

In this example the abbreviations for text style, text size, and user-elapsed time in minutes are TSTYLE:, TSIZE:, and ETM, respectively.

Command: **MODEMACRO**
Enter new value for MODEMACRO, or . for none<"">:
TSTYLE:$(GETVAR,TEXTSTYLE)TSIZE:$(GETVAR,TEXTSIZE)
ETM:$(FIX,$(*,60,$(*,24,$(GETVAR,TDUSRTIMER))))

Explanation

TSTYLE:$(GETVAR,TEXTSTYLE)
The **GETVAR** function obtains the name of the current textstyle from the system variable **TEXTSTYLE** and displays it next to TSTYLE: in the status bar.

TSIZE:$(GETVAR,TEXTSIZE)
The **GETVAR** function obtains the current size of the text from the system variable **TEXTSIZE** and then displays it next to TSIZE: in the status bar.

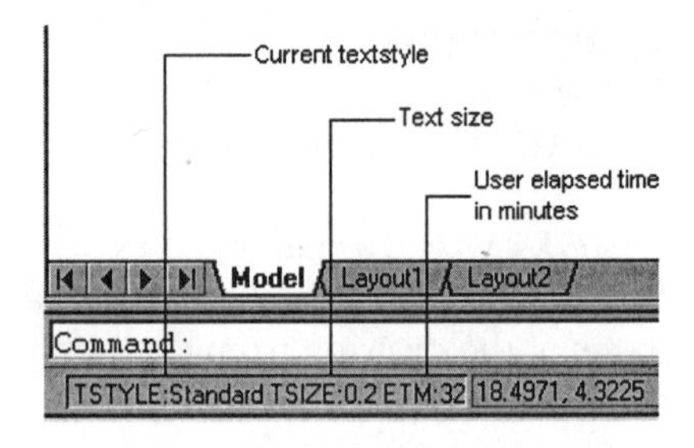

Figure 26-4 *Status bar for Example 3*

ETM:$(FIX,$(*,60,$(*,24,$(GETVAR,TDUSRTIMER))))
The **GETVAR** function obtains the user-elapsed time from the system variable **TDUSRTIMER** in the following format:

<Number of days>.<Fraction>

Example
0.03206400 (time in days)
To change this time into minutes, multiply the value obtained from the system variable **TDUSRTIMER** by 24 to change it into hours, and then multiply the product by 60 to change the time into minutes. To express the minutes value without a decimal, determine the integer value using the DIESEL string function FIX.

Example
Assume that the value returned by the system variable TDUSRTIMER is 0.03206400. This time is in days. Use the following calculations to change the time into minutes, and then express the time as an integer:

0.03206400 days x 24 = 0.769536 hr
0.769536 hr x 60 = 46.17216 min
integer of 46.17216 min = 46 min

DIESEL EXPRESSIONS IN MENUS

You can also define a DIESEL expression in the tablet, pull-down, or button menu. When you choose the menu item, it will automatically assign the value to the MODEMACRO system variable and then display the new status line. The following example illustrates the use of the DIESEL expression in the menu.

Example 4

Write a DIESEL macro for the menus that displays the following information in the status bar (Figure 26-5):

Macro-1	**Macro-2**	**Macro-3**
Project name	Pline width	Dimtad

Drawing name	Fillet radius	Dimtix
Current layer	Offset distance	Dimscale

The following file is a listing of the menu that contains the definition of three DIESEL macros for Example 4. The following .mnu file can be written using any Text editor. This menu is loaded using AutoCAD LT's **MENULOAD** command which displays the **Menu Customization** dialog box. In the **Menu Groups** tab use the **Browse** button to select the name of the .mnu file and then select the **Load** button. Also the menu title DIESEL is to be inserted before the Edit title (Figure 26-5) using the **Menu Bar** tab of the dialog box where the new .mnu file is selected from the drop-down list. Choose **Edit** from the **Menu** title list and then choose the **Insert** button. If you choose the second item, DIESEL2, it will automatically display the new status bar (Figure 26-6).

Figure 26-5 Inserting menu title

```
***POP1
[*DIESEL*]
[DIESEL1:]^C^CMODEMACRO;$M=Cust-ACLT N:$(GETVAR,DWGNAME)
L:$(GETVAR,CLAYER);
[DIESEL2:]^C^CMODEMACRO;$M=PLWID:$(GETVAR,PLINEWID)
FRAD:$(GETVAR,FILLETRAD) OFFSET:$(GETVAR,OFFSETDIST)
LTSCALE:$(GETVAR,LTSCALE);
[DIESEL3:]^C^CMODEMACRO;$M=DTAD:$(GETVAR,DIMTAD)
DTIX:$(GETVAR,DIMTIX) DSCALE:$(GETVAR,DIMSCALE);
```

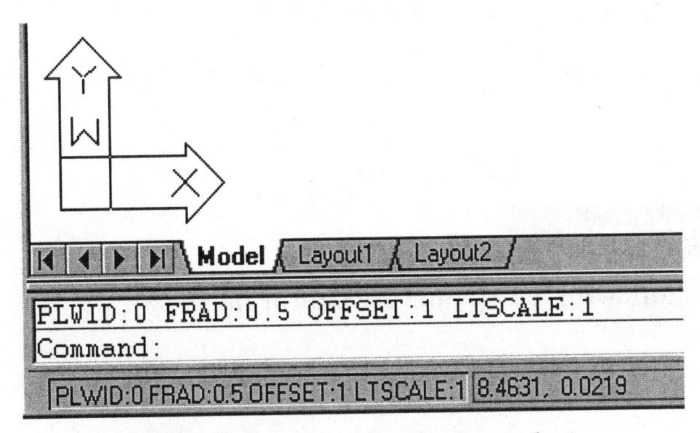

Figure 26-6 Status bar for Example 4

MACROTRACE SYSTEM VARIABLE

The MACROTRACE variable is an AutoCAD LT system variable that can be used to debug a DIESEL expression. The default value of this variable is 0 (off). It is turned on by assigning a value of 1 (on). When on, the MACROTRACE system variable will evaluate all DIESEL expressions and display the results in the command prompt area. For example, if you have defined several DIESEL expressions in a drawing session, all of them will be evaluated at the same time and the messages, if any, will be displayed in the command prompt area.

Example

Command: **MODEMACRO**
Enter new value for MODEMACRO, or . for none<“”>: **$(getvar,dwgname) $(getvar clayer)**

Note that in this DIESEL expression a comma is missing between getvar and clayer. If the MACROTRACE system variable is on, the following message will be displayed in the Command prompt area:

Eval: $(GETVAR,DWGNAME)
====>UNNAMED
Eval: $(GETVAR CLAYER)
Err: $(GETVAR CLAYER)??

This error message gives you an idea about the location of the error in a DIESEL expression. In the previous example, the first part of the expression successfully returns the name of the drawing (unnamed) and the second part of the expression results in the error message. This confirms that there is an error in the second part of the DIESEL expression. You can further determine the cause of the error by comparing it with the error messages in the following table:

Error Message	Description
$?	Syntax error
$?(func,??)	Incorrect argument to function
$(func)??	Unknown function
$(++)	Output string too long

Review Questions

Indicate whether the following statements are true or false.

1. The comment section in the ACLT.PGP file can contain any number of lines. (T/F)

2. AutoCAD LT ignores any line that is preceded by a semicolon. (T/F)

3. The command alias must not be an AutoCAD LT command. (T/F)

4. In the command alias section, the command alias must be preceded by a semicolon. (T/F)

5. The ACLT.PGP file does not come with AutoCAD LT software. (T/F)

6. The ACLT.PGP file is an ASCII file. (T/F)

7. DIESEL (direct interpretively evaluated string expression language) is a string expression language. (T/F)

8. The value assigned to the **MODEMACRO** variable is a string, and the output it generates is not a string. (T/F)

9. You cannot define a DIESEL expression in the menu. (T/F)

10. The coordinate information displayed in the status bar can be dynamic only. (T/F)

11. Once the value of the **MODEMACRO** variable is changed, it retains that value until you enter a new value, start a new drawing, or open an existing drawing file. (T/F)

12. The coordinate display field cannot be changed or edited. (T/F)

13. You can write a macro expression using DIESEL to assign a value to the **MODEMACRO** system variable. (T/F)

14. In DIESEL, the drawing name can be obtained by using the macro expression **$(getvar,dwgname)**. (T/F)

Exercises

Exercise 1 — *General*

Add the following AutoCAD LT command aliases to the AutoCAD LT program parameters file **(ACLT.PGP)**.

Abbreviation	Command	Abbreviation	Command
BL	**BLOCK**	LT	**LTSCALE**
INS	**INSERT**	EX	**EXPLODE**
DIS	**DISTANCE**	G	**GRID**
T	**TIME**		

Exercise 2 — *General*

Using the AutoCAD LT **MODEMACRO** command, redefine the status bar to display the following information in the status bar:

Your name
Name of drawing

Chapter 27

Pull-down, Shortcut, and Partial Menus and Customizing Toolbars

Learning Objectives

After completing this chapter, you will be able to:
- *Write pull-down menus.*
- *Load menus.*
- *Write cascading submenus in pull-down menus.*
- *Write shortcut menus.*
- *Pull-down menus.*
- *Write partial menus.*
- *Define accelerator keys.*
- *Write toolbar definitions.*
- *Write menus to access online help.*
- *Customize toolbars.*

AUTOCAD LT MENU

The AutoCAD LT menu provides a powerful tool to customize AutoCAD LT. The AutoCAD LT software package comes with a standard menu file named **ACLT.MNU**. When you start AutoCAD LT, the menu file **ACLT.MNU** is automatically loaded. The AutoCAD LT menu file contains AutoCAD LT commands, separated under different headings for easy identification. For example, all draw commands are under **Draw** and all editing commands are under **Modify** menu. The headings are named and arranged to make it easier for you to locate and access the commands. However, there are some commands that you may never use. Also, some users might like to regroup and rearrange the commands so that it is easier to access those most frequently used.

AutoCAD LT lets the user eliminate rarely used commands from the menu file and define new ones. This is made possible by editing the existing **ACLT.MNU** file or writing a new menu file. There is no limit to the number of files you can write. You can have a separate menu file for each application. For example, you can have separate menu files for mechanical, electrical, and architectural drawings. You can load these menu files any time by using the AutoCAD LT **MENULOAD** command. The menu files are text files with the extension **.MNU**. These files can be written by using any text editor like Wordpad or Notepad. The menu file can be divided into ten sections, each section identified by a.section label. AutoCAD LT uses the following labels to identify different sections of the AutoCAD LT menu file:

*****TABLET(n)**	**n** is from 1 to 4
*****IMAGE**	
*****POP(n)**	**n** is from 1 to 499 (For Shortcut menus n=0 and 500 to 999)
*****BUTTONS(n)**	**n** is from 1 to 4
*****AUX(n)**	**n** is from 1 to 4
*****MENUGROUP**	
*****TOOLBARS**	
*****HELPSTRING**	
*****ACCELERATORS**	
*****SCREEN**	

The tablet menu can have up to four different sections. The **POP** menu (pull-down and shortcut menu) can have up to 16 sections, and auxiliary and buttons menus can have up to four sections.

Tablet Menus	Pull-Down and Shortcut Menus
***TABLET1	***POP0
***TABLET2	***POP1
***TABLET3	***POP2
***TABLET4	***POP3
	***POP4
BUTTONS Menus	***POP5
***BUTTONS1	***POP6
***BUTTONS2	***POP7
***BUTTONS3	***POP8
***BUTTONS4	
Auxiliary Menus	
***AUX1	
***AUX2	***POP498
***AUX3	***POP499
***AUX4	

STANDARD MENUS

The menu is a part of the AutoCAD LT standard menu file, **ACLT.MNU**. The **ACLT.MNU** file is automatically loaded when you start AutoCAD LT, provided the standard configuration of AutoCAD LT has not been changed.

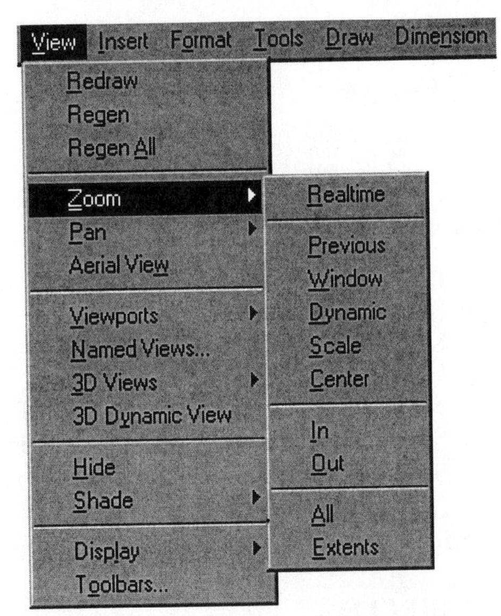

Figure 27-1 *Pull-down and cascading menu*

The menus can be selected by moving the crosshairs to the top of the screen, into the menu bar area. If you move the pointing device sideways, different menu bar titles are highlighted and you can select the desired item by pressing the pick button on your pointing device. Once the item is selected, the corresponding menu is displayed directly under the title (Figure 27-1). The menu can have 499 sections, named as POP1, POP2, POP3, . . ., POP499.

WRITING A MENU

Before you write a menu, you need to design a menu and to know the exact sequence of AutoCAD LT commands and the prompts associated with particular commands. To design a menu, you should select and arrange the commands in a way that provides easy access to the most frequently used commands. A careful design will save a lot of time in the long run. Therefore, consider several possible designs with different command combinations, and then select the one best suited for the job. Suggestions from other CAD operators can prove very valuable.

The second important thing in developing a menu is to know the exact sequence of the commands and the prompts associated with each command. To better determine the prompt entries required in a command, you should enter all the commands and the prompt entries at the keyboard. The following is a description of some of the commands and the prompt entries required for Example 1.

LINE Command
Command: **LINE**

Notice the command and input sequence:

LINE
<RETURN>
CIRCLE (C,R) Command
Command: **CIRCLE**
Specify center point for circle or [3P/2P/Ttr (tan tan radius)]: *Specify center point.*
Specify radius of circle or [Diameter]: *Enter radius.*

Notice the command and input sequence:

CIRCLE
<RETURN>
Center point
<RETURN>
Radius
<RETURN>

CIRCLE (C,D) Command
Command: **CIRCLE**
Specify center point for circle or [3P/2P/Ttr (tan tan radius)]: *Specify center point.*
Specify radius of circle or [Diameter]: **D**
Specify diameter of circle: *Enter diameter.*

Notice the command and input sequence:

CIRCLE
<RETURN>
Center Point
<RETURN>
D
<RETURN>
Diameter
<RETURN>

CIRCLE (2P) Command
Command: **CIRCLE**
Specify center point for circle or [3P/2P/Ttr (tan tan radius)]: **2P**
Specify first end point of circle's diameter: *Specify first point.*
Specify second end point of circle's diameter: *Specify second point.*

Notice the command and input sequence:

CIRCLE
<RETURN>
2P
<RETURN>
Select first point on diameter
<RETURN>
Select second point on diameter
<RETURN>

ERASE Command
Command: **ERASE**

Notice the command and input sequence:

ERASE
<RETURN>
MOVE Command
Command: **MOVE**

Notice the command and prompt entry sequence:

MOVE
<RETURN>

The difference between the Center-Radius and Center-Diameter options of the **CIRCLE** command is that in the first one the RADIUS is the default, whereas in the second one you need to enter D to use the **Diameter** option. This difference, although minor, is very important when writing a menu file. Similarly, the 2P (two-point) option of the **CIRCLE** command is different from the other two options. Therefore, it is important to know both the correct sequence of the AutoCAD LT commands and the entries made in response to the prompts associated with those commands.

You can use any text editor (like Notepad) to write the file. The filename can be up to eight characters long, and the file extension must be **.MNU**. If the filename exists, it will be automatically loaded; otherwise a new file will be created. To understand the process of developing a down menu, consider the following example.

Example 1

Write a menu for the following AutoCAD LT commands:

LINE	**ERASE**	**REDRAW**	**SAVE**
PLINE	**MOVE**	**REGEN**	**QUIT**
CIRCLE C,R	**COPY**	**ZOOM ALL**	**PLOT**
CIRCLE C,D	**STRETCH**	**ZOOM WIN**	
CIRCLE 2P	**EXTEND**	**ZOOM PRE**	
CIRCLE 3P	**OFFSET**		

The first step in writing any menu is to design the menu so that the commands are arranged in the desired configuration. Figure 27-2 shows one of the possible designs of this menu. This menu has four different groups of commands; therefore, it will have four sections: POP1, POP2, POP3, and POP4, and each section will have a section label. The following file is a listing of the pull-down menu file for Example 1. **The line numbers are not a part of the file; they are shown here for reference only.**

```
***POP1                  1
[DRAW]                   2
[LINE]*^C^CLINE          3
[PLINE]^C^CPLINE         4
```

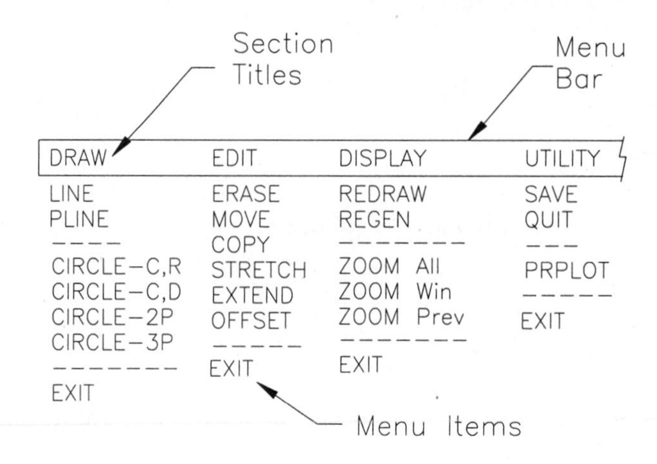

Figure 27-2 Design of menu

```
[—]                                      5
[CIR-C,R]^C^CCIRCLE                      6
[CIR-C,D]^C^CCIRCLE \D                   7
[CIR-2P]^C^CCIRCLE 2P                    8
[CIR-3P]^C^CCIRCLE 3P                    9
[—]                                      10
[Exit]^C                                 11
***POP2                                  12
[EDIT]                                   13
[ERASE]*^C^CERASE                        14
[MOVE]^C^CMOVE                           15
[COPY]^C^CCOPY                           16
[STRETCH]^C^CSTRETCH;C                   17
[OFFSET]^C^COFFSET                       18
[EXTEND]^C^CEXTEND                       19
[—]                                      20
[Exit]^C                                 21
***POP3                                  22
[DISPLAY]                                23
[REDRAW]'REDRAW                          24
[REGEN]^C^CREGEN                         25
[—]                                      26
[ZOOM-All]^C^C'ZOOM A                    27
[ZOOM-Window]'ZOOM W                     28
[ZOOM-Prev]'ZOOM PREV                    29
[—]                                      30
[~Exit]^C                                31
***POP4                                  32
[UTILITY]                                33
```

```
[SAVE]^C^CSAVE;                                                34
[QUIT]^C^CQUIT                                                 35
[——]                                                           36
[PLOT]^C^CPLOT                                                 37
[—]                                                            38
[Exit]^C                                                       39
```

Explanation

Line 1
*****POP1**
POP1 is the section label for the first pull-down menu. All section labels in the AutoCAD LT menu begin with three asterisks (***), followed by the section label name, such as POP1.

Line 2
[DRAW]
In this menu item **DRAW** is the menu bar title displayed when the cursor is moved in the menu bar area. The title names should be chosen so you can identify the type of commands you expect in that particular pull-down menu. In this example, all the draw commands are under the title **DRAW** (Figure 27-3), all edit commands are under **EDIT**, and so on for other groups of items. The menu bar title can be of any length. However, it is recommended to keep them short to accommodate other menu items. Some of the display devices provide a maximum of 80 characters. To have 16 sections in a single row in the menu bar, the length of each title should not exceed five characters. If the length of the menu bar items exceeds 80 characters, AutoCAD LT will wrap the items that cannot be accommodated in 80 characters space, and display them in the next line. This will result in two menu item lines in the menu bar.

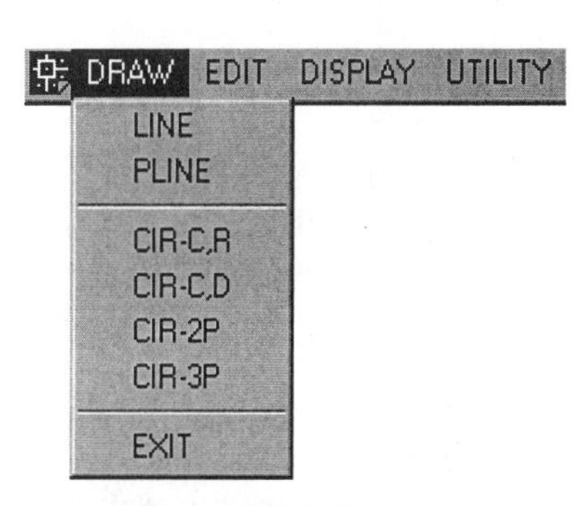

Figure 27-3 *Draw menu*

If the first line in a menu section is blank, the title of that section is not displayed in the menu bar area. Since the menu bar title is not displayed, you **cannot** access that menu. This allows you to turn off the menu section. For example, if you replace [**DRAW**] with a blank line, the **DRAW** section (POP1) of the menu will be disabled; the second section (POP2) will be displayed in its place.

Example

***POP1	Section label
	Blank line (turns off POP1)
[LINE:]^CLINE	Menu item
[PLINE:]^CPLINE	
[CIRCLE:]^CCIRCLE	

Line 3
***^C^CLINE**
In this menu item, the command definition starts with an asterisk (*). This feature allows the

command to be repeated automatically until it is canceled by pressing ESCAPE, entering CTRL C, or by selecting another menu command. ^C^C cancels the existing command twice; **LINE** is an AutoCAD LT command that generates lines.

***^C^CLINE**

Where ***** ------------------- Repeats the menu item (command)
^C^C ---------- Cancels existing command twice
LINE ------------- AutoCAD LT's **LINE** command

Line 5
[--]
To separate two groups of commands in any section, you can use a menu item that consists of two or more hyphens (--). This line automatically expands to fill the **entire width** of the menu. You can use a blank line in a menu. If any section of a menu (**POP section) has a blank line, it is ignored.

Line 11
[Exit]^C
In this menu item, **^C** command definition has been used to cancel the menu. This item provides you with one more option for canceling the menu. This is especially useful for new AutoCAD LT users who are not familiar with all AutoCAD LT features. The menu can also be canceled by any of the following actions:

1. Selecting a point.
2. Selecting an item in the screen menu area.
3. Selecting or typing another command.
4. Pressing ESC at the keyboard.
5. Selecting any menu title in the menu bar.

Line 28
[ZOOM-Window]'ZOOM W
In this menu item the single quote (') preceding the **ZOOM** command makes the **ZOOM** Window command transparent. When a command is transparent, the existing command is not canceled. After the **ZOOM** Window command (Figure 27-4), AutoCAD LT will automatically resume the current operation.

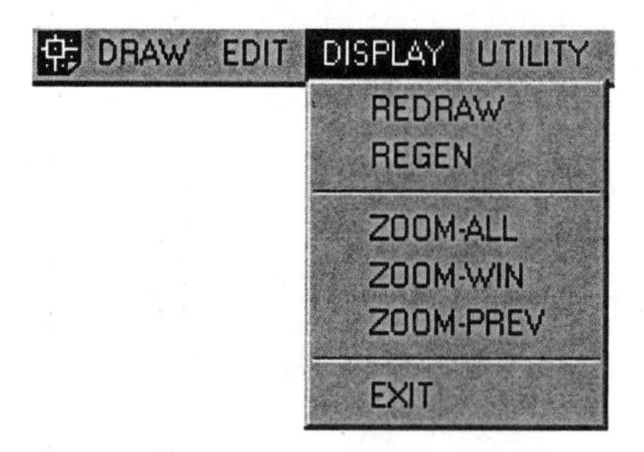

Figure 27-4 *Display menu*

[ZOOM-Window]'ZOOM W

Where **W** ----------------- Window option
ZOOM ----------- AutoCAD LT **ZOOM** command
' -------------------- Single quote makes **ZOOM** transparent

Line 31
[~Exit]^C
This menu item is similar to the menu item in line 11, except for the tilde (~). Since this menu item has a tilde (~), the menu item is not available (displayed grayed out), and if you select this item it will not cancel the menu. You can use this feature to disable a menu item or to

indicate that the item is not a valid selection. If there is an instruction associated with the item, the instruction will not be executed when you select the item. For example, [~OSNAPS]^C^C$S=OSNAPS will not load the OSNAPS submenu on the screen.

Line 34
[SAVE]^C^CSAVE;
In this menu item, the semicolon (;) that follows the **SAVE** command enters RETURN. The semicolon is not required; the command will also work without a semicolon.

[SAVE]^C^CSAVE;
Where **SAVE** ------------- AutoCAD LT **SAVE** command
; ------------------- Semicolon enters RETURN

Line 36
[----]
This menu item has four hyphens. The line will extend to fill the width of the menu. If there is only one hyphen [-], AutoCAD LT gives a syntex error when you load the menu.

Note

For all menus, the menu items are displayed directly beneath the menu title and are left-justified. If any menu [for example, the rightmost menu (POP16)] does not have enough space to display the entire menu item, the menu will expand to the left to accommodate the entire length of the longest menu item.

You can use // (two forward slashes) for comment lines. AutoCAD LT ignores the lines that start with //.

From this example it is clear that every statement in the menu is based on the AutoCAD LT commands and the information that is needed to complete these commands. This forms the basis for creating a menu file and should be given consideration. Following is a summary of the AutoCAD LT commands used in Example 1 and their equivalents in the menu file.

AutoCAD LT Commands	Menu File
Command: **LINE**	**[LINE]^C^CLINE**
Command: **CIRCLE** Specify center point for circle or [3P/2P/Ttr (tan tan radius)]: Specify radius of circle or [Diameter]:	**[CIR-C,R]^C^CCIRCLE**
Command: **CIRCLE** Specify center point for circle or [3P/2P/Ttr (tan tan radius)]: Specify radius of circle or [Diameter]: **D** Specify diameter of circle:	**[CIR-C,D]^C^CCIRCLE;\D**
Command: **CIRCLE** Specify center point for circle or [3P/2P/Ttr (tan tan radius)]: **2P**	

Specify first end point of circle's diameter: Specify second end point of circle's diameter:	**[CIR- 2P]^C^CCIRCLE;2P**
Command: **ERASE**	**[ERASE]^C^CERASE**
Command: **MOVE**	**[MOVE]^C^CMOVE**

LOADING MENUS

AutoCAD LT automatically loads the **ACLT.MNU** file when you get into the AutoCAD LT drawing editor. However, you can also load a different menu file by using the AutoCAD LT **MENULOAD** command. You can also load a menu file from the **Options** dialog box. Choose the **Files** tab and open the **Miscellaneous File Name** folder and then use the **Browse** button to choose the new menu file. The **MENULOAD** command is as follows:

Command: **MENULOAD**

When you enter the **MENULOAD** command, AutoCAD LT displays the **Menu Customization** dialog box (Menu Groups tab) (Figure 27-5) . ACLT is displayed as it is automatically loaded. Select ACLT and then choose the **Unload** button. Choose the **Browse** button to display the **Select Menu File** dialog box on the screen. Select the menu file (**.mnu**) that you want to load from the directory in which you have saved it, and then choose the **Open** button. The menu file will be displayed in the **File Name:** edit box. Choose the **Load** button to load the new menu file. Choose **Yes** in the AutoCAD LT Warning dialog box.

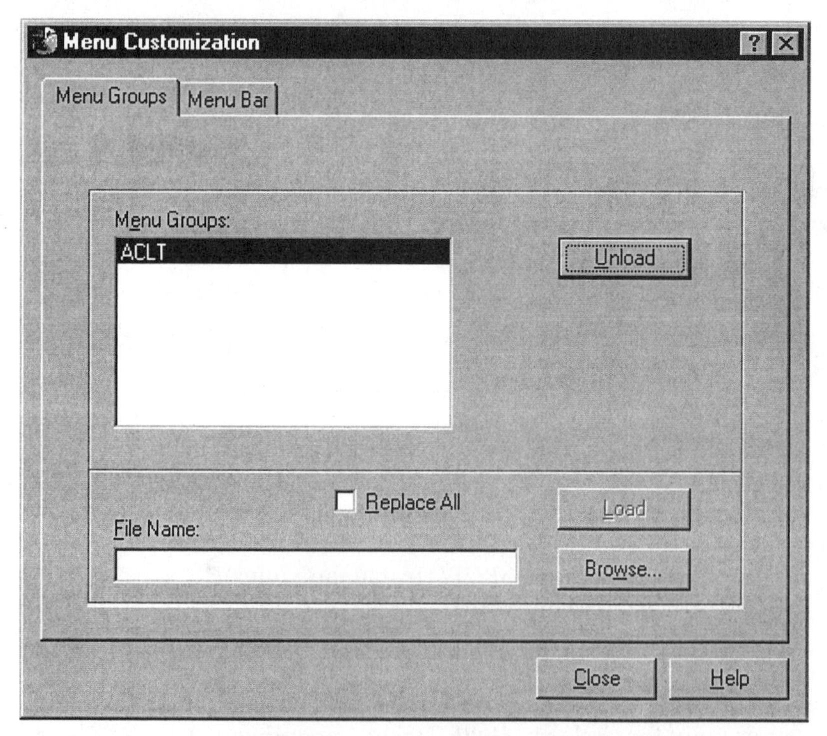

***Figure 27-5 Menu Customization** dialog box **(Menu Groups** tab)*

Choose the **Menu Bar** tab (Figure 27-6) in the dialog box. All the menu items are displayed in the Menus and the Menu Bar list boxes. Choose the **Close** button in the dialog box. The new menus are displayed in the menu bar.

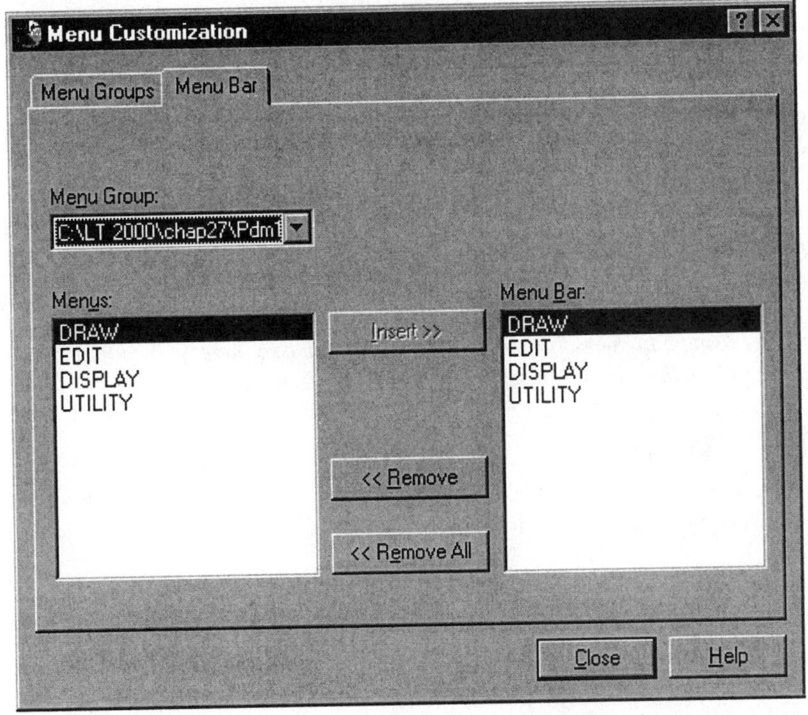

***Figure 27-6 Menu Customization** dialog box (**Menu Bar** tab)*

AutoCAD LT will automatically compile the menu file into **MNC** and **MNR** files. When you load a menu file in windows, AutoCAD LT creates the following files:

.mnc and **.mnr** files — When you load a menu file (.mnu), AutoCAD LT compiles the menu file and creates **.mnc** and **.mnr** files. The .mnc file is a compiled menu file. The **.mnr** file contains the bitmaps used by the menu.

.mns file — When you load the menu file, AutoCAD LT also creates an **.mns** file. This is an ASCII file that is same as the **.mnu** file when you initially load the menu file. Each time you make a change in the contents of the file, AutoCAD LT changes the **.mns** file.

Note

*After you load the new menu, you cannot use the buttons menu, or digitizer because the original menu, **ACLT.MNU**, is not present and the new menu does not contain these menu areas.*

*To activate the original menu, use the **MENULOAD** command. In the **Menu Customization** dialog box, unload the previous menu and load **ACLT.MNU** from the **Support** directory.*

If you need to use input from a keyboard or a pointing device, use the backslash (\). The system will pause for you to enter data.

There should be no space after the backslash (\).

The menu items, menu labels, and command definition can be uppercase, lowercase, or mixed.

You can introduce spaces between the menu items to improve the readability of the menu file. The blank lines are ignored and are not displayed on the screen.

If there are more items in the menu than the number of spaces available, the excess items are not displayed on the screen. For example, if the display device limits the number of items to 21, items in excess of 21 will not be displayed on the screen and are therefore inaccessible.

If you are using a high-resolution graphics board, you can increase the number of lines that can be displayed on the screen. On some devices this is 80 lines.

RESTRICTIONS

The menus are easy to use and provide a quick access to frequently used AutoCAD LT commands. However, the menu bar and the menus are disabled during the following commands:

DTEXT Command
After you assign the text height and the rotation angle to a **DTEXT** command, the menu is automatically disabled.

SKETCH Command
The menus are disabled after you set the record increment in the **SKETCH** command.

VPOINT Command
The menus are disabled while the axis tripod and compass are displayed on the screen.

ZOOM and DVIEW Commands
The menus are disabled when the dynamic zoom or dynamic view is in progress.

Exercise 1 — *General*

Write a menu for the following AutoCAD LT commands. (The layout of the menu is shown in Figure 27-7).

DRAW	**EDIT**	**DISP/TEXT**	**UTILITY**
LINE	**FILLET0**	**DTEXT,C**	**SAVE**
PLINE	**FILLET**	**DTEXT,L**	**QUIT**
ELLIPSE	**CHAMFER**	**DTEXT,R**	**END**
POLYGON	**STRETCH**	**ZOOM WIN**	**DIR**
DONUT	**EXTEND**	**ZOOM PRE**	**PLOT**
	OFFSET		

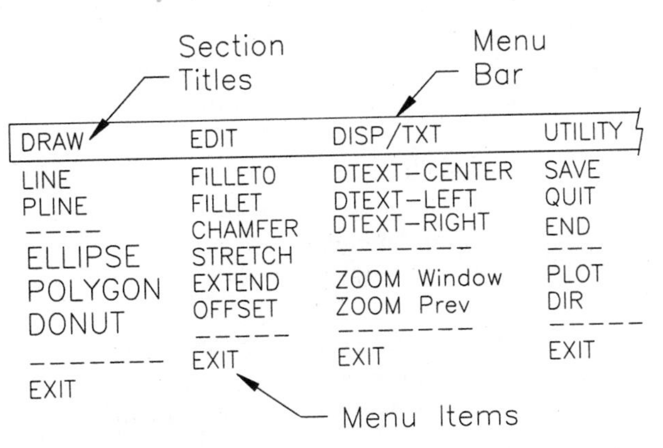

Figure 27-7 Menu display for Exercise 1

CASCADING SUBMENUS IN MENUS

The number of items in a menu or shortcut menu can be very large, and sometimes they cannot all be accommodated on one screen. For example, the maximum number of items that can be displayed on some display devices is 21. If the menu or the shortcut menu has more items than can be displayed, the excess menu items are not displayed on the screen and cannot be accessed. You can overcome this problem by using cascading menus that let you define smaller groups of items within a menu section. When an item is selected, it loads the cascading menu and displays the items, defined in the cascading menu, on the screen.

The cascading feature of AutoCAD LT allows pull-down and shortcut menus to be displayed in a hierarchical order that makes it easier to select submenus. To use the cascading feature in pull-down and shortcut menus, AutoCAD LT has provided some special characters. For example, -> defines a cascaded submenu and <- designates the last item in the menu. The following table lists some of the characters that can be used with the pull-down or shortcut menus.

Character	Character Description
--	The item label consisting of two hyphens automatically expands to fill the entire width of the menu. Example: [--]
+	Used to continue the menu item to the next line. This character has to be the last character of the menu item. Example: [Triang:]^C^CLine;1,1;+3,1;2,2;
->	This label character defines a cascaded submenu; it must precede the name of the submenu. Example: [**->Draw**]
<-	This label character designates the last item of the cascaded menu or shortcut menu. The character must precede the label item. Example: [**<-CIRCLE 3P**]**^C^CCIRCLE;3P**

<-<-... This label character designates the last item of the pull-down or shortcut menu and also terminates the parent menu. The character must precede the label item.
Example: [**<-<-Center Mark**]**^C^C_dim;_center**

$(This label character can be used with the pull-down and shortcut menus to evaluate a DIESEL expression. The character must precede the label item.
Example: **$**(if,**$**(getvar,orthomode),Ortho)

~ This item indicates the label item is not available (displayed grayed out); the character must precede the item.
Example: [~Application not available]

!. When used as a prefix, it displays the item with a check mark.

& When placed directly before a character, the character is displayed undescored. For Example, [W&Block] is displayed as WBlock. It also specifies the character as a menu accelerator key in the pull-down or shortcut menu.

\t The label text to the right of \t is displayed to the right side of the menu.

Some display devices provide space for a maximum of 80 characters. Therefore, if there are 10 menus, the length of each menu title should average eight characters. If the combined length of all menu bar titles exceeds 80 characters, AutoCAD LT automatically wraps the excess menu items and displays them on the next line in the menu bar. The following is a list of some additional features of the menu.

1. The section labels of the menus are ***POP1 through ***POP16. The menu bar titles are displayed in the menu bar.

2. The menus can be accessed by selecting the menu title from the menu bar at the top of the screen.

3. A maximum of 999 menu items can be defined in the menu. This includes the items that are defined in the submenus. The menu items in excess of 999 are ignored.

4. The number of menu items that can be displayed depends on the display device you are using. If the shortcut or the menu contains more items than can be accommodated on the screen, the excess items are truncated. For example, if your system can display only 21 menu items, the menu items in excess of 21 are automatically truncated.

Example 2

Write a menu for the commands shown in Figure 27-8. The menu must use the AutoCAD LT cascading feature.

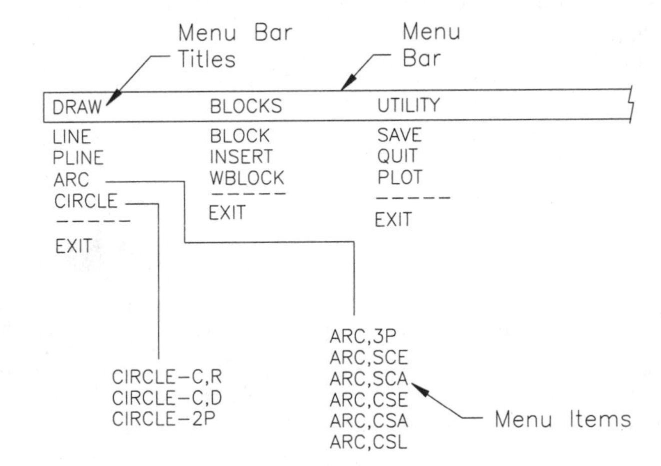

***Figure 27-8** Menu structure for Example 2*

The following file is a listing of the menu for Example 2. **The line numbers are not a part of the menu; they are shown here for reference only.**

```
***POP1                                  1
[DRAW]                                   2
[LINE]^C^CLINE                           3
[PLINE]^C^CPLINE                         4
[->ARC]                                  5
  [ARC]^C^CARC                           6
  [ARC,3P]^C^CARC;\\DRAG                 7
  [ARC,SCE]^C^CARC;\C;\DRAG              8
  [ARC,SCA]^C^CARC;\C;\A;DRAG            9
  [ARC,CSE]^C^CARC;C;\\DRAG              10
  [ARC,CSA]^C^CARC;C;\\A;DRAG            11
  [<-ARC,CSL]^C^CARC;C;\\L;DRAG          12
[->CIRCLE]                               13
  [CIRCLE C,R]^C^CCIRCLE                 14
  [CIRCLE C,D]^C^CCIRCLE;\D              15
  [CIRCLE 2P]^C^CCIRCLE;2P               16
  [<-CIRCLE 3P]^C^CCIRCLE;3P             17
[—]                                      18
[Exit]^C                                 19
***POP2                                  20
[BLOCKS]                                 21
[BLOCK]^C^CBLOCK                         22
[INSERT]*^C^CINSERT                      23
```

```
[WBLOCK]^C^CWBLOCK                                          24
[—]                                                         25
[Exit]^C                                                    26
***POP3                                                     27
[UTILITY]                                                   28
[SAVE]^C^CSAVE                                              29
[QUIT]^C^CQUIT                                              30
[PLOT]^C^CPLOT                                              31
[—]                                                         32
[Exit]^C                                                    33
```

Explanation

Line 5

[->ARC]

In this menu item, **ARC** is the menu item label that is preceded by the special label character **->**. This special character indicates that the menu item has a submenu. The menu items that follow it (Lines 6-12) are the submenu items.

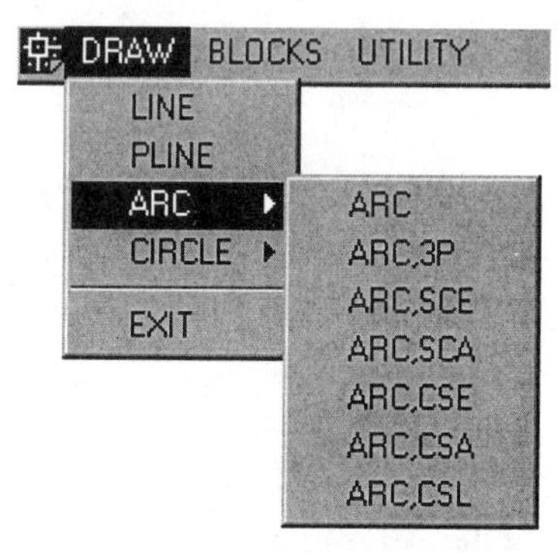

Figure 27-9 *Draw menu*

Line 12

[<-ARC,CSL]^C^CARC;C;\\L;DRAG

In this line, the menu item label **ARC,CSL** is preceded by another special label character, **<-**, which indicates the end of the submenu. The item that contains this character must be the last menu item of the submenu (Figure 27-9).

Lines 13 and 17

[->CIRCLE]

[<-CIRCLE 3P]^C^CCIRCLE;3P

The special character -> in front of **CIRCLE** indicates that the menu item has a submenu; the character <- in front of **CIRCLE 3P** indicates that this item is the last menu item in the submenu. When you select the menu item **CIRCLE** from the menu, it will automatically display the submenu on the side (Figure 27-10).

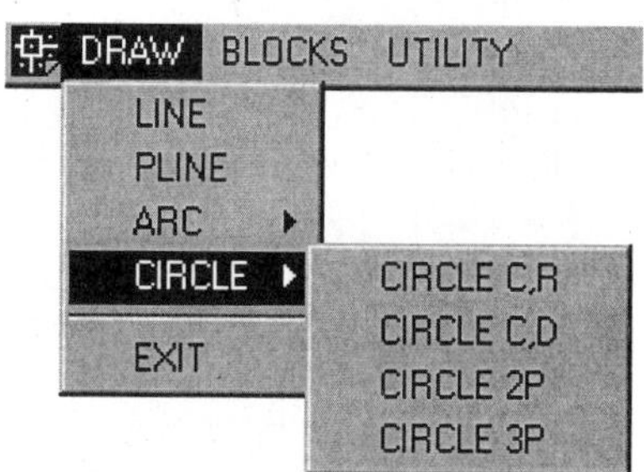

Figure 27-10 *Draw menu with CIRCLE submenu*

Example 3

Write a menu that has the cascading submenus for the commands shown in Figure 27-11.

The following file is a listing of the menu for Example 3. **The line numbers are not a part of the menu; they are shown here for reference only.**

```
***POP1                                                     1
[DRAW]                                                      2
[->CIRCLE]                                                  3
  [CIRCLE C,R]^C^C_CIRCLE                                   4
```

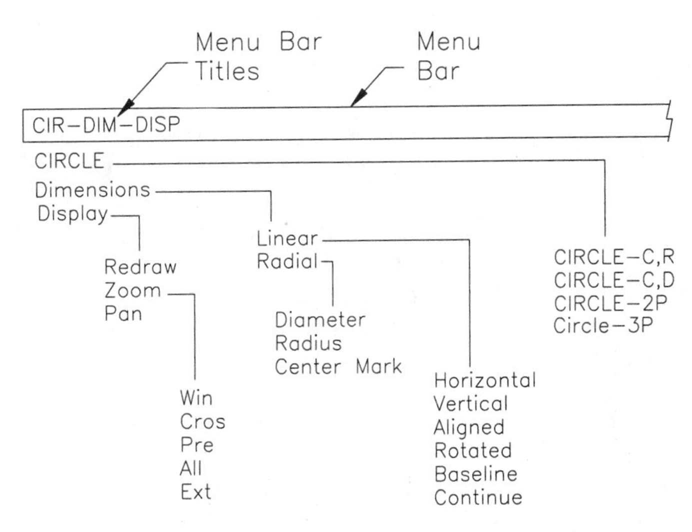

Figure 27-11 Menu structure for Example 3


```
  [CIRCLE C,D]^C^C_CIRCLE;\_D                 5
  [CIRCLE 2P]^C^C_CIRCLE;_2P                  6
  [<-CIRCLE 3P]^C^C_CIRCLE;_3P                7
[->DIMENSIONS]                                8
  [->LINEAR]                                  9
   [Horizontal]^C^C_dim;_horizontal           10
   [Vertical]^C^C_dim;_vertical               11
   [Aligned]^C^C_dim;_aligned                 12
   [Rotated]^C^C_dim;_rotated                 13
   [Baseline]^C^C_dim;_baseline               14
   [<-Continue]^C^C_dim;_continue             15
  [->RADIAL]                                  16
   [Diameter]^C^C_dim;_diameter               17
   [Radius]^C^C_dim;_radius                   18
   [<-<-Center Mark]^C^C_dim;_center          19
[->DISPLAY]                                   20
  [REDRAW]^C^CREDRAW                          21
  [->ZOOM]                                    22
   [...Win]^C^C_ZOOM;_W                       23
   [...Cros]^C^C_ZOOM;_C                      24
   [...Pre]^C^C_ZOOM;_P                       25
   [...All]^C^C_ZOOM;_A                       26
   [<-...Ext]^C^C_ZOOM;_E                     27
  [<-PAN]^C^C_Pan                             28
```

Explanation

Lines 8 and 9

[->DIMENSIONS]

[->LINEAR]

The special label character **->** in front of the menu item **DIMENSIONS** indicates that it has a submenu, and the character -> in front of **LINEAR** indicates that there is another submenu. The second submenu **LINEAR** is within the first submenu Dimensions (Figure 27-12). The menu items on lines 10 to 15 are defined in the **LINEAR** submenu, and the menu items Linear and **RADIAL** are defined in the submenu Dimensions.

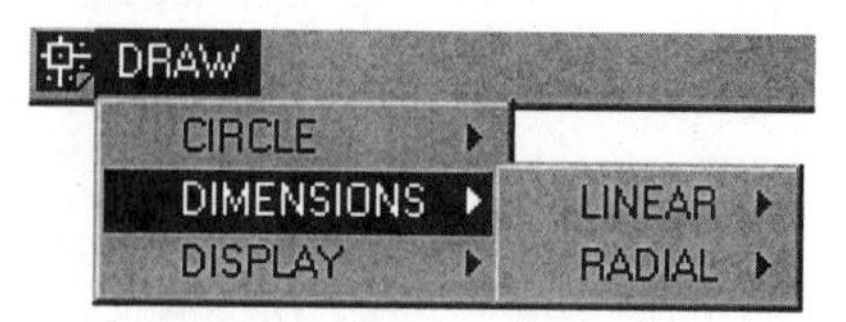

***Figure 27-12** Draw menu with cascading menus*

Line 16

[->RADIAL]

This menu item defines another submenu; the menu items on line numbers 17, 18, and 19 are part of this submenu.

Line 19

[<-<-Center Mark]^C^C_dim;_center

In this menu item the special label character **<-<-** terminates the **Radial** and **Dimensions** (parent submenu) submenus.

Lines 27 and 28

[<-...Ext]^C^C_ZOOM;_E

[<-PAN]^C^C_Pan

The special character **<-** in front of the menu item **...Ext** terminates the **ZOOM** submenu (Figure 27-13); the special character in front of the menu item **PAN** terminates the **DISPLAY** submenu.

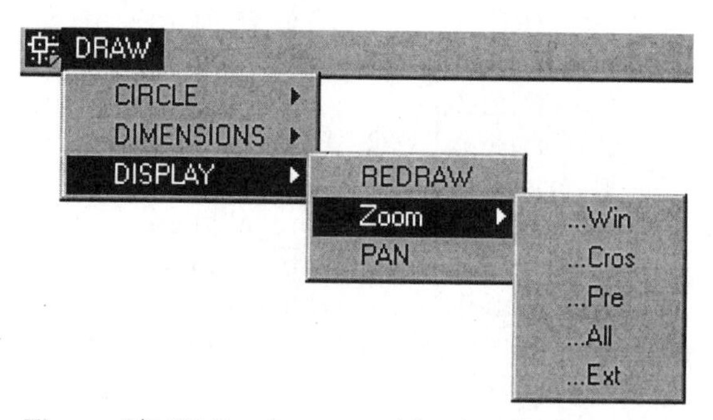

***Figure 27-13** Draw menu with cascading menus*

SHORTCUT AND CONTEXT MENUS

The shortcut menus are similar to the pull-down menus, except that it can contain only 499 menu items compared with 999 items in the pull-down menu. The section label of the shortcut menu can be ***POP0 and ***POP500 to ***POP 999. The shortcut menus are displayed near or at the cursor location. Therefore, they can be used to provide convenient and quick access to some of the frequently used commands. The shortcut menus in the upper range are also referred as context menus. The following are some of the features of shortcut menu.

1. The section label of the shortcut menu are ***POP0 and ***POP500 to ***POP999. The menu bar title defined under this section label is not displayed in the menu bar.

2. On most systems, the menu bar title is not displayed at the top of the shortcut menu. However, for compatibility reasons you should give a dummy menu bar title.

3. The POP0 menu can be accessed through the **$P0=*** menu command. The shortcut menus POP500 through POP999 must be referenced by their alias names. The reserved alias names for AutoCAD LT use are GRIPS, CMDEFAULT, CMEDIT, and CMCOMMAND. For example, to reference POP500 for grips, use **GRIPS command line under POP5000. This commands can be issued by a menu item in another menu, such as the button menu, or the auxiliary menu.

4. A maximum of 499 menu items can be defined in the shortcut menu. This includes the items that are defined in the shortcut submenus. The menu items in excess of 499 are ignored.

5. The number of menu items that can be displayed on the screen depends on the system you are using. If the shortcut or pull-down menu contains more items than your screen can accommodate, the excess items are truncated. For example, if your system displays 21 menu items, the menu items in excess of 21 are automatically truncated.

6. The system variable **SHORTCUTMENU** controls the availability of Default, Edit, and Command mode shortcut menus. If the value is 0, it disables the Default, Edit, and Command mode shortcut menus. The default value of this variable is 11.

Example 4

Write a shortcut menu for the following AutoCAD LT commands using cascading submenus. The menu should be compatible with foreign language versions of AutoCAD LT. Use the second button of the **BUTTONS** menu to display the shortcut menu.

Osnaps	Draw	DISPLAY
Center	Line	**REDRAW**
Endpoint	**PLINE**	**ZOOM**
Intersection	**CIR C,R**	...Win
Midpoint	**CIR 2P**	...Cen
Nearest	**ARC SCE**	...Prev
Perpendicular	**ARC CSE**	...All
Quadrant		...Ext
Tangent		PAN
None		

The following file is a listing of the menu file for Example 4. **The line numbers are not a part of the file; they are for reference only.**

```
***AUX1                                   1
$P0=*                                     2
***POP0                                   3
[Osnaps]                                  4
[Center]_Center                           5
[End point]_Endp                          6
[Intersection]_Int                        7
```

```
[Midpoint]_Mid                                          8
[Nearest]_Nea                                           9
[Perpendicular]_Per                                    10
[Quadrant]_Qua                                         11
[Tangent]_Tan                                          12
[None]_Non                                             13
[—]                                                    14
[->Draw]                                               15
 [Line]^C^C_Line                                       16
 [PLINE]^C^C_Pline                                     17
 [CIR C,R]^C^C_Circle                                  18
 [CIR 2P]^C^C_Circle;_2P                               19
 [ARC SCE]^C^C_ARC;\C                                  20
 [<-ARC CSE]^C^C_Arc;C                                 21
[—]                                                    22
[->DISPLAY]                                            23
  [REDRAW]^C^_REDRAW                                   24
  [->ZOOM]                                             25
   [...Win]^C^C_ZOOM;_W                                26
   [...Cen]^C^C_ZOOM;_C                                27
   [...Prev]^C^C_ZOOM;_P                               28
   [...All]^C^C_ZOOM;_A                                29
   [<-...Ext]^C^C_ZOOM;_E                              30
  [<-PAN]^C^C_Pan                                      31
***POP1                                                32
[SHORTCUTMENU]                                         33
[SHORTCUTMENU=0]^C^CSHORTCUTMENU;0                     34
[SHORTCUTMENU=11]^C^CSHORTCUTMENU;11                   35
```

Explanation

Line 1
*****AUX1**
AUX1 is the section label for the first auxiliary menu; *** designates the menu section. The menu items that follow it, until the second section label, are a part of this buttons menu.

Lines 2
$P0=*
The special command **$P0=*** is assigned to the second button of the pointing device.

Lines 3 and 4
*****POP0**
[Osnaps]
The menu label **POP0** is the menu section label for the shortcut menu; Osnaps is the menu bar title. The menu bar title is not displayed, but is required. Otherwise, the first item will be interpreted as a title and will be disabled.

Line 5
[Center]_Center
In this menu item, **_Center** is the center object Snap mode. The menu files can be used with foreign language versions of AutoCAD LT, if AutoCAD LT commands and the command options are preceded by the underscore (_) character.

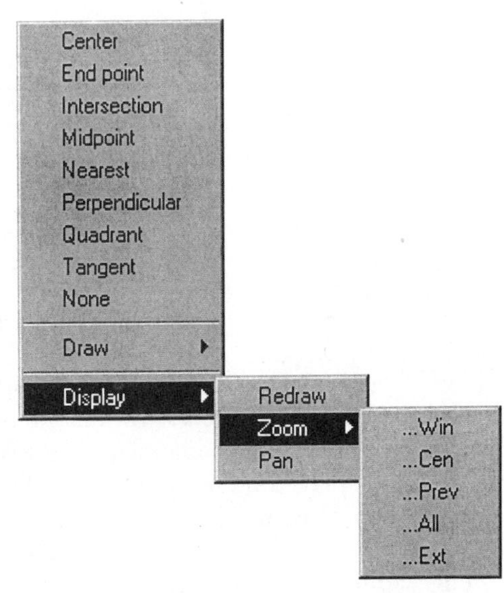

Figure 27-14 *Shortcut menu for Example 4*

After loading the menu, if you press the second button of your pointing device, the shortcut menu (Figure 27-14) will be displayed at the cursor (screen crosshairs) location. The system variable **SHORTCUTMENU** should be set to 0. If the cursor is close to the edges of the screen, the Shortcut menu will be displayed at a location that is closest to the cursor position. When you select a submenu, the items contained in the submenu will be displayed.

Lines 32 and 33
*****POP1**
[Draw]
*****POP1** defines the first pull-down menu. If no POPn sections are defined or the status bar is turned off, the Shortcut menu is automatically disabled.

Exercise 2

General

Write a menu for the following AutoCAD LT commands. Use a cascading menu for the **LINE** command options in the menu. (The layout of the menu is shown in Figure 27-15.)

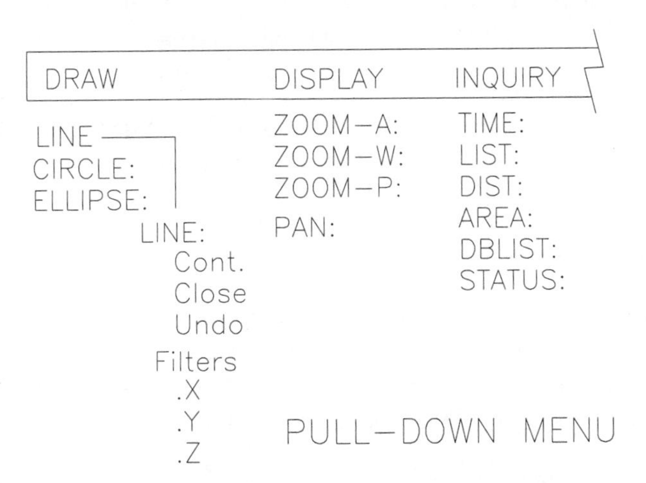

Figure 27-15 *Design of menu for Exercise 2*

LINE	**ZOOM All**	**TIME**
Continue	**ZOOM Win**	**LIST**
Close	**ZOOM Pre**	**DISTANCE**
Undo	**PAN**	**AREA**
.X		
.Y		
.Z		
CIRCLE		
ELLIPSE		

SUBMENUS

The number of items in a pull-down menu or shortcut menu can be very large and sometimes they cannot all be accommodated on one screen. For example, the maximum number of items that can be displayed on most of the display devices is 21. If the menu or the shortcut menu has more than 21 items, the menu excess items are not displayed on the screen and cannot be accessed. You can overcome this problem by using submenus that let you define smaller groups of items within a menu section. When a submenu is selected, it loads the submenu items and displays them on the screen.

The menus that use AutoCAD LT's cascading feature are the most efficient and easy to write. The submenus follow a logical pattern that are easy to load and use without causing any confusion. It is strongly recommended to use the cascading menus whenever you need to write the pull-down or the shortcut menus. However, AutoCAD LT provides the option to swap the submenus in the menus. These menus can sometimes cause distraction because the original menu is completely replaced by the submenu when swapping the menus.

Submenu Definition

A submenu definition consists of two asterisk signs (**) followed by the name of the submenu. A menu can have any number of submenus and every submenu should have a unique name. The items that follow a submenu, up to the next section label or submenu label, belong to that submenu. Following is the format of a submenu label:

****Name**

Where ****** ----------------- Two asterisk signs (**) designate a submenu
Name ------------ Name of the submenu

Note

The submenu name can be up to 31 characters long.

The submenu name can consist of letters, digits, and the special characters like: $ (dollar), - (hyphen), and _ (underscore).

The submenu name cannot have any embedded blanks.

The submenu names should be unique in a menu file.

Submenu Reference

The submenu reference is used to reference or load a submenu. It consists of a "$" sign followed by a letter that specifies the menu section. The letter that specifies a menu section is Pn, where n designates the number of the menu section. The menu section is followed by "=" sign and the name of the submenu that the user wants to activate. The submenu name should be without "**". Following is the format of a submenu reference:

$Section=Submenu

Where **$** ------------------ "$" sign
Section ---------- Menu section specifier
= ------------------ "=" sign
Submenu ------- Name of submenu

Example
$P1=P1A

Where **$P1** -------------- P1-Specifies pull-down menu section 1
P1A -------------- Name of submenu

Displaying a Submenu

When you load a submenu in a menu, the submenu items are not automatically displayed on the screen. For example, when you load a submenu P1A that has DRAW-ARC as the first item, the current title of POP1 will be replaced by the DRAW-ARC. But the items that are defined under DRAW-ARC are not displayed on the screen. To force the display of the new items on the screen, AutoCAD LT uses a special command $Pn=*.

$Pn=*

Where **P** ----------------- P for menu
Pn ---------------- Menu section number (1 to 10)
***** ------------------ Asterisk sign (*)

LOADING MENUS

From the menu, you can load any menu that is defined in the image tile menu sections by using the appropriate load commands. It may not be needed in most of the applications, but if you want to, you can load the menus that are defined in other menu sections.

Loading an Image Tile Menu

You can also load an image tile menu from the menu by using the following load command:

$I=IMAGE1 $I=*

Where **$I=Image1** ----- Load the submenu IMAGE1
$I=* ------------ To display the dialog box

This menu item consists of two load commands. The first load command $I=IMAGE1 loads the image tile submenu IMAGE1 that has been defined in the image tile menu section of the file. The second load command $I=* displays the new dialog box on the screen.

Chapter 27

Example 5

Write a menu for the following AutoCAD LT commands. Use submenus for the **ARC** and **CIRCLE** commands.

LINE	**BLOCK**	**QUIT**
PLINE	**INSERT**	**SAVE**
ARC	**WBLOCK**	——
ARC 3P	**PLOT**	
ARC SCE		
ARC SCA		
ARC CSE		
ARC CSA		
ARC CSL		
CIRCLE		
CIRCLE C,R		
CIRCLE C,D		
CIRCLE 2P		

The layout shown in Figure 27-16 is one of the possible designs for this menu. The **ARC** and **CIRCLE** commands are in separate groups that will be defined as submenus in the menu file.

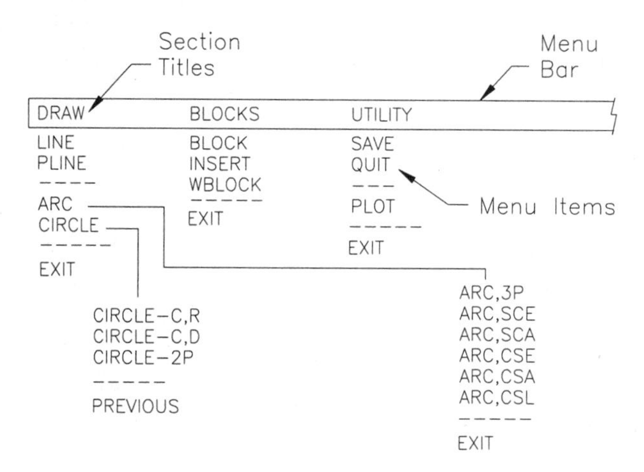

***Figure 27-16** Design of menu for Example 5*

The following file is a listing of the menu for Example 5. The line numbers are not a part of the menu file. They are given here for reference only.

```
***POP1                  1
**P1A                    2
[DRAW]                   3
[LINE]^C^CLINE           4
[PLINE]^C^CPLINE         5
[—]                      6
```

```
[ARC]^C^C$P1=P1B $P1=*                         7
[CIRCLE]^C^C$P1=P1C $P1=*                      8
[—]                                            9
[Exit]^C                                       10
                                               11
**P1B                                          12
[ARC]                                          13
[ARC,3P]^C^CARC \\DRAG                         14
[ARC,SCE]^C^CARC \C \DRAG                      15
[ARC,SCA]^C^CARC \C \A DRAG                    16
[ARC,CSE]^C^CARC C \\DRAG                      17
[ARC,CSA]^C^CARC C \\A DRAG                    18
[ARC,CSL]^C^CARC C \\L DRAG                    19
[—]                                            20
[PREVIOUS]$P1=P1A $P1=*                        21
                                               22
**P1C                                          23
[CIRCLE]                                       24
[CIRCLE C,R]^C^CCIRCLE                         25
[CIRCLE C,D]^C^CCIRCLE \D                      26
[CIRCLE 2P]^C^CCIRCLE 2P                       27
[—]                                            28
[PREVIOUS]$P1=P1A $P1=*                        29
                                               30
***POP2                                        31
[BLOCKS]                                       32
[BLOCK]^C^CBLOCK                               33
[INSERT]*^C^CINSERT                            34
[WBLOCK]^C^CWBLOCK                             35
[—]                                            36
[Exit]$P1=P1A $P1=*                            37
                                               38
***POP3                                        39
[UTILITY]                                      40
[SAVE]^C^CSAVE                                 41
[QUIT]^C^CQUIT                                 42
[~—]                                           43
[PLOT]^C^CPLOT                                 44
[~—]                                           45
[Exit]^C                                       46
                                               47
```

Line 2
****P1A**
**P1A defines the submenu P1A. All the submenus have two asterisk signs (**) followed by the name of the submenu. The submenu can have any valid name. In this example, P1A has been chosen because it is easy to identify the location of the submenu. P indicates that it is a menu, 1 indicates that it is in the first menu (POP1), and A indicates that it is the first submenu.

Line 6

[--]

The two hyphens enclosed in the brackets will automatically expand to fill the entire width of the menu. This menu item cannot be used to define a command. If it does contain a command definition, the command is ignored. For example, if the menu item is [--]^C^CLINE, the command ^C^CLINE will be ignored.

Line 7

[ARC]^C^C$P1=P1B $P1=*

In this menu item, $P1=P1B loads the submenu P1B and assigns it to the first menu section (POP1), but the new menu is not displayed on the screen. $P1=* forces the display of the new menu on the screen.

For example, if you select **CIRCLE** from the first menu (POP1), the menu bar title **DRAW** will be replaced by **CIRCLE**, but the new menu is not displayed on the screen. Now, if you select **CIRCLE** from the menu bar, the command defined under the **CIRCLE** submenu will be displayed in the menu. To force the display of the menu that is currently assigned to POP1, you can use AutoCAD LT's special command **$P1=***. If you select **CIRCLE** from the first menu (POP1), the **CIRCLE** submenu will be loaded and automatically displayed on the screen (Figure 27-17).

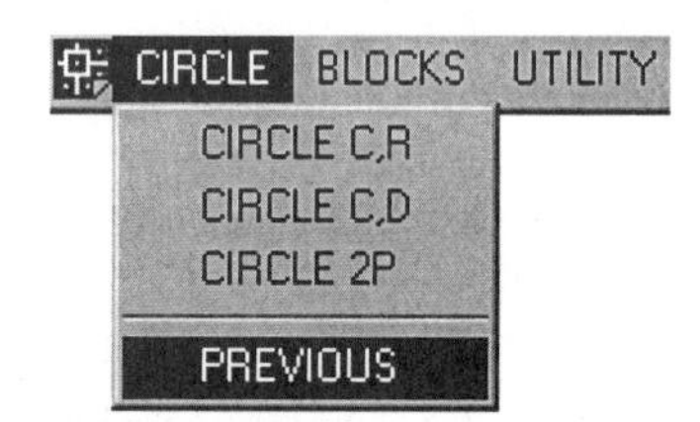

***Figure 27-17** CIRCLE menu replaces the DRAW menu*

Line 29

[PREVIOUS]$P1=P1A $P1=*

In this menu item $P1=P1A loads the submenu P1A, which happens to be the previous menu in this case. You can also use $P1= to load the previous menu. $P1=* forces the display of the submenu P1A.

PARTIAL MENUS

AutoCAD LT has provided a facility that allows users to write their own menus and load them in the menu bar. For example, in Windows you can write partial menus, toolbars, and definitions for accelerator keys. After you write the menu, AutoCAD LT lets you load the menu and use it with the standard menu. For example, you could load a partial menu and use it like a menu. You can also unload the menus that you do not want to use. These features make it convenient to use the menus that have been developed by AutoCAD LT users and developers.

Menu Section Labels

The following is a list of the additional menu section labels for Windows.

Section label	**Description**
***MENUGROUP	Menu file group name
***TOOLBARS	Toolbar definition
***HELPSTRING	Online help
***ACCELERATORS	Accelerator key definitions

Writing Partial Menus

The following example illustrates the procedure for writing a partial menu for Windows:

Example 6

In this example you will write a partial menu for Windows. The menu file has two menus, POP1 (MyDraw) and POP2 (MyEdit), as shown in Figure 27-18.

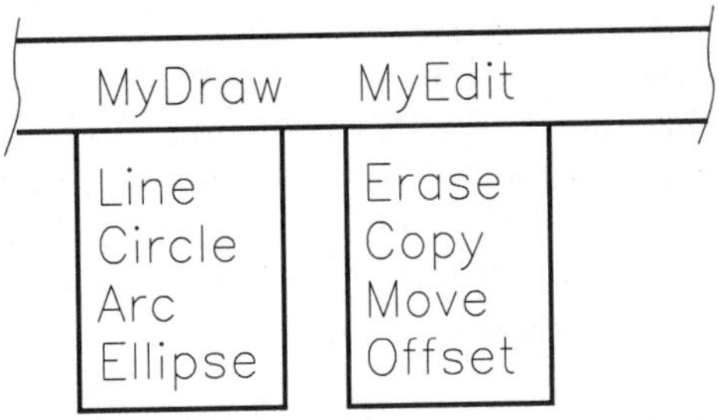

MyDraw MyEdit
Line Erase
Circle Copy
Arc Move
Ellipse Offset

Figure 27-18 Partial menus for Example 6

Step 1

Use a text editor to write the following menu file. The name of the file is assumed to be **MYMENU1.MNU**. The following is the listing of the menu file for this example:

```
***MENUGROUP=Menu1                         1
***POP1                                    2
[/MMyDraw]                                 3
[/LLine]^C^CLine                           4
[/CCircle]^C^CCircle                       5
[/AArc]^C^CArc                             6
[/EEllipse]^C^CEllipse                     7
***POP2                                    8
[/EMyEdit]                                 9
[/EErase]^C^CErase                         10
[/CCopy]^C^CCopy                           11
[/MMove]^C^CMove                           12
[/OOffset]^C^COffset                       13
```

]Explanation

Line 1

*****MENUGROUP=Menu1**

MENUGROUP is the section label and the Menu1 is the name tag for the menu group. The MENUGROUP label must precede all menu section definitions. The name of the MENUGROUP (Menu1) can be up to 32 characters long (alphanumeric), excluding spaces and punctuation marks. There is only one MENUGROUP in a menu file. All section labels must be preceded by *** (***MENUGROUP).

Line 2

*****POP1**

POP1 is the menu section label. The items on line numbers 3 through 7 belong to this section. Similarly, the items on line numbers 9 through 13 belong to the menu section **POP2**.

Line 3

[/MMyDraw]

/M defines the mnemonic key you can use to activate the menu item. For example, /M will display an underline under the letter M in the text string that follows it. If you enter the letter M, AutoCAD LT will execute the command defined in that menu item. MyDraw is the menu item label. The text string inside the brackets [ ] , except /M, has no function. It is used for displaying the function name so that the user can recognize the command that will be executed by selecting that item.

Line 4
[/LLine]^C^CLine
In this line, the /L defines the mnemonic key, and the Line that is inside the brackets is the menu item label. ^C^C cancels the command twice, and the Line is the AutoCAD LT **LINE** command. The part of the menu item statement that is outside the brackets is executed when you select an item from the menu. When you select Line 4, AutoCAD LT will execute the **LINE** command.

Step 2
Save the file and then load the partial menu file using the **MENULOAD** command.

Command: **MENULOAD**

When you enter the **MENULOAD** command, AutoCAD LT displays the **Menu Customization** dialog box (Figure 27-19) which displays the ACLT.mnu file. To load a partial menu do not unload the ACLT.mnu but the partial menu can be added. To load the menu file, enter the name of the menu file, **MYMENU1.MNU**, in the **File Name**: edit box. You can also use the

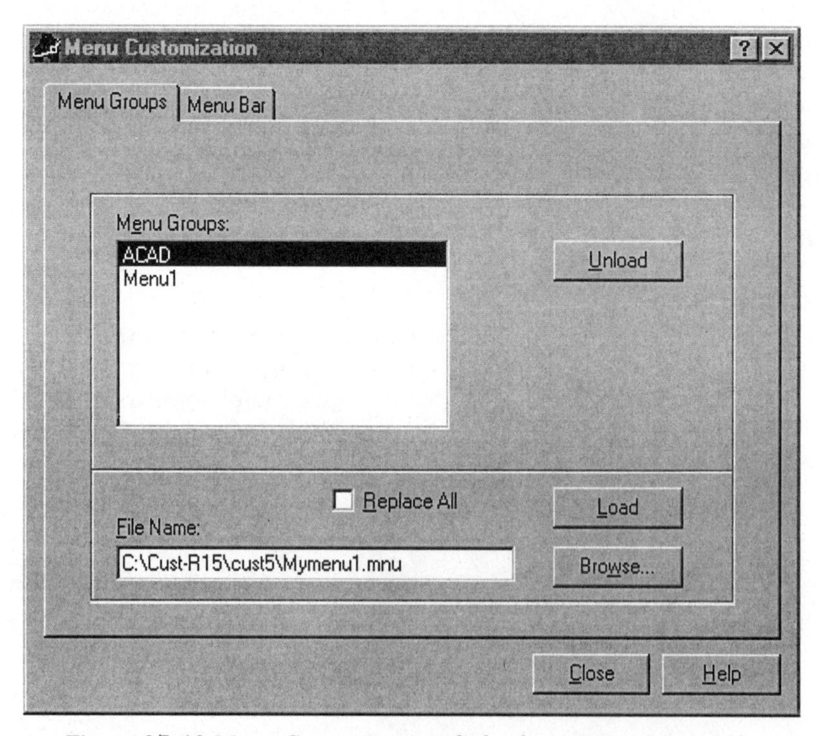

Figure 27-19 Menu Customization *dialog box (Menu Groups tab)*

Browse option to invoke the **Select Menu File** dialog box. Select the name of the file, and then use the **OK** button to return to the **Menu Customization** dialog box. To load the selected menu file, choose the **Load** button and choose the **Yes** button in the Warning dialog box. The name of the menu group (MENU1) will be displayed in the **Menu Groups** list box along with the original menu ACLT.

Step 3

In the **Menu Customization** dialog box, select the **Menu Bar** tab to display the menu bar options (Figure 27-20). In the **Menu Group** drop-down select **Menu1**; the menus defined in the menu group (Menu1) will be displayed in the **Menus** list box. In the **Menus** list box select the menu (MyDraw) that you want to insert in the menu bar. In the **Menu Bar** list box select the position where you want to insert the new menu. For example, if you want to insert the new menu in front of Format, select the **Format** menu in the **Menu Bar** list box. Choose the **Insert** button to insert the selected menu (MyDraw) in the menu bar. The menu (MyDraw) is displayed in the menu bar located at the top of your screen.

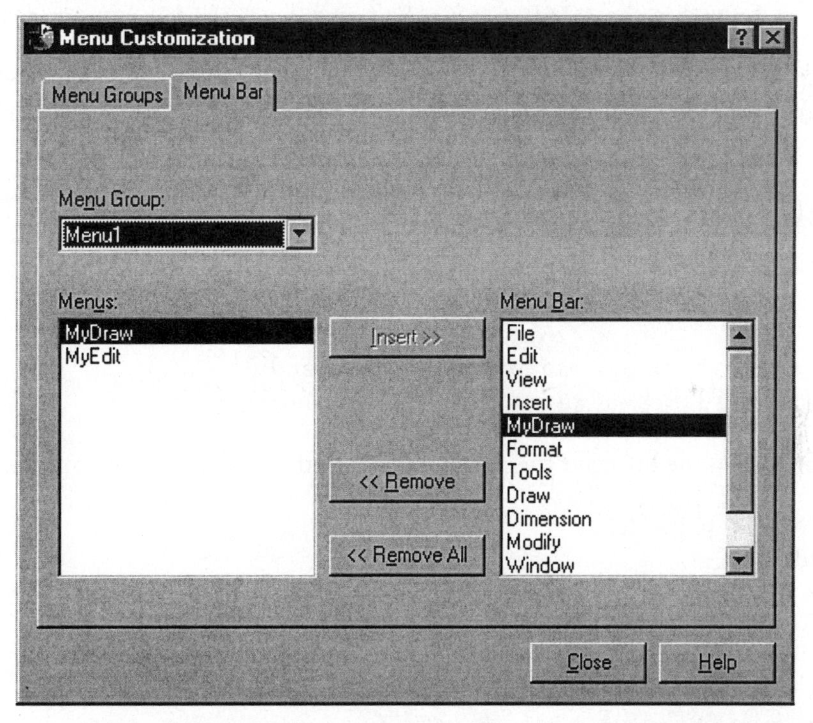

***Figure 27-20 Menu Customization** dialog box (Menu Bar tab)*

After inserting the menu items, AutoCAD LT will display the menu titles in the menu bar, as shown in Figure 27-21. If you select MyDraw, the corresponding menu as defined in the menu file will be displayed on the screen. Similarly, selecting MyEdit will display the corresponding edit menu.

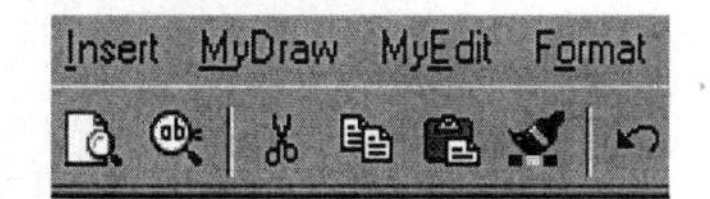

***Figure 27-21** Menu titles in the menu bar*

Step 4

If you want to unload the menu (MyDraw), select the Menu Bar tab of the **Menu Customization** dialog box. In the **Menu Bar** list box select the menu item that you want to unload. Choose the **Remove** button to remove the selected item.

Step 5

You can also unload the menu groups (Menu1) using the **Menu Customization** dialog box. Enter the **MENULOAD** or **MENUUNLOAD** command to invoke the **Menu Customization** dialog box. AutoCAD LT will display the names of the menu files in the **Menu Groups:** list box. Select the **Menu1** menu, and then choose the **Unload** button. AutoCAD LT will unload the menu group. Choose the **Close** button to exit the dialog box.

Accelerator Keys

AutoCAD LT for Windows also supports user-defined accelerator keys. For example, if you enter **C** at the Command prompt, AutoCAD LT draws a circle because it is a command alias for a circle as defined in the ACLT.PGP file. You cannot use the **C** key to enter the **COPY** command. To use the **C** key for entering the **COPY** command, you can define the accelerator keys. You can combine the SHIFT key with C in the menu file so that when you hold down the SHIFT key and then press the C key, AutoCAD LT will execute the **COPY** command. The following example illustrates the use of accelerator keys.

Example 7

In this example you will add the following accelerator keys to the partial menu of Example 6.

CONTROL+"E"	to draw an ellipse (**ELLIPSE** command)
SHIFT+"C"	to copy (**COPY** command)
[CONTROL"Q"]	to quit (**QUIT** command)

The following file is the listing of the partial menu file that uses the accelerator keys.

```
***MENUGROUP=Menu1
***POP1
**Alias
[/MMyDraw]
[/LLine]^C^CLine
[/CCircle]^C^CCircle
[/AArc]^C^CArc
ID_Ellipse [/EEllipse]^C^CEllipse

***POP2
[/EMyEdit]
[/EErase]^C^CErase
ID_Copy [/CCopy]^C^CCopy
[/OOffset]^C^COffset
[/VMove]^C^CMove
```

```
***ACCELERATORS
ID_Ellipse [CONTROL+"E"]
ID_Copy [SHIFT+"C"]
[CONTROL"Q"]^C^CQuit
```

This menu file defines three accelerator keys. The **ID_Copy [SHIFT+"C"]** accelerator key consists of two parts. The ID_Copy is the name tag, which must be the same as used earlier in the menu item definition. The SHIFT+"C" is the label that contains the modifier (SHIFT) and the keyname (C). The keyname or the string, such as "ESCAPE", must be enclosed in quotation marks. After you load the file, SHIFT+C will enter the **COPY** command and CTRL+E will draw an ellipse. Similarly, CTRL+Q will cancel the existing command and enter the **QUIT** command.

The accelerator keys can be defined in two ways. One way is to give the name tag followed by the label containing the modifier. The modifier is followed by a single character or a special virtual key enclosed in quotation marks [CONTROL+"E"] or ["ESCAPE"]. You can also use the plus sign (+) to concatenate the modifiers [SHIFT + CONTROL + "L"]. The other way of defining an accelerator key is to give the modifier and the key string, followed by a command sequence [CONTROL "Q"]^C^CQuit.

Special Virtual Keys

The following are the special virtual keys. These keys must be enclosed in quotation marks when used in the menu file.

String	Description	String	Description
"F1"	F1 key	"NUMBERPAD0"	0 key
"F2"	F2 key	"NUMBERPAD1"	1 key
"F3"	F3 key	"NUMBERPAD2"	2 key
"F4"	F4 key	"NUMBERPAD3"	3 key
"F5"	F5 key	"NUMBERPAD4"	4 key
"F6"	F6 key	"NUMBERPAD5"	5 key
"F7"	F7 key	"NUMBERPAD6"	6 key
"F8"	F8 key	"NUMBERPAD7"	7 key
"F9"	F9 key	"NUMBERPAD8"	8 key
"F10"	F10 key	"NUMBERPAD9"	9 key
"F11"	F11 key	"UP"	UP-ARROW key
"F12"	F12 key	"DOWN"	DOWN-ARROW key
"HOME"	HOME key	"LEFT"	LEFT-ARROW key
"END"	END key	"RIGHT"	RIGHT-ARROW key
"INSERT"	INS key	"ESCAPE"	ESC key
"DELETE"	DEL key		

Valid Modifiers. The following are the valid modifiers:

String	Description
CONTROL	The CTRL key on the keyboard
SHIFT	The SHIFT key (left or right)

COMMAND	The Apple key on Macintosh keyboards
META	The meta key on UNIX keyboards

Toolbars

The contents of the toolbar and its default layout can be specified in the Toolbar section (***TOOLBARS) of the menu file. Each toolbar must be defined in a separate submenu.

Toolbar Definition. The following is the general format of the toolbar definition:

```
***TOOLBARS
**MYTOOLS1
TAG1 [Toolbar ("tbarname", orient, visible, xval, yval, rows)]
TAG2 [Button ("btnname", id_small, id_large)]macro
TAG3 [Flyout ("flyname", id_small, id_large, icon, alias)]macro
TAG4 [control (element)]
[—]
```

*****TOOLBARS** is the section label of the toolbar, and **MYTOOLS1** is the name of the submenu that contains the definition of a toolbar. Each toolbar can have five distinct items that control different elements of the toolbar: TAG1, TAG2, TAG3, TAG4, and separator ([—]). The first line of the toolbar (TAG1) defines the characteristics of the toolbar. In this line, **Toolbar** is the keyword, and it is followed by a series of options enclosed in parentheses. The following describes the available options.

tbarname	This is a text string that names the toolbar. The tbarname text string must consist of alphanumeric characters with no punctuation other than a dash (-) or an underscore (_).
orient	This determines the orientation of the toolbar. The acceptable values are Floating, Top, Bottom, Left, and Right. These values are not case-sensitive.
visible	This determines the visibility of the toolbar. The acceptable values are Show and Hide. These values are not case-sensitive.
xval	This is a numeric value that specifies the X ordinate in pixels. The X ordinate is measured from the left edge of the screen to the right side of the toolbar.
yval	This is a numeric value that specifies the Y ordinate in pixels. The Y ordinate is measured from the top edge of the screen to the top of the toolbar.
rows	This is a numeric value that specifies the number of rows.

The second line of the toolbar (TAG2) defines the button. In this line the **Button** is the key word and it is followed by a series of options enclosed in parentheses. The following is the description of the available options.

btnname This is a text string that names the button. The text string must consist of alphanumeric characters with no punctuation other than a dash (-) or an underscore (_). This text string is displayed as ToolTip when you place the cursor over the button.

id_small This is a text string that names the ID string of the small image resource (16 by 15 bitmap). The text string must consist of alphanumeric characters with no punctuation other than a dash (-) or an underscore (_). The id_small text string can also specify a user-defined bitmap (Example: ICON_16_CIRCLE).

id_big This is a text string that names the ID string of the large image resource (24 by 22 bitmap). The text string must consist of alphanumeric characters with no punctuation other than a dash (-) or an underscore (_). The id_big text string can also specify a user-defined bitmap (Example: ICON_32_CIRCLE).

macro The second line (TAG2), which defines a button, is followed by a command string (macro). For example, the macro can consist of ^C^CLine. It follows the same syntax as that of any standard menu item definition.

The third line of the toolbar (TAG3) defines the flyout control. In this line the **Flyout** is the key word, and it is followed by a series of options enclosed in parentheses. The following describes the available options.

flyname This is a text string that names the flyout. The text string must consist of alphanumeric characters with no punctuation other than a dash (-) or an underscore (_). This text string is displayed as ToolTip when you place the cursor over the flyout button.

id_small This is a text string that names the ID string of the small image resource (16 by 15 bitmap). The text string must consist of alphanumeric characters with no punctuation other than a dash (-) or an underscore (_). The id_small text string can also specify a user-defined bitmap.

id_big This is a text string that names the ID string of the large image resource (24 by 22 bitmap). The text string must consist of alphanumeric characters with no punctuation other than a dash (-) or an underscore (_). The id_big text string can also specify a user-defined bitmap.

icon This is a Boolean key word that determines whether the button displays its own icon or the last icon selected. The acceptable values is **othericon**. The value is not case-sensitive.

alias The alias specifies the name of the toolbar submenu that is defined with the standard ****aliasname** syntax.

The fourth line of the toolbar (TAG4) defines a special control element. In this line the Control is the key word, and it is followed by the type of control element enclosed in parentheses. The following describes the available control element types.

element This parameter can have one of the following five values:
Layer: This specifies the layer control element.
Linetype: This specifies the linetype control element.
Color: This specifies the color control element.
Undo: This specifies the undo control element.
Redo: This specifies the redo control element.

Example 8

In this example you will write a menu file for a toolbar for the **LINE**, **PLINE**, **CIRCLE**, **ELLIPSE**, and **ARC** commands. The name of the toolbar is MyDraw1 (Figure 27-22).

The following is the listing of the menu file:

```
MENUGROUP=M1
***TOOLBARS
**TB_MyDraw1
ID_MyDraw1[_Toolbar("MyDraw1", _Floating, _Hide, 10, 200, 1)]
ID_Line [_Button("Line", ICON_16_LINE, ICON_32_LINE)]^C^C_line
ID_Pline [_Button("Pline", ICON_16_PLine, ICON_32_PLine)]^C^C_PLine
ID_Circle[_Button("Circle", ICON_16_CirRAD, ICON_32_CirRAD)]^C^C_Circle
ID_ELLIPSE[_Button("Ellipse",ICON_16_EllCEN,ICON_32_EllCEN)]^C^C_ELLIPSE
ID_Arc[_Button("Arc 3Point", ICON_16_Arc3Pt, ICON_32_Arc3Pt)]^C^C_Arc
```

Use the **MENULOAD** command to load the MyDraw1 menu group as discussed earlier. To display the new toolbar (MyDraw1) on the screen, choose **Toolbars** from the **View** menu to display the **Customize** dialog box (**Toolbars** tab) and then select **MyDraw1** in the **Menu Group** list box. Turn the MyDraw1 toolbar on in the **Toolbars** list box by checking the check box (Figure 27-23); MyDraw1 toolbar is displayed on the screen (27-22).

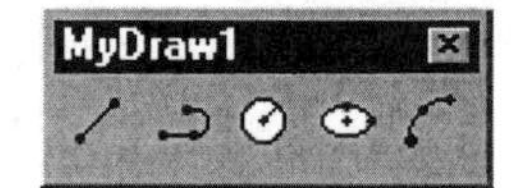

Figure 27-22
MyDraw1 toolbar

You can also load the new toolbar from the command line. After using the **MENULOAD** command to load the MyDraw1 menu group, use the **-TOOLBAR** command to display the MyDraw1 toolbar.

Command: **-TOOLBAR**
Enter toolbar name or [ALL]: **MYDRAW1**
Enter an option [Show/Hide/Left/Right/Top/Bottom/Float] <Show>: **S**

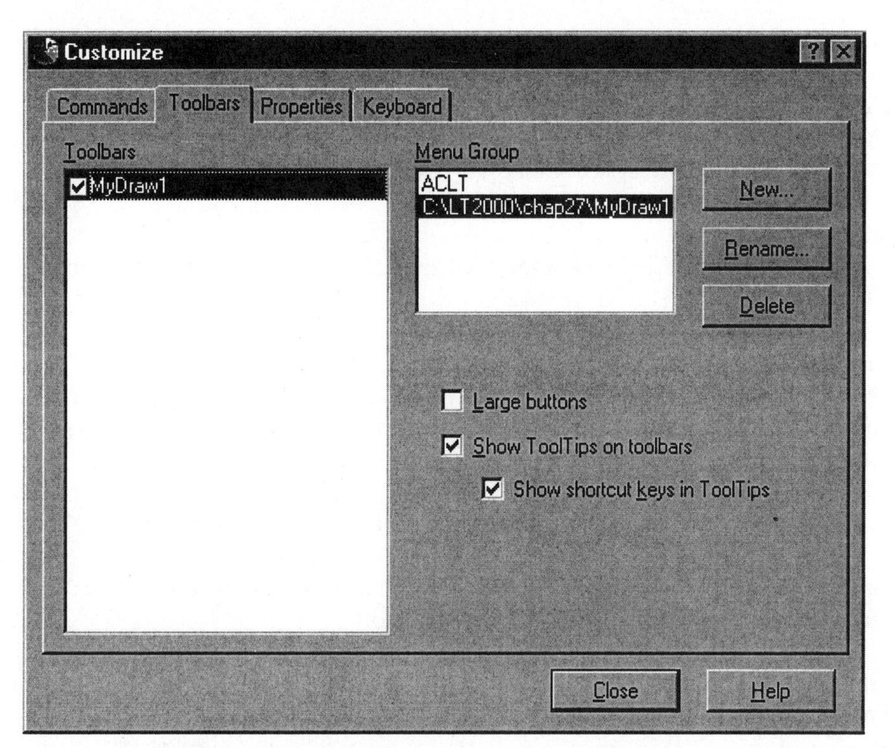

***Figure 27-23 Customize** dialog box (**Toolbars** tab)*

Example 9

In this example you will write a menu file for a toolbar with a flyout. The name of the toolbar is MyDraw2 (Figure 27-24), and it contains two buttons, Circle and Arc. When you select the Circle button, it should display a flyout with Radius, Diameter, 2P, and 3P buttons (Figure 27-25). Similarly, when you select the Arc button, it should display the 3Point, SCE, and SCA buttons.

The following is the listing of the menu file:

```
***Menugroup=M2
***TOOLBARS
**TB_MyDraw2
ID_MyDraw2[_Toolbar("MyDraw2", _Floating, _Show, 10, 100, 1)]
ID_TbCircle[_Flyout("Circle", ICON_16_Circle, ICON_32_Circle, _OtherIcon,
M2.TB_Circle)]
ID_TbArc[_Flyout("Arc", ICON_16_Arc, ICON_32_Arc, _OtherIcon, M2.TB_Arc)]

**TB_Circle
ID_TbCircle[_Toolbar("Circle", _Floating, _Hide, 10, 150, 1)]
ID_CirRAD[_Button("Circle C,R", ICON_16_CirRAD, ICON_32_CirRAD)]^C^C_Circle
ID_CirDIA[_Button("Circle C,D",ICON_16_CirDIA,ICON_32_CirDIA)]^C^C_Circle;\D
ID_Cir2Pt[_Button("Circle 2Pts", ICON_16_Cir2Pt, ICON_32_Cir2Pt)]^C^C_Circle;2P
ID_Cir3Pt[_Button("Circle 3Pts", ICON_16_Cir3Pt, ICON_32_Cir3Pt)]^C^C_Circle;3P
```

```
**TB_Arc
ID_TbArc[_Toolbar("Arc", _Floating, _Hide, 10, 150, 1)]
ID_Arc3PT[_Button("Arc,3Pts", ICON_16_Arc3PT, ICON_32_Arc3PT)]^C^C_Arc
ID_ArcSCE[_Button("Arc,SCE", ICON_16_ArcSCE, ICON_32_ArcSCE)]^C^C_Arc;\C
ID_ArcSCA[_Button("Arc,SCA", ICON_16_ArcSCA, ICON_32_ArcSCA)]^C^C_Arc;\C;\A
```

Explanation

ID_TbCircle[_Flyout("Circle", ICON_16_Circle, ICON_32_Circle, _OtherIcon, **M2.TB_Circle**)]

In this line M2 is the MENUGROUP name (***MENUGROUP=M2) and TB_Circle is the name of the toolbar submenu. **M2.TB_Circle** will load the submenu TB_Circle that has been defined in the M2 menugroup. If M2 is missing, AutoCAD LT will not display the flyout when you select the Circle button.

ID_CirDIA[_Button("Circle C,D", ICON_16_**CirDIA**, ICON_32_**CirDIA**)]^C^C_Circle;\D

CirDIA is a user-defined bitmap that displays the Circle-diameter button. If you use any other name, AutoCAD LT will not display the desired button.

To load the toolbar, use the **MENULOAD** command to load the menu file. The **MyDraw1** toolbar will be displayed on the screen.

Figure 27-24 *Toolbar for Example 9*

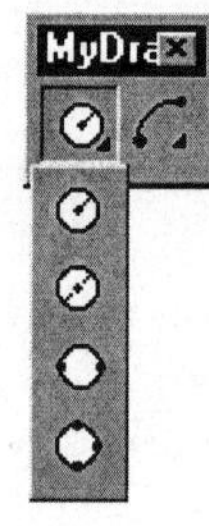

Figure 27-25 *Toolbar with flyout*

Menu-Specific Help

AutoCAD LT for Windows allows access to online help. For example, if you want to define a helpstring for the **CIRCLE** and **ARC** commands, the syntax is as follows:

***HELPSTRING
ID_Copy *(This command will copy the selected object.)*
ID_Ellipse *(This command will draw an ellipse.)*

The ***__*HELPSTRING** is the section label for the helpstring menu section. The lines defined in this section start with a name tag (ID_Copy) and are followed by the label enclosed in square brackets. If you press the F1 key when the menu item is highlighted, the help engine gets activated and AutoCAD LT displays the helpstring defined with the name tag.

CUSTOMIZING THE TOOLBARS

AutoCAD LT has provided several toolbars that should be sufficient for general use. However, sometimes you may need to customize the toolbars so that the commands that you use frequently are grouped in one toolbar. This saves time in selecting commands. It also saves the drawing space because you do not need to have several toolbars on the screen. The following example explains the process involved in creating and editing the toolbars.

Example 10

In this example you will create a new toolbar (MyToolbar1) that has Line, Polyline, Circle (Center, Radius option), Arc (Center, Start, End option), Spline, and Paragraph Text (**MTEXT**) commands. You will also change the image and tooltip of the Line button and perform other operations like deleting toolbars and buttons and copying buttons between toolbars.

1. Choose **Toolbars** in the **View** menu. The **Customize** dialog box with the **Toolbars** tab selected is displayed. You can also invoke this dialog box by entering **TOOLBAR** at AutoCAD LT's Command prompt.

2. Choose **New** button in the dialog box to display the **New Toolbar** dialog box.

3. Enter the name of the toolbar in the **Toolbar Name** edit box (Example **MyToolbar1**) (Figure 27-26) and then choose the **OK** button to exit the dialog box. The name of the

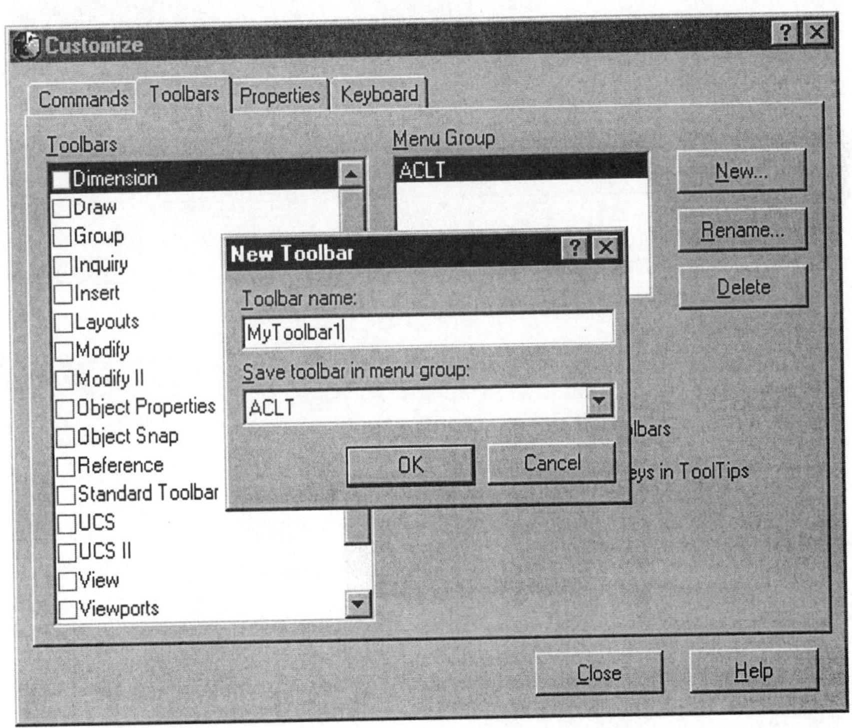

*Figure 27-26 **Customize** dialog box with the **New Toolbar** dialog box*

Chapter 27

new toolbar (MyToolbar1) is displayed in the **Toolbars** dialog box.

4. Select the new toolbar, if it is not already selected, by selecting the check box. The new toolbar (MyToolbar1) appears on the screen.

5. Choose the **Commands** tab in the **Customize** dialog box.

6. In the **Categories** list box select the **Draw** toolbar. AutoCAD LT displays the draw image tool buttons and names in the **Commands** list box (Figure 27-27).

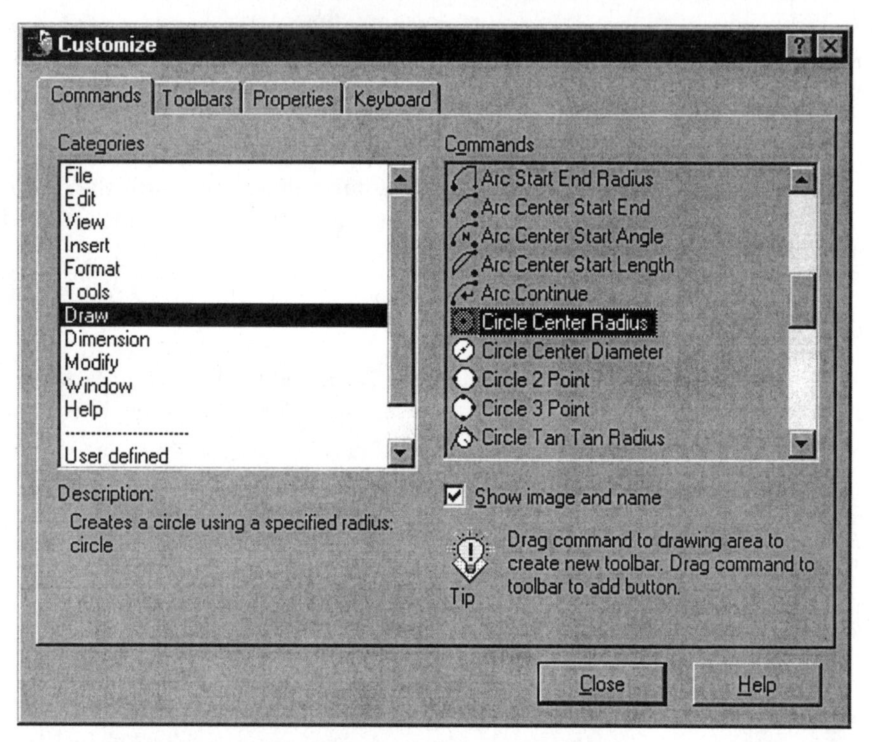

***Figure 27-27 Customize** dialog box (**Commands** tab)*

7. Choose and drag the Line button and position it in the MyToolbar1 toolbox. Repeat the same for Polyline, Circle (Center, Radius option), Arc (Center, Start, End option), Spline, and Multiline Text buttons (Figure 27-28).

***Figure 27-28** MyToolbar1 toolbar*

8. Now, select the Dimension category and then select and drag some of the dimensioning commands to MyToolbar1.

9. Similarly, you can open any toolbar category and add commands to the new toolbar.

10. When you are done defining the commands for the new toolbar, choose the **Close** button in the **Customize** dialog box.

11. Test the buttons in the new toolbar (MyToolbar1).

Creating a New Image and Tooltip for a Button

1. Make sure the button with the image you want to edit is displayed on the screen. In this example we want to edit the image of the **Line** button of MyToolbar1.

2. Select **Toolbars** from the **View** menu to display the **Customize** dialog box. Choose the **Properties** tab and then choose the **Line** button in the **MyToolbar1** toolbar. The different properties of the Line button are displayed (Figure 27-29).

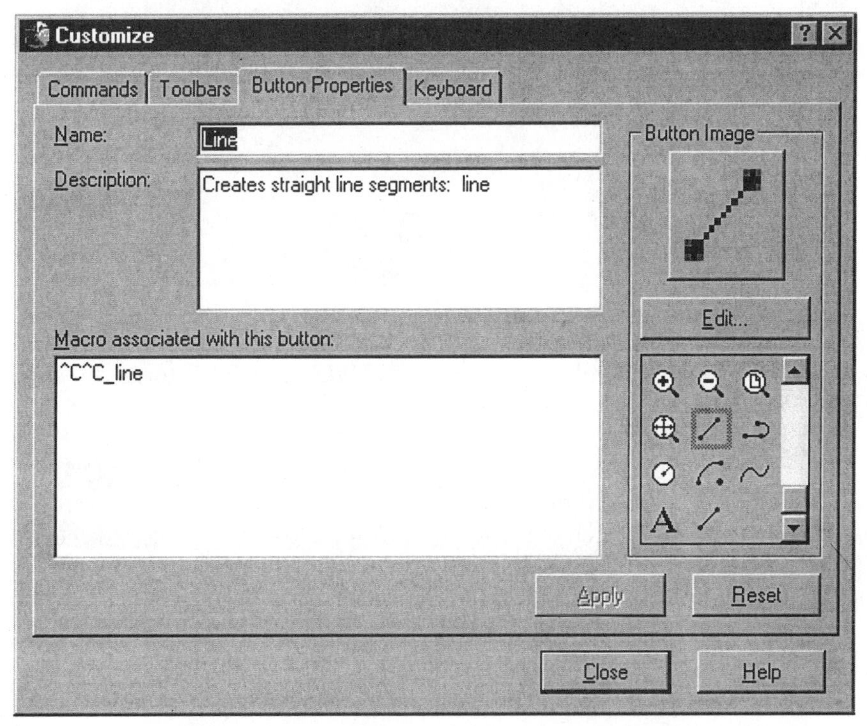

***Figure 27-29 Customize** dialog box (**Properties** tab)*

3. To edit the shape of the image, choose the **Edit** button to access the **Button Editor**. Select the Grid to display the grid lines.

   You can edit the shape by using different tools in the **Button Editor**. You can draw a Line by choosing the **Line** button and specifying two points. You can draw a circle or an ellipse by using the **Circle** button. The **Erase** button can be used to erase the image.

4. In this example, we want to change the color of Line image. To accomplish this, erase the existing line, and then select the color and draw a line. Also, create the shape L in the lower right corner (Figure 27-30).

Figure 27-30 ***Button Editor*** *dialog box*

5. Choose the **SaveAs** button and save the image as **MyLine** in the Tutorial directory. Choose the **Close** button to exit the **Button Editor**.

6. In the **Customize** dialog box, enter MyLine in the **Name** edit box. This changes the tooltip of this button. Now, choose the **Apply** button to apply the changes to the image. Close the dialog box to return to the AutoCAD LT screen. Notice the change in the button image and tooltip (Figure 27-31).

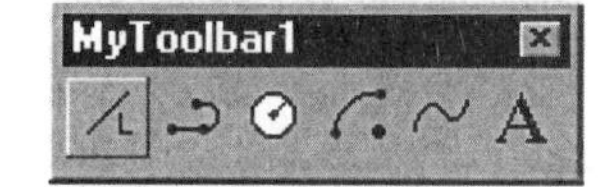

Figure 27-31 *Changed button image*

Deleting the Button from a Toolbar

1. Right-click on the button that you want to remove from the toolbox. Choose **Customize** from the Shortcut menu to display the **Customize** dialog box. In this example, we want to remove the **Spline** button from the MyToolbar1 toolbox.

2. Right-click again on the **Spline** button to display another Shortcut menu. Choose **Delete** to delete the **Spline** button from the MyToolbar1 toolbox.

3. Repeat the above step to delete other buttons, if needed. Close the **Customize** dialog box to return to the screen.

Deleting a Toolbar

1. Select **Toolbars** from the **View** menu.

2. Select the **Toolbar** tab in the **Customize** dialog box. Choose the toolbar that you want to delete and then choose the **Delete** button. The toolbar you selected is deleted.

Copying a Tool Button

In this example we want to copy the **Ordinate Dimension** button from the **Dimension** toolbar to **MyToolbar1**.

1. Choose **Toolbars** from the **View** menu or right-click on any button in any toolbar and choose **Customize** from the Shortcut menu to display the **Customize** dialog box.

2. In the **Customize** dialog box choose the **Toolbars** tab. Select the check box for Dimension and MyToolbar1. The Dimension and MyToolbar1 toolbars are displayed on the screen.

3. Hold the CTRL key down and then choose and drag the **Ordinate Dimension** button from the **Dimension** toolbar to **MyToolbar1**. The Ordinate dimensioning toolbar is copied to MyToolbar1.

Note

*If you do not hold down the CTRL key, the button you selected will be moved from the **Dimension** toolbar to MyToolbar1.*

Any changes made in the toolbars are saved in the ACLT.MNS and ACLT.MNR files. The following is the partial listing of ACLT.MNS file:

Creating Custom Toolbars with Flyout Icons

In this example you will create a custom toolbar with flyout buttons.

1. Choose **Toolbars** from the **View** menu to display the **Customize** dialog box.

2. In the **Customize** dialog box, choose the **New** button to display the **New Toolbar** dialog box. Enter the name of the new toolbar (for example, MYToolbar2) and then choose the **OK** button to exit the box.

3. In the **Customize** dialog box choose the **Commands** tab. Choose **User defined** from the **Categories** list. Choose and drag **User Defined Flyout** from the **Commands** list and place it in the **MyToolbar2**. An empty flyout button is created in the **MyToolbar2**.

4. Right-click on the **flyout** button of the **MyToolbar2** and choose **Properties** from the Shortcut menu. The **Flyout Properties** tab is opened in the **Customize** dialog box (Figure 27-32). To associate a predefined toolbar, select the toolbar name (for example, Draw) from the list.

5. Choose the **Apply** button and then close the dialog boxes. If you click on the **flyout** button of the new toolbar, the corresponding **flyout** buttons are displayed (Figure 27-32).

6. Now, you can add new buttons to the toolbar or delete the ones you do not want in this toolbar.

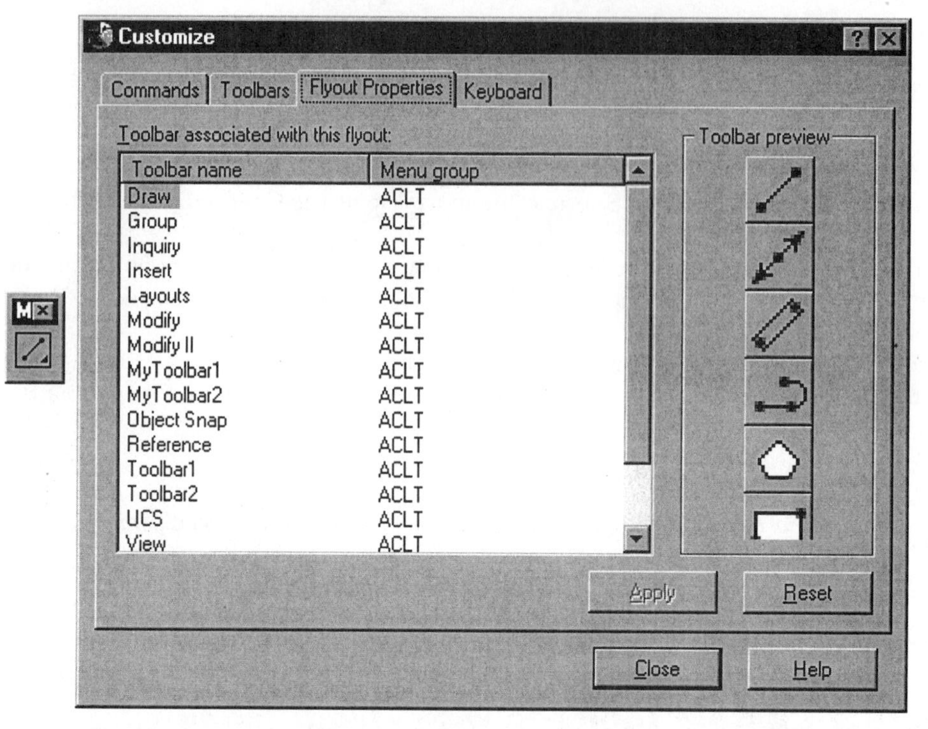

Figure 27-32 ***Customize*** *dialog box (**Flyout Properties** tab) and MyToolbar2*

Review Questions

1. A pull-down menu can have ______________ sections.
2. The length of the section title should not exceed ______________ characters.
3. The section titles in a menu are __________ justified.
4. In a pull-down menu, a line consisting of two hyphens ([--]) ______________ automatically to fill the ______________ of the menu.
5. If the menu item begins with a tilde (~), the items will be ______________ .
6. The pull-down menus are ______________ when the dynamic zoom is in progress.
7. Every cascading menu in the menu file should have a ______________ name.

8. The cascading menu name can be ________________ characters long.

9. The cascading menu names should not have any ________________ blanks.

10. In Windows you can write partial menus, toolbars, and accelerator key definitions. (T/F)

11. A menu file can contain only one MENUGROUP. (T/F)

12. You can load the partial menu file by using the AutoCAD LT ____________ command.

Exercises

Exercise 3 — General

Write a menu for the following AutoCAD LT commands. (The layout of the menu is shown in Figure 27-33).

LINE
CIRCLE C,R
CIRCLE C,D
ARC 3P
ARC SCE
ARC CSE

DIM HORZ
DIM VERT
DIM RADIUS
DIM DIAMETER
DIM ANGULAR
DIM LEADER

DTEXT LEFT
DTEXT RIGHT
DTEXT CENTER
DTEXT ALIGNED
DTEXT MIDDLE
DTEXT FIT

PULL-DOWN MENU

DRAW	DIM	DTEXT
LINE	DIM-HORZ	DTEXT-LEFT
CIRCLE C,R	DIM-VERT	DTEXT-RIGHT
CIRCLE C,D	DIM-RADIUS	DTEXT-CENTER
ARC 3P	DIM-DIAMETER	DTEXT-ALIGNED
ARC SCE	DIM-ANGULAR	DTEXT-MIDDLE
ARC CSE	DIM-LEADER	DTEXT-FIT

Figure 27-33 *Layout of menu*

Exercise 4 — General

Write a menu for the following AutoCAD LT commands.

LINE
PLINE
CIRCLE C,R
CIRCLE C,D

BLOCK
WBLOCK
INSERT
BLOCK LIST

ELLIPSE AXIS ENDPOINT
ELLIPSE CENTER
ATTDEF
ATTEDIT

Exercise 5 — *General*

Write a partial menu for Windows. The menu file should have two menus, POP1 (MyArc) and POP2 (MyDraw). The MyArc menu should contain all Arc options and must be displayed at the sixth position. Similarly, the MyDraw menu should contain **LINE**, **CIRCLE**, **PLINE**, **TEXT**, and **MTEXT** commands and should occupy the ninth position.

Exercise 6 — *General*

Write a menu file for a toolbar with a flyout. The name of the toolbar is MyDrawX1, and it contains two buttons, **Polygon** and **Ellipse**. When you select the **Polygon** button, it should display a flyout with **Rectangle** and **Polygon** button. Similarly, when you choose the **Ellipse** button, it should display the **Ellipse-Center** Option and **Ellipse-Edge** Option button.

Exercise 7 — *General*

Write a menu for the following AutoCAD LT commands. (The layouts of the menu is shown in Figure 27-34.)

LAYER NEW
LAYER MAKE
LAYER SET
LAYER LIST
LAYER ON
LAYER OFF

SNAP 0.25
SNAP 0.5
GRID 1.0
DRID 10.0
APERTURE 5
PICKBOX 5

UCS WORLD
UCS PREVIOUS
VPORTS 2
VPORTS 4
VPORTS SING.

PULL-DOWN MENU

LAYER	SETTINGS	UCS-PORT
LAYER-New	SNAP 0.25	UCS-World
LAYER-Make	SNAP 0.5	UCS-Pre
LAYER-Set	GRID 1.0	VPORTS-2
LAYER-List	GRID 10.0	VPORTS-4
LAYER-On	APERTURE 5	VPORTS-1
LAYER-Off	PICKBOX 5	

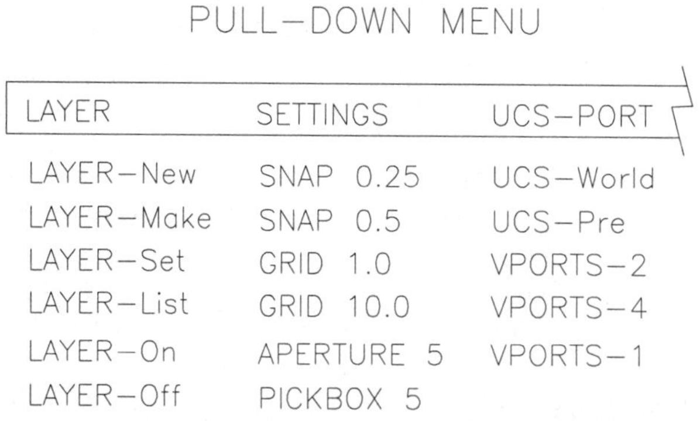

Figure 27-34 *Design of menu for Exercise 7*

Chapter 28

Tablet Menus

Learning Objectives

After completing this chapter, you will be able to:

- *Understand the advantages of tablet menus.*
- *Write and customize tablet menus.*
- *Load menus and configure tablet menus.*
- *Write tablet menus with different block sizes.*
- *Assign commands to tablet overlays.*

STANDARD TABLET MENU

The tablet menu provides a powerful alternative for entering commands. In the tablet menu, the commands are selected from the template that is secured on the surface of a digitizing tablet. To use the tablet menu you need a digitizing tablet and a pointing device. You also need a tablet template (Figure 28-1) that contains AutoCAD LT commands arranged in various groups for easy identification.

The standard AutoCAD LT menu file has four tablet menu sections: TABLET1, TABLET2, TABLET3, and TABLET4. When you start the AutoCAD LT and get into drawing editor, the tablet menu sections TABLET1, TABLET2, TABLET3, and TABLET4 are automatically loaded. The commands defined in these four sections are then assigned to different blocks of template.

The first tablet menu section (TABLET1) has 225 blank blocks that can be used to assign up to 225 menu items. The remaining tablet menu sections contain AutoCAD LT commands, arranged in functional groups that make it easier to identify and access the commands. The commands contained in the TABLET2 section include RENDER, SOLID MODELING, DISPLAY, INQUIRY, DRAW, ZOOM, and PAPER SPACE commands. The TABLET3 section

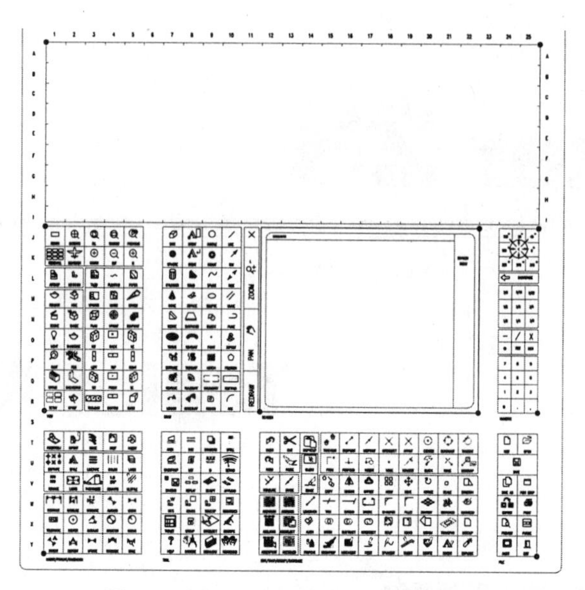

***Figure 28-1** Sample tablet template*

contains numbers, fractions, and angles. The commands contained in the TABLET4 section include **TEXT**, **DIMENSIONING**, **OBJECT SNAPS**, **EDIT**, **UTILITY**, **XREF**, and SETTINGS.

The AutoCAD LT tablet template has four tablet areas (Figure 28-2), which correspond to four tablet menu sections, TABLET1, TABLET2, TABLET3, and TABLET4.

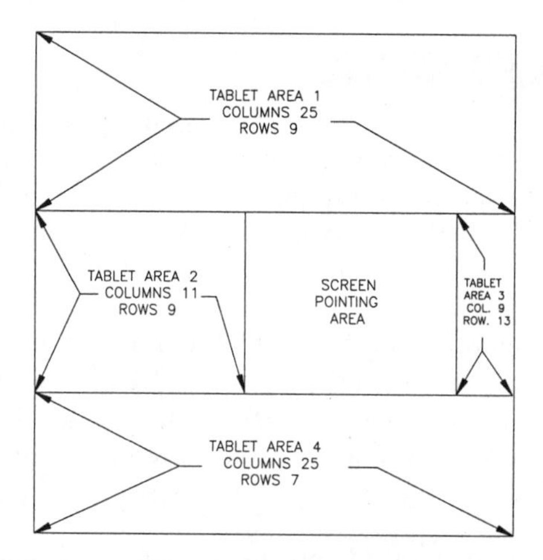

***Figure 28-2** Four tablet areas of the AutoCAD LT tablet template*

ADVANTAGES OF A TABLET MENU

A tablet menu has the following advantages over the pull-down menu, image menu, or keyboard.

1. In the tablet menu the commands can be arranged so that the most frequently used commands can be accessed directly. This can save considerable time in entering AutoCAD LT commands. In the pull-down menu some of the commands cannot be accessed directly. For example, to generate a horizontal dimension you have to go through several steps. You first select Dimension from the menu and then Linear. In the tablet menu you can select the Linear dimensioning command directly from the digitizer. This saves time and eliminates the distraction that takes place as you page through different screens.

2. You can have the graphical symbols of the AutoCAD LT commands drawn on the tablet template. This makes it much easier to recognize and select commands. For example, if you are not an expert in AutoCAD LT dimensioning, you may find Baseline and Continue dimensioning confusing. But if the command is supported by the graphical symbol illustrating what a command does, the chances of selecting a wrong command are minimized.

3. You can assign any number of commands to the tablet overlay. The number of commands you can assign to a tablet is limited only by the size of the digitizer and the size of the rectangular blocks.

CUSTOMIZING A TABLET MENU

Customizing tablet menus is a very powerful customizing tool to make AutoCAD LT more efficient.

The tablet menu can contain a maximum of four sections: TABLET1, TABLET2, TABLET3, and TABLET4. Each section represents a rectangular area on the digitizing tablet. These rectangular areas can be further divided into any number of rectangular blocks. The size of each block depends on the number of commands that are assigned to the tablet area. Also, the rectangular tablet areas can be located anywhere on the digitizer and can be arranged in any order. The AutoCAD LT **TABLET** command configures the tablet. The **MENULOAD** command loads and assigns the commands to the rectangular blocks on the tablet template.

Before writing a tablet menu file, it is very important to design the layout of the design of tablet template. A well-thought-out design can save a lot of time in the long run. The following points should be considered when designing a tablet template:

1. Learn the AutoCAD LT commands that you use in your profession.

2. Group the commands based on their function, use, or relationship with other commands.

3. Draw a rectangle representing a template so that it is easy for you to move the pointing device around. The size of this area should be appropriate to your application. It should not be too large or too small. Also, the size of the template depends on the active area of the digitizer.

4. Divide the remaining area into four different rectangular tablet areas for TABLET1, TABLET2, TABLET3, and TABLET4. It is not necessary to use all four areas; you can have fewer tablet areas, but four is the maximum.

5. Determine the number of commands you need to assign to a particular tablet area; then determine the number of rows and columns you need to generate in each area. The size of the blocks does not need to be the same in every tablet area.

6. Use the **TEXT** command to print the commands on the tablet overlay, and draw the symbols of the command, if possible.

7. Plot the tablet overlay on good-quality paper or a sheet of Mylar. If you want the plotted side of the template to face the digitizer board, you can create a mirror image of the tablet overlay and then plot the mirror image.

WRITING A TABLET MENU

When writing a tablet menu you must understand the AutoCAD LT commands and the prompt entries required for each command. Equally important is the design of the tablet template and the placement of various commands on it. Give considerable thought to the design and layout of the template, and, if possible, invite suggestions from AutoCAD LT users in your trade. To understand the process involved in developing and writing a tablet menu, consider Example 1.

Example 1

Write a tablet menu for the following AutoCAD LT commands. The commands are to be arranged as shown in Figure 28-3. Make a tablet menu template for configuration and command selection (filename **TM1.MNU**).

LINE	**CIRCLE**	**PLINE**
CIRCLE C,D	**ERASE**	**CIRCLE 2P**

Figure 28-3 represents one of the possible template designs, where the AutoCAD LT commands are in one row at the top of the template, and the screen pointing area is in the center. There is only one area in this template; therefore, you can place all these commands under the

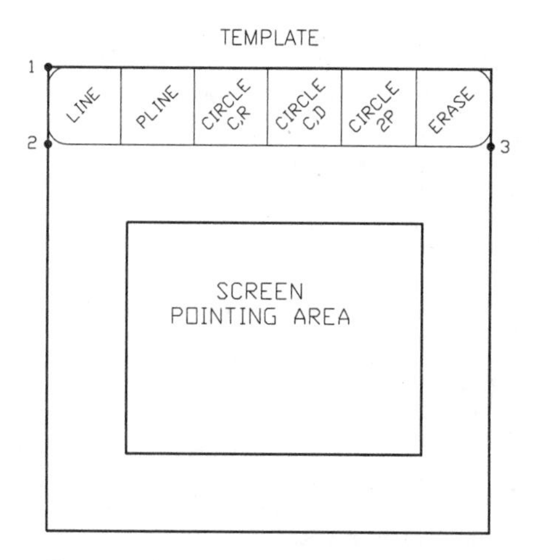

Figure 28-3 *Design of tablet template*

section label TABLET1. To write a menu file you can use any text editor like Notepad.

The name of the file is TM1, and the extension of the file is **.MNU. The line numbers are not part of the file. They are shown here for reference only.**

```
***TABLET1              1
^C^CLINE                2
^C^CPLINE               3
^C^CCIRCLE              4
^C^CCIRCLE;\D           5
^C^CCIRCLE;2P           6
^C^CERASE               7
```

Explanation

Line 1
*****TABLET1**
TABLET1 is the section label of the first tablet area. All the section labels are preceded by three asterisks (***).

*****TABLET1**

Where *** ---------------- Three asterisks designate section label.
TABLET1 ------ Section label for TABLET1

Line 2
^C^CLINE
^C^C cancels the existing command twice; **LINE** is an AutoCAD LT command. There is no space between the second ^C and **LINE.**

^C^C LINE

Where **^C^C** ---------- Cancels the existing command twice
Line -------------- AutoCAD LT **LINE** command

Line 3
^C^CPLINE
^C^C cancels the existing command twice; **PLINE** is an AutoCAD LT command.

Line 4
^C^CCIRCLE
^C^C cancels the existing command twice; **CIRCLE** is an AutoCAD LT command. The default input for the **CIRCLE** command is the center and the radius of the circle; therefore, no additional input is required for this line.

Line 5
^C^CCIRCLE;\D
^C^C cancels the existing command twice; **CIRCLE** is an AutoCAD LT command like the previous line. However, this command definition requires the diameter option of the circle command. This is accomplished by using \D in the command definition. There should be no space between the backslash (\) and the D, but there should always be a space **before** the backslash (\). The backslash (\) lets the user enter a point, and in this case it is the center point

of the circle. After you enter the center point, the diameter option is selected by the letter D, which follows the backslash (\).

^C^CCIRCLE;\D

Where **CIRCLE** --------- **CIRCLE** command
; ------------------ Space for RETURN
\------------------- Pause for input
D ----------------- Diameter option

Line 6
^C^CCIRCLE;2P
^C^C cancels the existing command twice; **CIRCLE** is an AutoCAD LT command. Semicolon is for ENTER or RETURN. The 2P selects the two-point option of the **CIRCLE** command.

Line 7
^C^CERASE
^C^C cancels the existing command twice; **ERASE** is an AutoCAD LT command that erases the selected objects.

Note

In the tablet menu, the part of the menu item that is enclosed in the brackets is used for screen display only. For example, in the following menu item, T1-6 will be ignored and will have no effect on the command definition.

[T1-6]^C^CCIRCLE;2P

Where **[T1-6]** ----------- For reference only and has no effect on the command definition
T1 ---------------- Tablet area 1
6 ------------------ Item number 6

The reference information can be used to designate the tablet area and the line number.

Before you can use the commands from the new tablet menu, you need to configure the tablet and load the tablet menu.

TABLET CONFIGURATION

To use the new template to select the commands, you need to configure the tablet so that AutoCAD LT knows the location of the tablet template and the position of the commands assigned to each block. This is accomplished by using the AutoCAD LT **TABLET** command. Secure the tablet template (Figure 28-3) on the digitizer with the edges of the overlay approximately parallel to the edges of the digitizer. Enter the AutoCAD LT **TABLET** command, select the **Configure option**, and respond to the following prompts. Figure 28-4 shows the points you need to select to configure the tablet.

Command: **TABLET** [Enter]
Enter an option [ON/OFF/CAL/CFG]:**CFG** [Enter]

Enter number of tablet menus desired (0–4) <current>: **1** Enter
Do you want to realign tablet menus? [Yes/No] <N>: **Y** Enter
Digitize upper left corner of menu area 1; **P1** Enter
Digitize lower left corner of menu area 1; **P2** Enter
Digitize lower right corner of menu area 1; **P3** Enter
Enter the number of columns for menu area 1: **6** Enter
Enter the number of rows for menu area 1: **1** Enter
Do you want to respecify the Fixed Screen Pointing Area? [Yes/No]<N>: **Y** Enter
Digitize lower left corner of Fixed Screen pointing area: **P4** Enter
Digitize upper right corner of Fixed Screen pointing area: **P5** Enter
Do you want to specify the Floating Screen Pointing Area? [Yes/No]<N>: **N** Enter
Digitize lower-left corner of the Floating Screen pointing area: *Digitize a point.*
Digitize upper-right corner of the Floating Screen pointing area: *Digitize a point.*

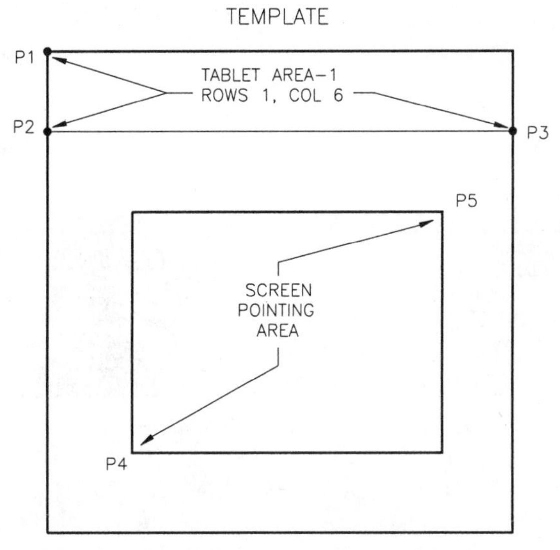

Figure 28-4 *Points that need to be selected to configure the tablet*

Note

The three points P1, P2, and P3 should form a 90-degree angle. If the selected points do not form a 90-degree angle, AutoCAD LT will prompt you to enter the points again until they do. The tablet areas should not overlap the screen pointing area.

The screen pointing area can be any size and located anywhere on the tablet as long as it is within the active area of the digitizer. The screen pointing area should not overlap other tablet areas. The screen pointing area you select will correspond to the monitor screen area. Therefore, the length-to-width ratio of the screen pointing area should be the same as that of the monitor, unless you are using the screen pointing area to digitize a drawing.

LOADING MENUS

AutoCAD LT automatically loads the **ACLT.MNU** file when you get into the AutoCAD LT drawing editor. However, you can also load a different menu file by using the AutoCAD LT

MENULOAD command.

Command: **MENULOAD** Digitize lower left corner of the Floating Screen pointing area: *Digitize a point.*

When you enter the **MENU** command, AutoCAD LT displays the **Menu Customization** dialog box. In this dialog box, choosing the **Browse** button displays the **Select Menu File** dialog box on the screen (Figure 28-5). Select the menu file that you want to load and then choose the **Open** button.

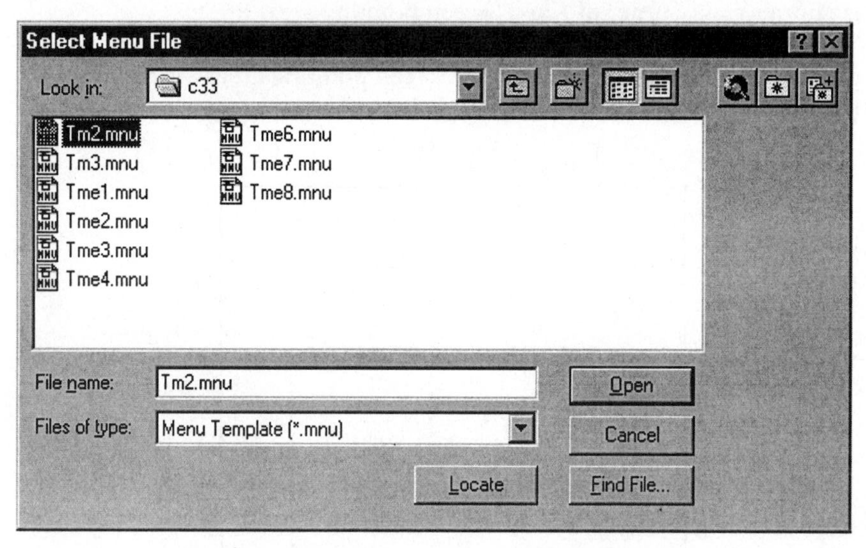

***Figure 28-5 Select Menu File** dialog box*

Exercise 1 *General*

Write a tablet menu for the following AutoCAD LT commands. Make a tablet menu template for configuration and command selection (filename **TME1.MNU**).

LINE	**TEXT-Center**
CIRCLE C,R	**TEXT-Left**
ARC C.S.E	**TEXT-Right**
ELLIPSE	**TEXT-Aligned**
DONUT	

Use the template in Figure 28-6 to arrange the commands. The draw and text commands should be placed in two separate tablet areas.

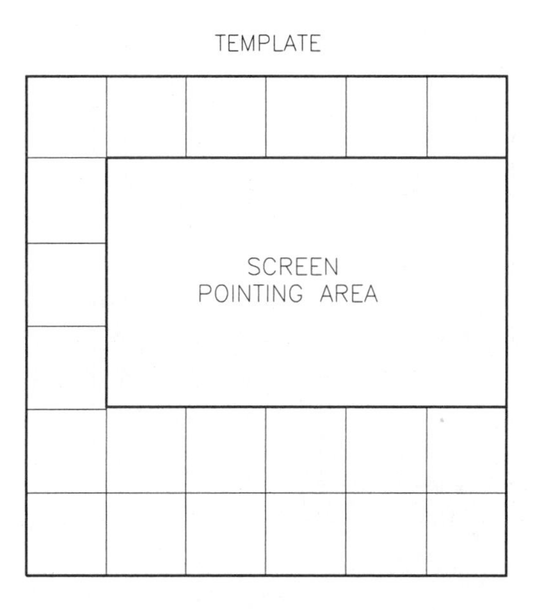

Figure 28-6 *Template for Exercise 1*

TABLET MENUS WITH DIFFERENT BLOCK SIZES

As mentioned earlier, the size of each tablet area can be different. The size of the blocks in these tablet areas can also be different. But the size of every block in a particular tablet area must be the same. This provides you with a lot of flexibility in designing a template. For example, you may prefer to have smaller blocks for numbers, fractions, or letters, and larger blocks for draw commands. You can also arrange these tablet areas to design a template layout with different shapes, such as L-shape and T-shape.

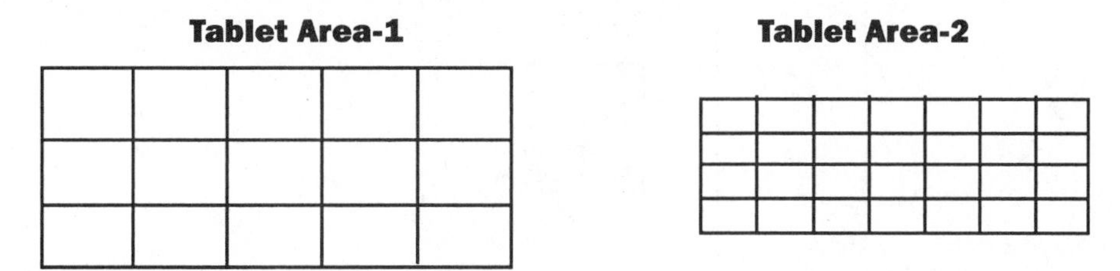

The following example illustrates the use of multiple tablet areas with different block sizes.

Example 2

Write a tablet menu for the tablet overlay shown in Figure 28-7(a). Figure 28-7(b) shows the number of rows and columns in different tablet areas (filename **TM2.MNU**).

Notice that this tablet template has four different sections in addition to the screen pointing area. Therefore, this menu will have four section labels: TABLET1, TABLET2, TABLET3,

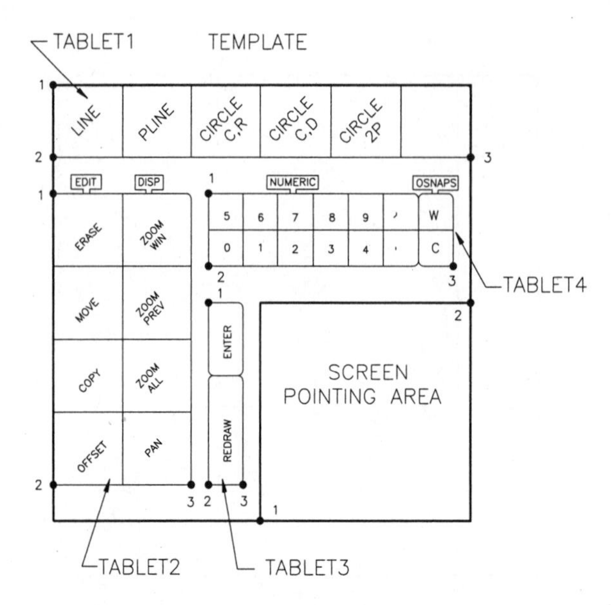

Figure 28-7(a) Tablet overlay for Example 2

and TABLET4. You can use any text editor (Notepad) to write the file.

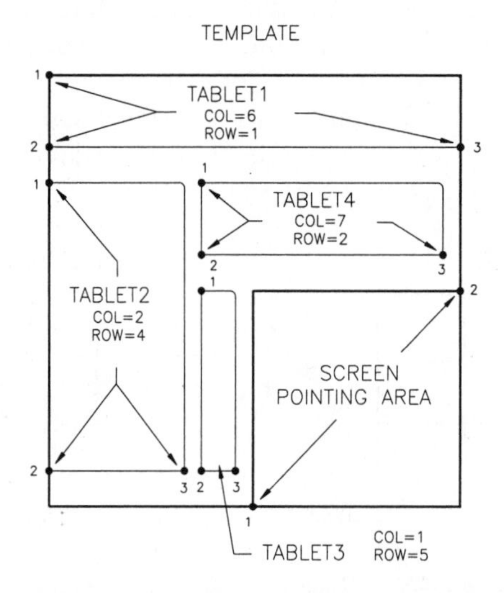

Figure 28-7(b) Number of rows and columns in different tablet

The following file is a listing of the tablet menu of Example 2.

```
***TABLET1                                              1
^C^CLINE                                                2
```

```
^C^CPLINE                3
^C^CCIRCLE               4
^C^CCIRCLE;\D            5
^C^CCIRCLE;2P            6
***TABLET2               7
^C^CERASE                8
^C^CZOOM;W               9
^C^CMOVE                10
^C^CZOOM;P              11
^C^CCCOPY               12
^C^CZOOM;A              13
^C^COFFSET              14
^C^CPAN                 15
***TABLET3              16
;                       17
;                       18
'REDRAW                 19
'REDRAW                 20
'REDRAW                 21
***TABLET4              22
5\                      23
6\                      24
7\                      25
8\                      26
9\                      27
,\                      28
WINDOW                  29
0\                      30
1\                      31
2\                      32
3\                      33
4\                      34
.\                      35
CROSSING                36
```


Explanation

Lines 1-6
The first six lines are identical to the first six lines of the tablet menu in Example 1.

Line 9
^C^CZOOM;W
ZOOM is an AutoCAD LT command; W is the window option of the **ZOOM** command.

^C^CZOOM;W

Where **ZOOM** ----------- AutoCAD LT **ZOOM** command
; ------------------- Space for Return
W ----------------- Window option of **ZOOM** command

This menu item could also be written as:

^C^CZOOM W

Space between ZOOM nad W causes a RETURN

Lines 17 and 18
The semicolon (;) is for RETURN. It has the same effect as entering RETURN at the keyboard.

Lines 19-21
'REDRAW
REDRAW is an AutoCAD LT command that redraws the screen. Notice that there is no ^C^C in front of the **REDRAW** command. If it had ^C^C, the existing command would be canceled before redrawing the screen. This may not be desirable in most applications because you might want to redraw the screen without canceling the existing command. The single quotation (') in front of REDRAW makes the **REDRAW** command transparent.

Line 23
5
The backslash (\) is used to introduce a pause for user input. Without the backslash you cannot enter another number or a character, because after you select the digit **5** it will automatically be followed by RETURN. For example, without the backslash (\), you will not be able to enter a number like **5.6**. Therefore, you need the backslash to enable you to enter decimal numbers or any characters. To terminate the input, enter RETURN at the keyboard or select RETURN from the digitizer.

5

Where \ --------- Backslash for user input

ASSIGNING COMMANDS TO A TABLET

After loading the menu by means of the AutoCAD LT **MENU** command, you must configure the tablet. At the time of configuration AutoCAD LT actually generates and stores the information about the rectangular blocks on the tablet template. When you load the menu, the commands defined in the tablet menu are assigned to various blocks. For example, when you select the three points for tablet area 4, Figure 28-7(a), and enter the number of rows and columns, AutoCAD LT generates a grid of seven columns and two rows, as shown in the following diagram.

After Configuration

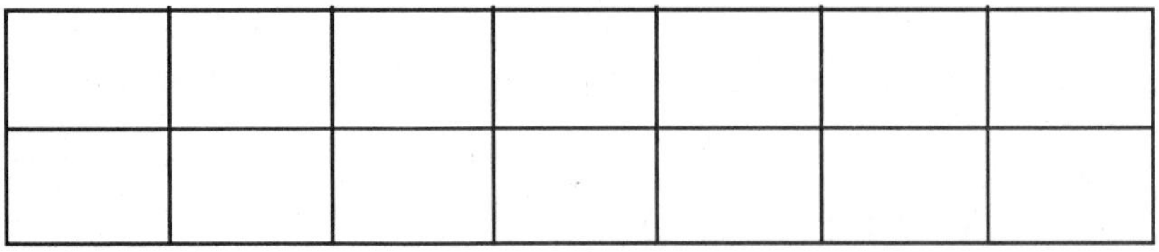

When you load the new menu, AutoCAD LT takes the commands under the section label TABLET4 and starts filling the blocks from left to right. That means "5", "6", "7", "8", "9", "," and "Window" will be placed in the top row. The next seven commands will be assigned to the next row, starting from the left, as shown in the following diagram.

After Loading Tablet Menu

5	6	7	8	9	.	Window
0	1	2	3	4	.	Cross

Similarly, tablet area 3 has been divided into five rows and one column. At first, it appears that this tablet area has only two rows and one column, as shown in Figure 28-7(a). When you configure this tablet area by specifying the three points and entering the number of rows and columns, AutoCAD LT divides the area into one column and five rows, as shown in the following diagrams.

After loading the menu, AutoCAD LT takes the commands in the TABLET3 section of the tablet menu and assigns them to the blocks. The first command (;) is placed in the first block. Since there are no more blocks in the first row, the next command (;) is placed in the second row. Similarly, the three **REDRAW** commands are placed in the next three rows. If you pick a point in the first two rows, you will select the **ENTER** command. Similarly, if you pick a point in the next three blocks, you will select the **REDRAW** command.

After Configuration	After Loading Menu
	;
	;
	Redraw
	Redraw
	Redraw

This process is carried out for all the of tablet areas, and the information is stored in the AutoCAD LT configuration file, **ACLT.CFG**. If for any reason the configuration is not right, the tablet menu may not perform the desired function.

Review Questions

1. The maximum number of tablet menu sections is __________.
2. A tablet menu area is __________ in shape.
3. The blocks in any tablet menu area are __________ in shape.
4. A tablet menu area can have __________ number of rectangular blocks.
5. You __________ assign the same command to more than one block on the tablet menu template.
6. The AutoCAD LT __________ command is used to configure the tablet menu template.
7. The AutoCAD LT __________ command is used to load a new menu.

Exercises

Exercise 2 *General*

Write a tablet menu for the commands shown in the tablet menu template of Figure 28-8. Make a tablet menu template for configuration and command selection.

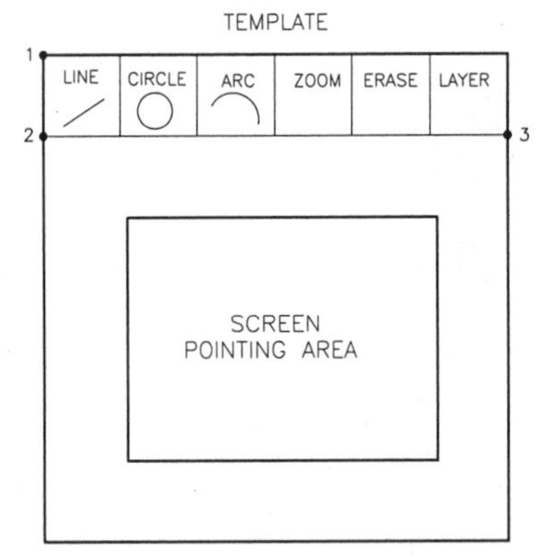

Figure 28-8 *Tablet menu template for Exercise 2*

Chapter 29

Image Tile, Button, and Auxiliary Menus

Learning Objectives

After completing this chapter, you will be able to:

- *Write image tile menus.*
- *Reference and display submenus.*
- *Make slides for image tile menus.*
- *Write Buttons menus.*
- *Learn special handling for button menus.*
- *Define and load submenus.*

IMAGE TILE MENUS

The image tile menus, also known as **icon menus**, are extremely useful for inserting a block, selecting a text font, or drawing a 3D object. You can also use the image tile menus to load an AutoLISP routine or a predefined macro. Thus, the image tile menu is a powerful tool for customizing AutoCAD LT.

The image tile menus can be accessed from the pull-down, tablet, button, or screen menu. However, the image tile menus **cannot** be loaded by entering the command from the keyboard. When you select an image tile, a dialog box that contains **20 image tiles** is displayed on the screen (Figure 29-1). The names of the slide files associated with image tiles below the image tiles with a scrolling bar that can be used to scroll the image tiles. The title of the image tile menu is displayed at the top of the dialog box (Figure 29-1). When you activate the image tile menu, an arrow that can be moved to select any image tile appears on the screen. You can select an image tile by selecting the slide file name from the dialog box and then choosing the **OK** button from the dialog box or by double-clicking on the slide file name.

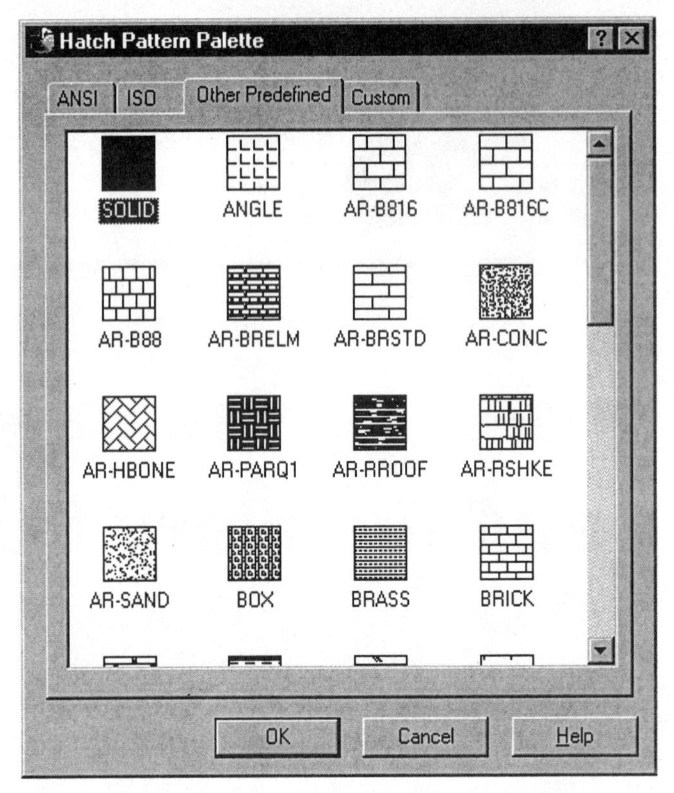

Figure 29-1 *Sample image tile menu display*

When you select the slide file, AutoCAD LT highlights the corresponding image tile by drawing a rectangle around the image tile (Figure 29-1). You can also select an image tile by moving the arrow to the desired image tile, and then pressing the PICK button of the pointing device. The corresponding slide file name will be automatically highlighted. And if you choose the **OK** button or double-click on the image tile, the command associated with that menu item will be executed. You can cancel an image tile menu by pressing ESCAPE on the keyboard, or selecting an image tile from the dialog box.

SUBMENUS

You can define an unlimited number of menu items in the image tile menu, but only 20 image tiles will be displayed at a time. If the number of items exceeds 20, you can use the **Next** and **Previous** buttons or the scroll bars of the dialog box to page through different pages of image tiles. You can also define submenus that let you define smaller groups of items within an image tile menu section. When you select a submenu, the submenu items are loaded and displayed on the screen.

Submenu Definition

A submenu label consists of two asterisks (**) followed by the name of the submenu. The image tile menu can have any number of submenus, and every submenu should have a unique name. The items that follow a submenu, up to the next section label or submenu label, belong to that submenu. The format of a submenu label is:

****Name**

Where ** ----------------- Two asterisks designate a submenu
Name ------------ Name of the submenu

Note

The submenu name can be up to 31 characters long.

The submenu name can consist of letters, digits, and special characters, like $ (dollar), - (hyphen), and _ (underscore).

The submenu name should not have any embedded blanks.

Submenu names should be unique in a menu file.

Submenu Reference

The submenu reference is used to reference or load a submenu. It consists of a $ sign followed by a letter that specifies the menu section. The letter that specifies an image tile menu section is I. The menu section is followed by an equal sign (=) and the name of the submenu you want to activate. The submenu name should be without the **. Following is the format of a submenu reference:

$Section=Submenu

Where **$** ------------------- "$" sign
Section ---------- Menu section specifier
= ----------------- "=" sign
Submenu ------- Name of submenu

$I=IMAGE1

Where **I** ------------------- I specifies image tile menu section
IMAGE1 -------- Name of submenu

Displaying a Submenu

When you load a submenu, the new dialog box and the image tiles are not automatically displayed. For example, if you load submenu IMAGE1, the items contained in this submenu will not be displayed. To force the display of the new image tile menu on the screen, AutoCAD LT uses the special command $I=*:

$I=*

Where **I** ------------------- I for image tile menu
* ------------------- Asterisk (*)

WRITING AN IMAGE TILE MENU

The image tile menu consists of the section label ***IMAGE followed by image tiles or image tile submenus. The menu file can contain only one image tile menu section (***IMAGE);

therefore, all image tiles must be defined in this section.

*****IMAGE**

Where ******* ---------------- Three asterisks designate a section label
IMAGE ---------- Section label for an image tile

You can define any number of submenus in the image tile menu. All submenus have two asterisks followed by the name of the submenu (**PARTS or **IMAGE1):

****IMAGE1**

Where ****** ----------------- Two asterisks designate a submenu
IMAGE1 -------- Name of submenu

The first item in the image tile menu is the title of the image tile menu, which is also displayed at the top of the dialog box. The image tile dialog box title has to be enclosed in brackets ([PLC-SYMBOLS]) and should not contain any command definition. If it does contain a command definition, AutoCAD LT ignores the definition. The remaining items in the image tile menu file contain slide names in the brackets and the command definition outside the brackets.

***IMAGE	Image tile menu section
**BOLTS	Image tile submenu (BOLTS)
[HEX-HEAD BOLTS]	Image tile title
[BOLT1]^C^CINSERT;B1	BOLT1 is slide file name; B1 is block name

Example 1

Write an image tile menu that will enable you to insert the block shapes from Figure 29-2 in a drawing by selecting the corresponding image tile from the dialog box. Use the **MENULOAD** to load the image tile menu.

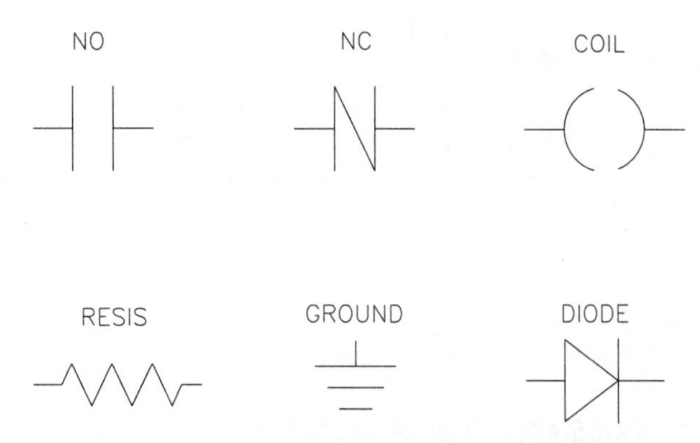

Figure 29-2 *Block shapes for the image tile menu*

PLC SYMBOLS	**ELECTRIC SYMBOLS**
NO (NORMALLY OPEN)	RESIS (RESISTANCE)
NC (NORMALLY CLOSED)	DIODE
COIL	GROUND

As mentioned in earlier chapters, the first step in writing a menu is to design the menu so that the commands are arranged in a desired configuration. Figure 29-3 shows one possible design for the menu and the image tile menu for Example 1.

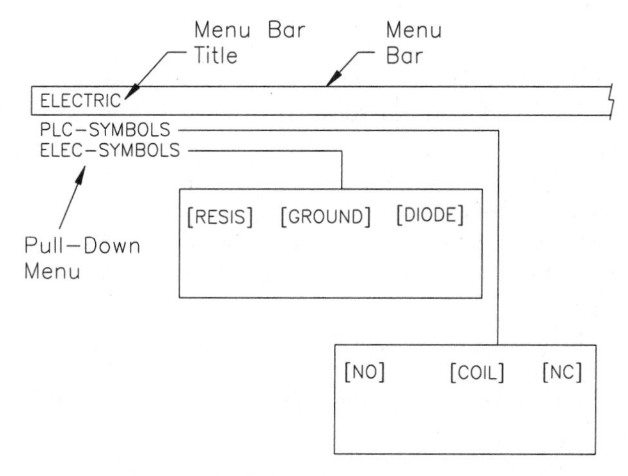

Figure 29-3 *Design of the menu and image tile menu*

You can use the text editors like **Notepad** to write the file. **The line numbers in the following file are for reference only and are not a part of the menu file.**

```
***POP1                                      1
[ELECTRIC]                                   2
[PLC-SYMBOLS]$I=IMAGE1 $I=*                  3
[ELEC-SYMBOLS]$I=IMAGE2 $I=*                 4
***IMAGE                                     5
**IMAGE1                                     6
[PLC-SYMBOLS]                                7
[NO]^C^CINSERT;NO;\1.0;1.0;0                 8
[NC]^C^CINSERT;NC;\1.0;1.0;0;                9
[COIL]^C^CINSERT;COIL                        10
[ No-Image]                                  11
[blank]                                      12
**IMAGE2                                     13
[ELECTRICAL SYMBOLS]                         14
[RESIS]^C^CINSERT;RESIS;\\\\                 15
[DIODE]^C^CINSERT;DIODE;\1.0;1.0;\           16
[GROUND]^C^CINSERT;GRD;\1.5;1.5;0;;          17
```

Explanation

Line 1
*****POP1**
In this menu item, ***POP1 is the section label and defines the first section of the menu.

Line 2
[ELECTRIC]
In this menu item [ELECTRIC] is the menu bar label for the POP1 menu. It will be displayed in the menu bar.

Line 3
[PLC-SYMBOLS]$I=IMAGE1 $I=*
In this menu item, $I=IMAGE1 loads the submenu IMAGE1; $I=* displays the current image tile menu on the screen.

[PLC-SYMBOLS]$I=IMAGE1 $I=*
Where **Image1** ---------- Loads submenu IMAGE1
$ ------------------- Forces display of current menu

Line 5
*****IMAGE**
In this menu item, ***IMAGE is the section label of the image tile menu. All the image tile menus have to be defined within this section; otherwise, AutoCAD LT cannot locate them.

Line 6
****IMAGE1**
In this menu item, **IMAGE1 is the name of the image tile submenu.

Line 7
[PLC-SYMBOLS]
When you select line 3 ([PLC-SYMBOLS]$I=IMAGE1 $I=*), AutoCAD LT loads the submenu IMAGE1 and displays the title of the image tile at the top of the dialog box (Figure 29-4). This title is defined in Line 7. If this line is missing, the next line will be displayed at the top of the dialog box. Image tile titles can be any length, as long as they fit the length of the dialog box.

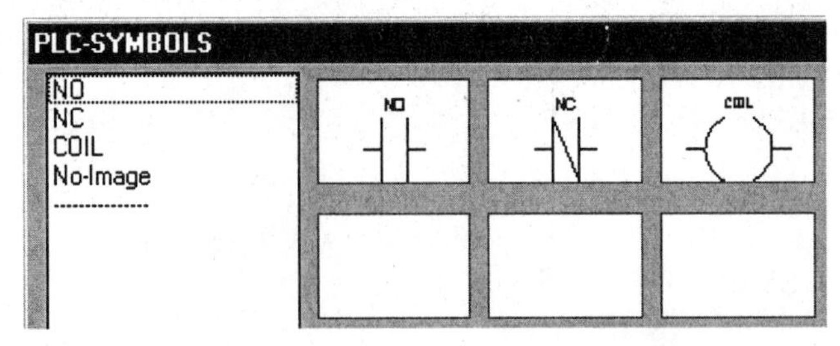

Figure 29-4 ***Image tile*** *box for PLC-SYMBOLS*

Line 8
[NO]^C^CINSERT;NO;\1.0;1.0;0
In this menu item, the first NO is the name of the slide and has to be enclosed within brackets. The name should not have any trailing or leading blank spaces. If the slides are not present, AutoCAD LT will not display any graphical symbols in the image tiles. However, the menu items will be loaded, and if you select this item, the command associated with the image tile will be executed, Figure 29-5. The second NO is the name of the block that is to be inserted. The backslash (\) pauses for user input; in this case it is the block insertion point. The first 1.0 defines the Xscale factor. The second 1.0 defines the Yscale factor, and the following 0 defines the rotation.

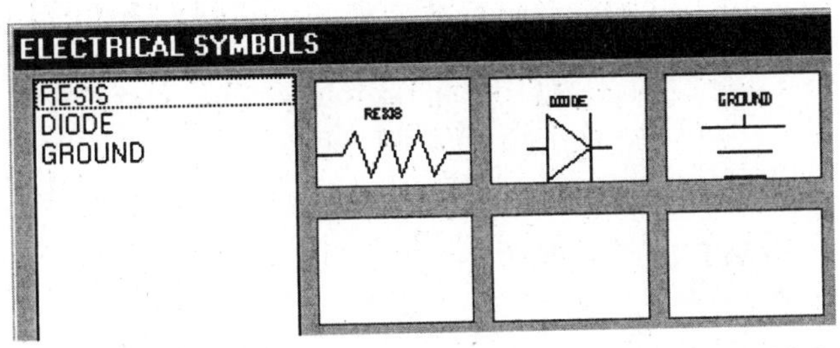

Figure 29-5 Image tile *box for ELECTRICAL-SYMBOLS*

[NO]^C^CINSERT;NO;\1.0;1.0;0
Where **NO** -------------- Block name
\\------------------- Pause for block insertion point
1.0 ---------------- Xscale factor
1.0 ---------------- Yscale factor
0 ------------------- Rotation angle

When you select this item, it will automatically enter all the prompts of the **INSERT** command and insert the NO block at the given location. The only input you need to enter is the insertion point of the block.

Line 10
[COIL]^C^CINSERT;COIL
In this menu item the block name is given, but you need to define other parameters when inserting this block.

Line 11
[No-Image]
Notice the **blank space before No-Image**. If there is a space following the open bracket, AutoCAD LT does not look for a slide. AutoCAD LT instead displays the text, enclosed within the brackets, in the slide file list box of the dialog box.

Line 12
[blank]
Line 12 consists of **blank**; this displays a separator line in the list box and a blank image (no image) in the image box.

Line 15
[RESIS]^C^CINSERT;RESIS;
This menu item inserts the block RESIS. The first backslash (\) is for the block insertion point. The second and third backslashes are for the Xscale and Yscale factors. The fourth backslash is for the rotation angle. This menu item could also be written as:

[RESIS]^C^CINSERT;RESIS;
or
[RESIS]^C^CINSERT;RESIS

Line 16
[DIODE]^C^CINSERT;DIODE;\1.0;1.0;
If you select this menu item, AutoCAD LT will prompt you to enter the block insertion point and the rotation angle. The first backslash is for the block insertion point; the second backslash is for the rotation angle.

[DIODE]^C^CINSERT;DIODE;\1.0;1.0;
Where \-------------------- Pause for insertion point
\-------------------- Pause for rotation angle

Line 17
[GROUND]^C^CINSERT;GRD;\1.5;1.5;0;;
This menu item has two semicolons (;) at the end. The first semicolon after 0 is for RETURN and completes the block insertion process. The second semicolon enters a RETURN and repeats the **INSERT** command. However, when the command is repeated you will have to respond to all of the prompts. It does not accept the values defined in the menu item.

Note
The menu item repetition feature cannot be used with the image tile menus. For example, if the command definition starts with an asterisk ([GROUND]^C^CINSERT;GRD;\1.5;1.5;0;;), the command is not automatically repeated, as is the case with a pull-down menu.*

A blank line in an image tile menu terminates the menu and clears the image tiles.

The menu command $I=, which displays the current menu, cannot be entered at the keyboard.*

If you want to cancel or exit an image tile menu, press the ESC (Escape) key on the keyboard. AutoCAD LT ignores all other entries from the keyboard.

You can define any number of image tile menus and submenus in the image tile menu section of the menu file.

SLIDES FOR IMAGE TILE MENUS

The idea behind creating slides for the image tile menus is to display graphical symbols in the image tiles. This symbol makes it easier for you to identify the operation that the image tile will perform. Any slide can be used for the image tile. However, the following guidelines should be kept in mind when creating slides for the image tile menu:

1. When you make a slide for an image tile menu; draw the object so that it fills the entire screen. The **MSLIDE** command makes a slide of the existing screen display. If the object is small, the picture in the image tile menu will be small. Use **ZOOM** Extents or **ZOOM** Window to display the object before making a slide.

2. When you use the image tile menu, it takes some time to load the slides for display in the image tiles. The more complex the slides, the more time it will take to load them. Therefore, the slides should be kept as simple as possible and at the same time give enough information about the object.

3. Do not fill the object, because it takes a long time to load and display a solid object. If there is a solid area in the slide, AutoCAD LT does not display the solid area in the image tile.

4. If the objects are too long or too wide, it is better to center the image with the AutoCAD LT **PAN** command before making a slide.

5. The space available on the screen for image tile display is limited. Make the best use of this small area by giving only the relevant information in the form of a slide.

6. The image tiles that are displayed in the dialog box have the length-to-width ratio (aspect ratio) of 1.5:1. For example, if the length of the image tile is 1.5 units, the width is 1 unit. If the drawing area of your screen has an aspect ratio of 1.5 and the slide drawing is centered in the drawing area, the slide in the image tile will also be centered.

LOADING MENUS

AutoCAD LT automatically loads the **ACLT.MNU** file when you get into the AutoCAD LT drawing editor. However, you can also load a different menu file by using the AutoCAD LT's **MENULOAD** command.

Command: **MENULOAD** Enter

When you enter the **MENULOAD** command, AutoCAD LT displays the **Menu Customization** dialog box (Figure 29-6). Before loading a new menu file, you have to unload the default **ACLT.MNU** file by choosing the **Unload** button in the **Menu Customization** dialog box. Now choose the **Browse** button in this dialog box to display the **Select Menu File** dialog box on the screen from where you can select the desired menu file. Select the menu file that you want to load and then choose the **Open** button.

To load the original menu, use the **MENULOAD** command again and enter the name of the default menu file (**ACLT.MNU**).

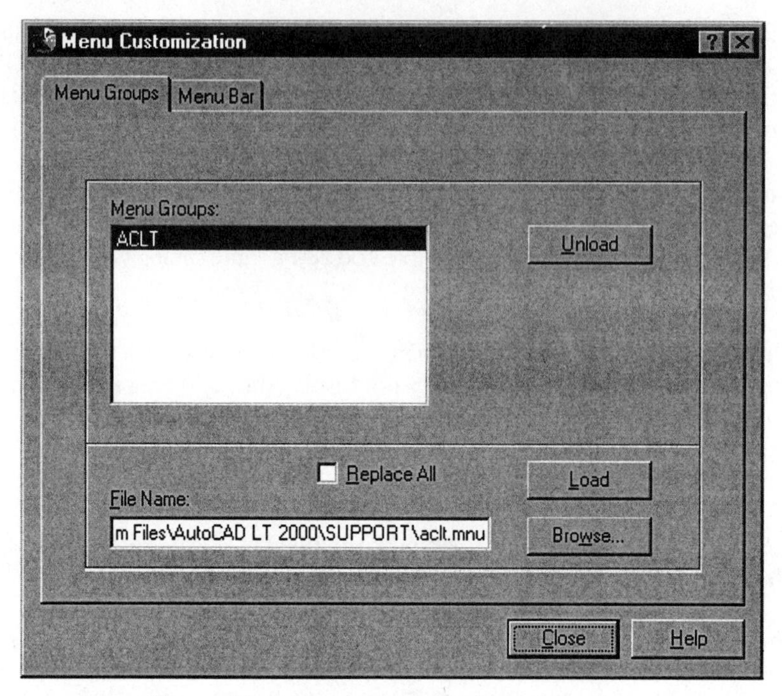

***Figure 29-6 Menu Customization** dialog box*

RESTRICTIONS

The pull-down and image tile menus are very easy to use and provide quick access to some of the frequently used AutoCAD LT commands. However, the menu bar, the menus, and the image tile menus are disabled during the following commands:

TEXT Command
After you assign the text height and the rotation angle to a **TEXT** command, the menu is automatically disabled.

VPOINT Command
The menus are disabled while the axis tripod and compass are displayed on the screen.

ZOOM Command
The menus are disabled when the dynamic zoom is in progress.

DVIEW
The menus are disabled when the dynamic view is in progress.

Exercise 1 *General*

Write an image tile menu for inserting the blocks shown in Figure 29-7. Arrange the blocks in two groups so that you have two submenus in the image tile menu.

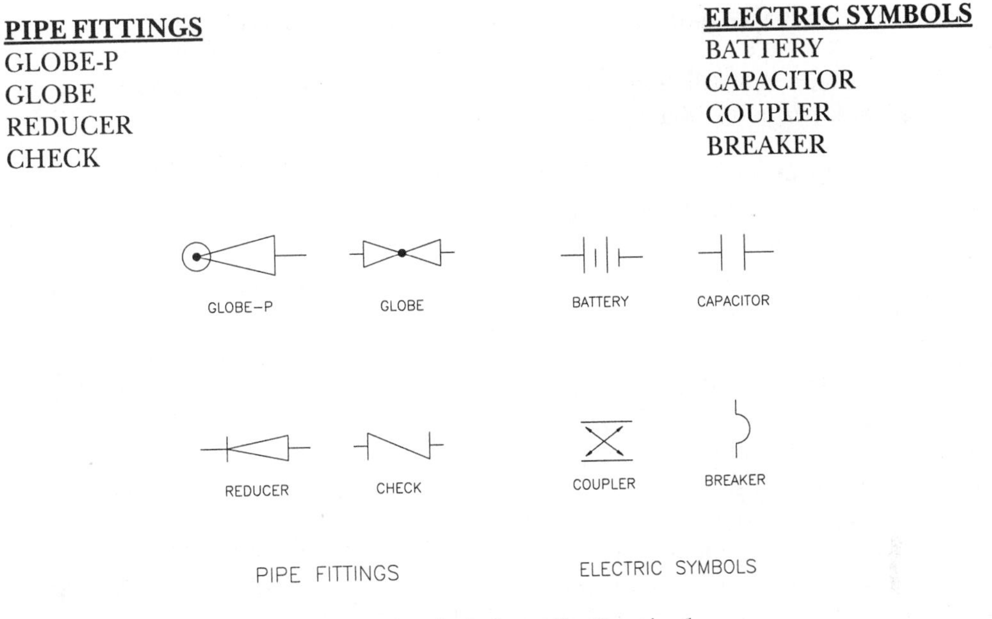

Figure 29-7 *Block shapes for Exercise 1*

IMAGE TILE MENU ITEM LABELS

As with menus, you can use menu item labels in the image menus. However, the menu item labels in the image tile menus use different formats, and each format performs a particular function in the image tile menu. The menu item labels appear in the slide list box of the dialog box. The maximum number of characters that can be displayed in this box is 23. The characters in excess of 23 are not displayed in the list box. However, this does not affect the command that is defined with the menu item.

Menu Item Label Formats

[slidename]. In this menu item label format, **slidename** is the name of the slide displayed in the image tile. This name (slidename) is also displayed in the list box of the corresponding dialog box.

[slidename,label]. In this menu item label format, **slidename** is the name of the slide displayed in the image tile. However, unlike the previous format, the **slidename** is **not** displayed in the list box. The **label** text is displayed in the list box. For example, if the menu item label is **[BOLT1,1/2-24UNC-3LG]**, **BOLT1** is the name of the slide and **1/2-24UNC-3LG** is the label that will be displayed in the list box.

[slidelib(slidename)]. In this menu item label format, **slidename** is the name of the slide in the slide library file **slidelib**. The slide (slidename) is displayed in the image tile, and the slide file name (slidename) is also displayed in the list box of the corresponding dialog box.

[slidelib(slidename,label)]. In this menu item label format, **slidename** is the name of the slide in the slide library file **slidelib**. The slide (slidename) is displayed in the image tile, and the label text is displayed in the list box of the corresponding dialog box.

[blank]. This menu item will draw a line that extends through the width of the list box. It also displays a blank image tile in the dialog box.

[label]. If the **label** text is preceded by a space, AutoCAD LT does not look for a slide. The label text is displayed in the list box only. For example, if the menu item label is [EXIT]^C, the label text (EXIT) will be displayed in the list box. If you select this item, the cancel command (^C) defined with the item will be executed. The **label** text is **not** displayed in the image tile of the dialog box.

BUTTON MENUS

You can use a multibutton pointing device to specify points, select objects, and execute commands. These pointing devices come with different numbers of buttons, but four-button and twelve-button pointing devices are very common. In addition to selecting points and objects, the multibutton pointing devices can be used to provide easy access to frequently used AutoCAD LT commands. The commands are selected by pressing the desired button; AutoCAD LT automatically executes the command or the macro that is assigned to that button. Figure 29-8 shows one such pointing device with 12 buttons.

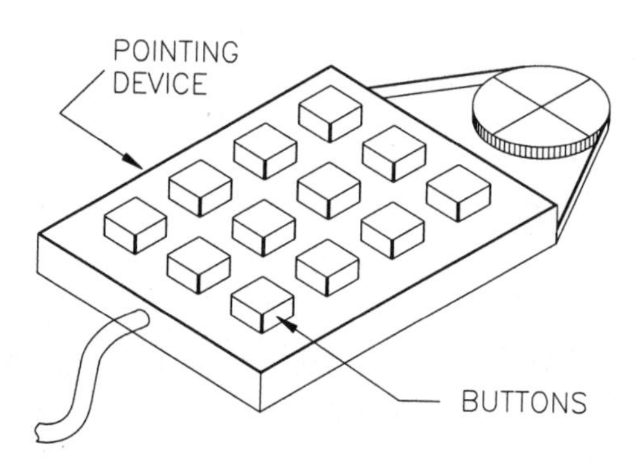

Figure 29-8 *Pointing device with 12 buttons*

The AutoCAD LT software package comes with a standard button menu that is part of the **ACLT.MNU** file. The **standard** menu is automatically loaded when you start AutoCAD LT and enter the drawing editor. You can write your own button menu and assign the desired commands or macros to various buttons of the pointing device.

AUXILIARY MENUS

In a menu file, you can have up to four auxiliary menu sections (AUX1 through AUX4). The auxiliary menu sections (***AUXn) are identical to the button menu sections. The difference is in the hardware. If your computer uses a system mouse, it will automatically use the auxiliary menu.

WRITING BUTTON AND AUXILIARY MENUS

In a menu file, you can have up to four button menus (BUTTONS1 through BUTTONS4) and four auxiliary menus (AUX1 through AUX4). The buttons and the auxiliary menus are identical. However, they are OS dependent. For example, in LT 97 the system mouse uses auxiliary menus. If your system has a pointing device (digitizer puck), AutoCAD LT automatically assigns the commands defined in the BUTTONS sections of the menu file to the buttons of the pointing device. When you load the menu file, the commands defined in the BUTTONS1 section of the menu file are assigned to the pointing device (digitizer puck) and if your computer has a system mouse, the mouse will use the auxiliary menus. You can also access other button menus (BUTTONS2 through BUTTONS4) by using the following keyboard and button (buttons of the pointing device-digitizer puck) combinations.

Aux Menu	Buttons Menu	Keyboard + Button Sequence
AUX1	BUTTONS1	Press the button of the pointing device.
AUX2	BUTTONS2	Hold down the SHIFT key and press the button of the pointing device.
AUX3	BUTTONS3	Hold down the CTRL key and press the button of the pointing device.
AUX4	BUTTONS4	Hold down the SHIFT and CTRL keys and press the button of the pointing device.

One of the buttons, generally the first, is used as a pick button to specify the coordinates of the screen crosshairs and send that information to AutoCAD LT. This button can also be used to select commands from various other menus, such as tablet menu, menu, and image tile menu. This button cannot be used to enter a command, but AutoCAD LT commands can be assigned to other buttons of the pointing device. Before writing a button menu, you should decide the commands and options you want to assign to different buttons, and know the prompts associated with those commands. The following example illustrates the working of the button menu and the procedure for assigning commands to different buttons.

Example 2

Write a buttons menu for the following AutoCAD LT commands. The pointing device has 12 buttons (Figure 29-9), and button number 1 is used as a pick button (filename **BM1.MNU**).

Button	Function	Button	Function
2	RETURN	3	CANCEL
4	CURSOR MENU	5	SNAP
6	ORTHO	7	AUTO
8	INT,END	9	LINE
10	CIRCLE	11	ZOOM Win
12	ZOOM Prev		

You can any text editor to write the menu file. The following file is a listing of the button menu for Example 1. **The line numbers are for reference only and are not a part of the menu file.**

```
***BUTTONS1        1
;                  2
^C^C               3
$P0=*              4
^B                 5
^O                 6
AUTO               7
```

```
INT,ENDP                                                    8
^C^CLINE                                                    9
^C^CCIRCLE                                                 10
'ZOOM;Win                                                  11
'ZOOM;Prev                                                 12
```

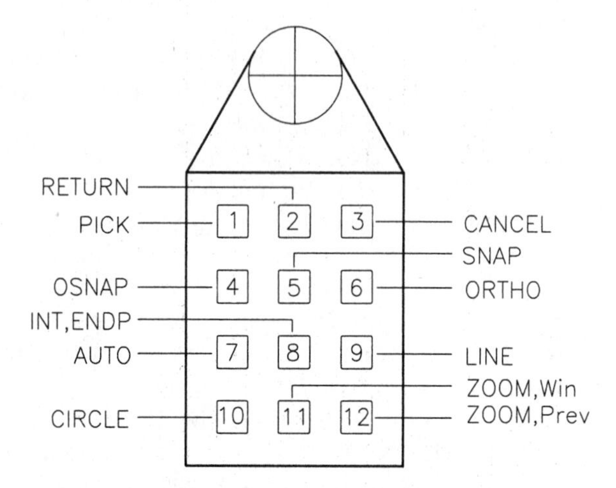

Figure 29-9 *Pointing device*

Explanation

Line 1
*****BUTTONS**

***BUTTONS1 is the section label for the first button menu. When the menu is loaded, AutoCAD LT compiles the menu file and assigns the commands to the buttons of the pointing device.

Line 2
;

This menu item assigns a semicolon (;) to button number 2. When you specify the second button on the pointing device, it enters a Return. It is like entering Return at the keyboard or the digitizer.

Line 3
^C^C

This menu item cancels the existing command twice (^C^C). This command is assigned to button number 3 of the pointing device. When you pick the third button on the pointing device, it cancels the existing command twice.

Line 4
$P0=*

This menu item loads and displays the cursor menu POP0, which contains various object snap modes. It is assumed that the POP0 menu has been defined in the menu file. This command is assigned to button number 4 of the pointing device. If you press this button, it will load and display the shortcut menu on the screen near the crosshairs location.

Line 5
^B
This menu item changes the Snap mode; it is assigned to button number 5 of the pointing device. When you pick the fifth button on the pointing device, it turns the Snap mode on or off. It is like holding down the CTRL key and then pressing the B key.

Line 6
^O
This menu item changes the ORTHO mode; it is assigned to button number 6. When you pick the sixth button on the pointing device, it turns the ORTHO mode on or off.

Line 7
AUTO
This menu item selects the AUTO option for creating a selection set; this command is assigned to button number 7 on the pointing device.

Line 8
INT,ENDP
In this menu item, INT is for the Intersection Osnap, and ENDP is for the Endpoint Osnap. This command is assigned to button number 8 on the pointing device. When you pick this button, AutoCAD LT looks for the intersection point. If it cannot find an intersection point, it then starts looking for the endpoint of the object that is within the pick box.

INT,ENDP
Where **INT** -------------- Intersection object snap
ENDP ------------ Endpoint object snap

Line 9
^C^CLINE
This menu item defines the **LINE** command; it is assigned to button number 9. When you select this button, AutoCAD LT cancels the existing command, and then selects the **LINE** command.

Line 10
^C^CCIRCLE
This menu item defines the **CIRCLE** command; it is assigned to button number 10. When you pick this button, AutoCAD LT automatically selects the **CIRCLE** command and prompts for the user input.

Line 11
'ZOOM;Win
This menu item defines a transparent **ZOOM** command with Window option; it is assigned to button number 11 of the pointing device.

'ZOOM;Win
Where ' -------------------- Single quote makes **ZOOM** command transparent
ZOOM ----------- AutoCAD LT **ZOOM** command

; ------------------- Semicolon for RETURN
Win --------------- Window option of **ZOOM** command

Line 12
'ZOOM;Prev
This menu item defines a transparent **ZOOM** command with **previous** option; it is assigned to button number 12 of the pointing device.

Note
If the button menu has more menu items than the number of buttons on the pointing device, the menu items in excess of the number of buttons are ignored. This does not include the pick button. For example, if a pointing device has three buttons in addition to the pick button, the first three menu items will be assigned to the three buttons (buttons 2, 3, and 4). The remaining lines of the button menu are ignored.

The commands are assigned to the buttons in the same order in which they appear in the file. For example, the menu item that is defined on line 3 will automatically be assigned to the fourth button of the pointing device. Similarly, the menu item that is on line 4 will be assigned to the fifth button of the pointing device. The same is true of other menu items and buttons.

Review Questions

Image Tile Menus

1. The image tiles are displayed in the ________________ box.

2. An **image tile** menu can be canceled by entering ____________ at the keyboard.

3. The **image title** dialog box can contain a maximum of __________ image tiles.

4. A blank line in an image tile menu __________ the image tile menu.

5. The menu item repetition feature ________________ be used with the image tile menu.

6. The drawing for a slide should be __________ on the entire screen before making a slide.

7. You __________________ fill a solid area in a slide for image tile menu.

8. An image tile menu _________ be accessed from a tablet menu.

Button and Auxiliary Menus

9. A multibutton pointing device can be used to specify ____________________________, or select ______________, or enter AutoCAD LT ______________.

10. AutoCAD LT receives the button ______________ and ______________ of screen crosshairs when a button is activated on the pointing device.

11. If the number of menu items in the button menu is more than the number of buttons on the pointing device, the excess lines are ______________.

12. Commands are assigned to the buttons of the pointing device in the ______________ order in which they appear in the buttons menu.

13. The format of referencing or loading a submenu that has been defined in the image menu is ______________.

14. The format of displaying a loaded submenu that has been defined in the image menu is ______________.

Exercises

Exercise 2

General

Write a tablet and image tile menu for inserting the following blocks. The layout of the template for the tablet menu is shown in Figure 29-10.

B1	B2	B3	C1	C2	C3
B4	B5	B6	C4	C5	C6
B7	B8	B9	C7	C8	C9

TEMPLATE

B1	B2	B3		C1	C2	C3
B4	B5	B6		C4	C5	C6
B7	B8	B9		C7	C8	C9

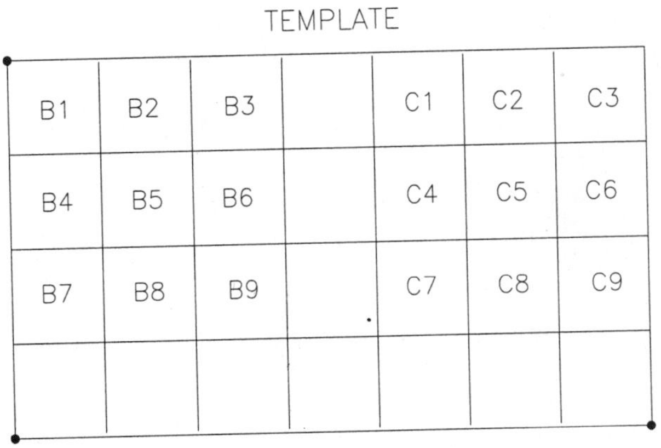

Figure 29-10 *Tablet template for Exercise 2*

Exercise 3 General

Write a button menu for the following AutoCAD LT commands. The pointing device has 10 buttons (Figure 29-11), and button number 1 is used for specifying the points. The blocks are to be inserted with a scale factor of 1.00 and a rotation of 0 degrees (file name **BME1.MNU**).

1. PICK BUTTON
2. RETURN
3. CANCEL
4. OSNAPS
5. INSERT B1
6. INSERT B2
7. INSERT B3
8. ZOOM Window
9. ZOOM All
10. ZOOM Previous

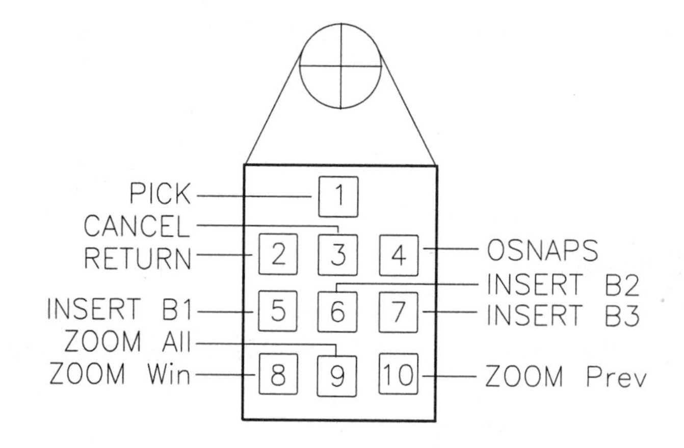

Figure 29-11 *Pointing device with 10 buttons*

Chapter 30

AutoCAD LT on the Internet

Learning Objectives

After completing this chapter, you will be able to:

- *Launch a Web browser from AutoCAD LT.*
- *Understand the importance of the uniform resource locator (URL).*
- *Open and save drawings to and from the Internet.*
- *Place hyperlinks in a drawing.*
- *Convert drawings to DWF file format.*
- *View DWF files with a Web browser.*
- *Learn about the Whip plug-in and the DWF file format.*

INTRODUCTION

The Internet has become the most important way to exchange information in the world. AutoCAD LT allows you to interact with the Internet in several ways. It can open and save files that are located on the Internet; it can launch a Web browser (AutoCAD LT 2000 includes a simple Web browser); and it can create drawing Web format (DWF) files for viewing as drawings on Web pages. Before you can use the Internet features in AutoCAD LT 2000, some components of Microsoft's Internet Explorer must be present in your computer. If Internet Explorer v4.0 (or later) is already set up on your computer, then you already have these required components. If not, then the components are installed automatically when during AutoCAD LT 2000's setup you select: (1) the **Full Install**; or (2) the **Internet Tools** option during the **Custom** installation.

This chapter introduces you to the following Web-related commands:

BROWSER:
Launches a Web browser from within AutoCAD LT.

HYPERLINK*:
Attaches and removes a URL to an object or an area in the drawing.

HYPERLINKFWD*:
Move to the next hyperlink (an undocumented command).

HYPERLINKBACK*:
Moves back to the previous hyperlink (an undocumented command).

HYPERLINKSTOP*:
Stops the hyperlink access action (an undocumented command).

PASTEASHYPERLINK*:
Attaches a URL to an object in the drawing from text stored in the Windows Clipboard (an undocumented command).

HYPERLINKBASE*:
A system variable for setting the path used for relative hyperlinks in the drawing.

In addition, all of AutoCAD LT 2000's file-related dialog boxes are "Web enabled."

CHANGES FROM AUTOCAD LT 97

If you used the Internet features in AutoCAD LT 97, then you should be aware of changes in AutoCAD LT 2000.

Attached URLs from LT 97

In AutoCAD LT 97, attached URLs were not active until the drawing was converted to a DWF file; in AutoCAD LT 2000, URLs are active in the drawing.

If you attached URLs to drawing in LT 97, they are converted to AutoCAD LT 2000-style hyperlinks the first time you save the drawing in the AutoCAD LT 2000 DWG format.

When you use the **SAVEAS** command to save an AutoCAD LT 2000 drawing in LT 97 format, hyperlinks are converted back to LT 97-style URLs. Hyperlink descriptions are stored as proxy data and not displayed in LT 97 (this ensures that hyperlink descriptions are available again when the drawing is brought back to AutoCAD LT 2000 format.)

WHIP v3.1 Limitations

AutoCAD LT 2000 can create DWF files compatible with the WHIP v3.1 plug-in. It is recommended, however, that you download WHIP v4.0 from the Autodesk Web site. Whip v3.1 does not display the following AutoCAD LT 2000-specific features:

- Hyperlink descriptions are not saved in v3.1 DWF format.
- All objects, except polylines and traces, are displayed at minimum lineweight.
- Linetypes defined by a plot style table are displayed as continuous lines.
- Nonrectangular viewports are displayed by rectangular viewports.

Changed Internet Commands

Many of LT 97 Internet-related commands have been discontinued and replaced. The following summary describes the status of all LT 97 Internet commands:

BROWSER (launches Web browser from within AutoCAD LT): continues to work in AutoCAD LT 2000; the default is now Home.Html, a Web page located on your computer.

ATTACHURL (attaches a URL to an object or an area in the drawing): continues to work in AutoCAD LT 2000, but has been superceded by the **-HYPERLINK** command's **Insert** option.

SELECTURL (selects all objects with attached URLs): continues to work in AutoCAD LT 2000, but has been superceded by the **QSelect** dialog box's **Hyperlink** option.

LISTURL (lists URLs embedded in the drawing): removed from AutoCAD LT 2000. As a replacement, use the **Properties** dialog box's **Hyperlink** option.

DETACHURL (removes the URL from an object): continues to work in AutoCAD LT 2000, but has been superceded by the **-HYPERLINK** command's **Remove** option.

DWFOUT (exports the drawing and embedded URLs as a DWF file): continues to work in AutoCAD LT 2000, but has been superceded by the **Plot** dialog box's **ePlot** option.

DWFOUTD (exports drawing in DWF format without a dialog box): removed from AutoCAD LT 2000. As a replacement, use the **-PLOT** command's **ePlot** option.

INETCFG (configures AutoCAD LT for Internet access): removed from AutoCAD LT 2000; it has been replaced by the **Internet** applet of the Windows Control Panel (choose the **Connection** tab).

INSERTURL (inserts a block from the Internet into the drawing): automatically executes the **INSERT** command and displays the **Insert** dialog box. You may type a URL for the block name.

OPENURL (opens a drawing from the Internet): automatically executes the **OPEN** command and displays the **Select File** dialog box. You may type a URL for the filename.

SAVEURL (saves the drawing to the Internet): automatically executes the **SAVE** command and displays the **Save Drawing As** dialog box. You may type a URL for the filename.

You are probably already familiar with the best-known uses for the Internet: email (electronic mail) and the Web (short for "World Wide Web"). Email lets users exchange messages and data at very low cost. The Web brings together text, graphics, audio, and movies in an easy-to-use format. Other uses of the Internet include FTP (file transfer protocol) for effortless binary file transfer, Gopher (presents data in a structured, subdirectory-like format), and USENET, a collection of more than 29,000 news groups.

AutoCAD LT allows you to interact with the Internet in several ways. AutoCAD LT is able to launch a Web browser from within AutoCAD LT with the **BROWSER** command. Hyperlinks can be inserted in drawings with the **HYPERLINK** command, which lets you link the drawing with other documents on your computer and the Internet. With the **PLOT** command's **ePlot** option (short for "electronic plot), AutoCAD LT creates DWF (short for "drawing Web format") files for viewing drawings in 2D format on Web pages. AutoCAD LT can open, insert, and save drawings to and from the Internet via the **OPEN**, **INSERT**, and **SAVEAS** commands.

UNDERSTANDING URLs

The uniform resource locator, known as the URL, is the file naming system of the Internet. The URL system allows you to find any resource (a file) on the Internet. Example resources include a text file, a Web page, a program file, an audio or movie clip—in short, anything you might also find on your own computer. The primary difference is that these resources are located on somebody else's computer. A typical URL looks like the following examples:

Example URL	**Meaning**
http://www.autodesk.com	Autodesk Primary Web Site
news://adesknews.autodesk.com	Autodesk News Server
ftp://ftp.autodesk.com	Autodesk FTP Server
http://www.autodeskpress.com	Autodesk Press Web Site
http://users.uniserve.com/~ralphg	Editor Ralph Grabowski's Web site

Note that the **http://** prefix is not required. Most of today's Web browsers automatically add in the *routing* prefix, which saves you a few keystrokes.

URLs can access several different kinds of resources—such as Web sites, email, news groups—but always take on the same general format, as follows:

scheme://netloc

The scheme accesses the specific resource on the Internet, including these:

Scheme	**Meaning**
file://	File is located on your computer's hard drive or local net work
ftp://	File Transfer Protocol (used for downloading files)
http://	Hyper Text Transfer Protocol (the basis of Web sites)
mailto://	Electronic mail (email)
news://	Usenet news (news groups)
telnet://	Telnet protocol
gopher://	Gopher protocol

The **://** characters indicate a network address. Autodesk recommends the following format for specifying URL-style filenames with AutoCAD LT:

Resource	**URL Format**
Web Site	**http://***servername/pathname/filename*
FTP Site	**ftp://***servername/pathname/filename*

Local File **file:**///*drive*:/*pathname*/*filename*
or *drive*:*pathname**filename*
or **file:**///*drive* | /*pathname*/*filename*
or **file:**//*localPC**pathname**filename*
or **file:**////*localPC*/*pathname*/*filename*
Network File **file:**//*localhost*/*drive*:/*pathname*/*filename*
or *localhost**drive*:*pathname**filename*
or **file:**//*localhost*/*drive* | /*pathname*/*filename*

The terminology can be confusing. The following definitions will help to clarify these terms.

Term	**Meaning**
servername	The name or location of a computer on the Internet, for example: *www.autodesk.com*
pathname	The same as a subdirectory or folder name
drive	The driver letter, such as C: or D:
localpc	A file located on your computer
localhost	The name of the network host computer

If you are not sure of the name of the network host computer, use Windows Explorer to check the Network Neighborhood for the network names of computers.

LAUNCHING A WEB BROWSER

The **BROWSER** command lets you start a Web browser from within AutoCAD LT. Commonly used Web browsers include Netscape's Navigator, Microsoft's Internet Explorer, and Operasoft's Opera. By default, the **BROWSER** command uses whatever brand of Web browser program is registered in your computer's Windows operating system. AutoCAD LT prompts you for the uniform resource locator (URL), such as *http://www.autodeskpress.com*. The **BROWSER** command can be used in scripts, toolbar or menu macros to automatically access the Internet.

Command: **BROWSER**
Browse <C:\Program Files\ACLT 2000\Home.htm>: *Enter the URL.*

The default URL is an HTML file added to your computer during AutoCAD LT's installation. After you type the URL and press *, AutoCAD LT launches the Web browser and contacts the Web site. Figure 30-1 shows the popular Internet Explorer with the Autodesk Web site at *http://www.autodesk.com*.

CHANGING DEFAULT WEB SITE

To change the default Web page that your browser starts with from within AutoCAD LT, change the setting in the **INETLOCATION** system variable. The variable stores the URL used by the **BROWSER** command and the **Browse the Web** dialog box. Make the change, as follows:

Command: **INETLOCATION**
Enter new value for INETLOCATION <"C:\Program Files\ACLT 2000\Home.htm">:
Type URL.

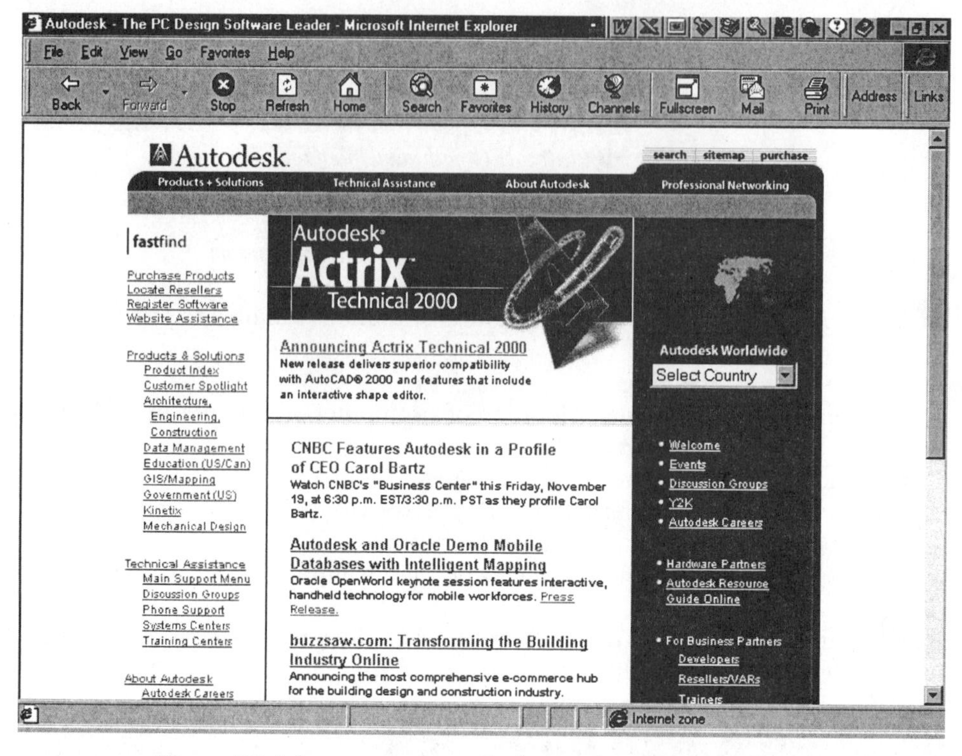

Figure 30-1 Internet Explorer displaying the Autodesk Web Site

DRAWINGS ON THE INTERNET

When a drawing is stored on the Internet, you access it from within AutoCAD LT 2000 using the standard **OPEN**, **INSERT**, and **SAVE** commands. (In AutoCAD LT 97, these commands were known as **OPENURL**, **INSERTURL**, and **SAVEURL**.) Instead of specifying the file's location with the usual drive-subdirectory-file name format, such as c:\program files\aclt\filename.dwg, use the URL format. (Recall that the URL is the universal file naming system used by the Internet to access any file located on any computer hooked up to the Internet.)

OPENING DRAWINGS FROM THE INTERNET

To open a drawing from the Internet (or your firm's intranet), use the **OPEN** command (**File** > **Open**). Notice the three buttons at the upper right of the **Select File** dialog box (see Figure 30-2). These are specifically for use with the Internet. From left to right, they are:

Search the Web: Displays the **Browse the Web** dialog box, a simplified Web browser.

Look in Favorites: Opens the **Favorites** folder, equivalent to "bookmarks" that store Web addresses.

Add to Favorites: Saves the current URL (hyperlink) to the **Favorites** folder.

When you click the **Search the Web** button, AutoCAD LT opens the **Browse the Web** dialog box (Figure 30-3). This dialog box is a simplified version of the Microsoft brand of Web browser.

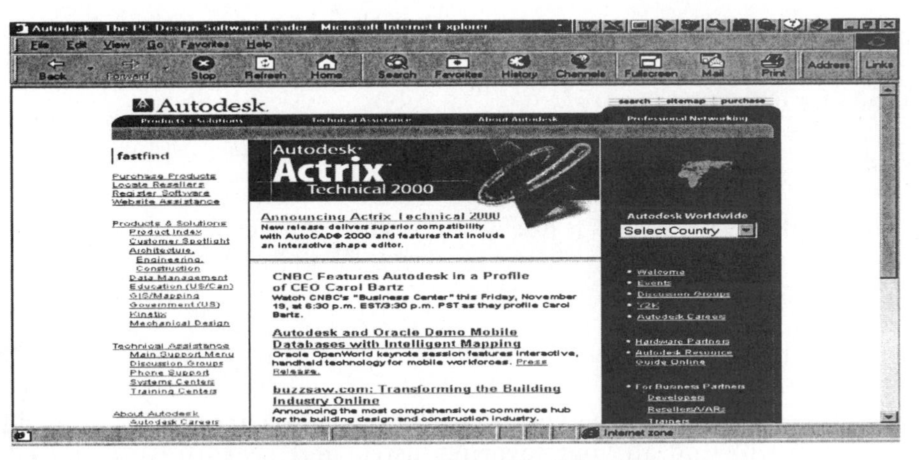

Figure 30-2 ***Select File*** *dialog box*

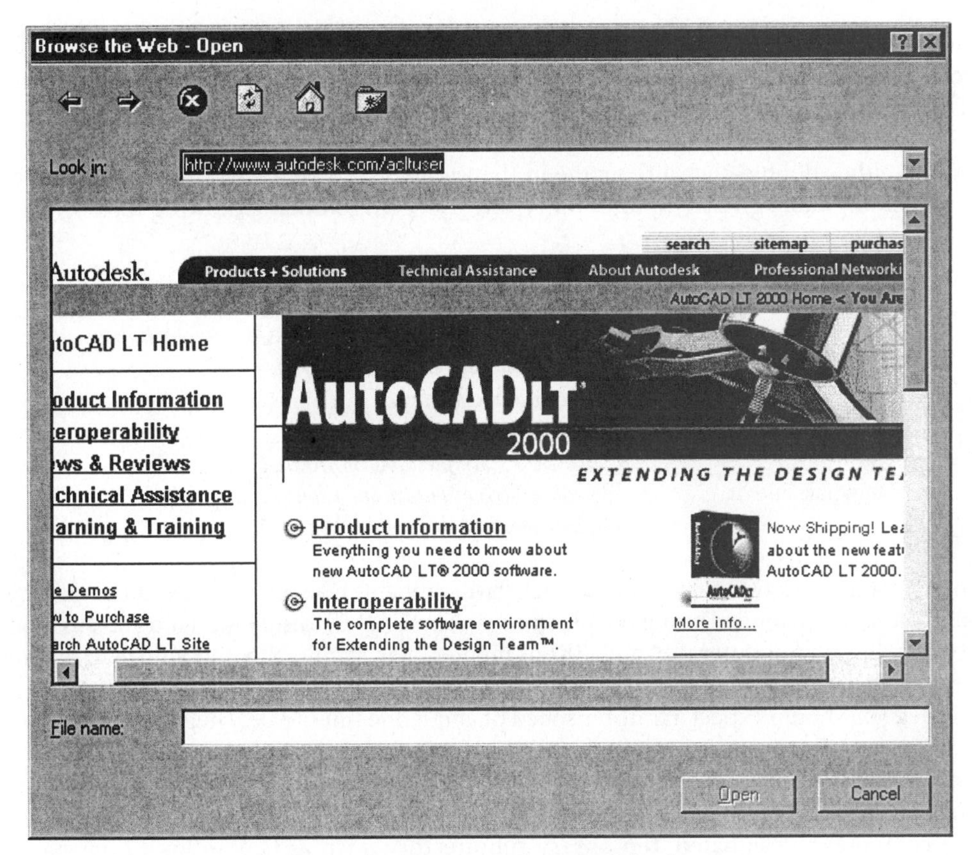

Figure 30-3 ***Browse the Web*** *dialog box*

The purpose of this dialog box is to allow you to browse files at a Web site.

By default, the **Browse the Web** dialog box displays the contents of the URL stored in **INETLOCATION** system variable. You can easily change this to another folder or Web site, as noted earlier.

Along the top, the dialog box has six buttons:

Back: Go back to the previous URL.

Forward: Go forward to the next URL.

Stop: Halt displaying the Web page (useful if the connection is slow or the page is very large.)

Refresh: Redisplay the current Web page.

Home: Return to the location specified by the **INETLOCATION** system variable.

Favorites: List stored URLs (hyperlinks) or bookmarks. If you have previously used Internet Explorer, you will find all your favorites listed here. Favorites are stored in the **\Windows\Favorites** folder on your computer.

The **Look in** field allows you to type the URL. Alternatively, click the down arrow to select a previous destination. If you have stored Web site addresses in the Favorites folder, then select a URL from that list.

You can either double-click a filename in the window, or type a URL in the **File name** field. The following table gives templates for typing the URL to open a drawing file:

Drawing Location	**Template URL**
Web or HTTP Site	*http://servername/pathname/filename.dwg*
	http://practicewrench.autodeskpress.com/wrench.dwg
FTP Site	*ftp://servername/pathname/filename.dwg*
	ftp://ftp.autodesk.com
Local File	*drive:\pathname\filename.dwg*
	c:\aclt 2000\sample\tablet2000.dwg
Network File	*\\localhost\drive:\pathname\filename.dwg*
	\\upstairs\e:\install\sample.dwg

When you open a drawing over the Internet, it will probably take much longer than opening a file found on your computer. During the file transfer, AutoCAD LT displays a dialog box to report the progress. If your computer uses a 28.8Kbps modem, you should allow about five to ten minutes per megabyte of drawing file size. If your computer has access to a faster T1 connection to the Internet, you should expect a transfer speed of about one minute per megabyte.

It may be helpful to understand that the **OPEN** command does not copy the file from the Internet location directly into AutoCAD LT. Instead, it copies the file from the Internet to your computer's designated **Temporary** subdirectory, such as C:\Windows\Temp (and then loads the drawing from the hard drive into AutoCAD LT). This is known as *caching*. It helps to speed up the processing of the drawing, since the drawing file is now located on your computer's fast hard drive, instead of the relatively slow Internet. Note that the **Locate** and **Find File** buttons in the **Select File** dialog box do not work for locating files on the Internet.

Example 1

The Autodesk Press Web site has an area that allows you to practice using the Internet with AutoCAD LT. In this example, you open a drawing file located at that Web site.

1. Start AutoCAD LT.

2. Ensure that you have a live connection to the Internet. If you normally access the Internet via a telephone (modem) connection, dial your Internet service provider now.

3. From the menu bar, select **File > Open.** Or, choose the **Open** button on the toolbar. Notice that AutoCAD LT displays the **Select File** dialog box.

4. Click the **Search the Web** button. Notice that AutoCAD LT displays the **Browse the Web** dialog box.

5. In the **Look in** field, type:

   practicewrench.autodeskpress.com

   and press ENTER. After a few seconds, the **Browse the Web** dialog box displays the Web site (Figure 30-4).

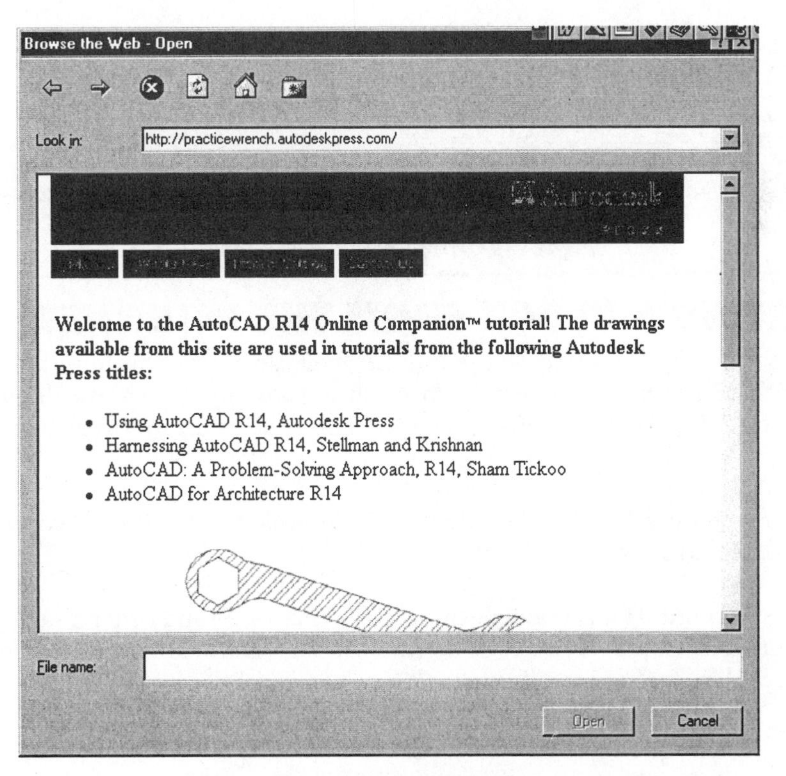

Figure 30-4 *Practice Wrench Web site*

6. In the Filename field, type:

   wrench.dwg

   and press ENTER. AutoCAD LT begins transferring the file. Depending on the speed of your Internet connection, this will take between a couple of seconds and a half-minute. Notice the drawing of the wrench in AutoCAD LT (Figure 30-5).

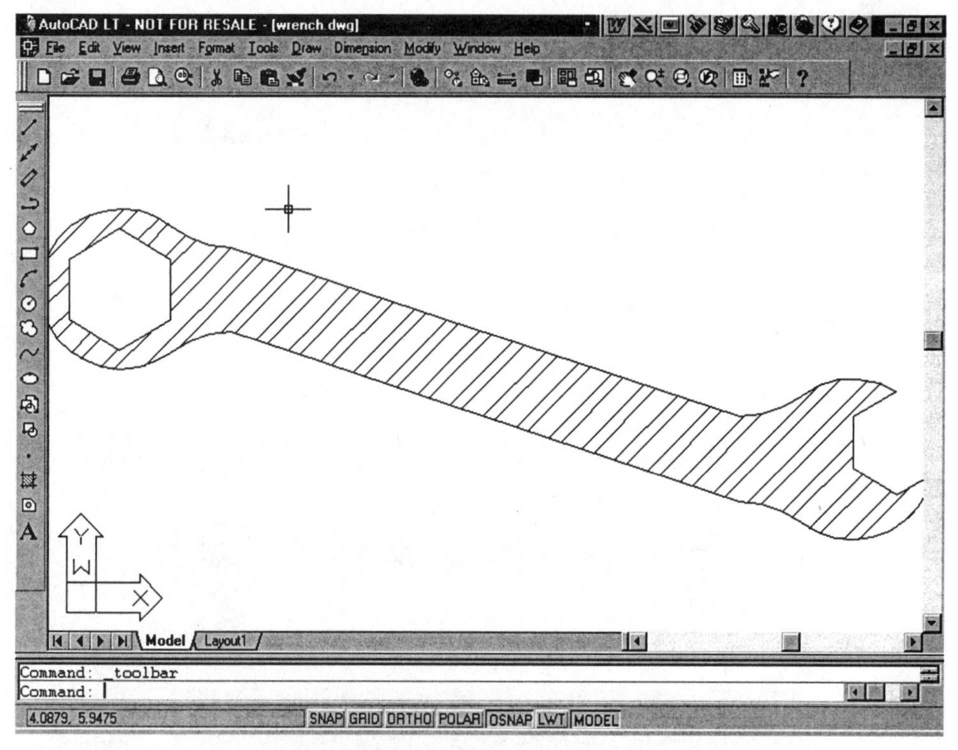

Figure 30-5 *Wrench drawing in AutoCAD*

INSERTING A BLOCK FROM THE INTERNET

When a block (symbol) is stored on the Internet, you can access it from within AutoCAD LT using the **INSERT** command. When the **Insert** dialog box appears, choose the **Browse** button to display the **Select Drawing File** dialog box. This is identical to the dialog box discussed previously.

After you choose the select the file, AutoCAD LT downloads the file and continues with the **INSERT** command's familiar prompts.

The process is identical for accessing external reference (xref) and raster image files. Another file that AutoCAD LT can access over the Internet include: WMF (Windows metafile). This option is found in the **Insert** menu on the menu bar.

ACCESSING OTHER FILES ON THE INTERNET

Most other file-related dialog boxes allow you to access files from the Internet or intranet. This allows your firm or agency to have a central location that stores drawing standards. When you need to use a linetype or hatch pattern, for example, you access the LIN or PAT file over the Internet. More than likely, you would have the location of those files stored in the Favorites list. Some examples include:

Linetypes: From the **Format** menu, choose **Linetype**. In the **Linetype Manager** dialog box, choose the **Load**, **File**, and **Look in Favorites** buttons.

Hatch Patterns: Use the Web browser to copy .PAT files from a remote location to your computer.

Layer Name: From the **Format** menu, choose **Layer> Layer Manager**. In the **Layer Manager** dialog box, choose the **Import** and **Look in Favorites** buttons.

Scripts: From the **Tools** menu, choose **Run Script**.

Menus: From the **Tools** menu, choose **Customize Menus**. In the **Menu Customization** dialog box, choose the **Browse** and **Look in Favorites** buttons.

You cannot access text files, text fonts (SHX and TTF), color settings, lineweights, dimension styles, plot styles, OLE objects, or named UCSs over the Internet.

SAVING THE DRAWING TO THE INTERNET

When you are finished editing a drawing in AutoCAD LT, you can save it to a file server on the Internet with the **SAVE** command. If you inserted the drawing from the Internet (using **INSERT**) into the default Drawing.Dwg drawing, AutoCAD LT insists that you first save the drawing to your computer's hard drive.

When a drawing of the same name already exists at that URL, AutoCAD LT warns you, just as it does when you use the **SAVEAS** command. Recall from the **OPEN** command that AutoCAD LT uses your computer system's Temporary subdirectory, hence the reference to it in the dialog box.

USING HYPERLINKS WITH AUTOCAD LT*

AutoCAD LT 2000 allows you to employ URLs in two ways: (1) directly within an AutoCAD LT drawing; and (2) indirectly in DWF files displayed by a Web browser. (URLs are also known as *hyperlinks*, the term we use from now on.)

Hyperlinks are created, edited, and removed with the **HYPERLINK** command (displays a dialog box) and the **-HYPERLINKS** command (for prompts at the command line). The **HYPERLINK** command (**Insert** > **Hyperlink**) prompts you to "Select objects" and then displays the **Insert Hyperlink** dialog box (Figure 30-6(b)). As a shortcut, you may press CTRL+K or choose the **Insert Hyperlink** button on the **Standard** toolbar, (Figure 30-6(a)).

As an alternative, you may use the **–HYPERLINK** command. This command displays its prompts at the command line, and is useful for scripts and AutoLISP routines. The **–HYPERLINK** command has the following syntax:

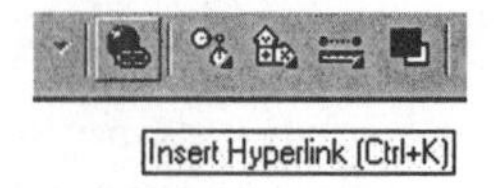

Figure 30-6(a) *Invoking the* ***HYPERLINK*** *command from the* ***Standard*** *toolbar*

Command: **-HYPERLINK**
Enter an option [Remove/Insert] <Insert>: *Press ENTER.*
Enter hyperlink insert option [Area/Object] <Object>: *Press ENTER.*
Select objects: *Select an object.*
1 found Select objects: *Press ENTER.*
Enter hyperlink <current drawing>: *Enter the name of the document or Web site.*
Enter named location <none>: *Enter the name of a bookmark or AutoCAD LT view.*
Enter description <none>: *Enter a description of the hyperlink.*

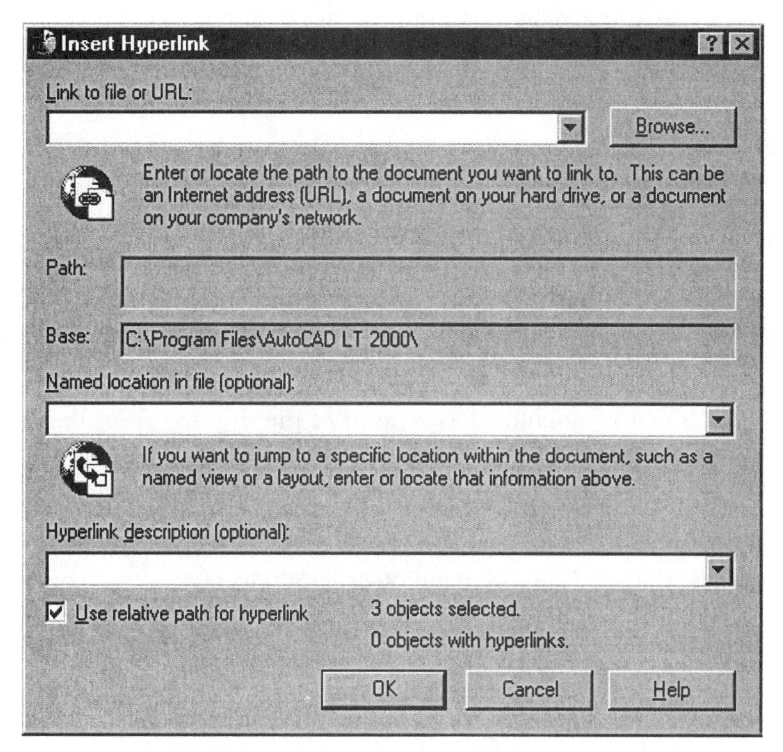

Figure 30-6(b) *The* ***Insert Hyperlink*** *dialog box*

Notice that the command also allows you to remove a hyperlink. It does not, however, allow you to edit a hyperlink. To do this, use the **Insert** option and respecify the hyperlink data. In addition, this command allows you to create hyperlink *areas*, which is a rectangular area that can be thought of as a 2D hyperlink (the dialog box-based **HYPERLINK** command does not create hyperlink areas). When you select the **Area** option, the rectangle is placed automatically on layer URLLAYER and colored red (Figure 30-7).

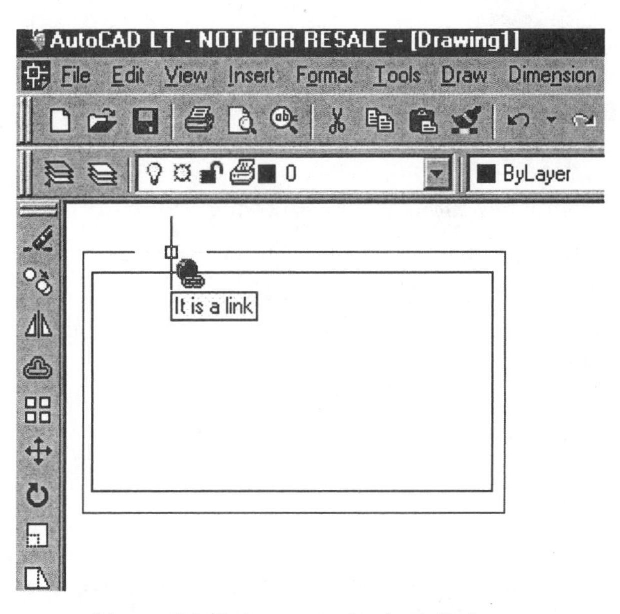

Figure 30-7 A rectangular hyperlink area

In the following sections, you learn how to apply and use hyperlinks in an AutoCAD LT drawing and in a Web browser via the dialog box-based **HYPERLINK** command.

Hyperlinks Inside AutoCAD LT*

AutoCAD LT allows you to add a hyperlink to any object in the drawing. An object is permitted just a single hyperlink; on the other hand, a single hyperlink may be applied to a selection set of objects.

You can tell an object has a hyperlink by passing the cursor over it. The cursor displays the "linked Earth" icon, as well as a tooltip describing the link. See Figure 30-8.

***Figure 30-8** The cursor reveals a hyperlink*

If, for some reason, you do not want to see the hyperlink cursor, you can turn it off. From the **Tools** menu, choose **Options**, then choose the **User Preferences** tab. The **Display hyperlink cursor and shortcut menu** item toggles the display of the hyperlink cursor, as well as the **Hyperlink** tooltip option on the shortcut menu.

Attaching Hyperlinks*

Let's see how this works with an example.

Example 2

In this example, we have the drawing of a floorplan. To this drawing, we will add hyperlinks to another AutoCAD LT drawing, a Word document, and a Web site. Hyperlinks must be attached to objects. For this reason, we will place some text in the drawing, then attach the hyperlinks to the text.

1. Start AutoCAD LT 2000.

2. Open the **Foundation Plan.dwg** file found in the AutoCAD LT 2000\ **Sample** folder. (If necessary, choose the **Model** tab to display the drawing in model space.)

3. Invoke the **TEXT** command and place the following text in the drawing (Figure 30-9):

   Command: **TEXT**
   Current text style: "Standard" Text height: 0.20000
   Specify start point of text or [Justify/Style]: *Specify a point in the drawing.*
   Specify height <0.2000>: **2**
   Specify rotation angle of text <0>: *Press ENTER.*
   Enter text: **Foundation Plan**
   Enter text: **Basic Design**
   Enter text: **Other Factors**
   Enter text: *Press ENTER.*

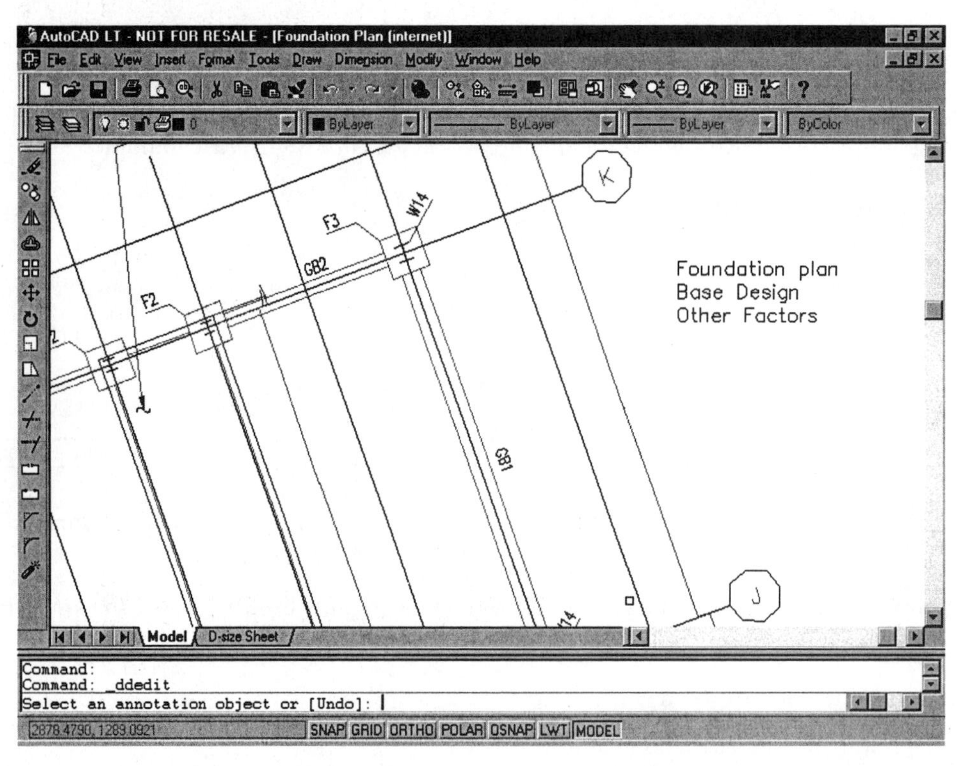

Figure 30-9 *Text placed in drawing*

4. Select the "Foundation Plan" text to select it.

5. Choose the **Insert Hyperlink** button on the toolbar. Notice the **Insert Hyperlinks** dialog box.

6. Next to the **Link to File or URL** field, choose the **Browse** button. Notice the **Browse the Web – Select Hyperlink** dialog box (Figure 30-10).

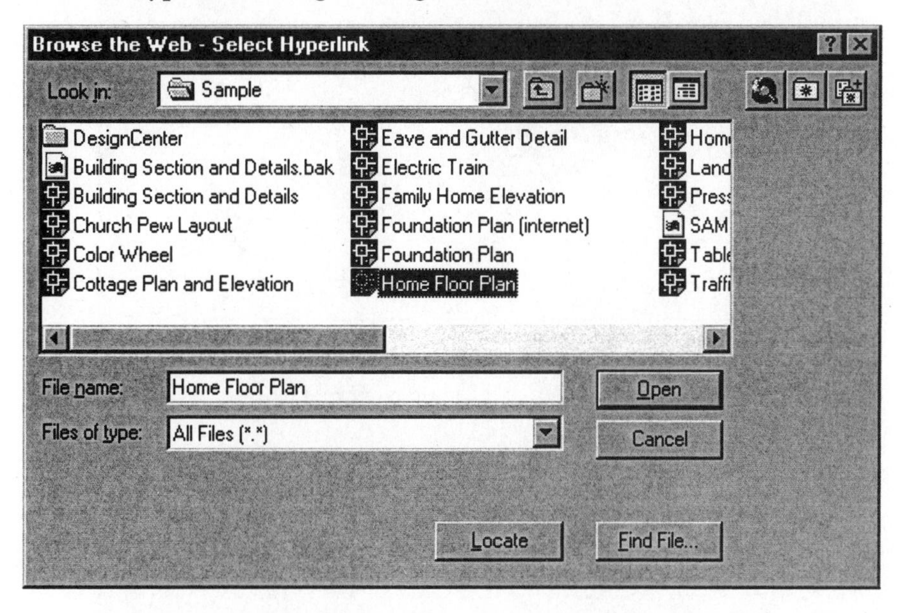

Figure 30-10 Browse the Web – Select Hyperlink dialog box

7. Go to AutoCAD LT 2000's **Sample** folder and select the **HomeFloor Plan.dwg** file. Choose **Open**. AutoCAD LT does not open the drawing; rather, it copies the file's name to the **Insert Hyperlinks** dialog box (Figure 30-11). Notice the name of the drawing you selected in the **Link to File or URL** field.

You can fill in two other fields:

Named Location in File: When the hyperlinked file is opened, it goes to the named location, called a "Bookmark." In AutoCAD LT, the bookmark is a named view (created with the **VIEW** command).

Hyperlink Description: You may type a description for the hyperlink, which is displayed by the tooltip. If you leave this blank, the URL is displayed by the tooltip.

8. Choose **OK** to dismiss the dialog box. Move the cursor over the Site Plan text. Notice the display of the "linked Earth" icon; a moment later, the tooltip displays "HomeFloor Plan.dwg"(Figure 30-12).

9. Repeat the **HYPERLINK** command twice more, attaching these files to the drawing text:

Text	URL
Basic Design	License.Rtf (found in the AutoCAD LT 2000 folder)
Other Factors	Home.Htm (also found in the AutoCAD LT 2000 folder)

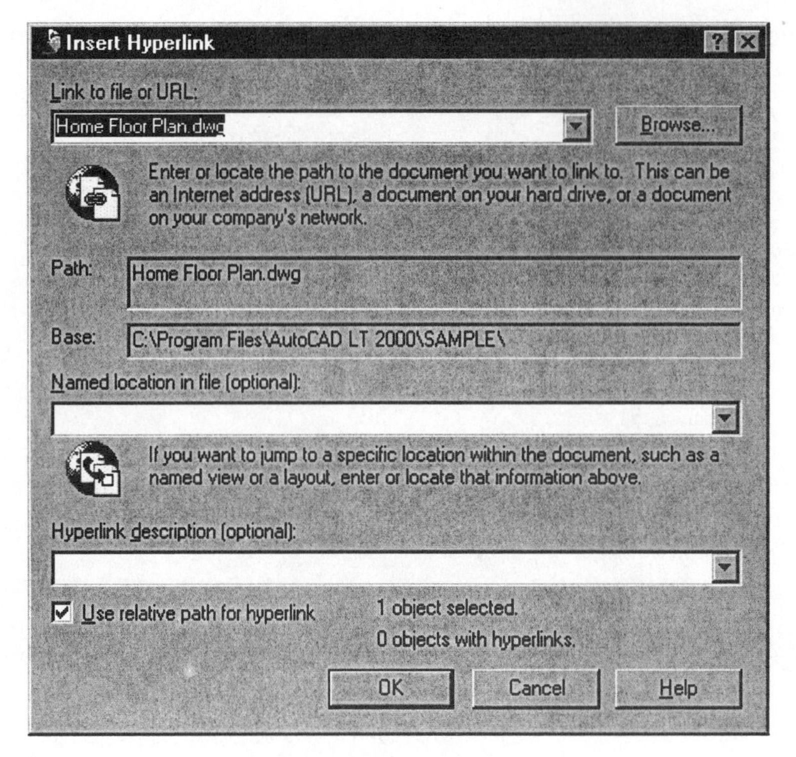

Figure 30-11 *The **Insert Hyperlinks** dialog box*

Figure 30-12 *The Hyperlink cursor and tooltip*

You have now attached a drawing, a text document, and a Web document to objects in the drawing. Let's try out the hyperlinks.

10. Click "Foundation plan" to select it. Right-click and select **Hyperlink** > **Open "HomeFloorPlan.dwg"** from the shortcut menu (Figure 30-13).

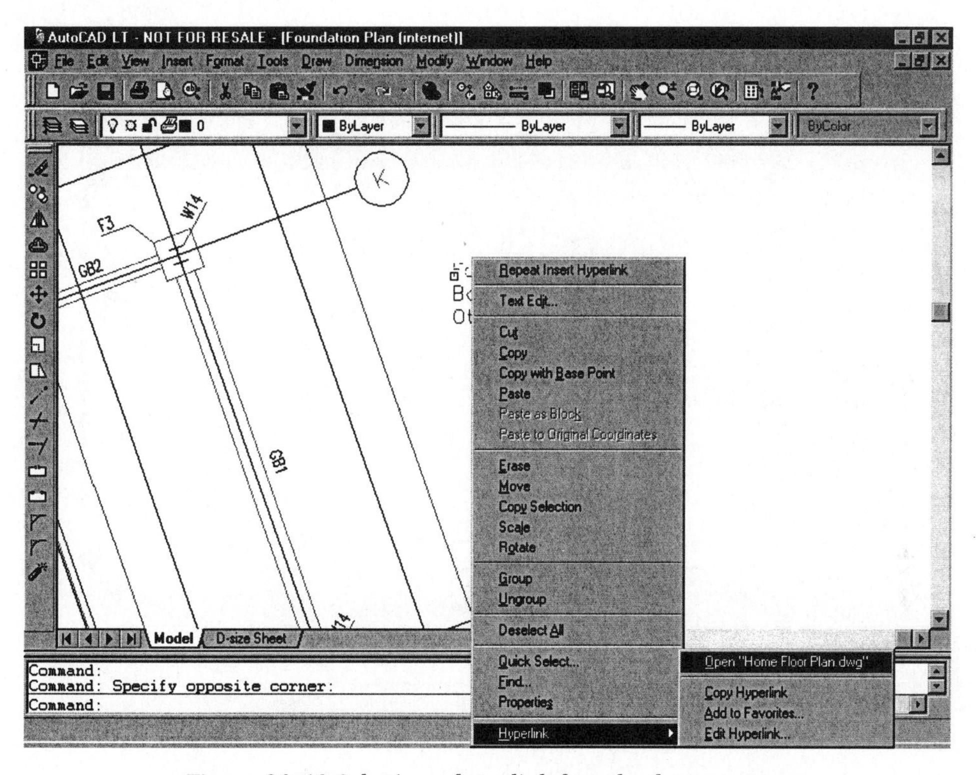

***Figure 30-13** Selecting a hyperlink from the shortcut menu*

Notice that AutoCAD LT opens "Home Floor Plan.dwg". To see both drawings, select **Window > Tile Vertically** from the menu bar (Figure 30-14).

11. Select, then right-click "Basic Design" hyperlink. Choose **Hyperlink > Open**. Notice that Windows starts a word processor and opens the License.Rtf file (Figure 30-15).

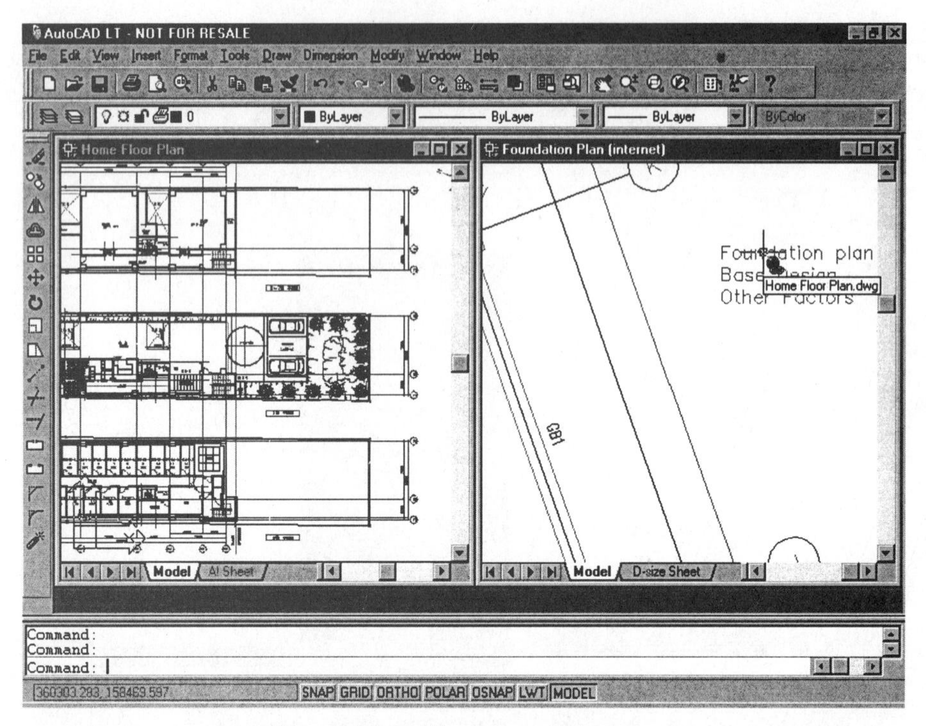

Figure 30-14 Viewing two drawings

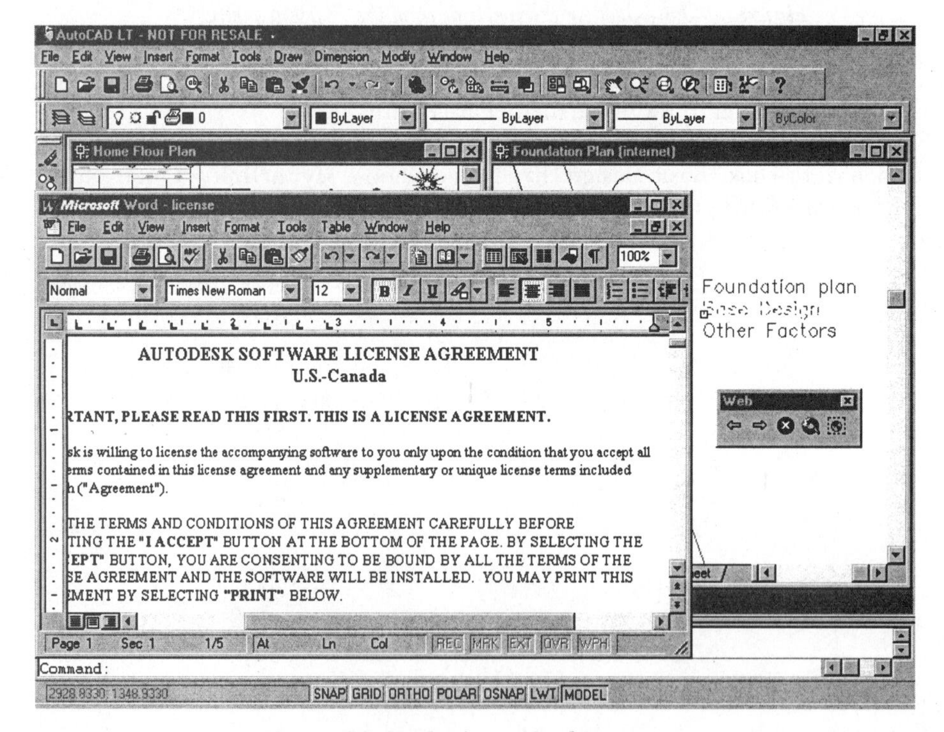

Figure 30-15 Viewing a Word Document

12. If your word processor has a **Web** toolbar, open it. Choose the **Back** button (the back arrow), as shown in Figure 30-16. Notice how that action sends you back to AutoCAD LT.

***Figure 30-16** AutoCAD's **Web** toolbar*

13. Select, then right-click "Other Factors" hyperlink. Select **Hyperlink > Open**. Notice that Windows starts your Web browser and opens the Home.Html file (Figure 30-17).

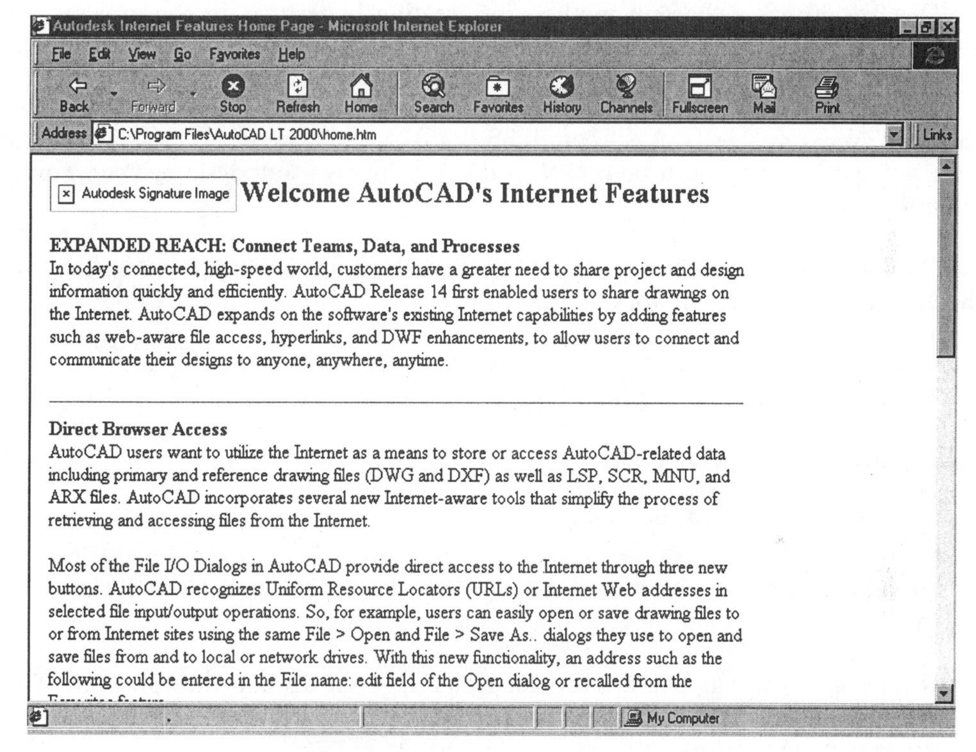

***Figure 30-17** Viewing a Web Site*

14. Back in AutoCAD LT, right-click any toolbar and choose **Web**. Notice that the **Web** toolbar has four buttons. From left to right, these are:

Button	**Command**	**Meaning**
Go Back	**HYPERLINKBACK**	Move back to the previous hyperlink.
Go Forward	**HYPERLINKFWD**	Move to the next hyperlink.
Stop Navigation	**HYPERLINKSTOP**	Stops the hyperlink access action.
Browse the Web	**BROWSER**	Launches the Web browser.

15. Try choosing the **Go Forward** and **Go Back** buttons. Notice how these allow you to navigate between the two drawings, the RTF document, and the Web page.
    When you work with hyperlinks in AutoCAD LT, you may come across these limitations:

    - AutoCAD LT does not check that the URL you type is valid.
    - If you attach a hyperlink to a block, be aware that the hyperlink data is lost when you scale the block unevenly, stretch the block, or explode it.
    - Wide polylines and rectangular hyperlink areas are only "sensitive" on their outline.

Pasting As Hyperlink*

AutoCAD LT 2000 has a shortcut method for creating hyperlinks in the drawing. The undocumented **PASTEASHYPERLINK** command pastes any text in the Windows Clipboard as a hyperlink to any object in the drawing. Here is how it works:

1. In a word processor, select some text and copy it to the Clipboard (via CTRL+C or choose **Edit > Copy**). The text can be a URL (such as http://www.autodeskpress.com) or any other text.

2. Switch to AutoCAD LT and select **Edit > Paste As Hyperlink** from the menu bar. Note that this command does not work (is grayed out) if anything else is in the Clipboard, such as a picture.

3. Select one or more objects, as prompted:

   Command: **_PASTEASHYPERLINK**
   Select objects: *Select an object.*
   1 found Select objects: *Press ENTER.*

4. Pass the cursor over the object and note the hyperlink cursor and tooltip. The tooltip displays the same text that you copied from the document.

If the text you copy to the Clipboard is very long, AutoCAD LT displays only portions of it in the tooltip, using ellipses (...) to shorten the text. You cannot select text in the AutoCAD LT drawing to paste as a hyperlink.

You can, however, copy the hyperlink from one object to another. Select the object, right-click, and select **Hyperlink > Copy Hyperlink** from the shortcut menu. The hyperlink is copied to the Clipboard. You can now paste the hyperlink into another document, or use AutoCAD LT's **Edit > Paste as Hyperlink** command to attach the hyperlink to another object in the drawing.

Note
*The **MATCHPROP** command does not work with hyperlinks.*

Highlighting Objects with URLs*

Although you can see the rectangle of area URLs, the hyperlinks themselves are invisible. For this reason, AutoCAD LT has the **QSELECT** command, which highlights all objects that match specifications. From the menu bar, choose **Tools > Quick Select**. AutoCAD LT displays the **Quick Select** dialog box (Figure 30-18).

In the fields, enter these specifications:

Apply to: **Entire drawing**
Object type: **Multiple**
Properties: **Hyperlink**
Operator: *** Wildcard Match**
Value: *

Choose **OK** and AutoCAD LT highlights all objects that have a hyperlink. Depending on your computer's display system, the highlighting shows up as dashed lines or as another color.

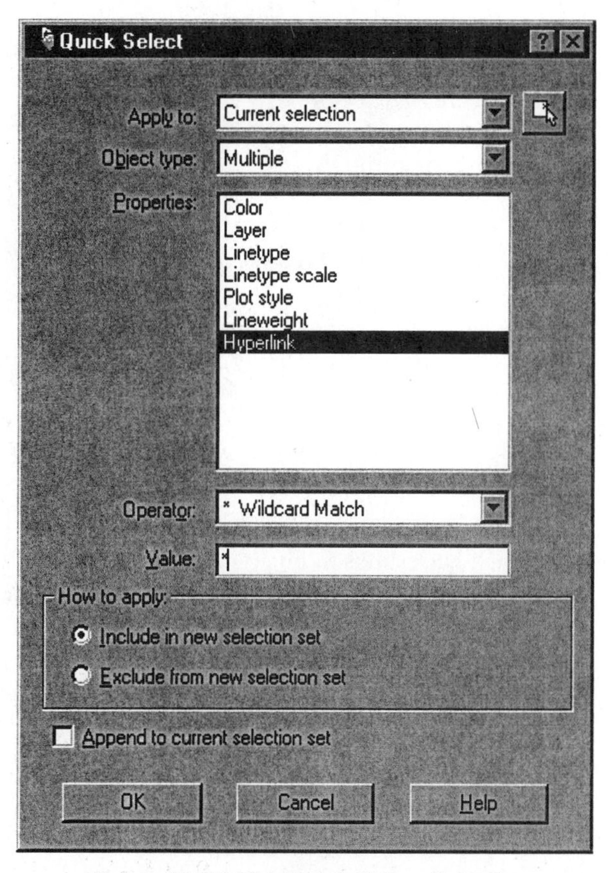

*Figure 30-18 The **Quick Select** dialog box*

Editing Hyperlinks*

Now that you know where the objects with hyperlinks are located, you can use the **HYPERLINK** command to edit the hyperlinks and related data. Select the hyperlinked object and invoke the **HYPERLINK** command. When the **Edit Hyperlink** dialog box appears (it looks identical to the **Insert Hyperlink** dialog box), make the changes and choose **OK**.

Removing Hyperlinks from Objects*

To remove a URL from an object, use the **HYPERLINK** command on the object. When the **Edit Hyperlink** dialog box appears, choose the **Remove Hyperlink** button.

To remove a rectangular area hyperlink, you can simply use the **ERASE** command; select the rectangle and AutoCAD LT erases the rectangle. (Unlike in LT 97, AutoCAD LT no longer purges the URLLAYER layer.)

As an alternative, you may use the **–HYPERLINK** command's **Remove** option, as follows:

Command: **-HYPERLINK**
Enter an option [Remove/Insert] <Insert>: **R**
Select objects: **ALL**

1 found Select objects: *Press ENTER.*
1. *www.autodesk.com*
2. *www.autodeskpress.com*
Enter number, hyperlink, or * for all: **1**
Remove, deleting the Area.
1 hyperlink deleted.

Hyperlinks Outside AutoCAD LT*

The hyperlinks you place in the drawing are also available for use outside of AutoCAD LT. The Web browser makes use of the hyperlink(s) when the drawing is exported in DWF format. To help make the process clearer, here are the steps that you need to follow:

Step 1

Open a drawing in AutoCAD LT.

Step 2

Attach hyperlinks to objects in the drawing with the **HYPERLINK** command. To attach hyperlinks to areas, use the **–HYPERLINK** command's **Area** option.

Step 3

Export the drawing in DWF format using the **PLOT** command's "DWF ePlot PC2" plotter configuration.

Step 4

Copy the DWF file to your Web site.

Step 5

Start your Web browser with the **BROWSER** command.

Step 6

View the DWF file and click on a hyperlink spot.

DRAWING WEB FORMAT

To display AutoCAD LT drawings on the Internet, Autodesk invented a new file format called Drawing Web Format (DWF). The DWF file has several benefits and some drawbacks over DWG files. The DWF file is compressed as much as eight times smaller than the original DWG drawing file so that it takes less time to transmit over the Internet, particularly with the relatively slow telephone modem connections. The DWF format is more secure, since the original drawing is not being displayed; another user cannot tamper with the original DWG file.

However, the DWF format has some drawbacks:

- You must go through the extra step of translating from DWG to DWF.
- DWF files cannot display rendered or shaded drawings.
- DWF is a flat 2D-file format; therefore, it does not preserve 3D data, although you can export a 3D view.

- AutoCAD LT itself cannot display DWF files.
- DWF files cannot be converted back to DWG format without using file translation software from a third-party vendor.
- Earlier versions of DWF did not handle paperspace objects (version 2.x and earlier), or linewidths and non-rectangular viewports (version 3.x and earlier).

To view a DWF file on the Internet, your Web browser needs a *plug-in*—a software extension that lets a Web browser handle a variety of file formats. Autodesk makes the DWF plug-in freely available from its Web site at *http://www.autodesk.com/whip*. It is a good idea to regularly check for updates to the DWF plug-in, which is updated about twice a year.

Other DWF Viewing Options

In addition to Whip with a Web browser, Autodesk provides two other options for viewing DWF files. **CADViewer Light** is designed to be a DWF viewer that works on all operating systems and computer hardware because it is written in Java. As long as your Windows, Mac, or Unix computer has access to Java (included with most Web browsers), you can view DWF files.

Volo View Express is a stand-alone viewer that views and prints DWG, DWF, and DXF files. Both products can be downloaded free from the Autodesk Web site.

Viewing DWG Files

To view AutoCAD LT DWG and DXF files on the Internet, your Web browser needs a DWG-DXF plug-in. At the time of publication, the plug-ins were available from the following vendors:

Autodesk: *http://www.autodesk.com/whip*
SoftSource: *http://www.softsource.com/*
California Software Labs: *http://www.cswl.com*
Cimmetry systems: *http://www.cimmetry.com*

Creating a DWF File*

To create a DWF file from AutoCAD LT 2000, follow these steps:

1. Type the **PLOT** command or choose **Plot** from the **File** menu. Notice that AutoCAD LT displays the **Plot** dialog box (Figure 30-19).

2. In the **Name** list box (found in the **Plotter Configuration** area), select **DWF ePlot pc3**.

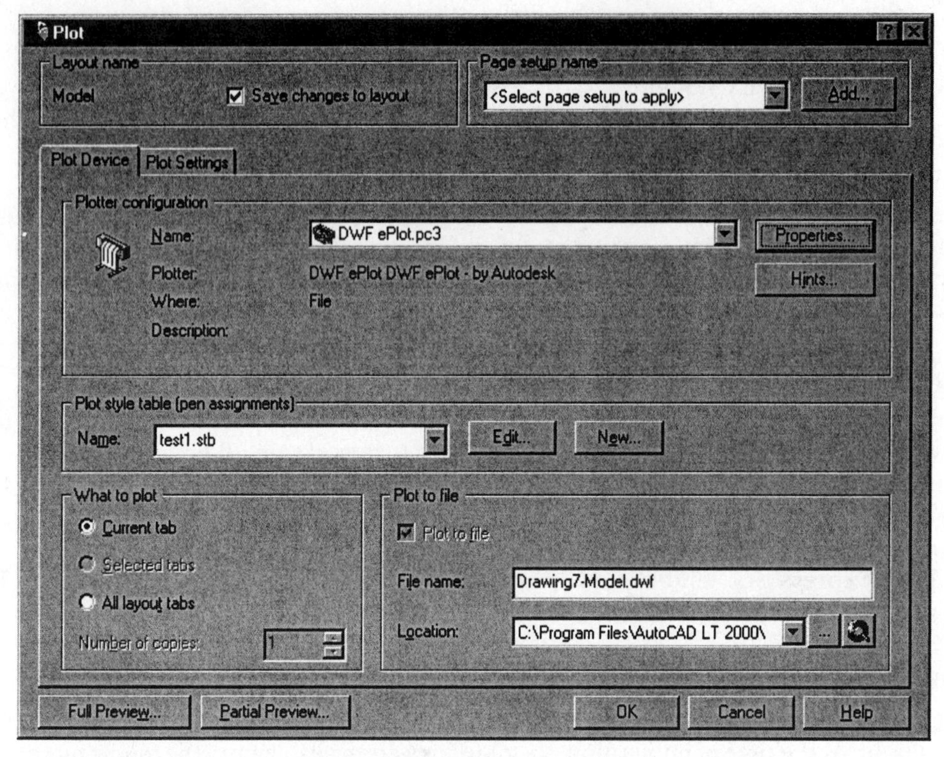

***Figure 30-19 Plot** dialog box*

3. Choose the **Properties** button. Notice the **Plotter Configuration Editor** dialog box (Figure 30-20).

4. Choose **Custom Properties** in the tree view. Notice the **Custom Properties** button, which appears in the lower half of the dialog box.

5. Choose the **Custom Properties** button. Notice the **DWF Properties** dialog box (Figure 30-21).

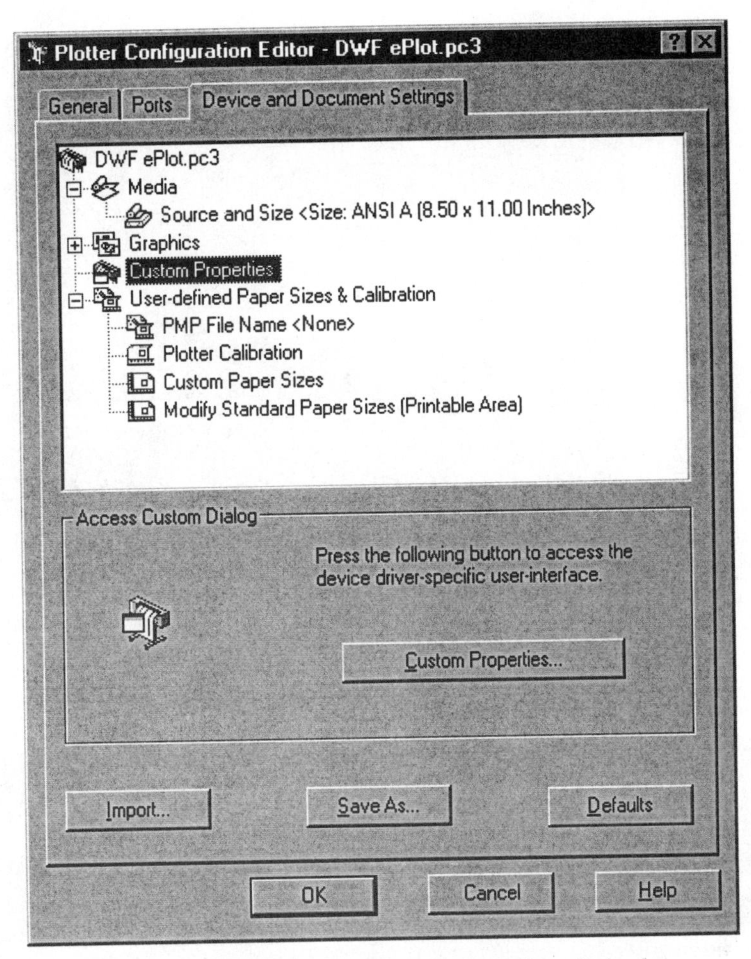

Figure 30-20 ***Plotter Configuration Editor*** *dialog box*

Resolution

Unlike AutoCAD LT DWG files, which are based on real numbers, DWF files are saved using integer numbers. The **Medium** precision setting saves the drawing using 16-bit integers, which is adequate for all but the most complex drawings. **High** resolution saves the DWF file using 20-bit integers, while **Extreme** resolution saves using 32-bit integers. Figure 30-22 shows an extreme close-up of a drawing exported in DWF format. The portion at left was saved at medium resolution and shows some bumpiness in the arcs. The portion at right was saved at extreme resolution.

More significant is the difference in file size. When a DWF file was created from a file of 342 KB, the medium resolution DWF file was just 27 KB, while the extreme resolution file became 1,100 KB – some 40 times larger. That means that the medium resolution DWF file transmits over the Internet 40 times faster, a significant savings in time.

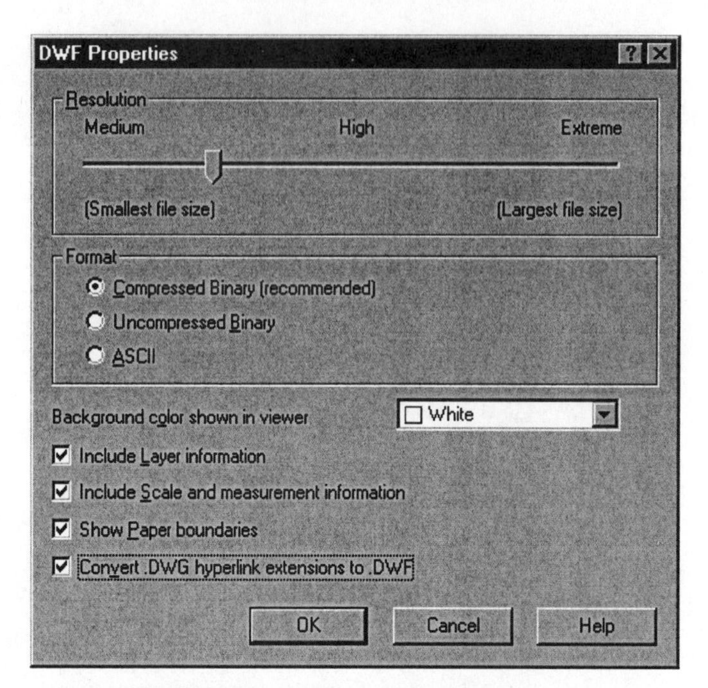

***Figure 30-21 DWF Properties** dialog box*

Figure 30-22 *DWF output in medium resolution (at left) and extreme resolution (at right)*

Format

Compression further reduces the size of the DWF file. You should always use compression, unless you know that another application cannot decompress the DWF file. Compressed binary format is 7 times smaller than ASCII format (7,700 KB). Again, that means the compressed DWF file transmits over the Internet 7 times faster than an ASCII DWF file.

Other Options

Background Color Shown in Viewer: While white is probably the best background color, you may choose any of AutoCAD LT's 255 colors.

Include Layer Information: This option includes layers, which allows you to toggle layers off and on when the drawing is viewed by the Web browser.

Include Scale and Measurement Information: This allows you to use the **Location** option in the Web browser's Whip plug-in to show scaled coordinate data.

Show Paper Boundaries: Includes a rectangular boundary at the drawing's extents.

Convert .DWG Hyperlink Extensions to .DWF: Includes hyperlinks in the DWF file.

In most cases, you would turn on all options.

7. Choose **OK** to exit the dialog boxes back to the **Plot** dialog box.

8. Accept the DWF file name listed in the **File name** text box, or type a new name. If necessary, change the location that the file will be stored in. Note the two buttons: one has an ellipsis (...), which displays the **Browse for Folder** dialog box. The second button brings up AutoCAD LT's internal Web browser, a simplified version of the Microsoft-branded product.

9. Choose **OK** to save the drawing in DWF format.

Viewing DWF Files

To view a DWF file, you need to use a Web browser with a special plug-in that allows the browser to correctly interpret the file. (Remember: you cannot view a DWF file with AutoCAD LT.) Autodesk has named their DWF plug-in "Whip!," short for "Windows HIgh Performance."

Autodesk updates the DWF plug-in approximately twice a year. Each update includes some new features. In summary, all versions of the DWF plug-in perform the following functions:

- Views DWF files created by AutoCAD LT within a browser.
- Right-clicking the DWF image displays a shortcut menu with commands.
- Real-time pan and zoom lets you change the view of the DWF file as quickly as a drawing file in AutoCAD LT.
- Embedded hyperlinks let you display other documents and files.
- File compression means that a DWF file appears in your Web browser much faster than the equivalent DWG drawing file would.
- Print the DWF file alone or along with the entire Web page.
- Works with Netscape Navigator or Microsoft Internet Explorer. A separate plug-in is required, depending upon which of the two browsers you use.
- Allows you to "drag and drop" a DWG file from a Web site into AutoCAD LT as a new drawing or as a block.
- Views a named view stored in the DWF file.
- Can specify a view using x,y- coordinates.
- Toggles layers off and on.

If you don't know whether the DWF plug-in is installed in your Web browser, select **Help > About Plug-ins** from the browser's menu bar.

If the plug-in is not installed, or is an older version, then you need to download it from Autodesk's Web site at:

http://www.autodesk.com/whip

For Netscape users, the file is quite large at 3.5 MB and takes about a half-hour to download using a typical 28.8 K baud modem. The file you download from the Autodesk Web site is either a self-installing file (has a .JAR extension) or a self-extracting installation file with a name such as Whip4.Exe. After the download is complete, follow the instructions on the screen.

DWF Plug-in Commands

To display the DWF plug-in's commands, move the cursor over the DWF image and press the mouse's right button. This displays a shortcut menu with options, such as **Pan**, **Zoom**, and **Named Views** (Figure 30-23). To select an option, move the cursor over the option name and press the left mouse button.

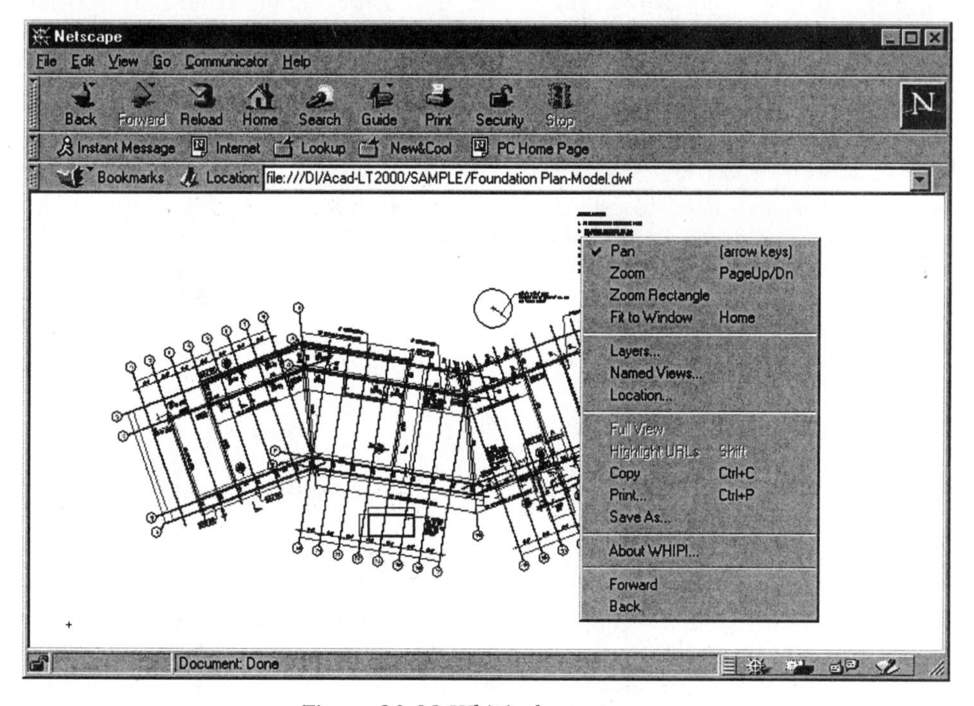

Figure 30-23 *Whip's shortcut menu*

Pan is the default option. Press the left mouse button and move the mouse. The cursor changes to an open hand, signaling that you can pan the view around the drawing. This is exactly the same as real-time panning in AutoCAD LT. Naturally, panning only works when you are zoomed in; it does not work in Full View mode.

Zoom is like the **ZOOM** command in AutoCAD LT. The cursor changes to a magnifying glass. Hold down the left mouse button and move the cursor up (to zoom in) and down (to zoom out).

Zoom Rectangle is the same as **ZOOM Window** in AutoCAD LT. The cursor changes to a plus sign. Click the left mouse button at one corner, then drag the rectangle to specify the size of the new view.

Fit to Window is the same as AutoCAD LT's **ZOOM Extents** command. You see the entire drawing.

Layers displays a nonmodal dialog box (Figure 30-24), which lists all layers in the drawing. (A *nonmodal* dialog box remains on the screen; unlike AutoCAD LT's modal dialog boxes, you do not need to dismiss a nonmodal dialog box to continue working). Click a layer name to toggle its visibility between on (yellow light bulb icon) and off (blue light bulb).

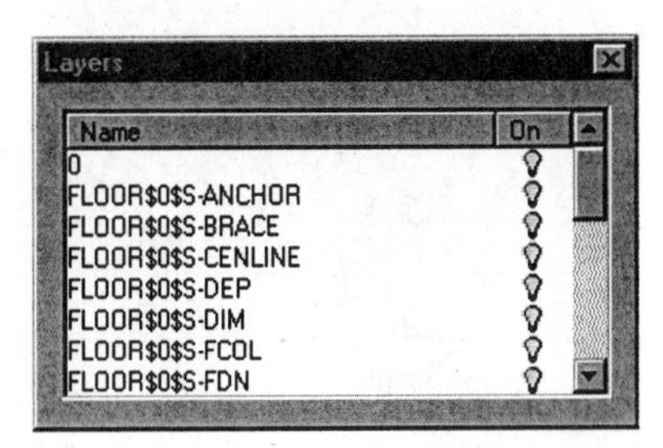

***Figure 30-24** Whip's **Layer** dialog box*

Named Views works only when the original DWG drawing file contained named views created with the **VIEW** command. Selecting **Named Views** displays a non- modal dialog box that allows you to select a named view (Figure 30-25). Click a named view to see it; click the small **x** in the upper-right corner to dismiss the dialog box.

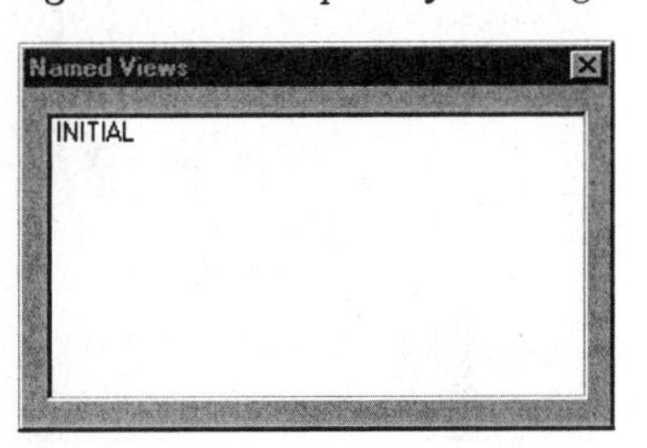

***Figure 30-25** Whip's **Named Views** dialog box*

Location displays a non-modal dialog box that shows the x,y,z-coordinates of the cursor location (Figure 30-26). Since DWF files are only 2D, the Z coordinate is always 0.

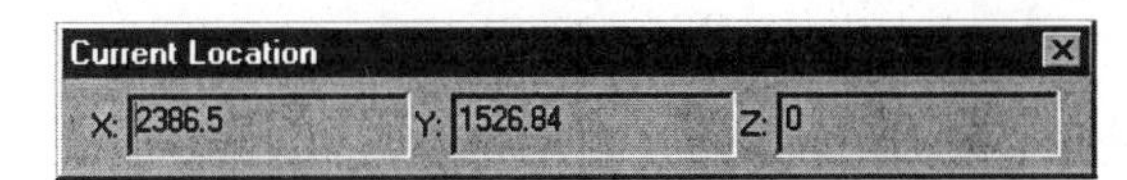

***Figure 30-26 Whip's Location** dialog box*

Full View causes the Web browser to display the DWF image as large as possible all by itself. This is useful for increasing the physical size of a small DWF image. When you have finished viewing the large image, right-click and choose the **Back** or choose the browser's **Back** button to return to the previous screen.

Highlight URLs displays a flashing gray effect on all objects and areas with a hyperlink (Figure 30-27). This helps you see where the hyperlinks are. To read the associated URL, pass the cursor over a highlighted area and look at the URL on the browser's status line. (A short-cut is to hold down the SHIFT key, which causes the URLs to highlight until you release the key). Choosing the **URL** opens the document with its associated application.

Copy copies DWF image to the Windows Clipboard in EMF (enhanced Metafile or "Picture") format.

Print prints the DWF image alone. To print the entire Web page (including the DWF image), use the browser's **Print** button.

Save As saves the DWF file in three formats to your computer's hard drive:

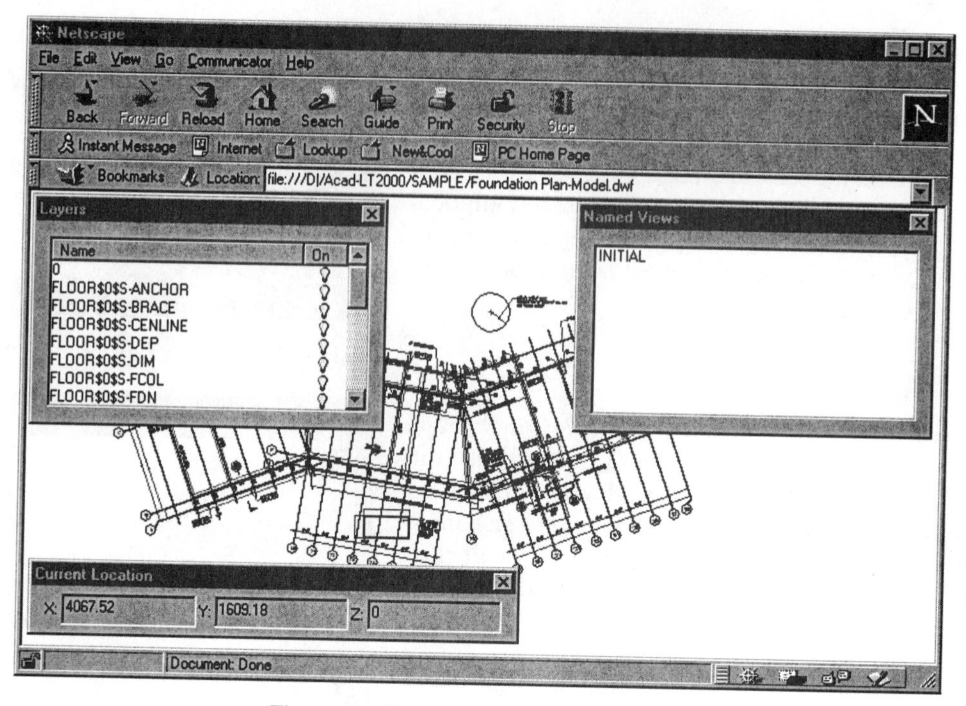

Figure 30-27 *Highlighting hyperlinks*

DWF
BMP (Windows bitmap)
DWG (AutoCAD LT drawing file) works only when a copy of the DWG file is available at the same subdirectory as the DWF file.

You cannot use the **Save As** option until the entire DWF file has been transmitted to your computer.

About WHIP displays information about the DWF file, including DWF file revision number, description, author, creator, source filename, creation time, modification time, source creation time, source modification time, current view left, current view right, current view bottom, and current view top (Figure 30-28).

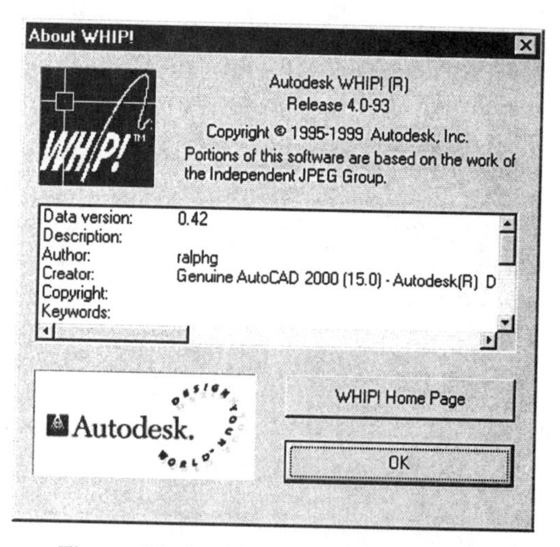

Figure 30-28 About WHIP *dialog box*

Forward is almost the same as choosing the browser's **Forward** button. When the DWF image is in a frame, only that frame goes forward.

Back is almost the same as choosing the browser's **Back** button. It works differently when the DWF image is displayed in a frame.

Drag and Drop from Browser to Drawing

The DWF plug-in allows you to perform several "drag and drop" functions. *Drag and drop* is when you use the mouse to drag an object from one application to another. Hold down the key to drag a DWF file from the browser into AutoCAD LT. Recall that AutoCAD LT cannot translate a DWF file into DWG drawing format. For this reason, this form of drag and drop only works when the originating DWG file exists in the same subdirectory as the DWF file. AutoCAD LT executes the **-INSERT** command, as follows:

Command: **-INSERT**
Enter block name or [?]: C:\CAD\ACAD 2000\SAMPLE\1st floor plan.dwg
Specify insertion point or [Scale/X/Y/Z/Rotate/PScale/PX/PY/PZ/PRotate]: **0,0**
Enter X scale factor, specify opposite corner, or [Corner/XYZ] <1>: *Press ENTER.*
Enter Y scale factor <use X scale factor>: *Press ENTER.*
Specify rotation angle <0>: *Press ENTER.*

To see the inserted drawing, you may need to use the **ZOOM Extents** command.

Another drag and drop function is to drag a DWF file from the Windows Explorer (or File Manager) into the Web browser. This causes the Web browser to load the DWF plug-in, then display the DWF file. Once displayed, you can execute all of the commands listed in the previous section.

Finally, you can also drag and drop a DWF file from Windows Explorer (in Windows 95/98) or File Manager (in Windows NT) into AutoCAD LT. This causes AutoCAD LT to launch another program that is able to view the DWF file. This does not work if you do not have other software on your computer system capable of viewing DWF files.

Embedding a DWF File

To let others view your DWF file over the Internet, you need to embed the DWF file in a Web page. Here are the steps to embedding a DWF file in a Web page.

Step 1. This hype text markup language (HTML) code is the most basic method of placing a DWF file in your Web page:

```
<embed src="filename.dwf">
```

The **<embed>** tag embeds an object in a Web page. The **src** option is short for "source." Replace **filename.dwf** with the URL of the DWF file. Remember to keep the quotation marks in place.

Step 2. HTML normally displays an image as large as possible. To control the size of the DWF file, add the **Width** and **Height** options:

```
<embed width=800 height=600 src="filename.dwf">
```

The **Width** and **Height** values are measured in pixels. Replace 800 and 600 with any appropriate numbers, such as 100 and 75 for a "thumbnail" image, or 300 and 200 for a small image.

Step 3. To speed up a Web page's display speed, some users turn off the display of images. For this reason, it is useful to include a description, which is displayed in place of the image:

<embed width=800 height=600 name=description src=”filename.dwf ”>

The **Name** option displays a textual description of the image when the browser does not load images. You might replace description with the DWF filename.

Step 4. When the original drawing contains named views created by the **VIEW** command, these are transferred to the DWF file. Specify the initial view for the DWF file:

<embed width=800 height=600 name=description namedview=”viewname”
src=”filename.dwf ”>

The **namedview** option specifies the name of the view to display upon loading. Replace **viewname** with the name of a valid view name. When the drawing contains named views, the user can right-click on the DWF image to get a list of all named views.

As an alternative, you can specify the 2D coordinates of the initial view:

<embed width=800 height=600 name=description view=”0,0 9,12"
src=”filename.dwf ”>

The **View** option specifies the x,y-coordinates of the lower-left and upper-right corners of the initial view. Replace 0,0 9,12 with other coordinates. Since DWF is 2D only, you cannot specify a 3D viewpoint. You can use **View** or **Namedview**, but not both.

Step 5. Before a DWF image can be displayed, the Web browser must have the DWF plug-in called “Whip!”. For users of Netscape Communicator, you must include a description of where to get the Whip plug-in when the Web browser is lacking it.

<embed pluginspage=http://www.autodesk.com/products/AutoCAD LT/whip/whip.htm
width=800 height=600 name=description view=”0,0 9,12" src=”filename.dwf ”>

The **pluginspage** option describes the page on the Autodesk Web site where the Whip-DWF plug-in can be downloaded.

The code listed above works for Netscape Navigator. To provide for users of Internet Explorer, the following HTML code must be added:

<object classid =”clsid:B2BE75F3-9197-11CF-ABF4-08000996E931" codebase = “ftp://
ftp.autodesk.com/pub/AutoCAD LT/plugin/whip.cab#version=2,0,0,0” width=800 height=600>
<param name=”Filename” value=”filename.dwf ”>
<param name=”View” value=”0,0 9,12">
<param name=”Namedview” value=”viewname”>
<embed pluginspage=http://www.autodesk.com/products/AutoCAD LT/whip/
whip.htm width=800 height=600 name=description view=”0,0 9,12"

src="filename.dwf">
</object>

The two **<object>** and three **<param>** tags are ignored by Netscape Navigator; they are required for compatibility with Internet Explorer. The **classid** and **codebase** options tell Explorer where to find the plug-in. Remember that you can use **View** or **Namedview** but not both.

Step 6: Save the HTML file.

Review Questions

1. Can you launch a Web browser from within AutoCAD LT?______________________
2. What does DWF mean?______________________
3. What is the purpose of DWF files?______________________
4. URL is an acronym for?______________________
5. Which of the following URLs are valid?
   www.autodesk.com
   http://www.autodesk.com
   Both of the above.
   None of the above.
6. FTP is an acronym for:______________________
7. What is a "local host"?______________________
8. Are hyperlinks active in an AutoCAD LT 2000 drawing?______________________
9. The purpose of URLs is to let you create ________ between files.______________________
10. When you attach a hyperlink to a block, the hyperlink data is ________ when you scale the block unevenly, stretch the block, or explode it.
11. Can you can attach a URL to any object?______________________
12. The **-HYPERLINK** command allows you to attach a hyperlink to ____________ and ________.
13. To see the location of hyperlinks in a drawing, use the ________ command.

14. Rectangular (area) hyperlinks are stored on layer:______________________________.

15. Compression in the DWF file causes it to take__________ (less, more, the same) time to transmit over the Internet.

16. A DWF is created from a __________ file using the __________ command.

17. What does a "plug-in" let a Web browser do?______________________________

18. Can a Web browser view DWG drawing files over the Internet?____________________

19. __________ is an HTML tag for embedding graphics in a Web page.

20. A file being transmitted over the Internet via a 28.8 Kpbs modem takes about __________ minutes per megabyte.

Appendix A

System Requirements and AutoCAD LT Installation

SYSTEM REQUIREMENTS

The following are the minimum system requirements for running AutoCAD LT 2000:

1. Operating systems: Windows® 95, Windows NT® 4.0, and Windows 98.
2. 32 MB of RAM
3. Pentium 133, or better/compatible processor
4. 130 MB of hard disk space
5. 64 MB of disk swap space (minimum)
6. 10 MB of additional memory (RAM) for each concurrent session of AutoCAD LT
7. 2.5 MB of free disk space required at the time of installation for temporary files that are removed when installation is complete
8. 4X speed or faster CD-ROM for initial installation
9. 800 by 600 VGA video display
10. Mouse or digitizer (with Wintab driver) pointing device
11. IBM compatible parallel port and hardware lock for international single user and educational versions only
12. Serial port (for digitizers and some plotters)
13. Sound card for multimedia learning

The following hardware is optional:

1. Printer
2. Plotter
3. Digitizing tablet

INSTALLING AUTOCAD LT

The installation process has now been simplified that has made it easier and faster to install AutoCAD LT. Also, the installation procedure is more consistent with other Windows installations. The following steps describe the installation process for AutoCAD LT 2000:

1. The first step is to insert the AutoCAD LT 2000 CD in the CD-ROM drive of your computer.
2. If AutoPlay is enabled, the **Setup** dialog box is automatically displayed and the **InstallShield Wizard** is installed on your system. (AutoPlay is a Windows 95, Windows 98, and Windows NT 4.0 feature that automatically runs an executable file. The AutoPlay feature can be turned off by holding down the Shift key when you insert the CD in the CD-ROM drive, or by changing the play setting of the CD-ROM.) If your system matches the minimum hardware requirements, the **Welcome** dialog box is displayed on the screen.
3. Choose the **Next** button in the **Welcome** dialog box; the **Software License Agreement** dialog box is displayed on the screen.
4. After reading the Software License Agreement, select **Accept** to continue installation; the **Serial Number** dialog box appears on the screen.
5. Enter the Serial number and CD Key and then choose the **Next** button. (The Serial Number and the CD Key can be found at the back of the box containing the AutoCAD LT 2000 CD.). If the numbers you entered are correct, the **Personal Information** dialog box is displayed.
6. Enter the requested information and then choose the **Next** button. Each field must contain at least one character to display the next screen that shows the personal information you just entered. You can edit the information or choose the **Next** button to confirm the data. The **Destination Location** dialog box is displayed on the screen.
7. In the **Destination Location** dialog box, use the **Browse** button to select the directory where you want to install AutoCAD LT and then choose the **Next** button to display the **Setup Type** dialog box.
8. In the **Setup Type** dialog box select the type of setup you prefer (Typical, Full, Compact, or Custom). If you select Custom, the **Custom Components** dialog box is displayed. From this dialog box clear the check boxes for the components that you do not want to install and then choose the **Next** button. If your computer has enough hard disk space, the **Folder Name** dialog box is displayed.
9. In the **Folder Name** dialog box, accept the AutoCAD LT 2000 (default) folder or select any other folder. Choose the **Next** button; the **Setup Confirmation** dialog box is displayed.
10. If the selection you made is OK, choose the **Next** button in the **Setup Confirmation** dialog box. The installation program installs the program files on your system and when the installation is complete, the **Setup Complete** dialog box is displayed on the screen.

Appendix B

AutoCAD LT Toolbars

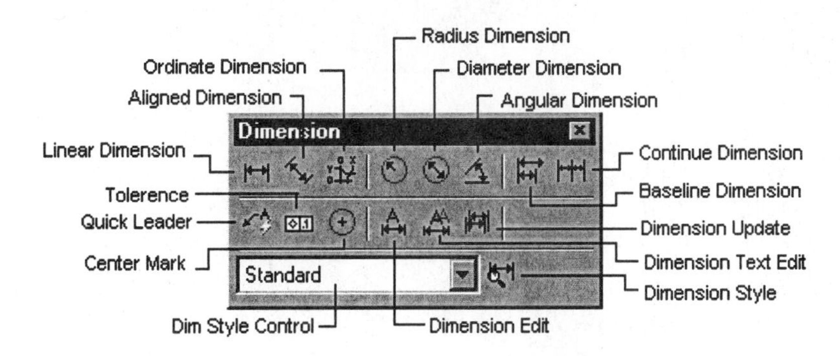

Figure B-1 Dimension toolbar

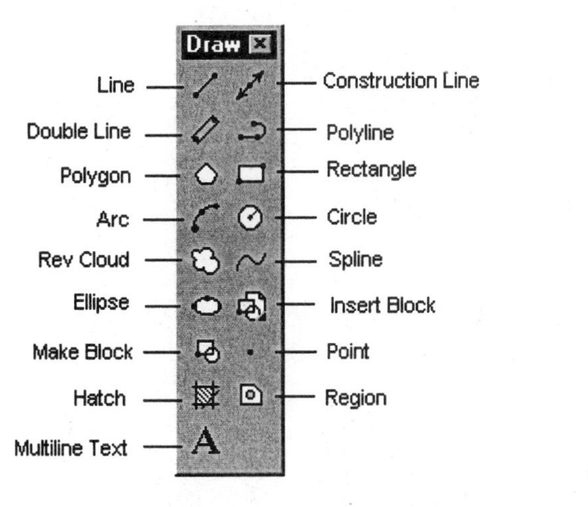

Figure B-2 Draw toolbar

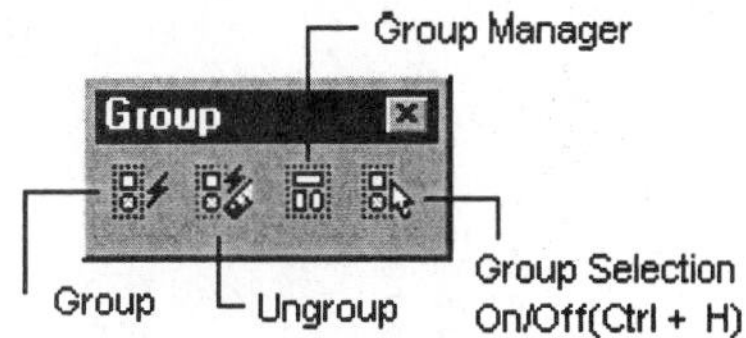

Figure B-3 Group toolbar

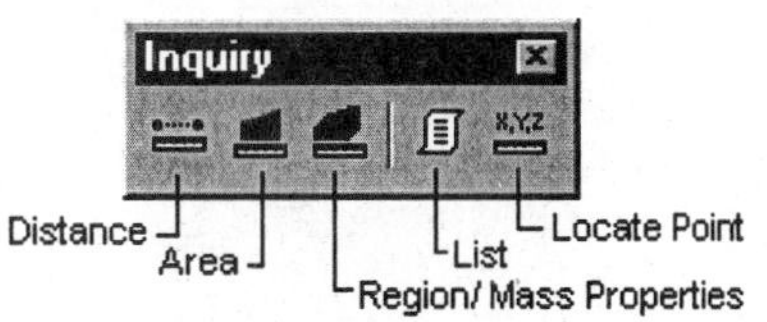

Figure B-4 Inquiry toolbar

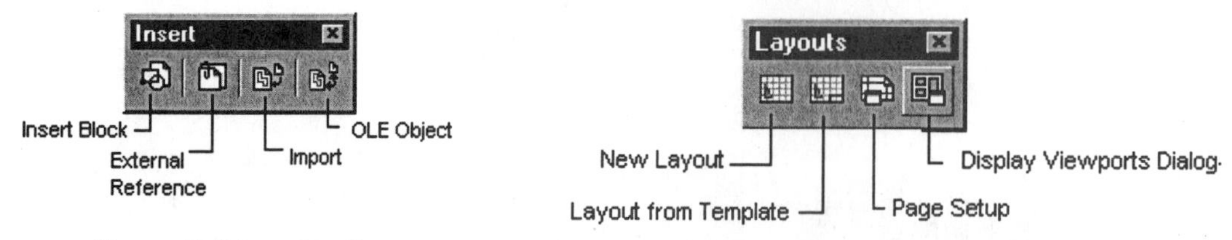

Figure B-5 *Insert toolbar*

Figure B-6 *Layout toolbar*

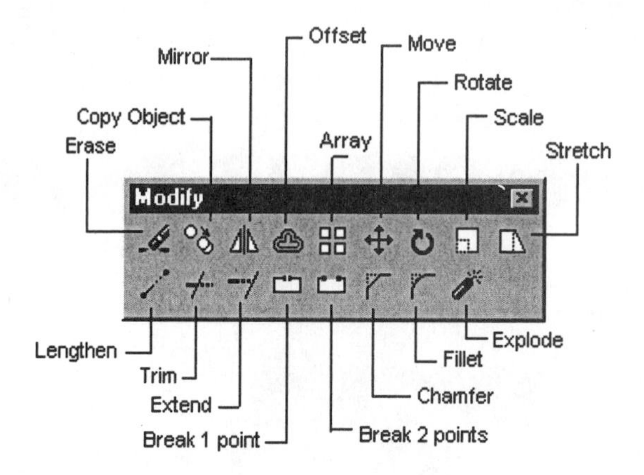

Figure B-7 *Modify toolbar*

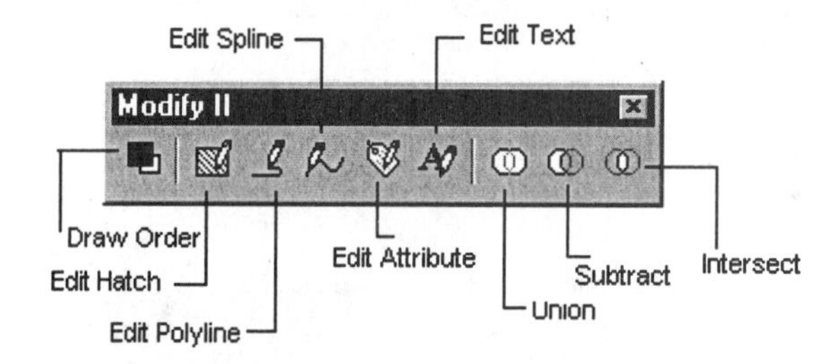

Figure B-8 *Modify II toolbar*

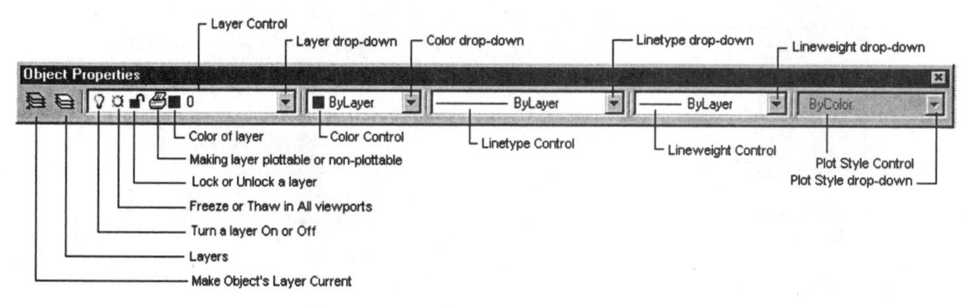

Figure B-9 *Object Properties toolbar*

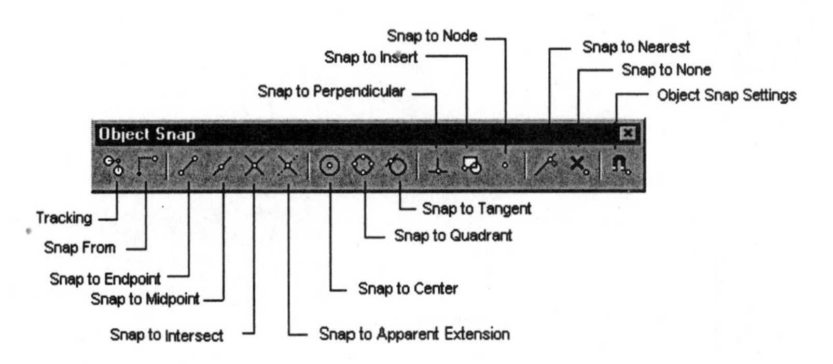

Figure B-10 Object Snap toolbar

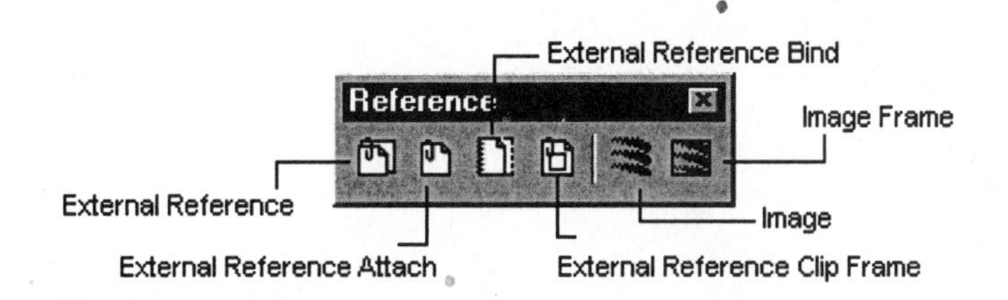

Figure B-11 Reference toolbar

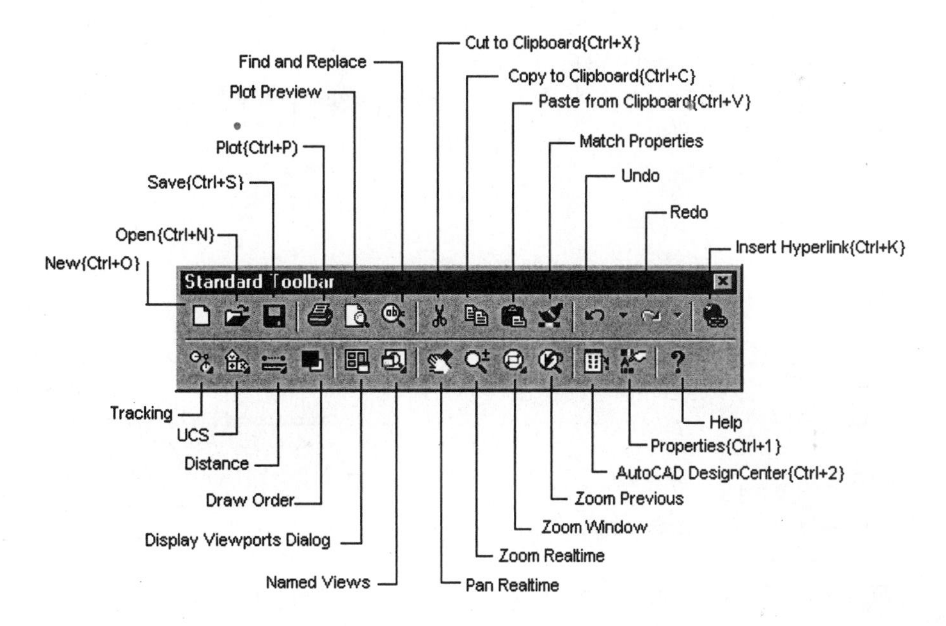

Figure B-12 Standard toolbar

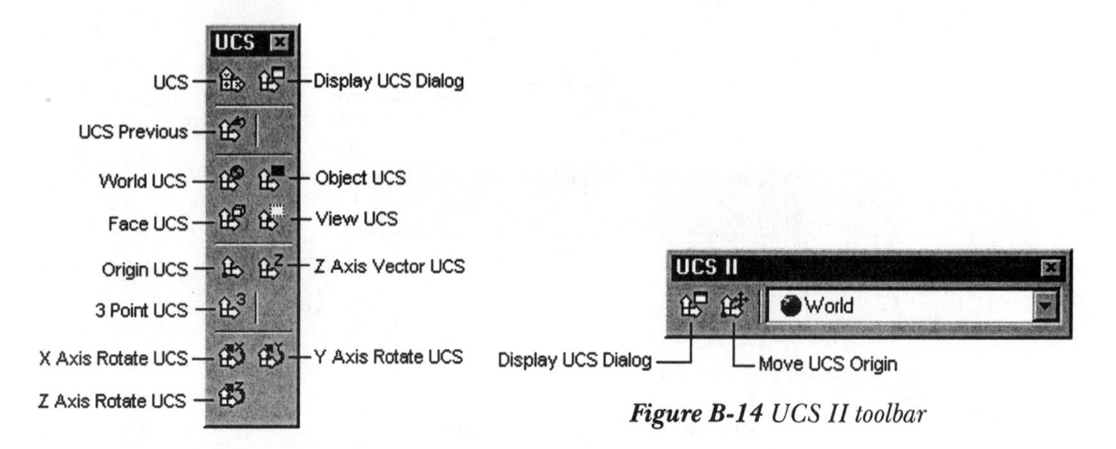

Figure B-14 UCS II toolbar

Figure B-13 UCS toolbar

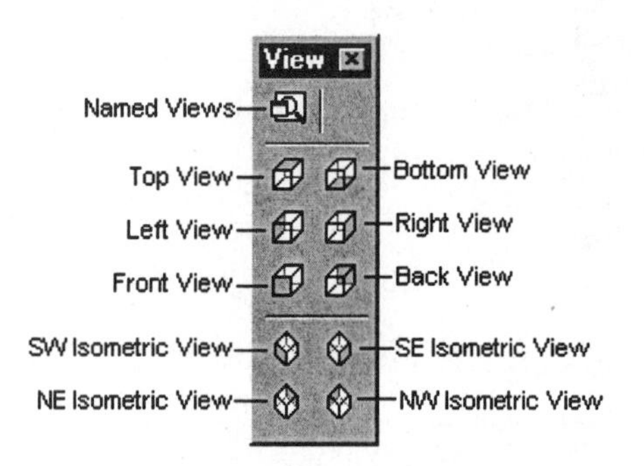

Figure B-15 View toolbar

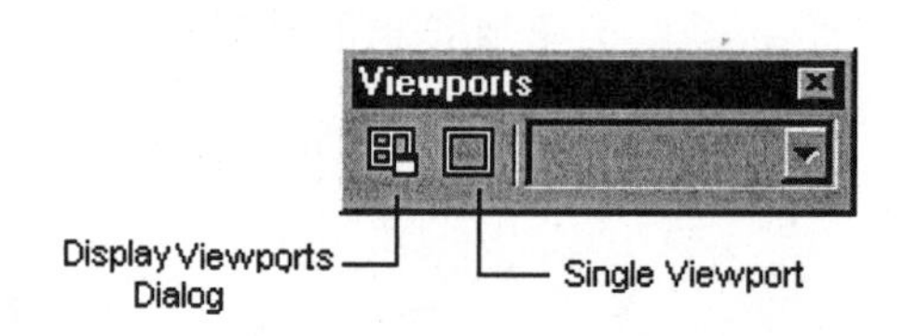

Figure B16 Viewports toolbar

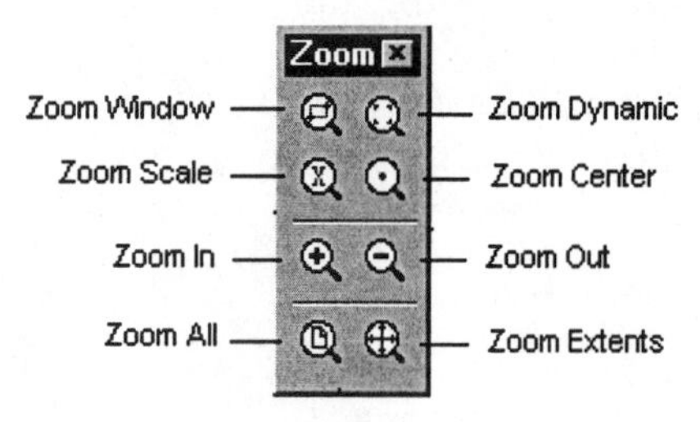

Figure B-17 Zoom toolbar

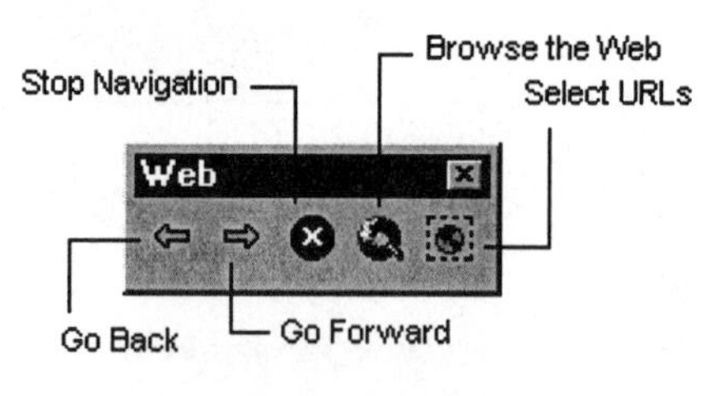

Figure B-18 Web toolbar

Appendix C

AutoCAD LT Commands

Command	Description and Options

3DPOLY Draws a 3D polyline having segments of a straight line. Options:

Point - Draws the 3D polyline to the specified point, C - Closes the 3D polyline by joining the last point with the first point, U - Deletes the last segment.

ABOUT Displays AutoCAD LT version and serial numbers and license and copyright information.

ADCCLOSE Closes the AutoCAD LT Design Center.

ADCENTER Starts and displays AutoCAD LT Design Center.

ADCNAVIGATE Directs the Desktop in AutoCAD LT DesignCenter.

APERTURE Controls the size of the object snap target box.

ARC Draws an arc of any size. The default method is to specify two endpoints and a point along the arc. Options:

A - Included angle, C - Center point, D - Starting direction, E - Endpoint, L - Length of chord, R - Radius.

AREA Computes the area and perimeter of different objects and of areas formed by specifying a sequence of points. Options:

A - Add mode, F - First point (Area by specifying points), O - Area of the object, S - Subtract mode.

ARRAY Creates specified number of copies of a selected object. Options:

P - Polar array, R - Rectangular array.

ATTDEF Creates an attribute definition (characteristics of an attribute). Options:

I - Invisible mode: Attribute remains invisible, C - Constant mode: Constant value of attribute, V - Verify mode: Verifies attribute value is correct, P - Preset mode: Default value to attribute.

ATTDISP Controls the visibility of the attribute globally. Options:

ON - Attributes made visible, OFF -

Attributes made invisible, N - Current visibility kept.

ATTEDIT Edits attributes within a block or blocks. Options:
Y - Yes (one attribute at a time), N - No (all attributes at a time).

ATTEXT Attribute information is extracted from the drawing. Options:
C - CDF: Comma-Delimited File, D - DXF: Drawing Interchange File, S - SDF: Space Delimited File, O - Drawing objects.

AUDIT Identifies errors in a drawing. Options:
Y - Corrects the errors, N - Informs about the error without correcting it.

BASE Sets the point of origin for inserting a drawing into another drawing.

BHATCH A specified enclosed area is filled with an associative hatch pattern through a dialog box. Previewing a hatch and adjusting the boundary is also possible. Options:
I - Internal point, P - Properties, S - Select, R - Remove islands, A - Advanced.

'BLIPMODE Controls the appearance of marker blip that is displayed on the screen when a point is picked. Options:
ON - Marker blip displayed, OFF - Marker blip not displayed.

BLOCK Creates a compound object as a block definition from a set of entities. Options:
? - Lists names of previously defined blocks.

BLOCKICON Generates preview images for the blocks created in earlier releases of AutoCAD LT.

BMPOUT Creates a bitmap image of the drawing and saves the screen to a file having a .bmp extension.

BOUNDARY Creates a polyline or region of a boundary that defines an enclosed area. Options:
A - Advanced Options.

BREAK Removes specified portions of an object or splits the object. Options:
F - Respecifies first point.

BROWSER Launches the default Web browser defined in the registry of the system.

CHAMFER Connects two nonparallel objects with a beveled line. Options:
A - Chamfer distance is set using angle and distance, D - Sets chamfer distance, P - Chamfers entire polyline, T - Controls the trimming of the edges to chamfer line endpoints, M - Trim method.

CHANGE Alters the properties of selected objects. Options:
C - Change point - Changes lines, circles, Text, Attribute Definitions, Blocks, P - Changes properties like Color, Elev, LAyer, LType, Thickness, ltScale, LWeight, PLotstyle.

CHPROP Alters the drawing properties of selected objects. Options:
C - Changes color, LA - Changes layer, LT - Changes linetype, S - Changes linetype scale factor, LW - Changes lineweight, T - Changes thickness, PL - Changes the plotstyle of the selected objects.

CIRCLE Draws a circle using any of the four methods available. Options:
C - Circle drawn on the basis of center point and diameter or radius, 3P - Drawn on the basis of 3 points on circumference, 2P - Drawn on the basis of 2 endpoints of the diameter, TTR - Circle drawn tangent to two objects with a specified radius.

CLOSE Closes the current drawing if no

change has occurred since last save or will prompt you to save the drawing.

COLOR Sets color for subsequent objects drawn. Options:
value - Sets color by number (1-255), name - Sets color by name, Byblock - The current color setting is inherited by the block at the time of insertion, Bylayer - Objects inherit the color of the layer in which they are drawn.

COPY Draws a copy of the selected object leaving the original object intact. The default method is to specify the base point. Options:
M - Multiple copies of object in single **COPY** command.

CONVERT Converts 2D polylines or hatches created in releases earlier than AutoCAD LT 97 into an optimized format. Options:
H - Hatch, P - polylines, A - All, S - Select Objects.

CONVERTCTB Converts a color-dependent plot style table (CTB) to a named plot style table (STB).

CONVERTPSTYLES Converts a currently open drawing to either named or color-dependent plot style, depending on which plot style method is being used.

COPYBASE Copies the drawing objects with a specified base point.

COPYCLIP Copies the selected objects to the Clipboard.

COPYHIST Copies the text in the command line history to the Clipboard.

COPYLINK Copies the current view to the Clipboard to link with other OLE applications.

CUSTOMIZE Customizes toolbars, button and shortcut keys.

CUTCLIP Copies objects to the Clipboard but erases them from the drawing.

DDEDIT Displays a dialog box that allows the user to edit text and attribute definitions. Options:
U - Undo - reverts the text to the pre edited value

DDPTYPE Displays a dialog box that sets the point style and also the size of the point object.

DDVPOINT Controls the direction of 3D views through a dialog box.

DELAY The execution of the next command is postponed for a specified time duration. In other words a specified pause is provided within the script.

DIM and DIM1 Dimensioning mode is invoked and permits the use of dimension subcommands from AutoCAD LT's previous releases.

DIMALIGNED Creates a linear dimension aligned to the specified points or object.

DIMANGULAR Creates an angular dimension.

DIMBASELINE Starts drawing from the baseline of the previous dimension. The new dimension can be linear, angular, or ordinate.

DIMCENTER Draws center mark or center lines in circles and arcs.

DIMCONTINUE Starts drawing a new dimension from the second extension line of the previous or selected dimension. The new dimension can be a linear, angular, or ordinate dimension.

DIMDIAMETER Draws diameter dimensions for different circles and arcs.

DIMEDIT Edits dimension text and extension lines. Options:

H - Dimension text moved back to default position, N - Dimension text is replaced, R - Dimension text is rotated, O - Extension lines placed at obliquing angle.

DIMLINEAR Draws linear dimensions.

DIMORDINATE Creates ordinate dimensions.

DIMOVERRIDE The settings of the dimensioning system variables concerning the dimension object are overridden. The current dimension style is not affected. Options:

C - Clears overrides - clears any overrides on a dimension object.

DIMRADIUS Draws radial dimensions for different circles and arcs.

DIMSTYLE New dimension styles are created and the existing ones are modified. Options:

R - Dimensioning system variable setting changed, S - Current settings of dimensioning variables saved, ST - Current values of dimensioning variables displayed, V - Dimensioning variable setting of a style is listed, A - Selected dimension objects are updated, ? - Named dimension styles are listed.

DIMTEDIT Dimension text is moved and rotated. Options:

A - Angle of dimension text changed, H - Dimension text moved to default position, L - Dimension text left justified, R - Dimension text right justified.

DIST Distance and angle between two points is measured.

DIVIDE Places blocks or points as markers at equal distance along the length or perimeter of an entity, thus dividing it into a specified number of equal parts. Options:

B - Places blocks as markers.

DLINE Creates double lines using line segments and arcs. Options:

B - Break - decides whether double lines should be broken when intersection takes place, C - Caps - determines the placement of end caps, D - Dragline - determines whether the double line will be drawn by offsetting to the left, right, or center of the located points, O - Offset - locates the start point of the double line, S - Snap - starts or ends a double line by snapping to an existing object, U - Undo - Removes in reverse order the points located for the current double line, W - Width - sets the width of a double line.

DONUT Draws wide polyline with specified inside and outside diameters, thus forming a ring.

DRAWORDER Changes the display order of images and other objects. Options:

A - above object, U - under object, F - Front, B - Back.

DSETTINGS Specifies the settings for Snap mode, grid, polar, and object snap tracking through a dialog box.

DSVIEWER Displays the Aerial View window.

DVIEW Parallel projection or perspective views are defined. Options:

CA - Sets camera position by rotating about the target, TA - Sets target position by rotating about the camera, D - Camera to target distance is set, PO - Locates target and camera points, PA - Pans image, Z - Zooms In/Out, TW - Tilts view around line of sight, CL - The view is clipped in front and back, H - Hidden lines removed on selected objects, OFF - Perspective viewing turned off, U - Last DVIEW operation reversed.

DWGPROPS Sets and displays all the properties of the current drawing through a dialog box.

ELEV The elevation and extrusion thickness is set for the new objects.

ELLIPSE Draws ellipses or elliptical arcs using different options. Specifying the axis endpoint is the default method. Options:
A - Draws elliptical arc, C - Specifies the center point of the ellipse, I - Draws isometric circle in current isometric plane.

ERASE Erases the selected objects from the drawing.

EXPLODE Compound objects (blocks, dimensions, polylines, 3D solids, regions, polygon meshes, multilines) are broken into their constituent parts.

EXPORT Objects are saved to other file formats via a dialog box.

EXTEND Lengthens a selected entity to meet another entity. Options:
P - Specifies projection mode like UCS and View, E - Controls the extension to implied or actual edge, U - Latest extension is undone.

FILL Controls whether multilines, traces, solids, or wide polylines are filled or not filled. Options:
ON - Fill mode is enabled, OFF - Fill mode is disabled.

FILLET The edges of two specified lines, arcs, or circles are filleted by construction of an arc of specified radius. The default method is to specify the two objects. Options:
P - Entire polyline is filleted, R - The radius of the fillet arc is specified, T - Controls the trimming of the edges to fillet arc endpoints.

FILTER Creates a list of properties on the basis of which the objects are selected.

FIND Finds, replaces, selects, or zooms to specified text through a dialog box.

GETENEV Displays values of specified system variables.

GRAPHSCR Flips to the graphics window from the text window.

GRID A grid of dots at specified spacing is displayed. Options:
Grid spacing(X) - Grid set to specified value, ON - Grid turned on at current spacing, OFF - Grid turned off, S - Grid spacing set to current Snap interval, A - Grid set to different spacing in X and Y.

GROUP Creates and changes object selection groups, which are sets of objects having specific names. Options:
? - Lists names and descriptions of all the group, A - Adds objects to the group, R - Remove objects from the group, U - Ungroup - removes the group name and the association of objects in the group, REN - Assigns a new name to an existing group, S - Specifies whether a group is selectable, C - Creates a new group.

HATCH A specified area is filled with a selected pattern. Options:

? - Lists the hatch patterns in acad.pat file, name - A pattern name as defined in acad.pat file is specified, U - User-defined hatch pattern is specified. U can be followed by a comma and a hatch style, S - Specifies a solid fill.

HATCHEDIT Edits a hatch block through a dialog box. It sets the pattern type and properties and then applies it to a block. Options:
D - Removes the associative quality from

an associative hatch, S - Changes the hatch style type, P - Specifies new hatch properties.

HELP(F1) Displays help for a specific command and also lists the commands and data entry options.

HIDE Regenerating of a 3D object is performed with the removal of hidden lines.

HYPERLINK Attaches a hyperlink to a graphical object or modifies an existing hyperlink through a dialog box.

HYPERLINKOPTIONS Controls the visibility of the hyperlink cursor and the display of hyperlink tooltips. Options:
Y - Shows the hyperlink cursor and tooltip, N - Does not display.

ID The coordinates of a specified point are displayed.

IMAGE Lists and modifies the path of inserted images into an AutoCAD LT drawing file through a dialog box.

IMAGEFRAME Controls the display of image frame on the screen. Options:
On - Displays image frames, Off - Hides image frames.

IMPORT Imports the different file formats into an AutoCAD LT drawing via a dialog box.

INSERT Places a previously defined named block or drawing into the current drawing.

INSERTOBJ Inserts a previously linked or embedded object.

INTERFERE Highlights all of the interfering solids and then creates new solids from the intersections of the interfering pairs of solids.

INTERSECT A new composite solid is created from the intersecting region of two or more solids.

ISOPLANE An isometric plane is selected to be the current plane for an orthogonal drawing. Options:
L - Left-hand plane, T - Top plane, R - Right-hand plane.

LAYER Creates layers and sets different properties for the specified layers. Options:
? - Lists defined layers, M - Creates a layer and makes it the current layer, S - Makes a specified already existing layer current, N - Creates one or more new layers, ON - Turns on the specified layers, OFF - Turns off the specified layers, C - Sets the color of the specified layer, L - Sets the linetype of the specified layer, LW - Changes the lineweight, P - Controls whether visible layers are plotted, PS - Sets the plot style assigned to a layer, F - Makes a layer invisible by freezing it, T - The frozen layer is thawed, LO - Locks layers, thus prevents editing on them, U - Unlocks specified locked layers.

LAYOUT Creates a new layout and renames, copies, saves, or deletes an existing layout. Options:
C - Copies a layout, D - Deletes a layout, N - Creates a new layout tab, T - Creates a new template, R - Renames a layout, SA - Saves a layout, S - Makes a layout current, ? - Lists all the available layouts.

LAYOUTWIZARD Starts the Layout wizard to designate page and plot settings for a new layout.

LEADER Creates a line segment with an arrowhead that connects the text to a feature. The leader is created from a specified point to another point depending upon the options. Options:
A - Annotation is inserted at the end of leader line, F - Controls the type of leader

(Spline, Straight, Arrow), U - The last vertex point is removed.

LENGTHEN Alters the length of specified entities and the included angle of arcs. Options:
DE - Lengthens the object by a specified incremental distance, P - Alters the length by a specified percentage of its total length, T - Alters the length by specified total absolute length, DY - The object is lengthened to where its endpoint is dragged.

LIMITS Sets the drawing boundaries (and WCS grid extents) for the current space. Options:
Lower - Specifies 2 points-lower left corner and the left-upper right corner, ON - Limits checking is enabled, OFF - Limits checking is disabled.

LINE Draws straight line segments of any length by specifying the endpoints. Options:
ENTER - Continues from end of previous line or arc, U - Removes the most recent segment, C - Closes polygon.

LINETYPE Defines line characteristics, loads linetypes and sets them for new entities. It also creates new linetype definitions to a library file. Options:
? - Lists linetypes in a file, C - Creates new linetype definition, L - Loads an already existing linetype definition, Sets linetype for new entities, S - Sets the current linetype.

LIST Lists database information (type, layer, X,Y,Z position, thickness, and so on) about the specified entity.

LOGFILEOFF The log file already opened is closed by this command.

LOGFILEON The subsequent contents of the text window are recorded into the log file.

LTSCALE Sets the global scale factor of the linetype so as to alter the relative length of dashes and dots.

LWEIGHT Sets the current lineweight, lineweight display options, and lineweight units.

MASSPROP Calculates and lists the mass characteristics of 2D and 3D objects. The properties displayed are Area, Perimeter, Bounding box, Centroid. For Coplanar regions, additional properties displayed are Moments of Inertia, Products of Inertia, Radii of Gyration, and Principal Moments.

MATCHPROP Copies the properties from one object to one or more objects.

MEASURE Places blocks or points as markers at measured intervals along the length or perimeter of an entity. Options:
B - Places blocks as markers.

MENULOAD Displays a dialog box that loads and permits you to add partial menu files to an already present base menu file.

MENUUNLOAD Displays the same dialog box as in the case of the **MENULOAD** command that can also be used to unload the partial menu files.

MIRROR Reflects objects so as to create their mirror images about a specified line.

MODEL Switches from a layout tab to the Model tab and makes it current.

MOVE Moves objects from one location to another by specifying a displacement.

MREDO Reverses the effect of multiple previous UNDO and U commands. Options:
A - All, L - Last

MSLIDE Creates a slide file from the current

display.

MSPACE Switches to model space in a floating viewport from paper space.

MTEXT Creates paragraph text within a specified text boundary. Displays a dialog box where you can specify the different options for the multiline text.

MULTIPLE Causes the repetition of the next command until it is cancelled.

MVIEW Creates viewports and controls the number and layout of paper space viewports. You specify diagonal corners of new viewport as the default option. Options:

ON - Viewport is turned on, OFF - Viewport is turned off, H - Hideplot-Hidden lines removed during plotting, F - Fit - Single viewport created that fills the display area completely, L - Locks the selected viewport, O - Specifies a closed polygon to convert it into viewport, P - Creates an irregularly shaped viewport using specified points,
2 - The specified area is divided into two viewports either horizontally or vertically, 3 - The specified area is divided into 3 viewports, 4 - The specified area is divided into 4 viewports, R - Restore - Viewport configurations changed into individual viewports.

NEW Displays a dialog box that creates a new drawing file.

OFFSET Creates offset curves, concentric circles, and parallel lines at a specified distance from the original object. Options:

value - Specify the offset distance, T - Through - The offset object passes through the specified point.

OLELINKS Updates, changes, and cancels existing OLE links through a dialog box.

OLESCALE Displays and modifies the OLE Properties through a dialog box.

OOPS Restores those entities that have been erased by the last **ERASE** command.

OPEN Displays a dialog box through which an existing drawing can be opened. The dialog box also displays the directory, files, preview, name of the file, and the pattern.

OPTIONS Customizes the AutoCAD LT settings through a dialog box.

ORTHO The movement of the cursor is restrained to only vertical or horizontal directions and aligned with the grid. Options:

ON - Constrains cursor movement, OFF - Does not constrain cursor movement.

OSNAP Specifies a point at an exact location on an object by setting the Object Snap modes. Options:

END - Closest endpoint of arc (arcs and lines include polyline segments), elliptical arc, ray, mline, line and closest corner of trace, solid, 3D face, MID - Midpoint of arc, elliptical arc, spline, ellipse, ray, solid, xline, mline, or line, INT - Intersection of line, arc, spline, elliptical arc, ellipse, ray, xline, mline, or circle, EXT - Snaps to the extension point of an object, APP - Apparent or extended (projected) intersection (which may not actually intersect in 3D space) of line, arc, spline, elliptical arc, ellipse, ray, xline, mline, or circle, CEN - Center of arc, elliptical arc, ellipse, or circle, QUA - Quadrant point of arc, elliptical arc, ellipse, solid, or circle, PER - Point perpendicular to arc, elliptical arc, ellipse, spline, ray, xline, mline, line, solid, or arc, TAN - Tangent to arc, elliptical arc, ellipse, or circle, NOD - Point object, INS - Insertion point of text, block, shape, or attribute, NEA - Nearest point of arc, elliptical arc, ellipse, spline, ray, xline, mline, line, circle, or point, QUI -

First snap point, NON - Turns Object Snap mode off.

PAGESETUP Specifies the layout page, paper size, plotting device, and settings for each new layout through a dialog box.

PAN Moves the drawing display by a specified displacement.

PASTEBLOCK Pastes a copied block in a new drawing.

PASTECLIP Inserts data from the Clipboard.

PASTEORIG Pastes a copied object in the new drawing using the geometric coordinates of the original drawing.

PASTESPEC Inserts data from the Clipboard and controls the format of the data.

PCINWIZARD Starts a wizard to import PCP and PC2 configuration file plot settings in current layout or model tab.

PEDIT Editing of 2D polyline, 3D polyline, or 3D mesh. Options:

2D polyline C - Closes polyline segment, O - Closing segment removed, J - Joins to polyline, W - Specifies uniform width, E - Edits the vertices. The first vertex is marked by placing an X. Editing includes moving the X to next or previous vertex, adding a new vertex, setting the first vertex for break, moving the vertex, regenerating, straightening and attaching a tangent direction to the current vertex, F - Fits arc curves smoothly to the polyline, replacing each line segment with a pair of arcs, S - Vertices are used as a frame for spline curve, L - Linetype generation in a continuous pattern, U - Reverses the previous operation, 3D polyline C - Closes polyline segment, O - Closing segment removed, E - Edits the vertices. Same suboptions as in 2D Edit except the tangent suboption, S - Vertices are used as a frame for spline curve, D - Removes a spline curve to its control frame, U - reverses the previous option, X - Exits **PEDIT**. 3D polygon mesh E - Edits vertices. The first vertex is marked by placing an X. Editing includes moving the X to next or previous vertex, moving the X marker to the next vertex or the previous vertex in the N direction, moving the marker to the next or previous vertex in the M direction, regenerating the mesh, S - Fits a smooth surface, D - The control point polygon mesh is restored, Mclose - M-direction polylines are closed, Mopen - M-direction polylines are opened, Nclose - N-direction polylines are closed, Nopen - N-direction polylines are opened, U - Reverses editing operations as far back as the beginning of the **PEDIT** session.

PLAN Allows you to view the drawing from plan view of a User Coordinate System. Options:

C - Plan view of the current UCS, U - Plan view of the specified UCS, W - Plan view of the World Coordinate System.

PLINE Draws 2D polylines. The default is to draw a polyline between two specified points. Options:

A - Arc mode - Arc segments can be added to polyline. The arc segment starts from the endpoint of the previous polyline segment and can be drawn by specifying the endpoint of the arc, the included angle, center of the arc, starting direction of the arc, halfwidth of the arc, radius of the arc, Width. You can also close the polyline with the arc segment, or reverse the previous operation or you can shift to the Line mode, C - Closes the polyline, H - Sets the halfwidth, L - Draws polyline of specified length, U - Last polyline segment is removed, W - The width of the next segment is specified.

PLOT Displays a dialog box that allows you to plot the drawing to the plotting device or file. Through a series of dialog boxes you can set the different parameters, device information, drawing extents and limits, plot size, paper size, orientation, plot scale, rotation, and origin. You can also plot a view or a specific portion of the drawing and also preview the plot.

PLOTSTYLE Sets the current plot style for new drawing objects, or the assigned plot style for selected objects through a dialog box.

PLOTTERMANAGER Opens the **Plotter Manager** and enables you to launch Add-a-plotter wizard and Plotter Configuration Editor.

POINT Draws a point object at a specified location.

POLYGON Draws a polygon (closed polyline object) having specified number of sides. Options:

C - Specifies the center of polygon. Suboptions:

I - Inscribed in the circle, C - Circumscribed about the circle.

E - Defines one edge of the polygon.

PREVIEW Shows how the drawing will look when it is printed or plotted.

PROPERTIES Controls properties of a drawing object and displays Properties window.

PROPERTIESCLOSE Closes the Properties window.

PSETUPIN Imports a user-defined page setup into a new drawing layout.

PSOUT Creates an Encapsulated PostScript file into which the current view of the drawing is exported through a dialog box.

PSPACE Switches from a model space viewport to paper space.

PURGE Removes those references from the database that are not being used. Options:

B - Removes unused blocks, D - Removes unused dimstyles, LA - Removes unused layers, LT - Removes unused linetypes, P - Removes unused plotstyle, SH - Removes unused shape files, ST - Removes unused text styles, M - Removes unused mline styles, A - Removes all unused objects.

QKUNGROUP Deletes the group definition from a drawing

QLEADER Quickly creates a leader and its annotation.

QSAVE Saves and backs up the drawing without asking for a filename.

QSELECT Quickly creates selection sets based on filtering criteria.

QTEXT Sets the text and the attribute objects to be displayed without drawing the text detail. Options:

ON - Text displayed as a bounding box.

OFF - Quick text mode off.

QUIT Exits AutoCAD LT without saving.

RAY Draws a semi-infinite line used as a construction line.

RECOVER Recovers a damaged and corrupted drawing.

RECTANG Creates a polyline rectangle by specifying the diagonally opposite corners.

REDO The effect of the previous command if it was **UNDO** is reversed.

REDRAW Cleans up the current viewport by

removing the blip marks and other stray pixels and redrawing missing portions of objects.

REGEN Regenerates the current viewport.

REGENALL Regenerates all the viewports.

REGION Region entities (2D enclosed areas) are created from a selection set.

REINIT Reinitializes the I/O ports, digitizer, display, or parameters file through a dialog box.

RENAME Alters the name of entities. Options:

B - Renames block, D - Renames dimstyle, LA - Renames layers, LT - Renames linetype, S - Renames style, U - Renames UCS, VI - Renames view, VP - Renames viewport configuration.

RESUME Resumes an interrupted script.

REVCLOUD Creates a cloud-shaped polyline containing sequential arcs. Options:

Start point, A - Arc length.

REVDATE Inserts or updates a block containing information on user name, current time and date and drawing name.

ROTATE Rotates specified entities about a base point. Options:

Angle - Rotates object through a specified angle, R - Rotates object with respect to the reference angle.

RSCRIPT Repeats a script continuously.

SAVE A name is requested under which the drawing is saved. If the drawing is already named, then it is saved under the current filename.

SAVEAS An unnamed drawing is saved with a filename or the current drawing is renamed.

SCALE The size of the existing objects is changed. The default is to specify a scale factor. Options:

R - The object is scaled according to the reference length and a new length.

SCRIPT Executes a command script.

SELECT Creates a selection set of specified group of objects. Options:

AU - Automatic selection, A - Add mode - Objects are added to the selection set, ALL - Selects all objects, BOX - Objects inside or crossing a rectangle are selected, C - Objects are selected that lie inside and crossing an area specified by two points, CP - Those objects are selected that lie inside and crossing the polygon created by specifying points around the objects, F - Those objects are selected that are crossing the specified fence, G - Objects within a group are selected, L - Recently created object is selected, M - Objects are picked without highlighting them, P - Recent selection set is selected, R - Remove mode - Objects can be removed from the selection set, SI - Selects first object or a set of objects, U - Removes the most recently added object from the selection set, W - Selects those objects that lie completely inside an area specified by two points, WP - Selects those objects that lie completely inside an area specified by picking points around the objects.

SELECTURL Selects all objects in the drawing which have URLs attached to them.

SETENV Sets values of specified registry variables from within AutoCAD LT.

SETUV Lets you map materials onto geometry.

SETVAR Sets values of the system variables.

Options:
? - Lists variables with their current values.

SHADE Removes hidden lines and displays a flat shaded image of the drawing.

SHADEMODE Displays a shaded picture of the drawing in the current viewport. Options:
2D - Displays the objects using lines and curves to represent the boundaries, 3D - Displays the objects using lines and curves to represent the boundaries, 3D wireframe - Displays the objects using 3D wireframe representation, H - Hides lines representing back faces, F - Shades the objects between the polygon faces, G - Shades the objects and smooths the edges, L - Combines the Flat Shaded and Wireframe options, O - Combines the Gouraud Shaded and Wireframe options.

'SNAP The movement of the cursor is constrained to the snap spacing. Options:
ON - Snap mode is turned on, OFF - Snap mode is turned off, A - Sets different X and Y spacings, R - Snap grid is rotated, S - Sets the style (Standard or Isometric) of the snap grid, T - Specifies the snap type (Polar or Grid).

SOLID Draws polygons that are solid-filled.

SPELL Allows spellcheck of text objects in a drawing. If an ambiguous word is found, then the dialog box is displayed that lists the alternatives for the word, or permits you to replace the current word with another one, or add the word to the dictionary.

SPLINE Draws smooth spline curves between points. Options:
Point - Specify points to define the spline curve. Suboptions:
Point - Adds spline curve segments by specifying points, C - Spline curve is closed, F - Fit, Tolerance - The tolerance for fitting is changed, O - 2D or 3D spline-fit polylines are changed to splines.

SPLINEDIT Allows you to edit a spline entity. Options:
F - Fit data is edited. Suboptions:
A - Fit points are added, C - An open spline is closed, O - A closed spline is opened, D - Fit points are removed, M - Fit points are moved, P - A spline fit data is removed from database, T - Beginning and end tangents are edited, L - Tolerance value for spline fit are changed, X - Exits fit data option, C - An open spline is closed, O - A closed spline is opened, M - Move Vertex - The position of the control vertices is changed, R - Refines a spline by adding control points, or by increasing its order, or by changing the weight, E - Spline direction is reversed, U - Reverses the previous operation of **SPLINEDIT**.

STRETCH Stretches lines, arcs, and polylines by moving the endpoints to another specified location, and moves the objects.

STYLE Creates new text styles or modifies the existing ones through a dialog box.

STYLESMANAGER Displays Plot style Manager dialog box.

SUBTRACT Subtracts the area of one set of regions from another and subtracts the volume of one set of solids from another, thus creating a new composite region or solid.

SYSWINDOWS Arranges windows and is equivalent to standard Window menu options in Windows applications.

TABLET Aligns the tablet with the coordinate system of a paper drawing. Options:
ON - Tablet mode is turned on, OFF - Tablet mode is turned off, CAL - Calibrates the tablet, CFG - Configures tablet menu area and screen pointing area.

TEXT Writes text using a variety of character patterns. Displays text on screen as it is entered. Options:

Start Point - Specifies a start point for the text object, J - Controls justification of the text, Suboptions:
A - Specifies both text height and text orientation, F - Specifies that text fits within an area, C - Center aligned, M - Horizontally aligned, R - Right justified, TL - Top left, TC - Top centered, TR - Top right, ML - Middle left, MC - Middle centered, MR - Middle right, BL - Bottom left, BC - Bottom centered, BR - Bottom right. S - Specifies the text style, which determines the appearance of the text characters.

TEXTSCR Flips to the text window from the graphics window.

TIME The date and time of drawing creation is displayed. It also displays the time and the date when the current drawing was last updated and controls an elapsed timer. Options:

D - Displays the updated times, O - Elapsed timer is turned on, OFF - Elapsed timer is turned off, R - Resets the user elapsed timer.

TOLERANCE Creates and adds geometric tolerances to a drawing through a dialog box.

TRACKING Locates a point in reference with a series of points.

TRIM Removes the extra portion of an entity which extends beyond a specified boundary. Options:

P - Sets projection mode, E - Controls trimming of objects until the implied edge, U - Reverses the previous operation of the **TRIM** command, O - Specifies the object to trim.

U Reverses the most recent operation.

UCS Sets and modifies the user coordinate system. Options:

W - Current UCS set to World Coordinate System, N - New UCS, M - Moves the origin, G - Specifies one of the six orthographic UCSs, P - Restores the previous UCS, R - Restores a saved UCS so that it becomes the current, S - Saves the current UCS to a specified name, D - Removes the specified UCS from the list, A - Applies the current UCS setting, ? - Lists names of user coordinate systems.

UCSICON Manages the location and the visibility of the UCS icon. Options:

ON - Coordinate system icon is enabled, OFF - Coordinate system icon is disabled, A - Icon is changed in all active viewports, N - Icon displayed at the lower left corner, OR - Icon displayed at the origin of current coordinate system.

UCSMAN Manages defined user coordinate systems through a dialog box.

UNDO Reverses the effect of commands. Options:

N - The effect of a specified number of previous commands used is reversed, A - The effect of the menu items is reversed by a single U command, C - The **UNDO** command is limited or is turned off, BE - A number of operations are grouped together and are treated as a single operation, E - The group is terminated, M - Mark - A marker is placed in the undo information, B - Back - Undoes all work until the marker is encountered.

UNION Combines the area of two or more regions, or the volume of two or more solids to create a composite region or solid.

UNITS Sets the coordinate and angle display formats and precision.

VIEW The graphics display is saved and

restored as a view with a specified name through a dialog box.

VIEWRES Controls the appearance of objects by setting their resolution in the current viewport.

VPLAYER Controls the visibility of layers in different viewports. Options:

? - Lists the frozen layers in a specified viewport, F - Layers are frozen in current, or all, or specified viewport, T - Layers are thawed in current, or all, or specified viewport, R - Rests the layers default visibility, N - New layers that are frozen in all viewports are created, V - Viewport Visibility Default - Controls thawing and freezing of layers.

VPOINT The viewing direction for 3D visualization. Options:

ENTER - Displays compass and axis tripod for controlling viewing direction, V - Specifies a point from which drawing can be viewed, R - New direction using two angles is specified.

VPORTS Divides the graphics display into a number of viewports through a dialog box.

VSLIDE Displays an existing raster image slide file in the current viewport.

WBLOCK Writes a block definition or specified objects to a new disk file through a dialog box.

WHOHAS Displays information about ownership of opened drawings.

WMFIN Imports a Windows metafile.

WMFOPTS Sets options for WMFIN.

WMFOUT Saves objects to a Windows metafile.

XATTACH Attaches an external reference to the current drawing.

XBIND Adds Xref 's dependent symbols to a drawing through a dialog box.

XLINE Creates a line of infinite length. Options:

point - Specifies the point through which the xline passes, H - Creates a horizontal xline, V - Creates a vertical xline, A - Creates a xline at an angle, B - Creates an xline through the vertex of two lines so that it bisects the angle between those two lines, O - Creates an xline parallel to another linear object.

XREF Manages external references to a drawing through Xref Manager.

ZOOM Changes the display of the entities in the current drawing. Options:

value - Scale(X/XP) - Changes the display by a specified scale factor, Scale X - Zoom relative to current scale, Scale XP - Scale relative to paper space, A - Zooms the entire drawing in current viewport, C - Displays at a specified center point, D - Displays the portion of the drawing with a view box, E - Displays the drawing extents, S - Zooms the display at a specified scale factor, W - Displays an area specified by two corners of the window, Realtime - Using the pointing device, zooms interactively to a logical extent.

Appendix D

AutoCAD LT System Variables

Variable Name	Type and Description

AFLAGS Integer
The AFLAGS variable establishes the attribute flags for the **ATTDEF** command bit-code. The initial value for this variable is 0. Basically the value of this variable is the addition of the following:
0 - No attribute mode selected, 1 - Invisible, 2 - Constant, 4 - Verify, 8 - Preset.

ANGBASE Real
The ANGBASE variable establishes the base angle 0 in relation to the current UCS. This variable is saved in the drawing and has an initial value of 0.0000.

ANGDIR Integer
The ANGDIR variable establishes the angle from angle 0 in relation to the prevailing UCS. This variable is saved in the drawing and has an initial value of 0.
0 - Direction is counterclockwise, 1 - Direction is clockwise.

APBOX Integer
The APBOX variable turns the AutoSnap aperture box on or off. Initial value is 1.
0 - AutoSnap aperture box is not displayed, 1 - AutoSnap aperture box is displayed.

APERTURE Integer
The APERTURE variable defines the object snap target height in pixels. This variable is saved in registry and has an initial value of 10.

AREA Real
The most recently calculated area with commands such as **AREA**, **LIST**, or **DBLIST** is stored in this variable. This is a read-only variable. You can examine this variable through the **SETVAR** command.

ATTDIA Integer
With the ATTDIA variable you can specify whether you want to enter the attribute

value through the INSERT dialog box or from the command line. This variable is saved in the drawing and has an initial value of 0.
0 - Attribute values can be specified on the command line, 1 - Attribute values can be specified in the dialog box.

ATTMODE Integer
The ATTMODE variable controls the Attribute Display mode and is saved in the drawing and its initial value is 1.
0 - Attribute Display mode is off, 1 - Normal, 2 - On.

ATTREQ Integer
The value contained in the ATTREQ variable determines whether the **INSERT** command uses the default attribute settings when the blocks are being inserted.
0 - The default values for all the attributes are used, 1 - This is also the initial value and enables prompts or dialog box for attribute values (depending on the value of ATTDIA variable).

AUDITCTL Integer
The AUDITCTL variable determines whether an .adt file (audit report file) will be created by AutoCAD LT. This variable is saved in registry.
0 - Does not allow writing of .adt files. This is also the initial value, 1 - Allows writing of .adt files.

AUNITS Integer
The AUNITS variable establishes the Angular Units mode and is saved in the drawing.
0 - Decimal degrees (initial value), 1 - Degrees/ minutes/ seconds, 2 - Gradians, 3 - Radians, 4 - Surveyor's units.

AUPREC Integer
The AUPREC variable establishes the angular units decimal places. This variable is saved in the drawing and has an initial value of 0.

AUTOSNAP Integer
The AUTOSNAP variable controls the display of AutoSnap marker, and Snap Tips and turns the AutoSnap magnet on or off. Initial value is 7.
0 - Turns off marker, Snap Tip and magnet, 1 - Turns on marker, 2 - Turns on Snap tooltips, 4 - Turns on magnet

BACKZ Real
The BACKZ variable contains the back clipping plane offset (in current drawing units) from the target plane for the current viewport. You can determine the distance between the back clipping plane and the camera point by subtracting the BACKZ value from the camera to target distance. This variable is saved in the drawing.

BLIPMODE Integer
The visibility of the blip marks is controlled by the BLIPMODE variable. This variable is saved in the drawing and its initial value is 0.
0 - Blip marks are not visible, 1 - Blip marks are visible.

CDATE Real
The calendar date and time is stored in this variable. This is a read-only variable.

CECOLOR String
The CECOLOR variable defines the color of new objects. This variable is saved in the drawing and its initial value is "BYLAYER" (256)

CELTSCALE Real
The CELTSCALE variable defines the current global linetype scale factor for objects. This variable is saved in the drawing and its initial value is 1.0000.

CELTYPE String
The CELTYPE variable defines the linetype that will be used in the new objects. This variable is saved in the drawing and its initial value is "BYLAYER."

CELWEIGHT Integer
The CELWEIGHT variable is stored in the drawing. It set the lineweight of the new objects. Its initial value is 1.
1 - Sets the lineweight to "ByLayer", 2 - Sets the lineweight to "ByBlock", 3 - Sets the lineweight to "Default" (controlled by LWDEFAULT system variable).

CHAMFERA Real
The CHAMFERA variable defines the first chamfer distance. This variable is saved in the drawing and its initial value is 0.5000.

CHAMFERB Real
The CHAMFERB variable defines the second chamfer distance. This variable is saved in the drawing and its initial value is 0.5000.

CHAMFERC Real
The CHAMFERC variable sets the chamfer length. This variable is saved in the drawing and its initial value is 1.0000.

CHAMFERD Real
The CHAMFERD variable sets the chamfer angle. This variable is saved in the drawing and its initial value is 0.0000.

CHAMMODE Integer
With the CHAMMODE variable you can specify the method that will be used to create chamfers.
0 - This is the initial value and in this case two chamfer distances are required, 1 - One chamfer length and an angle are required.

CIRCLERAD Real
The CIRCLERAD variable defines the default circle radius. The initial value of this variable is 0.0000.

CLAYER String
The CLAYER variable sets the current layer. This variable is saved in the drawing and its initial value is "0."

CMDACTIVE Integer
The CMDACTIVE variable contains the bit-code that signifies whether an ordinary command, transparent command, dialog box, or script is active. Basically the value of this variable is the addition of the following:
1 - Only ordinary command is active, 2 - Ordinary command as well as transparent command are active, 4 - Script is active, 8 - If this bit is set then Dialog box is active.

CMDNAMES String
The CMDNAMES variable displays the name of the presently active command and transparent command. This variable is read-only.

COORDS Integer
The COORDS variable determines when the coordinates are updated. This variable is saved in the drawing and its initial value is 1.
0 - Coordinates are updated only upon picking points, 1 - Absolute coordinates are continuously updated, 2 - Absolute continuously plus, when a distance or angle are requested, then the distance and angle from the last point are displayed.

CPLOTSTYLE String
Controls the current plot style for new objects. The AutoCAD LT defined values are ByLayer, ByBlock, Normal, User Defined.

CTAB String
The CTAB variable returns the name of the current (model or layout) tab in the

drawing. The value is saved in drawing and it is a read-only variable.

CURSORSIZE Integer
This variable determines the size of the crosshairs as a percentage of the screen size. Initial value is 5.

CVPORT Integer
The CVPORT variable establishes the identification number of the current viewport. When this value is changed, the current viewport is also changed in case the following conditions hold good:
1 - The specified identification number belongs to an active viewport, 2 - The cursor movement to the specified viewport is not locked by the command being executed, 3 - Tablet mode is off. The variable is saved in the drawing and its initial value is 2.

DATE Real
The DATE variable contains the current date and time as a Julian date and fraction in a real number. This variable is read-only.

DCTCUST String
The DCTCUST variable shows the current custom spelling dictionary path and filename. This variable is saved in registry and its initial value is "".

DCTMAIN String
The DCTMAIN variable shows the current main spelling dictionary filename. Normally this file is located in the \support directory. The default main spelling dictionary can be specified using the **SETVAR** command. This variable is saved in registry and its initial value is "".

DEFLPLSTYLE String
The DEFLPSTYLE variable specifies the default plot style for new layers. This variable is saved in registry and its initial value is "".

DEFPLSTYLE String
The DEFPLSTYLE variable specifies the default plot style for new objects. This variable is saved in registry and its initial value is "By Layer".

DELOBJ Integer
Thw DELOBJ variable determines whether objects used to draw other objects are kept or deleted from the drawing database. This variable is saved in the drawing and its initial value is 1.
1 - Objects are deleted from the drawing database, 0 - Objects are kept in the drawing database.

DIMADEC Integer
The DIMADEC variable controls the number of decimal places displayed for the angular dimension. Initial value for this variable is 0.
1 - Angular dimension is drawn using the number of decimal places corresponding to the DIMDEC setting.
0-8 - Angular dimension is drawn using the number of decimal places corresponding to the DIMADEC setting.

DIMALT Switch
The DIMALT variable controls the dimensioning in alternate units system. If the DIMALT variable is on (1), alternate unit dimensioning is facilitated. This variable is saved in the drawing and its initial value is Off (0).

DIMALTD Integer
The DIMALTD (DIMension ALTernate units Decimal places) variable controls the number of decimal places (decimal precision) of the dimension text in the alternate units if DIMALT variable is on. This variable is saved in the drawing and its initial value is 2.

DIMALTF Real
The DIMALTF variable (DIMension ALTernate units scale Factor) controls the alternate units scale factor. In case the DIMALT variable is enabled, all the linear dimensions will be multiplied with this factor to generate a value in an alternate units system. The initial value for DIMALTF is 25.4. This variable is saved in the drawing.

DIMALTRND Real
The DIMALTRND variable rounds off the alternate dimension units. The value is saved in drawing and initial value is 0.00.

DIMALTTD Integer
The DIMALTTD variable establishes the number of decimal places for the tolerance values of an alternate units dimension. This variable is saved in the drawing and has an initial value of 2.

DIMALTTZ Integer
The DIMALTTZ variable controls the suppression of zeros for alternate tolerance values. With this variable, the real-to-string transformation carried out by AutoLISP functions rtos and angtos is also influenced. This variable is saved in the drawing and has an initial value of 0. 0 - Suppresses zero feet and precisely zero inches, 1 - Includes zero feet and precisely zero inches, 2 - Includes zero feet and suppresses zero inches, 3 - Includes zero inches and suppresses zero feet. Value in the range of 0 and 3 influence only the feet and inches dimensions. However, you can add 4 to the above values to omit the leading zeros in all decimal dimensions. If you add 8, the trailing zeros are omitted. If 12 (both 4 and 8) is added, the leading and the trailing zeros are omitted.

DIMALTU Integer
The DIMALTU variable establishes the units format for alternate units of all dimensions except angular. This variable is saved in the drawing and has an initial value of 2.
1 - Scientific, 2 - Decimal, 3 - Engineering, 4 - Architectural (Stacked), 5 - Fractional (Stacked), 6 - Architectural, 7 - Fractional, 8 - Windows® Desktop settings.

DIMALTZ Integer
The DIMALTZ variable controls the suppression of zeros for alternate units dimension values. With this variable, the real-to-string transformation carried out by AutoLISP functions rtos and angtos is also influenced. This variable is saved in the drawing and has an initial value of 0. 0 - Suppresses zero feet and precisely zero inches, 1 - Includes zero feet and precisely zero inches, 2 - Includes zero feet and suppresses zero inches, 3 - Includes zero inches and suppresses zero feet. Value in the range of 0 and 3 influence only the feet and inches dimensions. However, you can add 4 to the above values to omit the leading zeros in all decimal dimensions. If you add 8, the trailing zeros are omitted. If 12 (both 4 and 8) is added, the leading and the trailing zeros are omitted.

DIMAPOST String
With the help of DIMAPOST variable, you can append a text prefix, suffix, or both to an alternate dimensioning measurement. This can be done in case of all the dimensions except angular dimensions. The variable is saved in the drawing and has an initial value of "". In order to disable an existing suffix or prefix, set the value of this variable to a single period.

DIMASO Switch
The DIMASO variable governs the creation of associative dimensions. This variable is saved in the drawing (not in the dimension style) and its initial value is set to on.

Off (0) - The dimension created are not associative in nature and hence in such dimensions no association exists between the dimension and the points on the object. All the dimensioning entities such as arrowheads, dimension lines, extension lines, dimension text, etc. are drawn as separate entities, On (1)- The dimension created are associative in nature and hence in such dimensions there exists an association between the dimension and the definition points. If you edit the object (editing like trimming or stretching), the dimensions associated with that object also change. Also, the appearance of associative dimensions can be preserved when they are edited by commands such as STRETCH or TEDIT. For example, a vertical associative dimension is retained as a vertical dimension even after an editing operation. The associative dimension is always generated with the same dimension variable settings as defined in the dimension style.

DIMASZ Real
The DIMASZ (Dimension arrowhead size) variable specifies the size of dimension line and leader line arrowheads when DIMTSZ is set to zero. The size of arrowhead blocks set by DIMBLK is also controlled by the DIMASZ variable. Multiples of this variable determine whether the dimension line and text will be located between the extension lines. This variable is saved in the drawing and has an initial value of 0.18 units.

DIMATFIT Integer
The DIMATFIT variable determines how dimension text and arrows are arranged when space is not sufficient to place both within the extension lines. The initial value is 3 and is saved in drawing.
0 - Places both text and arrows outside extension lines, 1 - Moves arrows first, then text, 2 - Moves text first, then arrows, 3 - Moves either text or arrows, whichever fits best.

DIMAUNIT Integer
The DIMAUNIT variable establishes the angle format for angular dimensions. This variable is saved in the drawing and its initial value is 0.
0 - Decimal degrees format, 1 - Degrees/minutes/seconds format, 2 - Gradians format, 3 - Radians format.

DIMAZIN Integer
The DIMAZIN variable suppresses zeros for angular dimensions. The initial value is 0 and saved in drawing.
0 - Displays all leading and trailing zeros, 1 - Suppresses leading zeros in decimal dimensions, 2 - Suppresses trailing zeros in decimal dimensions, 3 - Suppresses leading and trailing zeros.

DIMBLK String
DIMBLK variable replaces the default arrowheads at the end of the dimension lines with a user defined block. The user defined block that may replace the standard arrowhead can be a custom designed arrow or some other symbol. DIMBLK (DIMension BLocK) takes the name of the block as its string value. This variable is saved in the drawing and its initial value is no block (""). To discard an existing block name, set its value to a single period (.).

DIMBLK1 String
DIMBLK1 variable designates the user defined arrow block for the first end of the dimension line. This option can be used only if the DIMSAH (DIMension Separate Arrow blocks) variable is on. The value of this variable is the name of the earlier formulated block as in the case of DIMBLK. You can discard an existing block name by setting its value to a single period (.). This variable is saved in the

drawing and its initial value is no block ("").

DIMBLK2 String
DIMBLK2 variable designates a user defined arrow block for the second end of the dimension line. This option can be used only if the DIMSAH (DIMension Separate Arrow blocks) variable is on. The value of this variable is the name of the earlier formulated block as in the case of DIMBLK. You can discard an existing block name, by setting its value to a single period (.). This variable is saved in the drawing and its initial value is no block ("").

DIMCEN Real
The DIMCEN (DIMension CENter) variable governs the drawing of center marks and the center lines of circles and the arcs by the DIMCENTER, DIMDIAMETER, and DIMRADIUS commands. DIMCEN takes a distance as its argument. The value of the DIMCEN variable determines the result. This variable is saved in the drawing and its initial value is 0.0900.
0 - Center marks or center lines are not drawn, >0 - Center marks are drawn and their size is governed by the value of the DIMCEN. For example, a value of 0.250 displays center dashes that are 0.2500 units long, <0 - Center lines in addition to center marks are drawn and again the size of the mark portion is governed by the absolute value of the DIMCEN. The center lines extend beyond the circle or arc by the value entered. For example, a value of -0.2500 for the DIMCEN variable will draw center dashes eacs 0.25 units long and also the center lines will be extended beyond the circle/arc by a distance of 0.25 units. With the DIMRADIUS and DIMDIAMETER commands, center mark or center line is generated only when the dimension line is located outside the circle or arc.

DIMCLRD Integer
The DIMCLRD variable is used to assign colors to dimension lines, arrowheads, and the dimension leader lines. This variable can take any permissible color number or the special color labels BYBLOCK (0) or BYLAYER (256) as its value. If you use the SETVAR command, then you have to enter the integer number of the color you want to assign to the DIMCLRD variable. This variable is saved in the drawing and its initial value is 0.

DIMCLRE Integer
DIMCLRE variable is used to assign color to the dimension extension lines. Just as DIMCLRD, DIMCLRE (DIMension CoLOr Extension) can take any permissible color number or the special color labels BYBLOCK or BYLAYER. This variable is saved in the drawing and its initial value is 0.

DIMCLRT Integer
The DIMCLRT (DIMension CoLoR Text) variable is used to assign a color to the dimension text. DIMCLRT can take any permissible color number or the special color labels BYBLOCK (0) or BYLAYER (256). This variable is saved in the drawing and its initial value is 0.

DIMDEC Integer
The DIMDEC variable establishes the number for decimal places of a primary units dimension. This variable is saved in the drawing and its initial value is 4.

DIMDLE Real
By default the dimension lines meet the extension lines. But if you want the dimension line to continue past the extension lines, the DIMDLE (Dimension Line Extension) variable can be used for this function. DIMDLE is used only when

the DIMTSZ variable is nonzero (When the DIMTSZ variable is nonzero, ticks are drawn instead of arrows). The dimension line will extend past the extension line by the value of DIMDLE. This variable is saved in the drawing and its initial value is 0.0000.

DIMDLI Real
The DIMDLI variable governs the spacing between the successive dimension lines when dimensions are created with the DIMCONTINUE and DIMBASELINE commands. Successive dimension lines are offset by the DIMDLI value, if needed, to avert drawing over the previous dimension. This variable is saved in the drawing and its initial value is 0.38 units.

DIMDSEP Single character
The DIMDSEP variable specifies a single-character decimal separator to use when creating dimensions whose unit format is decimal. The initial value is a decimal point and it is saved in drawing.

DIMEXE Real
The extension of the extension line past the dimension line is governed by the DIMEXE (Dimension EXtension line Extension) variable. This variable is saved in the drawing and has an initial value of 0.18 units.

DIMEXO Real
There exists a small space between the origin points you specify and the start of the extension lines. The size of this gap is controlled by the DIMEXO (DIMension EXtension line Offset) variable. The offset distance is equal to the value of the DIMEXO variable. This variable is saved in the drawing and has an initial value of 0.0625 units.

DIMFIT Integer
Obsolete. Has no effect in AutoCAD LT 2000 except to preserve the integrity of pre-AutoCAD LT 2000 scripts and AutoLISP routines. This variable is saved in the drawing and its initial value is 3. DIMFIT is replaced by DIMATFIT and DIMTMOVE.

DIMFRAC Integer
The DIMFRAC variable controls the fraction format when DIMLUNIT is set to 4 (Architectural) or 5 (Fractional). It is saved in drawing and initial value is 0.
0 - Horizontal, 1 - Diagonal, 2 - Not stacked (for example, 3/5).

DIMGAP Real
The DIMGAP variable controls the space between the dimension line and the dimension text (distance maintained around the dimension text), when the dimension line is split into two for the placement of dimension text. The gap between the leader and annotation created with the LEADER command is also governed by DIMGAP variable. This variable is saved in the drawing and its initial value for DIMGAP is 0.0900 units. By entering a negative DIMGAP value, you can create a reference dimension, in which case you get the dimension text with a box drawn around it. DIMGAP value is also used by AutoCAD LT as the measure of minimum length needed for the segments of the dimension line. AutoCAD LT places the dimension text inside the extension lines only if the dimension line is split into two segments each of which is at least as long as DIMGAP. In case the text is positioned over or under the dimension line, it is placed inside the dimension line only if there is space for the arrows, dimension text, and a margin between them has a minimum value at least as much as DIMGAP: 2*(DIMGAP + DIMASZ).

DIMJUST Integer

The DIMJUST variable governs the horizontal dimension text position. This variable is saved in the drawing and its initial value is 0.

0 - The text is center justified between the extension lines, 1 - The text is placed next to the first extension line, 2 - The text is placed next to the second extension line, 3 - The text is placed above and aligned with the first extension line, 4 - The text is placed above and aligned with the second extension line.

DIMLDRBLK String

The DIMLDRBLK variable controls the arrow type for leaders. To turn off the arrowhead display, enter a single period (.).

DIMLFAC Real

The DIMLFAC (DIMension Length FACtor) variable acts as a global scale factor for all linear dimensioning measurements. The linear distances measured by dimensioning include coordinates, diameter, and radii. These linear distances are multiplied by the prevailing DIMLFAC value before they are projected as dimension text. In this manner DIMLFAC scales the contents of the default text. The angular dimensions are not scaled. Also DIMLFAC does not apply to the values held in DIMTM, DIMTP, or DIMRND. For example, if you want to scale the default dimension measurement by a value of 2, set the value of DIMLFAC to 2. When dimensioning in the paper space, if the value of DIMLFAC variable is not zero, then the distance measured is multiplied by the absolute value of DIMLFAC. In case of dimensioning in the model space, values less than zero are neglected; instead, the value of DIMLFAC is taken as 1.0. If in paper space you select the Viewport option and try to change DIMLFAC from the Dim: prompt, AutoCAD LT will compute a value for the DIMLFAC for you. This is illustrated as follows:

Dim: DIMLFAC, Current value <1.0000> New value (Viewport): V Select viewport to set scale: The scaling of model space to paper space is computed by AutoCAD LT and the negative of the computed value is assigned to DIMLFAC. This variable is saved in the drawing and its initial value is 1.0000.

DIMLIM Switch

The DIMLIM (DIMension LIMits) variable acts as a switch and creates the dimension limits as the default text if it is on (1). Also DIMTOL is forced to be off. This variable is saved in the drawing and its initial value is off.

DIMLUNIT Integer

The DIMLUNIT variable controls units for all dimension types except Angular. The initial value is Off (0) and it is saved in drawing.

1 - Scientific, 2 - Decimal, 3 - Engineering, 4 - Architectural, 5 - Fractional, 6 - Windows desktop settings.

DIMLWD Enum

The DIMLWD variable assigns lineweight to dimension lines. Values are standard lineweights. The initial value is "By Block" and it is saved in drawing.

DIMLWE Enum

The DIMLWD variable assigns lineweight to extension lines. Values are standard lineweights. The initial value is "By Block" and it is saved in drawing.

DIMPOST String

The DIMPOST variable is used to define prefix or suffix to the dimension measurement. The variable is saved in the drawing and has an initial value "" (empty string). DIMPOST takes a string value as

its argument. For example if you want to have a suffix for centimeters, set DIMPOST to "cm". A distance of 4.0 units will be displayed as 4.0 cm. In case tolerances are enabled, the suffix you have defined gets applied to the tolerances as well as to the main dimension.
To establish a prefix to a dimension text, type "<>" and then the prefix at the same prompt.

DIMRND Real
The DIMRND (DIMension RouND) variable is used for rounding all the dimension measurements to the specified value. For example if the DIMRND is set to 0.10, then all the measurements are rounded to the nearest 0.10 unit. Likewise a value of 1 for this variable will result in the rounding of all the measurements to the nearest integer. The angular measurements cannot be rounded. The variable is saved in the drawing and has an initial value of 0.0000.

DIMSAH Switch
The DIMSAH (DIMension Separate custom Arrow Head) variable governs the placement of user-defined arrow blocks instead of the standard arrows at the end of the dimension line. As explained before, the DIMBLK1 variable places a user defined arrow block at the first end of the dimension line and DIMBLK2 places a user defined arrow block at the other end of the dimension line. This variable is saved in the drawing and its initial value is off.
On - DIMBLK1 and DIMBLK2 specify different user-defined arrow blocks to be drawn at the two ends of the dimension line, Off - Ordinary arrowheads or user-defined arrowhead block defined by the DIMBLK variable is used.

DIMSCALE Real
The DIMSCALE variable controls the scale factor for all the size-related dimension variables such as those that affect text size, center mark size, arrow size, leader objects, etc. The DIMSCALE is not applied to the measured lengths, coordinates, angles, or tolerances. The default value for this variable is 1.0000; and in this case the dimensioning variables assume their preset values and the drawing is plotted at full scale. If the drawing is to be plotted at half the size, then the scale factor is the reciprocal of the drawing size. Hence the scale factor or the DIMSCALE value will be the reciprocal of 1/2 which is 2/1 = 2.
0.0 - A default value based on the scaling between the current model space viewport and paper space is calculated. In case you are not using the paper space feature, then the scale factor is 1.0, >0 - A scale factor is computed that makes the text sizes, arrowhead sizes, and scaled distances to plot at their face value.

DIMSD1 Switch
The DIMSD1 (DIMension Suppress Dimension line 1) variable suppresses the drawing of the first dimension line when it is on. This variable is saved in the drawing and its initial value is off.

DIMSD2 Switch
The DIMSD1 (DIMension Suppress Dimension line 2) variable suppresses the drawing of the second dimension line when it is on. This variable is saved in the drawing and its initial value is off.

DIMSE1 Switch
The DIMSE1 variable is used to suppress drawing of the first extension line. When DIMSE1 (DIMension Suppress Extension line 1) is on, the first extension line is not drawn. This variable is saved in the drawing and its initial value is off.

DIMSE2 Switch
The DIMSE2 variable is used to suppress drawing of the second extension line. When DIMSE2 (DIMension Suppress Extension line 2) is on, the second extension line is not drawn. This variable is saved in the drawing and its initial value is off.

DIMSHO Switch
DIMSHO variable governs the redefinition of dimension entities while dragging into some position. If DIMSHO (DIMension SHOw dragged dimensions) is on, associative dimensions will be computed dynamically as they are dragged. The DIMSHO value is saved in the drawing (not in a dimension style) and its initial value is on (1). Dynamic dragging reduces the speed of some computers and hence in such situations DIMSHO should be set off (0). However, when you are using the pointing device to specify the length of the leader in Radius and Diameter dimensioning, the DIMSHO setting is neglected and dynamic dragging is used.

DIMSOXD Switch
If you want to place text inside the extension lines, you will have to set the DIMTIX variable on. And if you want to suppress the dimension lines and the arrow heads you will have to set the DIMDSOXD (DIMension Suppress Outside eXtension Dimension lines) variable on. DIMSOXD suppresses the drawing of dimension lines and the arrow heads when they are placed outside the extension lines. If DIMTIX is on and DIMSOXD is off and there is not enough space inside the extension lines for drawing the dimension lines, then dimension lines will be drawn outside the extension lines. In such a situation, if both DIMTIX and DIMSOXD are on, then the dimension line will be totally suppressed. DIMSOXD works only when DIMTIX is on. The DIMSOXD variable is saved in the drawing and its initial value is off.

DIMSTYLE String
The DIMSTYLE variable is used for displaying the name of the present dimension style. DIMSTYLE is a read-only variable and is saved in the drawing. You can change the dimension style using the DDIM or DIMSTYLE command.

DIMTAD Integer
The DIMTAD (DIMension Text Above Dimension line) variable governs the vertical placement of the dimension text with respect to the dimension line. DIMTAD gets actuated when dimension text is drawn between the extension lines and is aligned with the dimension line, or when the dimension text is placed outside the extension lines. This variable is saved in the drawing and its initial value is 0.
0 - For this value the dimension text is placed at the center between the extension lines, 1 - The dimension text is placed above the dimension line and a single (unsplit) dimension line is drawn under it spanning between the extension lines. The exceptions to this arise when the dimension line is not horizontal and text inside the extension line is forced to be horizontal by making DIMTIH = 1. The space between the dimension line and the baseline of the lowest line of text is nothing but the prevailing DIMGAP value, 2 - The dimension text is placed on the side of the dimension line most remote from the defining points, 3 - The dimension text is placed to tune to a JIS representation.

DIMTDEC Integer
The DIMTDEC variable establishes the number of decimal places for the tolerance values for the primary units

dimension. This variable is saved in the drawing and its initial value is 4.

DIMTFAC Real

With the DIMTFAC (DIMension Tolerance scale FACtor) variable you can control the scaling factor of the text height of the tolerance values in relation to the dimension text height set by DIMTXT. Suppose DIMTFAC is set to 1.0 (the default value for DIMTFAC variable), then the text height of the tolerance text will be equal to the dimension text height. If DIMTFAC is set to a value of 0.50, the text height of the tolerance is half of the dimension text height. This variable is saved in the drawing and its initial value is 1.0000. It is important to remember that the scaling of tolerance text to any requirement is possible only when DIMTOL is on and DIMTM and DIMTP variable values are not identical, or when DIMLIM is on.

DIMTIH Switch

The DIMTIH (DIMension Text Inside Horizontal) variable controls the placement of the dimension text inside the extension lines for Linear, Radius, Angular, and Diameter dimensioning. DIMTIH is effective only when the dimension text fits between the extension lines.

On - If DIMTIH is on (the default setting), it forces the dimension text inside the extension lines to be placed horizontally, rather than aligned, Off - In case DIMTIH is off, the dimension text is aligned with the dimension line.

DIMTIX Switch

The DIMTIX variable draws the text between the extension lines. This variable is saved in the drawing and its initial value is Off.

On - When DIMTIX is set to on, the dimension text is placed amidst the extension lines even if it would normally be placed outside the extension lines, Off - If DIMTIX is off, the placement of the dimension text depends on the type of dimension. For example, if the dimensions are Linear or Angular, the text will be placed inside the extension lines by AutoCAD LT if there is enough space available. While as for the Radius and Diameter dimensions, the text is placed outside the object being dimensioned.

DIMTM Real

The DIMTM variable establishes the lower (minimum) tolerance limit for the dimension text. Tolerance is defined as the total amount by which a particular dimension is permitted to vary. The tolerance or limit values are drawn only if DIMTOL or DIMLIM variable is on. DIMTM (DImension Tolerance Minus) identifies the lower tolerance and DIMTP (DIMension Tolerance Plus) identifies the upper tolerance. You can specify signed values for the DIMTM and DIMTP variables. If DIMTOL is on and both DIMTM and DIMTP have same value, AutoCAD LT draws the "ñ" symbol followed by the tolerance value. If DIMTM and DIMTP hold different values, the upper tolerance is drawn above the lower tolerance. Also a positive (+) sign is appended to the DIMTP value if it is positive. For minus tolerance value (DIMTM), the negative of the value you enter (negative sign if you enter positive value and positive sign if you enter negative value) is displayed. Signs are not appended with zero. This variable is saved in the drawing and its initial value is 0.0000.

DIMTMOVE Integer

The DIMTMOVE variable controls the dimension text movement rules. Its initial value is 0 and it is saved in drawing.

0 - Moves the dimension line with dimension text, 1 - Adds a leader when

dimension text is moved, 2 - Allows text to be moved freely without a leader.

DIMTOFL Switch

If DIMTOFL variable is turned on, a dimension line is drawn between the extension lines even if the text is located outside the extension lines. When DIMTOFL is off, for radius and diameter dimensions, the dimension line and the arrowheads are drawn inside the arc or circle, while the text and the leader are placed outside. This variable is saved in the drawing and its initial value is Off.

DIMTOH Switch

The DIMTOH (DIMension Text Outside Horizontal) variable controls the orientation of the dimension text outside the extension lines. If DIMTOH is on, it forces the dimension text outside the extension lines to be placed horizontally, rather than aligned. In case DIMTOH is off, the dimension text is aligned with the dimension line. You must have noticed that the variable DIMTOH is the same as DIMTIH variable except it controls text drawn outside the extension lines. This variable is saved in the drawing and its initial value is On.

DIMTOL Switch

DIMTOL (DIMension with TOLerance) variable is used for controlling the appending of dimension tolerances to the dimension text. With DIMTM and DIMTP you can define the values of the lower and upper tolerances. If the DIMTOL variable is set on, the tolerances are appended to the default text. When DIMTOL is set on, DIMLIM variable is set off. This variable is saved in the drawing and its initial value is Off.

DIMTOLJ Integer

DIMTOLJ variable establishes the vertical justification for the tolerance values with respect to the normal dimension text. This variable is saved in the drawing and its initial value is 1.

0 - Bottom, 1 - Middle, 2 - Top.

DIMTP Real

The DIMTP (DIMension Tolerance Plus) variable establishes the upper (maximum) tolerance limit for the dimension text. Tolerance is defined as the total amount by which a particular dimension is permitted to vary. The tolerance or limit values are drawn only if the DIMTOL or DIMLIM variable is on. If DIMTOL is on and both DIMTM and DIMTP have same value, AutoCAD LT draws the "ñ" symbol followed by the tolerance value. If DIMTM and DIMTP hold different values, the upper tolerance is drawn above the lower tolerance. Also a positive (+) sign is appended to the DIMTP value if it is positive. This variable is saved in the drawing and its initial value is 0.0000.

DIMTSZ Real

The DIMTSZ variable defines the size of oblique strokes (ticks) instead of arrowheads at the end of the dimension lines (just as in architectural drafting), for Linear, Radius, and Diameter dimensioning. This variable is saved in the drawing and its initial value is 0.0000.

0 - Arrows are drawn, >0 - Oblique strokes instead of arrows are drawn. The size of the ticks is computed as DIMTSZ*DIMSCALE. Hence if the DIMSCALE factor is one, then the size of the tick is equal to the DIMTSZ value. This variable is also used to determine whether the dimension line and dimension text will get accommodated between the extension lines.

DIMTVP Real

The DIMTVP (DIMension Text Vertical Position) variable, controls the vertical placement of the dimension text over or under the dimension line. In certain cases DIMTVP is used as DIMTAD to control the vertical position of the dimension text. DIMTVP value holds good only when DIMTAD is off. The vertical placing of the text is done by offsetting the dimension text. The amount of the vertical offset of dimension text is a product of text height and DIMTVP value. If the value of DIMTVP is 1.0, DIMTVP acts as DIMTAD. However if the value of the DIMTVP is less than 0.70, the dimension line is broken into two segments to accommodate the dimension text. This variable is saved in the drawing and its initial value is 0.0000.

DIMTXSTY String
The DIMTXTSTY variable specifies the text style of the dimension. This variable is saved in the drawing and its initial value is "STANDARD".

DIMTXT Real
The DIMTXT variable is used to control the height of the dimension text except if the current text style has a fixed height. This variable is saved in the drawing and its initial value is 0.1800.

DIMTZIN Integer
With the DIMZIN variable you can control the suppression of the zeros for tolerance values. The variable is saved in the drawing and its initial value is 0.
0 - Suppresses zero feet and precisely zero inches, 1 - Includes zero feet and precisely zero inches, 2 - Includes zero feet and suppresses zero inches, 3 - Includes zero inches and suppresses zero feet. You can add 4 to the above values to omit the leading zeros in all decimal dimensions. If you add 8, the trailing zeros are omitted. If 12 (both 4 and 8) is added, the leading and the trailing zeros are omitted.

DIMUPT Switch
This variable governs the cursor functionality for User Positioned Text. This variable is saved in the drawing and its initial value is Off.
0 - The cursor controls the location of the dimension line only, 1 - The cursor controls the location of both the dimension text and the dimension line.

DIMZIN Integer
The DIMZIN (DIMension Zero INch) controls the suppression of the inches part of a feet-inches dimension when the distance is an integral number of feet or the suppression of the feet portion when the distance is less than one foot. This variable is saved in the drawing and its initial value is 0.
0 - Suppress zero feet and exactly zero inches, 1 - Include zero feet and, exactly zero inches, 2 - Include zero feet, suppress zero inches, 3 - Include zero inches, suppress zero feet. If the dimension has feet and a fractional inch part, the number of inches is included even if it is zero. This is independent of the DIMZIN setting. For example a dimension such as 1'-2/3" can never exist. It will be in the form 1'-0 2/3". The integer values 0 to 3 of the DIMZIN variable control the feet and inch dimension only, while as you can add 4 to omit the leading zeros in all decimal dimensions. For example 0.2600 becomes .2600. If you add 8, the trailing zeros are omitted. For example, 4.9600 becomes 4.96. If 12 (both 4 and 8) is added, the leading and the trailing zeros are omitted. For example, 0.2300 becomes .23.

DISTANCE Real
The DISTANCE variable holds the distance value determined by the DIST command. This command is read-only.

DONUTID Real
The DONUTID variable establishes the default inside diameter of a donut. The

initial value for this variable is 0.5000.

DONUTOD Real
The DONUTOD variable establishes the default outside diameter of a donut. It is important that the value of this variable be greater than zero. In case the value of DONUTID is greater than that of DONUTOD, then the two values are interchanged by the next command. The initial value for this variable is 1.0000.

DRAGMODE Integer
The DRAGMODE variable establishes the Object Drag mode while carrying out editing operations.
0 - Dragging disabled, 1 - Dragging enabled if invoked, 2 - Auto. This variable is saved in the drawing and is initially set to 2.

DWGCHECK Integer
The DWGCHECK variable determines whether a drawing was last edited by a product other than AutoCAD LT. Its initial value is 0 and it is stored in registry.
0 - The dialog box display is suppressed, 1 - The dialog box will be displayed.

DWGNAME String
The DWGNAME variable holds the name of the drawing as specified by the user. In case the drawing has not been assigned a name, the DWGNAME variable conveys that the drawing is unnamed. The drive and directory is also included if it was specified. This variable is a read-only variable.

DWGPREFIX String
The DWGPREFIX variable holds the drive and directory prefix for the drawing. This variable is a read-only variable.

DWGTITLED Integer
The DWGTITLED variable reflects whether the present drawing has been named.
0 - Indicates that the drawing has not been named, 1 - Indicates that the drawing has been named.

EDGEMODE Integer
With the EDGEMODE variable you can control how the EXTEND and TRIM commands determine boundary and cutting edges.
0 - In this case the selected edge is used without an extension, 1 - The object is trimmed or extended to an imaginary extension of the cutting or boundary edges. This is the initial value for this variable.

ELEVATION Real
The ELEVATION variable holds the current 3D elevation associated to the current UCS for the current space. This variable is saved in the drawing and has an initial value of 0.0000.

EXEDIR String
EXEDIR variable displays the folder path of the AutoCAD LT executable file.

EXPERT Integer
The issuance of some prompts is controlled with the EXPERT variable. The initial value for this variable is 0.
0 - All the prompts are issued, 1 - The "About to regen, proceed?" prompt and "Really want to turn the current layer off?" prompts are suppressed, 2 - The preceding prompts and "Block already defined. Redefine it?" (BLOCK command) and "A drawing with this name already exists. Overwrite it?" (SAVE or WBLOCK commands) are suppressed, 3 - The preceding prompts and the ones issued by LINETYPE if you try to load a linetype that is already loaded or create a new linetype in a file that already defines it are suppressed, 4 - The preceding prompts and the ones issued by UCS Save and VPORTS Save in case the name you

provide already exists are suppressed, 5 - The preceding prompts and the ones issued by the DIMSTYLE Save option, and DIMOVERRIDE in case the dimension style name you provide already exists, are suppressed. Whenever the **EXPERT** command suppresses a prompt, the corresponding operation is carried out as if you have entered Y as the response to the prompt. The **EXPERT** command can influence menu macros, scripts, AutoLISP, and the command functions.

EXPLMODE Integer

The EXPLMODE variable govern whether the **EXPLODE** command can explode nonuniformly scaled blocks. This variable is saved in the drawing and its initial value is 1.

0 - Nonuniformly scaled blocks cannot be exploded, 1 - Nonuniformly scaled blocks can be exploded.

EXTMAX 3D Point

The EXTMAX variable holds the upper-right point of the drawing extents and is saved in the drawing. The drawing extents increase outward when new objects are drawn and reduce only when ZOOM All or ZOOM Extents is used. The variable is reported in the World coordinates for the current space.

EXTMIN 3D Point

The EXTMIN variable holds the lower-left point of the drawing extents and is saved in the drawing. The drawing extents increase outward when new objects are drawn and reduce only when ZOOM All or ZOOM Extents is used. The variable is reported in the World coordinates for the current space.

EXTNAMES Integer

The EXTNAMES variable controls the parameters for named object names (such as linetypes, lineweights, and layers) stored in symbol tables. Its initial value is 1 and is saved in drawing.

0 - Limits the names to 255 in length which can include letters A to Z, numerals 0 to 9, and special characters like dollar sign ($), underscore (_), and hyphens (-), 1 - Limits the names to 255 in length which can include letters A to Z, numerals 0 to 9, and special characters not used by Microsoft Windows and AutoCAD LT for other puprose.

FACETRATIO Integer

The FACETRATIO variable sets the aspect ratio of faceting for cylindrical and conic ACIS solids. Its initial value is 0 and it is not saved.

0 - Creates an N by 1 mesh for cylindrical and conic ACIS solids, 1 - Creates an N by M mesh for cylindrical and conic ACIS solids.

FACETRES Real

The FACETRES variable adjusts the smoothness of shaded and objects whose hidden lines have been removed. This variable can be assigned values in the range of 0.010 to 10.0. The variable is saved in the drawing and has an initial value of 0.5.

FILEDIA Integer

The FILEDIA variable suppresses the display of file dialog boxes. This variable is saved in registry and has an initial value of 1.

0 - The file dialog boxes are disabled. However you can make AutoCAD LT to display the file dialog box by entering a tilde (~) as the response to the prompt. This applied for AutoLISP and ADS functions also, 1 - The file dialog boxes are enabled except when a script or AutoLISP/ADS program is active in which case only a prompt appears.

FILLETRAD Real
The FILLETRAD variable holds the current fillet radius and is saved in the drawing and its initial value is 0.5000.

FILLMODE Integer
FILLMODE variable indicates whether objects drawn with SOLID command are filled in. This variable is saved in the drawing and its initial value is 1.
0 - Objects are not filled, 0 - Objects are filled.

FONTALT String
The FONTALT variable specifies the alternate font to be used in case the specified font file cannot be found. In case you have not specified an alternate font, AutoCAD LT issues a warning. This variable is saved in registry and its initial value is “simplex.shx”.

FONTMAP String
The FONTMAP variable specifies the font mapping file to be used in case the specified font file cannot be found. This file holds one font mapping per line. The original font and the substitute font are separated by a semicolon (;). This variable is saved in registry and its initial value is “Aclt.fmp”.

FRONTZ Real
The FRONTZ variable contains the front clipping plane offset (in current drawing units) from the target plane for the current viewport. You can determine the distance between the front clipping plane and the camera point by subtracting the FRONTZ value from the camera to target distance. This variable is saved in the drawing and is read-only.

GRIDMODE Integer
The GRIDMODE variable specifies whether the grid is turned on or off. This variable is saved in the drawing and its initial value is 0.
0 - The grid is turned off, 1 - The grid is turned on.

GRIDUNIT 2D point
The GRIDUNIT variable specifies the X and Y grid spacing for the current viewport. The changes made to the grid spacing are manifested only after using the REDRAW or REGEN command. This variable is saved in the drawing and its initial value varies.

GRIPBLOCK Integer
The GRIPBLOCK variable controls the assignment of grips in blocks. This variable is saved in registry and its initial value is 0.
0 - The grip is assigned only to the insertion point of the block, 1 - Grips are assigned to objects within the block.

GRIPCOLOR Integer
The GRIPCOLOR variable controls the color of nonselected grips. It can take a value in the range of 1 to 255. This variable is saved in registry and its initial value is 5.

GRIPHOT Integer
The GRIPHOT variable controls the color of selected grips. It can take a value in the range of 1 to 255. This variable is saved in registry and its initial value is 1.

GRIPS Integer
With the GRIPS variable you can make use of selection set grips for the Stretch, Move, Rotate, Scale, and Mirror grip modes. This variable is saved in registry and its initial value is 1.
0 - Grips are disabled, 1 - Grips are enabled.

GRIPSIZE Integer
The GRIPSIZE variable allows you to assign a size to the box drawn to show the

grip. It sets its half height in pixels. This variable can be assigned a value in the range of 1 to 255. The variable is saved in registry and its initial value is 3.

HANDLES Integer
The HANDLES variable is always on (1), which states that object handles are enabled and can be accessed by applications. This variable is saved in the drawing and is read-only.

HIGHLIGHT Integer
The HIGHLIGHT variable governs object highlighting. Objects selected with grips are not influenced.
0 - Object selection highlighting is disabled, 1 - Object selection highlighting is enabled. This is the initial value for the variable.

HPANG Real
The HPANG variable specifies the angle of the hatch pattern. The initial value for this variable is 0.

HPBOUND Real
The HPBOUND variable governs the object type created by the **BHATCH** and **BOUNDARY** commands. This variable is saved in the drawing and its initial value is 1.
0 - A region is created, 1 - A polyline is created.

HPDOUBLE Integer
The HPDOUBLE variable governs the hatch pattern doubling for user-defined patterns. The initial value of this variable is 0.
0 - Hatch pattern doubling disabled, 1 - Hatch pattern doubling enabled.

HPNAME String
The default hatch pattern name is established with HPNAME variable. The name can be up to 34 characters and spaces are not allowed. Empty string ("") is returned if no default exists. To set no default, enter a period (.). The initial value of this variable is "ANSI31".

HPSCALE Real
The hatch pattern scale factor is specified with HPSCALE variable. This variable cannot assume zero value. The initial value of this variable is 1.0000.

HPSPACE Real
The hatch pattern line spacing for user-defined simple patterns is specified by HPSPACE variable. This variable cannot assume zero value. The initial value of this variable is 1.0000.

HYPERLINKBASE String
The HYPERLINKBASE variable specifies the path used for all relative hyperlinks in the drawing. Its initial value is "" and is saved in drawing.

INDEXCTL Integer
Controls whether layer and spatial indexes are created and saved in drawing. Initial value is 0.
0 - No indexes created, 1 - Layer index created, 2 - Spatial index is created, 3 - Layer and spatial are created.

INETLOCATION String
Stores the Internet location used by BROWSER. Initial value is "www autodesk.com/acltuser".

INSBASE 3D point
The insertion base point established by the **BASE** command is stored in this variable. This point is defined in the UCS coordinates for the current space. The variable is saved in the drawing and its initial value is 0.0000,0.0000,0.0000.

INSNAME String
The INSNAME variable establishes the

default block name for or INSERT commands. To set no default enter a period (.). The initial value of this variable is "".

INSUNITS Integer
The INSUNITS variable specifies a drawing units value, when you drag a block or image from AutoCAD LT DesignCenter. Its initial value is 0 and it is stored in drawing.
0 - Unspecified (No units), 1 - Inches, 2 - Feet, 3 - Miles, 4 - Millimeters, 5- Centimeters, 6 - Meters, 7 - Kilometers, 8 - Microinches, 9 - Mils, 10 - Yards, 11 - Angstroms, 12 - Nanometers, 13 - Microns, 14 - Decimeters, 15 - Decameters, 16 - Hectometers, 17 - Gigameters, 18 - Astronomical Units, 19 - Light Years, 20 - Parsecs.

INSUNITSDEFSOURCE Integer
The INSUNITSDEFSOURCE controls source content units value. Valid ranges between 0 to 20. Its initial value is 0 and it is stored in registry.

INSUNITSDEFTARGET Integer
The INSUNITSDEFTARGET controls source content units value. Valid ranges between 0 to 20. Its initial value is 0 and it is stored in registry.

ISAVEBAK Integer
Improves the speed of incremental saves, especially for large drawings on Windows. Initial value is 1.
0 - No BAK file is created, 1 - A BAK file is created.

ISAVEPERCENT Integer
Determines the amount of wasted space tolerated in a drawing file. Initial value is 50.

ISOLINES Integer
The ISOLINES variable specifies the number of isolines per surface on objects. The variable can accept a value in the range of 0 to 2047. This variable is saved in the drawing and its initial value is 4.

LASTANGLE Real
The LASTANGLE variable holds the end angle of the last arc entered, with respect to the XY plane of the current UCS for the current space. This variable is a read-only variable.

LASTPOINT 3D point
The LASTPOINT variable holds the UCS coordinates for the current space of the most recently entered point. This variable is saved in the drawing and its initial value is 0.0000,0.0000,0.0000.

LENSLENGTH Real
The LENSLENGTH variable holds the length of the lens (in mm) used in perspective viewing for the current viewport. This variable is saved in the drawing.

LIMCHECK Integer
This variable governs the drawing of objects outside the specified drawing limits. This variable is saved in the drawing and its initial value is 0.
0 - Object can be drawn outside the drawing limits, 1 - Object cannot be drawn outside the drawing limits.

LIMMAX 2D point
The upper-right drawing limits stated in World coordinates (for the current space) are held in the LIMMAX variable. This variable is saved in the drawing and its initial value varies.

LIMMIN 2D point
The lower-left drawing limits stated in World coordinates (for the current space) are held in the LIMMIN variable. This variable is saved in the drawing and its

initial value is 0.0000,0.0000.

LOCALE String
The LOCALE variable shows the ISO language code of the current AutoCAD LT version in use. The initial value of this variable is "enum" (varies by country).

LOGFILEMODE Integer
Specifies whether the contents of the text window are written to a log file. Its initial value is 0.
0 - Log file is not maintained, 1 - Log file is maintained.

LOGFILENAME String
Specifies the path for the log file. Initial value is "C:\ACAD2000\aclt.log".

LOGFILEPATH String
The LOGFILEPATH variable specifies the path for the log files for all drawings in a session. The initial value varies depending on where you installed AutoCAD LT and is saved in registry.

LOGINNAME String
The LOGINNAME variable shows the user's name as specified while configuring when AutoCAD LT is loaded.

LTSCALE Real
The LTSCALE variable establishes the global linetype scale factor. The variable is saved in the drawing and its initial value is 1.0000.

LUNITS Integer
The LUNITS variable establishes the Linear Units mode. The variable is saved in the drawing and its initial value is 2.
1 - Scientific units mode, 2 - Decimal units mode, 3 - Engineering units mode, 4 - Architectural units mode, 5 - Fractional units mode.

LUPREC Integer
The LUPREC variable establishes the linear units decimal places or denominator. The variable is saved in the drawing and its initial value is 4.

LWDEFAULT Enum
The LWDEFAULT variable controls the value for the default lineweight. The default lineweight can be set to any valid lineweight value in millimeters.

LWDISPLAY Integer
The LWDISPLAY variable controls whether the lineweight is displayed on the Model or Layout tab. Its initial value 0 and is saved in drawing.
0 - Lineweight is not displayed, 1 - Lineweight is displayed.

LWUNITS Integer
The LWUNITS variable controls whether lineweight units are displayed in inches or millimeters. Its initial value is 1 and is saved in registry.
0 - Inches, 1 - Millimeters.

MAXACTVP Integer
The MAXACTVP variable specifies the maximum number of viewports to regenerate at one time. The initial value of this variable is 64.

MAXSORT Integer
The MAXSORT variable sets the maximum number of symbol names of file names that are to be sorted by listing commands. In case the total number of items is greater than this number, then no items are sorted. This variable is saved in registry and its initial value is 200.

MBUTTONPAN Integer
The MBUTTONPAN variable controls the behavior of the third button or wheel on the pointing device. Its initial value is 1 and it is saved in registry.
0 - Supports the action defined in the

AutoCAD LT menu (.mnu) file, 1 - Supports panning by holding and dragging the button or wheel.

MEASUREINIT Integer
The MEASUREMENT variable sets the drawing units as English or metric and controls which hatch pattern and linetype files an existing drawing uses when it is opened. It is saved in drawing and the initial value is 0.
0 - English(AutoCAD LT uses ANSI hatch patterns and linetypes 0, 1 - Metric (AutoCAD LT uses ISOHatch and ISOLinetypes).

MEASUREMENT Integer
The MEASUREMENT variable sets the drawing units as English or metric and controls which hatch pattern and linetype files for the current drawing only. It is saved in drawing and the initial value is 0.
0 - English (AutoCAD LT uses ANSI hatch patterns and linetypes0, 1 - Metric (AutoCAD LT uses ISOHatch and ISOLinetypes).

MENUECHO Integer
The MENUECHO variable sets menu echo and prompt control bits. The initial value for this variable is 0. The variable is the addition of the following:
1 - The echo of menu items is suppressed, 2 - The display of system prompts during menu is suppressed, 4 - The ^P toggle of menu echoing is disabled, 8 - The input/output strings and debugging aid for DIESEL macros is displayed.

MIRRTEXT Integer
The MIRRTEXT variable governs how the MIRROR command mirrors text. This variable is saved in the drawing and its initial value is 1.
0 - The text direction is retained, 1 - The text is mirrored.

MODEMACRO String
The MODEMACRO variable shows a text string or DIESEL expression on the status line. This string reveals information like the name of the current drawing, time/date stamp, or special modes. The initial value of this variable is "".

MTEXTED String
The MTEXTED variable sets the name of the program to be used for the editing of mtext objects. This variable is saved in registry and its initial value is "Internal".

OFFSETDIST Real
This variable sets the default offset distance. If the value of this variable is less than zero, then the offset distance can be specified with the through mode. If the value of this variable is greater than zero, then the default offset distance is established. The initial value of this variable is 1.0000.

OLEHIDE Integer
Controls the display of OLE objects in AutoCAD LT. It is saved in registry and its initial value is 0.
0 - All OLE objects are visible, 1 - OLE objects are visible in paper space, 2 - OLE objects are visible in model space, 3 - No OLE objects are visible.

OLEQUALITY Integer
The OLEQUALITY variable controls the default quality level for embedded OLE objects. Its initial value is 1 and it is stored in registry.
0 - Line art quality, such as an embedded spreadsheet, 1 - Text quality, such as an embedded Word document, 2 - Graphics quality, such as an embedded pie chart, 3 - Photograph quality, 4 - High quality photograph.

OLESTARTUP Integer
The OLESTARTUP variable controls

whether the source application of an embedded OLE object loads when plotting. Its initial value is 0 and it is saved in drawing.
0 - Does not load the OLE source application, 1 - Loads the OLE source application when plotting.

ORTHOMODE Integer
The ORTHOMODE variable governs the orthogonal display of lines or polylines. This variable is saved in the drawing and its initial value is 0.
0 - The Ortho mode is turned off, 1 - The Ortho mode is turned on.

OSMODE Integer
The OSMODE variable sets the running Object Snap modes using the following bit-codes:
0 - NONe object snap, 1 - ENDpoint object snap, 2 - MIDpoint object snap, 4 - CENter object snap, 8 - NODe object snap, 16 - QUAdrant object snap, 32 - INTersection object snap, 64 - INSertion object snap, 128 - PERpendicular object snap, 256 - TANgent object snap, 512 - NEArest object snap, 1024 - QUIck object snap, 2048 - APPint object snap. If you want to specify more than one object snap, enter the sum of their values. For example, if you want to specify the node and center object snaps, enter 4+8 = 12 as the value for the OSMODE variable. This variable is saved in the drawing and its initial value is 0.

OSNAPCOORD Integer
Controls whether coordinates entered on the command line override running object snaps.
0 - Running object snap settings override keyboard entry, 1 - Keyboard entry overrides object snap settings, 2 - (Initial value) Keyboard entry overrides object snap setting except in scripts.

PAPERUPDATE Integer
The PAPERUPDATE variable controls the display of a warning dialog when attempting to print a layout with a paper size different from the paper size specified by the default for the plotter configuration file. Its initial value is 0 and it is saved in registry.
0 - Displays a warning dialog box if the paper size specified in the layout is not supported by the plotter, 1 - Sets paper size to the configured paper size of the plotter configuration file.

PDMODE Integer
The PDMODE variable sets Point Object Display mode. This variable is saved in the drawing and its initial value is 0.

PDSIZE Real
The PDSIZE variable sets the display size of the point object. This variable is saved in the drawing and its initial value is 0.0000.
0 - For this value, point is created at 5 percent of the graphics height, >0 - In this case the value entered specifies the absolute size, <0 - In this case the value entered specifies the percentage of the viewport size.

PELLIPSE Integer
The PELLIPSE controls the type of ellipse created with the **ELLIPSE** command. This variable is saved in the drawing and its initial value is 0.
0 - A true ellipse object is drawn, 1 - A polyline representation of an ellipse is drawn.

PERIMETER Real
The PERIMETER variable holds the most recently perimeter value computed by **AREA**, **DBLIST**, or **LIST** commands. This variable is a read-only variable.

PICKADD Integer
The PICKADD variable controls additive selection of objects. This variable is saved in registry and its initial value is 1.
0 - PICKADD variable is disabled, 1 - PICKADD variable is enabled. All the objects selected by any method are added to the selection set. If you want to remove objects from the selection set, hold down the Shift key and select the objects.

PICKAUTO Integer
PICKAUTO variable controls the automatic windowing feature when the "Select object" prompt appears. This variable is saved in registry and its initial value is 1.
0 - PICKAUTO variable is disabled, 1 - PICKAUTO variable is enabled and a selection window is automatically drawn at the Select objects prompt.

PICKBOX Integer
The PICKBOX variable sets the object selection target half height (in pixels). This variable is saved in registry and its initial value is 3.

PICKDRAG Integer
The PICKDRAG variable governs the method of drawing a selection window:
0 - For this value the selection window is drawn by clicking the pointing device at one corner and then clicking again at the other corner of the window, 1 - For this value the selection window is drawn by clicking the pointing device at one corner, holding down the pick button, dragging the cursor, and finally releasing the pick button of the pointing device at the other corner of the window. This variable is saved in registry and its initial value is 0.

PICKFIRST Integer
The PICKFIRST variable governs the method of object selection in such a manner that you can first select the object and then specify the desired edit or inquiry command. Its initial value is 1.
0 - PICKFIRST variable disabled, 1 - PICKFIRST variable enabled.

PICKSTYLE Integer
The PICKSTYLE variable controls the associative hatch selection and group selection. This variable is saved in the drawing and its initial value is 1.
0 - Associative hatch selection and group selection not possible, 1 - Group selection possible, 2 - Associative hatch selection possible, 3 - Associative hatch selection and group selection possible.

PLINEGEN Integer
The PLINEGEN variable sets the linetype pattern generation around the vertices of a 2D polyline. This variable does not affect polylines with tapered segments. This variable is saved in the drawing and its initial value is 0.
0 - Polylines are generated with a dash at each vertex, 1 - Linetype is created in a continuous pattern around the vertices of the polyline.

PLINETYPE Integer
Specifies whether AutoCAD LT uses optimized 2D polylines.
0 - PLINE creates old format polylines, 1 - PLINE creates optimized polylines, 2 - PLINE creates optimized polylines and the polylines in older drawings are converted on open.

PLINEWID Real
The default polyline width is stored in this variable. This variable is saved in the drawing and its initial value is 0.0000.

PLOTROTMODE Integer
The PLOTROTMODE variable controls the orientation of plots. This variable is saved in the drawing and its initial value is 1.

0 - The effective plotting area is rotated in order to align the corner with the Rotation icon with the paper at the lower-left for a rotation of 0, top-left for a rotation of 90, top-right for a rotation of 180, and lower left for a rotation of 270, 1 - The lower-left corner of the effective plotting area is aligned with the lower-left corner of the paper.

POLARADDANG String
The POLARADDANG variable contains all user-defined polar angles. You can add up to 10 angles. Each angle can be up to 25 characters, separated with semicolons (;). Its initial value is null and it is stored in registry.

POLARANG Real
The POLARANG variable controls the polar angle increment. Values are 90, 45, 30, 22.5, 18, 15,10, and 5. Its initial value is 90 and is saved in registry.

POLARDIST Real
The POLARDIST variable controls the snap increment when the SNAPSTYL system variable is set to 1 (polar snap). Its initial value is 0.0000 and it is stored in registry.

POLYSIDES Integer
The POLYSIDES variable establishes the default number of sides for a polygon. This variable can take values in the range of 3 to 1024. The initial value of this variable is 4.

PROJMODE Integer
The PROJMODE variable establishes the current Projection mode for Extend or Trim operations. Its initial value is 1.
0 - True 3D mode established (no projection), 1 - Projection to XY plane of the current UCS, 2 - Projection to current view plane.

PROXYGRAPHICS Integer
Specifies whether images of proxy objects are saved in the drawing. The initial value is 1.
0 - Image is not saved, 1 - Image is saved.

PROXYSHOW Integer
Controls the display of proxy objects in a drawing. The initial value is 1.
0 - Proxy objects are not displayed, 1 - Graphic images are displayed for all proxy objects, 2 - Only bounding box is displayed for all proxy objects.

PSLTSCALE Integer
The PSLTSCALE variable governs the paper space linetype scaling. This variable is saved in the drawing and its initial value is 1.
0 - Special linetype scaling not allowed. Linetype dash lengths depend on the drawing units of the space in which the objects were drawn, 1 - Linetype scaling governed by viewport scaling. In case TILEMODE is set to 0, dash lengths depend on the paper space drawing units, even if objects are in model space.

PSPROLOG String
The PSPROLOG variable assigns a name for a prologue section which is to be read from the aclt.psf file when PSOUT command is being used. This variable is saved in registry and its initial value is "".

PSTYLEMODE Read only
The PSTYLEMODE variable indicates whether the current drawing is in a Color-Dependent or Named Plot Style mode. Its initial value is 0 and it is saved in drawing.
0 - Uses named plot style tables in the current drawing, 1 - Uses color-dependent plot style tables in the current drawing.

PSTYLEPOLICY Integer
The PSTYLEPOLICY variable controls whether an object's color property is

associated with its plot style. The new value you assign affects only newly created drawings. Its initial value is 1 and it is saved in registry.
0 - No association is made between color and plot style.1 - An object's plot style is associated with its color.

PSVPSCALE Real
The PSVPSCALE variable controls the view scale factor for all newly created viewports. The view scale factor is defined by comparing the ratio of units in paper space to the units in newly created model space viewports. A value of 0 means the scale factor is "Scaled to Fit". A scale must be a positive real value. Its initial value is 0 and it is not saved.

PUCSBASE String
The PUCSBASE variable stores the name of the UCS that defines the origin and orientation of orthographic UCS settings in paper space only. Its initial value is "" and it is stored in drawing.

QTEXTMODE Integer
The QTEXTMODE controls the Quick Text mode. This variable is saved in the drawing and its initial value is 0.
0 - The Quick Text mode is turned off and characters are displayed, 1 - The Quick Text mode is turned on and a box instead of text is displayed.

RASTERPREVIEW Integer
The RASTERPREVIEW variable determines whether the drawing preview images are saved with the drawing and in which format they will be saved. This variable is saved in the registry and its initial value is 1.
0 - No preview image created, 1 - Preview image is created.

REGENMODE Integer
The REGENMODE variable controls the automatic regeneration of the drawing. This variable is saved in the drawing and its initial value is 1.
0 - REGENAUTO is turned off, 1 - REGENAUTO is turned on.

RTDISPLAY Integer
Controls the display of raster images during realtime zoom or pan.
0 - Displays raster image content, 1 - Displays raster image outline only (initial value).

SAVEFILE String
The present auto-save filename is held in the SAVEFILE variable. This variable is a read-only variable. The initial value is "auto.sv$".

SAVEFILEPATH String
The SAVEFILEPATH variable specifies the path to the directory for all automatic save files for the AutoCAD LT session. Its initial value is "C:\Temp\" and it is saved in registry.

SAVENAME String
You can save the current drawing to a different name and this name is held in the SAVENAME variable. This variable is a read-only variable.

SAVETIME Integer
AutoCAD LT has provided the facility of automatically saving your work at specific intervals. You can specify the automatic save time intervals (in minutes) with the SAVETIME variable. This variable is set to an initial value of 120.
0 - Automatic save facility is disabled, 0 - The drawing is saved according to the intervals specified. Once you make changes to the drawing, the SAVETIME timer starts. The **SAVE**, **SAVEAS**, or **QSAVE** commands reset and restart this timer. AutoCAD LT saves the drawing under the filename auto.sv$.

SDI Integer
This variable controls whether AutoCAD LT runs in single- or multiple-document interface. Its initial value is 0 and it is saved in registry.
0 - Turns on multiple-drawing interface, 1 - Turns off multiple-drawing interface, 2 - (Read-only) Multiple-drawing interface is disabled, 3 - (Read-only) Multiple-drawing interface is disabled.

SHADEDGE Integer
The SHADEDGE variable governs the shading of edges in rendering. This variable is saved in the drawing and its initial value is 3.
0 - Faces are shaded and edges are not highlighted, 1 - Faces are shaded and edges are drawn in background color, 2 - Faces are not filled and edges are in object color, 3 - Faces are in object color and edges are drawn in background color.

SHADEDIF Integer
The SHADEIF variable establishes the ratio of diffuse reflective light to ambient light (in percent of diffuse reflective light). This variable is saved in the drawing and its initial value is 70.

SHORTCUTMENU Integer
This variable Controls whether Default, Edit, and Command mode shortcut menus are available in the drawing area. Its initial value is 11 and it is stored in registry. The following bitcodes are used by SHORTCUTMENU:
0 - Disables all Default, Edit, and Command mode shortcut menus, restoring LT 97 legacy behavior, 1 - Enables Default mode shortcut menus, 2 - Enables Edit mode shortcut menus, 4 - Enables Command mode shortcut menus. In this case, the Command mode shortcut menu is available whenever a command is active, 8 - Enables Command mode shortcut menus only when command options are currently available from the command line.

SNAPANG Real
The SNAPANG variable specifies the snap/grid rotation angle relative to the UCS for the current viewport. This variable is saved in the drawing and its initial value is 0. Changes to this variable are manifested only after a redraw is performed.

SNAPBASE 2D point
The SNAPBASE variable specifies the snap/grid origin point (in UCS X, Y coordinates) for the current viewport. This variable is saved in the drawing and its initial value is 0.0000,0.0000. Changes to this variable are manifested only after a redraw is performed.

SNAPISOPAIR Integer
The SNAPISOPAIR variable controls the current isometric plane for the current viewport. This variable is saved in the drawing and its initial value is 0.
0 - Left, 1 - Top, 2 - Right.

SNAPMODE Integer
The SNAPMODE variable controls the Snap mode. This variable is saved in the drawing and its initial value is 0.
0 - Snap disabled, 1 - Snap enabled for the current viewport.

SNAPSTYL Integer
The SNAPSTYL variable establishes the snap style for the current viewport. This variable is saved in the drawing and its initial value is 0.
0 - Standard, 1 - Isometric.

SNAPTYPE Integer
This variable the snap style for the current viewport. Its initial value is 0 and it is saved in registry.

0 - Grid, or standard snap, 1 - Polar snap.

SNAPUNIT 2D point
The SNAPUNIT variable specifies the X and Y snap spacing for the current viewport. This variable is saved in the drawing and its initial value is 0.5000,0.5000. The changes to this variable are manifested only after a redraw is performed.

SORTENTS Integer
The SORTENTS variable governs the display of object sort order operations using the following values:
0 - SORTENTS is disabled, 1 - Sorts for object selection, 2 - Sorts for object snap, 4 - Sorts for redraw, 8 - Sorts for MSLIDE slide creation, 16 - Sorts for regens, 32 - Sorts for plotting, 64 - Sorts for PostScript output. More than one option can be selected by specifying the sum of the values of these options. Its initial value is 96. This value specifies sort operations for plotting and PostScript output.

SPLFRAME Integer
The SPLFRAME variable governs the display of spline-fit polylines. This variable is saved in the drawing and its initial value is 0.
0 - The control polygon for spline fit polylines is not displayed, 1 - The control polygon for spline fit polylines is displayed.

SPLINESEGS Integer
The SPLINESEGS variable governs the number of line segments used to construct each spline. Hence with this variable you can control the smoothness of the curve. This variable is saved in the drawing and its initial value is 8. With this value a reasonably smooth curve is generated which does not need much regeneration time. The greater the value of this variable, the smoother the curve and greater the regeneration time and the space occupied by the drawing file.

SPLINETYPE Integer
The SPLINETYPE variable specifies the type of spline curve that will be generated by the Spline option of **PEDIT** command. This variable is saved in the drawing and its initial value is 6.
5 - Quadratic B-spline is generated, 6 - Cubic B-spline is generated.

TABMODE Integer
The TABMODE variable governs the use of Tablet mode.
0 - Tablet mode disabled (initial value), 1 - Tablet mode enabled.

TARGET 3D point
The TARGET variable holds the position of the target point (in UCS coordinates) for the current viewport. This is a read-only variable and is saved in the drawing.

TDCREATE Real
The TDCREATE variable holds the creation time and date of a drawing. This is a read-only variable and is saved in the drawing.

TDINDWG Real
The TDINDWG variable holds the total editing time. This is a read-only variable and is saved in the drawing.

TDUCREATE Real
This variable stores the universal time and date the drawing was created. This is a read-only variable and is saved in drawing.

TDUPDATE Real
The TDUPDATE variable holds the time and date of the most recent update/save. This is a read-only variable and is saved in the drawing.

TDUSRTIMER Real
The TDUSRTIMER variable stores the user elapsed timer. This is a read-only variable and is saved in the drawing.

TDUUPDATE Real
This variable stores the universal time and date of the last update/save. This is a read-only variable and stored in drawing.

TEXTFILL Integer
The TEXTFILL variable governs the filling of TrueType fonts. This variable is saved in the registry and has an initial value of 1.
0 - The text is displayed as outlines, 1 - The text is displayed as filled images.

TEXTQLTY Real
The TEXTQLTY variable defines the resolution of TrueType, Bitstream, and Abode Type 1 fonts. The higher the value of this variable, the higher the resolution and lower the display and plotting speed. On the other hand the lower the value of this variable, the lower the resolution and higher the display and plotting speed. This variable is saved in the drawing and can take values in the range of 0 to 100.0. Its initial value is 50.

TEXTSIZE Real
The TEXTSIZE variable controls the text height of the text drawn with the current text style. But this is possible only if the style does not have a fixed height. This variable is saved in the drawing and its initial value is 0.2000.

TEXTSTYLE String
The TEXTSTYLE variable stores the name of the current text style. This variable is saved in the drawing and its initial value is STANDARD.

THICKNESS Real
The THICKNESS variable defines the current 3D thickness. This variable is saved in the drawing and its initial value is 0.0000.

TILEMODE Integer
The TILEMODE variable governs entry into paper space and also how the AutoCAD LT viewports act. This variable is saved in the drawing and its initial value is 1.
0 - The paper space and viewport objects are enabled. The graphics area is cleared and you are prompted to use the MVIEW command to define viewports, 1 - Release 10 Compatibility mode is enabled. Automatically you are taken into Tiled Viewport mode and previously active viewport configuration is restored on the screen. Paper space objects including viewport objects are not displayed. The **MSPACE**, **PSPACE**, **VPLAYER**, and **MVIEW** commands are disabled.

TOOLTIPS Integer
The TOOLTIPS variable is concerned with the Windows version of AutoCAD LT and determines the display of ToolTips. Its initial value is 1.
0 - The display of ToolTips is turned off, 1 - The display of ToolTips is turned on.

TRIMMODE Integer
The TRIMMODE variable determines whether selected edges for chamfers and fillets will be trimmed.
0 - Selected edges are not trimmed after chamfering and filleting, 1 - Selected edges are trimmed after chamfering and filleting (initial value).

TSPACEFAC Real
This variable controls the multiline text line spacing distance measured as a factor of text height. The valid values are 0.25 to 4.0. The initial value is 1 and it is not saved.

TSPACETYPE Integer
Controls the type of line spacing used in multiline text. At Least adjusts line spacing based on tallest characters in a line. Exactly uses the specified line spacing, regardless of individual character sizes. Its initial value is 1 and it is not saved.
1 - At Least, 2 - Exactly.

TSTACKALIGN Integer
This variable controls the vertical alignment of stacked text. Its initial value is 1 and it is saved in drawing.
0 - Bottom aligned, 1 - Center aligned, 2 - Top aligned.

TSTACKSIZE Integer
This variable controls the percentage of stacked text fraction height relative to selected text's current height. The valid value ranges from 1 to 127. Its initial value is 70 and it is saved in drawing.

UCSAXISANG Integer
This variable stores the default angle when rotating the UCS around one of its axes using the X, Y, or Z options of the UCS command. The values must be entered as an angle in degrees (valid values are: 5, 10, 15, 18, 22.5, 30, 45, 90, 180). Its initial value is 90 and it is stored in registry.

UCSBASE String
This variable stores the name of the UCS that defines the origin and orientation of orthographic UCS settings. The valid values include any named UCS. Its initial value is "World" and is stored in drawing.

UCSFOLLOW Integer
The UCSFOLLOW variable controls the automatic displaying of a plan view when you switch from one UCS to another. All the viewports have the UCSFOLLOW facility and hence you need to specify the UCSFOLLOW setting separately for each viewport. This variable is saved in the drawing and its initial value is 0.
0 - Switch from one UCS to another, does not alter the view, 1 - Plan view of the new UCS is automatically displayed when you switch from one UCS to another.

USERNAME String
USERNAME variable specifies the user name. Initially set by first name, last name, and organization name entered during installation of aAutoCAD LT.

UCSICON Integer
The UCSICON variable displays the present UCS icon using bit-code for the current viewport. Initial value for this variable is 3. The value of this variable is the sum of the following:
0 - The icon is not displayed, 1 - Icon is displayed, 2 - The icon moves to the origin if the icon display in enbled, 3 - Icon is displayed at the origin.

UCSNAME String
The UCSNAME variable contains the name of the current UCS. This is a read-only variable and is saved in the drawing. In case the current UCS is unnamed, then a null string is returned.

UCSORG 3D point
The coordinate value of the origin of the current UCS is held in the UCSORG variable. This is a read-only variable and is saved in the drawing. Initial value is 0.0000, 0.0000, 0.0000.

UCSORTHO Integer
This variable Determines whether the related orthographic UCS setting is restored automatically when an orthographic view is restored. Its initial value is 1 and it is stored in registry.
0 - Specifies that the UCS setting remains unchanged when an orthographic view is restored, 1 - Specifies that the related

orthographic UCS setting is restored automatically when an orthographic view is restored.

UCSVIEW Integer
This variable determines whether the current UCS is saved with a named view. Its initial value is 1 and is saved in registry.
0 - Does not save current UCS with a named view, 1 - Saves current UCS whenever a named view is created.

UCSXDIR 3D point
The X axis direction of the current UCS for the current space is held in the UCSXDIR variable. This is a read-only variable and is saved in the drawing.

UCSYDIR 3D point
The Y axis direction of the current UCS for the current space is held in UCSYDIR variable. This is a read-only variable and is saved in the drawing.

UNITMODE Integer
The UNITMODE variable governs the units display format. This variable is saved in the drawing and its initial value is 0.
0 - The fractional, feet and inches, and surveyor's angles are displayed as previously defined, 1 - The fractional, feet and inches, and surveyor's angles are displayed in the input format. This variable is saved in the drawing and its initial value is 0.

VIEWCTR 3D point
The VIEWCTR variable stores the center of view in the current viewport, defined in the UCS coordinates. This variable is a read-only variable and is saved in the drawing.

VIEWDIR 3D vector
The VIEWDIR variable contains the viewing direction in the current viewport expressed in the UCS coordinates. The camera position is expressed as a 3D offset from the target position. This variable is a read-only variable and is saved in the drawing.

VIEWMODE Integer
The VIEWMODE variable governs Viewing mode for the current viewport using bit-code. The value for this variable is the addition of the following bit values:
0 - Viewing mode disabled, 1 - Perspective view active, 2 - Front clipping on, 4 - Back clipping on, 8 - UCS Follow mode on, 16 - Front clip not at eye. In case it is on, the front clipping plane is determined by the front clip distance stored in the FRONTZ variable. If it is off, the front clipping plane passes through the camera point and in this case the FRONTZ variable is not taken into consideration. If the front clipping bit (2) is off, then this flag is neglected. This variable is a read-only variable and is saved in the drawing.

VIEWSIZE Real
The VIEWSIZE variable contains the view height in the current viewport and is defined in the drawing units. This variable is a read-only variable and is saved in the drawing.

VIEWTWIST Real
The VIEWTWIST variable contains the view twist angle for the current viewport. This variable is a read-only variable and is saved in the drawing.

VISRETAIN Integer
The VISRETAIN variable determines whether changes to the visibility, color, and linetype of xref dependent layers are saved in the current drawing.
0 - Changes to On/Off, Freeze/Thaw, color, and linetype settings for the xref-dependent layers are not saved in the current

drawing, 1 - Changes to the xref layer definitions in the current drawing are saved with the current drawing.

VSMAX 3D point
The VSMAX variable contains the upper-right corner of the virtual screen of the current viewport and is expressed in UCS coordinates. This variable is a read-only variable and is saved in the drawing.

VSMIN 3D point
The VSMIN variable contains the lower-left corner of the virtual screen of the current viewport and is expressed in UCS coordinates. This variable is a read-only variable and is saved in the drawing.

WHIPARC Integer
This variable controls whether the display of circles and arcs is smooth. Its initial value is 0 and it is saved in registry.
0 - Circles and arcs are not smooth, but rather are displayed as a series of vectors, 1 - Circles and arcs are smooth, displayed as true circles and arcs.

WMFBKGND Integer
The WMFBKGND variable controls whether the background display of AutoCAD LT objects is transparent in other applications when these objects are Output to a Windows metafile using the **WMFOUT** command, Copied to the Clipboard in AutoCAD LT and pasted as a Windows metafile or Dragged and dropped from AutoCAD LT as a Windows metafile. Its initial value is 1 and the values are not saved.
0 - The background is transparent, 1 - The background color is the same as the AutoCAD LT current background color.

WORLDVIEW Integer
The WORLDVIEW variable determines whether UCS changes to WCS during DVIEW or VPOINT commands. This variable is saved in the drawing and its initial value is 1.
0 - Current UCS is not changed, 1 - Current UCS is changed to WCS until the **DVIEW** or **VPOINT** commands are in progress. The **DVIEW** and **VPOINT** commands input are with respect to the current UCS, 2 - Current UCS is changed relative to UCS specified by the UCSBASE system variable.

XCLIPFRAME Integer
This variable controls visibility of xref clipping boundaries and its initial value is 0.
0 - Clipping boundary is not visible, 1 - Clipping boundary is visible.

XEDIT Integer
This variable controls whether the current drawing can be edited in-place when being referenced by another drawing. Its initial value is 1 and it is saved in drawing.
0 - Can not use in-place reference editing,
1 - Can use in-place reference editing.

XLOADCTL Integer
Turns demand load on and off and controls whether it loads the original drawing or a copy. Initial value is 1.
0 - Turns off demand loading; entire drawing is loaded, 1 - Turns on demand loading; reference file is kept open, 2 - Turns on demand loading; a copy of reference file is opened.

XLOADPATH String
Creates a path for storing temporary copies of demand-loaded xref files. Initial value is " ".

XREFCTL Integer
The XREFCTL variable determines whether AutoCAD LT writes .xlg files

(external reference log files). This variable is saved in registry and its initial value is 0. 0 - Xref log files are not written, 1 - Xref log files are written.

ZOOMFACTOR Integer
This value accepts an integer between 3 and 100 as valid values. The higher the number, the more incremental the change applied by each mouse-wheel's forward/ backward movement. Its initial value is 10 and it is stored in registry.

Index

D

E

F

G

H

I

J

L

P

Q

R

S

T

U

V

W

X

Y

Z